Electronic Devices and Circuits

Sixth Edition

Theodore F. Bogart, Jr.
University of Southern Mississippi

Jeffrey S. Beasley
New Mexico State University

Guillermo Rico
New Mexico State University

PEARSON
Prentice Hall

Upper Saddle River, New Jersey
Columbus, Ohio

Library of Congress Cataloging-in-Publication Data
Bogart Theodore F.
Electronic devices and circuits / Theodore F. Bogart, Jr., Jeffrey S. Beasley, Guillermo Rico.—6th ed.
p. cm.
Includes bibliographical references and index.
ISBN 0-13-111142-6
1. Electronic circuits. 2. Electronic apparatus and appliances. I. Beasley, Jeffrey S., 1995– II. Rico, Guillermo. III. Title.

TK7867.B57 2004
621.3815—dc21

2003051282

Editor in Chief: Stephen Helba
Acquisitions Editor: Dennis Williams
Development Editor: Kate Linsner
Editorial Assistant: Lara Dimmick
Production Editor: Stephen C. Robb
Production Supervision: Carlisle Publishers Services
Design Coordinator: Diane Y. Ernsberger
Cover Designer: Ali Mohrman
Cover art: Digital Vision
Production Manager: Pat Tonneman
Marketing Manager: Ben Leonard

This book was set in Times Europa by Carlisle Communications, Ltd. It was printed and bound by Courier Kendallville, Inc. The cover was printed by Phoenix Color Corp.

Pearson Education Ltd.
Pearson Education Singapore Pte. Ltd.
Pearson Education Canada, Ltd.
Pearson Education—Japan
Pearson Education Australia Pty. Limited
Pearson Education North Asia Ltd.
Pearson Educación de Mexico, S.A. de C.V.
Pearson Education Malaysia Pte. Ltd.

PEARSON
Prentice Hall

10 9 8 7 6 5 4 3 2 1
ISBN 0-13-111142-6

Preface

Electronic Devices and Circuits is a modern, thorough treatment of the topics traditionally covered during a two- or three-semester course in electronic device theory and applications. The minimum preparation for students beginning their study of this material is a course in dc circuit analysis. Students will need an understanding of impedance concepts and must be able to perform phasor computations before beginning Chapter 9, "Frequency Response." Therefore, students who begin their study *without* having completed a course in ac circuit analysis should be taking that course as a corequisite. Basic calculus is used as needed for the development of theoretical principles, such as in the discussion of electronic differentiators, integrators, and wave shaping.

Principal considerations in selecting topics for this edition were the significance of each topic in modem industrial applications and the impact that they are likely to have on emerging technologies. Consequently, integrated-circuit applications are presented in great detail, including coverage of analog and digital integrated circuit design, operational amplifier theory and applications, and specialized electronic devices and circuits such as switching regulators and optoelectronics.

Each new concept in *Electronic Devices and Circuits* is introduced from a systems or block-diagram approach. For example, the effect of the input and output resistance on the voltage gain of an amplifier is developed in Chapter 6 by regarding the amplifier as a functional block rather than as a particular circuit. Once the fundamental concept has been thoroughly discussed, it is applied to each amplifier studied thereafter. A similar approach is used to develop the theory of the feedback, frequency response, amplifier bias, distortion, linearity, oscillation, filters, voltage regulation, and analog and digital transistor circuit-building blocks.

Most chapters are accompanied by PSpice® examples and exercises. Appendix A contains general PSpice® instructional material that parallels the sequence in which electronic device theory is introduced in the book, so new simulation skills can be developed at the same pace that new theory is taught. Additionally, new sections on circuit analysis using Electronics Workbench Multisim® have been introduced for students to experience this widely used simulation program.

FEATURES OF THE SIXTH EDITION

1. Basic power supplies and voltage regulation are introduced in Chapter 3, along with a new section on power supply component specification.
2. Amplifier fundamentals are covered separately in the new Chapter 6.
3. More biasing circuits for bipolar junction transistors are included, along with their small-signal ac analyses.
4. The theory and usage of transistors in analog and digital VLSI circuits are presented in Chapters 17 and 18.
5. Multistage amplifiers are covered in a section in the new Chapter 6, and less emphasis is given to discrete multistage amplifiers in Chapter 7.
6. Frequency response topics have been expanded, including alternative approximations, more design, and more PSpice® exercises.
7. Op-amp theory and applications are presented with emphasis on design, based on meeting or exceeding desired levels of performance.
8. Active filters, part of op-amp applications, include wideband versions of both bandpass and band-stop filters. Design is very practical and limited to unity-gain filter blocks.
9. Each chapter concludes with a practical exercise using Electronics Workbench Multisim®.

SUPPLEMENTS

Supplements available for this textbook include:

- *CD with Circuit Files.* This CD is included with every copy of the text and contains an extensive number of circuit files prepared by the authors for students to experiment with using Electronics Workbench Multisim®. Also included on the CD is the Enhanced Textbook Edition of Multisim 2001®, which is all that is required to open and work with all of these circuits. However, anyone who wishes to order the full version of Multisim may do so through Prentice Hall by visiting http://www.prenhall.com, phoning 1-800-282-0693, or sending a fax request to 1-800-835-5327. Technical questions relating specifically to Multisim should be referred to Electronics Workbench by writing to *support@electronicsworkbench.com.*
- *Laboratory Manual.* Written by James Brown of Peavey Electronics Corporation, this manual contains experiments that coincide directly with the text.
- *Instructor's Resource Manual,* with worked-out solutions and a Test Item File.
- *PowerPoint® Transparencies.* This CD contains figures and tables from the text for presentation in the classroom.
- *Companion Website: www.prenhall.com/bogart.* This site is provided as a student study guide and consists of questions designed to assist students in the review of each chapter.
- *Electronics Supersite: www.prenhall.com/electronics.* This site pertains to many Prentice Hall Electronics Technology products and provides multiple resources for instructors and students.

ACKNOWLEDGMENTS

The authors gratefully acknowledge the following reviewers, who provided insightful suggestions for this edition: Sudqi H. Alayyan, Texas Tech University; Chris B. Effiong, University of Tennessee at Martin; Gerald Hayler, California State Polytechnic University, Pomona; and Peter Tampas, Michigan Technological University.

The authors also thank the following reviewers of previous editions: Ernie Abbott, Napa Valley College; Luis Acevedo, DeVry University; M. A. Baset, DeVry University; Dr. Ray Bellem, Embry-Riddle Aeronautical University; James Brown, Peavey Electronics Corp.; Wayne Brown, Dekalb Technical Institute; Richard Cliver, Rochester Institute of Technology; Nolan Coleman, Embry-Riddle Aeronautical University; Alan H. Czarapata; Tom Dingman, Rochester Institute of Technology; Joe Ennesser, DeVry University; Leonard Geis, DeVry University; Tom Grady, Western Washington University; Thurman Grass, Lima Technical College; Roger Hack, Indiana University—Purdue University Extension; Mohammed Hague, Kansas City Kansas Community College; Dave Krispinsky, Rochester Institute of Technology; David A. O'Brien; Joe O'Connell, DeVry University; Edward Peterson, Arizona State University; Roger Sash, University of Nebraska at Omaha; Earl Schoenwetter, California Polytechnic University—Pomona; Francis M. Turner; Dr. Mel Turner, Oregon Institute of Technology; Mike Wilson, Kansas State University—Salina; Paul Wojnowiak, Southern College of Technology; and Behbood Zoghi, Texas A&M University.

Contents

CHAPTER 1

INTRODUCTION

1–1 THE STUDY OF ELECTRONICS

The field of study we call *electronics* encompasses a broad range of specialty areas, including audio systems, digital computers, communications systems, instrumentation, and automatic controls. Each of these areas in turn has its own specialty areas. However, all electronic specialties are the same in one respect: They all utilize electronic *devices*—transistors, diodes, integrated circuits, and various special components. The electrical characteristics of these devices make it possible to construct circuits that perform useful functions in many different kinds of applications. For example, a transistor may be used to construct an audio amplifier in a high-fidelity system, a high-frequency amplifier in a television receiver, an electronic switch used in a computer, or a current controller in a dc power supply. Regardless of one's specialization, a knowledge of device theory is a vital prerequisite to understanding and applying developments in that area.

The study of electronic devices is now almost wholly synonymous with the study of *semiconductor* devices. Semiconductor material is widely abundant yet unique in terms of its electrical properties because it is neither a conductor nor an insulator. *Silicon,* an element found in ordinary sand, is now the most widely used semiconductor material. As we shall discover in a subsequent chapter, it must be subjected to many closely controlled manufacturing processes before it acquires the properties that make it useful in the fabrication of electronic devices.

If any one trend characterizes modern electronics, it is the quest for miniaturization. Each year we learn of new technological advances that result in ever more sophisticated circuitry being packaged in ever smaller integrated circuits. The primary reasons for this trend are improved reliability, reduced costs, and, in the case of high-frequency and digital computer circuitry, increased speed due to the shortening of interconnecting paths.

The effort to miniaturize has been so successful, and the complexity of the circuitry contained in a single package is now so great, that some students and educators despair, or question the validity, of studying the fundamental entities of which these devices are composed. There is a temptation to abandon the study of semiconductor structure and the properties of individual transistors, diodes, and similar building blocks of integrated circuits. However, what is now the "microscopic" nature of electronic circuits should be studied for numerous compelling reasons, not the least of which is the many career opportunities in the design and fabrication of the devices themselves. Furthermore, the intelligent application of integrated circuits depends on a knowledge of certain underlying limitations and special characteristics, as well as insights that can be gained only through a study of their structure. Finally, there are many applications in which ultraminiature electronic devices cannot and never will be used—applications involving heavy power dissipation. In these applications, only fundamental, building-block–type components, such as *discrete* transistors, can be used, and an understanding of their behavior is paramount.

1–2 BRIEF HISTORY OF ELECTRONICS

Although many electrical devices were in full use in the nineteenth century, it was not until the late 1800s when discoveries and contributions by Thomas Edison, Heinrich Hertz, Guglielmo Marconi, and Aleksandr Popov developed into what is generally accepted as the beginning of the electronics era: the radio. Marconi is considered the father of the radio after having been the first to send a radio transmission across the Atlantic. By the time of the Titanic tragedy, radiotelegraphy was the most common means of long-distance communication. By World War I, radio was used all over the world.

The first electronic device is considered to be Fleming's valve, a diode vacuum tube invented by J. A. Fleming that was based on Edison's early work in transmitting electron currents through a vacuum. Later on, Lee De Forest developed the triode, a three-element vacuum tube capable of voltage amplification. Edwin Armstrong's contributions to radio communication included the vacuum-tube oscillator and the superheterodine receiver, whose principle is used even today.

The 1920s saw a huge boom in commercial radio communication. Even during the Great Depression in the 1930s, people all around the world enjoyed many forms of radio entertainment: music, shows, news. Back then, radio speakers were not of the permanent-magnet–type. The voice coil was surrounded by an electromagnet that doubled as an inductor for the filter in the power supply. Today's baby boomers still remember those floor-standing radios in their grandparents' homes that sounded much better than the more modern table radios that flooded the market years later. After the big success of radio transmission of voice and music, the natural thing was to move towards transmitting images as well. In the 1930s, RCA developed television and officially started broadcasting in 1939 with President Franklin Roosevelt in front of the camera. The arrival of World War II, however, put TV broadcasting on hold until the end of the war. The 1950s saw the launch of the television era.

The vacuum tubes of the 1940s were much smaller than those of the 1920s, consumed less power, and were more efficient. But they still needed filaments for heating their cathodes to liberate electrons. In time, the first electronic devices not requiring heat would be introduced: the selenium diodes. But by 1947, based on the research of Shockley, Bardeen, and Brattain of Bell Telephone Laboratories, the first transistor was developed, and the "solid-state era" began. Transistor radios, however, did not flourish until the late 1950s, perhaps due to the existing huge stocks of vacuum tubes. Television sets would not have transistors until well into the 1960s.

Miniaturization of electronic circuits followed the development of the transistor. Jack Kilby of Texas Instruments developed the first integrated circuit in 1959. Integrated circuits contain many semiconductor devices, such as transistors and diodes, in a very small area. In 1961, Robert Noyce of the Fairchild Corporation was the developer of the first commercial integrated circuit: an operational amplifier. Today, almost every piece of electronic equipment is based on integrated circuits that can be manufactured with a high degree of accuracy and reliability at a relatively low cost. The most palpable example of this is the high performance and low cost of today's personal computers.

1–3 THE USE OF COMPUTERS

The study and understanding of electronics demands certain mathematical skills, because circuit behavior can be described in practical terms only by equations. Quantitative (numerical) results obtained from solving equations are the principal means we have for comparing, predicting, designing, and evaluating electronic circuits. In this book, the reader is given the opportunity to use computers as an aid in obtaining those results.

There are basically two ways that computers are used in the study of electronics. First, we can choose to write our own programs in one of the standard computer languages, such as BASIC or FORTRAN. Each program we write will be designed to solve a specific circuit problem. For example, we might wish to determine the magnitude of the output voltage in a certain single-transistor amplifier. Our program will be designed to produce that result, and only that result, for that circuit, and only that circuit. (However, it is usually very easy to change the *values* of the components in the circuit and to reuse the program with those changes.) When a computer is used this way, programmers must have a good knowledge of electronic circuit theory, because they must be able to select and rearrange the theoretical equations that apply to the circuit. At the same time, they must be skilled programmers, having a good knowledge of the computer language and the ability to use it to solve various kinds of equations.

On the other hand, in the second way of using computers, we rely on someone else's programming skills by acquiring a program specially designed to solve electronic circuit problems. A popular example of such a program is SPICE (Simulation Program with Integrated Circuit Emphasis), developed at the University of California, Berkeley. To use this kind of program, we need only specify the components in the circuit we wish to study, describe how they are interconnected, and "tell" the program what kind of results we want (output voltage, etc.). Very little programming skill is required because we simply supply *data* to a program that is already in computer memory. Also, very little knowledge of electronics theory is required: We need just enough to be able to describe the components in the circuit and how they are connected.

SPICE and PSpice

In this book, discussion of electronic circuit theory is accompanied by numerous examples and exercises using SPICE to solve circuit problems. For reference and/or review purposes, Appendix A contains a summary of SPICE programming techniques. It also contains material on the use of PSpice, a popular microcomputer version of the original (Berkeley) SPICE. With a few minor exceptions, any circuit simulation written for SPICE can be executed successfully by PSpice, although the reverse is not true. All SPICE programs in examples and exercises in this book have been checked and found to run successfully in PSpice and SPICE.

1–4 CIRCUIT ANALYSIS AND CIRCUIT DESIGN

In the context of electronic circuits, *analysis* generally means finding voltages, currents, and/or powers of given devices and the component values in a circuit. Circuit *design,* or *synthesis,* turns that process around by finding component values and selecting devices so that certain voltages, currents, and/or powers are developed at specific points in a circuit. As a simple example, we can analyze a voltage divider to determine the voltage it develops, given the voltage across it and the values of its resistors. A typical design problem is to select resistor values so that a specified voltage is developed.

In circuit analysis, we usually derive equations for voltage, current, or power in terms of component values. Thus, circuit design is often performed by solving such equations for component values in terms of voltage, current, or power. However, there are no hard-and-fast rules for teaching design. It is generally agreed that a thorough understanding of analysis techniques is a vital prerequisite to developing design skills. Accordingly, the principal thrust of this book is electronic circuit analysis. Nevertheless, because of the practical importance of design, numerous examples are given to show how analysis techniques can be turned around and used to create circuits having specific properties. In some cases, circuit design can be accomplished only by trial-and-error repetitions of the analysis procedure. Computers are very useful in that type of design activity, as we shall see in forthcoming examples.

CHAPTER 2

THE *pn* JUNCTION

OUTLINE

OBJECTIVES

- Develop an understanding of the *pn* junction.
- Perform basic diode circuit analysis.
- Use the basic diode equation to determine the current through a diode.
- Investigate temperature effects on a *pn* junction.
- Develop a strong understanding of the forward- and reverse-bias operating modes of a diode.
- Recognize that a *pn* junction has a small parasitic capacitance.

2–1 INTRODUCTION

Semiconductor devices are the fundamental building blocks from which all types of useful electronic products are constructed—amplifiers, high-frequency communications equipment, power supplies, computers, and control systems, to name only a few. It is possible to learn how semiconductor devices can be connected to create useful products with little or no knowledge of the physics and theory of the device. However, the person who truly understands integrated circuit behavior has a greater knowledge of the capabilities and limitations of the device and is therefore able to use it in more innovative and efficient ways than the person who only has a functional understanding. Furthermore, it is often the case that the success or failure of a complex electronic system can be traced to a certain peculiarity or operating characteristic of a single device, and an intimate knowledge of how and why that device behaves the way it does is the key to reliable designs or to practical remedies for substandard performance.

This chapter introduces the reader to the *pn* junction. The *pn* junction is a fundamental building block that exists naturally in all integrated circuit structures. The diode, the result of a *pn* junction, is a very useful semiconductor device because it can control the direction of current flow—called *rectification*—in an electronic circuit. For practical purposes, current flows in the diode when it is on (forward bias), and no current flows when the diode is off (reverse bias). This chapter examines the basic operational characteristics of the *pn* junction and provides an introduction to basic diode concepts such as the diode current equation, current curve, threshold voltage, forward- and reverse-bias current behavior, junction capacitance, device types, ratings, and specifications. Some *pn* junctions, created by adjacent *p* and *n* regions, are unwanted parasitics that can affect the circuit's performance. This is addressed in greater detail in Chapters 17 and 18. Chapter 3 will introduce the reader to more advanced applications and analysis of diode circuits and devices.

Readers who wish to study the fundamental theory underlying the flow of charged particles through semiconductor material are encouraged to refer to

Appendix D, "Semiconductor Theory." The theory presented there includes numerous equations that allow us to compute and assign quantitative values to important atomic-level properties of semiconductors.

2–2 CREATING THE *pn* JUNCTION

Silicon (Si), the primary material currently used for manufacturing semiconductor devices, is a Group IVB element, which means that it contains four valence electrons. (The IVB identifies the location on the periodic table.) Germanium (Ge), used in special cases, is also a Group IVB element containing four valence electrons. These electrons, located in the outer shell, are loosely attached to the nucleus and can be easily altered through a process known as *doping,* which consists of adding impurities to intrinsic (pure) silicon to alter its electrical characteristics. These impurities are known as *dopant atoms.* The *p* or *n* material is created through doping processes such as ion implantation and diffusion. The *p*-type (positive) silicon material is created by doping intrinsic (pure) silicon (Si) with a dopant from a Group IIIB element such as boron (B). (The III indicates that the material contains three valence electrons.) The *n*-type (negative) silicon material is created by doping intrinsic silicon with a dopant from a Group VB element such as phosphorous (P). (The V indicates that the material contains five valence electrons.) A portion of the periodic table is shown in Figure 2–1.

Silicon, in its intrinsic or pure state, contains an equal concentration of protons (positive charges) and electrons (negative charges). (See Appendix D for a detailed examination of semiconductor theory.) The doping process alters this concentration and creates *extrinsic* or unpure material. When a covalent bond in semiconductor material is ruptured as in the doping process, a "hole" is left by the loss of an electron.

A dopant from Group IIIB is used to create the *p* regions. Doping the silicon with a Group IIIB element results in material with a greater concentration of *holes.* The increase in the concentration of holes yields a decrease in the *electron* concentration. The substitution of Group III for Group IV material results in material with one less electron in the outer shell. Hence, the material will be more positive. This *p*-type material is called *acceptor* material because of its ability to accept electrons.

A dopant from Group V is used to create *n* regions. Doping the silicon with a Group VB element results in material with a greater concentration of electrons and a decreased concentration of holes. A Group V substitution for Group IV results in material with one additional electron in the outer shell.

FIGURE 2–1 The portion of the periodic table containing the semiconductor elements

IIIB	IVB	VB
B Boron	**C** Carbon	**N** Nitrogen
Al Aluminum	**Si** Silicon	**P** Phosphorous
Ga Gallium	**Ge** Germanium	**As** Arsenic

Hence, the material will be more negative. The *n*-type material is called *donor* material because of its ability to donate electrons.

A *pn* silicon bar (a diode) and the schematic symbol for a diode are shown in Figure 2–2(a). The doped silicon bar is shown with adjoining *p* doped (*p*) and *n*-doped (*n*) regions. The *p* region is called the *anode* of the diode, and the *n* region is called the *cathode*. In Figure 2–2(b), the positive terminal of a battery is attached to the *p* region (anode); the negative lead of the battery is connected to the *n* region (cathode). This results in the diode being connected in the forward-biased mode and current will flow. Recall that the *p* region is full of holes (positive charges—majority carriers for the *p* material); therefore, the holes will be repelled by the positive terminal and pushed toward the *pn* junction. The *n* region is connected to the negative lead; therefore, the electrons (negative charges—majority carriers for the *n* material) will be repelled by the negative battery terminal and pushed toward the *pn* junction. If sufficient voltage is applied to the diode, the electrons and holes will jump across the junction in opposite directions.

In Figure 2–2(c), the *pn* junction is connected in the reverse-biased mode. In this case, the positive lead of the battery is connected to the *n* region, while the negative lead of the battery is connected to the *p* region. This pulls the majority carriers in the *p* (holes) and *n* (electrons) regions away from the

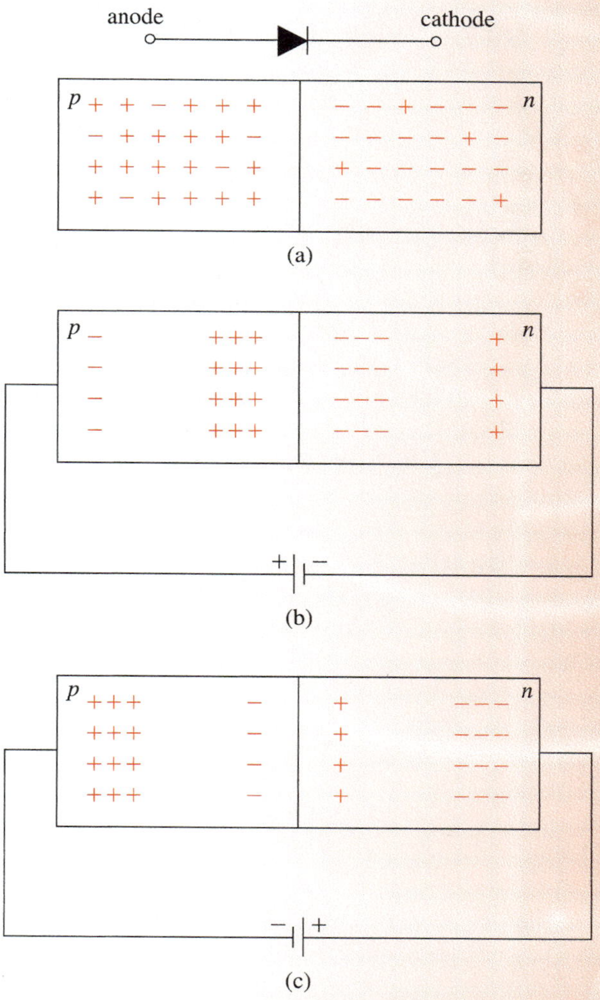

FIGURE 2–2 (a) The schematic symbol for a diode and the *pn* junction, (b) forward-biased *pn* junction, and (c) reverse-biased *pn* junction

junction leaving only the minority carriers at the junction, and essentially no current will flow. Recall that intrinsic silicon has equal concentrations of electrons and holes. When a region of the silicon is doped, this alters the *n* and *p* concentrations. The *p* region will have a heavy concentration of holes and a light concentration of electrons, while the *n* region will have a heavy concentration of electrons and a light concentration of holes.

2–3 BASIC DIODE OPERATION

A *discrete* diode is a single *pn* junction that has been fitted with an appropriate case (enclosure) and to which externally accessible leads have been attached. The leads allow us to make electrical connections to the anode (*p*) and cathode (*n*) sections of the diode. Semiconductor diodes are also formed inside *integrated circuits* as part(s) of larger, more complex networks and may or may not have leads that are accessible outside the package. (Integrated circuits have a large number of semiconductor components embedded and interconnected in a single wafer.)

In this section, we will investigate current and voltage relationships in circuits that contain diodes. We will learn how to analyze a circuit containing a diode as well as the behavior of the diode as a circuit element in considerable detail. A thorough analysis of this, the most fundamental building block in semiconductor electronics, is motivated by the following important facts:

1. The diode is an extremely useful device in its own right. It finds wide application in practical electronic circuits.
2. Many electronic devices that we will study later, such as transistors, contain *pn* junctions that behave like diodes. A solid understanding of diodes will help us understand and analyze these more complex devices.
3. A number of standard techniques used in the analysis of electronic circuits of all kinds will be introduced in the context of diode circuit analysis.

For a silicon diode, depending on small manufacturing variations and on the actual current flowing in it, the voltage drop is around 0.6 to 0.7 V. In practice, it is usually assumed to be 0.7 V. For *germanium* diodes, the drop is assumed to be 0.3 V. Therefore, for analysis purposes, we can replace the diode in a circuit by either a 0.7-V or a 0.3-V voltage source whenever the diode has a significant forward-biased current. Of course, the diode does not store energy and cannot produce current like a true voltage source, but the voltages and currents in the rest of a circuit containing the forward-biased diode are exactly the same as they would be if the diode were replaced by a voltage source. (The substitution theorem in network theory justifies this result.) Figure 2–3 illustrates these ideas.

In Figure 2–3(a), we assume that a forward-biased silicon diode has sufficient current to bias it and that it therefore has a voltage drop of 0.7 V. Using Ohm's law and Kirchhoff's voltage law, we can obtain

$$I = \frac{E - 0.7}{R} \tag{2–1}$$

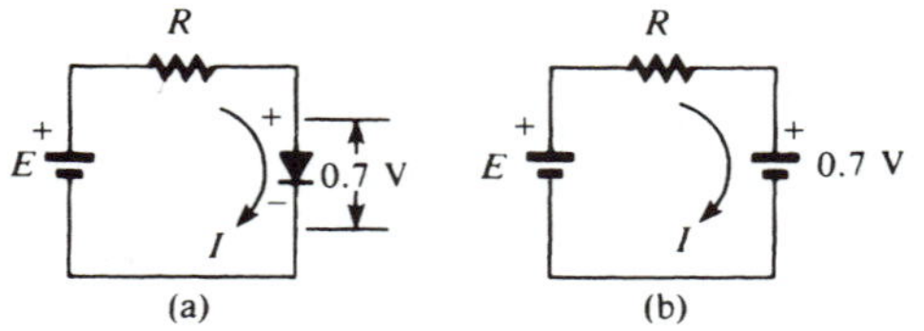

FIGURE 2–3 For analysis purposes, the forward-biased diode in (a) can be replaced by a voltage source, as in (b)

Figure 2–3(b) shows the equivalent circuit with the diode replaced by a 0.7-V source.

Our assumption that the voltage drop across the forward-biased diode is constant is usually accompanied by the additional assumption that the current through the diode is zero for all lesser voltages.

The idealized characteristic of the diode implies that it is an *open* circuit (infinite resistance, zero current) for all voltages less than 0.3 V or 0.7 V and becomes a short circuit (zero resistance) when one of those voltage values is reached. These approximations are quite valid in most real situations. Examples 2–1 to 2–3 examine basic diode circuit analysis.

EXAMPLE 2–1

Assume that a silicon diode is used in the circuit shown in Figure 2–4. Answer the following questions:

1. Is the diode forward or reverse biased?
2. What is the voltage drop across diode D_1?
3. Calculate the value of the current I in the circuit.

Solution

1. The anode of the diode is connected to the positive voltage terminal of the battery through the resistor R_1. The cathode of the diode is directly connected to the negative terminal of the battery. Therefore, the diode is forward biased.
2. The voltage drop across a forward-biased silicon (Si) diode is assumed to be 0.7 V.
3. Write the KVL for the circuit and solve for I.

$$12 - (1\text{k})I - 0.7 = 0$$
$$I = 11.3/1\text{k} = 11.3 \text{ mA}$$

EXAMPLE 2–2

Assume that a silicon diode is used in the circuit shown in Figure 2–5. Answer the following questions:

1. Is the diode forward or reverse biased?
2. What is the voltage drop across diode D_1?
3. Determine the value of the current I in the circuit.
4. How would the circuit analysis change if D_1 was changed to a germanium (Ge) diode?

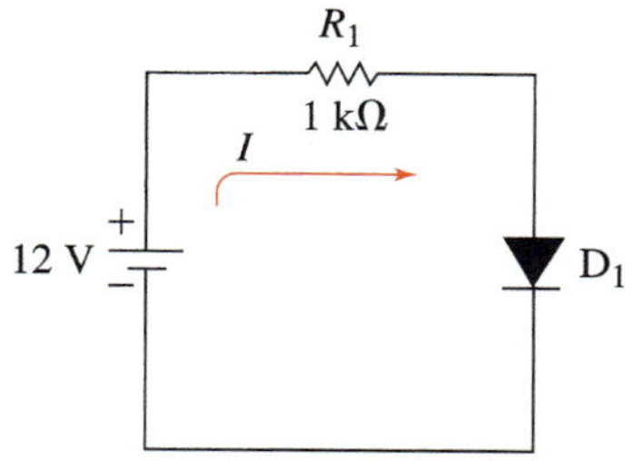

FIGURE 2–4 The circuit for Example 2–1

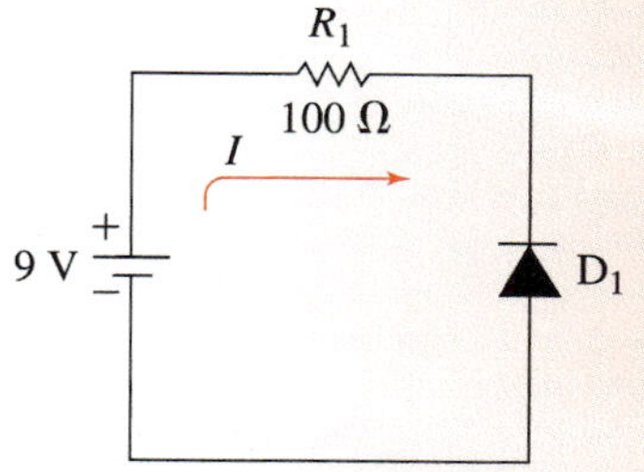

FIGURE 2–5 The circuit for Example 2–2

Solution

1. The anode of the diode is directly connected to the negative voltage terminal of the battery. The cathode of the diode is connected to the positive terminal of the battery through resistor R_1. Therefore, the diode is reverse biased.
2. In the reverse-bias mode, the diode can be approximated as an open circuit. This means that no current is flowing in the circuit and the voltage drop across R_1 is zero. The voltage across the diode is therefore 9.0 V.
3. Based on the answer to part 2, the current I is zero.
4. A forward-biased Ge diode has a voltage drop of 0.3 V but diode D_1 is reverse biased. The diode is still open and the value of the current I is zero.

EXAMPLE 2–3

In this example, diodes D_1 and D_2 are connected in series, as shown in Figure 2–6. Assume that D_1 and D_2 are silicon (Si) diodes. Answer the following questions:

1. Are the diodes forward or reverse biased?
2. Determine the value of the current I in the circuit.
3. What is the value of the voltage measured from point A to point B, V_{AB}?

Solution

1. The anode of diode D_1 is connected to the positive terminal of the battery through resistor R_1. The cathode of diode D_1 is connected to the cathode of diode D_2 while the anode of D_2 is directly connected to the negative terminal of the battery. For a diode to be forward biased, the anode lead of the diode must be more positive than the cathode lead. It can be seen that D_2 could never be forward biased since its anode lead is connected to the most negative point in the circuit. Therefore, diodes D_1 and D_2 are reverse biased (no current will flow).
2. In the reverse-bias mode, the diode is approximated by an open circuit. This means that no current is flowing in the circuit.
3. Because no current is flowing in the circuit, the voltage drop across R_1 will be zero. This means that the voltage at points A and B equals the voltage across the battery terminals. Therefore, the measured voltage V_{AB} is 6.0 V.

2–4 THE DIODE CURRENT EQUATION

A common application of a *pn* junction is in the construction of a *diode*. The relationship between the voltage across a *pn* junction and the current through it is given by the so-called diode equation:

$$I_D = I_s(e^{V_D/\eta V_T} - 1) \qquad (2–2)$$

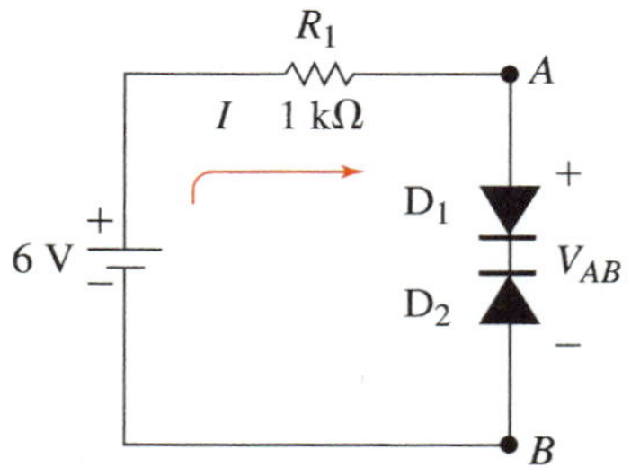

FIGURE 2–6 The circuit for Example 2–3

FIGURE 2–7 Current versus voltage in a typical forward-biased silicon junction. $I_s = 0.1$ pA

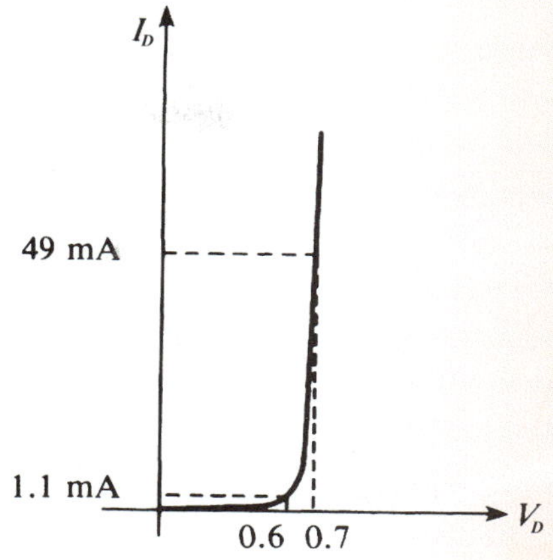

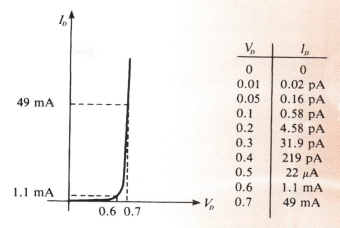

V_D	I_D
0	0
0.01	0.02 pA
0.05	0.16 pA
0.1	0.58 pA
0.2	4.58 pA
0.3	31.9 pA
0.4	219 pA
0.5	22 μA
0.6	1.1 mA
0.7	49 mA

where I_D = diode current, A
V_D = diode voltage, V (positive, for forward bias)
I_s = *saturation* current, A
η = *emission coefficient* (a function of V_D whose value depends also on the material; $1 \le \eta \le 2$)

$$V_T = \text{thermal voltage} = \frac{kT}{q} = \frac{[1.38 \times 10^{-23}][273 + T(°C)]}{1.6 \times 10^{-19}} = \frac{T}{11{,}600}$$

where k (Boltzmann constant) $= 1.38 \times 10^{-23}$ J/°K
T (Kelvin) $= 273 + T(°C)$; C (degrees centigrade)
q (charge of an electron) $= 1.6 \times 10^{-19}$ C; C (coulombs)

Equation 2–2 reveals some important facts about the nature of a forward-biased junction. The value of V_T at room temperature is about 0.026 V. The value of η for silicon is usually assumed to be 1 for $V_D \geq 0.5$ V and to approach 2 as V_D approaches 0. So when V_D is greater than $2V_T$, or about 0.05 V, the term $(e^{V_D/\eta V_T})$ begins to increase quite rapidly with increasing V_D. For $V_D > 0.2$, the exponential is much greater than 1. Consequently, equation 2–2 shows that the current I_D in a silicon *pn* junction increases dramatically once the forward-biasing voltage exceeds 200 mV or so. The *saturation current* I_s, in equation 2–2 is typically a very small quantity (we will have more to say about it when we discuss *reverse* biasing), but the fact that I_s is multiplied by the exponential $(e^{V_D/\eta V_T})$ means that I_D itself can become very large very quickly. For $V_D > 0.2$ V, $I_D \approx I_s e^{V_D/V_T}$ and $V_D \approx V_T \ln \frac{I_D}{I_s}$. Figure 2–7 shows a plot of I versus V for a typical forward-biased silicon junction. Note the rapid increase in the current I_D that is revealed by the plot and the accompanying table of values. For this figure we assumed that $I_s = 0.1$ pA $(= 10^{-13}$ A). Examples 2–4 and 2–5 examine more advanced diode circuit analysis.

Example 2–4 provides an example of using the diode equation (equation 2–2) to calculate the current through the diode when the exact value of the voltage drop across the diode is not known. This requires the use of an iterative process where the voltage across the diode must be guessed initially.

EXAMPLE 2–4

The silicon diode in the circuit shown in Figure 2–8 has a saturation current $I_s = 3 \times 10^{-14}$ A. Determine the current through and the voltage across the diode D_1. The temperature is 55°C. Assume $\eta = 1$.

Solution

Step 1. Diode D_1 is forward biased; therefore, assume that $V_D = 0.7$ V. The current I flowing through D_1 will be

$$I = (2 - 0.7)/3\text{ k} = 0.433\text{ mA}$$

Step 2. With this estimated current value of 0.433 mA, which is based on a guess of the voltage drop across D_1, we can calculate a more accurate diode voltage drop using the form

$$V_D \approx V_T \ln \frac{I_D}{I_s}$$

where $V_T = \dfrac{T}{11{,}600} = \dfrac{273 + 55}{11{,}600} = 28.3\text{ mV}$

$$V_D = V_T \ln(I_D / I_s) = (28.3\text{ mV})\ln(0.433\text{ mA}/3 \times 10^{-14}\text{A}) = 0.662\text{ V}$$

Step 3. Through an iterative process, a new current value for I can be calculated using the voltage value calculated in step 2.

$$I = (2 - 0.662)/3\text{ k} = 0.446\text{ mA}$$

Step 4. Calculate a new value for V_D.

$$V_D = (28.3\text{ mV})\ln(0.466\text{ mA}/3 \times 10^{-14}\text{ A}) = 0.662\text{ V}$$

This equals the value calculated in step 2, which means that the iterative process is complete and that the voltage converged to a solution very quickly.

Example 2–5 demonstrates a technique for calculating the actual voltage drop across a set of parallel diodes when each diode has a different value for I_s.

EXAMPLE 2–5

Diodes D_1 and D_2 in Figure 2–9 have saturation currents $I_{s1} = 2 \times 10^{-14}$A and $I_{s2} = 5 \times 10^{-14}$ A. Assuming room temperature, determine the current through and the voltage across each diode.

Solution

Because the diodes are connected in parallel, their voltages will be the same. Let us assume a diode voltage drop of 0.6 V since, by inspection, the current could be small. The current I will be $I = (3 - 0.6)/1\text{ k} = 2.4\text{ mA} = I_{D1} + I_{D2}$. In

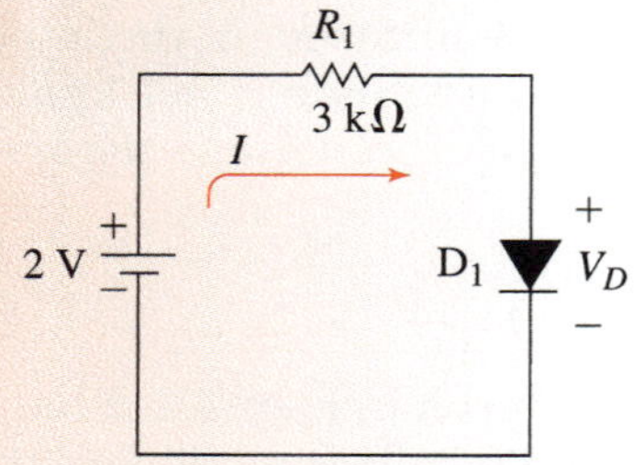

FIGURE 2–8 The circuit for Example 2–4

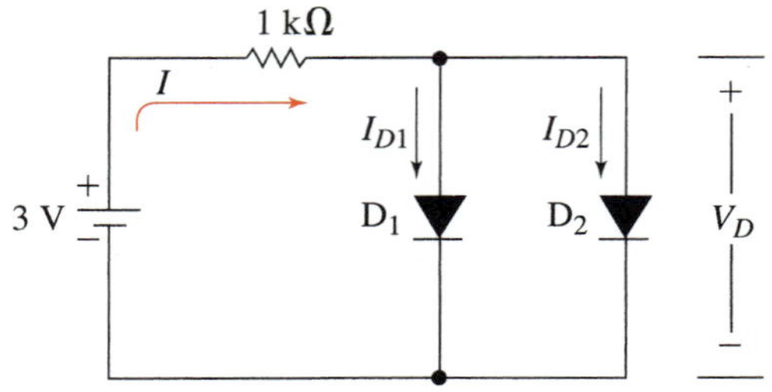

FIGURE 2–9 The circuit for Example 2–5

order to determine the current flowing through each device, equate their voltage, $V_T \ln(I_{D1}/I_{s1}) = V_T \ln(I_{D2}/I_{s2})$, and obtain $I_{D1} = I_{D2}(I_{s1}/I_{s2})$. But $I_{D1} =$ 2.4 mA $- I_{D2}$; substituting I_{D1} we can write

$$2.4 \text{ mA} - I_{D2} = I_{D2}\ (2/5)$$

or

$$I_{D2} = (2.4 \text{ mA})/1.4 = 1.714 \text{ mA}$$

and

$$I_{D1} = 2.4 - 1.714 = 0.686 \text{ mA}$$

With either of these two currents we can calculate a more approximated value for the voltage drop. Using I_{D2}, we obtain $V_D = 0.026 \ln(1.714 \text{ mA}/5 \times 10^{-14} \text{ A}) = 0.631$ V. The reader can verify that using I_{D1} yields the same result. With this new value for V_D, the current I can be recalculated as $I = (3 - 0.631)/1$ k $= 2.369$ mA. Iterating I_{D2} as well, we obtain $I_{D2} = 2.369 - 1.692 = 0.677$ mA. If we do one more iteration, V_D will turn out to be 0.630 V and $I = 2.370$ mA. Even if we had started with $V_D = 0.7$ V, we would have seen that the diode voltage converges with one or at most two iterations.

Suppose now that the connections between the *pn* junction and the external voltage source are reversed, so that the positive terminal of the source is connected to the *n* side of the junction and the negative terminal is connected to the *p* side. This connection *reverse biases* the junction and is illustrated in Figure 2–10.

The polarity of the bias voltage in this case *reinforces*, or strengthens, the internal barrier field at the junction. Consequently, diffusion current is inhibited to an even greater extent than it was with no bias applied. The increased field intensity must be supported by an increase in the number of ionized donor and acceptor atoms, so the depletion region widens under reverse bias.

The unbiased *pn* junction has a component of drift current (see Appendix D) consisting of minority carriers that cross the junction from the *p* to the *n* side. This reverse current is the direct result of the electric field across the depletion region. Because a reverse-biasing voltage increases the magnitude of that field, we can expect the reverse current to increase correspondingly. This is indeed the case. However, because the current is due to the flow of minority carriers only, its magnitude is very much smaller than the current that flows under forward bias (the forward current).

It is this distinction between the ways a *pn* junction reacts to bias voltage—very little current flow when it is reverse biased and substantial

FIGURE 2–10 A voltage source V connected to a reverse-bias *pn* junction. The depletion region (shown shaded) is widened

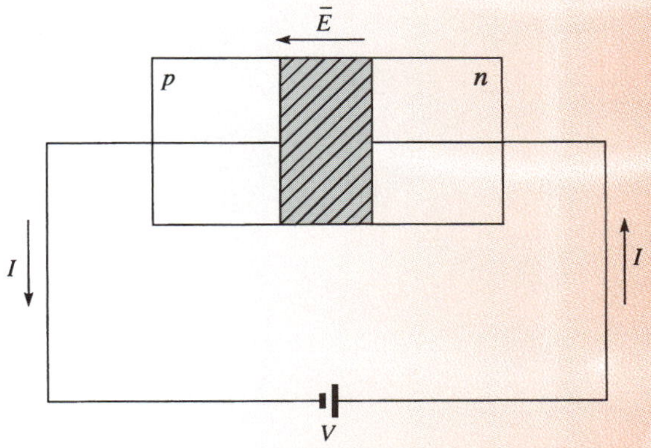

FIGURE 2–11 Diode symbols and bias circuits

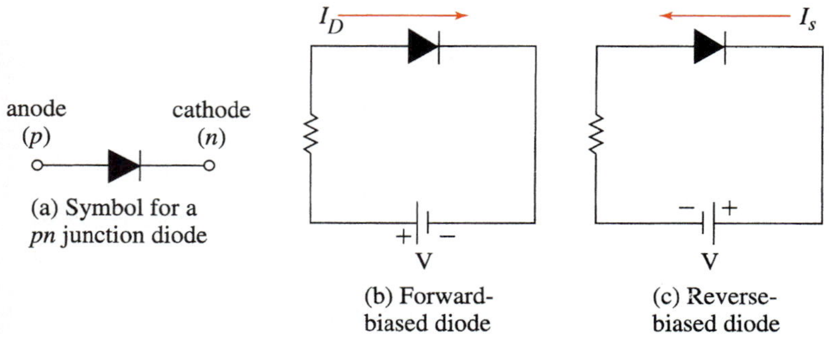

current flow when it is forward biased—that makes it a very useful device in many circuit applications. Figure 2–11(a) shows the diode connected to an external source for forward biasing, and 2–11(b) shows reverse biasing. Diode circuits will be studied in detail in Chapter 3.

Returning to our discussion of the reverse-biased junction, we should mention that it is conventional to regard reverse voltage and reverse current as *negative* quantities. When this convention is observed, equation 2–3 can be used to compute reverse current due to a reverse-biasing voltage:

$$I_D = I_s(e^{V_D/\eta V_T} - 1) \tag{2–3}$$

To illustrate, suppose $\eta V_T = 0.05$, $I_s = 0.01$ pA, and the reverse-biasing voltage is $V_D = -0.1$ V. Then

$$I_D = 10^{-14}\,(e^{-0.1/0.05} - 1) = -0.8647 \times 10^{-14}\text{ A}$$

From the standpoint of plotting the *I*-versus-*V* relationship in a *pn* junction, the sign convention makes further good sense. If forward current is treated as positive (upward), then reverse current should appear below the horizontal axis, i.e., downward, or negative. Similarly, forward voltage is plotted to the right of 0 and reverse voltage is plotted to the left of 0, i.e., in a negative direction. Figure 2–12 shows a plot of I_D versus V_D in which this convention is observed. Note that the current scale is exaggerated in the negative direction, because the magnitude of the reverse current is so very much smaller than that of the forward current.

When V_D is a few tenths of a volt negative in equation 2–3, the magnitude of the term $e^{V_D/\eta V_T}$ is negligible compared to 1. For example, if $V_D = -0.5$, then $e^{V_D/\eta V_T} \approx 4.5 \times 10^{-5}$. Of course, as V_D is made even more negative, the value of $e^{V_D/\eta V_T}$ becomes even smaller. As a consequence, when the junction is reverse biased beyond a few tenths of a volt,

FIGURE 2–12 Current–voltage relations in a *pn* junction under forward and reverse bias. The negative current scale in the reverse-biased region is exaggerated.

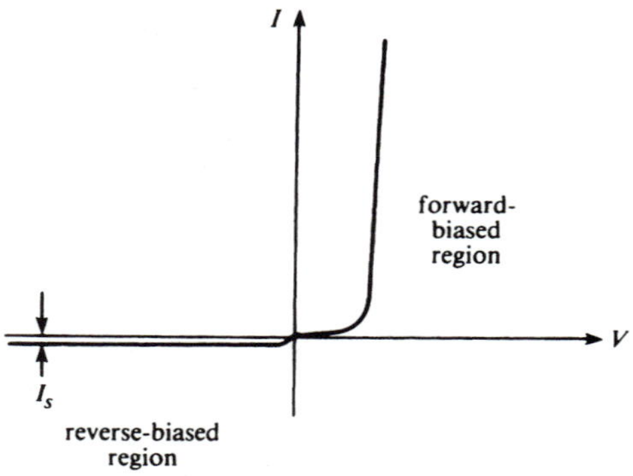

$$I_D \approx I_s\,(0 - 1) = -I_s \tag{2–4}$$

Equation 2–4 shows that the magnitude of the reverse current in the junction under these conditions is essentially equal to I_s, the saturation current. This result accounts for the name *saturation* current: The reverse current predicted by the equation never exceeds the magnitude of I_s.

Equation 2–3 is called the *ideal* diode equation. In real diodes, the reverse current can, in fact, exceed the magnitude of I_s. One reason for this deviation from theory is the existence of *leakage current*, current that flows along the surface of the diode and that obeys an Ohm's law relationship not accounted for in equation 2–3. In a typical silicon diode having $I_s = 10^{-14}$ A, the leakage current may be as great as 10^{-9} A, or 100,000 times the theoretical saturation value.

Breakdown

The reverse current also deviates from that predicted by the ideal diode equation if the reverse-biasing voltage is allowed to approach a certain value called the reverse *breakdown* voltage, V_{BR}. When the reverse voltage approaches this value, a substantial reverse current flows. Furthermore, a very small increase in the reverse-bias voltage in the vicinity of V_{BR} results in a very large increase in reverse current. In other words, the diode no longer exhibits its normal characteristic of maintaining a very small, essentially constant reverse current with increasing reverse voltage. Figure 2–13 shows how the current–voltage plot is modified to reflect breakdown. Note that the reverse current follows an essentially vertical line as the reverse voltage approaches V_{BR}. This part of the plot conveys the fact that large increases in reverse current result from very small increases in reverse voltage in the vicinity of V_{BR}.

In ordinary diodes, the breakdown phenomenon occurs because the high electric field in the depletion region imparts high kinetic energy (large velocities) to the carriers crossing the region, and when these carriers collide with other atoms they rupture covalent bonds. The large number of carriers that are freed in this way accounts for the increase in reverse current through the junction. The process is called *avalanching*. The magnitude of the reverse current that flows when V approaches V_{BR} can be predicted from the following experimentally determined relation:

$$I_D = \frac{I_s}{1 - \left(\dfrac{V_D}{V_{BR}}\right)^n} \tag{2–5}$$

where n is a constant determined by experiment and has a value between 2 and 6.

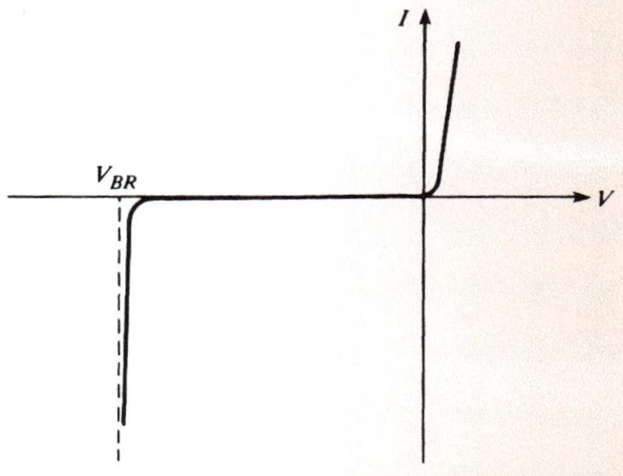

FIGURE 2–13 A plot of the *I*–*V* relation for a diode, showing the sudden increase in reverse current near the reverse breakdown voltage

Certain special kinds of diodes, called *zener* diodes, are designed for use in the breakdown region. The essentially vertical characteristic in the breakdown region means that the voltage across the diode remains constant in that region, independent of the (reverse) current that flows through it. This property is useful in many applications where the zener diode serves as a *voltage reference*, similar to an ideal voltage source. Zener diodes are more heavily doped than ordinary diodes, and they have narrower depletion regions and smaller breakdown voltages. The breakdown mechanism in zener diodes having breakdown voltages less than about 5 V differs from the avalanching process described earlier. In these cases, the very high electric field intensity across the narrow depletion region directly forces carriers out of their bonds, i.e., strips them loose. Breakdown occurs by avalanching in zener diodes having breakdown voltages greater than about 8 V, and it occurs by a combination of the two mechanisms when the breakdown voltage is between 5 V and 8 V. The characteristics and special properties of zener diodes are discussed in detail in Chapter 3.

Despite the name *breakdown*, nothing about the phenomenon is *inherently* damaging to a diode. On the other hand, a diode, like any other electronic device, is susceptible to damage caused by overheating. Unless there is sufficient current-limiting resistance connected in series with a diode, the large reverse current that would result if the reverse voltage were allowed to approach breakdown could cause excessive heating. Remember that the power dissipation of any device is

$$P = VI \text{ watts} \tag{2–6}$$

where V is the voltage across the device and I is the current through it. At the onset of breakdown, both V (a value near V_{BR}) and I (the reverse current) are likely to be large, so the power computed by equation 2–6 may well exceed the device's ability to dissipate heat. The value of the breakdown voltage depends on doping and other physical characteristics that are controlled in manufacturing. Depending on these factors, ordinary diodes may have breakdown voltages ranging from 10 or 20 V to hundreds of volts.

Temperature Effects

The ideal diode equation shows that both forward- and reverse-current magnitudes depend on temperature, through the thermal voltage term V_T (see equation 2–2). It is also true that the saturation current, I_s in equation 2–2, depends on temperature. In fact, the value of I_s is more sensitive to temperature variations than is V_T, so it can have a pronounced effect on the temperature dependence of diode current. A commonly used rule of thumb is that I_s doubles for every 10°C rise in temperature. The following example illustrates the effect of a rather wide temperature variation on the current in a typical diode.

EXAMPLE 2–6

A silicon diode has a saturation current of 0.1 pA at 20°C. Find its current when it is forward biased by 0.55 V. Find the current in the same diode when the temperature rises to 100°C.

Solution

At $T = 20°C$,

$$V_T = \frac{kT}{q} = \frac{1.38 \times 10^{-23}(273 + 20)}{1.6 \times 10^{-19}} = 0.02527 \text{ V}$$

From equation 2–3, assuming that $\eta = 1$,

$$I = I_s\,(e^{V_D/\eta V_T} - 1) = 10^{-13}\,(e^{0.55/0.02527} - 1) = 0.283 \text{ mA}$$

At $T = 100°\text{C}$,

$$V_T = \frac{1.38 \times 10^{-23}(273 + 100)}{1.6 \times 10^{-19}} = 0.03217 \text{ V}$$

In going from 20°C to 100°C, the temperature increases in 8 increments of 10°C each: $100 - 20 = 80$; $80/10 = 8$. Therefore, I_s doubles 8 times, i.e., increases by a factor of $2^8 = 256$. So, at 100°C, $I_s = 256 \times 10^{-13}$ A.

$$I = 256 \times 10^{-13}(e^{0.55/0.03217} - 1) = 0.681 \text{ mA}$$

In this example, we see that the forward current increases by a factor of 2.4 or 140% over the temperature range from 20°C to 100°C.

Example 2–6 illustrates that forward current in a diode increases with temperature when the forward voltage is held constant. This result is evident when the *I–V* characteristic of a diode is plotted at two different temperatures, as shown in Figure 2–14. At voltage V_1 in the figure, the current can be seen to increase from I_1 to I_2 as the temperature changes from 20°C to 100°C. (Follow the vertical line drawn upward from V_1—the line of constant voltage V_1.) Note that the effect of increasing temperature is to shift the *I–V* plot toward the left. Note also that when the *current* is held constant, the voltage *decreases* with increasing temperature. At the constant current I_2 in the figure, the voltage can be seen to decrease from V_2 to V_1 as temperature increases from 20°C to 100°C. (Follow the horizontal line drawn through I_2—the line of constant current I_2.) As a rule of thumb, the forward voltage decreases 2.5 mV for each 1°C rise in temperature when the current is held constant.

Of course, temperature also affects the value of reverse current in a diode, because the ideal diode equation (and its temperature-sensitive factors) applies to the reverse- as well as forward-biased condition. In many practical applications, the increase in reverse current due to increasing temperature is a more severe limitation on the usefulness of a diode than is the increase in forward current. This is particularly the case for germanium diodes, which have values of I_s that are typically much larger than those of silicon. In a germanium diode, the value of I_s may be as great as or greater than the reverse leakage current across the surface. Because I_s doubles for every 10°C rise in temperature, the total reverse current through a germanium junction can become quite large with a relatively small increase in temperature. For this reason, germanium devices are not so widely used as their silicon counterparts. Also, germanium devices can withstand temperatures up to only about 100°C, while silicon devices can be used up to 200°C.

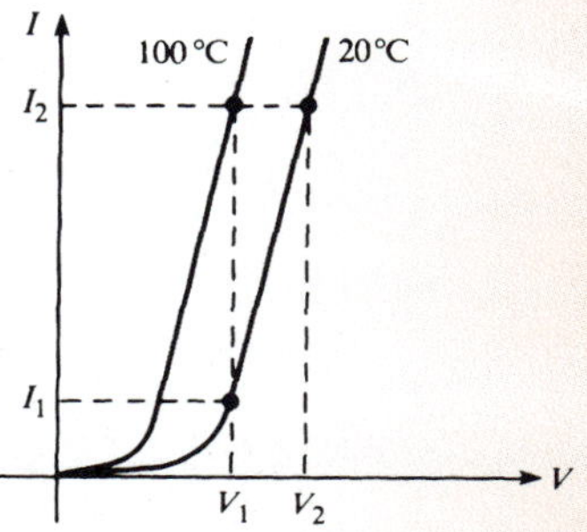

FIGURE 2–14 Increasing temperature causes the forward *I–V* characteristic to shift left

EXAMPLE 2–7

Use SPICE to obtain a plot of the diode current versus diode voltage for a forward-biasing voltage that ranges from 0.0 V to 0.7 V in 5-mV steps. The diode has a saturation current (IS) of 0.01 pA and an emission coefficient (N) of 1.0.

Solution

Figure 2–15(a) shows a diode circuit that can be used by SPICE to perform a .DC analysis of the circuit and the text listing the contents of the .CIR file used to describe the circuit. Although the .MODEL statement specifies the saturation current (IS) and emission coefficient (N), the values used are the same as the default values, so these values could have been omitted. Figure 2–15(b) shows the resulting plot. Note that the current varies from 0 mA to over 5.6 mA. Also note the dramatic increase in the diode current once the diode voltage drop exceeds 0.6 V.

FIGURE 2–15 (a) The circuit for the PSpice simulation used in Example 2–7, (b) SPICE simulation

```
* Forward Biased Diode Current
****************************************
* DEFINE THE CIRCUIT ELEMENTS AND NODES
****************************************
V1 1 0 DC 6.0V
R1 1 2 100
D1 2 0 DIODE
********************************************
* DEFINE THE MODEL PARAMETERS FOR THE DIODE
********************************************
.MODEL DIODE D IS = 1E-14 N = 1
***************************************************
* PERFORM A DC ANALYSIS ON THE CIRCUIT BY VARYING
* THE VOLTAGE OF V1 FROM ZERO VOLTS TO 0.7 VOLTS
* IN .005 VOLT INCREMENTS
***************************************************
.DC V1 0 1.5 .005
.PROBE
.END
```

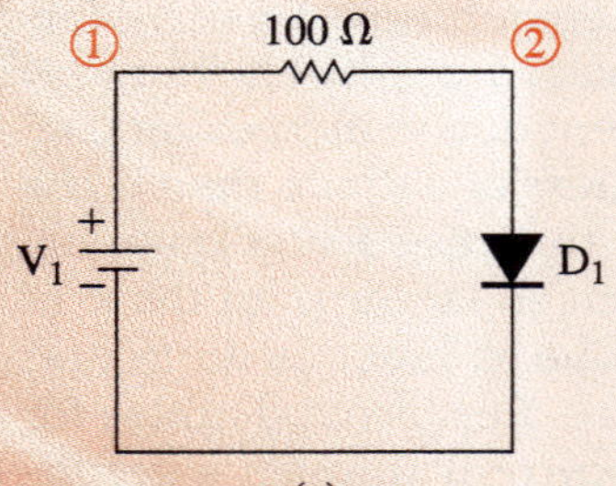

(a)

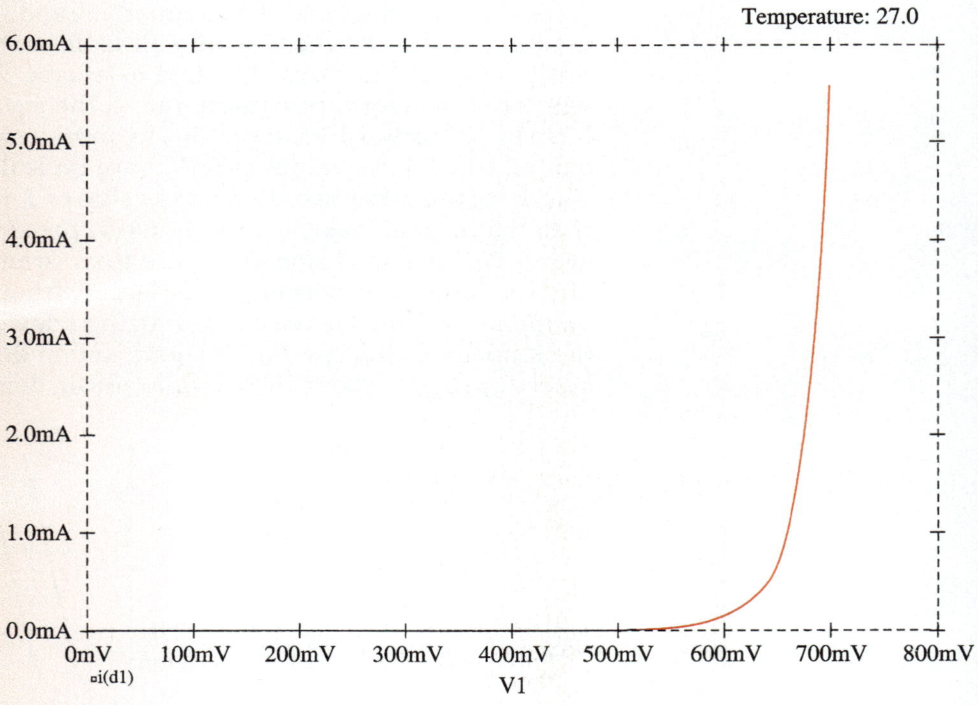

(b)

2–5 IDENTIFYING FORWARD- AND REVERSE-BIAS OPERATING MODES

In an important large-signal application of diodes, the devices are switched rapidly back and forth between their high-resistance and low-resistance states. In these applications, the circuit voltages are pulse-type waveforms, or square waves, that alternate between a "low" voltage, often 0 V, and a "high" voltage, such as +5 V. These essentially instantaneous changes in voltage between low and high cause the diode to switch between its "off" and "on" states. Figure 2–16 shows the voltage waveform that is developed across a resistor in series with a silicon diode when a square wave that alternates between 0 V and +5 V is applied to the combination. When $e(t) = +5$ V, the diode is forward biased, or ON, so current flows through the resistor and a voltage equal to $5 - 0.7 = 4.3$ V is developed across it. When $e(t) = 0$ V, the diode is in its high-resistance state, or OFF, and, because no current flows, the resistor voltage is zero. This operation is very much like rectifier action. However, we study digital logic circuits in just the extreme cases where the voltage is either low or high. In other words, we assume that every voltage in the circuit is at one of those two levels. Because the diode in effect performs the function of switching a high level into or out of a circuit, these applications are often called *switching circuits.*

Diode switching circuits typically contain two or more diodes, each of which is connected to an independent voltage source. Understanding the operation of a diode switching circuit depends first on determining which diodes, if any, are forward biased and which, if any, are reverse biased. The key to this determination is remembering that a diode is forward biased only if its anode is positive *with respect to* its cathode. The important words here (the ones that usually give students the most trouble) are "with respect to." Stated another way, the anode voltage (with respect to ground) must be more positive than the cathode voltage (with respect to ground) in order for a diode to be forward biased. This is, of course, the same as saying that the cathode voltage must be more negative than the anode voltage. Conversely, in order for a diode to be reverse biased, the anode must be negative with respect to the cathode, or, equivalently, the cathode positive with respect to the anode. The following example should help clarify these ideas.

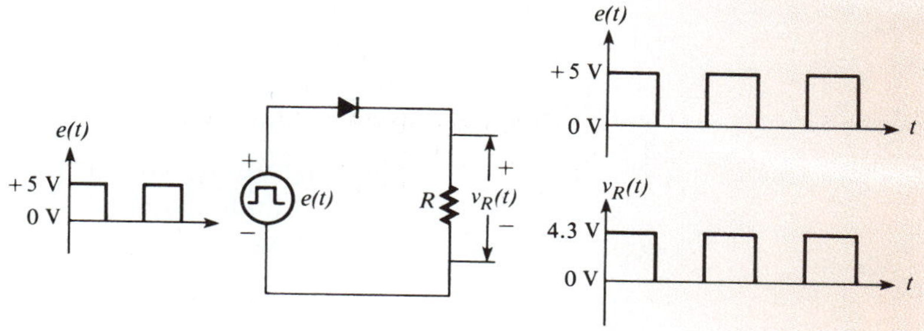

FIGURE 2–16 The diode is forward biased when the square-wave voltage is +5 V. Note that the (silicon) diode voltage is 0.7 V when it conducts.

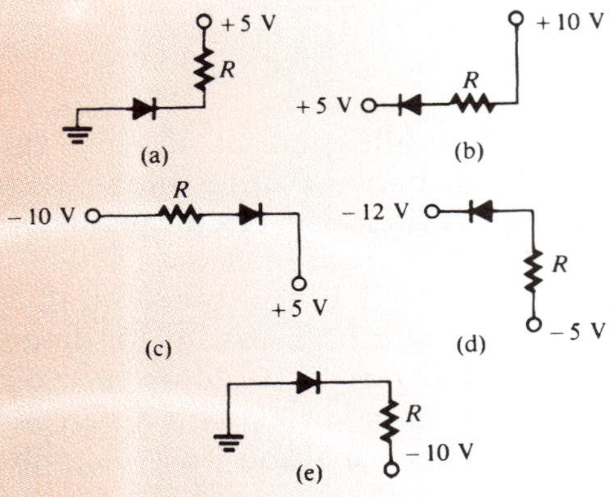

FIGURE 2–17 Forward- and reverse-biased diode configurations

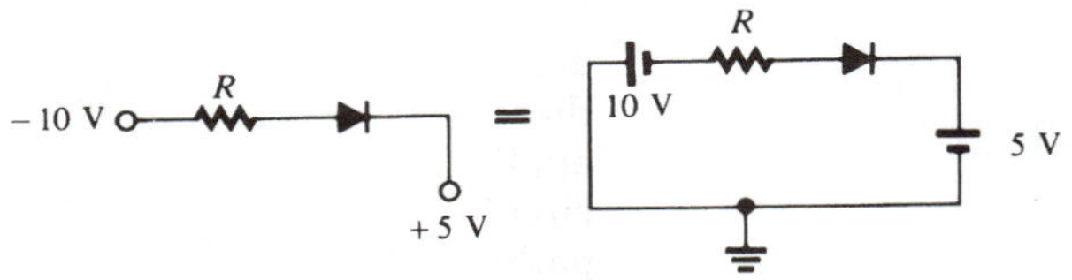

FIGURE 2–18 The circuit of Figure 2–17(c) is redrawn in an equivalent form that shows all ground connections

EXAMPLE 2–8

Determine which diodes are forward biased and which are reverse biased in each of the configurations shown in Figure 2–17. The schematic diagrams in each part of Figure 2–17 are drawn using the standard convention of omitting the connection line between one side of a voltage source and ground. In this convention, it is understood that the *opposite* side of each voltage source shown in the figure is connected to ground. If the reader is not comfortable with this convention, then he or she should begin the process now of becoming accustomed to it, for it is widely used in electronics. As an aid in understanding the explanations given below, redraw each circuit with all ground connections included. For example, Figure 2–18 shows the circuit that is equivalent to Figure 2–17(c).

Solution

1. In (a) the anode is grounded and is therefore at 0 V. The cathode side is positive by virtue of the +5-V source connected to it through resistor *R*. The cathode is therefore positive with respect to the anode; i.e., the anode is more negative than the cathode, so the diode is reverse biased.
2. In (b) the anode side is more positive than the cathode side ($+10\,\text{V} > +5\,\text{V}$), so the diode is forward biased. Current flows from the 10-V source, through the diode, and into the 5-V source.
3. In (c) the anode side is more negative than the cathode side, so the diode is reverse biased. Note that (essentially) no current flows in the circuit, so there is no drop across resistor *R*. Therefore, the total reverse-biasing voltage across the diode is 15 V. (See also Figure 2–18, and note that the sources are series-adding.)
4. In (d) the cathode side is more negative than the anode side ($-12\,\text{V} < -5$ V), so the diode is forward biased. Current flows from the −5-V source, through the diode, and into the −12-V source.
5. In (e) the anode is grounded and is therefore at 0 V. The cathode side is more negative than the anode side ($-10\,\text{V} < 0\,\text{V}$), so the diode is forward biased. Current flows from ground, through the diode, and into the − 10-V source.

Figure 2–19 shows a diode switching circuit. It consists of three diodes whose anodes are connected together and whose cathodes may be connected to independent voltage sources. The voltage levels connected to the cathodes are called *inputs* to the circuit, labeled *A*, *B*, and *C*, and the voltage developed

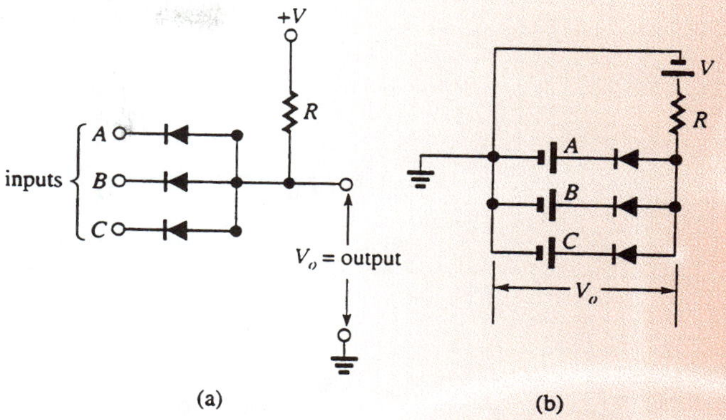

FIGURE 2–19 A typical diode switching circuit like those used in digital logic applications. The equivalent circuit, showing all ground paths, is shown in (b)

at the point where the anodes are joined is called the *output* of the circuit. All voltages are referenced to the circuit's common ground. The voltage source V is a fixed positive voltage called the *supply* voltage. The figure shows the conventional way of drawing this kind of circuit (a), and the complete equivalent circuit (b).

Let us assume that the inputs A, B, and C in Figure 2–19 can be either +5 V (high) or 0 V (low). Suppose further that the supply voltage is $V = +10$ V. If A, B, and C are all +5 V, then all three diodes are forward biased (+10 > +5) and are therefore conducting. Current flows from the 10-V source, through the resistor, and then divides through the three diodes. See Figure 2–20.

Typically, the dc resistance of the diode is not known and an approximate solution can be obtained by assuming a 0.7-V drop across each (silicon) diode. Under this assumption, the equivalent circuit appears as shown in Figure 2–21. Using this equivalent circuit, we can clearly see that $V_o = 5\text{ V} + 0.7\text{ V} = 5.7\text{ V}$.

Suppose now that input $A = 0$ V and $B = C = +5$ V, as shown in Figure 2–22(a). It is clear that the diode connected to input A is forward biased. If we temporarily regard the "ON" diode as a perfect closed switch, then we see that the anode side of all diodes will be connected through this closed switch to 0 V. Therefore, the other two diodes have +5 V on their cathodes and 0 V on their anodes, causing them to be reverse biased. In reality, the "ON" diode is not a perfect switch, so it has some small voltage drop across it and the

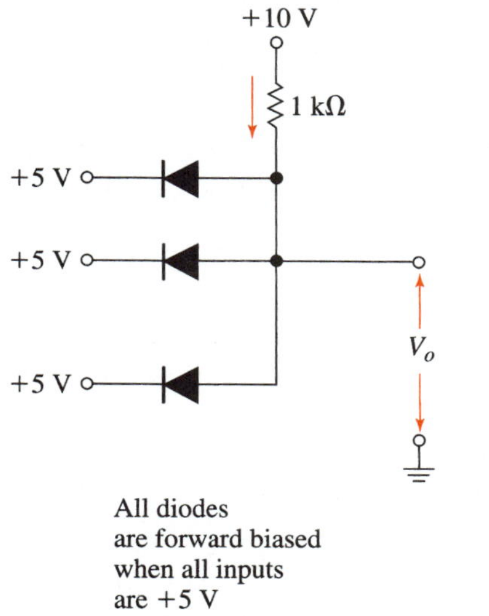

FIGURE 2–20 The diode circuit of Figure 2–19 when all inputs are +5 V

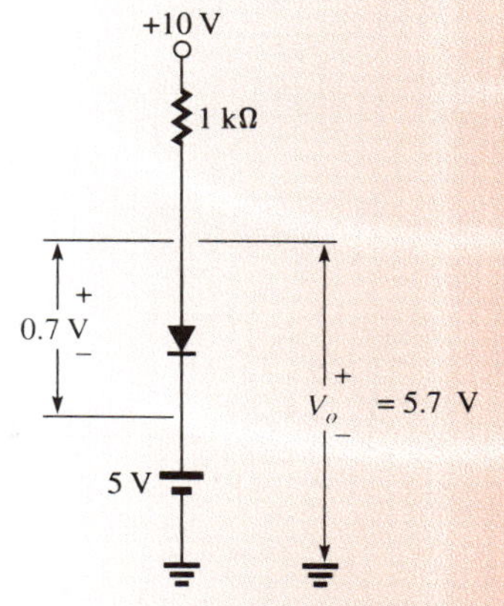

FIGURE 2–21 The circuit that is equivalent to Figure 2–20(a) when the diodes are assumed to have a fixed 0.7-V drop

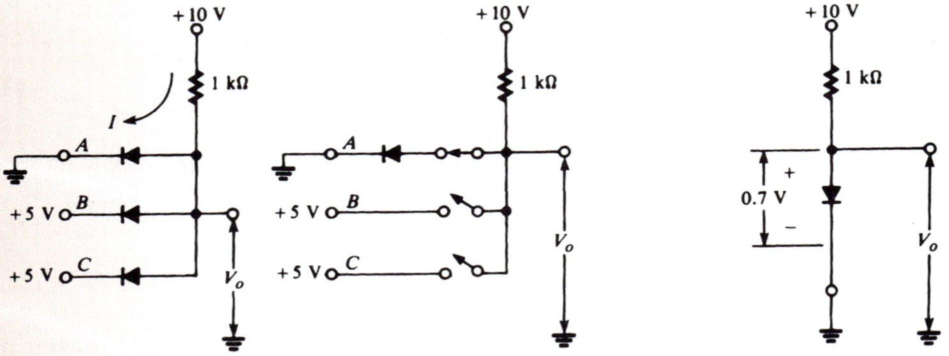

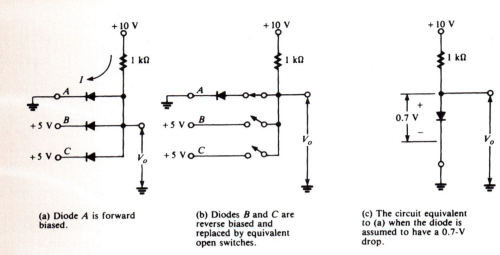

(a) Diode *A* is forward biased.

(b) Diodes *B* and *C* are reverse biased and replaced by equivalent open switches.

(c) The circuit equivalent to (a) when the diode is assumed to have a 0.7-V drop.

FIGURE 2–22 The circuit of Figure 2–19 when input *A* is 0 V and inputs *B* and *C* are +5 V

anodes are *near* 0 V rather than exactly 0 V. The net effect is the same: One diode is forward biased and the other two are reverse biased.

Figure 2-22(b) shows the equivalent circuit that results if we treat the reverse-biased diodes as open switches. Figure 2–22(c) shows the equivalent circuit that results when we assume that the diode voltage drop is 0.7 V. In this case, we see that $V_o = 0.7$ V.

If input *B* is at 0 V while $A = C = +5$ V, then it should be obvious that the output voltage V_o is exactly the same as in the previous case. Any combination of inputs that causes one diode to be forward biased and the other two to be reverse biased has the same equivalent circuits as shown in Figure 2–22.

When any two of the inputs are at 0 V and the third is at +5 V, then two diodes are forward biased and the third is reverse biased. It is left as an exercise to determine the output voltage in this case.

Finally, if all three inputs are at 0 V, then all three diodes are forward biased. The equivalent circuits are shown in Figure 2–23. If we regard the drop across each diode as 0.7 V, then $V_o = 0.7$ V, as shown in Figure 2–23(b). The diode circuit we have just analyzed is called a diode AND gate because the output is high if and only if inputs *A and B and C* are all high.

We have seen that the first step in this kind of analysis is to determine which diodes are forward biased and which are reverse biased. This determination is best accomplished by temporarily regarding each diode as a perfect, voltage-controlled switch. At this point, one might legitimately question how

FIGURE 2–23 The circuit of Figure 2–19 when all inputs are 0 V

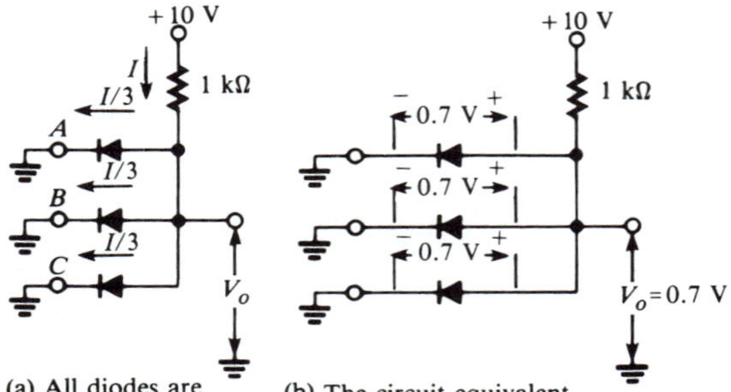

(a) All diodes are forward biased.

(b) The circuit equivalent to (a) when all diodes are assumed to have a 0.7-V drop.

we determined that the forward-biased diode shown in Figure 2–22 is the only one that is forward biased. After all, the other two diodes appear to have their anode sides more positive (10 V) than their cathode sides (5 V) and seem therefore to meet the criterion for forward bias. However, if D_1 is forward biased, its anode voltage will be +0.7 V forcing the other two diodes to be reverse biased, because their cathodes are more positive than their anodes.

A rule that is useful for determining which diode is truly forward biased is to determine which one has the greatest forward-biasing potential measured from the supply voltage to its input voltage. For example, in Figure 2–22, the net voltage between the supply and input *A* is 10 V − 0 V = 10 V, while the net voltage between the supply and inputs *B* and *C* is +10 V − 5 V = 5 V. Therefore, the first diode is forward biased and the other two diodes are reverse biased.

EXAMPLE 2–9

Determine which diodes are forward biased and which are reverse biased in the circuits shown in Figure 2–24. Assuming a 0.7-V drop across each forward-biased diode, determine the output voltage.

Solution

1. In (a) diodes D_1 and D_3 have a net forward-biasing voltage between supply and input of 5 V − 5 V = 0 V. Diodes D_2 and D_4 have a net forward-biasing voltage of 5 V − (−5 V) = 10 V. Therefore, D_2 and D_4 are forward biased and D_1 and D_3 are reverse biased. Figure 2–25 shows the equivalent circuit path between input and output. Writing Kirchhoff's voltage law around the loop, we determine $V_o = -5\text{ V} + 0.7\text{ V} = -4.3\text{ V}$.
2. In (b) the net forward-biasing voltage between supply and input for each diode is

   D_1: +15 V − (+5 V) = +10 V

   D_2: +15 V − 0 V = +15 V

FIGURE 2–24 (Example 2–9)

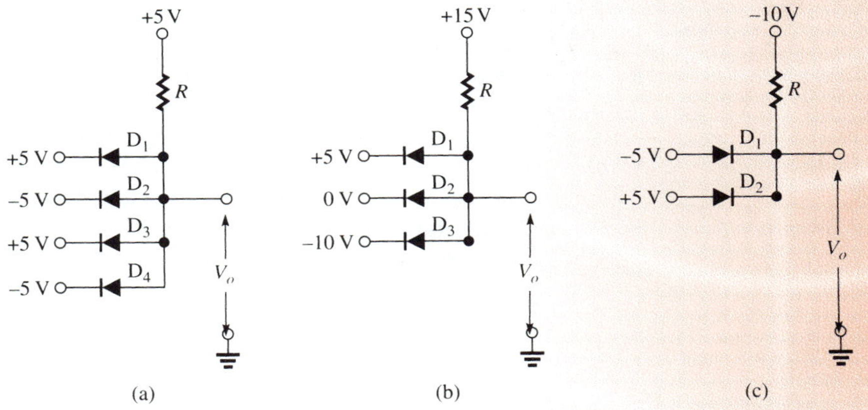

FIGURE 2–25 The circuit equivalent to Figure 2–24(a)

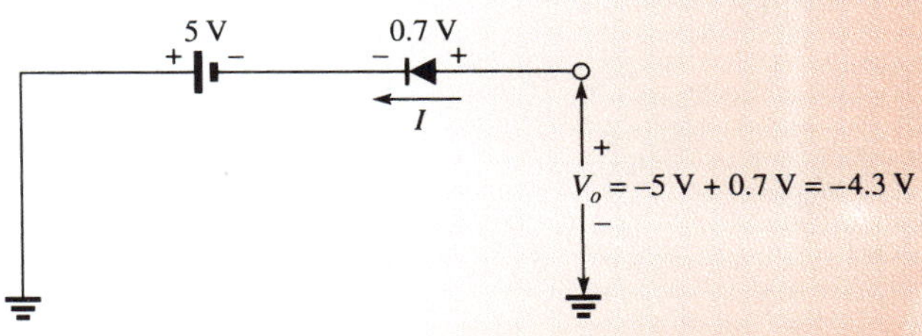

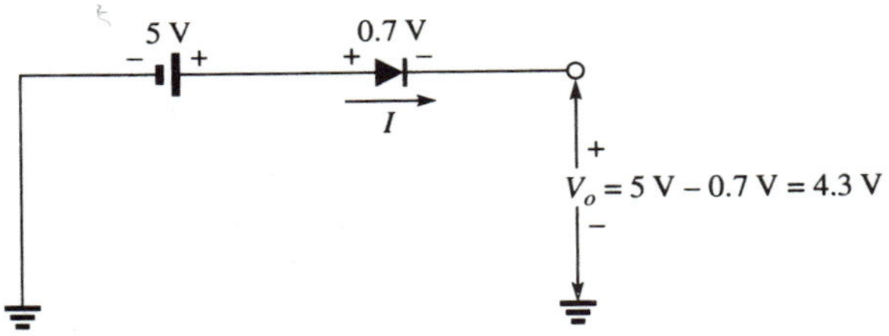

FIGURE 2–26 The circuit equivalent to Figure 2–24(c)

D_3: $+15\text{ V} - (-10\text{ V}) = +25\text{ V}$

Therefore, D_3 is forward biased and D_1 and D_2 are reverse biased.

$V_o = -10\text{ V} + 0.7\text{ V} = -9.3\text{ V}$

3. In (c) the net forward-biasing voltage between supply and input for each diode is

D_1: $-10\text{ V} - (-5\text{ V}) = -5\text{ V}$

D_2: $-10\text{ V} - (+5\text{ V}) = -15\text{ V}$

Notice that the diode positions are reversed with respect to those in (a) and (b), in the sense that the cathodes are joined together and connected through resistor R to a negative supply. Thus, the diode for which there is the greatest negative voltage between supply and input is the forward-biased diode. In this case, that diode is D_2. D_1 is reverse biased, by virtue of the fact that its cathode is near +5 V and its anode is at −5 V. Figure 2–26 shows the equivalent circuit path between input and output. Writing Kirchhoff's voltage law around the loop, we see that $V_o = 5\text{ V} - 0.7\text{ V} = 4.3\text{ V}$.

2–6 *pn* JUNCTION CAPACITANCE

In a *pn* junction, a small parasitic capacitance exists due to the charges associated with the ionized donor and acceptor atoms. This capacitance exists both in the reverse- and forward-bias modes. The capacitance is extremely important when studying *pn* junctions because it affects frequency behavior. For many high-frequency circuits, parasitics will limit the overall performance. The parasitic capacitance can also have beneficial behavior. For a reverse-biased diode, the width of the depletion region increases as the reverse-biasing voltage increases. Thus, increasing the reverse-bias voltage on a diode causes the distance (d) to increase in the capacitive plates. Recall that a capacitor is constructed by two plates separated by a dielectric. In this case, the plates are formed by the p and n regions, and the dielectric is provided by the depletion region. The capacitance equation is

$$C = \frac{\varepsilon A}{d} \qquad (2\text{–}7)$$

where ε is the permittivity of the dielectric, A is the cross-sectional surface area of the conducting region, and d is the distance separating the regions (the thickness, or width, of the dielectric). Thus, increasing the reverse bias on a diode causes d to increase and the capacitance $C = \varepsilon A/d$ to decrease. This behavior is the fundamental principle governing the operation of a *varactor diode*. The capacitance value obtained from a varactor diode is small, on the order of 100 pF or less and is used in practice only to alter the ac impedance it presents to a high-frequency signal. For example, it can be used in tuned LC networks (also called "tank" circuits). The equation for calculating the junction capacitance (C_j) is

$$C_j = \frac{C_{jo}}{\sqrt{1 + \dfrac{V_R}{\psi_o}}} \qquad (2\text{–}8)$$

where C_{jo} = zero-bias capacitance
V_R = *pn* junction reverse-voltage bias (magnitude)
ψ_o = zero-bias voltage potential

EXAMPLE 2–10

Given that the zero-bias capacitance of a *pn* junction is 2 pF and the zero-bias voltage potential is 0.55 V, calculate the capacitance of the *pn* junction if the reverse-bias voltage bias is 9.0 V.

Solution

$C_{jo} = 2$ pF

$V_R = 9.0$ V

$\psi_o = 0.55$ V

$$C_j = \frac{2.0}{\sqrt{1 + \dfrac{9.0}{.55}}} = 0.479 \text{ pF}$$

An important characteristic of a varactor diode is the ratio of its largest to its smallest capacitance when the voltage across it is adjusted through a specified range. Sometimes called the capacitance *tuning ratio*, this value is governed by the *doping profile* of the semiconductor material used to form the *p* and *n* regions, that is, the doping density in the vicinity of the junction. In diodes having an *abrupt* junction, the *p* and *n* sides are uniformly doped, and there is an abrupt transition from *p* to *n* at the junction. Abrupt-junction varactors exhibit capacitance ratios from about 2:1 to 3:1. In the *hyperabrupt* (extremely abrupt) junction, the doping level is increased as the junction is approached from either side, so the *p* material becomes more heavily *p* near the junction, and the *n* material becomes more heavily *n*. The hyperabrupt junction is very sensitive to changes in reverse voltage, and this type of varactor may have a capacitance ratio up to 20:1. Figure 2–27 shows typical plots of abrupt and hyperabrupt varactor capacitance versus reverse voltage. The figure also shows the schematic symbol for a varactor diode.

2–7 CIRCUIT ANALYSIS WITH ELECTRONICS WORKBENCH MULTISIM

This text present simulation examples of circuit analysis of fundamental electronic circuits using Electronics Workbench Multisim (EWB). Examples of key analog circuit concepts are presented in each chapter. EWB provides a unique opportunity for the student to examine electronic circuits and analog concepts in a way that reflects techniques used on both the bench and on the computer. The use of EWB provides the student with additional hands-on insight into many of the fundamental analog circuits, concepts, and test equipment while improving the student's ability to perform logical thinking when troubleshooting circuits and systems. The test equipment tools in EWB reflect the type of tools that are common on well-equipped test benches, and the analysis tools reflect the type of analytical tools available.

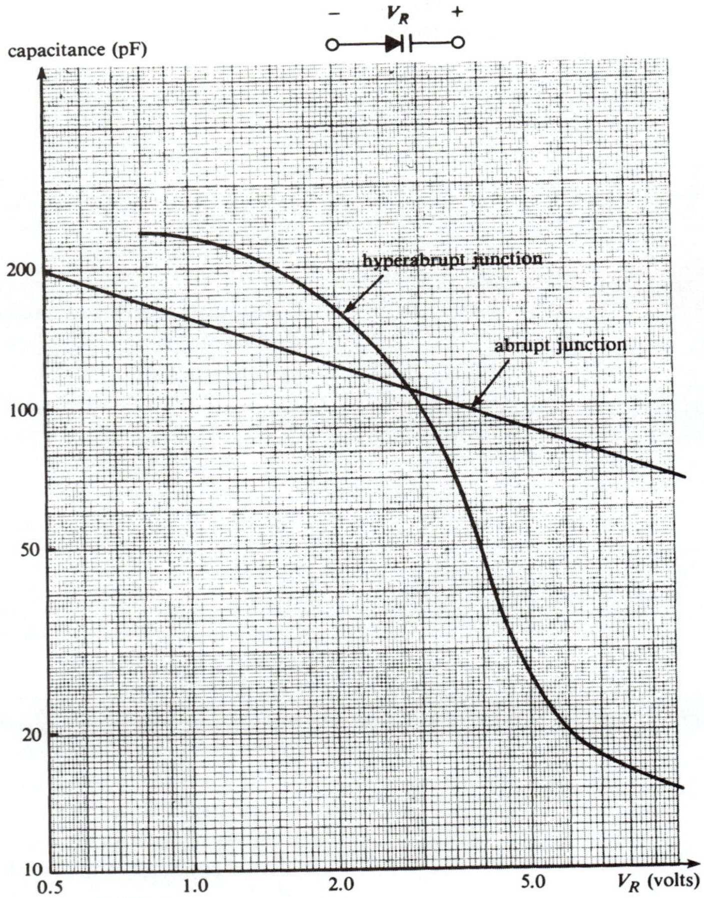

FIGURE 2–27 Capacitance versus reverse voltage for abrupt and hyperabrupt varactor diodes

Chapter 2 introduced the basic concept of the *pn* junction (a diode). The student learned that the voltage drop across a forward-biased ideal diode is 0.7 V. The following exercise demonstrates how a computer simulation package such as EWB can be used to test and analyze a basic diode circuit. The objectives of this exercise are as follows:

- Show the student how to construct and simulate a simple diode circuit.
- Use EWB to show that the forward voltage drop across an ideal diode is 0.7 V.
- Use EWB to measure the current flowing in the circuit.

Open the circuit, **Ch2_EWB.msm** that is found in the Electronics Workbench 2001 CD-ROM packaged with the text. This is a simple circuit (shown in Figure 2-28) containing a +12 V voltage source, a 1-kΩ resistor and an ideal diode. Two multimeters are being used in this circuit: One is being used to measure the diode voltage drop, and the other is used to measure the current in the circuit. The voltage measurement requires that the multimeter be set to DC and V selected. The current meter is set to measure DC current (A). Notice that the current meter is in series with the diode and the power supply.

Each multimeter has a settings button. This is shown in Figure 2–29. The user of the multimeter has the capability to adjust the settings for ammeter resistance, volmeter resistance, and ohmmeter current. Remember, real test gear has some resistance or requires some current to make a measurement. These settings simulate "real-world" conditions.

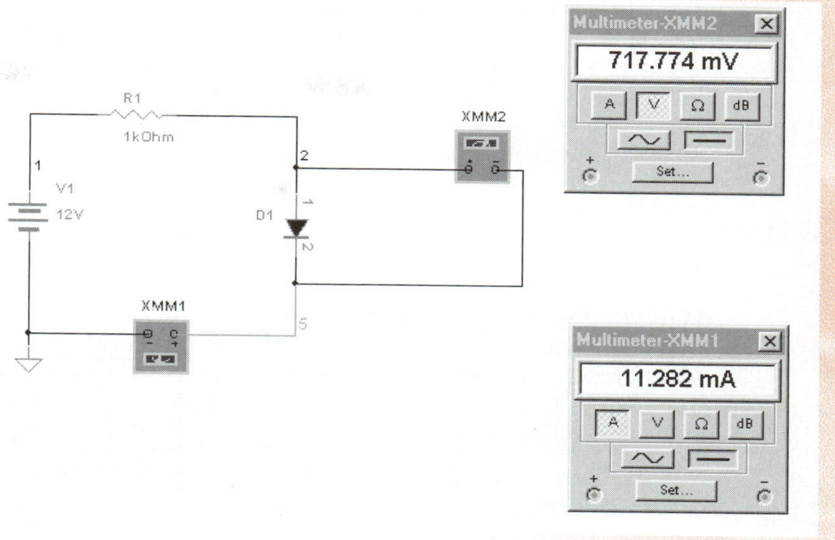

FIGURE 2–28 The simple diode circuit for the EWB exercise

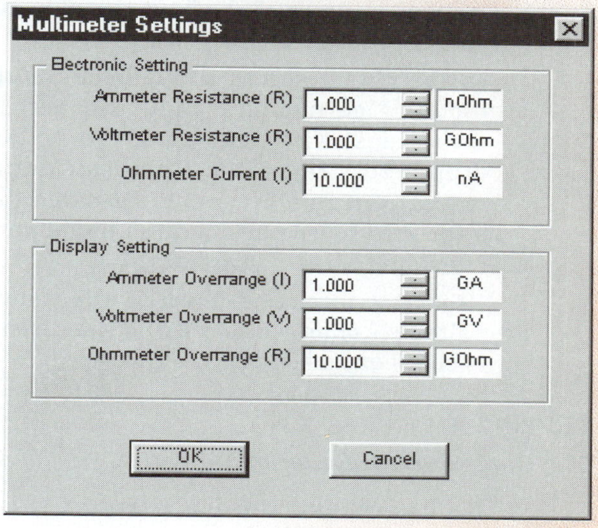

FIGURE 2–29 The setting options for the EWB multimeters

A key to success with computer simulation of electronic circuits is to have a reasonable ballpark estimate of what the expected voltages and currents should be. The expected voltage drop should be about 0.7 V and the current should be about $(12 - 0.7)/1\text{k} = 11.3$ mA. The computer simulation is started by clicking on the simulation start button. The multimeters will immediately display results. The voltmeter shows a value of 771.774 mV. The current meter shows 11.282 mA.

SUMMARY

This chapter has presented the basics of the *pn* junction. Students should have mastered the following concepts and skills:

- Identifying forward and reverse biased diodes.
- Understanding that current flows in a forward-bias diode and essentially no current flows in a reverse-bias diode.

- Knowing that the voltage drop across an ideal forward-biased diode is approximately 0.7 V.
- Analyzing a basic diode circuit to determine currents and voltages.

EXERCISES

SECTION 2–3
Basic Diode Operation

2–1. Assume that the voltage drop across a forward-biased silicon diode is 0.7 V and that across a forward-biased germanium diode is 0.3 V.

(a) If D_1 and D_2 are both silicon diodes in Figure 2–30, find the current I in the circuit.

(b) Repeat if D_1 is silicon and D_2 is germanium.

2–2. Repeat Exercise 2–1 when the constant source voltage is changed to 9 V.

2–3. In the circuit shown in Figure 2–31, the diode is germanium: Find the percent error caused by neglecting the voltage drop across the diode when calculating the current I in the circuit. (Assume that a forward-biased germanium diode has a constant voltage drop of 0.3 V.)

2–4. Repeat Exercise 2–3 when the source voltage is changed to 3 V and the resistor is changed to 470 Ω.

SECTION 2–4
The Diode Current Equation

2–5. A silicon *pn* junction has a saturation current of 1.8×10^{-14} A. Assuming that $\eta = 1$, find the current in the junction when the forward-biasing voltage is 0.6 V and the temperature is 27°C.

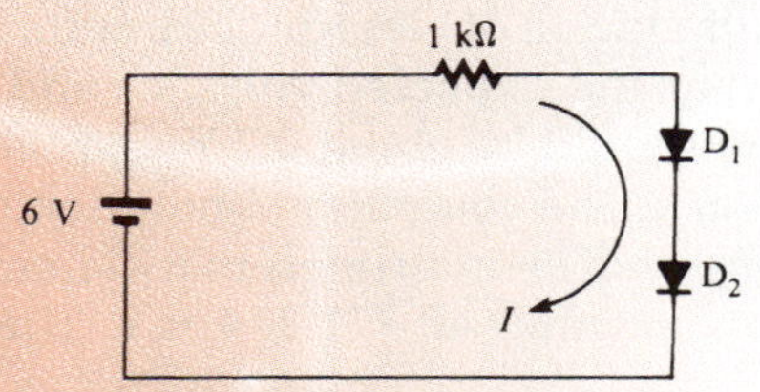

FIGURE 2–30 (Exercise 2–1)

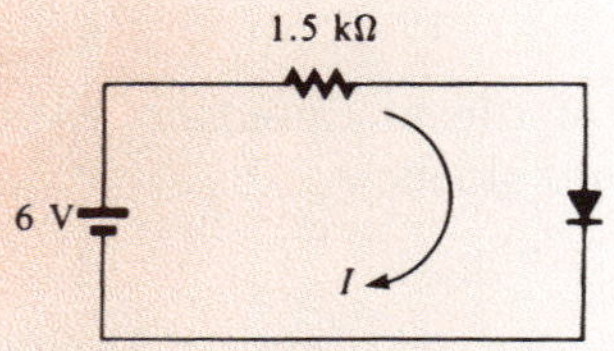

FIGURE 2–31 (Exercise 2–3)

2–6. Repeat Exercise 2–5 when the forward-biasing voltage is 0.65 V.

2–7. The forward current in a *pn* junction is 1.5 mA at 27°C. If $I_s = 2.4 \times 10^{-14}$ A and $\eta = 1$, what is the forward-biasing voltage across the junction?

2–8. The forward current in a *pn* junction is 22 mA when the forward-biasing voltage is 0.64 V. If the thermal voltage is 26 mV and $\eta = 1$, what is the saturation current?

2–9. A junction diode has an external voltage source of 0.15 V connected across it, with the positive terminal of the source connected to the cathode of the diode. The saturation current is 0.02 pA, the thermal voltage is 26 mV, and $\eta = 2$.

(a) Find the theoretical (ideal) diode current. Repeat for a source voltage of

(b) 0.3 V, and

(c) again for a source voltage of 0.5 V.

2–10. A junction diode is connected across an external voltage source so that the negative terminal of the source is connected to the anode of the diode. If the external voltage source is 5 V and the saturation current is 0.06 pA, what is the theoretical (ideal) diode current?

2–11. The reverse breakdown voltage of a certain diode is 150 V and its saturation current is 0.1 pA. Assuming that the constant n in equation 2–5 is 2, what is the current in the diode when the reverse-biasing voltage is 149.95 V?

2–12. In an experiment designed to investigate the breakdown characteristics of a certain diode, a reverse current of 9.3 nA was measured when the reverse voltage across the diode was 349.99 V. If the breakdown voltage of the diode was 350 V and its saturation current was known to be 1.0 pA, what value of the constant n in equation 2–5 is appropriate for this diode?

2–13. The manufacturer of a certain diode rates its maximum power dissipation as 0.1 W and its reverse breakdown voltage as 200 V. What maximum reverse current could it sustain at breakdown without damage?

2–14. A certain diode has a reverse breakdown voltage of 100 V and a saturation current of 0.05 pA. How much power does it dissipate when the reverse voltage is 99.99 V? Assume that n in equation 2–5 is 2.5.

2–15. A diode has a saturation current of 45 pA at a temperature of 373 K. What is the approximate value of I_s at $T = 273$ K?

2–16. When the voltage across a forward-biased diode at $T = 10°C$ is 0.621 V, the current is 4.3 mA. If the current is held constant, what is the voltage when,

(a) $T = 40°C$?

(b) Repeat for $T = -30°C$.

SECTION 2–5

Identifying Forward- and Reverse-Bias Operating Modes

2–17. Determine which of the diodes in Figure 2–32 are forward biased and which are reverse biased.

2–18. Determine which of the diodes shown in Figure 2-33 are forward biased and which are reverse biased.

2–19. The inputs A and B in Figure 2–34 can be either 0 V or +10 V. Each diode is silicon. Find V_o for each of the following cases. (Assume ideal silicon diodes.)

(a) $A = 0$ V, $B = 0$ V

(b) $A = 0$ V, $B = +10$ V

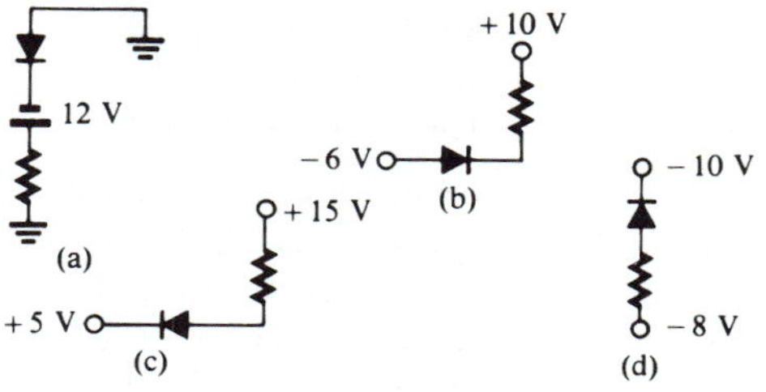

FIGURE 2–32 (Exercise 2–17)

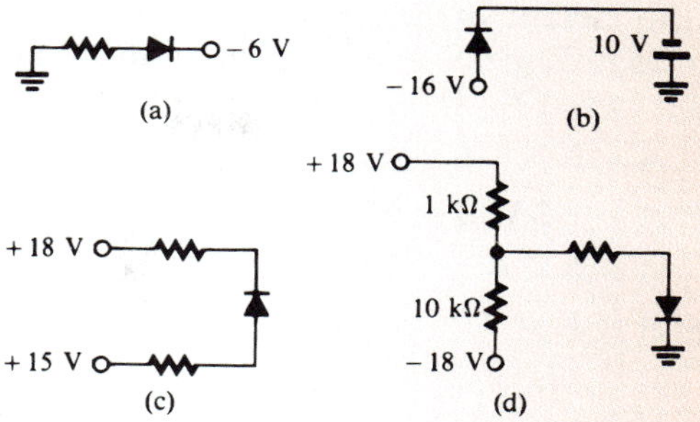

FIGURE 2–33 (Exercise 2–18)

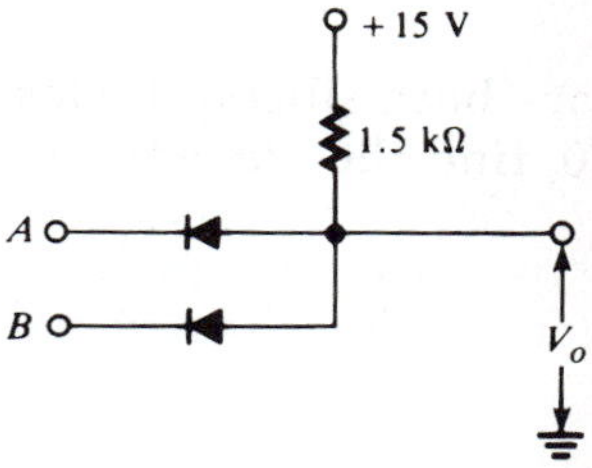

FIGURE 2–34 (Exercise 2–19)

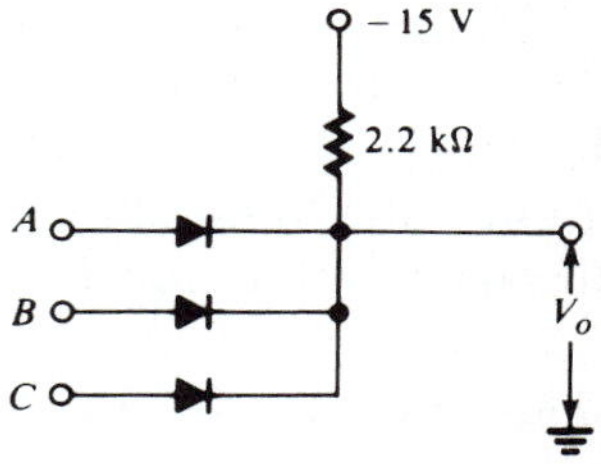

FIGURE 2–35 (Exercise 2–20)

(c) $A = +10$ V, $B = 0$ V

(d) $A = +10$ V, $B = +10$ V

2–20. The inputs A, B, and C in Figure 2–35 can be either +10 V or −5 V. Assume ideal silicon diodes. Find V_o when

(a) $A = B = C = -5$ V

(b) $A = C = -5$ V, $B = +10$ V

(c) $A = B = +10$ V, $C = -5$ V

(d) $A = B = C = +10$ V

2–21. In the circuit of Exercise 2–20. A, B, and C can be either 0 V or −5 V. Assuming that the forward $V_D = 0.7$ V, find V_o when

(a) $A = B = C = 0$ V

(b) $A = B = 0$ V, C $= -5$ V

(c) $A = C = -5$ V, B $= 0$ V

(d) $A = B = C = -5$ V

2–22. In the circuit of Exercise 2–19, A and B can be either 0 V or -5 V. Assuming that the forward $V_D = 0.7$ V, find V_o when

(a) $A = B = -5$ V
(b) $A = -5$ V, $B = 0$ V
(c) $A = 0$ V, $B = -5$ V
(d) $A = B = 0$ V

SECTION 2–6

pn Junction Capacitance

2–23. Given that the zero-bias capacitance of a *pn* junction is 5 pF and the zero-bias voltage potential is 0.62 V, calculate the capacitance of the *pn* junction if the reverse-bias voltage-bias is 10.0 V.

2–24. Given that a diode, with an abrupt junction, has the capacitance-versus-reverse-voltage curve shown in Figure 2–25, determine the capacitance range as the reverse-bias voltage is varied from 1.0 V to 5.0 V.

2–25. Given that $\varepsilon = 1.04 \times 10^{-12}$ F/cm, the plate area is 2 cm^2, and the width of the dielectric is 1 μ (1×10^{-6} meters), calculate the capacitance.

2-26. Given that the zero-bias capacitance of a *pn* junction is 7 pF and the zero-bias voltage potential is 0.65 V, calculate the capacitance of the *pn* junction if the reverse-bias voltage bias is

(a) 2.5 V,
(b) 5.0 V,
(c) 8.8 V.

SPICE EXERCISES

Note: In the exercises that follow, assume that all device parameters have their default values unless otherwise specified.

2–27. Use SPICE to simulate the circuit shown in Figure 2–4. Obtain a value for the current *I*.

2–28. Use SPICE to simulate the circuit shown in Figure 2–5. Obtain a value for the current *I* and the voltage across diode D_1.

2–29. Use SPICE to simulate the circuit shown in Figure 2–6. Obtain a value for the current *I* and the voltage V_{AB}.

2–30. Use SPICE to simulate the circuit shown in Figure 2–20. Obtain a value for the output voltage V_o.

CHAPTER 3

THE DIODE AS A CIRCUIT ELEMENT

OUTLINE

OBJECTIVES

- Develop an understanding of the operation of diodes in dc circuits.
- Differentiate between the concepts of ac and dc resistance in diodes.
- Understand the load line concept in a simple diode circuit.
- Graphically visualize the half-wave and full-wave rectification process.
- Recognize the effect of capacitive filtering in a rectifier.
- Specify components for a simple power supply.

3–1 INTRODUCTION

In Chapter 2 we studied the construction and properties of a *pn* junction and mentioned that a semiconductor diode is an example of an electronic device that contains such a junction. We learned about the *I–V* relationship (the diode equation) that describes mathematically the behavior of a diode and used it for obtaining the operating voltage and current in a simple diode circuit. We also learned to identify in a circuit when a diode is forward or reverse biased and used this fact in the analysis of diode logic circuits.

In this chapter we will study the diode as a circuit *device* or component of an electronic circuit with a specific function. We will also learn about modeling a diode as an ideal or semi-ideal device by replacing it with a simpler *equivalent* circuit element. This approach will serve as our introduction to this standard and widely used method of electronic circuit analysis: Replace actual devices by simpler equivalent circuits in order to obtain solutions that are sufficiently accurate for the application in which they are used. This *analysis* method allows us to use standard procedures of solution for dc and ac circuits. Understand that this "replacement" of the diode by its equivalent circuit is done on paper only, in order to simplify calculations.

Studying and understanding important concepts such as linearity, small- and large-signal operation, quiescent points, bias, load lines, and equivalent circuits are best accomplished by applying them to the analysis of the relatively simple diode circuit. In later chapters we will apply our knowledge of diode circuit analysis to the study of several practical circuits in which these versatile devices are used.

Finally, this chapter will cover elementary *power supplies* (electronic circuits used for converting ac voltage to dc voltage) where diodes play a very important role. A section on elementary *voltage regulation* is included with the aim of allowing students to build their own multi-voltage power supply early in the course.

3–2 THE DIODE AS A NONLINEAR DEVICE

Linearity is an exceptionally important concept in electronics. For our purposes now, we can best understand the practical implications of this rather broad concept by restricting ourselves to the following definition of a linear electronic device: *A device is linear if the graph relating the voltage across it to the current through it is a straight line.*

If we have experimental data that shows measured values of voltage and the corresponding values of current, then it is a simple matter to plot these and determine whether the linearity criterion is satisfied. Often we have an equation that relates the voltage across a device to the current through it (or that relates the current to the voltage). If the equation is in one of the general *forms*

$$V = a_1 I + a_2 \qquad (3\text{–}1)$$

or

$$I = b_1 V + b_2 \qquad (3\text{–}2)$$

where a_1, a_2, b_1, and b_2 are any constants, positive, negative, or 0, then the graph of V versus I is a straight line and the device is linear. In equation 3–1, a_1 is the *slope* of the line and has the units of ohms. In equation 3–2, b_1 is the slope and has the units siemens (formerly mhos).

In fundamental circuit analysis courses, we learn that resistors, capacitors, and inductors are all linear electrical devices because their voltage–current equations are of the form of equations 3–1 and 3–2, such as $V = IR$, $I = VY$, $V_L = I_L X_L$, or $I_C = V_C/X_C$, to name a few.

In these voltage–current equations, we of course assume that any other circuit characteristics such as frequency are held constant. In other words, only the magnitudes of voltage and current are regarded as variables. In each case, all else being equal, increasing or decreasing the voltage causes a proportional increase or decrease in the current. Figure 3–1 is a plot of the voltage V across a 200-Ω resistor versus the current I through it. The linearity property of the resistor is clearly evident and follows from the Ohm's law relation $V = 200I$. Note that the slope of the line equals the resistance, $r = \Delta V/\Delta I = 200\ \Omega$, and that the linear relation applies to *negative* voltages and currents as well as positive. Reversing the directions (polarities) of the voltage across and current through a linear device does not alter its linearity property. Note also that the slope of the line is everywhere the same: No matter where along the line the computation $\Delta V/\Delta I$ is performed, the result equals 200 Ω.

When displaying the voltage–current relationship of an electronic device on a graph, it is conventional to plot current along the vertical axis and voltage along the horizontal axis—the reverse of that shown in Figure 3–1. Of course, the graph of a linear device is still a straight line; reversal of axes is equivalent to expressing the V–I relation in the form of equation 3–2, with slope having the units of conductance, $G = \Delta I/\Delta V = 1/R$ (siemens) instead of resistance.

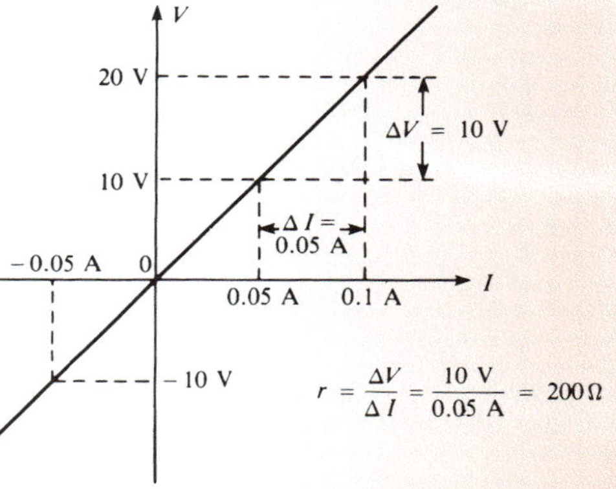

FIGURE 3–1 The graph of V versus I for a resistor is a straight line. A resistor is a linear device, and the value of $\Delta V/\Delta I$ is the same no matter where it is computed.

In Chapter 2 we stated that the current–voltage relation for a *pn* junction (and therefore for a diode) is

$$I_D = I_s(e^{V_D/\eta V_T} - 1) \tag{3–3}$$

where I_s = saturation current

V_T = thermal voltage ≈ 26 mV at room temperature

η = a function of V_D, whose value ranges between 1 and 2

Equation 3–3 is clearly not in the form of either equation 3–1 or 3–2, so the diode's voltage–current relation does not meet the criterion for a linear electronic device. We conclude that a diode is a *nonlinear* device. Figure 3–2 is a graph of the *I–V* characteristic of a typical silicon diode in its forward-biased region. The graph is most certainly not a straight line.

Figure 3–2 shows how identical ΔV values result in different ΔI values along the *I–V* curve, revealing that the resistance $\Delta V/\Delta I$ decreases (steeper slope) as diode current increases. Unlike a linear device, the resistance of a nonlinear device depends on the voltage across it (or current through it)—i.e., the resistance depends on the point where the values of ΔV and ΔI are calculated—specifically, at the biasing or operating point. In the case of a diode, we further note that the *I–V* characteristic becomes very nearly horizontal at low values of current and in the reverse-biased region (see Figure 2–12). Therefore, in these regions, large changes in voltage, ΔV, create very small changes in current, ΔI, so the value of $\Delta V/\Delta I$ is very large.

The region on the *I–V* curve where the transition from high resistance to low resistance takes place is called the *knee* of the curve. When the diode current is significantly greater or less than that in the vicinity of the knee, we will say that it is biased *above* or *below* the knee, respectively. The biasing point is also called the *quiescent* point or *Q*-point, for short.

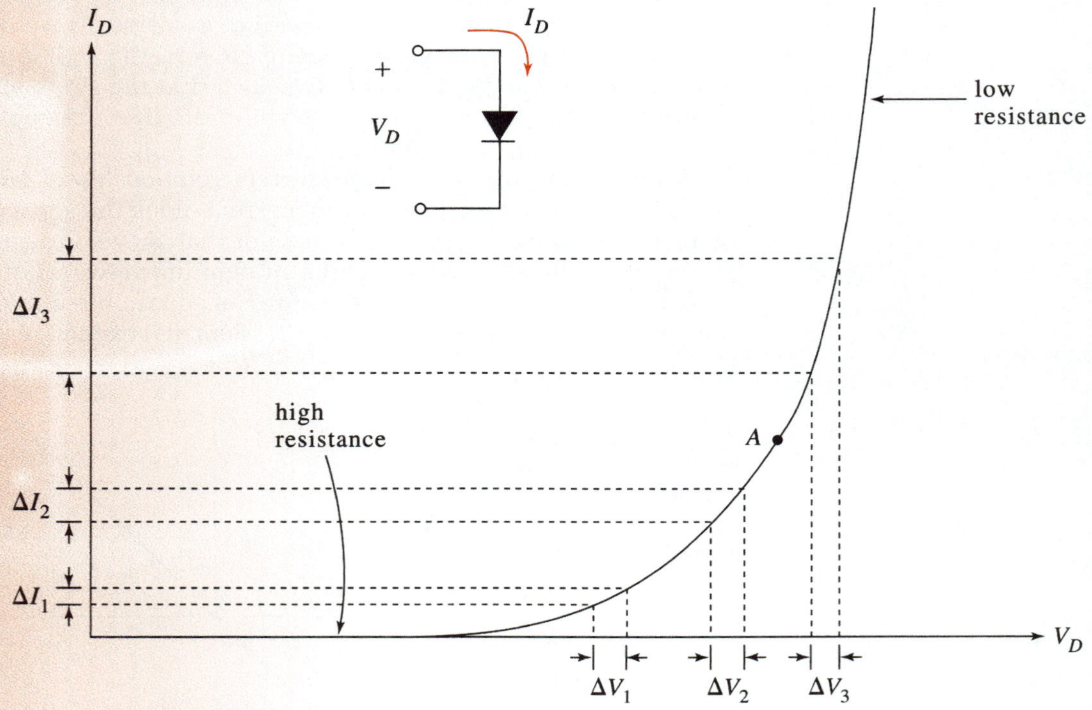

FIGURE 3–2 A forward-biased diode characteristic. The value of $\Delta I/\Delta V$ depends upon the location where it is computed.

3–3 ac AND dc RESISTANCE

The resistance $\Delta V/\Delta I$ is called the *ac* (or *dynamic*) *resistance* of the diode. It is called *ac* resistance because we consider the small *change* in voltage, ΔV, such as might be generated by an ac generator, causing a *change* in current, ΔI. In using this graphical method to calculate the ac resistance, the changes ΔV and ΔI must be kept small enough to avoid covering sections of the *I–V* curve over which there is an appreciable change in slope.

Henceforth we will refer to the ac resistance of the diode as r_D, where the lowercase r is in keeping with the convention of using lowercase letters for ac quantities. Thus, let us define r_D as

$$r_D \equiv \frac{\Delta V_D}{\Delta I_D} \tag{3–4}$$

for small ΔV_D and ΔI_D about the operating point. For very small variations, $\Delta I_D/\Delta V_D$ approaches the slope of the tangent at the *Q*-point.

When a dc voltage is applied across a diode, a certain dc current will flow through it. The *dc resistance* of a diode is found by dividing the dc voltage across it by the dc current through it. Thus the dc resistance, also called the *static* resistance, is found by direct application of Ohm's law. We will designate dc diode resistance by R_D:

$$R_D = \frac{V_D}{I_D} \tag{3–5}$$

Like ac resistance, the dc resistance of a diode depends on the point on the *I–V* curve at which it is calculated. For instance, the dc resistance of the diode at point *A* in Figure 3–2 is represented by a straight line that passes through the origin and point *A*. If the diode is biased at a larger current, it is apparent that the straight line will have a larger (steeper) slope, meaning that the dc resistance will be smaller. (The larger the slope, the smaller the resistance.) We see that the diode is nonlinear in both the dc and the ac sense; that is, both its dc and ac resistances change over a wide range.

When analyzing or designing diode circuits, it is often the case that the *I–V* characteristic curve is not available. In most practical work, the ac resistance of a diode is not calculated graphically but is found using a widely accepted approximation. It can be shown that the ac resistance is closely approximated by $r_D \cong V_T/I_D$, where V_T is the thermal voltage and I_D is the dc current in amperes. For $T = 300$ K, V_T is about 26 mV so at room temperature

$$r_D \approx \frac{26\text{ mV}}{I_D} \tag{3–6}$$

This approximation is valid for both silicon and germanium diodes and is obtained using calculus by differentiating the diode equation

$$I_D \approx I_S e^{V_D/V_T} \tag{3–7}$$

with respect to V_D; that is,

$$\frac{dI_D}{dV_D} = \frac{I_S}{V_T}e^{V_D/V_T} = \frac{I_D}{V_T} = \frac{1}{r_D} \quad \text{or} \quad r_D = \frac{V_T}{I_D} \tag{3–8}$$

There is one additional component of diode resistance that should be mentioned. The resistance of the semiconductor material and the contact resistance where the external leads are attached to the *pn* junction can be lumped together and called the *bulk resistance*, r_B, of the diode. Usually less than 1 Ω, the bulk resistance also changes with the dc current in the diode, becoming quite small at high current levels. The total ac resistance of the diode is $r_D + r_B$, but at low current levels r_D is so much greater than r_B that r_B can usually be neglected. At high current levels, r_B is typically on the order of 0.1 Ω.

When a diode is connected in a circuit in a way that results in the diode being forward biased, there should always be resistance in series with the diode to limit the current that flows through it. The following example illustrates a practical circuit that could be used to determine *I–V* characteristics.

EXAMPLE 3–1

The circuit shown in Figure 3–3 was connected to investigate the relation between the voltage and current in a certain diode. The adjustable voltage source was set to several different values in order to control the diode current, and the diode voltage was recorded at each setting. The results are tabulated in the table in Figure 3–3.

1. Find the dc resistance of the diode when the voltage across it is 0.56, 0.62, and 0.67 V.
2. Find the ac resistances presented by the diode to an ac signal generator that causes the voltage across the diode to vary between 0.55 V and 0.57 V, between 0.61 V and 0.63 V, and between 0.66 V and 0.68 V.
3. Find the approximate ac resistances when the diode voltages are 0.56 V, 0.62 V, and 0.67 V. Assume bulk resistances of 0.8 Ω, 0.5 Ω, and 0.1 Ω, respectively.

Solution

1. The dc diode resistances at the voltages specified are found from equation 3–5, $R_D = V_D/I_D$. At $V = 0.56$ V,

$$R_D = \frac{0.56\ \text{V}}{1.04 \times 10^{-3}\ \text{A}} = 538.5\ \Omega$$

At $V = 0.62$ V,

$$R_D = \frac{0.62\ \text{V}}{10.8 \times 10^{-3}\ \text{A}} = 57.4\ \Omega$$

FIGURE 3–3 (Example 3–1)

Measurement Number	*I* (mA)	V (volts)
1	0.705	0.55
2	1.04	0.56
3	1.54	0.57
4	7.33	0.61
5	10.8	0.62
6	15.9	0.63
7	51.1	0.66
8	75.3	0.67
9	110.8	0.68

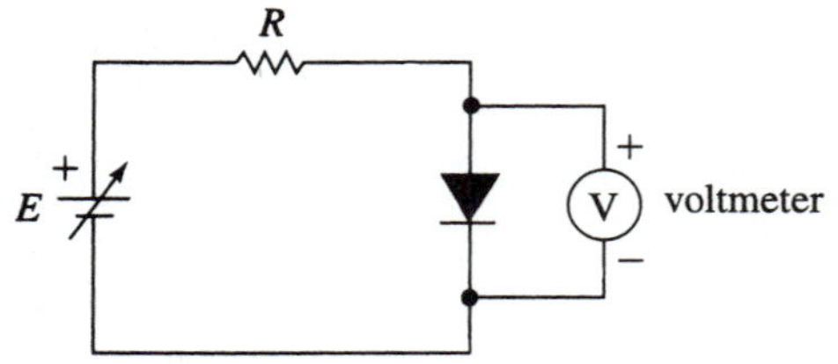

At $V = 0.67$ V,

$$R_D = \frac{0.67\text{ V}}{75.3 \times 10^{-3}\text{ A}} = 8.90\ \Omega$$

2. The ac diode resistances are found from equation 3–4, $r_D = \Delta V_D/\Delta I_D$, as follows:

$$r_D = \frac{(0.57 - 0.55)\text{ V}}{(1.54 - 0.705) \times 10^{-3}\text{ A}} = \frac{0.02\text{ V}}{0.835 \times 10^{-3}\text{ A}} = 23.95\ \Omega$$

$$r_D = \frac{0.02\text{ V}}{8.57 \times 10^{-3}\text{ A}} = 2.33\ \Omega$$

$$r_D = \frac{0.02\text{ V}}{59.7 \times 10^{-3}\text{ A}} = 0.34\ \Omega$$

3. The approximate ac resistances are found using relation (3–6), $r_D \approx 0.026/I_D$ and adding the bulk resistance r_B. At $V = 0.56$ V,

$$r_D = \frac{0.026}{I_2} + r_B = \frac{0.026}{1.04 \times 10^{-3}\text{ A}} + 0.8\ \Omega = 25.8\ \Omega$$

At $V = 0.62$ V,

$$r_D = \frac{0.026}{I_5} + r_B = \frac{0.026}{10.8 \times 10^{-3}\text{ A}} + 0.5\ \Omega = 2.91\ \Omega$$

At $V = 0.67$ V,

$$r_D = \frac{0.026}{I_8} + r_B = \frac{0.026}{75.3 \times 10^{-3}\text{ A}} + 0.1\ \Omega = 0.445\ \Omega$$

Note that each ac resistance calculated in part 3 is at a diode voltage in the middle of a range (ΔV) over which an ac resistance is calculated in part 2. We can therefore expect the approximations for r_D to agree reasonably well with the values calculated using $r_D = \Delta V_D/\Delta I_D$. The results bear out this fact.

3–4 ANALYSIS OF dc CIRCUITS CONTAINING DIODES

In virtually every practical dc circuit containing a diode, there is one simplifying assumption we can make when the diode current is beyond the knee. We have seen (Figure 3–2, for example) that the I–V curve is essentially a vertical line above the knee. *The implication of a vertical line on an I–V characteristic is that the voltage across the device remains constant, regardless of the current that flows through it.* Thus the voltage drop across a diode remains substantially constant for all current values above the knee. This fact is responsible for several interesting applications of diodes. For present purposes, it suggests that the diode is equivalent to another familiar device that has this same property of maintaining a constant voltage, independent of current: a voltage source! Indeed, our first simplified equivalent circuit of a diode is a voltage source having a potential equal to the (essentially) constant drop across it when the current is above the knee.

Figure 3–4(a) shows a simple circuit containing a forward-biased diode. Assuming the current is above the knee, the diode can be replaced by a 0.7-V source as shown in Figure 3–4(b).

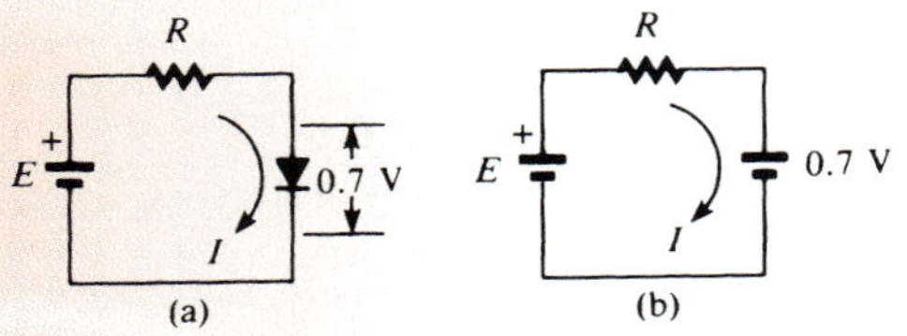

FIGURE 3–4 For analysis purposes, the forward-biased diode in (a) can be replaced by a voltage source, as in (b)

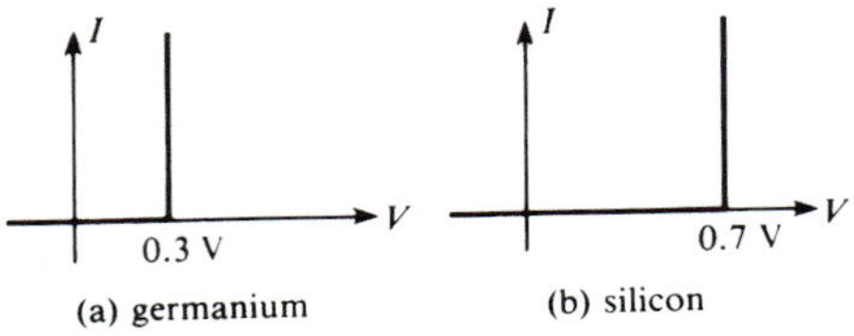

FIGURE 3–5 Idealized characteristic curves. The diodes are assumed to be open circuits until the forward-biasing voltages are reached.

The idealized characteristic curves in Figure 3–5 imply that the diode is an *open* circuit (infinite resistance, zero current) for all voltages less than 0.3 V or 0.7 V and becomes a short circuit (zero resistance) when one of those voltage values is reached. These approximations are quite valid in most real situations. Note that it is not possible to have, say, 5 V across a forward-biased diode. If a diode were connected directly across a +5-V source, it would act like a short circuit and damage either the source, the diode, or both. When troubleshooting a circuit that contains a diode that is supposed to be forward biased, a diode voltage measurement greater than 0.3 V or 0.7 V means that the diode has failed and is in fact open.

In some dc circuits, the voltage drop across a forward-biased diode may be so small in comparison to other dc voltages in the circuit that it can be neglected entirely. For example, suppose a circuit consists of a 25-V source in series with a 1-kΩ resistor in series with a germanium diode. Then $I = (25 - 0.3)/(1\ \text{k}\Omega) = 24.7$ mA. Neglecting the drop across the diode, we would calculate $I = 25/(1\ \text{k}\Omega) = 25$ mA, a result that in most practical situations would be considered close enough to 24.7 mA to be accurate.

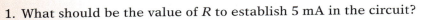

EXAMPLE 3–2

Assume that the silicon diode in Figure 3–6 requires a minimum current of 1 mA to be above the knee of its *I–V* characteristic.

1. What should be the value of *R* to establish 5 mA in the circuit?
2. With the value of *R* calculated in (1), what is the minimum value to which the voltage *E* could be reduced and still maintain diode current above the knee?

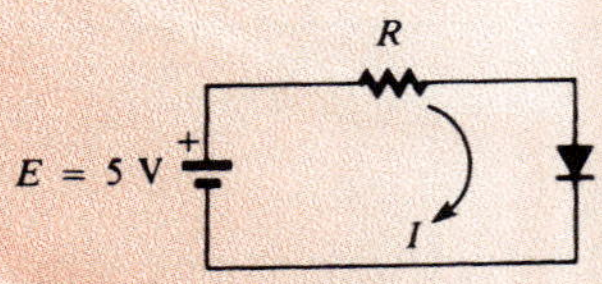

FIGURE 3–6 (Example 3–2)

Solution

1. If *I* is to equal 5 mA, we know that the voltage across the diode will be 0.7 V. Therefore, solving for *R*,

$$R = \frac{E - 0.7}{I} = \frac{(5 - 0.7)\ \text{V}}{5 \times 10^{-3}\ \text{A}} = 860\ \Omega$$

2. In order to maintain the diode current above the knee, *I* must be at least 1 mA. Thus,

$$I = \frac{E - 0.7}{R} \geq 10^{-3}\ \text{A}$$

Therefore, since $R = 860\ \Omega$,

$$\frac{E - 0.7}{860} \geq 10^{-3}\ \text{A}$$

or

$$E \geq (860 \times 10^{-3}) + 0.7$$
$$E \geq 1.56 \text{ V}$$

Circuits containing dc sources and two or more diodes can be analyzed through general circuit analysis by assuming a conducting or nonconducting state for each diode according to the polarity or orientation of the sources. Each diode assumed to be conducting is replaced by a 0.7-V or 0.3-V source. Even in cases where the conduction state is not obvious, one can assume an arbitrary state and solve for the currents. If a diode is assumed to be forward biased and the calculated forward current turns out to be negative, it means that the diode is not conducting and should be replaced by an open circuit. Obviously, the circuit needs to be reanalyzed with the new condition. An example illustrating the concept follows.

EXAMPLE 3–3

Determine the current through each branch in the following circuit (Figure 3–7). Assume silicon diodes.

Solution

Assuming both diodes are conducting, we replace both with 0.7-V sources as indicated in Figure 3–8 and write the two mesh equations according to the assigned directions for I_1 and I_2.

$$\text{Mesh 1:}\quad 3I_1 - 2I_2 = 13.6 \text{ (using V, k}\Omega\text{, and mA)}$$
$$\text{Mesh 2:}\quad -2I_1 + 5I_2 = -4.3$$

Using Kramer's rule or any of the other methods, we obtain

$$I_1 = 5.4 \text{ mA} = I_{D1} \text{ and } I_2 = 1.3 \text{ mA}$$

The (forward) current through D_2, according to KCL, is then

$$I_{D2} = 5.4 - 1.3 = 4.1 \text{ mA}$$

Because both diode currents turned out to be positive, both diodes are indeed conducting and the answers are all correct.

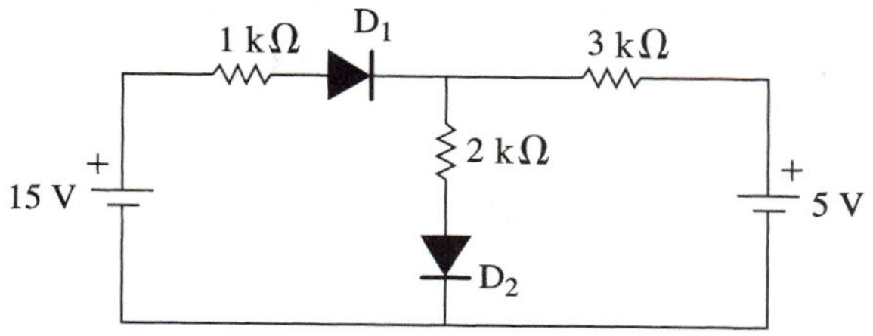

FIGURE 3–7 (Example 3–3)

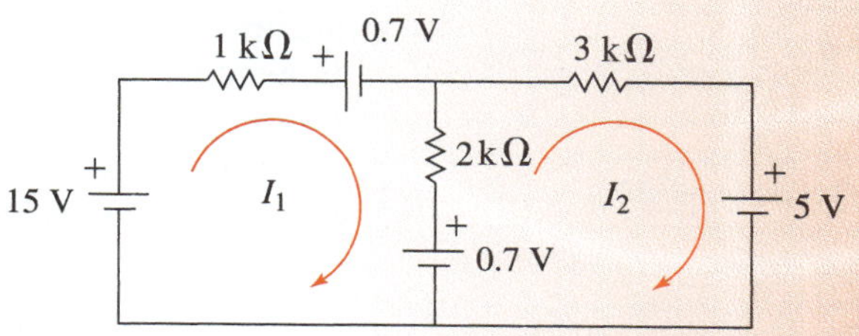

FIGURE 3–8 (Example 3–3)

EXAMPLE 3–4

Sketch the output voltage V_o in the following circuit for the variable input voltage V_S depicted in Figure 3–9.

Solution

When $V_S = 0$, the diode is reverse biased and is therefore an open circuit. The output voltage at this point is also zero. As V_S increases, the diode will continue being open and V_o will be equal to, or track, V_S since no drop occurs across the

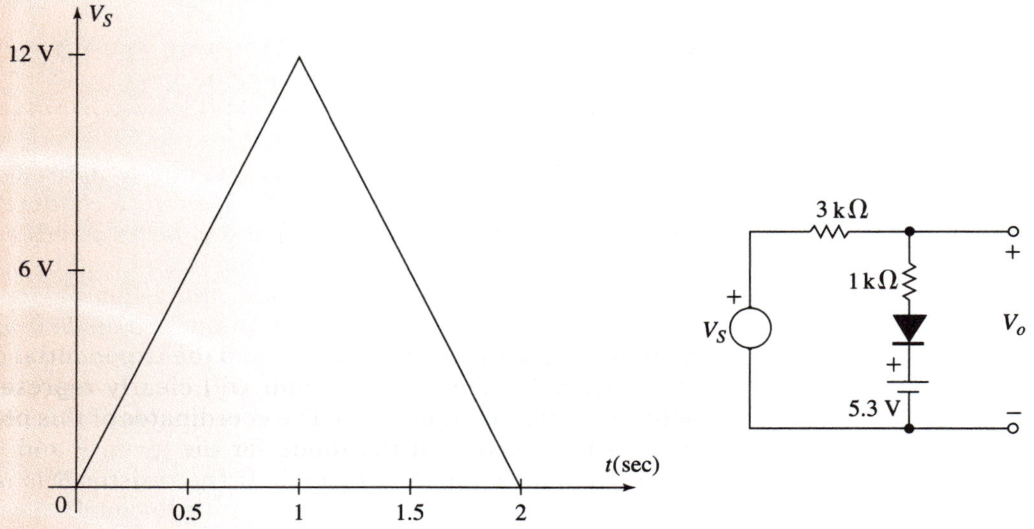

FIGURE 3–9 (Example 3–4)

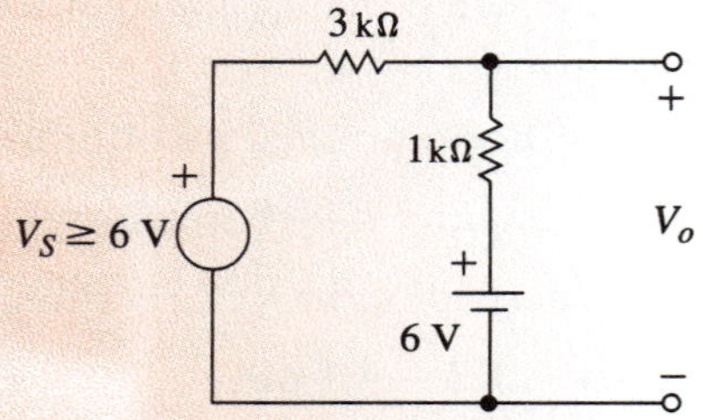

FIGURE 3–10 (Example 3–4)

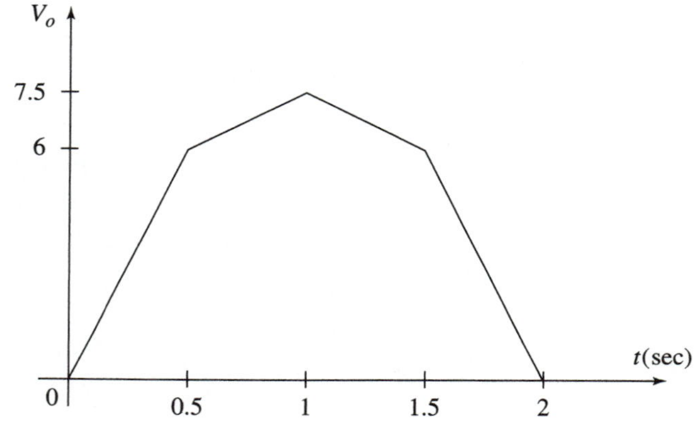

FIGURE 3–11 (Example 3–4)

series resistor. The diode starts conducting at $t = 0.5$ sec when $V_S = 6$ V, that is, when it overcomes the 5.3-V battery plus the 0.7-V diode drop. When the diode is conducting, the equivalent circuit in Figure 3–10 results.

Using superposition and the voltage-division rule, we can write (using V, mA, and kΩ) an expression for V_o as a function of V_S:

$$V_o = \left(\frac{1}{1+3}\right)V_S + \left(\frac{3}{1+3}\right)6 = 0.25\ V_S + 4.5\ \ (V_S \geq 6\text{ V})$$

Alternatively, we could have written the loop current $I = (V_S - 6)/4$, from which we obtain the voltage V_o as

$$V_o = I(1) + 6 = \frac{V_S - 6}{4} + 6 = 0.25\ V_S + 4.5\text{ V}\ \ (V_S \geq 6\text{ V})$$

According to this expression, when $V_S = 6$ V at $t = 0.5$ sec, $V_o = 0.25(6) + 4.5 = 6$ V, as expected. At $t = 1$ sec, when $V_S = 12$ V, V_o is $0.25(12) + 4.5 = 7.5$ V. Using symmetry, we can determine the behavior of the circuit for $1 \leq t \leq 2$. The resulting waveform for V_o is illustrated in Figure 3–11.

The Load Line

The concept of biasing a diode by means of a voltage source and a series resistor can also be seen from a graphical perspective. In later chapters, we will see that this graphical analysis applies to other devices as well. Consider the diode circuit of Figure 3–12. According to KVL, the loop equation yields $I_D = (E - V_D)/R = -(1/R)V_D + E/R$. Note that this equation is a straight line (recall $y = mx + b$) with slope $-1/R$ and y-intercept (I_D-intercept, in our case) E/R. The relationship between V_D and I_D also obeys the diode equation, namely:

$$I_D \approx I_s e^{V_D/V_T} \tag{3–9}$$

So if we plot both the straight line and the exponential diode curve as shown in Figure 3–12, the crossing point will clearly represent the simultaneous solution of the two equations. The coordinates of this point are the operating voltage and current of the diode for the given E and R values, that is, the biasing operating point (Q-point). If the resistor R is reduced in value, the slope of the straight line will be steeper (dotted line) with the crossing point at a higher current as expected.

To further illustrate the point, suppose we wanted to reduce the current back to its original value but now by reducing the source voltage E. This effect can also be seen graphically as the shifting of the straight line to the left while maintaining the same slope (same resistance) as shown in the figure.

The process just described dealt with changing the resistance or the source voltage to change the operating point. Now let us look at what happens if, for instance, the diode is exposed to higher temperatures. As we know, the diode drop decreases at a rate of approximately −2.2 mV/°C. Figure 3–12 shows how the operating point moves according to the shifting of the diode curve due to change in temperature. As temperature increases, V_{DQ} decreases and I_{DQ} increases. In this situation, of course, we are not dealing with changing the operating point for design purposes, but with the undesirable effect of having the operating point displaced due to changes in temperature.

3–5 ELEMENTARY POWER SUPPLIES

A typical application of diodes is in the construction of *dc power supplies,* which are electronic circuits that convert ac voltage to dc voltage, or dc voltage to a different dc voltage. Every electronic apparatus needs a power supply to operate. In this section, we will learn about the different basic schemes of obtaining dc voltage from an ac voltage source, such as a 120-V household outlet.

We will first look at the behavior of a diode when it is operated under *large-signal* conditions, that is, when the current and voltage changes it undergoes extend over a large portion of its characteristic curve, from full conduction to reverse biasing and vice versa. When this is the case, the diode resistance will change between very small and very large values and, for all practical purposes, the diode will behave very much like a *switch.*

An ideal (perfect) switch has zero resistance when closed and infinite resistance when open. Similarly, an *ideal* diode for large-signal applications is one whose resistance changes between these same two extremes. When analyzing such circuits, it is often helpful to think of the diode as a *voltage-controlled switch:* A forward-biasing voltage closes it, and a zero or reverse-biasing voltage opens it. Depending on the magnitudes of other voltages in the circuit, the 0.3-

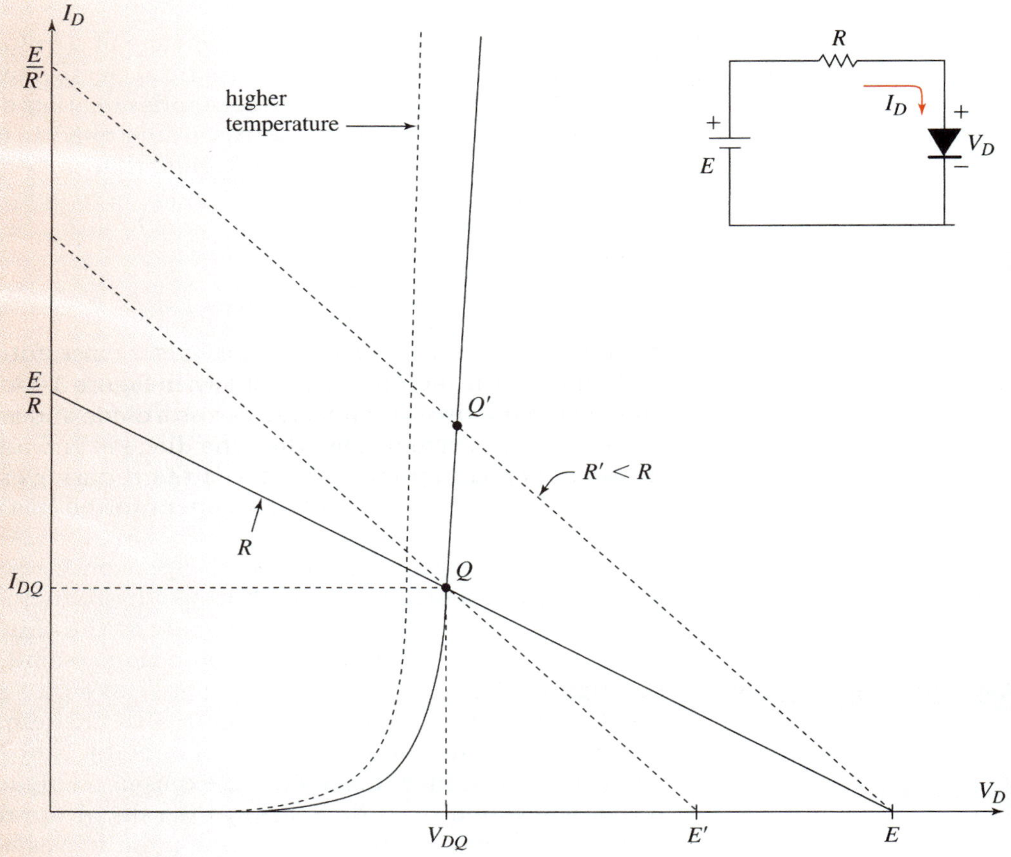

FIGURE 3–12 *Q* point displacement due to changes in load line and temperature

or 0.7-V drop across the diode when it is forward biased may or may not be negligible. Figure 3–13 shows the idealized characteristic curve for a silicon diode (a) when the 0.7-V drop is neglected and (b) when it is not. In case (a), the characteristic curve is the same as that of a perfect switch.

Half-Wave and Full-Wave Rectifiers

One of the most common uses of a diode in large-signal operation is in a *rectifier* circuit. A rectifier is a device that permits current to flow through it in one direction only. It is easy to see how a diode performs this function when we think of it as a voltage-controlled switch. When the anode voltage is positive with respect to the cathode, i.e., when the diode is forward biased, the "switch is closed" and current flows through it from anode to cathode. If the anode becomes negative with respect to the cathode, the "switch is open"

FIGURE 3–13 Idealized silicon diode characteristics, used for large-signal analysis

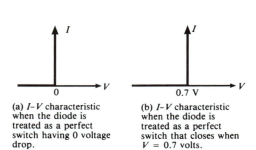

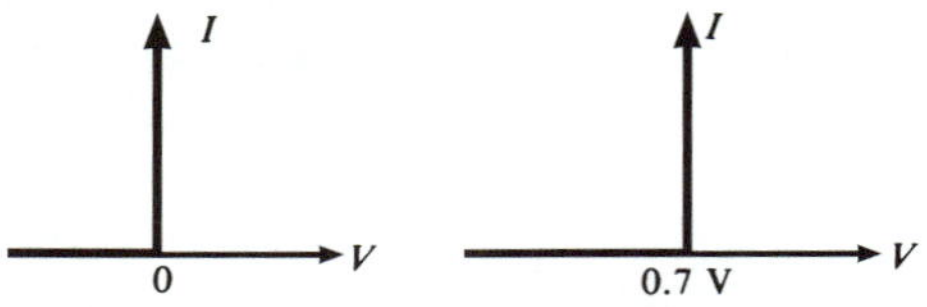

(a) I–V characteristic when the diode is treated as a perfect switch having 0 voltage drop.

(b) I–V characteristic when the diode is treated as a perfect switch that closes when V = 0.7 volts.

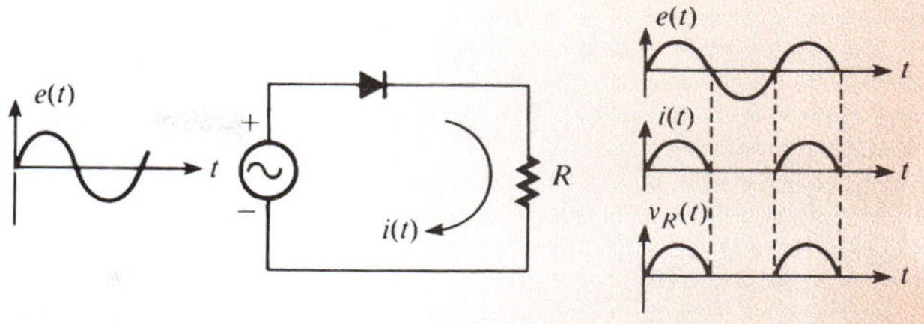

FIGURE 3–14 The diode used as a rectifier. Current flows only during the positive half-cycle of the input.

and no current flows. Of course, a *real* diode is not perfect, so there is in fact some very small reverse current that flows when it is reverse biased. Also, as we know, there is a nonzero voltage drop across the diode when it is forward biased (0.3 or 0.7 V), a drop that would not exist if it were a perfect switch.

Consider the rectifier circuit shown in Figure 3–14. We see in the figure that an ac voltage source is connected across a diode and a resistor, *R*, the latter designed to limit current flow when the diode is forward biased. Notice that no dc source is present in the circuit. Therefore, during each positive half-cycle of the ac source voltage $e(t)$, the diode is forward biased and current flows through it in the direction shown. During each negative half-cycle of $e(t)$ the diode is reverse biased and no current flows. The waveforms of $e(t)$ and $i(t)$ are sketched in the figure. We see that $i(t)$ is a series of positive current pulses separated by intervals of zero current. Also sketched is the waveform of the voltage $v_R(t)$ that is developed across *R* as a result of the current pulses that flow through it. Note that the net effect of this circuit is the conversion of an ac voltage into a pulsating dc voltage, a fundamental step in the construction of a dc *power supply*.

If the diode in the circuit of Figure 3–14 is turned around so that the anode is connected to the resistor and the cathode to the generator, then the diode will be forward biased during the negative half-cycles of the sine wave. The current would then consist of a sequence of pulses representing current flow in a counterclockwise, or negative, direction around the circuit.

EXAMPLE 3–5

Assume that the silicon diode in the circuit of Figure 3–15 has a characteristic like that shown in Figure 3–13(b). Find the peak values of the current $i(t)$ and the voltage $v_R(t)$ across the resistor when

1. $e(t) = 20 \sin \omega t$, and
2. $e(t) = 1.5 \sin \omega t$. In each case, sketch the waveforms for $e(t)$, $i(t)$, and $v_R(t)$.

Solution

1. When $e(t) = 20 \sin \omega t$, the peak positive voltage generated is 20 V. At the instant $e(t) = 20$ V, the voltage across the resistor is $20 - 0.7 = 19.3$ V, and the current is $i = 19.3/(1.5\ \text{k}\Omega) = 12.87$ mA. Figure 3–16 shows the resulting waveforms. Note that because of the characteristic assumed in Figure 3–13(b), the diode does not begin conducting until $e(t)$ reaches $+0.7$ V and ceases conducting when $e(t)$ drops below 0.7 V. The time interval between the point where $e(t) = 0$ V and $e(t) = 0.7$ V is very short in comparison to the half-cycle of conduction time. From a practical

FIGURE 3–15 (Example 3-5)

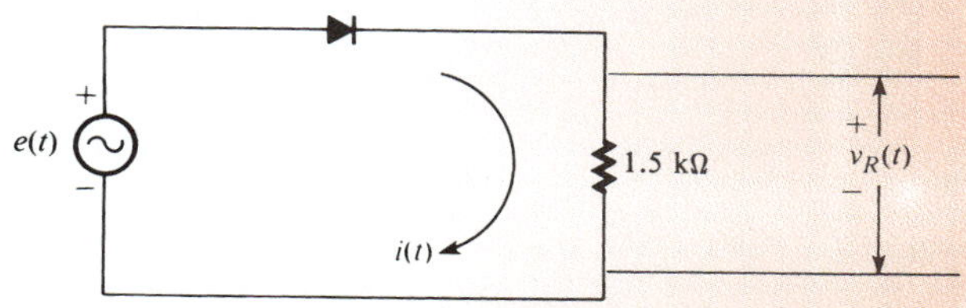

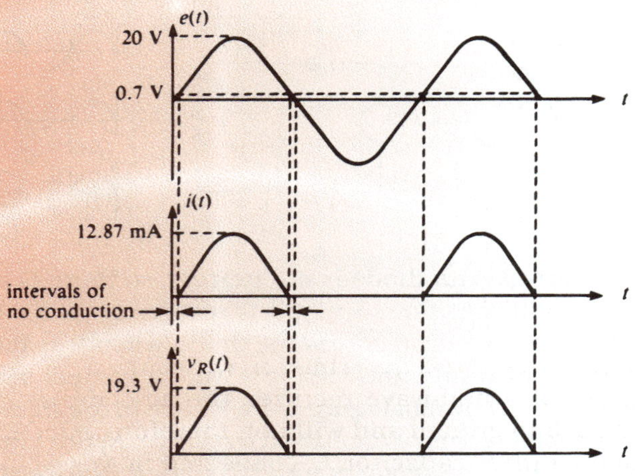

FIGURE 3–16 Diode current and voltage in the circuit of Figure 3–15. Note that the diode does not conduct until $e(t)$ reaches 0.7 V, so short intervals of nonconduction occur during each positive half-cycle.

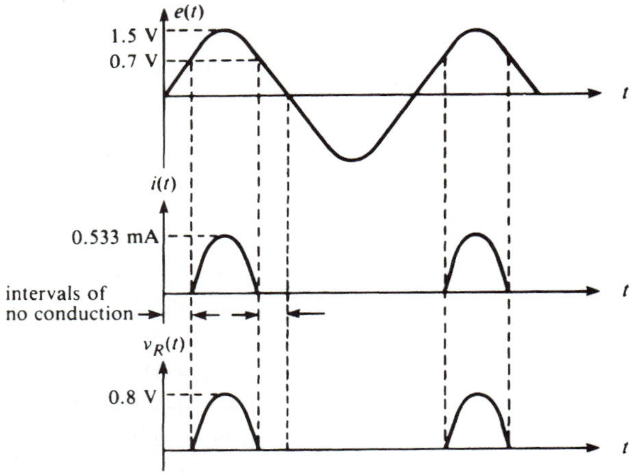

FIGURE 3–17 Diode current and voltage in the circuit of Figure 3–15 when the sine wave peak is reduced to 1.5 V. Note that the intervals of nonconduction are much longer than those in Figure 3–16.

standpoint, we could have assumed the characteristic in Figure 3–13(a), i.e., neglected the 0.7-V drop, and the resulting waveforms would have differed little from those shown.

2. When $e(t) = 1.5 \sin \omega t$, the peak positive voltage generated is 1.5 V. At that instant, $v_R(t) = 1.5 - 0.7 = 0.8\text{ V}$ and $i(t) = (0.8\text{ V})/(1.5\text{ k}\Omega) = 0.533\text{ mA}$. The waveforms are shown in Figure 3–17. Note once again that the diode does not conduct until $e(t) = 0.7$ V. However, in this case, the time interval between $e(t) = 0$ V and $e(t) = 0.7$ V is a significant portion of the conducting cycle. Consequently, current flows in the circuit for significantly less time than one-half cycle of the ac waveform. In this case, it clearly would *not* be appropriate to use Figure 3–13(a) as an approximation for the characteristic curve of the diode.

As already mentioned, an important application of diodes is in the construction of dc power supplies. It is instructive at this time to consider how diode rectification and waveform filtering, the first two operations performed by every power supply, are used to create an elementary dc power source.

The single-diode circuit in Figure 3–14 is called a *half-wave* rectifier, because the waveforms it produces ($i(t)$ and $v_R(t)$) each represent half a sine wave. These half-sine waves are a form of pulsating dc and by themselves are of little practical use. (They can, however, be used for charging batteries, an application in which a steady dc current is not required.)

Capacitive Filtering

Most practical electronic circuits require a dc voltage source that produces and maintains a *constant* voltage. For that reason, the pulsating half-sine waves must be converted to a steady dc level. This conversion is accomplished by *filtering* the waveforms. Filtering is a process in which selected frequency components of a complex waveform are *rejected* (filtered out) so that they do not appear in the output of the device (the filter) performing the filtering operation. The pulsating half-sine waves (like all periodic waveforms) can be regarded as waveforms that have both a dc component and ac components.

Our purpose in filtering these waveforms for a dc power supply is to reject *all* the ac components. It can be shown that the average (or dc) value of a half-wave rectified waveform is given by

$$V_{avg} = \frac{V_{PR}}{\pi} \quad \text{(half-wave)} \tag{3–10}$$

where V_{PR} = peak value of the rectified waveform $= \sqrt{2}V_{rms} - 0.7$ V.

This average or dc value is what a dc voltmeter would read when connected across the load resistor.

The simplest kind of filter that will perform the filtering task we have just described is a capacitor. Recall that a capacitor has reactance inversely proportional to frequency: $X_C = 1/2\pi fC$. Thus, if we connect a capacitor directly across the output of a half-wave rectifier, the ac components will "see" a low-impedance path to ground and will not, therefore, appear in the output. Figure 3–18 shows a filter capacitor, C, connected in this way. In this circuit the capacitor charges to the peak value of the rectified waveform, V_{PR}, so the output is the dc voltage V_{PR}. Note that $V_{PR} = V_P - V_D$, where V_P is the peak value of the sinusoidal input and V_D is the dc voltage drop across the diode (0.7 V for silicon).

In practice, a power supply must provide dc current to whatever load it is designed to serve, and this load current causes the capacitor to discharge and its voltage to drop. The capacitor discharges during the intervals of time between input pulses. Each time a new input pulse occurs, the capacitor recharges. Consequently, the capacitor voltage rises and falls in synchronism with the occurrence of the input pulses. These ideas are illustrated in Figure 3–19. The output waveform is said to have a *ripple voltage* superimposed on its dc level, V_{dc}.

When the peak-to-peak value of the output ripple voltage, V_{PP}, is small compared to V_{dc}, (a condition called *light loading*), we can assume that the load current is essentially constant and will discharge the capacitor linearly according to the basic equation

$$\Delta V = \frac{I\Delta t}{C} \tag{3–11}$$

where ΔV is the reduction in capacitor voltage over the time interval Δt and I is the current discharging the capacitor. Notice that Δt, within our discussion, is very close to the period of the rectified waveform; therefore, replacing ΔV with V_{PP}, we can write the following expression:

$$V_{PP} = \frac{I_L}{f_r C} = \frac{V_{dc}}{f_r R_L C} \tag{3–12}$$

where I_L = load current
f_r = frequency of the rectified waveform
C = filter capacitance

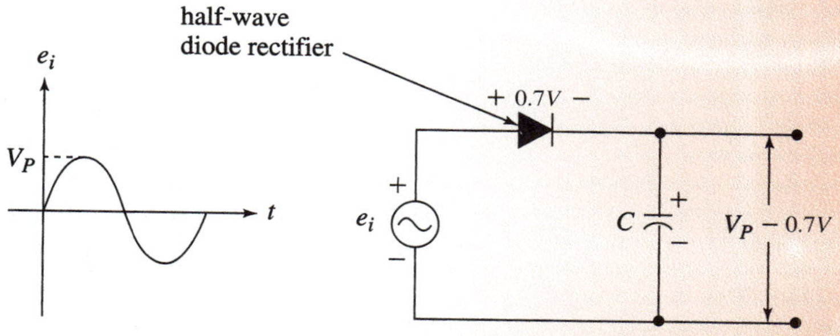

FIGURE 3–18 Filter capacitor C effectively removes the ac components from the half-wave rectified waveform.

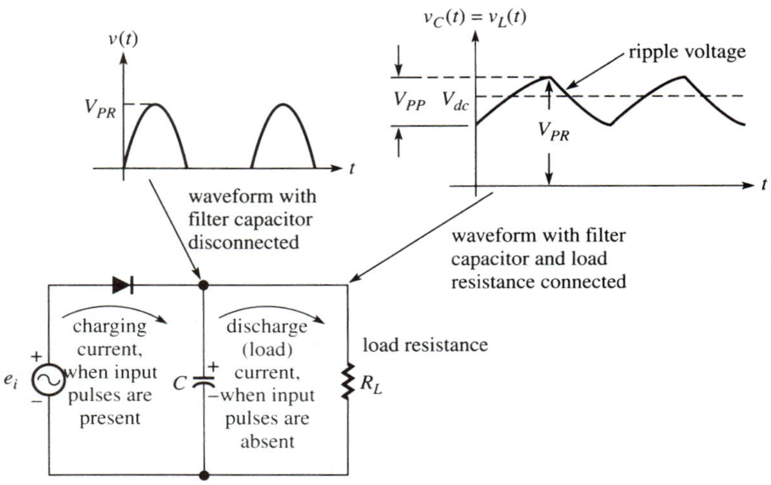

FIGURE 3–19 When load resistance R_L is connected across the filter capacitor, the capacitor charges and discharges, creating a load voltage that has a ripple voltage superimposed on a dc level.

The dc voltage across the load is the average of the maximum and minimum values caused by the ripple voltage. This can clearly be approximated as the maximum value minus one half of V_{PP}, expressly

$$V_{dc} = V_{PR} - \frac{V_{PP}}{2} \tag{3–13}$$

or

$$V_{dc} = V_{PR} - \frac{I_L}{2f_rC} \tag{3–14}$$

Note that this form is expressed in terms of I_L, which is the most general situation. However, in the case where the load is a fixed resistance R_L, I_L can be replaced with V_{dc}/R_L, yielding

$$V_{dc} = \frac{V_{PR}}{1 + \dfrac{1}{2f_r R_L C}} \tag{3–15}$$

Observing a rectified and filtered waveform, it is obvious that the smaller the variation V_{PP}, the more the waveform will resemble a pure dc voltage. The variation portion is known as *ripple* and the value V_{PP} is known as the *ripple voltage.* Furthermore, the ratio of the ripple voltage to the dc or average voltage is known as the *ripple factor* or *percent ripple* and represents a measure of how close the filtered waveform resembles a dc voltage. Obviously, low ripple factors are desirable and can be achieved by properly selecting the capacitor value. The ripple factor r is then expressed by

$$r = \frac{V_{PP}}{V_{dc}} \times 100\%$$

Ripple factors of up to 10% are typically acceptable in noncritical applications. However, precision electronic circuits could require supply voltages with very low ripple factors. Although ripple factors could be reduced arbitrarily by using large capacitor values, a more practical solution is to use special circuits called *voltage regulators* that not only reduce the ripple voltage substantially but also maintain a constant dc voltage under variable load current. These circuits will be addressed later in the chapter.

EXAMPLE 3–6

The sinusoidal input, e_i, in Figure 3–19 is 120 V rms and has frequency 60 Hz. The load resistance is 2 kΩ and the filter capacitance is 100 μF. Assuming light loading and neglecting the voltage drop across the diode,

1. find the dc value of the load voltage;
2. find the peak-to-peak value of the ripple voltage.

Solution

1. The peak value of the sinusoidal input voltage is $V_P = \sqrt{2}$ (120 V rms) = 169.7 V. Since the voltage drop across the diode can be neglected, $V_{PR} = V_P = 169.7$ V. From equation 3–15,

$$V_{dc} = \frac{169.7\text{ V}}{1 + \dfrac{1}{2(60\text{ Hz})(2\text{ k}\Omega)(100\ \mu\text{F})}} = 162.9\text{ V}$$

2. From equation 3–12,

$$V_{PP} = \frac{162.9\text{ V}}{(60\text{ Hz})(2\text{ k}\Omega)(100\ \mu\text{F})} = 13.57\text{ V}$$

Full-Wave Rectification

A *full-wave* rectifier effectively inverts the negative half-pulses of a sine wave to produce an output that is a sequence of positive half-pulses with no intervals between them. Figure 3–20 shows a widely used full-wave rectifier constructed from four diodes and called a full-wave diode *bridge*. Also shown is the full-wave rectified output. In this case, the average or dc value of the rectified waveform is

$$V_{avg} = \frac{2}{\pi}V_{PR} \quad \text{(full-wave)} \qquad (3\text{–}16)$$

Note that on each half-cycle of input, current flows through *two* diodes, so the peak value of the rectified output is $V_{PR} = V_P - 2V_D$ or $V_P - 1.4$ V for silicon.

As in the half-wave rectifier, the full-wave rectified waveform can be filtered by connecting a capacitor in parallel with load R_L. The advantage of the full-wave rectifier is that the capacitor does not discharge so far between input pulses, because a new charging pulse occurs every half-cycle instead of every full cycle. Consequently, the magnitude of the output ripple voltage is smaller. This fact is illustrated in Figure 3–21.

Equations 3–12 and 3–14 for V_{PP} and V_{dc} are valid for both half-wave and full-wave rectifiers. Note that f_r in those equations is the frequency of the *rectified* waveform, which, in a full-wave rectifier, is *twice* the frequency of the

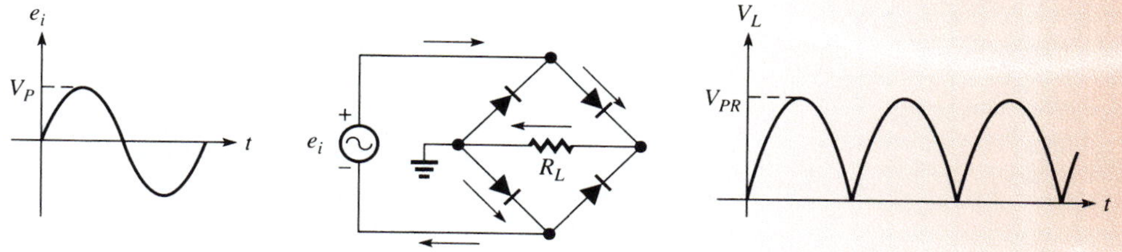

FIGURE 3–20 The full-wave bridge rectifier and output waveform. The arrows show the direction of current flow when e_i is positive.

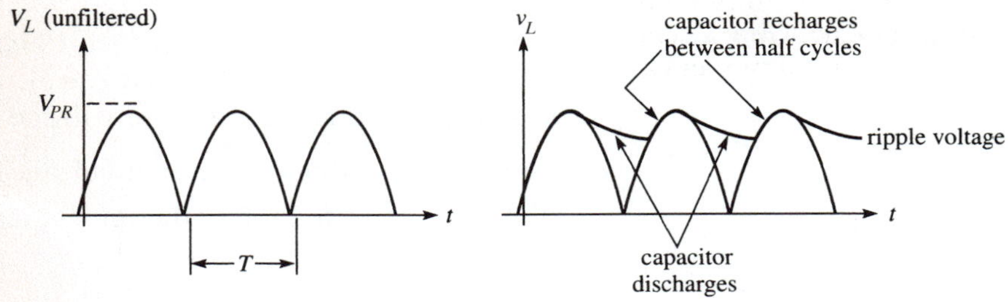

FIGURE 3–21 The ripple voltage in the filtered output of a full-wave rectifier is smaller than in the half-wave case because the capacitor recharges at shorter intervals.

unrectified sine wave (see Figure 3–21). If the same input and component values used in Example 3–6 are used to compute V_{dc} and V_{PP} for a full-wave rectifier (f_r = 120 Hz), we find

$$V_{dc} = \frac{169.7\text{ V}}{1 + \dfrac{1}{2(120\text{ Hz})(2\text{ k}\Omega)(100\ \mu\text{F})}} = 166.2\text{ V}$$

and

$$V_{PP} = \frac{166.2\text{ V}}{(120\text{ Hz})(2\text{ k}\Omega)(100\ \mu\text{F})} = 6.92\text{ V}$$

Note that the value of the ripple voltage is one-half that found for the half-wave rectifier.

Another means of obtaining full-wave rectification is through a center-tapped transformer and two diodes, as shown in Figure 3–22. Assume that the transformer is wound so that terminal A on the secondary is positive with respect to terminal B at an instant of time when v_{in} is positive, as signified by the polarity symbols (dot convention) shown in the figure. Then, with the center tap as reference (ground), v_A is positive with respect to ground and v_B is negative with respect to ground. Similarly, when v_{in} is negative, v_A is negative with respect to ground and v_B is positive with respect to ground.

Figure 3–22(b) shows that when v_{in} is positive, v_A forward biases diode D_1. As a consequence, current flows in a clockwise loop through R_L. Figure 3–22(c) shows that when v_{in} is negative, D_1 is reverse biased, D_2 is forward biased, and current flows through R_L in a counterclock-wise loop. Notice that the voltage developed across R_L has the same polarity in either case. Therefore, positive voltage pulses are developed across R_L during both the positive and negative half-cycles of v_{in}, and a full-wave–rectified waveform is created.

The peak rectified voltage is the secondary voltage in the transformer, between center tap and one side, less the diode drop:

$$V_{PR} = V_P - 0.7\text{ V} \qquad \textbf{(3–17)}$$

where V_P is the peak secondary voltage per side.

To determine the maximum reverse bias to which each diode is subjected, refer to the circuit in Figure 3–23. Here, we show the voltage drops in the rectifier when diode D_1 is forward biased and diode D_2 is reverse biased. Neglecting the 0.7-V drop across D_1, the voltage across R_L is v_A volts. Thus the cathode-to-ground voltage of D_2 is v_A volts. Now, the anode-to-ground voltage of D_2 is $-\ v_B$ volts, as shown in the figure. Therefore, the total reverse bias across D_2 is $v_A + v_B$ volts, as shown. When v_A is at its positive peak, v_B is at its negative peak, so the maximum reverse bias equals *twice* the peak value of either. We conclude that the PIV *(peak inverse voltage)* rating of each diode must be equal to at least twice the peak value of the rectified voltage:

$$\text{PIV} \geq 2V_{PR}$$

FIGURE 3–22 A full-wave rectifier employing a center-tapped transformer and two diodes

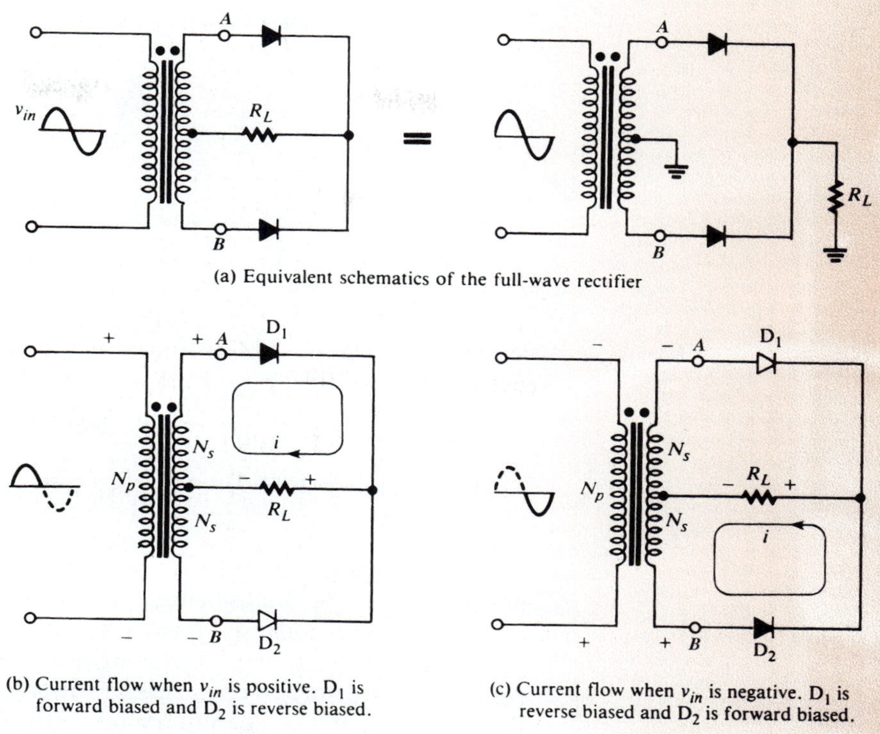

FIGURE 3–23 Diode D_2 is reverse biased by $v_A + v_B$ volts, which has a maximum value of $2V_{PR}$ volts

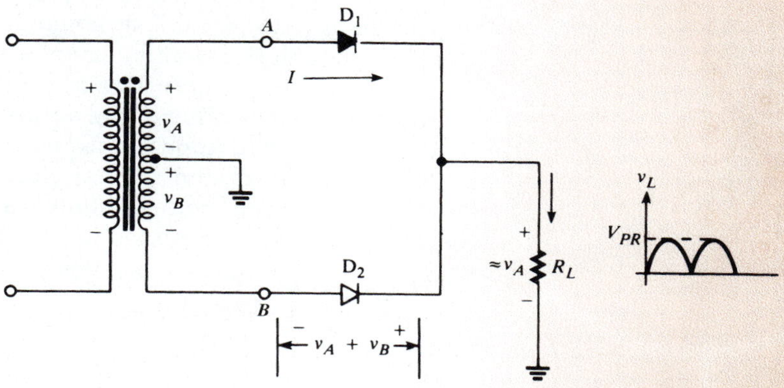

EXAMPLE 3–7

The primary voltage in the circuit shown in Figure 3–24 is 120 V rms, and the secondary voltage is 60 V rms from side to side (60 VCT). Find

1. the average value of the voltage across R_L;
2. the (approximate) average power dissipated by R_L; and
3. the minimum PIV rating required for each diode.

Solution

1. $V_P = \sqrt{2}(30) = 42.4$ V per side

   From equation 3–17, $V_{PR} = V_P - 0.7\text{ V} = 42.4 - 0.7\text{ V} = 41.7\text{ V}$.

   Although equation 3–16 does not strictly apply to a rectified waveform with 0.7-V nonconducting gaps, it is a good approximation when the peak value is so much greater than 0.7 V:

$$V_{avg} \approx \frac{2V_{PR}}{\pi} = \frac{2(41.7\text{ V})}{\pi} = 26.5\text{ V}$$

FIGURE 3–24 (Example 3–7)

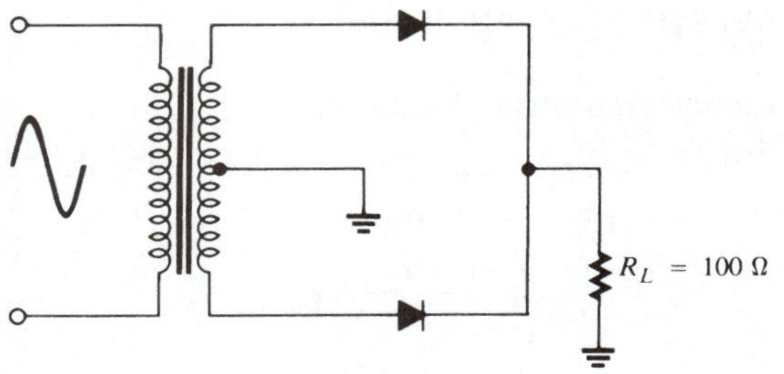

2. $$V_{rms} = \frac{V_{PR}}{\sqrt{2}} = \frac{41.7\text{ V}}{\sqrt{2}} = 29.5\text{ V rms}$$

$$P_{avg} = \frac{V_{rms}^2}{R_L} = \frac{(29.5\text{ V})^2}{100\ \Omega} = 8.7\text{ W}$$

3. $\text{PIV} \geq 2V_{PR} = 2(41.7) = 83.4\text{ V}$

Capacitive filtering with the center-tap configuration is identical to that described for the diode-bridge circuit. The only difference is the peak voltage of the rectified waveform which, in this case, involves only one diode drop. Remember that in this configuration, the circuit rectifies the voltage from the center tap to each side of the transformer in alternating half-cycles. For example, if the secondary of a transformer is rated 18 VCT, meaning 18 volts rms with a center tap, the peak voltage of the rectified waveform V_{PR} will be $9 \times 1.414 - 0.7$, or about 12 V.

Although the elementary power supplies we have described can be used in applications where the presence of some ripple voltage is acceptable, where the exact value of the output voltage is not critical, and where the load does not change appreciably, more sophisticated power supplies have more elaborate filters and special circuitry (voltage regulators) that maintain a constant output voltage under a variety of operating conditions. These refinements are discussed in detail in Chapter 13.

Voltage Multipliers

Diodes and capacitors can be connected in various configurations to produce filtered, rectified voltages that are integer multiples of the peak value of an input sine wave. By using a transformer to change the amplitude of an ac voltage before it is applied to a voltage multiplier, a wide range of dc levels can be produced using this technique. One advantage of a voltage multiplier is that high voltages can be obtained without using a high-voltage transformer.

HALF-WAVE VOLTAGE DOUBLER Figure 3–25(a) shows a half-wave voltage *doubler*. When v_{in} first goes positive, diode D_1 is forward biased and diode D_2 is reverse biased. Because the forward resistance of D_1 is quite small, C_1 charges rapidly to V_P (neglecting the diode drop), as shown in (b). During the ensuing negative half-cycle of v_{in}, D_1 is reverse biased and D_2 is forward biased, as shown in (c). Consequently, C_2 charges rapidly, with polarity shown. Neglecting the drop across D_2, we can write Kirchhoff's voltage law around the loop at the instant v_{in} reaches its negative peak, and obtain

$$V_P = -V_P + V_{C_2}$$

or

FIGURE 3–25 A half-wave voltage doubler

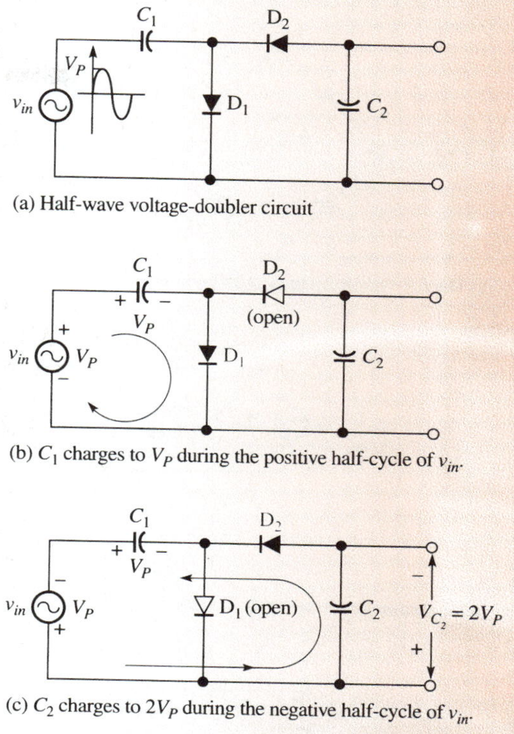

$$V_{C_2} = 2V_P \tag{3–18}$$

During the next positive half-cycle of v_{in}, D_2 is again reverse biased and the voltage across the output terminals remains at $V_{C_2} = 2V_P$ volts. Note carefully the polarity of the output. If a load resistor is connected across C_2, then C_2 will discharge into the load during positive half-cycles of v_{in} and will recharge to $2V_P$ volts during negative half-cycles, creating the usual ripple waveform. The PIV rating of each diode must be at least $2V_P$ volts.

FULL-WAVE VOLTAGE DOUBLER Figure 3–26(a) shows a full-wave voltage doubler. This circuit is the same as the full-wave bridge rectifier shown in Figure 3–20, with two of the diodes replaced by capacitors. When v_{in} is positive, D_1 conducts and C_1 charges to V_P volts, as shown in (b). When v_{in} is negative, D_2 conducts and C_2 charges to V_P volts, with the polarity shown in (c). It is clear that the output voltage is then $V_{C_1} + V_{C_2} = 2V_P$ volts. Since one or the other of the capacitors is charging during every half-cycle, the output is the same as that of a capacitor-filtered, full-wave rectifier. Note, however, that the effective filter capacitance is that of C_1 and C_2 in series, which is less than either C_1 or C_2. The PIV rating of each diode must be at least $2V_P$ volts.

VOLTAGE TRIPLER AND QUADRUPLER By connecting additional diode-capacitor sections across the half-wave voltage doubler, output voltages equal to three and four times the input peak voltage can be obtained. The circuit is shown in Figure 3–27. When v_{in} first goes positive, C_1 charges to V_P through forward-biased diode D_1. On the ensuing negative half-cycle, C_2 charges through D_2 and, as demonstrated earlier, the voltage across C_2 equals $2V_P$. During the next positive half-cycle, D_3 is forward biased and C_3 charges to the same voltage attained by C_2: $2V_P$ volts. On the next negative half-cycle, D_2 and D_4 are forward biased and C_4 charges to $2V_P$ volts. As shown in the figure, the voltage across the combination of C_1 and C_3 is $3V_P$ volts, and that across C_2 and C_4 is $4V_P$ volts.

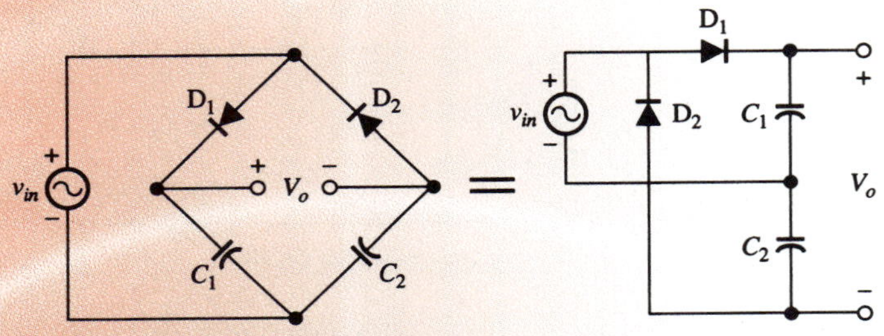

(a) The full-wave voltage-doubler circuit

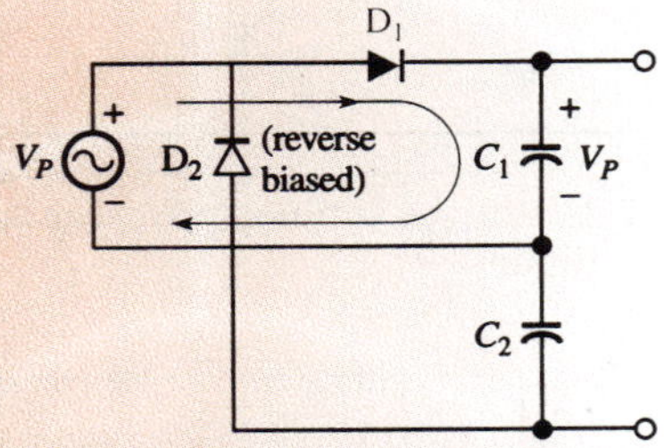

(b) C_1 charges to V_P during the positive half-cycle of v_{in}.

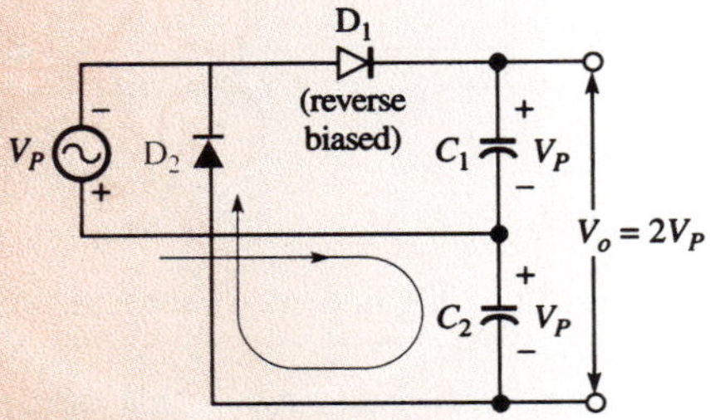

(c) C_2 charges to V_P during the negative half-cycle of v_{in}.

FIGURE 3–26 A full-wave voltage doubler

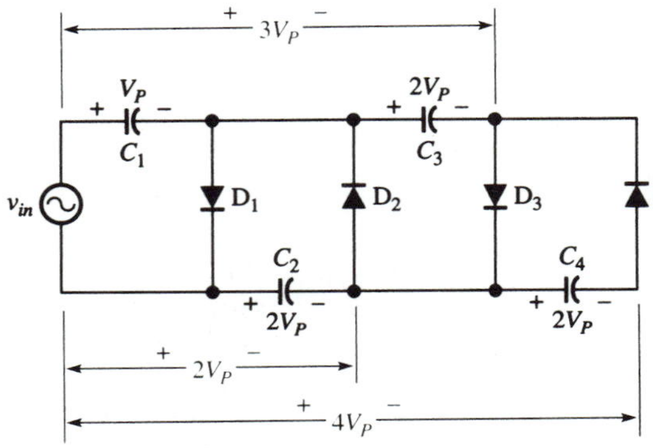

FIGURE 3–27 Voltage tripler and quadrupler

Additional stages can be added in an obvious way to obtain even greater multiples of V_P. The PIV rating of each diode in the circuit must be at least $2V_P$ volts.

3–6 ELEMENTARY VOLTAGE REGULATION

A basic power supply consisting of a transformer, one or more diodes, and a capacitor filter is subject to output voltage variations caused by changes in the load current and the ac line voltage to which the primary of the transformer is connected. If the dc voltage provided by the power supply needs to be constant regardless of the changes just mentioned, some form of voltage regulation must be employed. A simple form of voltage regulation can be obtained by using a *zener* diode, mentioned in Chapter 2, or by employing a three-terminal integrated-circuit (IC) voltage regulator with fixed or adjustable voltage. More advanced power supplies and voltage regulators will be presented later in this text where additional electrical parameters and temperature effects will also be addressed.

The Zener-Diode Voltage Regulator

Zener diodes are specifically designed for operating in their reverse-bias region and are fabricated with a specific zener (avalanche) voltage and power rating. Their forward-bias behavior is no different from common diodes.

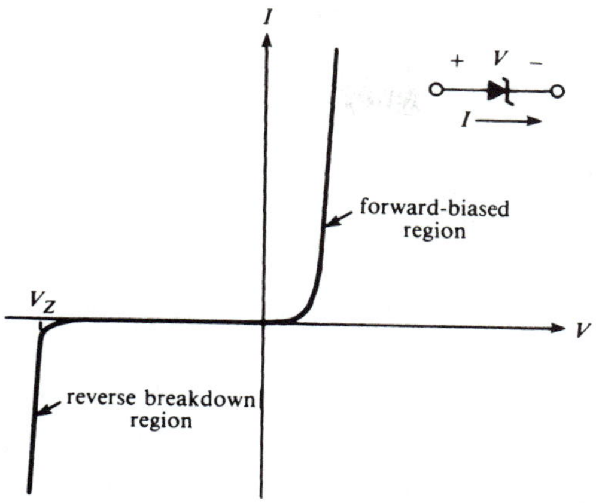

FIGURE 3–28 *I–V* characteristic of a zener diode

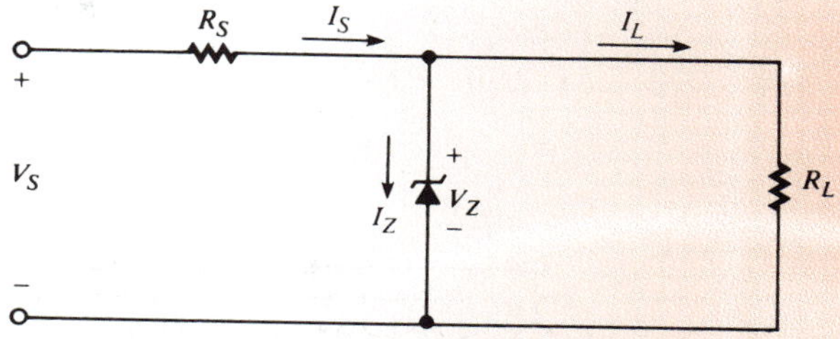

FIGURE 3–29 A simple voltage regulator using a zener diode

Figure 3–28 shows a typical *I–V* characteristic for a zener diode. The forward-biased characteristic is identical to that of a forward-biased silicon diode and obeys the same diode equation that we developed in Chapter 2 (equation 2–2). The zener diode is normally operated in its reverse-biased breakdown region, where the voltage across the device remains substantially constant as the reverse current varies over a large range. Like a fixed voltage source, this ability to maintain a constant voltage across its terminals, independent of current, is what makes the device useful as a voltage reference. The fixed breakdown voltage is called the *zener voltage*, V_Z, as illustrated in the figure.

To demonstrate how a zener diode can serve as a constant voltage reference, Figure 3–29 shows a simple but widely used configuration that maintains a constant voltage across a load resistor. Notice the orientation of V_Z and I_Z. The circuit is an elementary *voltage regulator* that holds the load voltage near V_Z volts as R_L and/or V_S undergo changes. So the voltage across the parallel combination of the zener and R_L remains at V_Z volts, the reverse current I_Z through the diode must at all times be large enough to keep the device in its breakdown region, as shown in Figure 3–28. The value selected for R_S is critical in that respect. As we shall presently demonstrate, R_S must be small enough to permit adequate zener current, yet large enough to prevent the zener current and power dissipation from exceeding permissible limits.

A couple of rules regarding the operation of the zener diode in Figure 3–29 should be addressed at this point:

- The zener current changes in direct proportion to input voltage variations.
- The zener current changes in inverse proportion to load current variations.

This is because the zener diode adjusts its current in order to increase or decrease the voltage drop across R_S and hence maintain a constant voltage V_Z. If V_S increases, I_Z increases, and vice versa; if I_L increases, I_Z decreases, and vice versa.

It is apparent in Figure 3–29 that

$$I_S = I_Z + I_L \tag{3–19}$$

Also, I_S is the voltage difference across R_S divided by R_S:

$$I_S = \frac{V_S - V_Z}{R_S} \tag{3–20}$$

The power dissipated in the zener diode is

$$P_Z = V_Z I_Z \tag{3–21}$$

Solving equation 3–20 for R_S, we find

$$R_S = \frac{V_S - V_Z}{I_S} \tag{3–22}$$

Substituting (3–19) into (3–22) gives

$$R_S = \frac{V_S - V_Z}{I_Z + I_L} \tag{3–23}$$

Let $I_Z(\text{min})$ denote the minimum zener current necessary to ensure that the diode is in its breakdown region. As mentioned earlier, R_S must be small enough to ensure that the $I_Z(\text{min})$ flows under worst-case conditions, namely, when V_S falls to its smallest possible value, $V_S(\text{min})$, and I_L reaches its largest possible value, $I_L(\text{max})$. Thus, from (3–23), we require

$$R_S = \frac{V_S(\text{min}) - V_Z}{I_Z(\text{min}) + I_L(\text{max})} \tag{3–24}$$

With the established value for R_S, we can now determine the actual power dissipation for the resistor and the zener diode. Obviously, maximum power will be dissipated by the resistor when the input voltage is maximum; that is,

$$P_{R_S}(\text{max}) = (V_S(\text{max}) - V_Z)^2/R_S \tag{3–25}$$

Because the zener voltage is constant, maximum power will be dissipated by the zener diode when I_Z is maximum. This happens when V_S is maximum and I_L is minimum according to the preceding rules. Therefore,

$$I_Z(\text{max}) = \frac{V_S(\text{max}) - V_Z}{R_S} - I_L(\text{min}) \tag{3–26}$$

and

$$P_Z(\text{max}) = V_Z I_Z(\text{max}) \tag{3–27}$$

Note that the power rating for R_S should be three or four times the actual maximum power dissipated by R_S. However, such a large safety factor is not necessary for the zener diode. A 50% safety margin is very adequate in this case. For example, if the maximum dissipated power in a zener diode is, say, 600 milliwatts, a 1-watt zener diode will do the job.

Regarding the minimum zener current necessary to keep the diode in its avalanche region (i.e., maintaining regulation), a good rule of thumb is to use 5% to 10% of the maximum load current but no less than a few milliamps in the case of small load currents. For instance, if the maximum load current is 150 mA, the minimum zener current can be set at about 10 mA or so. But if it is only about 10 mA, then the minimum zener current should probably not be set at about 1 mA but more like 3 or 4 mA. It is important to note that, as far as zener dissipation is concerned, the worst-case load condition in some applications may correspond to an open output; that is, $R_L = \infty$ and $I_L = 0$. In that case, all of the current through R_S flows in the zener.

EXAMPLE 3–8

In the circuit of Figure 3–29, $R_S = 20\ \Omega$, $V_Z = 18$ V, and $R_L = 200\ \Omega$. If V_S can vary from 20 V to 30 V, find

1. the minimum and maximum currents in the zener diode;
2. the minimum and maximum power dissipated in the diode; and
3. the rated power dissipation that R_S should have.

Solution

1. Assuming that the zener diode remains in breakdown, then the load voltage remains constant at $V_Z = 18\text{ V}$, and the load current therefore remains constant at

$$I_L = \frac{V_Z}{R_L} = \frac{18\text{ V}}{200\ \Omega} = 90\text{ mA}$$

From equation 3–20, when $V_S = 20\text{ V}$,

$$I_S = \frac{(20\text{ V}) - (18\text{ V})}{20\ \Omega} = 100\text{ mA}$$

Therefore, $I_Z = I_S - I_L = (100\text{ mA}) - (90\text{ mA}) = 10\text{ mA}$. When $V_S = 30\text{ V}$,

$$I_S = \frac{(30\text{ V}) - (18\text{ V})}{20\ \Omega} = 600\text{ mA}$$

and $I_Z = I_S - I_L = (600\text{ mA}) - (90\text{ mA}) = 510\text{ mA}$.

2. $P_Z(\text{min}) = V_Z I_Z(\text{min}) = (18\text{ V})(10\text{ mA}) = 180\text{ mW}$
$P_Z(\text{max}) = V_Z I_Z(\text{max}) = (18\text{ V})(510\text{ mA}) = 9.18\text{ W}$

3. $P_{R_S}(\text{max}) = I_S^2(\text{max})R_S = (0.6)^2(20) = 7.2\text{ W}$ (rated power should be 20 W)

EXAMPLE 3–9

The current in a certain 10-V, 2-W zener diode must be at least 5 mA to ensure that the diode remains in breakdown. The diode is to be used in the regulator circuit shown in Figure 3–30, where V_S can vary from 15 V to 20 V. Note that the load can be switched out of the regulator circuit in this application. Find a value for R_S. What power dissipation rating should R_S have?

Solution

$$V_S(\text{min}) = 15\text{ V}$$
$$I_Z(\text{min}) = 5\text{ mA}$$
$$R_L(\text{min}) = R_L = 500\ \Omega \text{ (when the switch is closed)}$$

Therefore, from equation 3–24,

$$R_S = \frac{V_S(\text{min}) - V_Z}{I_Z(\text{min}) + V_Z/R_L(\text{min})} = \frac{(15 - 10)\text{ V}}{(5\text{ mA}) + (10\text{ V})/(500\ \Omega)} = 200\ \Omega$$

FIGURE 3–30 (Example 3–9)

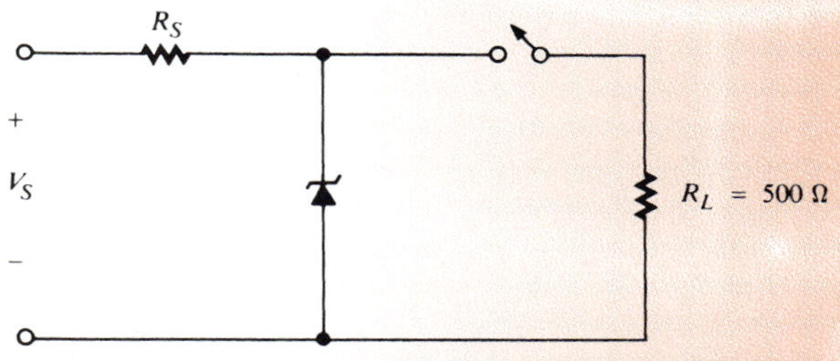

$$P_{R_S} = \frac{(20 - 10)^2}{200} = 0.5 \text{ W (use 2-watt rating)}$$

$$V_S(\text{max}) = 20 \text{ V}, \; I_L(\text{min}) = 0 \; \text{(switch open)}$$

$$I_Z(\text{max}) = \frac{V_S(\text{max}) - V_Z}{R_S} - I_L(\text{min}) = \frac{20 - 10}{200} - 0 = 50 \text{ mA}$$

$$P_Z(\text{max}) = (10 \text{ V})(50 \text{ mA}) = 0.5 \text{ W}$$

The zener diode is operating well under its power rating.

EXAMPLE 3–10

DESIGN

An unregulated dc power supply provides a dc voltage that can vary between 18 and 22 V. Design a 15-volt zener voltage regulator for a load having I_L (min) = 20 mA and I_L (max) = 120 mA. Specify resistor and zener diode values, including power ratings.

Solution

First, the resistor value is calculated based on worst-case conditions for the minimum zener current. Minimum zener current occurs when V_S is minimum and I_L is maximum. Using equation 3–24 with I_Z (min) = 5% of I_L (max), we obtain

$$R_S = \frac{V_S(\text{min}) - V_Z}{I_L(\text{max}) + I_Z(\text{min})} = \frac{18 - 15}{120 + 6} = 23.8 \; \Omega \; \text{(use 24 ohms)}$$

From equation 3–25,

$$P_{R_S}(\text{max}) = \frac{(22 - 15)^2}{24} = 2.04 \text{ W} \; \text{(use rating of 5 or 7 watts)}$$

From equation 3–26,

$$I_Z(\text{max}) = \frac{22 - 15}{24} - .02 \text{ A} = 272 \text{ mA}$$

$$P_Z(\text{max}) = (15 \text{ V})(272 \text{ mA}) = 4.08 \text{ W} \; \text{(use a 5-watt zener diode)}$$

Temperature Effects

The breakdown voltage of a zener diode is a function of the width of its depletion region, which is controlled during manufacturing by the degree of impurity doping. Recall that heavy doping increases conductivity, which narrows the depletion region and therefore decreases the voltage at which breakdown occurs. Zener diodes are available with breakdown voltages ranging from 2.4 V to 200 V. As noted in Chapter 2, the mechanism by which breakdown occurs depends on the breakdown voltage itself. When V_Z is less than about 5 V, the high electric field intensity across the narrow depletion region (around 3×10^7 V/m) strips carriers directly from their bonds, a phenomenon usually called *zener breakdown*. For V_Z greater than about 8 V, breakdown occurs as a result of collisions between high-energy carriers, the mechanism called *avalanching*. Between 5 V and 8 V, both the avalanching and zener mechanisms contribute to breakdown. The practical significance of these facts is that the breakdown mechanism determines how temperature variations affect the value of V_Z. Low-voltage zener diodes that break down by the zener mechanism have negative temperature coefficients (V_Z decreases with increasing temperature) and higher-voltage avalanche zeners have positive temperature coefficients. When V_Z is between about 3 V and 8 V, the temperature coefficient is also strongly influenced by the current in the diode: The coefficient may be positive or negative, depending on current, becoming more positive as current increases.

The temperature coefficient of a zener diode is defined to be its change in breakdown voltage per degree Celsius increase in temperature. For example, a temperature coefficient of +8 mV/°C means that V_Z will increase 8 mV for each degree Celsius increase in temperature. *Temperature stability* is the ratio of the temperature coefficient to the breakdown voltage. Expressed as a percent,

$$S(\%) = \frac{\text{T.C.}}{V_Z} \times 100\% \qquad (3\text{–}28)$$

where T.C. is the temperature coefficient. Clearly, small values of S are desirable.

In applications requiring a zener diode to serve as a highly stable voltage reference, steps must be taken to *temperature compensate* the diode. A technique that is used frequently is to connect the zener in series with one or more semiconductor devices whose voltage drops change with temperature in the opposite way that V_Z changes, i.e., devices having the opposite kind of temperature coefficient. If a temperature change causes V_Z to increase, then the voltage across the other components decreases, so the total voltage across the series combination is (ideally) unchanged. For example, the temperature coefficient of a forward-biased silicon diode is negative, so one or more of these can be connected in series with a zener diode having a positive temperature coefficient, as illustrated in Figure 3–31. The next example illustrates that several forward-biased diodes having relatively small temperature coefficients may be required to compensate a single zener diode.

EXAMPLE 3–11

A zener diode having a breakdown voltage of 10 V at 25°C has a temperature coefficient of +5.5 mV/°C. It is to be temperature compensated by connecting it in series with three forward-biased diodes, as shown in Figure 3–31. Each compensating diode has a forward drop of 0.65 V at 25°C and a temperature coefficient of −2 mV/°C.

1. What is the temperature stability of the uncompensated zener diode?
2. What is the breakdown voltage of the uncompensated zener diode at 100°C?
3. What is the voltage across the compensated network at 25°C? At 100°C?

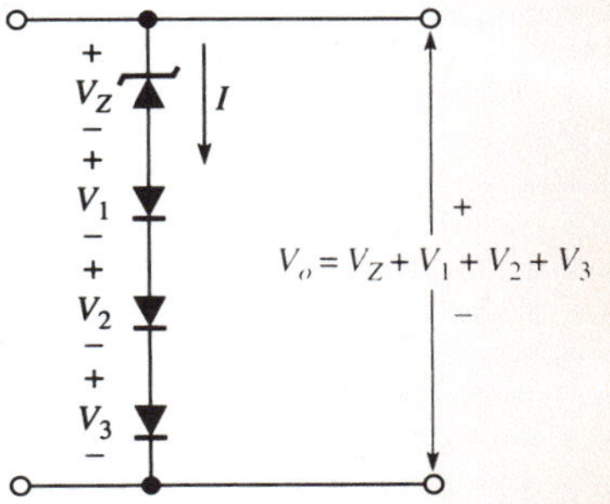

FIGURE 3–31 Temperature compensating a zener diode by connecting it in series with forward-biased diodes having opposite temperature coefficients

4. What is the temperature stability of the compensated network?

Solution

1. From equation 3–27,

$$S = \frac{\text{T.C.} \times 100\%}{V_Z} = \frac{5.5 \times 10^{-3}}{10\ \text{V}} \times 100\% = 0.055\%$$

2. $V_Z = (10\,\text{V}) + \Delta T(\text{T.C.}) = (10\,\text{V}) + (100°\text{C} - 25°\text{C})(5.5\ \text{mV/°C}) = 10.4125\,\text{V}$
3. As shown in Figure 3–31, $V_o = V_Z + V_1 + V_2 + V_3$. At 25°C, $V_o = 10 + 3(0.65) = 11.95\,\text{V}$. At 100°C, the drop V_D across each forward-biased diode is $V_D = (0.65\,\text{V}) + (100°\text{C} - 25°\text{C})(-2\ \text{mV/°C}) = 0.5\,\text{V}$. Therefore, at 100°C, $V_o = (10.4125\,\text{V}) + 3(0.5\,\text{V}) = 10.5625\,\text{V}$.
4. The temperature coefficient of the compensated network is T.C. = $(+5.5\ \text{mV/°C}) + 3(-2\ \text{mV/°C}) = (+5.5\ \text{mV/°C}) - (6\ \text{mV/°C}) = -0.5\ \text{mV/°C}$. The voltage drop across the network (at 25°C) was found to be 11.95 V, so

$$S = \frac{-0.5\ \text{mV/°C}}{11.95} \times 100\% = -0.00418\%$$

We see that compensation has improved the stability by a factor greater than 10.

Temperature-compensated zener diodes are available from manufacturers in single-package units called *reference diodes.* These units contain specially fabricated junctions that closely track and oppose variations in V_Z with temperature. Although it is possible to obtain an extremely stable reference this way, it may be necessary to maintain the zener current at a manufacturer's specified value in order to realize the specified stability.

Zener-Diode Impedance

The breakdown characteristic of an *ideal* zener diode is a perfectly vertical line, signifying zero change in voltage for any change in current. Thus, the ideal diode has zero impedance (or ac resistance) in its breakdown region. A practical zener diode has nonzero impedance, which can be computed in the usual way:

$$Z_Z = \frac{\Delta V_Z}{\Delta I_Z} \tag{3–29}$$

Z_Z is the reciprocal of the slope of the breakdown characteristic on an I_Z-V_Z plot. The slope is not constant, so the value of Z_Z depends on the point along the characteristic where the measurement is made. The impedance decreases as I_Z increases; that is, the breakdown characteristic becomes steeper at points farther down the line, corresponding to greater reverse currents. For this reason, the diode should be operated with as much reverse current as possible, consistent with rating limitations.

Manufactures' specifications for zener impedances are usually given for a specified ΔI_Z that covers a range from a small I_Z near the onset of breakdown to some percentage of the maximum rated I_Z. The values may range from a few ohms to several hundred ohms. There is also a variation in the impedance of zener diodes among those having different values of V_Z. Diodes with breakdown voltages near 7 V have the smallest impedances.

EXAMPLE 3–12

A zener diode has impedance 40 Ω in the range from $I_Z = 1$ mA to $I_Z = 10$ mA. The voltage at $I_Z = 1$ mA is 9.1 V. Assuming that the impedance is constant over the given range, what minimum and maximum zener voltages can be expected if the diode is used in an application where the zener current changes from 2 mA to 8 mA?

Solution

From equation 3–29, the voltage change between $I_Z = 1$ mA and $I_Z = 2$ mA is $\Delta V_Z = \Delta I_Z Z_Z = [(2\text{ mA}) - (1\text{ mA})](40\ \Omega) = 0.04$ V. Therefore, the minimum voltage is $V_Z(\text{min}) = (9.1\text{ V}) + \Delta V_Z = (9.1\text{ V}) + (0.04\text{ V}) = 9.14$ V. The voltage change between $I_Z = 2$ mA and $I_Z = 8$ mA is $\Delta V_Z = [(8\text{ mA}) - (2\text{ mA})](40\ \Omega) = 0.16$ V. Therefore, the maximum voltage is $V_Z(\text{max}) = V_Z(\text{min}) + \Delta V_Z = (9.14\text{ V}) + (0.16\text{ V}) = 9.3$ V.

Three-Terminal Integrated-Circuit Regulators

A three-terminal regulator is a compact, easy-to-use, fixed-voltage regulator packaged in a single integrated circuit. To use the regulator, it is necessary only to make external connections to the three terminals: V_{in}, V_o, and ground. These devices are widely used to provide *local regulation* in electronic systems that may require several different supply voltages. For example, a 5-V regulator could be used to regulate the power supplied to all the chips mounted on one printed circuit board, and a 12-V regulator could be used for a similar purpose on a different board. The regulators might well use the same unregulated input voltage, say, 20 V.

A popular series of three-terminal regulators is the 7800/7900 series, available from several manufacturers with a variety of output voltage ratings. Figure 3–32 shows National Semiconductor specifications for their 7800-series regulators, which carry the company's standard LM prefix and which are available with regulated outputs of +5 V, +12 V, and +15 V. The last two digits of the 7800 number designate the rated output voltage. For example, the 7805 is a +5-V regulator and the 7815 is a +15-V regulator. The 7900-series regulators provide negative output voltages. Notice that the integrated circuitry shown in the schematic diagram is considerably complex. It can be seen that the circuit incorporates a zener diode as an internal voltage reference. The 7800/7900 series also has internal current-limiting circuitry.

Important points to note in the 7800-series specifications include the following:

1. The output voltage of an arbitrarily chosen device might not *exactly* equal its nominal value. For example, with a 23-V input, the 7815 output may be anywhere from 14.4 V to 15.6 V. This specification does not mean that the output voltage of a single device will vary over that range, but that one 7815 chosen at random from a large number will hold its output constant at some voltage within that range.
2. The input voltage cannot exceed 35 V and must not fall below a certain minimum value, depending on type number, if output regulation is to be maintained. The minimum specified inputs are 7.3, 14.6, and 17.7 V for the 7805, 7812, and 7815, respectively.
3. Load regulation is specified as a certain output voltage change (ΔV_o) as the load current (I_o) is changed over a certain range. For example, the output of the 7805 will change a maximum of 50 mV as load current changes from 5 mA to 1.5 A, and will change a maximum of 25 mV as load current changes from 250 mA to 750 mA.

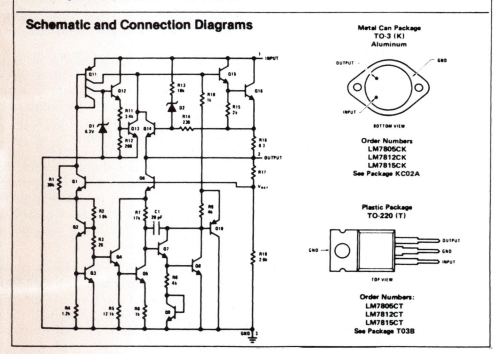

National Semiconductor

Voltage Regulators

LM78XX Series

LM78XX Series Voltage Regulators

General Description

The LM78XX series of three terminal regulators is available with several fixed output voltages making them useful in a wide range of applications. One of these is local on card regulation, eliminating the distribution problems associated with single point regulation. The voltages available allow these regulators to be used in logic systems, instrumentation, HiFi, and other solid state electronic equipment. Although designed primarily as fixed voltage regulators these devices can be used with external components to obtain adjustable voltages and currents.

The LM78XX series is available in an aluminum TO-3 package which will allow over 1.0A load current if adequate heat sinking is provided. Current limiting is included to limit the peak output current to a safe value. Safe area protection for the output transistor is provided to limit internal power dissipation. If internal power dissipation becomes too high for the heat sinking provided, the thermal shutdown circuit takes over preventing the IC from overheating.

Considerable effort was expended to make the LM78XX series of regulators easy to use and minimize the number of external components. It is not necessary to bypass the output, although this does improve transient response. Input bypassing is needed only if the regulator is located far from the filter capacitor of the power supply.

For output voltage other than 5V, 12V and 15V the LM117 series provides an output voltage range from 1.2V to 57V.

Features

- Output current in excess of 1A
- Internal thermal overload protection
- No external components required
- Output transistor safe area protection
- Internal short circuit current limit
- Available in the aluminum TO-3 package

Voltage Range

LM7805C	5V
LM7812C	12V
LM7815C	15V

Schematic and Connection Diagrams

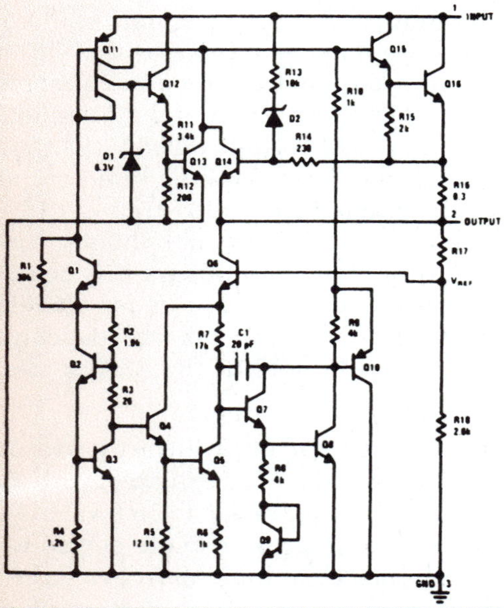

Metal Can Package
TO-3 (K)
Aluminum

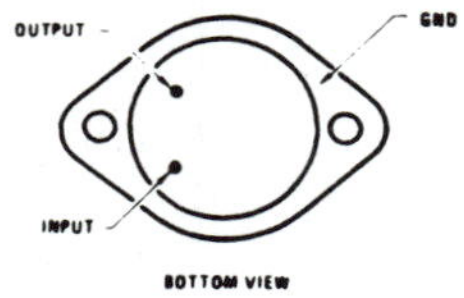

BOTTOM VIEW

Order Numbers
LM7805CK
LM7812CK
LM7815CK
See Package KC02A

Plastic Package
TO-220 (T)

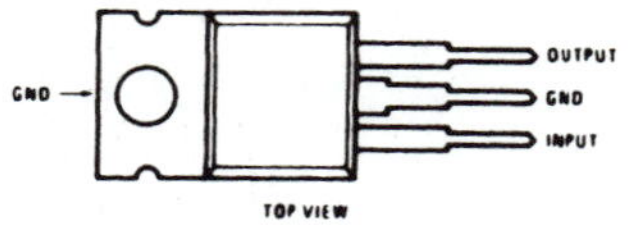

TOP VIEW

Order Numbers:
LM7805CT
LM7812CT
LM7815CT
See Package T03B

FIGURE 3–32 7800-series voltage-regulator specifications (Courtesy of National Semiconductor)

For negative voltage regulation, the 7900 series provides the same characteristics as the 7800 series but for negative input and output voltage.

Another popular three-terminal IC regulator is the LM317, which can provide adjustable output voltage by simply adding a few external components. The LM317 can supply up to 1.5 A of current to a load when mounted

LM78XX Series

Absolute Maximum Ratings

Input Voltage (V_O = 5V, 12V and 15V)	35V
Internal Power Dissipation (Note 1)	Internally Limited
Operating Temperature Range (T_A)	0°C to +70°C
Maximum Junction Temperature	
(K Package)	150°C
(T Package)	125°C
Storage Temperature Range	−65°C to +150°C
Lead Temperature (Soldering, 10 seconds)	
TO-3 Package K	300°C
TO-220 Package T	230°C

Electrical Characteristics LM78XXC (Note 2) 0°C ≤ Tj ≤ 125°C unless otherwise noted.

OUTPUT VOLTAGE			5V			12V			15V			UNITS
INPUT VOLTAGE (unless otherwise noted)			10V			19V			23V			
PARAMETER		CONDITIONS	MIN	TYP	MAX	MIN	TYP	MAX	MIN	TYP	MAX	
V_O	Output Voltage	T_j = 25°C, 5 mA ≤ I_O ≤ 1A	4.8	5	5.2	11.5	12	12.5	14.4	15	15.6	V
		P_D ≤ 15W, 5 mA ≤ I_O ≤ 1A	4.75		5.25	11.4		12.6	14.25		15.75	V
		V_{MIN} ≤ V_{IN} ≤ V_{MAX}			(7 ≤ V_{IN} ≤ 20)			(14.5 ≤ V_{IN} ≤ 27)			(17.5 ≤ V_{IN} ≤ 30)	V
ΔV_O	Line Regulation	I_O = 500 mA, T_j = 25°C		3	50		4	120		4	150	mV
		ΔV_{IN}			(7 ≤ V_{IN} ≤ 25)			(14.5 ≤ V_{IN} ≤ 30)			(17.5 ≤ V_{IN} ≤ 30)	V
		I_O = 500 mA, 0°C ≤ T_j ≤ +125°C			50			120			150	mV
		ΔV_{IN}			(8 ≤ V_{IN} ≤ 20)			(15 ≤ V_{IN} ≤ 27)			(18.5 ≤ V_{IN} ≤ 30)	V
		I_O ≤ 1A, T_j = 25°C			50			120			150	mV
		ΔV_{IN}			(7.3 ≤ V_{IN} ≤ 20)			(14.6 ≤ V_{IN} ≤ 27)			(17.7 ≤ V_{IN} ≤ 30)	V
		I_O ≤ 1A, 0° ≤ T_j ≤ +125°C			25			60			75	mV
		ΔV_{IN}			(8 ≤ IN ≤ 12)			(16 ≤ V_{IN} ≤ 22)			(20 ≤ V_{IN} ≤ 26)	V
ΔV_O	Load Regulation	T_j = 25°C, 5 mA ≤ I_O ≤ 1.5A		10	50		12	120		12	150	mV
		T_j = 25°C, 250 mA ≤ I_O ≤ 750 mA			25			60			75	mV
		5 mA ≤ I_O ≤ 1A, 0°C ≤ T_j ≤ +125°C			50			120			150	mV
I_Q	Quiescent Current	I_O ≤ 1A, T_j = 25°C			8			8			8	mA
		I_O ≤ 1A, 0°C ≤ T_j ≤ +125°C			8.5			8.5			8.5	mA
ΔI_Q	Quiescent Current Change	5 mA ≤ I_O ≤ 1A			0.5			0.5			0.5	mA
		T_j = 25°C, I_O ≤ 1A			1.0			1.0			1.0	mA
		V_{MIN} ≤ V_{IN} ≤ V_{MAX}			(7.5 ≤ V_{IN} ≤ 20)			(14.8 ≤ V_{IN} ≤ 27)			(17.9 ≤ V_{IN} ≤ 30)	V
		I_O ≤ 500 mA, 0°C ≤ T_j ≤ +125°C			1.0			1.0			1.0	mA
		V_{MIN} ≤ V_{IN} ≤ V_{MAX}			(7 ≤ V_{IN} ≤ 25)			(14.5 ≤ V_{IN} ≤ 30)			(17.5 ≤ V_{IN} ≤ 30)	V
V_N	Output Noise Voltage	T_A = 25°C, 10 Hz ≤ f ≤ 100 kHz		40			75			90		μV
$\frac{\Delta V_{IN}}{\Delta V_{OUT}}$	Ripple Rejection	f = 120 Hz, I_O ≤ 1A, T_j = 25°C or	62	80		55	72		54	70		dB
		f = 120 Hz, I_O ≤ 500 mA, 0°C ≤ T_j ≤ +125°C	62			55			54			dB
		V_{MIN} ≤ V_{IN} ≤ V_{MAX}			(8 ≤ V_{IN} ≤ 18)			(15 ≤ V_{IN} ≤ 25)			(18.5 ≤ V_{IN} ≤ 28.5)	V
R_O	Dropout Voltage	T_j = 25°C, I_{OUT} = 1A		2.0			2.0			2.0		V
	Output Resistance	f = 1 kHz		8			18			19		mΩ
	Short-Circuit Current	T_j = 25°C		2.1			1.5			1.2		A
	Peak Output Current	T_j = 25°C		2.4			2.4			2.4		A
	Average TC of V_{OUT}	0°C ≤ T_j ≤ +125°C, I_O = 5 mA		0.6			1.5			1.8		mV/°C
V_{IN}	Input Voltage Required to Maintain Line Regulation	T_j = 25°C, I_O ≤ 1A	7.3			14.6			17.7			V

NOTE 1: Thermal resistance of the TO-3 package (K, KC) is typically 4°C/W junction to case and 35°C/W case to ambient. Thermal resistance of the TO-220 package (T) is typically 4°C/W junction to case and 50°C/W case to ambient.

NOTE 2: All characteristics are measured with capacitor across the inut of 0.22 μF, and a capacitor across the output of 0.1 μF. All characteristics except noise voltage and ripple rejection ratio are measured using pulse techniques (t_W ≤ 10 ms, duty cycle ≤ 5%). Output voltage changes due to changes in internal temperature must be taken into account separately.

FIGURE 3–32 Continued

on a suitable heat sink. Figure 3–33 shows the specifications for this regulator. For negative voltage regulation, the LM337 is very much like the mirror image of the LM317 positive regulator.

Both regulators operate based on an internal fixed voltage reference of 1.25 V and a bias current of 100 μA, as shown in Figure 3–34. Resistor R_1 establishes a current of (1.25 V)/R_1, which, together with the bias current I_A, produces a voltage drop across R_2. The output voltage will then be

National Semiconductor

August 1999

LM117/LM317A/LM317 3-Terminal Adjustable Regulator

LM117/LM317A/LM317 3-Terminal Adjustable Regulator

General Description

The LM117 series of adjustable 3-terminal positive voltage regulators is capable of supplying in excess of 1.5A over a 1.2V to 37V output range. They are exceptionally easy to use and require only two external resistors to set the output voltage. Further, both line and load regulation are better than standard fixed regulators. Also, the LM117 is packaged in standard transistor packages which are easily mounted and handled.

In addition to higher performance than fixed regulators, the LM117 series offers full overload protection available only in IC's. Included on the chip are current limit, thermal overload protection and safe area protection. All overload protection circuitry remains fully functional even if the adjustment terminal is disconnected.

Normally, no capacitors are needed unless the device is situated more than 6 inches from the input filter capacitors in which case an input bypass is needed. An optional output capacitor can be added to improve transient response. The adjustment terminal can be bypassed to achieve very high ripple rejection ratios which are difficult to achieve with standard 3-terminal regulators.

Besides replacing fixed regulators, the LM117 is useful in a wide variety of other applications. Since the regulator is "floating" and sees only the input-to-output differential voltage, supplies of several hundred volts can be regulated as long as the maximum input to output differential is not exceeded, i.e., avoid short-circuiting the output.

Also, it makes an especially simple adjustable switching regulator, a programmable output regulator, or by connecting a fixed resistor between the adjustment pin and output, the LM117 can be used as a precision current regulator. Supplies with electronic shutdown can be achieved by clamping the adjustment terminal to ground which programs the output to 1.2V where most loads draw little current.

For applications requiring greater output current, see LM150 series (3A) and LM138 series (5A) data sheets. For the negative complement, see LM137 series data sheet.

Features

- Guaranteed 1% output voltage tolerance (LM317A)
- Guaranteed max. 0.01%/V line regulation (LM317A)
- Guaranteed max. 0.3% load regulation (LM117)
- Guaranteed 1.5A output current
- Adjustable output down to 1.2V
- Current limit constant with temperature
- P+ Product Enhancement tested
- 80 dB ripple rejection
- Output is short-circuit protected

Typical Applications

1.2V–25V Adjustable Regulator

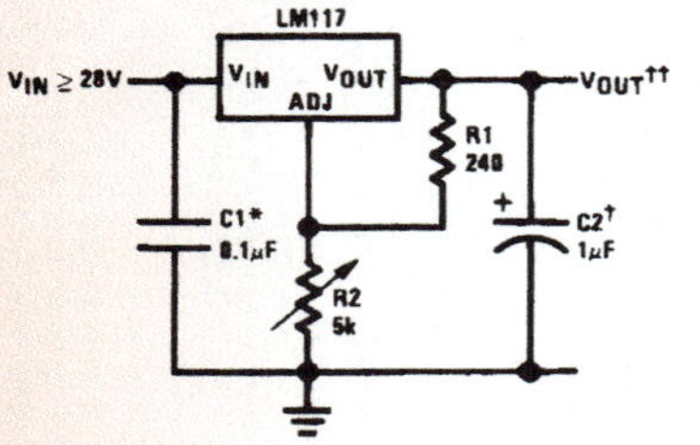

Full output current not available at high input-output voltages

*Needed if device is more than 6 inches from filter capacitors.

†Optional — improves transient response. Output capacitors in the range of 1 µF to 1000 µF of aluminum or tantalum electrolytic are commonly used to provide improved output impedance and rejection of transients.

$$††V_{OUT} = 1.25V \left(1 + \frac{R2}{R1}\right) + I_{ADJ}(R_2)$$

LM117 Series Packages

Part Number Suffix	Package	Design Load Current
K	TO-3	1.5A
H	**TO-39**	**0.5A**
T	TO-220	1.5A
E	**LCC**	**0.5A**
S	TO-263	1.5A
EMP	SOT-223	1A
MDT	**TO-252**	**0.5A**

SOT-223 vs D-Pak (TO-252) Packages

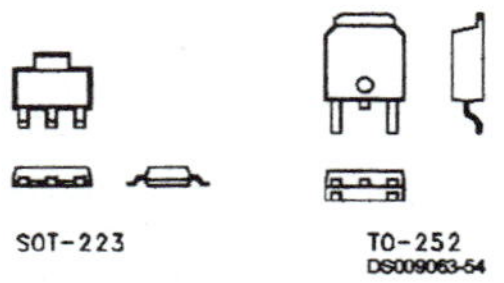

Scale 1:1

FIGURE 3–33 Specifications for the LM317/LM117 (Reprinted with permission of National Semiconductor Corporation)

$$V_o = V_{ref} + \left(I_A + \frac{V_{ref}}{R_1}\right)R_2$$

where I_A = 100 µA and V_{ref} = 1.25 V, nominally. The manufacturer recommends 240 ohms for R_1 to establish a current of about 5 mA through it. When this is the case, I_A can be neglected and V_o can be approximated by $V_o = V_{ref}(1 + R_2/R_1)$. The output voltage can be made adjustable by using

Electrical Characteristics (Note 3)

Specifications with standard type face are for $T_J = 25°C$, and those with **boldface type** apply over **full Operating Temperature Range**. Unless otherwise specified, $V_{IN} - V_{OUT} = 5V$, and $I_{OUT} = 10$ mA.

Parameter	Conditions	LM317A Min	LM317A Typ	LM317A Max	LM317 Min	LM317 Typ	LM317 Max	Units
Reference Voltage		1.238	1.250	1.262				V
	$3V \le (V_{IN} - V_{OUT}) \le 40V$, $10\ mA \le I_{OUT} \le I_{MAX}$, $P \le P_{MAX}$	**1.225**	**1.250**	**1.270**	**1.20**	**1.25**	**1.30**	V
Line Regulation	$3V \le (V_{IN} - V_{OUT}) \le 40V$ (Note 4)		0.005	0.01		0.01	0.04	%/V
			0.01	**0.02**		**0.02**	**0.07**	%/V
Load Regulation	$10\ mA \le I_{OUT} \le I_{MAX}$ (Note 4)		0.1	0.5		0.1	0.5	%
			0.3	**1**		**0.3**	**1.5**	%
Thermal Regulation	20 ms Pulse		0.04	0.07		0.04	0.07	%/W
Adjustment Pin Current			**50**	**100**		**50**	**100**	μA
Adjustment Pin Current Change	$10\ mA \le I_{OUT} \le I_{MAX}$, $3V \le (V_{IN} - V_{OUT}) \le 40V$		**0.2**	**5**		**0.2**	**5**	μA
Temperature Stability	$T_{MIN} \le T_J \le T_{MAX}$		**1**			**1**		%
Minimum Load Current	$(V_{IN} - V_{OUT}) = 40V$		**3.5**	**10**		**3.5**	**10**	mA
Current Limit	$(V_{IN} - V_{OUT}) \le 15V$							
	K, T, S Packages	**1.5**	**2.2**	**3.4**	**1.5**	**2.2**	**3.4**	A
	H Package	**0.5**	**0.8**	**1.8**	**0.5**	**0.8**	**1.8**	A
	MP Package	**1.5**	**2.2**	**3.4**	**1.5**	**2.2**	**3.4**	A
	$(V_{IN} - V_{OUT}) = 40V$							
	K, T, S Packages	0.15	0.4		0.15	0.4		A
	H Package	0.075	0.2		0.075	0.2		A
	MP Package	0.55	0.4		0.15	0.4		A
RMS Output Noise, % of V_{OUT}	$10\ Hz \le f \le 10\ kHz$		0.003			0.003		%
Ripple Rejection Ratio	$V_{OUT} = 10V$, f = 120 Hz, $C_{ADJ} = 0\ μF$		**65**			**65**		dB
	$V_{OUT} = 10V$, f = 120 Hz, $C_{ADJ} = 10\ μF$	**66**	**80**		**66**	**80**		dB
Long-Term Stability	$T_J = 125°C$, 1000 hrs		0.3	1		0.3	1	%
Thermal Resistance, Junction-to-Case	K Package					2.3	3	°C/W
	MDT Package					5		°C/W
	H Package		12	15		12	15	°C/W
	T Package		4	5		4		°C/W
	MP Package		23.5			23.5		°C/W
Thermal Resistance, Junction-to-Ambient (No Heat Sink)	K Package		35			35		°C/W
	MDT Package(Note 6)					92		°C/W
	H Package		140			140		°C/W
	T Package		50			50		°C/W
	S Package (Note 6)		50			50		°C/W

Note 1: Absolute Maximum Ratings indicate limits beyond which damage to the device may occur. Operating Ratings indicate conditions for which the device is intended to be functional, but do not guarantee specific performance limits. For guaranteed specifications and test conditions, see the Electrical Characteristics. The guaranteed specifications apply only for the test conditions listed.

Note 2: Refer to RETS117H drawing for the LM117H, or the RETS117K for the LM117K military specifications.

Note 3: Although power dissipation is internally limited, these specifications are applicable for maximum power dissipations of 2W for the TO-39 and SOT-223 and 20W for the TO-3, TO-220, and TO-263. I_{MAX} is 1.5A for the TO-3, TO-220, and TO-263 packages, 0.5A for the TO-39 package and 1A for the SOT-223 Package. All limits (i.e., the numbers in the Min. and Max. columns) are guaranteed to National's AOQL (Average Outgoing Quality Level).

Note 4: Regulation is measured at a constant junction temperature, using pulse testing with a low duty cycle. Changes in output voltage due to heating effects are covered under the specifications for thermal regulation.

Note 5: Human body model, 100 pF discharged through a 1.5 kΩ resistor.

Note 6: If the TO-263 or TO-252 packages are used, the thermal resistance can be reduced by increasing the PC board copper area thermally connected to the package. If the SOT-223 package is used, the thermal resistance can be reduced by increasing the PC board copper area (see applications hints for heatsinking).

3 www.national.com

FIGURE 3–33 Continued

a variable resistor for R_2. For better results, 1-μF tantalum capacitors can be connected across both the input and output sides (see data sheet).

3–7 DIODE TYPES, RATINGS, AND SPECIFICATIONS

Discrete diodes—those packaged in individual cases with externally accessible anode and cathode connections—are commercially available in a wide

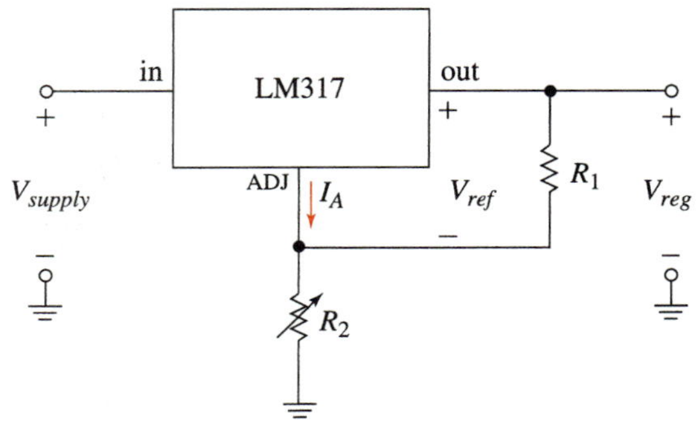

FIGURE 3–34 A very simple adjustable positive voltage regulator with the LM317

range of types designed for different kinds of service and for a variety of applications. We find, for example, *switching* diodes designed specifically for use in logic circuit applications, like those discussed in the last section. These diodes typically have low power-dissipation ratings, are small in size, and are designed to respond rapidly to pulse-type inputs, that is, to switch between their ON and OFF states with minimum delay. *Rectifier* or *power* diodes are designed to carry larger currents and to dissipate more power than switching diodes. They are used in power supply applications, where heavier currents and higher voltages are encountered. *Small-signal* diodes are general-purpose diodes used in applications such as signal detection in radio and TV.

Figure 3–35 illustrates the variety of sizes and shapes that commercially available diodes may have. Each of those shown has a designation that identifies the standard case size it has (DO-4, DO-7, etc.). Materials used for case construction include glass, plastic, and metal. Metal cases are used for large, rectifier-type diodes to enhance the conduction of heat and improve their power-dissipation capabilities.

There are two particularly important diode ratings that a designer using commercial, discrete diodes should know when selecting a diode for any application: the *maximum reverse voltage* (V_{RM}) and the *maximum forward current*. The maximum reverse voltage, also called the *peak inverse voltage* (PIV), is the maximum reverse-biasing voltage that the diode can withstand without breakdown. If the PIV is exceeded, the diode "breaks down" only in the sense that it readily conducts current in the reverse direction. As discussed in Chapter 2, breakdown *may* result in permanent failure if the power dissipation rating of the device is exceeded. The maximum forward current is the maximum current that the diode can sustain when it is forward biased. Exceeding this rating will cause excessive heat to be generated in the diode and will lead to permanent failure. Manufacturers' ratings for the maximum forward current will specify whether the rating is for continuous, peak, average, or rms current, and they may provide different values for each. The symbols I_o and I_F are used to represent forward current.

EXAMPLE 3–13

DESIGN

In the circuit of Figure 3–36, a rectifier diode is used to supply positive current pulses to the 100-Ω resistor load. The diode is available in the combinations of ratings listed in the table portion of the figure. Which is the least expensive diode that can be used for the application?

Solution

The applied voltage is 120 V rms. Therefore, when the diode is reverse biased by the *peak* negative value of the sine wave, it will be subjected to a maximum

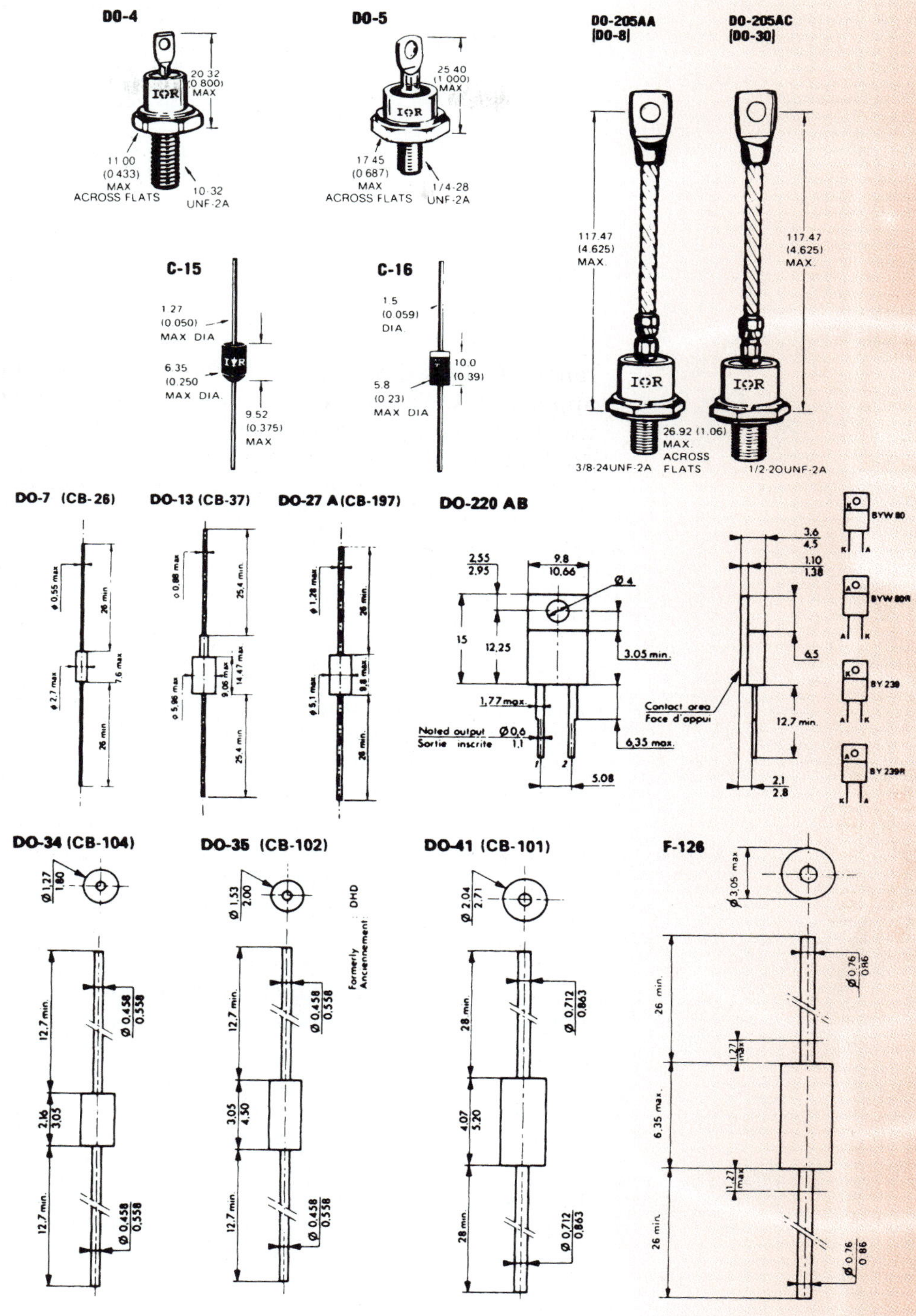

FIGURE 3–35 Discrete diode case styles (Courtesy of International Rectifier Corp. and Thomson-CSF Components Corp.)

reverse-biasing voltage of (1.414)(120) = 169.7 V. The V_{RM} rating must be greater than 169.7 V.

The average value of the current is one-half the average value of a single sinusoidal pulse: $I_{AVG} = (1/2)(0.637\ I_P)$ A, where I_P is the peak value of the pulse. (Note that the factor 1/2 must be used because the pulse is present for only one-half of each full cycle.) The peak forward current in the example (neglecting the

FIGURE 3–36
(Example 3–13)

V_{RM}	Max I_o (average)	Unit Cost
100 V	1.0 A	$0.50
150 V	2.0 A	1.50
200 V	1.0 A	2.00
200 V	2.0 A	3.00
500 V	2.0 A	3.50
500 V	5.0 A	5.00

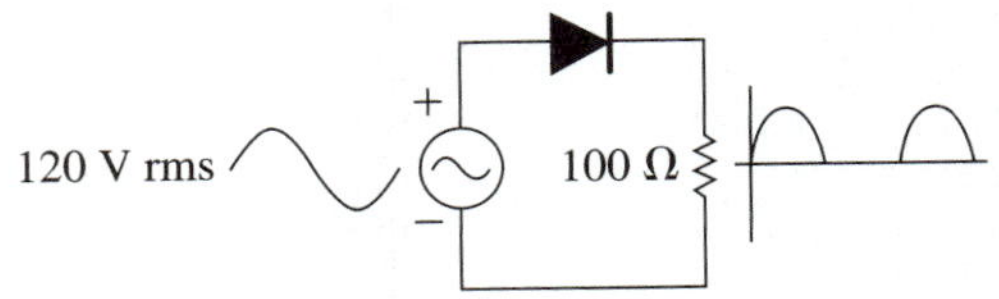

drop across the diode) is $I_P = (169.7\text{ V})/(100\ \Omega) = 1.697$ A. Therefore, the average forward current through the diode is $I_{AVG} = (1/2)(0.637 + 1.697) = 0.540$ A.

The least expensive diode having ratings adequate for the peak inverse voltage and average forward current values we calculated is the one costing $2.00.

Figure 3–37 shows a typical manufacturer's specification sheet for a line of silicon small-signal diodes. Like many other manufactured electronic components, diodes are often identified by a standard *type* number in accordance with JEDEC (Joint Electron Devices Engineering Council) specifications. Diode-type numbers have the prefix 1N, like those shown in the leftmost column of Figure 3–37. (Not all manufacturers provide JEDEC numbers; many use their own commercial part numbers.) The second column in the specification sheet shows the maximum reverse voltage, V_{RM}, for each of the diode types. Note that V_{RM} ranges from 20 V to 200 V for the diodes listed. The third column shows the rated average forward current, I_o, of each diode in mA, and these range from 0.1 mA to 200 mA. The next two pairs of columns list values of reverse current, I_R, for different values of reverse voltage, V_R, and ambient temperature, T_{amb}. The next column gives capacitance values in pF, an important specification in high-frequency and switching applications. The column headed t_{rr} lists the reverse recovery time of each diode, in nanoseconds. This specification relates to the time required for a diode to switch from its ON to its OFF state and is another important parameter in switching circuit design. Finally, the maximum rated power dissipation is given in mW. The product of diode voltage and diode current should never exceed this rating in any application (unless there is some auxiliary means for removing heat, such as a cooling fan).

Figure 3–38(a) shows a typical specification sheet for a line of silicon rectifier diodes. Note that the forward current ratings for these diodes are generally larger than those of the small-signal diodes. The current ratings are given as $I_{F(AV)}$ (average), and I_{FSM}, each in units of amperes. I_{FSM} is the maximum nonrepetitive forward current that the diode can sustain, that is, the maximum value of momentary or *surge* current it can conduct. Note that the I_{FSM} values are much larger than the $I_{F(AV)}$ values. The voltage ratings are specified by V_{RPM}, the maximum repetitive reverse voltage that each diode can sustain. Also note the large physical sizes and the metal cases of the stud-mounted rectifiers that are capable of conducting currents from 12 to 40 A.

Diode bridges are commonly available in single-package units. These packages have a pair of terminals to which the ac input is connected and

THOMSON-CSF

silicon signal diodes

Type	V_R-V_{RM} max (V)	I_O $V_F = 1V$ min (mA)	I_R / (µA)	V_R (V)	I_R (µA)	/ T_{amb} (°C)	C max (pF)	t_{rr} (ns)	P_{tot} (mW)
GENERAL PURPOSE AND HIGH SPEED SWITCHING									
1N 456	30	40	0,025	25	5	150			250
1N 456A	30	100	0,025	25	5	150			250
1N 457	70	20	0,025	60	5	150			250
1N 457A	70	100	0,025	60	5	150			250
1N 458	150	7	0,025	125	5	150			250
1N 458A	150	100	0,025	125	5	150			250
1N 461	30	15	0,5	25	30	150			250
1N 461A	30	100	0,5	25	30	150			250
1N 462	70	5	0,5	60	30	150			250
1N 462A	70	100	0,5	60	30	150			250
1N 463	200	1	0,5	175	30	150			250
1N 464	150	3	0,5	125	30	150			250
1N 464A	150	100	0,5	125	30	150			250
1N 482	40	100 *	0,25	30	30	150			250
1N 482A	40	100	0,025	30	15	150			250
1N 482B	40	100	0,025	30	5	150			250
1N 483	70	100 *	0,25	60	30	150			250
1N 484	130	100 *	0,25	125	30	150			250
1N 484A	130	100	0,025	125	15	150			250
1N 484B	130	100	0,025	125	5	150			250
1N 914	100	10	0,025	20	50	150	4	4	250
1N 914A	100	20	0,025	20	50	150	4	4	250
1N 914B	100	100	0,025	20	50	150	4	4	250
1N 916	100	10	0,025	20	50	150	2	4	250
1N 916A	100	20	0,025	20	50	150	2	4	250
1N 916B	100	30	0,025	20	50	150	2	4	250
1N 3062	75	20	0,1	50	100	150	1	2	250
1N 3063	75	10	0,1	50	100	150	2	4	250
1N 3064	75	10	0,1	50	100	150	2	4	250
1N 3066	75	50	0,1	50	100	150	6	50	250
1N 3070	200	100	0,1	150	100	150	5	50	250
1N 3071	200	200	0,1	125	500	125	8	3000	500
1N 3595	150	200	0,1	50	100	150	2,5	4	250
1N 3600	50	10	0,1	50	100	150	1	2	250
1N 3604		50	0,05	50	50	150	2	2	250
1N 3605		0,1**	0,05	30	50	150	2	2	250
1N 3606		0,1**	0,05	50	50	150	2	2	250
1N 4148	100	10	0,025	20	50	150	4	4	500
1N 4149	100	10	0,025	20	50	150	2	4	500
1N 4150	50	200	0,1	50	100	150	2,5	4	500
1N 4151	75	50	0,05	50	50	150	2	2	500
1N 4152	40	0,1**	0,05	30	50	150	2	2	500
1N 4153	75	0,1**	0,05	50	50	150	2	2	500
1N 4154	25	30	0,1	25	100	150	4	2	500
1N 4244	20	20	0,1	10	100	150	0,8	0,75	250
1N 4305	75	10	0,1	50	100	150	2	4	500
1N 4446	100	20	0,025	20	50	150	4	4	500
1N 4447	100	20	0,025	20	50	150	2	4	500
1N 4448	100	100	0,025	20	50	150	4	4	500
1N 4449	100	30	0,025	20	50	150	2	4	500
1N 4450	40	200	0,05	30	50	150	4	4	500
1N 4454	75	10	1	50	100	150	2	4	500

Case

T_{amb} = 25°C

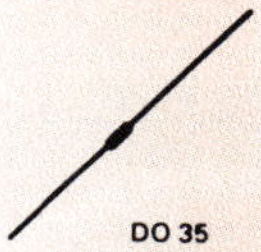

*V_F = 1,1V **V_F = 0,55V

FIGURE 3–37 A typical diode data sheet (Courtesy of Thomson-CSF)

another pair at which the full-wave–rectified output is taken. Figure 3-38(b) shows a typical manufacturer's specification sheet for a line of single-package bridges with current ratings from 1 to 100 A. The V_{RRM} specifications refer to the maximum repetitive reverse voltage ratings of each, i.e., the peak inverse voltage ratings when operated with repetitive inputs, such as sinusoidal voltages. These ratings range from 50 to 1200 V. Note the I_{FSM} specifications, which refer to maximum forward surge current.

Power Supply Component Specifications

Designing a simple power supply, as the reader might have already concluded, should not be a lengthy and difficult task. Nevertheless, the designer

General Purpose Stud Mounted Silicon Rectifiers

SILICON RECTIFIERS — GENERAL PURPOSE

Silicon rectifiers are available from less than 1 Amp to 3000 Amps, and voltages to 3000 volts. In addition to the standard industry packages, there are a number of other packages available as required. Two of these are shown below. These are ruggedly built devices with an excellent reputation for reliability and performance.

AXIAL LEAD SILICON RECTIFIERS — 750 mA TO 6 AMPS

$I_{F(AV)}$ (A) @ Max T_C (°C)	750 mA @ 25	1 @ 75	1.5 @ 40	1.5 @ 40	2 @ 100	3 @ 125	4 @ 120	6 @ 95
I_{FSM} (A)	22	30	50	50	50	150	200	400
Notes	(1)(2)	(1)(2)	(1)(2)	(1)(2)	(2)	(2)	(2)	(2)
Case Style	DO-41	DO-41	DO-41	DO-41	DO-41	C-12	C-16	C-15
PART NUMBERS								
V_{RRM}								
50 Volts	--	1N4001	10D05	1N4816	20D05	—	–	60S05
100 Volts	—	1N4002	10D1	1N4817	20D1	30S1	40D1	60S1
200 Volts	1N2069 A	1N4003	10D2	1N4818	20D2	30S2	40D2	60S2
300 Volts	—	—	10D3	1N4819	—	30S3	—	–
400 Volts	1N2070 A	1N4004	10D4	1N4820	20D4	30S4	40D4	60S4
500 Volts	—	—	10D5	1N4821	—	30S5	–	60S5
600 Volts	1N2071 A	1N4005	10D6	1N4822	20D6	30S6	40D6	60S6
700 Volts	--	—	–	1N5052	—	—	—	—
800 Volts	—	1N4006	10D8	1N5053	20D8	30S8	40D8	60S8
1000 Volts	--	1N4007	10D10	1N5054	20D10	30S10	—	60S10

(1) Temperature given is ambient temperature. (2) Also available on tape reel.

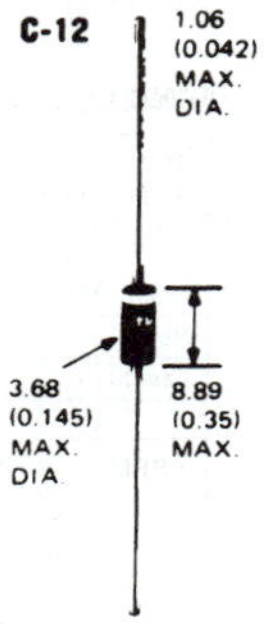

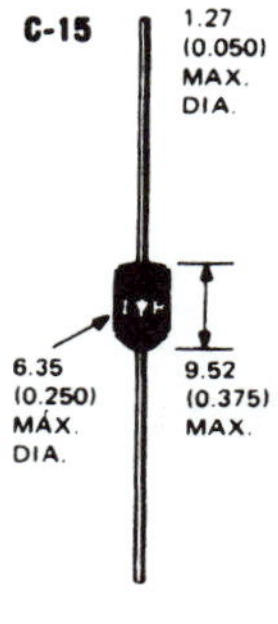

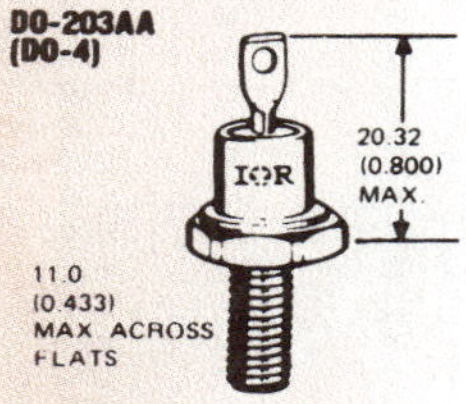

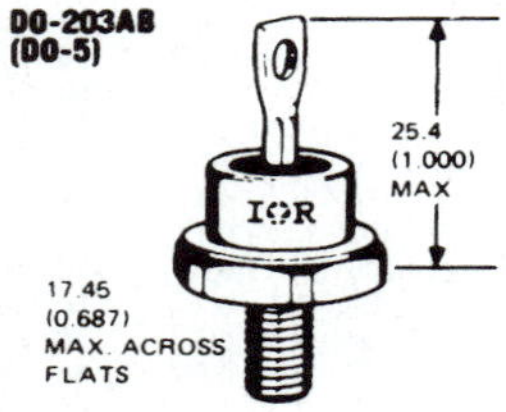

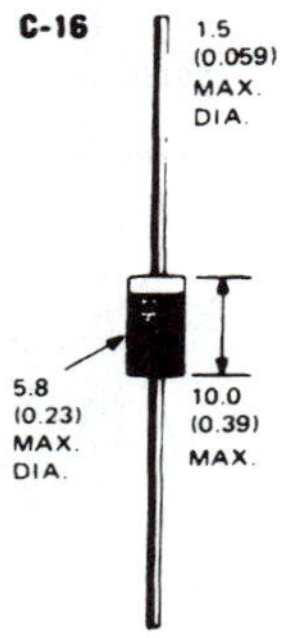

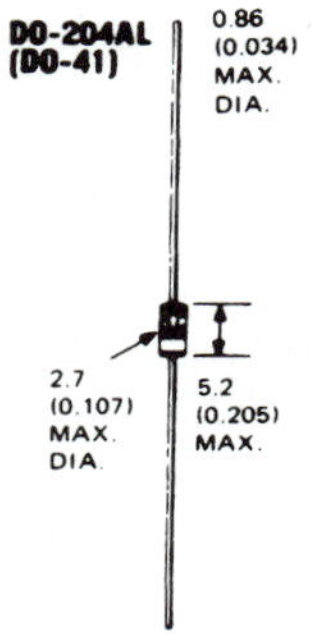

STUD MOUNTED[4] SILICON RECTIFIERS — 12 TO 40 AMPS

$I_{F(AV)}$ (A) @ Max T_C (°C)	12 @ 150	12 @ 150	12 @ 150	15 @ 150	16 @ 150	35 @ 140	40 @ 140	40 @ 150
I_{FSM} (A)	200	240	250	250	300	500	500	800
Notes	(5)	(5)	(5)	(5)	(5)	(5)	(5)	(5)
Case Style	DO-4	DO-4	DO-4	DO-5	DO-4	DO-5	DO-5	DO-5
PART NUMBERS								
V_{RRM}								
50 Volts	12F5	1N1199A	12F5B	1N3208	16F5	1N1183	40HF5	1N1183A
100 Volts	12F10	1N1200A	12F10B	1N3209	16F10	1N1184	40HF10	1N1184A
150 Volts	–	1N1201A	–	–	--	1N1185	–	1N1185A
200 Volts	12F20	1N1202A†	12F20B	1N3210	16F20	1N1186	40HF20	1N1186A
300 Volts	–	1N1203	–	1N3211	–	1N1187	40HF30	1N1187A
400 Volts	12F40	1N1204A†	12F40B	1N3212	16F40	1N1188	40HF40	1N1188A
500 Volts	–	1N1205A		1N3213		1N1189	40HF50	1N1189A
600 Volts	12F60	1N1206A†	12F60B	1N3214	16F60	1N1190	40HF60	1N1190A
700 Volts		1N3670A	--	–	--	1N3765	40HF80	
800 Volts	12F80	1N3671A	12F80B		16F80	1N3766	40HF100	–
900 Volts		1N3672A	–	–	--	1N3767	40HF120	-
1000 Volts	12F100	1N3673A	12F100B		16F100	1N3768		

(4) Metric threads available on some stud packages.
(5) Cathode-to-stud. For anode-to-stud, add "R" to base number (example 12FLR, 40HFLR, 1N3889R).
† JAN and/or JAN-TX types available.

OTHER CONNECTIONS

In addition to the flex leads shown in the case style drawings, IR also has threaded stud, threaded hole and flag terminal top connections available for some case styles. Contact your local IR Distributor, IR Field Office or IR El Segundo offices for more information.

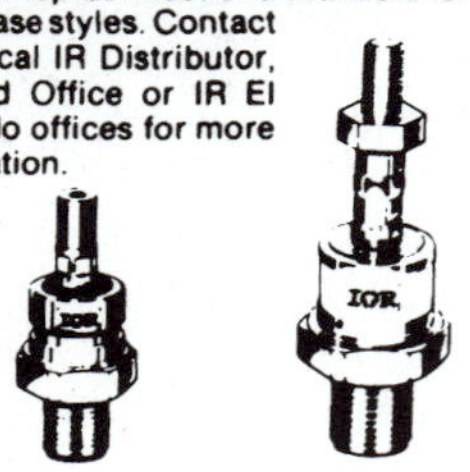

FIGURE 3–38(a) A typical rectifier data sheet (Courtesy of International Rectifier)

will need to perform a few calculations that will yield the required specifications and ratings for the power supply components.

Let us begin with the transformer. We will need to work backwards from the regulated output voltage, to the required unregulated voltage, to the secondary voltage from the transformer. An important point to remember is that if a diode-bridge rectifier is to be used, the secondary

INTERNATIONAL RECTIFIER IOR

Diode Bridges

Molded diode bridges are available from 1 to 100 Amps in a variety of space-saving, easy-to-use packages

Single ended bridges are available in three popular case styles for 1 to 6 Amps PC board-mount applications

The 10 to 35 Amp range is covered by the JB and MB series which are isolated base. **U.L. recognized bridges** with very high surge current capability

To make a bridge circuit into a center tap circuit, connect to terminals as shown

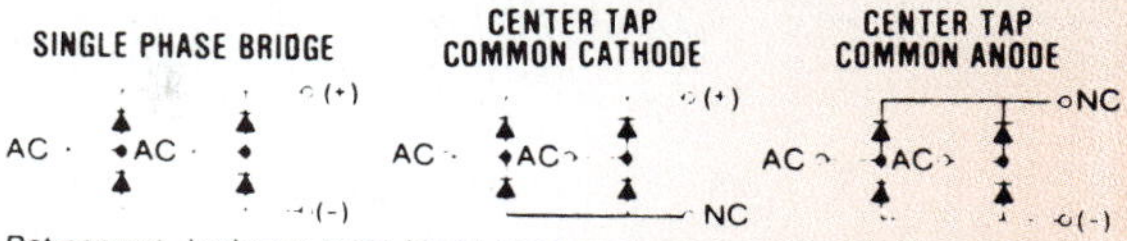

Ratings per diode are same for all three applications NC No Connection

SINGLE PHASE DIODE MOLDED BRIDGES — 1.0 TO 100 AMPS (I_O)

I_O (A) Output Current	1.0	1.2	1.8	1.9	2.0	3.0	6.0	10	25	35	80	100
@ T_C (°C)	25	45	50	45	50	50	50	65	65	55	85	80
I_{FSM} (A) (Surge)	30	52	52	52	60	50	125	130	350	420	800	1000
Notes	(1)	(1)	(1)	(1) (3)	(1)	(1)	(1)	(2)	(2)	(2)	(2)	(2)
Case Style	D-43	D-38	D-2	D-37	D-44	D-45	D-46	D-34	D-34	D-34	D-20-6	D-20-6
PART NUMBERS								UL	UL	UL		
V_{RRM} 50 Volts	–	–	18DB05A	–	2KBP005	KBPC1005	KBPC6005	100JB05L	250JB05L	35MB5A	800HB05U	1000HB05U
100 Volts	1DMB10	1KAB10E	18DB1A	2KBB10	–	–	–	100JB1L	250JB1L	35MB10A	800HB1U	1000HB1U
200 Volts	1DMB20	1KAB20E	18DB2A	2KBB20	2KBP02	KBPC102	KBPC602	100JB2L	250JB2L	35MB20A	800HB2U	1000HB2U
300 Volts	–	–	–	–	–	–	–	–	–	–	800HB3U	1000HB3U
400 Volts	1DMB40	1KAB40E	18DB4A	2KBB40	2KBP04	KBPC104	KBPC604	100JB4L	250JB4L	35MB40A	800HB4U	1000HB4U
600 Volts	–	1KAB60E	18DB6A	2KBB60	2KBP06	KBPC106	KBPC606	100JB6L	250JB6L	35MB60A	800HB6U	1000HB6U
800 Volts	–	1KAB80E	18DB8A	2KBB80	2KBP08	KBPC108	KBPC608	100JB8L	250JB8L	35MB80A		–
1000 Volts	–	1KAB100E	18DB10A	2KBB100	2KBP10	KBPC110	KBPC610	100JB10L	250JB10L	35MB100A		–
1200 Volts	–	–	–	–	–	–	–	100JB12L	250JB12L	35MB120A	–	

NOTES: (1) Ambient temperature (T_A). (2) Must be used with suitable heatsink. (3) For lead configuration - ~ ~ + add 'R' to part number.

* For detailed specifications, contact your local IR Field Office or IR Distributor.

D-2 D-37 D-38 D-43 D-44 D-45/D-46

THREE-PHASE BRIDGES—MOLDED

I_O (A) Output Current	60	100
@ T_C °C	70	100
I_{FSM} (A) (Surge)	500	800
Case Style	D-20-8	D-20-8
PART NUMBERS		
V_{RRM} (V) 100 Volts	600HT1U	1000HT1U
200 Volts	600HT2U	1000HT2U
300 Volts	600HT3U	1000HT3U
400 Volts	600HT4U	1000HT4U
600 Volts	600HT6U	1000HT6U
800 Volts	600HT8U	1000HT8U
1000 Volts	600HT10U	–

SINGLE PHASE BRIDGES — FINNED

I_O (A) Output Current	24	32
@ T_C °C	50	50
I_{FSM} (A) (Surge)	250	1500
Case Style	D10	D10
PART NUMBERS		
V_{RRM} (V) 400 Volts	B12F40	B70H40
600 Volts	B12F60	B70H60
800 Volts		B70H80
1000 Volts	B12F100	B70H100

D-34

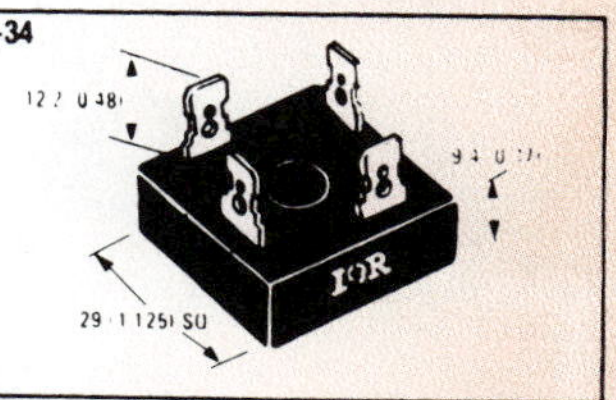

D-20-8

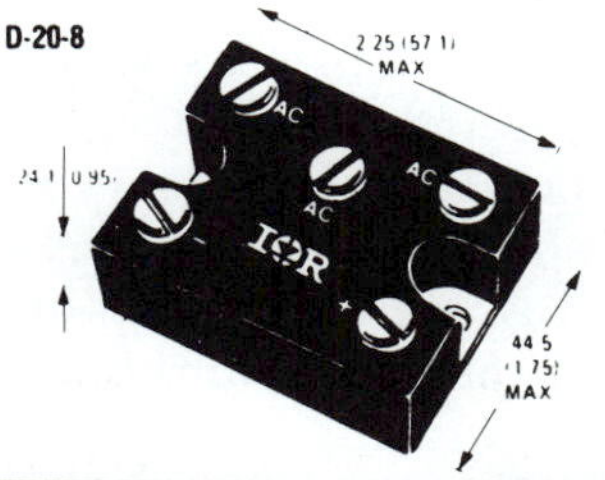

D-10

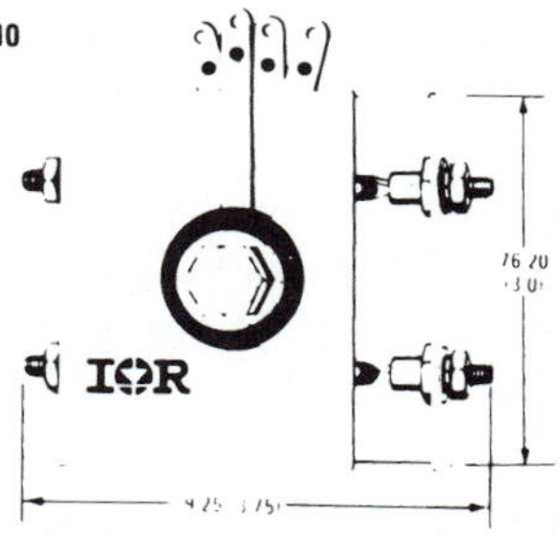

D-20-6

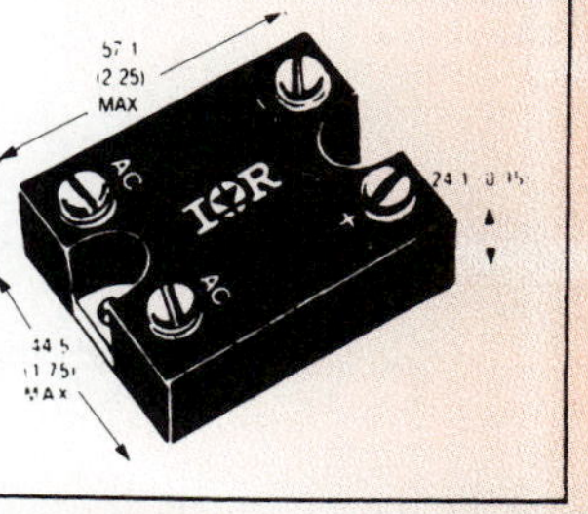

FIGURE 3–38(b) Diode bridge specifications (Courtesy of International Rectifier)

peak voltage from the transformer should be, in ballpark numbers, close to the required unregulated voltage. If a center-tapped transformer and two diodes are used, the secondary voltage must be twice as much. Remember that the filtering capacitor charges to the peak voltage of the rectified waveform.

Because the dc unregulated voltage has a ripple component, the peak voltage of the rectified waveform (V_{PR}) needs to be somewhat larger than the minimum required unregulated dc voltage. When a voltage regulator IC is used, the ripple factor does not have to be very low; a 10% figure would be adequate. Therefore, V_{PR} needs to be about 5% higher than the minimum unregulated dc voltage. We will also need to take into account one or two diode drops depending on whether we use a rectifier with center-tapped transformer or a diode bridge, respectively.

The current rating of both transformer and diodes is the same. The sum of all the load currents times a safety factor will determine the current rating. Additionally, the PIV rating of the diodes needs to be specified as well. Recall that the maximum reverse voltage per diode equals the peak voltage from the secondary. A safety factor can also be used for the PIV rating. In both cases, a safety factor of 1.5 to 2 is adequate.

For the capacitor value(s) and rating(s), we make use of the ripple voltage formula (equation 3–12) in terms of the load current. Again, if the power supply includes a regulator, a 10% ripple factor is acceptable. However, if a regulator is not required, the ripple factor should be less than 5%. The capacitor voltage should be somewhat larger than V_{PR}; a safety factor of 1.5 to 2 is also appropriate for capacitor voltage.

Finally, a fuse needs to be placed in series with one of the ac power lines to protect the power supply in case an excessive current develops anywhere in the circuit or the load. The power in the primary of the transformer is about the same as that in the secondary because the efficiency of commercial transformers is high. Therefore, the rms voltage-current product in the primary will equal the rms voltage-current product in the secondary. The fuse current rating should be about 50% higher than the calculated primary current. Additionally, if more than one regulator will be used, it would be a good idea to also protect the input to each regulator with a quick-acting fuse.

EXAMPLE 3–14

Design an experimenter's power supply that provides two regulated voltages: +12 V at 1 A and +5 V at 1 A.

Solution

We start by specifying the two regulators; the 7812 for the 12-V output, and the 7805 for the 5-V output. The 7812 requires a minimum unregulated input voltage of 14.6 V. Let us work with a target of 16 V.

Now we can determine the specs for the transformer. With a 10% ripple, V_{PR} needs to be 16.8 V, which is 5% above 16 V. If we choose to use a diode-bridge rectifier, then $16.8 = 1.41\ V_{rms} - 1.4$ V, from which $V_{rms} = 12.9$ V (secondary voltage). We can specify a 12-V transformer because the actual secondary voltage would be about 13 V to 14 V at normal loading. The transformer's current rating should be at least 2 A because the current from each regulator can be up to 1 A. However, it is not a good engineering practice to specify ratings based on strict minimum requirements; one should always use a safety margin. A rating of 2.5 or 3 A will do the job.

The filter capacitance value is determined by using equation 3–12 with $I_L = 2\text{A}$, $f_r = 120$ Hz, and $V_{PP} = 10\%$ of 16 V = 1.6 V. (We are using a 10% ripple factor.)

$$C = \frac{I_L}{f_r V_{PP}}$$

$$= \frac{2\ \text{A}}{(120\ \text{Hz})(1.6\ \text{V})} = 10{,}400\ \mu\text{F}$$

The voltage rating should be at least the value of V_{PR} or, in this case, about 17 V. A capacitor with a voltage rating of 25 or 30 V is appropriate. The final capacitor specification is: Electrolytic type, 10,000 μF at 25 or 30 V.

Diode-bridge specification is based on the total load current and the reverse voltage to which the diodes will be exposed. As stated previously, the diode rating is the same as for the transformer: 2 or 2.5 A. The reverse voltage per diode is the peak voltage from the secondary or about 17 V. We will use a PIV rating of 50 V, which is the lowest typically available.

The fuse is finally specified according to the current in the primary which can be obtained from equating primary and secondary powers, that is,

$$V_{pri}\, I_{pri} = V_{sec}\, I_{sec}$$

which yields

$$I_{pri} = \frac{V_{sec} I_{sec}}{V_{pri}}$$

and

$$I_{pri} = \frac{(2\text{ A})(12\text{ V})}{120\text{ V}} = 200\text{ mA}$$

A 250 mA, slow-blow type would be a good choice for the fuse. Slow-blow specification is necessary because when the power is first turned on, the initial current is much larger than the normal operating current. Additionally, the fuse should also protect the transformer if its maximum rating is exceeded. Therefore, the current rating of the fuse should not exceed the value of the primary current that results at the maximum rated secondary current. In our case, for example, if we use a 12–V, 3–A transformer, the primary current at the rated secondary current would be

$$I_{pri} = \frac{(3\text{ A})(12\text{ V})}{120\text{ V}} = 300\text{ mA}$$

The rating of the fuse must not exceed this value.

3–8 MULTISIM EXERCISE

Observing on the Oscilloscope the Filtered Output Voltage and the Diode Current in a Half-Wave Rectifier

Figure 3–39 shows a half-wave rectifier with capacitive filter and a resistive load. Notice that a 1-Ω resistor has been added in series with the ac voltage source to sense the diode current and display it on the oscilloscope. The resistor must be on the ground side because the oscilloscope must be grounded.

Before you start the simulation, double-click on the oscilloscope and set it up to 5 ms/div for the time axis, 1 V/div for channel A, and 100 V/div for channel B. Once the simulation is running, pause it and observe the waveform from channel A, which represents the diode current. Because the sensing resistor has a value of 1 Ω, the current will be numerically identical to the displayed voltage. Note that the current peaks reach almost 2 amperes. The current peaks appear negative on the oscilloscope because of the polarity of the voltage drop according to the direction of the diode current. Notice also that current flows through the diode only during the recharge of the capacitor.

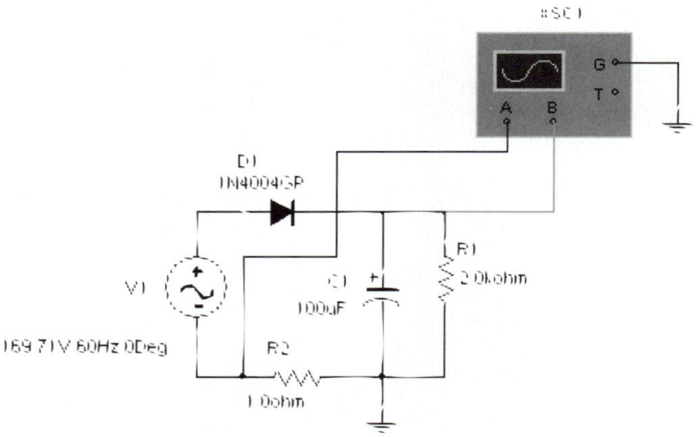

FIGURE 3–39 Half-wave rectifier with oscilloscope connected to observe diode current and output voltage

SUMMARY

This chapter has presented the use of diodes in circuits, particularly in rectifiers for basic power supplies. At the end of this chapter, the student should have an understanding of the following concepts:

- Diodes are used for allowing current to flow in only one direction.
- Diodes are an essential part in all power supplies.
- Rectified waveforms can be full-wave or half-wave.
- Full-wave rectification can be done with four diodes and an ac source, or with two diodes and a center-tapped transformer.
- Capacitive filtering substantially reduces the pulsating character of a rectified waveform.
- Zener diodes and special integrated circuits provide voltage regulation.

EXERCISES

SECTION 3–2

The Diode as a Nonlinear Device

3–1. Make a sketch of the *I–V* characteristic curve for a 10-kΩ resistor when current *I* is plotted along the vertical axis and voltage *V* along the horizontal axis. What is the slope of the characteristic? Be certain to include units in your answer.

3–2. Make a sketch of the *I–V* characteristic curve for a diode with $I_s = 0.02$ pA for V_D values from zero to 0.7 V. Use the approximation

$$I_D = I_s e^{V_D/V_T}$$

What is the slope of a line tangent to the curve at $I_D = 2$ mA?

SECTION 3–3

ac and dc Resistance

3–3. Using the diode *I–V* characteristic shown in Figure 3–40, find (graphically) the approximate ac resistance when the current in the diode is 0.1 mA. Repeat when the voltage across it is 0.64 V. Is the diode silicon or germanium?

3–4. Using the diode *I–V* characteristic shown in Figure 3–40, find (graphically) the approximate value of the dynamic resistance when the current in the diode is 0.2 mA. Repeat when the voltage across the diode is 0.62 V. What is the approximate maximum knee current?

3–5. Find the dc resistance of the diode at each point specified in Exercise 3–3.

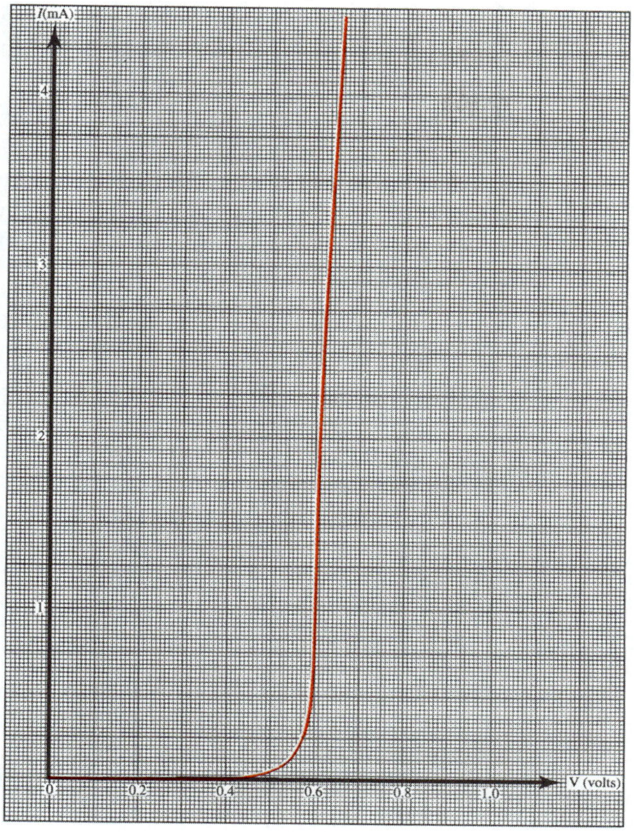

FIGURE 3–40 (Exercises 3–3 and 3–4).

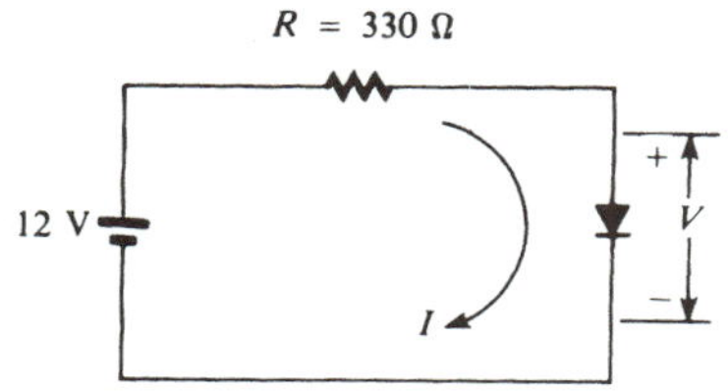

FIGURE 3–41 (Exercise 3–12)

3–6. Find the static resistance of the diode at each point specified in Exercise 3–4.

3–7. Neglecting bulk resistance, use equation 3–6 to find the approximate ac resistance of the diode at each point specified in Exercise 3–3.

3–8. Assume that the bulk resistance of the diode whose *I–V* characteristic is shown in Figure 3–41 is 0.1 Ω when the current is greater than 1.5 mA and 0.5 Ω when the current is less than 1.5 mA. Use equation 3–6 to find the approximate dynamic resistance of the diode at each point specified in Exercise 3–4.

3–9. A certain diode conducts a current of 440 nA from cathode to anode when the reverse-biasing voltage across it is 8 V. What is the diode's dc resistance under these conditions?

3–10. When the reverse-biasing voltage in Exercise 3–9 is increased to 24 V, the reverse current increases to 1.20 μA. What is its dc resistance in this case?

3–11. In the test circuit shown in Figure 3–3, a diode voltage of 0.69 V was measured when the diode current was 163 mA.

(a) What is the dc resistance of the diode at $V_D = 0.69$ V?

(b) What is the ac resistance of the diode when the voltage across it changes from 0.68 V to 0.69 V?

3–12. In the circuit shown in Figure 3–41, the current *I* is 34.28 mA. What is the voltage drop across the diode? What is its dc resistance?

3–13. Repeat Exercise 3–12 if the resistor *R* is 220 Ω and the current *I* is 51.63 mA.

SECTION 3–4

Analysis of dc Circuits Containing Diodes

3–14. Determine the currents I_1 and I_2 in Figure 3–42. Assume silicon diodes.

3–15. Repeat Problem 3–14 when diode D_1 is reversed.

3–16. The voltage V_S in Figure 3–43 can be varied between zero and 20 V. Assuming the diodes behave as shown in Figure 3–13(b):

(a) Determine the voltage levels of V_S at which D_1 and D_2 begin conducting.

(b) Find the currents through each diode when V_S reaches 20 V.

3–17. Find the current in each diode in the circuit of Figure 3–44. Assume diode drops of 0.7 V.

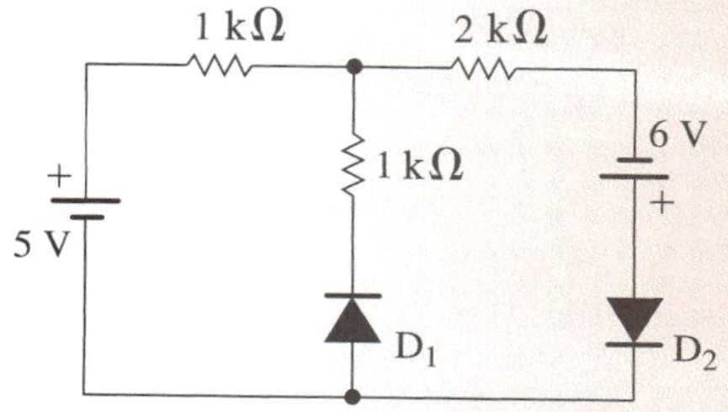

FIGURE 3–42 (Exercise 3–14)

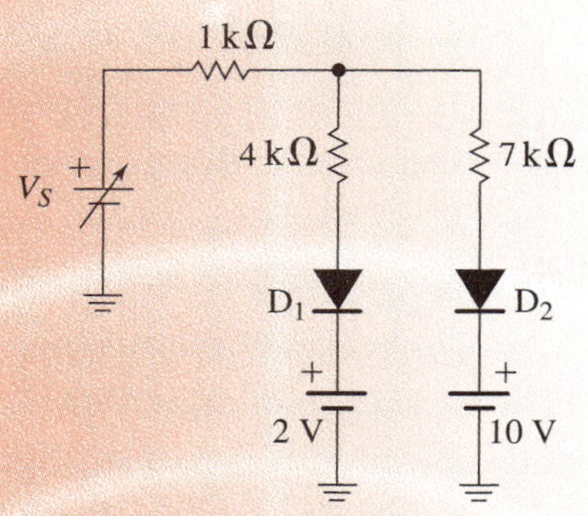

FIGURE 3–43 (Exercise 3–16)

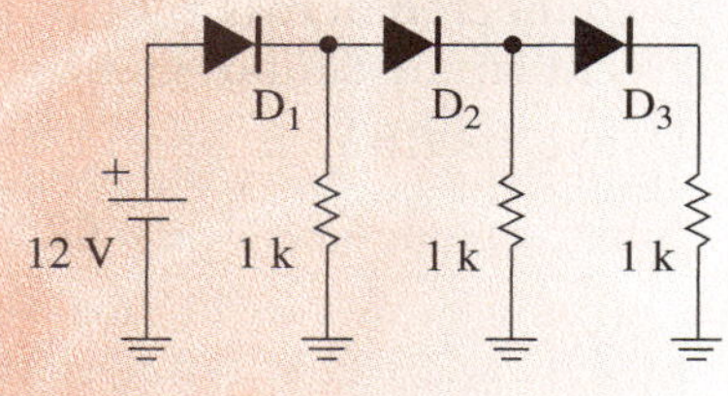

FIGURE 3–44 (Exercise 3–17)

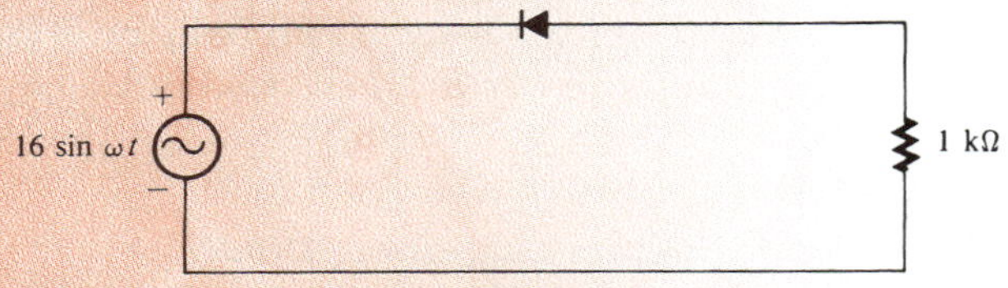

FIGURE 3–45 (Exercise 3–19)

SECTION 3–5

Elementary Power Supplies

3–18. In the circuit shown in Figure 3–15, the 1.5-kΩ resistor is replaced with a 2.2-kΩ resistor. Assume that the silicon diode has a characteristic curve like that shown in Figure 3–13(b). If $e(t) = 2 \sin \omega t$, find the peak value of the current $i(t)$ and the voltage $v_R(t)$ across the resistor. Sketch the waveforms for $e(t)$, $i(t)$, and $v_R(t)$.

3–19. The silicon diode in Figure 3–45 has a characteristic curve like that shown in Figure 3–13(b). Find the peak values of the current $i(t)$ and the voltage $v_R(t)$ across the resistor. Sketch the waveforms for $e(t)$, $i(t)$, and $v_R(t)$.

3–20. What peak-to-peak sinusoidal voltage must be connected to a half-wave rectifier if the rectified waveform is to have a dc value of 6 V? Assume that the forward drop across the diode is 0.7 V.

3–21. What should be the rms voltage of a sinusoidal wave connected to a full-wave rectifier if the rectified waveform is to have a dc value of 50 V? Neglect diode voltage drops.

3–22. The half-wave rectifier in Figure 3–19 has a 250-μF filter capacitor and a 1.5-kΩ load. The ac source is 120 V rms with frequency 60 Hz. The voltage drop across the silicon diode is 0.7 V. Assuming light loading, find

(a) the dc value of the load voltage;

(b) the peak-to-peak value of the ripple voltage.

3–23. The half-wave rectifier in Exercise 3–22 is replaced by a silicon full-wave rectifier, and a second 250–μF capacitor is connected in parallel with the filter capacitor. Assume light loading and do not neglect the voltage drop across the diodes. Find

(a) the dc value of the load voltage;

(b) the peak-to-peak value of the ripple voltage.

(c) If the load resistance is decreased by a factor of 2, determine (without recalculating) the approximate factor by which the ripple voltage is changed.

3–24. The primary voltage on the transformer shown in Figure 3–46 is 120 V rms and $R_L = 10\ \Omega$. Neglecting the forward voltage drops across the diodes, find

(a) the turns ratio $N_P : N_S$, if the average current in the resistor must be 1.5 A;

(b) the average power dissipated in the resistor, under the conditions of (a); and

(c) the maximum PIV rating required for the diodes, under the conditions of (a).

3–25. The primary voltage on the transformer in Figure 3–46 is 120 V rms and $N_P : N_S = 15{:}1$. Diode voltage drops are 0.7 V.

(a) What should be the value of R_L if the average current in R_L must be 0.5 A?

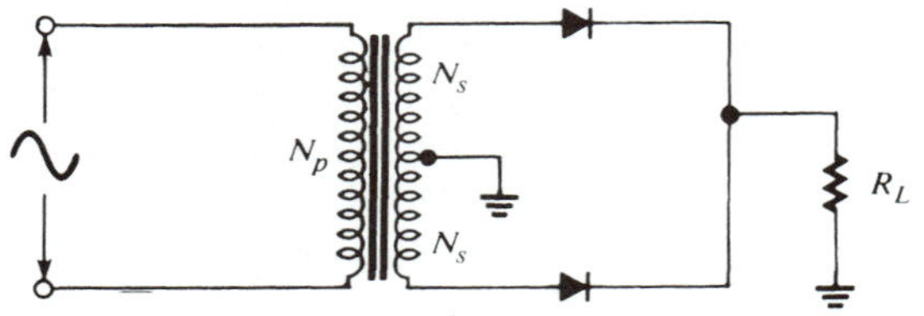

FIGURE 3–46 (Exercises 3–24 and 3–25)

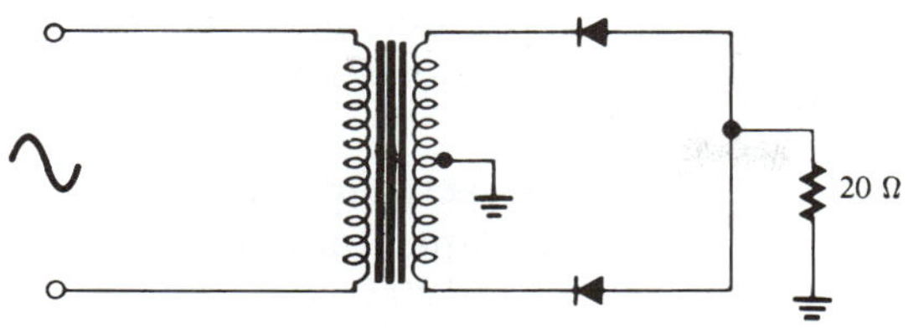

FIGURE 3–47 (Exercise 3–26)

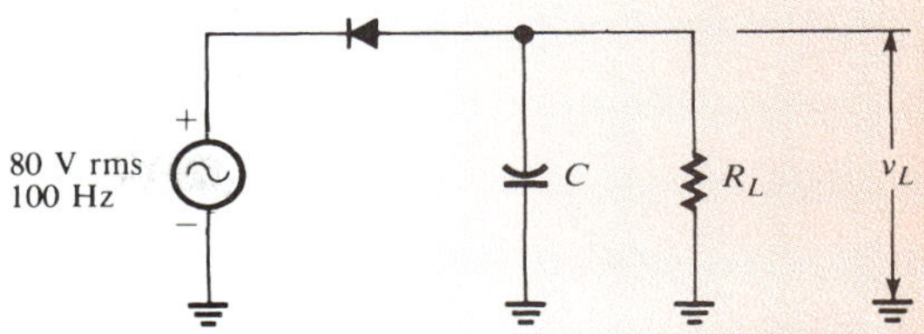

FIGURE 3–49 (Exercise 3–29)

(b) What power is dissipated in R_L under the conditions of (a)?

(c) What minimum PIV rating is required for the diodes under the conditions of (a)?

3–26. The secondary voltage on the transformer in Figure 3-47 is 30 VCT rms at 50 Hz. The diode voltage drops are 0.7 V. Sketch the waveforms of the voltage across and current through the 20-Ω resistor. Label peak values and the time points where the waveforms go to 0.

3–27. Each of the diodes in Figure 3–48 has a forward voltage drop of 0.7 V. Find

(a) the average voltage across R_L;

(b) the average power dissipated in the 1-Ω resistor; and

(c) the minimum PIV rating required for the diodes.

3–28. Repeat Exercise 3–27 if R_L is changed to 5 Ω and the transformer turns ratio is changed to 1:1.5.

3–29. Sketch the waveform of the voltage v_L in the circuit shown in Figure 3–49. Include the ripple and show the value of its period on the sketch. Also show the value of V_{PR}. Neglect the forward drop across the diode.

3–30. What is the percent ripple of a full-wave–rectified waveform having a peak value 75 V and frequency 120 Hz if $C = 220\ \mu F$ and $I_L = 80$ mA? What is the percent ripple if the frequency is halved?

3–31. A half-wave rectifier is operated from a 60-Hz line and has a 1000-μF filter capacitance connected across it. What is the minimum value of load resistance that can be connected across the capacitor if the percent ripple cannot exceed 5%?

3–32. A full-wave rectifier is operated from a 60-Hz, 50-V-rms source. It has a 500-μF filter capacitor and a 750-Ω load. Find

(a) the average value of the load voltage;

(b) the peak-to-peak ripple voltage; and

(c) the percent ripple.

3–33. A half-wave rectifier has a 1000-μF filter capacitor and a 500-Ω load. It is operated from a 60-Hz, 120-V-rms source. It takes 1 ms for the capacitor to recharge during each input cycle. For what minimum value of repetitive surge current should the diode be rated?

3–34. Repeat Exercise 3–33 if the rectifier is full-wave and the capacitor takes 0.5 ms to recharge.

3–35. Assuming negligible ripple, find the average current in the 100-kΩ resistor in Figure 3–50.

3–36. The transformer shown in Figure 3–51 has a *tapped secondary* each portion of the secondary winding having the number of turns shown. Assuming that the primary voltage is that shown in the figure, design two separate circuits that can be used with the transformer to obtain an (unloaded) dc

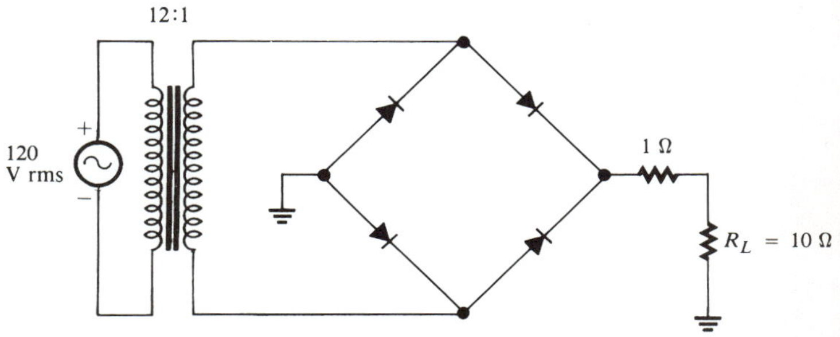

FIGURE 3–48 (Exercise 3–27)

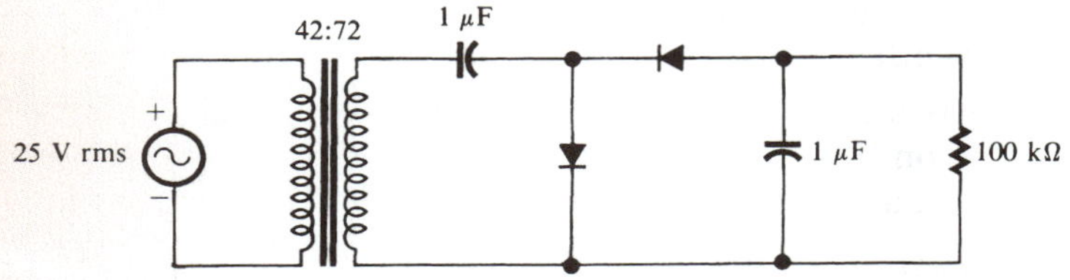

FIGURE 3–50 (Exercise 3-35)

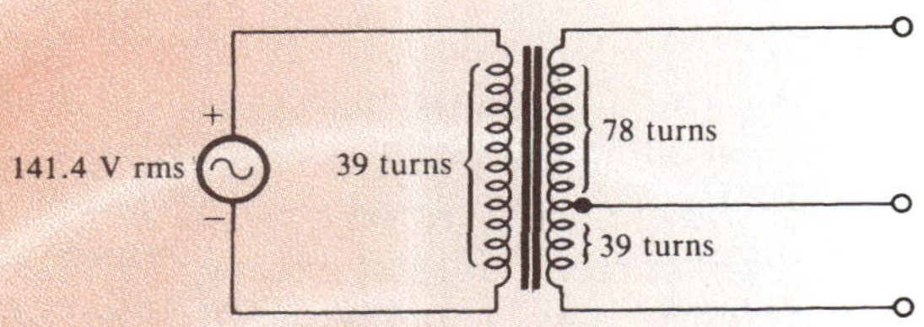

FIGURE 3–51 (Exercise 3-36)

voltage of 1200 V. It is not necessary to specify capacitor sizes. What minimum PIV ratings should the diodes in each design have?

SECTION 3–6

Elementary Voltage Regulation

3–37. In the circuit shown in Figure 3–52, the zener diode has a reverse breakdown voltage of 12 V. $R_S = 50\ \Omega$, $V_S = 20$ V, and R_L can vary from 100 Ω to 200 Ω. Assuming that the zener diode remains in breakdown, find

(a) the minimum and maximum current in the zener diode;

(b) the minimum and maximum power dissipated in the diode; and

(c) the minimum rated power dissipation that R_S should have.

3–38. Repeat Exercise 3–37 if, in addition to the variation in R_L, V_S can vary from 19 V to 30 V.

3–39. The 6-V zener diode in Figure 3–53 has a maximum rated power dissipation of 0.5 W. Its reverse current must be at least 5 mA to keep it in breakdown. Find a suitable value for R_S if V_S can vary from 8 V to 12 V and R_L can vary from 500 Ω to 1 kΩ.

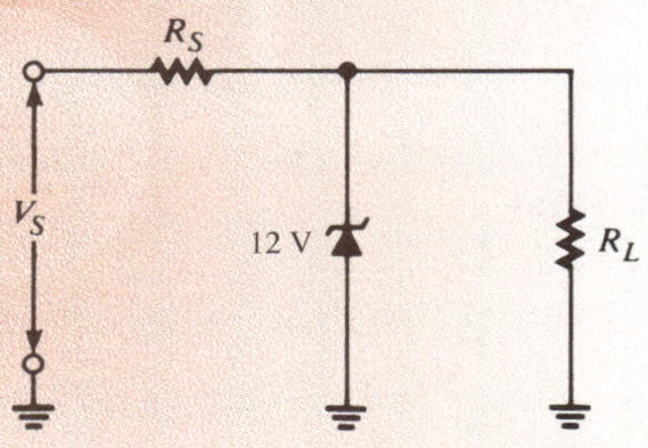

FIGURE 3–52 (Exercises 3-37 and 3-38)

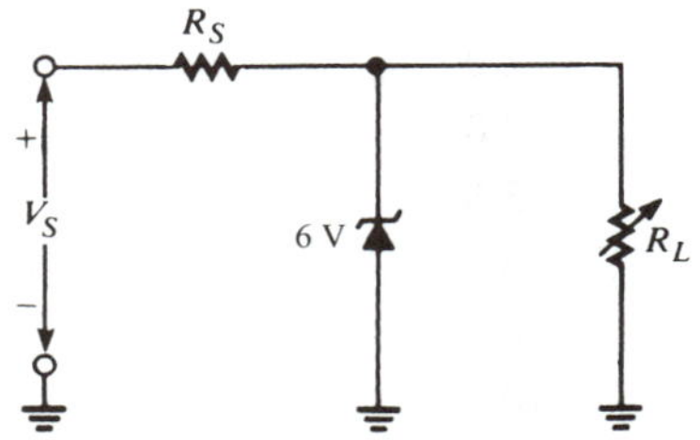

FIGURE 3–53 (Exercise 3-39)

3–40. (a) If R_S in Exercise 3–39 is set to its maximum permissible value, what is the maximum permissible value of V_S?

(b) If R_S in Exercise 3–39 is set equal to its minimum permissible value, what is the minimum permissible value of R_L?

3–41. A zener diode has a breakdown voltage of 12 V at 25°C and a temperature coefficient of +0.5 mV/°C.

(a) Design a temperature-stabilizing circuit using silicon diodes that have temperature coefficients of −0.21 mV/°C. The forward drop across each diode at 25°C is 0.68 V.

(b) Find the voltage across the stabilized network at 25°C and at 75°C.

(c) Find the temperature stability of the stabilized network.

3–42. A zener diode has a breakdown voltage of 15.1 V at 25°C. It has a temperature coefficient of +0.78 mV/°C and is to be operated between 25°C and 100°C. It is to be temperature stabilized in such a way that the voltage across the network is never less than its value at 25°C.

(a) Design a temperature-stabilizing network using silicon diodes whose temperature coefficients are −0.2 mV/°C. The forward drop across each diode at 25°C is 0.65 V.

(b) What is the maximum voltage across the stabilized network?

3–43. Following is a set of measurements that were made on the voltage across and current through a zener diode:

I_z*(mA)*	V_z*(volts)*
0.5	30.1
1.0	30.15
2.0	30.25
3.5	30.37
6	30.56
8	30.68
10	30.80
30	31.90
40	32.40
90	34.00

(a) Find the approximate zener impedance over the range from $I_Z = 3.5$ mA to $I_Z = 10$ mA.

(b) Show that the zener impedance decreases with increasing current.

3–44. The breakdown voltage of a zener diode when it is conducting 2.5 mA is 7.5 V. If the voltage must not increase more than 10% when the current increases 50%, what maximum impedance can the diode have?

SECTION 3–7

Diode Types, Ratings, and Specifications

3–45. In the circuit shown in Example 3–13 (Figure 3–36), suppose the load resistor R is changed to 47 Ω. What then is the least expensive of the diodes listed in the example that can be used in this application?

3–46. In the circuit shown in Example 3–13 (Figure 3–36), suppose the ac voltage is 100 V rms and the load resistor is changed to 68 Ω. What then is the least expensive of the diodes listed in the example that can be used in this application?

3–47. A small-signal diode is to be used in an application where it will be subjected to a reverse voltage of 35 V. It must conduct a forward current of 0.01 A when the forward-biasing voltage is 1.0 V. The reverse current must not exceed 30 nA when the reverse voltage is 30 V. Select a diode type number from Figure 3–37 that meets these requirements.

3–48. A silicon diode is to be used in an application where it will be subjected to a reverse-biasing voltage of 85 V. The forward current will not exceed 100 mA, but it must have a 0.5-W power dissipation rating. Select a diode type number from Figure 3–37 that meets these requirements.

3–49. A rectifier diode is to be used in a power supply design where it must repeatedly withstand sine wave reverse voltages of 250 V rms and must conduct 0.6 A (average) of forward current. The forward surge current through the diode when the supply is first turned on will be 25 A. It is estimated that the diode case temperature (T_C) will be 30°C. Select a diode type number from Figure 3–38 that meets these requirements.

3–50. A rectifier diode is to be used in a large power supply where it must be capable of withstanding repeated reverse voltages of 450 peak volts. The forward current in the diode will average 13.5 A. Select a diode type number from Figure 3–38 that meets these requirements.

3–51. A full-wave bridge is to be connected to a 240-V-rms power line. The output will be filtered and will supply an average voltage of 150 V to a 50-Ω load. The worst-case current that will flow through the bridge when power is first applied is 20 times the average load current. Using the International Rectifier specifications in Figure 3–38(b), select a bridge (give its part number) that can be used for this application.

PSPICE EXERCISES

3–52. Simulate Problem 3–14 using PSpice. Use the part DIN4002 in the "Eval" library for both diodes. Hint: Include in your circuit file the .OP statement in order to obtain a report on the diode currents in the output file.

3–53. Simulate Problem 3–16 using PSpice with the voltage source V_S swept from zero to 20 V. Plot both diode currents versus V_S. Hint: Use .DC analysis.

CHAPTER 4

BIPOLAR JUNCTION TRANSISTORS

OUTLINE

OBJECTIVES

- Explore the basic dc bias configurations for the bipolar junction transistor.
- Develop an understanding of the relationship between I_C, I_B, and I_E.
- Understand the transistor modes of operation and how they apply to circuit operation.
- Understand the characteristics of basic transistor curves.
- Explore the use of a BJT transistor switch and understand how to bias the transistor switch for use in a circuit.

4–1 INTRODUCTION

The workhorse of modern electronic circuits, both discrete and integrated analog and digital, is the *transistor*. The importance of this versatile device stems from its ability to produce *amplification,* or *gain,* in a circuit. We say that amplification has been achieved when a small variation in voltage or current is used to create a large variation in one of those same quantities, and this is the fundamental goal of most electronic circuits. As a means for creating gain, the transistor is in many ways analogous to a small valve in a large water system: By expending a small amount of energy (turning the valve), we are able to control—increase or decrease—a large amount of energy (in the flow of a large quantity of water). When a device such as a transistor is used to create gain, we supply a small signal to it and refer to that as the *input;* the large current or voltage variations that then occur at another point in the device are referred to as the *output*.

The two most important kinds of transistors are the *bipolar junction* type and the *field-effect* type. The bipolar junction transistor (BJT) is so named (*bi polar*—two polarities) because of its dependence on both holes and electrons as charge carriers. We will study the theory and applications of "bipolars" in this and the next chapter, and devote equal time to field-effect transistors (FETs), which operate under completely different principles, in Chapter 5.

We should note that bipolars are studied first for historical reasons, and because the theory of these devices follows naturally from a study of *pn* junctions. Their precedence in our study should in no way imply that they are of greater importance than field-effect transistors. Bipolars were the first kind of transistors to be widely used in electronics, and they are still an important segment of the semiconductor industry. Most people in the industry still use the word *transistor* (as we shall) with the understanding that a bipolar transistor is meant. However, field-effect technology has now evolved to the point where FETs are used in greater numbers than BJTs in integrated circuits for digital and analog applications.

4–2 THEORY OF BJT OPERATION

A bipolar junction transistor is a specially constructed, *three-terminal* semiconductor device containing *two pn* junctions. It can be formed from a bar of material that has been doped in such a way that it changes from *n* to *p* and back to *n,* or from *p* to *n* and back to *p*. In either case, a junction is created at each of the two boundaries where the material changes from one type to the other. Figure 4–1 shows the two ways that it is possible to alternate material types and thereby obtain two junctions.

FIGURE 4–1 *npn* and *pnp* transistor construction

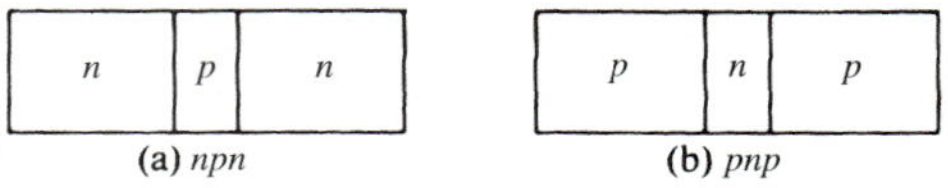

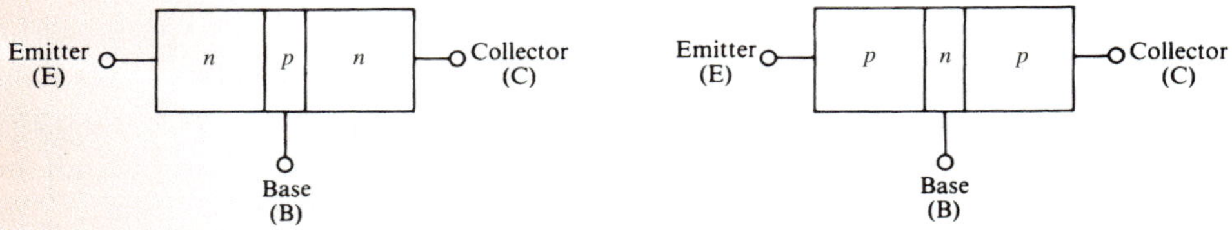

FIGURE 4–2 Base, emitter, and collector terminals of *npn* and *pnp* transistors

When a transistor is formed by sandwiching a single *p* region between two *n* regions, as shown in Figure 4–1(a), it is called an *npn* type. Figure 4–1(b) shows the *pnp* type, containing a single *n* region between two *p* regions.

The middle region of each transistor type is called the *base* of the transistor. Of the remaining two regions, one is called the *emitter* and the other is called the *collector* of the transistor. Let us suppose that terminals are attached to each region so that external electrical connections can be made between them. (In integrated circuits, no such accessible terminals may be provided, but it is still possible to identify the base, emitter, and collector and to form conducting paths between those regions and other internal components.) Figure 4–2 shows terminals attached to the regions of each transistor type. The terminals are labeled according to the region to which they connect.

The physical appearance of an actual transistor bears little resemblance to the figures we have shown so far. However, these kinds of diagrams are very helpful in understanding transistor theory, and Figures 4–1 and 4–2 are representative of actual transistors in at least one respect: The base region is purposely shown *thinner* than either the emitter or collector region. For reasons that will become evident soon, the base region in an actual transistor is made even thinner in proportion to the other regions than is depicted by the figures. Also, the base region is much more lightly doped than the other regions.* Both these characteristics of the base are important for the transistor to be a useful device and to perform what is called normal "transistor action."

For the sake of clarity, we will continue our discussion of transistor theory in terms of the *npn* type only. The underlying theory is equally applicable to the *pnp* type and can be "translated" for *pnp*s simply by changing each carrier type mentioned in connection with *npn*s and reversing each voltage polarity. To obtain normal transistor action, it is necessary to bias both *pn* junctions by connecting dc voltage sources across them. Figure 4–3 illustrates the correct bias for each junction in the *npn* transistor. In the figure, we show the result of biasing each junction separately, while in practice both junctions will be biased simultaneously by one external circuit, as will be described presently.

As shown in Figure 4–3(a), the emitter–base junction is forward biased by the dc source labeled V_{EE}. Note that the *n*egative terminal of V_{EE} is connected to the *n* side of the *np* junction, as required for forward bias. Consequently, there is a substantial flow of diffusion current across the junction due to the flow of the majority carriers (electrons) from the *n*-type emitter. This action is exactly that which we discussed in connection with a forward-biased *pn*

---

*In integrated circuits, the emitter is usually heavily doped (n^+) material. Most integrated-circuit BJTs are of the *npn* type.

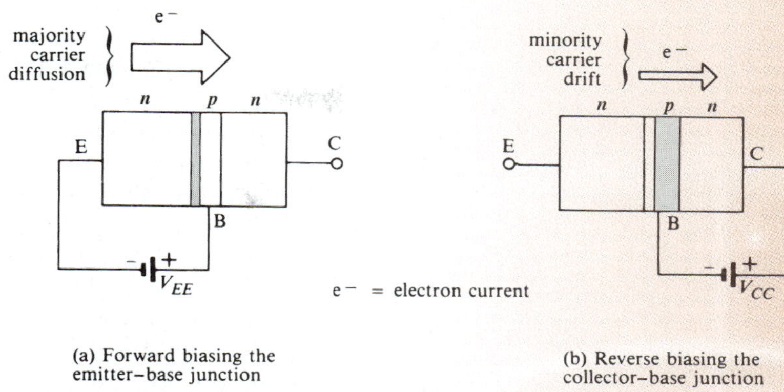

FIGURE 4–3 Biasing the two *pn* junctions in an *npn* transistor

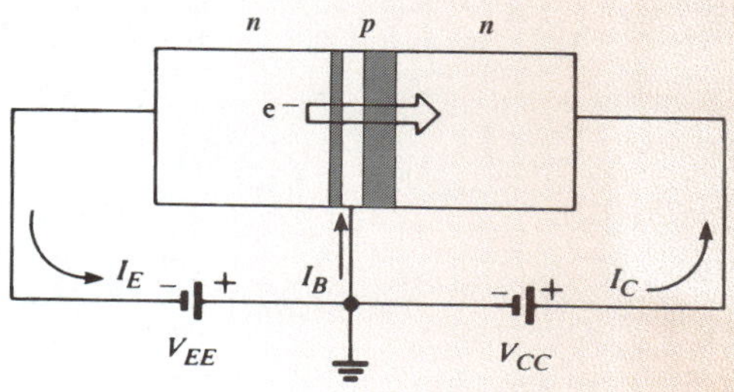

FIGURE 4–4 The *npn* transistor with both bias sources connected

junction in Chapter 2. The depletion region at this junction is made narrow by the forward bias, as also described in Chapter 2. The width of the base region is exaggerated in the figure for purposes of clarity. When the majority electrons diffuse into the base, they become minority carriers in that *p*-type region. We say that minority carriers have been *injected* into the base.

Figure 4–3(b) shows that the collector-base junction is reverse biased by the dc source labeled V_{CC}. The positive terminal of V_{CC} is connected to the *n*-type collector. As a result, the depletion region at this junction is widened, and the only current that flows from base to collector is due to the minority electrons crossing the junction from the *p*-type base. Recall from Chapter 2 that *minority* carriers readily cross a reverse-biased junction under the influence of the electric field, and they constitute the flow of reverse current in the junction:

Figure 4–4 shows the *npn* transistor the way it is biased for normal operation, with both dc sources V_{EE} and V_{CC} connected simultaneously. Note in the figure that the negative terminal of V_{CC} is connected to the positive terminal of V_{EE} and that both of these are joined to the base. The base is then the "ground," or common point of the circuit, and can therefore be regarded as being at 0 V. The emitter is negative *with respect to the base,* and the collector is positive *with respect to the base.* These are the conditions we require in order to forward bias the emitter–base junction and to reverse bias the collector–base junction.

Because the base region is very thin and is lightly doped relative to the heavily doped emitter (so there are relatively few holes in it), very few of the electrons injected into the base from the emitter recombine with holes. Instead, they diffuse to the reverse-biased base–collector junction and are swept across that junction under the influence of the electric field established by V_{CC}. Remember, again, that the electrons injected into the base are the minority carriers there, and that minority carriers readily cross the reverse-biased junction. We conclude that electron flow constitutes the dominant current type in an *npn* transistor. For a *pnp* transistor, in which everything is "opposite," hole current is the dominant type.

Despite the fact that most of the electrons injected into the base cross into the collector, a few of them do combine with holes in the base. For each electron that combines with a hole, an electron leaves the base region via the base terminal. This action creates a very small base current, about 2% or less of the electron current from emitter to collector. As we shall see, the smaller this percentage, the more useful the transistor is in practical applications.

Note in Figure 4–4 that arrows are drawn to indicate the direction of *conventional* current in the *npn* transistor. Of course, each arrow points in the opposite direction from the electron flows that we have described. Conventional current flowing from V_{CC} into the collector is called *collector current* and designated I_C. Similarly, current into the base is I_B, the *base current*, and current from V_{EE} into the emitter is *emitter current*, I_E. Figure 4–5(a) shows the standard electronic symbol for an *npn* transistor, with these currents labeled alongside. Figure 4–5(b) shows the same block form of the *npn* that we have shown earlier and is included as an aid for relating the physical device to the symbol. Figure 4–6 shows the standard symbol for a *pnp* transistor and its equivalent block form. Comparing Figures 4–5 and 4–6, we note first that the emitter of an *npn* transistor is represented by an arrow pointing out from the base, whereas the emitter of a *pnp* transistor is shown as an arrow pointing into the base. It is easy to remember this distinction by thinking of the arrow as pointing in the direction of conventional current flow, out of or into the emitter of each type of transistor. We further note that the polarities of the bias sources for the *pnp* transistor, V_{EE} and V_{CC}, are the opposite of those for the *npn* transistor. In other words, the positive and negative terminals of each source in Figure 4–6 are the reverse of those in Figure 4–5. These polarities are, of course, necessary in each case to maintain the forward and reverse biasing of the junctions, as we have described. Note, for example, that the negative terminal of V_{CC} is connected to the *p*-type collector of the *pnp* transistor. Recapitulating, here is the all-important universal rule for biasing transistors for normal operation (memorize it!): *The emitter–base junction must be forward biased, and the collector–base junction must be reverse biased.*

To emphasize and clarify an important point concerning transistor currents, Figure 4–7 replaces each type of transistor by a single block and shows the directions of currents entering and leaving each. Applying

FIGURE 4–5 Equivalent *npn* transistor diagrams

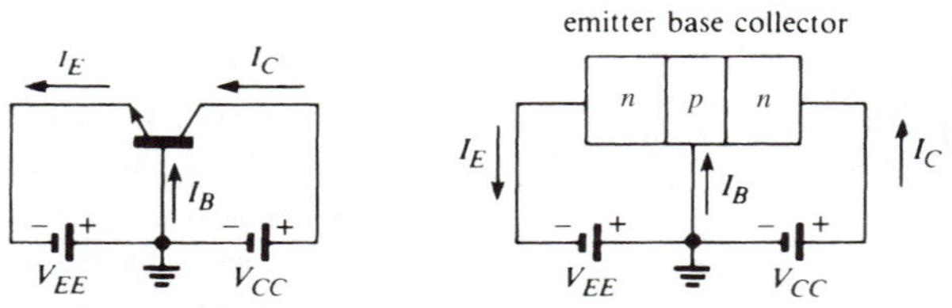

(a) *npn* transistor symbol and bias currents

(b) The block form of the *npn* transistor corresponding to (a)

FIGURE 4–6 Equivalent *pnp* transistor diagrams

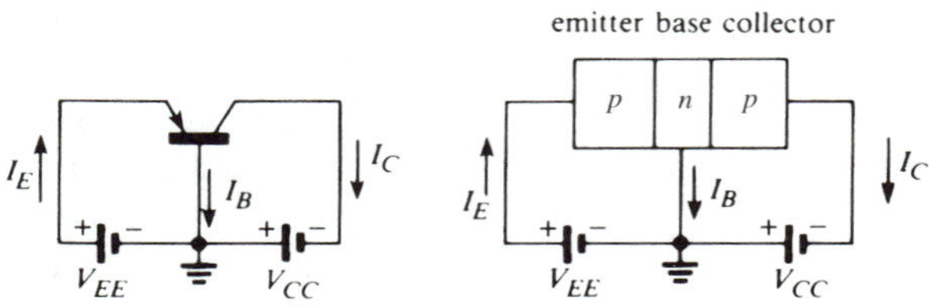

(a) *pnp* transistor symbol and bias currents

(b) The block form of the *pnp* transistor corresponding to (a)

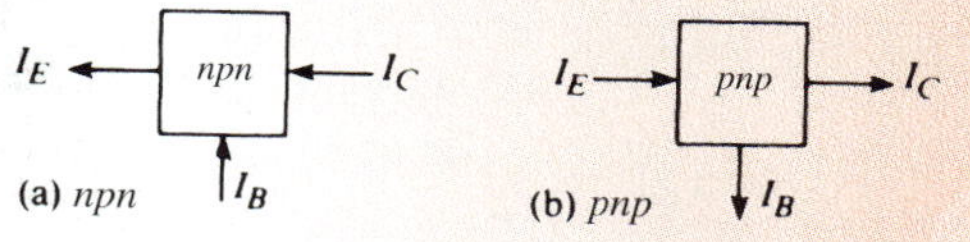

FIGURE 4–7 Each transistor type is replaced by a single block to highlight current flows in and out of the devices

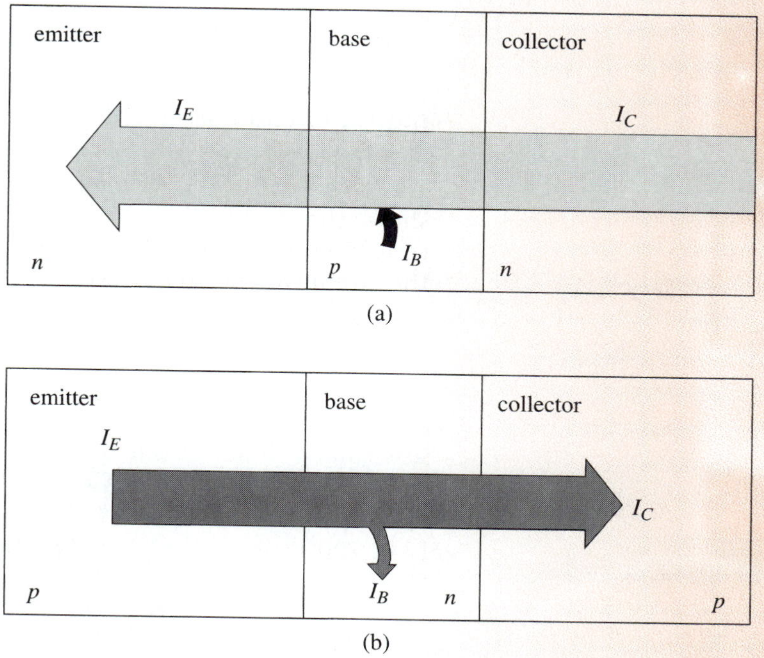

FIGURE 4–8 Graphical depiction of the relationships among the emitter, base, and collector currents

Kirchhoff's current law to each of Figures 4–7(a) and (b), we immediately obtain this important relationship, applicable to both *npn* and *pnp* transistors:

$$I_E = I_C + I_B \qquad (4\text{–}1)$$

that is graphically depicted in Figure 4–8.

I_{CBO} Reverse Current

Recall from Chapter 2 that a small reverse current flows across a *pn* junction due to *thermally* generated minority carriers that are propelled by the barrier potential. When the junction is reverse biased, this reverse current increases slightly. For moderate reverse-bias voltages, the reverse current reaches its saturation value, I_s. Because the collector–base junction of a transistor is reverse biased, there is likewise a reverse current due to thermally generated carriers. Of course this "reverse" current, in the context of a transistor, is in the same direction as the main (collector) current flowing through the device due to the injection of minority carriers into the base. The total collector current is, therefore, the *sum* of these two components: the injected minority carriers and the thermally generated minority carriers.

Suppose that the external connections between the base and emitter are left open and that the collector–base junction has its normal reverse bias, as shown in Figure 4–9 (see also Figure 4–3). Because the emitter is open, there can be no carriers injected into the base. Consequently, the only current that flows must be that "reverse" component due to thermally generated carriers. This current is designated I_{CBO}, the Collector-to-Base

FIGURE 4–9 I_{CBO} is the collector current that flows when the emitter is open

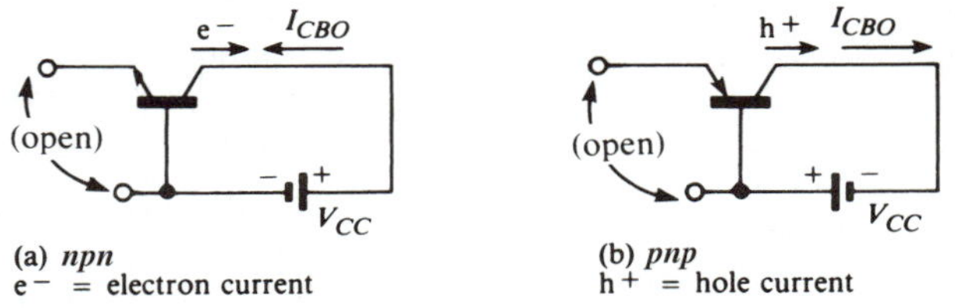

current, with the emitter *O*pen. Therefore, in normal operation, with the emitter circuit connected, the total collector current is expressed as

$$I_C = \alpha I_E + I_{CBO} \tag{4–2}$$

where the important transistor parameter *alpha* is defined as the ratio of the collector current resulting from carrier injection to the total emitter current:

$$\alpha = \frac{I_{C(INJ)}}{I_E} \tag{4–3}$$

Thus, α measures the portion of the emitter current that "survives," after passage through the base, to become collector current. Clearly, α will always be less than 1, because some of the emitter current is drained off in the base through recombinations. Generally speaking, the greater the value of α (the closer it is to 1), the better the transistor, from the standpoint of many practical applications that we will explore later. In other words, we want a transistor to be constructed so that its base current is as small as possible, because that makes I_C close to I_E and α close to 1. Typical transistors have values of α that range from 0.95 to 0.995.

Equation 4–2 states that the total collector current is that portion of the emitter current that makes it through the base (αI_E) plus the thermally generated collector current (I_{CBO}).

In modern transistors, particularly silicon, I_{CBO} is so small that it can be neglected for most practical applications. Remember that I_{CBO} is exactly the same as the reverse diode current we discussed in Chapter 2. We saw that its theoretical value, as a function of temperature and voltage, is given by equation 2–3, and that it is quite sensitive to temperature variations. Because the collector–base junction in a transistor is normally reverse biased by at least a volt or so, the theoretical value of I_{CBO} is for all practical purposes equal to its saturation value (I_s, in Chapter 2). Remember that I_s approximately doubles for every 10°C rise in temperature, so we can say the same about I_{CBO} in a transistor. This sensitivity to temperature can become troublesome in some circuits if high temperatures and large power dissipations are likely. We will explore those situations in more detail later.

We should also remember that the theoretical value of I_{CBO}, like reverse diode current, is usually much smaller than the reverse *leakage* current that flows across the surface. In silicon transistors, this surface leakage may so completely dominate the reverse current that temperature-related increases in I_s remain negligible. In fact, it is conventional in most texts and product literature to refer to I_{CBO} as the (collector-to-base) *leakage current*.

Because I_{CBO} is negligibly small in most practical situations, we can set it equal to 0 in equation 4–2 and obtain the good approximation

$$I_C \approx \alpha I_E \tag{4–4}$$

and

$$\alpha \approx \frac{I_C}{I_E} \tag{4–5}$$

EXAMPLE 4–1

The emitter current in a certain *npn* transistor is 8.4 mA. If 0.8% of the minority carriers injected into the base recombine with holes and the leakage current is 0.1 μA, find (1) the base current, (2) the collector current, (3) the exact value of α, and (4) the approximate value of α, neglecting I_{CBO}.

Solution

1. $I_B = (0.8\% \text{ of } I_E) = (0.008)(8.4 \text{ mA}) = 67.2\ \mu\text{A}$.
2. From equation 4–1, $I_C = I_E - I_B = 8.4 \text{ mA} - 0.0672 \text{ mA} = 8.3328 \text{ mA}$.
3. From equation 4–2, $\alpha I_E = I_C - I_{CBO} = 8.3328 \times 10^{-3}\text{A} - 10^{-7}\text{A} = 8.3327 \text{ mA}$. By equation 4–3, $\alpha = (8.3327 \text{ mA})/(8.4 \text{ mA}) = 0.9919881$.
4. By approximation 4–5, $\alpha \approx I_C/I_E = (8.3328 \text{ mA})/(8.4 \text{ mA}) = 0.992$.

For the conditions of this example, we see that the exact and approximate values of α are so close that the difference between them can be entirely neglected.

4–3 COMMON-BASE CHARACTERISTICS

In our introduction to the theory of transistor operation, we showed a bias circuit (Figure 4–4) in which the base was treated as the ground, or "common" point of the circuit. In other words, all voltages (collector-to-base and emitter-to-base) were *referenced* to the base. This bias arrangement results in what is called the *common-base* (CB) configuration for the transistor. It represents only one of three possible ways to arrange the external circuit to achieve a forward-biased base-to-emitter junction and a reverse-biased collector-to-base junction, because any one of the three terminals can be made the common point. We will study the other two configurations in later discussions.

The significance of having a common point in a transistor circuit is that it gives us a single reference for both the *input* voltage to the transistor and the *output* voltage. In the CB configuration, the emitter–base voltage is regarded as the input voltage and the collector–base voltage is regarded as the output voltage. See Figure 4–10. Notice the reference for V_{EB} and V_{CB} where standard double-subscript notation means "voltage at first subscript with respect to second subscript." For normal transistor action, V_{EB} is positive for *pnp* and negative for *npn*. On the other hand, V_{CB} is negative for *pnp* and positive for *npn*.

In our analysis of the CB configuration, the "input" voltage will be the emitter–base *bias* voltage (V_{EB}), and the "output" voltage will be the collector–base bias voltage (V_{CB}). In Chapter 7, we will adopt a more realistic viewpoint in which we will regard small (ac) variations in the emitter–base and collector–base voltages as the input and output, respectively. For the time being, we will concern ourselves only with the effects of changes in V_{EB} and V_{CB} on the behavior of the transistor. Do not be confused by the fact that the "input" current in the *npn* circuit (Figure 4–10(a)) flows *out* of the

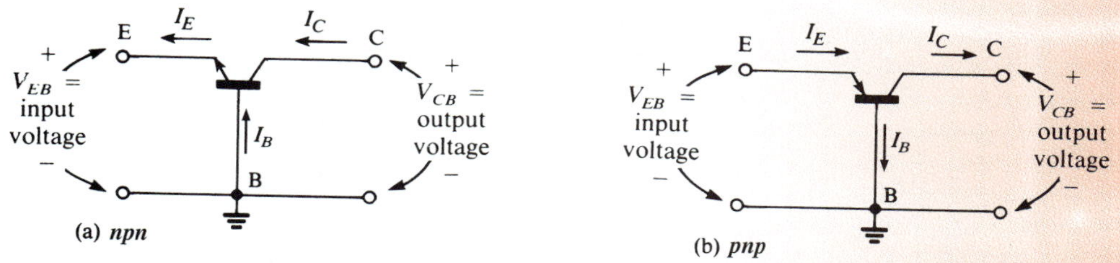

FIGURE 4–10 Input and output voltages in *npn* and *pnp* common-base transistors

emitter. Again, it will be small changes in the magnitude of I_E that we will ultimately regard as the "input."

Our objective now is to learn how the input and output voltages and the input and output currents are related to each other in a CB configuration. Toward that end, we will develop sets of characteristic curves called *input* characteristics and *output* characteristics. The input characteristics show the relation between input current and input voltage *for different values of output voltage,* and the output characteristics show the relation between output current and output voltage *for different values of input current.* As these statements suggest, there is in a transistor a certain *feedback* (the output voltage) that affects the input, and a certain "feedforward" (the input current) that affects the output.

Common-Base Input Characteristics

Let us begin with a study of the CB input characteristics of a typical *npn* transistor. Because the input is across the forward-biased base-to-emitter junction, we would expect a graph of input current (I_E) versus input voltage (V_{EB}) to resemble that of a forward-biased diode. That is indeed the case. However, the exact shape of this I_E–V_{EB} curve will, as we have already hinted, depend on the reverse-biasing output voltage, V_{CB}. The reason for this dependency is that the greater the value of V_{CB}, the more readily minority carriers in the base are swept through the base-to-collector junction. (Remember that the reverse-biasing voltage enhances such current.) The increase in emitter-to-collector current resulting from an increase in V_{CB} means that the (input) emitter current will be greater for a given value of (input) base-to-emitter voltage. Figure 4–11 shows a typical set of input characteristics in which this feedback effect can be discerned. Figure 4–11 is our first example of a *family* of transistor curves, a very useful way to display transistor behavior graphically and one that can provide rewarding insights if studied carefully. Although characteristic curves are seldom used in actual design or analysis

FIGURE 4–11 Common-base input characteristics (*npn*)

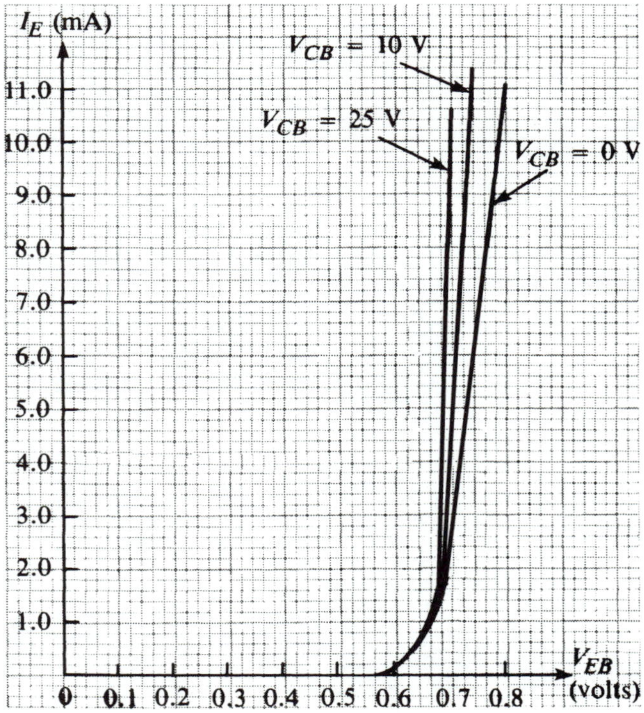

problems, they convey a wealth of information, and we will see many more of them in the future. Each set should be scrutinized and dwelled upon at length. Try to visualize how currents and/or voltages change when one quantity is held constant and the others are varied. Note that a family of curves can show the relationships among *three* variables: two represented by the axes, and the third represented by each curve. In Figure 4–11, each curve corresponds to a different value of V_{CB} and therefore each shows how emitter current varies with base-to-emitter voltage for a fixed value of V_{CB}. A good way to view this family is to think of an experiment in which the reverse-biasing voltage V_{CB} is fixed and a set of measurements of I_E is made for different settings of V_{EB}. Plot these results, and then set V_{CB} to a new value and repeat the measurements. Each time V_{CB} is set to a new value, a new curve is obtained.

Note in Figure 4–11 that each curve resembles a forward-biased diode characteristic, as expected. For a given value of V_{EB}, it can be seen that I_E increases with increasing V_{CB}. This variation has already been explained in terms of the way V_{CB} promotes minority carrier flow. We see in the figure that there is actually little difference in the shapes of the curves as V_{CB} is changed over a fairly wide range. For that reason, the effect of V_{CB} on the input is often neglected in practical problems. An "average" forward-biased diode characteristic is assumed.

The CB input characteristics for a *pnp* transistor will of course have the same general appearance as those shown for an *npn* in Figure 4–11. However, in a *pnp* transistor, a forward-biasing input voltage is positive when measured *from* emitter *to* base. Some data sheets show negative values for *pnp* voltages and/or currents because these quantities have directions that are the opposite of the corresponding *npn* quantities. For example, if the horizontal axis in Figure 4–11 were labeled V_{EB} for a *pnp* transistor, then all scale values would be negative. These sign conventions (rather, this *lack* of consistency) can be confusing but can always be resolved by remembering the fundamental rule for transistor bias: base–emitter forward and base–collector reverse. Finally, we should mention that some authors and some data sheets refer to the CB input characteristics as the *emitter characteristics* of a transistor.

EXAMPLE 4–2

The transistor shown in Figure 4–12 has the characteristic curves shown in Figure 4–11. When V_{CC} is set to 25 V, it is found that $I_C = 8.94$ mA.

1. Find the α of the transistor (neglecting I_{CBO}).
2. Repeat if $I_C = 1.987$ mA when V_{CC} is replaced by a short circuit to ground.

Solution

1. In Figure 4–12, we see that $V_{BE} = 0.7$ V. From Figure 4–11, the vertical line corresponding to $V_{BE} = 0.7$ V intersects the $V_{CB} = 25$ V curve at $I_E = 9.0$ mA. Therefore, $\alpha \approx I_C/I_E = (8.94\text{ mA})/(9.0\text{ mA}) = 0.9933$.
2. When V_{CC} is replaced by a short circuit, we have $V_{CB} = 0$. From Figure 4–11, $I_E = 2$ mA at $V_{CB} = 0$ V and $V_{BE} = 0.7$ V. Therefore, $\alpha \approx I_C/I_E = (1.987\text{ mA})/(2.0\text{ mA}) = 0.9935$.

FIGURE 4–12 (Example 4–2)

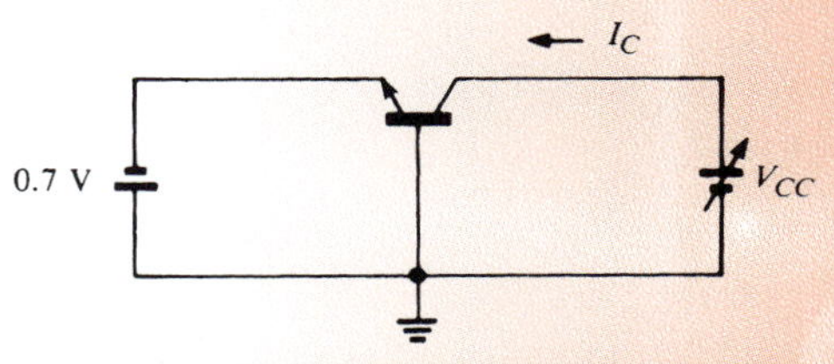

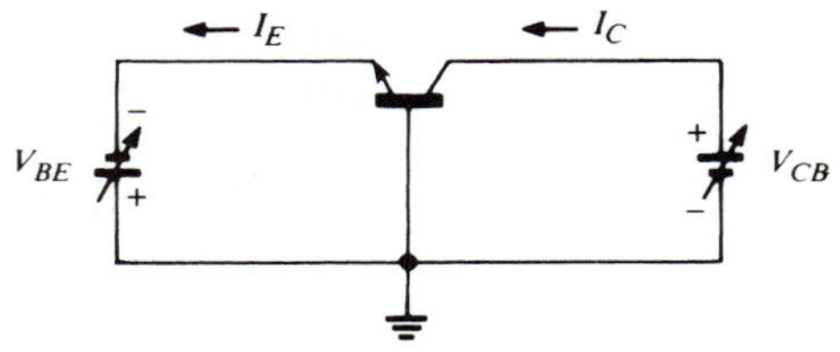

FIGURE 4–13 An experiment that could be used to produce the output characteristics shown in Figure 4–14

I_E	V_{CB}	I_C
1 mA	−1	
1 mA	0	
1 mA	5	
1 mA	10	
1 mA	15	
1 mA	20	

I_E	V_{CB}	I_C
2 mA	−1	
2 mA	0	
2 mA	5	
2 mA	10	
2 mA	15	
2 mA	20	

— etc. —

I_E	V_{CB}	I_C
9 mA	−1	
9 mA	0	
9 mA	5	
9 mA	10	
9 mA	15	
9 mA	20	

1. With V_{CB} set to −1 V, adjust V_{BE} to obtain $I_E = 1$ mA. Measure and record I_C.
2. Increase V_{CB} positively in small steps, each time measuring I_C. Adjust V_{BE} as necessary to maintain the original value of I_E. Continue until V_{CB} has reached 20 V. Plot I_C versus V_{CB}.
3. Repeat step 1, with V_{BE} adjusted to produce a new, slightly larger value of I_E. Then repeat step 2.

   Repeat step 3 until the fixed value of I_E has reached 9 mA.

Common-Base Output Characteristics

Consider now an experiment in which the collector (output) current is measured as V_{CB} (the output voltage) is adjusted for fixed settings of the emitter (input) current. Figure 4–13 shows a schematic diagram and a procedure that could be used to conduct such an experiment on an *npn* transistor. Understand that Figure 4–13 does not represent a practical circuit that could be used for any purpose other than investigating transistor characteristics. Practical transistor circuits contain more resistors and have input and output voltages that are different from the dc bias voltages. However, at this point in our study of transistor theory, we are interested in the *transistor* itself. We are using characteristic curves to gain insights into how the voltages and currents relate to each other in the *device*, rather than in the external circuit. Once we have gleaned all the device information we can from studying characteristic curves, we will have a solid understanding of what a transistor really is and can proceed to study practical circuits. When I_C is plotted versus V_{CB} for different values of I_E, we obtain the family of curves shown in Figure 4–14: the *output* characteristics for the CB configuration. A close examination of these curves will reveal some new facts about transistor behavior.

We note first in Figure 4–14 that each curve starts at $I_C = 0$ and rises rapidly for a small positive increase in V_{CB}. In other words, I_C increases rapidly just as V_{CB} begins to increase slightly beyond its initial negative value. Since each curve represents a fixed value of I_E, this means that while I_C is increasing, the ratio I_C/I_E must also be increasing. But I_C/I_E equals α, so the implication is that the value of α for a transistor is not constant. Alpha starts at 0 and increases as V_{CB} increases. The reason for this fact is that a very small portion of the emitter current is able to enter the collector region until the reverse-biasing voltage V_{CB} is allowed to reach a value large enough to propel all carriers across the junction. When V_{CB} is negative, the junction is actually

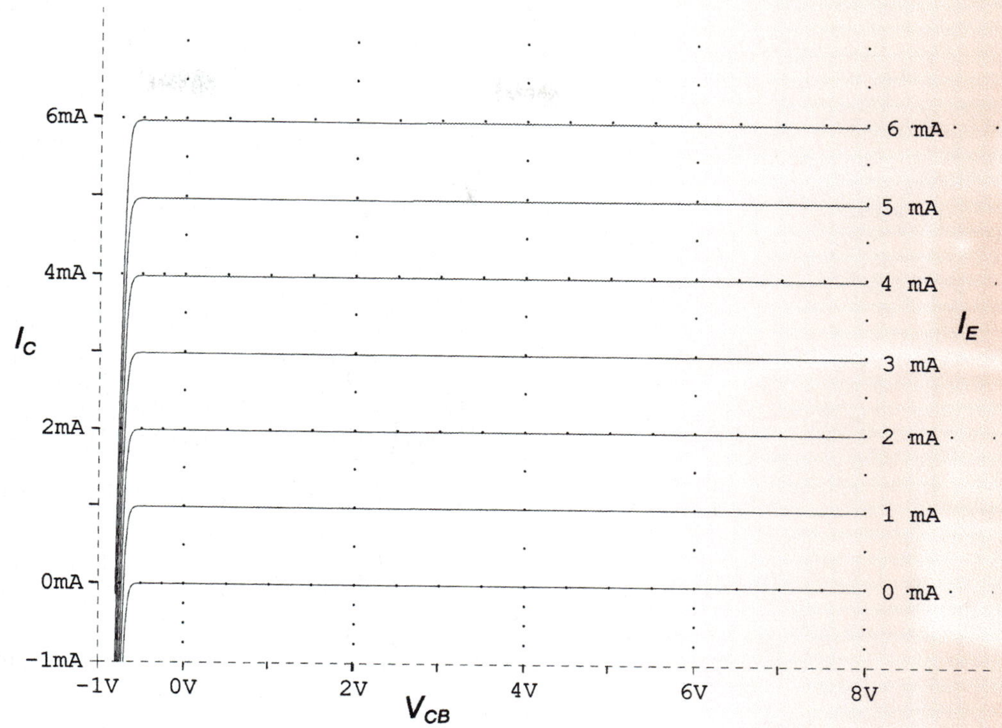

FIGURE 4–14 Common-base output characteristics (*npn*). Note that the negative V_{CB} scale is expanded.

forward biased, and minority carrier flow is inhibited. The proportion of the carriers that are swept across the junction (α) depends directly on the value of V_{CB} until V_{CB} no longer forward biases the junction. The portion of the plot where V_{CB} is negative is called the *saturation region* of the transistor. By definition (no matter what the transistor configuration), a transistor is *saturated* when *both* its collector-to-base junction *and* emitter-to-base junction are forward biased.

Once V_{CB} reaches a value large enough to ensure that a large portion of carriers enter the collector (close to 0 in Figure 4–14), we see that the curves more or less level off. In other words, for a fixed emitter current, the collector current remains essentially constant for further increases in V_{CB}. Note that this essentially constant value of I_C is, for each curve, very nearly equal to the value of I_E represented by the curve. In short, the ratio I_C/I_E, or α, is very close to 1 and is essentially constant. These observations correspond to what we had previously assumed about the nature of α, and the region of the plot where this is the case is called the *active region*. In its active region, a transistor exhibits those "normal" properties (transistor action) that we have associated with a forward-biased emitter–base junction and a reverse-biased collector–base junction. Apart from some special digital-circuit applications, a transistor is normally operated (used) in its active region. Note that we can detect a slight rise in the curves as they proceed to the right through the active region. Each curve of constant I_E approaches a horizontal line for which I_C is almost equal to I_E, implying that I_E approaches I_C, and that α approaches 1, for increasing V_{CB}. This we attribute to the increased number of minority carriers swept into the collector, which increases the collector current, as the reverse-biasing value of V_{CB} is increased.

There is one other region of the output characteristics that deserves comment. Note that the curve corresponding to $I_E = 0$ is very close to the $I_C = 0$

line. When the emitter current is made 0 (by opening the external emitter circuit), no minority carriers are injected into the base. Under those conditions, the only collector current that flows is the very small leakage current, I_{CBO}, as we have previously described (see Figure 4–9). With the scale used to plot the output characteristics in Figure 4–14, a horizontal line corresponding to $I_C = I_{CBO}$ coincides with the $I_C = 0$ line, for all practical purposes. The region of the output characteristics lying below the $I_E = 0$ line is called the *cutoff* region because the collector current is essentially 0 (cut off) there. A transistor is said to be in the cutoff state when *both* the collector–base and emitter–base junctions are reverse biased. Except for special digital circuits, a transistor is not normally operated in its cutoff region.

EXAMPLE 4–3

A certain *npn* transistor has the CB input characteristics shown in Figure 4–11 and the CB output characteristics shown in Figure 4–14.

1. Find its collector current when $V_{CB} = 10$ V and $V_{BE} = 0.7$ V.
2. Repeat when $V_{CB} = 5$ V and $I_E = 5.5$ mA.

Solution

1. From Figure 4–11, we find $I_E = 4$ mA at $V_{BE} = 0.7$ V and $V_{CB} = 10$ V. In Figure 4–14, the $I_E = 4$ mA curve runs just below $I_C = 4$ mA. The collector current under these conditions is practically 4 mA, independent of V_{CB}.
2. The conditions given require that we *interpolate* the output characteristics along the vertical line $V_{CB} = 5$ V between $I_E = 5$ mA and $I_E = 6$ mA. The value of I_C that is halfway between the $I_E = 5$ mA and $I_E = 6$ mA curves is approximately 5.5 mA. Note that high accuracy is not possible when using characteristic curves in this way. In most practical situations, we could simply assume that $I_C = I_E$ without seriously affecting the accuracy of other computations.

Breakdown

As is the case in a reverse-biased diode, the current through the collector–base junction of a transistor may increase suddenly if the reverse-biasing voltage across it is made sufficiently large. This increase in current is typically caused by the avalanching mechanism already described in connection with diode breakdown. However, in a transistor it can also be the result of a phenomenon called *punch through*. Punch through occurs when the reverse bias widens the collector–base depletion region to the extent that it meets the base-emitter depletion region. This joining of the two regions effectively shorts the collector to the emitter and causes a substantial current flow. Remember that the depletion region extends farther into the lightly doped side of a junction and that the base is more lightly doped than the collector. Furthermore, the base is made very thin, so the two junctions are already relatively close to each other. Punch through can be a limiting design factor in determining the doping level and base width of a transistor. Figure 4–15 shows how the CB output characteristics appear when the effects of breakdown are included. Note the sudden upward swing of each curve at a large value of V_{CB}. The collector-to-base breakdown voltage when $I_E = 0$ (emitter open) is designated BV_{CBO}. As can be seen in Figure 4–15, breakdown occurs at progressively lower voltages for increasing values of I_E.

Although the base–emitter junction is not normally reverse biased, there are practical applications in which it is periodically subjected to reverse bias.

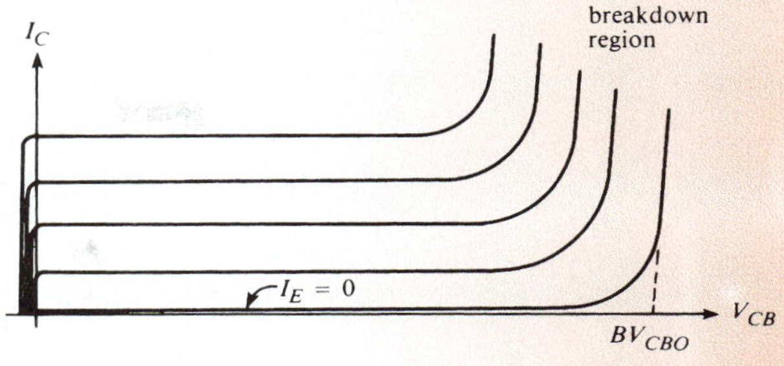

FIGURE 4–15 Common-base output characteristics showing the breakdown region

Of course, it too can break down, and its reverse breakdown voltage is usually much less than that of the collector–base junction. Base–emitter breakdown is often destructive, so designers must be aware of the manufacturer's specified maximum reverse base–emitter voltage.

As a final note on transistor operation, we should mention that some transistors can be (and occasionally are) operated in what is called an *inverted* mode. In this mode, the emitter is used as the collector and vice versa. Normally, the emitter is the most heavily doped of the three regions, so unless a transistor is specifically designed for inverted operation, it will not perform well in that mode. The α in the inverted mode, designated α_1, is generally smaller than the α that can be realized in conventional operation.

4–4 COMMON-EMITTER CHARACTERISTICS

The next transistor bias arrangement we will study is called the *common-emitter* (CE) configuration. It is illustrated in Figure 4–16. Note that the external voltage source V_{BB} is used to forward bias the base–emitter junction and the external source V_{CC} is used to reverse bias the collector–base junction. The magnitude of V_{CC} must be greater than V_{BB} to ensure that the collector–base junction remains reverse biased, because, as can be seen in the figure, $V_{CB} = V_{CC} - V_{BB}$. (Write Kirchhoff's voltage law around the loop from the collector, through V_{CC}, through V_{BB}, and back to the collector.) The emitter terminal is, of course, the ground, or common, terminal in this configuration.

Figure 4–17 shows that the input voltage in the CE configuration is the base–emitter voltage and the output voltage is the collector–emitter voltage. The input current is I_B and the output current is I_C. The common-emitter

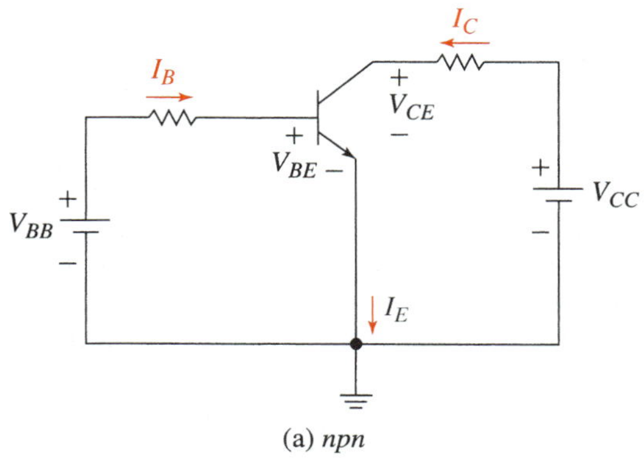

(a) *npn*

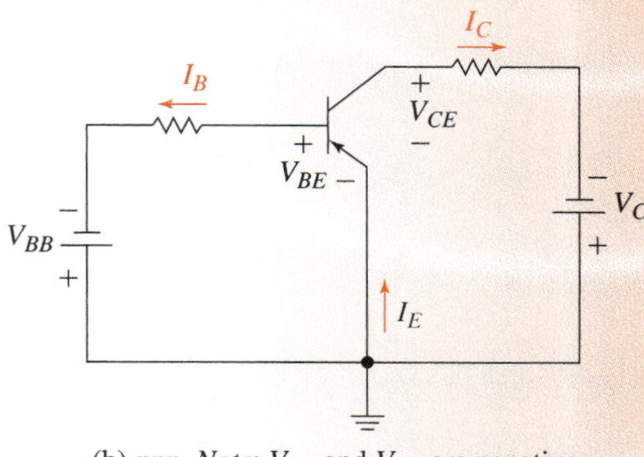

(b) *pnp*. *Note*: V_{BE} and V_{CE} are negative for *pnp* transistors.

FIGURE 4–16 Common-emitter (CE) bias arrangements

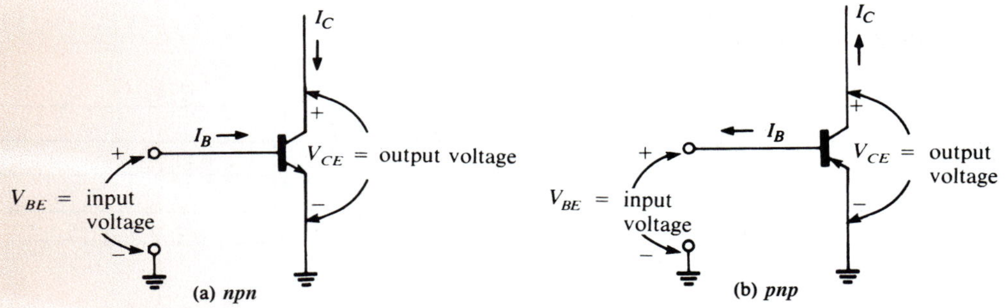

FIGURE 4–17 Input and output voltages and currents for *npn* and *pnp* transistors in the CE configuration

configuration is the most useful and most widely used transistor configuration, and we will study it in considerable detail. In the process we will learn some new facts about transistor behavior.

I_{CEO} and Beta

Before investigating the input and output characteristics of the CE configuration, we will derive a new relationship between I_C and I_{CBO}. Although this derivation does not depend in any way on the bias arrangement used, it will provide us with some new parameters that are useful for predicting leakage in the CE configuration and for relating CE input and output currents. Equation 4–2 states that

$$I_C = \alpha I_E + I_{CBO}$$

or

$$I_C - I_{CBO} = \alpha I_E$$

Dividing through by α,

$$\frac{I_C}{\alpha} - \frac{I_{CBO}}{\alpha} = I_E$$

Substituting $I_B + I_C$ for I_E on the right-hand side,

$$\frac{I_C}{\alpha} - \frac{I_{CBO}}{\alpha} = I_B + I_C$$

Collecting the terms involving I_C leads to

$$I_C\left(\frac{1}{\alpha} - 1\right) = I_B + \frac{I_{CBO}}{\alpha}$$

But $1/\alpha - 1 = (1 - \alpha)/\alpha$, so

$$I_C = \frac{\alpha I_B}{1 - \alpha} + \frac{I_{CBO}}{1 - \alpha} \qquad (4–6)$$

Using equation 4–6, we can obtain an expression for reverse "leakage" current in the CE configuration. Figure 4–18 shows *npn* and *pnp* transistors in which the base–emitter circuits are left open while the reverse-biasing voltage sources remain connected. As can be seen in Figure 4–18, the only current that can flow when the base is left open is reverse current across the collector–base junction. This current flows from the collector, through the base region, and into the emitter. It is designated I_{CEO}—*C*ollector-to-*E*mitter

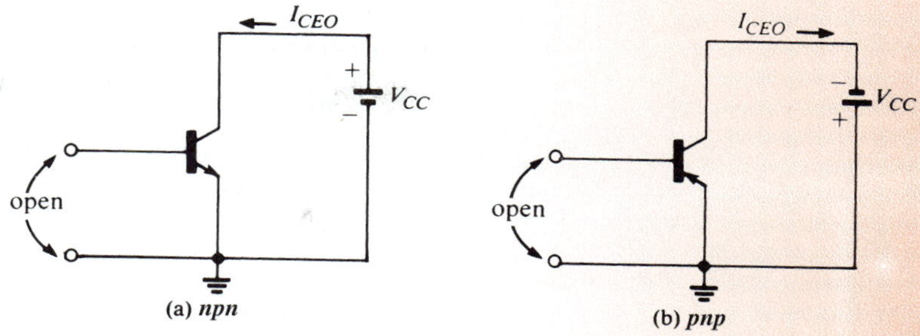

FIGURE 4–18 Collector–emitter leakage current I_{CEO}

current with the base Open. (Note, once again, that this "reverse" current is in the same direction as normal collector current through the transistor.) Because I_B must equal 0 when the base is open, we can substitute $I_B = 0$ in equation 4–6 to obtain

$$I_{CEO} = \frac{I_{CBO}}{1-\alpha} = \left(\frac{1}{1-\alpha}\right)I_{CBO} \tag{4–7}$$

Because the α of a transistor is close to 1, $1 - \alpha$ is close to 0, and so $1/(1 - \alpha)$ can be quite large. Therefore, equation 4–7 tells us that CE leakage current is much larger than CB leakage current. For example, if $I_{CBO} = 0.1$ μA and $\alpha = 0.995$, then $I_{CEO} = (0.1\ \mu\text{A})/0.005 = 20\ \mu\text{A}$. In effect, collector–base leakage current is amplified in the CE configuration, a result that can cause problems in high-temperature circuits, particularly those containing germanium transistors.

Returning to equation 4–6, let us focus on the factor $\alpha/(1 - \alpha)$ that multiplies I_B. This quantity is another important transistor parameter, called *beta:*

$$\beta = \frac{\alpha}{1-\alpha} \tag{4–8}$$

Beta is always greater than 1 and for typical transistors ranges from around 20 to several hundred. When α is close to 1, a small increase in α causes a large increase in the value of β. For example, if $\alpha = 0.99$, then $\beta = 0.99/(1 - 0.99) = 99$. If α is increased by 0.005 to 0.995, then $\beta = 0.995/(1 - 0.995) = 199$. Because a small change in α causes a large change in β, small manufacturing variations in transistors that are supposed to be of the same type cause them to have a wide range of β values. It is not unusual for transistors of the same type to have betas that vary from 50 to 200.

In terms of β, equation 4–6 becomes

$$I_C = \beta I_B + \frac{I_{CBO}}{1-\alpha} = \beta I_B + (\beta + 1)I_{CBO} \tag{4–9}$$

or

$$I_C = \beta I_B + I_{CEO} \tag{4–10}$$

Although I_{CEO} is much greater than I_{CBO}, it is generally quite small in comparison to βI_B, especially in silicon transistors, and it can be neglected in many practical circuits. Neglecting I_{CEO} in equation 4–10, we obtain the approximation $I_C \approx \beta I_B$. This approximation is widely used in transistor circuit analysis, and we will often write it as an equality in future discussions, with the understanding that I_{CEO} can be neglected:

$$I_C = \beta I_B \ (I_{CEO} = 0) \tag{4–11}$$

EXAMPLE 4–4

A transistor has $I_{CBO} = 48$ nA and $\alpha = 0.992$.

1. Find β and I_{CEO}.
2. Find its (exact) collector current when $I_B = 30\ \mu$A.
3. Find the approximate collector current, neglecting leakage current.

Solution

1. $$\beta = \frac{\alpha}{1-\alpha} = \frac{0.992}{0.008} = 124$$

$$I_{CEO} = \frac{I_{CBO}}{1-\alpha} = \frac{48 \times 10^{-9}}{0.008} = 6\ \mu\text{A} \quad \text{or} \quad I_{CEO} = (\beta + 1)I_{CBO} = 6\ \mu\text{A}$$

2. $$I_C = \beta I_B + I_{CEO} = (124)(30\ \mu\text{A}) + 6\ \mu\text{A} = 3726\ \mu\text{A} = 3.726\ \text{mA}$$

3. $$I_C \approx \beta I_B = 124(30\ \mu\text{A}) = 3.72\ \text{mA}$$

Equation 4–8 tells us how to find the β of a transistor, given its α. It is left as an exercise at the end of this chapter to show that we can find α, given β, by using the following relation:

$$\alpha = \frac{\beta}{\beta + 1} \tag{4–12}$$

EXAMPLE 4–5

As we will learn in a later discussion, the β of a transistor typically increases dramatically with temperature. If a certain transistor has $\beta = 100$ and an increase in temperature causes β to increase by 100%, what is the percent change in α?

Solution

A 100% increase in β means that β increases from 100 to 200. Let α_1 = value of α when $\beta = 100$ and α_2 = value of α when $\beta = 200$. From equation 4–12,

$$\alpha_1 = \frac{100}{100 + 1} = 0.990099$$

$$\alpha_2 = \frac{200}{200 + 1} = 0.995025$$

Thus, the percent change (increase) in α is

$$\frac{\alpha_2 - \alpha_1}{\alpha_1} \times 100\% = \frac{0.995025 - 0.990099}{0.990099} \times 100\% = 0.488\%$$

We see that a large change in β (100%) corresponds to a small change in α (0.488%).

Common-Emitter Input Characteristics

Because the input to a transistor in the CE configuration is across the base-to-emitter junction (see Figure 4–17), the CE input characteristics resemble a family of forward-biased diode curves. A typical set of CE input characteristics for an *npn* transistor is shown in Figure 4–19. Note that I_B increases as V_{CE} decreases, for a fixed value of V_{BE}. A large value of V_{CE} results in a large reverse bias of the collector–base junction, which widens the depletion region and makes the base smaller. When the base is smaller, there are fewer recombinations of injected minority carriers and there is a corresponding

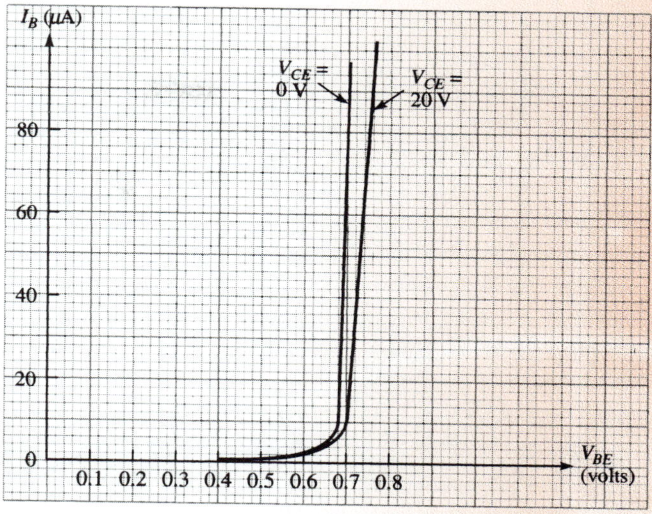

FIGURE 4–19 Common-emitter input characteristics

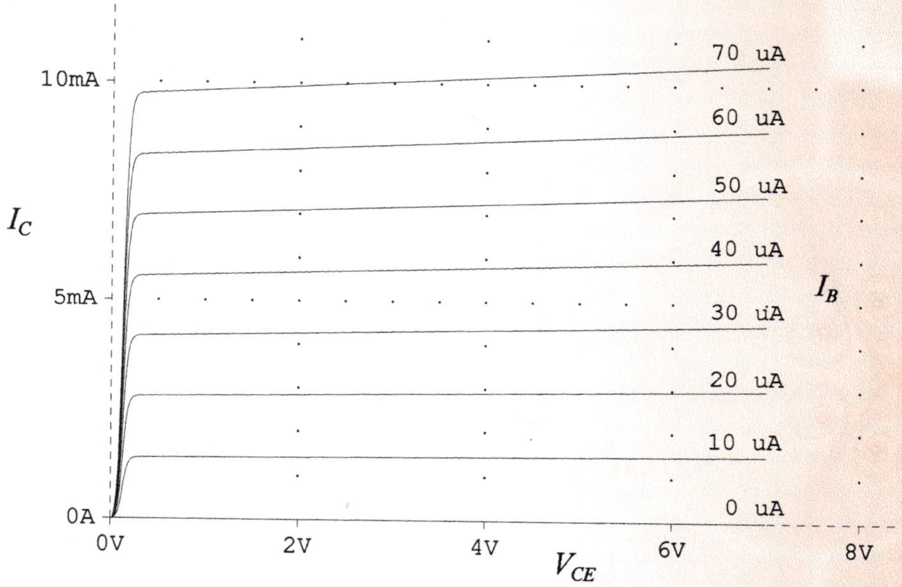

FIGURE 4–20 Common-emitter output characteristics

reduction in base current. In contrast to the CB input characteristics, note in Figure 4–19 that the input current is plotted in units of *micro*amperes. I_B is, of course, much smaller than either I_E or I_C ($I_B \approx I_C/\beta$). The CE input characteristics are often called the *base characteristics.*

Common-Emitter Output Characteristics

Common-emitter output characteristics show collector current I_C versus collector voltage V_{CE}, for different *fixed* values of I_B. These characteristics are often called *collector characteristics.* Figure 4–20 shows a typical set of output characteristics for an *npn* transistor in the CE configuration.

The approximate value of the β of the transistor can be determined at any point on the characteristic in Figure 4–20 simply by dividing the values I_C/I_B at the point. As illustrated in the figure, at $V_{CE} = 3$ V and $I_B = 50$ μA, the value of I_C is about 7 mA, and the value of the β at that point is therefore I_C/I_B = (7 mA)/(50 μA) = 140. Clearly β, like α, is not constant, but depends on the region of the characteristics where the transistor is operated. The region where the curves are approximately horizontal is the *active* region of the CE configuration. In this region, β is essentially constant but increases somewhat with V_{CE}, as can be deduced from the rise in each curve as V_{CE} increases to the right.

As in the CB configuration, the collector current in the CE configuration will increase rapidly if V_{CB} is permitted to become large. When $I_B = 0$ (base open), the collector-to-emitter current at which breakdown occurs is designated BV_{CEO}. The value of BV_{CEO} is always less than that of BV_{CBO} for a given transistor. BV_{CEO} is sometimes called the "sustaining voltage" and denoted LV_{CEO}.

When interpreting the characteristics of Figure 4–20, it is important to realize that each curve is drawn for a small, essentially constant value of V_{BE} (about 0.7 V for silicon). Figure 4–21 illustrates this point. Note in Figure 4–21 that the total collector-to-emitter voltage V_{CE} (which is the same as V_{CC} in our case) is the sum of the small, forward-biasing value of V_{BE} and the reverse-biasing value of V_{CB}. Thus $V_{CE} = V_{CB} + V_{BE}$, or, for silicon, $V_{CE} \approx V_{CB} + 0.7\text{V}$. Thus, if V_{CE} is reduced to about 0.7 V, V_{CB} must become 0, and the collector–base junction is no longer reverse biased. This effect can be seen in the characteristics of Figure 4–20. Notice that each curve is reasonably flat (in the active region) until V_{CE} is reduced to around 0.2 V to 0.3 V. As V_{CE} is reduced further, I_C starts to fall off. When V_{CE} is reduced below about 0.2 V or 0.3 V, the collector–base junction becomes well *forward* biased and collector current diminishes rapidly. Remember that V_{CB} is negative when the collector–base junction is forward biased. For example, if $V_{CE} = 0.2\text{ V}$, then $V_{CB} = V_{CE} - V_{BE} \approx 0.2\text{ V} - 0.7\text{ V} = -0.5\text{ V}$. In keeping with our previous definition, the transistor is said to be *saturated* when the collector–base junction is forward biased. The saturation region is shown on the characteristic curves. The saturation value of V_{CE}, designated $V_{CE(sat)}$, ranges from 0.1 V to 0.3 V, depending on the value of base current.

Notice in Figure 4–20 that when $I_B = 0$, the collector current is the same as that which flows when the base circuit is open (see Figure 4–18), that is, I_{CEO}. The region below $I_B = 0$ is the *cut-off* region.

Comparing the CB output characteristics in Figure 4–14 with the CE output characteristics in Figure 4–20, we note that the curves rise more steeply to the right in the CE case. This rise simply reflects the fact we have already discussed in connection with the CE input characteristics: The greater the value of V_{CE}, the smaller the base region and, consequently, the smaller the base current. But, because base current is constant along each curve in Figure 4–20, the effect appears as an increase in I_C. In other words, the fact that there are fewer recombinations occurring in the base means that a greater proportion of carriers cross the junction to become collector current.

It is apparent in Figure 4–20 that the characteristic curves corresponding to large values of I_B rise more rapidly to the right than those corresponding

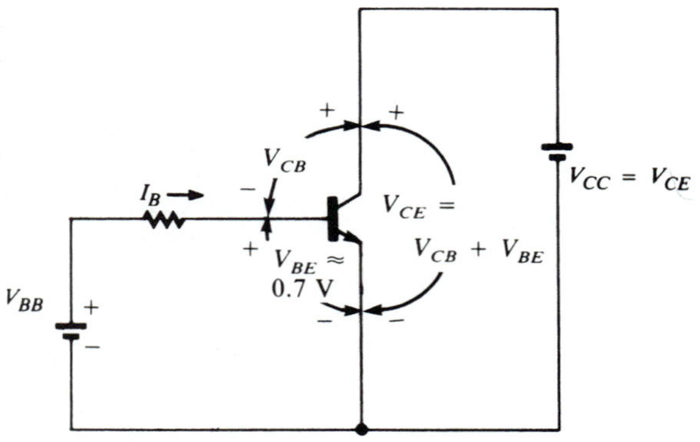

FIGURE 4–21 $V_{CE} \approx V_{CB} + 0.7\text{ V}$ for silicon. When V_{CE} is reduced to about 0.7 V, then $V_{CB} \approx 0$ and the collector–base junction is no longer reverse biased

FIGURE 4–22 The Early voltage, V_A, is the intersection of the characteristic curves with the horizontal axis

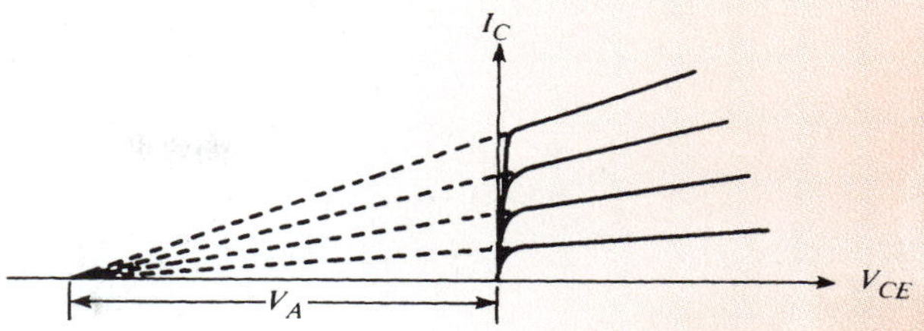

to small values of I_B. If these lines are projected to the left, as shown in Figure 4–22, it is found that they all intersect the horizontal axis at approximately the same point. The point of intersection, designated V_A in Figure 4–22, is called the *Early voltage,* after J. M. Early, who first investigated these relations. Of course, a transistor is never operated with V_{CE} equal to the Early voltage. V_A is simply another useful parameter characterizing a transistor's behavior. It is especially useful in computer simulation programs such as SPICE that analyze transistor circuits and require complete descriptions of transistor characteristics.

EXAMPLE 4–6

A certain transistor has the output characteristics shown in Figure 4–20.

1. Find the percent change in β as V_{CE} is changed from 1 V to 4 V while I_B is fixed at 40 μA.
2. Find the percent change in β as I_B is changed from 10 μA to 50 μA while V_{CE} is fixed at 3.5 V.

Neglect leakage current in each case.

Solution

1. At the intersection of the vertical line $V_{CE} = 1$ V with the curve $I_B = 40$ μA, we find $I_C \approx 5.6$ mA. Therefore, the β at that point is approximately (5.6 mA)/(40 μA) = 140.

   Traveling along the curve of constant $I_B = 40$ μA to its intersection with the vertical line $V_{CE} = 4$V, we find $I_C \approx 7.2$ mA. Therefore, $\beta \approx$ (7.2mA)/(40 μA) = 180. The percent change in β is

$$\frac{180 - 140}{140} \times 100\% = 28.57\%$$

2. At the intersection of the vertical line $V_{CE} = 3.5$ V with the curve $I_B =$ 10 μA, we find $I_C \approx 1.4$ mA. Therefore, $\beta \approx$ (1.4 mA)/(10 μA) = 140.

   Traveling up the vertical line of constant $V_{CE} = 3.5$ V to its intersection with the curve $I_B = 50$ μA, we find $I_C \approx 7.3$ mA. Therefore, $\beta \approx$ (7.3 mA)/(50 μA) = 146. The percent change in β is

$$\frac{146 - 140}{140} \times 100\% = 4.28\%$$

EXAMPLE 4–7

SPICE

(Computer Generation of Characteristic Curves) Use SPICE to obtain a set of output characteristics for an *npn* transistor in a CE configuration. The ideal maximum forward β is 100 and the Early voltage is 100 V. The characteristics should be plotted for V_{CE} ranging from 0 V to 15.0 V in 0.05-V steps and for I_B ranging from 0 to 40 μA in 10-μA steps. Use the results to determine the actual value of β at $V_{CE} = 10$ V and $I_B = 30$ μA.

FIGURE 4–23(A) (Example 4–7)

```
Transistor Curves (CE)
vce 2 0 dc 1
q1 2 1 0 tran
ib 0 3 dc 1
.model tran NPN(VAF=100 BF=100)
.dc vce 0 15 .05 ib 0 40u 10u
.probe
.end
```

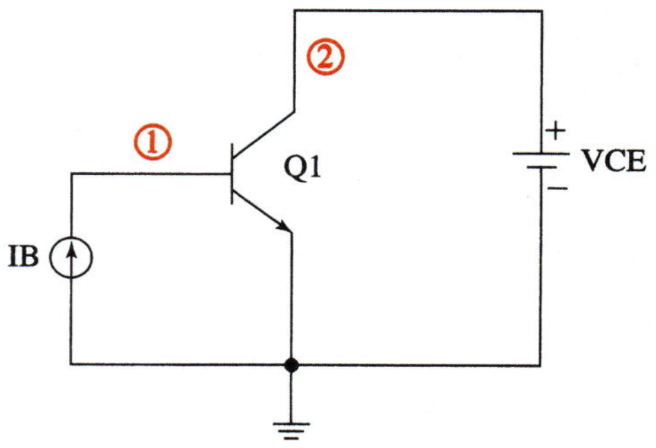

(a)

FIGURE 4–23(B) (Example 4–7)

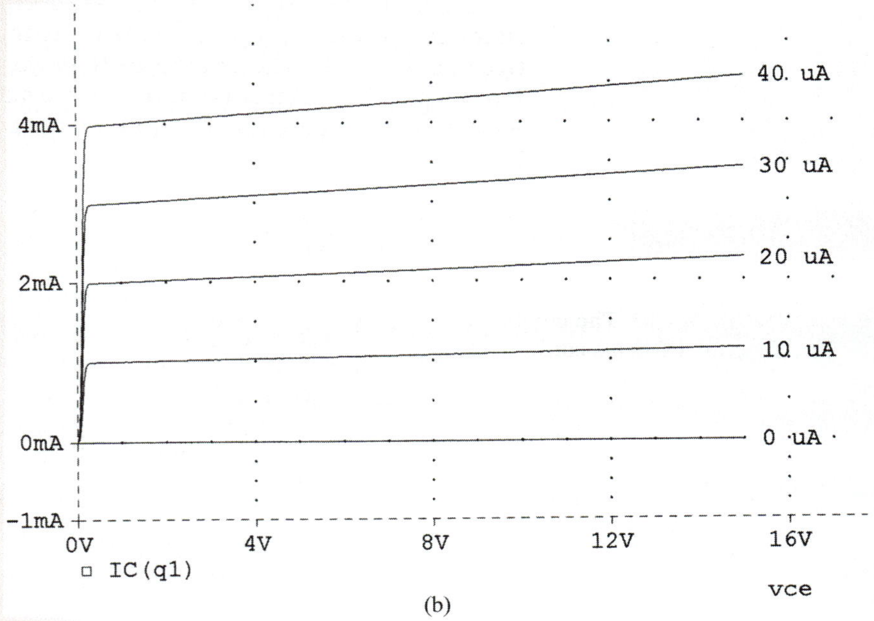

(b)

Solution

Figure 4–23(a) shows the circuit and the SPICE input file. Note that a constant-current source is used to supply base current. The transistor .MODEL statement specifies that $\beta = 100$ (BF = 100) and that the (forward) Early voltage VAF = 100 V.

The .DC statement is the *dc sweep* command that causes V_{CE} to be stepped in .05-V increments from 0 to 15.0 V and I_B to be stepped in 10-μA increments from 0 to 40 μA. The plot generated by the SPICE analysis and the Probe option is shown in Figure 4–23(b). Notice that at $V_{CE} = 15.0$ V and $I_B = 20$ μA, it can be seen from the graphical plot that $I_C = 2.3$ mA. Thus,

$$\beta = \frac{I_C}{I_B} = \frac{2.3\text{ mA}}{20\ \mu\text{A}} = 115$$

Note how the slope of the curves increase for an increasing voltage value of V_{CE}. This is due to the effect of the Early voltage, meaning that β increases with increasing V_{CE} (and with increasing I_B).

4–5 COMMON-COLLECTOR CHARACTERISTICS

In the third and final way to arrange the biasing of a transistor, the collector is made the common point. The result is called the *common-collector* (CC) configuration and is illustrated in Figure 4–24. It is apparent in Figure 4–24(a) that

$$V_{CE} = V_{CB} + V_{BE} \tag{4–13}$$

or $V_{CB} = V_{CE} - V_{BE}$. In our case, $V_{CE} = V_{CC}$ and $V_{CB} = V_{BB}$, so $V_{BB} = V_{CC} - V_{BE}$. Now V_{BE} is the small, essentially constant voltage across the forward-biased base-to-emitter junction (about 0.7 V for silicon). Thus,

$$V_{BB} = V_{CB} \approx V_{CC} - 0.7\text{ V} \tag{4–14}$$

Therefore, in order to keep the collector–base junction reverse biased ($V_{CB} > 0$), it is necessary that V_{BB} be larger than $V_{CC} - 0.7$ V.

Figure 4–25 shows that the base–collector voltage is the input voltage and the base current is the input current. The emitter–collector voltage is the output voltage, and the emitter current is the output current.

Figure 4–26 shows a typical set of input characteristics for an *npn* transistor in the CC configuration. It is clear that these are not the characteristics of a forward-biased diode, as they were in the CB and CE configurations. We can see that each curve is drawn for a different fixed value of V_{CE} and that each shows the base current going to 0 very quickly as V_{CB} increases slightly. This behavior can be explained by remembering that V_{BE} must remain in the neighborhood of 0.5 V to 0.7 V in order for any appreciable base current to flow. But, from equation 4–13,

$$V_{BE} = V_{CE} - V_{CB} \tag{4–15}$$

FIGURE 4–24 The common-collector (*CC*) bias configuration

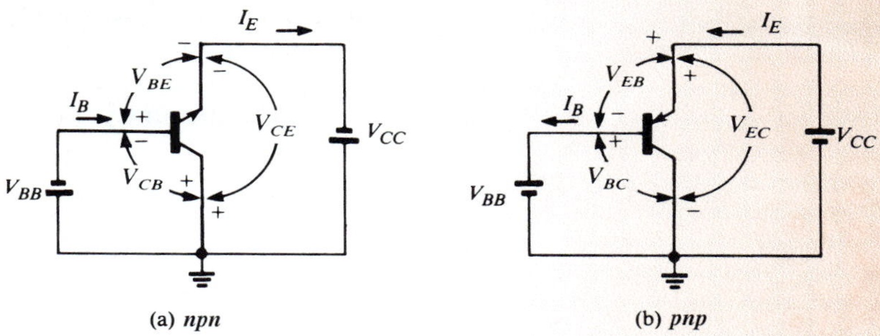

FIGURE 4–25 Input and output voltages and currents in the CC configuration

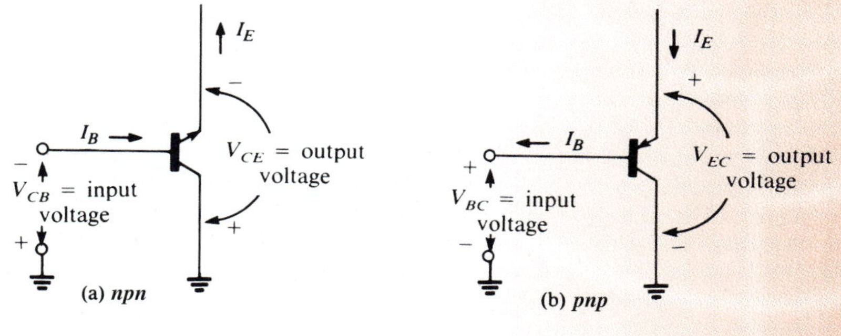

FIGURE 4–26 Common-collector input characteristics

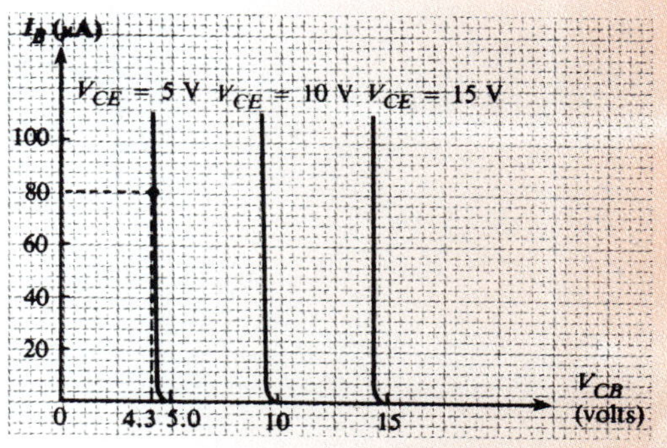

Therefore, if the value of V_{CB} is allowed to increase to a point where it is near the value of V_{CE}, the value of V_{BE} approaches 0, and no base current will flow. In Figure 4–26, refer to the curve that corresponds to $V_{CE} = 5$ V. When $V_{CB} = 4.3$ V, we have, from equation 4–15, $V_{BE} = 5 - 4.3 = 0.7$ V, and we therefore expect a substantial base current. In the figure, we see that the point $V_{CB} = 4.3$ V and $V_{CE} = 5$ V yields a base current of 80 μA. If V_{CB} is now allowed to increase to 5 V, then $V_{BE} = 5 - 5 = 0$ V, and the base–emitter junction is no longer forward biased. Note in the figure that $I_B = 0$ when $V_{CE} = V_{CB} = 5$ V.

Figure 4–27 shows a typical set of CC output characteristics for an *npn* transistor. These show emitter current, I_E, versus collector-to-emitter voltage, V_{CE}, for different fixed values of I_B. Note that these curves closely resemble the CE output characteristics shown in Figure 4–20. This resemblance is expected, because the only distinction is that I_E in Figure 4–27 is along the vertical axis instead of I_C, and $I_E \approx I_C$.

When leakage current is neglected, recall that $I_C = \beta I_B$. But $I_E = I_C + I_B = \beta I_B + I_B$. Therefore,

$$I_E = (\beta + 1)I_B \quad (4\text{–}16)$$

Equation 4–16 relates the input and output currents in the CC configuration.

4–6 BIAS CIRCUITS

In our discussion of BJT theory up to this point, we have been using the word *bias* simply to specify the polarity of the voltage applied across each of the *pn* junctions in a transistor. In that context, we have emphasized that the base–emitter junction must be forward biased and the collector–base junction must be reverse biased to achieve normal transistor action. We wish now

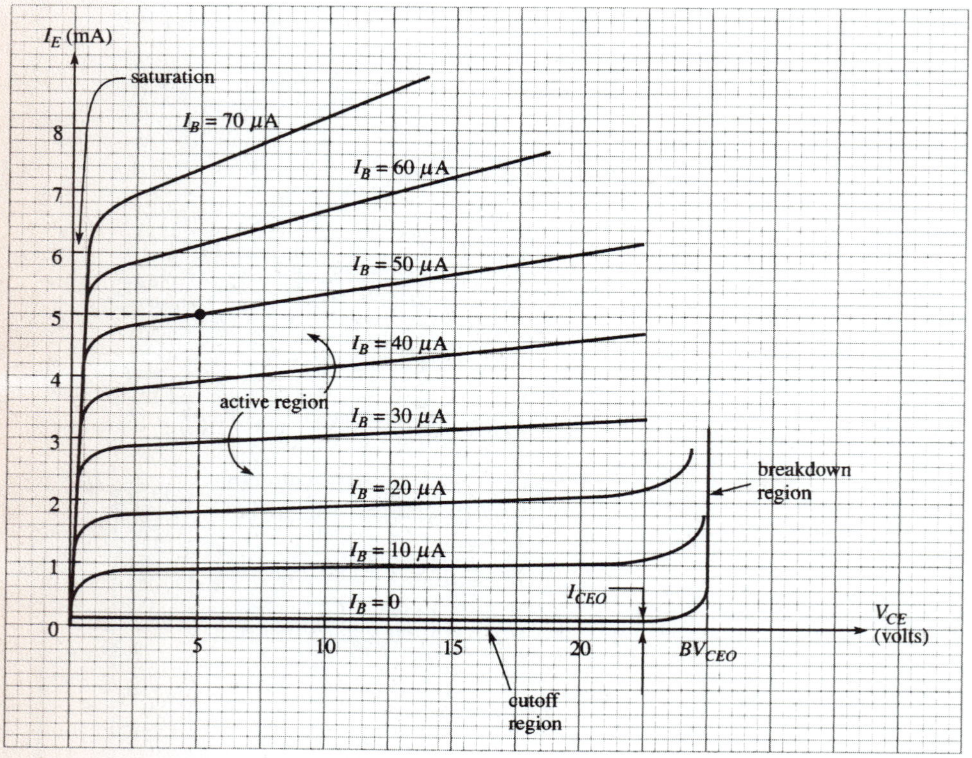

FIGURE 4–27 Common-collector output characteristics

to adopt a more restrictive interpretation of "bias." Henceforth, we will be concerned with adjusting the *value* of the bias, as needed, to obtain *specific* values of input and output currents and voltages. In other words, we accept the fact that both junctions must be biased in the proper *direction* and concentrate on a practical means for changing the *degree* of bias so that the output voltage, for example, is exactly the value we want it to be. When we have achieved a specific output voltage and output current, we say that we have set the *bias point* to those values.

Common-Base Bias Circuit

In practical circuits, we control the bias by connecting external resistors in series with the external voltage sources V_{CC}, V_{EE}, etc. We can then change resistor values instead of voltage source values to control the dc input and output voltages and currents. The circuit used to set the bias point this way is called a *bias circuit*. Figure 4–28 shows common-base bias circuits in which a resistor, R_E, is connected in series with the emitter and a resistor, R_C, is in series with the collector. Notice that we still regard emitter current as input current and base–emitter voltage as input voltage as in past discussions of the CB configuration (see Figure 4–10). Likewise, collector current and collector–base voltage are still outputs. The only difference is that the input voltage is no longer the same as V_{EE} because there is a voltage drop across R_E, and the output voltage is no longer the same as V_{CC} due to the drop across R_C. The external voltage sources V_{EE} and V_{CC} are called *supply* voltages. Of course, the characteristic curves are still perfectly valid for showing the relationships between input and output voltages and currents.

Writing Kirchhoff's voltage law around the collector–base loop in Figure 4–28(a), we have

$$V_{CC} = I_C R_C + V_{CB} \tag{4–17}$$

Rearranging equation 4–17 leads to

$$I_C = \frac{-1}{R_C} V_{CB} + \frac{V_{CC}}{R_C} \tag{4–18}$$

When we regard I_C and V_{CB} as variables and V_{CC} and R_C as constants, we see that equation 4–18 is the equation of a straight line. When plotted on a set of $I_C - V_{CB}$ axes, the line has slope $-1/R_C$, and it intercepts the I_C-axis at V_{CC}/R_C. Equation 4–18 is the equation for the (*npn*) CB *load line*. This load line has exactly the same interpretation as the diode load line we studied in Chapter 3: It is the line through *all possible* combinations of voltage (V_{CB}) and current (I_C). The actual bias point must be a point lying somewhere on the line. The precise location of the point is determined by the input current I_E.

We can find the point where the load line intercepts the V_{CB}-axis by setting $I_C = 0$ in equation 4–18 and solving for V_{CB}. Do this as an exercise and verify that the V_{CB}-intercept is $V_{CB} = V_{CC}$. Thus, the load line can be drawn simply by drawing a line through the two points $V_{CB} = 0$, $I_C = V_{CC}/R_C$ and $I_C = 0$, $V_{CB} = V_{CC}$.

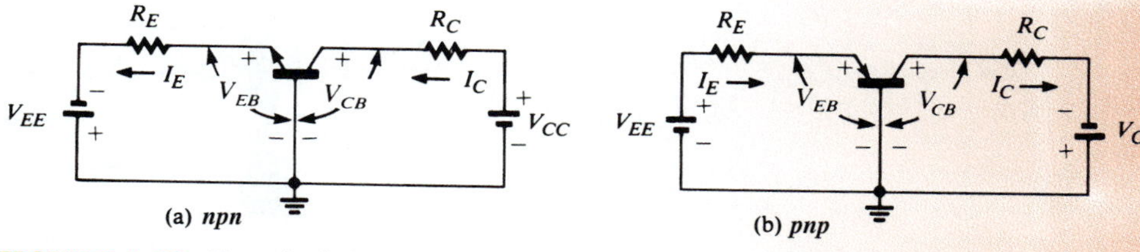

FIGURE 4–28 Practical CB bias circuits

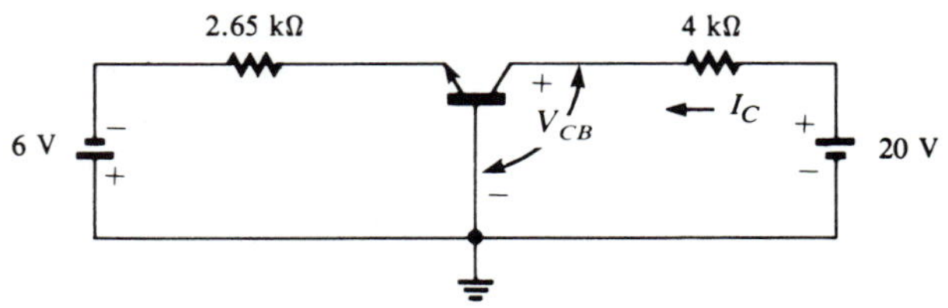

FIGURE 4–29 (Example 4–8)

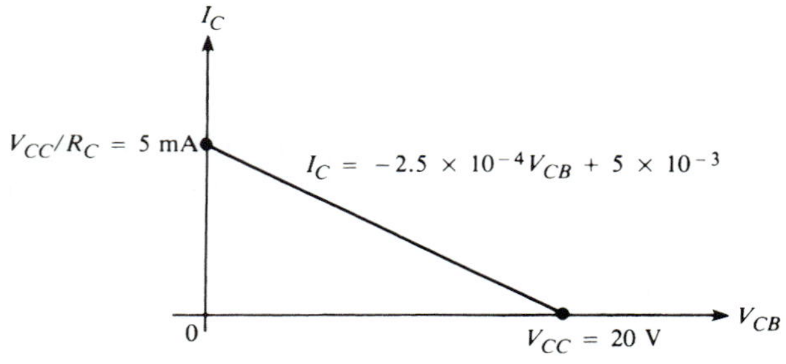

FIGURE 4–30 (Example 4–8) Load line for the bias circuit shown in Figure 4–29

EXAMPLE 4–8

Determine the equation of the load line for the circuit shown in Figure 4–29. Sketch the line.

Solution

$$\begin{aligned} I_C &= \frac{-1}{R_C}V_{CB} + \frac{V_{CC}}{R_C} \\ &= \frac{-1}{4 \times 10^3}V_{CB} + \frac{20}{4 \times 10^3} \\ &= -2.5 \times 10^{-4}V_{CB} + 5 \times 10^{-3}\text{A} \end{aligned}$$

The load line has slope -2.5×10^{-4} S, intercepts the I_C-axis at 5 mA, and intercepts the V_{CB}-axis at 20 V. It is sketched in Figure 4–30.

We can determine the bias point by plotting the load line on the output characteristics of the transistor used in the circuit. To illustrate, the load line determined in Example 4–8 is shown drawn on a set of CB output characteristics in Figure 4–31.

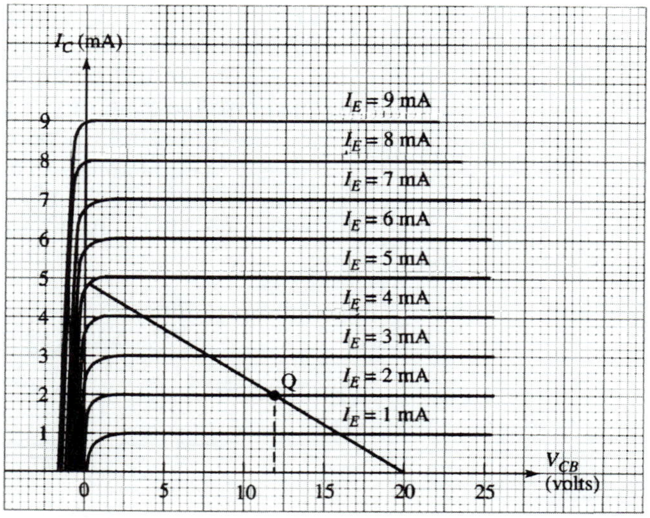

FIGURE 4–31 Load line plotted on CB output characteristics. The bias point (or quiescent point, labeled Q) is the intersection of the load line with the I_E = 2 mA curve.

To locate the bias point on the load line shown in Figure 4–31, we must determine the emitter current I_E in the circuit of Figure 4–29. One way to find I_E would be to draw an *input* load line on an input characteristic and determine the value of I_E where the line intersects the characteristic. This is the same technique used in Chapter 3 to find the dc current and voltage across a forward-biased diode in series with a resistor, which is precisely what the input side of the transistor circuit is. However, this approach is not practical for several reasons, not the least of which is the fact that input characteristics are seldom available.

The most practical way to determine I_E is to regard the base–emitter junction as a forward-biased diode having a fixed drop of 0.7 V (silicon) and solve for the diode current, the way we did in Chapter 3. Refer to Figure 4–32. In Figure 4–32, it is evident that

$$I_E = \frac{V_{EE} - 0.7}{R_E} \tag{4–19}$$

Note that we are neglecting the "feedback" effect of V_{CB} on the emitter current. Also note that the positive side of V_{EE} is connected to the circuit common, or ground, so it would normally be referred to as a *negative* voltage with respect to ground. However, in our equations, we treat V_{EE} as the absolute value of that voltage. Returning to our example circuit of Figure 4–29, we apply equation 4–19 to find

$$I_E = \frac{(6 - 0.7)\text{ V}}{2.65\text{ k}\Omega} = 2\text{ mA}$$

In Figure 4–31, the bias point, labeled Q, is seen to be the intersection of the load line with the curve $I_E = 2$ mA. At that point, $I_C \approx 2$ mA and $V_{CB} = 12$ V.

The bias point is often called the *quiescent* point, *Q*-point, or *operating* point. It specifies the dc output voltage and current when no ac voltage is superimposed on the input. As we shall discover in Chapter 7, the circuit is used as an *ac amplifier* by connecting an ac voltage source in series with the emitter. As the ac voltage alternately increases and decreases, the emitter current does the same. As a result, the output voltage and current change *along the load line* over a range determined by the change in I_E values.

Transistor input and output characteristics are useful for gaining insights into transistor behavior, and, when used with load lines, they help us visualize output current and voltage variations. However, they are seldom used to design or analyze transistor circuits. One reason they are not is that transistors of the same type typically have a wide variation in their characteristics. For that reason, manufacturers do not (cannot) publish a set of curves that could be used for every transistor of a certain type. Furthermore, the accuracy that can be obtained using approximations and purely algebraic methods of analysis (as opposed to graphical methods) is almost always adequate for practical applications. We have already seen an example of this

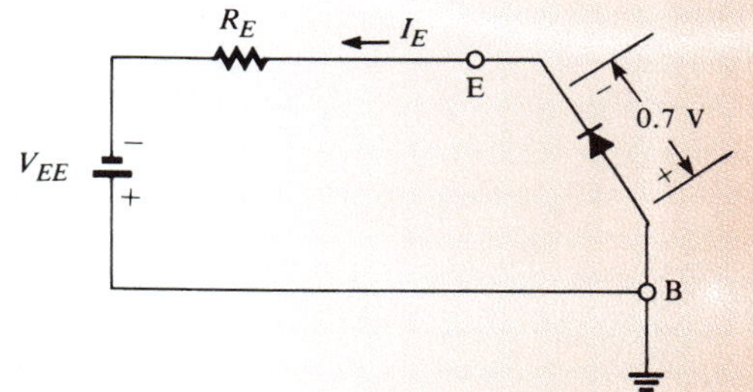

FIGURE 4–32 The input side of the transistor in a CB configuration can be regarded as a forward-biased diode for purposes of calculating I_E.

kind of algebraic approximation, when we regarded the input to the CB transistor as a forward-biased diode having a fixed voltage drop. We will now show how the entire bias circuit can be analyzed without the use of characteristic curves.

Because $\alpha \approx 1$ and $I_C = \alpha I_E$, it is true that $I_C \approx I_E$. Therefore, once we have determined I_E using equation 4–19, $I_E = (V_{EE} - 0.7)/R_E$ (silicon), we have a good approximation for I_C. We can then use equation 4–17 to solve for V_{CB}:

$$V_{CB} = V_{CC} - I_C R_C \tag{4–20}$$

EXAMPLE 4–9

Determine the bias point of the circuit shown in Figure 4–29 without using characteristic curves.

Solution

We have already shown (equation 4–19) that

$$I_E = \frac{V_{EE} - 0.7}{R_E} = \frac{(6 - 0.7)\text{ V}}{2.65\text{ k}\Omega} = 2\text{ mA}$$

Then, using $I_C \approx I_E$, we have from equation 4–20

$$\begin{aligned} V_{CB} &= V_{CC} - I_C R_C \\ &= 20\text{ V} - (2 \times 10^{-3}\text{ A})(4 \times 10^{3}\ \Omega) = 20\text{ V} - 8\text{ V} = 12\text{ V} \end{aligned}$$

Note that the bias point computed this way, $I_C = 2$ mA and $V_{CB} = 12$ V, is the same as that found graphically in Figure 4–31.

Summarizing, here are the four equations that can be used to solve for all input and output currents and voltages in the *npn* CB bias circuit of Figure 4–28(a):

$$\boxed{\begin{aligned} V_{BE} &= 0.7\text{ V (Si)}, \quad 0.3\text{ V (Ge)} \\ I_E &= \frac{V_{EE} - V_{EB}}{R_E} \\ I_C &\approx I_E \\ V_{CB} &= V_{CC} - I_C R_C \end{aligned}} \tag{4–21}$$

Equations 4–21 can be used for *pnp* transistors (Figure 4–28(b)) by substituting absolute values for V_{EB}, V_{CB}, and V_{CC}. For example, if we have $I_C = 1$ mA, $R_C = 1$ kΩ, and $V_{CC} = -15$ V in the *pnp* circuit of Figure 4–28(b), then $|V_{CB}| = 15 - (1\text{ mA})(1\text{ k}\Omega) = 5$ V. Since the base terminal is common in this circuit, the output voltage is expressed with the base as reference, i.e., as V_{CB}. Of course, $V_{CB} = -5$ V in this example.

Common-Emitter Bias Circuit

Figure 4–33 shows practical bias circuits for *npn* and *pnp* transistors in the common-emitter configuration. Notice that these bias circuits use only a single supply voltage (V_{CC}), which is a distinct practical advantage. The values of R_B and R_C must be chosen so that the voltage drop across R_B is greater than the voltage drop across R_C in order to keep the collector–base junction reverse biased. The selection of values for R_B and R_C is part of the procedure used to design a CE bias circuit, which we will cover in Chapter 7.

As discussed in Chapter 3, it is important to be able to visualize the operation of electronic circuits when their diagrams are drawn without the ground paths shown because that is the usual practice. The schematic

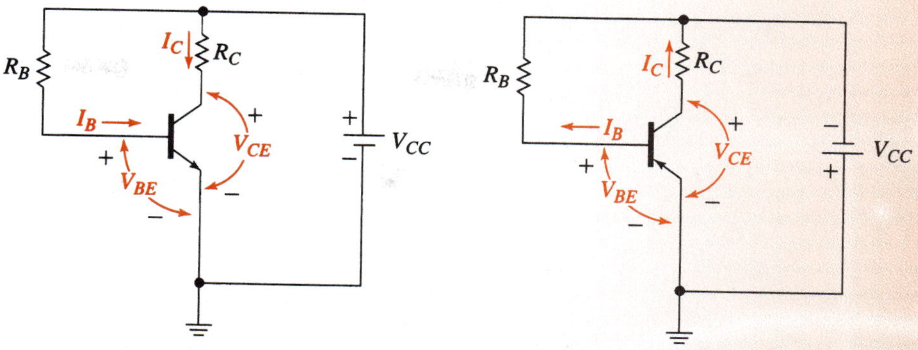

(a) *npn*

(b) *pnp*. *Note*: V_{BE} and V_{CE} are negative for *pnp* transistors.

FIGURE 4–33 Practical CE bias circuits. *Note:* V_{BE} and V_{CE} are negative for *pnp* transistors

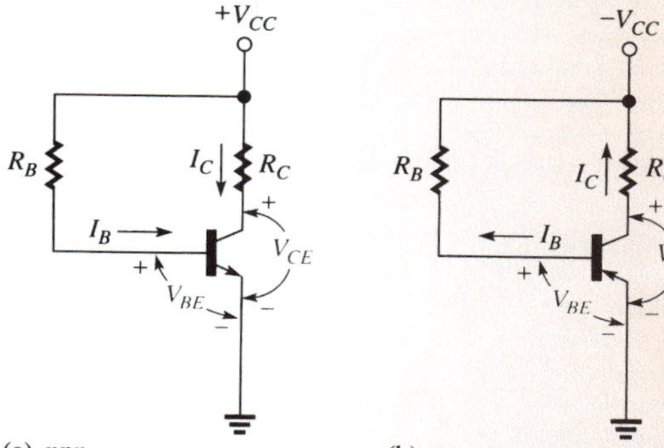

(a) *npn*

(b) *pnp*. *Note*: V_{BE} and V_{CE} are negative for *pnp* transistors.

FIGURE 4–34 Schematic diagrams of the CE bias circuits shown in Figure 4–33, with ground paths omitted

diagrams in Figure 4–33 are useful for identifying closed loops, around which Kirchhoff's voltage law can be written, but the diagrams are rarely drawn as shown. Figure 4–34 shows the conventional way of drawing schematic diagrams of the CE bias circuits. As we have done in Figure 4–34, we will hereafter omit the ground paths in transistor schematics. When studying an example or working an exercise, feel free to draw in any omitted ground paths whose presence would aid in understanding the circuit.

By writing Kirchhoff's voltage law around the output loop in Figure 4–33(a) or 4–34(a), we can obtain the equation for the load line of an *npn* transistor in a CE configuration: $V_{CC} = I_C R_C + V_{CE}$, or

$$I_C = \frac{-1}{R_C} V_{CE} + \frac{V_{CC}}{R_C} \qquad \text{(4–22)}$$

Note the similarity of equation 4–22 to the equation for the CB load line (4–18). The CE load line has slope $-1/R_C$, intercepts the I_C-axis at V_{CC}/R_C, and intercepts the V_{CE}-axis at V_{CC}. Figure 4–35 shows a graph of the load line plotted on $I_C - V_{CE}$-axes.

When the CE load line is plotted on a set of CE output characteristics, the bias point can be determined graphically, provided the value of I_B is known. To determine the value of I_B, we can again regard the input side of the transistor as a forward-biased diode having a fixed voltage drop, as shown in Figure 4–36. From Figure 4–36, we see that

$$I_B = \frac{V_{CC} - V_{BE}}{R_B} \qquad \text{(4–23)}$$

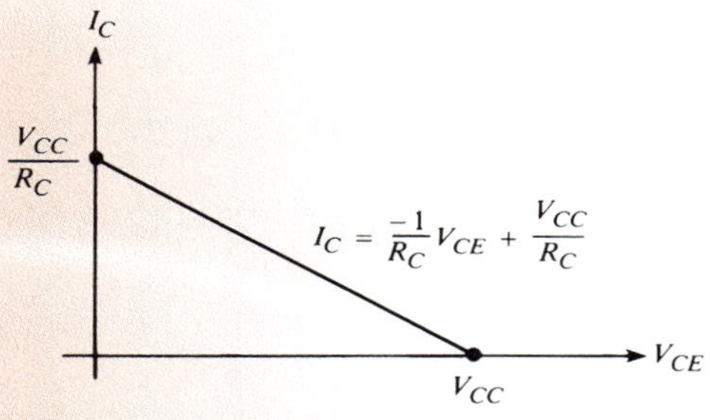

FIGURE 4–35 The CE load line for the bias circuit in Figure 4–34(a)

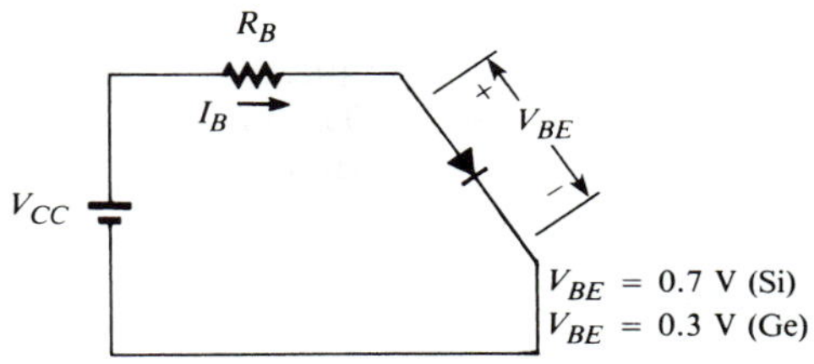

FIGURE 4–36 The input to an *npn* transistor in the CE configuration can be regarded as a forward-biased diode having a fixed voltage drop

where $V_{BE} = 0.7$ V for silicon and 0.3 V for germanium. Once again, we neglect the feedback effect of V_{CE} on I_B (the effect shown in Figure 4–19). As in the case of the CB bias circuit, the CE bias point can be determined algebraically. The equations for an *npn* circuit are summarized as follows:

$$\boxed{\begin{aligned} V_{BE} &= 0.7 \text{ V (Si)}, \;\; 0.3 \text{ V (Ge)} \\ I_B &= \frac{V_{CC} - V_{BE}}{R_B} \\ I_C &= \beta I_B \\ V_{CE} &= V_{CC} - I_C R_C \end{aligned}} \qquad (4\text{–}24)$$

Equations 4–24 can be applied to a *pnp* bias circuit (Figure 4–34(b)) as demonstrated in Example 4–10(b).

EXAMPLE 4–10

a. The silicon transistor in the CE bias circuit shown in Figure 4–37(a) has a β of 100.

   1. Assuming that the transistor has the output characteristics shown in Figure 4–38, determine the bias point graphically.
   2. Find the bias point algebraically.
   3. Repeat (1) and (2) when R_B is changed to 161.43 kΩ.

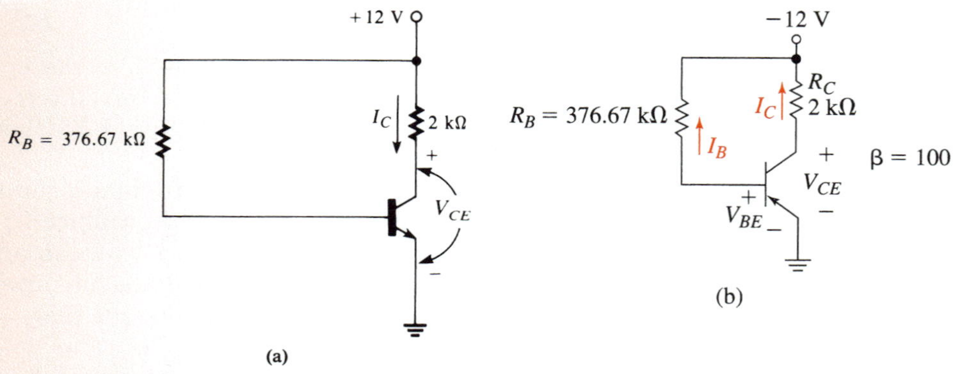

FIGURE 4–37 (Example 4–10)

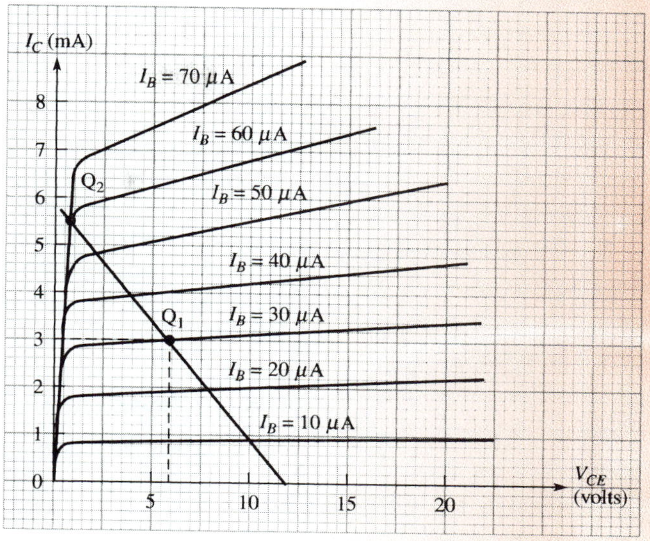

FIGURE 4–38 Load line plotted on CE output characteristics (Example 4–10). The bias point is shifted (into saturation) by the change in R_B.

Solution

1. The equation of the load line is

$$I_C = -\frac{1}{2 \times 10^3}V_{CE} + \frac{12}{2 \times 10^3}$$
$$= -0.5 \times 10^{-3}V_{CE} + 6 \times 10^{-3}$$

The load line intersects the I_C-axis at 6 mA and the V_{CE}-axis at 12 V. It is shown plotted on the output characteristics in Figure 4–38. To locate the bias point, we find I_B:

$$I_B = \frac{(12 - 0.7)\text{ V}}{376.67\text{ k}\Omega} = 30\ \mu\text{A}$$

At the intersection of the $I_B = 30\ \mu\text{A}$ curve with the load line, labeled Q_1 in Figure 4–38, we see that the bias point is $I_C \approx 2.95$ mA and $V_{CE} = 6$ V.

2. From equations 4–24, we find

$$V_{BE} = 0.7\text{ V}$$
$$I_B = \frac{(12 - 0.7)\text{ V}}{376.67\text{ k}\Omega} = 30\ \mu\text{A}$$
$$I_C = (100)(30\ \mu\text{A}) = 3\text{ mA}$$
$$V_{CE} = 12\text{ V} - (3\text{ mA})(2\text{ k}\Omega) = 6\text{ V}$$

These results are in good agreement with the bias values found graphically.

3. Changing R_B to 161.43 kΩ has no effect on the load line. Note that the load line equation (4–22) does not involve R_B. However, the value of I_B is changed to

$$I_B = \frac{(12 - 0.7)\text{ V}}{161.43\text{ k}\Omega} = 70\ \mu\text{A}$$

Thus, the bias point is shifted along the load line to the point labeled Q_2 in Figure 4–38. We see that Q_2 is now in the saturation region of the transistor. At Q_2, $I_C \approx 5.7$ mA and $V_{CE} \approx 0.5$ V. This result illustrates that the bias point can be changed by changing the value of external resistor(s) in the bias circuit.

Using equations 4–24 to find the new bias point, we have

$$I_C = \beta I_B = (100)(70\ \mu\text{A}) = 7\ \text{mA} \quad (!)$$

$$V_{CE} = 12\ \text{V} - (7\ \text{mA})(2\ \text{k}\Omega) = -2\ \text{V} \quad (!)$$

These are clearly erroneous results because the maximum value that I_C can have is 6 mA, and the minimum value that V_{CE} can have is 0 V. The reason equations 4–24 are not valid in this case is that the bias point is not in the active region. Remember that β decreases in the saturation region, and, in this example, can no longer be assumed to equal 100. (As an exercise, use Figure 4–38 to calculate the value of β at Q_2.)

b. For the *pnp* silicon transistor circuit shown in Figure 4–37(b), determine the following (assume $\beta = 100$):

1. Find the base current I_B.
2. Find the collector current I_C.
3. Find V_{CE}.

Solution

First examine Figure 4–37(b). Note that the currents and polarities of all voltages are indicated. V_{CE} and V_{BE} look the same as an *npn*. It is not necessary to write V_{CE} as V_{EC} and V_{BE} as V_{EB}. For a properly biased *pnp* circuit, these voltages will be negative (base–emitter junction forward biased and collector–base junction reverse biased).

1. To solve for I_B, write the KVL equation as follows:

$$-V_{CC} + I_B R_B - V_{BE} = 0$$

$$I_B = \frac{V_{CC} + V_{BE}}{R_B}$$

$$= \frac{12.0 - 0.7}{376.67\ \text{k}\Omega} = 30\ \mu\text{A}$$

2. To solve for I_C, use the equation

$$I_C = \beta I_B = 100 \times 30\ \mu\text{A} = 3\ \text{mA}$$

3. To solve for V_{CE}, write the KVL equation as follows:

$$-V_{CC} + I_C R_C - V_{CE} = 0 \rightarrow V_{CE} = -V_{CC} + I_C R_C$$
$$V_{CE} = -12.0 + (3\ \text{mA})(2\ \text{k}\Omega)$$
$$V_{CE} = -6\ \text{V}$$

Note: V_{CE} is a negative value as expected. Also note that I_B and V_{CE} have the same magnitude values as the *npn* circuit. This technique makes the analysis of *npn* and *pnp* circuits almost identical.

EXAMPLE 4–11

Use SPICE analysis to find the bias points of the CE circuit shown in Figure 4–37.

Solution

The circuit is redrawn with the nodes for the SPICE analysis as shown in Figure 4–39. Ground is always listed as node 0 (zero). The listing for the .CIR file follows. The .DC command specifies that the simulation will step the VCC

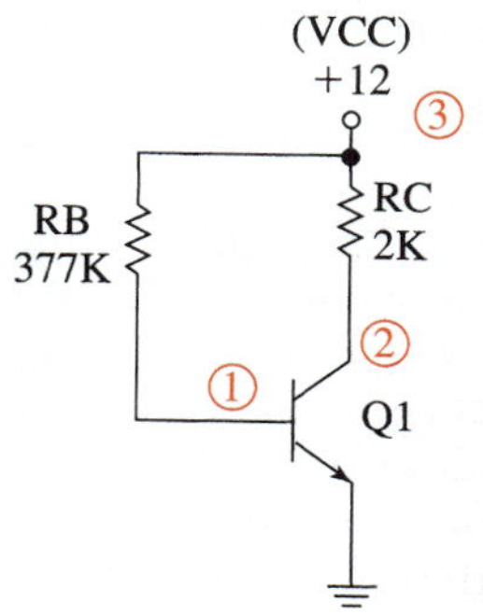

FIGURE 4–39 The CE circuit used in the simulation for Example 4–11

voltage from 0.0 V to 12.0 V in 0.1-V increments. The .OP command is used to tell the SPICE simulation to calculate the operating point values. This value is available in the output (.out) file.

```
Example 4-11
VCC 3 0 DC 12.0
Q1 2 1 0 NTRAN
RB 3 1 377k
RC 3 2 2k
.MODEL NTRAN NPN BF=100
.OP
.END
```

The following information was obtained by "Browsing" the output file (.out). Only a portion of the output file is listed.

```
**** SMALL SIGNAL BIAS SOLUTION TEMPERATURE = 27.000 DEG C
NODE VOLTAGE NODE VOLTAGE NODE VOLTAGE
(1) .8024 (2) 6.0596 (3) 12.0000
VOLTAGE SOURCE CURRENTS
NAME CURRENT
VCC −3.000E-03
OPERATING POINT INFORMATION TEMPERATURE = 27.000 DEG C
BIPOLAR JUNCTION TRANSISTORS
NAME Q1
MODEL NTRAN
IB 2.97E-05
IC 2.97E-03
VBE 8.02E-01
VBC −5.26E+00
VCE 6.06E+00
```

The SPICE simulation results show that $I_B = 29.7\ \mu\text{A}$, $I_C = 2.97$ mA, and $V_{CE} = 6.0596$ V. These values are in close agreement with the results obtained in Example 4–10.

The examples presented for common-emitter circuits use only a $+V_{CC}$ and ground connections. In some cases, it is desirable to power the circuit using a dual power supply (i.e., $+V_{CC}$ and $-V_{EE}$). Examples of these circuits are shown in Figures 4–40(a) and (b) for *npn* and *pnp* circuits. Analysis of these circuits is possible by writing KVL equations along input and output paths.

FIGURE 4–40 Common-emitter circuits using dual power supplies

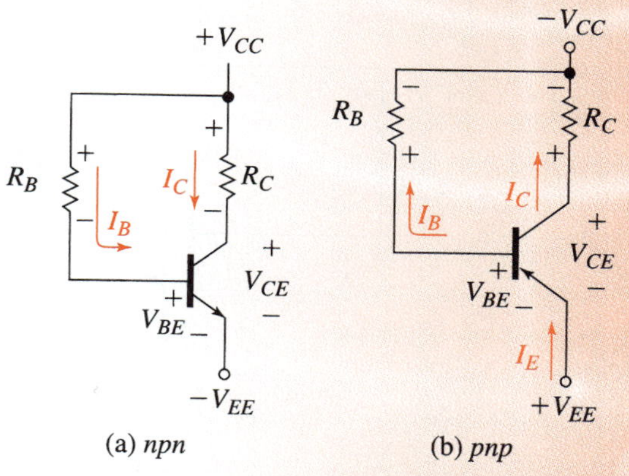

For the *npn* circuit in 4–40(a), I_B can be obtained by writing the KVL for the base current loop as follows: $V_{CC} - I_B R_B - V_{BE} + V_{EE} = 0$. Solving for V_{CE} requires that the KVL equation be written for the collector–emitter current loop as follows: $V_{CC} - I_C R_C - V_{CE} + V_{EE} = 0$.

For the *pnp* circuit in 4–40(b), I_B can be calculated in the same manner by writing the KVL for the base current loop as follows: $-V_{CC} + I_B R_B - V_{BE} - V_{EE} = 0$. Solving for V_{CE} requires that the KVL equation be written for the collector–emitter current loop as follows: $-V_{CC} + I_C R_C - V_{CE} - V_{EE} = 0$.

Common-Collector Bias Circuit

Figure 4–41 shows common-collector bias circuits for *npn* and *pnp* transistors. Once again, the load line for Figure 4–41(a) can be derived by writing Kirchhoff's voltage law around the output loop:

$$V_{CC} = I_E R_E + V_{CE}$$

$$I_E = \frac{-1}{R_E} V_{CE} + \frac{V_{CC}}{R_E} \qquad \textbf{(4–25)}$$

Recall that the output characteristics for the CC configuration are, for all practical purposes, the same as those for the CE configuration. Therefore, we will not present another example showing the load line plotted on output characteristics.

As in the previous configurations, we must find I_B in order to determine the bias point. Figure 4–42 shows a circuit that is equivalent to the loop in Figure 4–41(a) that starts at V_{CC}, passes through R_B, through the base–emitter junction, through R_E, and back to V_{CC}. Writing Kirchhoff's voltage law around the loop in Figure 4–42, we have

$$V_{CC} = I_B R_B + V_{BE} + I_E R_E \qquad \textbf{(4–26)}$$

By equation 4–16, $I_E = (\beta + 1)I_B$. Substituting for I_E in equation 4–26, we obtain

$$V_{CC} = I_B R_B + V_{BE} + (\beta + 1)I_B R_E$$

or

$$V_{CC} - V_{BE} = I_B[R_B + (\beta + 1)R_E]$$

Solving for I_B,

$$I_B = \frac{V_{CC} - V_{BE}}{R_B + (\beta + 1)R_E} \qquad \textbf{(4–27)}$$

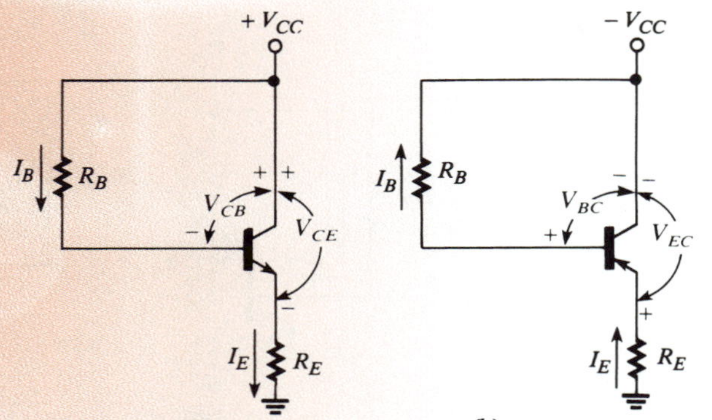

(a) *npn* (b) *pnp*

FIGURE 4–41 Common-collector bias circuits

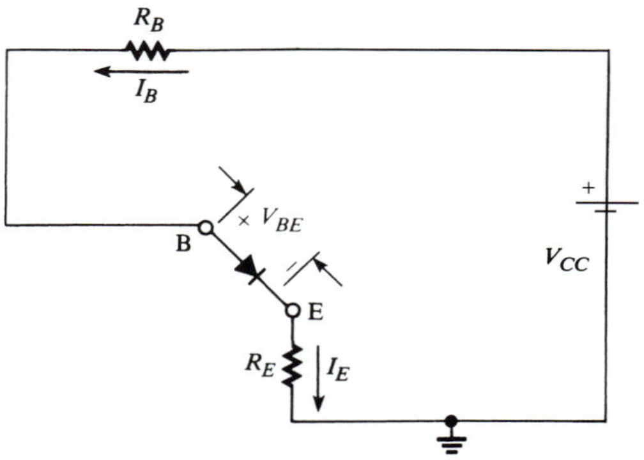

FIGURE 4–42 A circuit equivalent to the input side of Figure 4–41(a)

Summarizing, the equations for determining the bias point in an *npn* CC configuration are

$$
\begin{aligned}
V_{BE} &= 0.7\ \text{V (Si)}, \quad 0.3\ \text{V (Ge)} \\
I_B &= \frac{V_{CC} - V_{BE}}{R_B + (\beta + 1)R_E} \\
I_E &= (\beta + 1)I_B \approx \frac{V_{CC} - V_{BE}}{R_E + \dfrac{R_B}{\beta}} \approx I_C \\
V_{CE} &= V_{CC} - I_E R_E
\end{aligned}
\qquad (4\text{–}28)
$$

In most situations, we are normally concerned only with the collector (or emitter) current and the collector–emitter voltage. So the approximation in equations 4–28 for I_C or I_E can give a quick answer without need for calculating I_B. For a *pnp* CC bias circuit, use the absolute value of V_{CC}, V_{BE}, and V_{CE}.

EXAMPLE 4–12

a. Find the bias point of the germanium transistor in the circuit of Figure 4–43. Assume that $\beta = 120$.

Solution

From equations 4–28,

$$
\begin{aligned}
V_{BE} &= 0.3\ \text{V} \\
I_B &= \frac{16\ \text{V} - 0.3\ \text{V}}{116.5 \times 10^3\ \Omega + 121 \times 10^3\ \Omega} = 66.105\ \mu\text{A} \\
I_E &= (121)(66.105\ \mu\text{A}) = 8\ \text{mA}\ \ (7.97\ \text{mA using the approximation}) \\
V_{CE} &= 16\ \text{V} - (8 \times 10^{-3}\ \text{A})(10^3\ \Omega) = 8\ \text{V}
\end{aligned}
$$

b. Determine the bias point for the *pnp* circuit shown in Figure 4–43(b). Compare your answers to part (a).

Solution

Write the KVL equations for the circuit shown in Figure 4–43(b).

$$
\begin{aligned}
-V_{CC} + I_B R_B - V_{BE} - I_E R_E &= 0 \\
-V_{CC} + I_B R_B - V_{BE} - (\beta + 1)I_B R_E &= 0
\end{aligned}
$$

FIGURE 4–43
(Example 4–12)

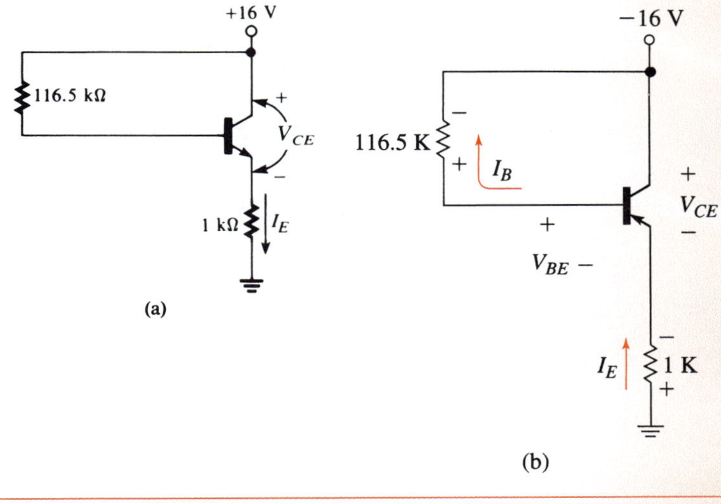

$$-V_{CC} + I_B[R_B + (\beta + 1)R_E] - V_{BE} = 0$$

$$I_B = \frac{V_{CC} + V_{BE}}{R_B + (\beta + 1)R_E} = \frac{16 - 0.3}{116.5\text{ k} + (121)(1\text{ k})}$$
$$= 66.11\ \mu\text{A}$$

To solve for V_{CE}:

$$I_E = (\beta + 1)I_B = (120 + 1)66.105\ \mu\text{A} = 8\text{ mA}$$

and

$$V_{CE} = -16 + (8\text{ mA})(1\text{ k}) = -8.0\text{ V}$$

V_{CE} is negative as expected.

Note that the values obtained for the *npn* and *pnp* transistors are the same. These values should be the same because the same component values are used and the betas for each transistor are the same. The process just described illustrates the use of KVL with the appropriate signs to solve for I_E and V_{CE}. In practice, we can simply write

$$I_C \approx I_E \approx \frac{|V_{CC}| - |V_{BE}|}{R_E + \dfrac{R_B}{\beta}}$$
$$\text{and } |V_{CE}| = |V_{CC}| - I_E R_E$$

4–7 DESIGN CONSIDERATIONS

CB Bias Design

Design is nothing more than "backwards" analysis. In other words, using equations and formulas derived through analysis in order to find items such as resistor values and power supply voltages. Although a bias circuit for the common-base configuration can be designed using a single dc power supply, we will consider only the two-source design (Figure 4–28) at this point in our discussion. In the usual design scenario, the supply voltages V_{EE} and V_{CC} are fixed, and we must choose values for R_E and R_C to obtain a specified bias current I_E and bias voltage V_{CB}. Letting $I_C = I_E$, equations 4–21 are easily solved for R_E and R_C in terms of I_E and V_{CB}:

$$\boxed{R_E = \frac{V_{EE} - V_{EB}}{I_E} \qquad R_C = \frac{V_{CC} - V_{CB}}{I_E}} \tag{4–29}$$

In practical discrete-circuit designs, it is often necessary to use standard-valued resistors. The standard values closest to the values calculated from equations 4–29 are used, and the circuit is analyzed to determine the resulting values of I_E and V_{CB}. If variation from the desired bias values is a critical consideration in a particular application, it may be necessary to use precision resistors or to calculate the total possible variation that could arise from using resistors that have a specified tolerance. The next example demonstrates these ideas.

EXAMPLE 4–13

A common-base bias circuit is to be designed for an *npn* silicon transistor to be used in a system having dc power supplies +15 V and −5 V. The bias point is to be $I_E = 1.5$ mA and $V_{CB} = 7.5$ V.

DESIGN

1. Design the circuit, using standard-valued resistors with 5% tolerance.
2. What are the actual bias values if the resistors selected have their nominal values?
3. What are the possible ranges of I_E and V_{CB}, taking the resistor tolerances into consideration?

Solution

1. From equations 4–29,

$$R_E = \frac{(5-0.7)\text{ V}}{1.5 \times 10^{-3}\text{ A}} = 2867\ \Omega$$
$$R_C = \frac{(15-7.5)\text{ V}}{1.5 \times 10^{-3}\text{ A}} = 5000\ \Omega$$

Appendix B contains a table of the standard values of resistors having 5% and 10% tolerances. The standard 5% resistors with values closest to those calculated for R_E and R_C are $E_E = 3\text{ k}\Omega$ and $R_C = 5.1\text{ k}\Omega$.

2. From equations 4–21,

$$I_E = \frac{(5-0.7)\text{ V}}{3\text{ k}\Omega} = 1.43\text{ mA}$$
$$V_{CB} = 15\text{ V} - (1.43\text{ mA})(5.1\text{ k}\Omega) = 7.69\text{ V}$$

3. The ranges of possible resistance values for R_E and R_C are

$$R_E = 3\text{ k}\Omega \pm 0.05(3\text{ k}\Omega) = 2850\text{–}3150\ \Omega$$
$$R_C = 5.1\text{ k}\Omega \pm 0.05(5.1\text{ k}\Omega) = 4845\text{–}5355\ \Omega$$
$$I_{E(min)} = \frac{V_{EE}-V_{BE}}{R_{E(max)}} = \frac{(5-0.7)\text{ V}}{3150\ \Omega} = 1.365\text{ mA}$$
$$I_{E(max)} = \frac{V_{EE}-V_{BE}}{R_{E(min)}} = \frac{(5-0.7)\text{ V}}{2850\ \Omega} = 1.509\text{ mA}$$
$$V_{CB(min)} = V_{CC} - I_{E(max)}R_{C(max)}$$
$$= 15\text{ V} - (1.509\text{ mA})(5355\ \Omega) = 6.92\text{ V}$$
$$V_{CB(max)} = V_{CC} - I_{E(min)}R_{C(min)}$$
$$= 15\text{ V} - (1.365\text{ mA})(4845\ \Omega) = 8.39\text{ V}$$

We see that considerable variation from the desired bias point is possible when using standard-valued resistors.

CE Bias Design

In Chapter 7 we discuss the design of a voltage-divider bias circuit that has certain properties superior to the design shown in Figure 4–33. However, when simplicity and minimization of the number of components are the primary considerations, the circuit of Figure 4–33 is used. Assuming the supply voltage V_{CC} is fixed, we solve equations 4–24 for R_B and R_C in terms of the desired bias values for I_C and V_{CE}:

$$R_B = \frac{V_{CC}-V_{BE}}{I_C/\beta}$$
$$R_C = \frac{V_{CC}-V_{CE}}{I_C} \qquad \textbf{(4–30)}$$

The practical difficulty with this design is that the bias point depends heavily on the value of β, which varies considerably with temperature. Also, there is typically a wide variation in the value of β among transistors of the same type. Consequently, this design is not recommended for applications where wide temperature variations may occur or for volume production (where different transistors are used). The next example demonstrates this point.

EXAMPLE 4–14

DESIGN

An *npn* silicon transistor having a nominal β of 100 is to be used in a CE configuration with $V_{CC} = 12$ V. The bias point is to be $I_C = 2$ mA and $V_{CE} = 6$ V.

1. Design the circuit, using standard-valued 5% resistors.
2. Find the range of possible bias values if the β of the transistor can change to any value between 50 and 150 (a typical range). Assume that the 5% resistors have their nominal values.

Solution

1. From equations 4–30,

$$R_B = \frac{(12 - 0.7)\text{ V}}{2 \times 10^{-3}\text{A}/100} = 565\text{ k}\Omega$$
$$R_C = \frac{(12 - 6)\text{ V}}{2 \times 10^{-3}\text{ A}} = 3\text{ k}\Omega$$

From Appendix B, the standard 5% resistors having values closest to those calculated are $R_B = 560$ kΩ and $R_C = 3$ kΩ.

2. From equations 4–24,

$$I_B = \frac{(12 - 0.7)\text{ V}}{560\text{ k}\Omega} = 20.18\ \mu\text{A}$$
$$I_{C(min)} = \beta_{(min)}I_B = 50(20.18\ \mu\text{A}) = 1.01\text{ mA}$$
$$I_{C(max)} = \beta_{(max)}I_B = 150(20.18\ \mu\text{A}) = 3.03\text{ mA}$$
$$V_{CE(min)} = V_{CC} - I_{C(max)}R_C = 12\text{ V} - (3.03\text{ mA})(3\text{ k}\Omega) = 2.92\text{ V}$$
$$V_{CE(max)} = V_{CC} - I_{C(min)}R_C = 12\text{ V} - (1.01\text{ mA})(3\text{ k}\Omega) = 8.97\text{ V}$$

In most practical applications, the possible variation of V_{CE} from 2.92 V to 8.97 V would be intolerable.

CC Bias Design

To obtain resistor values for the common-collector bias circuit (Figure 4–41), we solve equations 4-28 for R_E and R_B in terms of the desired bias values I_E and V_{CE}:

$$R_E = \frac{V_{CC} - V_{CE}}{I_E}$$
$$R_B = \frac{(\beta + 1)}{I_E}(V_{CC} - V_{BE} - I_E R_E) \approx \beta\left(\frac{V_{CC} - V_{BE}}{I_E} - R_E\right) \quad \textbf{(4–31)}$$

EXAMPLE 4–15

DESIGN

An *npn* silicon transistor having $\beta = 100$ is to be used in a CC configuration with $V_{CC} = 24$ V. The desired bias point is $V_{CE} = 16$ V and $I_E = 4$ mA.

1. Design the bias circuit using standard-valued 5% resistors.
2. Find the actual bias point when the standard resistors are used, assuming they have their nominal values.

Solution

1. From equation 4–31,

$$R_E = \frac{(24 - 16)\text{V}}{4 \text{ mA}} = 2 \text{ k}\Omega$$

$$R_B = \frac{101}{4 \times 10^{-3} \text{ A}}[24 \text{ V} - 0.7 \text{ V} - (4 \text{ mA})(2 \text{ k}\Omega)] = 386.3 \text{ k}\Omega$$

or

$$R_B \approx 100 \left(\frac{24 - 0.7}{4 \times 10^{-3}} - 2\text{k}\right) = 382.5 \text{ k}\Omega$$

From Appendix B, the standard 5% resistors having values closest to those calculated are $R_E = 2$ kΩ and $R_B = 390$ kΩ.

2. From equations 4–28,

$$I_B = \frac{(24 - 0.7) \text{ V}}{390 \text{ k}\Omega + 101(2 \text{ k}\Omega)} = 39.358 \text{ }\mu\text{A}$$

$$I_E = 101(39.358 \text{ }\mu\text{A}) = 3.98 \text{ mA}$$

$$V_{CE} = 24 \text{ V} - (3.98 \text{ mA})(2 \text{ k}\Omega) = 16.04 \text{ V}$$

4–8 THE BJT INVERTER (TRANSISTOR SWITCH)

Transistors are widely used in digital logic circuits and switching applications. Recall that the waveforms encountered in those applications periodically alternate between a "low" and a "high" voltage, such as 0 V and +5 V. The fundamental transistor circuit used in switching applications is called an *inverter*, the *npn* version of which is shown in Figure 4–44. Note in the figure that the transistor is in a common-emitter configuration, but there is no bias voltage connected to the base through a resistor, as in the CE bias circuits studied earlier. Instead, a resistor R_B is connected in series with the base and then directly to a square or pulse-type waveform that serves as the inverter's input. In the circuit shown, V_{CC} and the "high" level of the input are both +5 V. The output is the voltage between collector and emitter (V_{CE}), as usual.

When the input to the inverter is high (+5 V), the base–emitter junction is forward biased and current flows through R_B into the base. The values of

FIGURE 4–44 An *npn* transistor inverter, or switch

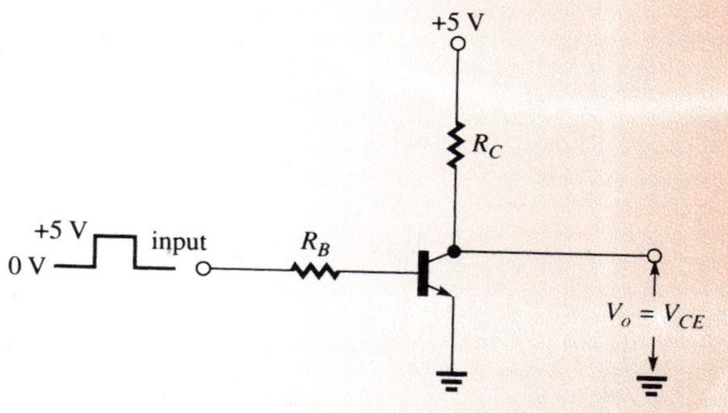

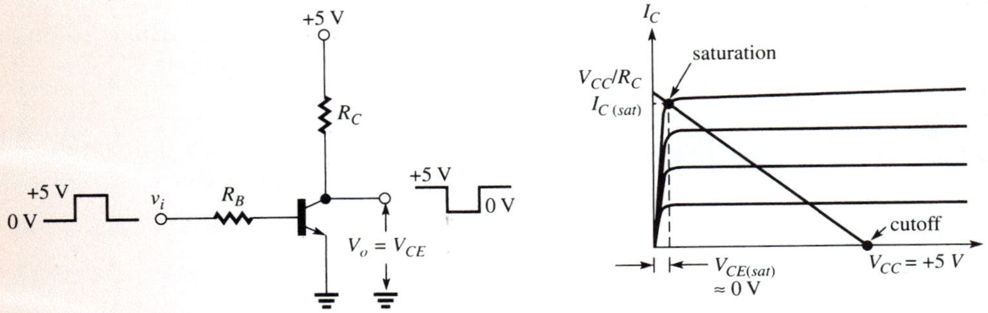

FIGURE 4–45 When the input to the inverter is high (+5 V), the transistor is saturated and its output is low (≈ 0 V). When the input to the inverter is low, the transistor is cut off and its output is high.

R_B and R_C are chosen (designed) so that the amount of base current flowing is enough to *saturate* the transistor, that is, to drive it into the saturation region of its output characteristics. Figure 4–45 shows a load line plotted on a set of CE output characteristics and identifies the point on the load line where saturation occurs. Note that the value of V_{CE} corresponding to this point, called $V_{CE(sat)}$, is very nearly 0 (typically about 0.1 V). The current at the saturation point is called $I_{C(sat)}$ and is very nearly equal to the intercept of the load line of the I_C-axis, namely, V_{CC}/R_C. When the transistor is saturated, it is said to be *ON*. This analysis has shown that a high input to the inverter (+5 V) results in a low output (≈ 0 V).

When the input to the transistor is low, that is, 0 V, the base–emitter junction has no forward bias applied to it, so no base current, and hence no collector current, flows. There is, therefore, no voltage drop across R_C, and it follows that V_{CE} must be the same as V_{CC}: +5 V. This fact is made evident by substituting $I_C = 0$ in the equation for V_{CE} (equation 4–24): $V_{CE} = V_{CC} - I_C R_C = V_{CC} - (0)(R_C) = V_{CC}$. In this situation, the transistor is in the cutoff region of its output characteristics, as shown in Figure 4–45, and is said to be *OFF*. A low input to the inverter results in a high output, and it is now obvious why this circuit is called an *inverter*.

In designing and analyzing transistor inverters, it is usually assumed that $I_{C(sat)} = V_{CC}/R_C$ and that $V_{CE(sat)} = 0$ V. These are very good approximations and lead to results that are valid for most practical applications. Under these assumptions, we can easily derive the voltage–current relations in a transistor inverter. Because the transistor is cut off when the input is low, regardless of the values of R_B and R_C, the equations we will study are those that apply when the input is high. Actually, these equations are precisely those we have already derived for a CE transistor, for the special case $I_C = I_{C(sat)}$. Thus, assuming saturation exists,

$$I_C = I_{C(sat)} = \frac{V_{CC}}{R_C} \qquad (4\text{–}32)$$

$$I_B = \frac{I_{C(sat)}}{\beta} = \frac{V_{CC}}{\beta R_C} \qquad (4\text{–}33)$$

$$I_B = \frac{V_{HI} - V_{BE}}{R_B} \qquad (4\text{–}34)$$

where V_{HI} is the high level of the input voltage, usually the same as V_{CC}.

FIGURE 4–46 (Example 4–16)

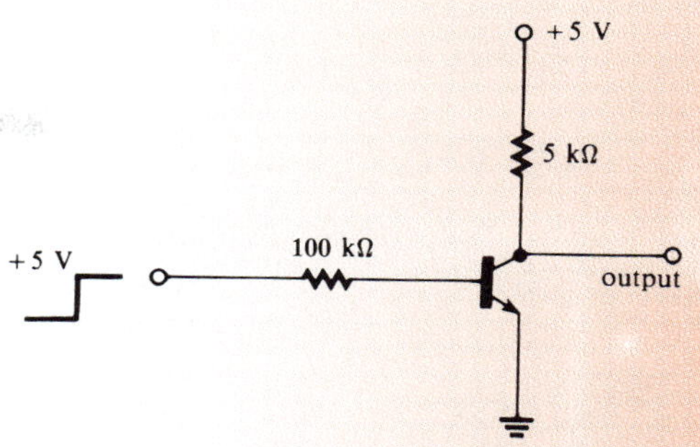

Regarding Equation 4–33, some books use the notation $I_{B(sat)} = \frac{I_{C(sat)}}{\beta}$ where $I_{B(sat)}$ is the minimum I_B necessary for saturation.

EXAMPLE 4–16

Verify that the circuit in Figure 4–46 behaves like an inverter when the input switches between 0 V and +5 V. Assume that the transistor is silicon and that $\beta = 100$.

Solution

It is only necessary to verify that the transistor is saturated when $V_i = +5\text{ V}$. From equation 4–34.

$$I_B = \frac{(5-0.7)\text{ V}}{100\text{ k}\Omega} = 43\ \mu\text{A}$$

Then

$$I_C = \beta I_B = 100(43\ \mu\text{A}) = 4.3\text{ mA}$$

and

$$V_{CE} = 5 - (4.3\text{ mA})(5\text{ k}\Omega) = 5 - 21.5 = -16.5\text{ V}$$

A negative V_{CE} indicates that the transistor is indeed in saturation and that $V_{CE} = V_{CE(sat)} \approx 0\text{ V}$, and $I_C = I_{C(sat)} \approx \frac{V_{CC}}{R_C} = 1\text{ mA}$. We could have concluded that the transistor was in saturation by calculating $I_{C(sat)} \approx V_{CC}/R_C$ and comparing it to the calculated I_C of 4.3 mA. When $I_C = \beta I_B$ is greater than $I_{C(sat)}$, saturation occurs. We can also determine saturation by comparing base currents. The minimum base current necessary to saturate is $I_B = \frac{V_{CC}}{\beta R_B}$. In this example, $\frac{V_{CC}}{\beta R_C} = \frac{5}{100(5\text{k})} = 10\ \mu\text{A}$. Since $I_B = 43\ \mu\text{A} > 10\ \mu\text{A}$, the transistor is in saturation. If we changed R_B to, say, 750 kΩ, the resulting base current would be 5.73 μA and the transistor would not be saturated. The collector current would then be 0.573 mA and V_{CE} would be 2.13 V.

Inverter Design

To design a transistor inverter we must have criteria for specifying the values of R_B and R_C. Typically, one of the two is known (or chosen arbitrarily), and the value of the other is derived from the first. Using equations 4–33 and 4–34, we can obtain the following relationships between R_B and R_C:

$$R_B = \frac{V_{HI} - V_{BE}}{I_B} = \frac{(V_{HI} - V_{BE})\beta R_C}{V_{CC}} \qquad (4\text{–}35)$$

$$R_C = \frac{V_{CC}R_B}{\beta(V_{HI} - V_{BE})} \tag{4–36}$$

Equation 4–35 can be used to find R_B when R_C is known, and equation 4–36 to find R_C when R_B is known. However, because these equations are valid only for a specific value of β, they are not entirely practical. We have already discussed the fact that the β of a transistor of a given type is likely to vary over a wide range. If the actual value of β is smaller than the one used in the design equations, the transistor will not saturate. For this reason, the β used in the design equations should always be the *smallest* possible value that might occur in a given application. In other words, equations 4–35 and 4–36 are more practical when expressed in the form of inequalities, as follows:

$$R_B \le \frac{(V_{HI} - V_{BE})\beta R_C}{V_{CC}} \tag{4–37}$$

$$R_C \ge \frac{V_{CC}R_B}{\beta(V_{HI} - V_{BE})} \tag{4–38}$$

These inequalities should hold for the entire range of β-values that transistors used in the inverter may have. This will be the case if the minimum possible β-value is used.

We should note that when a transistor has a higher value of β than the one for which the inverter circuit is designed, a high input simply drives it deeper into saturation. This *overdriving* of the transistor creates certain new problems, including the fact that it slows the speed at which the device can switch from ON to OFF, but the output is definitely low in the ON state. The results from Example 4–16 indicate an overdriving condition, because I_B was about four times larger than the minimum required. In practice, however, a moderate amount of overdriving is normally used in order to maintain a low $V_{CE(sat)}$.

EXAMPLE 4–17

DESIGN

An inverter having $R_C = 1.5\ \text{k}\Omega$ is to be designed so that it will operate satisfactorily with silicon transistors whose β-values range from 80 to 200. What value of R_B should be used? Assume that $V_{CC} = V_{HI} = +5\ \text{V}$.

Solution

Using equation 4–35 with $\beta = \beta_{(\text{min})} = 80$, we find

$$R_B \le \frac{(V_{HI} - V_{BE})\beta_{(\text{min})}R_C}{V_{CC}} = \frac{(4.3)(80)(1.5\ \text{k}\Omega)}{5} = 103.2\ \text{k}\Omega$$

An actual value of 100 kΩ would be appropriate.

The Transistor as a Switch

A transistor inverter is often called a transistor *switch*. This terminology is appropriate because the ON and OFF states of the transistor correspond closely to the closing and opening of a switch connected between the collector and the emitter. When the transistor is ON, or saturated, the voltage between collector and emitter is nearly 0, as it would be across a closed switch, and the current is the maximum possible, V_{CC}/R_C. When the transistor is OFF, no current flows from collector to emitter and the voltage is maximum, as it would be across an open switch. The switch is opened or closed by the input voltage: A high input closes it and a low input opens it. Figure 4–47 illustrates these ideas.

FIGURE 4–47 The transistor as a voltage-controlled switch. A high input closes the switch and a low input opens it.

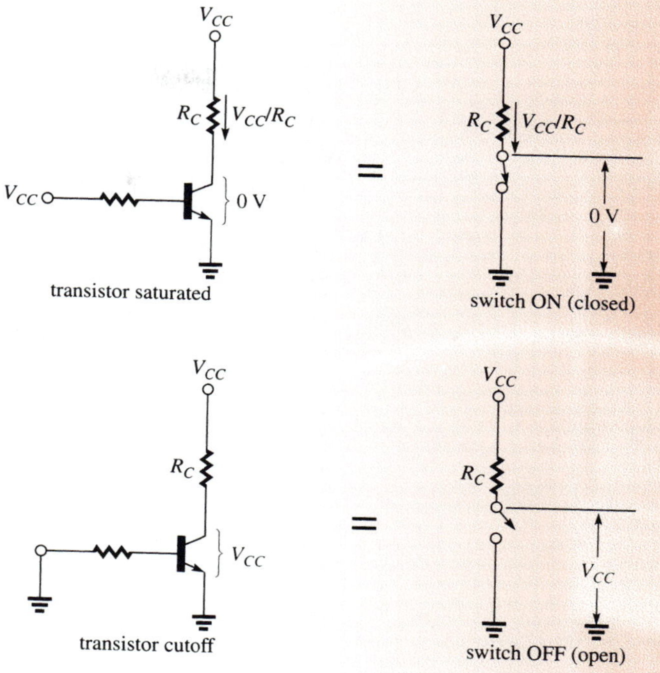

FIGURE 4–48 The BJT inverter interfaced to a relay. V_i is a TTL-level logic voltage.

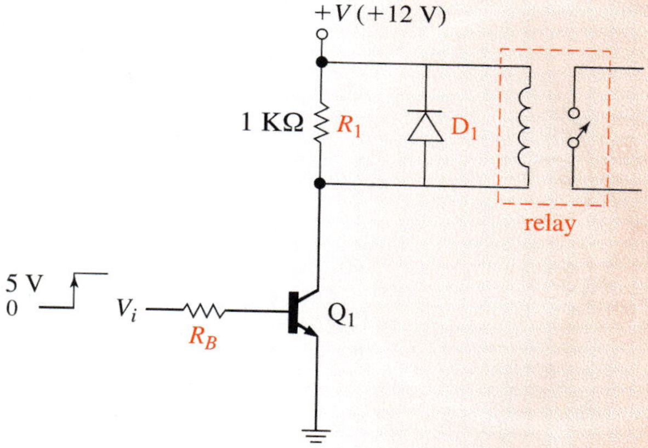

In many switching applications, the emitter may be connected to another circuit, or to another voltage source, instead of to ground. When analyzing such complex digital circuits, it is quite helpful to think of the transistor as simply a switch, for then it is easy to understand circuit operation in terms of the collector circuit being connected to or disconnected from the emitter circuit. For example, if the emitter in the basic inverter circuit were connected to −5 V instead of ground, then the output would clearly switch between +5 V and −5 V instead of between +5 V and 0 V.

A popular use of the BJT inverter is for interfacing logic signals to electromechanical devices such as relays. BJT transistors interface well to relays because they can easily sink the required current to fully turn on a relay's magnetic coil. A schematic of a BJT interfaced to a relay is shown in Figure 4–48.

The input to this circuit is a digital logic signal. The logical high-level voltage in digital signals varies considerably. Do not assume that a logical high is 5.0 V. Many of the new digital systems are using 3.3 V for a logical high. Also, for TTL 5.0-V logic systems, a logical high voltage can vary from 2.5 V to 5.0 V. When in doubt, measure the voltage level of your system. Second, the relay should turn on when the logic level is "high," and, accordingly, the relay will turn off when the input logic level is "low." The following is a summary of considerations for designing the relay interface:

1. TTL logic levels are not exact.
2. Transistor switching speed is not usually an issue when turning a relay on/off.
3. The β of the transistor is not usually known unless you take time to measure it.

Design Procedure

1. Determine the logic-level voltage you want for the transistor and relay to turn on. This should be some value that the logic level is guaranteed to reach. For a 5.0-V TTL system, 3.0 V is usually a guaranteed value. If you select a value such as 4.5 V, the transistor might not fully saturate, which could cause the relay coil to not fully energize. This could cause erratic behavior of the relay.
2. Recall that in many cases, the β of the transistor is not known. Of course, the user could set up a test or use a curve tracer to extract the β value for calculating R_B. For typical TTL logic levels, a resistor of about 1 kΩ to 5 kΩ for R_B works very well.
3. When the input logic level is low, the transistor will be placed in the cut-off condition ($I_C \cong 0$). This is called the *open collector* condition. To fully turn off the relay, a pull-up resistor (R_C) is connected across the relay.
4. When the transistor is ON, a current will be flowing in the magnetic coil. Recall that inductances store energy in terms of current. When the transistor is turned OFF, the coil will attempt to de-energize through the pull-up resistor R_C, which can lead to a significantly large voltage spike (transient). This transient can cause glitches in logic levels and can damage the circuits. This problem can be minimized by placing a reversed-biased diode (D_1) across the coil of the relay, as shown in Figure 4–48. D_1 will clamp any voltage on the collector to a maximum value of 0.7 V by providing a low-resistance path for the coil to de-energize. The diodes are also quick reacting, and the junction capacitance in the diode helps to suppress the back EMF induced by the de-energizing coil.

Q_1 is used to turn the relay on/off. Recall that when Q_1 is saturated, $V_{CE} \approx 0.1$ V (very close to 0). Typical transistors for relays requiring 50 mA or less are the 2N3904 and 2N2222, although many other transistors are suitable. The circuit is shown in Figure 4–48.

4–9 TRANSISTOR TYPES, RATINGS, AND SPECIFICATIONS

In modern electronic circuits, discrete transistors are used primarily for applications in which only one or a small number of devices are required and in applications where substantial power is dissipated. Although older designs, composed entirely of discrete devices, can still be found in large numbers, most new circuits containing a large number of transistors are constructed in integrated-circuit form. In many applications, both discrete and integrated components are used. In these applications, the integrated circuit typically performs complex, low-level *signal conditioning,* and a discrete transistor then drives a power-consuming load such as an indicator lamp or an audio speaker. This use of the transistor is an example of *interfacing;* it

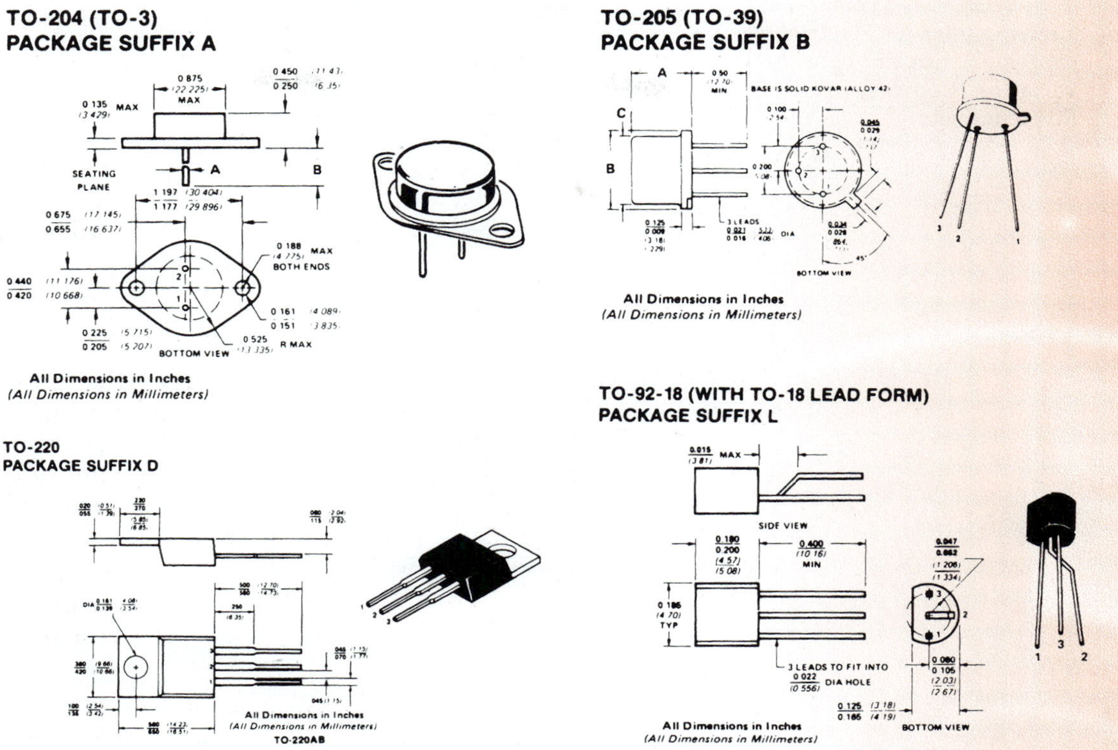

FIGURE 4–49 A few of the standard transistor case (enclosure) types, with TO-designations (old JEDEC) numbers in parentheses (Courtesy of Siliconix Inc.)

provides a link between a device having limited power capabilities and a load that requires large voltages or currents.

Discrete transistors are packaged in a wide variety of metal and plastic enclosures (cases). Figure 4–49 shows some of the standard case types, which are identified by standard TO numbers. Three leads are brought out through each enclosure to permit external connection to the transistor's emitter, base, and collector. In some power transistors, rated for high power dissipation, the collector is attached and electrically common to the metal case. (The majority of the power dissipated in a transistor occurs at the collector–base junction, since the collector voltage is usually the largest voltage in the device.) A transistor manufacturer uses a consistent scheme that can be followed to identify the base, emitter, and collector terminals for a given case type. For example, in the TO-39 case, the three leads are attached in a semicircular cluster and a metal tab on the case is adjacent to the emitter. The base is the center lead in the cluster, and the collector is the remaining lead.

A discrete transistor of a specific type, having registered JAN (military) specifications, is identified by a number with the prefix 2N. Although all transistors having the same number may not be identical, they are all designed to meet the same performance specifications related to voltage and current limits, power dissipation, operating temperature range, and parameter variations. More than one manufacturer may produce a transistor with a given 2N number. Many manufacturers also produce "commercial"-grade devices that do not have 2N designations.

Figure 4–50 shows parts of a typical set of manufacturer's transistor specifications. The maximum ratings show the maximum voltages that each device can sustain between different sets of transistor terminals and the maximum power dissipation, P_D, at 25°C. The effect of temperature on

FIGURE 4–50 Typical transistor specifications for electrical (dc) characteristics and beta (h_{FE}) variation (Courtesy of ON Semiconductor)

MPS2222, MPS2222A

MPS2222A is a Preferred Device

General Purpose Transistors

NPN Silicon

ON Semiconductor®

http://onsemi.com

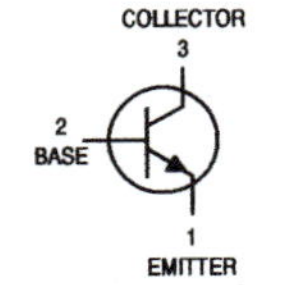

TO–92
CASE 29
STYLE 1

MARKING DIAGRAMS

Y = Year
WW = Work Week

MAXIMUM RATINGS

Rating	Symbol	Value	Unit
Collector–Emitter Voltage MPS2222 / MPS2222A	V_{CEO}	30 / 40	Vdc
Collector–Base Voltage MPS2222 / MPS2222A	V_{CBO}	60 / 75	Vdc
Emitter–Base Voltage MPS2222 / MPS2222A	V_{EBO}	5.0 / 6.0	Vdc
Collector Current – Continuous	I_C	600	mAdc
Total Device Dissipation @ T_A = 25°C; Derate above 25°C	P_D	625 / 5.0	mW / mW/°C
Total Device Dissipation @ T_C = 25°C; Derate above 25°C	P_D	1.5 / 12	Watts / mW/°C
Operating and Storage Junction Temperature Range	T_J, T_{stg}	–55 to +150	°C

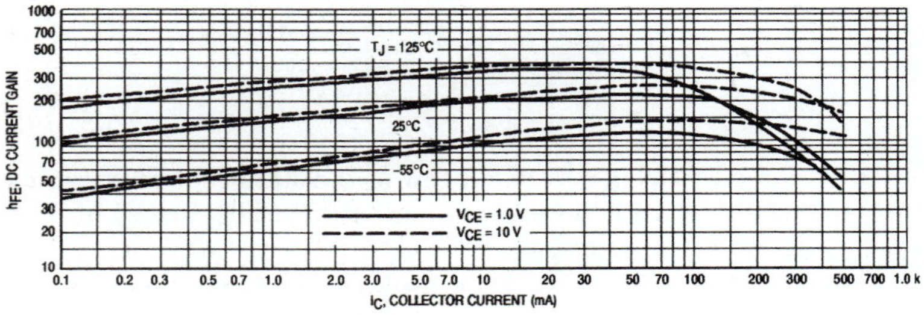

Figure 3. DC Current Gain

ELECTRICAL CHARACTERISTICS (T_A = 25°C unless otherwise noted)

Characteristic		Symbol	Min	Max	Unit
OFF CHARACTERISTICS					
Collector–Emitter Breakdown Voltage (I_C = 10 mAdc, I_B = 0)	MPS2222 / MPS2222A	$V_{(BR)CEO}$	30 / 40	– / –	Vdc
Collector–Base Breakdown Voltage (I_C = 10 μAdc, I_E = 0)	MPS2222 / MPS2222A	$V_{(BR)CBO}$	60 / 75	– / –	Vdc
Emitter–Base Breakdown Voltage (I_E = 10 μAdc, I_C = 0)	MPS2222 / MPS2222A	$V_{(BR)EBO}$	5.0 / 6.0	– / –	Vdc
Collector Cutoff Current (V_{CE} = 60 Vdc, $V_{EB(off)}$ = 3.0 Vdc)	MPS2222A	I_{CEX}	–	10	nAdc
Collector Cutoff Current		I_{CBO}			μAdc
(V_{CB} = 50 Vdc, I_E = 0)	MPS2222		–	0.01	
(V_{CB} = 60 Vdc, I_E = 0)	MPS2222A		–	0.01	
(V_{CB} = 50 Vdc, I_E = 0, T_A = 125°C)	MPS2222		–	10	
(V_{CB} = 50 Vdc, I_E = 0, T_A = 125°C)	MPS2222A		–	10	
Emitter Cutoff Current (V_{EB} = 3.0 Vdc, I_C = 0)	MPS2222A	I_{EBO}	–	100	nAdc
Base Cutoff Current (V_{CE} = 60 Vdc, $V_{EB(off)}$ = 3.0 Vdc)	MPS2222A	I_{BL}	–	20	nAdc
ON CHARACTERISTICS					
DC Current Gain		h_{FE}			–
(I_C = 0.1 mAdc, V_{CE} = 10 Vdc)			35	–	
(I_C = 1.0 mAdc, V_{CE} = 10 Vdc)			50	–	
(I_C = 10 mAdc, V_{CE} = 10 Vdc)			75	–	
(I_C = 10 mAdc, V_{CE} = 10 Vdc, T_A = –55°C)	MPS2222A only		35	–	
(I_C = 150 mAdc, V_{CE} = 10 Vdc) (Note 1.)			100	300	
(I_C = 150 mAdc, V_{CE} = 1.0 Vdc) (Note 1.)			50	–	
(I_C = 500 mAdc, V_{CE} = 10 Vdc) (Note 1.)	MPS2222		30	–	
	MPS2222A		40	–	
Collector–Emitter Saturation Voltage (Note 1.)		$V_{CE(sat)}$			Vdc
(I_C = 150 mAdc, I_B = 15 mAdc)	MPS2222		–	0.4	
	MPS2222A		–	0.3	
(I_C = 500 mAdc, I_B = 50 mAdc)	MPS2222		–	1.6	
	MPS2222A		–	1.0	
Base–Emitter Saturation Voltage (Note 1.)		$V_{BE(sat)}$			Vdc
(I_C = 150 mAdc, I_B = 15 mAdc)	MPS2222		–	1.3	
	MPS2222A		0.6	1.2	
(I_C = 500 mAdc, I_B = 50 mAdc)	MPS2222		–	2.6	
	MPS2222A		–	2.0	

1. Pulse Test: Pulse Width ≤ 300 μs, Duty Cycle ≤ 2%.

transistor specifications will be covered in detail in Chapter 16 (Section 16–3). A transistor circuit designer must be certain that a transistor used in a particular application will not be subjected to voltages or power dissipations that exceed the specified maximums; failure to do so may result in severe performance degradation or permanent damage.

The graph labeled "DC Current Gain" shows how β varies with V_{CE}, junction temperature (T_J), and collector current. (In the CE configuration, current *gain* is the ratio of output current to input current, or I_C/I_B, which is approximately β. As we shall learn later when *h* parameters are discussed, β is also designated by h_{FE}.) The chart illustrates a typical transistor characteristic: For a fixed collector current, *the value of* β *increases with increasing temperature.*

The *electrical characteristics* listed in Figure 4–50 show important transistor parameters associated with the dc operation of each device. Included are breakdown voltages and reverse leakage currents (called "cutoff" currents in the specifications). Notice that the breakdown voltages previously referred to as BV_{CBO} and BV_{CEO} are listed as $V_{BR(CBO)}$ and $V_{BR(CEO)}$, respectively.

Figure 4–50 does not show some other specifications that are usually furnished by a manufacturer, including small-signal characteristics. Small-signal characteristics are associated with the ac operation of a transistor, which we will cover in Chapter 7. Other specifications often furnished by the manufacturer include graphs showing additional parameter variations with temperature, voltage, and current.

4–10 TRANSISTOR CURVE TRACERS

We have mentioned that characteristic curves are seldom included in transistor specifications. These vary widely among transistors of a given type and are rarely used for circuit design purposes. However, in areas such as component testing, preliminary circuit development, and research, it is often useful to be able to study the characteristic curves of a single device and to obtain important parameter values from the curves. Recall that parameters such as α, β, BV_{CBO}, BV_{CEO}, leakage currents, saturation voltages, and the Early voltage can be discerned from appropriate sets of characteristic curves.

The most widely used method for obtaining a set of characteristic curves is by use of an instrument called a *transistor curve tracer.* A curve tracer is basically an oscilloscope equipped with circuitry that automatically steps the currents (or voltages) in a semiconductor device through a range of values and displays the family of characteristic curves that result. Selector switches allow the user to set the maximum value and the increment (step) value of each current or voltage applied to the device. For example, to obtain a family of transistor collector characteristics, the user might set the base current increment to be 10 μA, the maximum collector voltage to be 25 V, and the number of steps to be 10. The characteristics would then be displayed as a family of curves showing I_C versus V_{CE} for $I_B = 0$, 10 μA, 20 μA, ... Figure 4–51 shows two typical curve tracers.

Figure 4–52(a) is a photograph of a curve tracer display showing a typical set of *npn* collector characteristics. The horizontal sensitivity of the display was set for 2 V/division, so the horizontal axis (V_{CE}) extends from 0 to about 13 V. The vertical sensitivity was set for 1 mA/division, so the vertical axis (I_C) extends from 0 to about 6.5 mA. The base current increment is 10 μA. Using this display, we can determine, for example, that the β of the transistor at $V_{CE} = 4$ V and $I_B = 40$ μA is approximately

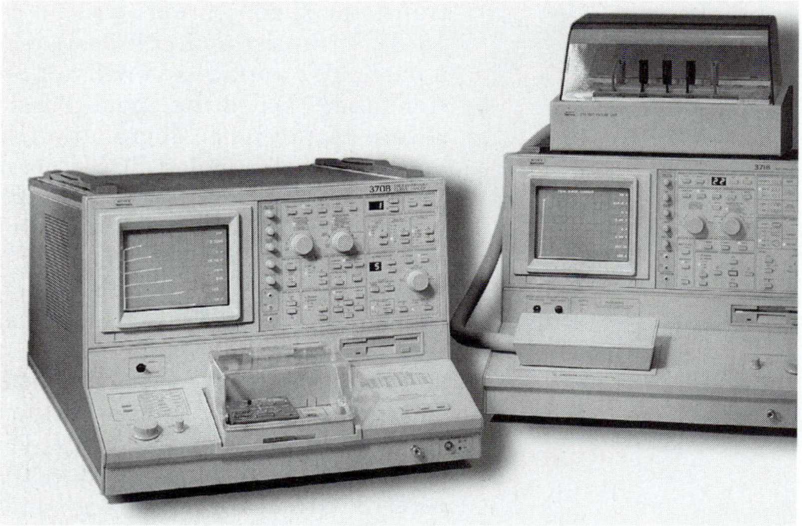

FIGURE 4–51 Two typical curve tracers (Tektronix Models 370B and 371B) (Courtesy of Tektronix, Inc.)

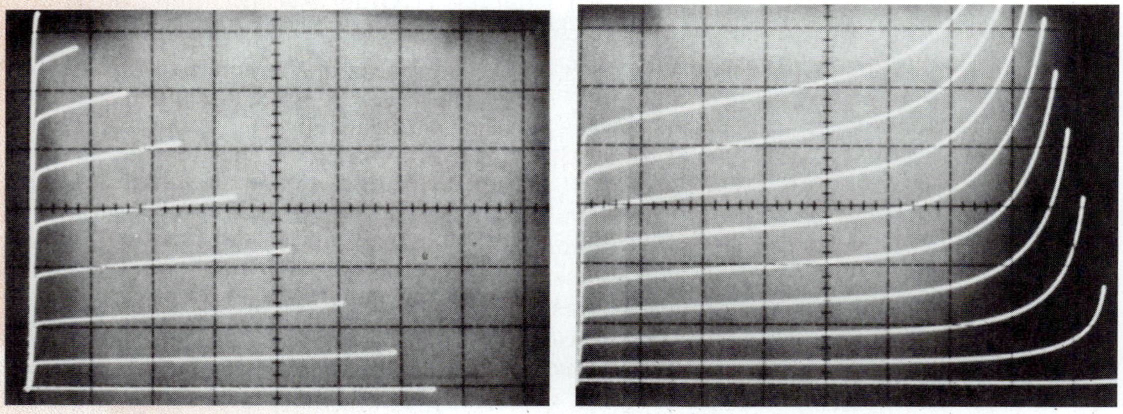

(a) Horizontal: 2 V/div
Vertical: 1 mA/div
$\Delta I_B = 10\ \mu A$

(b) Horizontal: 2 V/div
Vertical: 0.2 mA/div
$\Delta I_B = 2\ \mu A$

FIGURE 4–52 Photographs of curve tracer displays

$$\beta = \frac{I_C}{I_B} = \frac{3\text{ mA}}{40\ \mu\text{A}} = 75$$

The curve tracer from which this display was obtained permits the user to select a value of series collector resistance (R_C), which, for the display shown, was set to 2 kΩ. Notice that the base current curves become shorter with increasing current. *An imaginary line connecting the right-hand tips of each curve represents the load line for the circuit.* This load line is seen to intersect the V_{CE}-axis at 13 V and the I_C-axis at 6.5 mA. Thus, the value of V_{CC} used in this circuit is 13 V, and the load line intersects the I_C-axis at the value expected:

$$I_C = \frac{V_{CC}}{R_C} = \frac{13\text{ V}}{2\text{ k}\Omega} = 6.5\text{ mA}$$

One convenient feature of a curve tracer is that it permits a user to expand or contract the display in different regions of the characteristic curves by adjusting the sensitivity and range controls.

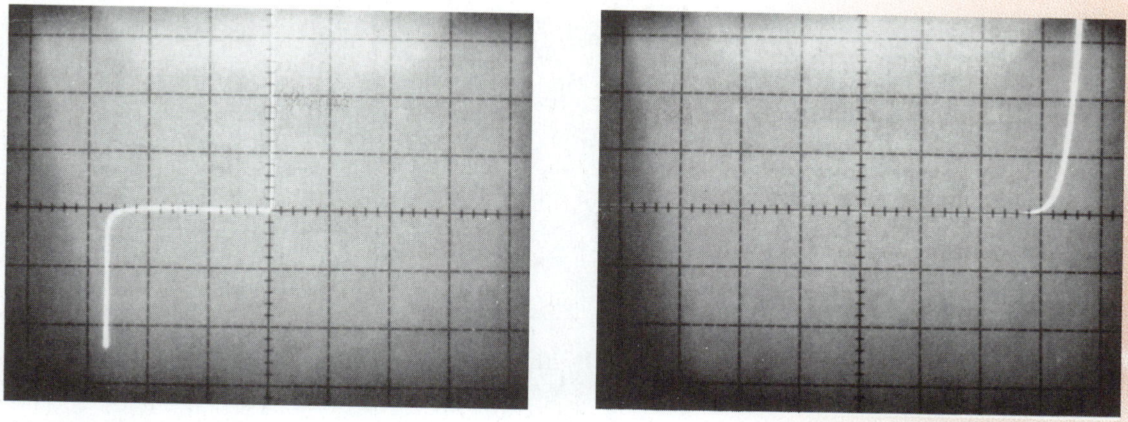

(a) Horizontal: 10 V/div (b) Horizontal: 0.2 V/div

FIGURE 4–53 Curve tracer displays of diode characteristics

Figure 4–52(b) shows collector characteristics of the same transistor when the curve tracer settings are adjusted to generate larger values of V_{CE}. In this example, the horizontal sensitivity is 2 V/division, the vertical sensitivity is 0.2 mA/division, and the base current increment is 2 μA. With these settings, the breakdown characteristics are clearly evident. For example, at $I_B = 12$ μA and $V_{CE} = 12$ V, it can be seen that the transistor is in its breakdown region and that the collector current is approximately 0.84 mA.

Most curve tracers can be used to obtain characteristic curves for devices other than transistors. Some even have special adapters that allow the testing of integrated circuits. Figure 4–53 shows photographs of diode characteristics that were obtained from a curve tracer display. The forward and reverse characteristics are shown in Figure 4–53(a), with a horizontal sensitivity of 10 V/division. It can be seen that the diode enters breakdown at a reverse voltage of about 25 V. With this scale, the forward characteristic essentially coincides with the vertical axis. However, when the sensitivity is set to 0.2 V/division, the forward characteristic appears as shown in Figure 4–53(b) and can be examined in detail. (The origin of the axes is at the center of the display.) We see that the knee of the characteristic occurs at about 0.62 V, and there is sufficient detail to compute dc and ac resistances in the forward region.

4–11 BJT CIRCUIT ANALYSIS WITH ELECTRONICS WORKBENCH MULTISIM

This section examines the use of EWB to simulate a transistor switch circuit. The transistor switch circuit is probably one of the most useful functions of a discrete BJT transistor. The transistor has plenty of drive current capability for driving most relays or even providing level conversion. For example, a signal that switches between +24 and ground can easily be converted to a +3.3/0.0 switching voltage level using a discrete BJT transistor. This section demonstrates how to simulate this circuit with EWB.

This chapter introduced the basic concepts of transistor operation. The student learned that the BJT transistor has three modes of operation: cutoff, active, and saturation. The BJT switch operates primarily in either the cutoff or saturation regions. In the cutoff region, $I_C \approx 0$ and $V_{CE} = V_{CC}$. In the saturation mode, $I_C = V_{CC}/R_C$ and $V_{CE} = 0.0$ V. You also learned that resistor values of 1 kΩ work well for R_B and R_C in most applications. The objective of this section are as follows:

FIGURE 4–54 The EWB Multisim circuit

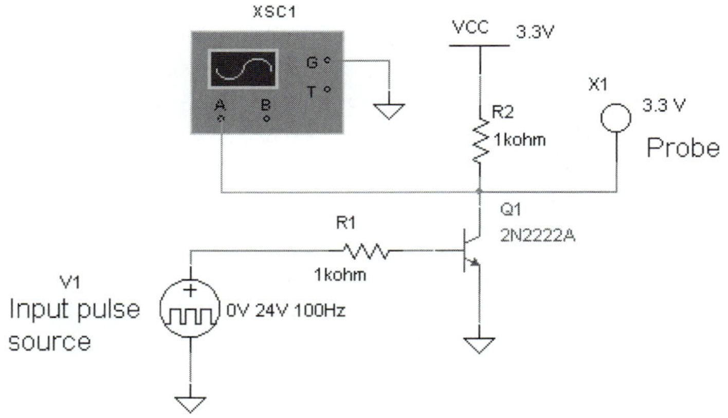

- Use EWB to construct a BJT switch circuit and verify that the circuit switches between the specified voltages.
- Develop an understanding on how to set the probe threshold voltage value and set the pulse parameters on a pulse voltage source.

Open the circuit **Ch4_EWB.msm** found in the Electronics Workbench CD-ROM packaged with the text. This file is the simple switch circuit (shown in Figure 4–54) containing a +24-V input switching signal, a 2N2222 BJT-NPN transistor, +3.3-V voltage source, two 1-kΩ resistors, and a Multisim probe (X1) that has been triggered to turn on at 3.3 volts.

The settings for the probe are obtained by double-clicking on the probe icon, selecting the value tab, and then setting the threshold voltage to 3.3 V (Figure 4–55). The menu for setting the probe values is provided in Figure 4–56.

FIGURE 4–55 The threshold voltage menu

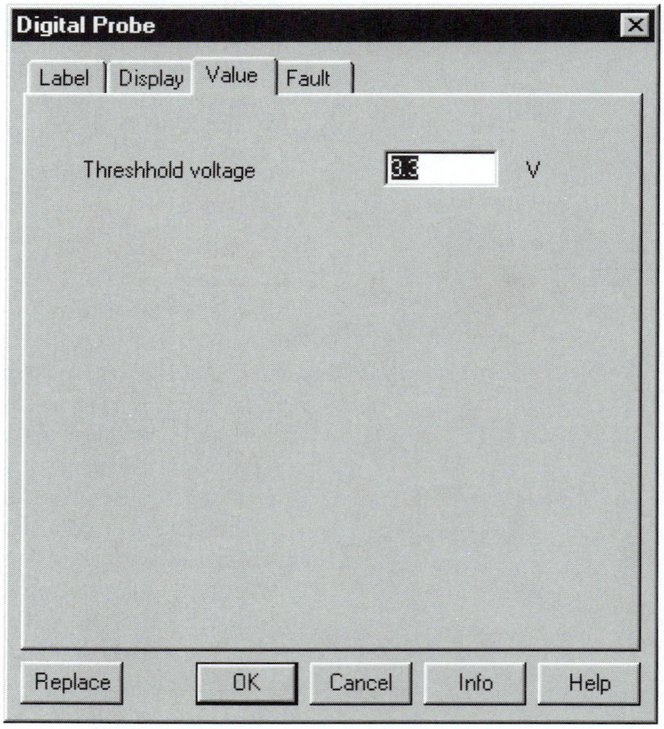

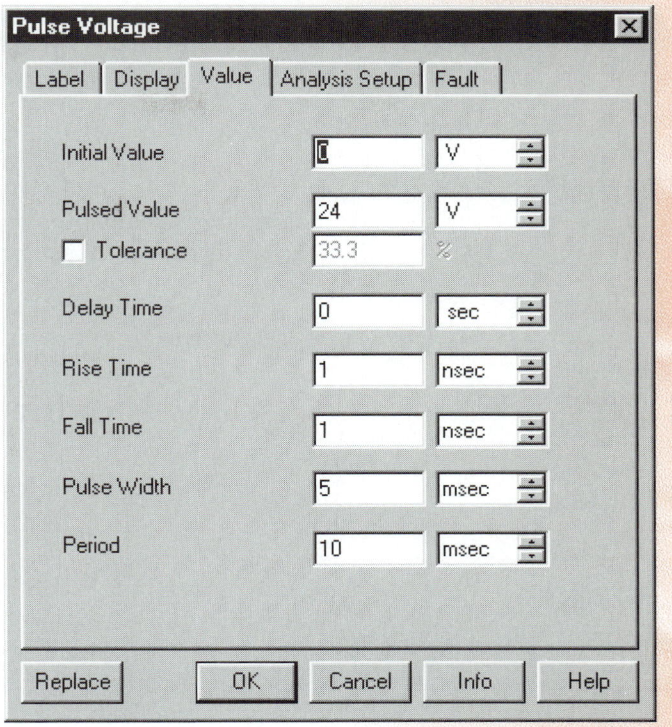

FIGURE 4–56 The menu for setting the pulse voltage values

The input pulse signal is set by double-clicking on the input pulse source. Next, click on value and set the initial and pulsed values. In this example, the initial value is 0 V and the pulsed value is 24 V. EWB also provides settings for the rise and fall time, pulse width, and period.

An oscilloscope (XSC1) is attached to the output to show that the signal is indeed switching. An example oscilloscope trace is provided in Figure 4–57.

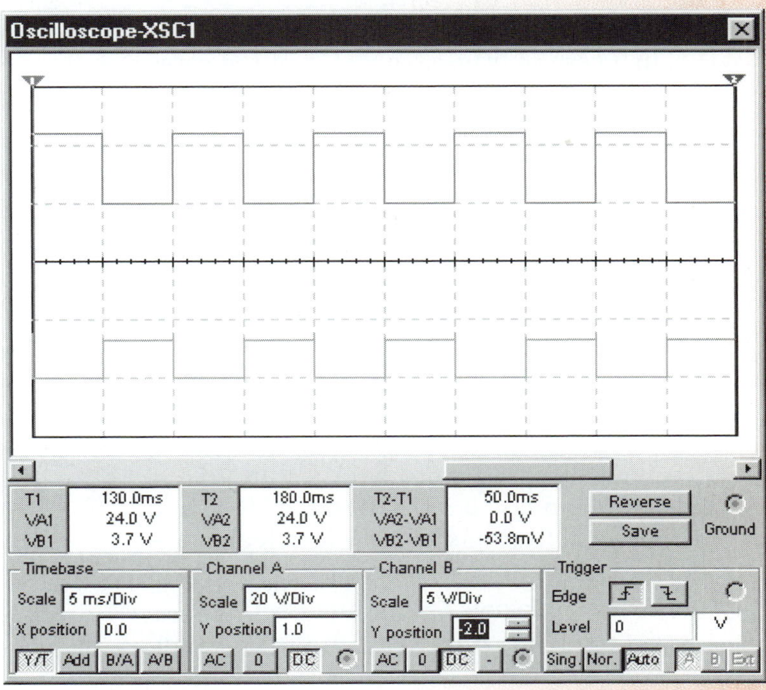

FIGURE 4–57 The oscilloscope trace for the input signal and the output of the switching circuit

Is this the expected result? The input voltage is switching between 0 and +24 V. The output is switching between 0 and 3.3 V. It appears that the design objective has been met.

SUMMARY

This chapter has presented the basics of the BJT transistor. Students should have mastered those concepts and skills:

- Basic circuit analysis techniques.
- Identifying the transistor mode of operation.
- The basics of transistor curves.
- How to use the BJT switch in a circuit.

EXERCISES

SECTION 4–2

Theory of BJT Operation

4–1. In a certain transistor, the emitter current is 1.01 times as large as the collector current. If the emitter current is 12.12 mA, find the base current.

4–2. A (conventional) current of 26 μA flows out of the base of a certain transistor. The emitter current is 0.94 mA. What is the collector current and what kind of transistor is it (*npn* or *pnp*)? Draw a transistor symbol and label all current flows, showing directions and magnitudes.

4–3. In a certain transistor, 99.5% of the carriers injected into the base cross the collector–base junction. If the leakage current is 5.0 μA and the collector current is 22 mA, find

(a) the exact α,

(b) the emitter current, and

(c) the approximate α when I_{CBO} is neglected.

4–4. A germanium transistor has a surface leakage current of 1.4 μA and a reverse current due to thermally generated minority carriers of 1.2 nA at 10°C. If $\alpha = 0.992$ and $I_E = 0.8$ mA, find I_C at 10°C and at 90°C. (Assume that surface leakage is independent of temperature.)

4–5. Using equation 4–2 and neglecting I_{CBO}, derive the following approximation: $I_B \approx (1 - \alpha)I_E$.

SECTION 4–3

Common-Base Characteristics

4–6. A certain transistor has the CB input characteristics shown in Figure 4–11. It is desired to hold I_E constant at 9.0 mA while V_{CB} is changed from 0 V to 25 V. What change in V_{BE} must accompany the change in V_{CB}?

4–7. A transistor has the CB input characteristics shown in Figure 4–11. If $\alpha = 0.95$, find I_C when $V_{BE} = 0.72$ V and $V_{CB} = 10$ V.

SECTION 4–4

Common-Emitter Characteristics

4–8. A transistor has an α of 0.98 and a collector-to-base leakage current of 0.02 μA.

(a) Find its collector-to-emitter leakage current.

(b) Find the β of the transistor.

(c) Find I_C when $I_B = 0.04$ mA.

(d) Find the approximate I_C, neglecting leakage current.

4–9. A transistor has $I_{CBO} = 0.1$ μA and $I_{CEO} = 16$ μA. Find its α.

4–10. Derive the relation $\alpha = \beta/(\beta + 1)$. (*Hint:* Solve equation 4–8 for α.)

4–11. Under what condition is the approximation $I_{CEO} \approx \beta\, I_{CBO}$ valid?

4–12. A transistor has the CE output characteristics shown in Figure 4–20.

(a) Find the emitter current at $V_{CE} = 5$ V and $I_B = 50$ μA.

(b) Find the α at that point (neglecting leakage current).

4–13. An *npn* transistor has the CE input characteristics shown in Figure 4–19 and the CE output characteristics shown in Figure 4–20.

(a) Find I_B when $V_{BE} = 0.7$ V and $V_{CE} = 20$ V.

(b) Find the β of the transistor at $V_{CE} = 6.0$ V and $I_B + 20\mu$A (neglecting leakage current).

4–14. Using graphical methods, determine the approximate value of the Early voltage for the transistor whose CE output characteristics are shown in Figure 4–58.

4–15. In a certain experiment, the collector current of a transistor was measured at different values of collector-emitter voltage, while the base current was held constant. The results of the experiment are summarized in the following table:

$I_B = 100\ \mu$A	
V_{CE}	I_C
5 V	15 mA
10 V	16 mA
15 V	17 mA
20 V	18 mA

$I_B = 200\ \mu$A	
V_{CE}	I_C
5 V	30 mA
10 V	32 mA
15 V	34 mA
20 V	36 mA

$I_B = 300\ \mu$A	
V_{CE}	I_C
5 V	45 mA
10 V	48 mA
15 V	51 mA
20 V	54 mA

Plot the experimental data and graphically determine approximate values for the following:

(a) β at $V_{CE} = 8$ V and $I_B = 100\ \mu$A,

(b) β at $V_{CE} = 14$ V and $I_B = 250\ \mu$A, and

(c) the Early voltage.

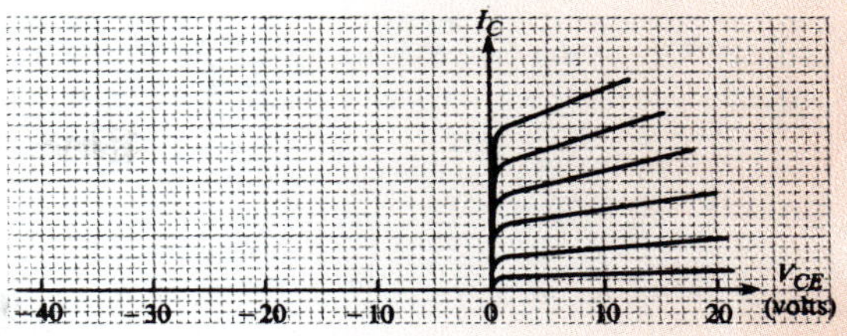

FIGURE 4–58 (Exercise 4–14)

SECTION 4–5

Common-Collector Characteristics

4–16. A certain transistor has the common-collector output characteristics shown in Figure 4–27. Neglecting leakage current, find approximate values for

(a) β at $V_{CE} = 12.5$ V and $I_B = 20\ \mu$A,

(b) I_E at $V_{CE} = 12.5$ V and $I_B = 45\ \mu$A, and

(c) α at $V_{CE} = 2.5$ V and $I_B = 70\ \mu$A.

4–17. Prove that equation 4–16 is equivalent to $I_E = I_B/(1 - \alpha)$.

SECTION 4–6

Bias Circuits

4–18. Determine the equation for the load line of the circuit shown in Figure 4–59. Sketch the line and label the values of its intercepts.

4–19. In the circuit of Figure 4–59 find

(a) I_C when $V_{CB} = 10$ V, and

(b) V_{CB} when $I_C = 1$ mA.

4–20. In the circuit shown in Figure 4–60, find

(a) I_C when $V_{BC} = 20$ V, and

(b) V_{BC} when $I_C = 4.2$ mA.

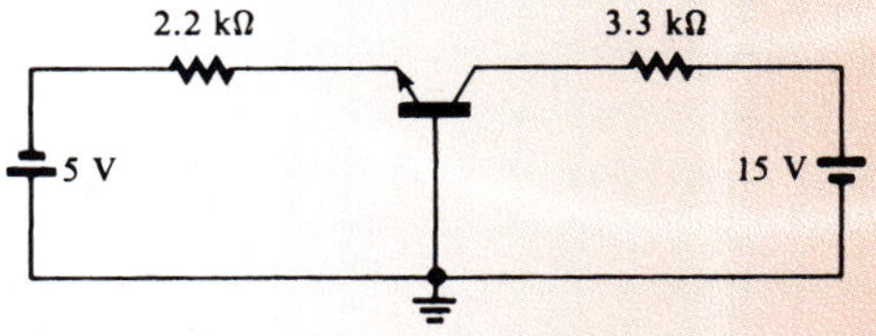

FIGURE 4–59 (Exercise 4–18)

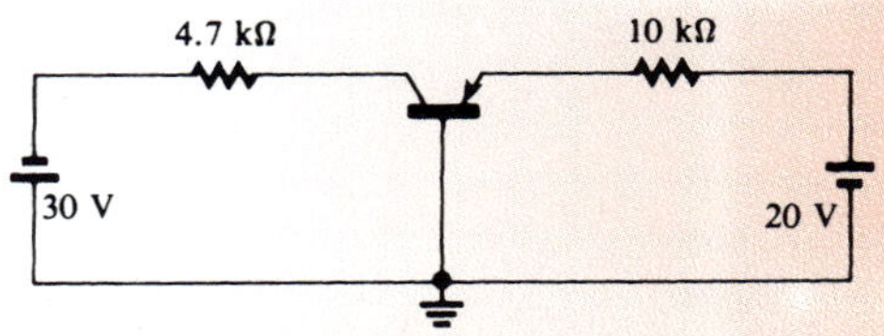

FIGURE 4–60 (Exercise 4–20)

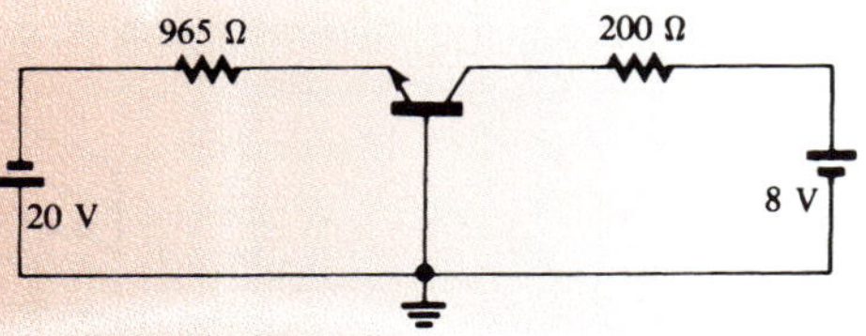

FIGURE 4–61 (Exercise 4–21)

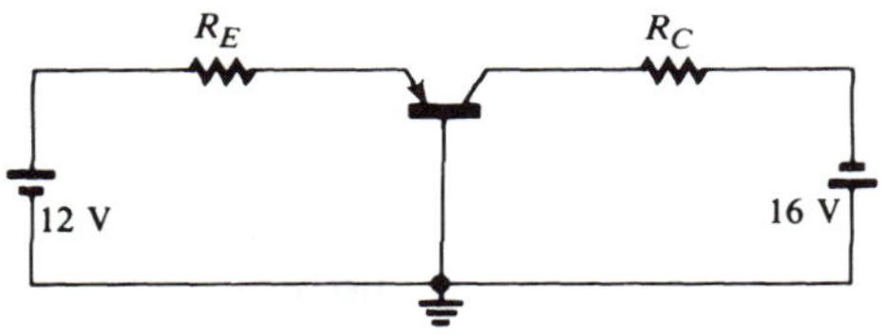

FIGURE 4–63 (Exercise 4–22)

4–21. The silicon transistor shown in Figure 4–61 has the CB output characteristics shown in Figure 4–62

(a) Draw the load line on the characteristics and graphically determine V_{CB} and I_C at the bias point.

(b) Determine the bias point without using the characteristic curves.

4–22. The transistor shown in Figure 4–63 is germanium.

(a) If $R_C = 1\ \text{k}\Omega$, what value of R_E will cause V_{BC} to equal 0 V?

(b) If $R_E = 1.5\ \text{k}\Omega$, what value of R_C will cause V_{BC} to equal 0 V?

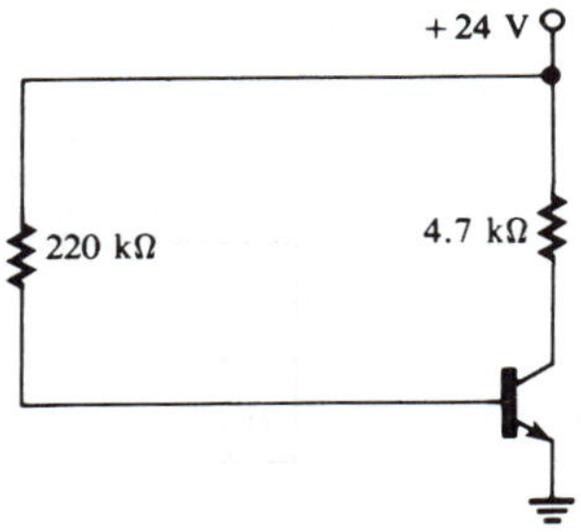

FIGURE 4–64 (Exercise 4–23)

4–23. In the circuit shown in Figure 4–64, find

(a) V_{CE} when $I_C = 1.5$ mA,

(b) I_C when $V_{CE} = 12$ V, and

(c) V_{CE} when $I_C = 0$.

4–24. In the circuit shown in Figure 4–65, find

(a) V_{EC} when $I_C = 12$ mA,

(b) I_C when $V_{EC} = 2.5$ V, and

(c) V_{EC} when $I_C = 0$.

4–25. The silicon transistor shown in Figure 4–66 has the CE output characteristics shown in Figure 4–67. Assume that $\beta = 105$.

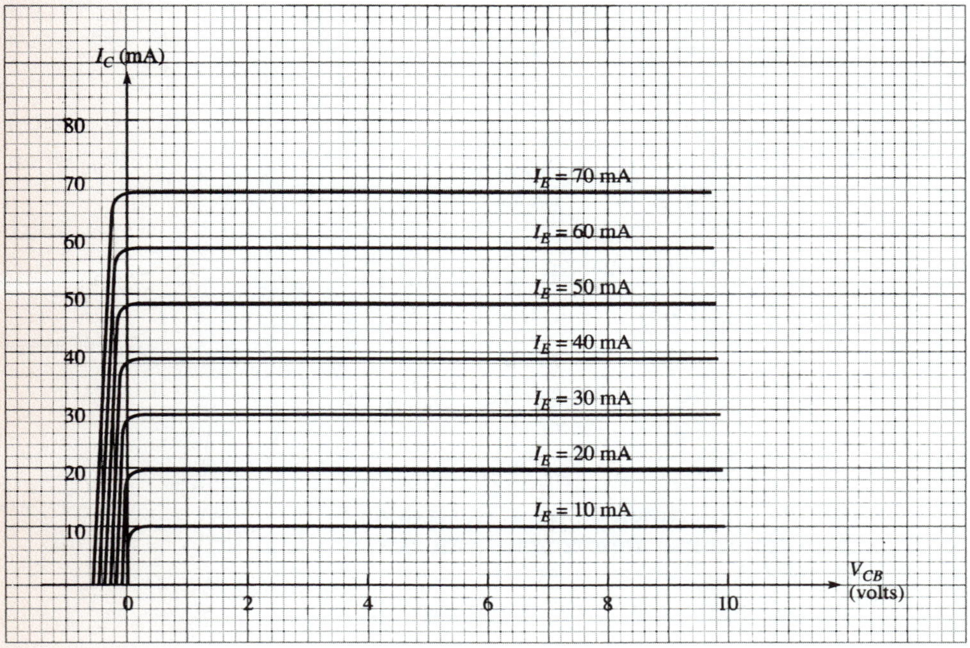

FIGURE 4–62 (Exercise 4–21)

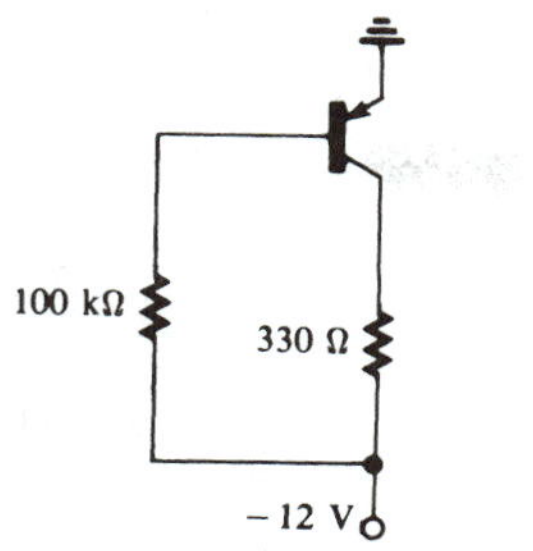

FIGURE 4–65 (Exercise 4–24)

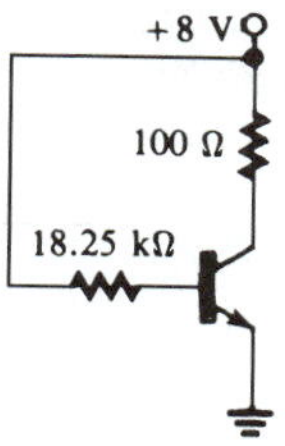

FIGURE 4–66 (Exercise 4–25)

(a) Draw the load line on the characteristics and graphically determine V_{CE} and I_C at the bias point.

(b) What is the approximate value of I_{CEO} for this transistor?

(c) Calculate V_{CE} and I_C at the bias point without using the characteristic curves.

4–26. Assuming that $\beta = 150$, find the bias point of the germanium transistor shown in Figure 4–68.

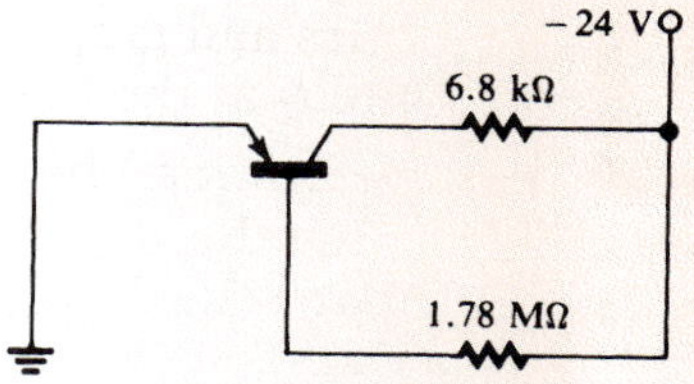

FIGURE 4–68 (Exercise 4–26)

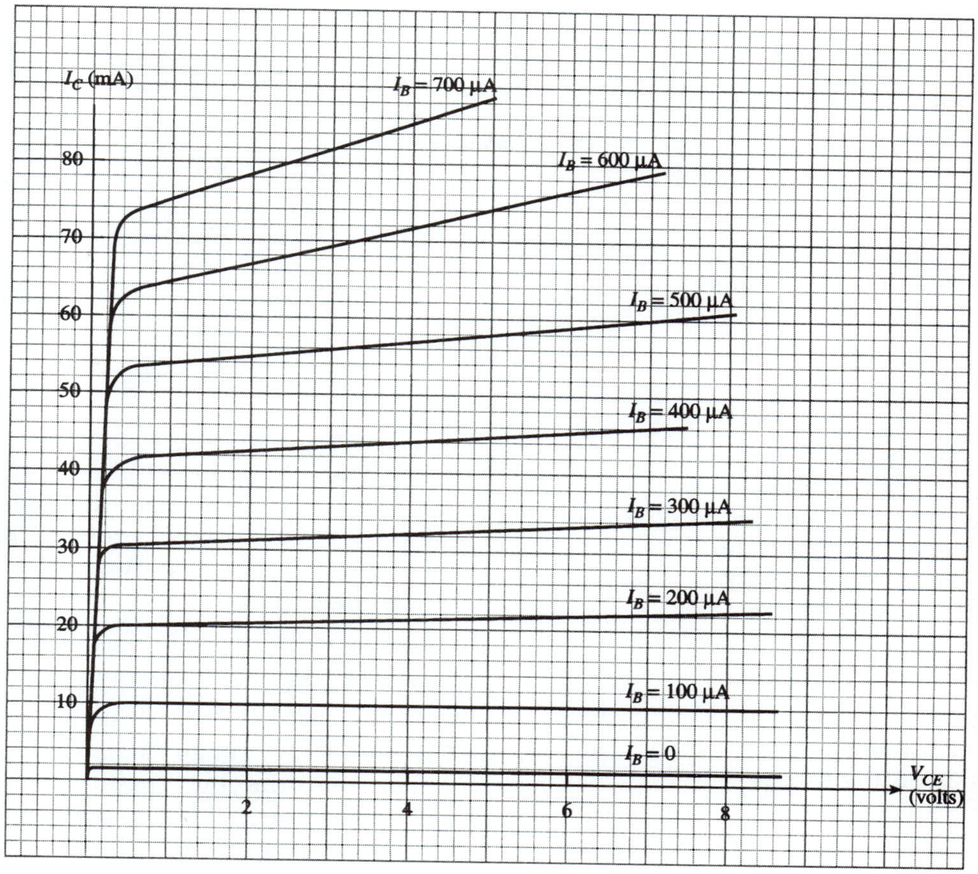

FIGURE 4–67 (Exercise 4–25)

4–27. What value of R_B in the circuit shown in Figure 4–69 will just cause the silicon transistor to be saturated, assuming that $\beta = 100$ and $V_{CE(sat)} = 0.3$ V?

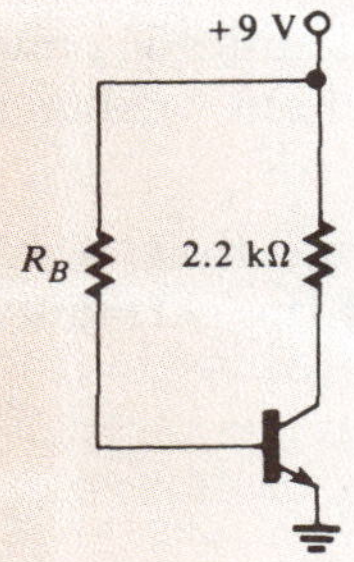

FIGURE 4–69 (Exercise 4–27)

4–28. Calculate the bias point of the silicon transistor shown in Figure 4–70. Assume that $\beta = 80$.

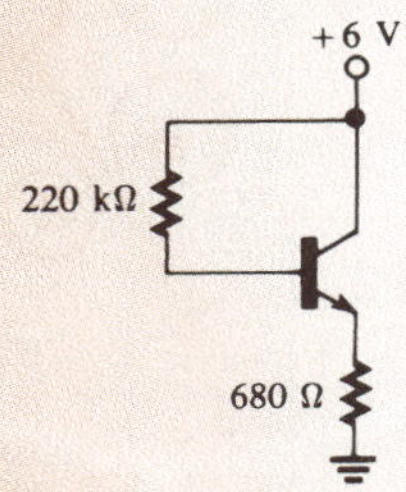

FIGURE 4–70 (Exercise 4–28)

4–29. Calculate the bias point of the silicon transistor shown in Figure 4–71. Assume that $\beta = 100$.

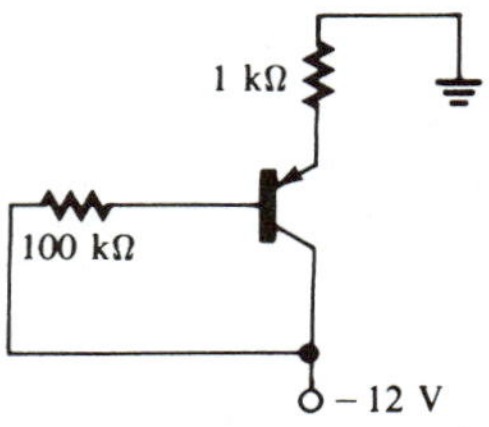

FIGURE 4–71 (Exercise 4–29)

4–30. Determine the bias configuration (CB, CE, or CC) of the transistor in each part of Figure 4–72.

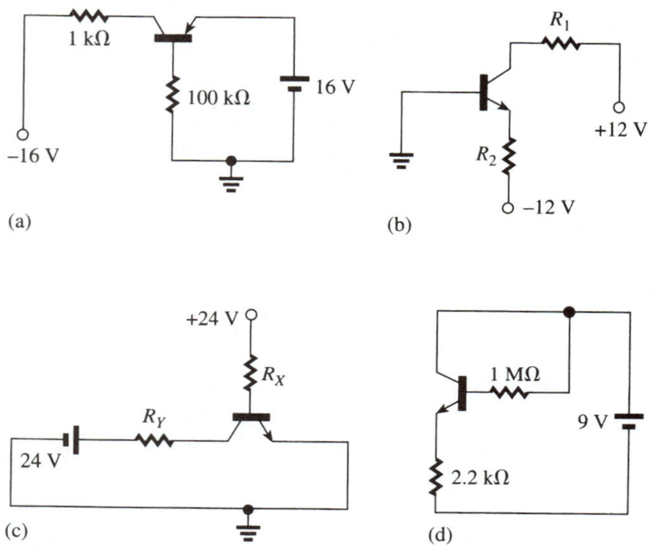

FIGURE 4–72 (Exercise 4–30)

DESIGN EXERCISES

4–31. (a) Design a bias circuit for an *npn* silicon transistor in a common-base configuration. The bias point should be $I_E = 2$ mA and $V_{CB} = 9$ V. Supply voltages are +20 V and −10 V. Use standard-valued resistors with 5% tolerance and draw a schematic diagram of your design.

(b) Calculate the range of possible values that the bias point could have, taking the resistor tolerances into consideration.

4–32. (a) Design a bias circuit for a *pnp* silicon transistor in a common-emitter configuration. The nominal β of the transistor is 80 and the supply voltage is −24 V. The bias point is to be $I_C = 5$ mA, and $V_{CE} = -10$ V. Use standard-valued resistors with 10% tolerance and draw a schematic diagram of your design.

(b) Calculate the actual bias point assuming the 10% resistors have their nominal values.

(c) Calculate the range of possible values that the bias point could have if the value of β changed over the range from 50 to 100. Assume the resistors have their nominal values.

4–33. (a) Design a bias circuit for an *npn* silicon transistor in a common-collector configuration. The nominal β for the transistor is 100, and the supply voltage is 30 V. The bias point is to be I_C = 10 mA, and V_{CE} = 12 V. Use standard-valued resistors having 5% tolerance and draw a schematic diagram of your design.

(b) Calculate the *minimum* value that V_{CE} could have if *both* the resistor tolerances and a variation in β from 60 to 120 are taken into account. *Hint*: Use equations 4–28 to derive the expression

$$V_{CE} = V_{CC} - \frac{(V_{CC} - V_{BE})}{\dfrac{R_B}{(\beta + 1)R_E} + 1}$$

4–34. (a) Design a bias circuit for an *npn* silicon transistor having a nominal β of 100, to be used in a common-emitter configuration. The bias point is to be I_C = 1 mA, and V_{CE} = 5 V. The supply voltage is 15 V. Use standard-valued 5% resistors and draw a schematic diagram of your design.

(b) Calculate the possible range of values of the bias point taking into consideration *both* the resistor tolerances and a possible variation in β from 30 to 150. Interpret and comment on your results.

SECTION 4–8

The BJT Inverter (Transistor Switch)

4–35. The input to the circuit shown in Figure 4–73 alternates between 0 V and 10 V. If the silicon transistor has a β of 120, verify that the circuit operates as an inverter.

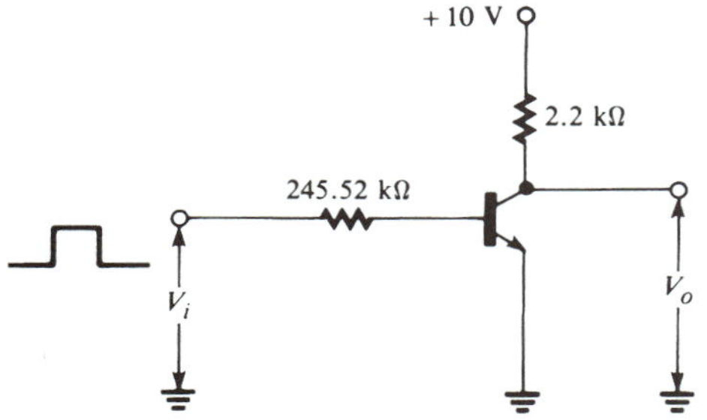

FIGURE 4–73 (Exercise 4–35)

4–36. What would be the output voltages from the inverter in Figure 4–73 if

(a) the input voltage levels were changed to −5 V and +10 V?

(b) the input voltage levels were changed to 0 V and +15 V?

(c) the β of the transistor were changed to 150?

4–37. A transistor inverter is to be designed using a silicon transistor whose β may vary from 60 to 120. If the series base resistance is to be 100 kΩ, what should be the value of R_C? Assume that $V_{CC} = V_{HI}$ = 4.5 V.

4–38. What is the minimum value of β for which the silicon transistor in Figure 4–74 will operate satisfactorily as an inverter?

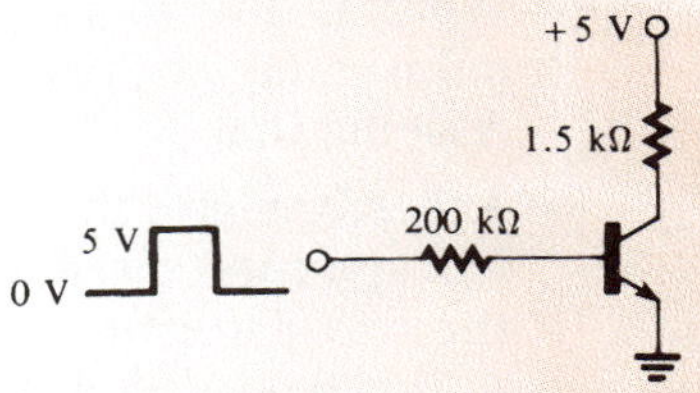

FIGURE 4–74 (Exercise 4–38)

SECTION 4–9

Transistor Types, Ratings, and Specifications

Note: In Exercises 4–39 through 4–40, refer to the manufacturer's specification sheets given in Section 4–9.

4–39. The input to the 2N2222,A, transistor in Figure 4–75 is a square wave that alternates between ±*V* volts. Assuming that no base current flows when the input is at −*V* volts (so there is no drop across R_B), what is the maximum safe value for *V*? Assume that a safe value for *V* is one that does not exceed 80% of the manufacturer's rated maximum.

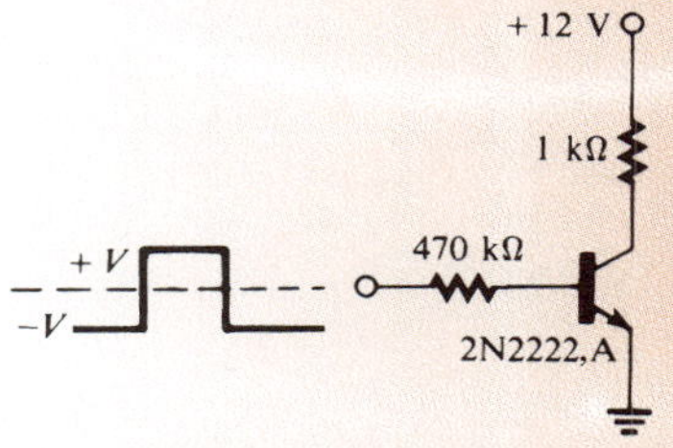

FIGURE 4–75 (Exercise 4–39)

4–40. A 2N2221, A, transistor is to be operated at 75°C. What is its maximum-rated power dissipation at that temperature?

4–41. The β of a 2N2218, A, transistor is 40 at $T = 25°C$, $V_{CE} = 10$ V, and $I_C = 3$ mA. What typical value for β could be expected at

(a) $T = -55°C$, $I_C = 5$ mA, and $V_{CE} = 10$ V?

(b) $T = 175°C$, $I_C = 80$ mA, and $V_{CE} = 10$ V?

4–42. The β of a 2N2219, A, transistor is 140 at $T = 175°C$, $V_{CE} = 10$ V, and $I_C = 80$ mA. What typical value for β could be expected at

(a) $T = 25°C$, $I_C = 3$ mA, and $V_{CE} = 10$ V?

(b) $T = -55°C$, $I_C = 5$ mA, and $V_{CE} = 10$ V?

4–43. What is the manufacturer's rated maximum value for the current I in Figure 4–76. Assume that $T = 25°C$.

4–44. A certain circuit is designed so that it will operate satisfactorily at $T = 25°C$ if the transistor has a β of at least 50 when $I_C = 10$ mA and $V_{CE} = 10$ V. Can the 2N2222, A series transistor be used for the application?

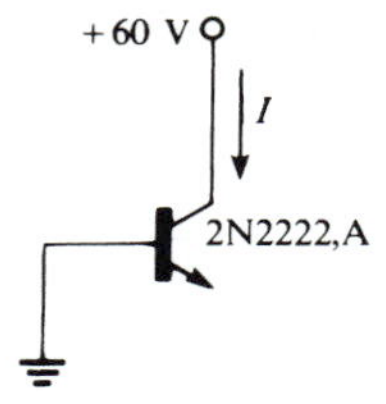

FIGURE 4–76 (Exercise 4–43)

SECTION 4–10

Transistor Curve Tracers

4–45. Using the curve tracer display of the collector characteristics shown in Figure 4–52(b), find the approximate β of the transistor when

(a) $V_{CE} = 7$ V and $I_B = 8\ \mu$A, and

(b) $V_{CE} = 8$ V and $I_B = 12\ \mu$A.

4–46. Using the curve tracer display of the diode characteristic shown in Figure 4–53, find the approximate dc resistance of the diode when it is forward biased by 0.64 V.

SPICE EXERCISES

Note: In the exercises that follow, assume all device parameters have their default values unless otherwise specified.

4–47. Use SPICE to obtain a set of output characteristics for an *npn* transistor in the CE configuration. The ideal maximum β is 150 and the Early voltage is 180 V. The characteristics should be plotted for V_{CE} ranging from 0 V to 12 V in 0.1 V steps and for I_B ranging from 0 to 50 μA in 10-μA steps.

4–48. Use SPICE to determine the bias point for the circuit shown in Figure 4–43 (Example 4–12). Set the ideal maximum forward β(BF) to 120 and then repeat the analysis with BF = 240. Comment on the effect the change in β has on the bias point. Use a silicon transistor.

4–49. Use SPICE to simulate the common-emitter circuit shown in Figure 4–37(a) (Example 4–10). Compare the change in the bias point when the temperature is changed from 27°C to 0°C.

4–50. Use SPICE to simulate a BJT inverter that is being driven by a 100-Hz TTL 5-V clock. Let the β of the transistor equal 100, $R_C = 1\ k\Omega$, $R_S = 1\ k\Omega$. Provide a plot of the output.

CHAPTER 5

FIELD-EFFECT TRANSISTORS

OUTLINE

OBJECTIVES

- Understand the basic operation of a JFET transistor.
- Explore the basic dc bias configurations for the Junction Field Transistor.
- Understand the modes of operation of the JFET transistor.
- Apply the JFET transfer characteristic equation in circuit analysis.
- Use the JFET transistor as an analog switch.
- Understand the basic operation of the MOSFET transistor.
- Use the JFET transistor in computer simulation.

5–1 INTRODUCTION

The field-effect transistor (FET), like the bipolar junction transistor, is a three-terminal semiconductor device. However, the FET operates under principles completely different from those of the BJT. A field-effect transistor is called a *unipolar* device because the current through it results from the flow of only one of the two kinds of charge carriers: holes or electrons. The name *field effect* is derived from the fact that the current flow is controlled by an electric field set up in the device by an externally applied voltage.

There are two main types of FETs: the junction field-effect transistor (JFET) and the metal-oxide-semiconductor FET (MOSFET). We will study the theory and some practical applications of each. Both types are fabricated as discrete components and as components of integrated circuits. The MOSFET is the most important component in modern digital integrated circuits, such as microprocessors and computer memories.

5–2 JUNCTION FIELD-EFFECT TRANSISTORS

Figure 5–1 shows a diagram of the structure of a JFET and identifies the three terminals to which external electrical connections are made. As shown in the figure, a bar of n-type material has regions of p material embedded in each side. The two p regions are joined electrically, and the common connection between them is called the *gate* (G) terminal. A terminal at one end of the n-type bar is called the *drain* (D), and a terminal at the other end is called the *source* (S). The region of n material between the two opposing p regions is called the *channel*. The transistor shown in the figure is therefore called an *n-channel* JFET, the type that we will study initially, and a device constructed from a p-type bar with embedded n regions is called a *p-channel* JFET. As we develop the theory of the JFET, it may be helpful at first to think of the drain as corresponding to the collector of a BJT, the source as corresponding to the emitter, and the gate as corresponding to the base. As we shall see, the voltage applied to the gate controls the flow of current between drain and source, just as the signal applied to the base of a BJT controls the flow of current between collector and emitter.

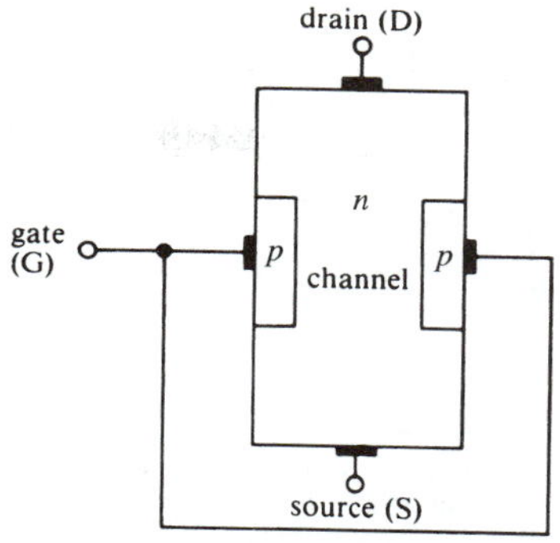

FIGURE 5–1 Structure of an n-channel JFET

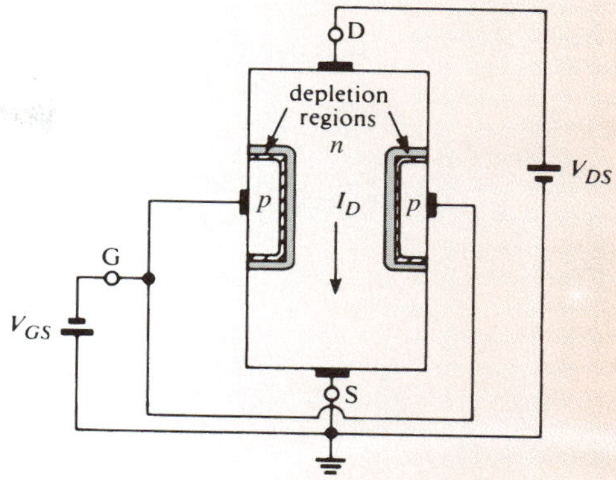

FIGURE 5–2 Reverse biasing the gate-to-source junctions causes the formation of depletion regions. V_{GS} is a small reverse-biasing voltage for the case illustrated.

When an external voltage is connected between the drain and the source of an n-channel JFET, so that the drain is positive with respect to the source, current is established by the flow of electrons through the n material from the source to the drain. (The source is so named because it is regarded as the origin of the electrons.) Thus, *conventional* current flows from drain to source and is limited by the resistance of the n material. In normal operation, an external voltage is applied between the gate and the source so that the pn junctions on each side of the channel are reverse biased. Thus, the gate is made negative with respect to the source, as illustrated in Figure 5–2. Note in the figure that the reverse bias causes a pair of depletion regions to form in the channel. The channel is more lightly doped than the gate, so the depletion regions penetrate more deeply into the n-type channel than into the p material of the gate.

The width of the depletion regions in Figure 5–2 depends on the magnitude of the reverse-biasing voltage V_{GS}. The figure illustrates the case where V_{GS} is only a few tenths of a volt, so the depletion regions are relatively narrow. (V_{DS} is also assumed to be relatively small; we will investigate the effect of a large V_{DS} presently.) As V_{GS} is made more negative, the depletion regions expand and the width of the channel decreases. The reduction in channel width increases the resistance of the channel and thus decreases the flow of current I_D from drain to source.

To investigate the effect of increasing V_{DS} on the drain current I_D, let us suppose for the moment that the gate is shorted to the source ($V_{GS} = 0$). As V_{DS} is increased slightly above 0, we find that the current I_D increases in direct proportion to it, as shown in Figure 5–3(a). This is as we would expect, because increasing the voltage across the fixed-resistance channel simply causes an Ohm's law increase in the current through it. As we continue to increase V_{DS}, we find that noticeable depletion regions begin to form in the channel, as illustrated in Figure 5–3(b). Note that the depletion regions are broader near the drain end of the channel (in the vicinity of point A) than they are near the source end (point B). This is explained by the fact that current flowing through the channel creates a voltage drop along the length of the channel. Near the top of the channel, the channel voltage is very nearly equal to V_{DS}, so there is a large reverse-biasing voltage between the n channel and the p gate. As we proceed down the channel, less voltage is available because of the drop that accumulates through the resistive n material. Consequently, the reverse-biasing potential between channel and gate

FIGURE 5–3 Effects of increasing V_{DS} while the gate is shorted to the source ($V_{GS} = 0$)

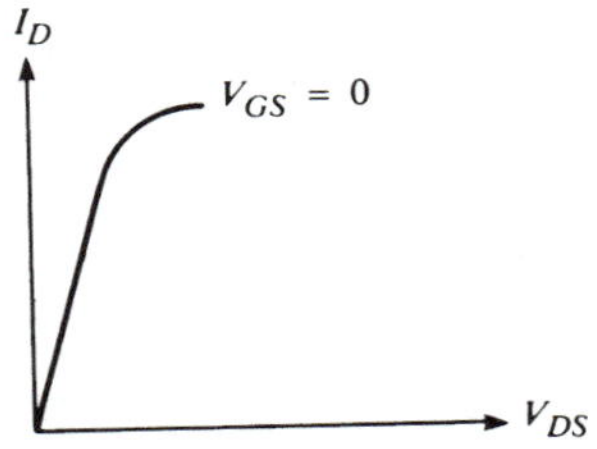

(a) The drain current rises linearly with V_{DS} until significant channel narrowing causes it to level off

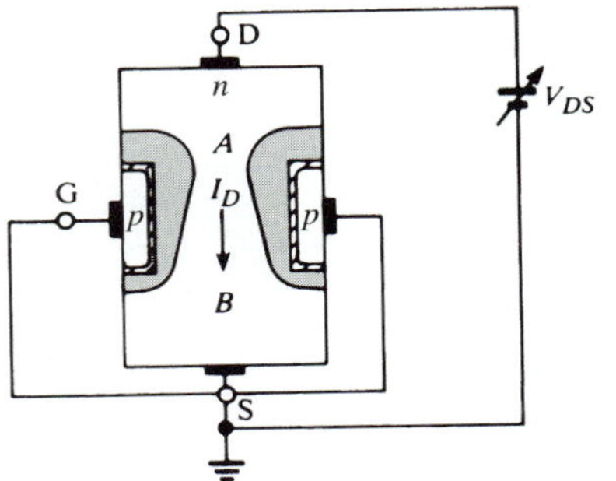

(b) Increasing V_{DS} creates depletion regions that narrow the channel width near the drain (point *A*)

becomes smaller and the depletion regions become narrower as we approach the source. When V_{DS} is increased further, the depletion regions expand and the channel becomes very narrow in the vicinity of point A, causing the total resistance of the channel to increase. As a consequence, the rise in current is no longer directly proportional to V_{DS}. Instead, the current begins to level off, as shown by the curved portion of the plot in Figure 5–3(a).

Figure 5–4(a) shows what happens when V_{DS} is increased to a value large enough to cause the depletion regions to meet at a point in the channel near the drain end. This condition is called *pinch-off*. At the point where pinch-off occurs, the gate-to-channel junction is reverse biased by the value of V_{DS}, so (the negative of) this value is called the *pinch-off voltage*, V_p. The pinch-off voltage is an important JFET parameter, whose value depends on the doping and geometry of the device. V_p is always a negative quantity for an *n*-channel JFET and a positive quantity for a *p*-channel JFET. Figure 5–4(b) shows that the current reaches a maximum value at pinch-off and that it remains at that value as V_{DS} is increased beyond $|V_p|$. This current is called the *saturation* current and is designated I_{DSS}—the *D*rain-to-*S*ource current with the gate *S*horted.

Despite the implication of the name *pinch-off*, note again that current continues to flow through the device when V_{DS} exceeds $|V_p|$. The value of the current remains constant at I_{DSS} because of a kind of self-regulating or equilibrium process that controls the current when V_{DS} exceeds $|V_p|$: Suppose that an increase in V_{DS} *did* cause I_D to increase; then there would be in the channel an increased voltage drop that would expand the depletion regions further and reduce the current to its original value. Conversely, if current ceased to flow at pinch-off, the depletion region would shrink and current flow would resume. Of course, this change in current never actually occurs: I_D simply remains constant at I_{DSS}.

FIGURE 5–4 The *n*-channel JFET at pinch-off

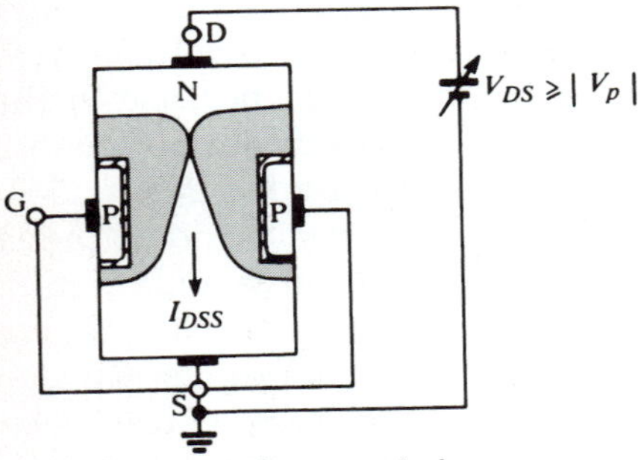

(a) When V_{DS} equals or exceeds the pinch-off voltage $| V_p |$, the depletion regions meet in the channel

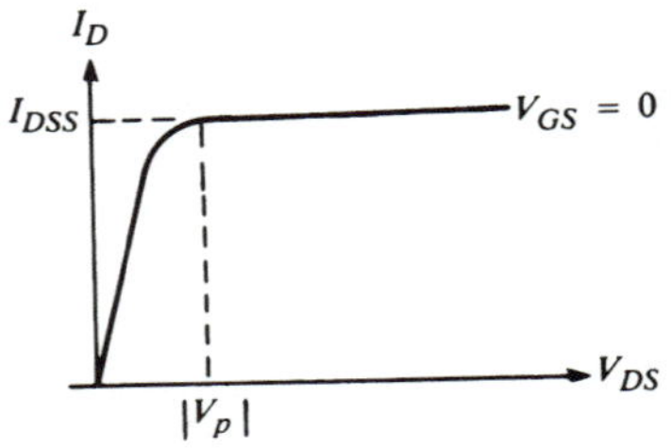

(b) The drain current remains at its saturation value I_{DSS} as V_{DS} is increased beyond pinch-off

A typical set of values for V_p and I_{DSS} are -4 V and 12 mA, respectively. Suppose we connect a JFET having those parameter values in the circuit shown in Figure 5–5(a). Note that the gate is no longer shorted to the source, but a voltage $V_{GS} = -1$ V is connected to reverse bias the gate-to-source junctions. The reverse bias causes the depletion regions to penetrate the channel farther along the entire length of the channel than they did when V_{GS} was 0. If we now begin to increase V_{DS} above 0, we find that the current I_D once more begins to increase linearly, as shown in Figure 5–5(b). Note that the slope of this line is not as steep as that of the $V_{GS} = 0$ line because the total resistance of the narrower channel is greater than before. As we continue to increase V_{DS}, we find that the depletion regions again approach each other in the vicinity of the drain. This further narrowing of the channel increases its resistance, and the current again begins to level off. Because there is already a 1–V reverse bias between the gate and the channel, the pinch-off condition, where the depletion regions meet, is now reached at $V_{DS} = 3$ V instead of 4 V ($V_{DS} = V_{GS} - V_p$). As shown in Figure 5–5(b), the current saturates at the lower value of 6.75 mA as V_{DS} is increased beyond 3 V.

If the procedure we have just described is repeated with V_{GS} set to -2 V instead of -1 V, we find that pinch-off is reached at $V_{DS} = 2$ V and that the current saturates at $I_D = 3$ mA. It is clear that increasing the reverse-biasing value of V_{GS} (making V_{GS} more negative) causes the pinch-off condition to occur at smaller values of V_{DS} and that smaller saturation currents result. Figure 5–6 shows the family of characteristic curves, the *drain characteristics*, obtained when the procedure is performed for $V_{GS} = 0, -1, -2, -3$, and -4 V. The dashed line, which is parabolic, joins the points on each curve where pinch-off occurs. A value of V_{DS} on the parabola is called a *saturation voltage* $V_{DS(sat)}$. At any value of V_{GS}, the corresponding value of $V_{DS(sat)}$ is the difference between V_{GS} and V_p: $V_{DS(sat)} = V_{GS} - V_p$, as we have already described. The equation of the parabola is

$$I_D = I_{DSS}\left(\frac{V_{DS(sat)}}{V_p}\right)^2 \tag{5–1}$$

To illustrate, we have, in our example, $V_p = -4$ V and $I_{DSS} = 12$ mA, so at $V_{DS(sat)} = 3$ V we find

$$I_D = (12\text{ mA})\left(\frac{3}{-4}\right)^2 = 6.74\text{ mA}$$

which is the saturation current at the $V_{GS} = -1$ V line (see Figure 5–5(b)). Note in Figure 5–6 that the region to the right of the parabola is called the *pinch-off region*. This is the region in which the JFET is normally operated

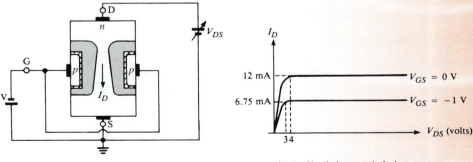

FIGURE 5–5 Effects of increasing V_{DS} when $V_{GS} = -1$ V

(a) A reverse-biasing voltage $V_{GS} = -1$ V creates, along the length of the channel, a depletion region that is wider than when $V_{GS} = 0$ V

(b) As V_{DS} is increased, I_D increases linearly until pinch-off occurs at $V_{DS} = 3$ V

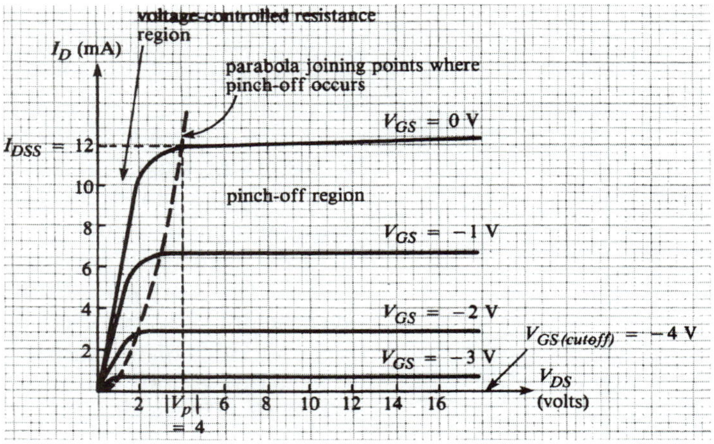

FIGURE 5–6 Drain characteristics of an *n*-channel JFET

when used for small-signal amplification. It is also called the *active* region, or the *saturation* region. The region to the left of the parabola is called the *voltage-controlled–resistance* region, the *ohmic* region, or the *triode* region. In this region, the resistance between drain and source is controlled by V_{GS}, as we have previously discussed, and we can see that the lines become less steep (implying larger resistance) as V_{GS} becomes more negative. The device acts like a voltage-controlled resistor in this region, and there are some practical applications that exploit this characteristic.

The line drawn along the horizontal axis in Figure 5–6 shows that $I_D = 0$ when $V_{GS} = -4$ V, regardless of the value of V_{DS}. When V_{GS} reverse biases the gate-to-source junction by an amount equal to V_p, depletion regions meet along the entire length of the channel and the drain current is cut off. Because the value of V_{GS} at which the drain current is cut off is the same as V_p, the pinch-off voltage is also called the *gate-to-source cutoff voltage*. Thus, there are two ways to determine the value of V_p from a set of drain characteristics: It is the value of V_{DS} where I_D saturates when $V_{GS} = 0$, and it is the value of V_{GS} that causes all drain current to cease, i.e., $V_p = V_{GS(cutoff)}$.

One property of a field-effect transistor that makes it especially valuable as a voltage amplifier is the very high input resistance at its gate. Because the path from gate to source is a reverse-biased *pn* junction, the only current that flows into the gate is the very small leakage current associated with a reverse-biased junction. Therefore, very little current is drawn from a signal source driving the gate, and the FET input looks like a very large resistance. A dc input resistance of several hundred megohms is not unusual. Although the gate of an *n*-channel JFET can be driven slightly positive, this action causes the input junction to be forward biased and radically decreases the gate-to-source resistance. In most practical applications, the sudden and dramatic decrease in resistance when the gate is made positive would not be tolerable to a signal source driving a FET.

Figure 5–7 shows the structure and drain characteristics of a typical *p*-channel JFET. Since the channel is *p* material, current is due to hole flow, rather than electron flow, between drain and source. The gate material is, of course, *n*-type. Note that all voltage polarities are opposite those in the *n*-channel JFET. Figure 5–7(b) shows that positive values of V_{GS} control the amount of saturation current in the pinch-off region.

Figure 5–8 shows the schematic symbols used to represent *n*-channel and *p*-channel JFETs. Note that the arrowhead on the gate points into an

FIGURE 5–7 Structure and characteristics of a *p*-channel JFET

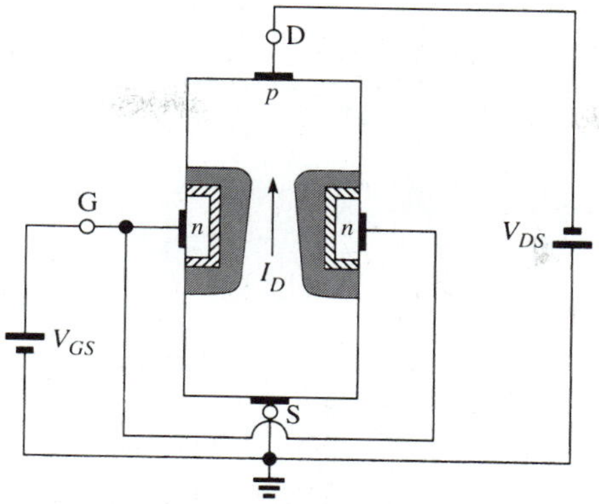

(a) The structure of a *p*-channel JFET

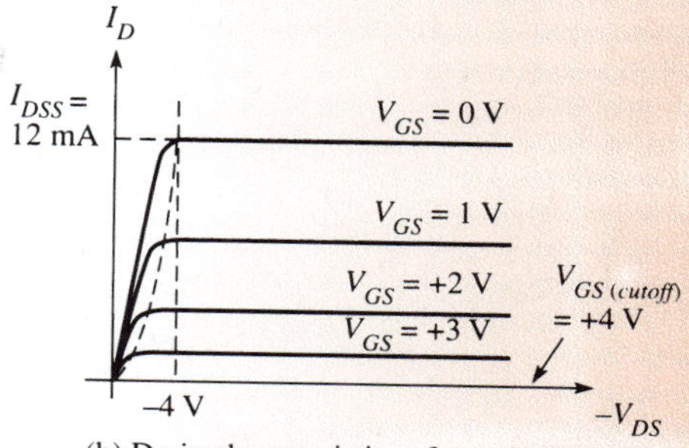

(b) Drain characteristics of a *p*-channel JFET. (Note that values of V_{DS} are negative and increase negatively to the right.)

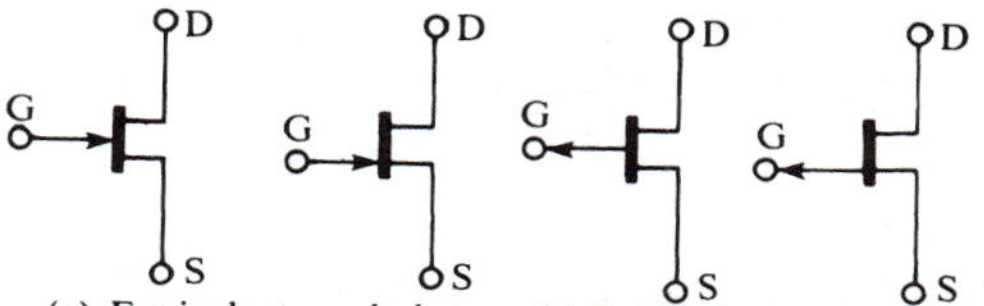

(a) Equivalent symbols for an *n*-channel JFET (b) Equivalent symbols for a *p*-channel JFET

FIGURE 5–8 Schematic symbols for JFETs

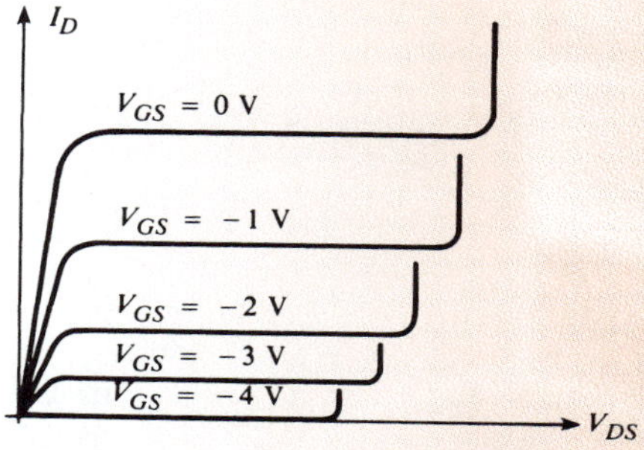

FIGURE 5–9 Breakdown characteristics of an *n*-channel JFET

n-channel JFET and outward for a *p*-channel device. The symbols showing the gate terminal off-center are used as a means of identifying the source: The source is the terminal drawn closest to the gate arrow. Some JFETs are manufactured so that the drain and source are interchangeable, and the symbols for these devices have the gate arrow drawn in the center.

Figure 5–9 shows the breakdown characteristics of an *n*-channel JFET. Breakdown occurs at large values of V_{DS} and is caused by the avalanche mechanism described in connection with BJTs. Note that the larger the magnitude of V_{GS}, the smaller the value of V_{DS} at which breakdown occurs.

Transfer Characteristics

The *transfer characteristic* of a JFET is a plot of output current versus input voltage, for a fixed value of output voltage. When the input to a JFET is the gate-to-source voltage and the output current is drain current (common-source configuration), the transfer characteristic can be derived from the drain characteristics. It is only necessary to construct a vertical line on the drain characteristics (a line of constant V_{DS}) and to note the value of I_D at each intersection of the line with a line of constant V_{GS}. The values of I_D can then be plotted against the values of V_{GS} to construct the transfer characteristic. Figure 5–10 illustrates the process.

In Figure 5–10, the transfer characteristic is shown for $V_{DS} = 8$ V. As can be seen in the figure, this choice of V_{DS} means that all points are in the pinch-off region. For example, the point of intersection of the $V_{DS} = 8$ V line and the $V_{GS} = 0$ V line occurs at $I_D = I_{DSS} = 12$ mA. At $V_{DS} = 8$ V and $V_{GS} = -1$ V, we find $I_D = 6.75$ mA. Plotting these combinations of I_D and V_{GS} produces the parabolic transfer characteristic shown. The nonlinear shape

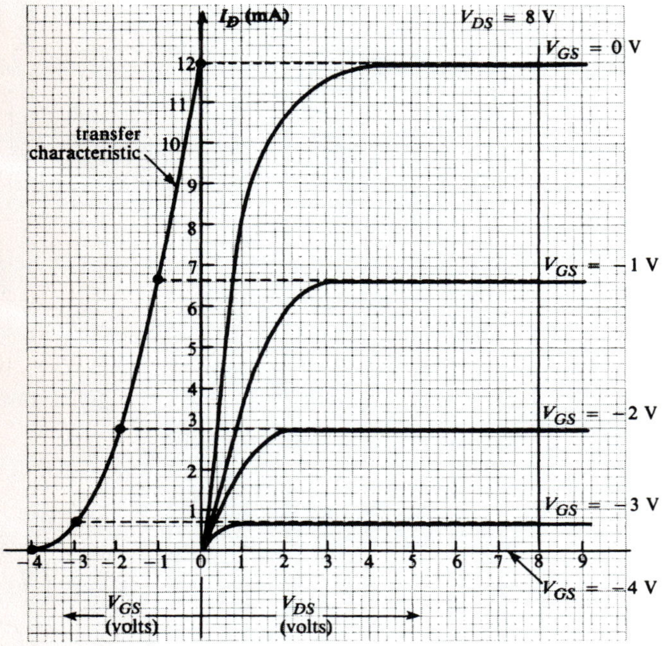

FIGURE 5–10 Construction of an *n*-channel transfer characteristic from the drain characteristics

of the transfer characteristic can be anticipated by observing that equal increments in the values of V_{GS} on the drain characteristics ($\Delta V_{GS} = 1\,\text{V}$) do not produce equally spaced lines. (Recall from our discussion of BJT output characteristics that this situation creates output signal distortion when the device is used as an ac amplifier; practical JFET circuits incorporate a means for reducing distortion, at the expense of gain.) Note that the intercepts of the transfer characteristic are I_{DSS} on the I_D-axis and V_p on the V_{GS}-axis.

The equation for the transfer characteristic *in the pinch-off region* is, to a close approximation,

$$I_D = I_{DSS}\left(1 - \frac{V_{GS}}{V_p}\right)^2 \tag{5–2}$$

Note that equation 5–2 correctly predicts that $I_D = I_{DSS}$ when $V_{GS} = 0$ and that $I_D = 0$ when $V_{GS} = V_p$. The transfer characteristic is often called the *square-law* characteristic of a JFET and is used in some interesting applications to produce outputs that are nonlinear functions of inputs.

EXAMPLE 5–1

An n-channel JFET has a pinch-off voltage of $-4.5\,\text{V}$ and $I_{DSS} = 9\,\text{mA}$.

1. At what value of V_{GS} in the pinch-off region will I_D equal 3 mA?
2. What is the value of $V_{DS(sat)}$ when $I_D = 3\,\text{mA}$?

Solution

1. We must solve equation 5–2 for V_{GS}:

$$(1 - V_{GS}/V_p)^2 = I_D/I_{DSS}$$
$$1 - V_{GS}/V_p = \sqrt{I_D/I_{DSS}}$$
$$V_{GS} = V_p(1 - \sqrt{I_D/I_{DSS}})$$
$$V_{GS} = -4.5[1 - \sqrt{(3\,\text{mA})/(9\,\text{mA})}] = -1.9\,\text{V}$$

2. Equation 5–1 relates I_D and $V_{DS(sat)}$. Solving for $V_{DS(sat)}$, we find

$$V_{DS(sat)} = \sqrt{(V_p)^2 I_D / I_{DSS}} = \sqrt{(4.5)^2(3\text{ mA})/(9\text{ mA})} = 2.6\text{ V}$$

Note that we use the positive square root, because V_{DS} is positive for an n-channel JFET. For a p-channel JFET, we would use the negative root. The value of $V_{DS(sat)}$ could also have been determined from the fact that $V_{DS(sat)} = V_{GS} - V_p = -1.9 - (-4.5) = 2.6$ V.

EXAMPLE 5–2

SPICE

Use SPICE to obtain a plot of the transfer characteristic of an n-channel JFET having $I_{DSS} = 10$ mA and $V_p = -2$ V. The characteristic should be plotted for $V_{DS} = 10$ V.

Solution

Figure 5–11(a) shows a SPICE circuit that can be used to obtain the desired characteristic. The value of BETA in the .MODEL statement is found from

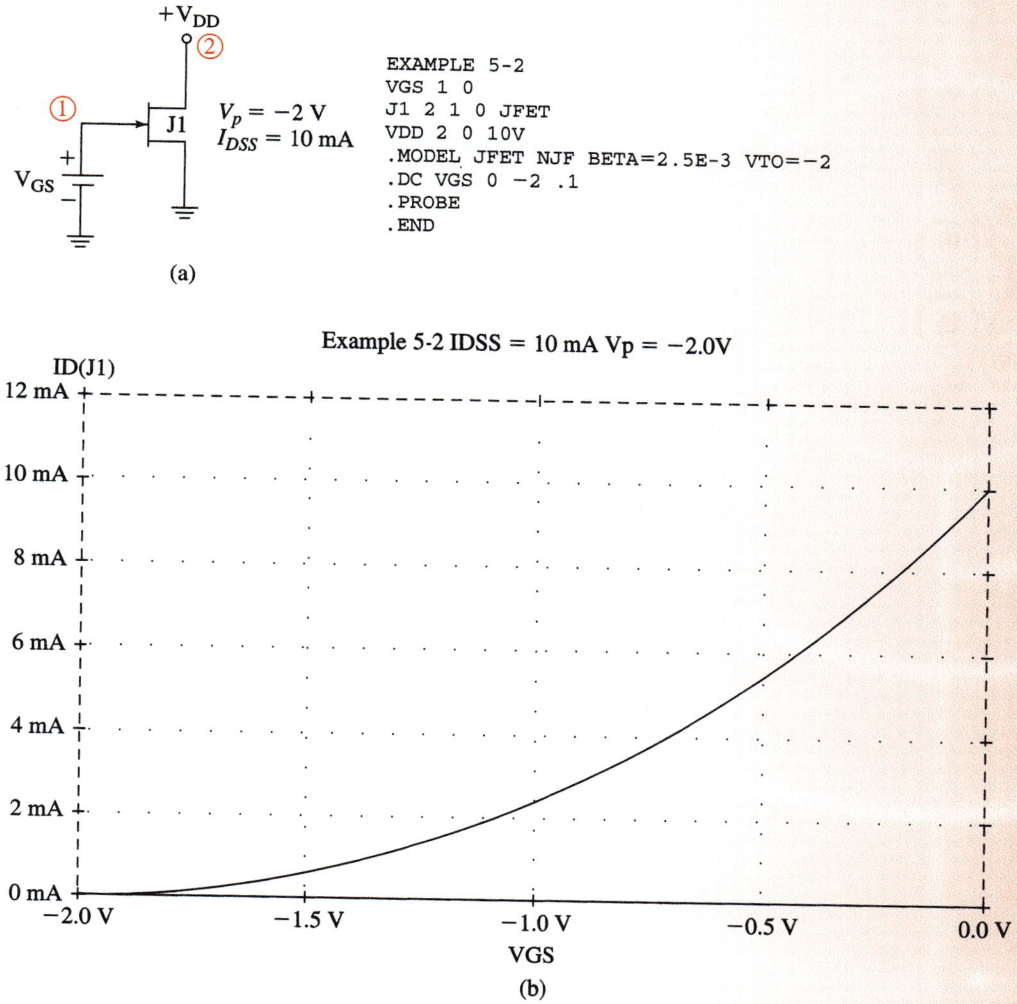


```
EXAMPLE 5-2
VGS 1 0
J1 2 1 0 JFET
VDD 2 0 10V
.MODEL JFET NJF BETA=2.5E-3 VTO=-2
.DC VGS 0 -2 .1
.PROBE
.END
```

FIGURE 5–11 (Example 5–2)

$$\beta = \frac{I_{DSS}}{V_p^2} = \frac{10 \times 10^{-3}\text{ A}}{(-2\text{ V})^2} = 2.5 \times 10^{-3}\text{ A/V}^2$$

Note that we step VGS from 0 to − 2 V in 0.1-V increments. The voltage source labeled VGS has its positive terminal connected to the gate of the FET, but the stepped voltages should all be negative. (Alternatively, we could reverse the polarity of VGS and step it through positive voltages.)

Figure 5–11(b) shows the plot produced by SPICE. (Rotate it 180° to obtain the orientation shown in Figure 5–10.) Note that $I_D = 10$ mA $= I_{DSS}$ when $V_{GS} = 0$ and that $I_D \approx 0$ when $V_{GS} = V_p$.

5–3 JFET BIASING

Fixed Bias

Like a bipolar transistor, a JFET used as an ac amplifier must be biased in order to create a dc output voltage around which ac variations can occur. When a JFET is connected in the *common-source* configuration, the input voltage is V_{GS} and the output voltage is V_{DS}. Therefore, the bias circuit must set dc (quiescent) values for the drain-to-source voltage V_{DS} and drain current I_D. Figure 5–12 shows one method that can be used to bias *n*-channel and *p*-channel JFETs.

Notice in Figure 5–12 that a dc supply voltage V_{DD} is connected to supply drain current to the JFET through resistor R_D and that another dc voltage is used to set the gate-to-source voltage V_{GS}. This biasing method is called *fixed bias* because the gate-to-source voltage is fixed by the constant voltage applied across those terminals. Writing Kirchhoff's voltage law around the output loops in Figure 5–12, we find

$$\begin{aligned} V_{DS} &= V_{DD} - I_D R_D \quad &(n\text{-channel}) \\ V_{DS} &= -V_{DD} + I_D R_D \quad &(p\text{-channel}) \end{aligned} \qquad (5\text{–}3)$$

When using these equations, always substitute a positive value for V_{DD} to ensure that the correct sign is obtained for V_{DS}. V_{DS} should always turn out to be a positive quantity in an *n*-channel JFET and a negative quantity in a *p*-channel JFET. For example, in an *n*-channel device where V_{DD} is +15 V from drain to ground, if $I_D = 10$ mA, and $R_D = 1$ kΩ, we have $V_{DS} = 15\text{ V} - (10\text{ mA})(1\text{ k}\Omega) = +5$ V. For a *p*-channel device where V_{DD} is −15 V from drain to ground, $V_{DS} = -15 + (10\text{ mA}) \times (1\text{ k}\Omega) = -5$ V. Equations 5–3 can be rewritten in the form

$$\begin{aligned} I_D &= -(1/R_D)V_{DS} + V_{DD}/R_D \quad &(n\text{-channel}) \\ I_D &= (1/R_D)V_{DS} + V_{DD}/R_D \quad &(p\text{-channel}) \end{aligned} \qquad (5\text{–}4)$$

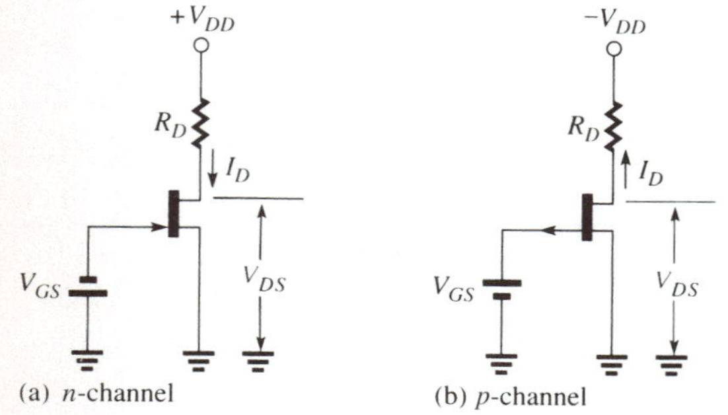

FIGURE 5–12 Fixed-bias circuits for *n*- and *p*-channel JFETs

Equations 5–4 are the equations of the dc load lines for *n*- and *p*-channel JFETs, and each can be plotted on a set of drain characteristics to determine a Q-point. This technique is the same as the one we used to determine the Q-point in a BJT bias circuit. The load line intersects the V_{DS}-axis at V_{DD} and the I_D-axis at V_{DD}/R_D.

EXAMPLE 5–3

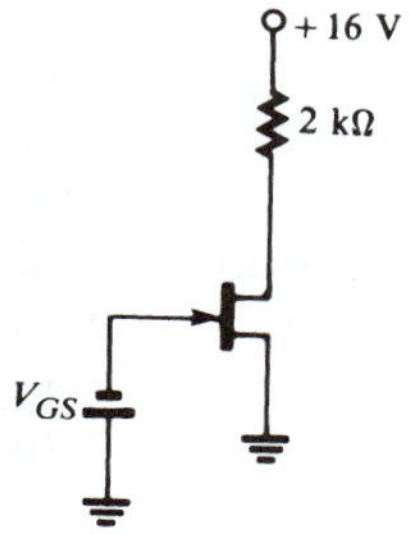

FIGURE 5–13 (Example 5–3)

The JFET in the circuit of Figure 5–13 has the drain characteristics shown in Figure 5–14. Find the quiescent values of I_D and V_{DS} when (1) $V_{GS} = -1.5$ V and (2) $V_{GS} = -0.5$ V.

Solution

1. The load line intersects the V_{DS}-axis at $V_{DD} = +16$ V and the I_D-axis at $I_D = (16\text{ V})/(2\text{ k}\Omega) = 8$ mA. It is plotted on Figure 5–14.
   At the intersection of the load line with $V_{GS} = -1.5$ V (labeled Q_1), we find the quiescent values $I_D \approx 4$ mA and $V_{DS} \approx 8$ V.
2. The load line is, of course, the same as in part (1). Changing V_{GS} to -0.5 V moves the Q-point to the point labeled Q_2 in Figure 5–14. Here we see that $I_D \approx 6.8$ mA and $V_{DS} \approx 2.4$ V.

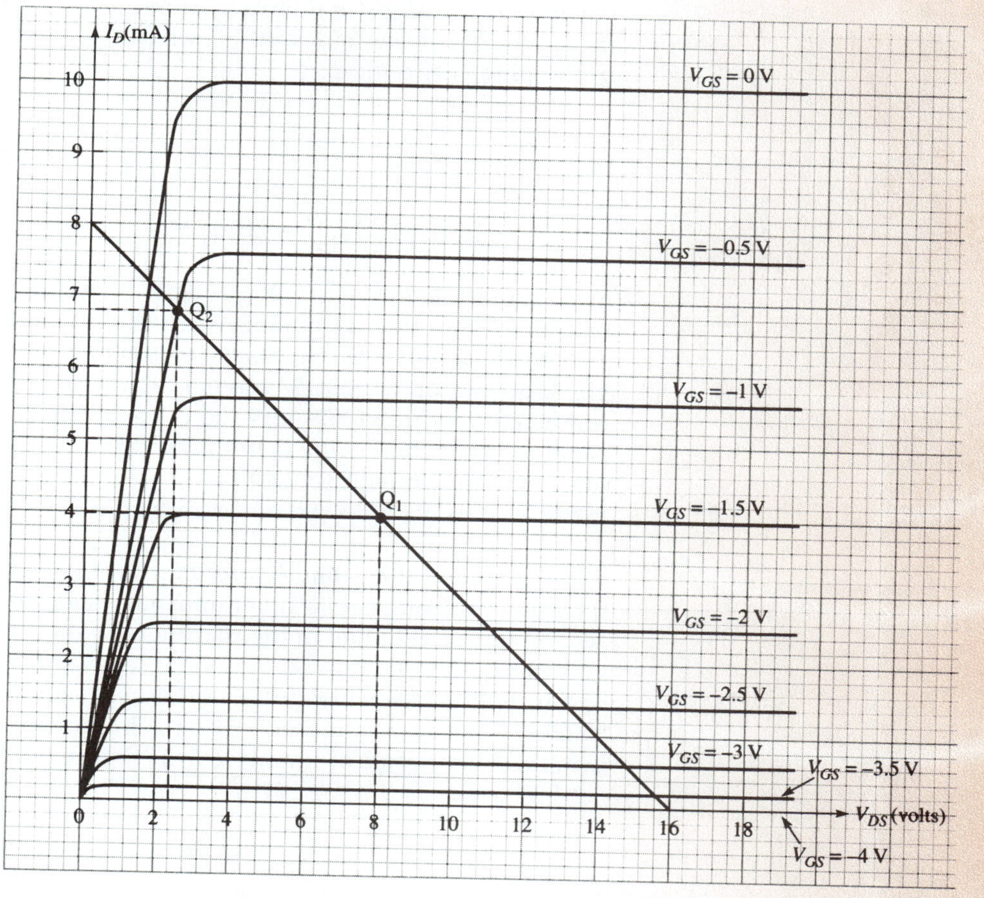

FIGURE 5–14 (Example 5–3)

Part 2 of the preceding example illustrates an important result. Note that changing V_{GS} to -0.5 V in the bias circuit of Figure 5–13 caused the Q-point to move out of the pinch-off region and into the voltage-controlled–resistance region. As we have already mentioned, the Q-point must be located in the pinch-off region for normal amplifier operation. *To ensure that the Q-point is in the pinch-off region, the quiescent value of* $|V_{DS}|$ *must be greater than* $|V_p|-|V_{GS}|$. The pinch-off voltage for the device whose characteristics are given in Figure 5–14 can be seen to be approximately -4 V. Because $|V_{GS}| = 0.5$ V and the quiescent value of V_{DS} at Q_2 is only 2.4 V, we do not satisfy the requirement $|V_{DS}| > |V_p| - |V_{GS}|$. Q_2 is therefore in the variable-resistance region.

Of course, the quiescent value of I_D can also be determined using the transfer characteristic of a JFET. Because the transfer characteristic is a plot of I_D versus V_{GS}, it is only necessary to locate the V_{GS} coordinate and read the corresponding value of I_D directly. The value of V_{DS} can then be determined using equation 5–3. Although graphical techniques for locating the bias point are instructive and provide insights into the way in which the circuit variables affect each other, the quiescent values of I_D and V_{DS} can be calculated using a straightforward computation, if the values of V_p and I_{DSS} are known. The next example illustrates that the square-law characteristic is used in this computation.

EXAMPLE 5–4

Given that the JFET in Figure 5–13 has $I_{DSS} = 10$ mA and $V_p = -4$ V, compute the quiescent values of I_D and V_{DS} when $V_{GS} = -1.5$ V. Assume that it is biased in the pinch-off region.

Solution

From equation 5–2,

$$I_D = I_{DSS}(1 - V_{GS}/V_p)^2 = (10\text{ mA})\left(1-\frac{-1.5}{-4}\right)^2 = 3.9\text{ mA}$$

From equation 5–3, $V_{DS} = V_{DD} - I_D R_D = 16 - (3.9\text{ mA})(2\text{ k}\Omega) = 8.2$ V. These results are in close agreement with those obtained graphically in Example 5–3. Note that it was necessary to assume that the JFET is biased in the pinch-off region to justify the use of equation 5–2. If the computation had produced a value of V_{DS} less than $|V_p| - |V_{GS}| = 2.5$ V, we would have had to conclude that the device is not biased in pinch-off and would then have had to use another means to find the Q-point.

The values of I_{DSS} and V_p are likely to vary widely among JFETs of a given type. A variation of 50% is not unusual. When the fixed-bias circuit is used to set a Q-point, a change in the parameter values of the JFET for which the circuit was designed (caused, for example, by substitution of another JFET) can result in an intolerable shift in quiescent values. Suppose, for example, that a JFET having parameters $I_{DSS} = 13$ mA and $V_p = -4.3$ V is substituted into the bias circuit of Example 5–3 (Figure 5–13), with V_{GS} once again set to -1.5 V. Then

$$I_D = (13\text{ mA})\left(1-\frac{-1.5}{-4.3}\right)^2 = 5.51\text{ mA}$$

$$V_{DS} = 16-(5.51\text{ mA})(2\text{ k}\Omega) = 4.98\text{ V}$$

These results show that I_D increases 41.3% over the value obtained in Example 5–3 and that V_{DS} decreases 68.7%. Note also that the value of

FIGURE 5–15 Self-bias circuits

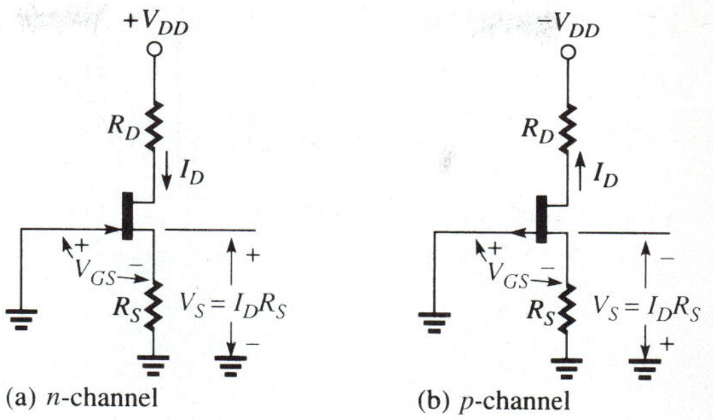

V_{DS} (4.98 V) is now perilously near the pinch-off voltage (4.3 V). We conclude that the fixed-bias circuit does not provide good Q-point stability against changes in JFET parameters.

Figure 5–15 shows a bias circuit that provides improved stability and requires only a single supply voltage. This bias method is called *self-bias* because the voltage drop across R_S due to the flow of quiescent current determines the quiescent value of V_{GS}. We can understand this fact by realizing that the current I_D in resistor R_S creates the voltage $V_S = I_D R_S$ at the source terminal, with respect to ground. For the n-channel JFET, this means that the source is positive with respect to the gate, because the gate is grounded. In other words, the gate is negative (by $I_D R_S$ volts) with respect to the source, as required for biasing an n-channel JFET: $V_{GS} = -I_D R_S$. For the p-channel device, the gate is positive by $I_D R_S$ volts, with respect to the source: $V_{GS} = I_D R_S$.

The equations

$$V_{GS} = -I_D R_S \quad (n\text{-channel}) \tag{5–5}$$

$$V_{GS} = I_D R_S \quad (p\text{-channel}) \tag{5–6}$$

describe straight lines when plotted on V_{GS}–I_D-axes. (Verify these equations by writing Kirchhoff's voltage law around each gate-to-source loop in Figure 5–15.) Each line is called the *bias line* for its respective type. The quiescent value of I_D in the self-bias circuit can be determined graphically by plotting the bias line on the same set of axes with the transfer characteristic. The intersection of the two locates the Q-point. In effect, we solve the bias-line equation and the square-law equation simultaneously by finding the point where their graphs intersect. The quiescent value of V_{DS} can be found by summing voltages (writing Kirchhoff's voltage law) around the output loops in Figure 5–15:

$$\begin{aligned} V_{DS} &= V_{DD} - I_D(R_D + R_S) \quad (n\text{-channel}) \\ V_{DS} &= -V_{DD} + I_D(R_D + R_S) \quad (p\text{-channel}) \end{aligned} \tag{5–7}$$

The next example illustrates the graphical procedure.

EXAMPLE 5–5

The transfer characteristic of the JFET in Figure 5–16 is given in Figure 5–17. Determine the quiescent values of I_D and V_{DS} graphically.

Solution

Because $R_S = 600\ \Omega$, the equation of the bias line is $V_{GS} = -600 I_D$. It is clear that the bias line always passes through the origin ($I_D = 0$ when $V_{GS} = 0$),

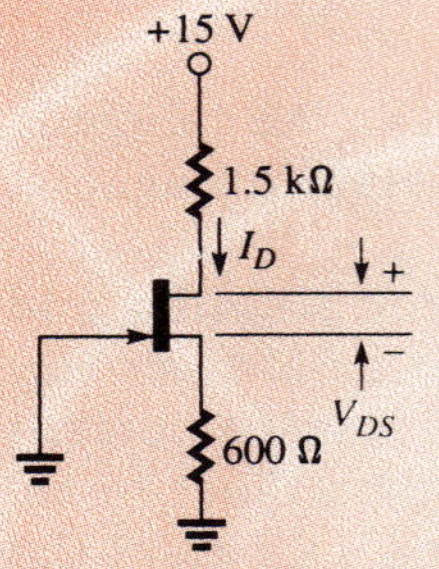

FIGURE 5–16
(Example 5–5)

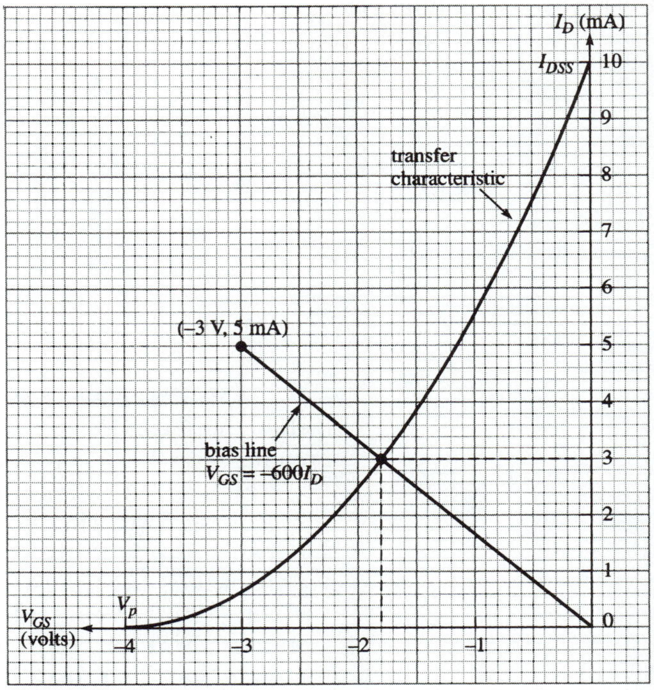

FIGURE 5–17 (Example 6–5).

so (0,0) is one point on the line. To determine another point on the line, choose a convenient value of V_{GS} and solve for I_D. In this example, if we let $V_{GS} = -3$, then

$$I_D = \frac{-V_{GS}}{600\ \Omega} = \frac{-(-3\text{ V})}{600\ \Omega} = 5\text{ mA}$$

Thus, (−3 V, 5 mA) is another point on the bias line. We can then draw a straight line between the two points (0,0) and (−3 V, 5 mA) and note where that line intersects the transfer characteristic. The line is plotted on the transfer characteristic shown in Figure 5–17. We note that it intersects the characteristic at $I_D \approx 3$ mA, which is the quiescent drain current. The corresponding value of V_{GS} is seen to be approximately −1.8 V. The quiescent value of V_{DS} is found from equation 5–7.

$$V_{DS} = 15\text{ V} - (3\text{ mA})\,[(1.5\text{ k}\Omega) + (0.6\text{ k}\Omega)] = 8.7\text{ V}$$

General Algebraic Solution—Self-Bias

The quiescent values of I_D and V_{GS} in the self-bias circuit can also be computed algebraically by solving the bias-line equation and the square-law equation simultaneously. To perform the computation, we must know the values of I_{DSS} and V_p. As in the fixed-bias case, the results are valid only if the Q-point is in the pinch-off region, i.e., if $|V_{DS}| > |V_p| - |V_{GS}|$. We must therefore assume that to be the case, but discard the results if the computation reveals the quiescent value of $|V_{DS}|$ to be less than $|V_p| - |V_{GS}|$. Equations 5–8 give the general form of the algebraic solution for the quiescent values of I_D, V_{DS}, and V_{GS} in the self-bias circuit. Because absolute values are used in the computations, the equations are valid for both p-channel and n-channel devices.

General algebraic solution for the bias point of self-biased JFET circuits

$$I_D = \frac{-B - \sqrt{B^2 - 4AC}}{2A} \quad (5\text{–}8)$$

where

$$A = R_S^2$$

$$B = -\left(2|V_p|R_S + \frac{V_p^2}{I_{DSS}}\right)$$

$$C = V_p^2$$

$|V_{DS}| = |V_{DD}| - I_D(R_D + R_S)$ See note 1.

$|V_{GS}| = I_D R_S$ See note 2.

Note 1. V_{DS} is positive for an n-channel JFET and negative for a p-channel JFET.

Note 2. V_{GS} is negative for an n-channel JFET and positive for a p-channel JFET.

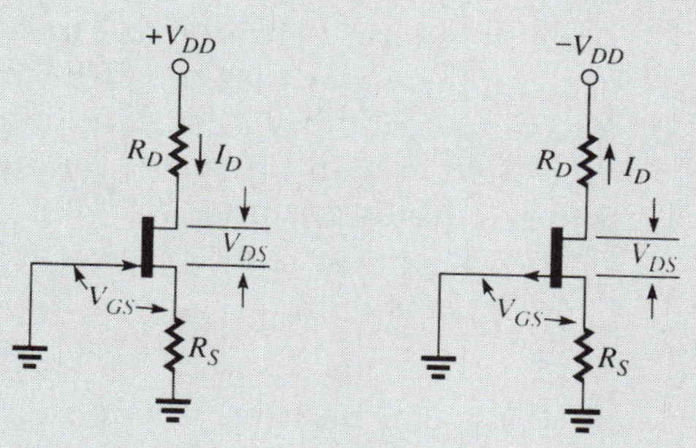

EXAMPLE 5–6

Use equations 5–8 to find the bias point that was determined graphically in Example 5–5.

Solution

As shown in Figure 5–16, $R_S = 600\ \Omega$ and $R_D = 1.5\ \text{k}\Omega$. Also, the transfer characteristic in Figure 5–17 shows that $I_{DSS} = 10$ mA and $V_p = -4$ V. Thus, with reference to equations 5–8, we find

$$A = R_S^2 = (600)^2 = 3.6 \times 10^5$$

$$B = -\left(2|V_p|R_S + \frac{V_p^2}{I_{DSS}}\right) = -\left[2(4)(600) + \frac{(-4)^2}{10 \times 10^{-3}}\right] = -6.4 \times 10^3$$

$$C = V_p^2 = (-4)^2 = 16$$

$$I_D = \frac{-B - \sqrt{B^2 - 4AC}}{2A}$$

$$= \frac{6.4 \times 10^3 - \sqrt{40.96 \times 10^6 - 4(3.6 \times 10^5)(16)}}{2(3.6 \times 10^5)} = 3.0 \text{ mA}$$

$$|V_{DS}| = |V_{DD}| - I_D(R_D + R_S) = 15\text{ V} - 3\text{ mA}(1.5\text{ k}\Omega + 600\ \Omega) = 8.7\text{ V}$$

$$|V_{GS}| = I_D R_S = (3\text{ mA})(600\ \Omega) = 1.8\text{ V}$$

Because the JFET is n-channel, $V_{GS} = -1.8$ V. These results agree well with those found in Example 5–5. Because $|V_{DS}| = 8.7\text{ V} > |V_p| - |V_{GS}| = 4\text{ V} - 1.8\text{ V} = 2.2$ V, we know the bias point is in the pinch-off region and the results are valid.

To demonstrate that the self-bias method provides better stability than the fixed-bias method, let us compare the shift in the quiescent value of I_D that occurs using each method when the JFET parameters of the previous example are changed to $I_{DSS} = 12$ mA and $V_p = -4.5$ V. In each case, we will assume that the initial bias point (using a JFET with $I_{DSS} = 10$ mA and $V_p = -4$ V) is set so that $I_D = 3$ mA and that a JFET having the new parameters is then substituted in the circuit. We have already seen that $I_D = 3$ mA when $V_{GS} = -1.8$ V, so let us suppose that a fixed-bias circuit has V_{GS} set to -1.8 V. When I_{DSS} changes to 12 mA and V_p to -4.5 V, with V_{GS} fixed at -1.8 V, we find that the new value of I_D in the fixed-bias circuit is

$$I_D = I_{DSS}\left(1 - \frac{V_{GS}}{V_p}\right)^2 = (12 \times 10^{-3}\text{ A})\left(1 - \frac{1.8}{4.5}\right)^2 = 4.32\text{ mA}$$

This change in I_D from 3 mA to 4.32 mA represents a 44% increase.

Suppose now that the JFET parameters in the self-bias circuit change by the same amount: $I_{DSS} = 12$ mA and $V_p = -4.5$ V. Using equations 5–8, we find $I_D = 3.46$ mA. In this case, the increase in I_D is 15.3%, less than half that of the fixed-bias design.

5–4 THE JFET CURRENT SOURCE

A JFET can be used to supply constant current to a variable load by connecting its gate directly to its source, as illustrated in Figure 5–18. Here, the resistor R_D is regarded as the (variable) load resistance. To be able to supply

FIGURE 5–18 JFET constant-current sources

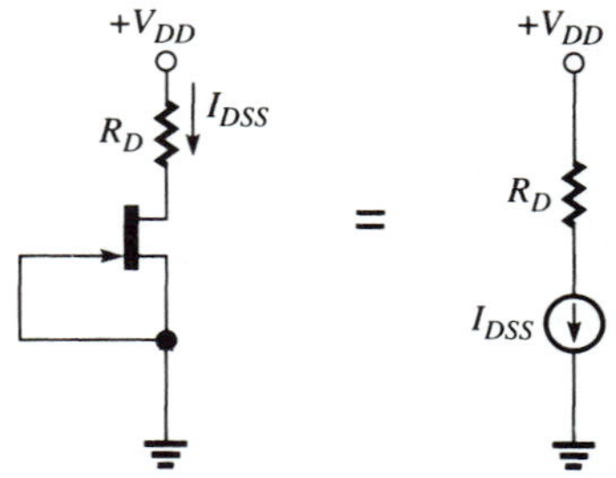

(a) n-channel JFET current source

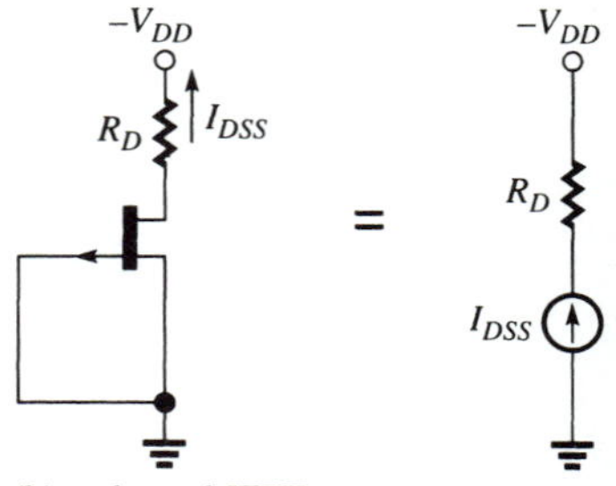

(b) p-channel JFET current source

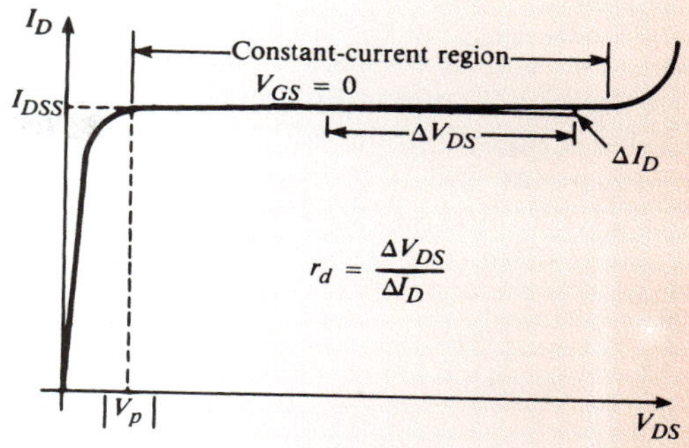

FIGURE 5–19 A JFET current source maintains an essentially constant current equal to I_{DSS} in the pinch-off region. If the characteristic were perfectly flat, then ΔI_D would equal zero and r_d would be infinite.

a current that is independent of R_D, the JFET must remain in its pinch-off region. Recall that the condition for pinch-off is $|V_{DS}| > |V_p| - |V_{GS}|$. Because $V_{GS} = 0$ in this case, the condition reduces to

$$|V_{DS}| > |V_p| \tag{5–9}$$

The constant current produced by the JFET is then $I_D = I_{DSS}$, because that is the drain current in the pinch-off region when $V_{GS} = 0$.

So long as the JFET is in its pinch-off region, the line corresponding to $V_{GS} = 0$ is essentially horizontal, meaning that the same current flows regardless of V_{DS}. See Figure 5–19. In reality, the line rises slightly to the right, so the current source is not perfect. Of course, no current source is perfect. The JFET current source would be perfect if r_d were infinite, which would be the case if the line were horizontal: $r_d = \Delta V_{DS}/\Delta I_D$ with $\Delta I_D = 0$. The JFET can also be used to supply a constant current equal to some value less than I_{DSS} by biasing it appropriately.

EXAMPLE 5–7

The JFET shown in Figure 5–20 has $V_p = -4$ V and $I_{DSS} = 14$ mA. What is the maximum value of R_L for which the circuit can be used as a constant-current source?

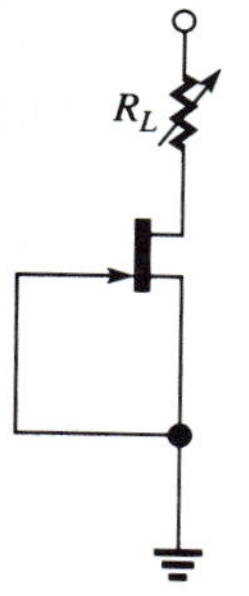

FIGURE 5–20 (Example 5–7)

Solution

To keep the JFET operating in the pinch-off region, we require

$$|V_{DS}| > |V_p|$$
$$18 - (14\text{ mA})R_L > 4$$
$$-14 \times 10^{-3} R_L > -14$$
$$R_L < \frac{14}{14 \times 10^{-3}} = 1\text{ k}\Omega$$

Thus, R_L must be less than 1 kΩ.

5–5 THE JFET AS AN ANALOG SWITCH

An *analog switch* is an electronically controlled device that will either pass or shut off a continuously varying analog-type signal. Figure 5–21 illustrates the concept. By way of contrast, a *digital switch* is one whose output switches between only two possible levels (low or high), such as the BJT inverter we discussed in Chapter 4. As illustrated in Figure 5–21, the analog switch is "opened" or "closed" by a digital-type input. Depending on the nature of the device, a high input may close the switch and a low input may open it, or vice versa. An analog switch is also called a *digital*

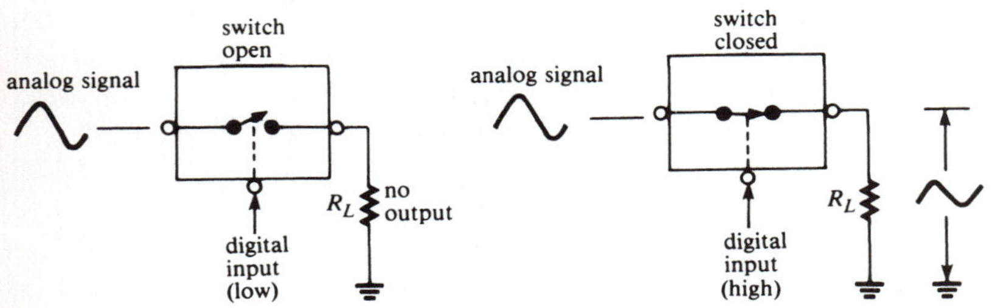

FIGURE 5–21 An analog switch connects or disconnects a variable (analog) signal, depending on the level of a digital input. In the arrangement shown, the switch is in series with the load R_L.

analog switch (DAS) because a digital input controls the switching of an analog signal.

A JFET can be used as an analog switch by connecting it as shown in Figure 5–22. Note that the analog signal (v_d) is connected to R_D, where a fixed supply voltage (V_{DD}) would normally be connected. The digital signal that opens and closes the switch is the gate-to-source voltage V_{GS}. V_{GS} is either 0 V, which causes the JFET to conduct, or V_p, a negative voltage (for an *n*-channel JFET) that cuts the JFET off. The output voltage of the switch, v_o, is the drain-to-source voltage, which will be either v_d (when the JFET is cut off) or close to 0 (when the JFET is conducting). Note that the switching arrangement is somewhat different from that shown in Figure 5–21, because the switch is now in *parallel* with the load resistance, R_L. When the switch is closed (JFET on), it effectively shorts out R_L; when the switch is open (JFET cut off), the short is removed.

When used as an analog switch, the JFET is operated in its voltage-controlled–resistance region rather than in pinch-off. As an aid in understanding how the JFET operates as a switch, refer to Figure 5–23, which shows a portion of the drain characteristics for $V_{GS} = 0$ and for $V_{GS} = V_p$. Only the rising portion of the $V_{GS} = 0$ curve in the voltage-controlled–resistance region is shown. The line corresponding to $V_{GS} = V_p$ coincides with the horizontal axis, because $I_D = 0$ in this case. Imagine that the variation in v_d creates a series of parallel load lines, each intersecting the V_{DS}-axis at an instantaneous value of v_d, just as a load line would intersect at V_{DD} if a fixed drain supply voltage were present. Thus, when $V_{GS} = V_p$, the values of V_{DS} are the same as the variations in v_d. This condition corresponds to that shown in Figure 5–22(c). When $V_{GS} = 0$, the operating point moves to the $V_{GS} = 0$ curve and the output voltage (V_{DS}) is very small. This corresponds to Figure 5–22(b). As long as V_{GS} remains at 0, the variations in I_D and V_{DS} are traced by a point that moves up and down the $V_{GS} = 0$

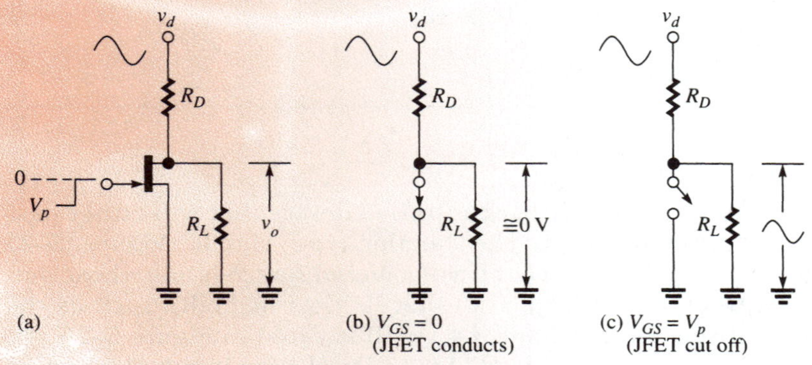

FIGURE 5–22 The JFET as an analog switch. Note that the switch is in parallel with R_L

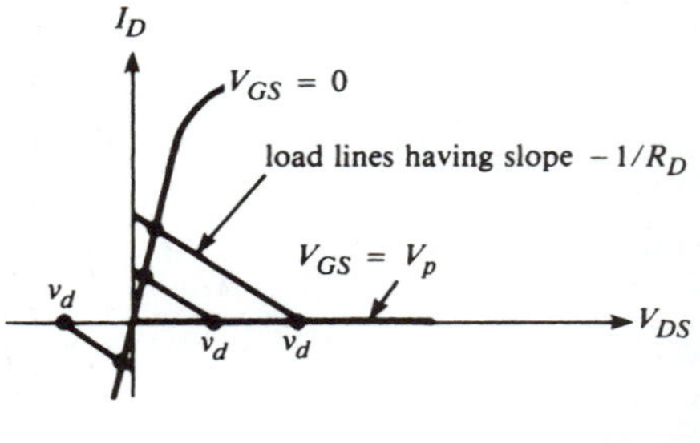

FIGURE 5–23 Operation of the JFET as an analog switch can be viewed as a variation in $V_{DS} = v_d$ along the horizontal axis (when $V_{GS} = V_p$) or as a variation along the $V_{GS} = 0$ curve when the JFET is conducting

curve. For small variations, the curve is nearly linear and can be seen to be quite steep. The resistance ($\Delta V/\Delta I$) in this region is therefore very small and approximates the short-circuit condition we discussed for the case when the FET is "on" (conducting).

Note in Figure 5–23 that the $V_{GS} = 0$ curve extends into the third quadrant: the region where V_{DS} is negative and I_D is negative. This is the region of operation when the analog signal v_d goes negative and the current through the channel reverses direction. The reversal of polarity causes the gate-to-source junction to be forward biased but does not affect the *channel* resistance, so the slope of the $V_{GS} = 0$ curve is unchanged. The total variation in v_d must be small so that operation takes place over a small, nearly linear portion of the $V_{GS} = 0$ curve on either side of the origin. Also, R_D must be large enough to ensure that the variation occurs along the lower portion of the $V_{GS} = 0$ curve; that is, the load line should not be steep.

When the JFET is conducting, the small resistance $V_{DS}/I_D \approx v_{ds}/i_d$ in the region around the origin is called the *ON resistance*, $R_{D(ON)}$. Typical values range from 20 to 100 ohms. The smaller the value of $R_{D(ON)}$, the more nearly ideal the switch. Although a BJT switch has a lower ON resistance, the JFET switch has the advantage that $i_d = 0$ when $v_d = 0$.

EXAMPLE 5–8

The JFET in Figure 5–24 has $R_{D(ON)} = 50\ \Omega$. If $v_d = 100$ mV, what is the load voltage v_L (1) when $V_{GS} = V_p$ and (2) when $V_{GS} = 0$ V?

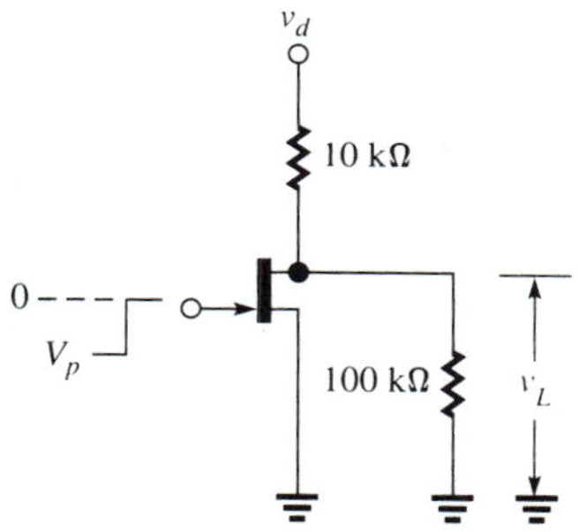

FIGURE 5–24 (Example 5–8)

Solution

1. When $V_{GS} = V_p$, the JFET is cut off, and the circuit is equivalent to that shown in Figure 5–25. By the voltage-divider rule,

$$v_L = \left[\frac{100\ \text{k}\Omega}{(100\ \text{k}\Omega) + (10\ \text{k}\Omega)}\right](100\ \text{mV}) = 90.9\ \text{mV}$$

2. When $V_{GS} = 0$, the circuit is equivalent to that shown in Figure 5–26.

$$R_L \| R_{D(ON)} = (100\ \text{k}\Omega) \| (50\ \Omega) \approx 50\ \Omega$$

Therefore,

$$v_L = \left(\frac{50\ \Omega}{50\ \Omega + 10\ \text{k}\Omega}\right)(100\ \text{mV}) = 0.497\ \text{mV}$$

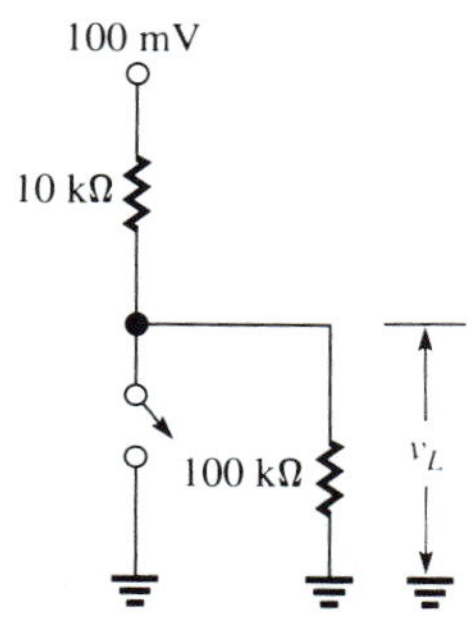

FIGURE 5–25 (Example 5–8) The circuit equivalent to Figure 5–24 when $V_{GS} = V_p$ and the JFET is cut off

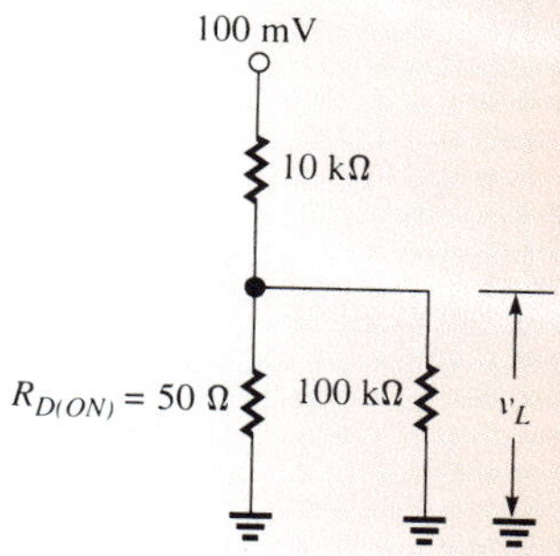

FIGURE 5–26 (Example 5–8) The circuit equivalent to Figure 5–24 when $V_{GS} = 0$ and the JFET is conducting

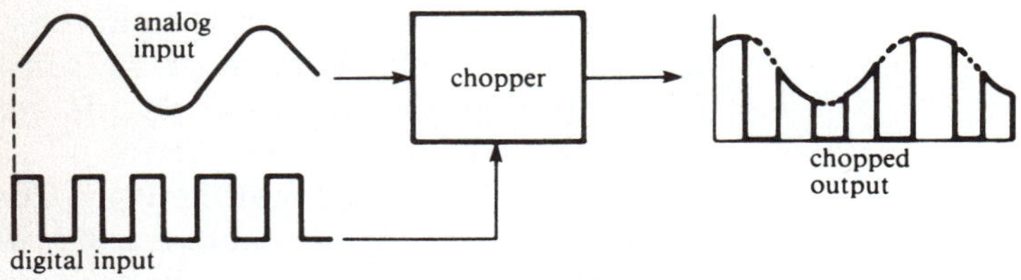

FIGURE 5–27 A chopper produces a series of pulses whose amplitudes follow the variations of an analog input signal.

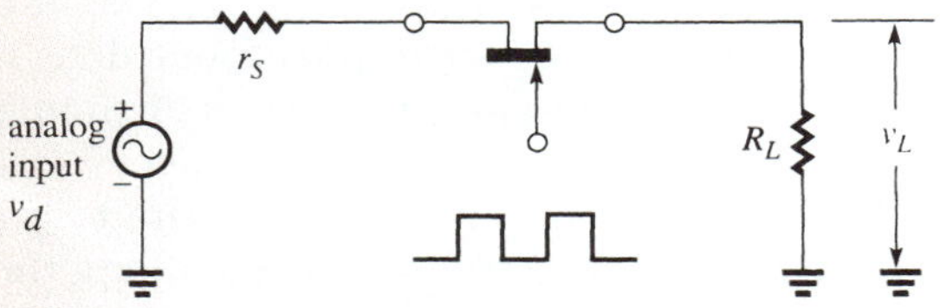

FIGURE 5–28 The JFET connected as a chopper. Note that the switch is in series with the load.

The JFET Chopper

A *chopper* is an analog switch that is turned on and off at a rapid rate by a periodic sequence of pulses, such as a square wave. It is used to convert a slowly varying signal into a series of pulses whose amplitudes vary slowly in the same way as the signal. Figure 5–27 illustrates the concept. A chopper is an example of a *modulator*—in this case, a pulse-amplitude modulator.

Figure 5–28 shows a JFET connected as a chopper. In this variation, the analog switch is in series with the load resistor, R_L, across which the chopped waveform is developed. When the switch is closed (JFET on), current flows from the analog signal source and into R_L. When the switch is open (JFET cut off), no current flows and the output voltage is 0.

Applying the voltage-divider rule to the circuit of Figure 5–28 when the JFET is on, we find

$$v_L = \left(\frac{R_L}{R_L + R_{D(ON)} + r_s}\right) v_d \qquad (5\text{–}10)$$

If R_L is much greater than $R_{D(ON)} + r_s$, then v_L is approximately the same as v_d. Thus, the amplitude of the output (pulse) follows the analog input during each interval when the JFET is conducting.

5–6 MANUFACTURERS' DATA SHEETS

Figures 5–29 and 5–30 show typical data sheets for a series of *n*-channel JFETs: the 2N4220, 2N4221, and 2N4222. Note in particular the specification for the range of values of I_{DSS} for each device, as shown in Figure 5–29. We see, for example, that I_{DSS} for the 2N4222 can range from 5 mA to 15 mA. The pinch-off voltage (designated $V_{GS(off)}$) for the 2N4222 is seen to have a maximum value of −8 V. Devices having small values of I_{DSS} will have smaller values of $V_{GS(off)}$. Another important static characteristic shown in Figure 5–29 is I_{GSS}, the *gate reverse current*, which is the gate current when the gate–source junction is reverse biased, the normal mode of operation. This current provides a measure of the dc input resistance of the device, from gate to source. We see that the maximum specified value for the magnitude of I_{GSS} is 0.1 nA when $V_{GS} = -15$ V and $V_{DS} = 0$. Thus, the minimum gate-to-source resistance under those conditions is $R = (15\text{ V})/(0.1 \times 10^{-9}\text{ A}) = 150 \times 10^9\ \Omega$.

Figure 5–30 shows typical "performance curves" for the 2N4220 series of JFETs. The curves are also applicable to a number of similar JFETs

FIGURE 5–29 Manufacturer's specifications for a series of *n*-channel JFETs (Courtesy of Siliconix, Inc.)

n-channel JFETs
designed for . . .

Siliconix

2N4220 2N4220A 2N4221 2N4221A 2N4222 2N4222A

- **Small-Signal Amplifiers**
- **VHF Amplifiers**
- **Oscillators**
- **Mixers**

Performance Curves NRL
See Section 4

BENEFITS
- High Gain
- Low Receiver Noise Figure

*ABSOLUTE MAXIMUM RATINGS (25°C)

Gate-Drain or Gate-Source Voltage (Note 1) . . . . -30 V
Gate Current . . . . 10 mA
Drain Current . . . . 15 mA
Total Device Dissipation at (or below) 25°C
Free-Air Temperature (Note 2) . . . . 300 mW
Storage Temperature Range . . . . -65 to +200°C
Lead Temperature
(1/16" from case for 10 seconds) . . . . 300°C

TO-72
See Section 6

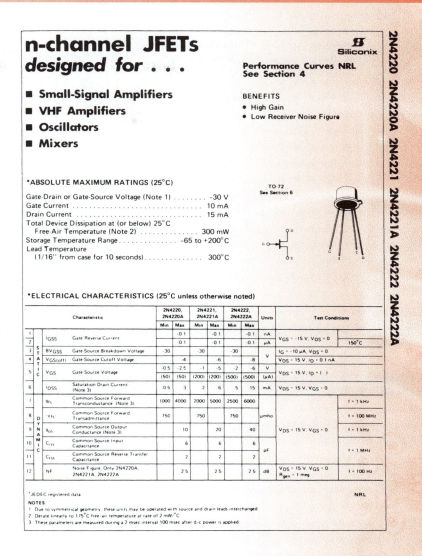

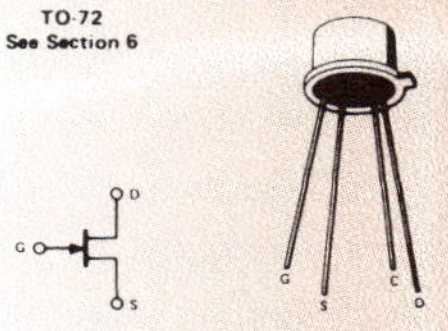

*ELECTRICAL CHARACTERISTICS (25°C unless otherwise noted)

		Characteristic		2N4220, 2N4220A Min	2N4220, 2N4220A Max	2N4221, 2N4221A Min	2N4221, 2N4221A Max	2N4222, 2N4222A Min	2N4222, 2N4222A Max	Units	Test Conditions	
1	STATIC	I_{GSS}	Gate Reverse Current		-0.1		-0.1		-0.1	nA	V_{GS} = -15 V, V_{DS} = 0	
2					-0.1		-0.1		-0.1	μA		150°C
3		BV_{GSS}	Gate-Source Breakdown Voltage	-30		-30		-30		V	I_G = -10 μA, V_{DS} = 0	
4		$V_{GS(off)}$	Gate-Source Cutoff Voltage		-4		-6		-8		V_{DS} = 15 V, I_D = 0.1 nA	
5		V_{GS}	Gate-Source Voltage	-0.5	-2.5	-1	-5	-2	-6	V	V_{DS} = 15 V, I_D = ( )	
				(50)	(50)	(200)	(200)	(500)	(500)	(μA)		
6		I_{DSS}	Saturation Drain Current (Note 3)	0.5	3	2	6	5	15	mA	V_{DS} = 15 V, V_{GS} = 0	
7	DYNAMIC	g_{fs}	Common-Source Forward Transconductance (Note 3)	1000	4000	2000	5000	2500	6000	μmho	V_{DS} = 15 V, V_{GS} = 0	f = 1 kHz
8		$\|Y_{fs}\|$	Common-Source Forward Transadmittance	750		750		750				f = 100 MHz
9		g_{os}	Common-Source Output Conductance (Note 3)		10		20		40			f = 1 kHz
10		C_{iss}	Common-Source Input Capacitance		6		6		6	pF		f = 1 MHz
11		C_{rss}	Common-Source Reverse Transfer Capacitance		2		2		2			
12		NF	Noise Figure, Only 2N4220A, 2N4221A, 2N4222A		2.5		2.5		2.5	dB	V_{DS} = 15 V, V_{GS} = 0, R_{gen} = 1 meg	f = 100 Hz

*JEDEC registered data

NRL

NOTES
1. Due to symmetrical geometry, these units may be operated with source and drain leads interchanged.
2. Derate linearly to 175°C free-air temperature at rate of 2 mW/°C.
3. These parameters are measured during a 2 msec interval 100 msec after d-c power is applied.

produced by the manufacturer and show how wide a range of characteristics is possible. Note that two sets of output characteristics (drain characteristics) are shown, one representing a device having a small I_{DSS} ($\approx$ 1.4 mA) and another having a larger I_{DSS} ($\approx$ 4.2 mA).

Two sets of transfer characteristics are also shown in the figure. These show the range that can be expected in the characteristic among devices of the same type, as well as variations due to temperature. The transfer characteristic in the center of the figure is applicable to the 2N4222 JFET. At 25°C, we see that any given 2N4222 could have a value V_p between about -2 V and -5 V. An interesting temperature phenomenon of JFETs is revealed by these characteristics. Note that the (maximum) value of I_{DSS} at 25°C is less than the value at -40°C, but greater than the value at 85°C. Although it is difficult to see in the figure, this result is accounted for by the fact that the three temperature curves intersect and cross through each other near the V_p end of the characteristics. For the 2N4222, this point can be seen to occur at about $V_{GS} = -4.5$ V. Thus, *the 2N4222 characteristics have zero temperature coefficient when the bias point is set at (about) -4.5 V.* For every JFET, there is a value of V_{GS} near V_p that results in a zero temperature coefficient.

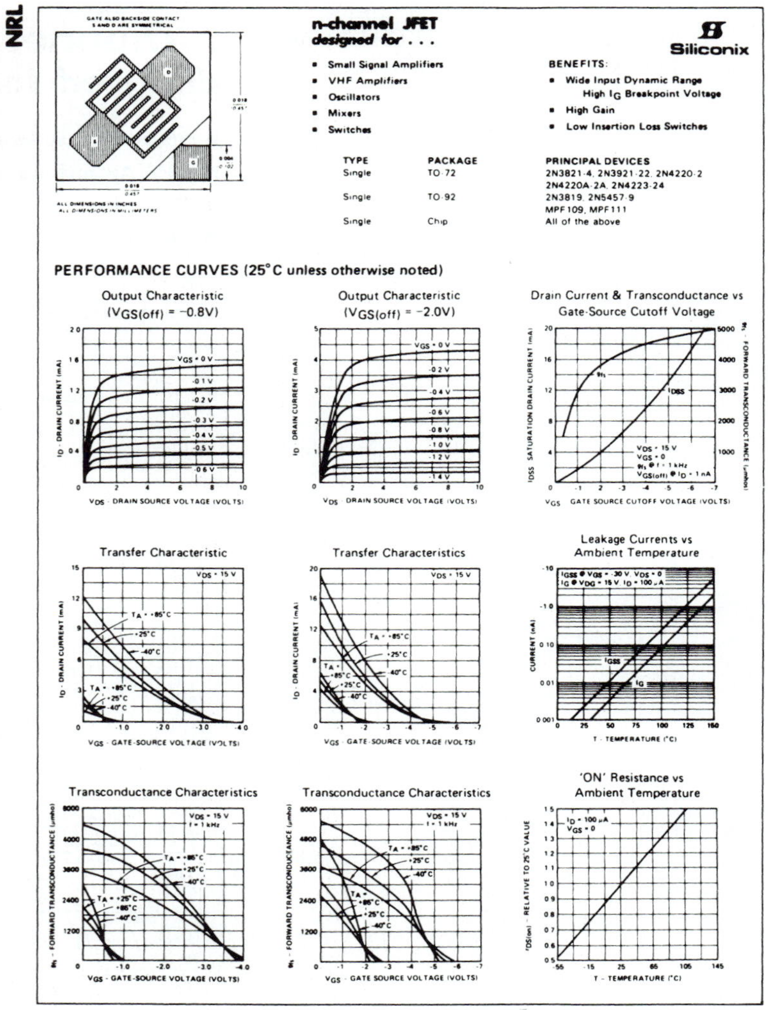

FIGURE 5–30 Typical manufacturer's performance curves for *n*-channel JFETs (Courtesy of Siliconix, Inc.)

5–7 METAL-OXIDE-SEMICONDUCTOR FETS

The metal-oxide-semiconductor FET (MOSFET) is similar in many respects to its JFET counterpart, in that both have drain, gate, and source terminals and both are devices whose channel conductivity is controlled by a gate-to-source voltage. The principal feature that distinguishes a MOSFET from a JFET is the fact that the gate terminal in a MOSFET is *insulated* from its channel region. For this reason, a MOSFET is often called an *insulated-gate* FET, or IGFET.

Enhancement-Type MOSFETs

In the *n*-channel enhancement-type MOSFET, a *p*-type substrate extends all the way to an insulating SiO_2 layer adjacent to a metallic gate. This structure is shown in Figure 5–31.

Figure 5–32 shows the normal electrical connections between drain, gate, and source; the substrate is usually connected to the source. Notice that V_{GS} is connected so that *the gate is positive with respect to the source.* The positive gate voltage attracts electrons from the substrate to the region along the insulating layer opposite the gate. If the gate is made sufficiently positive, enough electrons will be drawn into that region to

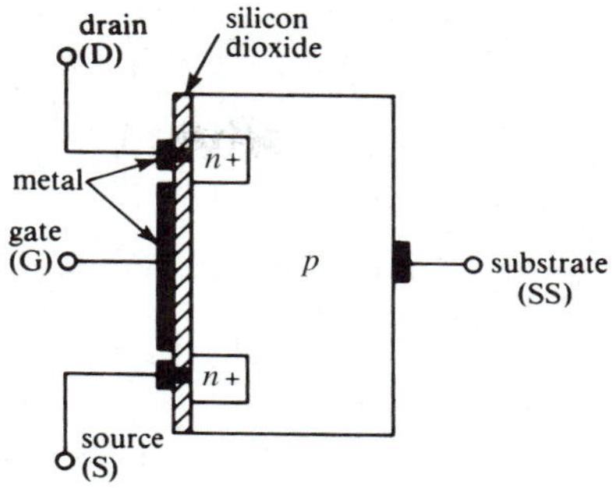

FIGURE 5–31 Enhancement-type MOSFET

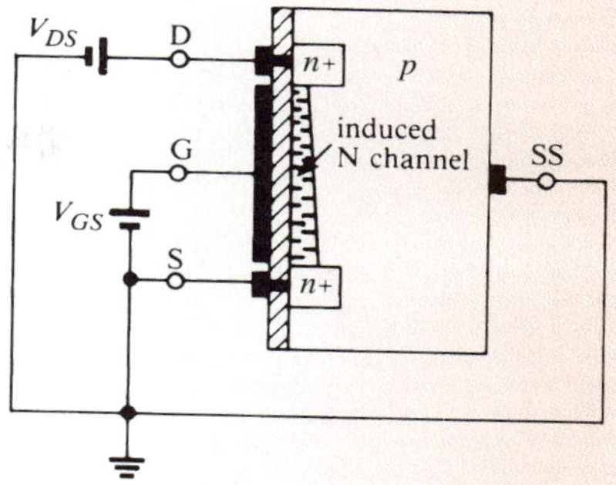

FIGURE 5–32 The positive V_{GS} induces an n-type channel in the substrate of an enhancement MOSFET

convert it to n-type material. Thus, an n-type channel will be formed between drain and source. The p material is said to have been *inverted* to form an n-type channel. If the gate is made still more positive, more electrons will be drawn into the region and the channel will widen, making it more conductive. In other words, making V_{GS} more positive *enhances* the conductivity of the channel and increases the flow of current from drain to source. Since electrons are induced into the channel to convert it to n-type material, the MOSFET shown in Figures 5–31 and 5–32 is often called an *induced n-channel* enhancement-type MOSFET. When this device is referred to simply as an n-channel enhancement MOSFET, it is understood that the n channel exists only when it is induced from the p substrate by a positive V_{GS}.

The induced n channel in Figure 5–32 does not become sufficiently conductive to allow drain current to flow until V_{GS} reaches a certain *threshold voltage,* V_T. In modern silicon MOSFETs, the value of V_T is typically in the range from 1 to 3 V. Suppose that $V_T = 2$ V and that V_{GS} is set to some value greater than V_T, say, 10 V. We will consider what happens when the drain-to-source voltage is gradually increased above 0 V. As V_{DS} increases, the drain current increases because of normal Ohm's law action. The current rises linearly with V_{DS}, as shown in Figure 5–33. As V_{DS} continues to increase, we find that the channel becomes narrower at the drain end, as illustrated in Figure 5–32. This narrowing occurs because the *gate-to-drain* voltage becomes smaller when V_{DS} becomes larger, thus reducing the positive field at the drain end. For example, if $V_{GS} = 10$ V and $V_{DS} = 3$ V, then $V_{GD} = 10 - 3 = 7$ V. When V_{DS} is increased to 4 V, $V_{GD} = 10 - 4 = 6$ V. The positive gate-to-drain voltage decreases by the same amount that V_{DS} increases, so the electric field at the drain end is reduced and the channel is narrowed. As a consequence, the resistance of the channel begins to increase, and the drain current begins to level off. This leveling off can be seen in the curve of Figure 5–33. When V_{DS}

FIGURE 5–33 The drain current in an n-channel enhancement MOSFET increases with V_{DS} until $V_{DS} = V_{GS} - V_T$ (= 10 − 2 = 8 V in this example)

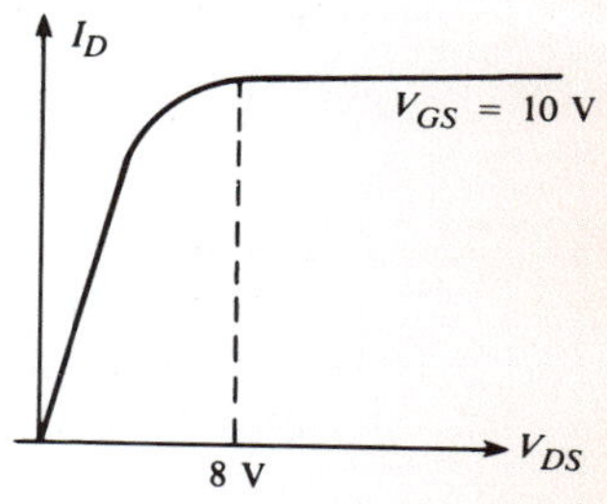

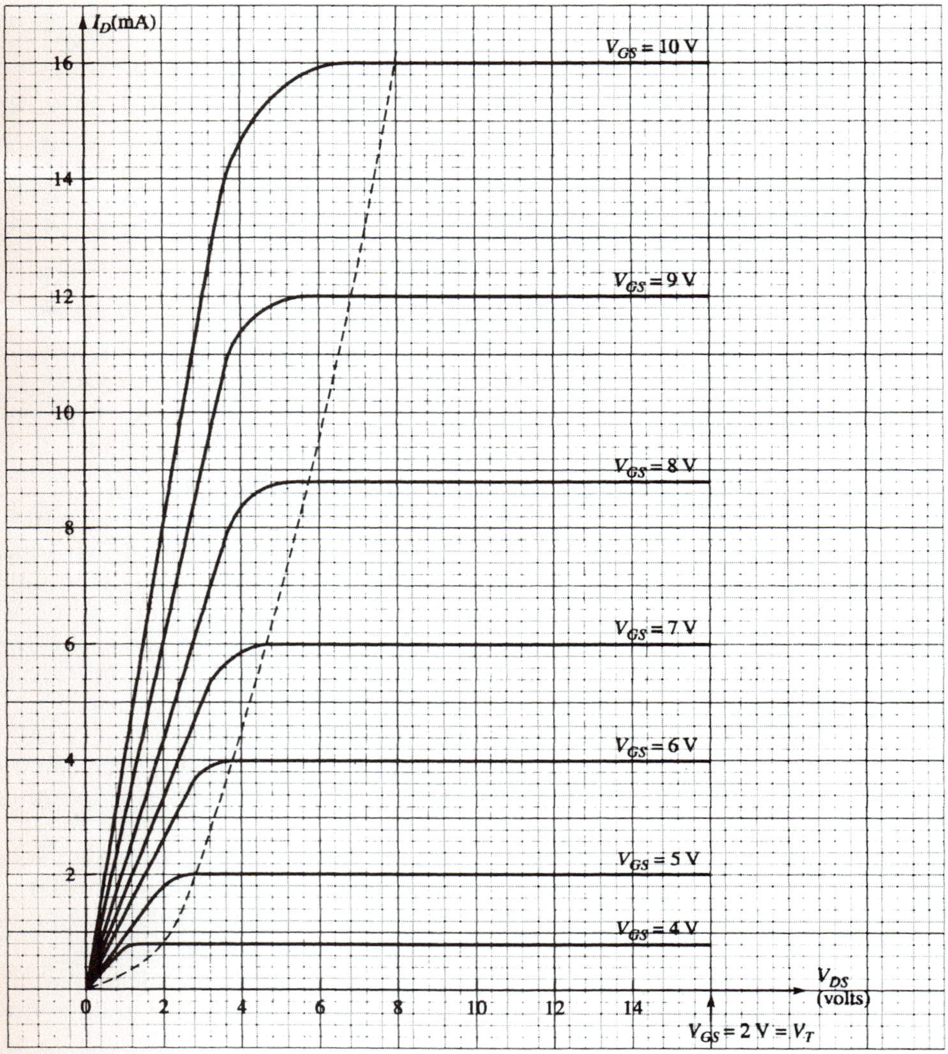

FIGURE 5–34 Drain characteristics of an induced *n*-channel enhancement MOSFET. Note that all values of V_{GS} are positive.

reaches 8 V, then $V_{GD} = 10 - 8 = 2\text{ V} = V_T$. That is, the positive voltage at the drain end reaches the threshold voltage, and the channel width at that end shrinks to zero. Further increases in V_{DS} do not change the shape of the channel, and the drain current does not increase any further; i.e., I_D saturates. This action is quite similar to the saturation that occurs at pinch-off in a junction FET.

When the process we have just described is repeated with V_{GS} fixed at 12 V, we find that saturation occurs at $V_{DS} = 12 - 2 = 10\text{ V}$. Letting $V_{DS(sat)}$ represent the voltage at which saturation occurs, we have, in the general case,

$$V_{DS(sat)} = V_{GS} - V_T \qquad \textbf{(5–11)}$$

Figure 5–34 shows a set of drain characteristics resulting from repetitions of the process we have described, with V_{GS} set to different values of positive voltage. When V_{GS} is reduced to the threshold voltage $V_T = 2\text{ V}$, notice that I_D is reduced to 0 for all values of V_{DS}. The drain characteristics are similar to those of an *n*-channel JFET, except that all values of V_{GS} are positive in the case of the enhancement MOSFET. The enhancement MOSFET can be operated only in an enhancement mode, unlike the depletion MOSFET, which can be operated in both depletion and enhancement modes. The dashed, parabolic line shown on the characteristics in Figure 5–34 joins the saturation voltages, i.e., those satisfying equation 5–11. As in JFET characteristics, the region to the left of the parabola is called the *voltage-controlled–resistance* region where the drain-to-source resistance changes with V_{GS}. We will refer to the region to the right of the parabola as the *active region*. The device is normally operated in the active region for small-signal amplification.

Figure 5–35(a) shows the structure of a *p*-channel enhancement MOSFET and its electrical connections. Note that the substrate is *n*-type material and that a *p*-type channel is induced by a negative V_{GS}. The field produced by V_{GS} drives electrons away from the region near the insulating layer and inverts it to *p* material. Figure 5–35(b) shows a typical set of drain characteristics for the *p*-channel enhancement MOSFET. Note that all values of V_{GS} are negative and that the threshold voltage V_T is negative. *n*-channel and *p*-channel MOSFETs are often called *NMOS* and *PMOS* devices for short.

Figure 5–36 shows the schematic symbols typically used to represent *n*-channel and *p*-channel enhancement MOSFETs. Symbols I (a) and (b) are typically used when the bulk substrate (B) connection needs to be shown. The broken line symbolizes the fact that the channel is induced rather than being an inherent part of the structure. The broken-line symbol is not always used and is replaced with a solid line. Symbols II (a) and (b) are typically used when the source and bulk connections are shorted together, $V_{SB} = 0$. Both sets of symbols are commonly used when preparing schematics that contain MOSFETs.

FIGURE 5–35 The induced *p*-channel enhancement MOSFET

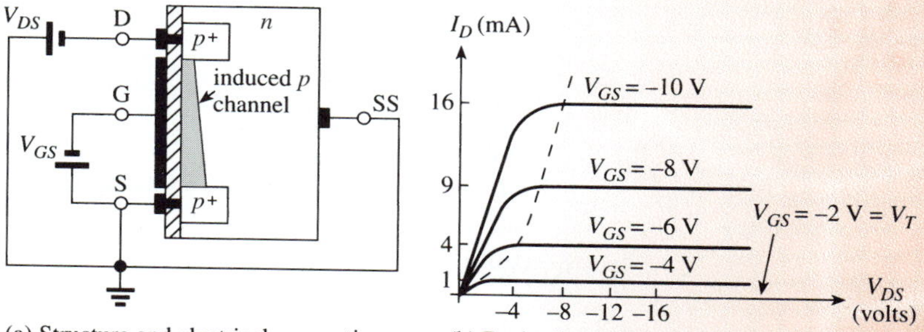

(a) Structure and electrical connections

(b) Drain characteristics

FIGURE 5–36 Symbols for enhancement-type MOSFETs

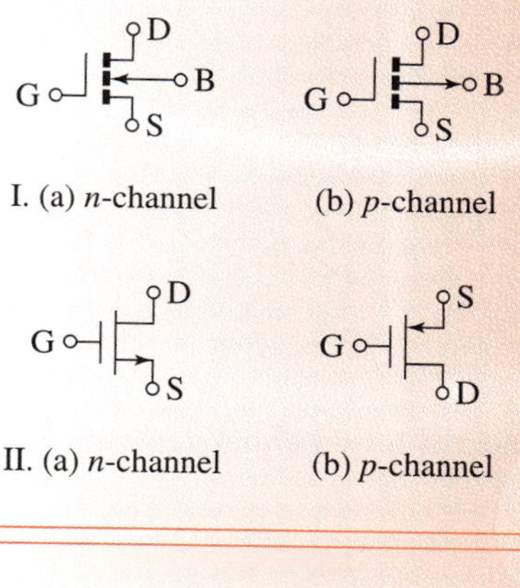

I. (a) *n*-channel

(b) *p*-channel

II. (a) *n*-channel

(b) *p*-channel

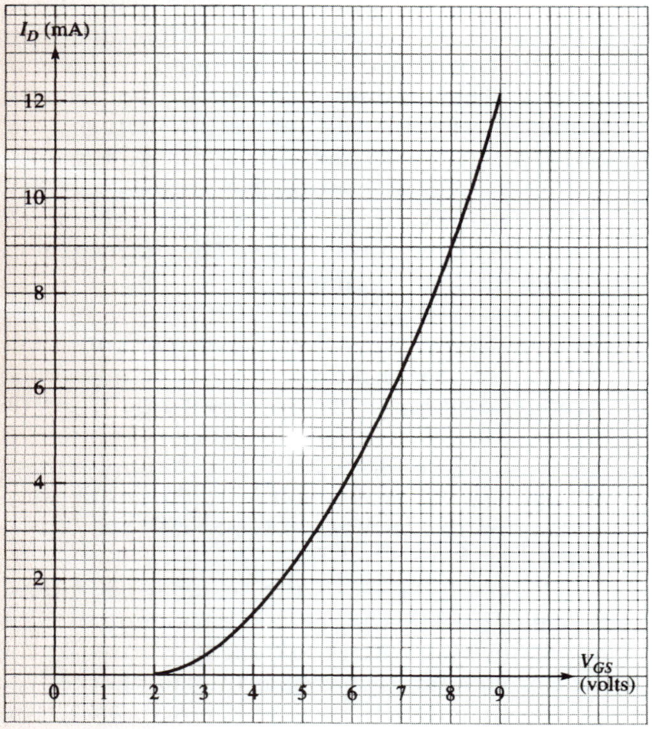

FIGURE 5–37 Transfer characteristic for an enhancement NMOS FET. $\beta = 0.5 \times 10^{-3}$; $V_T = 2$ V

Enhancement MOSFET Transfer Characteristic

In the active region, the drain current and gate-to-source voltage are related by

$$I_D = 0.5\beta(V_{GS} - V_T)^2 \qquad V_{GS} \geq V_T \tag{5–12}$$

where β is a constant whose value depends on the geometry of the device, among other factors. A typical value of β is 0.5×10^{-3} A/V^2. Figure 5–37 shows a plot of the transfer characteristic of an n-channel enhancement MOSFET for which $\beta = 0.5 \times 10^{-3}$ A/V^2 and $V_T = 2$ V.

Enhancement MOSFET Bias Circuits

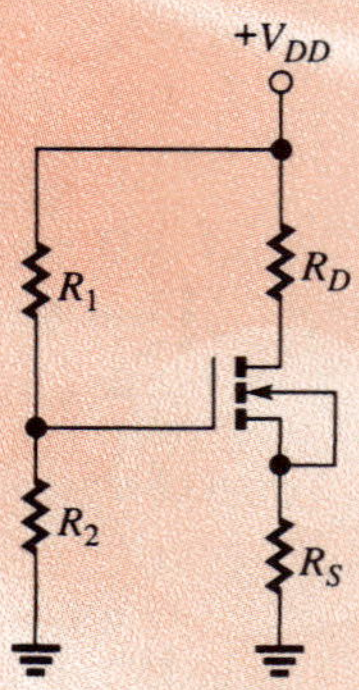

FIGURE 5–38 A bias circuit for an enhancement MOSFET

Enhancement MOSFETs are widely used in digital integrated circuits (and require no bias circuitry in those applications). They also find applications in discrete- and integrated-circuit small-signal amplifiers. Figure 5–38, shows one way to bias a discrete enhancement NMOS for such an application. The resistor R_S does not provide self-bias as it does in the JFET circuit. Self-bias is not possible with enhancement devices. In Figure 5–38 the resistor R_S is used to provide feedback for bias stabilization, in the same way that the emitter resistor does in a BJT bias circuit. The larger the value of R_S, the less sensitive the bias point is to changes in MOSFET parameters caused by temperature changes or by device replacement. Recall that R_S in a JFET self-bias circuit also provides this beneficial effect.

Figure 5–39 shows the voltage drops in the enhancement MOSFET bias circuit. R_1 and R_2 form a voltage divider that determines the gate-to-ground voltage V_G:

FIGURE 5–39 Voltage drops in the enhancement NMOS bias circuit

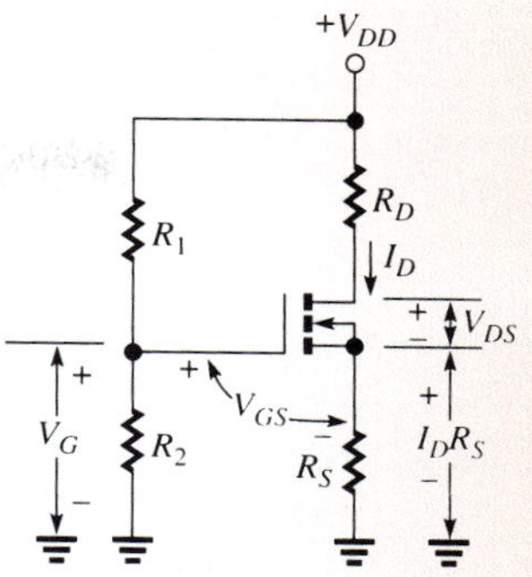

$$V_G = \left(\frac{R_2}{R_1 + R_2}\right)V_{DD} \tag{5–13}$$

The voltage divider is not loaded by the very large input resistance of the MOSFET, so the values of R_1 and R_2 are usually made very large to keep the ac input resistance of the stage large. Writing Kirchhoff's voltage law around the gate-to-source loop, we find

$$V_{GS} = V_G - I_DR_S \qquad \text{(NMOS)} \tag{5–14}$$

For a PMOS device, V_G and V_{GS} are negative, so equation 5–14 would be written

$$V_{GS} = V_G + I_DR_S \qquad \text{(PMOS)} \tag{5–15}$$

(Note that I_D is considered positive in both equations.) Writing Kirchhoff's voltage law around the drain-to-source loop, we find

$$V_{DS} = V_{DD} - I_D(R_D + R_S) \qquad \text{(NMOS)} \tag{5–16}$$

Again regarding I_D as positive in both the NMOS and PMOS devices, the counterpart of equation 5–16 for a PMOS device is

$$V_{DS} = -|V_{DD}| + I_D(R_D + R_S) \qquad \text{(PMOS)} \tag{5–17}$$

V_{DS} is negative in a PMOS circuit; note that the absolute value of V_{DD} must be used in equation 5–17 to obtain the correct sign for V_{DS}.

Equation 5–14 can be rewritten in the form

$$I_D = -\,(1/R_S)V_{GS} + V_G/R_S \tag{5–18}$$

Equation 5–18 is seen to be the equation of a straight line on the $I_D - V_{GS}$-axes. It intercepts the I_D-axis at V_G/R_S and the V_{GS}-axis at V_G. The line can be plotted on the same set of axes as the transfer characteristics of the device, and the point of intersection locates the bias values of I_D and V_{GS}.

General Algebraic Solution

We can obtain general algebraic expressions for the bias points in PMOS and NMOS circuits by solving equation 5–12 simultaneously with equation 5–14 or 5–15 for I_D. The results are shown as equation 5–19 and are valid for both NMOS and PMOS devices.

General algebraic solution for the bias point of NMOS and PMOS circuits

$$|V_G| = \frac{R_2}{R_1 + R_2}|V_{DD}|$$

$$I_D = \frac{-B - \sqrt{B^2 - 4AC}}{2A}$$

where

$$A = R_S^2$$

$$B = -2\left((|V_G| - |V_r|)R_S + \frac{1}{\beta}\right)$$

$$C = (|V_G| - |V_r|)^2$$

$$|V_{DS}| = |V_{DD}| - I_D(R_D + R_S) \quad \text{See note 1.}$$

$$|V_{GS}| = |V_G| - I_D R_S \quad \text{See note 2.}$$

Note 1. V_{DS} is positive for an NMOS FET and negative for a PMOS FET.

Note 2. V_{GS} is positive for an NMOS FET and negative for a PMOS FET.

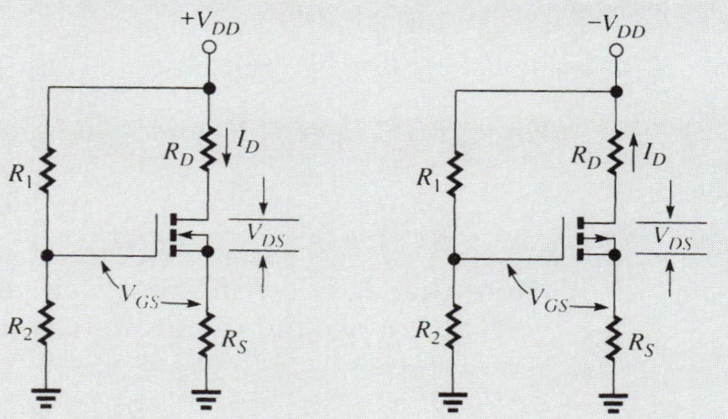

EXAMPLE 5–9

The transfer characteristic of the NMOS FET in Figure 5–40 is given in Figure 5–41 ($\beta = 0.5 \times 10^{-3}$ and $V_T = 2$ V). Determine values of V_{GS}, I_D, and V_{DS} at the bias point (1) graphically and (2) algebraically.

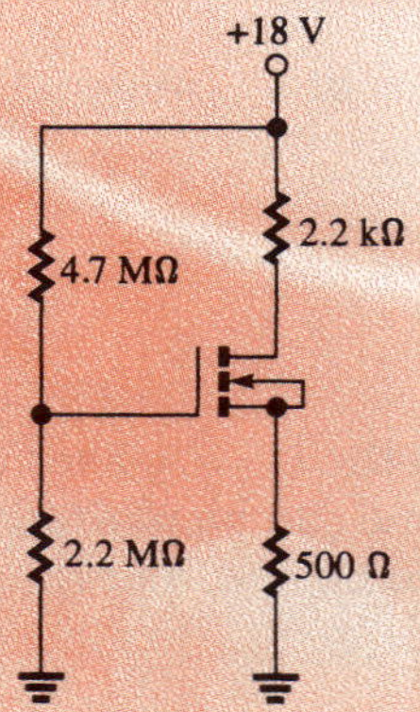

FIGURE 5–40 (Example 5–9).

Solution

1. From equation 5–13,

$$V_G = \left(\frac{22 \times 10^6}{47 \times 10^6 + 22 \times 10^6}\right) 18\text{ V} = 5.74\text{ V}$$

Substituting in equation 5–18, we have

$$I_D = -2 \times 10^{-3} V_{GS} + 11.48 \times 10^{-3}$$

This equation intersects the I_D-axis at 11.48 mA and the V_{GS}-axis at $V_G = 5.74$ V. It is shown plotted with the transfer characteristic in Figure 5–41. The two plots intersect at the quiescent point, where the values of I_D and V_{GS} are approximately $I_D = 2.0$ mA and $V_{GS} = 4.6$ V. The corresponding quiescent value of V_{DS} is found from equation 5–16:

$$V_{DS} = 18 - (2.0\text{ mA})[(2.2\text{ k}\Omega) + (0.5\text{ k}\Omega)] = 12.60\text{ V}$$

In order for this analysis to be valid, the Q-point must be in the saturation region: that is, we must have $V_{DS} > V_{GS} - V_T$. In our example, we have

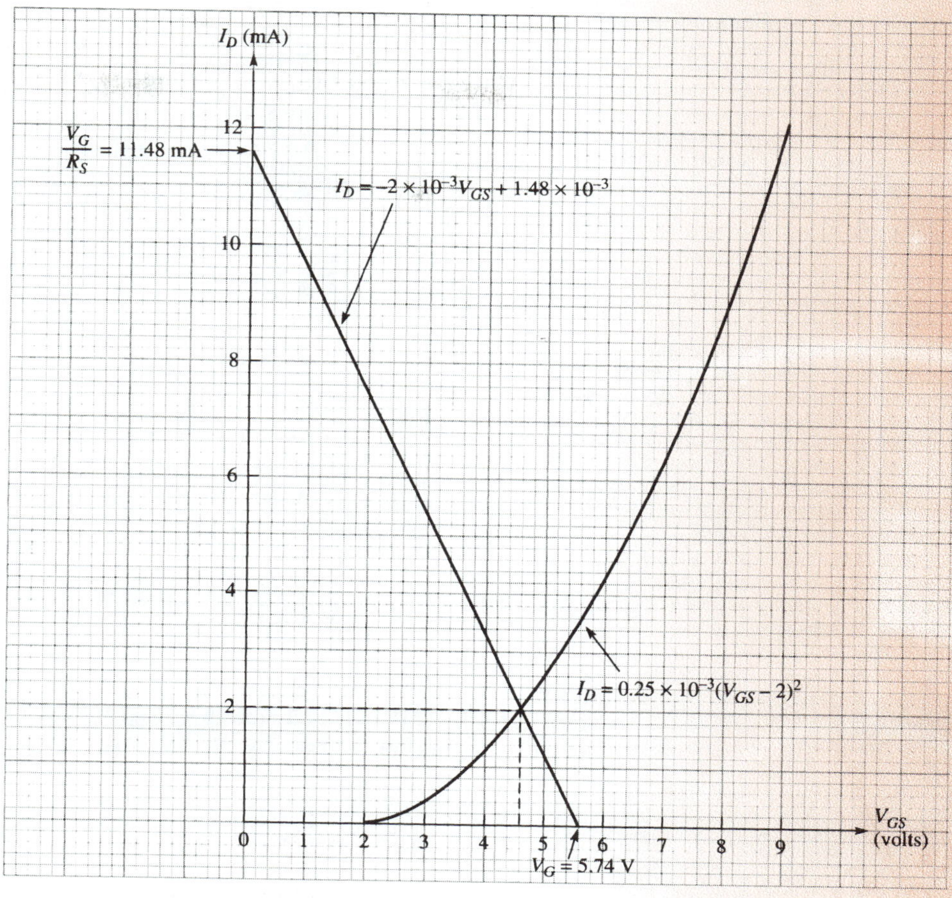

FIGURE 5–41 (Example 5–9)

V_{DS} = 12.60 V and $V_{GS} - V_T$ = 2.6 V, so we know the results are valid. The validity criterion can be expressed for both NMOS and PMOS FETs as $|V_{DS}| > |V_{GS} - V_T|$.

2. We have already found V_G = 5.74 V. Using R_S = 500 Ω, R_D = 2.2 kΩ, V_{DD} = 18 V, V_T = 2 V, and $\beta = 0.5 \times 10^{-3}$, we have, with reference to equations 5–19.

$$A = (500)^2 = 2.5 \times 10^5$$
$$B = -2[(5.74-2)500 + 1/(0.5 \times 10^{-3})] = -7.74 \times 10^3$$
$$C = (5.74-2)^2 = 13.9876$$

Substituting these values into the equation for I_D, we find I_D = 1.927 mA. Then, V_{DS} = 18 V − (1.927 mA)(2.2 kΩ + 500 Ω) = 12.8 V and V_{GS} = 5.74 V − (1.927 mA)(500 Ω) = 4.78 V. These results agree well with those obtained graphically in part 1.

EXAMPLE 5–10

SPICE

Use SPICE to find I_D, V_{DS} and V_{GS} for the figure shown.

Solution

The SPICE circuit and input data file are shown in Figure 5–42(a). Note that the parameter V_T is entered in the .MODEL statement as VTO = 2 and that β is entered as KP = 0.5E−3 (see Section A–12). All other parameter values are

FIGURE 5–42 (Example 5–10)

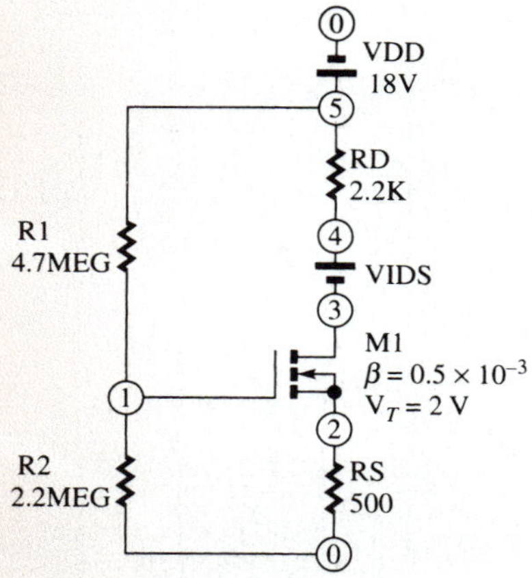

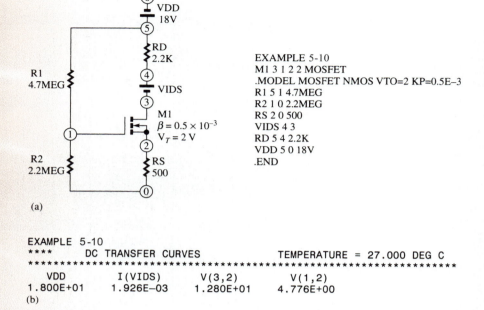

```
EXAMPLE 5-10
M1 3 1 2 2 MOSFET
.MODEL MOSFET NMOS VTO=2 KP=0.5E-3
R1 5 1 4.7MEG
R2 1 0 2.2MEG
RS 2 0 500
VIDS 4 3
RD 5 4 2.2K
VDD 5 0 18V
.END
```

(a)

```
EXAMPLE 5-10
****      DC TRANSFER CURVES                TEMPERATURE = 27.000 DEG C
**********************************************************************
   VDD          I(VIDS)        V(3,2)         V(1,2)
1.800E+01     1.926E-03     1.280E+01      4.776E+00
```

(b)

allowed to default. The results of the analysis are shown in Figure 5–42(b). We see that I_D = I(VIDS) = 1.926 mA, V_{DS} = V(3, 2) = 12.8 V and V_{GS} = V(1, 2) = 4.776 V, in good agreement with the previous example.

5–8 INTEGRATED-CIRCUIT MOSFETS

By far the greatest number of MOSFETs manufactured today are in integrated circuits. The enhancement-type MOSFET has a very simple structure (Figure 5–31) that makes its fabrication in a crystal substrate a straightforward and economical procedure. Furthermore, a very great number of devices can be fabricated in a single chip. Enhancement MOSFETs account for the vast majority of very large scale integrated (VLSI) circuits manufactured, and they are the primary ingredients of digital ICs such as microprocessors and computer memories.

The fabrication of integrated-circuit MOSFETs is accomplished using photolithographic techniques and batch production methods. Because millions of components may be fabricated in a single VLSI chip, the techniques we described for producing very fine masks and for direct writing of patterns using electron beams are particularly appropriate to VLSI technology. Ion implantation, which allows close control of impurity concentration and layer depth, is widely used to control the values of threshold voltages and other MOSFET characteristics.

Figure 5–43 shows cross-sectional views of PMOS and NMOS FETs embedded in crystal substrates. Note that a layer of polycrystalline silicon is deposited over the gate of an NMOS device to form the gate terminal. This layer improves device performance but adds to the complexity of the manufacturing procedure. PMOS devices are less expensive to produce but do not perform as well as NMOS circuits, primarily because the mobility of the majority carriers (holes) in p material is smaller than that of the majority carriers (electrons) in n material. NMOS circuits are generally preferred and can be produced with the greatest number of components per chip for a given performance capability.

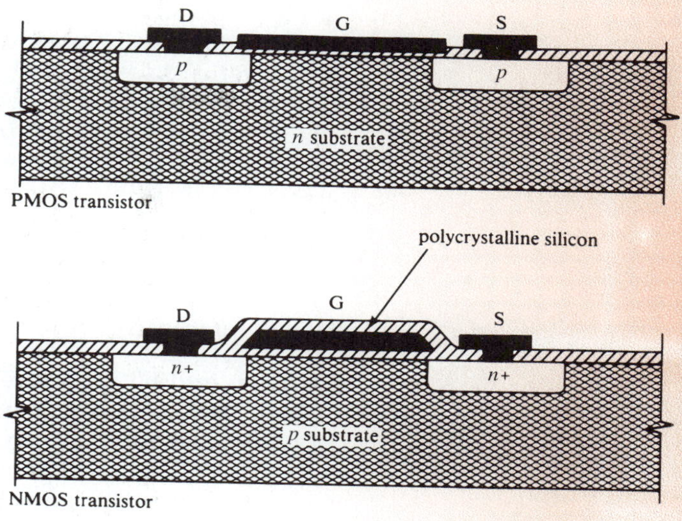

FIGURE 5–43 Cross-sectional views of integrated-circuit MOSFETs

Another type of digital integrated circuit using enhancement MOSFETs has both PMOS and NMOS devices embedded in the same substrate. These circuits are called *complementary* MOS, or CMOS, circuits. They are more difficult to construct than either PMOS or NMOS circuits, but they have the best performance characteristics, especially in terms of switching speed. We will discuss applications of CMOS circuitry in Chapter 18. Figure 5–44 shows a cross-sectional view of a CMOS circuit containing one PMOS and one NMOS transistor. Note that it is necessary to embed a *p*-type layer in the *n* substrate for the NMOS transistor. This *p* layer, frequently called a "tub" or "*p*-well," is necessary for the formation of the *induced n* channel of the NMOS transistor. Also note the n^+ and p^+ regions used to isolate the transistors. The CMOS structure is made with a polycrystalline silicon gate electrode. The more complex structure of a CMOS IC is evident in the figure.

5–9 VMOS AND DMOS TRANSISTORS

VMOS Transistors

Still another variation in MOS structure is called VMOS, which is used to produce both *n*-channel and *p*-channel enhancement MOSFETs. The name is derived from the appearance of the cross-sectional view (Figure 5–45), in which it can be seen that a V-shaped groove penetrates alternate *n* and *p* layers. (In reality, the device is formed in a crater shaped like an inverted pyramid.) As can be seen in the *n*-channel VMOS transistor shown in Figure 5–45, the length of the induced *n* channel is determined by the *thickness* of the *p* layer. The layers

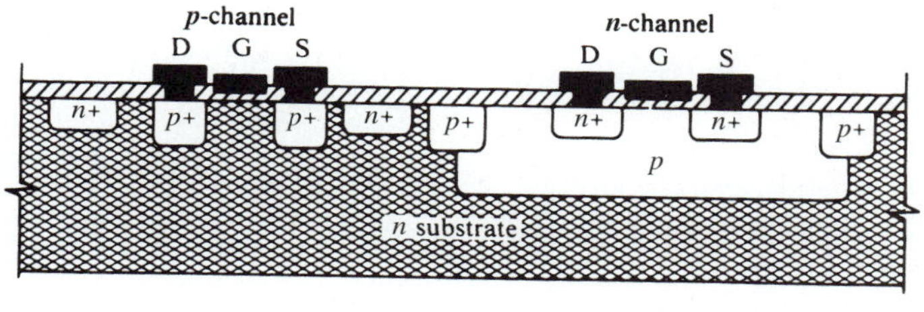

FIGURE 5–44 Complementary MOS (CMOS) integrated circuit containing both NMOS and PMOS devices

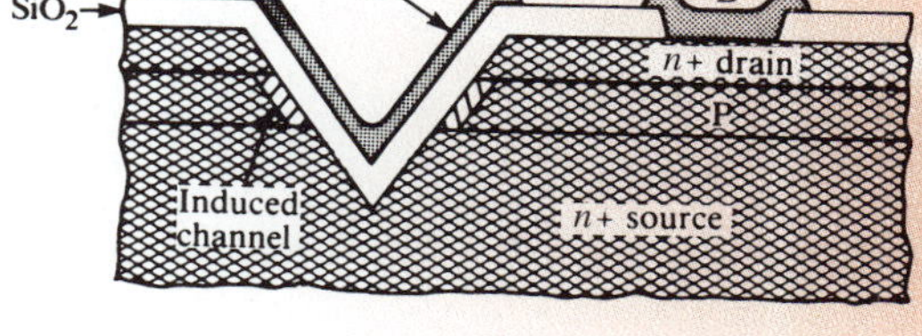

FIGURE 5–45 VMOS structure. Note that the length of the channel depends on the thickness of the diffused *p* layer.

are formed using epitaxial growth and diffusion methods, which provide good control over thickness and therefore good control of channel length. Since the aspect ratio of the channel determines some important properties of the FET, this control is a valuable feature of the method. Also, the technique conserves space on the chip surface because the channel can be made longer simply by making the *p* region thicker. As a result, a greater number of devices can be created in one chip using conventional photolithographic methods. Finally, VMOS transistors have greater current-handling capabilities than their planar counterparts and are finding use in power-amplifier applications.

DMOS Transistors

DMOS is an FET structure created specifically for high-power applications. It is a planar transistor whose name is derived from the double-diffusion process used to construct it. Figure 5–46(a) shows a cross-sectional view. Note that drain-to-source current flows through a *p*-type channel region (that is inverted by a positive gate-to-source voltage), an n^- (lightly doped) epitaxial region, and n^+ substrate. The channel length can be closely controlled in the diffusion process and is typically very short (a few micrometers). For this reason, it is called a *short-channel* MOS transistor. DMOS transistors are widely used for *switching* heavy currents and high voltages and switching regulators.

The DMOS transistor has a *parasitic* diode between its drain and source. A parasitic component of a semiconductor device is one that exists as a result of the structure of the device rather than by design. In other words, the parasitic diode in a DMOS transistor is inherent in its structure of *p* and *n* layers: Its existence is inevitable. In Figure 5–46(a), the n^+ source and the *p* channel (called the *body*) form an *electrical bond* rather than a *pn* junction. The parasitic diode is the *pn* junction between the body and drain. Because the body is electrically connected to the source, the parasitic diode is effectively connected across the drain and source terminals, with the anode connected to

FIGURE 5–46 The DMOS transistor

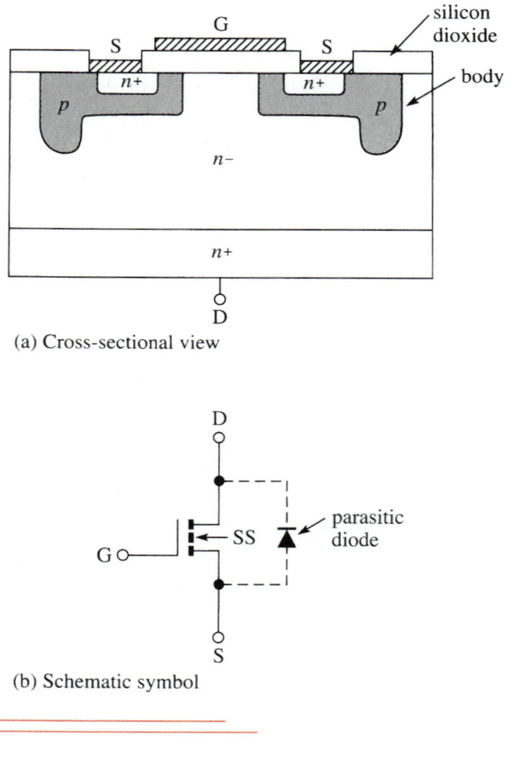

source and the cathode connected to drain. The schematic symbol for the DMOS transistor often includes the parasitic diode, as shown in Figure 5–46(b).

One property shared by short-channel power FETs is that they have a *linear* transfer characteristic. That is, when the gate-to-source voltage is greater than the threshold voltage, the drain current is a linear function of the gate-to-source voltage rather than the nonlinear function illustrated in Figure 5–37. The linear transfer characteristic of short-channel devices is the result of a phenomenon called *velocity saturation*, whereby the velocity of charge carriers reaches but does not exceed a certain value as drain-to-source current increases.

5–10 FET CIRCUIT ANALYSIS WITH ELECTRONICS WORKBENCH MULTISIM

This section examines the use of EWB to simulate a JFET transistor switch. The JFET can be used as an analog switch as described in Section 5–5. To begin the exercise, open the file Ch5-EWB.msm that is found in the Electronics Workbench Multisim 2001 CD-ROM packaged with the text. The circuit being demonstrated is shown in Figure 5–47. This is a JFET chopper circuit. Q1 is an ideal JFET transistor being driven by the function generator (XFG1). The resistor R1 is simulating the load, the input signal is a 2 $V_{p\text{-}p}$ 1000 Hz signal. The EWB oscilloscopes have been connected to monitor the input signal (XSC2), the function generator input signal, and the circuit output (XSC1).

This application will demonstrate the operation of a JFET chopper. The JFET transistor will be set up to switch from the cutoff region ($V_{GS} = V_P = -2$ V) to the pinch-off region ($V_{GS} = 0$). The EWB Multisim ideal transistor will be used. The model parameters for the JFET transistor can be viewed by double-clicking on the JFET. This opens the menu shown in Figure 5–48.

Click on Edit Model to view the model parameters used by EWB Multisim. This opens the list shown in Figure 5–49. Notice that VTO = −2. This is the setting for the pinch-off voltage for the transistor.

Double-click on the function generator (XFG1) to set the signal levels to those shown in Figure 5–50. The function generator is set to produce a square-wave with a sample frequency of 50 Hz. The offset voltage is set to −2 V and the amplitude is 2 V. The function generator will produce a square-wave that

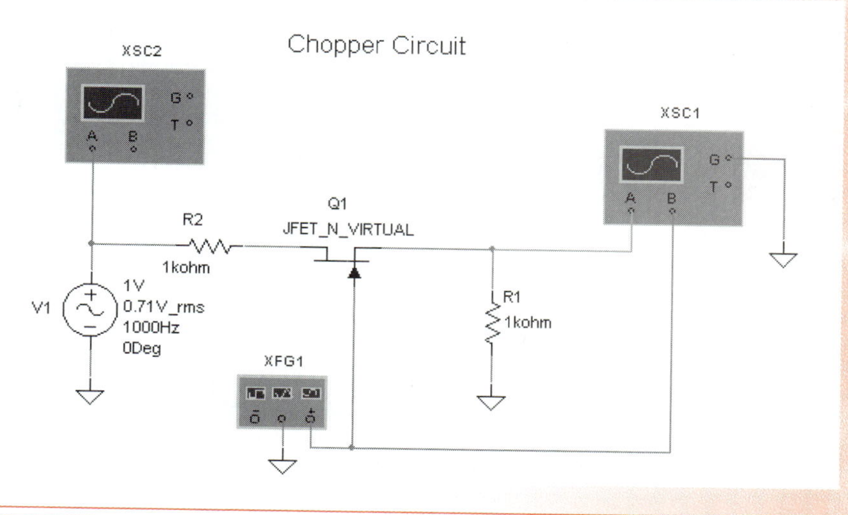

FIGURE 5–47 The JFET chopper circuit used in the EWB exercise

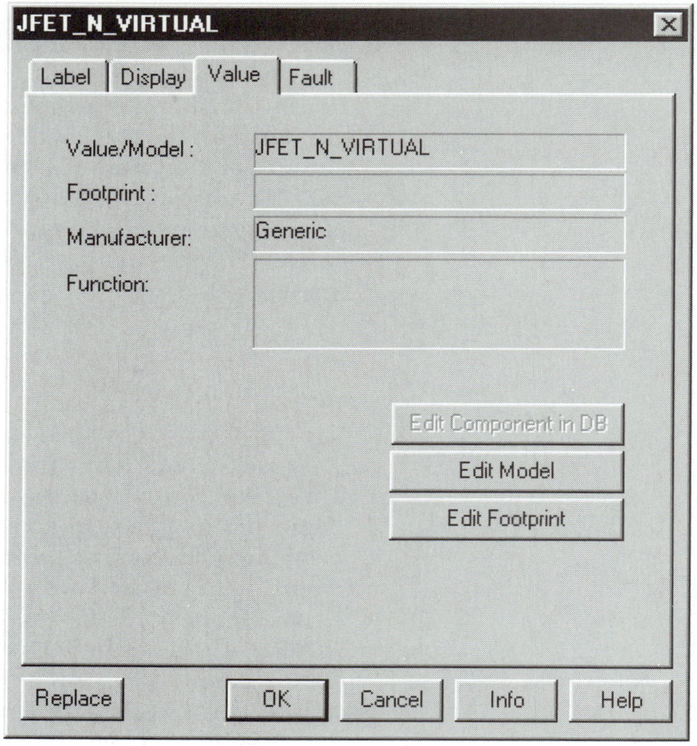

FIGURE 5–48 The menu for selecting the transistor model parameters

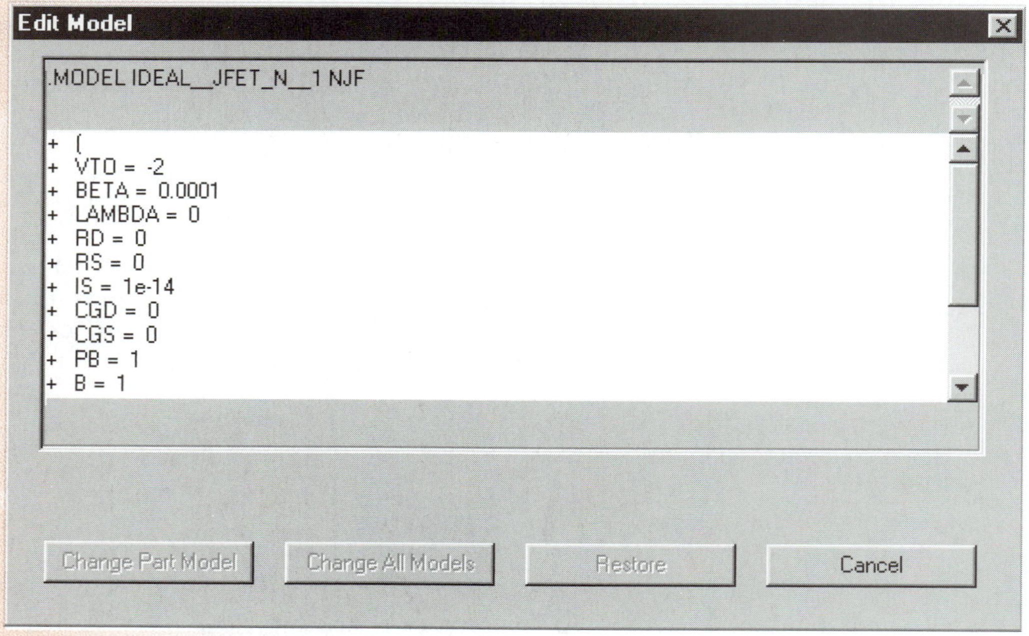

FIGURE 5–49 The list of the JFET model parameters

switches between −4 V and 0 V. This is adequate to place the transistor in the pinch-off and cutoff mode.

Start the simulation by clicking on the simulation start button and click on the oscilloscope XSC1 to examine the output of the circuit. You should observe signals similar to that shown in Figure 5–51. Notice that the output

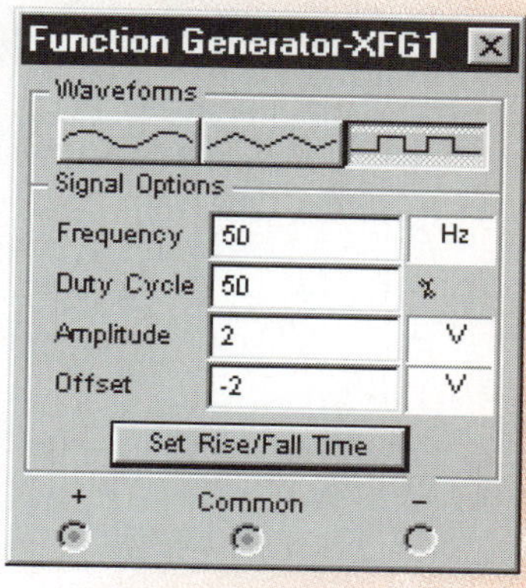

FIGURE 5–50 The settings for the EWB function generator

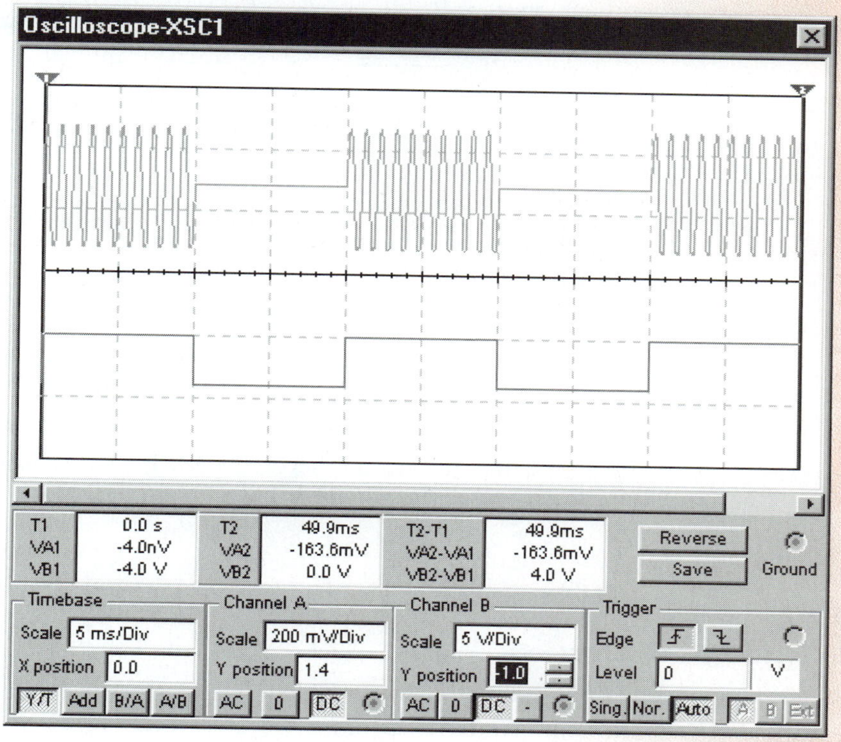

FIGURE 5–51 The output of the JFET switch

signal (shown on top) displays a sine-wave except when the input gate voltage goes to 0 V. At that time, the output goes to zero volts.

The JFET can be used as a chopper by setting the frequency of the function generator to a larger frequency than the input frequency. In this example, the input frequency remains at 1 kHz and the frequency of the function generator is set to 6 kHz. The output of the chopper circuit is shown in Figure 5–52.

This exercise has demonstrated the operation of the JFET switch and the JFET chopper. The setup is simple and the circuit is easy to implement using the simulator and also in the lab.

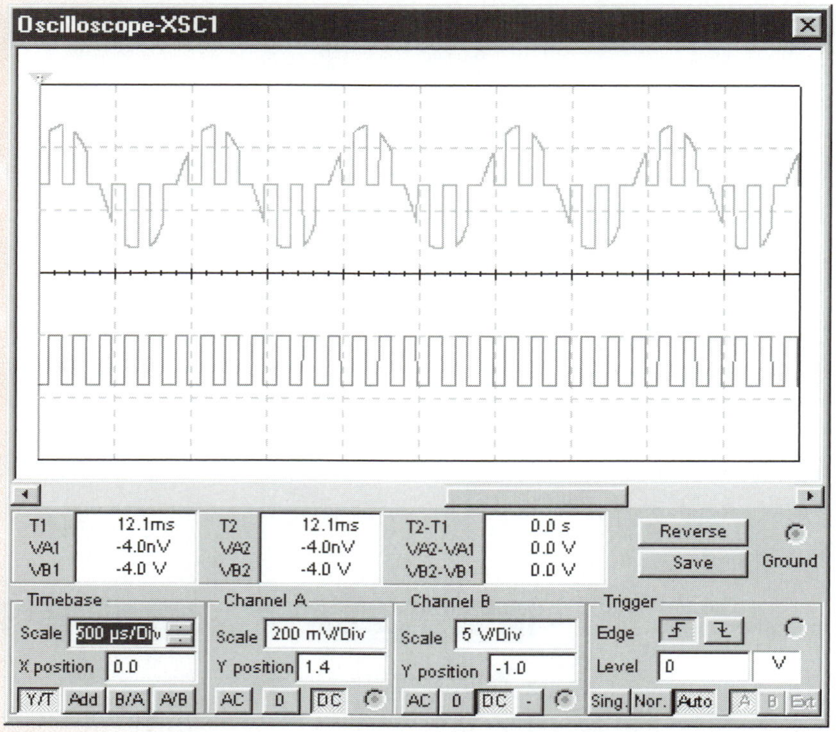

FIGURE 5–52 The output of the JFET chopper

SUMMARY

This chapter has presented the basics of the FET transistor. Students should have mastered these concepts and skills:

- Basic circuit analysis techniques.
- JFET transistor circuit analysis to identify the cutoff, pinch-off, and voltage-controlled–resistance of operation.
- Using the JFET transistor as an analog switch.
- How to bias a discrete MOSFET transistor.

EXERCISES

SECTION 5–2

Junction Field-Effect Transistors

5–1. An *n*-channel JFET has the drain characteristics shown in Figure 5–6. Find the approximate dc resistance between drain and source when $V_{DS} = 1$ V and (a) $V_{GS} = 0$ V, (b) $V_{GS} = -1$ V, and (c) $V_{GS} = -2$ V. Explain why these results confirm that these points lie in the voltage-controlled–resistance region of the characteristics.

5–2. An *n*-channel JFET has $I_{DSS} = 16$ mA and $V_p = -6$ V.

(a) What is the value of $V_{DS(sat)}$ when $V_{GS} = -4$ V?

(b) What is the saturation current at $V_{GS} = -4$ V?

5–3. A *p*-channel JFET has a pinch-off voltage of 8 V. At what value of V_{GS} does $V_{DS(sat)} = -3$ V?

5–4. Using the transfer characteristic shown in Figure 5–10, find approximate values for (a) I_D when $V_{DS} = 8$ V and $V_{GS} = -1.6$ V, and (b) V_{GS} when $V_{DS} = 8$ V and $I_D = 10$ mA.

5–5. Using the transfer characteristic shown in Figure 5–10, find the total change in I_D ($\Delta\ I_D$) when (a) V_{GS} changes from -2.8 V to -1.8 V and (b) V_{GS} changes from −1.8 V to −0.8 V. What do these results tell you about the linearity of the device?

5–6. An n-channel JFET has a pinch-off voltage of −5.8 V and I_{DSS} = 15 mA. Assuming that it is operated in its pinch-off region, find the value of I_D when (a) V_{GS} = 0 V, (b) V_{GS} = −2 V, and (c) V_{GS} = −6.5 V.

5–7. An n-channel JFET having a pinch-off voltage of −3.5 V has a saturation current of 2.3 mA when V_{GS} = −1 V. What is its saturation current when (a) V_{GS} = 0 V and (b) V_{GS} = −2 V?

5–8. A p-channel JFET has a pinch-off voltage of 6 V and I_{DSS} = 18 mA. At what value of V_{GS} in the pinch-off region will I_D equal 6 mA? What is the value of V_{DS} at the boundary of the pinch-off and voltage-controlled–resistance regions when I_D = 6 mA?

SECTION 5–3

JFET Biasing

5–9. The JFET in the circuit of Figure 5–53 has the drain characteristics shown in Figure 5–14. Find the quiescent values of I_D and V_{DS} when (a) V_{GS} = −2 V and (b) V_{GS} = 0 V. Which, if either, of the Q-points is in the pinch-off region?

5–10. Using Figure 5–14, determine the value of V_{GS} in Exercise 5–9 that would be required to obtain V_{DS} = 8 V.

5–11. The JFET shown in Figure 5–54 has I_{DSS} = 14 mA and V_p = −5 V. Algebraically determine the quiescent values of I_D and V_{DS} for (a) V_{GS} = −3.6 V, (b) V_{GS} = −3 V, and (c) V_{GS} = −1.7 V. In each case, check the validity of your results by verifying that the quiescent point is in the pinch-off region. Identify any cases that do not meet that criterion and for which the results are therefore not valid.

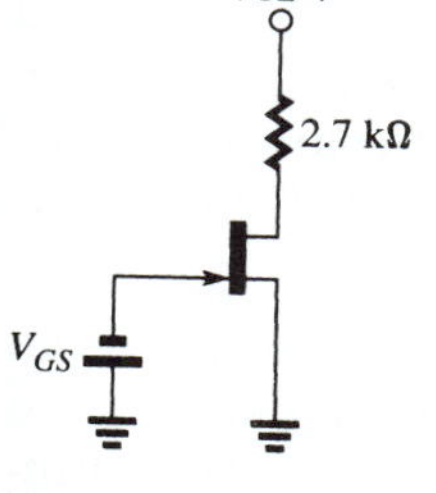

FIGURE 5–53 (Exercise 5–9)

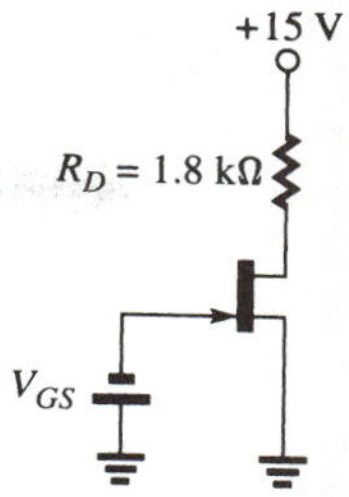

FIGURE 5–54 (Exercise 5–11)

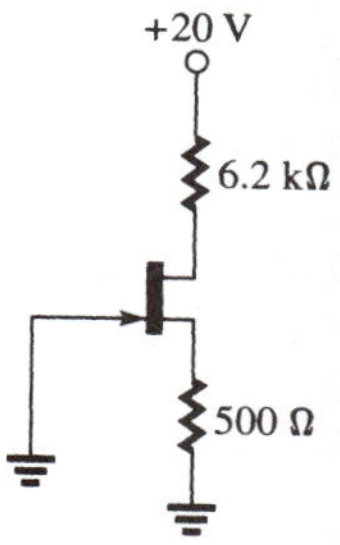

FIGURE 5–55 (Exercise 5–13)

5–12. Repeat Exercise 5–11 when R_D is changed to 1 kΩ. Does this modification affect the validity of the results in any of the three cases (a), (b), or (c)? Explain.

5–13. Figure 5–56 shows the transfer characteristics of the JFET in the circuit of Figure 5–55. Graphically determine the quiescent values of I_D and V_{GS}. Compute the quiescent value of V_{DS} based on your results and verify their validity.

5–14. Algebraically determine the quiescent values of I_D, V_{GS}, and V_{DS} in the circuit of Exercise 5–13. (Refer to Figure 5–56 to obtain values for I_{DSS} and V_p.)

5–15. Figure 5–58 shows the transfer characteristic for the JFET in the circuit of Figure 5–57.

(a) Graphically determine the quiescent values of I_D and V_{GS}. Compute the quiescent value of V_{DS} based on your results and verify their validity.

(b) Determine the quiescent values of I_D, V_{GS}, and V_{DS} algebraically.

SECTION 5–4

The JFET Current Source

5–16. The JFET shown in Figure 5–59 has V_p = −5 V and I_{DSS} = 12 mA. What is the

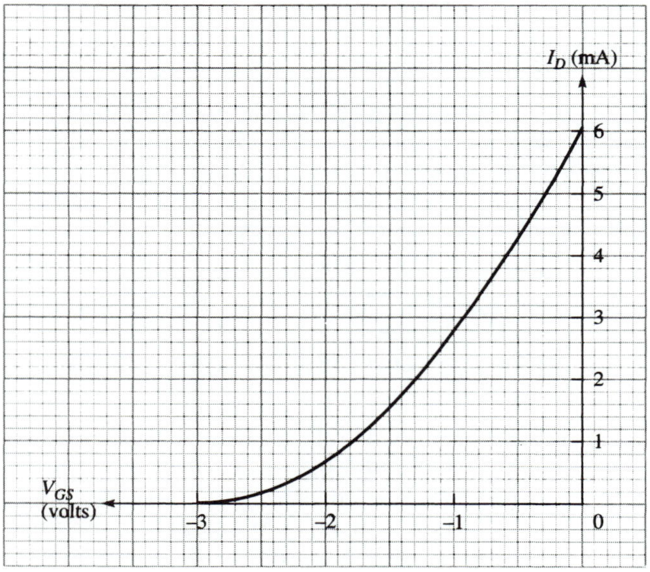

FIGURE 5–56 (Exercise 5–13)

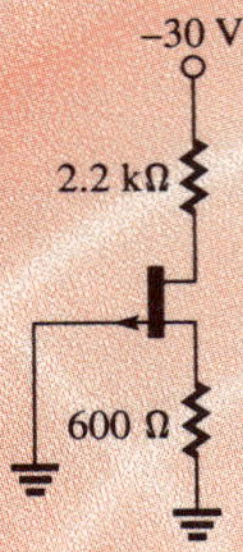

FIGURE 5–57 (Exercise 5–15)

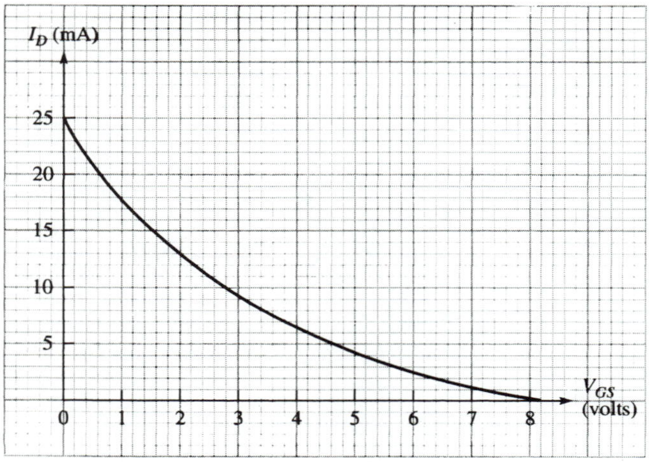

FIGURE 5–58 (Exercise 5–15)

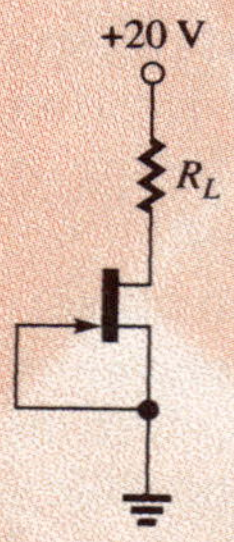

FIGURE 5–59 (Exercise 5–16)

maximum value of R_L for which the circuit can be used as a constant-current source?

5–17. When $V_{GS} = 0$ V, the JFET shown in Figure 5–60 begins to break down at $V_{DS} = 15$ V. If $V_p = -3.5$ V and $I_{DSS} = 10$ mA, over what range of R_L can the circuit be used as a constant-current source?

5–18. When $V_{GS} = 0$ V, the JFET shown in Figure 5–61 begins to break down at $V_{DS} = 18$ V. If $V_p = 4$ V and $I_{DSS} = 12$ mA, over what range of R_L can the circuit be used as a constant-current source?

SECTION 5–5

The JFET as an Analog Switch

5–19. The drain-to-source resistance when the JFET in Figure 5–62 is conducting is 80 Ω. If $v_d = 0.16$ V, find the load voltage v_L (a) when $V_{GS} = V_p$ and (b) when $V_{GS} = 0$ V.

5–20. The JFET in Figure 5–63 has $R_{D(ON)} = 100\ \Omega$ and $V_p = 4$ V. If $v_i = 0.1 \sin 1000t$, write the expression for v_L (a) when $V_{GS} = 4$ V and (b) when $V_{GS} = 0$ V.

5–21. The square wave shown in Figure 5–64 turns the chopper ON when high and OFF when low. If it has frequency 1 kHz, sketch v_L over a time period of 3 ms. Assume that $R_{D(ON)}$ for the JFET is 50 Ω.

5–22. The JFET shown in Figure 5–65 has $R_{D(ON)} = 40\ \Omega$. If the output voltage is to be no less than 90% of the signal voltage when the JFET is ON, what is the minimum permissible value of R_L?

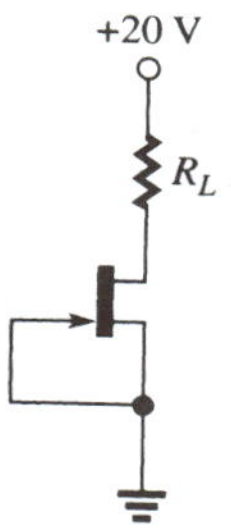

FIGURE 5–60 (Exercise 5–17)

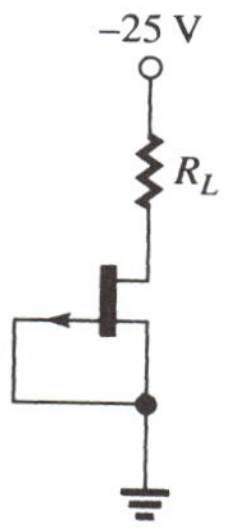

FIGURE 5–61 (Exercise 5–18)

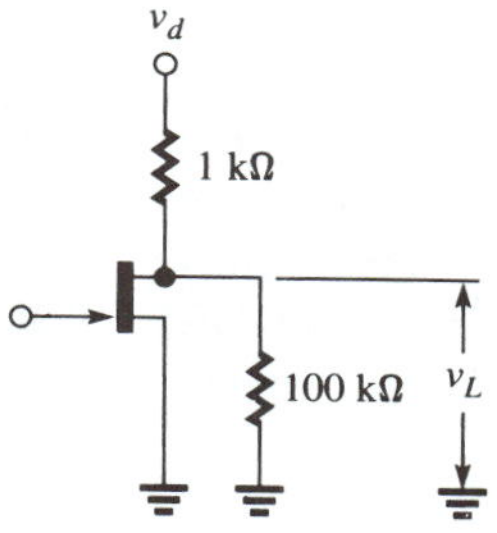

FIGURE 5–62 (Exercise 5–19)

SECTION 5–6
Manufacturers' Data Sheets

5–23. By referring to the manufacturer's data sheets for the 2N4220–2N4222 series of JFETs and assuming that $T = 25°C$, determine the following:

(a) What is the maximum permissible drain current in each device?

(b) What is the maximum pinch-off voltage for 2N4221 transistors?

(c) What is the maximum permissible value of V_{GS} for each device?

(d) For which device is the variation in possible values of I_{DSS} the greatest (in terms of the ratio of maximum to minimum values)?

5–24. Refer to the manufacturer's data sheet in Figure 5–30 to determine the following:

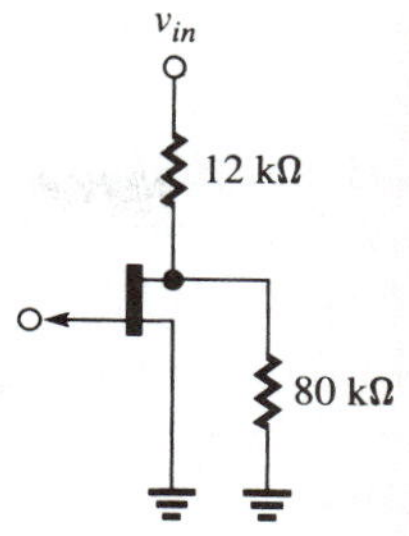

FIGURE 5–63 (Exercise 5–20)

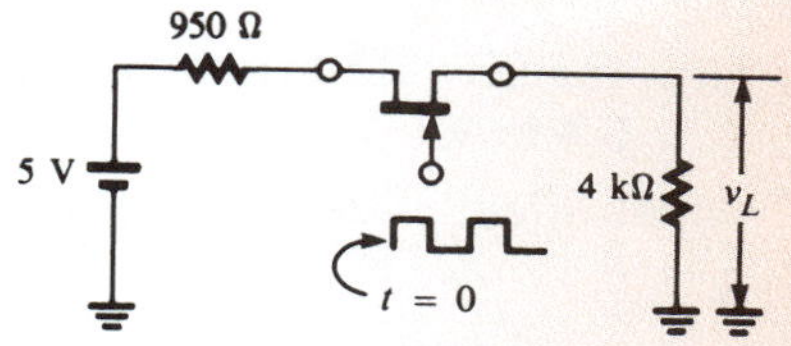

FIGURE 5–64 (Exercise 5–21)

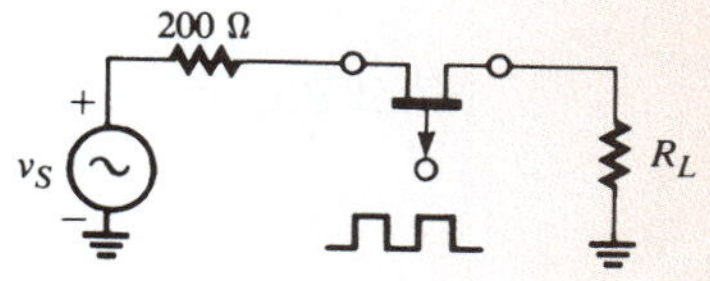

FIGURE 5–65 (Exercise 5–22)

(a) Locate the transfer characteristic that shows the maximum value of I_{DSS} to be 12 mA at 40°C. What is the approximate value of V_{GS} at the point of zero temperature coefficient? What is the approximate maximum value of I_{DSS} at 25°C?

(b) Locate the drain characteristics for which I_{DSS} is approximately 1.4 mA. What is the approximate dc resistance between drain and source when $V_{DS} = 0.5$ V and $V_{GS} = 0$ V? At what value of V_{DS} will $I_D = 1.2$ mA when $V_{GS} = -0.1$ V?

SECTION 5–7
Metal-Oxide-Semiconductor FETs

5–25. An induced *n*-channel enhancement MOSFET has the drain characteristics shown in Figure 5–34. At what value of V_{DS} would a curve corresponding to $V_{GS} = 7.35$ V intersect the parabola?

5–26. The induced *n*-channel enhancement MOSFET whose drain characteristics are shown in Figure 5–34 is to be operated in its active region with a drain current of 10.4 mA. What value should V_{DS} exceed?

5–27. An enhancement NMOS FET has $\beta = 0.5 \times 10^{-3}$ and $V_T = 2.5$ V. Find the value of I_D when (a) $V_{GS} = 6.14$ V, and (b) $V_{GS} = 0$ V.

5–28. An enhancement PMOS FET has $\beta = 0.5 \times 10^{-3}$ and $V_T = -2$ V. What is the value of V_{GS} when $I_D = 10.32$ mA?

5–29. An enhancement NMOS FET has the transfer characteristic shown in Figure 5–37.

(a) Graphically determine V_{GS} when $I_D = 6.4$ mA.

(b) Algebraically determine V_{GS} when $I_D = 6.4$ mA.

5–30. In the bias circuit of Figure 5–30, $R_1 = 2.2$ mΩ, $R_2 = 1$ mΩ, $V_{DD} = 28$ V, $R_D = 2.7$ kΩ, and $R_s = 600$ Ω . If $V_{GS} = 5.5$ V, find (a) I_D and (b) V_{DS}.

5–31. In the bias circuit of Figure 5-38, $R_1 = 470$ kΩ , $V_{DD} = 20$ V, $R_D = 1.5$ kΩ , $R_s = 220$ Ω, $I_D = 6$ mA, and $V_{GS} = 6$ V. Find (a) V_{DS} and (b) R_2.

5–32. The MOSFET in Figure 5–66 has $\beta = 0.62 \times 10^{-3}$ and $V_T = -2.4$ V. Algebraically determine the quiescent values of I_D, V_{GS}, and V_{DS}. Verify the validity of your results.

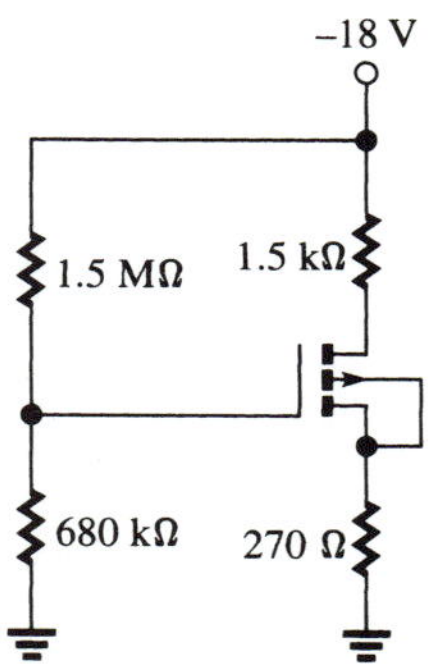

FIGURE 5–66 Exercise 5–32

SECTION 5–8

Integrated-Circuit MOSFETs

5–33. What does CMOS stand for? Why is it so named?

SECTION 5–9

VMOS and DMOS Transistors

5–34. List and briefly describe three advantages of VMOS technology.

5–35. What process is used to construct DMOS transistors? What is their primary application? What characteristics make them different from conventional MOSFETs?

SPICE EXERCISES

5–36. Verify the answers to Example 5–3 using SPICE analysis. BETA = .62SE−3, VTO = −4 for $V_{GS} = -1.5$ V. Why do the answers in Example 5–3 differ from the SPICE results?

5–37. Use SPICE to simulate the JFET chopper circuit shown in Figure 5–28. Use a 1–V p-p 1-kHz sinusoid for the input, a source resistance of 600 Ω, and a gate pulse of 0 to −5 V with a pulse width of 100 μs and a period of 200 μs. The BETA of the transistor is 7.5 E−4 and VTO = −4.

```
VS 5 0 SIN(0.5 1kHz)
RS 5 2 600
VG 3 0 PULSE(0 −5 0 1μs 1μs 100μs 200μs)
.MODEL NFET NJF BETA = 7.5E-4 VTO=−4
```

Provide a plot of the input sinusoid, the gate pulse, and the output waveform. What is the output waveform called?

CHAPTER 6

AMPLIFIER FUNDAMENTALS

OUTLINE

OBJECTIVES

- Develop an understanding of the amplification function.
- Visualize an amplifier in its different components.
- Rationalize the concepts of voltage, current, and power gains.
- Perform circuit analysis using amplifier models.
- Understand the difference between single-ended and differential amplifiers.
- Recognize and analyze cascading amplifier configurations.

6–1 INTRODUCTION

Within an electronics context, the term amplifier refers to a circuit used for boosting electric signals in a prescribed manner. These signals will typically be generated by *transducers* such as microphones, tape heads, thermocouples, and electrochemical cells. They can also come from self-contained electronic components or instruments such as tape decks, CD players, pressure sensors, among many others.

Let us consider a basic stereo amplifier with a turntable (remember that?) connected to it. The signal stamped in the grooves of the vinyl record (disc) is picked up by a needle and a magnetic cartridge (also called magnetic pickup) and converted to a very small ac signal voltage that represents music and voice. The corresponding sound will eventually come from a pair of loudspeakers connected to the output of the amplifier. The few millivolts of signal produced by the turntable's magnetic pickup is not enough for producing a sound on the loudspeaker. The stereo amplifier provides the necessary boosting of the signal and delivers the appropriate voltage level to the speakers.

Signal amplification in the stereo amplifier is accomplished by two identical multistage amplifiers which process the left and right signals that will eventually drive the speakers to produce sound. The initial stages boost the signal and provide additional processing such as tone control, balance, volume, and other functions. This section of the stereo amplifier is called the *pre-amplifier*. The signal from the pre-amplifier, although several hundred times larger than that from the turntable, is not yet capable of driving a pair of speakers. It is only strong enough to drive a set of headphones. The stage that can finally drive a speaker is a *power amplifier* that can produce several volts across the speaker load, which is typically 4 Ω to 8 Ω, and produce sound at relatively high levels.

Other types of amplifiers used in fields such as industry and medicine can process different types of signals. In this chapter, we will talk about amplifiers in the general sense without regard to the types of signals they will process. We will also look at the different ways of representing an amplifier using equivalent electric circuits. In this manner, amplifier analysis can easily be done using basic circuit theory.

6–2 AMPLIFIER CHARACTERISTICS

Consider the block illustrated in Figure 6–1(a), which shows two pairs of terminals. It is common in network analysis to refer to those pairs of terminals as *ports*. In this context, the block is said to be a *two-port network*. For our discussion, let port 1 be the input port and port 2 the output port. If we apply a voltage across the input port, as illustrated in Figure 6–1(b), and obtain a larger voltage across the output port, then we regard the block as being a voltage amplifier. Additionally, if upon connecting a load resistance across the output port, as shown in Figure 6–1(c), the load or output current obtained is larger than the input current, then we say that the block is also a current amplifier. An amplifier block can amplify voltage, or current, or both. And as a consequence, it can also amplify power.

There are many types of amplifiers; some can amplify not only ac or signal voltages but also dc voltages. Some others will produce sign inversion in addition to amplification. Amplifiers that are designed to produce large amounts of output current to drive low-impedance loads are regarded as power amplifiers. But there are several circuit parameters that are shared by all amplifiers, regardless of their type. These are

- input resistance,
- voltage gain,
- current gain,
- output resistance, and
- power gain.

Input Resistance

By definition, the input resistance of any network is the total equivalent resistance at its input terminals. In other words, it is the resistance "seen" by the input signal source when connected to the amplifier. Input resistance can also have different values depending on whether the input voltage or current is dc or ac. Furthermore, it can also vary with the frequency of the applied signal. But regardless of these facts, input resistance will always be defined as the ratio of input voltage to input current, as prescribed by Ohm's law. At this point, however, let us make the (possible) distinction between dc and ac input resistance, namely,

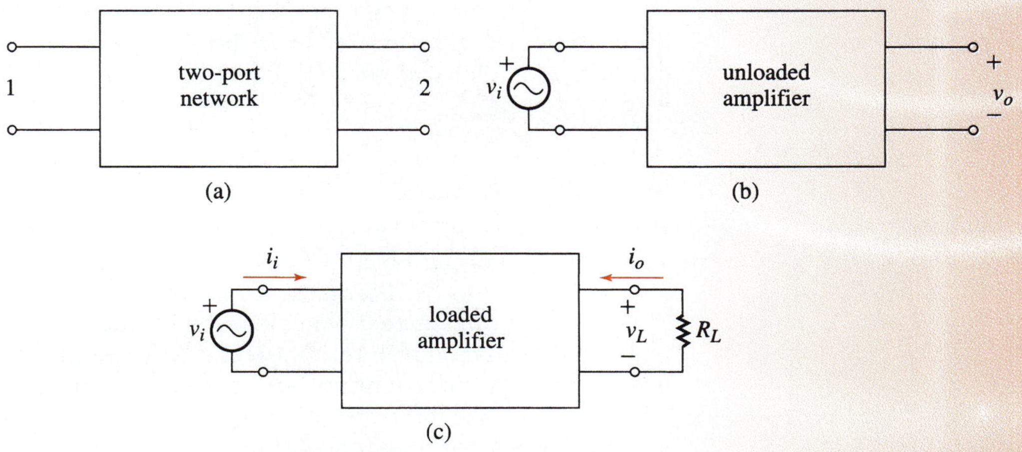

FIGURE 6–1 An amplifier pictured as a two-port network

$$R_i = V_i/I_i \text{ (dc input resistance)} \tag{6–1}$$

and

$$r_i = v_i/i_i \text{ (ac input resistance)} \tag{6–2}$$

where uppercase notation is used for dc quantities and lowercase notation for ac quantities.

Voltage Gain

Voltage gain is the voltage amplification factor from input to output. In words, output voltage equals input voltage *times* voltage gain. The voltage gain is therefore the *ratio* of output voltage to input voltage. Symbolized A_v, voltage gain is then defined by

$$A_v \equiv \frac{v_o}{v_i} \tag{6–3}$$

Let us stress the fact that voltage gain is the ratio, and not the difference, between output and input voltages. Also, if the amplifier produces sign inversion, the gain is said to be inverting and its value will be preceded by a negative sign.

Current Gain

Like voltage gain, current gain is an amplification factor defined as the ratio of output current to input current. Obviously, in order for an output current to actually take place, there must be a path for it, typically a load resistance, as seen in Figure 6–1(c). If the path for the output current is a short circuit across the output terminals, the current gain would be regarded as *short-circuit current gain*. Current gain, symbolized by A_i, is therefore defined as

$$A_i \equiv \frac{i_o}{i_i} \tag{6–4}$$

We will specify the output current as "entering" the positive output terminal. Note that this is an arbitrary reference and does not imply the actual direction of the output current. If the output current actually enters the positive output terminal when the input current enters (assumed reference) the positive input terminal, we say that the current gain is positive or noninverting. It is important to clarify that the reference direction for the output current is not universal and that technical literature and textbooks use it in either way. As long as there is consistency, the results will be identical one way or the other.

Output Resistance

The amplifier's output resistance is the resistance "seen" by the load resistance when "looking back" into the output terminals. Output resistance can also be interpreted as being the Thévenin resistance at the output terminals. This way, the output resistance can be obtained by driving the output port with an external source to obtain the *v*/*i* ratio at the output terminals. This has to be done with the input source deactivated, that is, replaced by a short circuit if it is a voltage source or by an open circuit if a current source, but leaving the source resistance, if any, in the circuit. The output resistance in this case is defined by

$$r_o \equiv \frac{v_{ext}}{i_{ext}} \tag{6–5}$$

where v_{ext} is the applied external voltage and i_{ext} is the resulting current.

Alternately, the Thévenin or output resistance can be obtained by driving the block with its normal input source to obtain the open-circuit voltage v_{oc} and short-circuit current i_{sc} at the output terminals. The output resistance can then be expressed as

$$r_o \equiv \frac{v_{oc}}{i_{sc}} \tag{6–6}$$

Incidentally, you may recall that the short-circuit current, by definition, is the Norton current I_N from Norton's theorem and that the ratio V_{Th}/I_N is the Thévenin (or Norton) resistance.

Power Gain

Power gain, symbolized A_p, is defined as the ratio of output power to input power, that is,

$$A_p \equiv \frac{p_o}{p_i} \tag{6–7}$$

But since $p_i = v_i i_i$ and $p_o = v_o i_o$, current gain can also be expressed as the product of the voltage and current gains, namely,

$$A_p = \frac{v_o i_o}{v_i i_i} = A_v A_i \tag{6–8}$$

Input power can also be computed using any of the other familiar forms:

$$p_i = i_i^2 r_i \quad \text{or} \quad p_i = v_i^2/r_i \tag{6–9}$$

An important point: Voltage and current gain, as well as input and output resistance, can be determined using rms, peak, and peak-to-peak units, as long as consistency is observed. However, power calculations using the standard formulas cited above must be carried out using rms units for both voltage and current in order to obtain correct results. Alternatively, the following modified formulas can be used for peak and peak-to-peak units.

$$p = \frac{v_p i_p}{2} = \frac{i_p^2 R}{2} = \frac{v_p^2}{2R} \quad \text{(peak units)} \tag{6–10}$$

$$p = \frac{v_{pp} i_{pp}}{8} = \frac{i_{pp}^2 R}{8} = \frac{v_{pp}^2}{8R} \quad \text{(peak-to-peak units)} \tag{6–11}$$

Unless it is necessary to emphasize that we are using a particular form of voltage (peak, p-p, rms), we will simply use lowercase notation for ac quantities with no units in particular. Again, as long as we are consistent, all calculations will be correct.

The Voltage Gain Formula

Because input voltage and input current in an amplifier are related by the input resistance, and the output voltage is the product of output current and load resistance, we can obtain an expression for the voltage gain as a function of the current gain as follows:

Observing Figure 6–1(c), input and output voltages are

$$v_i = i_i r_i \quad \text{and} \quad v_L = -\,i_o R_L$$

The *loaded* voltage gain is then

$$A_v = \frac{v_L}{v_i} = \frac{-i_o R_L}{i_i r_i}$$

which can be written as

$$A_v = -A_i \frac{R_L}{r_i} \tag{6–12}$$

Equation 6–12 is a general expression for the loaded voltage gain that will always hold true regardless of the type or structure of the amplifier, provided both input and output currents are defined entering the positive terminals. In case the output current is defined leaving the positive output terminal, equation 6–12 would have to be written as a positive relationship. Later in the chapter, we will make use of this expression to perform gain calculations and for developing other important formulas.

EXAMPLE 6–1

Figure 6–2 shows the conventional symbol for an amplifier: a triangular block with output at the vertex. As shown in the figure, the input voltage to the amplifier is $v_i(t) = 0.7 + 0.008 \sin 10^3 t$ V. The amplifier has an ac current gain of 80. If the input current is $i_i(t) = 2.8 \times 10^{-5} + 4 \times 10^{-6} \sin 10^3 t$ A, and the ac component of the output voltage is 0.4 V rms, find (1) A_v, (2) R_i, (3) r_i, (4) i_o (rms), (5) R_L, and (6) A_p.

Solution

1. $v_i\,(\text{rms}) = 0.707(0.008\ \text{V-pk}) = 5.66 \times 10^{-3}$ V rms

$$A_v = \frac{v_o\,(\text{rms})}{v_i\,(\text{rms})} = \frac{0.4\ \text{V}}{5.66 \times 10^{-3}\ \text{V}} = 70.7$$

2. The dc input resistance is the ratio of the dc component of the input voltage to the dc component of the input current:

$$R_i = \frac{V_i}{I_i} = \frac{0.7\ \text{V}}{2.8 \times 10^{-3}\ \text{A}} = 25\ \text{k}\Omega$$

3. The ac input resistance is the ratio of the ac components of the input voltage and current:

$$r_i = \frac{v_i}{i_i} = \frac{0.008\ \text{V-pk}}{4 \times 10^{-6}\ \text{A-pk}} = 2\ \text{k}\Omega$$

4. $i_o\,(\text{rms}) = A_i i_i\,(\text{rms}) = 80(0.707)(4 \times 10^{-6}\ \text{A-pk}) = 0.226$ mA rms

5. $$R_L = \frac{v_o\,(\text{rms})}{i_o\,(\text{rms})} = \frac{0.4\ \text{V}}{0.226 \times 10^{-3}\ \text{A}} = 1770\ \Omega$$

FIGURE 6–2 (Example 6–1)

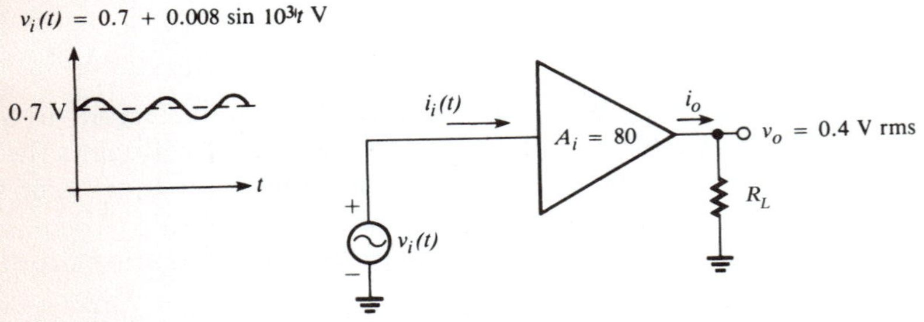

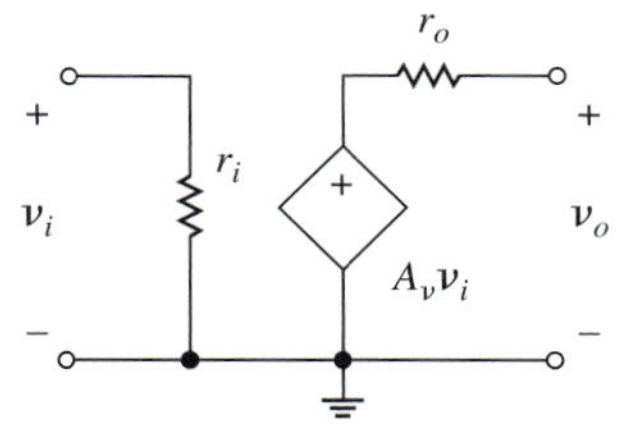

FIGURE 6–3 Amplifier model with a voltage-controlled voltage source

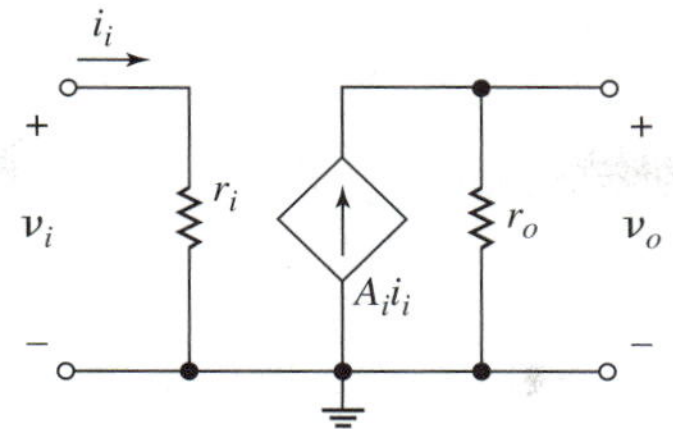

FIGURE 6–4 Amplifier model with a current-controlled current source

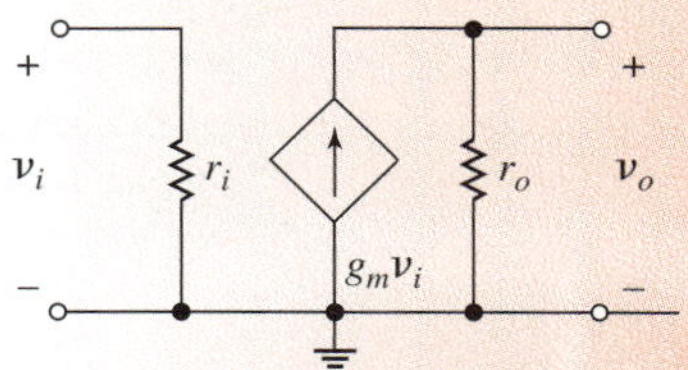

FIGURE 6–5 Amplifier model with a voltage-controlled current source

6. $$p_i = \frac{v_i^2(\text{rms})}{r_i} = \frac{(5.66 \times 10^{-3}\ \text{V})^2}{2 \times 10^3\ \Omega} = 1.6 \times 10^{-8}\ \text{W}$$

$$p_o = \frac{v_o^2(\text{rms})}{r_o} = \frac{(0.4\ \text{V})^2}{1770\ \Omega} = 9.04 \times 10^{-5}\ \text{W}$$

$$A_p = \frac{p_o}{p_i} = \frac{9.04 \times 10^{-5}\ \text{W}}{1.6 \times 10^{-8}\ \text{W}} = 5650$$

Note that the power gain can also be computed in this example as the product of the voltage and current gains: $A_p = A_v A_i = (70.7)80 = 5656$. The small difference between the two results is due to roundoff error.

6–3 AMPLIFIER MODELS

Now that we have a good idea of how an amplifier behaves with respect to input and output terminals, let us look at equivalent circuits, called models, that emulate the internal behavior of an amplifier. We will use special kinds of sources called dependent, or controlled, voltage and current sources. Additionally, we will adopt the standard symbol for dependent sources: a tilted square.

The model in Figure 6–3 employs a *voltage-controlled voltage source* (VCVS) on the output circuit. Notice that the output resistance, r_o, appears in series with the VCVS. Notice also the labeling of the VCVS: $A_v v_i$; it means that the voltage generated by the VCVS is the input voltage times the dimensionless factor A_v. The input voltage is said to be the controlling variable. As you can see, this model is based on voltage gain.

The model shown in Figure 6–4 makes use of a current-controlled current source (CCCS) and a parallel output resistance, that is, a Norton equivalent circuit. Note that this model is based on current gain because the current generated by the CCCS is the input current times the dimensionless factor A_i.

The last model we will consider is shown in Figure 6–5. It employs a *voltage-controlled current source* (VCCS) on the output. The VCCS generates a current that is proportional to the input voltage. The factor g_m is called the *transcounductance* of the amplifier and has units of siemens.

These models are used with all types of amplifiers, whether they employ vacuum tubes or semiconductor devices such as transistors. Furthermore, they can be used in amplifiers based on a single electronic device as well as in those built with multiple devices. In a later chapter we will study *operational amplifiers*, which are multitransistor structures in integrated circuit form. Operational amplifiers are the basic and most important building blocks in all modern amplifier designs.

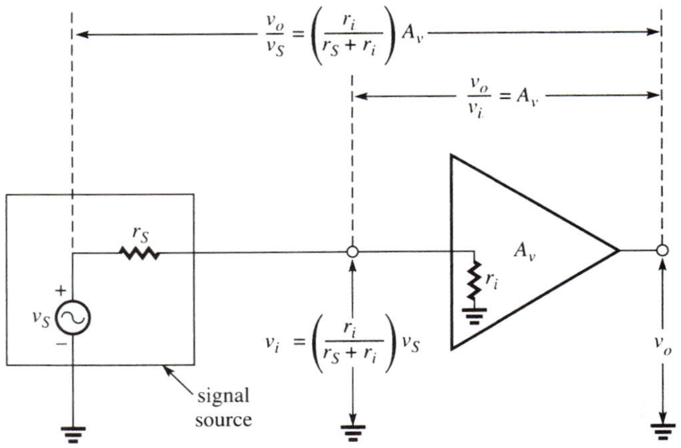

FIGURE 6–6 r_S and r_i form a voltage divider across the amplifier input. The voltage gain from source to output is reduced by the factor $r_i/(r_S + r_i)$.

Source Resistance

Every signal source has internal resistance (its Thévenin equivalent resistance), which we will refer to as *source resistance,* r_S. When a signal source is connected to the input of an amplifier, the source resistance is in series with the input resistance, r_i, of the amplifier. Notice in Figure 6–6 that r_S and r_i form a voltage divider across the input to the amplifier. The input voltage at the amplifier is

$$v_i = v_S\left(\frac{r_i}{r_S + r_i}\right) \tag{6–13}$$

Now,

$$v_o = A_v v_i = A_v v_S\left(\frac{r_i}{r_S + r_i}\right)$$

So

$$\frac{v_o}{v_S} = A_v\left(\frac{r_i}{r_S + r_i}\right) \tag{6–14}$$

Equation 6–14 shows that the overall voltage gain *between source voltage and amplifier output,* v_o/v_S, equals the amplifer voltage gain *reduced* by the factor $r_i/(r_S + r_i)$.

If r_i is much larger than r_S, then $r_i/(r_S + r_i) \approx 1$, so there is little reduction in the overall voltage gain caused by the voltage-divider effect. It is therefore desirable, in general, for a voltage amplifier to have as large an input resistance as possible.

On the other hand, if current amplification is desired, the amplifier should have as *small* an input resistance as possible. When r_i is small, the majority of the current generated at the signal source will be delivered to the amplifier input. This fact is illustrated in Figure 6–7, where the signal is a current source.

As shown in Figure 6–7, the current delivered to the amplifier input is the source current i_S reduced by the factor $r_S/(r_S + r_i)$. Therefore, r_i should be much less than r_S to make the quantity $r_S/(r_S + r_i)$ close to 1. The overall current gain from source to output is

$$\frac{i_o}{i_S} = A_i\left(\frac{r_S}{r_S + r_i}\right) \tag{6–15}$$

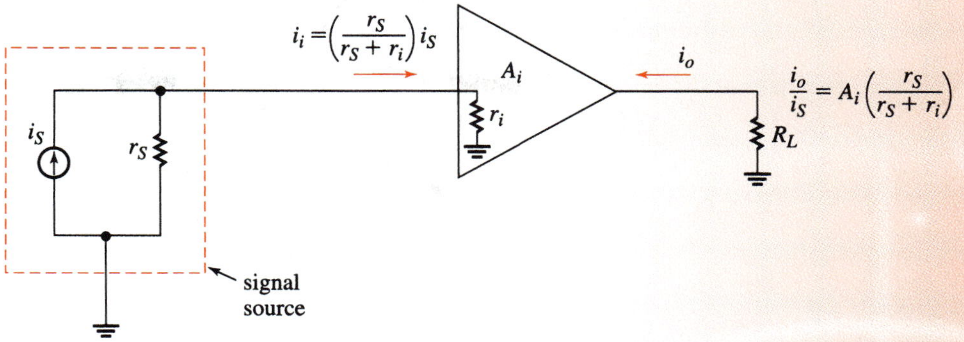

FIGURE 6–7 A current amplifier should have a small input resistance in order to make the quantity $r_S/(r_S + r_i)$ close to 1.

The amplifier shown in Figure 6–8 has $A_i = 10$ and $r_i = 10\ \text{k}\Omega$. It is driven by a signal source that has source resistance r_S. Find the overall voltage gain from source to output when (1) $r_S = 1\ \text{k}\Omega$ and (2) $r_S = 10\ \text{k}\Omega$. Note: Observe the orientation of i_o.

Solution

1. $r_S = 1\ \text{k}\Omega\ (r_S = 0.1\ r_i) \quad A_v = A_i\frac{R_L}{r_i} = 10\left(\frac{2\ \text{k}}{1\ \text{k}}\right) = 20$

$$\frac{v_o}{v_S} = A_v\left(\frac{r_i}{r_S + r_i}\right) = 20\left(\frac{10\ \text{k}}{1\ \text{k} + 10\ \text{k}}\right) = 18.2$$

2. $r_S = 10\ \text{k}\Omega \quad (r_i = 0.1 r_S)$

$$\frac{v_o}{v_S} = 20\left(\frac{10\ \text{k}}{10\ \text{k} + 10\ \text{k}}\right) = 10$$

This example shows that when $r_S = 0.1\ r_i$, the voltage gain is reduced by about 10%; when $r_i = r_S$, the voltage is reduced by 50%.

Load Resistance

An ac amplifier is always used to supply voltage, current, and/or power to some kind of *load* connected to its output. The load may be a speaker, an antenna, a siren, an indicating instrument, an electric motor, or any one of a large number of other useful devices. Often the load is the input to another amplifier. Amplifier performance is analyzed by representing its load as an equivalent load resistance (or impedance). When a load resistance R_L is connected to the output of an amplifier, there is again a voltage division

FIGURE 6–8 (Example 6–2).

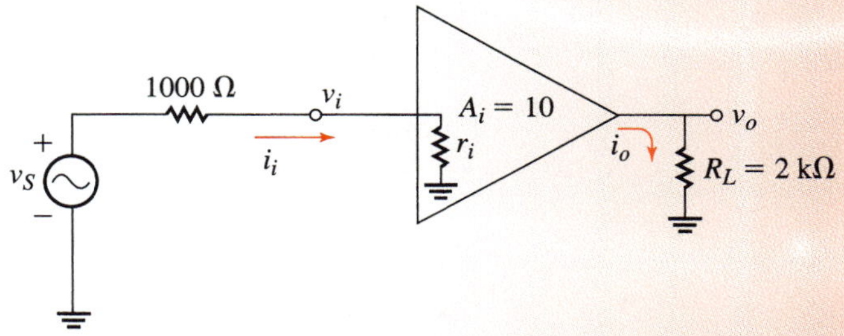

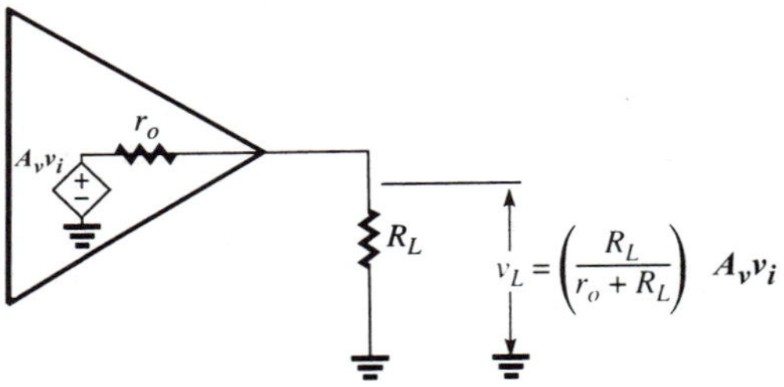

FIGURE 6–9 The output voltage from an ac amplifier divides between r_o and the load resistance R_L.

between the output resistance of the amplifier and the load. Figure 6–9 shows a Thévenin equivalent circuit of the output of an ac amplifier in which the output circuit is modeled by a VCVS in series with r_o. As can be seen in the figure, the load voltage v_L is determined by

$$v_L = \left(\frac{R_L}{r_o + R_L}\right) A_v v_i \tag{6–16}$$

Note that $A_v v_i$ should be regarded as the open-circuit (or unloaded) output voltage. For a voltage amplifier, r_o should be much smaller than R_L in order to maximize the portion of $A_v v_i$ that appears across the load.

When the effects of both r_S and R_L are taken into account, the overall voltage gain A_{vs} from source to load becomes

$$A_{vs} = \frac{v_L}{v_S} = A_v \left(\frac{r_i}{r_S + r_i}\right)\left(\frac{R_L}{r_o + R_L}\right) \tag{6–17}$$

where A_v is the unloaded voltage gain.

If a signal source has fixed resistance r_S, then from the *maximum power transfer theorem*, maximum power is transferred from the source to an amplifier when $r_i = r_S$. Similarly, if the amplifier has fixed output resistance r_o, then maximum power is transferred from the amplifier to a load when $R_L = r_o$. The amplifier is said to be *matched* to its source when $r_i = r_S$ and matched to its load when $R_L = r_o$. (We should also note that if the values of r_S and r_o can be controlled, maximum power transfer occurs when $r_S = 0$ and $r_o = 0$, irrespective of the values of r_i and R_L.)

As we have seen, the output of an amplifier can also be modeled using a current source in parallel with the output resistance, as shown in Figure 6–10. This is the Norton equivalent, which can be very useful for devices that are more current oriented such as transistors.

Loading effects under this situation are opposite to those for the Thévenin model. It is apparent that in this case more current will flow through the load if R_L is much less than r_o. Therefore, a well-designed current amplifier should have relatively large output resistance, as opposed to a voltage amplifier, whose output resistance should be as small as possible. The direction shown for the current depends on the particular device and configuration used in the amplifier. The open-circuit output voltage is clearly

$$v_o = i_o r_o \tag{6–18}$$

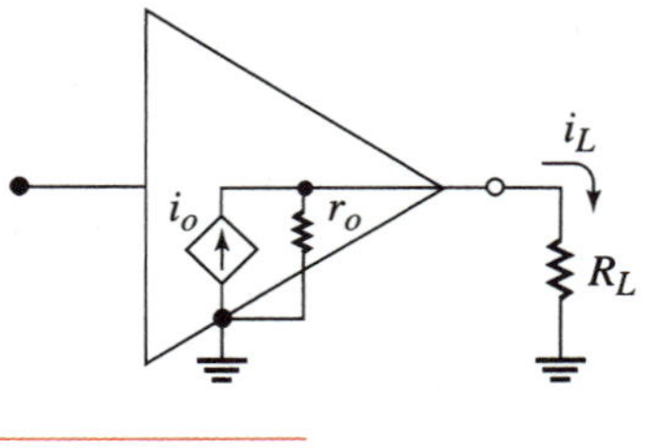

FIGURE 6–10 Norton equivalent of the output. The current i_o can be a function of v_i or i_i.

where i_o can either be $g_m v_i$ or $A_i i_i$. With the load resistance connected, the load voltage v_L can be obtained by multiplying i_o times the parallel combination $r_o \| R_L$, that is,

$$v_L = i_o(r_o \| R_L) \tag{6–19}$$

or by using voltage division on the open-circuit output voltage $(i_o r_o)$ in series with the output resistance r_o, that is, "Thévenizing" the parallel circuit. Thus,

$$v_L = (i_o r_o)\frac{R_L}{R_L + r_o} \tag{6–20}$$

which is equivalent to (6–19).

Alternatively, we could have performed current division to obtain i_L in terms of i_o and then multiplied by R_L to obtain the load voltage v_L, that is,

$$v_L = i_o\left(\frac{r_o}{r_o + R_L}\right)R_L$$

which yields (6–19) or (6–20).

Inverting and Noninverting Amplification

We had already learned that when an amplifier produces sign inversion it is regarded as an inverting amplifier. The models of Figures 6–3, 6–4, and 6–5 are all of the noninverting type, because a positive input voltage produces a positive output voltage. An inverting model can be obtained by either flipping over the dependent source or by indicating a negative voltage or current gain, or a negative transconductance factor.

As an example, Figure 6–11 shows an inverting amplifier modeled with the transconductance function g_m. Note the orientation of the VCCS, which produces a negative voltage across the output terminals, according to the indicated reference for v_o.

Let us obtain the overall voltage gain of this circuit when driven by a voltage source with a characteristic resistance r_S.

First, the voltage across the input, using voltage division is

$$v_i = v_S \frac{r_i}{r_S + r_i} \tag{6–21}$$

The loaded output voltage is clearly

$$v_L = -g_m v_i (r_o \| R_L) \tag{6–22}$$

that yields

$$v_L = -g_m v_S \frac{r_i}{r_S + r_i}(r_o \| R_L) \tag{6–23}$$

FIGURE 6–11 Inverting amplifier with the transconductance model

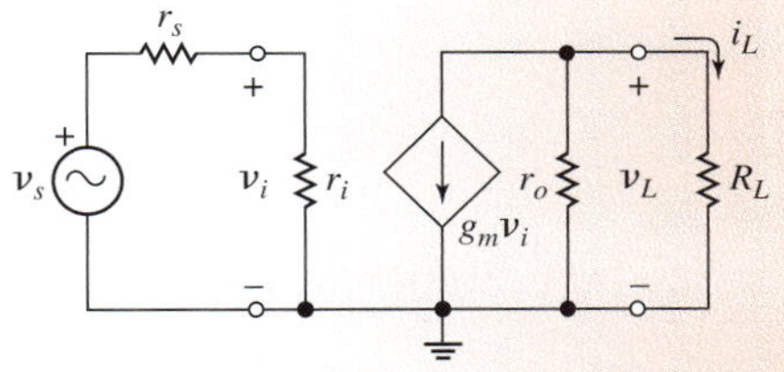

from which we solve for the overall gain A_{vs}:

$$A_{vs} = \frac{v_L}{v_S} = -g_m(r_o \| R_L)\frac{r_i}{r_S + r_i} \tag{6–24}$$

To further illustrate analysis with models, let us also find the loaded current gain from the input source to the load resistance, that is, i_L/i_S. Using the current-division rule on the parallel output circuit we can write

$$i_L = -\frac{r_o}{r_o + R_L} g_m v_i \tag{6–25}$$

But $v_i = i_S r_i$; therefore,

$$\frac{i_L}{i_S} = -g_m r_i \frac{r_o}{r_o + R_L} \tag{6–26}$$

EXAMPLE 6–3

The amplifier shown in Figure 6–11 has the following parameters: $r_i = 2.4$ kΩ, $g_m = 48$ mS, $r_o = 75$ kΩ. The amplifier has an external resistor $R_1 = 12$ kΩ connected across the input terminals, and the load resistance is 10 kΩ. If the input source resistance is 1 kΩ, find:

1. The overall current gain i_L/i_S.
2. The overall voltage gain A_{vs}.
3. The power gain from the input to the load resistance.

Solution

1. Using (6–26) and taking into account the current-division effect at the input, we write

$$\frac{i_L}{i_S} = -g_m r_i \left(\frac{r_o}{r_o + R_L}\right)\left(\frac{R_1}{R_1 + r_i}\right) = -84.7$$

2. To take into account R_1 in the calculation of the overall voltage gain, we simply put it in parallel with r_i in (6–24):

$$A_{vs} = -g_m(r_o \| R_L)\frac{r_i \| R_1}{r_s + r_i \| R_1} = -282.4$$

3. The power gain from input to load can be obtained as the product of the voltage and current gains, provided the voltage gain is with respect to v_i. From (6–22), this gain is

$$\frac{v_L}{v_i} = -g_m(r_o \| R_L) = -423.5$$

and the power gain is

$$A_p = \frac{i_L}{i_S} \times \frac{v_L}{v_i} = 3.587 \times 10^4$$

Differential Amplifiers

The models of Figures 6–3, 6–4, and 6–5 are referred to as single-ended amplifiers, meaning that both input and output voltages are referenced to ground. Another form, called *differential amplifier*, employs a double-ended

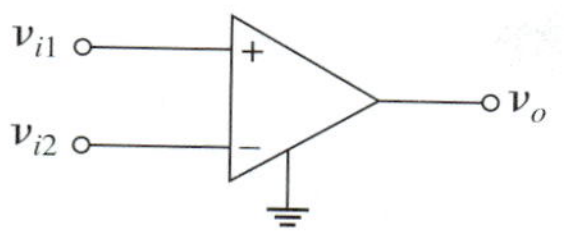

FIGURE 6–12 Basic differential amplifier

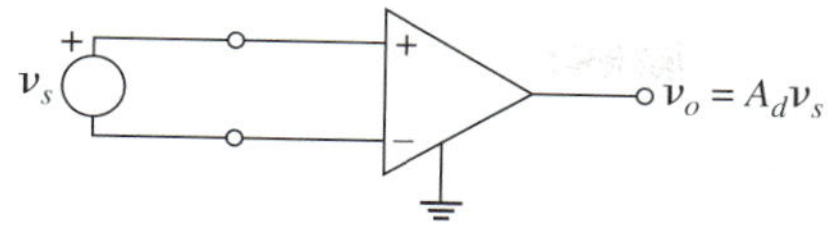

FIGURE 6–13 An input signal connected in the differential mode

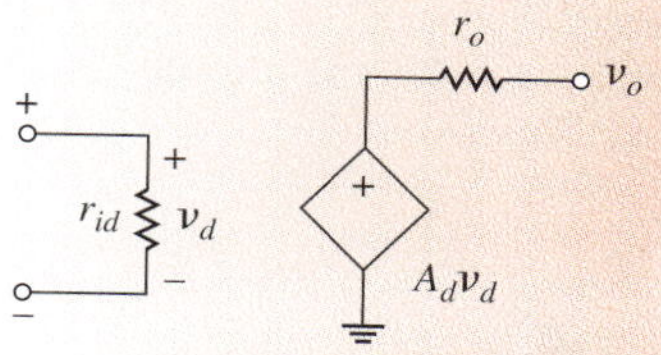

FIGURE 6–14 Basic differential amplifier model

input. The output can either be single- or double-ended. A double-ended or differential input consists of two terminals as shown in Figure 6–12.

The basic relationship between input and output in a differential amplifier is

$$v_o = A_d(v_{i1} - v_{i2}) \tag{6–27}$$

where A_d is the differential voltage gain of the amplifier. Note that the output voltage is the differential gain times the *difference* between the two input voltages. This difference is called the differential input voltage; however, in most practical situations there will only be one input signal which will be connected in a "floating" mode as shown in Figure 6–13.

A very common application of differential amplifiers is in microphone signal amplification. Due to the normally long cables associated with microphones, noise and interference will be picked up by the conductors. If a pair of twisted wires with an overall grounded shield is used for connecting the microphone to the differential amplifier, each of the conductors of the twisted pair will carry identical noise voltages, thereby cancelling at the differential amplifier. The signal voltage across the twisted pair appears as a differential input voltage and will be amplified by the differential gain. In a later chapter we will look at noise calculations in differential amplifiers.

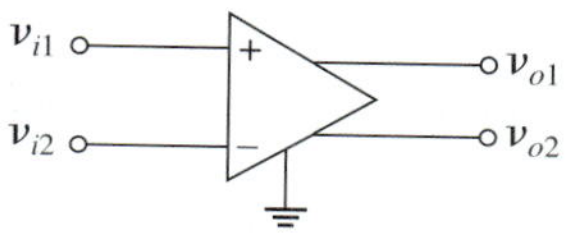

FIGURE 6–15 A differential amplifier with double-ended output

An important issue regarding differential amplifiers is that, if needed, they can be used as conventional (signal-ended) inverting or noninverting amplifiers. A noninverting amplifier can be obtained by applying the input signal to the noninverting (+) input while connecting the inverting (−) input to ground. Similarly, an inverting amplifier is obtained by using the inverting (−) input terminal and grounding the noninverting (+) one.

A basic model for a differential amplifier is shown in Figure 6–14. The single-ended output includes the output resistance r_o.

In certain applications, there is need for the output to be differential as well. Figure 6–15 shows a differential amplifier with double-ended output. In this case the input-output relationship is expressed as

$$v_{o1} - v_{o2} = A_d(v_{i1} - v_{i2}) \tag{6–28}$$

where the output signal is also a differential voltage that appears across the two output terminals. Double-ended outputs are useful when there is need for sending an output signal from a differential amplifier, through a long cable, to another differential amplifier. This way, the noise-cancelling property of differential amplifiers is preserved throughout the amplification process.

6–4 MULTISTAGE AMPLIFIERS

In many applications, a single amplifier cannot furnish all the gain that is required to drive a particular kind of load. For example, a speaker represents a

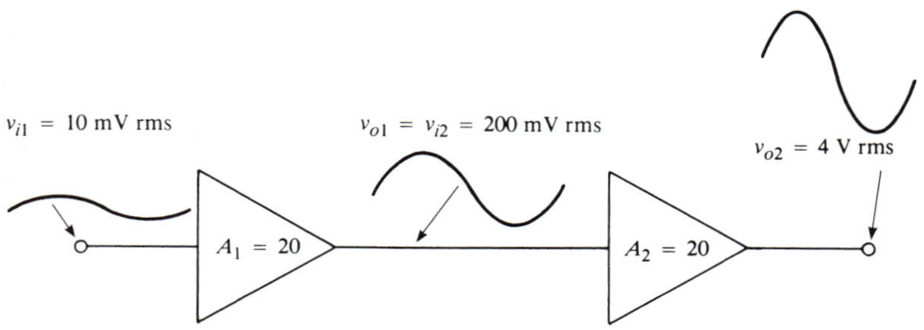

FIGURE 6–16 Two amplifier stages connected in cascade

"heavy" load in an audio amplifier system, and several amplifier stages may be required to "boost" a signal originating at a microphone or magnetic tape head to a level sufficient to provide a large amount of power to the speaker. We hear of *preamplifiers, power amplifiers*, and *output amplifiers*, all of which constitute stages of amplification in such a system. Actually, each of these components may itself consist of a number of individual transistor amplifier stages. Amplifiers that create voltage, current, and/or power gain through the use of two or more stages are called multistage amplifiers.

When the output of one amplifier stage is connected to the input of another, the amplifier stages are said to be in *cascade*. Figure 6–16 shows two stages connected in cascade. To illustrate how the overall voltage gain of the combination is computed, let us assume that the input to the first stage is 10 mV rms and that the voltage gain of each stage is $A_1 = A_2 = 20$, as shown in the figure. The output of the first stage is $A_1 v_{i1} = 20$ (10 mV rms) = 200 mV rms. Thus, the input to the second stage is 200 mV rms. The output of the second stage is, therefore, $A_2 v_{i2} = 20(200$ mV rms$) = 4$ V rms. Therefore, the overall voltage gain is

$$A_v = \frac{v_{o2}}{v_{i1}} = \frac{4 \text{ V rms}}{10 \text{ mV rms}} = 400$$

Notice that $A_v = A_1 A_2 = (20)(20) = 400$.

Figure 6–17 shows an arbitrary number (n) of stages connected in cascade. Note that the output of each stage is the input to the succeeding one ($v_{o1} = v_{i2}$, $v_{o2} = v_{i3}$, etc.). We will derive an expression for the overall voltage gain $v_{o,n}/v_{i1}$ in terms of the individual stage gains $A_1, A_2, \ldots, A_n$. We assume that each stage gain $A_1, A_2 \ldots, A_n$ is the value of the voltage gain between input and output of a stage *with all other stages connected* (more about that important assumption later).

By definition,

$$v_{o1} = A_1 v_{i1} \tag{6–29}$$

Also,

$$v_{o2} = A_2 v_{i2} = A_2 v_{o1} \tag{6–30}$$

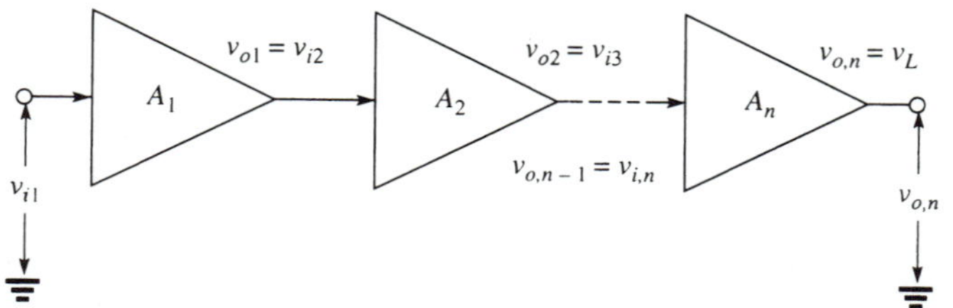

FIGURE 6–17 *n* amplifier stages connected in cascade. The output voltage of each stage is the input voltage to the next stage.

Substituting v_{o1} from (6–29) into (6–30) gives

$$v_{o2} = (A_2A_1)v_{i1} \tag{6–31}$$

Similarly,

$$v_{o3} = A_3v_{i3} = A_3v_{o2}$$

and, from (6–31),

$$v_{o3} = (A_3A_2A_1)v_{i1}$$

Continuing in this manner, we eventually find

$$v_{o,n} = v_L = (A_n\,A_{n-1}\ldots A_2A_1)v_{i1}$$

Therefore,

$$\frac{v_{o,n}}{v_{i1}} = A_nA_{n-1}\ldots A_2A_1 \tag{6–32}$$

Equation 6–32 shows that the overall voltage gain of n cascaded stages is the *product* of the individual stage gains (not the sum!). In general, any one or more of the stage gains can be negative, signifying, as usual, that the stage causes a 180° phase inversion. It follows from equation 6–32 that the cascaded amplifiers will cause the output of the last stage ($v_{o,n}$) to be out of phase with the input to the first stage (v_{i1}) if there is an *odd* number of inverting stages, and will cause $v_{o,n}$ to be in phase with v_{i1} if there is an even (or zero) number of inversions.

Our derivation of equation 6–32 did not include the effect of source or load resistance on the overall voltage gain. Source resistance r_S causes the usual voltage division to take place at the input to the first stage, and load resistance r_L causes voltage division to occur between r_L and the output resistance of the last stage. Under those circumstances, the overall voltage gain between load and signal source becomes

$$\frac{v_L}{v_S} = \left(\frac{r_{i1}}{r_S + r_{i1}}\right)A_n\,A_{n-1}\ldots A_2A_1\left(\frac{r_L}{r_{o,n} + r_L}\right) \tag{6–33}$$

where r_{i1} = input resistance to first stage and $r_{o,n}$ = output resistance of last stage.

EXAMPLE 6–4

Figure 6–18 shows a three-stage amplifier and the ac rms voltages at several points in the amplifier. Note that v_1 is the input voltage delivered by a signal source having zero resistance and that v_3 is the output voltage with no load connected.

1. Find the voltage gain of each stage and the overall voltage gain v_3/v_1.
2. Find the overall voltage gain v_L/v_S when the multistage amplifier is driven by a signal source having resistance 2000 Ω and the load is 25 Ω. Stage 1 has input resistance 1 kΩ and stage 3 has output resistance 50 Ω.
3. What is the power gain under the conditions of (2) (measured between the input to the first stage and the load)?
4. What is the overall current gain i_L/i_1 under the conditions of (2)?

Solution

1. $A_1 = (36\text{ mV})/(900\ \mu\text{V}) = 40$
   $A_2 = (1.25\text{ V})/(36\text{ mV}) = 34.722$
   $A_3 = (21\text{ V})/(1.25\text{ V}) = 16.8$

FIGURE 6–18 (Example 6–4)

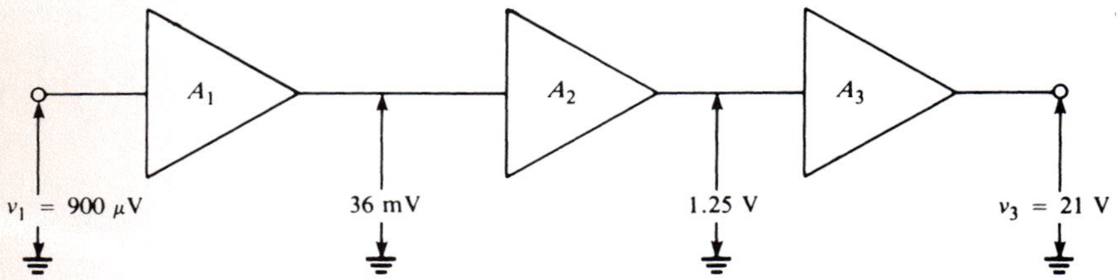

$$v_3/v_1 = A_1 A_2 A_3 = (40)(34.722)(16.8) = 23{,}333$$

Note that the product of the voltage gains equals the overall voltage gain, which, in this example, can also be calculated directly: $v_3/v_1 = (21 \text{ V})/(900 \ \mu\text{V}) = 23{,}333$.

2. From equation 6–33,

$$\frac{v_L}{v_S} = \left(\frac{1000}{2000 + 1000}\right)(23{,}333)\left(\frac{25}{50 + 25}\right) = 2592.5$$

3. When a signal-source resistance of 2000 Ω is inserted in series with the input, v_1 becomes

$$v_1 = \left(\frac{1000}{2000 + 1000}\right)(900 \ \mu\text{V}) = 300 \ \mu\text{V}$$

The input power is, therefore,

$$P_i = \frac{v_1^2}{r_{i1}} = \frac{(300 \times 10^{-6})^2}{1000} = 90 \text{ pW}$$

The voltage across the 25-Ω load is then

$$v_L = v_1(A_1A_2A_3)\left(\frac{R_L}{r_{o3} + R_L}\right)$$

$$= (300 \ \mu\text{V})(40)(34.722)(16.8)\left(\frac{25}{50 + 25}\right) = 2.33 \text{ V}$$

The output power developed across the load resistance is, therefore,

$$P_o = \frac{v_L^2}{R_L} = \frac{(2.33)^2}{25} = 0.217 \text{ W} = 217 \text{ mW}$$

Finally,

$$A_p = \frac{217 \text{ mW}}{90 \text{ pW}} = 2.41 \times 10^9$$

4. Recall that $A_p = A_vA_i$.

The voltage gain between the input to the first stage and the load is

$$A_v = \frac{v_L}{v_1} = \frac{2.33 \text{ V}}{300 \ \mu\text{V}} = 7766$$

Therefore,

$$A_i = \frac{A_p}{A_v} = \frac{2.41 \times 10^9}{7766} = 3.1 \times 10^5$$

Interstage Loading

It is important to remember that the gain equations we have derived are based on the *in-circuit* values of $A_1, A_2, \ldots$, that is, on the stage gains that result when all other stages are connected. Thus, we have assumed that each value of stage gain takes into account the loading the stage causes on the previous stage and the loading presented to it by the next stage (except we assumed that A_1 did not include loading by r_S and A_n did not include loading by r_L). If we know the *open-circuit* (unloaded) voltage gain of each stage and its input and output resistances, we can calculate the overall gain by taking into account the loading effects of each stage on another. Theoretically, the load presented to a given stage may depend on *all* of the succeeding stages lying to its right, since the input resistance of any one stage depends on its output load resistance, which in turn is the input resistance to the next stage, and so forth. In practice, we can usually ignore this cumulative loading effect of stages beyond the one immediately connected to a given stage, or assume that the input resistance that represents the load of one stage to a preceding one is given for the condition that all succeeding stages are connected.

To illustrate the ideas we have just discussed, Figure 6–19 shows a three-stage amplifier for which the individual open-circuit voltage gains A_{o1}, A_{o2}, and A_{o3} are assumed to be known, as well as the input and output resistances of each stage. From the voltage division that occurs at each node in the system, it is apparent that the following relations hold:

$$v_1 = \left(\frac{r_{i1}}{r_S + r_{i1}}\right) v_S$$

$$v_2 = A_{o1} v_1 \left(\frac{r_{i2}}{r_{o1} + r_{i2}}\right)$$

$$v_3 = A_{o2} v_2 \left(\frac{r_{i3}}{r_{o2} + r_{i3}}\right)$$

$$v_L = A_{o3} v_3 \left(\frac{r_L}{r_{o3} + r_L}\right)$$

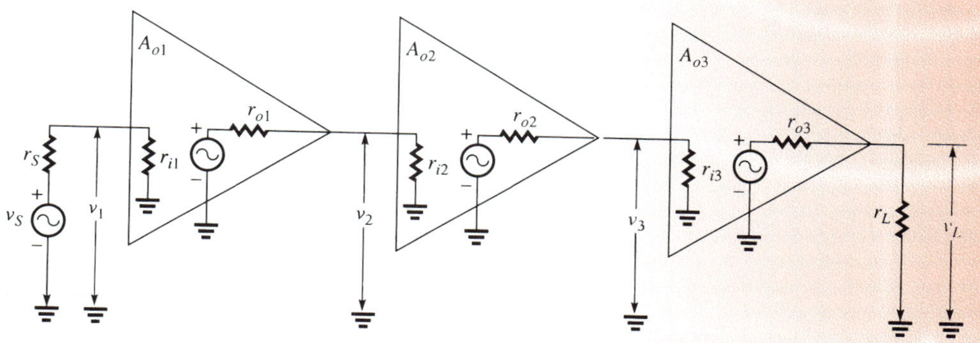

FIGURE 6–19 A three-stage amplifier. A_{o1}, A_{o2}, and A_{o3} are the open-circuit (unloaded) voltage gains of the respective stages.

Combining these relations leads to

$$\frac{v_L}{v_S} = \left(\frac{r_{i1}}{r_S + r_{i1}}\right)A_{o1}\left(\frac{r_{i2}}{r_{o1} + r_{i2}}\right)A_{o2}\left(\frac{r_{i3}}{r_{o2} + r_{i3}}\right)A_{o3}\left(\frac{r_L}{r_{o3} + r_L}\right) \tag{6–34}$$

As might be expected, equation 6–34 shows that the overall voltage gain of the multistage amplifier is the product of the open-circuit stage gains multiplied by the voltage-division ratios that account for the loading of each stage. Notice that a *single* voltage-division ratio accounts for the loading between any pair of stages. In other words, it is *not* correct to compute loading effects twice: once by regarding an input resistance as the load on a previous stage and again by regarding the output resistance of that previous stage as the source resistance for the next stage.

In amplifiers modeled with an output current source (CCCS or VCCS), the interstage loading is manifested as current division between the output resistance of one stage and the input resistance of the following stage. Figure 6–20 illustrates two cascaded amplifier stages. Let us analyze the circuit to find the overall transconductance i_L/v_S and the overall voltage gain v_L/v_S.

Starting on the input side, we can write an expression for the input current as

$$i_{i1} = \frac{v_S}{r_S + r_{i1}}$$

Using the current division rule, the input current to the second stage can be expressed as

$$\begin{aligned} i_{i2} &= \frac{r_{o1}}{r_{o1} + r_{i2}}A_{i1}i_{i1} \\ &= \frac{r_{o1}}{r_{o1} + r_{i2}}A_{i1}\frac{v_S}{r_S + r_{i1}} \end{aligned}$$

Similarly, the load current is

$$\begin{aligned} i_L &= \frac{r_{o2}}{r_{o2} + R_L}A_{i2}i_{i2} \\ &= \frac{r_{o2}}{r_{o2} + R_L}A_{i2}\frac{r_{o1}}{r_{o1} + r_{i2}}A_{i1}\frac{v_S}{r_S + r_{i1}} \end{aligned}$$

which yields the overall transconductance in terms of current gains:

$$\frac{i_L}{v_S} = \frac{A_{i1}A_{i2}}{r_S + r_{i1}}\left(\frac{r_{o1}}{r_{o1} + r_{i2}}\right)\left(\frac{r_{o2}}{r_{o2} + R_L}\right) \tag{6–35}$$

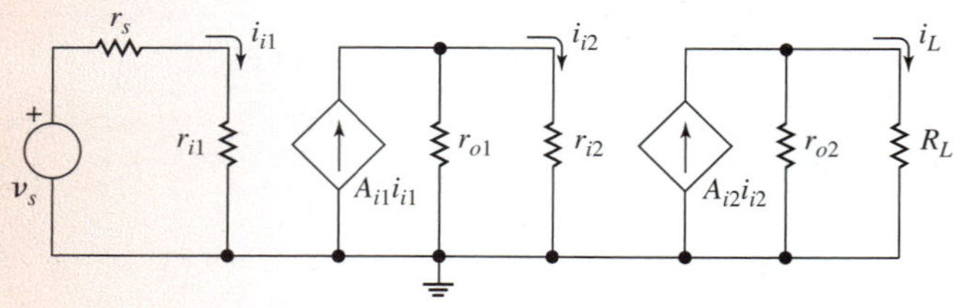

FIGURE 6–20 Two amplifier stages with interstage loading in the form of current division

To obtain an equation for the overall voltage gain, it suffices to recognize that the output voltage is the product of i_L and R_L, which yields

$$\frac{v_L}{v_S} = \frac{A_{i1}A_{i2}R_L}{r_S + r_{i1}}\left(\frac{r_{o1}}{r_{o1} + r_{i2}}\right)\left(\frac{r_{o2}}{r_{o2} + R_L}\right) \tag{6–36}$$

The last two terms in equation 6–36 are the two current division fractions. If both output resistances approach infinity, which is the case for ideal amplifiers, then both current division fractions will approach unity. Therefore, the first term in equation 6–36 represents the overall voltage gain under ideal conditions.

Invoking the general form $A_v = A_i(R_L/r_i)$ we can easily establish in equation 6–36 that the overall current gain must be

$$A_i(\text{ovr}) = A_{i1}A_{i2}\left(\frac{r_{o1}}{r_{o1} + r_{i2}}\right)\left(\frac{r_{o2}}{r_{o2} + R_L}\right) \tag{6–37}$$

Note also that $r_S + r_{i1}$ would be the input resistance immediately following v_S, even though r_S is the internal resistance of the source v_S. Therefore, if we remove the r_S term in equation 6–36, the resulting expression would be the voltage gain v_L/v_i, referenced to the voltage across the input terminals of the amplifier.

SUMMARY

This chapter has presented the basics of the amplification function in general terms, that is, without regard to the actual devices employed in the construction of amplifiers. At the end of this chapter, the student should be able to understand the following concepts:

- Amplifiers are used for boosting electrical signals.
- Amplifiers are characterized by a set of functional parameters.
- Amplifiers can be analyzed through equivalent electric circuits called models.
- There are different approaches for the modeling of amplifiers.
- The amplification of signals can be inverting or noninverting.
- Amplifiers can be single-ended or differential.
- Differential amplifiers can reduce noise and interference.
- Two or more amplifiers can be cascaded to form a multistage amplifier.

EXERCISES

SECTION 6–2

Amplifier Characteristics

6–1. An ac amplifier has a voltage gain of 55 and a power gain of 456.5. The ac output current is 24.9 mA rms and the ac input resistance is 200 Ω. Find

(a) the current gain,

(b) the rms value of the ac input current,

(c) the rms value of the ac input voltage,

(d) the rms value of the ac output voltage,

(e) the ac output resistance, and

(f) the output power.

6–2. An ac amplifier has a current gain of 0.95 and a voltage gain of 100. The ac input voltage is 120 mV rms and the ac input resistance is 25 Ω. Find the output power.

6–3. The signal source connected to the input of an ac amplifier has an internal resistance of 1.2 kΩ. The voltage gain of the amplifier from its input to its output is 140. What minimum value of input resistance should the amplifier have if the voltage gain from signal source to amplifier output is to be at least 100?

6–4. The amplifier in Figure 6–21 has current gain $A_i = 80$ from amplifier input to amplifier output. Find the load current i_L.

6–5. An ac amplifier is driven by a 20-mV-rms signal source having internal resistance 1 kΩ. The output resistance of the amplifier is 50 Ω. The voltage gain of the amplifier from its input to its (open-circuit) output (A_v) is 150. What power is delivered to the load if the amplifier is matched to its source and matched to its source and load?

SECTION 6–3

Amplifier Models

6–6. Using the model of Figure 6–3 and placing a short circuit across the output terminals, show that the short-circuit current gain, A_{isc} (current entering output terminal), is given by

$$A_{isc} = \frac{-A_v r_i}{r_o}$$

6–7. An amplifier modeled as in Figure 6–4 has $r_i = 5$ kΩ, $A_i = 130$, and $r_o = 90$ kΩ. A current source i_S with internal resistance of 15 kΩ is connected to the input of the amplifier, and a load resistance of 2 kΩ is connected across the output terminals. If $i_S = 1$ mA rms, what is the voltage across and the current through the load resistance?

6–8. An amplifier modeled as in Figure 6–11 has input resistance of 3 kΩ, its transconductance is 55 mS, and the output resistance is 90 kΩ. If $v_S = 60$ mV rms, $r_S = 600$ Ω, and $R_L = 1.5$ kΩ, find the power delivered to the load.

SECTION 6–4

Multistage Amplifiers

6–9. The in-circuit voltage gains of the stages in a multistage amplifier are shown in Figure 6–22. Find

(a) the overall voltage gain, v_o/v_{in}; and

(b) the voltage gain that would be necessary in a fifth stage, which, if added to the cascade, would make the overall voltage gain 10^5.

6–10. The in-circuit voltage gains of the stages in a multistage amplifier are shown in Figure 6–22. (The gain of the first stage does not include loading by the signal source and that of the fourth stage does not include loading by a load resistor.) The input resistance to the first stage is 20 kΩ, and the output resistance of the fourth stage is 20 Ω. The amplifier is driven by a signal source having resistance 25 kΩ, and a 12-Ω load is connected to the output of the fourth stage. If the source voltage is $v_s = 5$ mV rms, find

(a) the load voltage, v_L;

(b) the power gain, between the input to the first stage and the load.

6–11. It is desired to construct a three-stage amplifier whose overall voltage gain is 500. The in-circuit voltage gains of the first two stages are to be equal, and the voltage

FIGURE 6–21 (Exercise 6–4)

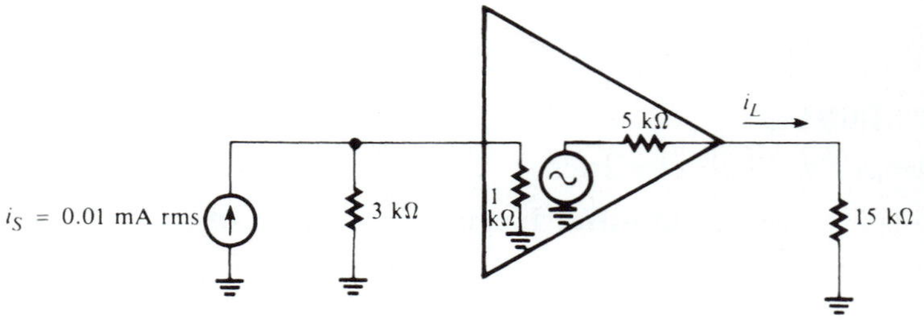

FIGURE 6–22 (Exercises 6–9 and 6–10)

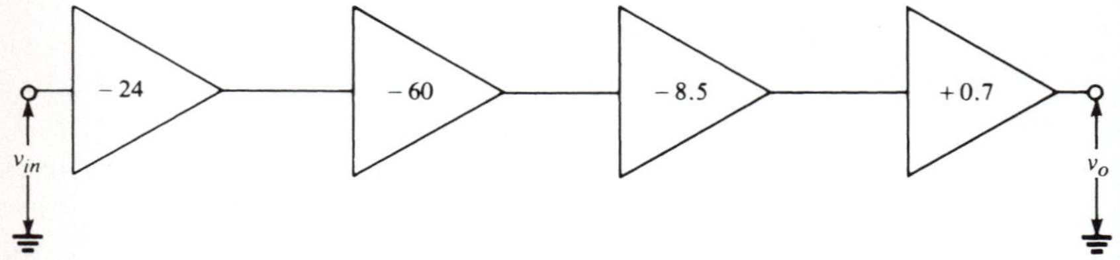

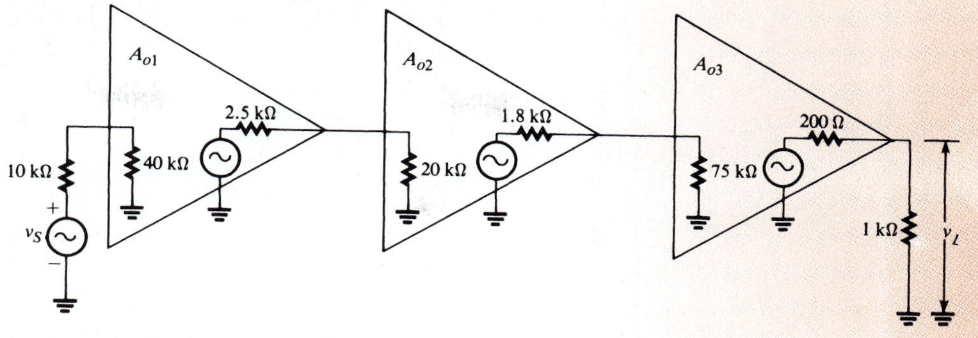

FIGURE 6–23 (Exercise 6–12)

gain of the third stage is to be one-half that of each of the first two. What should be the voltage gain of each stage?

6–12. The open-circuit (unloaded) voltage gains of the stages in the multistage amplifier shown in Figure 6–23 are $A_{o1} = -42$, $A_{o2} = -26$, and $A_{o3} = 1.8$. Find the overall voltage gain v_L/v_S.

CHAPTER 7

SMALL-SIGNAL TRANSISTOR AMPLIFIERS

OUTLINE

OBJECTIVES

- Understand how a BJT, a JFET, or a MOSFET can amplify a signal.
- Visualize why a transistor can clip or limit a signal.
- Differentiate between the common-emitter, common-base, and common-collector configurations in BJT amplifiers.
- Differentiate between the common-source, common-gate, and common-drain configurations in field-effect transistor amplifiers.
- Understand the difference between capacitively-coupled stages and direct-coupled stages.
- Recognize the hybrid-π model for a common-emitter amplifier.
- Perform analysis of BJT amplifiers using h-parameters.
- Develop an understanding of amplification based on field-effect transistors.

7–1 INTRODUCTION

In Chapters 4 and 5 we learned about the characteristics of bipolar and field-effect transistors and their behavior in dc circuits. We discussed their regions of operation (cutoff, active, saturation) and how to bias them in order to establish a desired quiescent condition.

In Chapter 6 we covered amplifier fundamentals in the general sense, that is, without regard to what device or devices are used for the amplification function.

In this chapter we will look at amplifiers based on transistors of both bipolar and field-effect types in discrete form. We will make extensive use of the models we learned in Chapter 6 and will expand more on concepts and characteristics that are specific for the particular device and configuration being considered.

At this point it is important that the reader understands that in modern electronics, amplification of low-level signals is seldom done with discrete components. That approach was the only choice several decades ago. Nevertheless, it is important to understand the behavior of discrete transistors in amplification as foundation material for more advanced topics in integrated-circuit amplification as well as in power amplifiers.

Bipolar junction transistors can be used in three distinct amplifying configurations: common emitter (CE), common base (CB), and common collector (CC). Similarly, field-effect transistor amplifiers, whether junction type or MOSFET, can be configured as common source (CS), common gate (CG), and common drain (CD). Each configuration must be properly biased in the center of its active region and capacitively coupled to the input signal source and to the load resistance.

The Purpose of Bias

In most single-transistor amplifiers, the output voltage must always be positive or must always be negative. When that is the case, it is not possible for the output to be a pure ac waveform. By definition, an ac waveform is alternately positive and negative. The purpose of bias in a transistor amplifier is to set a dc output level somewhere in the middle of the total range of possible output voltages so that an ac waveform can be superimposed on it. Figure 7–1 illustrates the point. The ac input causes the output voltage to vary above and below the bias voltage, but the instantaneous values of the output are always positive (in this example). In other words, the output is of the form

$$v_o(t) = V_B + A \sin \omega t$$

where V_B is the bias voltage, or dc component, of the output, and A is the peak value of the sinusoidal, ac component.

It is apparent that the values of V_B and A must be such that $V_B + A$ is not greater than the maximum possible output voltage and $V_B - A$ is not less than the minimum possible output voltage. If these conditions are not satisfied, then the output voltage will reach its minimum or maximum extremes before the total ac variation can take place. The result is a flattening of the output waveform called *clipping.* Figure 7–2 illustrates positive and negative clipping caused by values of V_B that are too large or too small and by values of A that are too large. When clipping is caused by the amplitude A being too large, as shown in Figure 7–2(c), the amplifier is said to be *overdriven.*

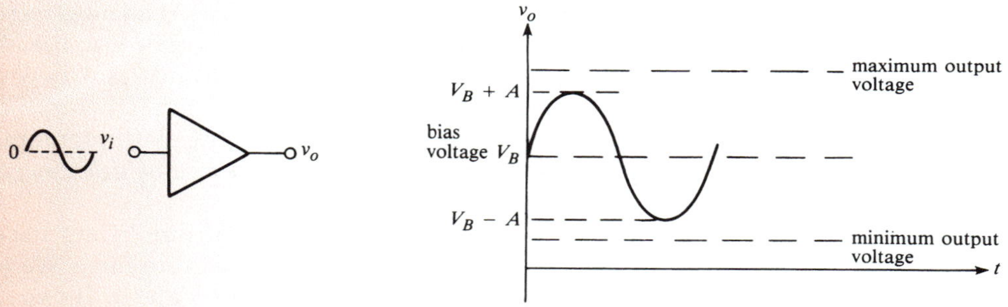

FIGURE 7–1 The purpose of bias is to provide a dc level about which ac variations can occur

FIGURE 7–2 Clipping caused by improper setting of the bias voltage and by overdriving

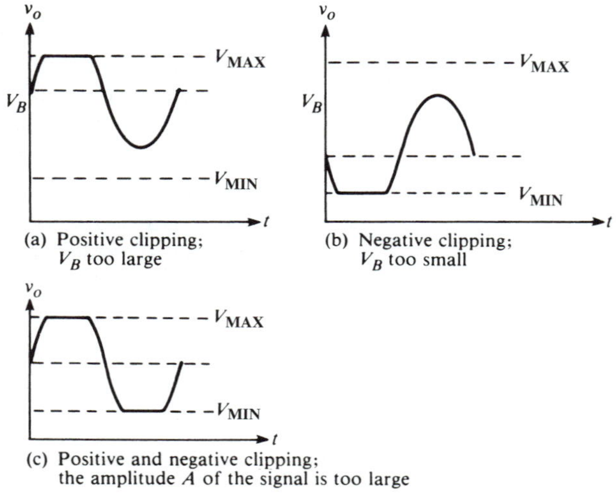

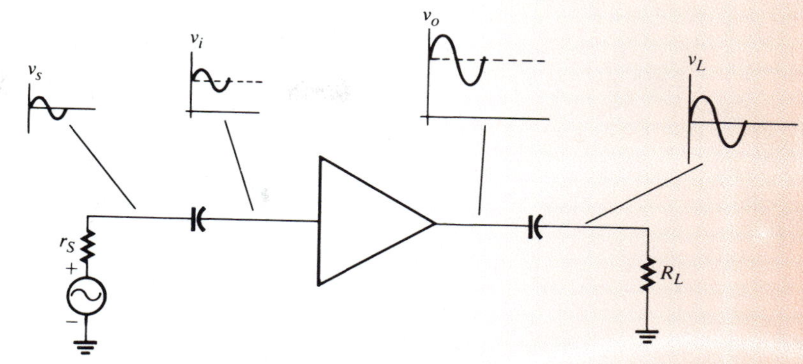

FIGURE 7–3 Coupling capacitors are used to block the flow of dc current between the amplifier and the signal source and between the amplifier and load

FIGURE 7–4 A common-emitter amplifier

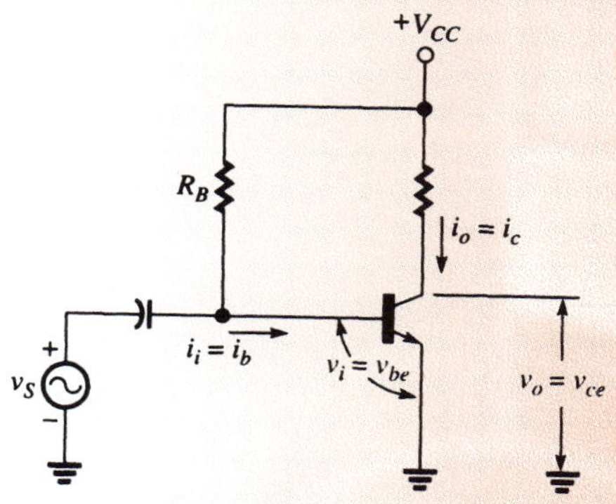

The purpose of an ac amplifier is to produce an output waveform that is an amplified version of the input waveform. Because clipping defeats this purpose, it is said to *distort* the signal, and clipping is an example of amplitude *distortion.* In a transistor amplifier, the minimum and maximum output voltages are typically the voltages at saturation and cutoff, respectively. Thus, the minimum output may be a saturation voltage of a few tenths of a volt, and the maximum output may be a cutoff voltage equal to the supply voltage.

Coupling Capacitors

In many amplifier applications, the source or the load, or both, cannot be subjected to a dc voltage or be permitted to conduct dc current. For example, an electromagnetic speaker is designed to respond to ac fluctuations only and may not operate properly if it conducts a dc current. To prevent the dc component of an amplifier's output voltage from producing dc current in the load, a capacitor is connected in series with the load. Similarly, to prevent the flow of dc current from the amplifier into the signal source (or vice versa), a capacitor is connected in series with the source. These connections are shown in Figure 7–3. The capacitors are called *coupling* capacitors, or *blocking* capacitors, because they block the flow of dc current. The capacitors must be large enough to present negligible impedance to the ac signals.

7–2 ANALYSIS OF THE CE AMPLIFIER

Our study of transistor amplifiers begins with an analysis of the common-emitter circuit because that is the most widely used configuration. Figure 7–4 shows the CE bias circuit we studied in Chapter 4 modified by the inclusion of an ac signal source in series with a capacitor connected to the base. As we will see later, this simple configuration has several drawbacks, but it is a good start for illustrating the amplification principle. The coupling capacitor is used for preventing the flow of dc current from the transistor into the signal source.

Notice in Figure 7–4 that we now designate input and output voltages and currents with lowercase letters to represent ac quantities. The signal source causes small variations in the transistor input voltage, which in turn cause small variations in the base current. As the base current decreases and increases, the collector current does the same. Because the collector current is approximately β times larger than the base current, we achieve current gain between input and output. The transistor's ability to produce gain can be attributed to the difference in impedance levels at input and output: A small current into a low-resistance, forward-biased junction controls current flow across a high-resistance, reverse-biased junction.

It is instructive to view ac operation in terms of the variation in output voltage and output current that occurs along a load line plotted on a set of transistor output characteristics. We will perform a graphical analysis of the circuit shown in Figure 7–4.

To observe the total variation in base current caused by the signal source in Figure 7–4, we use the input characteristic shown in Figure 7–5. We see in Figure 7–5 that the input voltage variation causes I_B to vary sinusoidally between a minimum and a maximum value producing ΔI_B as indicated.

The load line for the circuit in Figure 7–4 intersects the V_{CE}-axis at V_{CC} and the I_C-axis at V_{CC}/R_C. It is shown in Figure 7–6. The Q-point is located at the intersection of the load line with the base current curve corresponding to *the base current that flows when* $V_S = 0$, or I_{BQ}. By definition, the Q-point locates the *dc* bias values of V_{CEQ} and I_{CQ}, which are the output values when no ac signal is present.

Recall that the load line is a plot of all *possible* combinations of I_C and V_{CE}. Therefore, as the base current alternates between its extreme values, the values of I_C and V_{CE} change along the load line producing ΔV_{CE} and ΔI_C, as indicated in the figure.

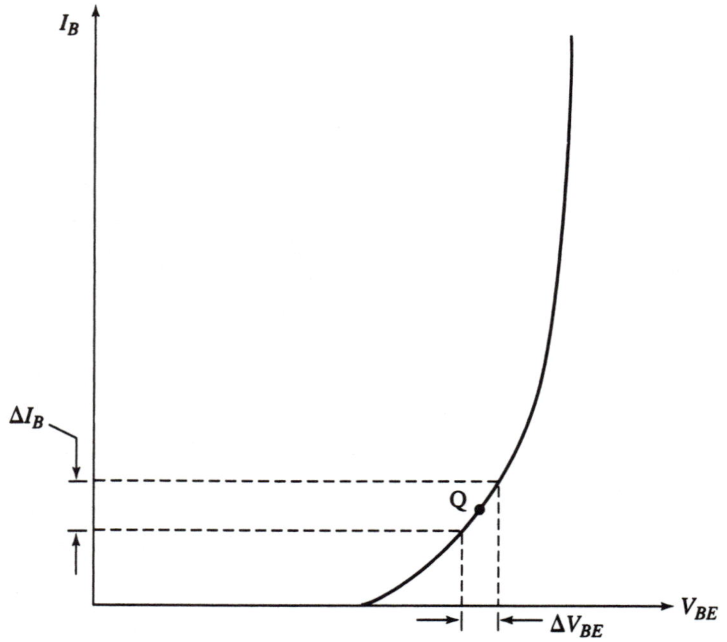

FIGURE 7–5 A signal voltage produces the change ΔV_{BE}, which translates into ΔI_B

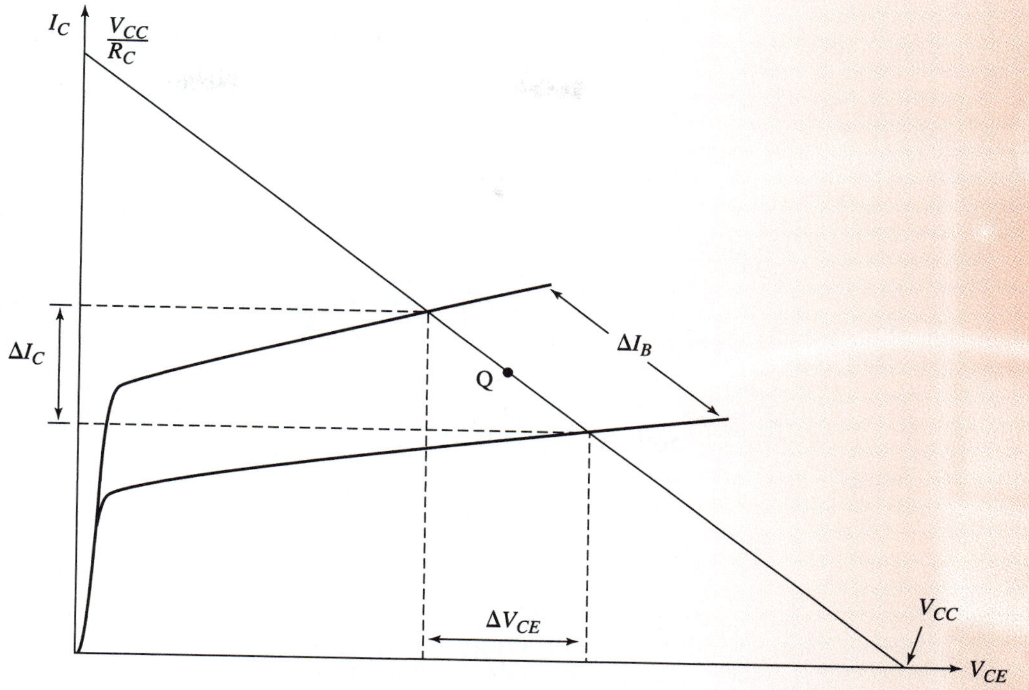

FIGURE 7–6 The base current change ΔI_B produces changes ΔI_C and ΔV_{CE}

Note carefully that V_{CE} *decreases* when I_B and I_C are increasing, and vice versa. Therefore, the sinusoidal voltage v_{ce} is 180° out of phase with each of the sinusoidal currents i_b and i_c. Because i_b is in phase with v_{be}, we conclude that v_{ce} is also 180° out of phase with v_{be}. In other words, *the ac output voltage from a common-emitter amplifier is 180° out of phase with the ac input voltage.* This fact can also be deduced from the load line equation:

$$I_C = \frac{-1}{R_C} V_{CE} + \frac{V_{CC}}{R_C} \tag{7–1}$$

or

$$V_{CE} = V_{CC} - I_C R_C$$

Because the term $I_C R_C$ in equation 7–1 subtracts from the constant V_{CC}, we observe that an increase in I_C causes a decrease in V_{CE}. Thus, as input voltage increases, I_B increases, I_C increases, and V_{CE} decreases. The common-emitter voltage amplifier is said to cause *phase inversion,* or to *invert* voltage.

Concluding, we can also relate this graphical analysis to gains and resistances. The current gain is

$$A_i = \frac{\Delta I_c}{\Delta I_B} \tag{7–2}$$

the voltage gain is

$$A_v = \frac{\Delta V_{CE}}{\Delta V_{BE}} \tag{7–3}$$

and the input resistance looking into the base-emitter junction is

$$r_i = \frac{\Delta V_{BE}}{\Delta I_B} \tag{7–4}$$

Bias-Stabilized CE Amplifier

The simple biasing configuration we have used so far for analyzing the common-emitter amplifier is commonly referred to as a *fixed* bias circuit. The term is somewhat misleading because the Q-point, far from being fixed, depends directly on the value of β. Actually, the only "fixed" variable is the base current, given by

$$I_B = \frac{V_{CC} - V_{BE}}{R_B}$$

Both the collector current and collector-emitter voltage depend on β, which in turn can vary substantially from transistor to transistor, even within the same model.

A well-designed biasing configuration should establish a collector current whose value would not be affected greatly by variations in β. In other words, I_C should not change more than a small percentage even if β were to change over a 2-to-1 range or more. The variations in β we are mostly concerned with are those that occur from transistor to transistor. However, there are some other factors, such as temperature, that can make β change within the same transistor.

Variation of the Q-point is not caused exclusively by variations in β. Although to a much lesser extent, fluctuations in V_{BE} and I_{CBO}, primarily caused by temperature variations, do contribute to instability of the bias operating point. Fortunately, the criterion used for reducing the effect of β on the Q-point applies also to both V_{BE} and I_{CBO}.

Figure 7–7(a) shows the biasing scheme known as the voltage-divider biasing circuit, which can be made arbitrarily insensitive to β variations. The smaller R_1 and R_2 are made, the "stiffer" the base voltage and collector current will be. However, making R_1 and R_2 too small will result in an impractically low input resistance. With this trade-off in mind, a good design should allow for some, albeit small, variation in collector current in order to avoid an excessively low input resistance to the amplifying stage.

To simplify the dc analysis of the circuit, it is convenient to use Thévenin's theorem at the voltage divider, resulting in the equivalent circuit shown in Figure 7–7(b), where

$$V_{Th} = \frac{R_2}{R_1 + R_2} V_{CC} \quad \text{and} \quad R_{Th} = R_1 \| R_2 = R_B$$

To obtain an expression for the base current we first write the KVL equation

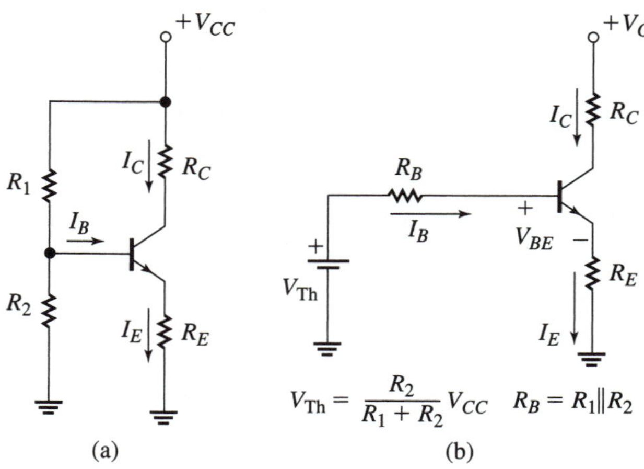

FIGURE 7–7 Voltage-divider biasing circuit and its Thévenin equivalent

$$V_{Th} = I_B R_B + V_{BE} + I_E R_E \tag{7–5}$$

Substituting $I_E = (\beta + 1)I_B$ and solving for I_B, we obtain

$$I_B = \frac{V_{Th} - V_{BE}}{R_B + (\beta + 1)R_E} \tag{7–6}$$

To obtain I_E, we multiply the above expression times $(\beta + 1)$ to obtain

$$I_E = \frac{V_{Th} - V_{BE}}{R_E + \dfrac{R_B}{\beta + 1}} \tag{7–7}$$

But because I_C is very close in value to I_E, we can approximate both with the same formula, that is,

$$I_E \approx \frac{V_{Th} - V_{BE}}{R_E + \dfrac{R_B}{\beta}} \approx I_C \tag{7–8}$$

Incidentally, equation 7–8 conveniently yields a value that is between the true I_C and I_E values, thus spreading the small error between the two.

Let us now focus on the design procedure. Let

β_{min} = minimum expected value of β

β_{max} = maximum expected value of β

I_{Cmin} = collector current due to β_{min}

I_{Cmax} = collector current due to β_{max}

The extreme values for the collector current according to the extreme values of β are

$$I_{Cmax} = \frac{V_{Th} - V_{BE}}{R_E + \dfrac{R_B}{\beta_{max}}} \tag{7–9}$$

and

$$I_{Cmin} = \frac{V_{Th} - V_{BE}}{R_E + \dfrac{R_B}{\beta_{min}}} \tag{7–10}$$

If we solve simultaneously these two equations by equating the $V_{Th} - V_{BE}$ terms, we obtain the expression

$$\frac{R_B}{R_E} = \frac{I_{Cmax} - I_{Cmin}}{I_{Cmin}/\beta_{min} - I_{Cmax}/\beta_{max}} \tag{7–11}$$

which can be used for determining the required R_B/R_E ratio according to the desired allowable variation for I_C. Once R_E and R_B are obtained, V_{Th}, R_1, and R_2 will follow.

The following steps summarize the design procedure for the voltage-divider biasing circuit.

1. Given a nominal value for I_{CQ} and a maximum acceptable variation, determine I_{Cmax} and I_{Cmin}.
2. With the expected extreme values of β, determine the required R_B/R_E ratio.
3. Determine R_E using the nominal collector current. A good rule of thumb is to establish a voltage drop across R_E equal to 10% of V_{CC}.

4. Determine R_B from the previously calculated R_B/R_E ratio.
5. Calculate the required V_{Th} using either expression for I_{Cmin} or I_{Cmax}.
6. Solve simultaneously the expressions for V_{Th} and R_B, that is,

$$V_{Th} = \frac{R_2}{R_1 + R_2} V_{CC} \quad \text{and} \quad R_B = \frac{R_1 R_2}{R_1 + R_2}$$

to obtain

$$R_1 = \frac{V_{CC}}{V_{Th}} R_B \tag{7–12}$$

and then R_2 from

$$\frac{1}{R_2} = \frac{1}{R_B} - \frac{1}{R_1}$$

7. Adjust R_1 and R_2 to the nearest commercial value and calculate the collector current for the two extreme values of β to make sure the design satisfies the requirements.

EXAMPLE 7–1

Design a voltage-divider biasing circuit with $V_{CC} = 18$ V, $I_{CQ} = 2$ mA ± 3%, and $V_{CEQ} = 7$ V. Expected β range is 80 to 200. Specify commercial values for resistors.

Solution

Using a voltage drop across R_E of 10% of V_{CC}, $R_E = 1.8$ V/2 mA = 900 Ω. The closest commercial value is 910 Ω.

The collector resistor is $R_C = V_{R_C}/I_{CQ}$, where $V_{R_C} = V_{CC} - V_{CE} - V_{R_E}$ according to KVL. Then $V_{R_C} = 18 - 7 - 1.8 = 9.2$ V and $R_C = 9.2$ V/2 mA = 4.6 kΩ. We will use 4.7 kΩ.

With the ±3% allowable collector current variation, we obtain I_{Cmax} = 2.06 mA and I_{Cmin} = 1.94 mA. The ratio R_B/R_E can now be calculated using equation 7–11:

$$\frac{R_B}{R_E} = \frac{2.06 - 1.94}{1.94/80 - 2.06/200} = 8.6$$

from which we obtain $R_B = 8.6(910\ \Omega) = 7.83$ kΩ. Solving for V_{Th} from the expression for I_{Cmax}, we obtain $V_{Th} = I_{Cmax}(R_E + R_B/\beta_{max}) + V_{BE} = 2.06(0.910 + 7.83/200) + 0.7 = 2.66$ V. (Note that the numbers were entered in mA and kΩ, consistent with volts.) You can also verify that the same result is obtained by using the expression for I_{Cmin}. Using equation 7–12, we obtain $R_1 = (V_{CC}/V_{Th})R_B = (18/2.66)7.83$ kΩ = 53 kΩ, from which it follows $R_2 = (1/7.83\text{ k} - 1/53\text{ k})^{-1} = 9.2$ kΩ. The closest commercial values are $R_1 = 51$ kΩ and $R_2 = 9.1$ kΩ.

To verify our design, we calculate I_{Cmax} and I_{Cmin} using the commercial values for resistance we have specified. First, $V_{Th} = (9.1/60.1)18 = 2.73$ V, and $R_B = 51\|9.1 = 7.72$ kΩ. Then,

$$I_{Cmin} = \frac{2.73 - 0.7}{0.91 + 7.72/80} = 2.02\text{ mA} \quad \text{and} \quad I_{Cmax} = \frac{2.73 - 0.7}{0.91 + 7.72/200} = 2.14\text{ mA}$$

You can see that even though the resulting currents are slightly off due to using commercial resistor values, the total variation in collector current is 0.12 mA, as specified.

In order to use the bias-stabilized circuit as a common-emitter amplifier, besides the input and output coupling capacitors, there may be a need

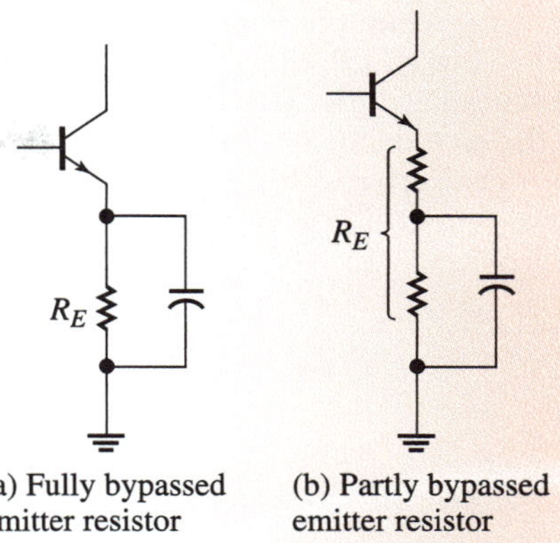

FIGURE 7–8 Using a bypass capacitor for the emitter

for using a bypass capacitor across the emitter resistor in order to place the emitter at *zero ac potential* or *ac ground*, as shown in Figure 7–8(a). It is also possible to partially bypass the emitter resistor as shown in Figure 7–8(b) in order to have control on the gain, as we will see shortly, and also for reducing distortion. The circuit could be used with no bypass capacitor at all, but at the cost of substantially reduced gain.

We will examine in full detail this amplifying configuration along with other configurations later in the chapter using *small-signal models*. However, a few issues regarding the relationship between dc biasing and ac performance need to be addressed beforehand.

The Effect of Q-Point Location on ac Operation

Let us consider now how the location of the Q-point affects the ac operation of the amplifier. Suppose the value of the base resistor R_B in Figure 7–4 is increased. The quiescent value of the base current will then become smaller. Figure 7–9 shows that the Q-point in this case is shifted down the load line to the point where it intersects the I_{BQ} curve. When the base current increases beyond the Q-point to I_{Bmax}, it can be seen in Figure 7–9 that the collector current increases to I_{Cmax} and V_{CE} decreases to V_{CEmin}. However, when the base current decreases below the Q-point, the transistor enters its cutoff region. Clearly I_C cannot be less than 0 and V_{CE} cannot be greater than V_{CC}. As shown in the figure, the output current prematurely becomes 0 in the sine-wave cycle, and clipping results. At the same time, V_{CE} reaches its limit of V_{CC}, and the output waveform shows positive clipping. With the Q-point in its new location, the output voltage change cannot exceed $V_{CC} - V_Q$ without positive clipping occurring. Thus, the maximum peak-to-peak output voltage is $2(V_{CC} - V_Q)$, and it represents the *maximum output swing*. This reduced swing limits the usefulness of the amplifier and illustrates the importance of locating the bias point somewhere near the center of the load line.

If the Q-point is located too far *up* the load line, the output swing will be limited by the onset of saturation. This fact is illustrated in Figure 7–10(a). In this case, a substantial increase in base current beyond its quiescent value causes the transistor to saturate. The collector current cannot exceed its saturation value and V_{CE} cannot be less than 0. Consequently, the output voltage is a negatively clipped waveform, as shown in the figure.

Even if the Q-point is located at the center of the load line, positive and negative clipping can occur if the input signal is too large. Figure 7–10(b) shows what happens when the total change in base current is so great that the transistor is driven into saturation at one end and cutoff at the other. We see that both positive and negative clipping occur due to the amplifier being *overdriven*.

FIGURE 7–9 When the base resistance is increased, the Q-point moves down the load line and the signal causes the amplifier to cut off, resulting in clipping

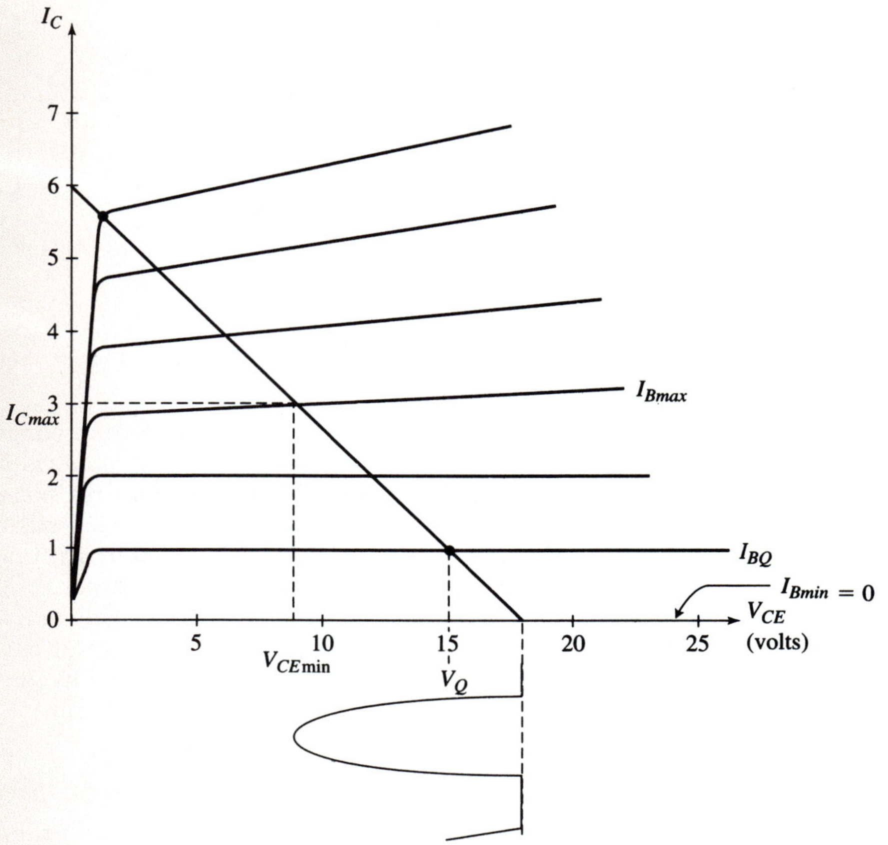

FIGURE 7–10

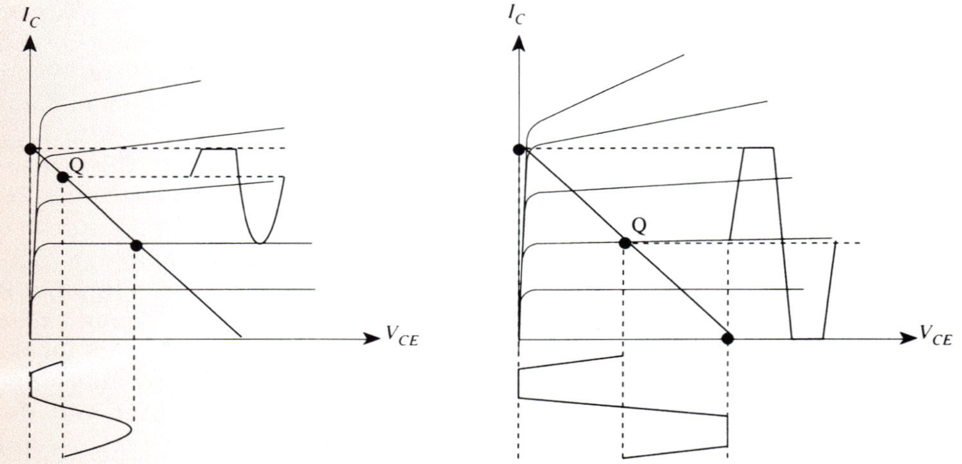

(a) The Q-point is located too close to saturation and the output voltage shows negative clipping.

(b) Too large an input signal causes both positive and negative clipping.

Linearity and Distortion

To be useful, an amplifier's output waveform must be a faithful replica of the input waveform (or a phase-inverted replica of the input). That clearly is not the case when clipping occurs. Apart from clipping, the degree to which the output waveform has the same shape as the input depends upon the amplifier's *linearity*. To be linear, any change in output voltage must be *directly proportional* to the change in input voltage that created it. For example, if $\Delta V_o = 1$ V when $\Delta V_i = 0.01$ V, then ΔV_o must equal 2 V when $\Delta V_i = 0.02$ V, and ΔV_o must equal 0.5 V when $\Delta V_i = 0.005$ V. The linearity of a transistor can be determined by examining the extent to which equal increments of base current correspond to equally spaced curves on the CE output characteristics. If we assume that input current is directly proportional to input voltage (i.e., that the base–emitter junction is linear, in the sense discussed in Chapter 3), then changes in input voltage should cause the output voltage to vary to a proportional extent along the load line. This will be the case only if the curves of constant base current are equally spaced. Figure 7–11 shows a set of CE output characteristics that have been intentionally distorted to exaggerate nonuniform spacing. Notice that the distance between the curves increases for larger values of base current.

The active region of a transistor's output characteristics is the region where the base current curves are generally found to have equal or nearly equal spacing. For this reason, the active region is often called the *linear* region. Of course, the characteristics shown in Figure 7–11 are decidedly nonlinear. In this example, the nonlinearity is due to the fact that device parameters (such as β) change significantly over the region of operation.

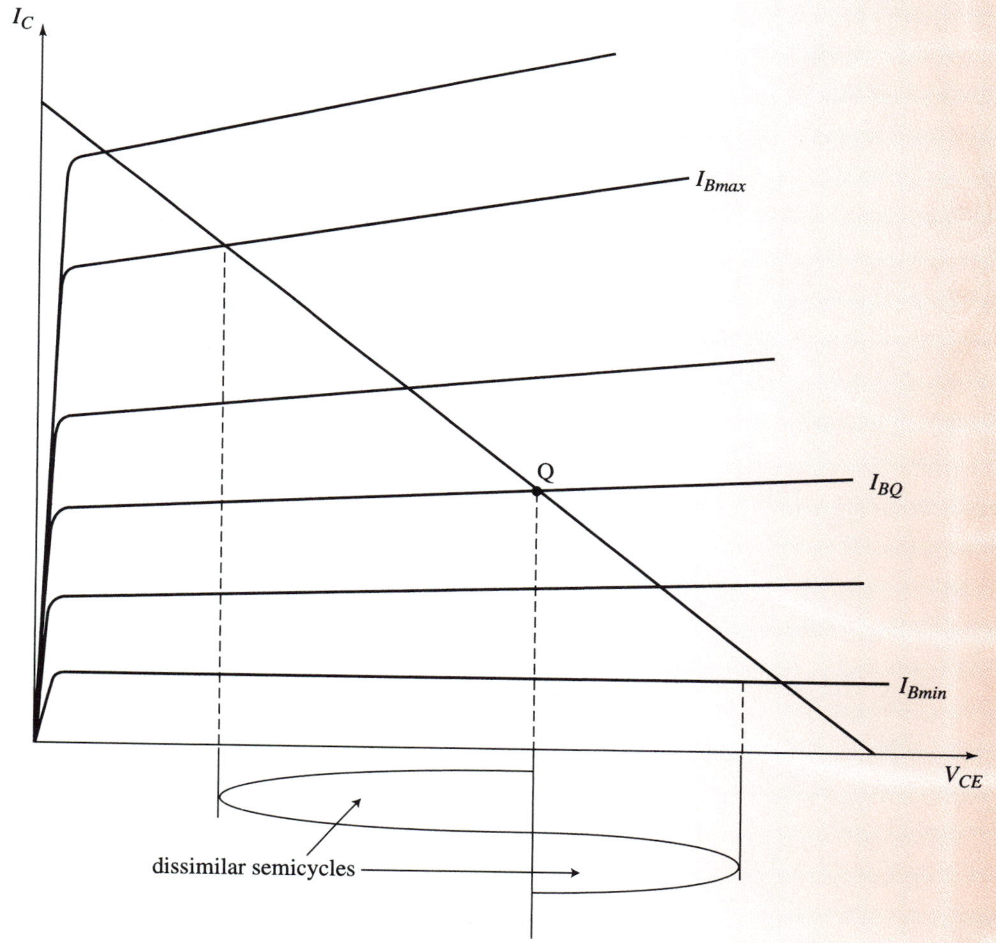

FIGURE 7–11 Unequal spacing between equal intervals of base current represents a nonlinear characteristic that causes a distorted output

Small-signal analysis of this device would thus be restricted to a very small range of operation along the load line.

The Effect of Load Resistance on ac Operation

Let us now consider the effect of connecting a load resistor R_L across the output of the CE amplifier, as shown in Figure 7–12. It is important to realize that *as far as ac performance is concerned, R_L is in parallel with R_C. A dc source is a short circuit to ac signals,* so resistor R_C in Figure 7–12 is effectively grounded through V_{CC}, and the ac voltage at the collector "sees" R_C in parallel with R_L. Another way of viewing this is from the standpoint of analysis by superposition: If we are interested in the output voltage due only to the ac source, v_S, we replace all other voltage sources by short circuits.

The ac load resistance, designated r_L, is the parallel combination of R_C and R_L:

$$r_L = R_C \| R_L = \frac{R_C R_L}{R_C + R_L} \tag{7–13}$$

The dc load resistance is, of course, still equal to R_C because the coupling capacitor blocks the flow of dc current into the load resistor.

The existence of an ac load that differs from the dc load means that the output voltage is no longer determined by variations along a load line based on R_C and V_{CC}. Instead, the output is determined by variations along an *ac load line,* based on $r_L = R_C \| R_L$. The load line based only on the value of R_C will hereafter be called the *dc load line.* Because the ac load line represents all possible combinations of collector voltage and current, it must include the point where the ac input goes through 0. That point is, of course, the Q-point on the dc load line, so we conclude that the dc and ac load lines intersect at the Q-point. Figure 7–13 shows both the dc and ac load lines of this example plotted on the output characteristics.

Note that the ac load line is steeper than the dc load line. Remember that the slope of the dc load line is $-1/R_L$, whereas that of the ac load line is $-1/r_L$. Since $r_L < R_L$, the latter slope is greater than the former. The intercepts of the ac load line are

$$I_o = \frac{V_Q}{r_L} + I_Q \tag{7–14}$$

and

$$V_0 = V_Q + I_Q r_L \tag{7–15}$$

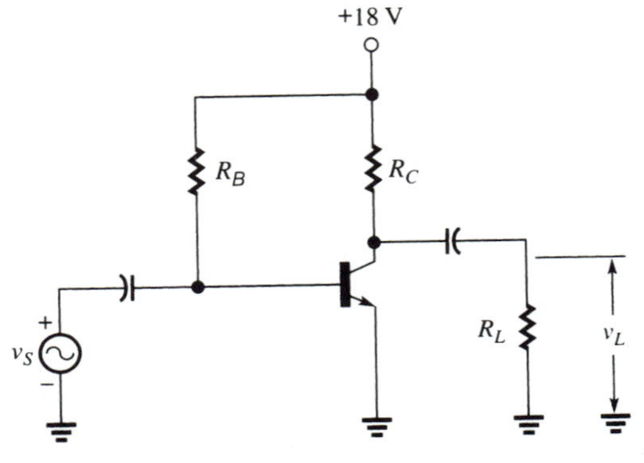

FIGURE 7–12 The CE amplifier circuit of Figure 7–4 modified to include a capacitor-coupled load resistor

FIGURE 7–13 dc and ac load lines

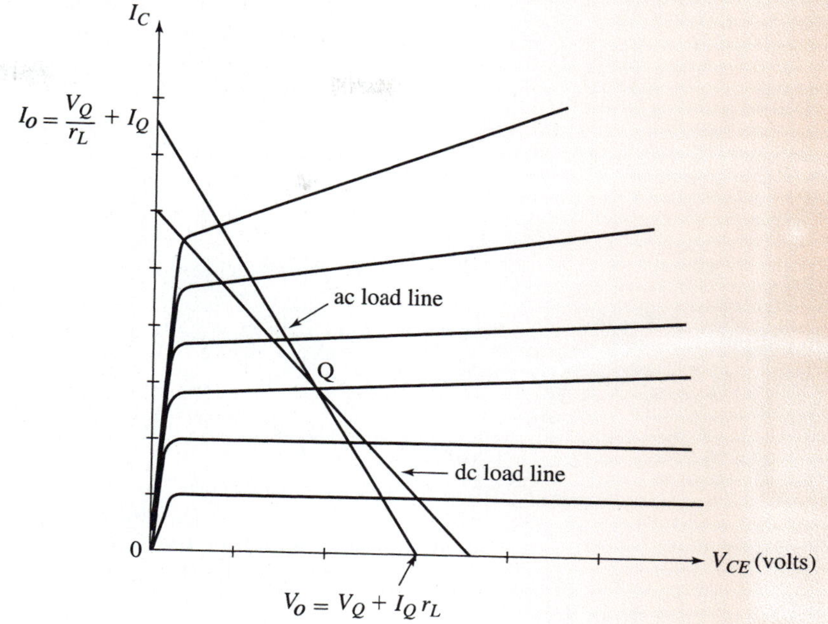

where I_0 = intercept of the ac load line on the I_C-axis
V_0 = intercept of the ac load line on the V_{CE}-axis
V_Q = quiescent value of V_{CE}
I_Q = quiescent value of I_C

It is an exercise at the end of this chapter to show that equations 7–14 and 7–15 also apply to the dc load line when R_C is substituted for r_L (i.e., for the case $R_L = \infty$).

It must be emphasized again that the ac load line represents all possible combinations of collector–emitter voltage and collector current, and that the dc load line no longer applies. It is a common mistake to believe that the dc load line governs the voltage across R_C while the ac load line governs the voltage across R_L. Remember that the current through and voltage across R_L are pure ac waveforms that go both positive and negative, because the capacitor blocks the dc component of the collector waveform. The only difference between v_L and the collector voltage is the dc component in the latter.

The practical implication of the ac load line is that it makes the magnitude of the ac output voltage smaller than it would be if the output variations were determined by the dc load line. This fact is illustrated in Figure 7–14, where the output voltages determined by both dc and ac load lines are plotted. The same base-current variation is assumed for both, and it can be seen that the steeper slope of the ac load line results in a smaller output. The connection of a load across the output of an amplifier always reduces the amplitude of its ac output.

If the base resistance R_B is changed, the Q-point will shift to a new location on the *dc* load line. Because the ac load line passes through the Q-point, it too will shift. As illustrated in Figure 7–15, ac load lines corresponding to different Q-points are parallel to each other, because all have the same slope, $-1/r_L$.

Another consequence of a capacitor-coupled load R_L is that it reduces the maximum permissible voltage swing at the output of the amplifier. Recall from equations 7–14 and 7–15 that the ac load line intersects the V_{CE}-axis at $V_0 = V_Q + I_Q r_L$ and intersects the I_C-axis at $I_0 = I_Q + V_Q/r_L$, where I_Q and V_Q are the Q-point coordinates. It is therefore clear that the positive-going output voltage cannot exceed the bias voltage by more than $I_Q r_L$ volts. This fact is illustrated

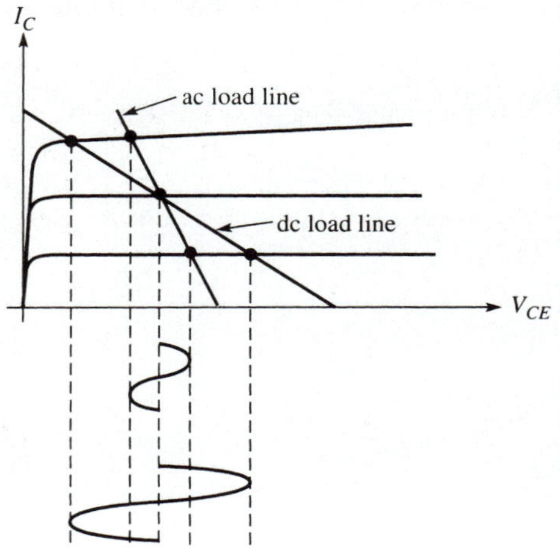

FIGURE 7–14 The output voltage determined by the ac load line is smaller than it would be if it were determined by the dc load line

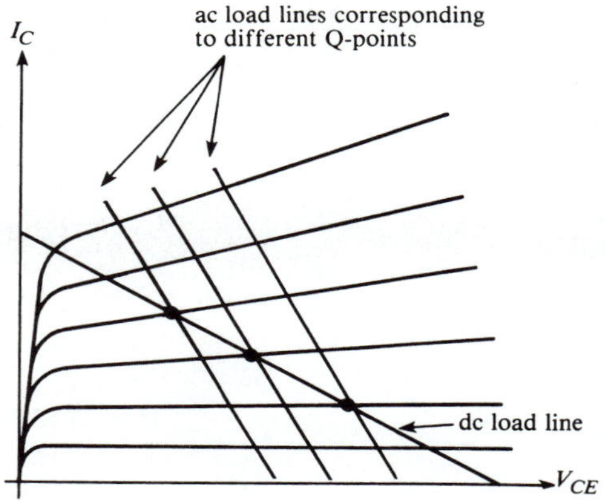

FIGURE 7–15 If the Q-point shifts along the dc load line, the ac load line shifts to a parallel location

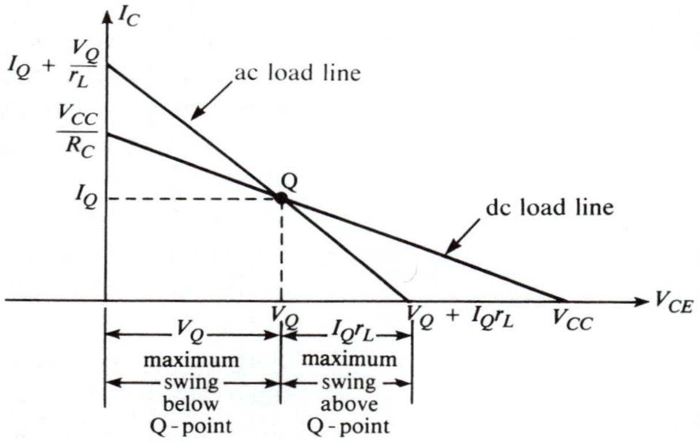

FIGURE 7–16 The ac load line limits the output voltage swing to the minimum of V_Q and $I_Q r_L$

in Figure 7–16. The output voltage can swing $I_Q r_L$ volts above the Q-point and V_Q volts below the Q-point. Thus the maximum peak output voltage is the *minimum* of the two values V_Q and $I_Q r_L$. If the output attempts to exceed V_Q by more than $I_Q r_L$ volts, its positive peak will be clipped; if it attempts to swing more than V_Q volts below V_Q, its negative peak will be clipped.

Figure 7–15 in our discussion of graphical amplifier analysis shows that the ac load line shifts with changes in the Q-point. It is obvious that shifting the ac load line causes it to have different intercepts on the horizontal and vertical axes. It should therefore be possible to locate the Q-point in such a way that the maximum positive swing equals the maximum negative swing, and thereby attain the maximum possible peak-to-peak output swing. As we can see in Figure 7–16, this condition occurs when

$$V_Q = I_Q r_L \tag{7–16}$$

When the dc load line equation is rewritten and evaluated at $V_{CE} = V_Q$ and $I_C = I_Q$, we obtain

$$V_Q = V_{CC} - I_Q R_C \tag{7–17}$$

Solving equations 7–16 and 7–17 simultaneously (see Exercise 7–14) leads to the following equations for the optimum Q-point coordinates:

$$\text{(optimum) } I_Q = \frac{V_{CC}}{R_C + r_L} \tag{7–18}$$

$$\text{(optimum) } V_Q = V_{CC} - I_Q R_C \tag{7–19}$$

EXAMPLE 7–2

Find the maximum peak-to-peak output voltage that the amplifier in Figure 7–17 can produce without causing clipping at the output.

FIGURE 7–17 (Example 7–2)

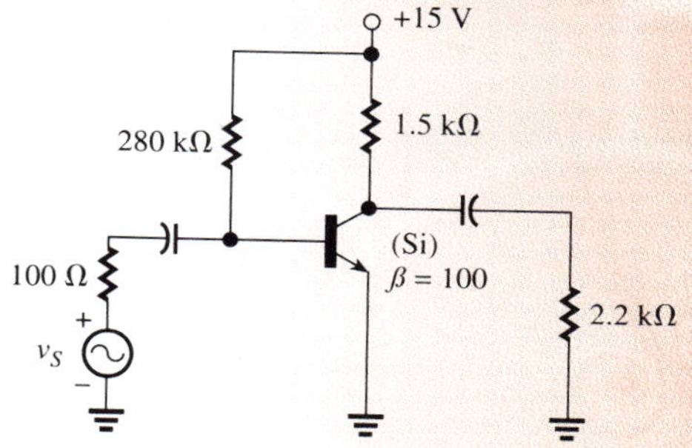

Solution

Solving first for the quiescent values, we have

$$I_B = \frac{(15 - 0.7)\text{V}}{280 \times 10^3\ \Omega} = 51.1\ \mu\text{A}$$

$$I_C = I_Q = \beta I_B = 100(51.1\ \mu\text{A}) = 5.11\text{ mA}$$

$$V_Q = V_{CC} - I_Q R_C = 15\text{ V} - (5.11\text{ mA})(1.5\text{ k}\Omega) = 7.34\text{ V}$$

$$r_L = R_C \| R_L = (1.5\text{ k}\Omega) \| (2.2\text{ k}\Omega) = 892\ \Omega$$

Therefore, the maximum output swing *above* $V_Q = 7.34$ V is $I_Q r_L = (5.11\text{ mA})(892\ \Omega) = 4.56$ V. Because $V_Q = 7.34$ V, the maximum output variation from the Q-point is the minimum of 4.56 V and 7.34 V, or 4.56 V. Thus, the maximum peak-to-peak ac output is (2)(4.56) = 9.12 V p–p.

EXAMPLE 7–3

PSPICE

Use PSpice to obtain a plot of v_{ce} versus time in the circuit shown on the next page. The plot should cover two full cycles of output. Also find the quiescent values of V_{CE} and I_C.

Solution

To obtain a plot of v_{ce} versus time, we must use a SIN source and perform a .TRAN analysis (see Appendix Sections A–4 and A–5). In order for the plot to cover two full cycles, we set *TSTOP* in the .TRAN statement to two times the period of the sine wave: 2(1/10 kHz) = 0.2 ms.

Also shown is the PSpice input circuit file. Note that we use the PSpice library to access the model statement for the 2N2222A transistor (see Appendix Section A–17). Also note that the Q-point can be obtained in a PSpice .PRINT statement by requesting outputs IC(Q1) (the dc collector current of Q1) and VCE(Q1) (the dc collector-to-emitter voltage of Q1).

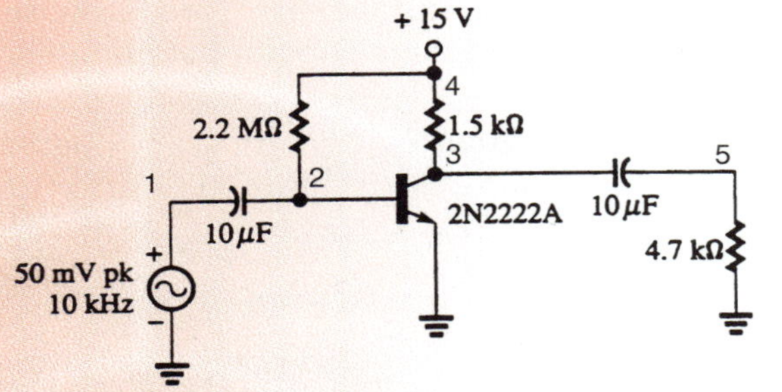


```
EXAMPLE 7-3
VS 1 0 SIN(0 50MV 10KHZ 0 0)
VCC 4 0 15V
C1 1 2 10UF
C2 3 5 10UF
RB 2 4 2.2MEG
RC 3 4 1.5K
RL 5 0 4.7K
Q1 3 2 0 Q2N2222
.LIB EVAL.LIB
.TRAN 500NS 0.2MS 0 500NS
.PROBE
.END
```

(EVAL.LIB for Evaluation version of PSpice only)

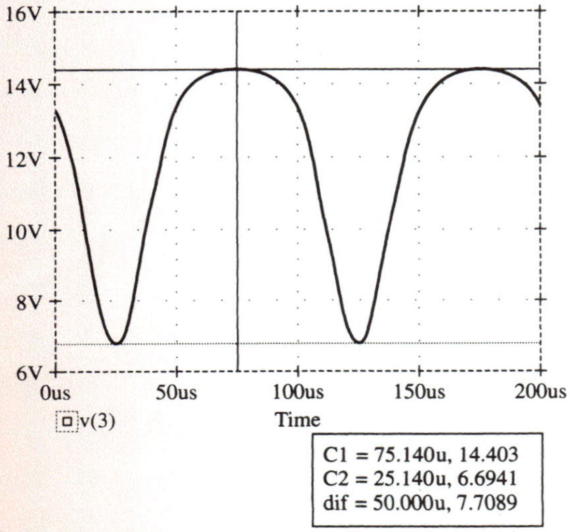

Execution of the program reveals that IC(Q1) = 1.125 mA and VCE(Q1) = 13.31 V. Since V_{CC} = 15 V, this quiescent point is close to the cutoff region of the transistor. We might therefore expect positive clipping to occur if the input to the amplifier is sufficiently large (see Figure 7–9). The plot produced by the PSpice Probe option (Section A–16) is shown, and we see that clipping does indeed occur. The Probe cursors are set at the minimum and maximum values of v_{ce}, C1 = 14.403 V and C2 = 6.6941 V. Thus, positive clipping occurs at v_{ce} = 14.403 V.

7–3 AMPLIFIER ANALYSIS USING SMALL-SIGNAL MODELS

Small-Signal Parameters

Because transistor circuits are usually analyzed using algebraic rather than graphical methods, it is convenient to have an equivalent circuit that can be substituted for the transistor wherever it appears. Many different kinds of equivalent circuits have been developed for transistors, each of which has special features that make it more useful or more accurate than others for a particular kind of analysis. The form that an equivalent circuit takes depends on the transistor *parameters* that are chosen as the basis for the circuit. A transistor parameter is simply a transistor characteristic or property that can be given a numerical value. For example, α and β are transistor parameters. The latter are examples of *derived* parameters: They are computed from a numerical relationship between two quantities (the ratio of two currents, in this case). Transistor parameters can also specify inherent physical characteristics, such as the resistance of the base region or the width of the collector–base depletion region.

Small-signal parameters are parameters whose values are determined under small-signal (ac) operating conditions and at the operating bias point. For example, the small-signal value of β is defined to be

$$\beta = \left.\frac{i_c}{i_b}\right|_{V_{CEQ}} \tag{7–20}$$

Equation 7–20 states that small-signal β is the ratio of ac collector current to ac base current at the quiescent value of V_{CE}. Small-signal β can be determined from a set of collector characteristics by constructing a vertical line at V_{CEQ} and finding $\Delta I_C/\Delta I_B$ along that line. Up to now, we have computed the (approximate) value of β by taking the ratio of two *dc* currents: $\beta \approx I_C/I_B$. To distinguish this value from the small-signal value, many authors use the notation β_{DC} ("dc beta") when referring to the ratio of dc currents. In most practical applications, the small-signal and dc values of β are close enough to be assumed equal, and we will hereafter use the notation β_{DC} only when it is necessary to emphasize that we are referring strictly to the dc value. Like small-signal β, small-signal α is defined in terms of ac currents:

$$\alpha = \left.\frac{i_c}{i_e}\right|_{V_{CBQ}} \tag{7–21}$$

An important physical parameter of a transistor is its small-signal resistance from emitter to base, called the *dynamic emitter resistance* and designated r'_e. This resistance is the same as the small-signal input resistance of the transistor in its common-base configuration. It is defined as

$$r'_e = \left.\frac{v_{eb}}{i_e}\right|_{V_{CEQ}} \tag{7–22}$$

Because the emitter–base junction can be regarded as a forward-biased diode, an approximate value for r'_e can be found in the same way in which we found the dynamic resistance of a diode (Chapter 3). Recall that $r_D \approx V_T/I \approx 0.026/I$ at room temperature, where I is the dc current in the diode. Similarly,

$$r'_e \approx \frac{26 \text{ mV}}{I_{EQ}} \tag{7–23}$$

where I_{EQ} is the Q-point emitter current.

The small-signal collector resistance r_c is the ac resistance from collector to base. It is the same as the output resistance of a transistor in its common-base configuration and typically has a value of several megohms, because it is across a reverse-biased junction.

$$r_c = \left.\frac{v_{cb}}{i_c}\right|_{I_{EQ}} \tag{7–24}$$

It is important to emphasize that small-signal parameters, in general, depend directly on the dc biasing conditions of the circuit.

Small-Signal CB Amplifier Model

An equivalent circuit for an electronic device is called a *model.* Using just the parameters we have discussed so far, we can construct a simple but reasonably accurate small-signal model for a transistor. Figure 7–18 shows a transistor in the common-base configuration and its approximate small-signal model. Remember that all voltages and currents are ac quantities, so all polarities periodically alternate. The polarities shown in the figure should be interpreted as reference directions for instantaneous values. For example, Figure 7–18(a) shows that an increase of current into the emitter terminal is accompanied by an increase of current out of the collector terminal. Models are the same for *npn* and *pnp* devices.

Notice that the model in Figure 7–18 includes a *controlled* ac current source that produces a current equal to αi_e. (Notice also the standard symbol for such a controlled source.) Thus, the collector current i_c is approximately equal to αi_e, so the model accurately reflects the relationship between i_e and i_c. The model does *not* show the feedback effect we discussed in Chapter 4; that is, it does not reflect the fact that the value of i_e depends on the value of V_{CB}. For most practical design and analysis problems, the feedback effect is insignificant and can be ignored.

It is clear that the transistor input resistance equals r'_e in our approximate CB model. This is a relatively small value, usually less than 50 Ω.

$$r_i = r'_e \quad \text{(common base)} \tag{7–25}$$

The transistor output resistance, on the other hand, can be seen to equal the large collector-base resistance r_c:

$$r_o = r_c \quad \text{(common base)} \tag{7–26}$$

The dynamic resistance r_c can also be described as

$$r_c = \left.\frac{\Delta V_{CB}}{\Delta I_C}\right|_{I_{EQ}} \quad \text{(at the operating emitter current)} \tag{7–27}$$

and can be seen graphically on the characteristic CB output curves as the reciprocal of the slope of the curve corresponding to I_{EQ} in the active region.

To illustrate how the small-signal model can be used to analyze a practical amplifier, we will incorporate it into the CB amplifier circuit shown in Figure 7–19(a). Figure 7–19(b) shows the ac equivalent of just that part of (a) that is external to the transistor. Note that all dc sources are treated as ac short circuits to ground, in accordance with our previous discussion. Also, the coupling capacitor is assumed to have negligible impedance and is replaced by a short in the ac equivalent circuit. Finally, Figure 7–19(c) shows the complete ac equivalent when the transistor is replaced by its small-signal model.

In Figure 7–19(c), we see that R_E is in parallel with r'_e and that r_c is in parallel with R_C. In practical circuits, R_E is usually much greater than r'_e ($R_E \gg r'_e$), so

FIGURE 7–18 A CB transistor and its small-signal model

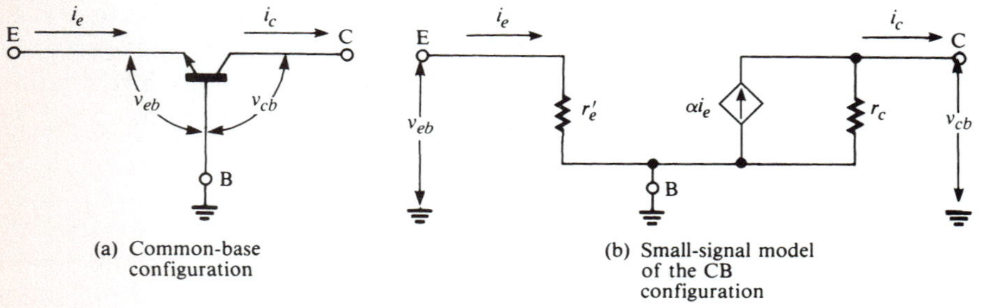

FIGURE 7–19 Evolution of an ac equivalent circuit for a common-base amplifier

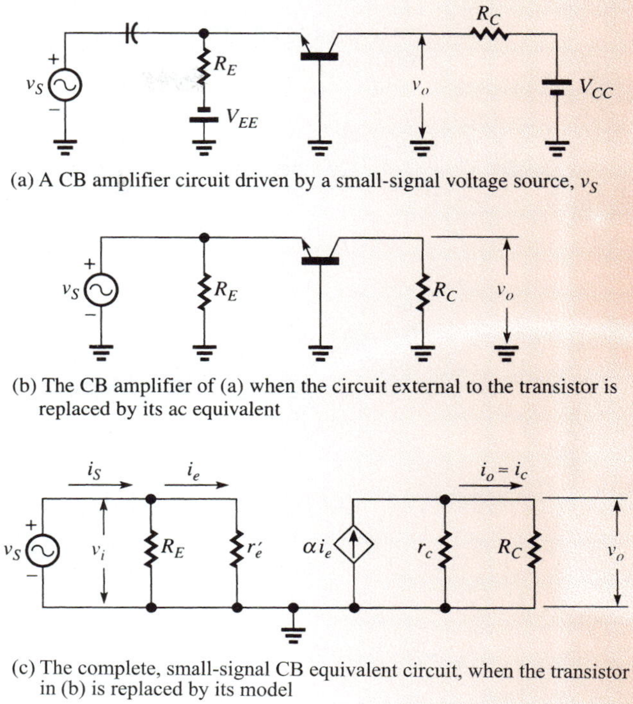

(a) A CB amplifier circuit driven by a small-signal voltage source, v_S

(b) The CB amplifier of (a) when the circuit external to the transistor is replaced by its ac equivalent

(c) The complete, small-signal CB equivalent circuit, when the transistor in (b) is replaced by its model

FIGURE 7–20 A practical CB equivalent circuit that incorporates the (usual) condition that $r_e' \parallel R_E \approx r_e'$ and $R_C \parallel r_c \approx R_C$

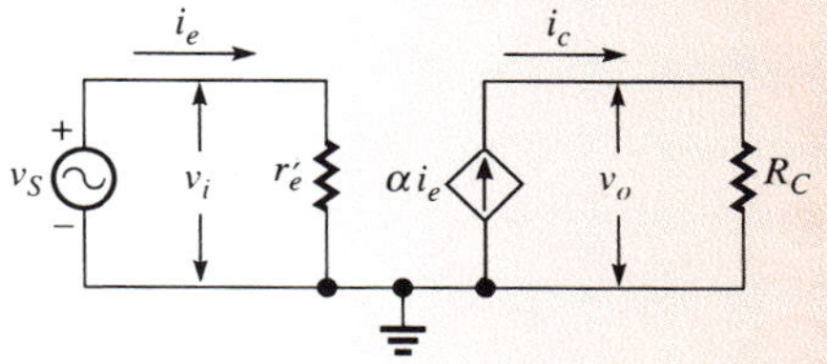

the parallel combination $R_E \parallel r_e'$ essentially equals r_e'. Also, $r_c \gg R_C$ in practical circuits, so $r_c \parallel R_C$ essentially equals R_C. Thus, Figure 7–19(c) can be replaced by the practical equivalent shown in Figure 7–20.

In Figure 7–20, it is clear that

$$v_S = v_i = i_e r_e' \quad \text{and} \quad v_o = i_c R_C = \alpha\, i_e R_C$$

or, since $\alpha \approx 1$, $v_o \approx i_e R_C$. Therefore, the voltage gain is

$$A_v = \frac{v_o}{v_i} \approx \frac{i_e R_C}{i_e r_e'} = \frac{R_C}{r_e'} \tag{7–28}$$

The current gain in Figure 7–20 is

$$A_i = \frac{i_o}{i_i} = \frac{i_c}{i_e} = \alpha \tag{7–29}$$

Thus, the current gain of a CB amplifier is always less than 1.

Figure 7–21(a) shows a CB amplifier that is driven by a source having internal resistance r_S. The amplifier load is the resistor R_L. Figure 7–21(b) shows the ac equivalent circuit that results when the assumptions $R_E \parallel r_e' \approx r_e'$ and $r_c \parallel R_C \approx R_C$ are once again imposed. Figure 7–21(c) shows the amplifier circuit when the transistor is replaced by a single block having the parameters described previously. Note that the voltage source in Figure 7–21(c) is the (Thévenin) equivalent of the current source in Figure 7–20.

FIGURE 7–21 The CB amplifier with source and load resistances included

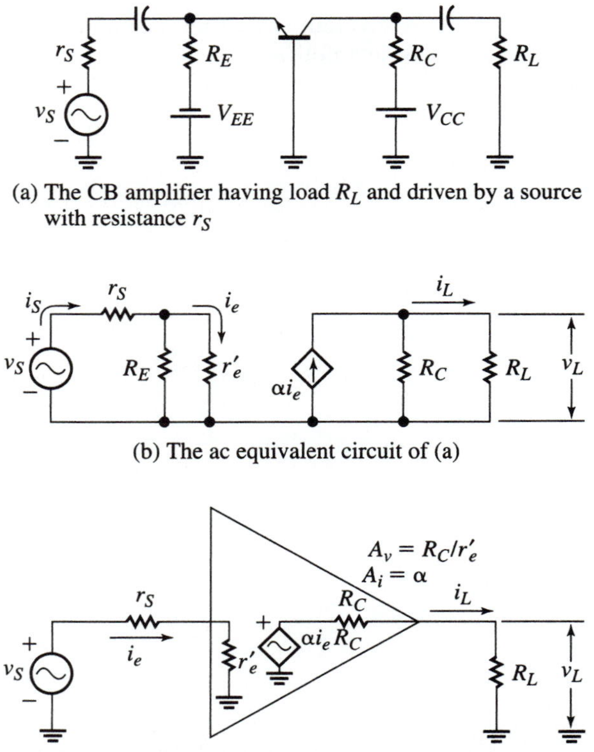

(a) The CB amplifier having load R_L and driven by a source with resistance r_S

(b) The ac equivalent circuit of (a)

(c) The circuit of (b) when the transistor is replaced by a single amplifier block

Equation 6–17 in Section 6–2 gives the overall voltage gain of an ac amplifier, from source to load. Applying the equation to Figure 7–21(c), we have

$$A_{vs} = \frac{v_L}{v_S} = \frac{R_C}{r'_e}\left(\frac{r'_e}{r_S + r'_e}\right)\left(\frac{R_L}{R_C + R_L}\right) \tag{7–30}$$

which reduces to

$$A_{vs} = \frac{R_C \| R_L}{r'_e + r_S} = \frac{r_L}{r'_e + r_S} \tag{7–31}$$

The ac output voltage from a CB amplifier is *in phase* with the ac input voltage since the gain is positive. We can also deduce this fact by the following: An increase in emitter-to-base (input) voltage reduces the forward bias on the emitter-base junction and thus reduces the emitter current. But a decrease in emitter current causes a decrease in collector current, since $I_C = \alpha I_E$. Decreasing I_C causes the voltage drop $I_C R_C$ to decrease and therefore causes V_{CB} to increase. Recapitulating, an increase in emitter (input) voltage causes an increase in collector (output) voltage, and we conclude that input and output are in phase.

When a transistor is connected in a circuit to produce gain, the transistor and all the associated external components it needs to operate properly (such as bias resistors) are referred to collectively as an amplifier *stage*. It is important to distinguish between the input and output resistances of the transistor alone and those of the stage of which it is a part. Hereafter, we will use the subscripts *in* and *out* when we wish to emphasize that it refers to a stage characteristic. Figure 7–22 illustrates these distinctions in a general amplifier stage.

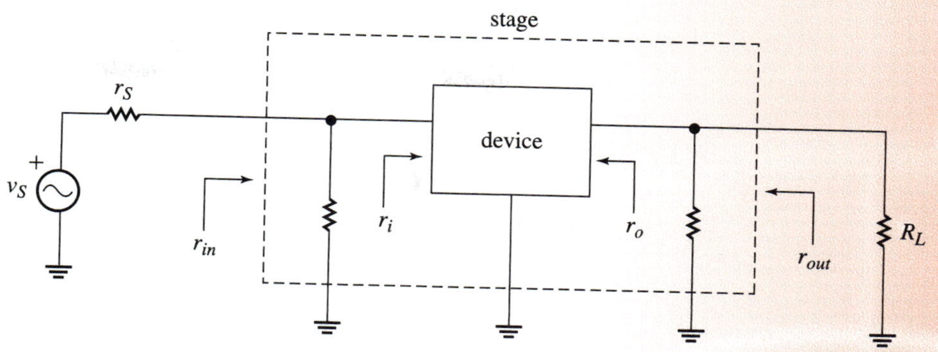

FIGURE 7–22 r_i and r_o are input and output resistance into the device, whereas r_{in} and r_{out} refer to input and output resistances of the amplifier stage

We note that for a CB amplifier,

$$r_i = r'_e$$

$$r_{in} = r'_e \| R_E \approx r'_e \qquad (7\text{–}32)$$

$$r_o = r_c$$

$$r_{out} = r_c \| R_C \approx R_C \qquad (7\text{–}33)$$

EXAMPLE 7–4

For the circuit in Figure 7–23, find (1) r_i, (2) r_{in}, (3) A_v, (4) v_L, (5) i_L, (6) i_L/i_S (assume that $\alpha = 1$), and (7) i_L, using the result of (6).

Solution

1. $I_E = \dfrac{|V_{EE}| - |V_{EB}|}{R_E} = \dfrac{6 - 0.7}{2\text{ k}\Omega} = 2.65\text{ mA}$

   Therefore,

$$r_i = r'_e = \frac{26\text{ mV}}{2.65\text{ mA}} = 9.81\ \Omega$$

2. $r_{in} = r'_e \| R_E = \dfrac{(9.81)(2000)}{2009.81} = 9.76\ \Omega$

   This result confirms equation 7–32, because for all practical purposes,

$$r_{in} = r'_e = 9.81\ \Omega$$

3. $A_v = R_C/r'_e = 1000/9.81 = 101.9$
4. $A_{vS} = 101.9\left(\dfrac{r'_e}{r_S + r'_e}\right)\left(\dfrac{R_L}{R_L + R_C}\right)$

$$= 101.9\left(\frac{9.81}{50 + 9.81}\right)\left(\frac{4\text{ k}}{1\text{ k} + 4\text{ k}}\right) = 13.37$$

FIGURE 7–23 (Example 7–4)

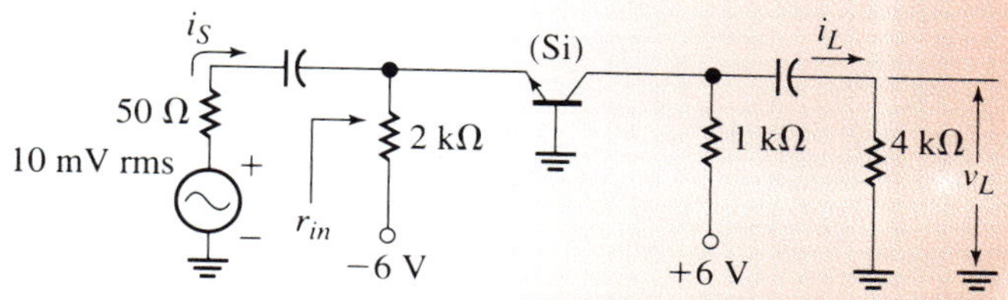

Therefore,

$$v_L = A_{vs}v_S = 13.37(10 \text{ mV rms}) = 133.7 \text{ mV rms}$$

5. $i_L = \dfrac{v_L}{R_L} = \dfrac{133.7 \text{ mV rms}}{4000\ \Omega} = 33.4\ \mu\text{A rms}$

6. $$\frac{i_L}{i_S} = \alpha\left(\frac{R_E}{R_E + r'_e}\right)\left(\frac{R_C}{R_C + R_L}\right)$$
$$= 1\left(\frac{2\text{ k}}{2\text{ k} + 9.81}\right)\left(\frac{1\text{ k}}{1\text{ k} + 4\text{ k}}\right) = 0.199$$

7. To compute i_L using the value 0.199 for i_L/i_S, we must first find the source current i_S.

$$i_S = \frac{10 \text{ mV}}{50\ \Omega + 9.81\ \Omega} = 167\ \mu\text{A}$$

Then, $i_L = 0.199\ (167\ \mu\text{A}) = 33.2\ \mu\text{A}$, which agrees with the value previously found.

Small-Signal CE Amplifier Model

To develop a model for the transistor in its common-emitter configuration, we will first investigate the input resistance in that configuration. Figure 7–24 shows the CE input circuit with r'_e drawn inside the emitter terminal, to emphasize that it is an internal transistor parameter. The ac input resistance is $r_i = v_{be}/i_b$. Recalling that $i_e = (\beta + 1)i_b$, we have

$$r_i = \frac{v_{be}}{i_e/(\beta + 1)} = (\beta + 1)\frac{v_{be}}{i_e}$$

But $v_{be}/i_e = r'_e$, so

$$r_i = (\beta + 1)r'_e \approx \beta r'_e \quad \text{(common emitter)} \qquad \textbf{(7–34)}$$

Equation 7–34 shows that the input resistance in the CE configuration is approximately β times greater than that in the CB configuration. The quantity $(\beta + 1)r'_e$ is often designated as r_π in texts and data sheets. It can also be shown that the output resistance of a transistor in its CE configuration is approximately β times *smaller* than it is in the CB configuration: $r_o \approx r_c/\beta$. Because the CE input resistance is β times greater and the output resistance is β times smaller than the corresponding values in the CB configuration, the common-emitter amplifier is inherently better suited for voltage amplification than its CB counterpart.

In a manner analogous to the CB model, the CE output resistance $r_o = r_c/\beta$ is the reciprocal of the slope of the curve for I_{BQ} in the active region of the CE output characteristic, that is,

$$r_o = \left.\frac{\Delta V_{CE}}{\Delta I_C}\right|_{I_{BQ}}$$

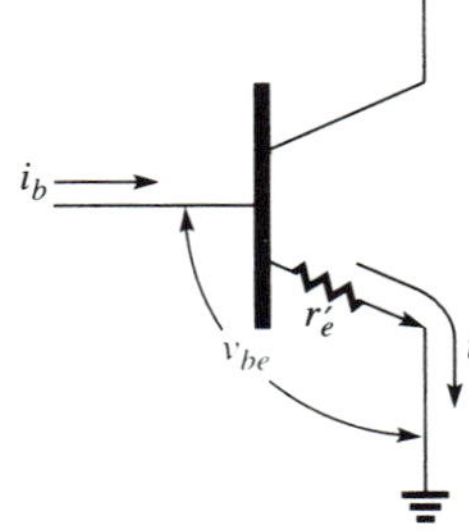

FIGURE 7–24 The input circuit for a transistor in the CE configuration

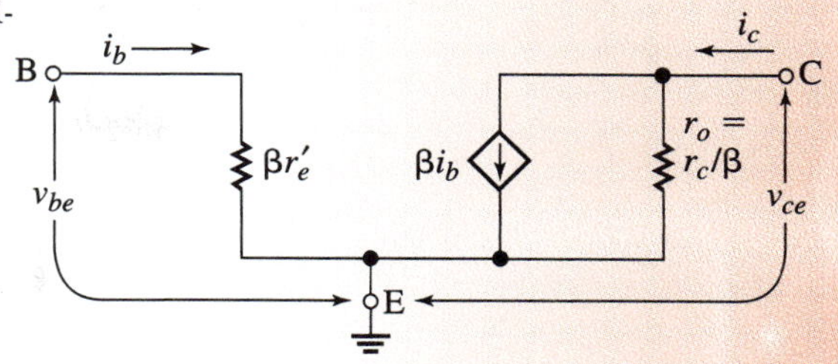

FIGURE 7–25 Approximate small-signal model for a transistor in the common-emitter configuration

which can also be determined from

$$r_o = \frac{V_A + V_{CEQ}}{I_{CQ}} \quad \textbf{(7–35)}$$

where V_A is the transistor's Early voltage. For example, if $V_A = 80$ V, $V_{CEQ} = 8$ V, and $I_{CQ} = 2$ mA, which can be fairly typical values, the resulting r_o is 44 kΩ.

Figure 7–25 shows an approximate model for the common-emitter transistor. Note that the controlled current source has value βi_b, reflecting the fact that $i_c = \beta i_b$. Once again, we neglect the feedback effect, whereby the value of i_b depends somewhat on V_{CE}.

Figure 7–26 shows a common-emitter amplifier stage with voltage-divider bias and its ac equivalent circuit. Note that the dc supply voltages are treated as ac short circuits, as before. In Figure 7–26(b), it can be seen that

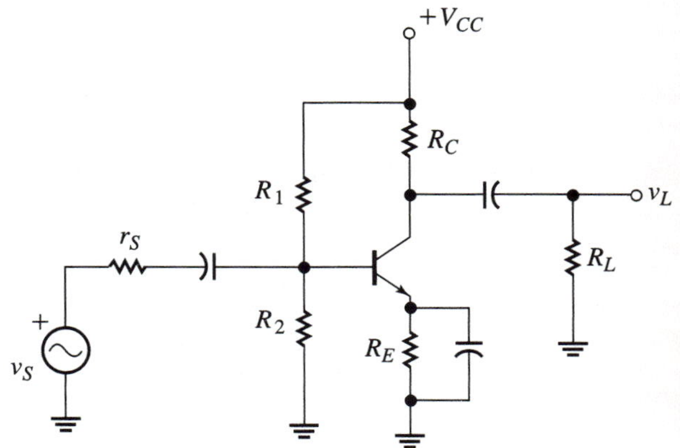

(a) Complete CE amplifier

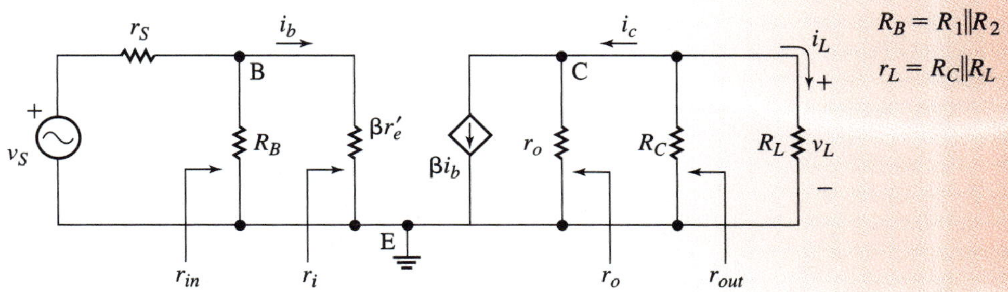

(b) Equivalent ac circuit with small-signal CE model

FIGURE 7–26 A common-emitter amplifier

r_c/β is in parallel with R_C, which implies that $r_{out} = r_o \| R_C$. Note also the ac load resistance, $r_L = R_C \| R_L$, defined in the figure. It is clear that

$$i_c = \frac{r_o}{r_o + r_L}(\beta i_b) \quad \text{(current division)}$$

and

$$A_i = \frac{i_c}{i_b} = \beta\left(\frac{r_o}{r_o + r_L}\right) \tag{7–36}$$

The input resistance in this case is clearly

$$r_i = \beta r_e' \tag{7–37}$$

Invoking the general form for the voltage gain in terms of A_i, namely $A_v = -A_i(R_L/r_i)$, we can write

$$A_v = -\beta\left(\frac{r_o}{r_o + r_L}\right)\frac{r_L}{\beta r_e'} \quad (r_L = R_C \| R_L \text{ is the total ac load resistance})$$

which can further be reduced to

$$A_v = \frac{-(r_L \| r_o)}{r_e'} \tag{7–38}$$

Note that if $r_o \gg r_L$, the current gain becomes $A_i \approx \beta$, and A_v reduces to

$$A_v \approx \frac{-r_L}{r_e'} \qquad (r_o \gg r_L) \tag{7–39}$$

The overall gain, $A_{vs} = v_L/v_S$, can be written as

$$A_{vs} = A_v\left(\frac{r_{in}}{r_s + r_{in}}\right) \tag{7–40}$$

where $r_{in} = R_B \| \beta r_e'$ is the stage input resistance.

The overall current gain i_L/i_S can be written in terms of A_i and the current-division terms as follows:

$$\frac{i_L}{i_S} = A_i\left(\frac{R_C}{R_C + R_L}\right)\left(\frac{R_B}{r_{in} + R_B}\right) \tag{7–41}$$

CE Amplifier with Partly Bypassed Emitter Resistance

Let us now do the ac analysis of the more general CE amplifier with a partly bypassed emitter resistor. When the emitter resistor is fully bypassed, maximum gain is obtained but at the cost of high distortion, unless the input signal level is no more than about 20 mV. There are also instances where there is no need for maximum gain, but rather for having a specific gain. This can be accomplished easily by bypassing a portion of R_E, as was shown in Figure 7–8(b). Refer to the equivalent circuit of Figure 7–27 where R_f is the portion of R_E not bypassed by the emitter capacitor. Notice that the transistor ac model is shown in a slightly different way with both input and output circuits drawn horizontally.

Because the controlled current source and r_o are no longer grounded, the current gain cannot be obtained directly as before; however, it can be shown that the current gain can be closely approximated by the same expression:

$$A_i \approx \beta\left(\frac{r_o}{r_o + r_L}\right) \tag{7–42}$$

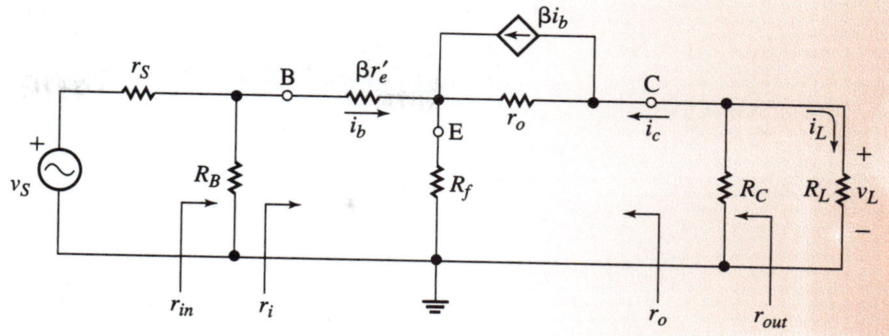

FIGURE 7–27 Equivalent ac circuit of the CE amplifier with emitter resistance R_f

To obtain the input resistance looking into the base, we first write the equation of the input voltage at the base terminal as follows:

$$v_b = i_b\beta r_e' + (i_b + A_i i_b)R_f = i_b[\beta r_e' + (1 + A_i)R_f]$$

then solve for v_b/i_b, which is the input resistance r_i:

$$r_i = \beta r_e' + (1 + A_i)R_f \tag{7–43}$$

The voltage gain can now be expressed by means of the general form $-A_i(R_L/r_i)$ but using the total ac resistance r_L:

$$A_v = -A_i\frac{r_L}{r_i} \tag{7–44}$$

where A_i and r_i are given by equations 7–42 and 7–43. Notice, however, that the expression for r_i contains also A_i, which in turn contains the term $r_o/(r_o + r_L)$. So even though these terms do not cancel algebraically, their effect on the expression closely cancel due to the fact that the $\beta r_e'$ term would be the smaller term in r_i. Therefore, with $A_i \gg 1$, it can easily be shown that the voltage gain, current gain, and input resistance can be approximated by the following:

$$A_v \approx \frac{-r_L}{r_e' + R_f} \tag{7–45}$$

$$A_i \approx \beta \tag{7–46}$$

$$r_i \approx \beta\,(r_e' + R_f) \tag{7–47}$$

The overall gain $A_{vs} = v_L/v_S$ is obtained by multiplying A_v by the voltage-division fraction of the input circuit:

$$A_{vs} = \frac{-r_L}{r_e' + R_f}\left(\frac{r_{in}}{r_{in} + r_s}\right) \tag{7–48}$$

where $r_{in} = R_B \| \beta(r_e' + R_f)$ = stage input resistance. Notice that all these equations reduce to those derived for the grounded-emitter case by simply making $R_f = 0$. This is why this configuration can be considered a general case.

The output resistance (r_o' in this case) can be shown to be several times larger than that for grounded emitter and can be ignored when compared to $R_C \| R_L$, further justifying the preceding approximations. An expression for r_o' can be obtained by the methods described in Section 6–2, yielding

$$r_o' \approx r_o\left(1 + \frac{\beta R_f}{R_f + \beta r_e' + r_s \| R_B}\right) \tag{7–49}$$

EXAMPLE 7–5

Figure 7–28 shows a CE amplifier in which $R_1 \| R_2 = 8\ \text{k}\Omega$. If the quiescent collector current is 2 mA, the Early voltage is 72 V, and $\beta = 150$, find

1. A_i, r_i, r_{in}, A_v, and A_{vs} without approximations.
2. A_i, r_i, r_{in}, A_v, and A_{vs} using approximate expressions. Determine percent errors.

FIGURE 7–28 (Example 7–5)

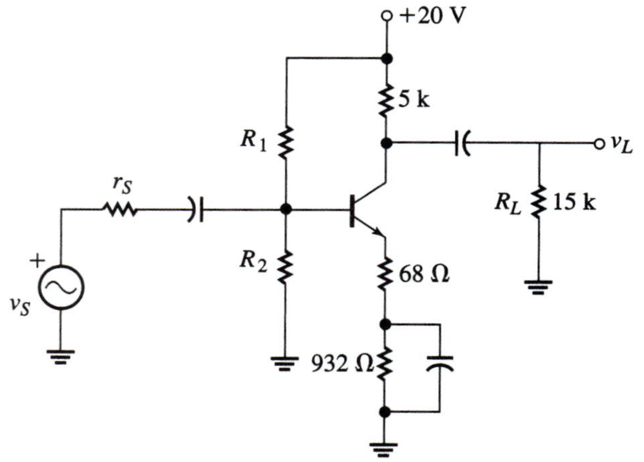

Solution

Use equations 7–42 through 7–49.

1. The ac emitter resistance is $r_e' = 26\ \text{mV}/2\ \text{mA} = 13\ \Omega$.
   The quiescent C-E voltage is $V_{CEQ} = 20 - (2\ \text{mA})(5\ \text{k} + 1\ \text{k}) = 8\ \text{V}$ (KVL).
   The ac output resistance is $r_o = (72 + 8)/2\ \text{mA} = 40\ \text{k}\Omega$ (equation 7–35).
   The total ac load resistance is $r_L = 5\ \text{k} \| 15\ \text{k} = 3.75\ \text{k}\Omega$.
   Current gain is $A_i = 150(40\ \text{k}/43.75\ \text{k}) = 137.1$.
   Input resistance is $r_i = 150(13) + 138.1(68) = 11.34\ \text{k}\Omega$.
   Stage input resistance is $r_{in} = 11.34\ \text{k} \| 8\ \text{k} = 4.69\ \text{k}$.
   Voltage gain is $A_v = -137.1(3.75\ \text{k}/11.34\ \text{k}) = -45.34$.
   Overall voltage gain is $A_{vs} = -45.34(4.69\ \text{k}/5.69\ \text{k}) = -37.4$.
2. $A_i \approx \beta = 150$ (error = 9.4%)
   $r_i = 150(13 + 68) = 12.2\ \text{k}\Omega$ (error = 7.1%)
   $r_{in} = 8\ \text{k} \| 12.2\ \text{k} = 4.83\ \text{k}$ (error = 3%)
   $A_v = -(3.75\ \text{k})/(13 + 68) = -46.3$ (error = 2.1%)
   $A_{vs} = -46.3(4.83\ \text{k}/5.83\ \text{k}) = -38.4$ (error = 2.7%)

Notice that the most significant errors were those for A_i and r_i because they are moderately influenced by the output resistance r_o. The other errors are not significant from a practical standpoint.

In order to obtain a particular overall gain from a CE amplifier, we must determine a value for R_f that will satisfy the requirement. Note that equation 7–48 includes r_{in} in addition to R_f. But r_{in} depends also on R_f. So in order to solve for R_f relatively easily, we must derive an expression for the overall gain A_{vs} in a more closed form. If we expand $r_{in} = R_B \| \beta(r_e' + R_f)$ as "product over sum" and substitute it into the term $r_{in}/(r_s + r_{in})$, we can obtain

$$\frac{r_{in}}{r_s + r_{in}} = \frac{r_e' + R_f}{\dfrac{r_s}{\beta} + (r_e' + R_f)\left(1 + \dfrac{r_s}{R_B}\right)}$$

which after substituted into (7–48) yields

$$A_{vs} = \frac{-r_L}{\dfrac{r_s}{\beta} + (r_e' + R_f)\left(1 + \dfrac{r_s}{R_B}\right)} \quad (7\text{–}50)$$

As a verification, observe that the preceding expression reduces to A_v if we make $r_s = 0$. Also notice that while not totally independent of β, as is the case for A_v, the overall gain, A_{vs}, is slightly influenced by β.

EXAMPLE 7–6

Change both R_f and the bypassed emitter resistance in the previous example to establish an overall gain of -50 while maintaining the same dc bias conditions.

Solution

Solve for R_f from equation 7–50:

$$R_f = \left(\frac{r_L}{|A_{vs}|} - \frac{r_s}{\beta}\right)\left(\frac{R_B}{r_s + R_B}\right) - r_e'$$

Substitute values to obtain $R_f = 47.7\ \Omega$ (47 Ω is a commercial value). In order to maintain the original R_E, the bypassed emitter resistance needs to be changed to 953 Ω.

Helpful hint: Try solving for R_f by substituting the values directly into equation 7–50 and then using the calculator to extract the value of R_f by "undoing" the equation's arithmetic. Do this without taking down any partial results. You will notice that with practice and confidence the answer can be obtained very quickly, with fewer chances of making "Math 101" errors. This is not to discourage you from practicing good math (a must in engineering), but just to illustrate another powerful use of your calculator.

We now turn again to the issue of obtaining maximum unclipped output voltage through proper location of the Q-point on the load line. Equation 7–18 is used for determining the optimum value of the collector current when the emitter is connected directly to ground. This equation can be modified for the general CE amplifier with fully or partly bypassed emitter resistance by an emitter capacitor. Basically, there are two resistive components in equation 7–18: The dc component, R_C, and the ac component, r_L. In the general CE amplifier circuit, however, there are two additional components: The emitter resistance, R_E, and the unbypassed ac resistance, R_f. A general expression for I_{CQ} (optimum) can then be written as

$$I_{CQ}(\text{optimum}) = \frac{V_{CC}}{R_C + R_E + R_f + r_L} \quad (7\text{–}51)$$

Again, we say that this is a general expression because it can be used in all other instances: Emitter directly grounded ($R_E = 0, R_f = 0$), emitter resistor totally bypassed ($R_f = 0$), and emitter resistor without bypass capacitor ($R_f = R_E$). Note that the dc emitter resistor R_E is the sum of R_f and the bypassed resistor. An example illustrating these issues follows.

EXAMPLE 7–7

Design a CE amplifier for a load resistance of 10 kΩ using V_{CC} = 20 V and R_C = 3.3 kΩ. Determine the optimum collector current for obtaining maximum output swing. The input source resistance is 600 ohms and the required overall gain is $A_{vs} = -45$. Specify R_E, R_1, and R_2 so that the Q-point collector current will not change more than ±4% over a β range of 70 to 180.

Solution

In order to determine the optimum collector current, we need to know R_E and R_f. R_E is normally determined based on a voltage drop of about 10% of V_{CC}. But since I_C is not known yet, we can assume that the drop across R_C will be about 50% of V_{CC}. This would leave 40% of V_{CC} for V_{CE}, which is consistent with the fact that the Q-point should be above the middle point of the dc load line (higher I_C, lower V_{CE}) for maximum output voltage swing. Therefore, R_E needs to be about one-fifth of R_C, or 660 ohms. Note that with 10 volts across R_C, the collector current will be about 3 mA, according to Ohm's law.

We also need to estimate the value of R_f based on the gain of −45. Here again we are faced with another unknown: $R_B = R_1 \| R_2$. However, since the magnitude of the gain from base to collector is approximately $r_L/(r_e' + R_f)$, we can calculate R_f based on a magnitude of about 50, since the specified overall gain is −45. With $r_L = 3.3\text{ k} \| 10\text{ k} = 2.5\text{ k}\,\Omega$ and r_e' about 8 ohms, we have $R_f = 2500/50 - 8 = 42$ Ω. The optimum collector current according to equation 7–51 is then

$$I_{CQ}(\text{optimum}) = 20/(3.3\text{ k} + 660 + 2.5\text{ k} + 42) \approx 3.1\text{ mA}$$

With the ±4% allowed variation of collector current we establish I_{Cmax} = 3.22 mA and I_{Cmin} = 2.98 mA. Using equation 7–10, we determine

$$R_B/R_E = (3.22 - 2.98)/(2.98/70 - 3.22/180) = 9.7$$

and $R_B = 9.7\,R_E = 6.4$ kΩ

From equation 7–9, $V_{Th} = I_C(R_E + R_B/\beta) + V_{BE} = 3.22(660 + 6.4\text{ k}/180) + 0.7 =$ 2.94 V.

From equation 7–12, $R_1 = (20/2.94)6.4\text{ k} = 43.5$ kΩ, from which $R_2 = 7.5$ kΩ follows.

Commercial values for R_1 and R_2 are 43 kΩ and 7.5 kΩ, respectively.

A more exact value for R_f can now be calculated with equation 7–50 using a mid β-value of 120.

$$45 = \frac{2500}{\dfrac{600}{120} + (8 + R_f)\left(1 + \dfrac{600}{6400}\right)} \longrightarrow R_f = 38.2\ \Omega\ \ (39\ \Omega\text{ commercial}).$$

You can verify that by using the extreme values of β, the resulting value for R_f does not change much. Now, because the sum of R_f and the bypassed emitter resistance equals R_E, or 660 Ω, then the bypassed emitter resistance must be 621 Ω, or 620 Ω commercial. Exercise 7–15 at the end of the chapter verifies this design thoroughly.

Small-Signal CC Amplifier Model

Figure 7–29(a) shows a common-collector amplifier and Figure 7–29(b) shows the corresponding small-signal equivalent circuit. Note that the collector in Figure 7–29(b) is shown grounded, since V_{CC} in Figure 7–29(a) is connected directly to the collector and dc sources are once again treated as ac short circuits. Notice again that the transistor model was drawn slightly differently, with the three terminals clearly identified. Some authors refer to the CC transistor as a "grounded collector" configuration, which is correct in the ac sense.

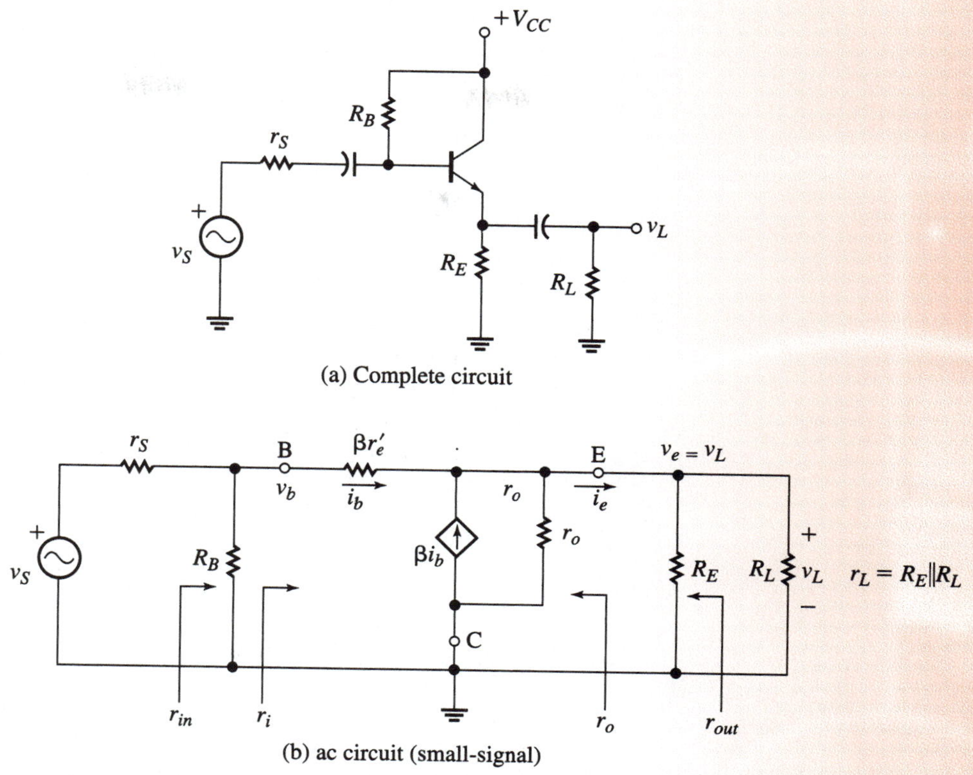

FIGURE 7–29 A common-collector amplifier and its small-signal equivalent circuit

As in the CE model, the resistance between base and emitter is seen to be $\beta r'_e$, which is again an acceptable approximation for $r_\pi = (\beta + 1)r'_e$. The total input resistance at the base of the transistor (between base and ground) is derived as follows:

$$r_i = \frac{v_i}{i_i} = \frac{v_i}{i_b}$$

Remember that in a CC circuit v_i *is the ac base-to-collector voltage, which the figure clearly shows is the same as the ac base-to-ground voltage.* The input current i_i is the same as the base current i_b.

For purposes of this derivation, let us temporarily replace $\beta r'_e$ in Figure 7–29(b) by the more accurate value $(\beta + 1)r'_e$. Then, writing Kirchhoff's voltage law from B to ground in Figure 7–29(b), we obtain

$$\begin{aligned} v_i = v_b &= i_b(\beta + 1)r'_e + i_e r_L \\ &= i_b(\beta + 1)r'_e + i_b(\beta + 1)r_L \\ &= i_b(\beta + 1)(r'_e + r_L) \end{aligned} \tag{7–52}$$

Solving for v_b/i_b from (7–52), we obtain

$$\frac{v_b}{i_b} = r_i = (\beta + 1)(r'_e + r_L) \tag{7–53}$$

Again invoking the approximation $\beta + 1 \approx \beta$, we have

$$r_i \approx \beta(r'_e + r_L) \tag{7–54}$$

It is apparent from Figure 7–29(b) that

$$r_{in} = R_B \| r_i \tag{7–55}$$

Equations 7–54 and 7–55 reveal the single most important feature of the CC amplifier in practical applications: Its input resistance can be made very large in comparison to the other configurations. For example, using the typical values $r_L = 500\ \Omega$ and $\beta = 100$, we have $r_i \approx 50\ \text{k}\Omega$. Later in the chapter we will discuss an application in which this feature is very valuable.

Remember from Chapter 4 that the output voltage in the CC configuration is the collector–emitter voltage. Because the collector is grounded, the ac output voltage is the same as the ac emitter-to-ground voltage (see Figure 7–29(b)). We can therefore derive the voltage gain of the CC transistor as follows:

$$A_v = \frac{v_o}{v_i} = \frac{i_e r_L}{v_i}$$

Substituting from equation 7–52 for v_i, we obtain

$$A_v = \frac{i_e r_L}{i_b(\beta + 1)(r_e' + r_L)}$$

$$= \frac{i_b(\beta + 1)r_L}{i_b(\beta + 1)(r_e' + r_L)} = \frac{r_L}{r_e' + r_L} \qquad \textbf{(7–56)}$$

Equation 7–56 shows that *the CC transistor always has a voltage gain less than 1.* As mentioned before, it is usually the case that $r_L \gg r_e'$, so the gain is very close to unity, from which it follows that $v_o \approx v_i$. It is not difficult to visualize why the ac output voltage is approximately the same as the ac input voltage in a CC configuration: The input and output are "separated" only by the small ac resistance of the forward-biased base–emitter junction. Note that there is no phase inversion between input and output. As the base-to-ground voltage increases, so does the emitter-to-ground voltage. Because the output voltage is essentially the same as the input voltage, in magnitude and phase, the emitter is said to *follow* the base. In that context, a transistor in the CC configuration is often called an *emitter follower* (more often, in fact, than it is called a *common-collector* circuit).

EXAMPLE 7–8

The common-collector amplifier in Figure 7–29(a) has $V_{CC} = 15\ \text{V}$, $R_B = 75\ \text{k}\Omega$, and $R_E = 910\ \Omega$. The β of the silicon transistor is 100 and the load resistor is 600 Ω. Find

1. r_{in},
2. A_v.

Solution

1. From equations 4–28,

$$I_B = \frac{V_{CC} - V_{BE}}{R_B + (\beta + 1)R_E} = \frac{15\ \text{V} - 0.7\ \text{V}}{75\ \text{k}\Omega + 101(910\ \Omega)} = 85.7\ \mu\text{A}$$

$$I_E = (\beta + 1)I_B = (101)(85.7\ \mu\text{A}) = 8.57\ \text{mA}$$

Then,

$$r_e' = \frac{0.026\ \text{V}}{I_E} = \frac{0.026\ \text{V}}{8.57\ \text{mA}} = 3.03\ \Omega$$

$$r_L = 910 \| 600 = 362\ \Omega$$

From equation 7–56,

$$r_i = (\beta + 1)(r_e + r_L) = 101(3.03\ \Omega + 362\ \Omega) = 36.9\ \text{k}\Omega$$

$$r_{in} = R_B \| r_i = 36.9\ \text{k} \| 75\ \text{k} = 24.7\ \text{k}\Omega$$

2. From equation 7–56,

$$A_v = \frac{r_L}{r'_e + r_L} = \frac{362\ \Omega}{3.03\ \Omega + 362\ \Omega} = 0.992$$

Note that $A_v \approx 1$.

A note regarding the output swing: In this example, the maximum output swing is $2I_{EQ}r_L$ or about 6.2 V(p-p). The optimum I_{EQ} would be $V_{CC}/(R_E + r_L) = 11.8$ mA, which would yield about 8.5 V(p-p) of available maximum output swing.

Since the common-collector input and output currents are i_b and i_e, respectively, we find the current gain for the transistor in this configuration to be

$$A_i = \frac{i_e}{i_b} = \frac{(\beta + 1)i_b}{i_b} = \beta + 1 \approx \beta \qquad (7\text{–}57)$$

Thus, while the CC voltage gain is somewhat less than 1, the current gain is substantially greater than 1, and so, therefore, is the power gain:

$$A_p \approx A_v A_i \approx A_i \qquad (7\text{–}58)$$

Rather than present a lengthy derivation for the output resistance of the CC transistor, let us make an intuitive observation, from which we can formulate a general rule for determining output resistance of CC amplifiers. We have already seen how the relationship $i_e = (\beta + 1)i_b$ leads to the result that the resistance looking into the base is $(\beta + 1)$ times greater than the total resistance from emitter to ground (equation 7–54). Conversely, that same relationship implies that resistance looking into the emitter is $(\beta + 1)$ times *smaller* than the resistance from the base back to the signal source. Now, the output resistance of the CC stage in Figure 7–29, r_{out}, is the resistance looking into the emitter in parallel with R_E. Therefore, applying the foregoing rule, we find

$$r_{out} = R_E \left\| \left(r'_e + \frac{R'_B}{\beta + 1} \right) \right. \qquad (7\text{–}59)$$

where R'_B is the resistance looking from the base toward the signal source. Figure 7–30 shows how R'_B is defined in the ac equivalent of the base circuit. R'_B is computed by shorting the signal source to ground, so

$$R'_B = R_B \| r_S \qquad (7\text{–}60)$$

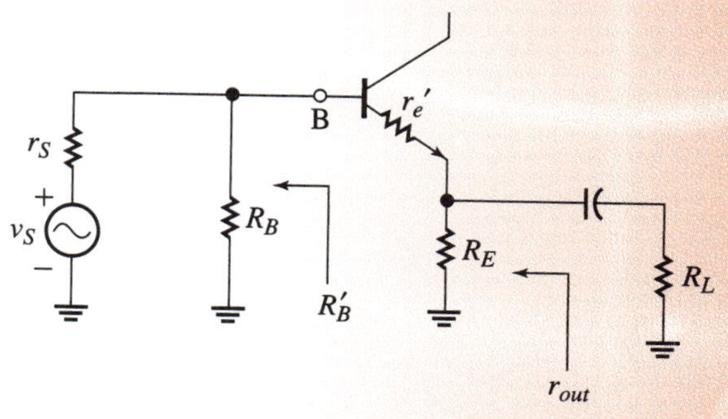

FIGURE 7–30 R'_B is the (Thévenin) equivalent resistance looking from the base toward the source. $R'_B = R_B \| r_s$

and with $\beta + 1 \approx \beta$,

$$r_{out} \approx R_E \left\| \left(r'_e + \frac{R_B \| r_S}{\beta} \right) \right. \tag{7–61}$$

Equation 7–61 shows that the output resistance of an emitter follower can be quite small. For example, for the typical values $R_E = 1\ \text{k}\Omega$, $r'_e = 5\ \Omega$, $R_B = 100\ \text{k}\Omega$, $r_S = 50\ \Omega$, and $\beta = 100$, we find

$$r_{out} \approx 1\ \text{k} \left\| \left(5 + \frac{100\ \text{k} \| 50}{100} \right) \right. \approx 5.5\ \Omega$$

When $r_S = 0$, notice that $R_B \| r_S = 0$ and the equation becomes

$$r_{out} \approx R_E \| r'_e \approx r'_e \quad (r_S = 0) \tag{7–62}$$

The overall voltage gain of the emitter-follower stage, taking source resistance into account, can be determined as follows:

$$A_{vs} = \frac{v_L}{v_S} = \frac{r_L}{r'_e + r_L}\left(\frac{r_{in}}{r_S + r_{in}}\right) \tag{7–63}$$

where $r_L = R_E \| R_L$. For small r_S, equation 7–63 is approximately equivalent to equation 7–56.

Although the emitter follower by itself has a voltage gain less than 1, it can be used to improve the voltage gain of a larger amplifier system. Because of its large input resistance, it does not "load" the output of another amplifier. In other words, the load presented by an emitter follower to another amplifier does not appreciably reduce the voltage gain of that amplifier. Also, because it has a small output resistance, the emitter follower can drive a "heavy" load (small resistance) whose presence would otherwise reduce voltage gain. For these reasons, an emitter follower is valuable as an intermediate stage between an amplifier and a load. When an emitter follower is used this way, it is called a *buffer* amplifier, or an *isolation* amplifier, because it effectively isolates another amplifier from the loading effect of R_L. This use is illustrated in the next example.

EXAMPLE 7–9

An amplifier having an output resistance of 1 kΩ is to drive a 100-Ω load, as shown in Figure 7–31(a). Assuming that the amplifier has a voltage gain of $A_v = 140$ (with no load connected), find

1. the voltage gain with the load connected, and
2. the voltage gain when an emitter follower is inserted between the amplifier and the load, as shown in Figure 7–31(b).

Solution

1. The voltage gain with the 100-Ω load connected is

$$\frac{v_L}{v_S} = A_v\left(\frac{R_L}{r_o + R_L}\right) = 140\left(\frac{100}{1000 + 100}\right) = 12.7$$

The 100-Ω load severely reduces the voltage gain.

2. To find the input resistance of the emitter follower, we must find r'_e, which means that we must first find the dc bias current I_E:

$$I_E \approx \frac{V_{CC} - 0.7}{R_E + R_B/\beta} = 92\ \text{mA}$$

Therefore,

$$r'_e = \frac{26\ \text{mV}}{92\ \text{mA}} = 0.28\ \Omega$$

FIGURE 7–31 (Example 7–9)

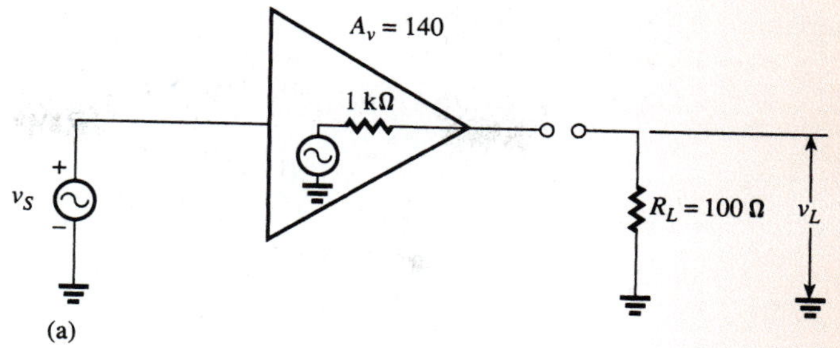

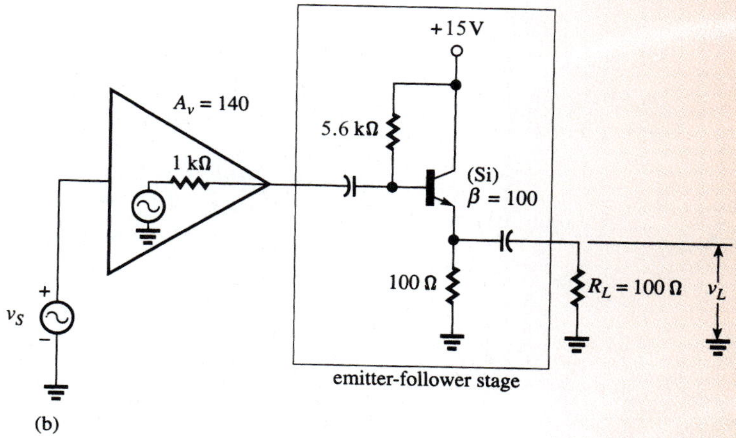

The ac load resistance for the emitter follower is

$$r_L = R_E \| R_L = (100\ \Omega)\|(100\ \Omega) = 50\ \Omega$$

From equation 7–56,

$$r_{in} = 5.6\ \text{k}\|100(0.28 + 50) = 2.65\ \text{k}\Omega$$

As far as the emitter follower is concerned, the source resistance is the output resistance (1 kΩ) of the amplifier driving it. Therefore, from equation 7–63, the overall gain of the emitter follower is

$$A_{vs} = \frac{50}{0.28 + 50}\left(\frac{2.65\ \text{k}}{1\ \text{k} + 2.65\ \text{k}}\right) = 0.726$$

The system gain from amplifier to load is then $(0.726)(140) = 102$. We see that insertion of the emitter follower improved the voltage gain from 12.7 to 102, an increase of 700%.

Table 7–1 is a summary of all important formulas for BJT amplifiers in the three configurations, including approximated expressions. Note that the common-emitter configuration is given as the general case from which all particular cases can be obtained. Table 7–2 compares, in a very general way, the important small-signal characteristics of the three configurations.

Capacitively Coupled BJT Amplifiers

Recall that the primary reason for employing capacitor coupling is to block the flow of dc current. We have observed that it is often necessary to prevent the flow of dc current between the input of an amplifier and its signal source,

TABLE 7–1 Summary of BJT amplifier equations

	CB	CE (general)	CC
Current Gain A_i	α	$\beta\left(\frac{r_o}{r_o + r_L}\right) \approx \beta$	$(\beta + 1)\frac{r_o}{r_o + r_L} \approx \beta$
Input Resistance r_i	r'_e	$\beta r'_e + (1 + A_i)R_f \approx \beta(r'_e + R_f)$	$\beta r'_e + (1 + A_i)r_L \approx \beta(r'_e + r_L)$
Voltage Gain A_v	$\frac{r_L}{r'_e}$	$-A_i\frac{r_L}{r_i} \approx \frac{-r_L}{r'_e + R_f}$	$\frac{r_L}{r'_e + r_L} \approx 1$
Stage Input Resistance r_{in}	$R_E \Vert r_i$	$R_B \Vert r_i$	$R_B \Vert r_i$
Overall Voltage Gain A_{vs}	$\frac{r_L}{r'_e}\left(\frac{r_{in}}{r_S + r_{in}}\right) \approx \left(\frac{r'_L}{r_S + r'_e}\right)$	$A_v\left(\frac{r_{in}}{r_S + r_{in}}\right) \approx \frac{-r_L}{\frac{r_S}{\beta} + (r'_e + R_f)\left(1 + \frac{r_s}{R_B}\right)}$	$A_v\left(\frac{r_{in}}{r_S + r_{in}}\right) \approx \frac{r_{in}}{r_S + r_{in}}$
Output Resistance r_o	$r_c \approx$ open circuit	$\approx r_o\left(1 + \frac{\beta R_f}{\beta r'_e + R_f + r_S \Vert R_B}\right)$	$r'_e + \frac{r_S \Vert R_B}{\beta + 1} \approx r'_e + \frac{r_S \Vert R_B}{\beta}$

TABLE 7–2

	CB	CE	CC
Voltage Gain	Large (noninverting)	Large (inverting)	≈ 1 (noninverting)
Current Gain	≈ 1	Large	Large
Power Gain	Moderate	Large	Small
Input Resistance	Low	Moderate	High
Output Resistance	High	Moderate	Low

as well as between the amplifier's output and its load. Similarly, capacitor coupling is used to prevent dc current from flowing between the output of one amplifier stage and the input of the next stage. The capacitor connected in the path between amplifier stages makes it possible to have a dc bias voltage at the output of one stage that is different from the dc bias voltage at the input to the next stage. This idea is illustrated in Figure 7–32, which shows the output of a BJT amplifier stage connected through a coupling capacitor

FIGURE 7–32 The capacitor used in the capacitor coupling method makes it possible to have different bias voltages on the amplifier stages. Note that the (electrolytic) capacitor has 6 V dc across it and has its positive lead connected to the more positive bias (9 V).

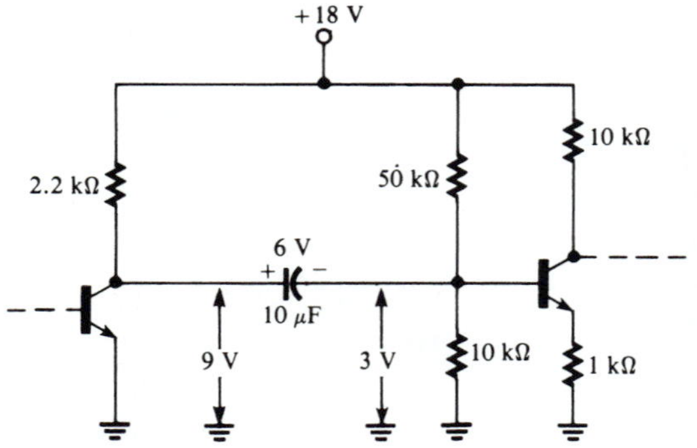

to the input of another BJT amplifier stage. Notice that the collector of the first stage is at +9 V and that the base of the second stage is at +3 V. The dc voltage across the capacitor is therefore 9 − 3 = 6 V, so the capacitor should have a dc-working-voltage (DCWV) rating somewhat greater than 6 V. If the 10-μF coupling capacitor is of the electrolytic type, it *must* be connected with its positive terminal to the more positive bias voltage: the 9-V collector voltage in this example.

Of course, a coupling capacitor permits the flow of *ac* signal current between stages, provided the frequency is high enough to keep the capacitive reactance small. The disadvantage of the RC coupling method is that it affects the low-frequency response of the amplifier; we must sometimes choose between an impractically large capacitor value and an unreasonably large lower cutoff frequency. RC coupling is not used in integrated circuits because it is difficult and uneconomical to fabricate capacitors on a chip.

EXAMPLE 7–10

Figure 7–33 shows two capacitor-coupled, common-emitter amplifier stages. Notice that the ac signal developed at the output of the first stage (the collector of Q_1) is coupled through the 0.85-μF capacitor to the input of the second stage (the base of Q_2). Assuming that the transistors are identical and have $\beta = 100$ and $r_c = 1\ \text{M}\Omega$, find the small signal, midband (1) voltage gain v_L/v_S and (2) current gain i_L/i_S.

Solution

We will use the common-emitter, small-signal relations that are summarized in Table 7–1.

1. The collector current for Q_1 is $I_{C1} = \beta\left(\dfrac{V_{CC} - V_{BE}}{R_B}\right) = 1.82\ \text{mA}$ and $r'_{e1} = \dfrac{26\ \text{mV}}{1.82\ \text{mA}} = 14.3\ \Omega$. The collector current for Q_2 is found using Thévenin's theorem: $V_{Th} = 12(10/78) = 1.54\ \text{V}$, $R_B = 10 \| 68 = 8.72\ \text{k}\Omega$, and $I_C = (1.54 - 0.7)/(220 + 8.72\ \text{k}/100) = 2.73\ \text{mA}$, yielding $r'_{e2} = 9.5\ \Omega$.

   The input resistance to the first stage is

$$r_{in1} = R_{B1} \| \beta r'_e = (620\ \text{k}\Omega) \| (1.43\ \text{k}\Omega) \approx 1.43\ \text{k}\Omega$$

   The output resistance of stage 1 (at the collector of Q_1) is

$$r_{out1} = R_{C1} \| (r_c/\beta) = (3.3\ \text{k}\Omega) \| [(1\ \text{M}\Omega)/100] = 2.48\ \text{k}\Omega$$

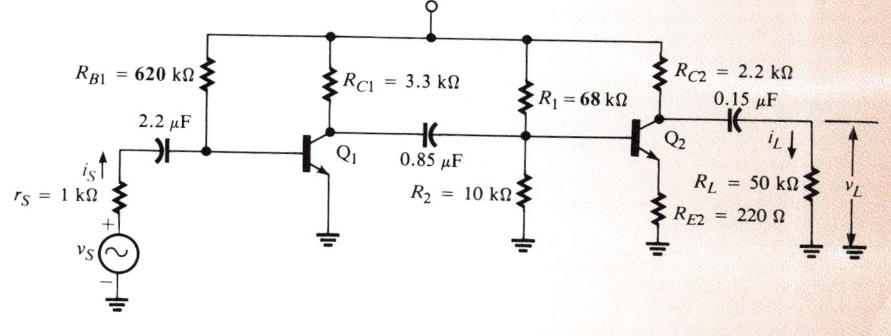

FIGURE 7–33 (Example 7–10)

The *unloaded* voltage gain of stage 1 is, therefore,

$$A_{v1} = \frac{-r_{out1}}{r'_e} = \frac{-2.48\ \text{k}\Omega}{14.3\ \Omega} = -173.4$$

The input resistance to the second stage is

$$\begin{aligned} r_{in2} &= R_1 \| R_2 \| \beta(r'_e + R_{E2}) \\ &= (68\ \text{k}\Omega)\|(10\ \text{k}\Omega)\|100[(9.5\ \Omega) + (220\ \Omega)] \\ &= (8.72\ \text{k}\Omega)\|(23\ \text{k}\Omega) = 6.32\ \text{k}\Omega \end{aligned}$$

The output resistance of stage 2 is

$$r_{out2} = R_{C2}\|(r_c/\beta) = (2.2\ \text{k}\Omega)\|(1\ \text{M}\Omega/100) = 1.8\ \text{k}\Omega$$

The *unloaded* voltage gain of stage 2 is, therefore,

$$A_{v2} = \frac{-r_{out2}}{R_{E2} + r'_e} = \frac{-1.8\ \text{k}\Omega}{(220\ \Omega) + (9.5\ \Omega)} = -7.84$$

The two-stage amplifier can now be represented as shown in Figure 7–34. No capacitors are shown in this figure because we are assuming operation in the midband frequency range. Using (the two-stage form of) equation 7–61, we find

$$\frac{v_L}{v_S} = \left[\frac{1.43\ \text{k}\Omega}{(1\ \text{k}\Omega) + (1.43\ \text{k}\Omega)}\right](-173.4)\left[\frac{6.32\ \text{k}\Omega}{(2.48\ \text{k}\Omega) + (6.32\ \text{k}\Omega)}\right](-7.84) \times \left[\frac{50\ \text{k}\Omega}{(1.8\ \text{k}\Omega) + (50\ \text{k}\Omega)}\right] = 554$$

The positive result shows that v_L is in phase with v_S.

Notice that an alternative approach to finding the overall voltage gain is to find the voltage gains A_1 and A_2 *with loads connected.* Taking this approach, we compute the ac load resistance r_L of each stage:

$$r_{L1} = r_{out1}\|r_{in2} = (2.48\ \text{k}\Omega)\|(6.32\ \text{k}\Omega) = 1.78\ \text{k}\Omega$$

$$r_{L2} = r_{out2}\|R_L = (1.8\ \text{k}\Omega)\|(50\ \text{k}\Omega) = 1.74\ \text{k}\Omega$$

The voltage gains with these loads connected are then

$$A_1 \approx \frac{-r_{L1}}{r'_{e1}} = \frac{-1.78\ \text{k}\Omega}{14.3\ \Omega} = -124.5$$

$$A_2 \approx \frac{-r_{L2}}{r'_{e2} + R_{E2}} = \frac{-1.74\ \text{k}\Omega}{(9.5\ \Omega) + (220\ \Omega)} = -7.58$$

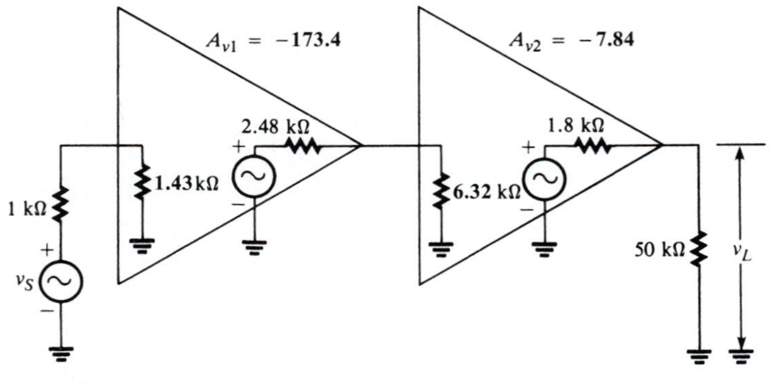

FIGURE 7–34 (Example 7–10) The two-stage amplifier of Figure 7–33

The overall voltage gain is then

$$\frac{v_L}{v_S} = \left[\frac{r_{in1}}{r_{in1} + r_S}\right] A_1 A_2 = \left[\frac{1.43\text{ k}\Omega}{(1.43\text{ k}\Omega) + (1\text{ k}\Omega)}\right](-124.5)(-7.58) = 555$$

Except for a small rounding error, this approach produces the same result as before.

2. To determine the overall current gain, it is helpful to draw the small-signal equivalent circuit of each stage and then trace the flow of ac current through the amplifier. We will use the current-divider rule at each node to determine the portion of signal current that continues to flow toward the load. Figure 7–35 shows the small-signal equivalent of the first stage. Applying the current-divider rule to the input side of stage 1 in Figure 7–35, we obtain

$$i_{b1} = \left[\frac{620\text{ k}\Omega}{1.43\text{ k}\Omega + 620\text{ k}\Omega}\right] i_S = 0.9977 i_S$$

This result shows that essentially all of the source current enters the base of Q_1 ($0.0023\ i_S$ is shunted to ground through R_B). At the output side of stage 1, we have

$$i_{c1} = \beta i_{b1} = 100 i_{b1} = 100(0.9977 i_S) = 99.77 i_S$$

To find the portion of βi_{b1} that flows into the base of Q_2, we must consider all the parallel paths to ground in the interstage circuitry between Q_1 and Q_2. Figure 7–36 shows the equivalent circuit of the second stage and the output side of the first stage. Note that r_{out1} is the parallel combination of r_c/β and R_{C1}, and this combination is in parallel with the bias resistors R_1 and R_2 at the input to the second stage. The total equivalent resistance r_{SH} shunting the input (base) of Q_2 is, therefore,

$$r_{SH} = r_{out1} \| R_1 \| R_2 = (2.48\text{ k}\Omega) \| (68\text{ k}\Omega) \| (10\text{ k}\Omega) = 1.93\text{ k}\Omega$$

The current i_{b2} into the base of Q_2 is then found by the current-divider rule:

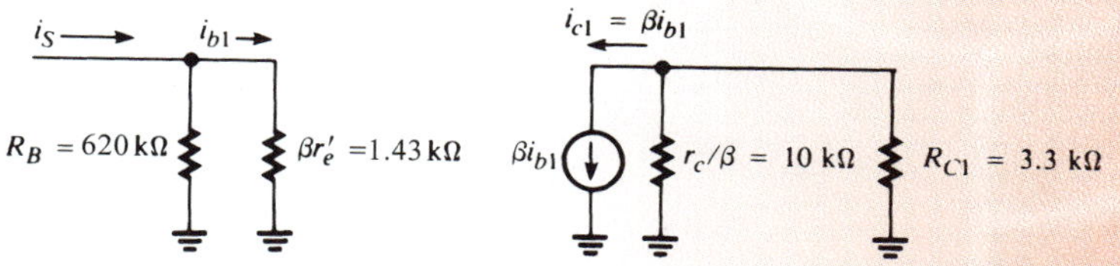

FIGURE 7–35 (Example 7–10) The small-signal equivalent circuit for the first stage of the amplifier in Figure 7–33. This circuit is used to compute the current i_{b1} in terms of i_S.

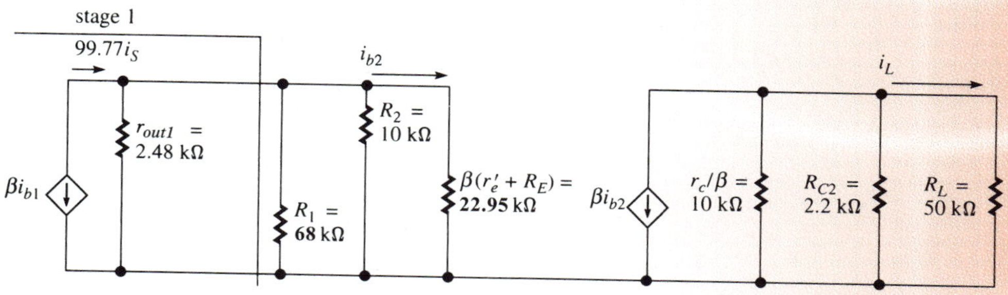

FIGURE 7–36 (Example 7–10) The small-signal equivalent circuit of the output side of stage 1 joined to amplifier stage 2. This circuit is used to find i_L in terms of i_S.

$$i_{b2} = -99.77 i_S \left[\frac{r_{SH}}{r_{SH} + \beta(r'_e + R_E)} \right]$$

$$= -99.77 i_S \left[\frac{1.93\ \text{k}\Omega}{(1.93\ \text{k}\Omega) + (22.95\ \text{k}\Omega)} \right] = -7.74 i_S$$

Notice that a significant portion of signal current is lost in the interstage circuitry due to the shunt paths to ground: Less than $\frac{1}{10}$ of i_{c1} reaches the base of Q_2. Now,

$$i_{c2} = \beta i_{b2} = 100(-7.74 i_S) = -774 i_S$$

The parallel combination of r_c/β and R_{C2} shunts R_L, so one more application of the current-divider rule gives

$$i_L = -\left[\frac{(r_c/\beta) \parallel R_{C2}}{R_L + (r_c/\beta) \parallel R_{C2}} \right] \beta i_{b2} = -\left[\frac{(10\ \text{k}\Omega) \parallel (2.2\ \text{k}\Omega)}{(50\ \text{k}\Omega) + (10\ \text{k}\Omega) \parallel (2.2\ \text{k}\Omega)} \right] (-774 i_S)$$

$$= 26.94 i_S$$

Therefore, the current gain from source to load is

$$\frac{i_L}{i_S} = 26.94$$

7–4 DIRECT COUPLING

We now examine *direct coupling,* the coupling method in which the output of one stage is electrically connected directly to the input of the next stage. In other words, both the dc and ac voltages at the output of one stage are identical to those at the input of the next stage. The method is often referred to as *dc,* which, in the context of signal coupling, means both direct coupling and direct current. Clearly, any change in the dc voltage at the output of one stage produces an identical change in dc voltage at the input to the next stage, so a direct-coupled amplifier behaves like a direct-current amplifier. Direct coupling is used in differential and operational amplifiers, which we will study extensively in Chapter 17, and in integrated circuits. Two practical configurations that employ direct coupling are introduced next.

The CE–CC Configuration

We already know that a CC stage is used for driving low-resistance loads. If the output of a CE stage is to be used with a relatively low load resistance, a CC should be used as a buffer. The CC stage can be directly coupled to the CE stage as shown in Figure 7–37. The collector resistance of Q_1 also serves for biasing the base of Q_2. Let us perform a dc analysis of the circuit.

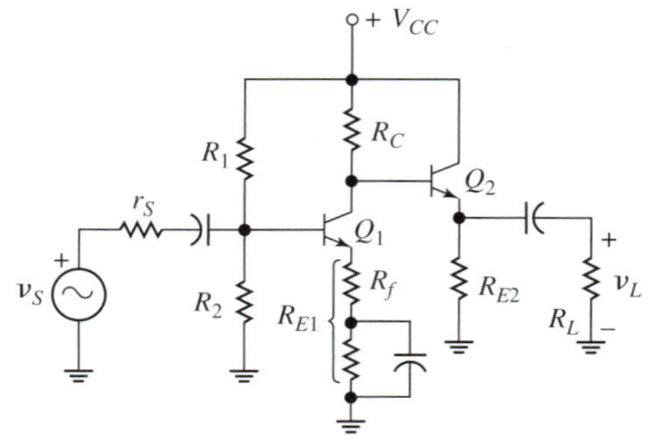

FIGURE 7–37 The CE–CC combination circuit

As we know, the collector current of Q_1 is

$$I_{C1} \approx \frac{V_{Th} - V_{BE}}{R_{E1} + R_B/\beta_1}$$

where $V_{Th} = V_{CC}\left(\dfrac{R_2}{R_1 + R_2}\right)$ and $R_B = R_1 \| R_2$

The emitter current of Q_2 is

$$I_{E2} = \frac{V_{B2} - V_{BE}}{R_{E2}} = \frac{V_{CC} - (I_{C1} + I_{B2})R_C - V_{BE}}{R_{E2}}$$

Substituting $I_{B2} = \dfrac{I_{E2}}{\beta_2 + 1}$ and reducing we obtain

$$I_{E2} \approx \frac{V_{CC} - V_{BE} - I_{C1}R_C}{R_{E2} + R_C/\beta_2} \tag{7–64}$$

To analyze the circuit in ac, we first draw the corresponding ac-equivalent circuit as shown in Figure 7–38. The overall gain v_L/v_S can be expressed as the product of the in-circuit gains of both stages, that is, considering the input resistance of the CC stage as part of the load resistance of the CE stage. Using the equations from Table 7–1 we can write

$$\frac{v_L}{v_S} = \frac{-r_{L1}}{\dfrac{r_S}{\beta_1} + (r'_{e1} + R_f)\left(1 + \dfrac{r_S}{R_B}\right)}\left(\frac{r_{L2}}{r'_{e2} + r_{L2}}\right) \tag{7–65}$$

where

$$r_{L1} = R_C \| [\beta_2(r'_{e2} + r_{L2})]$$

and

$$r_{L2} = R_{E2} \| R_L$$

The Darlington Pair

When the collectors of two BJTs are tied together and the emitter of one is direct-coupled to the base of the other, as shown in Figure 7–39, we obtain an important and highly useful configuration called a *Darlington pair.* The combination is used in amplifier circuits as if it were a single transistor having the base, collector, and emitter terminals labeled B, C, and E in the figure. We will analyze the Darlington pair to discover the effective beta (β_{DP}) of the single transistor it represents, as well as some of its small-signal characteristics.

Let β_1 and β_2 be the dc β-values of Q_1 and Q_2, respectively. Then, by definition,

$$I_{C1} = \beta_1 I_{B1} \qquad \text{and} \qquad I_{E1} = (\beta_1 + 1)I_{B1}$$

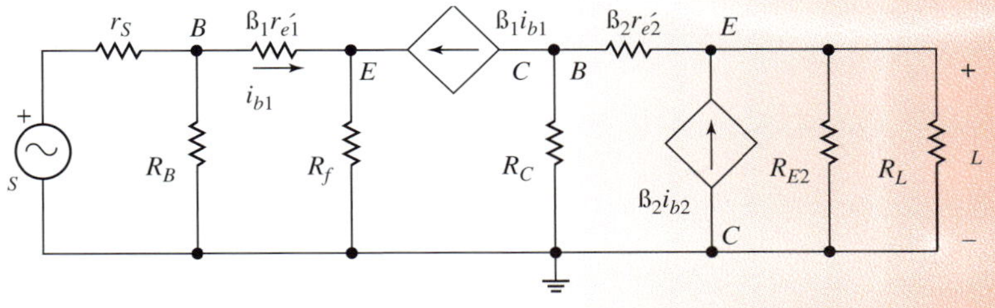

FIGURE 7–38 The ac-equivalent circuit of a CE-CC amplifier

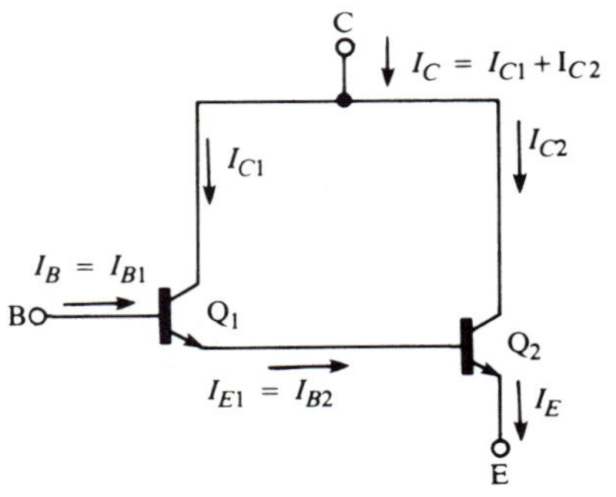

FIGURE 7–39 The Darlington pair is used as a single transistor having the collector, base, and emitter terminals labeled C, B, and E

But $I_{E1} = I_{B2}$, so

$$I_{C2} = \beta_2 I_{B2} = \beta_2(\beta_1 + 1)I_{B1}$$

Now,

$$I_C = I_{C1} + I_{C2} = \beta_1 I_{B1} + \beta_2(\beta_1 + 1)I_{B1} = [\beta_1\beta_2 + (\beta_1 + \beta_2)]I_{B1}$$

Since $I_{B1} = I_B$, we have $I_C = [\beta_1\beta_2 + (\beta_1 + \beta_2)]I_B$, or

$$\beta_{DP} = I_C/I_B = \beta_1\beta_2 + \beta_1 + \beta_2 \quad \text{(7–66)}$$

Equation 7–66 shows that the effective β of the Darlington pair is the product plus the sum of the βs of the individual transistors. It is usually true that $\beta_1\beta_2 \gg \beta_1 + \beta_2$, so

$$\beta_{DP} = \beta_1\beta_2 + \beta_1 + \beta_2 \approx \beta_1\beta_2 \quad \text{(7–67)}$$

Darlington pairs are often fabricated on a single chip to achieve matched Q_1 and Q_2 characteristics. When $\beta_1 = \beta_2 = \beta$, we have an effective β of

$$\beta_{DP} = \beta^2 + 2\beta \approx \beta^2 \quad \text{(7–68)}$$

For example, if $\beta_1 = \beta_2 = 100$, then $\beta_{DP} = 10{,}000 + 200 \approx 10{,}000$. We see that the Darlington pair can be regarded as a "super-β" transistor, and therefore enjoys all the advantages that high-β transistors have.

While the foregoing analysis was performed for dc currents and the dc β-values, an identical small-signal analysis shows that the small-signal value of β_{DP} is the product plus the sum of the small-signal values of β_1 and β_2. Hereafter we will make the usual assumption that the dc and small-signal values of β are approximately equal and will not distinguish between the two.

We wish now to determine the effective small-signal input resistance from B to E, $r_{i(DP)}$, and the emitter resistance $r'_{e(DP)}$, of the composite transistor. Recall the general relationship (equation 7–22):

$$r'_e \approx \frac{V_T}{I_E} \approx \frac{26 \text{ mV}}{I_E} \quad \text{(at room temperature)}$$

Therefore

$$r'_{e2} = \frac{26 \text{ mV}}{I_{E2}} \quad \text{(7–69)}$$

Also, recall that the ac resistance looking into the base of Q_2, r_{i2}, is

$$r_{i2} \approx \beta_2 r'_{e2} \quad \text{(7–70)}$$

Now

$$r'_{e1} \approx \frac{26 \text{ mV}}{I_{E1}} \qquad (7\text{–}71)$$

Since $I_{E2} \approx \beta_2 I_{B2} = \beta_2 I_{E1}$, we have

$$I_{E1} \approx \frac{I_{E2}}{\beta_2} \qquad (7\text{–}72)$$

Substituting (7–72) into (7–71) gives

$$r'_{e1} \approx \beta_2\left(\frac{26 \text{ mV}}{I_{E2}}\right) = \beta_2 r'_{e2} \qquad (7\text{–}73)$$

The total effective resistance looking into the base of Q_1 (across the composite B–E terminals), i.e., the effective small-signal input resistance of the Darlington pair, is

$$r_{i(DP)} = \beta_1(r'_{e1} + r_{i2}) \approx \beta_1(r'_{e1} + \beta_2 r'_{e2}) \qquad (7\text{–}74)$$

Substituting from equation 7–73, we have

$$r_{i(DP)} \approx \beta_1(\beta_2 r'_{e2} + \beta_2 r'_{e2}) = 2\beta_1\beta_2 r'_{e2} \qquad (7\text{–}75)$$

Since $\beta_{DP} \approx \beta_1\beta_2$, the effective emitter resistance $r'_{e(DP)}$, is

$$r'_{e(DP)} \approx \frac{r_{in(DP)}}{\beta_{DP}} = \frac{2\beta_1\beta_2 r'_{e2}}{\beta_1\beta_2} = 2r'_{e2} \qquad (7\text{–}76)$$

EXAMPLE 7–11

A Darlington pair is biased so that the total collector current is 2 mA. If $\beta_1 = 110$ and $\beta_2 = 100$, find the room-temperature values of (1) β_{DP}, (2) $r_{i(DP)}$, and (3) $r'_{e(DP)}$.

Solution

1. $\beta_{DP} = \beta_1\beta_2 + \beta_1 + \beta_2 = (110)(100) + 110 + 100 = 11{,}210$
2. From equation 7–69,

$$r'_{e2} \approx \frac{0.026}{2 \times 10^{-3}} = 13\ \Omega$$

From equation 7–75, $r_{i(DP)} \approx 2(110)(100)13 = 286\ \text{k}\Omega$.

3. From equation 7–76, $r'_{e(DP)} \approx 2(13) = 26\ \Omega$.

This example shows that the Darlington pair can be used to obtain a significant increase in base-to-emitter input resistance compared to that obtainable from a conventional BJT: 286 kΩ versus 1.3 kΩ.

The Darlington pair is most often used in an emitter-follower configuration because of the excellent buffering it provides between a high-impedance source and a low-impedance load. With an ac load resistance r_L connected to the emitter, the total input resistance to the follower is

$$r_i = r_{i(DP)} + \beta_{DP} r_L = \beta_{DP}(r'_{e(DP)} + r_L) \qquad (7\text{–}77)$$

When operated as an emitter follower, the current gain from the base of Q_1 to the emitter of Q_2 is $A_i = i_{e2}/i_{b1}$. Since $i_{e1} = i_{b2}$, we have

$$A_i = \frac{i_{e2}}{i_{b1}} = \left(\frac{i_{e2}}{i_{b2}}\right)\frac{i_{e1}}{i_{b1}} = (\beta_1 + 1)(\beta_2 + 1) \approx \beta_1\beta_2 \qquad (7\text{–}78)$$

The next example illustrates an extreme case of the need for buffering, where the source resistance is 5 kΩ and the load resistance is 100 Ω.

EXAMPLE 7–12

Figure 7–40 shows a common-emitter stage driving a Darlington pair connected as an emitter follower. The β-values for the silicon transistors are $\beta_1 = 200$, $\beta_2 = 100$, and $\beta_3 = 100$.

1. Find v_L/v_S.
2. Find v_L/v_S if the Darlington pair is removed and the 100-Ω load is capacitor-coupled to the collector of Q_1.

Solution

1. $$V_{Th} \approx \left(\frac{R_2}{R_1 + R_2}\right)V_{CC} = \left[\frac{10\text{ k}\Omega}{(47\text{ k}\Omega) + (10\text{ k}\Omega)}\right](15\text{ V}) = 2.6\text{ V}$$

$$R_B = 10\text{ k}\,\|\,47\text{ k} = 8.25\text{ k}$$

$$I_{C1} \approx I_{E1} \approx \frac{2.6 - 0.7}{978 + \dfrac{8.25\text{ k}}{200}} \approx 1.9\text{ mA}$$

$$r'_{e1} \approx \frac{26\text{ mV}}{I_{E1}} = \frac{26\text{ mV}}{1.9\text{ mA}} = 13.7\ \Omega$$

$$r_{in} = R_B\,\|\,\beta_1(r'_{e1} + R_f) = (8.25\text{ k}\Omega)\,\|\,200(13.7 + 68) = 5.48\text{ k}\Omega$$

Notice that the collector of Q_1 is direct-coupled to the base of Q_2 in the Darlington pair, so

$$V_{B2} = V_{C1} \approx V_{CC} - I_{C1}R_{C1} = 15\text{ V} - (1.9\text{ mA})(3.3\text{ k}\Omega) = 8.7\text{ V}$$

The dc emitter voltage of Q_3 is about 1.4 V less than the base voltage of Q_2, since there are *two* forward-biased base-emitter junctions between those two points:

$$V_{E3} = V_{B2} - 1.4 = 8.7 - 1.4 = 7.3\text{ V}$$

$$I_{E3} = \frac{V_{E3}}{R_{E3}} = \frac{7.3\text{ V}}{47\ \Omega} = 155\text{ mA}$$

Noting that Q_1 and Q_2 in our previous analysis of the Darlington pair are now designated Q_2 and Q_3, we have, from equation 7–69,

$$r'_{e3} = \frac{26\text{ mV}}{155\text{ mA}} = 0.17\ \Omega$$

From equation 7–76,

$$r'_{e(DP)} \approx 2(0.17\ \Omega) = 0.34\ \Omega$$

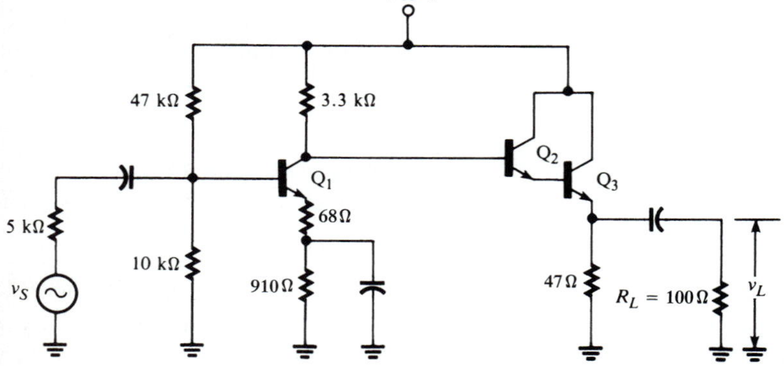

FIGURE 7–40 (Example 7–12)

From equation 7–68,

$$\beta_{DP} = (100)^2 + 200 = 10{,}200$$

From equation 7–77, the total resistance looking into the Darlington pair, with load connected, is

$$r_i = 10{,}200(0.34 + 47\|100) = 330\text{ k}\Omega$$

This value of input resistance is so large in comparison to the 3.3-kΩ collector resistance of Q_1 that virtually no loading occurs. Thus,

$$A_{v1} \approx \frac{-R_{C1}}{r'_{e1} + R_f} = \frac{-3.3\text{ k}\Omega}{81.7\ \Omega} = -40.4$$

The voltage gain of the emitter-follower stage is (from equation 7–56)

$$A_{v(DP)} \approx \frac{r_L}{r'_{e(DP)} + r_L} = \frac{32}{0.34 + 32} \approx 0.99$$

Taking into account the gain drop caused by r_S, the overall voltage gain is

$$\frac{v_L}{v_S} = \left[\frac{r_{in}}{r_S + r_{in}}\right] A_{v1}A_{v(DP)}$$

$$= \left[\frac{5.48\text{ k}\Omega}{(5\text{ k}\Omega) + (5.48\text{ k}\Omega)}\right](-40.4)(0.99) = -20.9$$

2. With the 100-Ω load connected to the collector of Q_1, the ac load on Q_1 is $r_{L1} = (3.3\text{ k}\Omega)\|(100\ \Omega) \approx 97\ \Omega$. Therefore,

$$A_{v1} \approx \frac{-r_{L1}}{r'_{e1} + R_f} = \frac{-97}{81.7} = -1.2$$

The overall gain is then

$$\frac{v_L}{v_S} = \left[\frac{5.48\text{ k}\Omega}{(5\text{ k}\Omega) + (5.48\text{ k}\Omega)}\right](-1.2) = -0.63$$

Without the buffering provided by the Darlington pair, the overall voltage gain turns out to be much less than the original value, not to mention the severe reduction of available output swing.

7–5 OTHER SMALL-SIGNAL MODELS

Transconductance

We have seen that one of the important parameters used in the transistor models of the last section, the ac emitter resistance r'_e, depends on the dc bias current by way of the relation

$$r'_e = \frac{V_T}{I_E} \approx \frac{26\text{ mV}}{I_{EQ}} \quad \text{(at room temperature)}$$

In fact, all small-signal parameters depend on dc operating conditions, some to a greater extent than others. Our goal now is to develop a transistor model that reflects that dependency better than the models of the previous section. One of the advantages of the model is that it permits us to perform (approximate) small-signal analysis based entirely on a knowledge of the dc characteristics of a transistor. We begin with a discussion of a new small-signal parameter, *transconductance,* whose approximate value can be found using only dc quantities.

Transconductance is another *derived* parameter that is widely used in the analysis of electronic devices of all kinds. It is designated g_m and is defined as the ratio of a small-signal *output* current to a small-signal *input* voltage, with dc output voltage held constant:

$$g_m = \left.\frac{i_o}{v_i}\right|_{V_o = \text{constant}} \tag{7–79}$$

Because g_m is the ratio of a current to a voltage, its units are those of conductance, that is, siemens. It is called *trans*conductance because it relates input and output quantities (*across* the device).

For a BJT, transconductance is defined in terms of the common-emitter configuration; that is, input voltage is v_{be} and output current is i_c:

$$g_m = \left.\frac{i_c}{v_{be}}\right|_{V_{CEQ}} \tag{7–80}$$

For v_{be} less than about 10 mV, it can be shown that the value of g_m is closely approximated by the following:

$$g_m \approx \frac{I_C}{V_T} \approx \frac{I_C}{26 \text{ mV}} \quad \text{(at room temperature)} \tag{7–81}$$

where I_C is the dc collector current and V_T is the thermal voltage kT/q. Because $I_E \approx I_C$, we see that $g_m \approx I_E/0.026 = 1/r'_e$ (from equation 7–23).

Output Resistance

The output resistance of a CE transistor can also be determined using dc bias values. Recall that

$$r_o = \left.\frac{v_{ce}}{i_c}\right|_{I_{BQ}} \tag{7–82}$$

Equation 7–82 states that the output resistance is the reciprocal of the slope of the Q-point base current on a set of CE output characteristics. See Figure 7–41.

From Figure 7–41, it is apparent that

$$r_o = \frac{V_A + V_{CEQ}}{I_{CQ}} \tag{7–83}$$

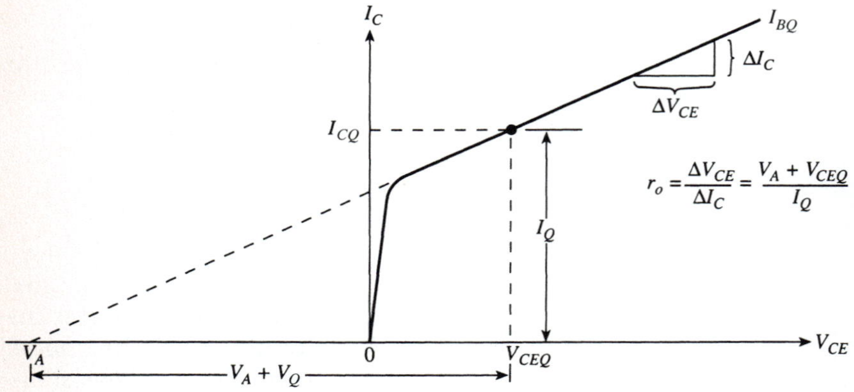

FIGURE 7–41 Common-emitter output resistance is the reciprocal of the slope of a line of constant base current on the CE output characteristics

where V_A = Early voltage
V_{CEQ} = quiescent collector-emitter voltage
I_{CQ} = quiescent collector current

Instead of output resistance, the output *conductance* g_o is used in some transistor models:

$$g_o = \frac{1}{r_o} \tag{7–84}$$

The Hybrid-π Model

Figure 7–42 shows a CE transistor model based on the parameters we have discussed in this section called the hybrid-π model. Note that the current source representing collector current is now labeled $g_m v_{be}$. From equation 7–80, we have $i_c = g_m v_{be}$, so the current source is correctly labeled. Because the hybrid-π model utilizes g_m, it is often called a *transconductance model.*

Analysis of a CE amplifier using the hybrid-π model is very similar to our previous analysis because the two structures are identical. Refer to the basic CE amplifier whose ac equivalent circuit is shown in Figure 7–43. The input resistance is clearly r_π; whereas the current gain, which is the ratio of collector current to base current, can be obtained using the current-division rule as

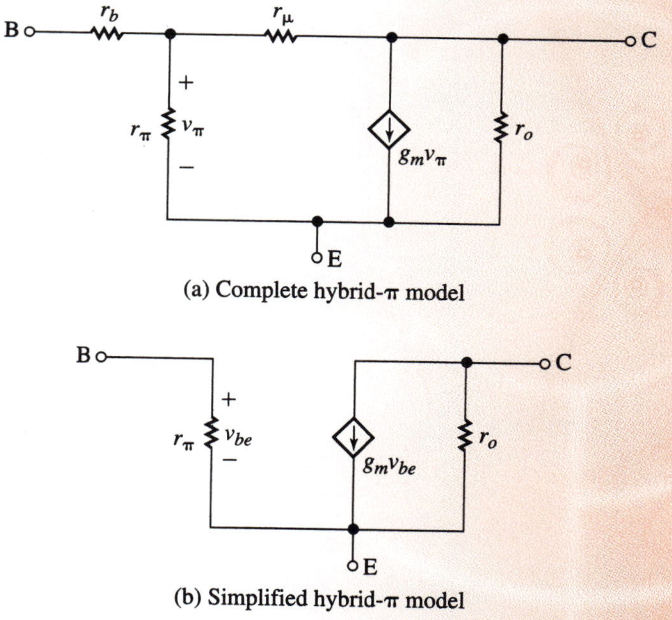

FIGURE 7–42 The common-emitter hybrid-π model for mid-band frequencies

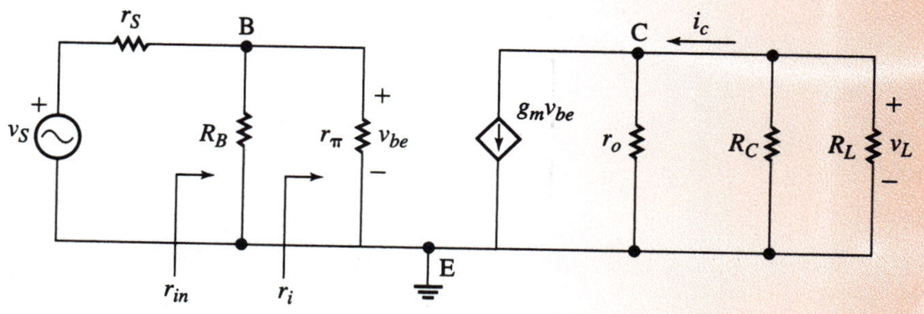

FIGURE 7–43 A basic CE amplifier with the hybrid-π model

$$A_i = g_m r_\pi \left(\frac{r_o}{r_o + r_L} \right) \qquad \textbf{(7–85)}$$

Compare this expression to the one previously derived, namely, $A_i = \beta[r_o/(r_o + r_L)]$. It is clear that the two models are then related by the following expressions:

$$r_\pi = \beta r_e' \qquad \textbf{(7–86)}$$

and

$$\beta = g_m r_\pi \qquad \textbf{(7–87)}$$

The voltage gain, which is the ratio of the load voltage to the base-emitter voltage can be written as

$$A_v = \frac{v_L}{v_{be}} = \frac{-g_m v_{be}(r_o \| r_L)}{v_{be}} = -g_m(r_o \| r_L) \qquad \textbf{(7–88)}$$

which, when $r_o \gg r_L$, reduces to

$$A_v = -g_m r_L \quad (r_L \gg r_o) \qquad \textbf{(7–89)}$$

The reader can verify that the same results can be obtained by invoking the general voltage gain expression $A_v = -A_i(r_L/r_i)$.

The overall gain A_{vs}, which takes into account the source resistance r_s, is written as

$$A_{vs} = -g_m(r_L \| r_o)\left(\frac{r_{in}}{r_s + r_{in}} \right) \qquad \textbf{(7–90)}$$

where $r_{in} = R_B \| r_\pi$ is the stage input resistance.

EXAMPLE 7–13

Use the simplified hybrid-π model of Figure 7–42(b) to determine the load voltage in the circuit of Figure 7–44. Assume that $\beta = 80$ and $V_A = 140$ V.

Solution

$$I_{CQ} = \beta \frac{V_{CC} - 0.7}{R_B} = (80)\frac{9.3 \text{ V}}{360 \text{ k}} = 2.07 \text{ mA}$$

$$r_e' = \frac{0.026 \text{ V}}{I_E} = \frac{0.026}{2.07 \text{ mA}} = 12.56 \ \Omega$$

$$r_\pi \approx \beta r_e' = (80)(12.56 \ \Omega) \approx 1 \text{ k}\Omega$$

From equation 7–83,

$$g_m = \frac{2.07 \text{ mA}}{26 \text{ mV}} = 79.6 \text{ mS}$$

FIGURE 7–44 (Example 7–13)

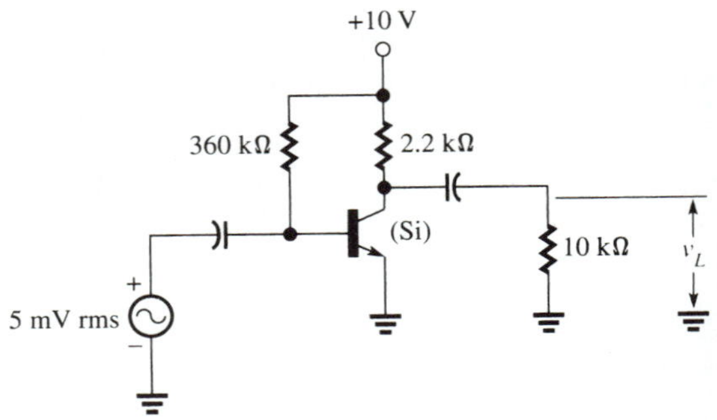

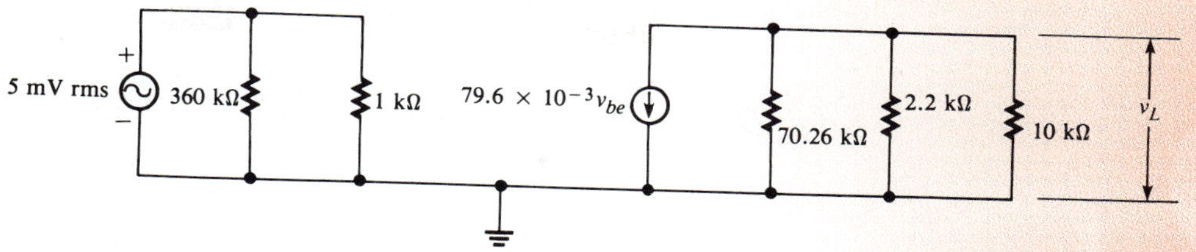

FIGURE 7–45 The ac equivalent circuit of Figure 7–44 (Example 7–13)

$$V_{CEQ} = V_{CC} - I_{CQ}R_C = 10\text{ V} - (2.07\text{ mA})(2.2\text{ k}\Omega) = 5.45\text{ V}$$

From equation 7–92,

$$r_o = \frac{V_A + V_{CEQ}}{I_Q} = \frac{(140 + 5.45)\text{ V}}{2.07 \times 10^{-3}\text{ A}} = 70.26\text{ k}\Omega$$

Figure 7–45 shows the ac equivalent circuit of the amplifier with the transconductance model incorporated. The total equivalent resistance across the output is $(70.26\text{ k}\Omega)\|(2.2\text{ k}\Omega)\|(10\text{ k}\Omega) \approx 1.8\text{ k}\Omega$. Therefore, the voltage gain, using equation 7–89, is

$$A_v = -(79.6\text{ mS})(1.8\text{ k}) = -143$$

The magnitude of the load voltage is therefore (5 mV)(143) = 715 mV.

The *h* Parameter Model

Another scheme of modeling transistors makes use of a two-port network structure with four different parameters, called *hybrid* or *h parameters.* The h-parameter circuit structure is shown in Figure 7–46, where we can identify the four parameters: h_{11}, h_{12}, h_{21}, and h_{22}. Their values are such that the following two network equations are always satisfied:

$$v_1 = h_{11}i_1 + h_{12}v_2 \tag{7–91}$$

$$i_2 = h_{21}i_1 + h_{22}v_2 \tag{7–92}$$

By observing the variables in equations 7–91 and 7–92, we can conclude that h_{11} must be resistance or impedance, h_{12} dimensionless, h_{21} dimensionless, and h_{22} must be a conductance or admittance. We can also observe that, according to the circuit structure, equation 7–91 is a KVL equation around the input loop, whereas (7–92) is a KCL equation for the parallel output circuit.

The formal definition of the four h parameters is given in Table 7–3. By observing carefully the circuit structure and the definitions of the four parameters, and by analogy with the previous CE models, we can identify h_{11} as being $\beta r_e'$ or r_π. Similarly, the parameter h_{21} is equivalent to β, and h_{22} is

FIGURE 7–46 The h-parameter two-port network model

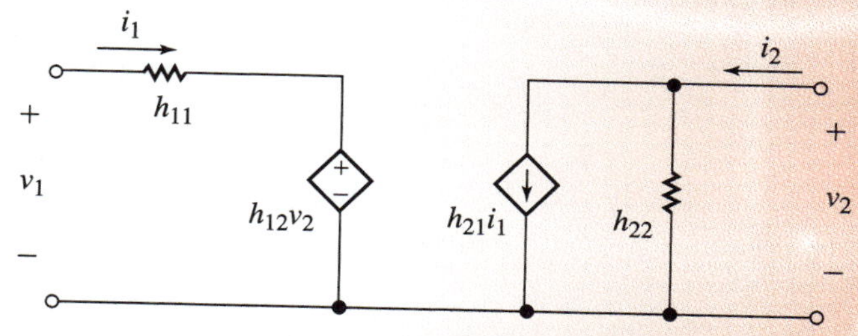

TABLE 7–3

Parameter	Definition	Name
h_{11}	$\left.\frac{v_1}{i_1}\right\rvert_{v_2 = 0}$	Short-circuit input impedance
h_{12}	$\left.\frac{v_1}{v_2}\right\rvert_{i_1 = 0}$	Open-circuit reverse voltage ratio
h_{21}	$\left.\frac{i_2}{i_1}\right\rvert_{v_2 = 0}$	Short-circuit forward current ratio
h_{22}	$\left.\frac{i_2}{v_2}\right\rvert_{i_1 = 0}$	Open-circuit output admittance

simply the reciprocal of r_o. Up to now, we had not used a parameter equivalent to h_{12}. This parameter is useful for modeling the small effect that the output voltage has on the input behavior, an effect that was mentioned at the beginning of the chapter but was considered negligible.

Even though these parameters can be used for describing any two-port network, we will concentrate on their use for characterizing BJTs as small-signal amplifiers. In this application, h parameters are used with two-letter subscripts that not only identify the specific parameter but also the particular configuration: CB, CE, or CC. The equivalencies for the first subscript are as follows:

$h_{11} = h_i$ (from input)

$h_{12} = h_r$ (from reverse)

$h_{21} = h_f$ (from forward)

$h_{22} = h_o$ (from output)

The second subscript is simply b, e, or c to indicate CB, CE, or CC, respectively. For example, the parameter h_{ie} is the input resistance in CE configuration, that is, $\beta r_e'$ or r_π; h_{fb} is the forward current transfer ratio in CB configuration, which would be equivalent to α (actually $-\alpha$ because of the direction of the collector current in the CB configuration). Thus, the 12 h parameters for a bipolar transistor are designated

$h_{ie}, h_{fe}, h_{re}, h_{oe}$: common-emitter h parameters

$h_{ic}, h_{fc}, h_{rc}, h_{oc}$: common-collector h parameters

$h_{ib}, h_{fb}, h_{rb}, h_{ob}$: common-base h parameters

If all of the h-parameter values in one configuration are known, then the values corresponding to any other configuration can be determined. The common-emitter values are the ones most often given in data sheets.

The values specified for transistor h parameters are almost always small-signal quantities. Occasionally, a dc value will be given, in which case it is common practice to use capital-letter subscripts, as, for example, h_{FE}. It is important to remember that all small-signal BJT parameter values are affected by dc (quiescent) current, as discussed previously. Each small-signal parameter is the ratio of certain ac voltages and currents, and each is valid only in a small region over which there is negligible change in device characteristics.

Now let us look at how the basic definitions of the h parameters relate to input and output characteristics of a BJT in the CE modality. Figure 7–47 shows how common-emitter h-parameter values can be determined graphically using characteristic curves. In the common-emitter configuration, note that

FIGURE 7–47 Graphical determination of common-emitter h parameters. Notice that the ac quantities v_{ce} and i_b are 0 when the dc quantities V_{CE} and I_B are held constant

$$h_{11} = \left.\frac{v_1}{i_1}\right|_{v_2 = 0} \qquad h_{ie} = \left.\frac{\Delta V_{BE}}{\Delta I_B}\right|_{V_{CE} = \text{constant}}$$

$$h_{21} = \left.\frac{i_2}{i_1}\right|_{v_2 = 0} \qquad h_{fe} = \left.\frac{\Delta I_C}{\Delta I_B}\right|_{V_{CE} = \text{constant}} = \beta$$

$$h_{12} = \left.\frac{v_1}{v_2}\right|_{i_1 = 0} \qquad h_{re} = \left.\frac{\Delta V_{BE}}{\Delta V_{CE}}\right|_{I_B = \text{constant}}$$

$$h_{22} = \left.\frac{i_2}{v_2}\right|_{i_1 = 0} \qquad h_{oe} = \left.\frac{\Delta I_C}{\Delta V_{CE}}\right|_{I_B = \text{constant}}$$

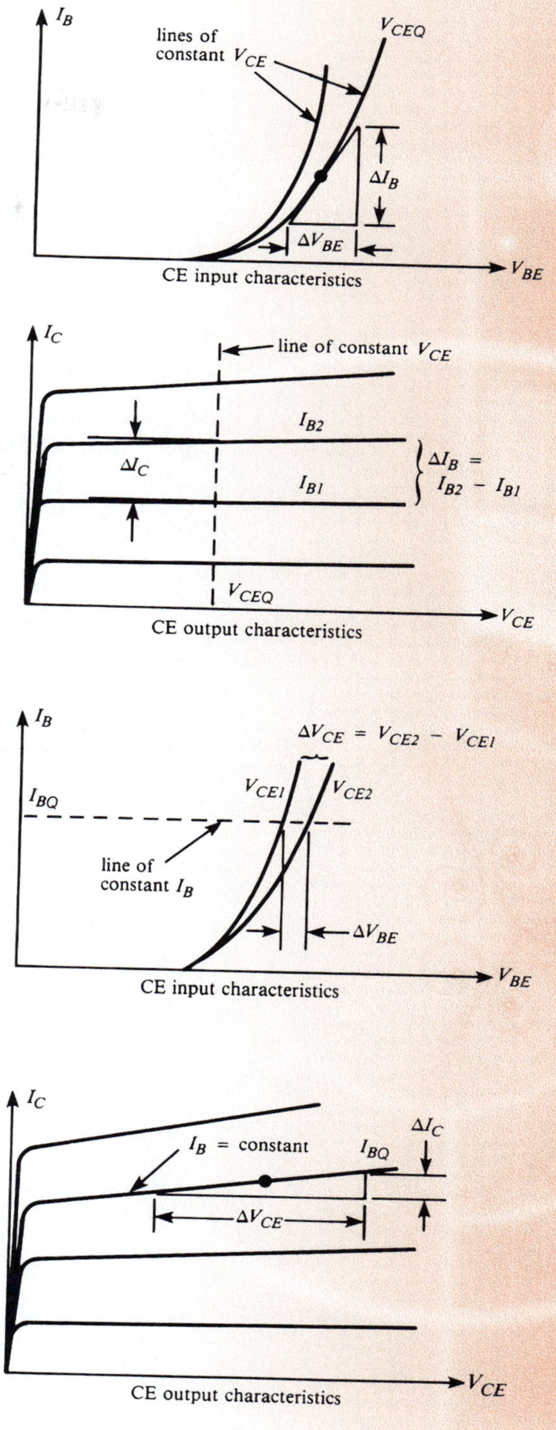

$$v_1 = v_{be} \qquad i_1 = i_b$$

and

$$v_2 = v_{ce} \qquad i_2 = i_c$$

Therefore, from the definitions of the h parameters, the computation of h_{ie} and h_{fe} must be performed with $v_{ce} = 0$. Note that requiring the *ac* quantity v_{ce} to be 0 is the same as requiring that the *dc* quantity V_{CE} be held constant.

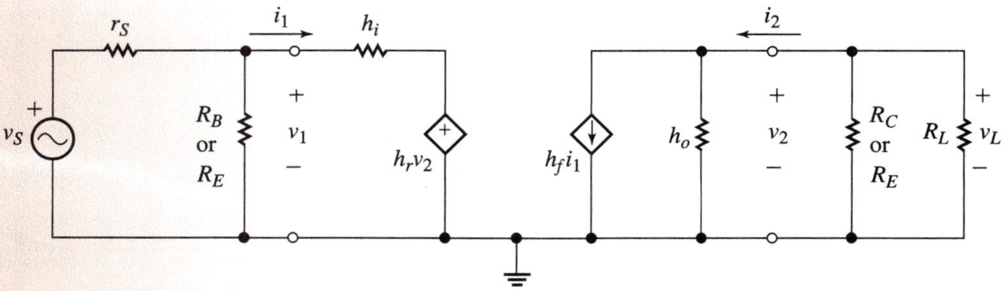

FIGURE 7–48 A general amplifier using the h-parameter small-signal model

Similarly, the computation of h_{re} and h_{oe} requires that $i_b = 0$, which is satisfied by requiring that I_B remain constant. Furthermore, besides being held constant, the conditional variables should be set at their Q-point values, such as I_{CQ}, V_{CEQ}, etc. A typical set of *h*-parameter values for a silicon *npn* transistor is

$h_{ie} = 1200\ \Omega$ $\qquad h_{re} = 2 \times 10^{-4}$
$h_{fe} = 100$ $\qquad h_{oe} = 20 \times 10^{-6}$ S

Figure 7–48 shows a general amplifier modeled with *h* parameters. Notice that the parameters are shown with the first subscript only because the analysis is identical for the three configurations. The resulting equations can be used for any particular case by simply adding the second subscript to all the expressions. The presence of the *h* parameter h_r makes the analysis slightly different but it remains straightforward.

Let us begin with the current gain A_i. Using the current-division principle in the output circuit, we can solve for the ratio i_2/i_1 and obtain

$$\frac{i_2}{i_1} = A_i = \frac{h_f}{1 + h_o r_L} \tag{7–93}$$

where r_L is the total ac load resistance, including any biasing resistor such as R_C or R_E.

To obtain the input resistance, we write the KVL equation

$$v_1 = h_i i_1 + h_r v_2$$

but $v_2 = -A_i i_1 r_L$. Therefore, substituting v_2 and solving for the ratio v_1/i_1, we obtain

$$\frac{v_1}{i_1} = r_i = h_i - A_i h_r r_L = h_i - \frac{h_r h_f r_L}{1 + h_o r_L} \tag{7–94}$$

Invoking again the general form for the voltage gain $A_v = -A_i \dfrac{r_L}{r_i}$, we can obtain

$$A_v = \frac{v_2}{v_1} = \frac{-h_f r_L}{h_i(1 + h_o r_L) - h_r h_f r_L} \tag{7–95}$$

Overall gain, A_{vs}, is again expressed as A_v times the voltage-division fraction of the input circuit; that is,

$$A_{vs} = A_v\left(\frac{r_{in}}{r_S + r_{in}}\right) \tag{7–96}$$

where r_{in} is the stage input resistance given by $r_{in} = R_B \| r_i$ for CE, or $r_{in} = R_E \| r_i$ for CB.

As you can see, there are more similarities between the three small-signal models we have so far analyzed than there are differences. To better use these models, we should concentrate on understanding the definitions of the different parameters and what they represent, try to memorize at least their typical values, understand the differences between the amplifying configurations, and avoid looking at these models as sets of magic formulas for substituting numbers. Of course, it is important to memorize a few important equations, but understanding where they came from is by far more important.

Conversion Formulas

If we have the h-parameter values for a particular transistor in a given configuration, the full set of parameters for another configuration can be obtained through conversion formulas. Let us derive the conversion formulas for two cases to illustrate the process. The reader is encouraged to derive a few more formulas because this is an excellent means to gain more good skills in amplifier analysis.

Consider deriving the formulas for converting parameters from common emitter to common collector. Although it is not strictly necessary, rearranging the original circuit can help in making the derivation less difficult. Figure 7–49 shows the CE h-parameter model with the base as input, the emitter as output, and the collector grounded.

Let us derive the formula for h_{fc}. The first step is to write the definition of the parameter we want to obtain. Using the definitions from Table 7–3,

$$h_{fc} = \left.\frac{i_e}{i_b}\right|_{v_{ec} = 0}$$

To satisfy the condition $v_{ec} = 0$, we place a short circuit across the output terminals, and then we write an equation that includes i_e and i_b. Depending on the circuit structure, we can use KVL, KCL, Ohm's law, or whatever is appropriate.

Using KCL we write

$$i_b + h_{fe}i_b + i_e = 0$$

Solving for $\frac{i_e}{i_b}$ yields

$$\frac{i_e}{i_b} = -(1 + h_{fe}) = h_{fc} \tag{7–97}$$

To obtain the conversion formulas for CB parameters from CE parameters, we use the circuit shown in Figure 7–50.

We will derive the formula for h_{ob} in terms of the CE parameters. The definition is

$$h_{ob} = \left.\frac{i_c}{v_{cb}}\right|_{i_e = 0}$$

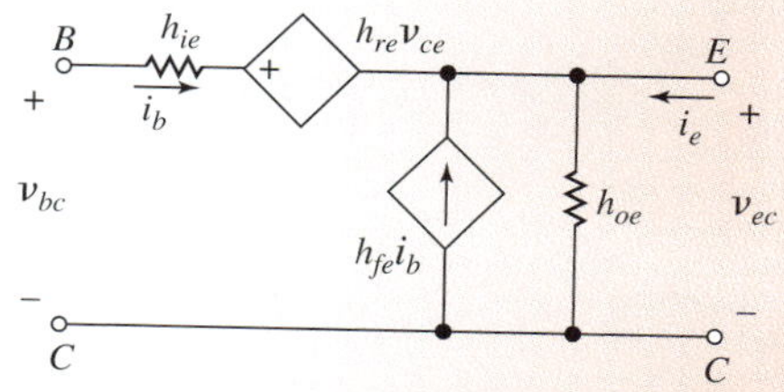

FIGURE 7–49 Circuit for obtaining CC parameters from CE parameters

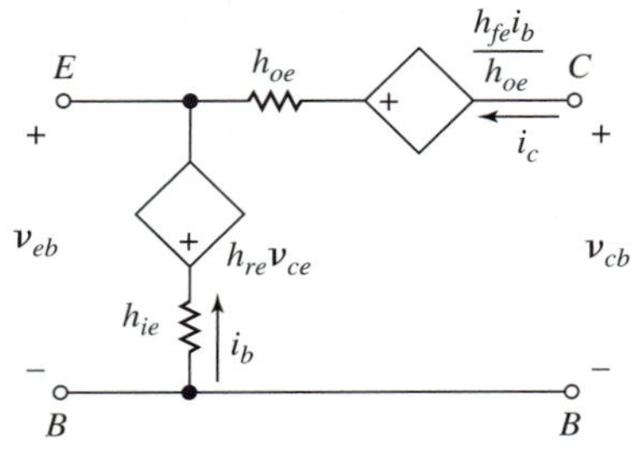

FIGURE 7–50 Circuit for obtaining CB parameters from CE parameters. Note the source transformation

Leaving the input port open to satisfy the condition $i_e = 0$, we write an expression that relates v_{cb} and i_c as follows:

$$i_c = \frac{v_{cb} + h_{fe}i_b/h_{oe} + h_{re}v_{ce}}{h_{ie} + 1/h_{oe}}$$

Noting that $i_b = -i_c$ and $v_{ce} = \dfrac{i_c}{h_{oe}} + \dfrac{h_{fe}i_c}{h_{oe}}$,

$$i_c\left(h_{ie} + \frac{1}{h_{oe}} + \frac{h_{fe}}{h_{oe}} - \frac{h_{re}}{h_{oe}} - \frac{h_{re}h_{fe}}{h_{oe}}\right) = v_{cb}$$

Solving for i_c/v_{cb} yields

$$\frac{i_c}{v_{cb}} = \frac{h_{oe}}{h_{ie}h_{oe} + (1 + h_{fe})(1 - h_{re})} = h_{ob} \qquad \textbf{(7–98)}$$

Table 7–4 shows the conversion formulas between the three BJT configurations along with typical values for the four parameters in each configuration. Note that these conversion formulas can be simplified substantially by neglecting terms that are very small compared to other terms in the expression. For example, h_{re} and h_{rb} are very small and therefore $(1 - h_{re})$ and $(1 - h_{rb})$ become simply 1. Other instances are the products $h_{ie}h_{oe}$ and $h_{ib}h_{ob}$ which are much less than 1, as well as the fact that $h_{rc} \approx 1$. It is left to the reader to develop a set of approximate conversion formulas based on these facts.

Figure 7–51 shows the manufacturer's small-signal specifications for the 2N2218A–2N2222A series transistors whose dc specifications were given in Chapter 4. The graphs showing how the *h* parameters vary with collector current clearly demonstrate the dependence of parameter values on dc conditions. Note also that a significant variation is possible among devices of the same type. Each graph shows the *h*-parameter variation for a typical "high-gain" unit (labeled 1) and for a typical "low-gain" unit (labeled 2). Table 7–5 shows approximate parameter equivalents and conversions.

EXAMPLE 7–14

Using Table 7–4 and the common-emitter output characteristics shown in Figure 7–52, find approximate values for h_{fe}, h_{oe}, h_{fb}, h_{ob}, h_{fc}, and h_{oc} when $I_C = 4.2$ mA and $V_{CE} = 15$ V. The transistor whose characteristics are shown has $h_{ie} = 1500\ \Omega$ and $h_{re} = 2 \times 10^{-4}$ at the operating point.

Solution

The operating point where $I_C = 4.2$ mA and $V_{CE} = 15$ V is labeled Q in Figure 7–52. To find h_{fe}, we determine ΔI_C along the vertical line through Q (a line of constant V_{CE}) and the corresponding value of ΔI_B. As shown in the

TABLE 7–4 *h*-Parameter conversion equations

		CE	CB	CC
CE	h_{ie}	1400 Ω	$\dfrac{h_{ib}}{(1+h_{fb})(1-h_{rb})+h_{ib}h_{ob}}$	h_{ic}
	h_{re}	2×10^{-4}	$\dfrac{h_{ib}h_{ob}-h_{rb}(1+h_{fb})}{(1+h_{fb})(1-h_{rb})+h_{ib}h_{ob}}$	$1-h_{rc}$
	h_{fe}	100	$\dfrac{-h_{fb}(1-h_{rb})-h_{ob}h_{ib}}{(1+h_{fb})(1-h_{rb})+h_{ib}h_{ob}}$	$-(1+h_{fc})$
	h_{oe}	2×10^{-5} S	$\dfrac{h_{ob}}{(1+h_{fb})(1-h_{rb})+h_{ib}h_{ob}}$	h_{oc}
CB	h_{ib}	$\dfrac{h_{ie}}{(1+h_{fe})(1-h_{re})+h_{ie}h_{oe}}$	14 Ω	$\dfrac{h_{ic}}{h_{ic}h_{oc}-h_{fc}h_{rc}}$
	h_{rb}	$\dfrac{h_{ie}h_{oe}-h_{re}(1+h_{fe})}{(1+h_{fe})(1-h_{re})+h_{ie}h_{oe}}$	4×10^{-5}	$\dfrac{h_{fc}(1-h_{rc})+h_{ic}h_{oc}}{h_{ic}h_{oe}-h_{fc}h_{rc}}$
	h_{fb}	$\dfrac{-h_{fe}(1-h_{re})-h_{ie}h_{oe}}{(1+h_{fe})(1-h_{re})+h_{ie}h_{oe}}$	−0.99	$\dfrac{h_{rc}(1+h_{fc})-h_{ic}h_{oc}}{h_{ic}h_{oc}-h_{fc}h_{rc}}$
	h_{ob}	$\dfrac{h_{oe}}{(1+h_{fe})(1-h_{re})+h_{ie}h_{oe}}$	2×10^{-7} S	$\dfrac{h_{oc}}{h_{ic}h_{oc}-h_{fc}h_{rc}}$
CC	h_{ic}	h_{ie}	$\dfrac{h_{ib}}{(1+h_{fb})(1-h_{rb})+h_{ib}h_{ob}}$	1400 Ω
	h_{rc}	$1-h_{re}$	$\dfrac{1+h_{fb}}{(1+h_{fb})(1-h_{rb})+h_{ib}h_{ob}}$	1
	h_{fc}	$-(1+h_{fe})$	$\dfrac{h_{rb}-1}{(1+h_{fb})(1-h_{rb})+h_{ib}h_{ob}}$	−101
	h_{oc}	h_{oe}	$\dfrac{h_{ob}}{(1+h_{fb})(1-h_{rb})+h_{ib}h_{ob}}$	2×10^{-5} S

figure, $\Delta I_C \approx (5.6\text{ mA}) - (3.1\text{ mA}) = 2.5\text{ mA}$, and $\Delta I_B = (50\ \mu\text{A}) - (30\ \mu\text{A}) = 20\ \mu\text{A}$. Therefore,

$$h_{fe} = \left.\frac{\Delta I_C}{\Delta I_B}\right|_{V_{CE}=15\text{ V}} \approx \frac{2.5\text{ mA}}{20\ \mu\text{A}} = 125$$

TABLE 7–5

Model Parameter	Hybrid-π Model (CE)	Approximate *h*-Parameter Equivalent	Typical Value
r'_e	$1/g_m$	h_{ib}	25 Ω
α	—	h_{fb}	0.99
r_c	$(\beta+1)r_o$	$1/h_{ob}$	5 MΩ
$(\beta+1)r'_e$	r_π	h_{ie}	2.5 kΩ
β	$g_m r_\pi$	h_{fe}	100
$r_c/(\beta+1)$	r_o	$1/h_{oe}$	50 kΩ
$(\beta+1)/r_c$	—	h_{oe}	20 μS
$1/r_c$	$1/(\beta+1)r_o$	h_{ob}	0.2 μS
$1/r'_e$	g_m	$1/h_{ib}$	40 mS

FIGURE 7–51 Typical small-signal specifications (Courtesy of Motorola Corporation)

SMALL-SIGNAL CHARACTERISTICS

Current-Gain — Bandwidth Product(2) (I_C = 20 mAdc, V_{CE} = 20 Vdc, f = 100 MHz)	All Types, Except 2N2219A, 2N2222A, 2N5582	f_T	250 300	— —	MHz
Output Capacitance(3) (V_{CB} = 10 Vdc, I_E = 0, f = 100 kHz)		C_{obo}	—	8.0	pF
Input Capacitance(3) (V_{EB} = 0.5 Vdc, I_C = 0, f = 100 kHz)	Non-A Suffix A-Suffix, 2N5581, 2N5582	C_{ibo}	— —	30 25	pF
Input Impedance (I_C = 1.0 mAdc, V_{CE} = 10 Vdc, f = 1.0 kHz)	2N2218A, 2N2221A 2N2219A, 2N2222A	h_{ie}	1.0 2.0	3.5 8.0	kohms
(I_C = 10 mAdc, V_{CE} = 10 Vdc, f = 1.0 kHz)	2N2218A, 2N2221A 2N2219A, 2N2222A		0.2 0.25	1.0 1.25	
Voltage Feedback Ratio (I_C = 1.0 mAdc, V_{CE} = 10 Vdc, f = 1.0 kHz)	2N2218A, 2N2221A 2N2219A, 2N2222A	h_{re}	— —	5.0 8.0	$\times 10^{-4}$
(I_C = 10 mAdc, V_{CE} = 10 Vdc, f = 1.0 kHz)	2N2218A, 2N2221A 2N2219A, 2N2222A		— —	2.5 4.0	
Small-Signal Current Gain (I_C = 1.0 mAdc, V_{CE} = 10 Vdc, f = 1.0 kHz)	2N2218A, 2N2221A 2N2219A, 2N2222A	h_{fe}	30 50	150 300	—
(I_C = 10 mAdc, V_{CE} = 10 Vdc, f = 1.0 kHz)	2N2218A, 2N2221A 2N2219A, 2N2222A		50 75	300 375	
Output Admittance (I_C = 1.0 mAdc, V_{CE} = 10 Vdc, f = 1.0 kHz)	2N2218A, 2N2221A 2N2219A, 2N2222A	h_{oe}	3.0 5.0	15 35	μmhos
(I_C = 10 mAdc, V_{CE} = 10 Vdc, f = 1.0 kHz)	2N2218A, 2N2221A 2N2219A, 2N2222A		10 25	100 200	

h PARAMETERS

V_{CE} = 10 Vdc, f = 1.0 kHz, T_A = 25°C

This group of graphs illustrates the relationship between h_{fe} and other "h" parameters for this series of transistors. To obtain these curves, a high-gain and a low-gain unit were selected and the same units were used to develop the correspondingly numbered curves on each graph.

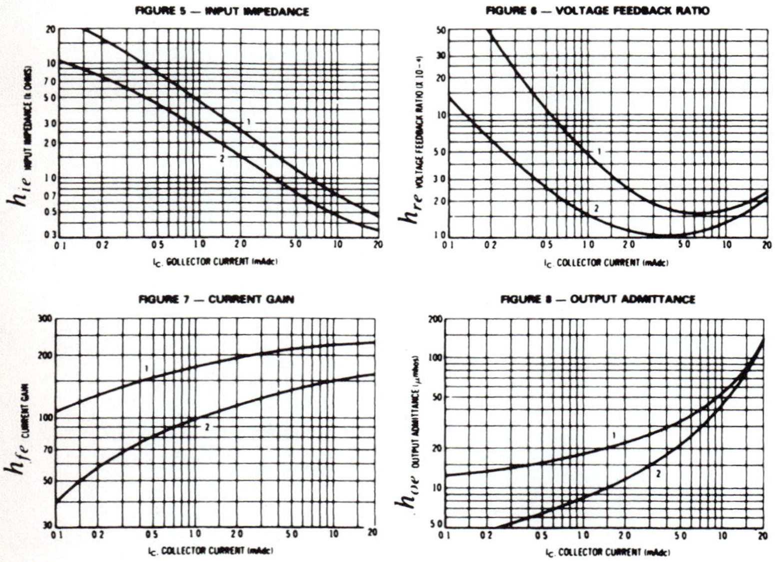

To find h_{oe}, we determine the slope of the line corresponding to $I_B = 40\ \mu A$ (a line of constant I_B). As shown in the figure, $\Delta I_C \approx (4.4\text{ mA}) - (4\text{ mA}) = 0.4$ mA, and $\Delta V_{CE} \approx (28\text{ V}) - (2\text{ V}) = 26$ V. Therefore,

$$h_{oe} = \left.\frac{\Delta I_C}{\Delta V_{CE}}\right|_{I_B = 40\ \mu A} \approx \frac{0.4 \times 10^{-3}}{26} = 15.38\ \mu S$$

From Table 7–4,

$$h_{fb} = \frac{-h_{fe}(1 - h_{re}) - h_{ie}h_{oe}}{(1 + h_{fe})(1 - h_{re}) + h_{ie}h_{oe}}$$

$$= \frac{-125(1 - 2 \times 10^{-4}) - 1500(15.38 \times 10^{-6})}{126(1 - 2 \times 10^{-4}) + 1500(15.38 \times 10^{-6})} = -0.992$$

$$h_{ob} = \frac{h_{oe}}{(1 + h_{fe})(1 - h_{re}) + h_{ie}h_{oe}}$$

FIGURE 7–52 (Example 7–14)

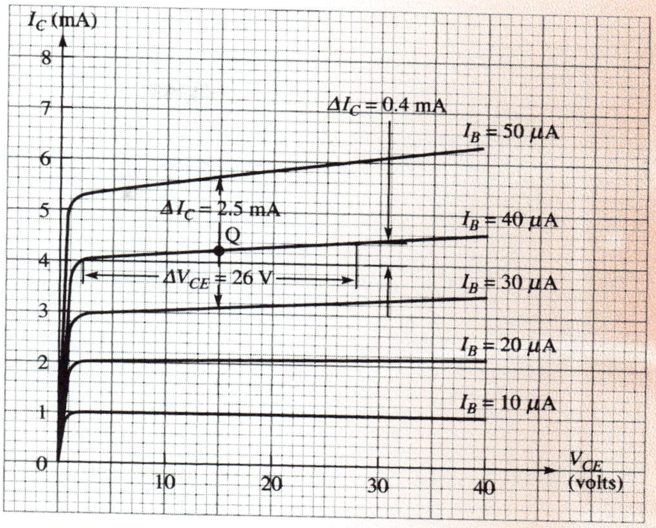

$$= \frac{15.38 \times 10^{-6}}{126(1 - 2 \times 10^{-4}) + 1500(15.38 \times 10^{-6})} = 1.22 \times 10^{-7}\ \text{S}$$

$$h_{fc} = -(1 + h_{fe}) = -126$$

$$h_{oc} = h_{oe} = 15.38\ \mu\text{S}$$

EXAMPLE 7–15

The transistor shown in Figure 7–53 has the following h-parameter values: $h_{ie} = 1600\ \Omega$, $h_{fe} = 80$, $h_{re} = 2 \times 10^{-4}$, $h_{oe} = 20\ \mu$S. Find (1) A_i, (2) r_i, (3) r_{in}, (4) A_v, (5) A_{vs}, (6) r_o, and (7) r_{out}.

Solution

1. $r_L = 2.2\text{ k} \| 10\text{ k} = 1.8\text{ k}$

$$A_i = \frac{h_{fe}}{1 + h_{oe}r_L} = \frac{80}{1 + (20\ \mu\text{S})(1.8\text{ k})} = 77.2$$

2. $r_i = h_{ie} - A_i h_{re} r_L = 1.6\text{ k} - 77.2(2 \times 10^{-4})(1.8\text{ k}) = 1.57\text{ k}\Omega$
3. $R_B = 33\text{ k} \| 12\text{ k} = 8.8\text{ k}\Omega$

FIGURE 7–53 (Example 7–15)

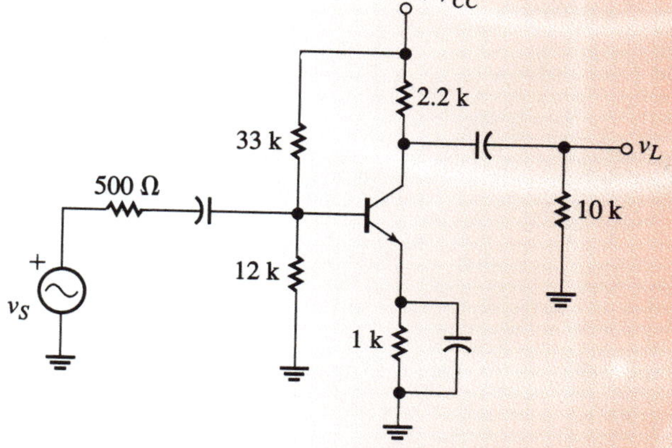

$$r_{in} = 1.57\text{ k}\|8.8\text{ k} = 1.33\text{ k}\Omega$$

4. $A_v = -A_i\dfrac{r_L}{r_i} = -77.2\left(\dfrac{1.8\text{ k}}{1.57\text{ k}}\right) = -88.5$

5. $A_{vs} = A_v\left(\dfrac{r_{in}}{r_S + r_{in}}\right) = -88.5\left(\dfrac{1.33\text{ k}}{500 + 1.33\text{ k}}\right) = -64.3$

6. $r_o = \dfrac{1}{h_{oe} - \dfrac{h_{fe}h_{re}}{h_{ie} + r_S\|R_B}} = \dfrac{1}{1.6\text{ k} - \dfrac{80(2\times 10^{-4})}{1.6\text{ k} + 500\|8.8\text{ k}}} = 81.4\text{ k}\Omega$

7. $r_{out} = r_o\|R_C = 81.4\text{ k}\|2.2\text{ k} = 2.14\text{ k}\Omega \approx R_C$

7–6 THE COMMON-SOURCE JFET AMPLIFIER

Like bipolar transistors, field-effect transistors can be connected as ac amplifiers to achieve voltage, power, and/or current gain. For example, in the most useful JFET configuration, the *common-source* amplifier, a small variation in the gate-to-source (input) voltage creates a large variation in the drain-to-source (output) voltage, when the device is operated in its pinch-off region. We will develop equations that allow us to compute the magnitude of the gain in this and other FET configurations when the input variation is small compared to its total possible range, i.e., under small-signal conditions. We begin with a discussion of the FET small-signal parameters that directly affect gain magnitude computations.

Small-Signal JFET Parameters

Recall that transconductance is defined to be the ratio of a small-signal output current (i_o) to the small-signal input voltage (v_i) that produced it. The transconductance of a JFET is defined in the context of a *common-source* configuration, for which the output current is i_d and the input voltage is v_{gs}:

$$g_m = \frac{i_d}{v_{gs}} = \left.\frac{\Delta I_D}{\Delta V_{GS}}\right|_{V_{DS} = \text{constant}} \tag{7–99}$$

where ΔI_D is the *change* in drain current caused by ΔV_{GS}, the *change* in gate-to-source voltage. Note that we resume the use of lowercase letters in keeping with the usual convention for representing small-signal quantities. The value of g_m can be found graphically using a JFET transfer characteristic, since the latter is a plot of drain current versus gate-to-source voltage. Transconductance is thus the *slope* of the transfer characteristic at the operating point, as illustrated in Figure 7–54.

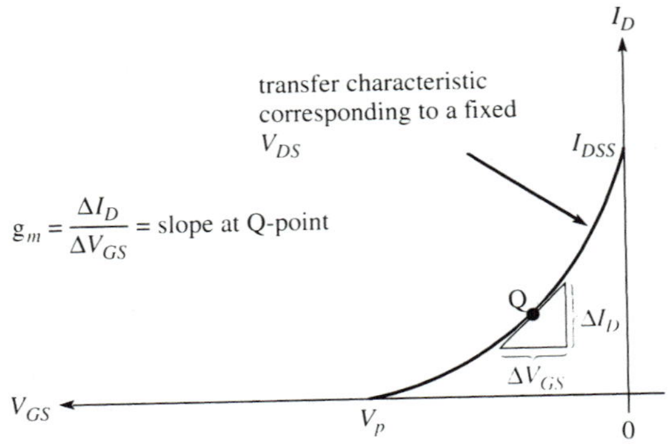

FIGURE 7–54 Transconductance is the slope of the transfer characteristic. Note that the slope, and hence the value of g_m increases with increasing I_D.

As shown in the figure, g_m is calculated graphically by drawing a line tangent to the characteristic at the quiescent point and then measuring its slope. Note that transconductance increases with increasing values of I_D, that is, at Q-points lying farther up the curve toward I_{DSS}, because the curve becomes steeper in that direction. Because the current in a JFET amplifier varies above and below the Q-point, and because the transconductance changes at every point along the characteristic, small-signal analysis is valid only if ΔI_D is small enough to make the change in g_m negligible.

It can be shown that under small-signal conditions the value of g_m is approximately given by

$$g_m = \frac{2I_{DSS}}{|V_p|}\left(1 - \left|\frac{V_{GS}}{V_p}\right|\right) \text{ siemens} \qquad \textbf{(7–100)}$$

where V_{GS} is the quiescent value of the gate-to-source voltage. The use of absolute values in equation 7–100 makes it valid for both n-channel and p-channel devices. Equation 7–100 can be used in conjunction with the equation for the transfer characteristic to express g_m in terms of I_D, the quiescent value of the drain current:

$$g_m = \frac{2I_{DSS}}{|V_p|}\sqrt{\frac{I_D}{I_{DSS}}} \text{ siemens} \qquad \textbf{(7–101)}$$

Equation 7–101 clearly shows that the value of g_m increases with increasing I_D. It is a maximum when $I_D = I_{DSS}$, for which it is given the special symbol g_{mO}. From equation 7–101 with $I_D = I_{DSS}$, we have

$$g_{mO} = \frac{2I_{DSS}}{|V_p|} \text{ siemens} \qquad \textbf{(7–102)}$$

Although a large value of g_m is desirable, and results in a large voltage gain (as we shall presently see), we would never bias a JFET at $I_D = I_{DSS}$ for small-signal operation. Obviously, no signal variation above that Q-point could occur if that were the case.

EXAMPLE 7–16

An n-channel JFET has the transfer characteristic shown in Figure 7–55. Determine, both (1) graphically and (2) algebraically, the value of its transconductance when $I_D = 4$ mA. Also determine the value of g_{mO}.

Solution

1. As shown in Figure 7–55, a line is drawn tangent to the transfer characteristic at the point where $I_D = 4$ mA. Using values obtained from the figure, the slope of the tangent line is determined to be

$$g_m = \frac{\Delta I_D}{\Delta V_{GS}} = \frac{(6\text{ mA}) - (2\text{ mA})}{(2.1\text{ V}) - (0.8\text{ V})} = \frac{4 \times 10^{-3}\text{ A}}{1.3\text{ V}} = 3.077 \times 10^{-3}\text{ S}$$

2. From Figure 7–55, it is evident that $V_p = -4$ V and $I_{DSS} = 10$ mA. Using equation 7–101 with $I_D = 4$ mA, we find

$$g_m = \frac{2(10 \times 10^{-3}\text{ A})}{4\text{ V}}\sqrt{\frac{4 \times 10^{-3}}{10 \times 10^{-3}}} = 3.16 \times 10^{-3}\text{ S}$$

We see that there is good agreement between the graphical and algebraic solutions.

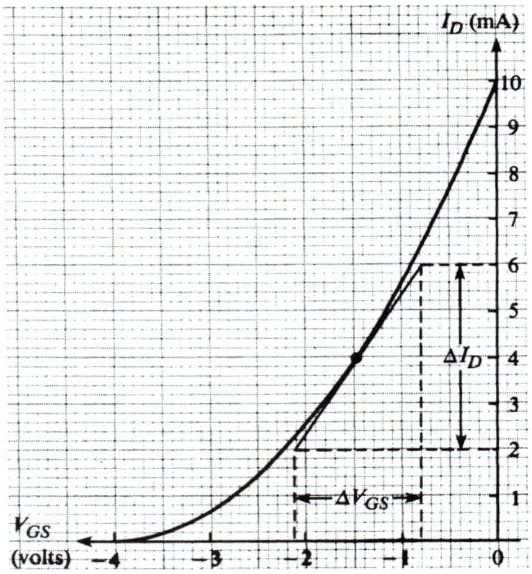

FIGURE 7–55 (Example 7–16)

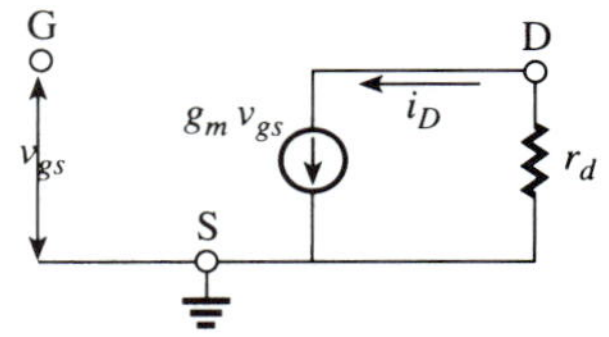

FIGURE 7–56 Small-signal equivalent circuit of a common-source JFET

The value of g_{mO} is computed using equation 7–102:

$$g_{mO} = \frac{2(10 \times 10^{-3}\text{ A})}{4\text{ V}} = 5 \times 10^{-3}\text{ S}$$

The small-signal output resistance of a common-source JFET is defined by

$$r_o = \frac{v_{ds}}{i_d} = \frac{\Delta V_{DS}}{\Delta I_D}\bigg|_{V_{GS} = \text{ constant}} \tag{7–103}$$

This parameter is also called the *drain resistance*, r_d or r_{ds}. It can be determined graphically from a set of drain characteristics, but the lines are so nearly horizontal in the pinch-off region that it is difficult to obtain accurate values. In any case, the value is generally so large that it has little effect on the computation of voltage gain in practical circuits. Values of r_d range from about 50 kΩ to several hundred kΩ in the pinch-off region.

Figure 7–56 shows the small-signal equivalent circuit of the common-source JFET, incorporating the parameters we have discussed. The current source having value $g_m v_{gs}$ is a voltage-controlled current source because the current it produces depends on the (input) voltage v_{gs}. It is apparent in the figure that $i_d = g_m v_{gs}$, in agreement with the definition of g_m (equation 7–99):

$$g_m = \frac{i_d}{v_{gs}} \longrightarrow i_d = g_m v_{gs}$$

Notice that there is an open circuit shown between the gate and source terminals. Because the gate-to-source junction is reverse biased in normal operation, the extremely large resistance between those terminals can be assumed to be infinite in most practical situations.

The Common-Source JFET Configuration

Figure 7–57 shows a common-source JFET amplifier with fixed-bias V_{GG}. R_G is a large resistance connected in series with V_{GG} to prevent the dc source from shorting the ac signal to ground. (Recall that a dc source is an ac short circuit.) The input resistance of the JFET is so large that there is negligible dc voltage division at the gate; that is, the major part of V_{GG} appears from gate-to-source instead of across R_G, so $V_{GS} \approx V_{GG}$. From another viewpoint, the dc gate current is so small that there is negligible drop across R_G. The input coupling capacitor serves the same purpose it does in a BJT amplifier, namely, to provide dc isolation between the signal source and the FET. For the moment, we ignore any signal-source resistance (r_S) and assume that the output is open ($R_L = \infty$).

The total gate-to-source voltage is the sum of the small-signal source voltage, v_S, and the bias voltage V_{GG}. For example, if $V_{GG} = -2$ V and v_S is a sine wave having peak value 0.1 V, then $v_{gs} = -2 + 0.1 \sin \omega t$ volts. Thus, v_{gs} varies between the extreme values $-2 - 0.1 = -2.1$ V and $-2 + 0.1 = -1.9$ V. When v_{gs} goes more positive (toward -1.9 V), the drain current increases. This increase in i_d causes the output voltage v_{ds} to *decrease*, since

$$V_{DS} = V_{DD} - I_D R_D \qquad (7\text{–}104)$$

We conclude that an increase in the input signal voltage causes a decrease in the output voltage and that the output is therefore 180° out of phase with the input.

Figure 7–58 shows the equivalent circuit of the common-source amplifier in Figure 7–57, with the JFET replaced by its small-signal equivalent. The coupling capacitor is assumed to have negligible impedance at the signal frequency we are considering (for now), so it is replaced by a short circuit. As usual, all dc sources are treated as ac short circuits to ground. Notice that the arrow in the controlled current source can be shown reversed (from Figure 7–56) with a minus sign attached. Note: If the arrow is reversed, the minus sign *must* be attached (not "can be"). The minus sign denotes the phase inversion between input and output, as we have described.

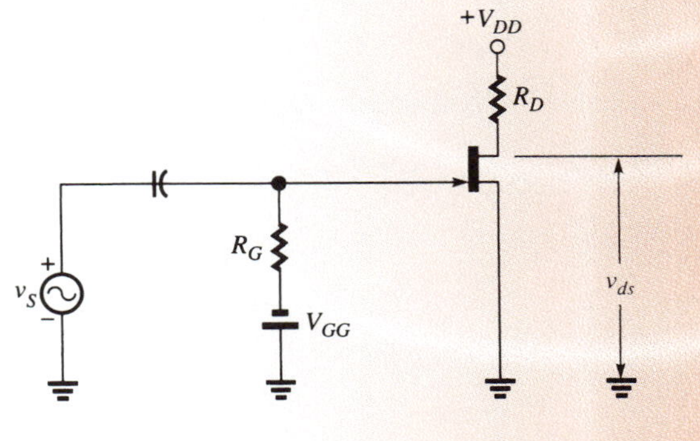

FIGURE 7–57 A common-source amplifier with fixed bias

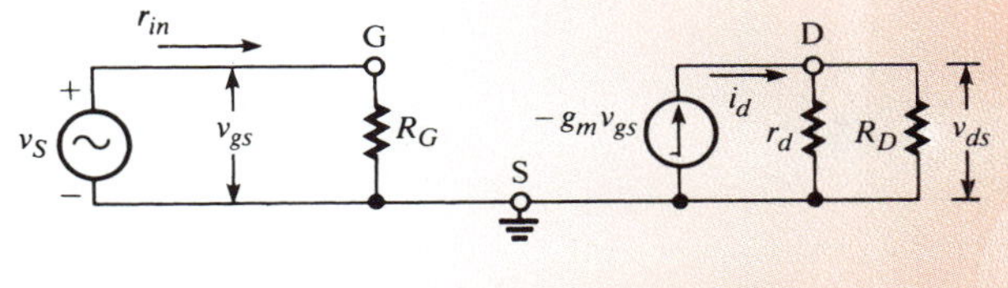

FIGURE 7–58 The small-signal equivalent circuit of the common-source amplifier shown in Figure 7–57

It is clear from Figure 7–58 that

$$v_{ds} = i_d(r_d \| R_D) \quad \textbf{(7–105)}$$

Since $i_d = -g_m v_{gs}$, we have

$$v_{ds} = -g_m v_{gs}(r_d \| R_D) \quad \textbf{(7–106)}$$

There is no signal-source resistance, so $v_{gs}=v_S$, and the voltage gain is therefore

$$A_v = \frac{v_{ds}}{v_S} = \frac{v_{ds}}{v_{gs}} = -g_m(r_d \| R_D) \quad \textbf{(7–107)}$$

In most practical amplifier circuits, the value of r_d is much greater than that of R_D, so $r_d \| R_D \approx R_D$, and a good approximation for the voltage gain is

$$\frac{v_{ds}}{v_{gs}} \approx -g_m R_D \quad \textbf{(7–108)}$$

It is important to remember that a JFET amplifier is operated in its pinch-off region. Therefore, the bias point and the voltage variations around the bias point must always satisfy $|V_{DS}| \geq |V_p| - |V_{GS}|$.

So long as the signal source does not drive the gate positive with respect to the source, the reverse-biased gate-to-source resistance is very large and the input resistance to the amplifier is essentially R_G. The equivalent circuit in Figure 7–58 is based on this assumption, and it can be seen that $r_{in} = R_G$.

EXAMPLE 7–17

The JFET in the amplifier circuit shown in Figure 7–59 has $I_{DSS} = 12$ mA, $V_p = -4$ V, and $r_d = 100$ kΩ.

1. Find the quiescent values of I_D and V_{DS}.
2. Find g_m.
3. Draw the ac equivalent circuit.
4. Find the voltage gain.

Solution

1. I_D is found using the equation of the transfer characteristic:

$$I_D = I_{DSS}\left(1 - \frac{V_{GS}}{V_p}\right)^2 = (12\text{ mA})\left(1 - \frac{-2\text{ V}}{-4\text{ V}}\right)^2 = 3\text{ mA}$$

From equation 7–104, $V_{DS} = 15 - (3\text{ mA})(2.2\text{ k}\Omega) = 8.4$ V. Since $V_{DS} = 8.4$ V and $|V_p| - |V_{GS}| = 4 - 2 = 2$ V, we have $|V_{DS}| > |V_p| - |V_{GS}|$, confirming that the JFET is biased in its pinch-off region.

FIGURE 7–59 (Example 7–17)

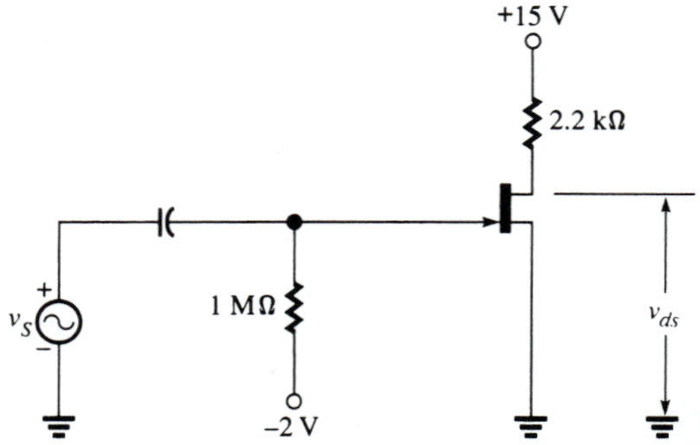

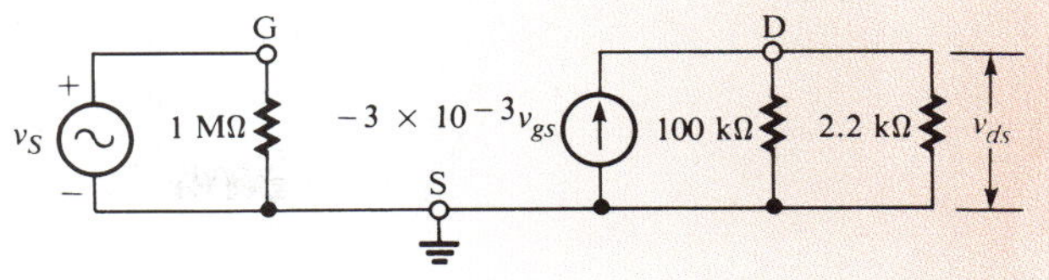

FIGURE 7–60 (Example 7–17)

2. From equation 7–101,

$$g_m = \frac{2(12\text{ mA})}{4\text{ V}}\sqrt{\frac{3\text{ mA}}{12\text{ mA}}} = 3 \times 10^{-3}\text{ S}$$

3. The ac equivalent circuit is shown in Figure 7–60.
4. Since $r_d \| R_D = (100\text{ k}\Omega) \| (2.2\text{ k}\Omega) \approx 2.2\text{ k}\Omega$, we have, from equation 7–108,

$$\frac{v_{ds}}{v_s} = \frac{v_{ds}}{v_{gs}} \approx (-3 \times 10^{-3}\text{ S})(2.2 \times 10^3\ \Omega) = -6.6$$

As the preceding example illustrates, the voltage gain obtainable from a JFET amplifier is much smaller than that which can be obtained from its BJT counterpart. The principal advantage of the JFET amplifier is its very large input resistance.

The Self-Biased JFET Amplifier

When the JFET amplifier is biased using only the self-biasing resistor, the voltage gain is again computed using $A_v = -g_m(r_d \| R_D)$. A resistor R_G is connected between gate and ground to provide dc continuity for the gate circuit. This resistor can be made very large to maintain a large input resistance to the amplifier. The reverse current through the reverse-biased gate-source junction is very small, so the voltage drop across R_G is negligible. However, if an extremely large resistance is used with a JFET having excessive leakage current, the drop should be taken into account. For example, if $R_G = 10\text{ M}\Omega$ and $I_{DS(reverse)} = 0.1\ \mu\text{A}$, the gate voltage will be increased by $(10\text{ M}\Omega)(0.1\ \mu\text{A}) = 1\text{ V}$ with respect to ground. Figure 7–61 shows the self-biased JFET amplifier and its ac equivalent circuit.

To avoid confusion, we will hereafter refer to resistance associated with the signal source as *signal-source resistance*, which we will designate by r_S, and we will continue to refer to resistance in series with the source terminal of the JFET as simply *source resistance*, R_S. When signal-source resistance is present, there is the usual voltage division between r_S and the input resistance of the amplifier, r_i. When a load resistor R_L is capacitor-coupled to the output, there is also a voltage division between the amplifier output resistance $r_d \| R_D$ and R_L. Thus,

$$\frac{v_L}{v_s} = \left(\frac{r_{in}}{r_S + r_{in}}\right) A_v \left(\frac{R_L}{R_L + r_d \| R_D}\right) \tag{7–109}$$

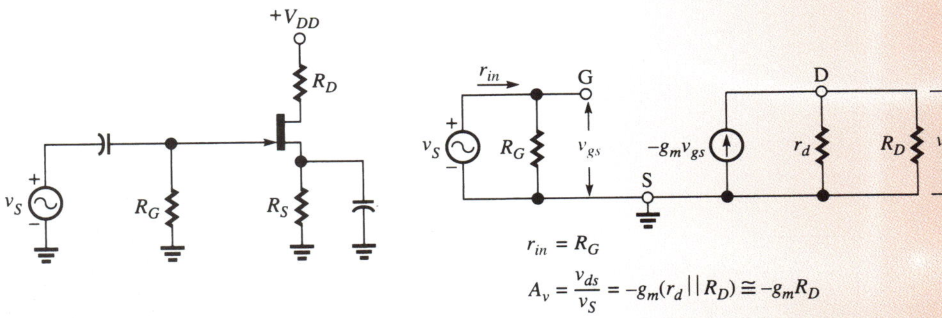

FIGURE 7–61 The self-biased JFET amplifier and its ac equivalent circuit

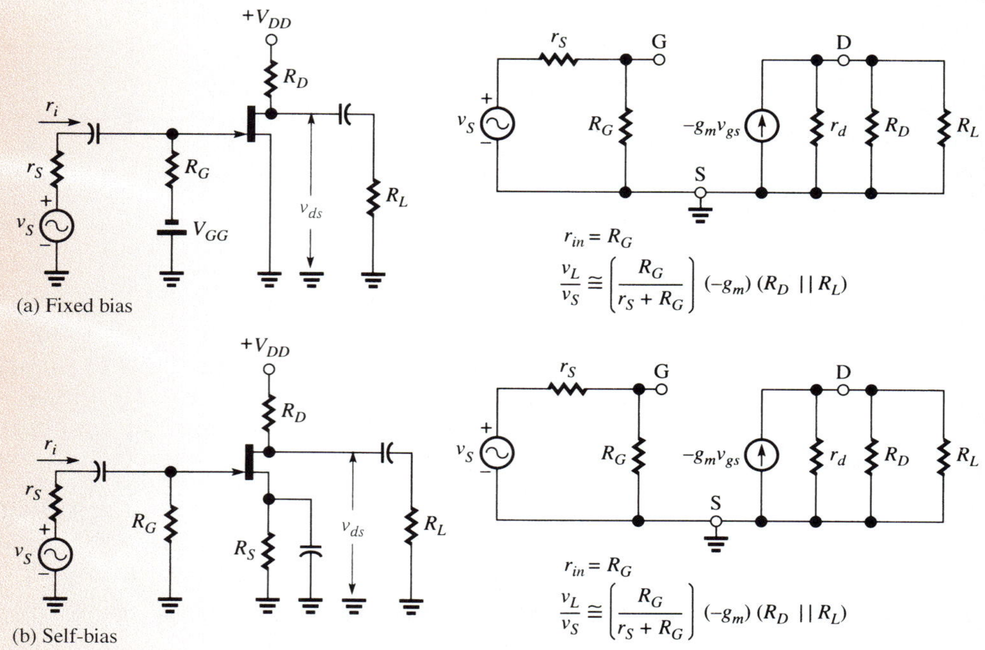

FIGURE 7–62 Common-source amplifiers with load and signal-source resistances included and their ac equivalent circuits

Figure 7–62 shows the common-source amplifier in a fixed and self-bias arrangement with load and signal-source resistances included. Also shown are the ac equivalent circuits of each. Note in each case that r_S is in series with the amplifier input and R_L is in parallel with the amplifier output.

For the fixed-bias amplifier (Figure 7–62(a)), equation 7–109 becomes

$$\frac{v_L}{v_S} = \left(\frac{R_G}{r_S + R_G}\right)[-g_m(r_d \| R_D)]\left(\frac{R_L}{R_L + r_d \| R_D}\right) \qquad \textbf{(7–110)}$$

$$\approx \left(\frac{R_G}{r_S + R_G}\right)(-g_m R_D)\left(\frac{R_L}{R_L + R_D}\right)$$

$$= \left(\frac{R_G}{r_S + R_G}\right)(-g_m)(R_D \| R_L) \qquad \textbf{(7–111)}$$

For the self-biased amplifier (Figure 7–62(b)), equation 7–109 becomes

$$\frac{v_L}{v_S} = \left(\frac{R_G}{r_S + R_G}\right)(-g_m)(r_d \| R_D)\left(\frac{R_L}{R_L + r_d \| R_D}\right) \qquad \textbf{(7–112)}$$

$$\approx \left(\frac{R_G}{r_S + R_G}\right)(-g_m)(R_D \| R_L) \qquad \textbf{(7–113)}$$

Of course, all of the gain relations just derived are exactly the same for *p*-channel JFET amplifiers.

If the amplifier is used with an unbypassed source resistor, it can be shown that the overall gain v_L/v_S becomes

$$\frac{v_L}{v_S} = \left(\frac{R_G}{r_S + R_G}\right)\left(\frac{-g_m}{1 + g_m R_S}\right)(R_D \| R_L) \quad \text{(unbypassed } R_S\text{)} \qquad \textbf{(7–114)}$$

We see that the effect of omitting the bypass capacitor is to reduce gain by the factor $1/(1 + g_m R_S)$. Using the typical values $g_m = 4000\ \mu S$, $R_D = 3\ k\Omega$, and $R_S = 500\ \Omega$, the gain with R_S bypassed is $-g_m R_D = -(4000\ \mu S)(3\ k\Omega) = -12$. This value is reduced in the unbypassed case by the factor $1 + g_m R_S = 1 + (4000\ \mu S)(500\ \Omega) = 3$ to the value $-\frac{12}{3} = -4$.

7–7 THE COMMON-DRAIN AND COMMON-GATE JFET AMPLIFIERS

Common-Drain Amplifier

Figure 7–63 shows a JFET connected as a common-drain amplifier. Note that the drain terminal is connected directly to the supply voltage V_{DD}, so the drain is at ac ground. Because the input and output signals are taken with respect to ground, the drain is common to both, which accounts for the name of the configuration. Because the ac output signal is between the source terminal and ground, it is the same signal measured from source to drain as measured across R_S. Note that the JFET is biased using the combined self-bias/voltage-divider method, though either of the other two bias methods we studied for the common-source configuration could also be used. As can be seen from the figure,

$$v_i = v_{gs} + v_L \tag{7–115}$$

Therefore, the gate-to-source voltage is

$$v_{gs} = v_i - v_L \tag{7–116}$$

Figure 7–64 shows the ac equivalent circuit of the common-drain amplifier, v_i is related to v_{gs} by equations 7–115 and 7–116. Neglecting the signal-source resistance r_S for the moment, we derive the voltage gain v_L/v_i as follows:

$$v_L = g_m v_{gs}(r_d \| R_S \| R_L) \tag{7–117}$$

Substituting $v_{gs} = v_i - v_L$ (equation 7–116) into (7–117), we obtain

$$v_L = g_m(v_i - v_L)(r_d \| R_S \| R_L)$$

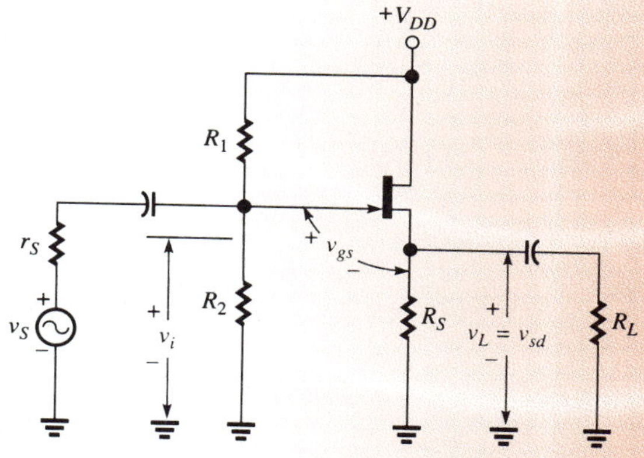

FIGURE 7–63 A common-drain JFET amplifier

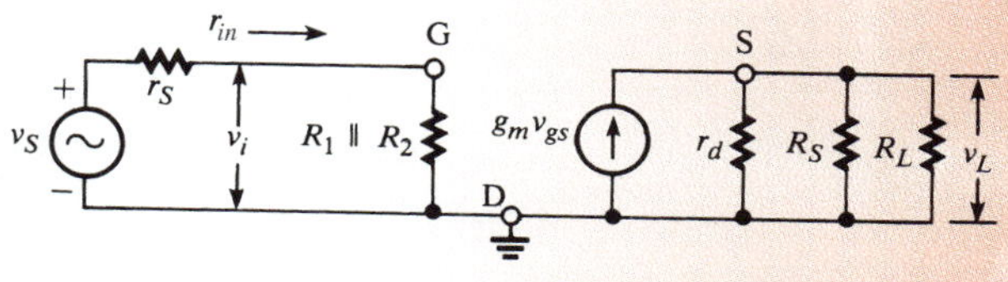

FIGURE 7–64 The small-signal equivalent circuit for the common-drain amplifier

$$= g_m v_i(r_d \| R_S \| R_L) - g_m v_L(r_d \| R_S \| R_L)$$

$$v_L + g_m v_L(r_d \| R_S \| R_L) = g_m v_i(r_d \| R_S \| R_L)$$

Solving for v_L/v_i, we obtain

$$\frac{v_L}{v_i} = \frac{g_m(r_d \| R_S \| R_L)}{1 + g_m(r_d \| R_S \| R_L)} \tag{7–118}$$

It is clear from equation 7–118 that the voltage gain of the common-drain configuration is always less than 1. Note that there is no phase inversion between input and output. When $g_m(r_d \| R_S \| R_L) \gg 1$ (often the case), equation 7–118 shows

$$\frac{v_L}{v_i} \approx 1 \tag{7–119}$$

In other words, the load voltage is approximately the same as the input voltage, in magnitude and phase, and we say that the output *follows* the input. For this reason, the common-drain amplifier is called a *source follower*.

The input resistance of the source follower can be seen from Figure 7–63 to be the same as it is for a common-source amplifier:

$$r_{in} = R_1 \| R_2 \tag{7–120}$$

Similarly, r_{in} is the same as it is for a common-source amplifier when the other bias arrangements are used. When the voltage division between signal-source resistance and r_i is taken into account, we find the overall voltage gain to be

$$\frac{v_L}{v_S} = \left(\frac{r_{in}}{r_S + r_{in}}\right)\left[\frac{g_m(r_d \| R_S \| R_L)}{1 + g_m(r_d \| R_S \| R_L)}\right] \tag{7–121}$$

The output resistance of the source follower (r_{out}), looking from R_L toward the source terminal, is R_S in parallel with the resistance "looking into" the JFET at the source terminal. The resistance looking into the JFET at the source terminal can be found using equation 7–118. Because we are looking to the left of R_S and R_L, the parallel combination of those resistances is not relevant in the equation. Substituting r_d for $r_d \| R_S \| R_L$, equation 7–118 becomes

$$A_v = \frac{v_o}{v_i} = \frac{g_m r_d}{1 + g_m r_d}$$

Now,

$$r_o = \frac{v_o}{i_o} = \frac{A_v v_i}{i_o} = \frac{A_v}{g_m} = \frac{r_d}{1 + g_m r_d}$$

Thus,

$$r_{out} = R_S \left\| \left(\frac{r_d}{1 + g_m r_d}\right)\right. \tag{7–122}$$

It is almost always true that $g_m r_d \gg 1$, so (7–122) reduces to

$$r_{out} \approx R_S \left\| \left(\frac{1}{g_m}\right)\right. \tag{7–123}$$

Like the BJT emitter follower, the source follower is used primarily as a buffer amplifier because of its large input resistance and small output resistance.

EXAMPLE 7–18

The JFET in Figure 7–65 has $g_m = 5 \times 10^{-3}$ S and $r_d = 100$ kΩ. Find

1. the input resistance,
2. the voltage gain, and
3. the output resistance of the amplifier stage.

Solution

1. $r_{in} = R_1 \| R_2 = (1.8 \text{ M}\Omega) \| (470 \text{ k}\Omega) = 372.7 \text{ k}\Omega.$
2. Since $r_d \| R_S \| R_L = (100 \text{ k}\Omega) \| (1.5 \text{ k}\Omega) \| (3 \text{ k}\Omega) = 990\ \Omega$, we have, from equation 7–121

$$\frac{v_L}{v_S} = \left(\frac{372.7 \text{ k}\Omega}{10 \text{ k}\Omega + 372.7 \text{ k}\Omega}\right)\left[\frac{(5 \times 10^{-3} \text{ S})(990\ \Omega)}{1 + (5 \times 10^{-3} \text{ S})(990\ \Omega)}\right) = 0.81$$

3. From equation 7–122,

$$r_{out} = (1.5 \text{ k}\Omega) \Bigg\| \left(\frac{100 \text{ k}\Omega}{1 + (5 \times 10^{-3} \text{ S})(100 \text{ k}\Omega)}\right)$$

$$= (1.5 \text{ k}\Omega) \| (199.6\ \Omega) = 176.2\ \Omega$$

Note that the approximation for r_o (equation 7–123) gives a nearly equal result:

$$r_{out} \approx R_S \| (1/g_m) = (1.5 \text{ k}\Omega) \| 200 = 176.5\ \Omega$$

Common-Gate Amplifier

Figure 7–66 shows a common-gate amplifier and its small-signal equivalent circuit. We see that the load voltage is developed across the parallel combination of r_d, R_D, and R_L:

$$v_L = -\ i_d r_L = g_m v_{gs} r_L \qquad (7\text{–}124)$$

where $r_L = R_D \| R_L$. The voltage gain v_L/v_{gs} is, therefore,

$$\frac{v_L}{v_{gs}} = \frac{g_m v_{gs} r_L}{v_{gs}} = g_m r_L \qquad (7\text{–}125)$$

We see that the magnitude of this gain is calculated in the same way as it is for the common-source amplifier. Note, however, that the gain is a positive quantity, so there is no phase inversion between input and output.

When signal-source resistance r_S is taken into account, voltage division occurs at the input, and the gain v_L/v_S becomes

$$\frac{v_L}{v_S} = \left(\frac{r_{in}}{r_S + r_{in}}\right) g_m r_L \qquad (7\text{–}126)$$

FIGURE 7–65 (Example 7–18)

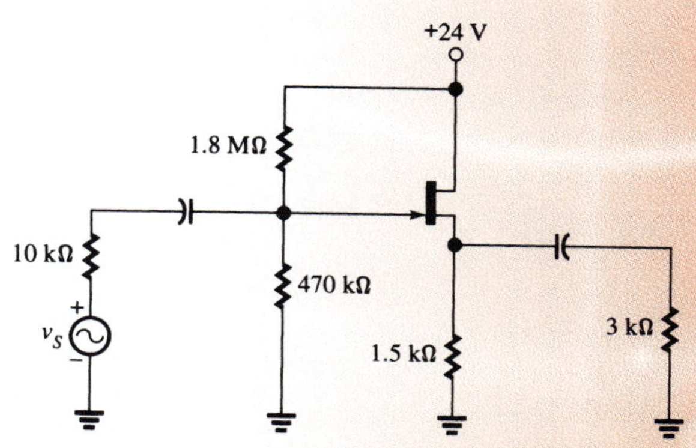

FIGURE 7–66 The common-gate amplifier

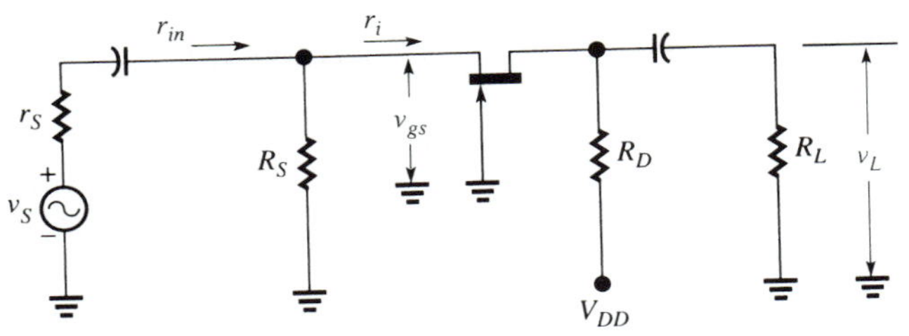

(a) Amplifier circuit

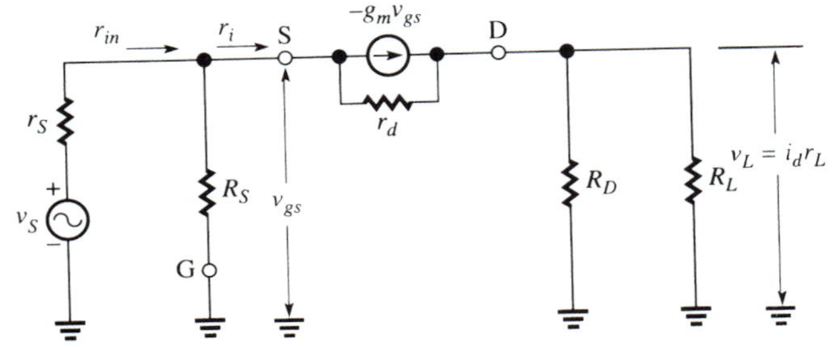

(b) Small-signal equivalent circuit

The input resistance, r_{in}, is the parallel combination of R_S and the resistance r_i looking into the source terminal. As can be seen in Figure 7–66(b),

$$r_i = \frac{v_{gs}}{i_s} = \frac{v_{gs}}{i_d} = \frac{v_{gs}}{g_m v_{gs}} = \frac{1}{g_m} \tag{7–127}$$

Therefore,

$$r_{in} = r_i \| R_S = \frac{1}{g_m} \Big\| R_S \tag{7–128}$$

The input resistance of the common-gate amplifier is much smaller than its common-source and common-drain counterparts. Using the typical values $g_m = 4000\ \mu\text{S}$ and $R_S = 500\ \Omega$, we find $r_{in} = (1/4000\ \mu\text{S}) \| (500\ \Omega) = (250\ \Omega) \| (500\ \Omega) = 166.7\ \Omega$. To avoid the large reduction in voltage gain predicted by equation 7–126 when such a small value of r_{in} is used, it is clear that signal-source resistance r_S must be very small.

7–8 SMALL-SIGNAL MOSFET AMPLIFIERS

The most convenient small-signal model for a MOSFET, like that of a JFET, incorporates the transconductance of the device.

The transconductance of the *enhancement*-type MOSFET can be found graphically using the transfer characteristic and the definition:

$$g_m = \frac{\Delta I_D}{\Delta V_{GS}} \Big|\, V_{DS} = \text{constant} \tag{7–129}$$

Figure 7–67 shows how g_m is computed as the slope of a line drawn tangent to the characteristic at the operating or Q-point. It is clear from the figure that the slope of the characteristic, and hence the value of g_m, changes as the Q-point is changed. Therefore, small-signal analysis requires that the signal variation around the Q-point be confined to a limited range over which there is negligible change in g_m, i.e., to an essentially linear segment of the characteristic. Also, to ensure that the device is operated within its active region, the variation must be such that the following inequality is always satisfied:

FIGURE 7–67 The transconductance of an enhancement MOSFET is the slope of a line drawn tangent to the transfer characteristic at the Q-point

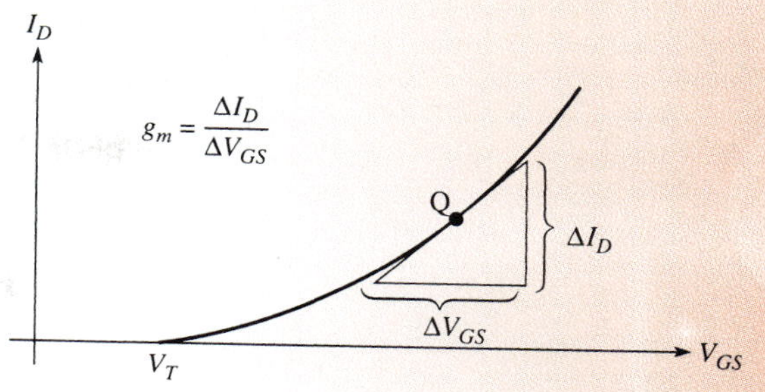

FIGURE 7–68 A common-source NMOS amplifier and its equivalent circuit

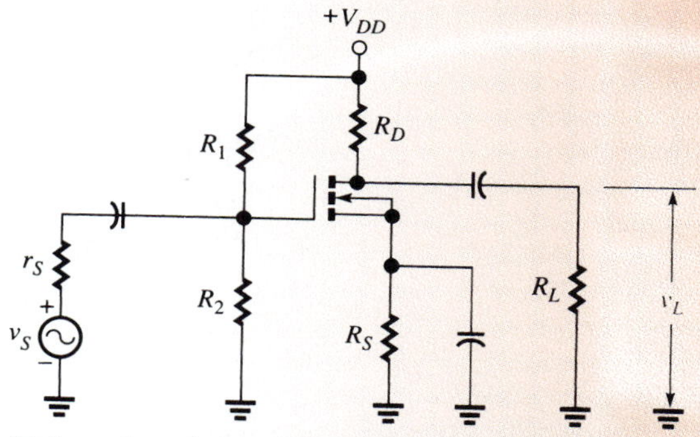

(a) An *n*-channel enhancement MOSFET (NMOS) amplifier with voltage-divider bias

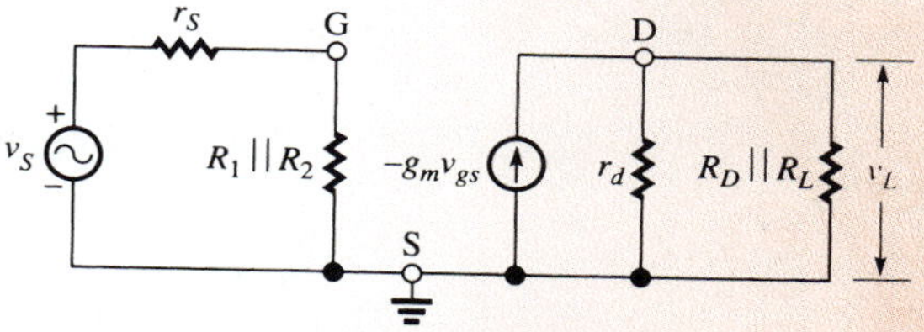

(b) The small-signal equivalent circuit of (a)

$$|V_{DS}| > |V_{GS} - V_T| \tag{7–130}$$

It can be shown that under small-signal conditions the transconductance of an enhancement MOSFET can be determined from

$$g_m = \beta(V_{GS} - V_T) \text{ siemens} \tag{7–131}$$

Figure 7–68 shows a common-source enhancement MOSFET amplifier and its small-signal equivalent circuit. The MOSFET is biased using the voltage-divider method (Figure 5–38), where

$$r_i = R_1 \parallel R_2 \tag{7–132}$$

$$\frac{v_L}{v_S} = \left(\frac{R_1 \parallel R_2}{r_S + R_1 \parallel R_2}\right)(-g_m)(r_d \parallel R_D \parallel R_L) \tag{7–133}$$

EXAMPLE 7–19

The MOSFET shown in Figure 7–69 has the following parameters: $V_T = 2$ V, $\beta = 0.5 \times 10^{-3}$, $r_d = 75$ kΩ. It is biased at $I_D = 1.93$ mA.

1. Verify that the MOSFET is biased in its active region.
2. Find the input resistance.
3. Draw the small-signal equivalent circuit and find the voltage gain, v_L/v_S.

FIGURE 7–69 (Example 7–19)

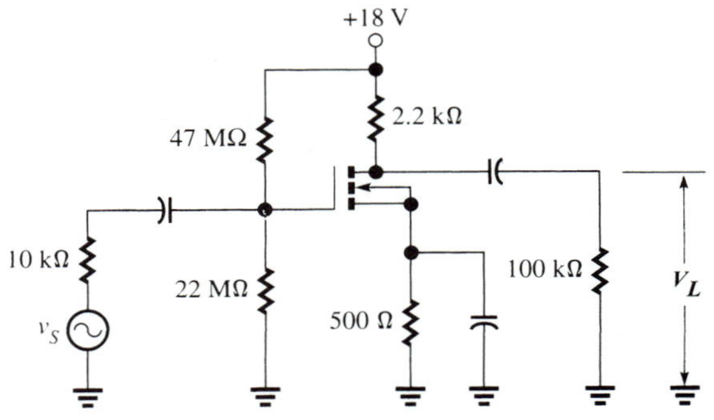

Solution

1. $V_{DS} = V_{DD} - I_D(R_D + R_S) = 18\text{ V} - (1.93\text{ mA})(2.2\text{ k}\Omega + 500\ \Omega) = 12.78\text{ V}$

$$V_G = \left(\frac{2.2\text{ M}\Omega}{4.7\text{ M}\Omega + 2.2\text{ M}\Omega}\right)18\text{ V} = 5.74\text{ V}$$

Using equation 5–14 to find V_{GS}, we have

$$V_{GS} = 5.74\text{ V} - (1.93\text{ mA})(500\ \Omega) = 4.78\text{ V}$$
$$|V_{GS} - V_T| = |4.78 - 2| = 2.78\text{ V}$$

Therefore, condition 7–130 is satisfied:

$$12.78 = |V_{DS}| > |V_{GS} - V_T| = 2.78$$

and we conclude that the MOSFET is biased in its active region.

2. $r_i = R_1 \parallel R_2 = (4.7\text{ M}\Omega) \parallel (2.2\text{ M}\Omega) = 1.5\text{ M}\Omega.$
3. From equation 7–131

$$g_m = 0.5 \times 10^{-3}(4.78\text{ V} - 2\text{V}) = 1.4 \times 10^{-3}\text{ S}$$

The small-signal equivalent circuit is shown in Figure 7–70. From equation 7–133

$$\frac{v_L}{v_S} = \left(\frac{15 \times 10^6}{10 \times 10^3 + 15 \times 10^6}\right)$$
$$\times (-1.4 \times 10^{-3})[(75 \times 10^3) \parallel (2.2 \times 10^3) \parallel (100 \times 10^3)]$$
$$= (0.999)(-1.4 \times 10^{-3})(2.09 \times 10^3) = -2.92$$

FIGURE 7–70 (Example 7–19) The small-signal equivalent circuit of Figure 7–69

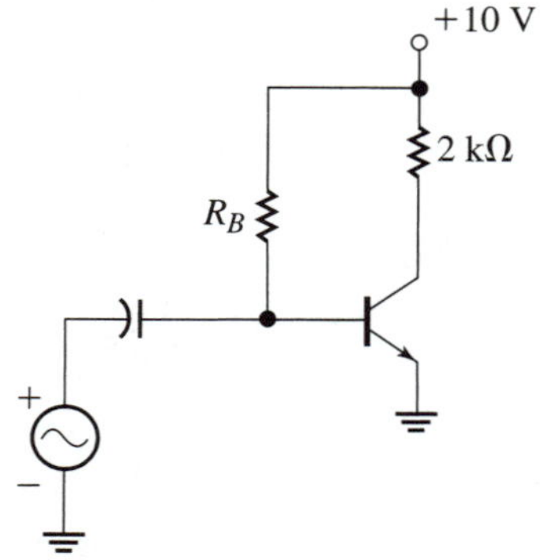

FIGURE 7–71 Common-emitter stage from Example 7–7

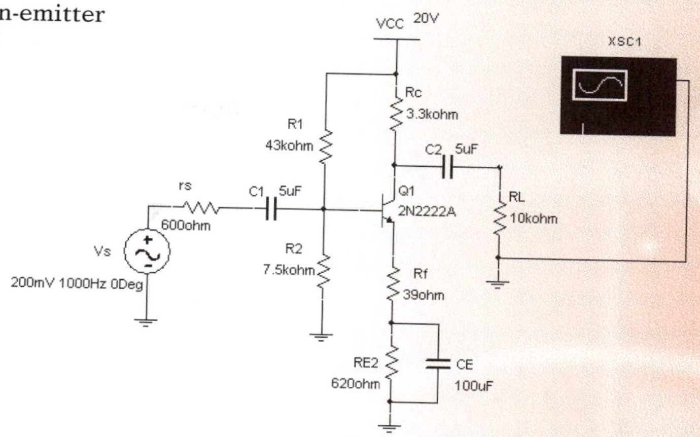

Like JFETs, MOSFETs can be operated in common-drain and common-gate configurations. Because the JFET and MOSFET small-signal models are identical, the gain and impedance equations are also identical, MOSFET amplifier circuits are discussed in greater detail in Chapter 8.

7–9 MULTISIM EXERCISE

We will test the design described in Example 7–7. Figure 7–71 shows the circuit as drawn on the EWB Multisim schematic editor. Note that the input voltage is 10 mV peak. Before you run the simulation, set up the oscilloscope to 200 mV/div for the vertical scale and 1 ms/div for the horizontal scale.

Observe the output waveform on the oscilloscope. You should read about 440 mV of peak voltage, which corresponds to an overall gain of 44 in magnitude. We had designed the amplifier stage for a gain of 45.

To obtain the maximum output voltage swing, change the peak value of the input voltage to 200 mV peak, the vertical scale to 5 V/div, run the simulation again, and observe the clipping of the waveform occurring on the lower side only. This means that the Q-point is not on the center of the ac line as expected; it is actually slightly above the center. Increase the input voltage to 220 mV and you will see that the upper side just begins to clip. Increase the input voltage to 300 mV, and both sides will be clipped. Calculate the expected maximum output voltage swing and verify that it agrees closely with the measured value.

If you measure the actual collector current, you will see that it is slightly larger than the nominal value used in the design. By changing R_2 to 6.8 kΩ, the collector current will be reduced slightly and the clipping will occur with more symmetry.

SUMMARY

This chapter presented detailed information regarding transistor amplifiers of both bipolar junction and field-effect types. Emphasis in the use of models was consistently present throughout the chapter. At the end of this chapter the student should have learned the following concepts:

- Transistors must be properly biased in the center of their active region before they can be used as amplifiers.
- Small-signal voltage or current refers to a signal level that will confine the operation of an amplifier to a relatively small portion of its linear operation region.
- The localization of the Q-point has a direct effect on the linearity and maximum output swing of a transistor amplifier stage.

- Each transistor configuration (CE, CB, CC, CS, CG, CD) has its own model and parameters, and presents unique characteristics that distinguishes it from the others.
- Directly coupling stages reduces the number of required components as compared to capacitively coupled stages.
- The Darlington configuration can be seen as a super-beta transistor with higher common-emitter input resistance.
- The different transistor models have many characteristics in common, and their equivalencies can be easily established.

EXERCISES

SECTION 7–2

Analysis of the CE Amplifier

7–1. Suppose the signal source voltage v_S in Figure 7–72 causes the transistor base current to vary sinusoidally between 10 μA and 50 μA. Assuming that the transistor has the output characteristics shown in Figure 7–76, what is the total (peak-to-peak) variation in V_{CE} and the total variation in I_C?

7–2. Assuming that the transistor shown in Figure 7–73 has the input characteristics shown in Figure 7–75 and the output characteristics shown in Figure 7–76, find

(a) V_{CE} and I_C at the Q-point; and

(b) the voltage gain.

7–3. If R_B in Figure 7–73 is changed to 1.934 MΩ, what is the maximum peak-to-peak output voltage without clipping?

7–4. At V_{CE} = 10 V, the CE output characteristics of a certain transistor show that I_C has the values listed in the following table for different values of I_B:

I_B	I_C
10 μA	0.5 mA
15 μA	0.75 mA
20 μA	1.00 mA
25 μA	1.25 mA
30 μA	2.10 mA
35 μA	3.15 mA
40 μA	4.00 mA

At the quiescent voltage V_{CE} = 10 V, what maximum range of I_B values results in the output being a linear reproduction of the input?

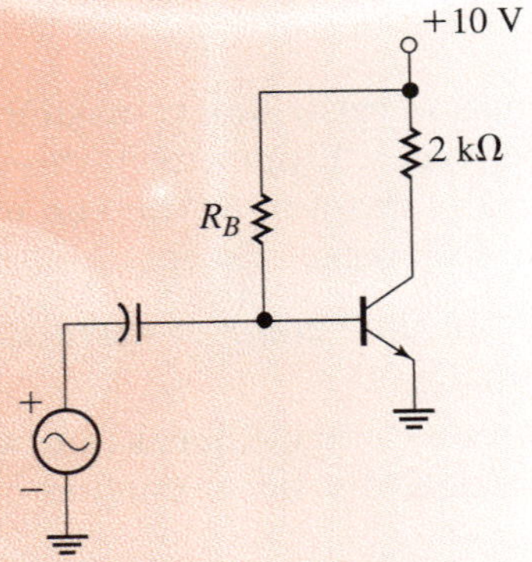

FIGURE 7–72 (Exercise 7–1)

7–5. Show that equations 7–14 and 7–15 correctly give the intercepts of a dc load line when R_C is substituted for r_L. (This shows that the dc load line is a special case of the ac load line, the case where $R_L = \infty$.)

7–6. The transistor shown in Figure 7–74 has the CE output characteristics shown in Figure 7–76.

(a) Plot the dc and ac load lines on the output characteristics.

(b) Determine the voltage gain of the circuit if a 24-mV p-p input voltage causes a 20-μA p-p variation in base current.

7–7. Repeat Exercise 7–6 when R_L is changed to 10.5 kΩ.

SECTION 7–3

Amplifier Analysis Using Small-Signal Models

7–8. The silicon transistor in the amplifier stage shown in Figure 7–77 has an α of 0.99 and collector resistance r_c = 2.5 MΩ. Find

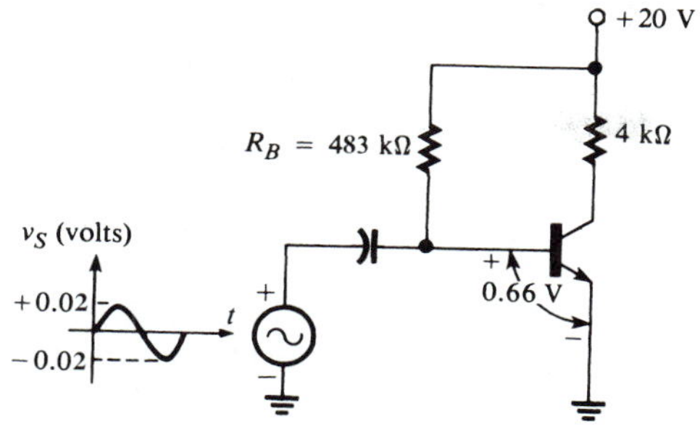

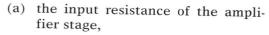

FIGURE 7–73 (Exercises 7–2 and 7–3)

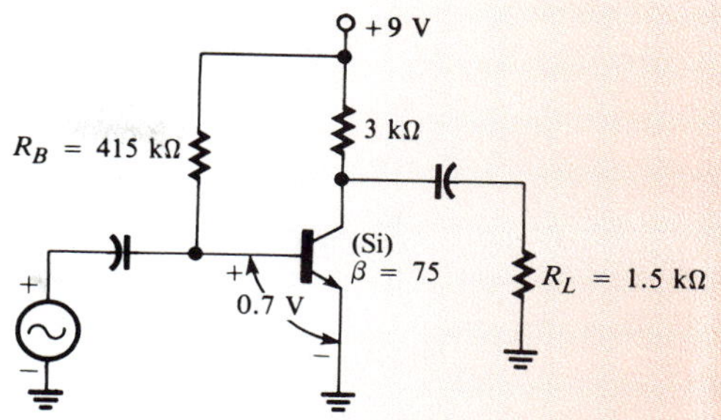

FIGURE 7–74 (Exercise 7–6)

(a) the input resistance of the amplifier stage,

(b) the output resistance of the amplifier stage,

(c) the voltage gain of the amplifier, and

(d) the current gain of the amplifier.

7–9. The silicon transistor in the amplifier stage shown in Figure 7–78 has a collector resistance $r_c = 1.5$ MΩ. Find

(a) the input resistance of the amplifier stage,

(b) the output resistance of the amplifier stage, and

(c) the rms load voltage v_L.

7–10. Find the voltage gain of the amplifier shown in Figure 7–79. The transistor is germanium.

7–11. What value of R_L in the amplifier of Exercise 7–10 would result in a voltage gain of 100?

7–12. Find the rms load voltage v_L in the amplifier of Figure 7–80 when R_L = (a) 1 kΩ, (b) 10 kΩ, and (c) 100 kΩ. Assume that β = 100.

7–13. Find the maximum peak-to-peak output voltage swing for each value of R_L in Exercise 7–12.

FIGURE 7–75 (Exercise 7–2)

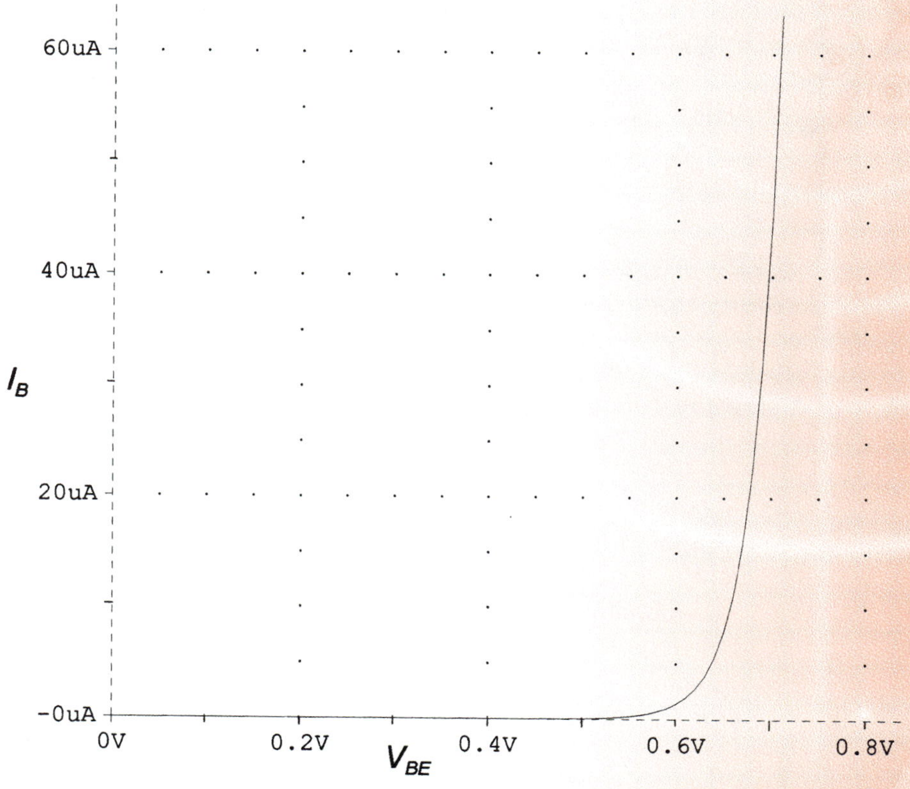

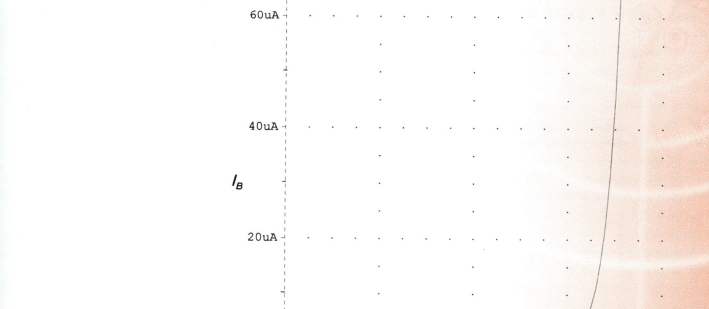

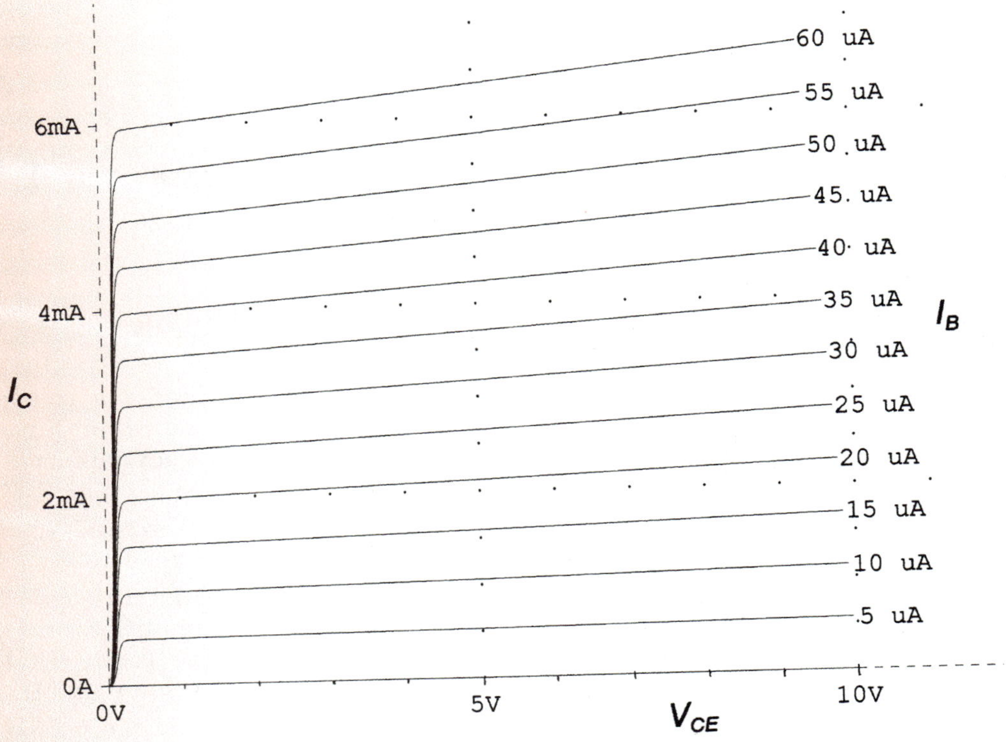

FIGURE 7–76 (Exercise 7–6)

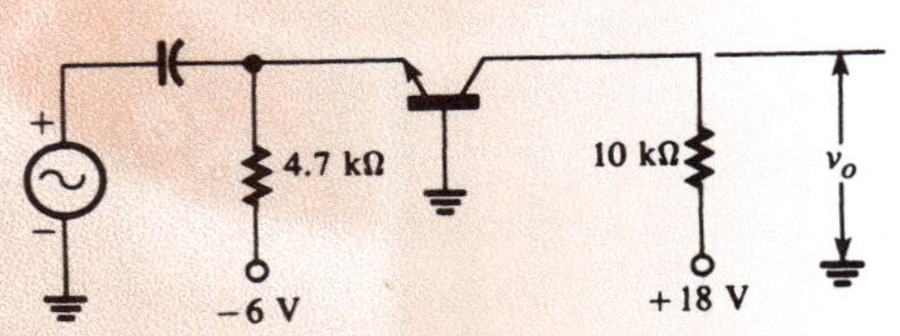

FIGURE 7–77 (Exercise 7–8)

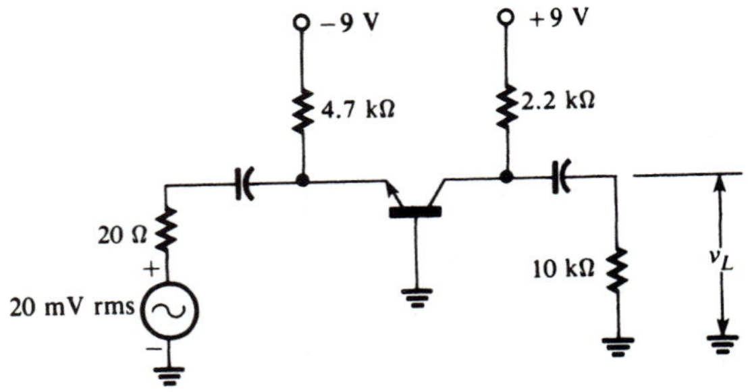

FIGURE 7–78 (Exercise 7–9)

7–14. Solve equations 7–16 and 7–17 simultaneously to derive the expressions for I_Q and V_Q at the bias point where maximum peak-to-peak output can be achieved.

7–15. Using the commercial values for resistance determined in Example 7–7, determine the actual

(a) Maximum Q-point collector current,

(b) Minimum Q-point collector current, and

(c) Maximum unclipped p-p output voltage across the load.

7–16. For the amplifier in Figure 7–81, find the value of R_1 and R_2 that results in maximum peak-to-peak output swing. Assume $R_B = R_1 \| R_2 = 8\ \text{k}\Omega$.

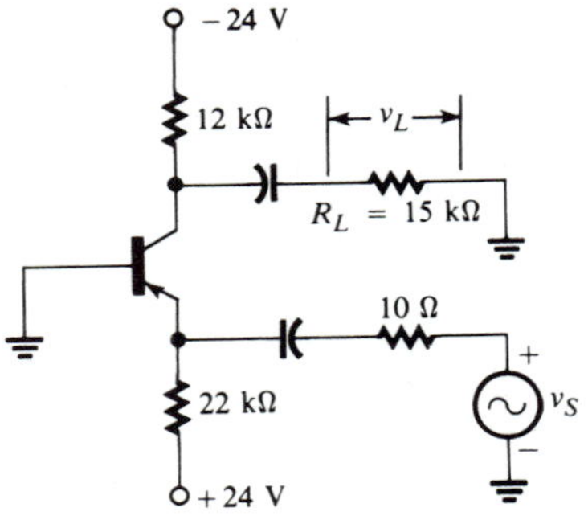

FIGURE 7–79 (Exercise 7–10)

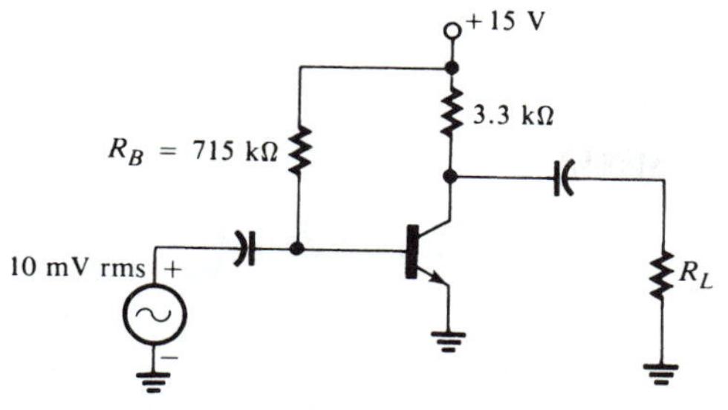

FIGURE 7–80 (Exercise 7–12)

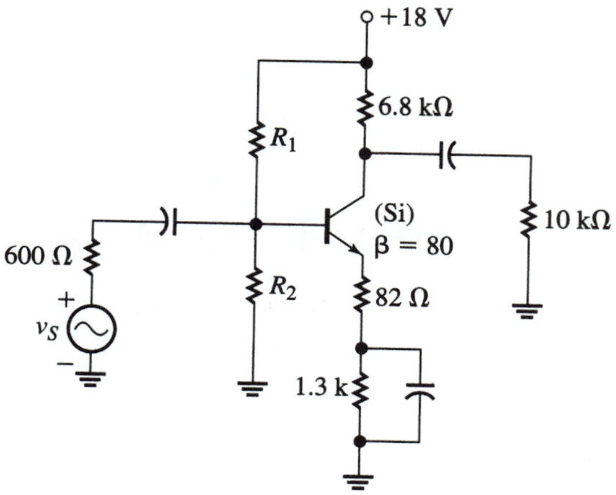

FIGURE 7–81 (Exercise 7–16)

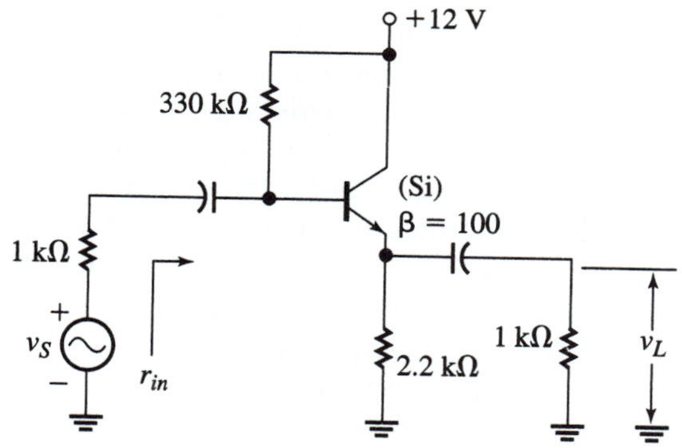

FIGURE 7–82 (Exercise 7–17)

7–17. For the amplifier circuit shown in Figure 7–82, find
(a) r_{in}
(b) r_{out}
(c) A_v and
(d) A_{vs}

7–18. Find the maximum unclipped peak-to-peak load voltage for the circuit in Figure 7–82.

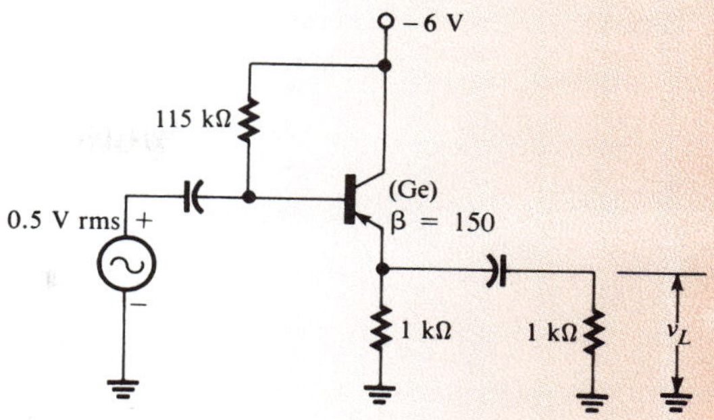

FIGURE 7–83 (Exercise 7–19)

7–19. Find the output voltage v_L in Figure 7–83.

7–20. The transistors in Figure 7–84 have the following parameter values:

Q_1: $r'_{e1} = 12\ \Omega$, $\beta_1 = 200$, $r_{c1} = 2\ \text{M}\Omega$

Q_2: $r'_{e2} = 10\ \Omega$, $\beta_2 = 100$

Find the midband values of
(a) the voltage gain v_L/v_S; and
(b) the current gain i_L/i_S.

SECTION 7–4

Direct Coupling

7–21. The silicon transistors in Figure 7–85 have $\beta_1 = 120$ and $\beta_2 = 110$. Find (a) r_{in}, (b) v_L/v_S, and (c) i_L/i_S.

7–22. The silicon transistors in Figure 7–86 have $\beta_1 = \beta_2 = 50$. The effective collector resistance (at Q_2) is 10 kΩ. Find approximate values for (a) r_{in} and (b) V_L/V_S.

SECTION 7–5

Other Small-Signal Models

7–23. Find the (approximate) transconductance of the transistor in the circuit shown in Figure 7–87 at room temperature, when
(a) $R_B = 330\ \text{k}\Omega$ and $\beta = 50$,
(b) $R_B = 330\ \text{k}\Omega$ and $\beta = 150$, and
(c) $R_B = 220\ \text{k}\Omega$ and $\beta = 50$.

7–24. Assuming that the Early voltage of the transistor in Exercise 7–23 is 100 V, find the output resistance of (a) the transistor and (b) the stage for each of the combinations of parameters given in Exercise 7–23.

7–25. Draw the ac equivalent circuit of the amplifier, shown in Figure 7–88, using the hybrid-π model for the transistor. Label all component values in the circuit

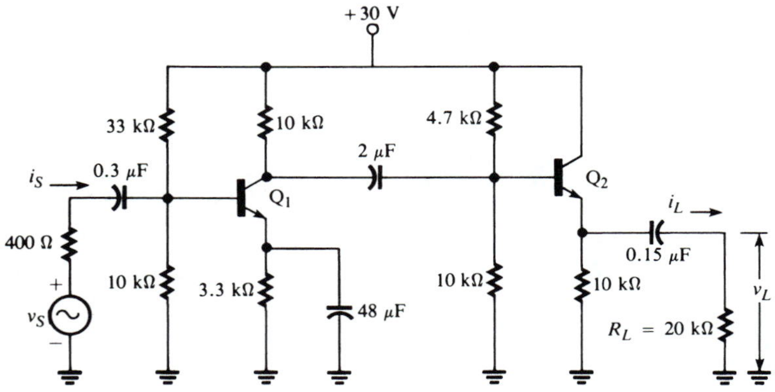

FIGURE 7–84 (Exercise 7–20)

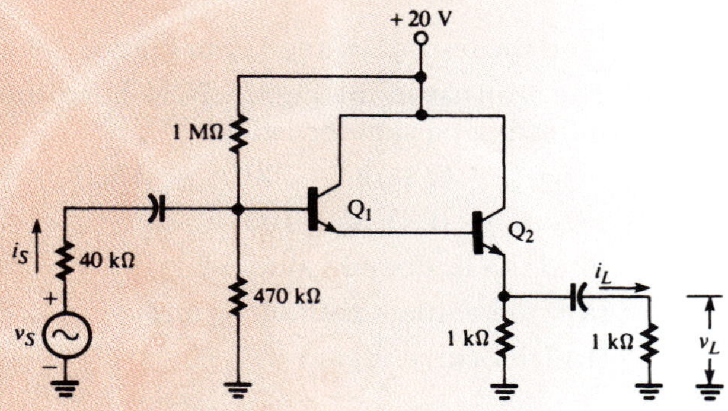

FIGURE 7–85 (Exercise 7–21)

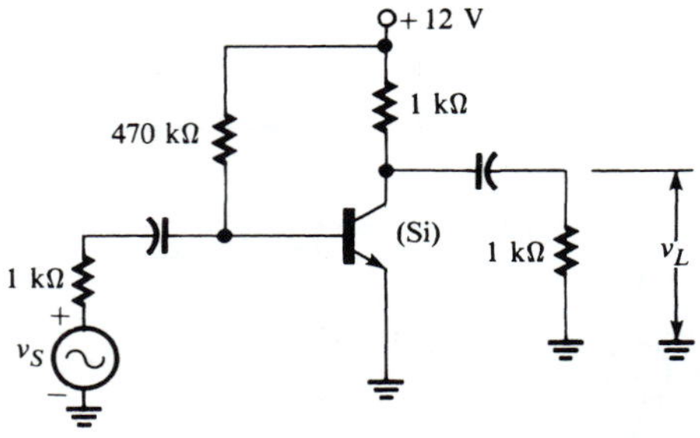

FIGURE 7–88 (Exercise 7–25)

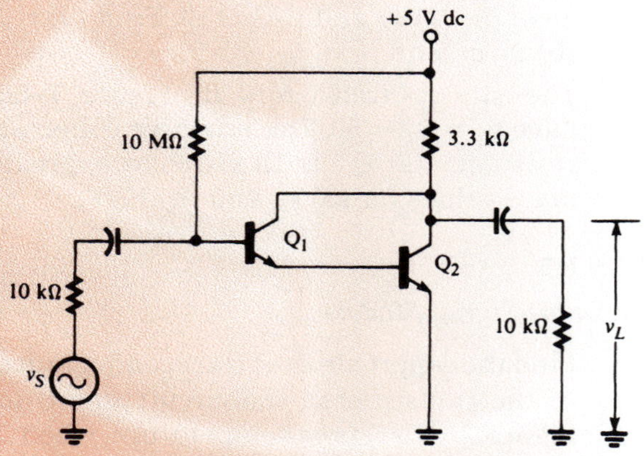

FIGURE 7–86 (Exercise 7–22)

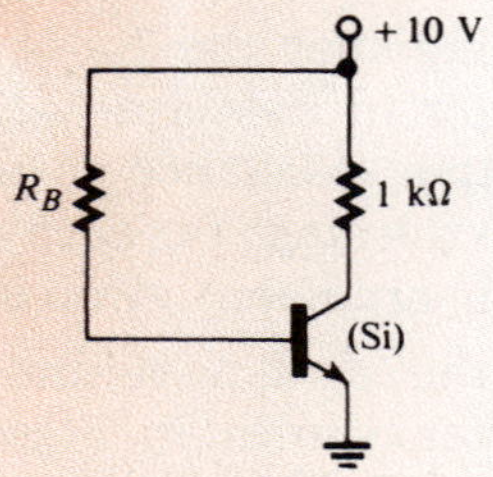

FIGURE 7–87 (Exercise 7–23)

and the model. Assume that $\beta = 100$ and $V_A = 100$ V.

7–26. Use the equivalent circuit drawn for Exercise 7–25 to find v_L/v_S.

7–27. Using the manufacturer's small-signal specifications furnished in Section 7–5 find the percent change in h_{fe}, h_{re}, h_{ie}, and h_{oe} for the 2N2222A transistor when the dc collector current is changed from 0.5 mA to 2.0 mA. Assume a "high-gain" unit.

7–28. Find the percent change in each of the following parameters of the transistor in Exercise 7–27 when I_C is changed from 0.5 mA to 2.0 mA: β, g_m, r_e', r_π, g_o.

7–29. A circuit has been designed using the 2N2221A transistor to amplify 1-kHz signals. After each unit is manufactured, the bias point of the transistor is adjusted so that $I_C = 1$ mA and $V_{CE} = 10$ V. What are the ranges of values of β and g_m that could be expected for the transistors that are used in the circuit?

7–30. The transistor shown in Figure 7–89 has the following h parameters: $h_{fe} = 120$, $h_{ie} = 2.2\ \text{k}\Omega$, $h_{oe} = 40\ \mu\text{S}$, and $h_{re} = 6 \times 10^{-4}$. Draw the complete ac equivalent circuit and find the value of the load voltage v_L.

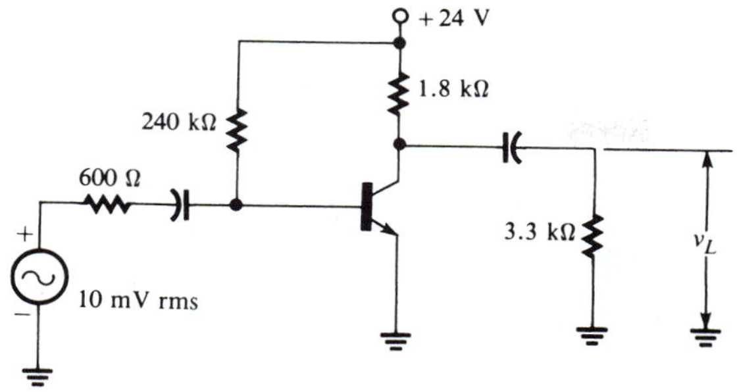

FIGURE 7–89 (Exercise 7–30)

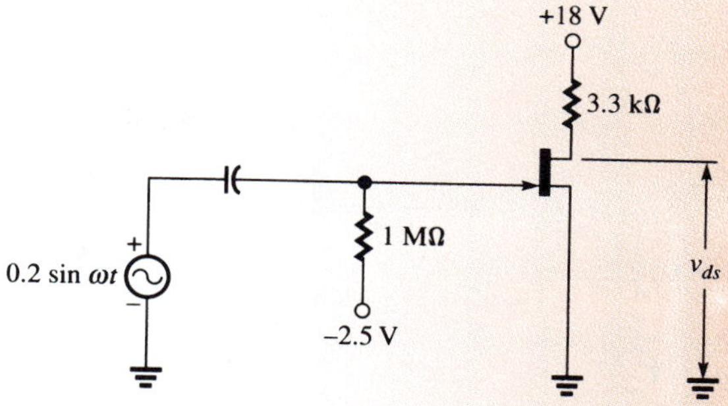

FIGURE 7–90 (Exercise 7–37)

SECTION 7–6

The Common-Source JFET Amplifier

7–31. (a) What is the value of the transconductance of a JFET when its gate-to-source voltage equals its pinch-off voltage?

(b) At what value of I_D does $g_m = 0$?

7–32. Find the transconductance of the JFET whose transfer characteristic is shown in Figure 7–55, when $I_D = 1$ mA both (a) graphically and (b) algebraically.

7–33. Find the transconductance of the JFET whose transfer characteristic is shown in Figure 7–55, when $V_{GS} = -0.5$ V both (a) graphically and (b) algebraically.

7–34. Use equation 7–100 and the equation of the transfer characteristic of a JFET to derive equation 7–101.

7–35. The maximum transconductance of a certain n-channel JFET is 9.8×10^{-3} S. If $I_{DSS} = 18$ mA, what is its pinch-off voltage?

7–36. The following measurements were taken from a curve-tracer display of the output characteristics of a p-channel JFET along the line $V_{GS} = +2.5$ V: $I_D = 4.2$ mA at $V_{DS} = -4.5$ V; $I_D = 4.3$ mA at $V_{DS} = -12$ V. What is the drain resistance at $V_{GS} = 2.5$ V?

7–37. The JFET in the amplifier shown in Figure 7–90 has $I_{DSS} = 16$ mA, $V_p = -4.5$ V, and $r_d = 80$ kΩ.

(a) Find the quiescent values of I_D and V_{DS}.

(b) Find g_m.

(c) Draw the ac equivalent circuit.

(d) What is the peak value of the ac component of v_{ds}?

7–38. The JFET in the amplifier shown in Figure 7–91 has $I_{DSS} = 14$ mA, $V_p = 3$ V, and $r_d = 120$ kΩ.

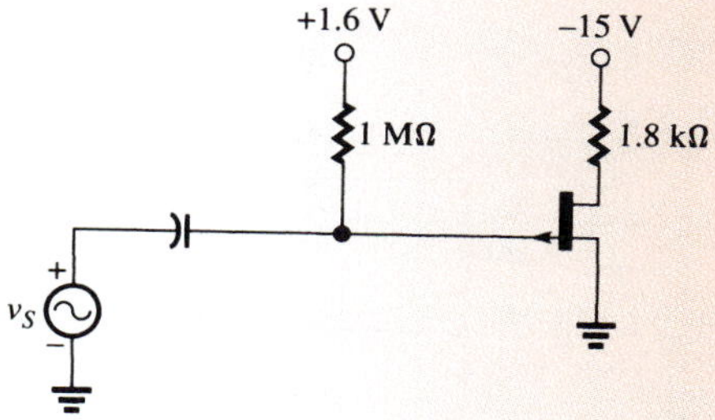

FIGURE 7–91 (Exercise 7–38)

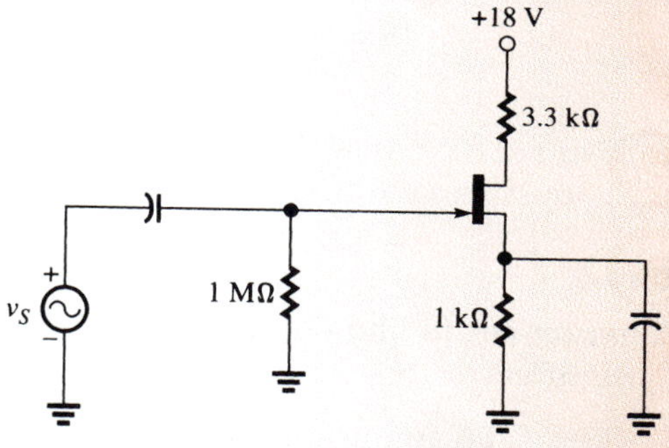

FIGURE 7–92 (Exercise 7–39)

(a) Find the quiescent values of I_D and V_{DS}.

(b) Find g_m.

(c) Draw the ac equivalent circuit.

(d) Find the voltage gain.

7–39. The JFET in the amplifier shown in Figure 7–92 has $r_d = 75$ kΩ, $V_p = -3.6$ V, and $I_{DSS} = 9$ mA. The dc voltage drop across the 3.3-kΩ resistor is 6.38 V. Find the voltage gain of the amplifier.

7–40. The JFET in the amplifier shown in Figure 7–93 has $r_d = 60$ kΩ, $V_p = 3.9$ V, and $I_{DSS} = 10$ mA. The quiescent value of V_{DS} is -9.24 V. Find the voltage gain v_L/v_S.

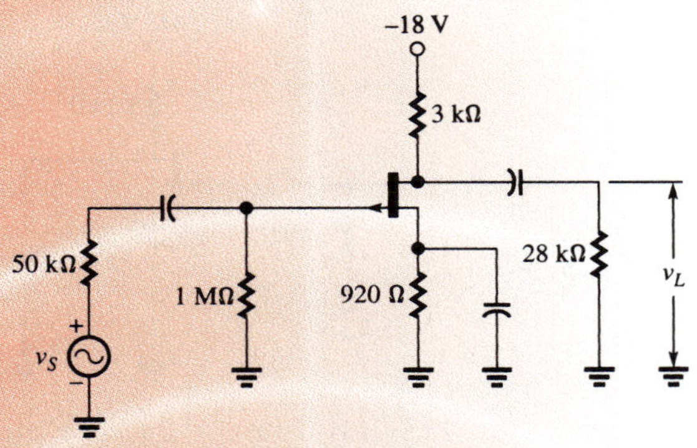

FIGURE 7–93 (Exercise 7–40)

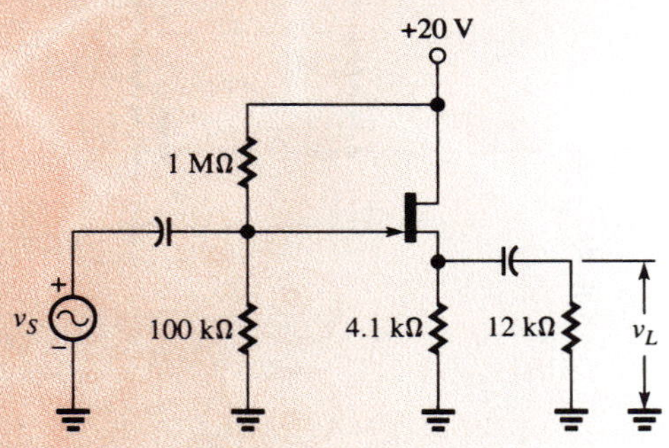

FIGURE 7–94 (Exercise 7–42)

7-41. Repeat Exercise 7–39 when the source bypass capacitor is removed.

SECTION 7–7

The Common-Drain and Common-Gate JFET Amplifiers

7–42. The JFET in the amplifier shown in Figure 7–94 has $g_m = 0.004$ S and $r_d = 90$ kΩ.

(a) Find the voltage gain v_L/v_S.

(b) Find the input resistance.

7–43. The JFET in the amplifier shown in Figure 7–95 has $g_m = 5200$ μS and $r_d = 80$ kΩ.

(a) Find the voltage gain v_L/v_S.

(b) Find the output resistance.

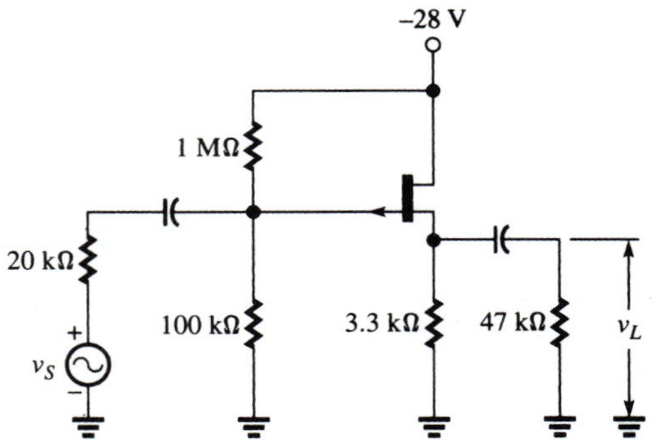

FIGURE 7–95 (Exercise 7–43)

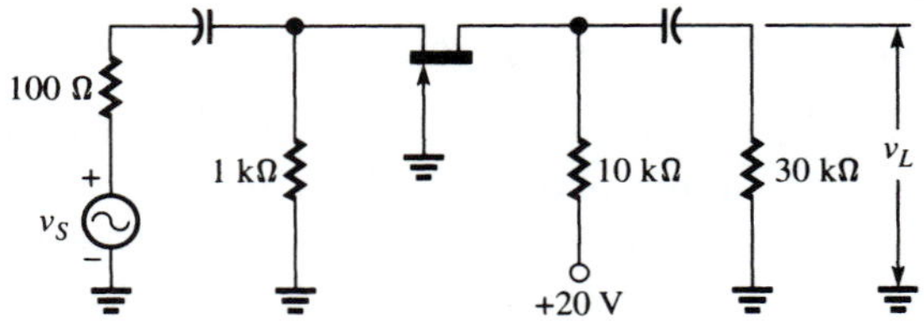

FIGURE 7–96 (Exercise 7–44)

7–44. The JFET in the amplifier in Figure 7–96 has $g_m = 3.58$ mS and $r_d = 100$ kΩ. Find the voltage gain v_L/v_S.

7–45. The amplifier in Figure 7–66(a) has $r_S = 50$ Ω, $R_S = 680$ Ω, $R_D = 2$ kΩ, $R_L = 4$ kΩ, and $V_{DD} = 15$ V. The pinch-off voltage of the JFET is −3.9 V and the saturation current is 15 mA. The dc voltage drop across R_D is 6.24 V. Find the voltage gain v_L/v_S. Drain resistance r_d can be neglected.

SECTION 7–8

Small-Signal MOSFET Amplifiers

7–46. Graphically determine the transconductance of the MOSFET whose transfer characteristic is shown in Figure 7–97 when $V_{GS} = 7$ V.

7–47. Algebraically determine the transconductance of the MOSFET in Exercise 7–46 given that its β is 0.5×10^{-3}.

SPICE EXERCISES

Note: In the exercises that follow, assume that all devices have their default values unless otherwise specified. The capacitor values and frequencies specified in these exercises are such that capacitive reactances are negligible and have no effect on amplifier performance.

7–48. Use SPICE to determine the quiescent point, voltage gain v_L/v_S, and current gain i_L/i_S of the common-emitter amplifier shown in Figure 7–26. The transistor has a beta of 200. Component values are: $r_S = 1$ kΩ, $R_1 = 47$ kΩ, $R_2 = 8.2$ kΩ $R_E = 470$ Ω, $R_C = 2.2$ kΩ, and $R_L = 10$ kΩ. Both coupling capacitors are 10 μF, and $V_{CC} = 18$ V. The analysis should be performed at 10 kHz.

FIGURE 7–97 (Exercise 7–46)

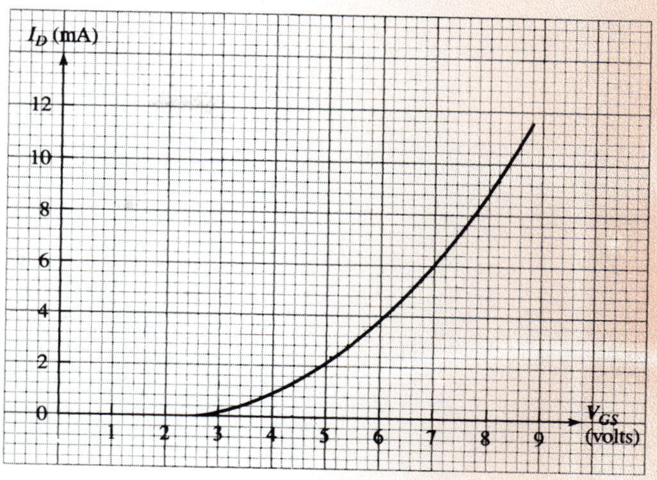

7–49. Use SPICE to determine the quiescent point, voltage gain v_L/v_S, and current gain i_L/i_S of the common-base amplifier shown in Figure 7–21(a). The transistor has a beta of 80. Component values are: r_S = 100 Ω, R_E = 2.2 kΩ, R_C = 6.2 kΩ, and R_L = 10 kΩ. The input coupling capacitor is 100 μF, and the output coupling capacitor is 10 μF. V_{CC} = 36 V and V_{EE} = 6 V. The analysis should be performed at 10 kHz.

7–50. Use SPICE to determine the quiescent point, voltage gain v_L/v_S, and current gain i_L/i_S of the common-collector amplifier shown in Figure 7–29(a). The transistor has a beta of 150. Component values are: r_S = 1 kΩ, R_B = 180 kΩ, R_E = 1 kΩ, and R_L = 300 Ω. The input coupling capacitor is 10 μF, and the output coupling capacitor is 100 μF. V_{CC} = 9 V.

7–51. Use SPICE to determine r_i and r_{in} (see Figure 7–29(b)) of the common-collector amplifier in Exercise 7–50. (*Hint*: Find ac voltages and currents at appropriate points in the circuit.)

CHAPTER 8

IDEAL OPERATIONAL AMPLIFIER CIRCUITS AND ANALYSIS

OUTLINE

OBJECTIVES

- Establish the ideal characteristics of an ideal operational amplifier.
- Establish the main amplifier configurations.
- Develop an understanding of inverting and noninverting amplification.
- Differentiate between the different summing and subtracting configurations.
- Recognize the various methods for implementing controlled sources.
- Distinguish the use of dual power supply and single power supply.

8–1 THE IDEAL OPERATIONAL AMPLIFIER

An operational amplifier is a direct-coupled amplifier with two (differential) inputs and a single output. It normally requires to be powered by a dual power supply ($+V$ and $-V$ with respect to ground) although later in the chapter we will look at how to connect an operational amplifier to a single power supply. We will define an *ideal* operational amplifier to be one that has the following attributes:

1. It has infinite gain.
2. It has infinite input impedance.
3. It has zero output impedance.
4. It has infinite bandwidth.

Although no real amplifier* can satisfy any of these requirements, we will see that most modern amplifiers have such large gains and input impedances, and such small output impedances, that a negligibly small error results from assuming ideal characteristics. A detailed study of the ideal amplifier will therefore be beneficial in terms of understanding how practical amplifiers are used as well as in building some important theoretical concepts that have broad implications in many areas of electronics.

Figure 8–1 shows the standard symbol for an operational amplifier. Note that the two inputs are labeled "+" and "−" and the input signals are correspondingly designated v_i^+ and v_i^-. In relation to our previous discussion of differential amplifiers, these inputs correspond to v_{i1} and v_{i2}, respectively, when the single-ended output is v_{o2}. The + input is called the *noninverting* input and the − input is called the *inverting input.* In many applications, one of the amplifier inputs is grounded, so v_o is in phase with the input if the signal is connected to the noninverting terminal, and v_o is out of phase with the input if the signal is connected to the inverting input. These ideas are summarized in the table accompanying Figure 8–1.

At this point, a legitimate question that may have already occurred to the reader is this: If the gain is infinite, how can the output be anything

*In this chapter, we will hereafter use the word *amplifier* with the understanding that operational amplifier is meant. We will also use the term *op-amp,* which is widely used in books, papers, and technical literature.

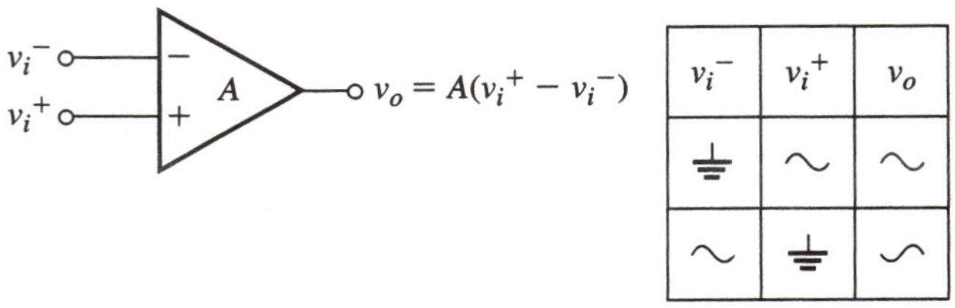

FIGURE 8–1 Operational amplifier symbol, showing inverting (−) and noninverting (+) inputs

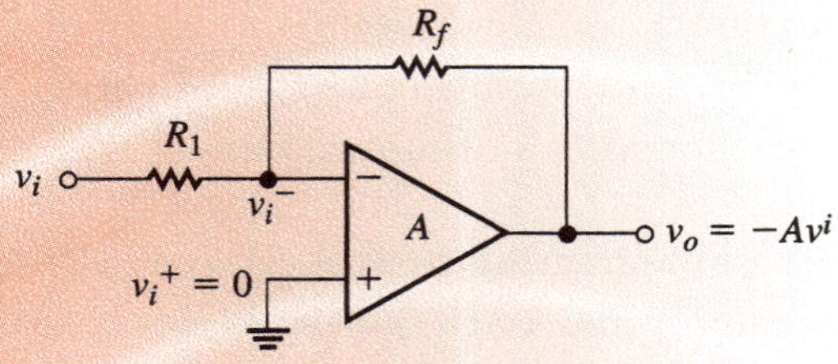

FIGURE 8–2 An operational-amplifier application in which signal v_i is connected through R_1. Resistor R_f provides feedback. $v_o/v_i^- = -A$.

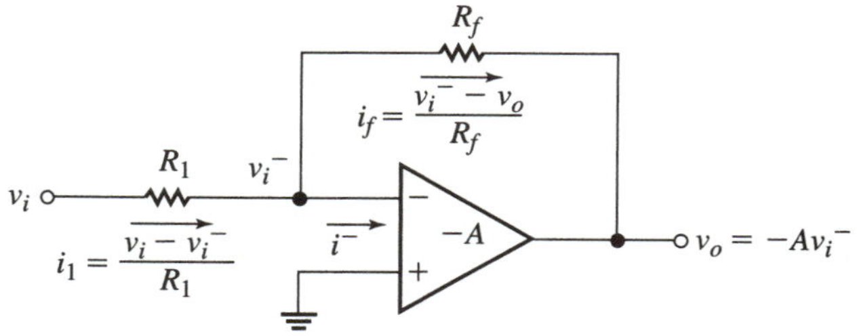

FIGURE 8–3 Voltages and currents resulting from the application of the signal voltage v_i

other than a severely clipped waveform? Theoretically, if the amplifier has infinite gain, an infinitesimal input voltage must result in an infinitely large output voltage. The answer, of course, is that the gain is not truly infinite, just very large. Nevertheless, it *is* true that a very small input voltage will cause the amplifier output to be driven all the way to its extreme positive or negative voltage limit. The practical answer is that an operational amplifier is seldom used in such a way that the full gain is applied to an input. Instead, external resistors are connected to and around the amplifier in such a way that the signal undergoes vastly smaller amplification. The resistors cause gain reduction through signal *feedback*, which we will soon study in considerable detail.

The Inverting Amplifier

Consider the configuration shown in Figure 8–2. In this very useful application of an operational amplifier, the noninverting input is grounded, v_i is connected through R_1 to the inverting input, and feedback resistor R_f is connected between the output and v_i^-. Let A denote the voltage gain of the amplifier: $v_o = A(v_i^+ - v_i^-)$. Since $v_i^+ = 0$, we have

$$v_o = -Av_i^- \tag{8–1}$$

(Note that $v_i \neq v_i^-$.) We wish to investigate the relation between v_o and v_i when the magnitude of A is infinite.

Figure 8–3 shows the voltages and currents that result when signal v_i is connected. From Ohm's law, the current i_1 is simply the difference in voltage across R_1, divided by R_1:

$$i_1 = (v_i - v_i^-)/R_1 \tag{8–2}$$

Similarly, the current i_f is the difference in voltage across R_f, divided by R_f:

$$i_f = (v_i^- - v_o)/R_f \tag{8–3}$$

Writing Kirchhoff's current law at the inverting input, we have

$$i_1 = i_f + i^- \tag{8–4}$$

where i^- is the current entering the amplifier at its inverting input. However, the ideal amplifier has infinite input impedance, which means i^- must be 0. So (8–4) is simply

$$i_1 = i_f \tag{8–5}$$

Substituting (8–2) and (8–3) into (8–5) gives

$$\frac{v_i - v_i^-}{R_1} = \frac{v_i^- - v_o}{R_f}$$

or

$$\frac{v_i}{R_1} - \frac{v_i^-}{R_1} = \frac{v_i^-}{R_f} - \frac{v_o}{R_f} \tag{8–6}$$

From equation 8–1,

$$v_i^- = -\frac{v_o}{A} \tag{8–7}$$

If we now invoke the assumption that $A = \infty$, we see that $-v_o/A = 0$ and, therefore,

$$v_i^- = 0 \text{ (ideal amp, with } A = \infty) \tag{8–8}$$

Substituting $v_i^- = 0$ into (8–6) gives

$$\frac{v_i}{R_1} = \frac{-v_o}{R_f}$$

or

$$\frac{v_o}{v_i} = \frac{-R_f}{R_1} \tag{8–9}$$

We see that the gain is negative, signifying that the configuration is an *inverting* amplifier. Equation 8–9 also reveals the exceptionally useful fact that the magnitude of v_o/v_i *depends only on the ratio of the resistor values* and not on the amplifier itself. Provided the amplifier gain and impedance remain quite large, variations in amplifier characteristics (due, for example, to temperature changes or manufacturing tolerance) do not affect v_o/v_i. For example, if $R_1 = 10$ kΩ and $R_f = 100$ kΩ, we can be certain that $v_o = -[(100\text{ k}\Omega)/(10\text{ k}\Omega)]v_i = -10\, v_i$, i.e., that the gain is as close to -10 as the resistor precision permits. The gain v_o/v_i is called the *closed-loop gain* of the amplifier, and A is called the *open-loop gain*. In this application, we see that an extremely large open-loop gain, perhaps 10^6, is responsible for giving us the very predictable, though much smaller, closed-loop gain equal to 10. This is the essence of most operational-amplifier applications: Trade the very large gain that is available for less spectacular but more precise and predictable characteristics.

In our derivation, we used the infinite-gain assumption to obtain $v_i^- = 0$ (equation 8–8). In real amplifiers, having very large, but finite, values of A, v_i^- is a very small voltage, near zero. For that reason, the input terminal where the feedback resistor is connected is said to be at *virtual ground*. For *analysis* purposes, we often assume that $v_i^- = 0$, but we cannot actually ground that point. Because v_i^- is at virtual ground, the impedance seen by the signal source generating v_i is R_1 ohms.

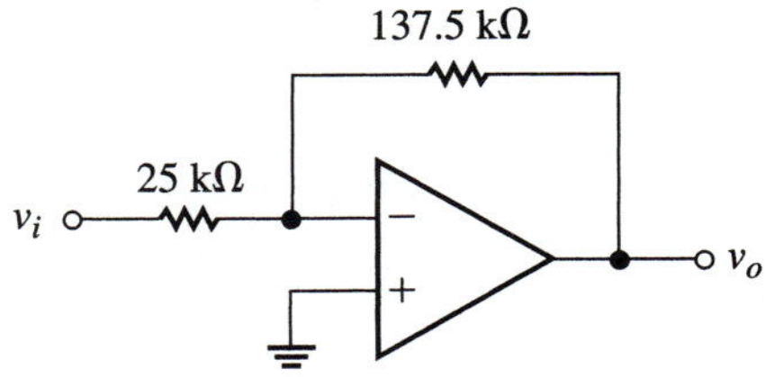

FIGURE 8–4 (Example 8–1)

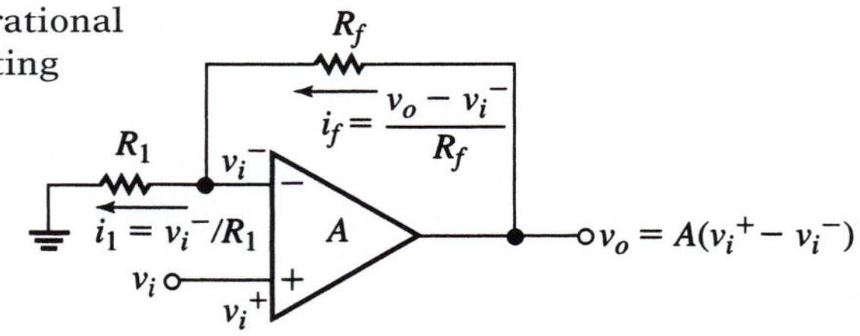

FIGURE 8–5 The operational amplifier in a noninverting configuration

EXAMPLE 8–1

Assuming that the operational amplifier in Figure 8–4 is ideal, find

1. the rms value of v_o when v_i is 1.5 V rms;
2. the rms value of the current in the 25-kΩ resistor when v_i is 1.5 V rms; and
3. the output voltage when $v_i = -0.6$ V dc.

Solution

1. From equation 8–9,

$$\frac{v_o}{v_i} = \frac{-R_f}{R_1} = \frac{-137.5\text{ k}\Omega}{25\text{ k}\Omega} = -5.5$$

Thus, $|v_o| = 5.5|v_i| = 5.5(1.5\text{ V rms}) = 8.25\text{ V rms}$.

2. Since $v_i \approx 0$ (virtual ground), the current in the 25-kΩ resistor is

$$i = \frac{v_i}{R_1} = \frac{1.5\text{ V rms}}{25\text{ k}\Omega} = 60\ \mu\text{A rms}$$

3. $v_o = (-5.5)v_i = (-5.5)(-0.6\text{ V}) = 3.3$ V dc. Notice that the output is a positive dc voltage when the input is a negative dc voltage, and vice versa.

The Noninverting Amplifier

Figure 8–5 shows another useful application of an operational amplifier, called the *noninverting* configuration. Notice that the input signal v_i is connected directly to the noninverting input and that resistor R_1 is connected from the inverting input to ground. Under the ideal assumption of infinite input impedance, no current flows into the inverting input, so $i_1 = i_f$. Thus,

$$\frac{v_i^-}{R_1} = \frac{v_o - v_i^-}{R_f} \tag{8–10}$$

Now, as shown in the figure,

$$v_o = A(v_i^+ - v_i^-) \tag{8–11}$$

Solving (8–11) for v_i^- gives

$$v_i^- = v_i^+ - v_o/A \tag{8–12}$$

Letting $A = \infty$, the term v_o/A goes to 0, and we have

$$v_i^- = v_i^+ \tag{8–13}$$

Substituting v_i^+ for v_i^- in (8–10) gives

$$\frac{v_i^+}{R_1} = \frac{v_o - v_i^+}{R_f} \tag{8–14}$$

Solving for v_o/v_i^+ and recognizing that $v_i^+ = v_i$ lead to

$$\frac{v_o}{v_i} = \frac{R_1 + R_f}{R_1} = 1 + \frac{R_f}{R_1} \tag{8–15}$$

We saw (equation 8–8) that when an operational amplifier is connected in an inverting configuration, with $v_i^+ = 0$, the assumption $A = \infty$ gives $v_i^- = 0$ (virtual ground), i.e., $v_i^- = v_i^+$. Also, in the noninverting configuration, the same assumption gives the same result: $v_i^- = v_i^+$ (equation 8–13). Thus, we reach the important general conclusion that *feedback in conjunction with a very large voltage gain forces the voltages at the inverting and noninverting inputs to be approximately equal.*

Equation 8–15 shows that the closed-loop gain of the noninverting amplifier, like that of the inverting amplifier, depends only on the values of external resistors. A further advantage of the noninverting amplifier is that the input impedance seen by v_i is infinite, or at least extremely large in a real amplifier. The inverting and noninverting amplifiers are used in voltage *scaling* applications, where it is desired to multiply a voltage precisely by a fixed constant, or scale factor. The multiplying constant in the inverting amplifier is R_f/R_1 (which may be less than 1), and it is $1 + R_f/R_1$ (which is always greater than 1) in the noninverting amplifier. A wide range of constants can be realized with convenient choices of R_f and R_1 when the gain is R_f/R_1, which is not so much the case when the gain is $1 + R_f/R_1$. For that reason, the inverting amplifier is more often used in precision scaling applications.

The reader may wonder why it would be desirable or necessary to use an amplifier to multiply a voltage by a number less than 1, since this can also be accomplished using a simple voltage divider. The answer is that the amplifier provides power gain to drive a load. Also, the ideal amplifier has zero output impedance, so the output voltage is not affected by changes in load impedance.

The Voltage Follower

Figure 8–6 shows a special case of the noninverting amplifier used in applications where power gain and impedance isolation are of primary concern. Notice that $R_f = 0$ and $R_1 = \infty$, so, by equation 8–15, the closed-loop gain is $v_o/v_i = 1 + R_f/R_1 = 1$. This configuration is called a *voltage follower* because v_o has the same magnitude and phase as v_i. Like a BJT emitter follower, it has large input impedance and small output impedance and is used as a buffer amplifier between a high-impedance source and a low-impedance load.

FIGURE 8–6 The voltage follower

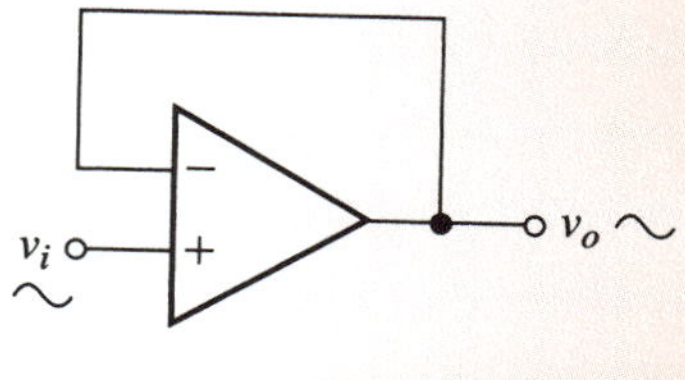

EXAMPLE 8–2

DESIGN

In a certain application, a signal source having 60 kΩ of source impedance produces a 1-V-rms signal. This signal must be amplified to 2.5 V rms and drive a 1-kΩ load. Assuming that the phase of the load voltage is of no concern, design an operational-amplifier circuit for the application.

Solution

Because phase is of no concern and the required voltage gain is greater than 1, we can use either an inverting or noninverting amplifier. Suppose we decide to use the inverting configuration and arbitrarily choose $R_f = 250\ \text{k}\Omega$. Then,

$$\frac{R_f}{R_1} = 2.5 \Rightarrow R_1 = \frac{R_f}{2.5} = \frac{250\ \text{k}\Omega}{2.5} = 100\ \text{k}\Omega$$

Note, however, that the signal source sees an impedance equal to $R_1 = 100\ \text{k}\Omega$ in the inverting configuration, so the usual voltage division takes place and the input to the amplifier is actually

$$v_i = \left(\frac{R_1}{R_1 + r_S}\right)(1\ \text{V rms}) =$$

$$\left[\frac{100\ \text{k}\Omega}{(100\ \text{k}\Omega) + (60\ \text{k}\Omega)}\right](1\ \text{V rms}) = 0.625\ \text{V rms}$$

Therefore, the magnitude of the amplifier output is

$$v_o = \frac{R_f}{R_1}(0.625\ \text{V rms}) = \frac{250\ \text{k}\Omega}{100\ \text{k}\Omega}(0.625\ \text{V rms}) = 1.5625\ \text{V rms}$$

Clearly, the large source impedance is responsible for a reduction in gain, and it is necessary to redesign the amplifier circuit to compensate for this loss. (Do this, as an exercise.)

In view of the fact that the source impedance may not be known precisely or may change if a replacement source is used, a far better solution is to design a noninverting amplifier. Because the input impedance of this design is extremely large, the choice of values for R_f and R_1 will not depend on the source impedance. Letting $R_f = 150\ \text{k}\Omega$, we have

$$1 + \frac{R_f}{R_1} = 2.5$$

$$\frac{R_f}{R_1} = 1.5$$

$$R_1 = \frac{R_f}{1.5} = \frac{150\ \text{k}\Omega}{1.5} = 100\ \text{k}\Omega$$

The completed design is shown in Figure 8–7. We can assume that the amplifier has zero output impedance, so we do not need to be concerned with voltage division between the amplifier output and the 1-kΩ load.

The Compensating Resistor R_c

Later in our study of operational amplifiers we will learn about certain nonideal characteristics. One of them is the fact that both inputs in an op-amp take a finite, albeit very small, current called *input bias current*. These two finite input currents can produce a small dc output voltage even when the input voltage is zero. The easiest way to minimize this problem is by including a compensating resistor, R_c, in series with the noninverting input, as shown in Figure 8–8. The value of this resistor, as will be shown in Chapter 10, should

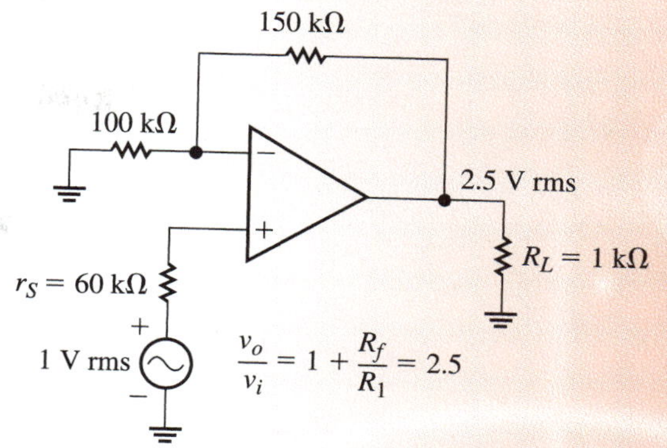

FIGURE 8–7 (Example 8–2)

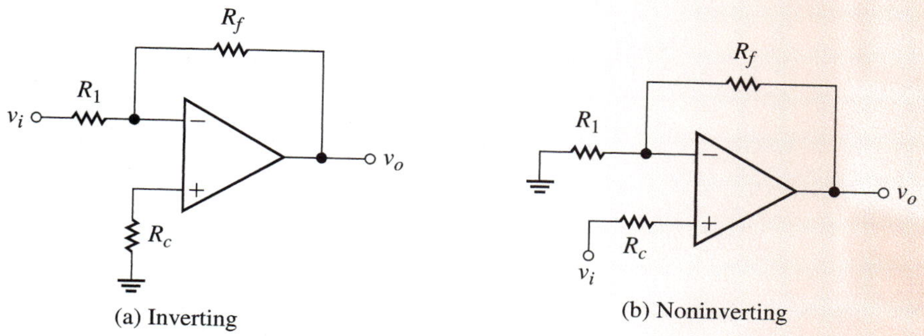

FIGURE 8–8 Using a compensating resistor in inverting and noninverting amplifiers

be approximately equal to the parallel combination of R_f and R_1. Any source resistance present in the circuit should be taken into account as well. For instance, in an inverting amplifier, the compensating resistor should be

$$R_c = R_f \| (r_S + R_1) \tag{8–16}$$

since r_S appears in series with R_1.

In the case of a noninverting amplifier, where the source is connected directly to the + input, the *sum* of r_S and R_c should be approximately equal to $R_1 \| R_f$,which means

$$R_c = (R_f \| R_1) - r_S \tag{8–17}$$

Single Power Supply Operation

When an amplifier is to be used with a single power supply, the $-V$ terminal is connected directly to ground and the supply voltage to the $+V$ terminal. The + input must be biased to one-half the supply voltage for proper linear operation. The resulting dc output voltage will also be one-half the supply voltage. Because of this, single power supply operation requires capacitive coupling for both the input signal source and the load resistance or subsequent stage.

Figure 8–9 shows the typical inverting amplifier configuration for single power supply operation. Note the voltage divider that provides the biasing for the + input. The gain formulas remain the same as for normal operation, but the coupling capacitors should be able to pass the lowest frequencies present in the signal; this topic will be covered in detail in a later chapter.

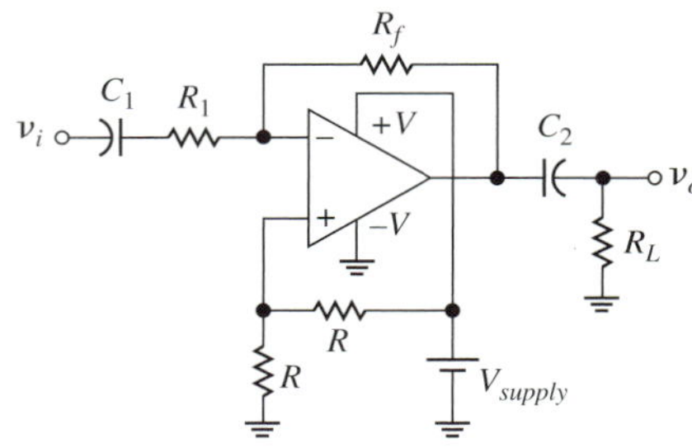

FIGURE 8–9 An inverting amplifier connected to a single power supply

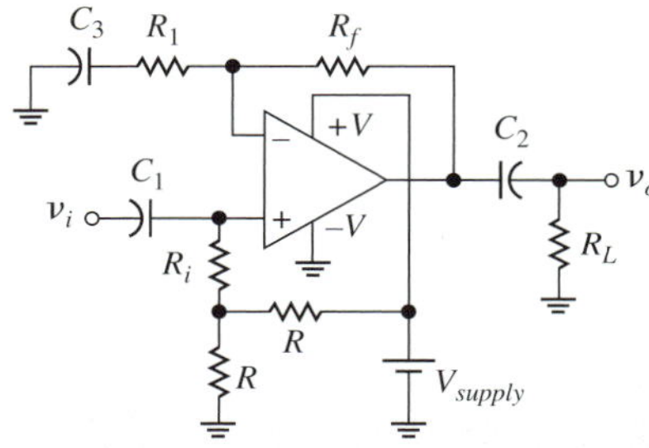

FIGURE 8–10 A noninverting amplifier connected to a single power supply

Figure 8–10 shows the noninverting version of the circuit. Note the addition of resistor R_i, which can be used to rise the input resistance seen by the signal source v_i to a particular desired level.

Because these configurations must be used with coupling capacitors, the biasing-compensating resistor is not much of an issue. In other words, any deviation of the output dc voltage from one-half the supply voltage due to bias currents will be irrelevant as far as ac operation is concerned. In any event, deviation can be maintained very small by making the parallel combination of the voltage-divider resistors (plus R_i, if any) match the value of R_f.

8–2 VOLTAGE SUMMATION, SUBTRACTION, AND SCALING

Voltage Summation

We have seen that it is possible to *scale* a signal voltage, that is, to multiply it by a fixed constant, through an appropriate choice of external resistors that determine the closed-loop gain of an amplifier circuit. This operation can be accomplished in either an inverting or noninverting configuration. It is also possible to sum several signal voltages in one operational-amplifier circuit and at the same time scale each by a different factor. For example, given inputs v_1, v_2, and v_3, we might wish to generate an output equal to $2v_1 + 0.5v_2 + 4v_3$. The latter sum is called a *linear combination* of v_1, v_2, and v_3, and the circuit that produces it is often called a *linear combination circuit*.

Figure 8–11 shows an inverting amplifier circuit that can be used to sum and scale three input signals. Note that input signals v_1, v_2, and v_3 are applied through separate resistors R_1, R_2, and R_3 to the summing junction of the amplifier and that there is a single feedback resistor R_f. Resistor R_c is the offset compensation resistor discussed previously.

Following the same procedure we used to derive the output of an inverting amplifier having a single input, we obtain for the three-input (ideal) amplifier

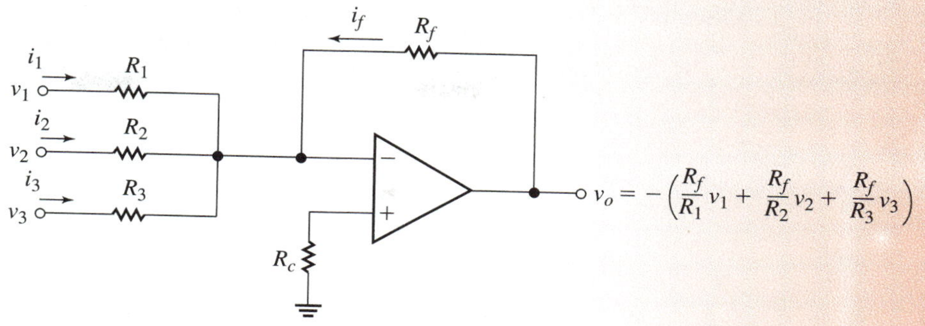

FIGURE 8–11 An operational-amplifier circuit that produces an output equal to the (inverted) sum of three separately scaled input signals

$$i_1 + i_2 + i_3 = -i_f$$

Or, since the voltage at the summing junction is ideally 0,

$$\frac{v_1}{R_1} + \frac{v_2}{R_2} + \frac{v_3}{R_3} = \frac{-v_o}{R_f}$$

Solving for v_o gives

$$v_o = -\left(\frac{R_f}{R_1}v_1 + \frac{R_f}{R_2}v_2 + \frac{R_f}{R_3}v_3\right) \quad (8\text{–}18)$$

Equation 8–18 shows that the output is the inverted sum of the separately scaled inputs, i.e., a *weighted* sum, or linear combination, of the inputs. By appropriate choice of values for R_1, R_2, and R_3, we can make the scale factors equal to whatever constants we wish, within practical limits. If we choose $R_1 = R_2 = R_3 = R$, then we obtain

$$v_o = \frac{-R_f}{R}(v_1 + v_2 + v_3) \quad (8\text{–}19)$$

and, for $R_f = R$,

$$v_o = -(v_1 + v_2 + v_3) \quad (8\text{–}20)$$

The theory can be extended in an obvious way to two, four, or any reasonable number of inputs. In this case, the compensating resistor is obtained from

$$R_c = R_f \| R_1 \| R_2 \| \ldots$$

EXAMPLE 8–3

DESIGN

1. Design an operational-amplifier circuit that will produce an output equal to $-(4v_1 + v_2 + 0.1v_3)$.
2. Write an expression for the output and sketch its waveform when $v_1 = 2 \sin \omega t$ V, $v_2 = +5$ V dc, and $v_3 = -100$ V dc.

Solution

1. We arbitrarily choose $R_f = 60$ kΩ. Then

$$\frac{R_f}{R_1} = 4 \Rightarrow R_1 = \frac{60 \text{ k}\Omega}{4} = 15 \text{ k}\Omega$$

$$\frac{R_f}{R_2} = 1 \Rightarrow R_2 = \frac{60 \text{ k}\Omega}{1} = 60 \text{ k}\Omega$$

FIGURE 8–12 (Example 8–3)

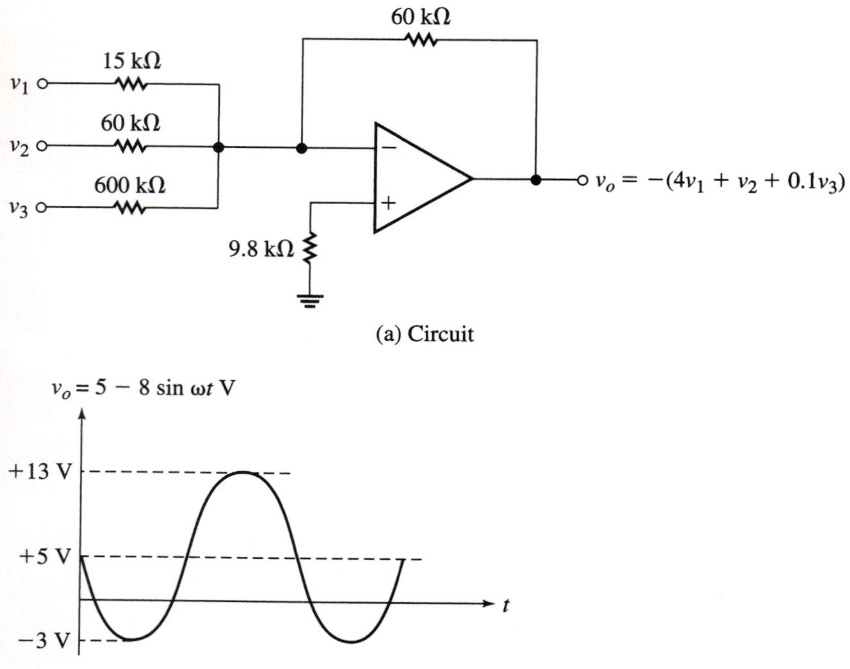

(b) Output waveform

$$\frac{R_f}{R_3} = 0.1 \Rightarrow R_3 = \frac{60\ \text{k}\Omega}{0.1} = 600\ \text{k}\Omega$$

By equation 8–21, the optimum value for the compensating resistor is $R_c = R_f \| R_1 \| R_2 \| R_3 = (60\ \text{k}\Omega) \| (15\ \text{k}\Omega) \| (60\ \text{k}\Omega) \| (600\ \text{k}\Omega) = 9.8\ \text{k}\Omega$. The circuit is shown in Figure 8–12(a).

2. $v_o = -[4(2 \sin \omega t) + 1(5) + 0.1(-100)] = -8 \sin \omega t - 5 + 10 = 5 - 8 \sin \omega t$. This output is sinusoidal with a 5-V offset and varies between $5 - 8 = -3$ V and $5 + 8 = 13$ V. It is sketched in Figure 8–12(b).

Figure 8–13 shows a noninverting version of the linear combination circuit. In this case, it can be shown (Exercise 8–13) that

$$v_o = \left(1 + \frac{R_f}{R_g}\right)\left(\frac{R_p}{R_1}v_1 + \frac{R_p}{R_2}v_2 + \frac{R_p}{R_3}v_3\right) \tag{8–21}$$

where $R_p = R_1 \| R_2 \| R_3$

Although this circuit does not invert the scaled sum, it is somewhat more cumbersome than the inverting circuit in terms of selecting resistor values to provide precise scale factors. Phase inversion is often of no consequence, but in those applications where a noninverted sum is required, it can also be obtained using the inverting circuit of Figure 8–11, followed by a unity-gain inverter.

Voltage Subtraction

Suppose we wish to produce an output voltage that equals the mathematical difference between two input signals. This operation can be performed by using the amplifier in a *differential* mode, where the signals are connected through appropriate resistor networks to the inverting and noninverting terminals. Figure 8–14 shows the configuration. We can use the superposition

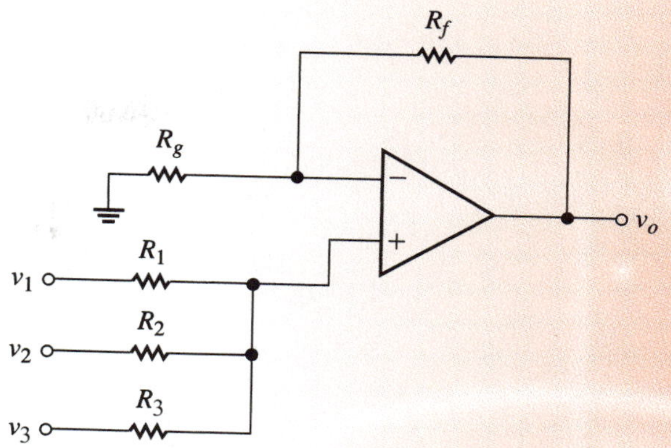

FIGURE 8–13 A noninverting linear combination circuit

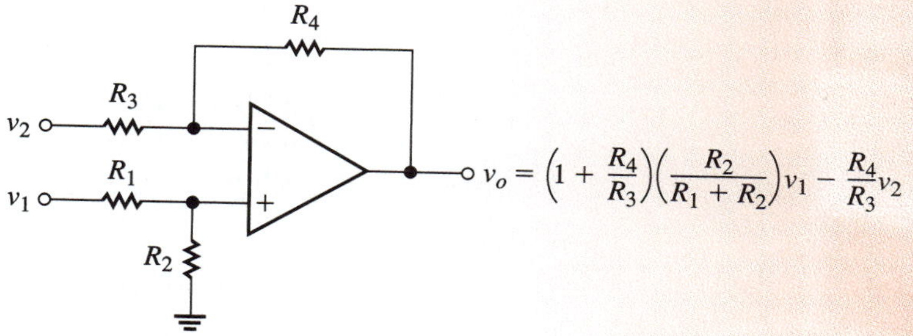

FIGURE 8–14 Using the amplifier in a differential mode to obtain an output proportional to the difference between two scaled inputs

principle to determine the output of this circuit. First, assume that v_2 is shorted to ground. Then

$$v^+ = \frac{R_2}{R_1 + R_2} v_1 \tag{8–22}$$

so

$$v_{o1} = \left(1 + \frac{R_4}{R_3}\right)v^+ = \left(1 + \frac{R_4}{R_3}\right)\left(\frac{R_2}{R_1 + R_2}\right)v_1 \tag{8–23}$$

Assuming now that v_1 is shorted to ground, we have

$$v_{o2} = \frac{-R_4}{R_3} v_2 \tag{8–24}$$

Therefore, with both signal inputs present, the output is

$$v_o = v_{o1} + v_{o2} = \left(1 + \frac{R_4}{R_3}\right)\left(\frac{R_2}{R_1 + R_2}\right) v_1 - \left(\frac{R_4}{R_3}\right)v_2 \tag{8–25}$$

Equation 8–25 shows that the output is proportional to the difference between scaled multiples of the inputs.

To obtain a difference or differential amplifier for which

$$v_o = A(v_1 - v_2) \tag{8–26}$$

where A is the differential gain, select the resistor values in accordance with the following:

$$R_1 = R_3 = R \quad \text{and} \quad R_2 = R_4 = AR \tag{8–27}$$

Substituting these values into (8–25) gives

$$\left(\frac{R + AR}{R}\right)\left(\frac{AR}{R + AR}\right)v_1 - \frac{AR}{R}v_2 = \frac{AR}{R}v_1 - \frac{AR}{R}v_2 = A(v_1 - v_2)$$

as required. When resistor values are chosen in accordance with (8–27), the bias compensation resistance ($R_1 \| R_2$) is automatically the correct value ($R_3 \| R_4$), namely $R \| AR$.

Let the general form of the output of Figure 8–14 be

$$v_o = a_1 v_1 - a_2 v_2 \tag{8–28}$$

where a_1 and a_2 are positive constants. Then, by equation 8–25, we must have

$$a_1 = \left(1 + \frac{R_4}{R_3}\right)\left(\frac{R_2}{R_1 + R_2}\right) \tag{8–29}$$

and

$$a_2 = \frac{R_4}{R_3} \tag{8–30}$$

Substituting (8–30) into (8–29) gives

$$a_1 = (1 + a_2)\frac{R_2}{R_1 + R_2} \tag{8–31}$$

But the quantity $R_2/(R_1 + R_2)$ is always less than 1. Therefore, equation 8–31 shows that in order to use the circuit of Figure 8–12 to produce $v_o = a_1 v_1 - a_2 v_2$, we must have

$$(1 + a_2) > a_1 \tag{8–32}$$

This restriction limits the usefulness of the circuit.

EXAMPLE 8–4

DESIGN

Design an op-amp circuit that will produce the output $v_o = 0.5v_1 - 2v_2$.

Solution

Note that $a_1 = 0.5$ and $a_2 = 2$, so $(1 + a_2) > a_1$. Therefore, it is possible to construct a circuit in the configuration of Figure 8–14.

Comparing v_o with equation 8–25, we see that we must have

$$\left(1 + \frac{R_4}{R_3}\right)\left(\frac{R_2}{R_1 + R_2}\right) = 0.5$$

and

$$\frac{R_4}{R_3} = 2$$

Let us arbitrarily choose $R_4 = 100\ \text{k}\Omega$. Then $R_3 = R_4/2 = 50\ \text{k}\Omega$. Thus

$$\left(1 + \frac{R_4}{R_3}\right)\left(\frac{R_2}{R_1 + R_2}\right) = \frac{3R_2}{R_1 + R_2} = 0.5$$

Arbitrarily choosing $R_2 = 20\ \text{k}\Omega$, we have

$$\frac{3(20\ \text{k}\Omega)}{R_1 + (20\ \text{k}\Omega)} = 0.5$$

$$60\ \text{k}\Omega = 0.5R_1 + (10\ \text{k}\Omega)$$

$$R_1 = 100\ \text{k}\Omega$$

The completed design is shown in Figure 8–15.

In Example 8–4, we note that the compensation resistance ($R_1 \| R_2 = (100\ \text{k}\Omega) \| (20\ \text{k}\Omega) = 16.67\ \text{k}\Omega$) is not equal to its optimum value ($R_3 \| R_4 = (50\ \text{k}\Omega) \| (100\ \text{k}\Omega) = 33.33\ \text{k}\Omega$). With some algebraic complication,

FIGURE 8–15 (Example 8–4)

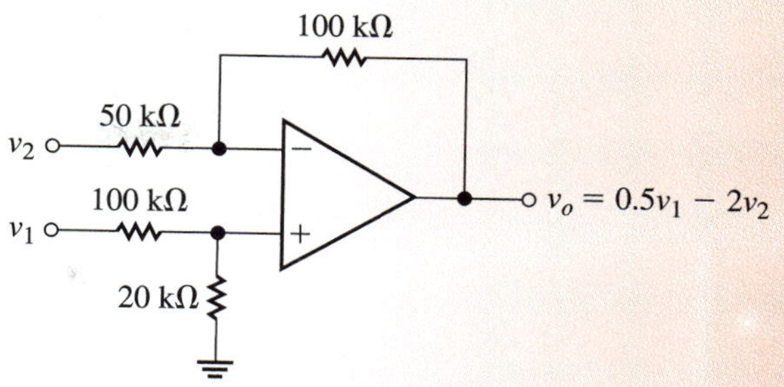

we can impose the additional condition $R_1 \| R_2 = R_3 \| R_4$ and thereby force the compensation resistance to have its optimum value. With $v_o = a_1 v_1 - a_2 v_2$, it can be shown (Exercise 8–17) that the compensation resistance $(R_1 \| R_2)$ is optimum when the resistor values are selected in accordance with

$$R_4 = a_1 R_1 = a_2 R_3 = R_2(1 + a_2 - a_1) \tag{8–33}$$

To apply this design criterion, choose R_4 and solve for R_1, R_2, and R_3. In Example 8–4, $a_1 = 0.5$ and $a_2 = 2$. If we choose $R_4 = 100$ kΩ, then $R_1 = (100\text{ k}\Omega)/0.5 = 200$ kΩ, $R_2 = (100\text{ k}\Omega)/2.5 = 40$ kΩ, and $R_3 = (100\text{ k}\Omega)/2 = 50$ kΩ. These choices give $R_1 \| R_2 = 33.3\text{ k}\Omega = R_3 \| R_4$, as required.

Although the circuit of Figure 8–14 is a useful and economical way to obtain a difference voltage of the form $A(v_1 - v_2)$, our analysis has shown that it has limitations and complications when we want to produce an output of the general form $v_o = a_1 v_1 - a_2 v_2$. An alternative way to obtain the difference between two scaled signal inputs is to use *two* inverting amplifiers, as shown in Figure 8–16. The output of the first amplifier is

$$v_{o1} = \frac{-R_2}{R_1} v_1 \tag{8–34}$$

and the output of the second amplifier is

$$v_{o2} = -\left(\frac{R_5}{R_3} v_{o1} + \frac{R_5}{R_4} v_2\right) = \frac{R_5 R_2}{R_3 R_1} v_1 - \frac{R_5}{R_4} v_2 \tag{8–35}$$

This equation shows that there is a great deal of flexibility in the choice of resistor values necessary to obtain $v_o = a_1 v_1 - a_2 v_2$, because a large number of combinations will satisfy

$$\frac{R_5 R_2}{R_3 R_1} = a_1 \quad \text{and} \quad \frac{R_5}{R_4} = a_2 \tag{8–36}$$

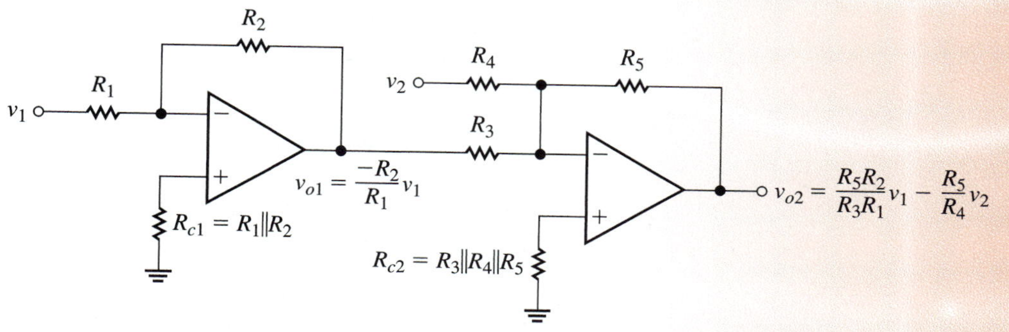

FIGURE 8–16 Using two inverting amplifiers to obtain the output $v_o = a_1 v_1 - a_2 v_2$

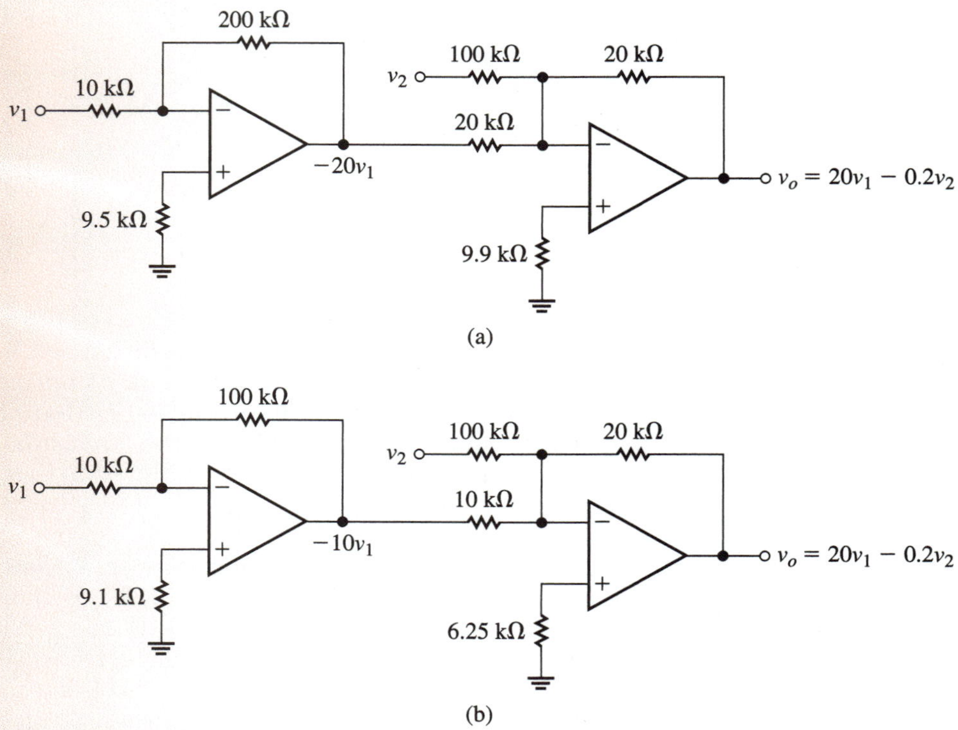

FIGURE 8–17 (Example 8–5) Two (of many) equivalent methods for producing $20v_1 - 0.2v_2$ using two inverting amplifiers

Furthermore, there are no restrictions on the choice of values for a_1 and a_2, nor any complications in setting R_c to its optimum value.

EXAMPLE 8–5

DESIGN

Design an operational-amplifier circuit using two inverting configurations to produce the output $v_o = 20v_1 - 0.2v_2$. (Note that $1 + a_2 = 1.2 < 20 = a_1$, so we cannot use the differential circuit of Figure 8–14.)

Solution

We have so many choices for resistance values that the best approach is to implement the circuit directly, without bothering to use the algebra of equation 8–33. We can, for example, begin the process by designing the first amplifier to produce $-20v_1$. Choose $R_1 = 10$ kΩ and $R_2 = 200$ kΩ. Then, the second amplifier need only invert $-20v_1$ with unity gain and scale the v_2 input by 0.2. Choose $R_5 = 20$ kΩ. Then $R_5/R_3 = 1 \Rightarrow R_3 = 20$ kΩ and $R_5/R_4 = 0.2 \Rightarrow R_4 = 100$ kΩ.

The completed design is shown in Figure 8–17(a). Figure 8–17(b) shows another solution, in which the first amplifier produces $-10v_1$ and the second multiplies that by the constant -2. The compensation resistors have values calculated as shown in Figure 8–16.

Although there are a large number of ways to choose resistor values to satisfy equation 8–34, there may, in practice, be constraints on some of those choices imposed by other performance requirements. For example, R_1 may have to be a certain minimum value to provide adequate input resistance to the v_1 signal source.

The method used to design a subtractor circuit in Example 8–5 can be extended in an obvious way to the design of circuits that produce a linear combination of voltage sums and differences. The most general form of a

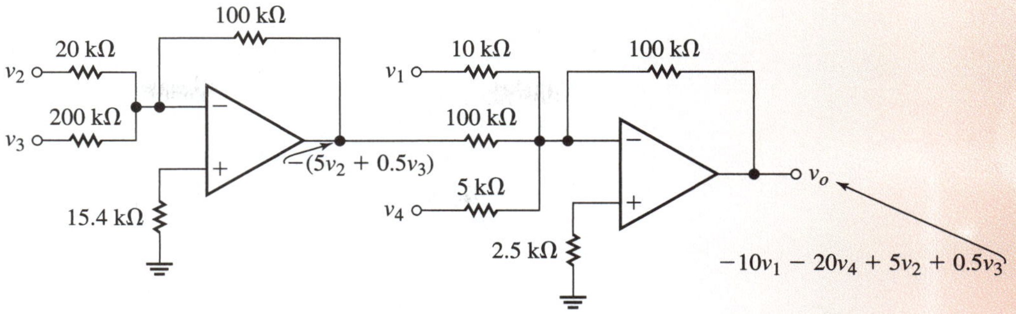

FIGURE 8–18 (Example 8–6)

linear combination is $v_o = \pm a_1 v_1 \pm a_2 v_2 \pm a_3 v_3 \pm \ldots \pm a_n v_n$. Remember that the input signal corresponding to any term that appears in the output with a positive sign must pass through two inverting stages.

EXAMPLE 8–6

DESIGN

Design an operational-amplifier circuit using two inverting configurations to produce the output $v_o = -10v_1 + 5v_2 + 0.5v_3 - 20v_4$.

Solution

Since v_2 and v_3 appear with positive signs in the output, those two inputs must be connected to the first inverting amplifier. We can produce $-(5v_2 + 0.5v_3)$ at the output of the first inverting amplifier and then invert and add it to $-(10v_1 + 20v_4)$ in the second amplifier. One possible solution is shown in Figure 8–18.

8–3 CONTROLLED VOLTAGE AND CURRENT SOURCES

Recall that a *controlled* source is one whose output voltage or current is determined by the magnitude of another, independent voltage or current. We have used controlled sources extensively in our study of transistor-circuit models, but those were, in a sense, fictitious devices that served mainly to simplify the circuit analysis. We wish now to explore various techniques that can be used to construct controlled voltage and current sources using operational amplifiers. As we shall see, some of these sources are realized simply by studying already-familiar circuits from a different viewpoint.

Voltage-Controlled Voltage Sources

An ideal, voltage-controlled voltage source (VCVS) is one whose output voltage V_o (1) equals a fixed constant (k) times the value of another, controlling voltage: $V_o = kV_i$; and (2) is independent of the current drawn from it. Notice that the constant k is dimensionless. Both the inverting and noninverting configurations of an ideal operational amplifier meet the two criteria. In each case, the output voltage equals a fixed constant (the closed-loop gain, determined by external resistors) times an input voltage. Also, since the output resistance is (ideally) zero, there is no voltage division at the output and the voltage is independent of load. We have studied these configurations in detail, so we will be content for now with the observation that they do belong to the category of voltage-controlled voltage sources.

Voltage-Controlled Current Sources

An ideal, voltage-controlled current source (VCCS) is one that supplies a current whose magnitude (1) equals a fixed constant (k) times the value of an independent, controlling voltage: $I_o = kV_i$; and (2) is independent of the load to which the current is supplied. Notice that the constant k has the dimensions of conductance (siemens). Because it relates output current to input voltage, it is called the *transconductance,* g_m, of the source.

Figure 8–19 shows two familiar amplifier circuits: the inverting and noninverting configurations of an operational amplifier. Note, however, that we now regard the feedback resistors as *load resistors* and designate each by R_L. We will show that each circuit behaves as a voltage-controlled current source, where the load current is the current I_L in R_L.

In Figure 8–19(a), v^- is virtual ground, so $I_1 = V_i/R_1$. Because no current flows into the inverting terminal of the ideal amplifier, $I_L = I_1$, or

$$I_L = \frac{V_i}{R_1} \tag{8–37}$$

Equation 8–37 shows that the load current is the constant $1/R_1$ times the controlling voltage V_i. Thus, the transconductance is $g_m = 1/R_1$ siemens. *Note that R_L does not appear in the equation, so the load current is independent of load resistance.* Like any constant-current source, the load voltage (voltage across R_L) will change if R_L is changed, but the current remains the same. The direction of the current through the load is controlled by the polarity of V_i. This version of a controlled current source is said to have a *floating load,* because neither side of R_L can be grounded. Thus, it is useful only in applications where the load is not required to have the same ground reference as the controlling voltage, V_i.

In Figure 8–19(b), $v^- = V_i$, so $I_1 = V_i/R_1$. Once again, no current flows into the inverting terminal, so $I_L = I_1$. Therefore,

$$I_L = \frac{V_i}{R_1} \tag{8–38}$$

As in the inverting configuration, the load current is independent of R_L and the transconductance is $1/R_1$ siemens. The load is also floating in this version.

Of course, there is a practical limit on the range of load resistance R_L that can be used in each circuit. If R_L is made too large, the output voltage of the amplifier will approach its maximum limit, as determined by the power supply voltages. For successful operation, the load resistance in each circuit must obey

$$R_L < \frac{R_1|V_{max}|}{V_i} \quad \text{(inverting circuit)} \tag{8–39}$$

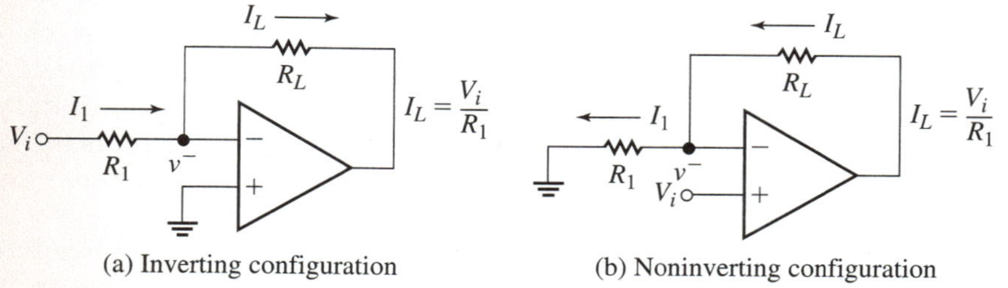

FIGURE 8–19 Floating-load, voltage-controlled current sources

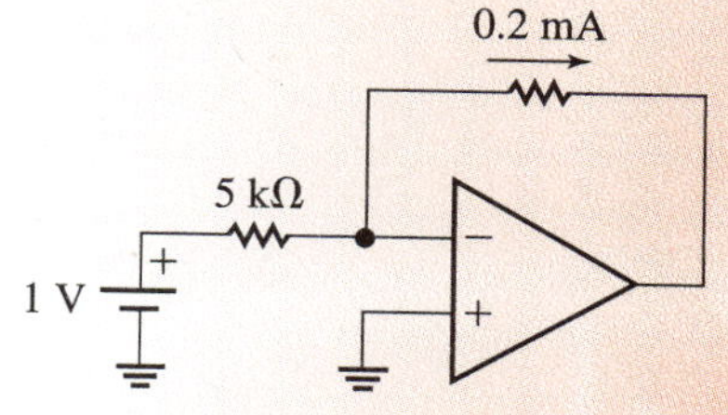

FIGURE 8–20 (Example 8–7)

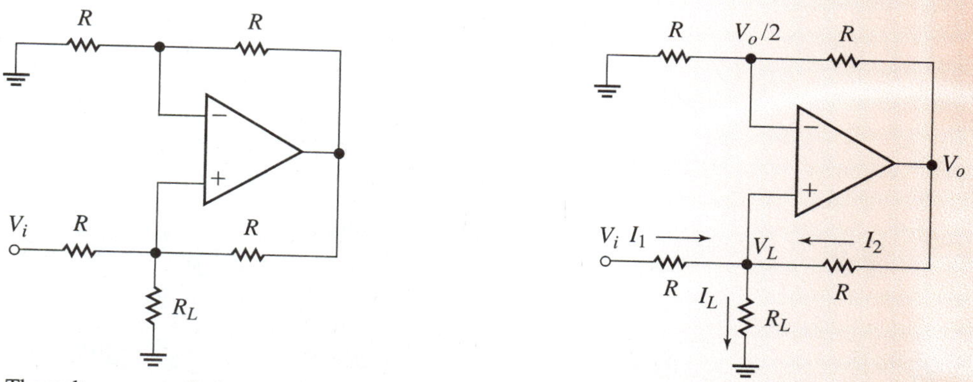

(a) The voltage-controlled current source

(b) Voltages and currents in the circuit of (a)

FIGURE 8–21 A voltage-controlled current source with a grounded load

$$R_L < R_1\left(\frac{|V_{max}|}{V_i} - 1\right) \quad \text{(noninverting circuit)} \tag{8–40}$$

where $|V_{max}|$ is the magnitude of the maximum output voltage of the amplifier.

EXAMPLE 8–7

DESIGN

Design an inverting, voltage-controlled current source that will supply a constant current of 0.2 mA when the controlling voltage is 1 V. What is the maximum load resistance for this supply if the maximum amplifier output voltage is 20 V?

Solution

The transconductance is $g_m = (0.2\text{ mA})/(1\text{ V}) = 0.2 \times 10^{-3}$ S. Therefore, $R_1 = 1/g_m = 5\text{ k}\Omega$. By (8–39),

$$R_L < \frac{R_1|V_{max}|}{V_i} = \frac{(5\text{ k}\Omega)(20\text{ V})}{1\text{ V}} = 100\text{ k}\Omega$$

The required circuit is shown in Figure 8–20.

Figure 8–21(a) shows a voltage-controlled current source that can be operated with a grounded load. To understand its behavior as a current source, refer to Figure 8–21(b), which shows the voltages and currents in the circuit. Because there is (ideally) zero current into the + input, Kirchhoff's current law at the node where R_L is connected to the + input gives

$$I_L = I_1 + I_2 \tag{8–41}$$

or

$$I_L = \frac{V_i - V_L}{R} + \frac{V_o - V_L}{R} \tag{8–42}$$

FIGURE 8–22 (Example 8–8)

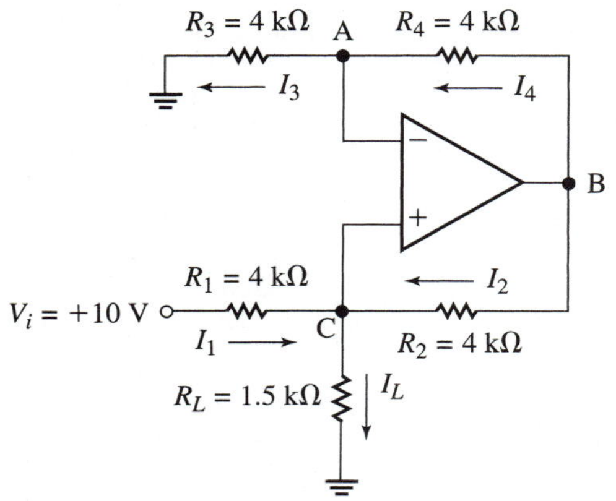

By voltage-divider action,

$$v^- = \left(\frac{R}{R+R}\right)V_o = V_o/2 \tag{8–43}$$

Since $v^- = v^+ = V_L$, we have $V_L = V_o/2$, which, upon substitution in (8–42), gives

$$I_L = \frac{V_i}{R} - \frac{V_o}{2R} + \frac{V_o}{R} - \frac{V_o}{2R}$$

or

$$I_L = \frac{V_i}{R} \tag{8–44}$$

This equation shows that the load current is controlled by V_i and that it is independent of R_L. Note that these results are valid to the extent that the four resistors labeled R are matched, i.e., truly equal in value. For successful operation, the loading condition must obey

$$2R_L|I_L| < |V_{max}| \tag{8–45}$$

EXAMPLE 8–8

Find the current through each resistor and the voltage at each node of the voltage-controlled current source in Figure 8–22. What is the transconductance of the source?

Solution

From equation 8–44, $I_L = V_i/R = (10\text{ V})/(4\text{ k}\Omega) = 2.5\text{ mA}$. Therefore, the voltage at node C (V_L) is $V_C = I_LR_L = (2.5\text{ mA})(1.5\text{ k}\Omega) = 3.75\text{ V}$. We know that the voltage at node B is twice V_C ($V_o = 2V_L$): $V_B = 2V_C = 2(3.75) = 7.5\text{ V}$. The voltage at node A is one-half that at node B ($v^- = V_o/2$): $V_A = (\frac{1}{2})(V_B) = (\frac{1}{2})(7.5) = 3.75\text{ V}$. The currents I_1, I_2, I_3, and I_4 in R_1, R_2, R_3, and R_4 can then be found:

$$I_1 = (V_i - V_C)/R_1 = (10 - 3.75)/(4 \times 10^3) = 1.5625\text{ mA}$$

$$I_2 = (V_B - V_C)/R_2 = (7.5 - 3.75)/(4 \times 10^3) = 0.9375\text{ mA}$$

$$I_3 = V_A/R_3 = 3.75/(4 \times 10^3) = 0.9375\text{ mA}$$

$$I_4 = (V_B - V_A)/R_4 = (7.5 - 3.75)/(4 \times 10^3) = 0.9375\text{ mA}$$

The transconductance of the source is $g_m = 1/R = 1/(4\text{ k}\Omega) = 0.25\text{ mS}$.

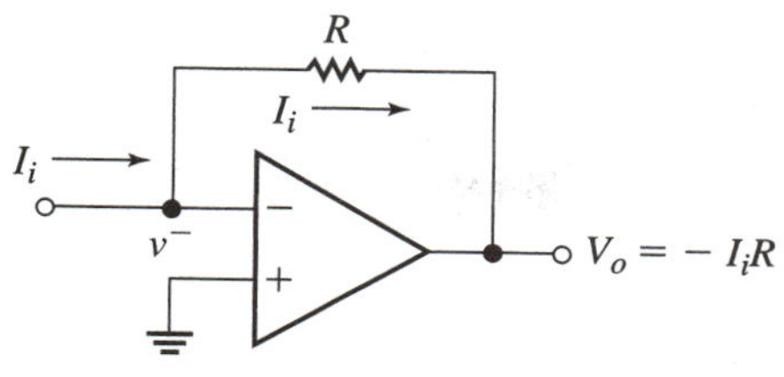

FIGURE 8–23 A current-controlled voltage source

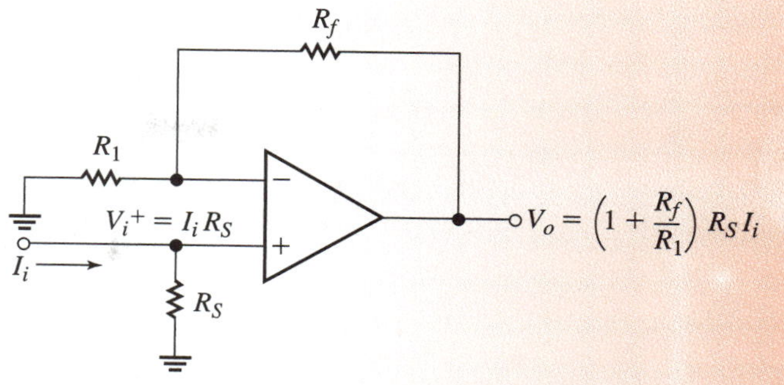

FIGURE 8–24 A current-controlled voltage source whose controlling current, I_i, has a return path to ground

Current-Controlled Voltage Sources

An ideal current-controlled voltage source (CCVS) has an output voltage that (1) is equal to a constant (k) times the magnitude of an independent current: $I_o = kI_i$, and (2) is independent of the load connected to it. Here, the constant k has the units of ohms. A current-controlled voltage source can be thought of as a *current-to-voltage converter,* since output voltage is proportional to input current. It is useful in applications where current measurements are required, because it is generally more convenient to measure voltages.

Figure 8–23 shows a very simple current-controlled voltage source. Because no current flows into the − input, the controlling current I_i is the same as the current in feedback resistor R. Since v^- is virtual ground,

$$V_o = -I_iR \tag{8–46}$$

Once again, the fact that the amplifier has zero output resistance implies that the output voltage will be independent of load.

Figure 8–24 shows a noninverting, current-controlled voltage source in which the controlling current has a return path to ground. Since $V_i^+ = I_iR_S$, we have

$$V_o = \left(1 + \frac{R_f}{R_1}\right)V_i^+ = \left(1 + \frac{R_f}{R_1}\right)R_S I_i \tag{8–47}$$

Current-Controlled Current Sources

An ideal current-controlled current source (CCCS) is one that supplies a current whose magnitude (1) equals a fixed constant (k) times the value of an independent controlling current: $I_o = kI_i$, and (2) is independent of the load to which the current is supplied. Note that k is dimensionless, since it is the ratio of two currents.

Figure 8–25 shows a current-controlled current source with floating load R_L. Because no current flows into the − input, the current in R_2 must equal I_i. Since v^- is at virtual ground, the voltage V_2 is

$$V_2 = -I_iR_2$$

Therefore, the current I_1 in R_1 is

$$I_1 = (0 - V_2)/R_1 = I_iR_2/R_1 \tag{8–48}$$

Writing Kirchhoff's current law at the junction of R_1, R_2, and R_L, we have

$$I_L = I_1 + I_i \tag{8–49}$$

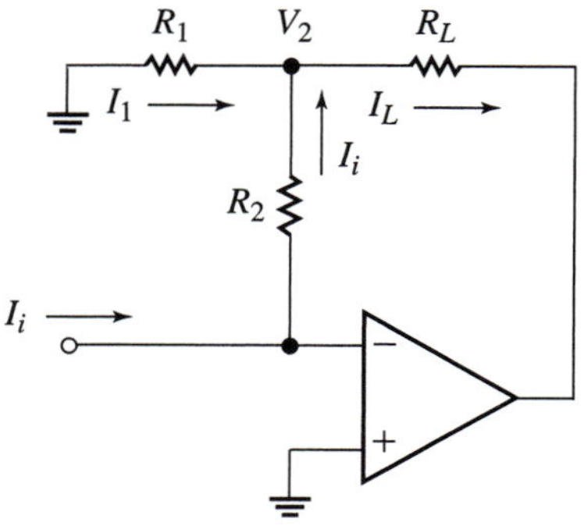

FIGURE 8–25 A current-controlled current source with floating load

or

$$I_L = \frac{R_2}{R_1}I_i + I_i = \left(\frac{R_2}{R_1} + 1\right)I_i \tag{8–50}$$

This equation shows that the load current equals the constant $(1 + R_2/R_1)$ times the controlling current and that I_L is independent of R_L. For successful operation, R_L must obey

$$R_L < \left(\frac{|V_{max}|}{I_i} - R_2\right)\left(\frac{R_1}{R_1 + R_2}\right) \tag{8–51}$$

Note that the circuit of Figure 8–25 may be regarded as a current *amplifier*, the amplification factor being

$$k = I_L/I_i = 1 + R_2/R_1 \tag{8–52}$$

The next example demonstrates the utility of current amplification and illustrates an application where a floating load may be used.

EXAMPLE 8–9

DESIGN

It is desired to measure a dc current that ranges from 0 to 1 mA using an ammeter whose range is 0 to 10 mA. To improve the measurement accuracy, the current to be measured should be amplified by a factor of 10.

1. Design the circuit.
2. Assuming that the meter resistance is 150 Ω and the maximum output voltage of the amplifier is 15 V, verify that the circuit will perform properly.

Solution

1. Figure 8–26 shows the required circuit. I_X is the current to be measured and the ammeter serves as the load through which the amplified current flows. From equation 8–52, the current amplification is $I_L/I_X = 1 + R_2/R_1 = 10$. Letting $R_1 = 1\ \text{k}\Omega$, we find $R_2 = (10 - 1)\ 1\ \text{k}\Omega = 9\ \text{k}\Omega$.
2. Inequality 8–51 must be satisfied for the smallest possible value of the right-hand side, which occurs when $I_i = 1$ mA:

$$R_L < \left[\frac{15\ \text{V}}{1\ \text{mA}} - (9\ \text{k}\Omega)\right]\left[\frac{1\ \text{k}\Omega}{(1\ \text{k}\Omega) + (9\ \text{k}\Omega)}\right] = 600\ \Omega$$

Since the meter resistance is 150 Ω, the circuit operates satisfactorily.

FIGURE 8–26 (Example 8–9) The current-controlled current source acts as a current amplifier, so a 0–1-mA current can be measured by a 0–10-mA ammeter

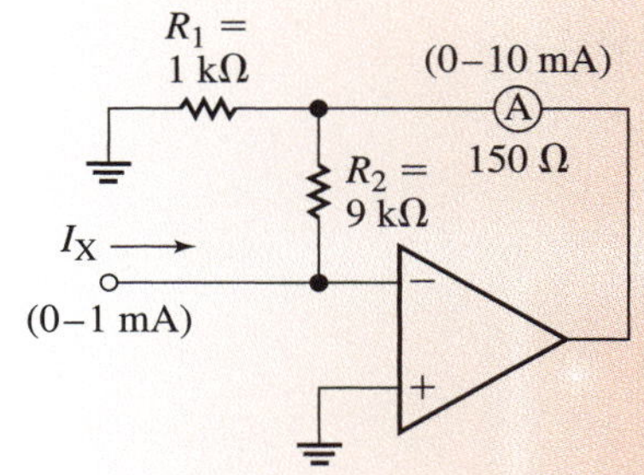

FIGURE 8–27 A noninverting amplifier with a single power supply

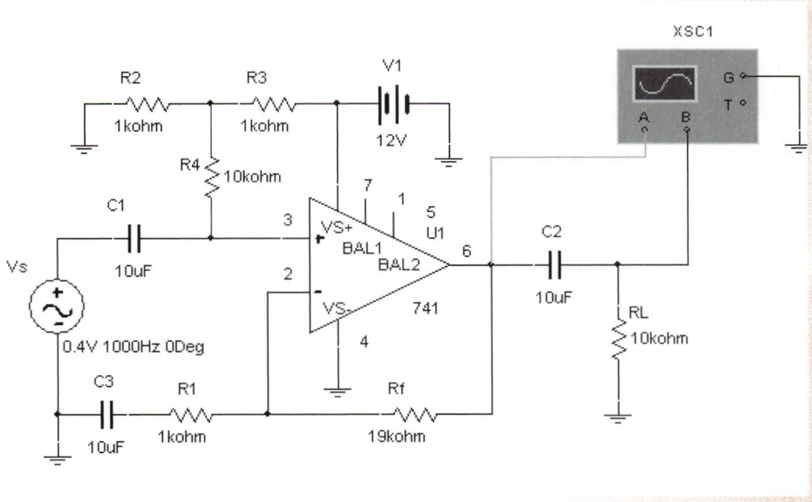

8–4 MULTISIM EXERCISE

We will simulate a noninverting amplifier operated with a single power supply. Figure 8–27 shows the Multisim Schematic which employs the 741 operational amplifier. Channel A from the oscilloscope will be connected to the op-amp's output, whereas channel B will be connected across the load resistor.

The objective of this exercise is to observe that the waveform at the output of the op-amp is centered at one-half the supply voltage, but the waveform at the load resistor is centered at zero volts. In other words, the op-amp's output waveform has a dc component but the waveform across the load resistance does not.

Before you start the simulation, double-click on the oscilloscope and set up the vertical scale to 5 V/div for both channels. The horizontal scale can be set to 0.5 or 1.0 ms/div.

SUMMARY

This chapter has established the foundation for the study of operational amplifiers. It was done by treating them as ideal devices and by introducing a number of basic op-amp applications. After completing this chapter, the reader should have a good understanding of the following concepts:

- An ideal op-amp has infinite input resistance, infinite differential gain, and zero output resistance.
- There are two basic amplifier configurations: inverting and noninverting.

- Voltage gain is established by two external resistors.
- A voltage follower is used as a buffer between a high-resistance source and a low-resistance load.
- Op-amps can be used for summing and/or subtracting scaled signals in inverting and noninverting modes.
- Controlled sources can be implemented with operational amplifiers and can be configured for floating or grounded loads.
- Op-amps are normally powered by dual power supplies, but with the addition of a few components, they can also be powered by a single power supply.

EXERCISES

SECTION 8–1

The Ideal Operational Amplifier

8–1. Find the output of the ideal operational amplifier shown in Figure 8–28 for each of the following input signals:

(a) $v_i = 120$ mV dc

(b) $v_i = 0.5 \sin \omega t$ V

(c) $v_i = -2.5$ V dc

(d) $v_i = 4 - \sin \omega t$ V

(e) $v_i = 0.8 \sin (\omega t + 75°)$ V

8–2. Assume that the feedback resistance in Exercise 8–1 is doubled and the input resistance is halved. Find the output for each of the following input signals:

(a) $v_i = -60.5$ mV dc

(b) $v_i = 500 \sin \omega t$ μV

(c) $v_i = -0.16 + \sin \omega t$ V

(d) $v_i = -0.2 \sin (\omega t - 30°)$ V

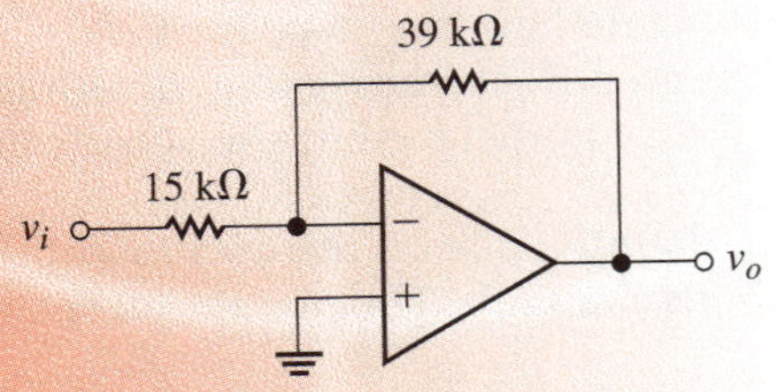

FIGURE 8–28 (Exercise 8–1)

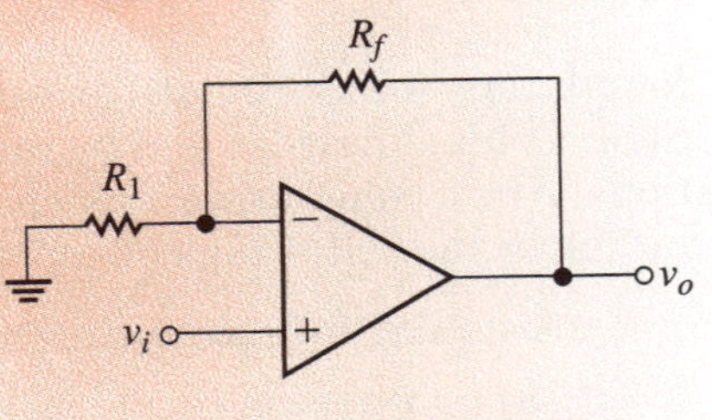

FIGURE 8–29 (Exercise 8–6)

8–3. Find the current in the feedback resistor for each part of Exercise 8–2.

8–4. The amplifier in Exercise 8–1 is driven by a signal source whose output resistance is 40 kΩ. The source voltage is 2.2 V rms. What is the rms value of the amplifier's output voltage?

8–5. Design an inverting operational-amplifier circuit that will provide an output of 10 V rms when the input is a 1-V-rms signal originating at a source having 10 kΩ source resistance.

8–6. The input to the ideal operational amplifier shown in Figure 8–29 is 0.5 V rms. Find the rms value of the output for each of the following combinations of resistor values:

(a) $R_1 = R_f = 10$ kΩ

(b) $R_1 = 20$ kΩ, $R_f = 100$ kΩ

(c) $R_1 = 100$ kΩ, $R_f = 20$ kΩ

(d) $R_f = 10R_1$

8–7. Repeat Exercise 8–6 for each of the following resistor combinations:

(a) $R_1 = 125$ kΩ, $R_f = 1$ MΩ

(b) $R_1 = 220$ kΩ, $R_f = 47$ kΩ

(c) $R_1/R_f = 0.1$

(d) $R_1/R_f = 10$

8–8. Assuming ideal operational amplifiers, find the load voltage v_L in Figure 8–30.

8–9. Assuming ideal operational amplifiers, find the load voltage v_L in each part of Figure 8–31.

FIGURE 8–30 (Exercise 8–8)

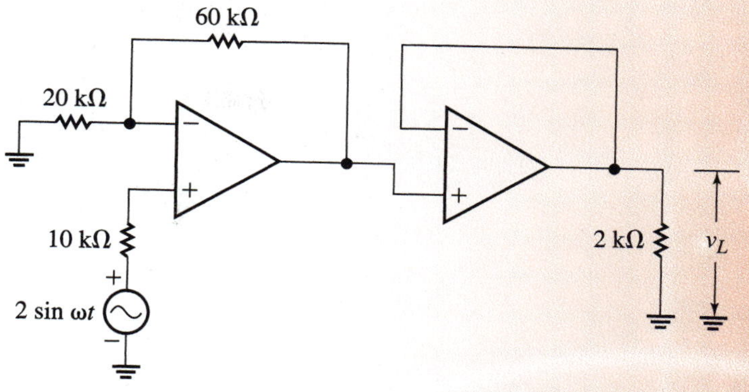

FIGURE 8–31 (Exercise 8–9)

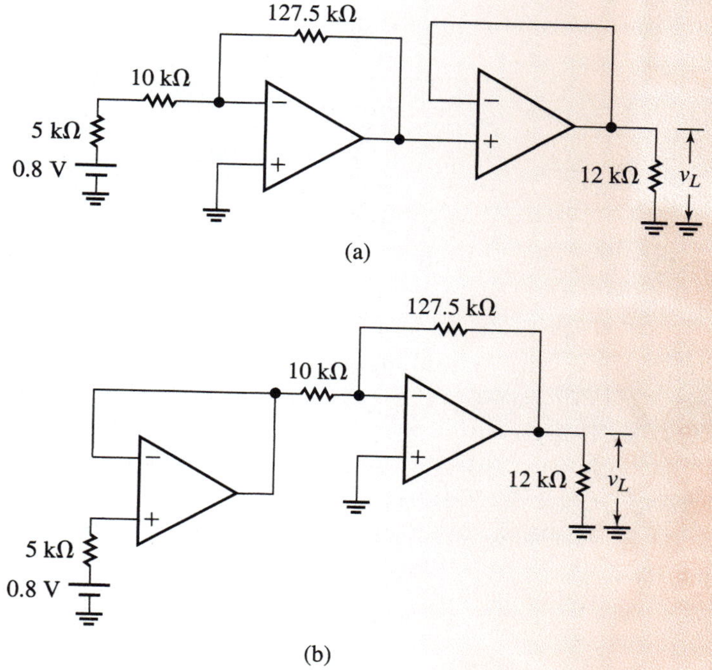

SECTION 10–2

Voltage Summation, Subtraction, and Scaling

8–10. (a) Write an expression for the output of the amplifier in Figure 8–32 in terms of v_1, v_2, v_3, and v_4.

(b) What is the output when $v_1 = 5 \sin \omega t$, $v_2 = -3$ V dc, $v_3 = -\sin \omega t$, and $v_4 = 2$ V dc?

(c) What value should R_c have?

8–11. (a) Design an operational-amplifier circuit that will produce the output $v_o = -10v_1 - 50v_2 + 10$. Use only one amplifier. (*Hint:* One of the inputs is a dc source.)

(b) Sketch the output waveform when $v_1 = v_2 = -0.1 \sin \omega t$ volts.

8–12. The operational amplifier in Exercise 8–10 has unity-gain frequency 1 MHz and input offset voltage 3 mV. Find

(a) the closed-loop bandwidth of the configuration, and

(b) the magnitude of the output offset voltage due to V_{io}.

8–13. Derive equation 8–21 for the output of the circuit shown in Figure 8–13. (*Hint:* Using source transformation, write an expression for v^+.)

8–14. (a) Write an expression, in terms of v_1 and v_2, for the output of the amplifier shown in Figure 8–33.

(b) Write an expression for the output in the special case in which v_1 and v_2

FIGURE 8–32 (Exercise 8–10)

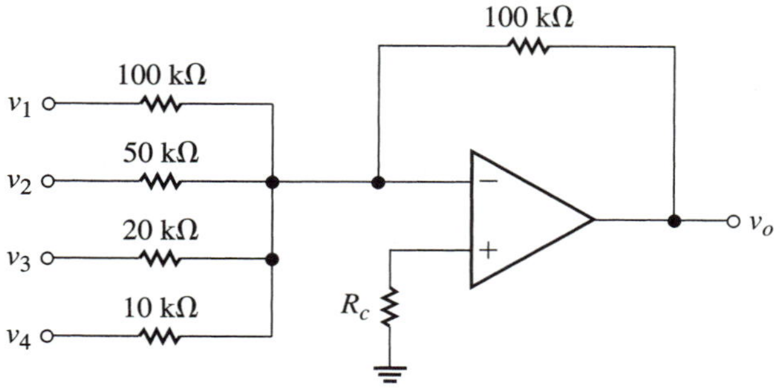

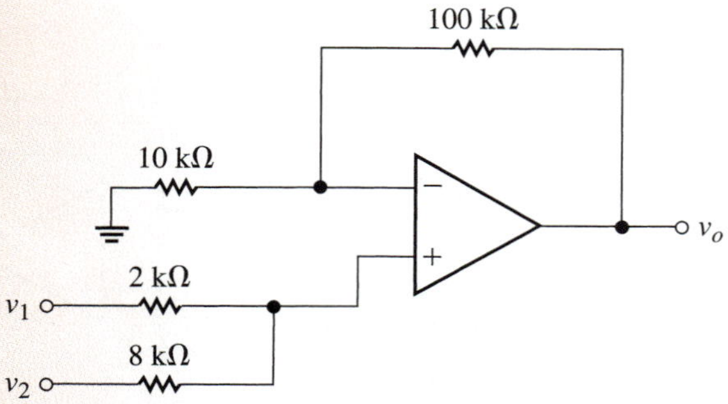

FIGURE 8–33 (Exercises 8–14 and 8–19)

are equal-magnitude, out-of-phase signals.

8–15. Design a noninverting circuit using a single operational amplifier that will produce the output $v_o = 4v_1 + 6v_2$.

8–16. Using a single operational amplifier in each case, design circuits that will produce the following outputs:

(a) $v_o = 0.1v_1 - 5v_2$

(b) $v_o = 10(v_1 - v_2)$

Design the circuits so that the compensation resistance has an optimum value.

8–17. If the resistor values in Figure 8–14 are chosen in accordance with $R_4 = a_1R_1 = a_2R_3 = R_2(1 + a_2 - a_1)$, then, assuming that $1 + a_2 > a_1$, show that

(a) $v_o = a_1v_1 - a_2v_2$, and

(b) the compensation resistance $(R_1 \| R_2)$ has its optimum value $(R_3 \| R_4)$.

8–18. Design operational-amplifier circuits to produce each of the following outputs:

(a) $v_o = 0.4v_2 - 10v_1$

(b) $v_o = v_1 + v_2 - 20v_3$

8–19. Change the two input resistors in Exercise 8–14 so that the same output expression is obtained but with proper bias compensation.

SECTION 8-3

Controlled Voltage and Current Sources

8–20. (a) Design an inverting voltage-controlled current source that will supply a current of 1 mA to a floating load when the controlling voltage is 2 V.

(b) If the source designed in (a) must supply its current to loads of up to 20 kΩ, what maximum output voltage should the amplifier have?

8–21. (a) Design a voltage-controlled current source that will supply a current of 2 mA to a floating load when the controlling voltage is 10 V. The input resistance seen by the controlling voltage source would have to be greater than 10 kΩ.

(b) If the maximum output voltage of the amplifier is 15 V, what is the maximum load resistance for which your design will operate properly?

8–22. (a) Design a voltage-controlled current source that will supply a current of 0.5 mA to a grounded load when the controlling voltage is 5 V.

(b) What will be the value of the amplifier output voltage if the load resistance is 12 kΩ?

8–23. The voltage-controlled current source in Figure 8–21 is to be used to supply current to a grounded 10-kΩ load when the controlling voltage is 5 V. If the maximum output voltage of the amplifier is 20 V, what is the maximum current that can be supplied to the load?

8–24. A certain temperature-measuring device generates current in direct proportion to temperature, in accordance with the relation $I = 2.5T$ μA, where T is in degrees Celsius. It is desired to construct a current-to-voltage converter for use with this device so that an output of 20 mV/°C can be obtained. Design the circuit.

8–25. The circuit shown in Figure 8–23 is used with the temperature-measuring device described in Exercise 8–24. If $R = 10$ kΩ, what is the output voltage when the temperature is 75°C?

8–26. Find the currents I_1, I_2, and I_3, and the voltages V_A and V_B, in the circuit of Figure 8–34. Assume an ideal operational amplifier.

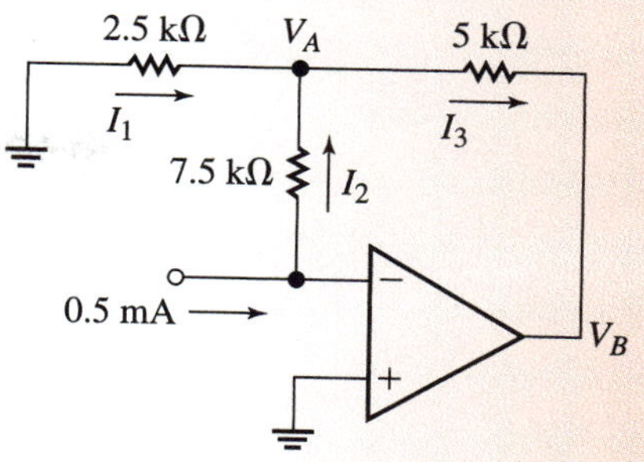

FIGURE 8–34 (Exercise 8–26)

8–27. (a) Design a current amplifier that will produce, in a 1-kΩ load, five times the current supplied to it.

(b) If the input current supplied to the amplifier is 2 mA, what should be the magnitude of the maximum output voltage of the amplifier?

CHAPTER 9

FREQUENCY RESPONSE

OUTLINE

OBJECTIVES

- Develop an understanding of the concept of frequency response in amplifiers.
- Investigate how RC circuits behave at different frequencies.
- Express gains and attenuations in decibels.
- Plot gain magnitude and phase versus frequency.
- Derive expressions for magnitude and phase in 2-port RC networks.
- Understand the Miller effect in inverting amplifiers.
- Differentiate between frequency response and transient response.

9–1 DEFINITIONS AND BASIC CONCEPTS

The *frequency response* of an electronic device or system is the variation it causes, if any, in the level of its output signal when the frequency of the signal is changed. In other words, it is the manner in which the device *responds* to changes in signal frequency. Variation in the level (amplitude, or rms value) of the output signal is usually accompanied by a variation in the *phase angle* of the output relative to the input, so the term *frequency response* also refers to phase shift as a function of frequency. (Phase shift versus frequency is sometimes called *phase response.*) Figure 9–1 shows an amplifier whose frequency response causes small output amplitudes at both low and high frequencies. Notice that the input signal amplitude is the same at each frequency, but the output signal amplitude changes with frequency. Thus, the *gain* of the amplifier is a function of frequency. In this example, the gain is small at the low frequency and small at the high frequency.

The frequency response of an amplifier is usually presented in the form of a graph that shows output amplitude (or, more often, voltage gain) plotted versus frequency. Phase-angle variation is sometimes plotted on the same graph. Figure 9–2 shows a typical plot of the voltage gain of an ac amplifier versus frequency. Notice that the gain is 0 at dc (zero frequency), then rises as frequency increases, levels off for further increases in frequency, and then begins to drop again at high frequencies.

The frequency range over which the gain is more or less constant ("flat") is called the *midband range,* and the gain in that range is designated A_m. As shown in Figure 9–2, the low frequency at which the gain equals $(\sqrt{2}/2)\,A_m \approx 0.707A_m$ is called the *lower cutoff frequency* and is designated f_L. The high frequency at which the gain once again drops to $0.707A_m$ is called the *upper cutoff frequency* and is designated f_H. The *bandwidth* of the amplifier is defined to be the difference between the upper and lower cutoff frequencies:

$$\text{bandwidth} = \text{BW} = f_H - f_L \tag{9–1}$$

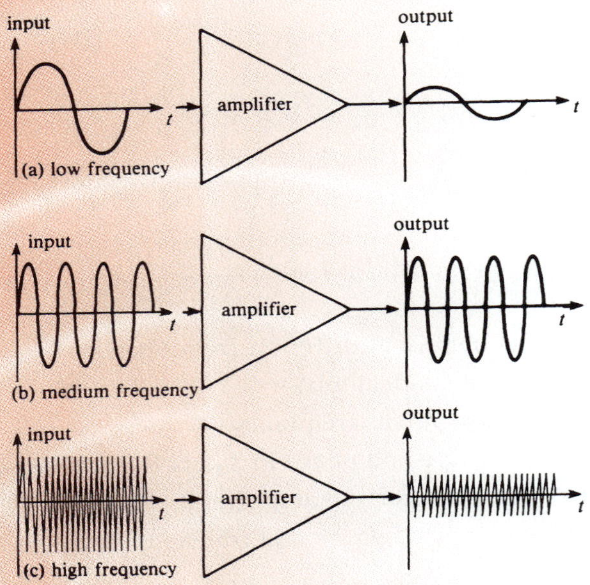

FIGURE 9–1 An amplifier whose frequency response is such that the output signal amplitude is small at both low and high frequencies

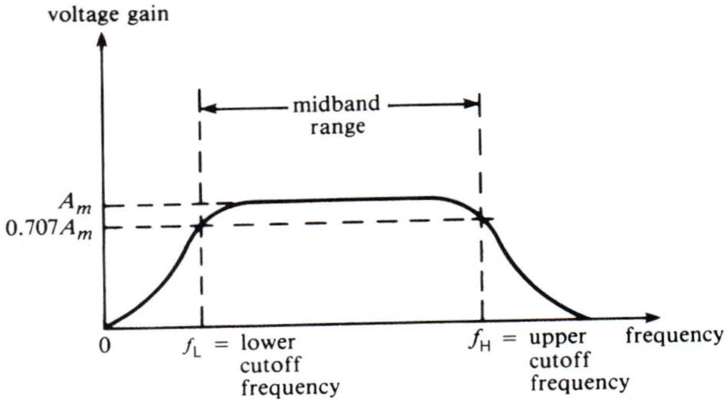

FIGURE 9–2 A typical amplifier's frequency response, showing gain versus frequency

The points on the graph in Figure 9–2 where the gain is $0.707A_m$ are often called *half-power points,* and the cutoff frequencies are sometimes called *half-power frequencies,* because the output power of the amplifier at cutoff is one-half of its output power in the midband range. To demonstrate this fact, suppose that an rms output voltage v is developed across R ohms in the midband range. Then the output power in midband is

$$P_{(midband)} = \frac{v^2}{R} \text{ watts} \qquad \textbf{(9–2)}$$

At each of the cutoff frequencies, the output voltage is $(\sqrt{2}/2)v$, so the power is

$$P_{(at\ cutoff)} = \frac{\left(\frac{\sqrt{2}}{2}v\right)^2}{R} = \frac{0.5\ v^2}{R} = 0.5\ P_{(midband)}$$

EXAMPLE 9–1

An audio amplifier has a lower cutoff frequency of 20 Hz and an upper cutoff frequency of 20 kHz. (This is the frequency range of sound waves—the audio frequency range.) The amplifier delivers 20 W to a 12-Ω load at 1 kHz.

1. What is the bandwidth of the amplifier?
2. What is the rms load voltage at 20 kHz?
3. What is the rms load voltage at 2 kHz?

Solution

1. $BW = f_H - f_L = 20 \times 10^3 \text{ Hz} - 20 \text{ Hz} = 19{,}980 \text{ Hz}$
2. Since 1 kHz is in the midband range, the midband power is 20 W. At the 20-kHz cutoff frequency, the power is $0.5(20) = 10$ W, so

$$\frac{v^2}{12} = 10$$
$$v^2 = 120$$
$$v = \sqrt{120} = 10.95 \text{ V rms}$$

3. Assuming that the load voltage is exactly the same throughout the midband range (not always the case in practice), the output at 2 kHz will be the same as that at 1 kHz, and we can use equation 9–2 with $P = 20$ W to solve for v. Alternatively, the midband voltage equals the voltage at cutoff *divided* by 0.707:

$$v_{(midband)} = \frac{10.95 \text{ V rms}}{0.707} = 15.49 \text{ V rms}$$

Amplitude and Phase Distortion

The signal passed through an ac amplifier is usually a complex waveform containing many different frequency components rather than a single-frequency ("pure") sine wave. For example, audio-frequency signals such as speech and music are combinations of many different sine waves occurring simultaneously with different amplitudes and different frequencies, in the range from 20 Hz to 20 kHz. As another example, any *periodic* waveform, such as a square wave or a triangular wave, can be shown to be the sum of a large number of sine waves whose amplitudes and frequencies can be determined mathematically. In previous discussions, we have analyzed ac amplifiers driven by single-frequency, sine-wave sources. This approach is justified by the fact that signals of all kinds can be regarded as sums of sine waves, as just described, so it is enough to know how an amplifier treats any single sine wave to know how it treats sums of sine waves (by the superposition principle).

In order for an output waveform to be an amplified version of the input, *an amplifier must amplify every frequency component in the signal by the same amount.* For example, if an input signal is the sum of a 0.5-V-rms, 100-Hz sine wave, a 0.2-V-rms, 1-kHz sine wave, and a 0.7-V-rms, 10-kHz sine wave, then an amplifier having gain 10 must amplify each frequency component by 10, so that the output consists of a 5-V-rms, 100-Hz sine wave, a 2-V-rms, 1-kHz sine wave, and a 7-V-rms, 10-kHz sine wave. If the frequency response of an amplifier is such that the gain at one frequency is different than it is at another frequency, the output will be *distorted,* in the sense that it will not have the same shape as the input waveform. This alteration in waveshape is called *amplitude distortion.* Figure 9–3 shows the distortion that results when a triangular waveform is passed through an amplifier having an inadequate frequency response. In this example, the high-frequency components in the waveform fall beyond the upper cutoff frequency of the amplifier, so they are not amplified by the same amount as low-frequency components.

It can be seen that knowledge of the frequency response of an amplifier is important in determining whether it will distort a signal having known frequency components. The bandwidth must cover the entire range of frequency

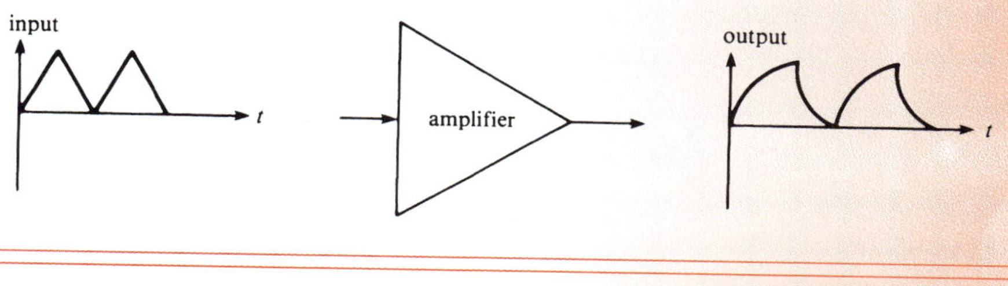

FIGURE 9–3 The output waveform is a distorted version of the input. The amplifier has a frequency response that is inadequate for the frequency components of the input waveform.

components in the signal if undistorted amplification is to be achieved. In general, "jagged" waveforms and signals having abrupt changes in amplitude, such as square waves and pulses, contain very broad ranges of frequencies and require wide bandwidth amplifiers.

An amplifier will also distort a signal if it causes components having different frequencies to be shifted by different *times*. For example, if an amplifier shifts one component by 1 ms, it must shift every component by 1 ms. This means that the *phase* shift at each frequency must be proportional to frequency. Distortion caused by failure to shift phase in this way is called *phase distortion*. In most amplifiers, phase distortion occurs at the same frequencies where amplitude distortion occurs because phase shifts are not proportional to frequency outside the midband range.

Amplitude and phase distortion should be contrasted with *nonlinear* distortion, which we discussed in an earlier chapter in connection with the nonuniform spacing of characteristic curves. Nonlinear distortion results when an amplifier's gain depends on signal *amplitude* rather than frequency. The effect of nonlinear distortion is to create frequency components in the output that were not present in the input signal. These new components are integer multiples of the frequency components in the input and are called *harmonic* frequencies. For example, if the input signal were a pure 1-kHz sine wave, the output would be said to contain third and fifth harmonics if it contained 3-kHz and 5-kHz components in addition to the 1-kHz *fundamental*. Such distortion is often called *harmonic distortion*.

9–2 DECIBELS AND LOGARITHMIC PLOTS

Decibels

Frequency-response data are often presented in *decibel* form. Recall that decibels (dB) are the units used to compare two power levels in accordance with the definition

$$\text{dB} = 10\log_{10}\frac{P_2}{P_1} \tag{9–3}$$

The two power levels, P_1 and P_2, are often the input and output power of a system, respectively, in which case equation 9–3 defines the power gain of the system in decibels. If $P_2 > P_1$, then equation 9–3 gives a *positive* number, and if $P_2 < P_1$, the result *is negative*, signifying a reduction in power. If $P_2 = P_1$, the result is 0 dB, since $\log_{10}(1) = 0$.

Let R_1 be the resistance across which the power P_1 is developed and R_2 be the resistance across which P_2 is developed. Then, since $P = v^2/R$, we have, from equation 9–3,

$$\text{dB} = 10\log_{10}\frac{(v_2^2/R_2)}{(v_1^2/R_1)} \tag{9–4}$$

where v_2 is the rms voltage across R_2 and v_1 is the rms voltage across R_1. If the resistance values are the *same* at the two points where the power comparison is made ($R_1 = R_2 = R$), then equation 9–4 becomes

$$\text{dB} = 10\log_{10}\left(\frac{v_2^2/\cancel{R}}{v_1^2/\cancel{R}}\right) = 10\log_{10}\left(\frac{v_2}{v_1}\right)^2 = 20\log_{10}\left(\frac{v_2}{v_1}\right) \tag{9–5}$$

Equation 9–5 gives power gain (or loss) in terms of the voltage levels at two points in a circuit, but it must be remembered that the equation is valid for power comparison only if the resistances at the two points are equal. *The*

FIGURE 9–4 (Example 9–2)

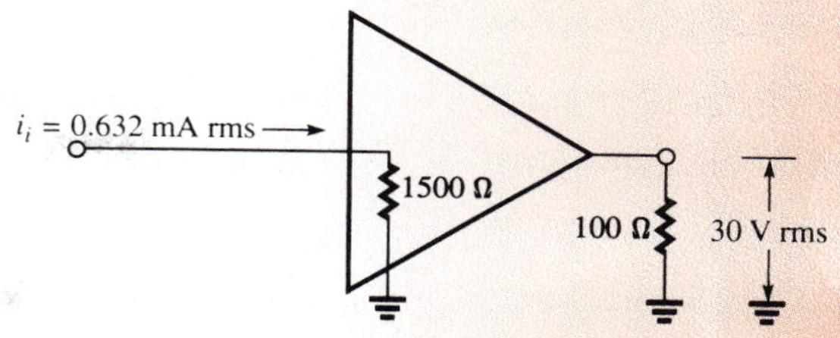

same equation is used to compare voltage *levels regardless of the resistance values at the two points.* In other words, it is common practice to compute voltage gain as

$$\text{dB(voltage gain)} = 20 \log_{10}\left(\frac{v_2}{v_1}\right) \qquad (9\text{–}6)$$

If the resistances R_1 and R_2 are equal, then the power gain in dB equals the voltage gain in dB.

EXAMPLE 9–2

The amplifier shown in Figure 9–4 has input resistance of 1500 Ω and drives a 100-Ω load. If the input current is 0.632 mA rms and the load voltage is 30 V rms, find

1. the power gain in dB, and
2. the voltage gain in dB.

Solution

1. $P_i = i_i^2 r_i = (0.632 \times 10^{-3}\text{ A rms})^2(1500\ \Omega) = 0.6\text{ mW}$

$$P_L = \frac{v_L^2}{R_L} = \frac{(30\text{ V rms})^2}{100\ \Omega} = 9\text{ W}$$

From equation 9–3,

$$\text{power gain} = 10 \log_{10}\left(\frac{9\text{ W}}{0.6 \times 10^{-3}\text{ W}}\right) = 10 \log_{10}(15 \times 10^3) = 41.76\text{ dB}$$

2. Since $v_i = i_i r_i = (0.632 \times 10^{-3}\text{ A})(1500\ \Omega) = 0.948$ V rms, we have from equation 9–6,

$$\text{voltage gain} = 20 \log_{10}\left(\frac{30}{0.948}\right) = 20 \log_{10}(31.645) = 30\text{ dB}$$

It is helpful to remember that a 2-to-1 change in voltage corresponds to approximately ± 6 dB, and a 10-to-1 change corresponds to ±20 dB. The sign (±) depends on whether the change represents an increase or a decrease in voltage. Suppose, for example, that $v_1 = 8$ V rms. If this voltage is doubled ($v_2 = 16$ V rms), then

$$20 \log_{10}\left(\frac{16}{8}\right) = 20 \log_{10}(2) \approx 6\text{ dB}$$

If v_1 is halved ($v_2 = 4$ V rms), then

$$20 \log_{10}\left(\frac{4}{8}\right) = 20 \log_{10}(0.5) \approx -6\text{ dB}$$

TABLE 9–1

(v_2/v_1)	$\text{dB} = 20\log_{10}(v_2/v_1)$	(v_2/v_1)	$\text{dB} = 20\log_{10}(v_2/v_1)$
0.001	− 60	2	6
0.002	− 54	4	12
0.005	− 46	8	18
0.008	− 42	10	20
0.01	− 40	20	26
0.02	− 34	40	32
0.05	− 26	80	38
0.08	− 22	100	40
0.1	− 20	200	46
0.2	− 14	400	52
0.5	− 6	800	58
0.8	− 2	1000	60
1.0	0		

If v_1 is increased by a factor of 10 ($v_2 = 80$ V rms), then

$$20\log_{10}\left(\frac{80}{8}\right) = 20\log_{10}(10) = 20 \text{ dB}$$

If v_1 is reduced by a factor of 10 ($v_2 = 0.8$ V rms), then

$$20\log_{10}\left(\frac{0.8}{8}\right) = 20\log_{10}(0.1) = -20 \text{ dB}$$

Every time a voltage is doubled, an additional 6 dB is *added* to the voltage gain, and every time it is increased by a factor of 10, an additional 20 dB is added to the voltage gain. For example, a gain of $100 = 10 \times 10$ corresponds to 40 dB, and a gain of $4 = 2 \times 2$ corresponds to 12 dB. As another example, a gain of $400 = 2 \times 2 \times 10 \times 10$ corresponds to $(6 + 6 + 20 + 20)$ dB = 52 dB. Similarly, a reduction in voltage by a factor of $0.05 = (1/2)(1/10)$ corresponds to $-6 - 20 = -26$ dB. Common logarithms (base 10) can be computed on scientific-type calculators, and the reader should become familar with the calculator's use for that purpose and for computing inverse logarithms. For reference and comparison purposes, Table 9–1 shows the decibel values corresponding to some frequently encountered ratios between 0.001 and 1000.

EXAMPLE 9–3

The input voltage to an amplifier is 4 mV rms. At point 1 in the amplifier, the voltage gain with respect to the input is − 4.2 dB, and at point 2 the voltage gain with respect to point 1 is 18.5 dB. Find

1. the voltage at point 1,
2. the voltage at point 2, and
3. the voltage gain in dB at point 2 with respect to the input.

Solution

Let v_i = the input voltage (4×10^{-3} V rms), v_1 = the rms voltage at point 1, and v_2 = the rms voltage at point 2.

1. $20\log_{10}\left(\dfrac{v_1}{4 \times 10^{-3}}\right) = -4.2$

$$\log_{10}\left(\frac{v_1}{4\times10^{-3}}\right) = -0.21$$

Taking the inverse log (antilog) of both sides,

$$\frac{v_1}{4\times10^{-3}} = 0.617$$

$$v_1 = 2.46 \text{ mV rms}$$

On most scientific calculators, the inverse log of $-$ 0.21 can be computed directly by entering a sequence such as -0.21, *inverse, log;* it can also be found on a calculator having the y^x function by computing $10^{-0.21}$. Many calculators have the function 10^x in addition to y^x.

2. $$20\log_{10}\left(\frac{v_2}{2.466\times10^{-3}}\right) = 18.5$$

$$\log_{10}\left(\frac{v_2}{2.466\times10^{-3}}\right) = 0.925$$

$$v_2 = 2.466\times10^{-3}\text{ antilog }(0.925) = 20.75 \text{ mV rms}$$

3. $$\text{voltage gain} = 20\log_{10}\left(\frac{v_2}{v_i}\right) = 20\log_{10}\left(\frac{20.75\times10^{-3}}{4\times10^{-3}}\right) = 14.3 \text{ dB}$$

Notice that this result is the same as -4.2 dB + 18.5 dB; the overall gain in dB is the *sum* of the intermediate dB gains.

It must be remembered that decibels are derived from a *ratio* and therefore represent a comparison of one voltage or power level to another. It is correct to speak of voltage or power *gain* in terms of decibels, but it is meaningless to speak of output *level* in dB, unless the reference level is specified. Popular publications and the broadcast media frequently abuse the term *decibel* because no reference level is reported. Do not be confused by this practice.

It is common practice in some technical fields to use one standard reference level for all decibel computations. For example, the power level 1 mW is used extensively as a reference. When the reference is 1 mW, the decibel unit is written dBm:

$$\text{dBm} = 10\log_{10}\left(\frac{P}{10^{-3}}\right) \tag{9–7}$$

Note that 0 dBm corresponds to a power level of 1 mW. Another standard reference is 1 W:

$$\text{dBW} = 10\log_{10}\left(\frac{P}{1}\right) = 10\log_{10}P \tag{9–8}$$

When the voltage reference is 1 V, voltage gain in decibels is written dBV:

$$\text{dBV} = 20\log_{10}\left(\frac{V}{1}\right) = 20\log_{10}V \tag{9–9}$$

The *neper* is a logarithmic unit based on the natural log (ln) of a ratio:

$$A_p\text{ (nepers)} = \frac{1}{2}\ln\left(\frac{P_2}{P_1}\right) \tag{9–10}$$

$$A_v\,(\text{nepers}) = \ln\left(\frac{v_2}{v_1}\right)$$

Semilog and Log-Log Plots

It is a convenient and widely followed practice to plot the logarithm of frequency-response data rather than actual data values. If the logarithm of frequency is plotted along the horizontal axis, a wide frequency range can be displayed on a convenient size of paper without losing resolution at the low-frequency end. For example, if it were necessary to scale frequencies directly on average-sized graph paper over the range from 1 Hz to 10 kHz, each small division might represent 100 Hz. It would then be impossible to plot points in the range from 1 Hz to 10 Hz, where the lower cutoff frequency might well occur. When the horizontal scale represents logarithms of frequency values, the low-frequency end is expanded and the high-frequency end is compressed.

One way in which a logarithmic frequency scale can be obtained is to compute the logarithm of each frequency and then label conventional graph paper with those logarithmic values. An easier way is to use specially designed *log paper,* on which coordinate lines are logarithmically spaced. It is then necessary only to label each line directly with an actual frequency value. If only one axis of the graph paper has a logarithmically spaced scale and the other has conventional linear spacing, the paper is said to be *semilog* graph paper. If both the horizontal and vertical axes are logarithmic, it is called *log-log* paper.

Any 10-to-1 range of values is called a *decade.* For example, each of the frequency ranges 1 Hz to 10 Hz, 10 kHz to 100 kHz, 500 Hz to 5 kHz, and 0.02 Hz to 0.2 Hz is a decade. Figure 9–5 shows a sample of log-log graph paper on which two full decades can be plotted along each axis. This sample is called *2-cycle-by-2-cycle,* or simply 2×2, log-log paper. Log-log graph paper is available with different numbers of decades along each axis, including 4×2, 5×3, 3×3, and so forth. Notice that each decade along each axis occupies the same amount of space. Several horizontal and vertical decades are identified on the figure. Notice also that the graph paper is printed with identical scale values along each decade. The user must relabel the divisions in accordance with the actual decade values that are appropriate for the data to be plotted. Suppose, for example, that the gain of an amplifier varies from 2 to 60 over the frequency range from 150 Hz to 80 kHz. Then the frequency axis must cover the three decades 100 Hz to 1 kHz, 1 kHz to 10 kHz, and 10 kHz to 100 kHz, and the gain axis must cover the two decades 1 to 10 and 10 to 100. 3×2 graph paper would be required. In Figure 9–5, the axes are arbitrarily labeled with decades 0.1–1 and 1–10 (vertical) and 10–100 and 100–1000 (horizontal).

An *octave* is any 2-to-1 range of values, such as 5–10, 80–160, and 1000–2000. Notice in Figure 9–5 that every octave occupies the same length. Notice also that the value *zero* does not appear on either axis of the figure. Zero can never appear on a logarithmic scale, no matter how many decades are represented, because $\log(0) = -\infty$.

Semilog graph paper is used to plot gain in dB versus the logarithm of frequency. When the frequency axis is logarithmic and the vertical axis is linear with its divisions labeled in decibels, the graph paper is essentially the same as log-log paper. Thus, a plot of an amplifier's frequency response will have the same general shape when constructed on either type of graph paper. Semilog graph paper is also used to plot phase shift on a linear scale versus the logarithm of frequency. Graphs of frequency

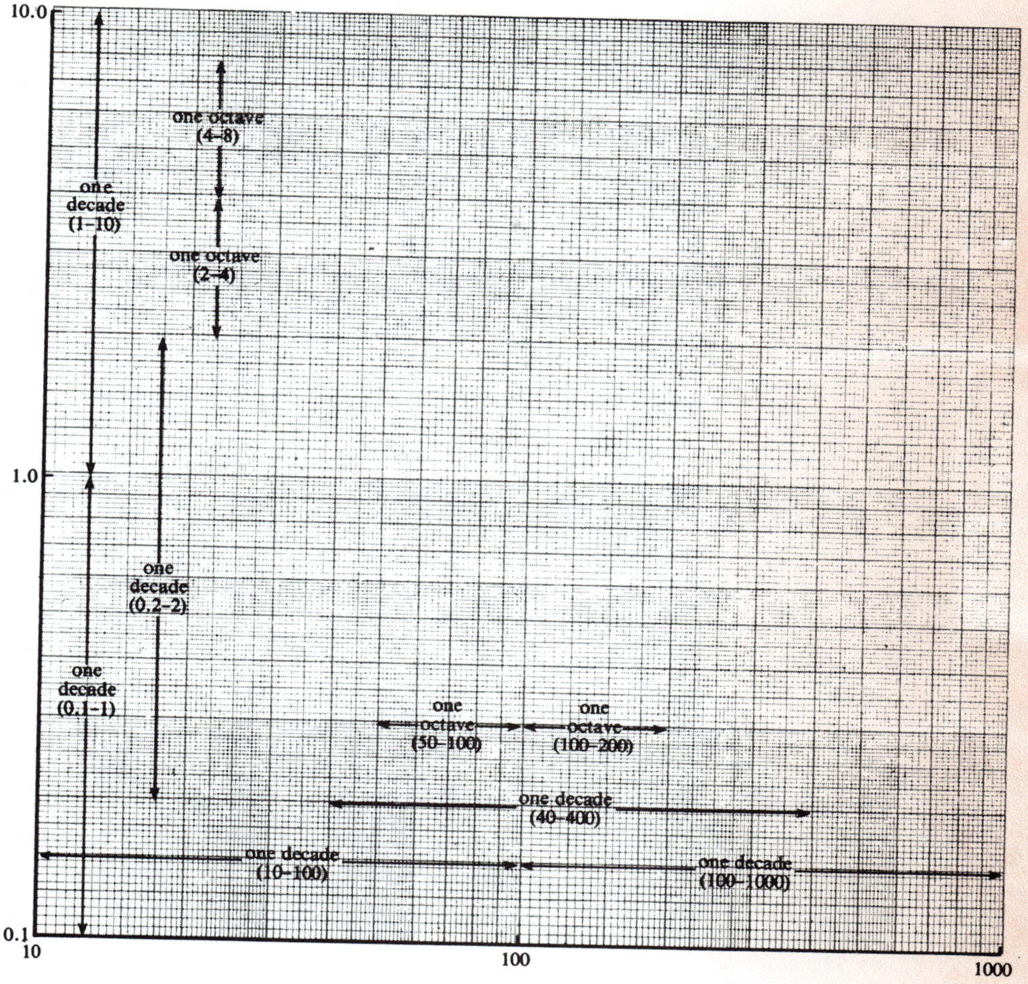

FIGURE 9–5 2-cycle-by-2-cycle (2 × 2) log-log graph paper, showing some typical octaves and decades along each axis

response plotted against the logarithm of frequency are called *Bode* (pronounced bō-dē) *plots.*

The cutoff frequency is the frequency at which the gain on a frequency-response plot is 3 dB less than the midband gain. At cutoff, the gain is said to be "3 dB down," and the cutoff frequencies are often called *3-dB frequencies.* The value is 3 dB because the output voltage is 0.707 times its value at midband, and

$$20 \log_{10}\left[\frac{(0.707)v_m}{v_m}\right] = 20 \log_{10}(0.707) \approx -3 \text{ dB} \qquad (9\text{–}11)$$

where v_m = the midband voltage.

EXAMPLE 9–4

Figure 9–6 shows the voltage gain of an amplifier plotted versus frequency on log-log paper. Find

1. the midband gain, in dB,
2. the gain in dB at the cutoff frequencies,

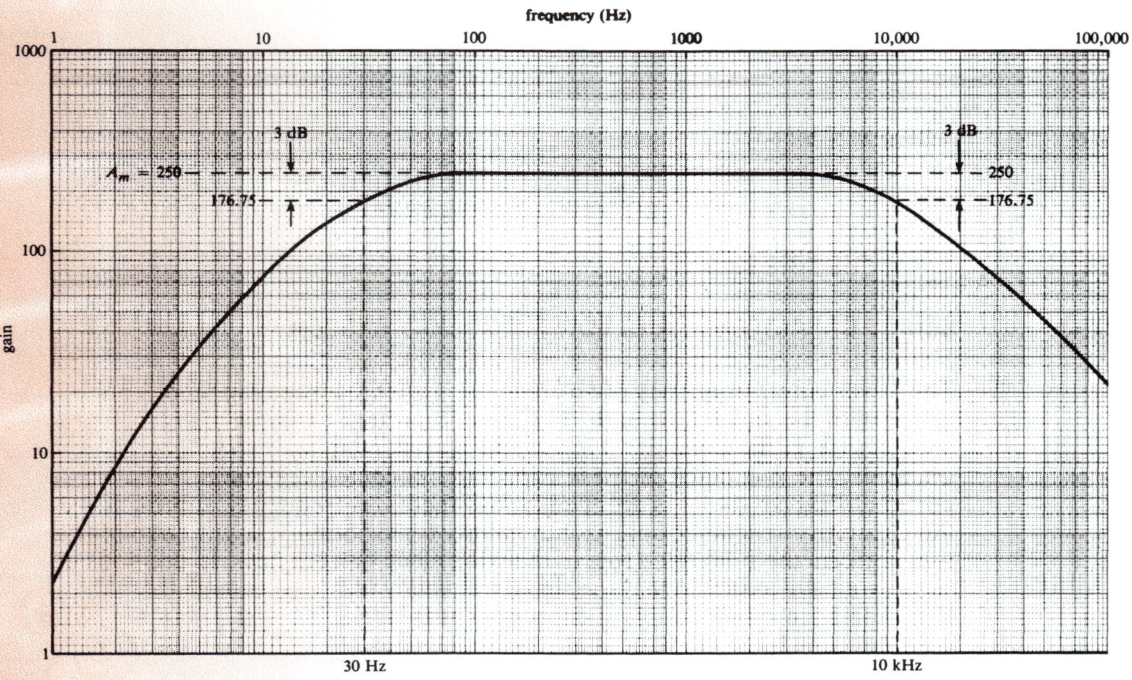

FIGURE 9–6 (Example 9–4)

3. the bandwidth,
4. the gain in dB at a frequency 1 decade below the lower cutoff frequency,
5. the gain in dB at a frequency 1 octave above the upper cutoff frequency, and
6. the frequencies at which the gain is down 15 dB from its midband value.

Solution

1. As shown in Figure 9–6, the magnitude of the midband gain is 250. Therefore, $A_m = 20 \log_{10}250 = 47.96 \text{ dB} \approx 48 \text{ dB}$.
2. Because the gain at each cutoff frequency is 3 dB less than the gain at midband, $A_v(\text{at cutoff}) = 48 - 3 = 45$ dB.
3. The magnitude of the gain at each cutoff frequency is $(0.707)250 = 176.8$. As shown in Figure 9–6, this value of gain is reached at $f_L = 30$ Hz and at $f_H = 10$ kHz. Therefore, $\text{BW} = f_H - f_L = (10 \text{ kHz}) - (30 \text{ Hz}) = 9970$ Hz.
4. The frequency 1 decade below f_L is $(0.1)f_L = 3$ Hz. From Figure 9–6, the magnitude of the gain at this frequency is approximately 15, so $A_v = 20 \log_{10}16 = 24.1$ dB.
5. The frequency 1 octave above f_H is $2 f_H = 20$ kHz. From Figure 9–6, the magnitude of the gain at 20 kHz is approximately 105, so $A_v = 20 \log_{10} 105 = 40.4$ dB.
6. The gain 15 dB below midband is (48 dB) − (15 dB) = 33 dB.

$$20 \log_{10}A_v = 33$$
$$\log_{10}A_v = 1.65$$
$$A_v = \text{antilog}(1.65) = 44.7$$

From Figure 9–6, the frequencies at which $A_v = 44.7$ are approximately 6.5 Hz and 50 kHz.

One-*n*th Decade and Octave Intervals

In many practical investigations, including computer-generated frequency-response data, it is necessary to specify logarithmic frequency intervals within one decade or within one octave. For example, we may want ten logarithmically spaced frequencies within one decade. These frequencies are said to be at *one-tenth decade* intervals. Similarly, three logarithmically spaced frequencies in one octave are said to be at *one-third octave* intervals. The frequencies in one-*n*th decade intervals beginning at frequency f_1 are

$$10^x,\ 10^{x+(1/n)},\ 10^{x+(2/n)},\ 10^{x+(3/n)},\ldots$$

where $x = \log_{10} f_1$.

The frequencies in one-*n*th octave intervals beginning at frequency f_1 are

$$2^x,\ 2^{x+(1/n)},\ 2^{x+(2/n)},\ 2^{x+(3/n)},\ldots$$

where $x = \log_2 f_1$.

9–3 SERIES CAPACITANCE AND LOW-FREQUENCY RESPONSE

The lower cutoff frequency of an amplifier is affected by capacitance connected in *series* with the signal flow path. The most important example of series-connected capacitance is the amplifier's input and output coupling capacitors. At low frequencies, the reactance of these capacitors becomes very large, so a significant portion of the ac signal is dropped across them. As frequency approaches 0 (dc), the capacitive reactance approaches infinity (open circuit), so the coupling capacitors perform their intended role of blocking all dc current flow. In previous discussions, we have assumed that the signal frequency was high enough that the capacitive reactance of all coupling capacitors was negligibly small, but we will now consider how large reactances at low frequencies affect the overall voltage gain.

Figure 9–7 shows the capacitor-resistor combination formed by the coupling capacitor and the input resistance at the input side of an amplifier. We omit consideration of any signal-source resistance for the moment. Notice that r_i and the capacitive reactance of C_1 form a voltage divider across the amplifier input. The amplifier input voltage, v_i, is found from the voltage-divider rule:

$$v_i = \left(\frac{r_i}{r_i - jX_{C_1}}\right) v_S \qquad \textbf{(9–12)}$$

where $\quad X_{C_1} = \dfrac{1}{\omega C_1}$ ohms

$\omega = 2\pi f$ radians/second

From equation 9–12, we can determine the *magnitude* (amplitude) of v_i, which we will designate $|v_i|$, as a function of ω:

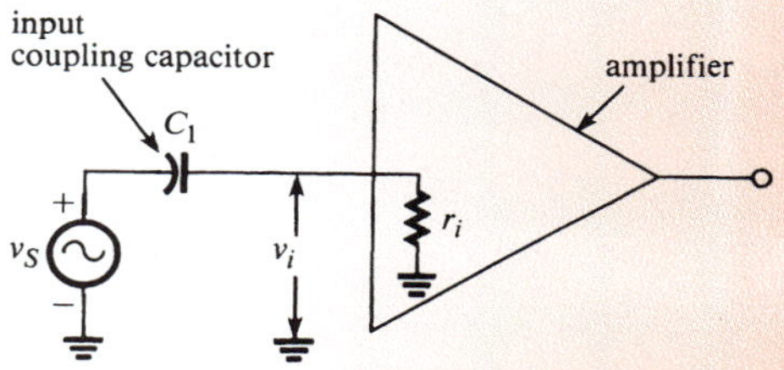

FIGURE 9–7 The input resistance of the amplifier and the coupling capacitor form an RC network that reduces the amplifier's input signal at low frequencies

$$|v_i| = \frac{r_i}{\sqrt{r_i^2 + X_{C_1}^2}}|v_S| = \frac{r_i}{\sqrt{r_i^2 + \left(\frac{1}{\omega C_1}\right)^2}}|v_S| = \frac{1}{\sqrt{1 + \left(\frac{1}{\omega r_i C_1}\right)^2}}|v_S| \quad (9\text{–}13)$$

Equation 9–13 shows that $|v_i| = 0$ when $\omega = 0$ (dc) and that $|v_i|$ approaches $|v_S|$ in value as ω becomes very large. At the frequency $\omega = 1/(r_iC_1)$ rad/s, we have, from equation 9–13,

$$|v_i| = \frac{1}{\sqrt{1+1}}|v_S| = \frac{1}{\sqrt{2}}|v_S| = 0.707|v_S|$$

This result shows that the amplifier input voltage falls to 0.707 times the source voltage when the frequency is reduced to $1/(r_iC_1)$ rad/s. Therefore, if there are no other frequency-sensitive components affecting the signal level, the overall gain from source to output is 0.707 times its midband value, meaning that $1/(r_iC_1)$ rad/s is the lower cutoff frequency. It is an exercise at the end of this chapter to show that the lower cutoff frequency is the frequency at which the capacitive reactance, X_{C_1}, equals the resistance, r_i.

If source resistance r_S is present, then equation 9–13 becomes

$$|v_i| = \left(\frac{r_i}{r_S + r_i}\right)\frac{1}{\sqrt{1 + \left[\frac{1}{\omega(r_S + r_i)C_1}\right]^2}} \quad (9\text{–}14)$$

and the lower cutoff frequency is

$$\omega_L = \frac{1}{(r_i + r_S)C_1} \quad (9\text{–}15)$$

or

$$f_L = \frac{1}{2\pi(r_i + r_S)C_1} \quad (9\text{–}16)$$

The cutoff frequency defined by equations 9–15 and 9–16 is the frequency at which the ratio $|v_i|/|v_S|$ is 0.707 times its *midband* value, namely, 0.707 times $r_i/(r_S + r_i)$. The ratio $|v_i|/|v_S|$ can be written in terms of the cutoff frequency f_L and the signal frequency f as follows:

$$\frac{|v_i|}{|v_S|} = K\left(\frac{1}{\sqrt{1 + (f_L/f)^2}}\right) \quad (9\text{–}17)$$

where

$$K = \frac{r_i}{r_S + r_i}$$

Let us now use v_S as a phase angle reference ($\angle v_S = 0°$) and compute the phase of v_i as a function of frequency. From equation 9–12,

$$\angle v_i = -\arctan\left(\frac{-X_{C_1}}{r_i}\right) + \angle v_S = \arctan\left(\frac{X_{C_1}}{r_i}\right) = \arctan\left(\frac{1}{\omega r_i C_1}\right) \quad (9\text{–}18)$$

Again, if source resistance r_S is present, we find

$$\angle v_i = \arctan\left[\frac{1}{\omega(r_i + r_S)C_1}\right] \quad (9\text{–}19)$$

FIGURE 9–8 (Example 9–5)

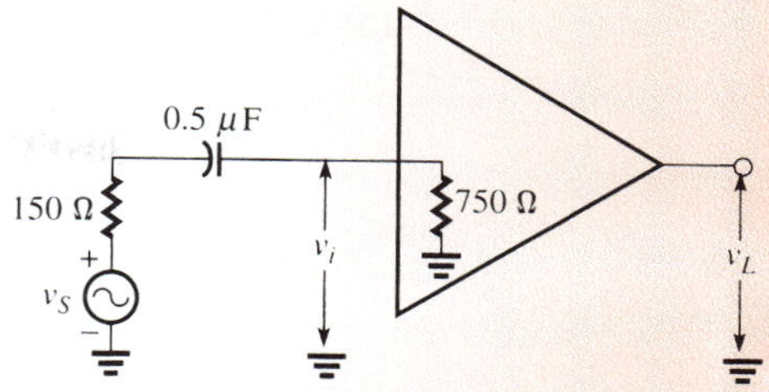

At $\omega = 1/(r_i + r_S)C_1$, equation 9–19 becomes

$$\angle v_i = \arctan(1) = 45°$$

Thus, v_i leads v_S by 45° at cutoff. The phase shift in terms of f_L and the signal frequency f is

$$\angle v_i = \arctan(f_L/f) \tag{9–20}$$

Note that $\angle v_i$ approaches 90° as f approaches 0.

EXAMPLE 9–5

The amplifier shown in Figure 9–8 has midband gain $|v_L|/|v_S|$ equal to 90. Find

1. the voltage gain $|v_L|/|v_i|$,
2. the lower cutoff frequency, and
3. the voltage gain $|v_L|/|v_S|$, in dB, at the cutoff frequency.

Solution

1. At midband, the reactance of the coupling capacitor is negligible, so the overall voltage gain $|v_L|/|v_S|$ is

$$\frac{|v_L|}{|v_S|} = \frac{|v_i|}{|v_S|}\frac{|v_L|}{|v_i|} = \left(\frac{r_i}{r_S + r_i}\right)\frac{|v_L|}{|v_i|}$$

Thus,

$$90 = \left(\frac{750\ \Omega}{150\ \Omega + 750\ \Omega}\right)\frac{|v_L|}{|v_i|}$$

$$\frac{|v_L|}{|v_i|} = 108$$

2. From equation 9–16,

$$f_L = \frac{1}{2\pi(150 + 750)(0.5\ \mu\text{F})} = 354\ \text{Hz}$$

3. At midband, the gain in dB is $20 \log_{10}(90) = 39.1$ dB. At cutoff, the gain is 3 dB less than its midband value: $39.1 - 3 = 36.1$ dB.

Figure 9–9 shows *normalized* plots of the gain of an RC network connected so that the capacitor is in series with the signal flow and the output is taken across the resistor. This is the configuration at the input of the capacitor-coupled amplifier, as shown in Figure 9–7. The gain of the network approaches 1 (0 dB) at high frequencies, which corresponds to the

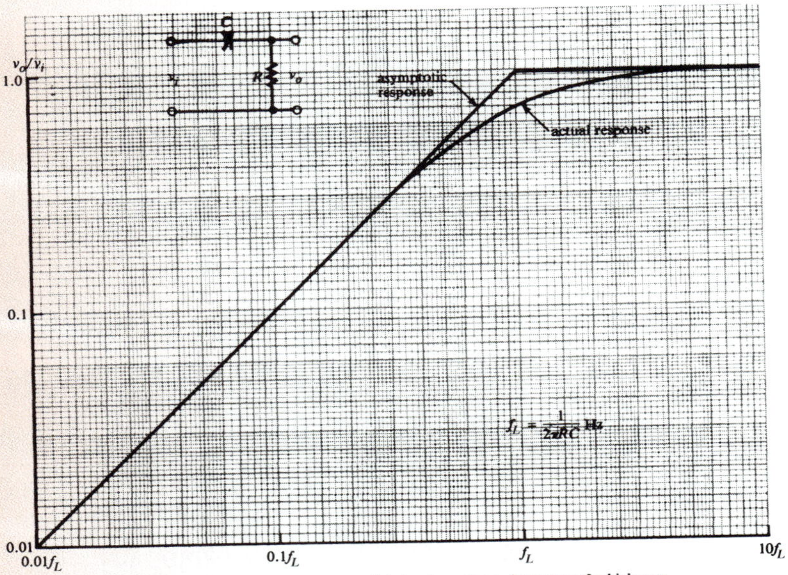

(a) Normalized frequency-response plot showing gain vs. frequency of a high-pass RC network. f_L = lower cutoff frequency. Both gain and frequency are plotted on logarithmic scales.

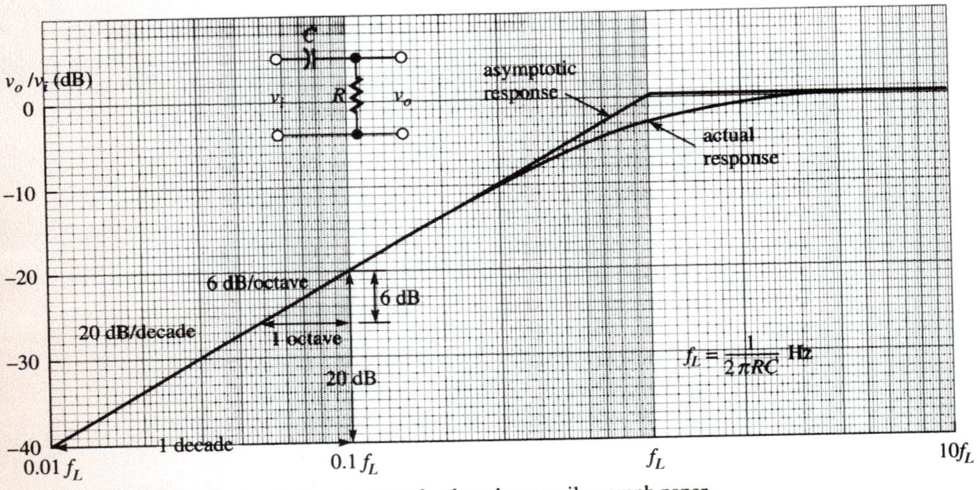

(b) Gain vs. frequency for the high-pass network, plotted on semilog graph paper. Note that the vertical axis is linear and is scaled in dB.

FIGURE 9–9 Normalized gain and phase plots for a high-pass RC network

condition $|v_i|/|v_S| = 1$, when the capacitive reactance is negligibly small. Of course, the *overall* gain of an amplifier having a capacitor-coupled input may be greater than 1 at frequencies above cutoff, but the shape of the frequency response is the same as that shown in Figure 9–9. The only difference is that the gain above cutoff is A_m—the amplifier's midband gain—rather than 1 (0 dB).

Note that the plots in Figure 9–9 extend two decades below and one decade above the cutoff frequency, which is labeled f_L. Figure 9–9(b) shows that the gain is "down" approximately 6 dB at a frequency one octave below cutoff ($0.5f_L$) and is down 20 dB one decade below cutoff (at $0.1f_L$). Figure 9–9(c) shows that the phase shift is 63° one octave below f_L and 86° one decade below f_L. The gain is down 0.04 dB and the phase shift is 6° at a frequency one decade above cutoff ($10f_L$). The gain plots also show the straight line *asymptotes* that the gain approaches at frequencies below cutoff. An

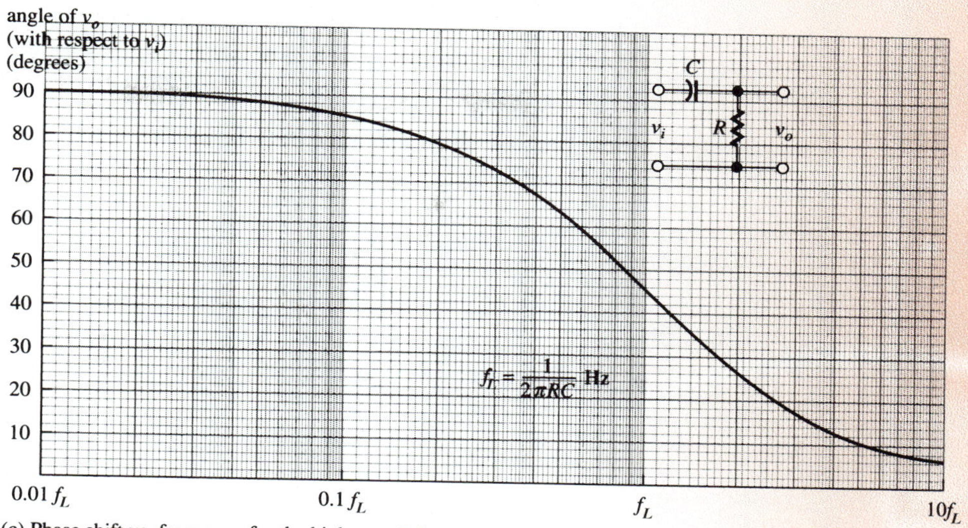

(c) Phase shift vs. frequency for the high-pass RC network. v_o leads v_i.

FIGURE 9–9 (Continued)

asymptote is often used to approximate the gain response. Note that it "breaks" downward at f_L, where its deviation from the actual response curve is the greatest (3 dB). The frequency at which an asymptote breaks (f_L in this example) is called a *break* frequency. *The asymptote has a slope of 6 dB/octave, or 20 dB/decade.*

When using these plots, remember that they are valid for only a *single* resistor–capacitor (RC) combination. We will study the effects of multiple RC combinations in the signal flow path in a later discussion. The plots in Figure 9–9 apply to any RC network in which the capacitor is in series with the signal path and the output is taken across the resistor. Such a network is called a *high-pass filter,* because, as the plots show, the gain is constant at high frequencies (above cutoff) and "falls off," at a constant rate, below cutoff. To find the overall gain of an RC network in which part of the resistance (r_S) is in the signal source, the gain determined from the plot must be multiplied by the factor $K = r_i/(r_S + r_i)$.

EXAMPLE 9–6

The amplifier shown in Figure 9–10 has voltage gain $|v_L|/|v_i| = 120$. Calculate

1. the lower cutoff frequency due to the input coupling capacitor,
2. the gain $|v_L|/|v_S|$ one octave below cutoff, and
3. the phase shift one decade above cutoff.

FIGURE 9–10 (Example 9–6)

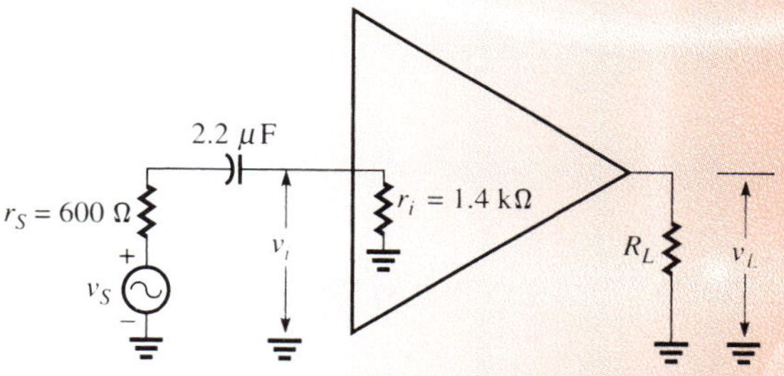

Use Figure 9–9 to find approximate values for

4. the asymptotic gain $|v_L|/|v_S|$ one octave below cutoff,
5. the gain $|v_L|/|v_S|$, in dB, at $f = 10.85$ Hz, and
6. the frequency at which the phase shift is 20°.

Solution

1. From equation 9–16,

$$f_L = \frac{1}{2\pi(1400 + 600)(2.2\ \mu\text{F})} = 36.2\ \text{Hz}$$

2. The frequency one octave below cutoff is $f_L/2 = 18.1$ Hz. From equation 9–17, with $f_L/f = 2$,

$$\frac{|v_i|}{|v_S|} = \left(\frac{r_i}{r_S + r_i}\right)\frac{1}{\sqrt{1 + (f_L/f)^2}} = \left(\frac{1400\ \Omega}{2000\ \Omega}\right)\frac{1}{\sqrt{1 + 2^2}} = 0.313$$

Therefore, $\dfrac{|v_L|}{|v_S|} = 0.313\,(120) = 37.6$

3. At one decade above cutoff, $f = 10f_L$, or $f_L/f = 0.1$. Assuming that the amplifier does not cause any phase shift beyond that due to the coupling capacitor, the phase shift is, from equation 9–20, $\arctan(0.1) = 5.71°$.
4. One octave below cutoff, $f = 0.5f_L$. From Figure 9–9(a), the gain at the intersection of the asymptote and the $0.5f_L$ coordinate line is 0.5. Therefore, the overall *asymptotic* gain is

$$\frac{|v_L|}{|v_S|} = 0.5\left(\frac{r_i}{r_S + r_i}\right)120 = 0.5\left(\frac{1400\ \Omega}{2000\ \Omega}\right)120 = 42$$

(Compare with the actual gain of 37.6, computed in part 2.)

5. At $f = 10.85$ Hz, $f/f_L = 10.85/36.2 = 0.3$, so $f = 0.3\,f_L$. From Figure 9–9(b), the gain curve intersects the 0.3 coordinate line at approximately −11 dB. The midband gain $|v_L|/|v_S|$ in decibels is

$$20\log_{10}\left[\left(\frac{1400\ \Omega}{600\ \Omega + 1400\ \Omega}\right)(120)\right] = 38.5\ \text{dB}$$

Therefore, the gain at 10.85 Hz is 38.5 − 11 = 27.5 dB.

6. From Figure 9–9(c), the 20° phase coordinate intersects the curve at approximately 2.8 f_L. Therefore, the phase shift is 20° at (2.8)(36.2 Hz) = 101.4 Hz.

The smaller the desired lower cutoff frequency of an amplifier, the larger the input coupling capacitor must be. If the input resistance of the amplifier is small, the required capacitance may be impractically large. For example, if an audio amplifier having input resistance of 100 Ω is to have a lower cutoff frequency of 15 Hz, then, assuming that $r_S = 0$, the coupling capacitor must have value

$$C_1 = \frac{1}{2\pi(100\ \Omega)(15\ \text{Hz})} = 106\ \mu\text{F}$$

On the other hand, if the input resistance is 10 kΩ, the required capacitance is 1.06 μF, a much more reasonable value.

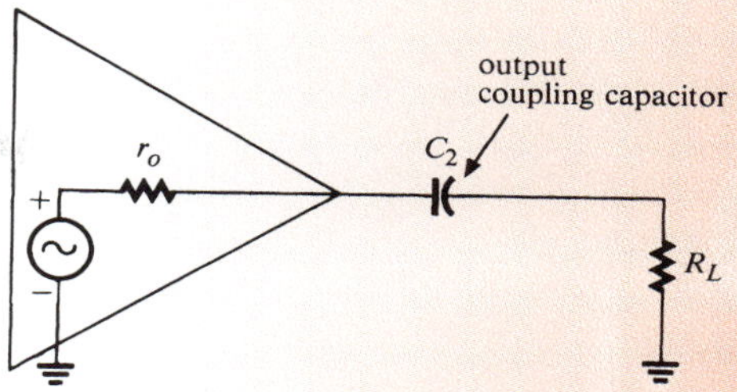

FIGURE 9–11 The output and load resistance of the amplifier and the output coupling capacitor form an RC network that reduces the signal reaching the load at low frequencies

Figure 9–11 shows the resistance–capacitance combination formed by the output resistance, output coupling capacitor, and load resistance of an amplifier. The effect of the output coupling capacitor on the frequency response is the same as that of the input coupling capacitor: At low frequencies, there is a significant voltage drop across the capacitive reactance, so less signal is delivered to the load. The break (or corner) frequency due to the output coupling capacitor (C_2) is found in the same way in which we found f_L due to C_1. It is the frequency at which the capacitive reactance, X_{C_2}, equals the resistance, $r_o + R_L$. To distinguish between the two corner frequencies, we will hereafter write f_1 and f_2:

$$f_1 = \frac{1}{2\pi(r_i + r_S)C_1} \qquad \text{(9–21)}$$

$$f_2 = \frac{1}{2\pi(r_o + R_L)C_2} \qquad \text{(9–22)}$$

If the lower cutoff frequency were determined by C_2 alone, then the frequency response in the vicinity of cutoff would have the same appearance as the normalized plots shown in Figure 9–9. However, when both input and output coupling capacitors are present, the low-frequency response is significantly different because both capacitors contribute *simultaneously* to a reduction in gain. Let us first consider the case where the frequencies f_1 and f_2 are considerably different, at least a decade apart. Then, to a good approximation, the overall frequency response will have a lower cutoff frequency, f_L, equal to the *larger* of f_1 and f_2. Figure 9–12 shows the frequency response for the case $f_1 = 10$ Hz and $f_2 = 100$ Hz. The midband gain is assumed to be 20 dB. It is clear that $f_L = 100$ Hz because the gain at that frequency is $20 - 3 = 17$ dB. The gain at 10 Hz is much lower (approximately -3 dB) because at frequencies below 100 Hz *both* C_1 and C_2 cause gain reduction.

Note that in Figure 9–12(a) the gain asymptote has slope 20 dB/decade (6 dB/octave) between 10 Hz and 100 Hz but breaks downward at 10 Hz with a slope of 40 dB/decade (12 dB/octave). There are, therefore, two break frequencies in this example, i.e., two frequencies where the asymptote changes slope: 10 Hz and 100 Hz. The phase shift is approximately 51° at $f_L = 100$ Hz and 129° at 10 Hz and approaches 180° as frequency approaches 0. The response characteristics shown in Figure 9–12 are valid for two high-pass RC networks connected in series, provided the networks are *isolated* from each other, in the sense that one does not load the other. In our illustration, the two networks are assumed to be isolated by an amplifier having gain 20 dB.

If the frequencies f_1 and f_2 are closer than one decade to each other, then the overall lower cutoff frequency is somewhat higher than the larger of the two. In such cases, it is usually adequate to assume that f_L equals the larger of the two. The exact value of f_L can be found by solving the following equation (derived in Appendix C):

$$f_L^4 - (f_1^2 + f_2^2)f_L^2 - f_1^2 f_2^2 = 0 \qquad \text{(9–23)}$$

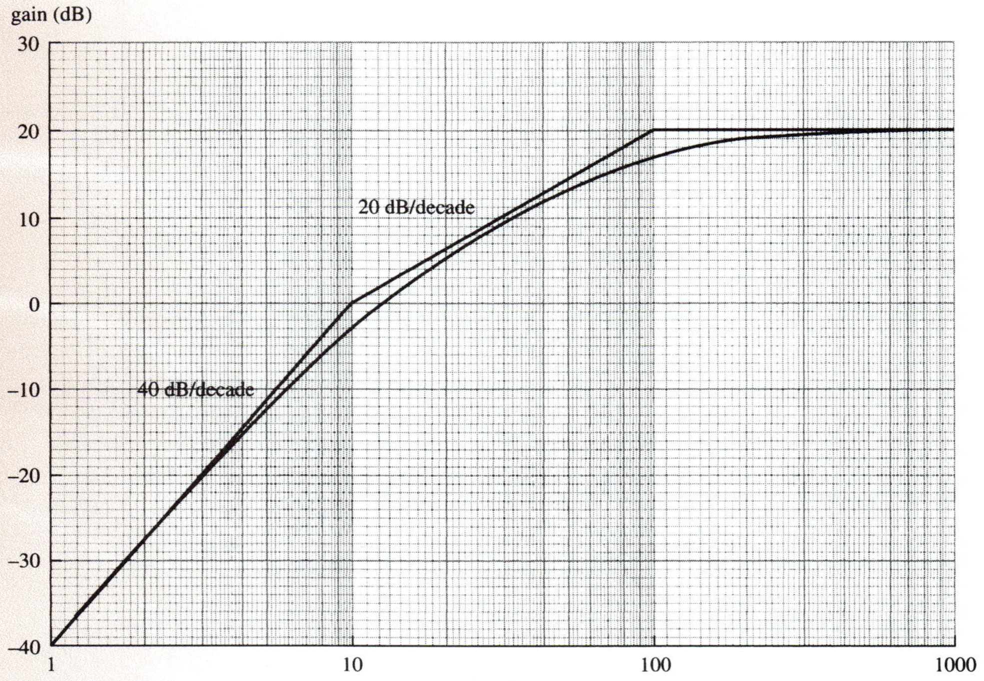

(a) Gain vs. frequency for an amplifier having $f_1 = 10$ Hz and $f_2 = 100$ Hz. The lower cutoff frequency in this case is the larger of the two, $f_L = 100$ Hz. Note that there are two "break" frequencies. $A_m = 20$ dB.

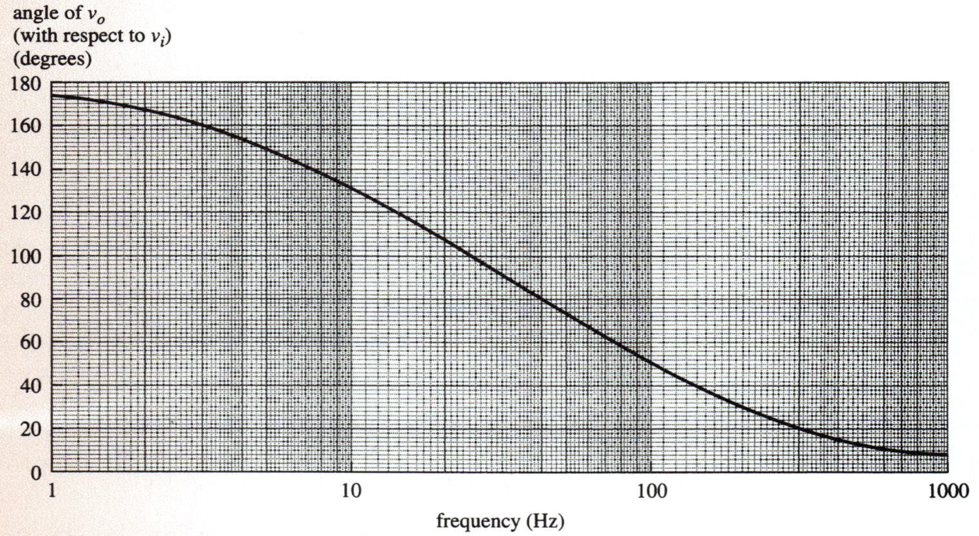

(b) Phase shift vs. frequency for the amplifier whose gain plot is shown in Figure 9-12 (a). Note that the phase angle approaches 180° as frequency approaches 0.

FIGURE 9–12 Gain and phase versus frequency for an amplifier having two break frequencies

Notice that this equation is a quadratic in f_L^2. The solution using the general quadratic formula yields, then, f_L^2.

If $f_1 = f_2$, equation 9–23 can be used to show that the overall lower cutoff frequency is 1.55 times the value of either. For example, if $f_1 = f_2 = 100$ Hz, then $f_L = 155$ Hz. Figure 9–13 shows the normalized gain and phase response

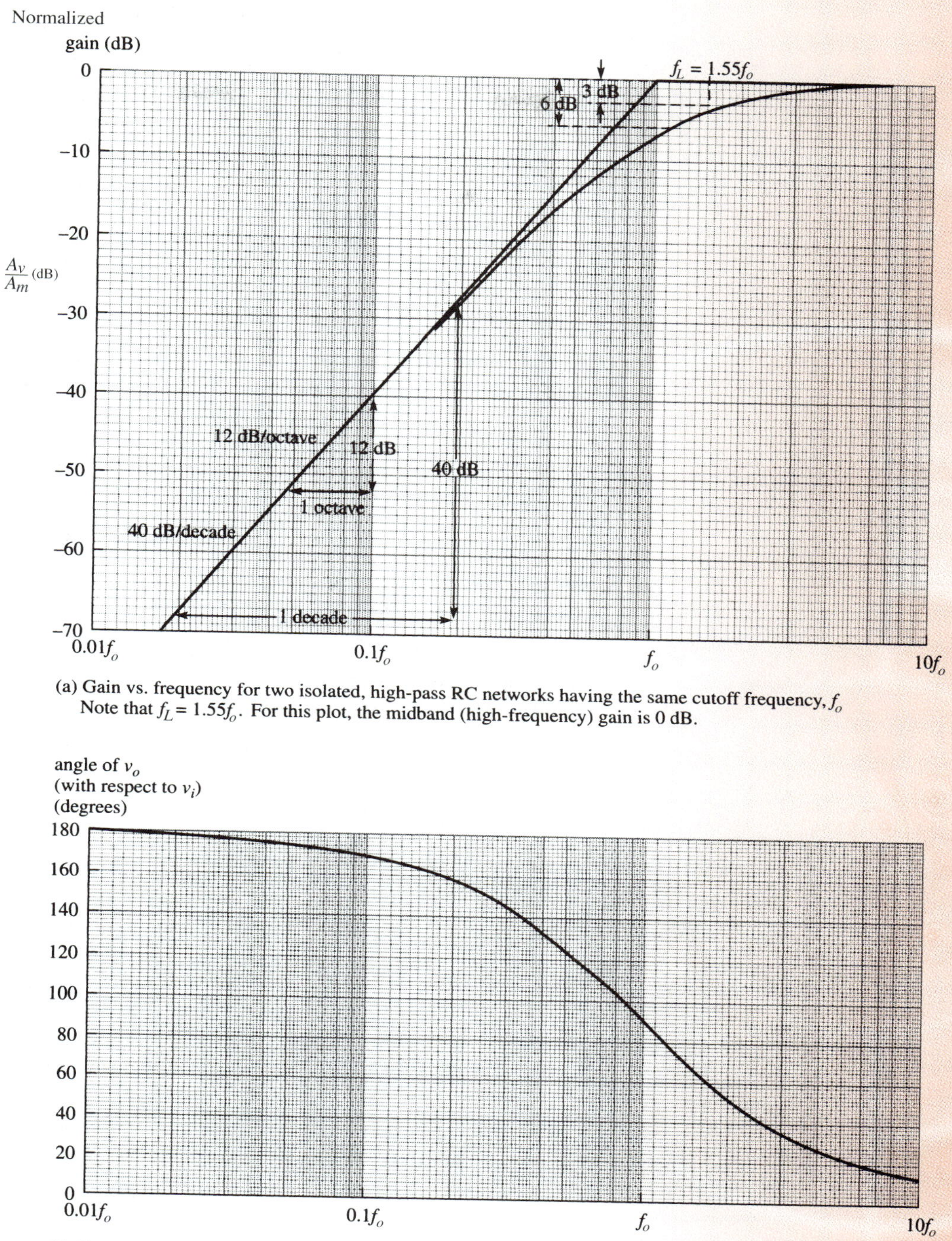

(a) Gain vs. frequency for two isolated, high-pass RC networks having the same cutoff frequency, f_o. Note that $f_L = 1.55f_o$. For this plot, the midband (high-frequency) gain is 0 dB.

(b) Phase shift vs. frequency for the two high-pass RC networks whose gain is shown in Figure 9-13(a). Note that the output leads the input by 90° at frequency f_o.

FIGURE 9–13 Normalized gain and phase response for the case $f_1 = f_2 = f_0$

for the special case $f_1 = f_2$. Note that the asymptote breaks downward at $f_o = f_1 = f_2$ and has a slope of 40 dB/decade, or 12 dB/octave. The actual response is 6 dB below the asymptote at that frequency. The cutoff frequency is $f_L = 1.55\,f_o$, where the gain is down 3 dB. Note that the break frequency (f_o) is not the same as the cutoff frequency (f_L) in this case. The phase shift is 90° at f_o and approaches 180° as frequency approaches 0. Once again, the plots shown in Figure 9–13 are valid only when the two high-pass RC networks whose

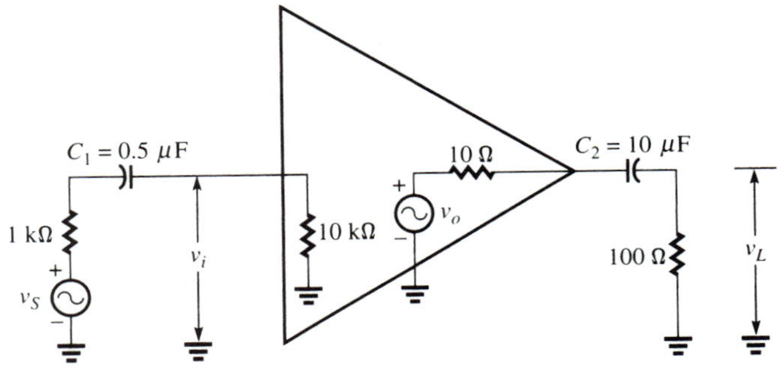

FIGURE 9–14 (Example 9–7)

response they represent are isolated from each other. In our case, we assume that the isolation is provided by an amplifier with gain A_m at midband.

EXAMPLE 9–7

The amplifier shown in Figure 9–14 has midband gain $|v_L|/|v_S| = 140$. Find

1. the approximate lower cutoff frequency,
2. the gain $|v_o|/|v_i|$,
3. the lower cutoff frequency when C_2 is changed to 50 μF,
4. the approximate gain $|v_L|/|v_S|$, in dB, at 2.9 Hz with $C_2 = 50$ μF, and
5. the value that C_2 would have to be in order to obtain a lower cutoff frequency of approximately 200 Hz.

Solution

1. From equation 9–21,

$$f_1 = \frac{1}{2\pi(1 \times 10^3 + 10 \times 10^3)(0.5 \times 10^{-6})} = 29 \text{ Hz}$$

From equation 9–22,

$$f_2 = \frac{1}{2\pi(10 + 100)(10 \times 10^{-6})} = 145 \text{ Hz}$$

Therefore, $f_L \approx 145$ Hz. (Actually 150 Hz using equation 9–23.)

2. $$\frac{|v_L|}{|v_S|} = \frac{|v_i|}{|v_S|}\frac{|v_o|}{|v_i|}\frac{|v_L|}{|v_o|}$$

$$140 = \left[\frac{10\text{ k}\Omega}{(10\text{ k}\Omega) + (1\text{ k}\Omega)}\right]\frac{|v_o|}{|v_i|}\left[\frac{100\ \Omega}{(10\ \Omega) + (100\ \Omega)}\right]$$

$$\frac{|v_o|}{|v_i|} = \frac{140}{\left(\dfrac{1 \times 10^4}{1.1 \times 10^4}\right)\left(\dfrac{1 \times 10^2}{1.1 \times 10^2}\right)} = 169.4$$

3. $$f_2 = \frac{1}{2\pi(10 + 100)(50\ \mu\text{F})} = 29 \text{ Hz}$$

Therefore, $f_1 = f_2 = 29$ Hz, so $f_L = 1.55(29) = 45$ Hz.

4. 2.9 Hz is one decade below the break frequency of 29 Hz. Because the gain falls at the rate of 40 dB/decade below the break frequency, at 2.9 Hz it will be 40 dB less than A_m:

$$A_m = 20\log_{10}140 = 42.9 \text{ dB}$$
$$A_v \text{ (at 2.9 Hz)} = 42.9 - 40 = 2.9 \text{ dB}$$

5. Using the approximation $f_L \approx f_1 = 200$ Hz,

$$200 = \frac{1}{2\pi(10 + 100)C_2}$$
$$C_2 = 7.2\ \mu\text{F}$$

9–4 SHUNT CAPACITANCE AND HIGH-FREQUENCY RESPONSE

Capacitance that provides an ac path between an amplifier's signal flow path and ground is said to *shunt* the signal. The most common form of shunt capacitance is that which exists between the terminals of an electronic device due to its structural characteristics. Recall, for example, that a *pn* junction has capacitance between its terminals because the depletion region forms a dielectric separating the two conductive regions of *p* and *n* material. Capacitance between device terminals is called *interelectrode* capacitance. Shunt capacitance is also created by wiring, terminal connections, solder joints, and any other circuit structure where conducting regions are close to each other. This type of capacitance is called *stray* capacitance.

Shunt capacitance affects the high-frequency performance of an amplifier because at high frequencies the small capacitive reactance diverts the signal to ground or to some other point besides the load. A semiconductor device has inherent frequency limitations that may impose more severe restrictions on high-frequency operation than the effect of shunt capacitance we will study now, but these *device* considerations will be temporarily postponed. Figure 9–15 shows shunt capacitance C_A between the input side of an amplifier and ground (in parallel with r_i). Notice that we can now neglect series coupling capacitance because we are considering only high-frequency operation.

In the midband frequency range, the effect of C_A in Figure 9–15 can be neglected because the frequency is not high enough to make the capacitive reactance small. In other words, the midband frequency range is that range of frequencies that are high enough to neglect coupling capacitance and low enough to neglect shunt capacitance. It is apparent in the figure that, in the midband frequency range,

$$\left|\frac{v_i}{v_S}\right| = \frac{r_i}{r_S + r_i} \qquad (9\text{–}24)$$

When the frequency is high enough to consider the effect of C_A, we must replace r_i in (9–24) by the parallel combination of r_i and $-jX_{C_A}$:

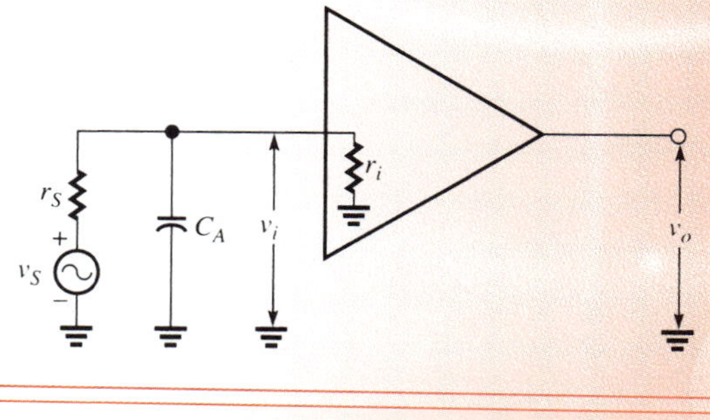

FIGURE 9–15 Shunt capacitance C_A in parallel with the input of an amplifier affects its high-frequency response

$$\frac{v_i}{v_S} = \frac{-jX_{C_A} \| r_i}{r_S + (-jX_{C_A} \| r_i)} \tag{9–25}$$

At very high frequencies, where X_{C_A} becomes very small, the parallel combination of X_{C_A} and r_i becomes very small, and the net effect is the same as if the input impedance of the amplifier were made small. As we know, the consequence of that result is that the overall gain is reduced due to the voltage division between r_S and the input impedance. Our goal now is to find the frequency at which that reduction in gain equals 0.707 times the midband value given by equation 9–24. From equation 9–25,

$$\frac{v_i}{v_S} = \frac{r_i/(1 + j\omega r_i C_A)}{r_S + r_i/(1 + j\omega r_i C_A)} = \frac{r_i}{r_S + j\omega r_S r_i C_A + r_i}$$
$$= \frac{r_i}{r_i + r_S} \cdot \frac{1}{1 + j\omega r_S r_i C_A/(r_S + r_i)} = \frac{r_i}{r_S + r_i} \cdot \frac{1}{1 + \dfrac{j\omega}{\omega_A}}$$

and

$$\left|\frac{v_i}{v_S}\right| = \frac{r_i}{r_S + r_i} \cdot \frac{1}{\sqrt{1 + \left(\dfrac{\omega}{\omega_A}\right)^2}}$$

where $\omega_A = 1/(r_i \| r_S)C_A$ rad/sec. Notice that when $\omega = \omega_A$, $|v_i/v_S| = 0.707 r_i/(r_i + r_S)$, meaning that ω_A is the break (or 3-dB) frequency for the input circuit. In terms of the cyclic frequency f in Hz, the expression can be written as

$$\left|\frac{v_S}{v_i}\right| = \frac{r_i}{r_S + r_i} \cdot \frac{1}{\sqrt{1 + \left(\dfrac{f}{f_A}\right)^2}} \tag{9–26}$$

where

$$f_A = \frac{1}{2\pi(r_S \| r_i)C_A} \tag{9-27}$$

Equation 9–27 shows that the break frequency due to C_A is inversely proportional to the parallel combination of r_S and r_i. Thus, to achieve a large bandwidth it is necessary to make $r_S \| r_i$ as small as possible, preferably by making r_S small. Theoretically, if $r_S = 0$, then $r_S \| r_i = 0$, and $f_A = \infty$. In practice, it is not possible to have $r_S = 0$, but a very small value of r_S can increase the upper cutoff frequency significantly (not always a desirable practice, as we shall see in later discussions).

An RC network in which the resistor is in series with the signal flow path and the output is taken across the shunt capacitor is called a *low-pass filter.* The cutoff frequency f_c is determined by $f_c = 1/(2\pi RC)$ Hz. Comparing with equation 9–27, we see that the input of an amplifier having shunt capacitance C_A is the same as a low-pass filter having $R = r_S \| r_i$ and $C = C_A$. Normalized gain and phase plots for the low-pass RC network are shown in Figure 9–16. Note that the gain plot is the mirror image of that of the high-pass RC network shown in Figure 9–9. The asymptote shows that the gain falls off at the rate of 6 dB/octave, or 20 dB/decade, at frequencies above f_c. The phase shift approaches −90° at frequencies above f_c. Note that negative phase means that the output voltage *lags* the input voltage, in contrast to the high-pass RC network, where the output leads the input. The gain and phase equations as functions of the ratio f/f_c are

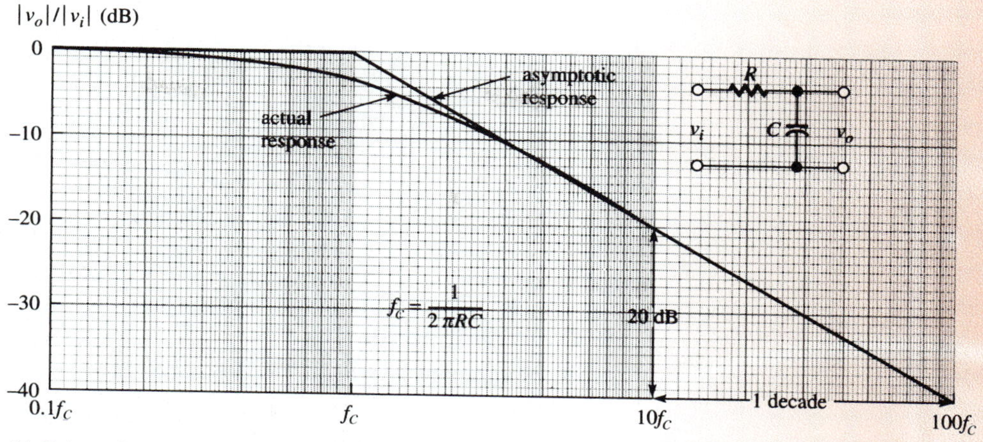

(a) Gain vs. frequency for the low-pass RC network. f_C = upper cutoff frequency.

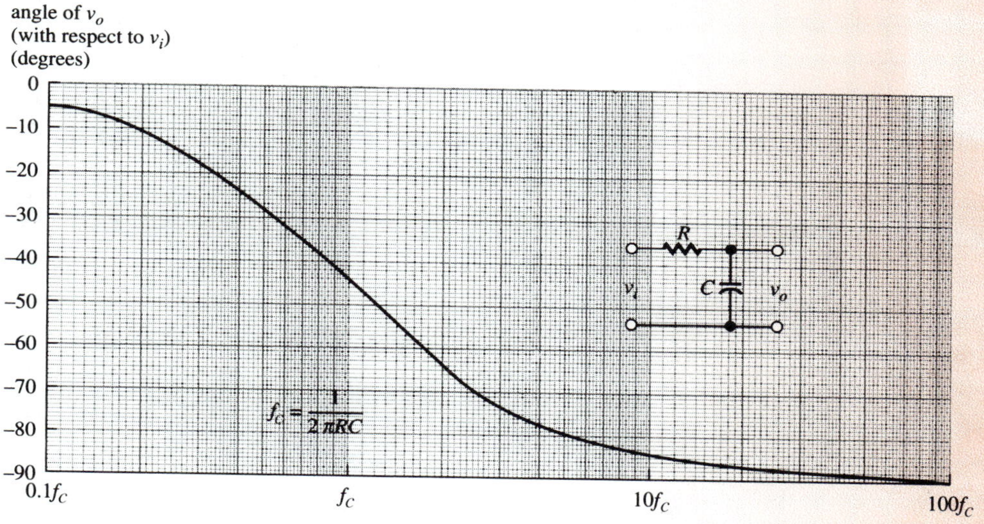

(b) Phase shift vs. frequency for the low-pass RC network. Note that v_o lags v_i.

FIGURE 9–16 Gain and phase versus frequency for a low-pass RC network

$$|A| = \frac{1}{\sqrt{1 + (f/f_c)^2}} = \left|\frac{v_o}{v_i}\right| \tag{9–28}$$

and

$$\angle A = -\arctan(f/f_c) \tag{9–29}$$

where $f_c = 1/(2\pi RC)$, $R = r_S \| r_i$, and $C = C_A$, for an amplifier having input shunt capacitance C_A.

EXAMPLE 9–8

The amplifier shown in Figure 9–17 has midband gain $|v_L/v_S| = 40$ dB. Calculate

1. the upper cutoff frequency f_H,
2. the gain $|v_L/v_S|$, in dB, at $f = 20$ MHz, and
3. the phase shift at $f = f_H$

FIGURE 9–17 (Example 9–8)

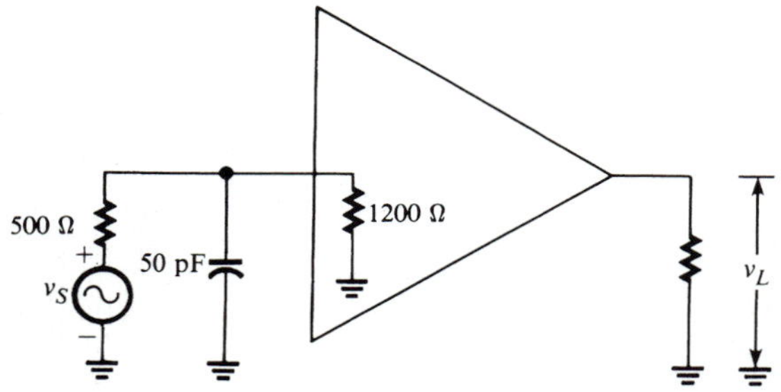

Use Figure 9–16 to find

4. the frequency at which $|v_L|/|v_S|$ is 35 dB, and
5. the frequency at which the output voltage lags the input voltage by 30°.

Solution

1. Since $r_S \| r_i = 500\ \Omega \| 1200\ \Omega = 353\ \Omega$, we have, from equation 9–27,

$$f_A = f_H = \frac{1}{2\pi(353)(50\text{ pF})} = 9.02\text{ MHz}$$

2. From equation 9–28, the gain of the equivalent RC network at the amplifier input when $f = 20$ MHz is

$$|A| = \frac{1}{\sqrt{1 + \left(\dfrac{20\text{ M}}{9.02\text{ M}}\right)^2}} = 0.441$$

$$20 \log_{10}(0.441) = -7.1\text{ dB}$$

Therefore, the overall gain $|v_L|/|v_S|$ at $f = 20$ MHz is (40 dB) − (7.1 dB) = 32.9 dB.

3. From equation 9–29, when $f = f_H$, $\angle A = -\arctan(f_H/f_H) = -\arctan(1) = -45°$. We conclude that *the output of a low-pass RC network lags the input by 45° at the cutoff frequency.* (See Figure 9–16(b).)

4. An overall gain of 35 dB corresponds to a 5-dB drop in gain from the midband value of 40 dB. From Figure 9–16(a), the gain of the equivalent RC network at the amplifier input is down 5 dB at approximately $f = 1.5f_c = 1.5(9.02\text{ MHz}) = 13.5$ MHz.

5. From Figure 9–16(b), the frequency at which the phase shift is −30° is approximately $f = 0.57f_c = 0.57(9.02\text{ MHz}) = 5.14$ MHz.

Figure 9–18 shows shunt capacitance C_B connected across the output of an amplifier. The output resistance r_o of the amplifier, the load resistance R_L, and the capacitance C_B form a low-pass RC network that has the same effect on the high-frequency response as the low-pass network at the input: The gain falls off because the impedance to ground decreases with increasing frequency. The upper corner frequency due to C_B is derived in the same way as that due to C_A.

$$f_A = \frac{1}{2\pi(r_S \| r_i)C_A} \tag{9–30}$$

$$f_B = \frac{1}{2\pi(r_o \| R_L)C_B} \tag{9–31}$$

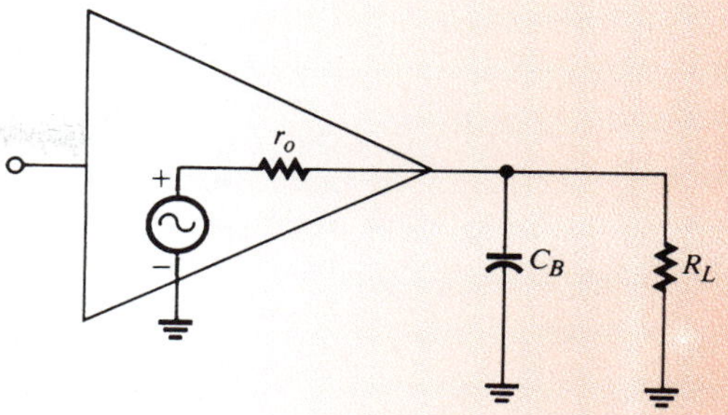

FIGURE 9–18 An amplifier having shunt capacitance C_B across its output

When there is shunt capacitance at both the input and the output of an amplifier, the frequency response is different than it would be if only one were present. If f_A and f_B are not close in value, then the actual upper cutoff frequency f_H is approximately equal to the *smaller* of f_A and f_B. The gain is asymptotic to a line that falls off at 20 dB/decade (6 dB/octave) between f_A and f_B and is asymptotic to a line that falls off at 40 dB/decade (12 dB/octave) at frequencies above the larger of the two. See Figure 9–19(a). At the higher frequencies, the gain falls off at twice the rate it would for a single low-pass network, because both C_A and C_B contribute *simultaneously* to gain reduction. The total phase shift approaches −180° at high frequencies. The exact value of the upper cutoff frequency in terms of f_A and f_B can be found by solving (Appendix C)

$$f_H^4 + (f_A^2 + f_B^2)\, f_H^2 - f_A^2 f_B^2 = 0 \tag{9–32}$$

in the same way as equation 9–23.

If $f_A = f_B$, equation 9–32 can be used to show that the cutoff frequency is $f_H = 0.645\, f_A$. In that case, there is but one break frequency, and the single asymptote has slope −40 dB/decade, or −12 dB/octave. See Figure 9–19(b). When the two frequencies are equal, the total phase shift is −90° at that frequency.

Thévenin Equivalent Circuits at Input and Output

We note that the same *form* of equation is used to compute both the lower and upper corner frequencies of an amplifier: $f = 1/2\pi RC$. In the case of f_L, C is either input or output coupling capacitance, and in the case of f_H, C is input or output shunt capacitance. In *both* cases, R is the Thévenin equivalent resistance seen by the capacitance. This observation provides an easy way to remember the equations for calculating corner frequencies:

$$f = \frac{1}{2\pi r_{\mathrm{TH}} C} \tag{9–33}$$

where r_{TH} is the Thévenin equivalent resistance with respect to the capacitor terminals at input or output. Recall that the Thévenin equivalent resistance is found by open-circuiting the capacitor terminals and computing the total equivalent resistance looking into those terminals when all voltage sources are replaced by short circuits. As an exercise, verify that the equations for f_1, f_2, f_A, and f_B can also be found using the Thévenin equivalent resistance with respect to the capacitor terminals.

The Overall Picture

When we look at the overall frequency response of a capacitively coupled amplifier, we see two main segments: the low-frequency behavior due to the coupling capacitors C_1 and C_2, and the high-frequency behavior due to the

FIGURE 9–19 High-frequency gain and phase response when shunt capacitance is present at both the input and the output of an amplifier. f_A = break frequency due to input shunt capacitance, f_B = break frequency due to output shunt capacitance.

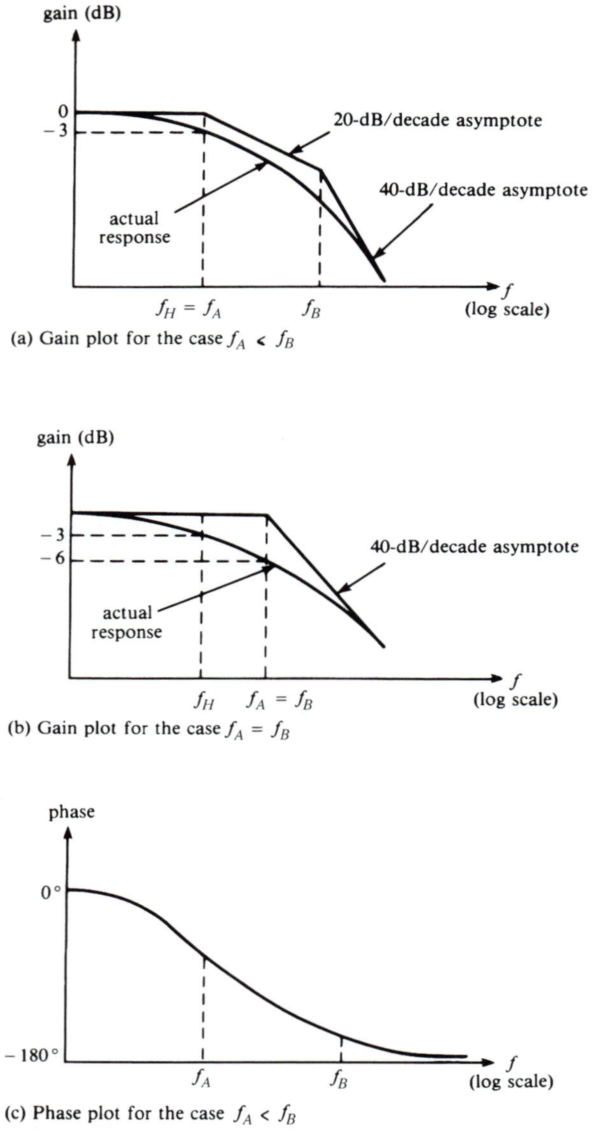

shunt capacitances C_A and C_B. Each of these capacitances produces a corner frequency, thus resulting in two corner frequencies on the lower side (f_1 and f_2) and two on the upper side (f_A and f_B).

On the lower side, if the corner (or break) frequencies f_1 and f_2 are separated by close to a decade or more, then the lower cutoff frequency f_L will be approximately equal to the larger of the two. But what do we mean by "close to a decade"? It is certainly not a factor of 3 or 4, but we can say that a factor of 7 or 8 is fairly close to a decade. The more separated f_1 and f_2 are, the closer the cutoff frequency will be to the larger of the two. For example if $f_1 = 4$ Hz and $f_2 = 50$ Hz, then f_L is 50 Hz without a doubt, because they are separated by (slightly) more than a decade. In that case, we say that the larger corner frequency is the "dominant" corner frequency because it influences the cutoff frequency, f_L, the most.

When f_1 and f_2 are identical (coincident corner frequencies), the cutoff frequency f_L is 1.55 times either frequency. However, if the corner frequencies are different but not sufficiently separated, the cutoff frequency will be "pushed up" somewhat above the larger of the two. In this case, we obtain f_L by solving equation 9–23 using the general quadratic formula. Note that when $a = 1$, the general solution for $ax^2 + bx + c = 0$ can be simplified to

$$x = -\frac{b}{2} \pm \sqrt{(b/2)^2 - c}$$

Regardless of the relative location of the lower corner frequencies, the equation for the magnitude of the gain as a function of frequency on the lower side is

$$|A_{vs}| = \frac{A_m}{\sqrt{1 + \left(\frac{f_1}{f}\right)^2}\sqrt{1 + \left(\frac{f_2}{f}\right)^2}} \tag{9–34}$$

where A_m is the magnitude of the overall gain in the midband. Remember that the response on the lower side is of high-pass nature.

The phase response on the lower side, which approaches $+180°$ as f approaches zero and goes to $0°$ in the midband, is given by

$$\phi_L(f) = \arctan(f_1/f) + \arctan(f_2/f) \tag{9–35}$$

On the upper side of the frequency response, the two upper corner frequencies are f_A (due to C_A) and f_B (due to C_B). The upper cutoff frequency (or 3-dB frequency) depends on their relative values (location on the f-axis). If they are sufficiently separated, the "dominant" corner frequency will be the smaller of the two and its value will approximate the upper cutoff frequency, f_H. When they coincide, the cutoff frequency f_H is 0.645 times either frequency. If they are not sufficiently separated, however, the cutoff frequency is "pushed down" from the smaller of the two corner frequencies, requiring the use of equation 9–32 to solve for f_H.

The equation for the magnitude of the gain as a function of frequency on the upper side, which is of low-pass character, is

$$|A_{vs}| = \frac{A_m}{\sqrt{1 + \left(\frac{f}{f_A}\right)^2}\sqrt{1 + \left(\frac{f}{f_B}\right)^2}} \tag{9–36}$$

and the overall phase response on the upper side is

$$\phi_H(f) = -\arctan(f/f_A) - \arctan(f/f_B) \tag{9–37}$$

Note that the phase on the upper side approaches zero when f is much smaller than the smaller of the two upper corner frequencies, that is, well into the midband. For f values much larger than the larger of the two, the phase approaches $-180°$. Figure 9–20 shows the overall response for gain magnitude and phase (Bode plot) of an amplifier having distinct corner frequencies on both ends. Note that the midband, shown broken in the figure, is typically three to four decades wide for an audio amplifier. On the other hand, video amplifiers have a much wider bandwidth, from about 15 Hz to more than 5 MHz.

Miller-Effect Capacitance

Figure 9–21 shows an amplifier having capacitance C_C connected between its input and output terminals. The most common example of such capacitance is interelectrode capacitance, as, for example, between the base and the collector of a common-emitter amplifier or between the gate and the drain of

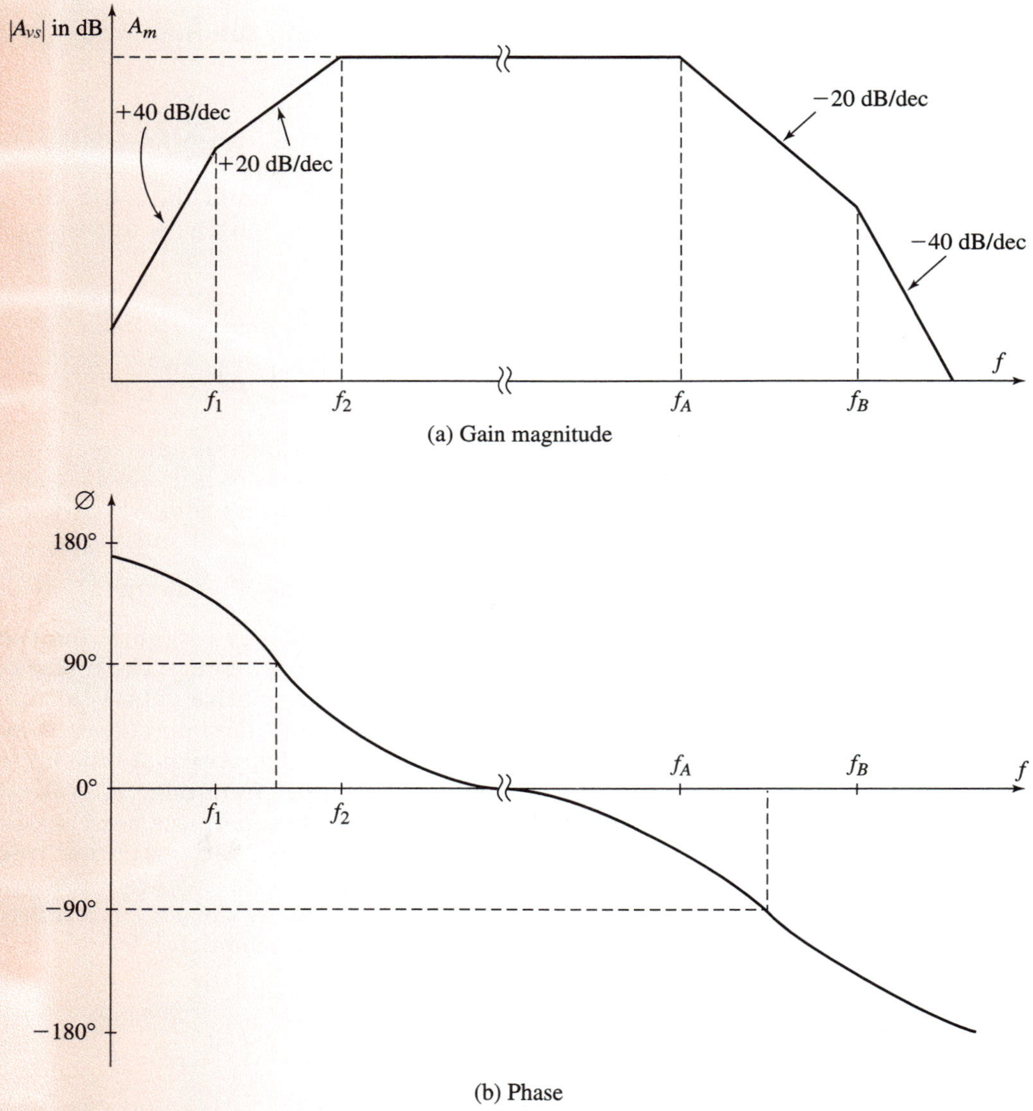

FIGURE 9–20 Overall frequency response of an amplifier with distinct corner frequencies

a common-source amplifier. This capacitance forms a *feedback* path for ac signals, and it can have a significant influence on the high-frequency response of an amplifier.

The conclusion in the derivation that follows is based on the method used in circuit analysis to determine the total admittance (y) of components connected in parallel, which we will review briefly now. Recall that admittance is the sum of conductance g and susceptance b: $y = g + jb$ siemens. Also recall that the total admittance of parallel-connected components is the *sum* of their individual admittances. The susceptance of capacitance C is $b = j\omega C$ S. In particular, the total admittance of resistance R in parallel with capacitance C at ω rad/s is

$$y = g + jb = 1/R + j\omega C$$

Thus, for example, if the *input admittance* to a circuit were $0.01 + j0.02$ S at frequency $\omega = 1000$ rad/s, we would conclude that the input circuit consisted

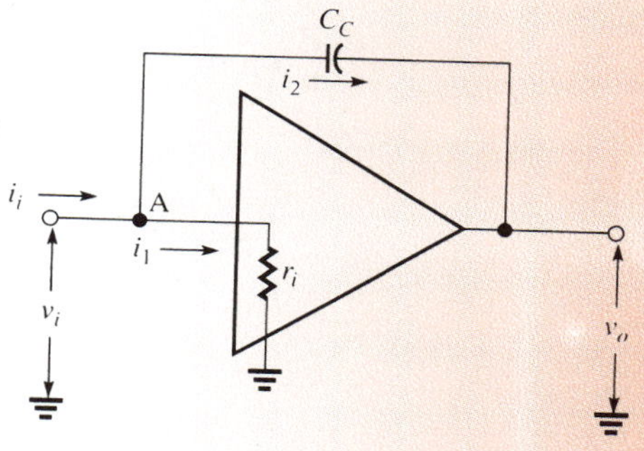

FIGURE 9–21 Capacitance C_C connected between the input and the output of an amplifier affects its high-frequency response due to the Miller effect

of resistance $R = 1/g = 1/0.01 = 100\ \Omega$ in parallel with capacitance $C = b/\omega = 0.02/1000 = 20\ \mu F$.

Writing Kirchhoff's current law at the node labeled A in Figure 9–21, we have

$$i_i = i_1 + i_2$$

The current i_2 that flows in the capacitor is the difference in voltage across it divided by the capacitor's impedance:

$$i_2 = \frac{v_i - v_o}{-jX_{C_C}}$$

Therefore,

$$i_i = \frac{v_i}{r_i} + \frac{v_i - v_o}{-jX_{C_C}} \qquad (9\text{–}38)$$

Let the amplifier gain be $A_v = v_o/v_i$. Substituting $v_o = A_v v_i$ in (9–38) gives

$$i_i = \frac{v_i}{r_i} + \frac{v_i - A_v v_i}{-jX_{C_C}} = v_i\left(\frac{1}{r_i} + \frac{1 - A_v}{-jX_{C_C}}\right)$$

Then

$$i_i = v_i\left[\frac{1}{r_i} + j\omega C_C(1 - A_v)\right] \qquad (9\text{–}39)$$

Equation 9–39 shows that the input admittance consists of the conductance component $1/r_i$ and the capacitive susceptance component $\omega C_C(1 - A_v)$. Note that *this input admittance is exactly the same as it would be if a capacitance having value $C_C(1 - A_v)$ were connected between input and ground* instead of capacitance C_C connected between input and output. In other words, as far as the input signal is concerned, capacitance connected between input and output has the same effect as that capacitance "magnified" by the factor $(1 - A_v)$ and connected so that it shunts the input. This magnification of feedback capacitance, reflected to the input, is called the *Miller effect,* and the magnified value $C_C(1 - A_v)$ is called the *Miller capacitance,* C_M. Miller capacitance is relevant only for an *inverting* amplifier, so A_v is a negative number and the magnification factor $(1 - A_v)$ equals one *plus* the magnitude of A_v.

By a derivation similar to the foregoing, it can be shown that capacitance in the feedback path is also reflected to the output side of an amplifier. In this case, the effective shunt capacitance at the output is $(1 - 1/A_v)C_C$. Once again, A_v is negative, so the magnitude of the reflected capacitance is $(1 + 1/|A_v|)C_C$. Because the increase in capacitance is inversely proportional to gain, the effect is much less significant than that of the capacitance reflected to the input.

The total shunt capacitance at the input is the sum of the Miller capacitance and any other input-to-ground capacitance that may be present. Also, the total shunt capacitance at the output is the sum of the reflected capacitance and any other output-to-ground capacitance present. Let us define the composite capacitances C'_A and C'_B as follows:

$$C'_A = C_A + (1 - A_v)C_C \tag{9–40}$$

and

$$C'_B = C_B + (1 - 1/A_v)C_C \tag{9–41}$$

Note that C'_A and C'_B are the total equivalent capacitances across the input and output, respectively, after reflecting the capacitance C_C. Therefore, the corresponding frequencies f'_A and f'_B are:

$$f'_A = \frac{1}{2\pi(r_S \| r_i)C'_A} \tag{9–42}$$

and

$$f'_B = \frac{1}{2\pi(r_o \| R_L)C'_A} \tag{9–43}$$

An important concept to understand at this point is that the frequencies f'_A and f'_B that result from the Miller effect when C_C is present will not be the actual upper corner frequencies of the amplifier. Furthermore, in this case a third corner that causes a positive slope change to the gain will occur at a very high frequency; however, its location is so far apart that it has practically no effect on the cutoff frequency. It can be shown that the lowest or dominant corner frequency, that is, the upper cutoff frequency f_H, can be closely approximated by

$$f_H \approx \frac{1}{\dfrac{1}{f'_A} + \dfrac{1}{f'_B}} \tag{9–44}$$

EXAMPLE 9–9

The inverting amplifier shown in Figure 9–22 has a gain $v_L/v_i = -200$. Find its upper cutoff frequency.

Solution

We will use the gain v_L/v_i to determine the Miller capacitance. Notice that this is the gain *between the points where the 20-pF capacitance is connected.* The Miller capacitance is, therefore, $C_M \approx (20\text{ pF})\,[1 - (-200)] = 4020\text{ pF}$. The total capacitance C'_A shunting the input is then $C'_A = (40\text{ pF}) + (4020\text{ pF}) = 4060\text{ pF}$. The frequency f'_A is

FIGURE 9–22 (Example 9–9)

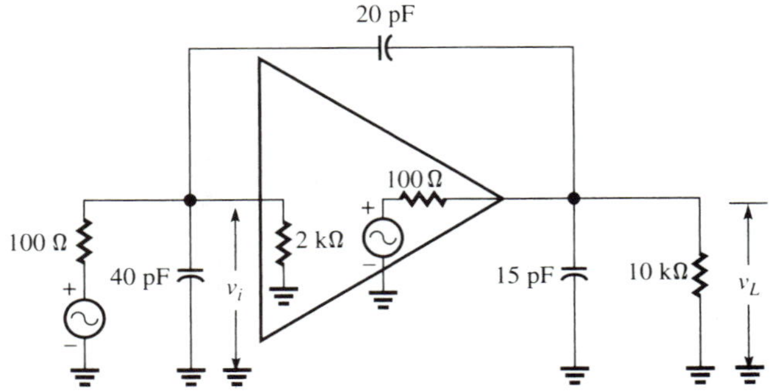

$$f'_A = \frac{1}{2\pi[(2\text{ k})\,\|\,100]4060\text{ pF}} = 412\text{ kHz}$$

The total capacitance shunting the output is $C'_B = 15\text{ pF} + (1 + 1/200)\,20\text{ pF} \approx 35\text{ pF}$. The frequency f'_B is

$$f'_B = \frac{1}{2\pi[(10\text{ k})\,\|\,100]35 \times 10^{-12}} = 46\text{ MHz}$$

The 3-dB frequency f_H is then

$$f_H \approx \frac{1}{\dfrac{1}{f'_A} + \dfrac{1}{f'_B}} \approx 408\text{ kHz}$$

9–5 TRANSIENT RESPONSE

The *transient response* of an electronic amplifier or system is the output waveform that results when the input is a pulse or a sudden change in level. Because the transient response is a waveform, it is presented as a plot of voltage versus time, in contrast to frequency response, which is plotted versus frequency. Figure 9–23 shows a typical transient response.

The transient response of an amplifier is completely dependent on its frequency response, and vice versa. In other words, if two amplifiers have identical frequency responses, they will have identical transient responses, and vice versa. A pulse can be regarded as consisting of an infinite number of frequency components, so the transient waveform represents the amplifier's ability (or inability) to amplify all frequency components equally and to phase-shift all components equally. Theoretically, if an amplifier had infinite bandwidth, its transient response would be an exact duplicate of the input pulse. It is not possible for an amplifier to have infinite bandwidth, so the transient response is always a distorted version of the input pulse. It is necessary for an amplifier to have a wide bandwidth (to be a *wideband,* or *broadband,* amplifier) in order for it to amplify pulse or square-wave signals with a minimum of distortion. *Square-wave testing* is sometimes used to check the frequency response of an amplifier, as shown in Figure 9–24. The figure shows typical waveforms that result when a square wave is applied to an amplifier whose frequency response causes attenuation of either low- or high-frequency components. "Low" or "high" frequency in any given case means low or high in relation to the square-wave frequency, f_S.

As a concrete example of how frequency response is related to transient response, consider the low-pass RC network shown in Figure 9–25. The figure shows the output transient when the input is an abrupt change in level (called a *step* input), such as might occur when a dc voltage is switched into the network. Recall that the time constant, $\tau = RC$ seconds, is the time required for the transient output to reach 63.2% of its final value. Also recall that the cutoff frequency of the network is $f_c = 1/(2\pi RC)$ Hz. Thus, $f_c = 1/(2\pi\tau)$ Hz. Since

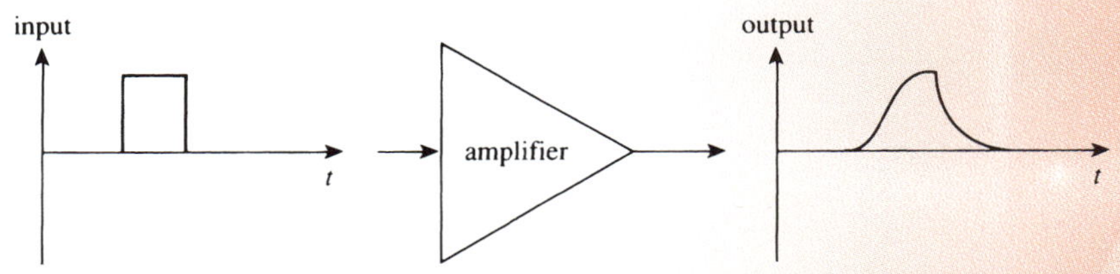

FIGURE 9–23 The output waveform is the transient response of the amplifier to a pulse-type input

FIGURE 9–24 Typical outputs resulting from square-wave testing of an amplifier. f_S = square-wave frequency. In (a) and (b), only low frequencies are attenuated, and in (c) and (d), only high frequencies are attenuated. In both cases, the attenuation outside cutoff is 20 dB/decade

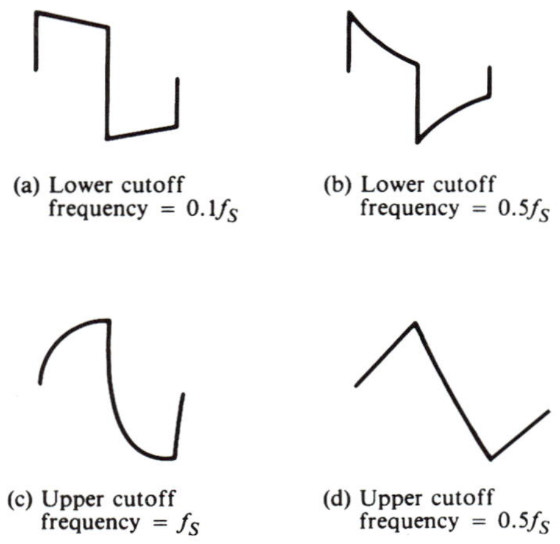

FIGURE 9–25 Relationship between transient response and frequency response. Note that the bandwidth is inversely proportional to the time constant of the transient

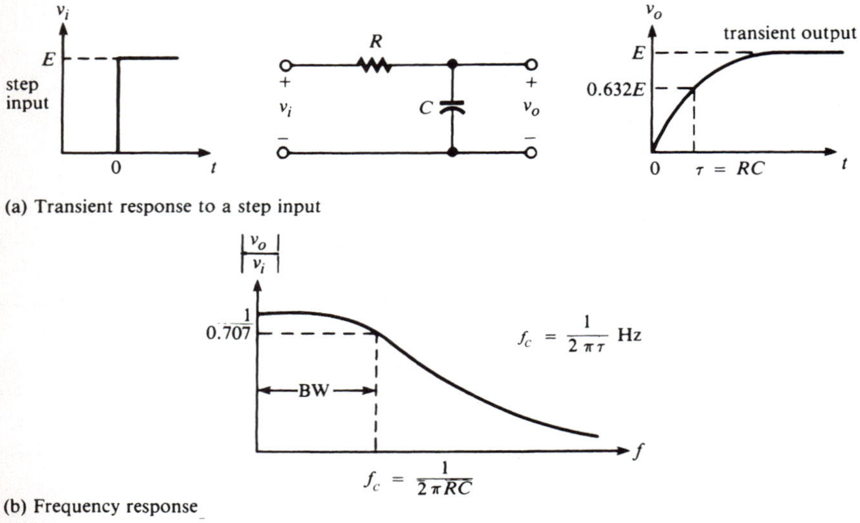

the network passes all frequencies below f_c, down to dc, its bandwidth equals $f_c - 0 = f_c$. Summarizing,

$$\text{BW} = f_c = \frac{1}{2\pi\tau}$$

This equation shows that the cutoff frequency and the bandwidth, which are frequency-response characteristics, are *inversely* proportional to the time constant, which is a characteristic of the transient response. It is generally true for all electronic devices that the time required for the transient response to rise to a certain level is inversely proportional to bandwidth.

Recall that rise time, t_r, is the time required for a waveform to change from 10% of its final value to 90% of its final value. A widely used approximation that relates the rise time of the transient response of an amplifier to its bandwidth is

$$t_r \approx \frac{0.35}{\text{BW}} \qquad \textbf{(9–45)}$$

where BW is the bandwidth, in hertz. Equation 9–45 is used when the lower cutoff frequency is 0 (dc) or very small so that the bandwidth is essentially the same as the upper cutoff frequency. The relationship is exact if the high-frequency response beyond cutoff is the same as that of the single RC low-pass network (Figure 9–25).

EXAMPLE 9–10

The specifications for a certain oscilloscope state that the rise time of the vertical amplifier is 8.75 ns. What is the approximate bandwidth of the amplifier?

Solution

From equation 9–45,

$$\text{BW} \approx \frac{0.35}{t_r} = \frac{0.35}{8.75 \times 10^{-9}} = 40 \text{ MHz}$$

A practical application of equation 9–45 is in the measurement of the bandwidth of an amplifier. A square waveform is applied to the input, and the output voltage is observed on the oscilloscope. The rise time is then measured with the oscilloscope and bandwidth calculated with equation 9–45. An interesting aspect to note is that the rise time does not depend on the frequency of the applied square waveform, only on the bandwidth of the circuit.

Transient response due to low-frequency characteristics of the amplifier is manifested as "sagging" of the output waveform as shown in Figure 9–26. The amount of sagging depends of course on the frequency of the square waveform (larger period, larger ΔV) as well as on the lower cutoff frequency of the amplifier, f_L. If the amplifier can pass dc (zero frequency), the output waveform will not sag. If we define the sag as the ratio of sagging voltage to peak voltage, that is,

$$\text{sag} = \frac{\Delta V}{E_P} \qquad \text{(9–46)}$$

we can derive an expression for the sag in terms of the square-wave frequency f and the lower cutoff frequency f_L. Using an analogy with a discharging RC circuit, the expression can be shown to be

$$\text{sag} = 1 - e^{-\pi f_L/f} \qquad \text{(9–47)}$$

This transient characteristic can be used for quickly obtaining the lower cutoff frequency of an amplifier by driving it with a square wave and measuring E_P and ΔV with the oscilloscope to obtain the sag. For example, if the sag of an amplifier when driven by a square wave is 0.18, then substituting into (9–47) yields

$$0.82 = e^{-\pi f_L/300}$$

from which we obtain $f_L = 19$ Hz.

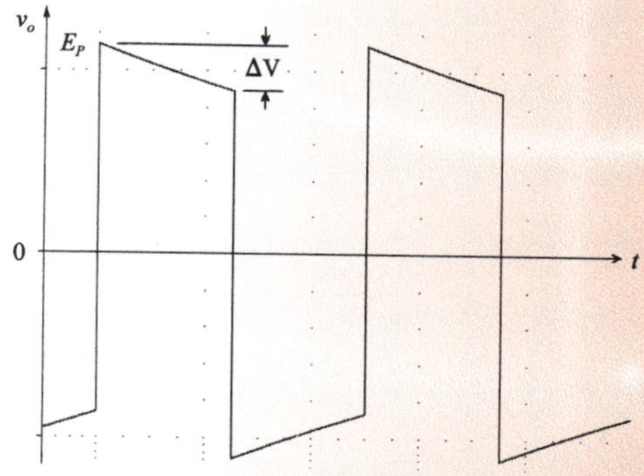

FIGURE 9–26 Transient response due to low-frequency response of an amplifier

9–6 FREQUENCY RESPONSE OF BJT AMPLIFIERS

Low-Frequency Response of BJT Amplifiers

We have learned that the lower cutoff frequency of an amplifier is approximately equal to the larger of f_1 and f_2, where

$$f_1 = \frac{1}{2\pi(r_S + r_i)C_1} \tag{9–48}$$

and

$$f_2 = \frac{1}{2\pi(r_o + R_L)C_2} \tag{9–49}$$

In a BJT amplifier, the term r_i appearing in equation 9–48 is r_{in}, the resistance seen by the source when looking into the amplifier. Its value depends on the transistor configuration, the bias resistors, and the values of the transistor parameters. Similarly, r_o in equation 9–49 is r_{out}, and its value depends on particular amplifier characteristics. The next example illustrates the application of equations 9–48 and 9–49 to determine the lower cutoff frequency of a fixed-bias common-emitter amplifier.

EXAMPLE 9–11

Find the lower cutoff frequency of the amplifier shown in Figure 9–27. Assume that the resistance looking into the base of the transistor is 1500 Ω and that the transistor output resistance at the collector, r_o, is 100 kΩ.

Solution

Since $r_{in} = (150\text{ k}\Omega)\|(1.5\text{ k}\Omega) \approx 1.5\text{ k}\Omega$, we have, from equation 9–46,

$$f_1 = \frac{1}{2\pi(50 + 1500)(1\ \mu\text{F})} \approx 103\text{ Hz}$$

Since $r_{out} = (100\text{ k}\Omega)\|(1\text{ k}\Omega) \approx 1\text{ k}\Omega$, we have, from equation 9–49,

$$f_2 = \frac{1}{2\pi(1\text{ k} + 3\text{ k})(1\ \mu\text{F})} \approx 40\text{ Hz}$$

Because f_1 and f_2 are not sufficiently separated, we use equation 9–23 to solve for f_L. Substituting the values for f_1 and f_2:

$$f_L^4 - (103^2 + 40^2)\,f_L^2 - 103^2\,40^2 = 0 \qquad \text{or} \qquad f_L^4 - 12209\,f_L^2 - 17 \times 10^6 = 0$$

Using the simplified quadratic formula, we solve for the positive solution for f_L^2 as follows:

FIGURE 9–27 (Example 9–11)

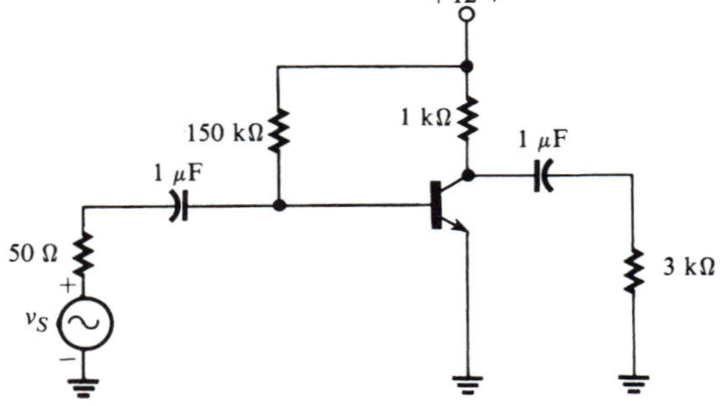

$$f_L^2 = 6104 + (6104^2 + 17 \times 10^6)^{1/2} = 13.47 \times 10^3$$

from which $f_L = 116$ Hz follows.

Figure 9–28 shows a common-emitter amplifier having an emitter bypass capacitor C_E designed to bring the emitter closer to ac ground, as discussed in Chapter 7. At low frequencies, the reactance of C_E can become significant, to the extent that the voltage gain is reduced. Thus, we can define another corner frequency due to C_E that affects the amplifier's low-frequency response in the same way as the coupling capacitors, C_1 and C_2. If the coupling capacitors are large enough to have negligible effect, then the gain will fall 3 dB below its midband value at the frequency where the reactance of C_E equals the resistance R_e looking into node A in Figure 9–28:

$$R_e = R_E \left\| \left(\frac{r_S \| R_B}{\beta} + r_e' + R_f \right) \right. \tag{9–50}$$

where $R_B = R_1 \| R_2$. Then,

$$f_L = \frac{1}{2\pi R_e C_E} \quad (C_1, C_2 \text{ large}) \tag{9–51}$$

Making C_1 and C_2 very large can be impractical. In general, C_1 and C_E produce two interdependent corner frequencies in addition to the corner frequency produced by C_2. Furthermore, it can also be shown that the parallel combination of R_E and C_E produces one more corner frequency given by

$$f_Z = \frac{1}{2\pi R_E C_E} \tag{9–52}$$

which causes an asymptotic break in a direction opposite to the other three corner frequencies. Figure 9–29 shows a typical asymptotic plot of the frequency response caused by C_1, C_2, and C_E. Note the break point at f_Z producing an increase in slope, opposite to the other three break points. In a practical design, C_2 can be chosen so that f_2 coincides with f_Z, thus mutually canceling their effect, because their asymptotic breaks are of opposite nature.

Although it is possible to obtain the exact value of the two corner frequencies caused by C_1 and C_E acting together and from there obtain the lower cutoff frequency f_L, the required analysis would be very extensive and elaborate. Instead, it can be shown that an acceptable approximation for the lower cutoff frequency is the sum of the short-circuit *natural* frequencies of C_1 and C_E using the general form $1/2\pi RC$, where R is the equivalent resistance across each capacitor C with the other capacitor

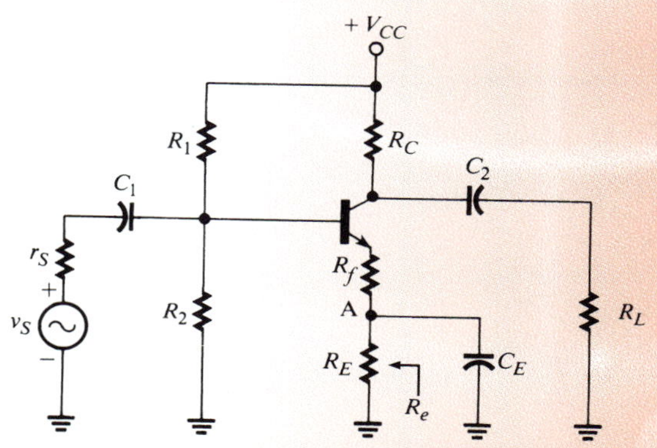

FIGURE 9–28 The corner frequency due to the bypass capacitor is the frequency where the reactance of C_E equals the resistance R_e

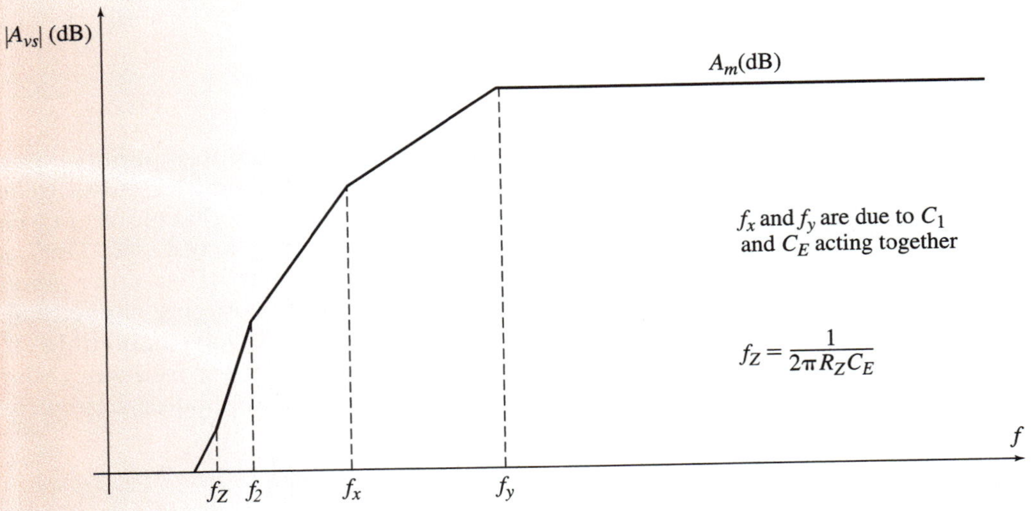

FIGURE 9–29 Asymptotic plot of $|A_{vs}|$ showing the corner frequencies produced by C_1 and C_E

shorted. A *natural frequency* in an RC circuit is associated with a particular time constant.

Let f_b be the natural frequency from C_1 and f_e the natural frequency from C_E. Then,

$$f_b = \frac{1}{2\pi[r_S + R_B \| \beta(r'_e + R_f)]C_1} \tag{9–53}$$

and

$$f_e = \frac{1}{2\pi R_e C_E} \tag{9–54}$$

Notice that the term $R_B \| \beta(r'_e + R_f)$ in equation 9–53 is the amplifier's stage input resistance, r_{in}.

The approximate lower cutoff frequency is then

$$f_L \approx f_b + f_e \tag{9–55}$$

This approximation is valid if the corner frequencies f_Z and f_2 cancel each other or if at least their values are smaller than both f_b and f_e. The condition $f_2 = f_Z$ can be satisfied by simply making $C_2(r_{out} + R_L) = C_E R_Z$.

An example illustrating these concepts follows.

EXAMPLE 9–12

Determine the approximate lower cutoff frequency for the CE amplifier shown in Figure 9–30. Assume $\beta = 150$ and $r'_e = 13\ \Omega$. Note that $C_2 = 7.5\ \mu\text{F}$, making the corner frequency f_2 coincide with the corner frequency f_Z, mutually canceling each other.

Solution

$$R_B = 40\text{ k} \| 10\text{ k} = 8\text{ k}\Omega$$
$$r_{in} = 8\text{ k} \| [150(82 + 13)] = 6.12\text{ k}\Omega$$
$$R_e = 680 \| [82 + 13 + (8\text{ k} \| 1\text{ k})/150] = 88\ \Omega$$

The two natural frequencies are then

FIGURE 9–30 (Example 9–12)

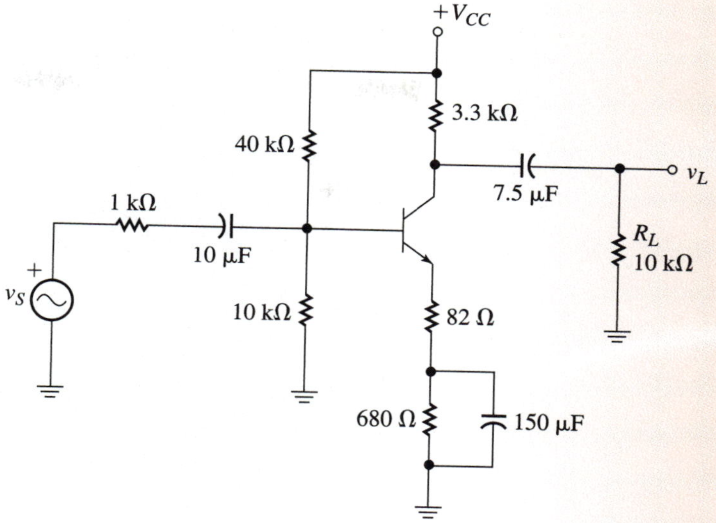

$$f_b = 1/[2\pi(6.12\ \text{k})10\ \mu] = 2.6\ \text{Hz}$$
$$f_e = 1/[2\pi(88)150\ \mu] = 12\ \text{Hz}$$

Therefore, the cutoff frequency is

$$f_L \approx 2.6 + 12 \approx 14.6\ \text{Hz}$$

EXAMPLE 9–13

SPICE

Use SPICE to simulate the circuit from Example 9–12 to obtain the plot of the frequency response over the frequency range from 1 Hz through 1 kHz. Use the transistor 2N3904 with $V_{CC} = 12$ volts.

Solution

Shown here is the circuit's file using PSpice. Note that the .AC statement specifies 50 frequency points per decade in order to get a smooth plot. The plot of the gain in Figure 9–31 was obtained in dB using the Probe interactive program. Note that an input voltage of 1 V may seem too large. Recall that SPICE does not consider practical voltage limitations and clipping when performing an AC analysis. Also note at the bottom of the plot the expression entered to obtain the gain in dB: db (v(7)/v(1)).

```
CE Amplifier with Emitter Feedback
*Nodes: Input=1, CBE=4, 3, 5, Vcc=8, Output=7
vs 1 0 ac 1
rs 1 2 1k
r1 8 3 40k
r2 3 0 10k
rc 8 4 3.3k
q1 4 3 5 q2n3904
rf 5 6 82
rz 6 0 680
rl 7 0 10k
cb 2 3 10uf
ce 6 0 150uf
cc 4 7 7.5uf
vcc 8 0 12
.lib eval.lib
```

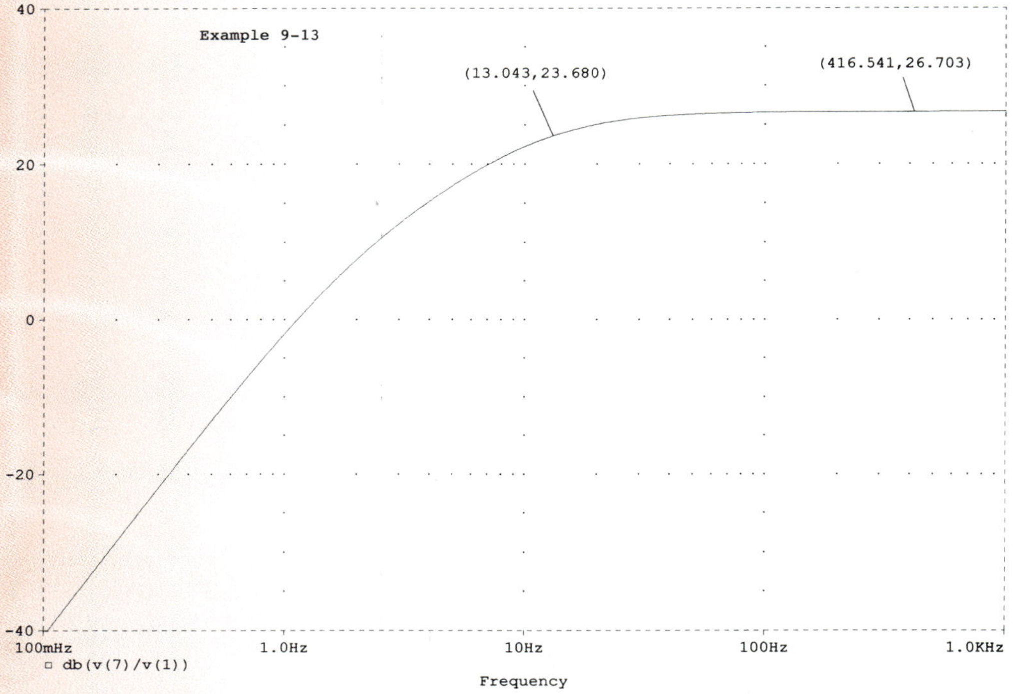

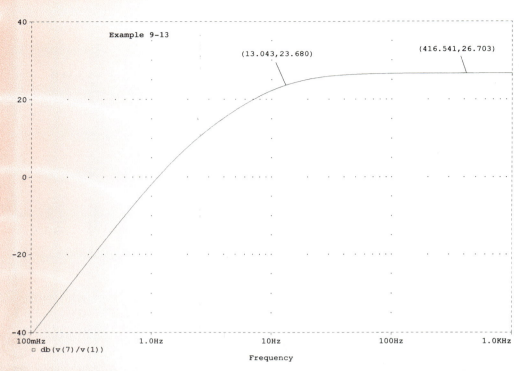

FIGURE 9–31 (Example 9–13)

```
.op
.ac dec 50 1 1k
.probe
.end
```

The Probe plot shows a value-labeled point in the midband and another at the cutoff frequency. The value of about 13 Hz from the plot agrees well with the calculated value of 14.6 Hz.

The principal consideration in the design and analysis of amplifiers in the low-frequency region is the (Thévenin equivalent) resistance across each capacitor used in the circuit. We have seen that a break frequency occurs whenever the frequency becomes low enough to make the capacitive reactance equal to the equivalent resistance at the point of connection. Therefore, in any BJT amplifier, the capacitor that is most critical in determining the lower cutoff frequency is the one that "sees" the smallest resistance. In the case of the common-emitter amplifier, C_E is that capacitor. In a common-base amplifier, the stage input resistance ($r_{in} \approx r_e'$) is quite small, so the input coupling capacitor, C_1, is the most crucial. In the common-collector amplifier, the output resistance is quite small, so the output coupling capacitor, C_2, is crucial. Of course, a large source resistance in the case of a CB amplifier, or a large load resistance in the case of a CC amplifier, will mitigate these circumstances, since each is in series with the affected capacitor.

In the particular case of the CE amplifier with emitter resistor R_f, in which the lower cutoff frequency is $f_L \approx f_b + f_e$, a practical design criterion is to make f_e 90% of f_L, f_b 10% of f_L, and $f_2 = f_Z$. If we assigned 90% of the cutoff frequency to f_b and 10% to f_e, the result would be the same but the value of C_E would be much larger, making it less cost-effective. It can be, however, a good SPICE simulation problem for verification.

FIGURE 9–32 CB and CC amplifiers and the equations for f_1 and f_2 for each

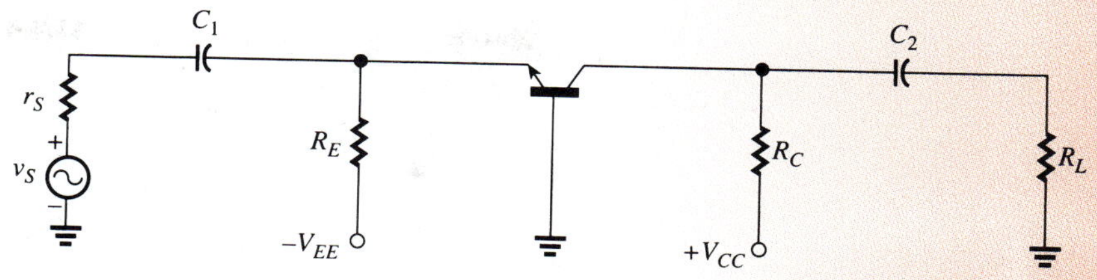

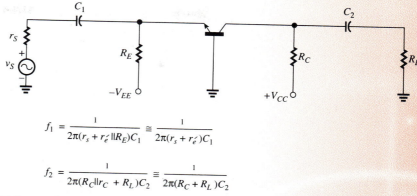

$$f_1 = \frac{1}{2\pi(r_s + r_e' \| R_E)C_1} \cong \frac{1}{2\pi(r_s + r_e')C_1}$$

$$f_2 = \frac{1}{2\pi(R_C \| r_C + R_L)C_2} \cong \frac{1}{2\pi(R_C + R_L)C_2}$$

(a) Common base

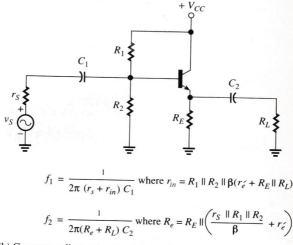

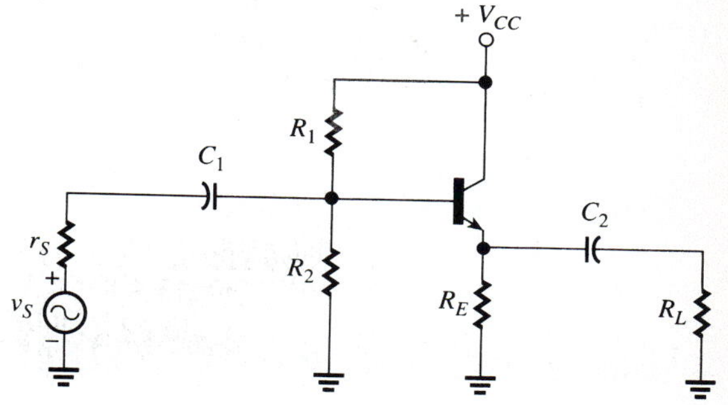

$$f_1 = \frac{1}{2\pi\ (r_s + r_{in})\ C_1} \text{ where } r_{in} = R_1 \| R_2 \| \beta(r_e' + R_E \| R_L)$$

$$f_2 = \frac{1}{2\pi(R_e + R_L)\ C_2} \text{ where } R_e = R_E \| \left(\frac{r_S \| R_1 \| R_2}{\beta} + r_e'\right)$$

(b) Common collector (emitter follower)

Figure 9–32 shows common-base and common-collector amplifiers and gives the equations for the break frequencies due to each coupling capacitor. These are derived in a straightforward manner by solving for the frequency at which the capacitive reactance of each capacitor equals the Thévenin equivalent resistance across it.

EXAMPLE 9–14

A certain transistor has $\beta = 100$, $r_C = 100\ \text{k}\Omega$, and $r_e' = 25\ \Omega$. The transistor is used in each of the circuits shown in Figure 9–32. Find the approximate lower cutoff frequency in each circuit, given the following component values:

1. (Common-base circuit)

$r_S = 100\ \Omega$	$R_C = 1\ \text{k}\Omega$	$C_1 = 2.2\ \mu\text{F}$
$R_E = 10\ \text{k}\Omega$	$R_L = 15\ \text{k}\Omega$	$C_2 = 1\ \mu\text{F}$

2. (Common-collector circuit)

$r_S = 100\ \Omega$	$R_E = 1\ \text{k}\Omega$	$C_1 = 2\ \mu\text{F}$
$R_1 = 33\ \text{k}\Omega$	$R_L = 50\ \Omega$	$C_2 = 10\ \mu\text{F}$
$R_2 = 10\ \text{k}\Omega$		

Solution

1. $$f_1 = \frac{1}{2\pi(100 + 25 \| 10 \times 10^3)(2.2 \times 10^{-6})} = 579\ \text{Hz}$$

$$f_2 = \frac{1}{2\pi(1 \times 10^3 \| 100 \times 10^3 + 15 \times 10^3)(1 \times 10^{-6})} = 9.9 \text{ Hz}$$

Therefore, $f_L = f_1 = 579$ Hz.

2. $r_{in} = (33 \text{ k}\Omega) \| (10 \text{ k}\Omega) \| (100[(25 \ \Omega) + (1 \text{ k}\Omega) \| (50 \text{ k}\Omega)]$
$= (7.67 \text{ k}\Omega) \| (7.26 \text{ k}\Omega) = 3.73 \text{ k}\Omega$

$$f_1 = \frac{1}{2\pi(100 + 3.73 \times 10^3)(2 \times 10^{-6})} = 20.8 \text{ Hz}$$

$$R_e = (1 \text{ k}\Omega) \left\| \left[\frac{(100 \ \Omega) \| (33 \text{ k}\Omega) \| (10 \text{ k}\Omega)}{100} + (25 \ \Omega) \right] \approx 25 \ \Omega \right.$$

$$f_2 = \frac{1}{2\pi(25 + 50)(10 \times 10^{-6})} = 212 \text{ Hz}$$

Therefore, $f_L \approx f_2 = 212$ Hz.

Design Considerations

Capacitors in practical discrete circuits are generally bulky and costly, or else they are unreliable if quality is sacrificed for cost. Therefore, a principal objective in the design of discrete BJT amplifiers is the selection of the smallest capacitors possible consistent with the low-frequency response desired. Equations 9–56 through 9–58 can be used to find minimum capacitor values for CE, CB, and CC amplifiers in terms of break frequencies and circuit parameters. As noted earlier, each configuration has one capacitor whose value is the most critical: The one connected to the point where the equivalent resistance is smallest. This capacitor should be selected first to ensure that the lower cutoff frequency is at least as low as the break frequency it produces. The other capacitor(s) can then be selected to break frequencies a decade or so below cutoff. (With two identical break frequencies occurring one decade below the third, the gain is actually reduced by a factor of 0.7, rather than 0.707, at the third frequency.)

EXAMPLE 9–15

DESIGN

A CE amplifier having $R_1 = 330$ kΩ, $R_2 = 47$ kΩ, $R_C = 3.3$ kΩ, and $R_E = 1.8$ kΩ is to have a lower cutoff frequency of 50 Hz. The amplifier drives a 10-kΩ load and is driven from a signal source whose resistance is 600 Ω. The transistor has $\beta = 90$ and $r_o = 100$ kΩ and is biased at $I_E = 0.5$ mA. Find coupling and bypass capacitors necessary to meet the low-frequency response requirement. Assume that capacitors must be selected from a line having standard values of 1 μF, 1.5 μF, 3.3 μF, 4.7 μF, 6.8 μF, 10 μF, 47 μF, 68 μF, and 100 μF.

Solution

To obtain a value for C_E, we must first find the value of r'_e:

$$r'_e \approx \frac{0.026}{I_E} = \frac{0.026}{0.5 \text{ mA}} = 52 \ \Omega$$

Then, from equations 9–56,

$$R_e = 1.8 \text{ k}\Omega \left\| \left[\frac{(600 \ \Omega) \| (330 \text{ k}\Omega) \| (47 \text{ k}\Omega)}{90} + 52 \ \Omega \right] = 56.7 \ \Omega \right.$$

and

$$C_E = \frac{1}{2\pi(56.7 \ \Omega)(50 \text{ Hz})} = 56.1 \ \mu\text{F}$$

Capacitor values necessary to achieve specified break frequencies in a common emitter (CE) amplifier

$$C_E^{*} = \frac{1}{2\pi R_e f_L} \qquad (9\text{–}56)$$

where $R_e = R_E \left\| \left(\frac{r_S \| R_1 \| R_2}{\beta} + r_e' \right) \right.$.

$$C_1 = \frac{1}{2\pi[r_{in} + r_S]f_1}$$

(make $f_1 = f_L/10$)

where $r_{in} = R_1 \| R_2 \| \beta r_e'$.

$$C_2 = \frac{1}{2\pi[r_{out} + R_L]f_z}$$

where $r_{out} = r_o \| R_C$ and $f_Z = \frac{1}{2\pi R_E C_E}$

*Select first.

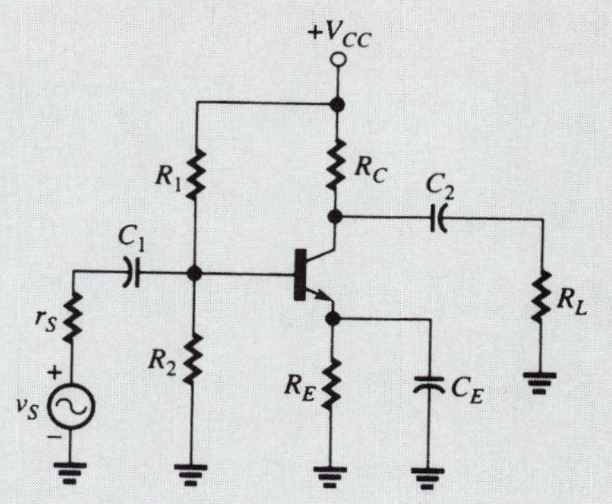

To ensure that the lower cutoff frequency is no greater than 50 Hz, we choose the standard-value capacitor with the next *higher* capacitance. Thus, we choose $C_E = 68\ \mu F$.

To find C_1 and C_2, we let $f_1 = 50\text{ Hz}/10 = 5\text{ Hz}$ and use equations 9–56 to calculate

$$r_{in} = 330\text{ k}\Omega \| 47\text{ k}\Omega \| (90)(52\ \Omega) = 4.2\text{ k}\Omega$$

$$C_1 = \frac{1}{2\pi(4.2\text{ k}\Omega + 600\ \Omega)5\text{ Hz}} = 6.63\ \mu F$$

$$r_{out} = 100\text{ k}\Omega \| 3.3\text{ k}\Omega = 3.19\text{ k}\Omega, \quad f_2 = f_Z = \frac{1}{2\pi(1.8\text{ k})68\ \mu F} = 1.3\text{ Hz}$$

$$C_2 = \frac{1}{2\pi(3.19\text{ k}\Omega + 10\text{ k}\Omega)1.3\text{ Hz}} = 9.3\ \mu F$$

Capacitor values necessary to achieve specified break frequencies in a common base (CB) amplifier

$$C_1^* = \frac{1}{2\pi(r_S + r_e' \| R_E)f_L} \quad (9\text{–}57)$$

$$C_2 = \frac{1}{2\pi(R_C \| r_C + R_L)f_2}$$

(make $f_2 = f_L/10$)

*Select first.

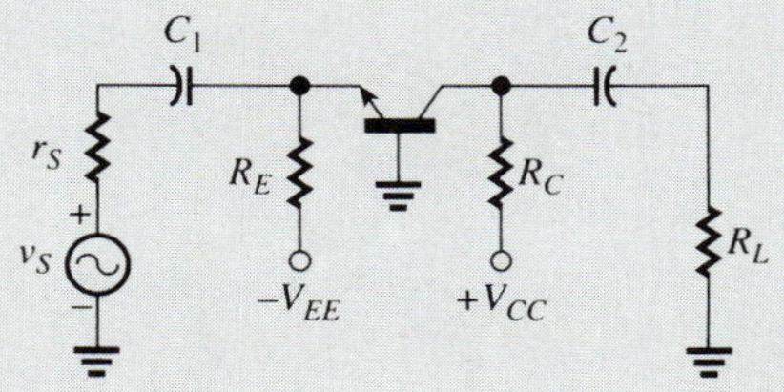

Capacitor values necessary to achieve specified break frequencies in a common collector (CC) amplifier

$$C_2^* = \frac{1}{2\pi(R_e + R_L)f_L}$$

where $R_e = R_E \| \left(\frac{r_s \| R_1 \| R_2}{\beta} + r_e'\right)$.

$$C_1 = \frac{1}{2\pi[r_{in} + r_S]f_1} \quad (9\text{–}58)$$

where $r_{in} = R_1 \| R_2 \| \beta(r_e' + R_E \| R_L)$.

(make $f_1 = f_L/10$)

*Select first.

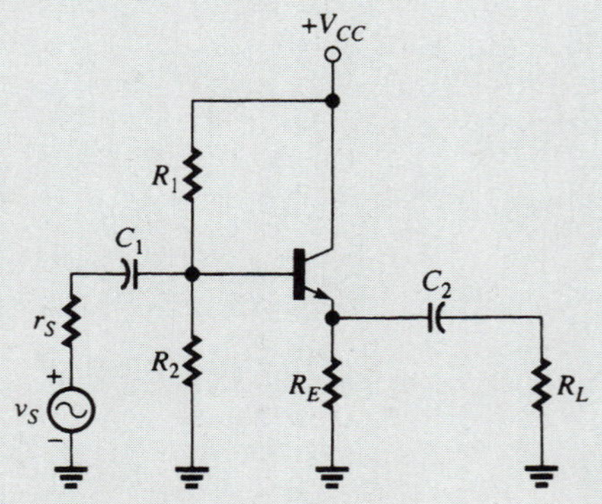

Again choosing standard capacitors with the next highest values, we select $C_1 = 6.8\ \mu F$ and $C_2 = 10\ \mu F$.

It is a SPICE programming exercise at the end of this chapter to find the actual reduction in gain at 50 Hz when these standard-value capacitors are used. Our choices result in an actual lower cutoff frequency of about 52 Hz.

High-Frequency Response of BJT Amplifiers

Figure 9–33 shows a common-emitter amplifier having interelectrode capacitance designated C_{be}, C_{bc}, and C_{ce}. Because we are now considering high-frequency performance, the emitter bypass capacitor effectively shorts the emitter terminal to ground, so C_{be} and C_{ce} are input-to-ground and output-to-ground capacitances, respectively. We can apply the general equations developed earlier to determine the corner frequency due to the interelectrode capacitances:

$$f_A = \frac{1}{2\pi(r_S \| r_i)C_A} \tag{9–59}$$

where $C_A = C_{be} + C_M = C_{be} + C_{bc}(1 - A_v)$ and $r_{in} = R_1 \| R_2 \| (\beta r'_e)$. Note that the Miller-effect capacitance, C_M, is due to C_{bc} because the latter is connected between the input (base) and output (collector) of the common-emitter configuration. The corner frequency due to the output shunt capacitance is

$$f_B = \frac{1}{2\pi(r_{out} \| R_L)C_B} \tag{9–60}$$

where $C_B = C_{ce} + C_{bc}(1 - 1/A_v)$ and $r_{out} = r_o \| R_C$. Any wiring or stray capacitance that shunts the input or output must be added to C_A or C_B. It is important to note that f_B is typically much larger than f_A. Therefore, the upper cutoff frequency will be approximately equal to f_A.

As mentioned in an earlier discussion, the high-frequency performance of a transistor is affected by the interelectrode capacitances that cause β to be frequency dependent. Its value drops at the rate of 20 dB/decade at frequencies beyond a certain frequency called the *beta cutoff frequency*, f_β. By definition, f_β is the frequency at which the β of a transistor is 0.707 times its low-frequency value. Figure 9–34 shows a plot of β versus frequency as it would appear on log-log paper. The low- (or mid-) frequency value of β is designated β_m.

If the frequency is increased beyond f_β, β continues to decrease until it eventually reaches a value of 1. As shown in Figure 9–34, the frequency at which $\beta = 1$ is designated f_T. Note that the asymptote's slope of -20 dB/decade is the same as a slope of -1 on log-log scales, since 20 dB is a 10-to-1 change along the vertical axis and a decade is a 10-to-1 change

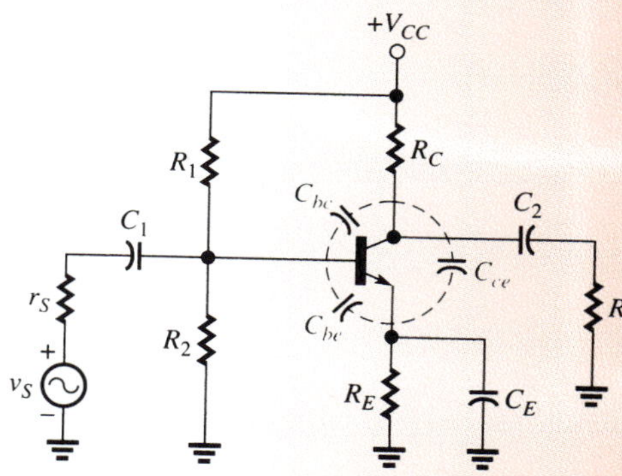

FIGURE 9–33 A common-emitter amplifier showing the interelectrode capacitances of the transistor

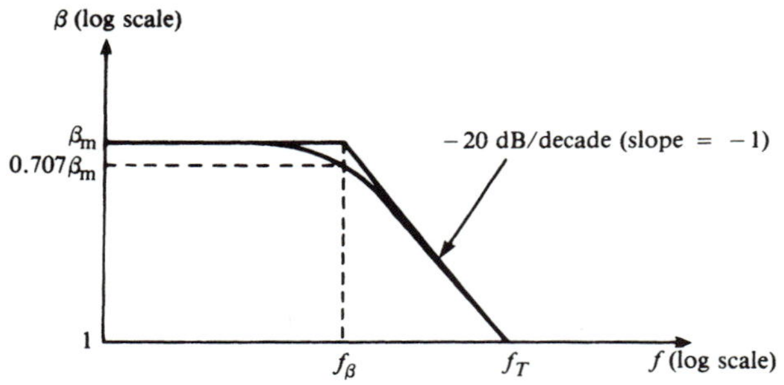

FIGURE 9–34 The β of a transistor becomes smaller at high frequencies. At frequencies above f_β, its value falls off at the rate of 20 dB/decade, and reaches value 1 at the frequency designated f_T

along the horizontal axis. Therefore, in Figure 9–34, the difference between the logarithms of β_m and 1 must be the same as the difference between the logarithms of f_T and f_β:

$$\log \beta_m - \log(1) = \log f_T - \log f_\beta$$
$$\log(\beta_m/1) = \log(f_T/f_\beta)$$

Taking the antilog of both sides leads to the important result

$$f_\beta = \frac{f_T}{\beta_m} \tag{9–61}$$

The frequency dependency of β is due to the base–emitter junction capacitance C_{be} and to the base–collector capacitance C_{bc}. To demonstrate this fact, consider the CE configuration modeled with the hybrid-π model shown in Figure 9–35. Notice that the output is shorted to satisfy the definition for β or h_{fe}. Note also that because of the short circuit, the capacitor C_μ is connected in parallel with C_π. The short-circuit collector current can clearly be written as $i_c = g_m v_\pi$. But $v_\pi = i_b Z_i$, where Z_i is the parallel combination of r_π, C_π, and C_μ, which can be shown to be

$$Z_i = \frac{r_\pi}{1 + j\omega r_\pi(C_\pi + C_\mu)} \tag{9–62}$$

Therefore, from $i_c = g_m i_b Z_i$ and substituting Z_i, we can solve for the ratio i_c/i_b, which yields

$$\frac{i_c}{i_b} = \beta = \frac{g_m r_\pi}{1 + j\omega r_\pi(C_\pi + C_\mu)} \tag{9–63}$$

This expression, as expected, is a low-pass function whose magnitude approaches $g_m r_\pi$ when ω approaches zero. Also, observe that the value of ω that makes the denominator of equation 9–63 equal to 1 + j1 is the 3-dB frequency ω_β, since $g_m r_\pi/|1 + j1| = 0.707\, g_m r_\pi$ at that point. Therefore, by letting $\beta_m = g_m r_\pi$, we can write an expression for the magnitude of β as a function of ω or f as follows:

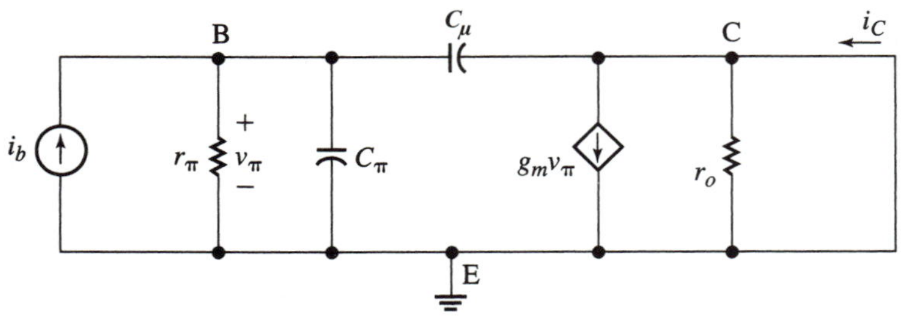

FIGURE 9–35 The hybrid-π model used for obtaining f_β and f_T

$$|\beta| = \frac{\beta_m}{\sqrt{1 + \left(\frac{\omega}{\omega_\beta}\right)^2}} = \frac{\beta_m}{\sqrt{1 + \left(\frac{f}{f_\beta}\right)^2}} \qquad (9\text{–}64)$$

where

$$\omega_\beta = \frac{1}{r_\pi(C_\pi + C_\mu)} \quad \text{and} \quad f_\beta = \frac{1}{2\pi r_\pi(C_\pi + C_\mu)} \qquad (9\text{–}65)$$

To obtain an expression for ω_T (or f_T) directly from equation 9–63, we note that for large ω values, the *magnitude* of β reduces to

$$|\beta| = \frac{g_m}{\omega(C_\pi + C_\mu)}$$

We now simply make $|\beta| = 1$ and solve for ω to obtain

$$\omega_T = \frac{g_m}{C_\pi + C_\mu} = \frac{1}{r'_e(C_\pi + C_\mu)} \qquad (9\text{–}66)$$

or

$$f_T = \frac{1}{2\pi r'_e(C_\pi + C_\mu)} \qquad (9\text{–}67)$$

Although the information given by manufacturers in data sheets regarding f_T is important, we have to understand that, by definition, β (or h_{fe}) is a short-circuit current gain, which is equivalent to the ratio of ac collector current to ac base current with $R_L = 0$. When the transistor operates normally in a circuit with a load resistance R_L, the frequency response of the current gain will have a 3-dB frequency significantly lower than f_β. The reader can easily verify this fact by simulating a CE amplifier with a normal R_L value to obtain the current gain and then giving R_L a value of, say, 0.1 Ω, and comparing results.

It can be shown that the overall voltage gain response of a CE amplifier contains three corner frequencies: one dominant (the smallest) and the other two at much higher frequencies, typically beyond f_T. The only possible way of bringing one of those two corner frequencies relatively close to the dominant corner frequency is by physically adding shunt capacitance across the output. Normal stray and wiring capacitance across the output will not have much effect on the dominant corner frequency. On the other hand, because of the "amplified" C_{bc} (or C_μ) due to the Miller effect, the dominant corner frequency will be mostly influenced by that capacitance. As pointed out earlier, the frequencies f'_A and f'_B due to Miller-effect capacitances are not the true corner frequencies of the amplifier. Recall that the dominant corner frequency, which is the upper cutoff frequency f_H, can be closely approximated by

$$f_H \approx \frac{1}{\frac{1}{f'_A} + \frac{1}{f'_B}} \qquad (9\text{–}68)$$

Equation 9–68 is another example of *duality* that we find so frequently in electric and electronic circuits. The upper cutoff frequency is found from a sum of frequency reciprocals, in contrast to the lower cutoff frequency which is the simple sum of the natural frequencies f_e and f_b (see equation 9–55).

Because β decreases with frequency, we would expect α to do the same. This is, in fact, the case, and the frequency at which α falls to 0.707 times its low-frequency value is called the *alpha cutoff frequency,* f_α. Because f_α is generally much larger than f_β, the common-base amplifier is not as frequency limited by parameter cutoff as is the common-emitter amplifier. A commonly used approximation is

$$f_\alpha \approx f_T$$

In specification sheets, f_T is often called the *gain-bandwidth product.* We will discuss this interpretation of a unity-gain frequency in Chapter 10. Recall that junction capacitance depends on the width of the depletion region, which in turn depends on the bias voltage. Because r'_e also depends on bias conditions, f_T is clearly a function of bias and is usually so indicated on specification sheets.

EXAMPLE 9–16

The transistor in Figure 9–36 has a low-frequency β of 120, $r'_e = 20\ \Omega$ and $r_o = 100\ \text{k}\Omega$. The interelectrode capacitances are $C_{be} = 40$ pF, $C_{bc} = 1.5$ pF, and $C_{ce} = 1$ pF. There is wiring capacitance equal to 4 pF across the input and 2 pF across the output. Find the approximate upper cutoff frequency.

Solution

To determine the Miller-effect capacitance, we must find the midband voltage gain of the transistor. Note that the *overall* voltage gain v_L/v_S is *not* used in this computation, because the gain reduction due to voltage division at the input does not affect the Miller capacitance. (It would, if the feedback capacitance were connected all the way back to v_S.)

$$A_v \approx \frac{-r_L}{r'_e} = \frac{-(4\ \text{k}\Omega)\,\|\,(20\ \text{k}\Omega)}{20\ \Omega} = -166.67$$

We use the midband gain for determining C_M:

$$C_M \approx C_{bc}\,(1 - A_v) = (1.5\ \text{pF})\,[1 - (-166.67)] = 251.5\ \text{pF}$$

The total capacitance shunting the input is, therefore,

$$C_A = C_{be} + C_M + C_{wiring} = (40\ \text{pF}) + (251.5\ \text{pF}) + (4\ \text{pF}) = 295.5\ \text{pF}$$

Then, since

$$r_{in} = R_1 \| R_2 \| \beta r'_e = (100\ \text{k}\Omega)\,\|\,(22\ \text{k}\Omega)\,\|\,(2.4\ \text{k}\Omega) = 2.1\ \text{k}\Omega$$

we have, from equation 9–59,

$$f_A = \frac{1}{2\pi(200\,\|\,2.1\times10^3)295.5\times10^{-12}} = 2.95\ \text{MHz}$$

Since $\quad r_{out} = r_o \| R_C = (100\ \text{k}\Omega)\,\|\,(4\ \text{k}\Omega) = 3.8\ \text{k}\Omega$

FIGURE 9–36 (Example 9–16)

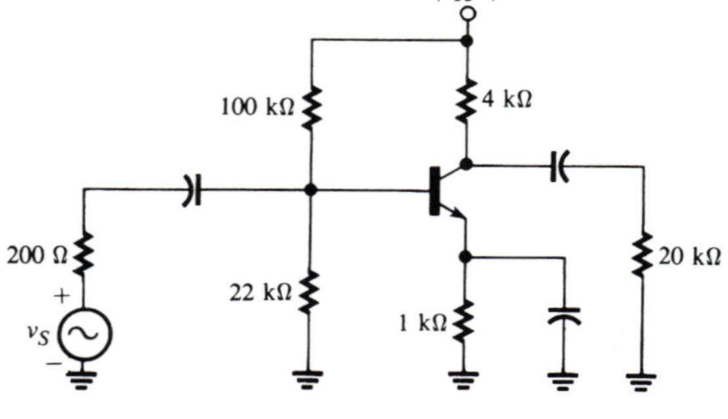

and $C_B = C_{ce} + C_{be}(1 - 1/A_v) + C_{wiring}$

$$= (1 \text{ pF}) + (1.5 \text{ pF})(1 + 1/167) + (2 \text{ pF}) = 4.5 \text{ pF}$$

we have, from equation 9–63,

$$f_B = \frac{1}{2\pi(3.8 \times 10^3 \| 20 \times 10^3)(4.5 \times 10^{-12})} = 11.1 \text{ MHz}$$

Therefore, from equation 9–68,

$$f_H \approx \frac{1}{\dfrac{1}{2.95} + \dfrac{1}{11.1}} = 2.33 \text{ MHz.}$$

Because the common-base amplifier is noninverting, the shunt capacitance at its input does not have a Miller-effect component. Therefore, the CB amplifier is not as severely limited as its CE counterpart in terms of the upper cutoff frequency, and CB amplifiers are often used in high-frequency applications. We will discuss one such application, the *cascode* amplifier, below. We have also noted that f_α is generally much higher than f_β. The common-collector amplifier (emitter follower) is also noninverting and unaffected by Miller capacitance. It, too, is generally superior to the CE configuration for high-frequency applications.

Cascode Amplifier

A simple example of a *cascode* amplifier is shown in Figure 9–37. Note that Q_1 is a common-emitter stage that uses R_{B1} for "fixed" bias. Because capacitor C_B grounds the base of Q_2 to ac signals, Q_2 is a common-base stage. It serves as the load on the collector of Q_1. Recall that the input to a common-base stage is at its emitter and the output is taken at its collector. Thus Q_1 is direct-coupled to the input of Q_2 and the output of the cascode amplifier is at the collector of Q_2. Resistors R_1 and R_2 form a voltage-divider bias circuit for Q_2.

The principal advantage of the cascode arrangement is that it has a small input capacitance, an important consideration in high-frequency amplifiers. The input capacitance is small because the voltage gain of Q_1 is small (near unity), which means that the Miller capacitance is minimized. Most of the voltage gain is achieved in the common-base stage. The voltage gain of Q_1 is small because the effective load resistance in its collector circuit is the small input resistance of the common-base stage.

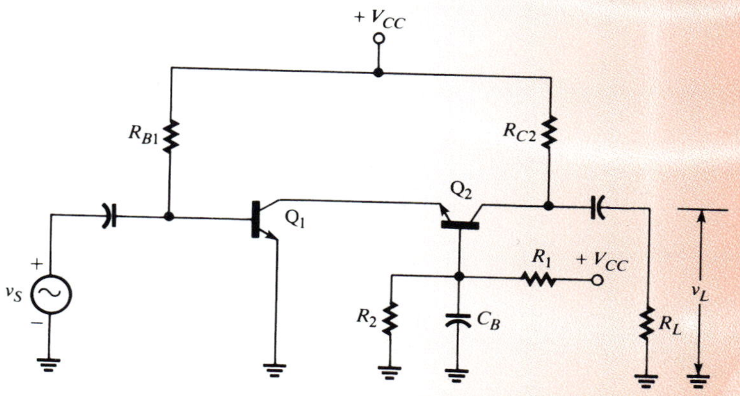

FIGURE 9–37 An example of a cascode amplifier. Q_1 is a common-emitter stage that is direct-coupled to the common-base stage formed by Q_2

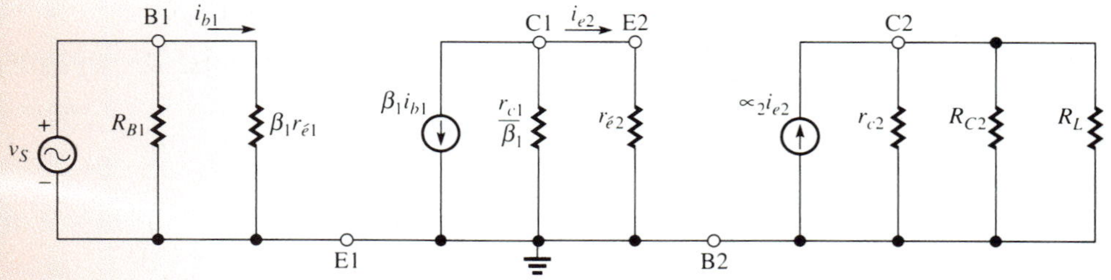

FIGURE 9–38 Small-signal equivalent circuit of a cascode amplifier

Calculation of the bias currents and voltages is straightforward:

$$I_{B1} = \frac{V_{CC} - V_{BE}}{R_{B1}}$$
$$I_{C1} = I_{E2} \approx I_{C2} = \beta_1 I_{B1}$$
$$V_{B2} = \left(\frac{R_2}{R_1 + R_2}\right) V_{CC}$$
$$V_{C1} = V_{E2} = V_{B2} - V_{BE}$$
$$V_{C2} = V_{CC} - I_{C2} R_{C2}$$
$$V_{CE2} = V_{C2} - V_{E2}$$

Q_1, like any other common-emitter stage, can also be biased using the voltage-divider method and an emitter resistor.

Figure 9–38 shows a small signal equivalent circuit of the cascode amplifier. The collector, base, and emitter terminals of each transistor are labeled in the diagram (C1, B1, etc.).

Referring to the equivalent circuit, we see that

$$r_{in} = R_{B1} \| \beta_1 r'_{e1}$$

Note that the input resistance of Q_2 is the small emitter resistance r'_{e2} of the common-base stage. The parallel combination of r_{c1}/β_1 and r'_{e2} form the ac load on the collector of Q_1. Since $r_{c1}/\beta_1 \gg r'_{e2}$,

$$r_{L1} = \left(\frac{r_{c1}}{\beta_1}\right) \Bigg\| \; r'_{e2} \approx r'_{e2}$$

Thus,

$$A_{v1} = \frac{-r_{L1}}{r'_{e1}} \approx \frac{-r_{e2}}{r'_{e1}} \approx -1 \qquad \text{(9–69)}$$

Equation 9–69 states that the voltage gain of Q_1 is approximately unity, which follows from the fact that $I_{E2} \approx I_{E1}$, making $r'_{e2} \approx r'_{e1}$. The ac load on the collector of Q_2 is seen to be

$$r_{L2} = r_{c2} \| R_{C2} \| R_L \approx R_{C2} \| R_L$$

Therefore, the voltage gain of the second stage is

$$A_{v2} \approx \frac{r_{L2}}{r'_{e2}} \approx \frac{R_{C2} \| R_L}{r'_{e2}} \qquad \text{(9–70)}$$

The overall voltage gain of the cascode is thus

$$A_v = A_{v1} A_{v2} \approx -\frac{R_{C2} \| R_L}{r'_{e2}} \qquad \text{(9–71)}$$

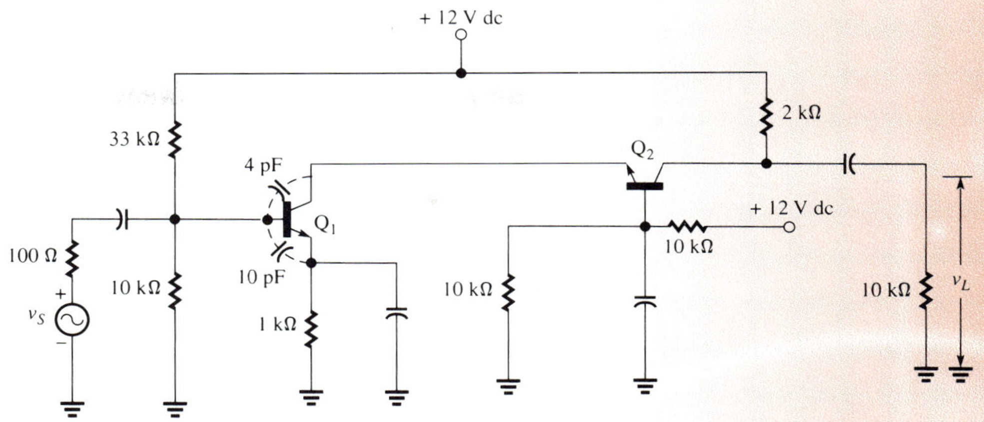

FIGURE 9–39 (Example 9–17)

Equation 9–71 shows that the common-base stage provides most of the voltage gain. The current gain of the cascode amplifier, not including current division at the input or output, is approximately

$$A_i \approx \beta_1\alpha_2 \approx \beta_1 \qquad \text{since } \alpha_2 \approx 1$$

The output resistance of the amplifier is

$$r_{out} = r_{c2} \| R_{C2} \approx R_{C2}$$

EXAMPLE 9–17

The silicon transistors in Figure 9–39 have the following parameter values:

$$Q_1: \beta_1 = 100, r_{c1} \approx \infty, C_{bc} = 4 \text{ pF}, C_{be} = 10 \text{ pF}$$
$$Q_2: \alpha_2 \approx 1, r_{c2} \approx \infty$$

Find approximate values for

1. the dc currents and voltages I_{C1}, I_{C2}, V_{C1}, and V_{C2};
2. the small-signal voltage gain v_L/v_S; and
3. the break frequency f_A due to shunt capacitance at the input of Q_1.

Solution

1. The base-to-ground bias voltage at the input stage (Q_1) is determined by the voltage-divider there.

$$V_{B1} \approx \left[\frac{10 \text{ k}\Omega \| 100 \text{ k}\Omega}{(10 \text{ k}\Omega \| 100 \text{ k}\Omega) + (33 \text{ k}\Omega)}\right](12 \text{ V}) = 2.6 \text{ V}$$

Therefore,

$$V_{E1} = V_{B1} - 0.7 = 2.6 - 0.7 = 1.9 \text{ V}$$
$$I_{C1} \approx I_{E1} = \frac{V_{E1}}{R_{E1}} = \frac{1.9 \text{ V}}{1 \text{ k}\Omega} = 1.9 \text{ mA} = I_{E2} \approx I_{C2}$$

The base-to-ground bias voltage of Q_2 is determined by its voltage divider:

$$V_{B2} = \left[\frac{10 \text{ k}\Omega}{(10 \text{ k}\Omega) + (10 \text{ k}\Omega)}\right](12 \text{ V}) = 6 \text{ V}$$

Therefore,

$$V_{C1} = V_{E2} = V_{B2} - 0.7 = 6 - 0.7 = 5.3 \text{ V}$$
$$V_{C2} = V_{CC} - I_{C2}R_{C2} = 12 - (1.9 \text{ mA})(2 \text{ k}\Omega) = 8.2 \text{ V}$$

2. Since $I_{E1} \approx I_{E2}$,

$$r'_{e2} \approx r'_{e1} = \frac{0.026}{I_{E1}} = \frac{0.026}{1.9\text{ mA}} = 13.7\ \Omega$$

$$r_{in} = (33\text{ k}\Omega) \parallel (10\text{ k}\Omega) \parallel 100(13.7\ \Omega) = 1.16\text{ k}\Omega$$

$$A_{v1} = \frac{-r'_{e2}}{r'_{e1}} \approx -1$$

$$A_{v2} = \frac{r_{L2}}{r'_{e2}} = \frac{(2\text{ k}\Omega) \parallel (10\text{ k}\Omega)}{13.7\ \Omega} = 121.6$$

$$\frac{v_L}{v_S} = \left[\frac{r_{in}}{r_S + r_{in}}\right] A_{v1}A_{v2}$$

$$= \left[\frac{1.16\text{ k}\Omega}{(100\ \Omega) + (1.16\text{ k}\Omega)}\right](-1)(121.6) = -112$$

3. The Miller capacitance at the input to Q_1 is determined by the voltage gain of stage 1 alone:

$$C_M = C_{bc}(1 - A_{v1}) = (4\text{ pF})(2) = 8\text{ pF}$$

The total input capacitance is, therefore,

$$C_A = C_M + C_{be} = (8\text{ pF}) + (10\text{ pF}) = 18\text{ pF}$$

Then

$$f_A = \frac{1}{2\pi[r_S \parallel r_{in}(\text{stage 1})]C_A} = \frac{1}{2\pi[100 \parallel (1.16 \times 10^3)](18 \times 10^{-12})}$$

$$= 96\text{ MHz}$$

We see that the break frequency due to input capacitance is quite high. While there may be break frequencies elsewhere in the circuit, the cascode amplifier effectively eliminates the most troublesome source of high-frequency loss—the input Miller capacitance.

9–7 FREQUENCY RESPONSE OF FET AMPLIFIERS

Low-Frequency Response of FET Amplifiers

Figure 9–40 shows a common-source JFET amplifier biased using the combination of self-bias and a voltage divider, as studied in Chapter 5. For our purposes now, the FET could also be a MOSFET. The capacitors that affect the low-frequency response are the input coupling capacitor, C_1, the

FIGURE 9–40 The low-frequency response of an FET amplifier is affected by capacitors C_1, C_2, and C_S

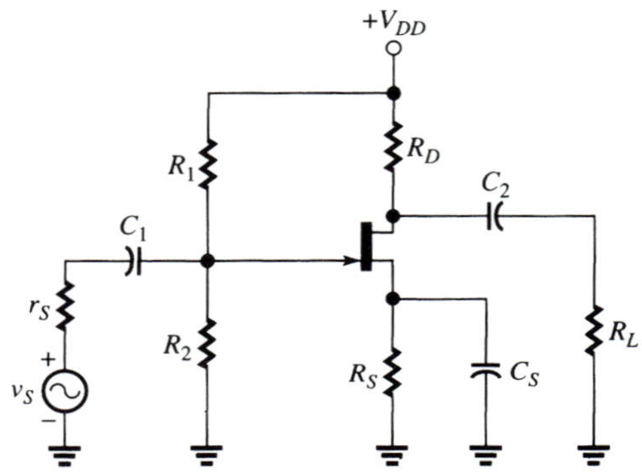

output coupling capacitor, C_2, and the source bypass capacitor, C_S. Applying equations 9–48 and 9–49 to this amplifier configuration, we find

$$f_1 = \frac{1}{2\pi(r_S + r_{in})C_1} \tag{9–72}$$

where $r_{in} \approx R_1 \| R_2$, and

$$f_2 = \frac{1}{2\pi(r_{out} + R_L)C_2} \tag{9–73}$$

where $r_{out} = r_d \| R_D$. The bypass capacitor C_S affects low-frequency response because at low frequencies its reactance is no longer small enough to eliminate degeneration. The corner frequency due to C_S is the frequency at which the reactance of C_S equals the resistance looking into the junction of R_S and C_S at the source terminal, namely, $R_S \| (1/g_m)$. Thus,

$$f_S = \frac{1}{2\pi[R_S \| (1/g_m)]C_S} \tag{9–74}$$

Since the bias resistors R_1 and R_2 are usually very large in an FET amplifier, the input coupling capacitor C_1 can be smaller than its counterpart in a BJT CE amplifier, and C_1 does not generally determine the low-frequency cutoff. The small value of $R_S \| (1/g_m)$ makes the source bypass capacitor the most troublesome, in that a large amount of capacitance is required to achieve a small value of f_S.

EXAMPLE 9–18

The FET shown in Figure 9–41 has $g_m = 3.4$ mS and $r_d = 100$ kΩ. Find the approximate lower cutoff frequency.

Solution

$$r_{in} \approx (1.5\ \text{M}\Omega) \| (330\ \text{k}\Omega) = 270\ \text{k}\Omega$$

$$f_1 = \frac{1}{2\pi(20 \times 10^3 + 270 \times 10^3)(0.02 \times 10^{-6})} = 27.4\ \text{Hz}$$

$$r_{out} = (100\ \text{k}\Omega) \| (2\ \text{k}\Omega) = 1.96\ \text{k}\Omega$$

$$f_2 = \frac{1}{2\pi(1.96 \times 10^3 + 40 \times 10^3)(0.02 \times 10^{-6})} = 189.6\ \text{Hz}$$

$$R_S \| (1/g_m) = 820 \| 294.1 = 216.5\ \Omega$$

$$f_S = \frac{1}{2\pi(216.5)(1 \times 10^{-6})} = 735.1\ \text{Hz}$$

The actual lower cutoff frequency is approximately equal to the largest of the three corner frequencies, namely, $f_H \approx 735.1$ Hz.

FIGURE 9–41 (Example 9–18)

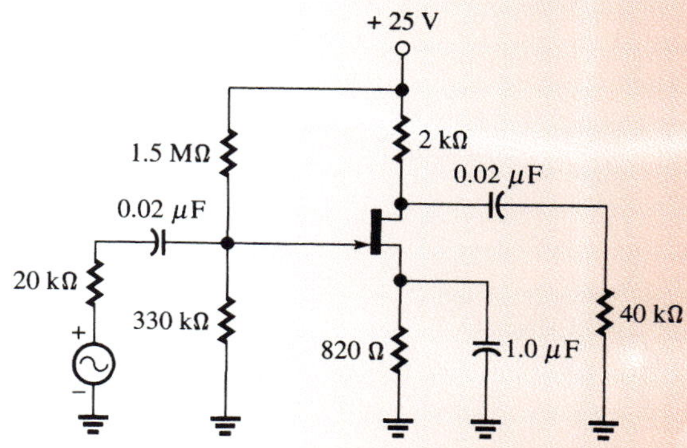

$$f_1 = \frac{1}{2\pi(r_S + R_1 \| R_2)C_1}$$

$$f_2 = \frac{1}{2\pi[R_S \|(1/g_m) + R_L]C_2}$$

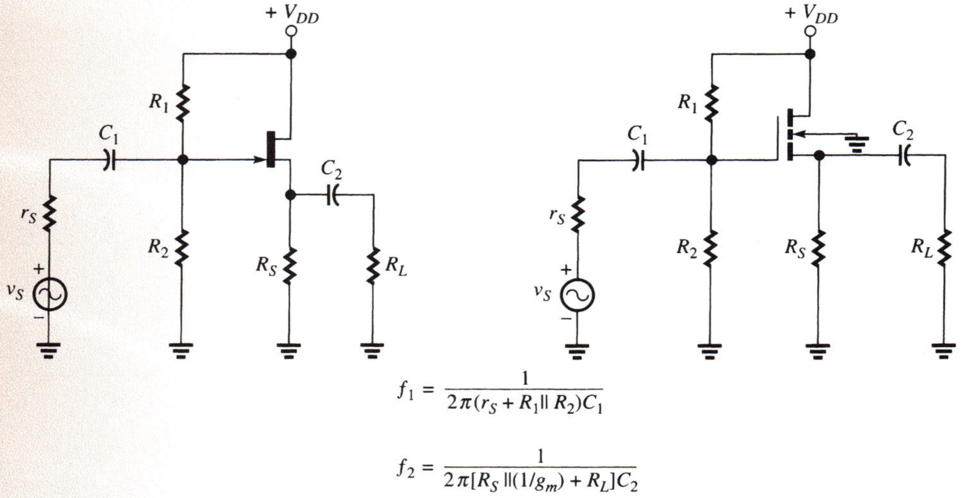

FIGURE 9–42 Low-frequency cutoff equations for the common-drain amplifier

Figure 9–42 shows JFET and MOSFET amplifiers in the common-drain (source-follower) configuration and gives the equations for f_1 and f_2. These equations are obtained in a straightforward manner using the same low-frequency cutoff theory we have discussed for general amplifiers.

For design purposes, each of the equations for f_1, f_2, or f_S in each of the three configurations is easily solved for C_1, C_2, or C_S, giving a set of capacitor equations similar to equations 9–56 through 9–58.

High-Frequency Response of FET Amplifiers

Like the BJT common-emitter amplifier, the FET common-source amplifier is affected by Miller capacitance that often determines the upper cutoff frequency. Figure 9–43 shows a common-source amplifier and the interelectrode capacitance that affects its high-frequency performance. Because R_S is completely bypassed at high frequencies, C_{gs} and C_{ds} are effectively shunting the input and output, respectively, to ground. The equations for f_A' and f_B' are obtained directly from the general high-frequency cutoff theory we have studied:

$$f_A' = \frac{1}{2\pi(r_S \| r_{in})C_A'} \tag{9–75}$$

where $C_A' = C_{gs} + C_M = C_{gs} + C_{gd}(1 - A_v)$ and $r_i = r_{in} \approx R_1 \| R_2$, and

FIGURE 9–43 Common-source amplifier with interelectrode capacitance that affects high-frequency response

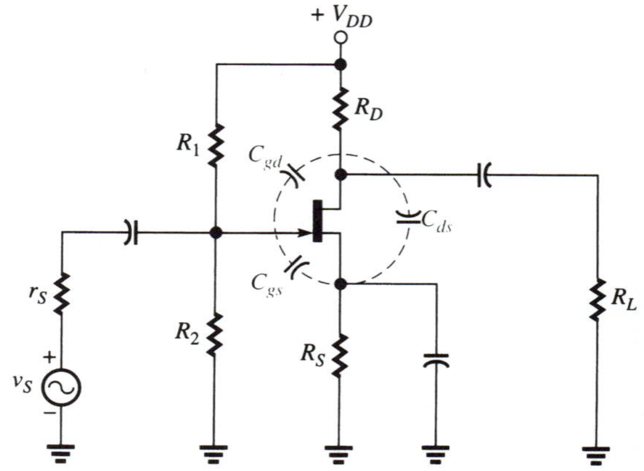

$$f'_B = \frac{1}{2\pi(r_{out} \| R_L)C'_B} \qquad (9\text{–}76)$$

where $C'_B = C_{ds} + C_{gd}(1 - 1/A_v)$ and $r_{out} = r_d \| R_D$.

EXAMPLE 9–19

The JFET in Figure 9–43 has $g_m = 3.2$ mS, $r_d = 100$ kΩ, $C_{gs} = 4$ pF, $C_{ds} = 0.5$ pF, and $C_{gd} = 1.2$ pF. The wiring capacitance shunting the input is 2.5 pF and that shunting the output is 4 pF. The component values in the circuit are

$$r_S = 20\text{ k}\Omega \qquad R_D = 2\text{ k}\Omega$$
$$R_1 = 1.5\text{ M}\Omega \qquad R_S = 820\ \Omega$$
$$R_2 = 330\text{ k}\Omega \qquad R_L = 40\text{ k}\Omega$$

Find the approximate upper cutoff frequency.

Solution

$$A_v = -g_m(r_d \| R_D \| R_L) = -3.2 \times 10^{-3}[(100\text{ k}\Omega) \| (2\text{ k}\Omega) \| (40\text{ k}\Omega)] = -6$$
$$C_M = (1.2\text{ pF})[1 - (-6)] = 8.4\text{ pF}$$
$$C_A = C_{gs} + C_M + C_{(wiring)} = (4\text{ pF}) + (8.4\text{ pF}) + (2.5\text{ pF}) = 14.9\text{ pF}$$
$$r_{in} = (1.5\text{ M}\Omega) \| (330\text{ k}\Omega) = 270\text{ k}\Omega$$
$$f_A = \frac{1}{2\pi(20 \times 10^3 \| 270 \times 10^3)(14.9 \times 10^{-12})} = 573.6\text{ kHz}$$
$$C_B = C_{ds} + C_{gd}(1 - 1/A_v) + C_{(wiring)}$$
$$= (0.5\text{ pF}) + (1.2\text{ pF})(1 + 1/6) + (4\text{ pF}) = 5.9\text{ pF}$$
$$r_{out} = (100\text{ k}\Omega) \| (2\text{ k}\Omega) = 1.96\text{ k}\Omega$$
$$f_B = \frac{1}{2\pi(1.96 \times 10^3 \| 40 \times 10^3)(5.9 \times 10^{-12})} = 14.43\text{ MHz}$$

Therefore, using equation 9–68, $f_H \approx 552$ kHz.

EXAMPLE 9–20

PSPICE

Use PSpice and Probe to determine the lower and upper cutoff frequencies of the MOSFET amplifier shown in Figure 9–44. The capacitances C_{gs}, C_{gd}, and C_{ds} shown in the figure represent stray capacitance and should be included in the input circuit file. The MOSFET transistor has VTO = 2 and KP = 0.5E − 3. Let other parameters default.

FIGURE 9–44 (Example 9–20)

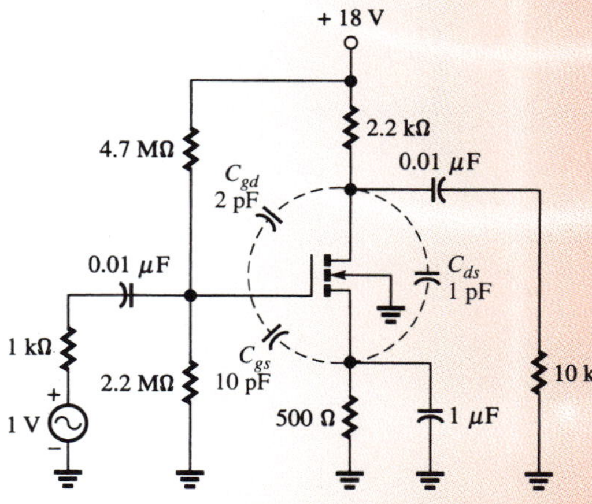

Solution

The PSpice circuit and input circuit file are shown in Figure 9–45(a). A plot of the voltage across the load resistance, V(7), as displayed by Probe, is shown in Figure 9–45(b). We see that the maximum load voltage is 2.5 V. Therefore, the lower and upper cutoff frequencies occur where the load voltage is 0.707(2.5 V) = 1.767 V. As shown in the Probe display, cursor C1 is moved to a point on the plot where the voltage equals 1.768 V, and we see that the lower cutoff frequency is 142.5 Hz. C2 is moved to 1.768 V at the high-frequency end of the plot, and we see that the upper cutoff frequency is 8.052 MHz.

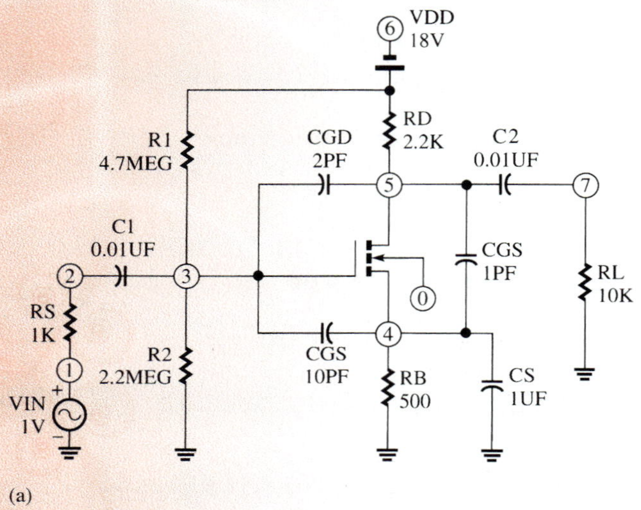


```
EXAMPLE 9-20
VIN 1 0 AC 1V
VDD 6 0 18V
RS 1 2 1K
C1 2 3 0.01UF
R1 6 3 4.7MEG
R2 3 0 2.2MEG
RD 5 6 2.2K
RB 4 0 500
CGD 3 5 2PF
CGS 3 4 10PF
CDS 5 4 1PF
CS 4 0 1UF
CL 5 7 0.01UF
RL 7 0 10K
M1 5 3 4 0 MOSFET
.MODEL MOSFET NMOS VTO=2 KP=0.5E-3
.AC DEC 10 1K 10MEG
.PROBE
.END
```

FIGURE 9–45(a) (Example 9–20)

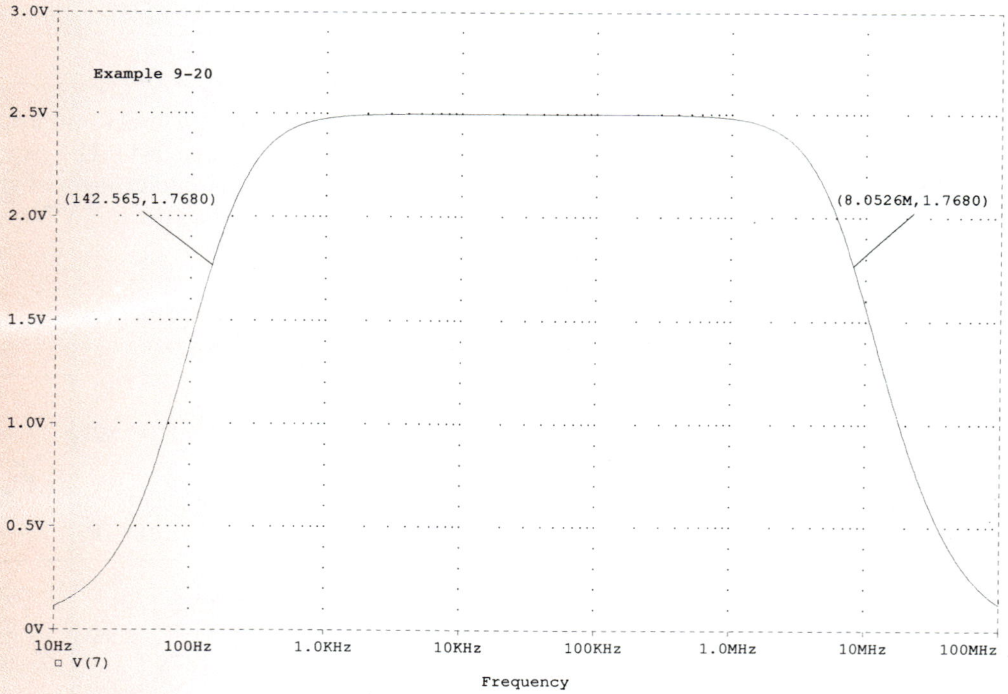

FIGURE 9–45(b) (Example 9–20)

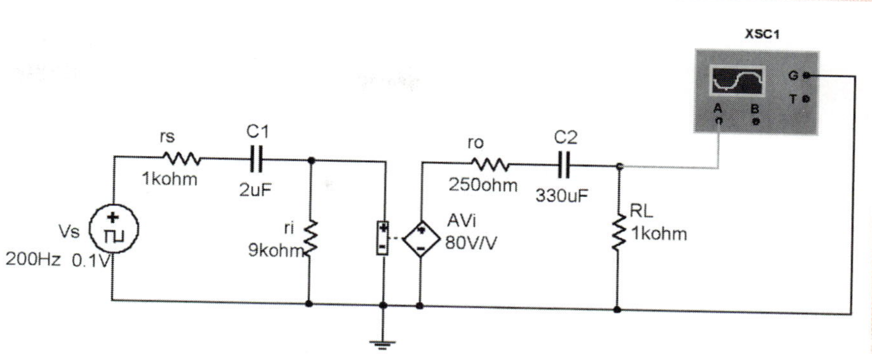

FIGURE 9–46 Amplifier circuit for experimenting with the sag of a square waveform

9–8 MULTISIM EXERCISE

Our objective for this exercise is to reinforce the concept of transient response due to the low-frequency response of an amplifier. The sagging in an amplified square waveform is caused by lack of low-frequency response. No sagging will occur, however, if the amplifier can respond to dc or zero frequency.

Figure 9–46 shows an amplifier modeled with a VCVS. It is capacitively coupled to the input source and to the load. The input signal is a square wave from a clock source. Before you start the simulation, double-click on the oscilloscope and set up the vertical scale to 2 V/div and the horizontal scale to 2 ms/div.

Start the simulation and wait until the output waveform reaches a steady-state condition. With the cursors, measure both the peak (maximum) voltage and the sagging voltage, and then compute the actual sag of the waveform. With the sag, determine the expected lower cutoff frequency f_L and compare to the value calculated from the circuit, which is approximately 8 Hz.

SUMMARY

This chapter has presented a general concept of how amplifiers respond to ac signals of different frequencies. Their behavior can be fully described by graphs of gain magnitude and phase versus frequency called Bode plots. At the end of this chapter the reader should have an understanding of the following concepts:

- An amplifier has a voltage gain whose magnitude and phase depend on the frequency of the input signal.
- Amplifiers have a fairly wide frequency range for which the magnitude of the gain is essentially constant. This region of the frequency spectrum is called the mid-frequency range.
- Gains and attenuations in amplifiers or circuits can be expressed in decibels. Decibels are logarithmic quantities that can be positive, zero, or negative.
- Series capacitance in circuits and amplifiers affect their low-frequency response, which is of high-pass and phase-lead nature.

- Shunt capacitance in circuits and amplifiers affect their high-frequency response, which is of low-pass and phase-lag nature.
- A dominant corner frequency is one whose value has the most influence on a cutoff frequency.
- The dominant corner frequency on the lower side of the frequency response is that with the largest value.
- The dominant corner frequency on the upper side of the frequency response is that with the smallest value.
- The transient response of an amplifier depends on its frequency response and vice versa.

EXERCISES

SECTION 9–1

Definitions and Basic Concepts

9–1. The lower cutoff frequency of a certain amplifier is 120 kHz and its upper cutoff frequency is 1 MHz. The peak-to-peak output voltage in the midband frequency range is 2.4 V p–p and the output power at 120 kHz is 0.4 W. Find

(a) the bandwidth,

(b) the rms output voltage at 1 MHz,

(c) the output power in the midband frequency range, and

(d) the output power at 1 MHz.

9–2. The voltage at the input of an amplifier is 15 mV rms. The amplifier delivers 0.02 A rms to a 12-Ω load at 1 kHz. The input resistance of the amplifier is 1400 Ω. Its lower cutoff frequency is 50 Hz and its bandwidth is 9.95 kHz. Find

(a) the upper cutoff frequency, f_H,

(b) the output power at 50 Hz,

(c) the midband power gain, and

(d) the power gain at f_H.

9–3. The input signal to a certain amplifier has a 1-kHz component with amplitude 0.2 V rms and a 4-kHz component with amplitude 0.05 V rms. In the amplifier's output, the 1-kHz component has amplitude 0.5 V rms and is delayed with respect to the 1-kHz input component by 0.25 ms. If the output is to be an undistorted version of the input, what should be the amplitude and delay of the 4-kHz output component?

SECTION 9–2

Decibels and Logarithmic Plots

9–4. Find the power gain P_2/P_1 in dB corresponding to each of the following:

(a) $P_1 = 4$ mW, $P_2 = 1$ W;

(b) $P_1 = 0.2$ W, $P_2 = 80$ mW;

(c) $P_1 = 5000\ \mu$W, $P_2 = 5$ mW; and

(d) $P_1 = 0.4\ P_2$.

9–5. A certain amplifier has a power gain of 42 dB. The output power is 16 W and the input resistance is 1 kΩ. What is the rms input voltage?

9–6. Find the voltage gain v_2/v_1 in dB corresponding to each of the following:

(a) $v_1 = 120\ \mu$V rms, $v_2 = 40$ mV rms;

(b) $v_1 = 1.8$ V rms, $v_2 = 18$ V peak-to-peak;

(c) $v_1 = 0.707$ V rms, $v_2 = 1$ V rms;

(d) $v_1 = 5 \times 10^3 v_2$; and

(e) $v_1 = v_2$.

9–7. The amplifier shown in Figure 9–47 has $v_S = 30$ mV rms and $v_L = 1$ V rms. Find

(a) the power gain in decibels, and

(b) the voltage gain in decibels.

9–8. Repeat Exercise 9–7 if the load resistor is 2 kΩ.

9–9. The voltage gain of an amplifier is 18 dB. If the output voltage is 6.8 V rms, and the input and load resistances both equal 800 Ω, find

(a) the input voltage, and

(b) the power gain in decibels.

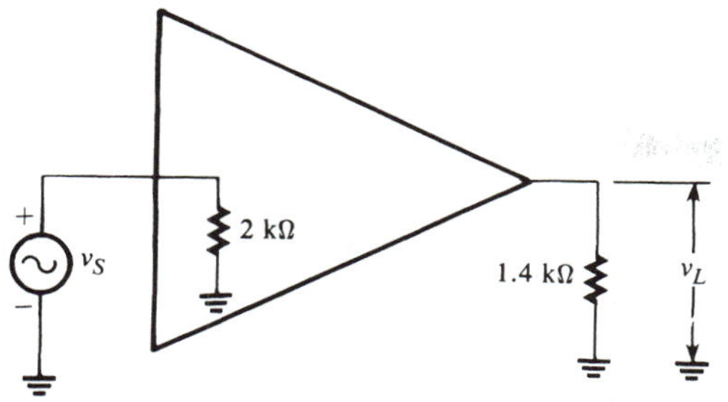

FIGURE 9–47 (Exercise 9–7)

9–10. The output voltage of an amplifier is 12.5 dB above its level at point A in the amplifier. The amplifier input voltage is 6.4 dB below the voltage at point A. If the output voltage is 2.4 V rms, find

(a) the voltage at point A,

(b) the input voltage, and

(c) the amplifier voltage gain, in decibels.

9–11. The input voltage of a certain amplifier is 0.36 V rms. Without performing any computations, estimate the output voltage that would correspond to each of the following gains:

(a) 20 dB,

(b) 46 dB,

(c) −6 dB, and

(d) 60 dB.

9–12. An amplifier has output power 28 dBm and output voltage 34 dBV. Find

(a) its output power in watts, and

(b) its output voltage in V rms.

9–13. In frequency response measurements made on an amplifier, the gain was found to vary from 0.06 to 75 over a frequency range from 7 Hz to 3.3 kHz. What kind of log-log graph paper would be required to plot this data (how many cycles on each axis)?

9–14. Frequency response measurements were made on an amplifier (see next column):

Frequency	*Voltage Gain*	*Phase Shift*
2 Hz	4.0	84.3°
4 Hz	7.8	78.6°
8 Hz	14.9	68.2°
15 Hz	24.0	53.1°
30 Hz	33.3	33.7°
40 Hz	35.8	14.0°
100 Hz	39.2	6.0°
200 Hz	39.8	3.4°
500 Hz	40.0	−3.4°
1 kHz	39.2	−11.3°
2 kHz	37.1	−21.8°
4 kHz	31.2	−38.6°
8 kHz	21.2	−58.0°
10 kHz	17.9	−63.4°

Plot the gain data on log-log graph paper and the phase data on semilog graph paper. Use your plots to estimate the following:

(a) the lower and upper cutoff frequencies,

(b) the (positive) phase shift when the gain is 20,

(c) the gain when the phase shift is −30°, and

(d) the gain two octaves below the lower cutoff frequency.

9–15. The following frequency-response measurements were made on an amplifier:

Frequency	*Voltage Gain*	*Phase Shift*
150 Hz	23.2 dB	89.6°
200 Hz	25.4 dB	68.2°
400 Hz	29.9 dB	51.3°
600 Hz	31.7 dB	39.8°
800 Hz	32.6 dB	29.0°
1 kHz	33.0 dB	22.8°
2 kHz	33.7 dB	6.4°
5 kHz	34.0 dB	−12.7°
8 kHz	32.9 dB	−28.1°
10 kHz	32.4 dB	−33.7°
12 kHz	31.9 dB	−38.7°
18 kHz	30.1 dB	−50.2°
20 kHz	29.6 dB	−53.1°
40 kHz	24.9 dB	−69.4°
100 kHz	17.4 dB	−81.5°
200 kHz	11.5 dB	−85.7°
500 kHz	3.5 dB	−88.3°
1 MHz	−2.5 dB	−89.1°

Plot the gain and phase data on semilog graph paper and use your plots to estimate the following:

(a) the lower and upper cutoff frequencies,

(b) the magnitude of the output voltage when the phase shift is $-60°$ (the input voltage is 20 mV rms),

(c) the (positive) phase shift when the input voltage is 15 mV rms and the output voltage is 0.336 V rms, and

(d) the magnitude of the voltage gain one decade above the upper cutoff frequency.

SECTION 9–3

Series Capacitance and Low-Frequency Response

9–16. Show that the lower cutoff frequency f_L in Figure 9–7 is the frequency at which the capacitive reactance X_{C_1} equals the resistance r_i.

9–17. The amplifier shown in Figure 9–48 has midband voltage gain $|v_L|/|v_S|$ equal to 80. Calculate

(a) the lower cutoff frequency,

(b) the gain $|v_L|/|v_S|$ at 300 Hz,

(c) the phase shift one decade below cutoff, and

(d) the frequency at which the gain $|v_L|/|v_S|$ is 10 dB down from its midband value.

9–18. Repeat Exercise 9–17 using the normalized plots in Figure 9–9 for (b), (c), and (d).

9–19. The amplifier shown in Figure 9–49 has voltage gain $|v_L|/|v_i|$ equal to 200. Calculate

(a) the lower cutoff frequency,

(b) the frequency at which v_L leads v_S by 75°, and

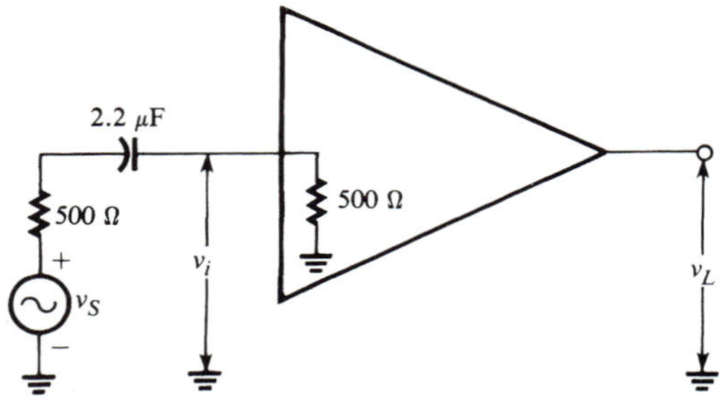

FIGURE 9–49 (Exercise 9–19)

(c) the voltage gain $|v_L|/|v_S|$ two octaves above cutoff.

9–20. The amplifier shown in Figure 9–50 has voltage gain $|v_L|/|v_i|$ equal to 180. What should be the value of the coupling capacitor C in order that the gain $|v_L|/|v_S|$ be no less than 100 at 20 Hz?

9–21. The amplifier shown in Figure 9–51 has midband gain $|v_L|/|v_S|$ equal to 160. Find the (approximate) values of

(a) the lower cutoff frequency,

(b) the gain $|v_o|/|v_i|$, and

(c) a value for the input coupling capacitor that would cause the asymptotic gain plot to have a single break frequency.

9–22. The amplifier shown in Figure 9–52 has a very large (essentially infinite) input

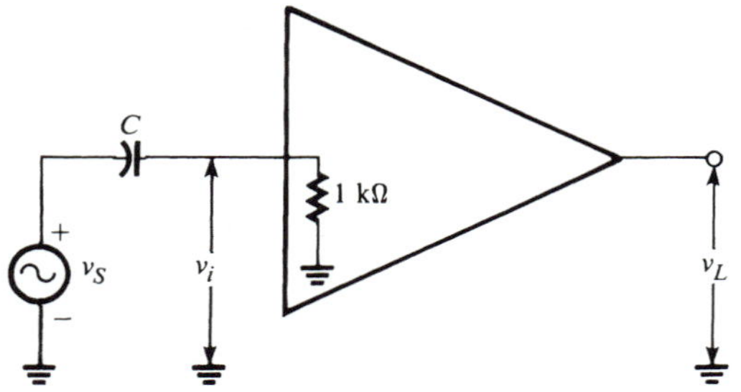

FIGURE 9–50 (Exercise 9–20)

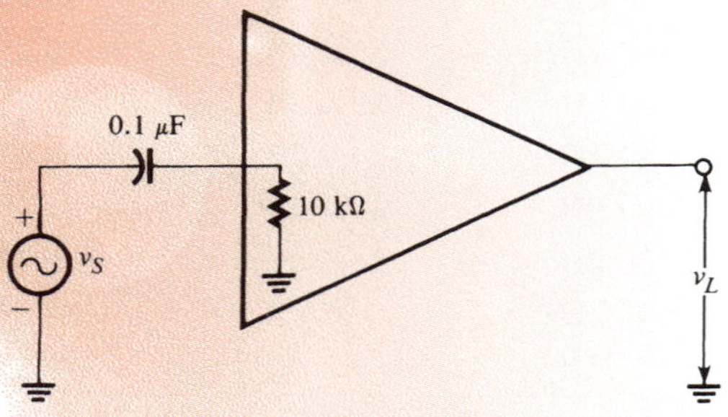

FIGURE 9–48 (Exercise 9–17)

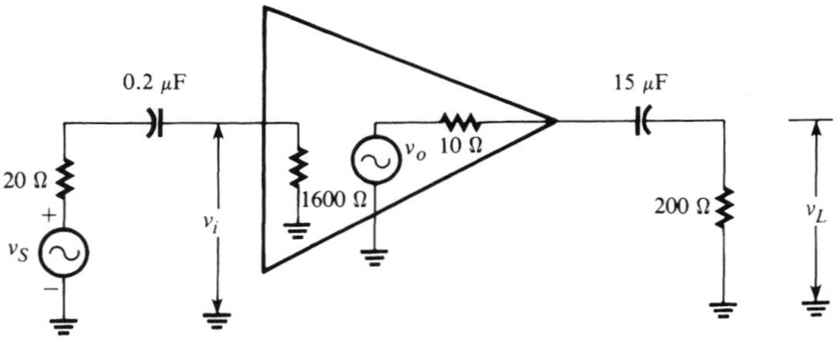

FIGURE 9–51 (Exercise 9–21)

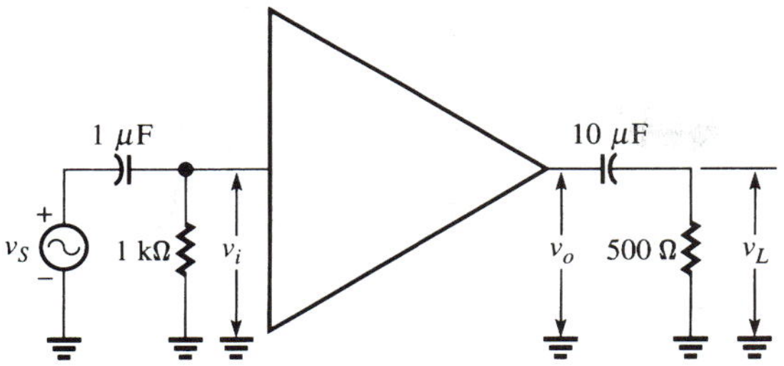

FIGURE 9–52 (Exercise 9–22)

impedance and a very small (essentially 0) output impedance. What should be the gain $|v_o|/|v_i|$ if it is desired to have $|v_L| = 1$ V rms when $|v_S| = 20$ mV rms at 100 Hz?

9–23. The amplifier shown in Figure 9–53 has midband gain $|v_L|/|v_S| = 100$. Calculate

(a) the approximate lower cutoff frequency;

(b) the approximate gain $|v_L|/|v_S|$, in decibels, at $f = 120$ Hz; and

(c) the approximate gain $|v_L|/|v_S|$, in decibels, at $f = 77.6$ Hz.

Use Figure 9–13 to find

(d) the phase shift one octave below the cutoff frequency, and

(e) the frequency at which v_L leads v_S by 160°.

9–24. Find values for the input and output coupling capacitors in Exercise 9–23 that will result in a single break frequency and a lower cutoff frequency at 40 Hz.

SECTION 9–4

Shunt Capacitance and High-Frequency Response

9–25. The amplifier shown in Figure 9–54 has midband gain $|v_L|/|v_S| = 150$. Calculate

(a) the upper cutoff frequency,

(b) $|v_L|/|v_i|$ at midband,

(c) $|v_L|/|v_S|$ at 10 MHz, and

(d) the phase angle of the output with respect to v_S, at 10 MHz.

9–26. For the amplifier of Exercise 9–25, use Figure 9–16 to find approximate values of

(a) $|v_L|/|v_S|$ one octave above cutoff,

(b) the frequency at which $|v_L|/|v_S| = 75$, and

(c) the frequency at which v_L lags v_S by 60°.

9–27. What is the maximum permissible value of the source resistance for the amplifier of Exercise 9–25 if the upper cutoff frequency must be at least 1 MHz?

9–28. Design a low-pass RC network, consisting of a single resistor and a single capacitor, whose output voltage is 0.5 V rms when the input is a 1-V-rms signal having frequency 120 kHz. Draw the schematic diagram, label input and output terminals, and show the component values of your design.

9–29. The amplifier shown in Figure 9–55 has midband gain $|v_L|/|v_S| = 84$.

(a) Find the approximate upper cutoff frequency.

(b) Find the midband value of $|v_o|/|v_i|$.

(c) Sketch the asymptotic Bode plot of $|v_L|/|v_S|$ as it would appear if plotted on log-log graph paper. (It is not necessary to use log-log graph paper for your sketch.) Label important frequency and gain coordinates on your sketch.

9–30. What is the approximate upper cutoff frequency of the amplifier shown in Figure 9–56?

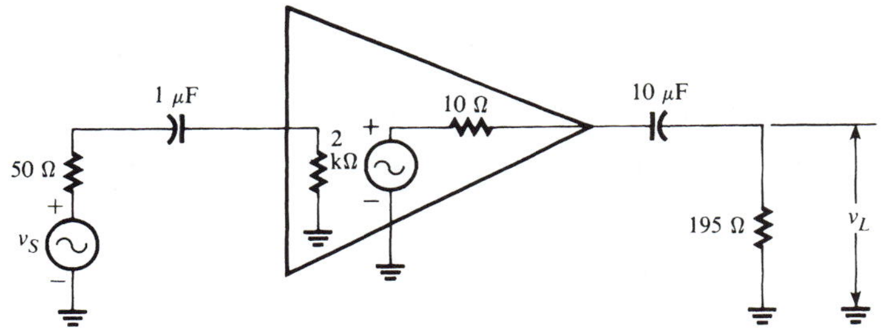

FIGURE 9–53 (Exercise 9–23)

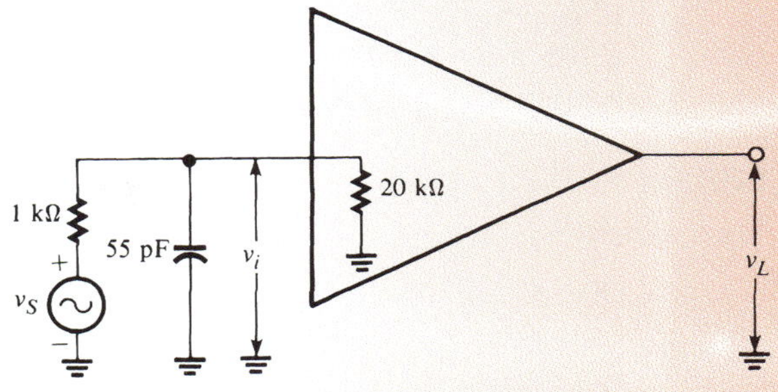

FIGURE 9–54 (Exercise 9–25)

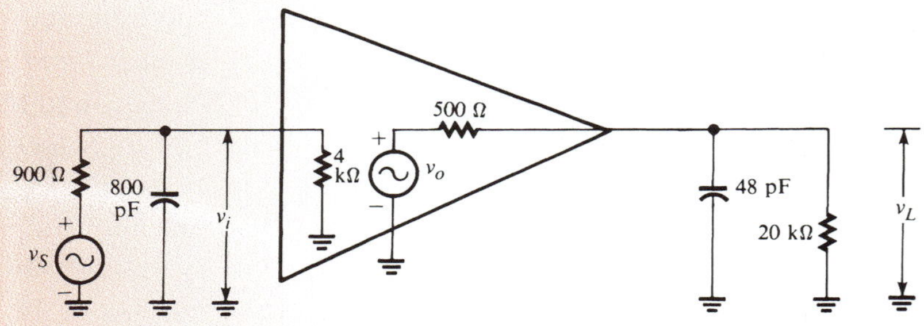

FIGURE 9–55 (Exercise 9–29)

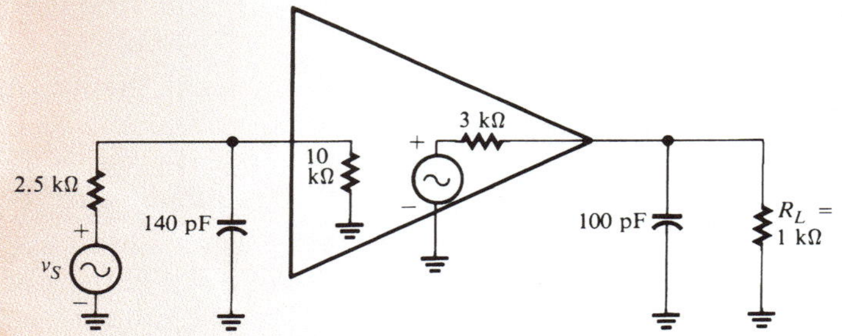

FIGURE 9–56 (Exercise 9–30)

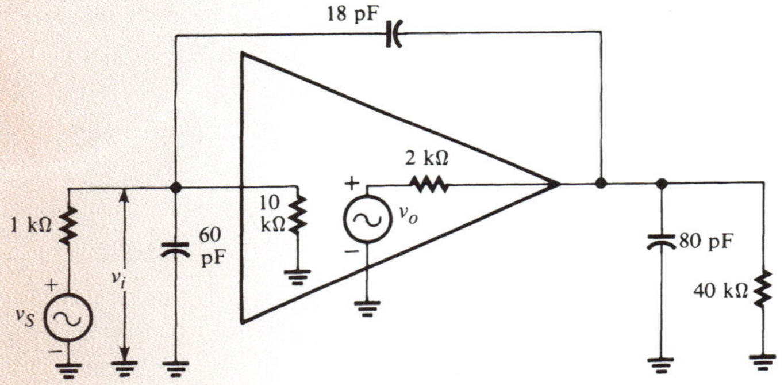

FIGURE 9–57 (Exercise 9–31)

9–31. The amplifier shown in Figure 9–57 has midband gain $|v_L|/|v_i|$ equal to -160. Find the approximate upper cutoff frequency.

9–32. For what value of load resistance in Exercise 9–30 would the frequency response have a single break frequency?

SECTION 9–5

Transient Response

9–33. (a) What is the bandwidth of the low-pass network shown in Figure 9–58?

(b) What is the rise time at the output when the input is a step voltage?

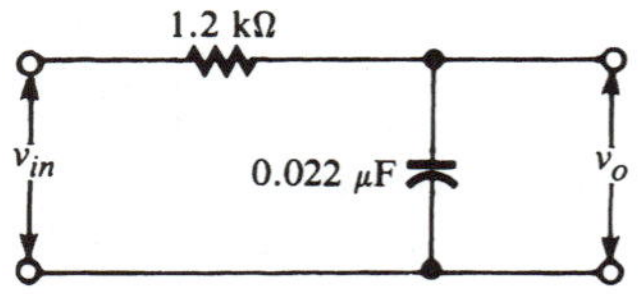

FIGURE 9–58 (Exercise 9–33)

(c) How would the bandwidth be affected by an increase in capacitance? In resistance?

(d) How would the rise time be affected by an increase in capacitance? In resistance?

9–34. A capacitively-coupled amplifier is required to pass a 500-Hz square waveform with no more than 0.1 of sag. What is the maximum lower cutoff frequency that the amplifier can have?

SECTION 9–6

Frequency Response of BJT Amplifiers

9–35. The transistor shown in Figure 9–59 has $\beta = 90$, $r'_e = 15\ \Omega$, and $r_o = 125$ kΩ. Find the approximate lower cutoff frequency.

9–36. The transistor shown in Figure 9–60 has $r'_e = 10\ \Omega$, $\beta = 100$, and $r_o = 100$ kΩ. Find the approximate lower cutoff frequency. Also $r'_e = 18\ \Omega$. Find the approximate lower cutoff frequency.

9–37. The transistor shown in Figure 9–61 has $r'_e = 20\ \Omega$ and $\beta = 150$. Find the approximate lower cutoff frequency.

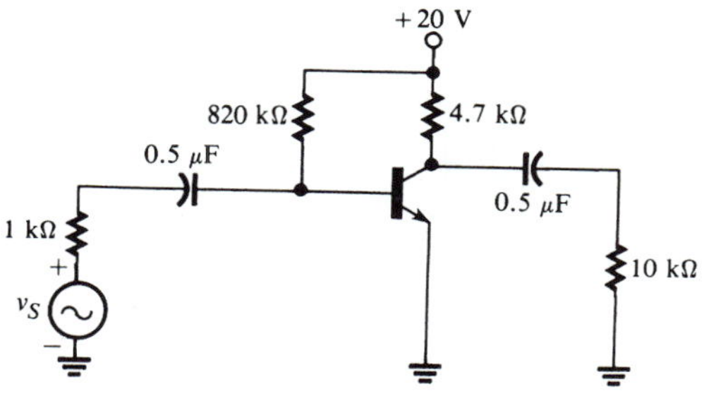

FIGURE 9–59 (Exercise 9–35)

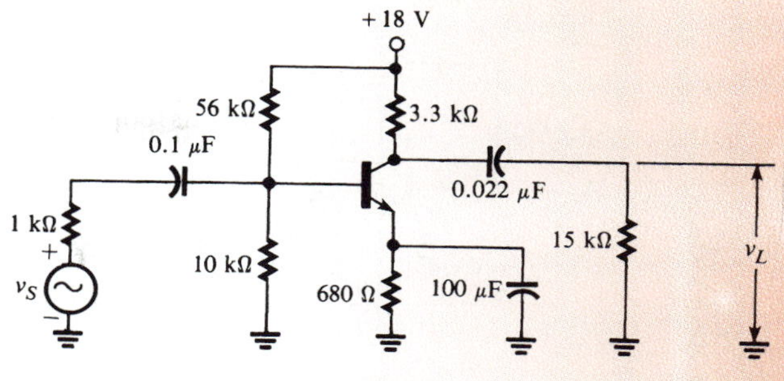

FIGURE 9–60 (Exercise 9–36)

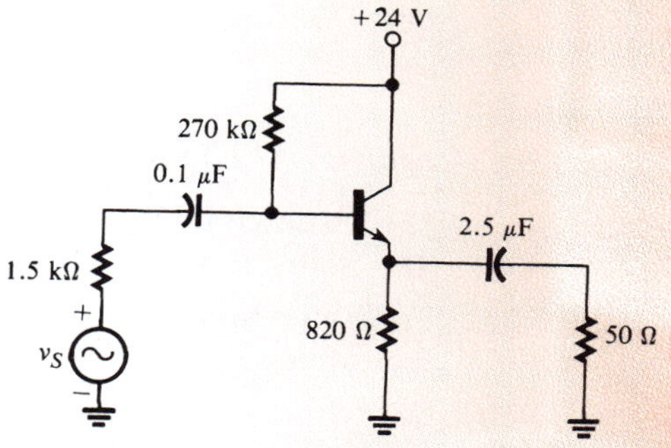

FIGURE 9–61 (Exercise 9–37)

9–38. Following the design procedure given in Section 9–6, redesign the amplifier in Exercise 9–36 (i.e., find new capacitor values) so that the lower cutoff frequency is approximately 100 Hz.

9–39. The common-base amplifier in Figure 9–32(a) has $r_S = 100\ \Omega$, $R_E = 1$ kΩ, $R_C = 4.7$ kΩ, $R_L = 10$ kΩ, $V_{EE} = -2$ V, and $V_{CC} = 15$ V. Select capacitor values so that the lower cutoff frequency is approximately 250 Hz. Assume that capacitors must be selected from a line having standard values 0.1 μF, 0.5 μF, 1 μF, 5 μF, 7.5 μF, 10 μF, 15 μF, 50 μF, and 100 μF.

9–40. The emitter follower in Figure 9–32(b) has $r_S = 600\ \Omega$, $R_1 = 220$ kΩ, $R_2 = 47$ kΩ, $R_E = 1$ kΩ, $R_L = 500\ \Omega$, and $V_{CC} = 15$ V. The β of the transistor is 100.

(a) Find capacitor values so that the lower cutoff frequency is approximately 50 Hz. Assume that capacitors must be selected from a line having standard values 1 μF, 1.5 μF, 2.2 μF, 3.3 μF, 4.7 μF, 10 μF, 15 μF, 22 μF, and 33 μF.

(b) Find the approximate lower cutoff frequency when the standard-value capacitors are used.

(c) Find the approximate lower cutoff frequency when the standard-value capacitors are used and the β of the transistor changes to 200.

9–41. The transistor shown in Figure 9–62 has $r'_e = 25\ \Omega$, $r_o = 50$ kΩ, and a midband β equal to 100. Find the approximate upper cutoff frequency.

9–42. Repeat Exercise 9–41 using the following parameters: $r'_e = 25\ \Omega$, $r_o = 50$ kΩ, $\beta_{(\text{midband})} = 100$, $C_{bc} = 2$ pF, $C_{be} = 22$ pF, $C_{ce} = 15$ pF.

9–43. How much input-to-ground wiring capacitance could be tolerated in the amplifier

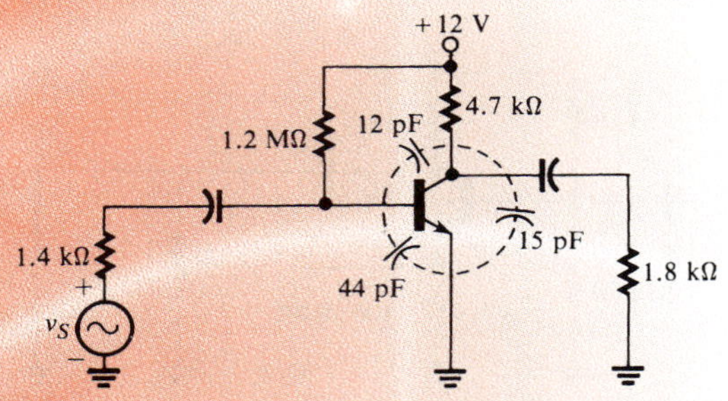

FIGURE 9–62 (Exercise 9–41)

of Exercise 9–41 without the upper cutoff frequency falling below 100 kHz?

SECTION 9–7

Frequency Response of FET Amplifiers

9–44. The JFET shown in Figure 9–63 has $g_m = 4 \times 10^{-3}$ S and $r_d = 100$ kΩ.

(a) Find the approximate lower cutoff frequency.

(b) Sketch asymptotic gain versus frequency as it would appear if plotted on log-log graph paper. Label the break frequencies on your sketch. It is not necessary to show actual gain values, but indicate the rate of gain reduction corresponding to each asymptote.

9–45. The MOSFET shown in Figure 9–64 has $g_m = 3 \times 10^{-3}$ S and $r_d = 75$ kΩ. Find the approximate lower cutoff frequency.

9–46. What is the minimum value of R_L in Exercise 9–45 that could be used without the lower cutoff frequency being greater than 10 Hz?

9–47. The MOSFET shown in Figure 9–65 has $g_m = 4000$ μS and $r_d = 60$ kΩ. Find the approximate upper cutoff frequency.

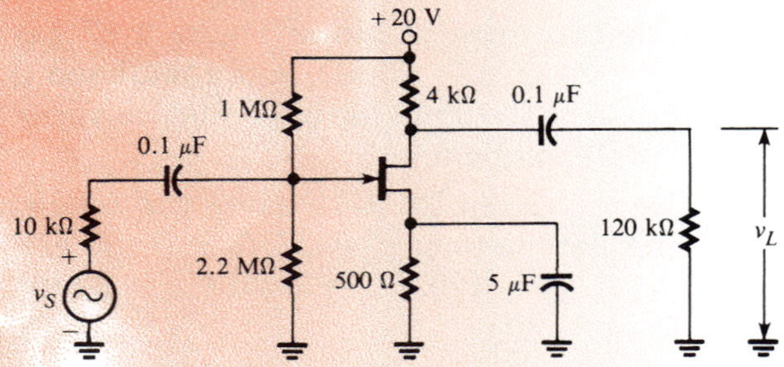

FIGURE 9–63 (Exercise 9–44)

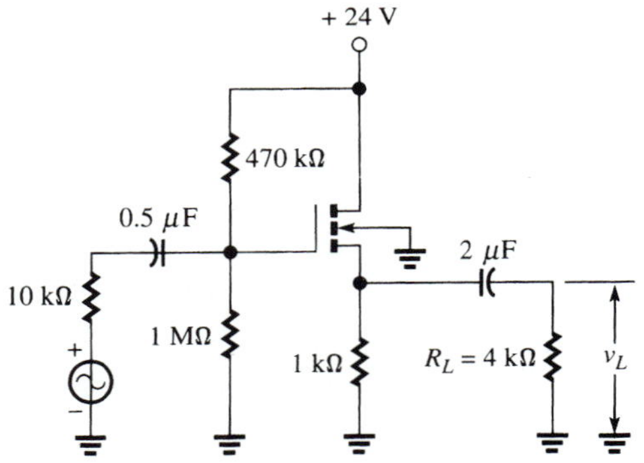

FIGURE 9–64 (Exercise 9–45)

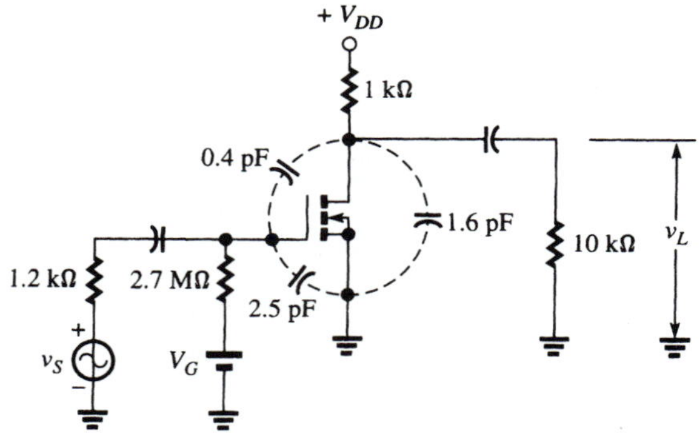

FIGURE 9–65 (Exercise 9–47)

9–48. What is the maximum input wiring capacitance (to ground) that could be tolerated in the amplifier of Exercise 9–47 if the upper cutoff frequency must be at least 1 MHz?

9–49. The JFET in Figure 9–66 has $C_{iss} = 6$ pF, $C_{rss} = 2$ pF, $C_{oss} = 2.8$ pF, $g_m = 3.8 \times 10^{-3}$ S, and $r_d = 100$ kΩ. The wiring and stray capacitance shunting input and output are 0.6 pF and 15 pF, respectively. Find the approximate upper cutoff frequency.

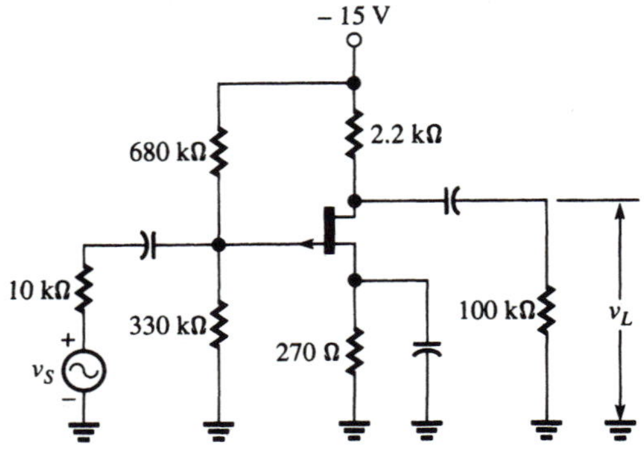

FIGURE 9–66 (Exercise 9–49)

SPICE EXERCISES

9–50. The amplifier in Figure 9–28 has $r_s = 1$ kΩ, $R_1 = 100$ kΩ, $R_2 = 22$ kΩ, $R_C = 3$ kΩ, $R_E = 2$ kΩ, $R_L = 20$ kΩ, $C_1 = 0.22$ μF, $C_2 = 0.22$ μF, $C_E = 15$ μF, and $V_{CC} = 24$ V. The β of the transistor is 120. Use SPICE to find the lower cutoff frequency and to obtain a plot of the low-frequency response.

9–51. Use SPICE to determine the actual factor by which the gain of the amplifier in Example 9–15 is reduced at 50 Hz. The supply voltage is $V_{CC} = 15$ V. Also find the actual lower cutoff frequency.

9–52. Use SPICE to solve Exercise 9–37.

9–53. The amplifier in Figure 9–43 has $r_S = 10$ kΩ, $R_1 = 1.5$ MΩ, $R_2 = 330$ kΩ, $R_D = 2$ kΩ, $R_S = 820$ Ω, and $R_L = 20$ kΩ. The input and output coupling capacitors are each 1 μF and the source bypass capacitor is 10 μF. The JFET has interelectrode capacitance $C_{gd} = 1.5$ pF, $C_{gs} = 6$ pF, and $C_{ds} = 5$ pF. The JFET parameters are $I_{DSS} = 10$ mA and $V_p = -2$ V. Use SPICE to obtain a plot of the high-frequency response of the amplifier and to determine its upper cutoff frequency. (Use discrete capacitors to represent interelectrode capacitance in the SPICE model.)

CHAPTER 10

OPERATIONAL AMPLIFIER THEORY AND PERFORMANCE

OUTLINE

OBJECTIVES

- Transition from ideal to actual op-amp characteristics.
- Analyze op-amp characteristics using feedback concepts.
- Understand the basic components of the open-loop frequency response.
- Relate the closed-loop frequency response of an amplifier to its open-loop frequency response and the feedback ratio.
- Understand the difference in frequency response between inverting and noninverting configurations.
- Learn how slew rate in an op-amp can distort an amplified waveform and how to avoid it.
- Recognize that the input offset parameters in an op-amp cause a dc offset on the output voltage.
- Read and correctly interpret information provided in op-amp data sheets.

10–1 MODELING AN OPERATIONAL AMPLIFIER

In Chapter 8 we examined several op-amp configurations under ideal op-amp characteristics, such as infinite input resistance, infinite gain, and zero output resistance. In this chapter we will study the behavior of the operational amplifier working under its actual, nonideal parameters. We will also study the effect that op-amp parameters have on the overall circuit performance.

In order to simplify the analysis of the different op-amp configurations under nonideal parameters, it is convenient to work with a circuit model that incorporates the nonideal aspects of operational amplifiers. The circuit of Figure 10–1 represents a relatively simple model for an op-amp that includes three important parameters: differential voltage gain A, differential input resistance r_{id}, and output resistance r_o. To give the reader an idea of the values for these parameters, the 741 operational amplifier has a typical differential gain of about 200,000, around 2 megohms of differential input resistance, and an output resistance of about 75 ohms. This model can also be used when simulating circuits containing several op-amps using student-version or evaluation software. Normally, these packages substantially limit the size of the simulated circuit in terms of the number of nodes and components. Op-amp models included with the software, or models provided by manufacturers (called *macromodels*) contain a significant number of nodes and components, thereby preventing the simulation of circuits with several op-amps unless a full version of the software is available.

10–2 FEEDBACK THEORY

We have seen that we can control the closed-loop gain v_o/v_i of an operational amplifier by introducing feedback through external resistor combinations.

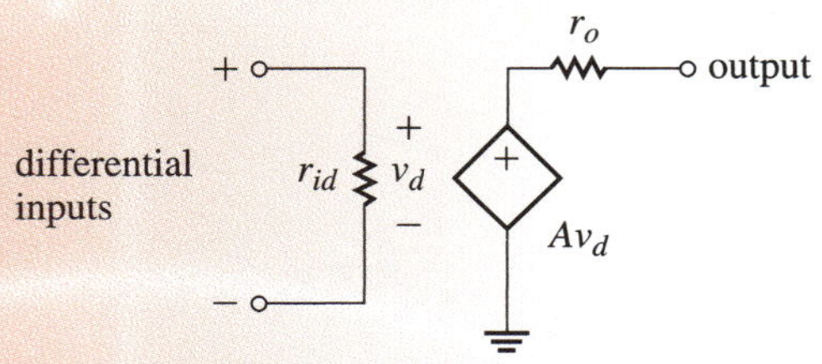

FIGURE 10–1 A simple operational amplifier model with three components: differential input resistance, differential gain, and output resistance

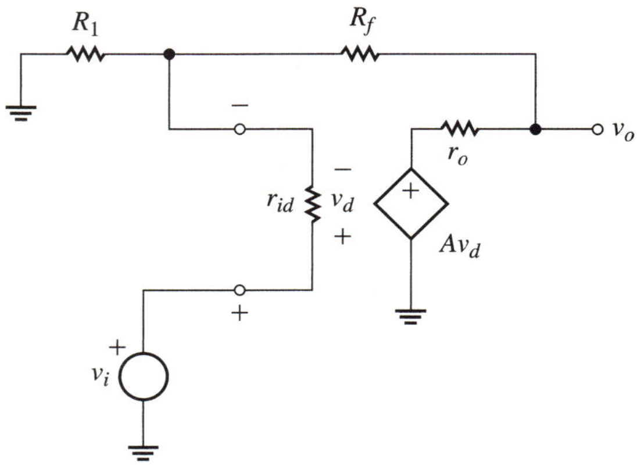

FIGURE 10–2 The noninverting amplifier using the simplified op-amp model

We wish now to examine the feedback mechanism in detail and discover some other important consequences of its use. Feedback theory is widely used to study the behavior of electronic components as well as complex systems in many different technical fields, so it is important to develop an appreciation and understanding of its underlying principles.

Feedback in the Noninverting Amplifier

Consider the noninverting amplifier circuit of Figure 10–2, where the op-amp has been replaced by the model from Figure 10–1. To simplify the analysis, we will assume that both R_1 and R_f are much smaller that the differential input resistance r_{id} (i.e., the voltage divider formed by R_1 and R_f will not be loaded by r_{id}). Similarly, we assume that the output resistance r_o is much smaller than R_1 and R_f and hence we can neglect any voltage drop across it.

We had seen in Chapter 8 that the op-amp's output voltage is $v_o = A(v_i^+ - v_i^-)$. In the noninverting configuration, the input voltage source is connected directly to the + input, making $v_i^+ = v_i$. The inverting (−) input is connected to the voltage divider formed by R_1 and R_f, that is,

$$v_i^- = \left(\frac{R_1}{R_1 + R_f}\right) v_o$$

Note that the voltage division ratio represents the *fraction* of v_o that is fed back to the inverting input. For example, if the fraction is 0.25, then a voltage equal to one-fourth of the output level will appear at the inverting input. Let us define this fraction as the *feedback ratio* β, that is,

$$\beta = \frac{R_1}{R_1 + R_f} \quad \text{(dimensionless)} \qquad \textbf{(10–1)}$$

which yields

$$v_i^- = \beta v_o \qquad \textbf{(10–2)}$$

The output voltage can then be expressed as

$$v_o = A(v_i - \beta v_o) \qquad \textbf{(10–3)}$$

Before we solve for the voltage gain v_o/v_i, let us analyze the preceding expression and derive an equivalent block diagram. The expression reads: The output voltage equals the open-loop gain A times the difference between v_i and the feedback voltage βv_o. Figure 10–3 shows a block diagram that

FIGURE 10–3 A block diagram representation of the noninverting amplifier

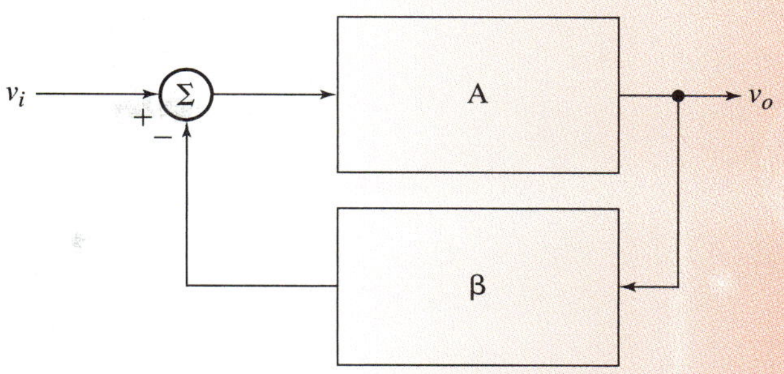

satisfies this relationship. Note the directed arrows interconnecting the blocks. The relationship for *each* block is very simple: Signal going out equals signal going in *times* the block function. Block diagrams are very useful for better visualizing signal paths and the overall structure of circuits and systems. Observe that the block diagram for the noninverting amplifier depicts more clearly the forward and feedback signal paths represented by A and β, respectively. Notice also the standard symbol for the *summing junction:* a circle identified by the Greek letter sigma and signs indicating the nature of the operation, which in our case is a difference.

Going back to our analysis, we want to solve for the ratio v_o/v_i, which is the closed-loop voltage gain of the noninverting amplifier. Grouping the v_o terms, we obtain

$$\frac{v_0}{v_i} = \frac{A}{1 + A\beta} = A_{CL} \qquad (10\text{–}4)$$

where we have used A_{CL} to symbolize closed-loop gain.

Equation 10–4 is the expression for the noninverting closed-loop gain as a function of open-loop gain A and feedback ratio β. This equation gives us the means for investigating how significant the value of open-loop gain A is in the determination of the closed-loop gain A_{CL}. First, note that when $A = \infty$, (10–4) reduces to $A_{CL} = 1/\beta$, which is exactly the same result we obtained in Section 8–1 for the ideal noninverting amplifier: $(R_1 + R_f)/R_1 = 1 + R_f/R_1$.

EXAMPLE 10–1

Find the closed-loop gain of the amplifier in Figure 10–4 when (1) $A = \infty$, (2) $A = 10^6$, and (3) $A = 10^3$.

Solution

1. The feedback ratio is

$$\beta = \frac{R_1}{R_1 + R_f} = \frac{10\text{ k}\Omega}{(10\text{ k}\Omega) + (90\text{ k}\Omega)} = 0.1$$

Therefore, the closed-loop gain when $A = \infty$ is $v_o/v_i = 1/\beta = 1/0.1 = 10$.

FIGURE 10–4 (Example 10–1)

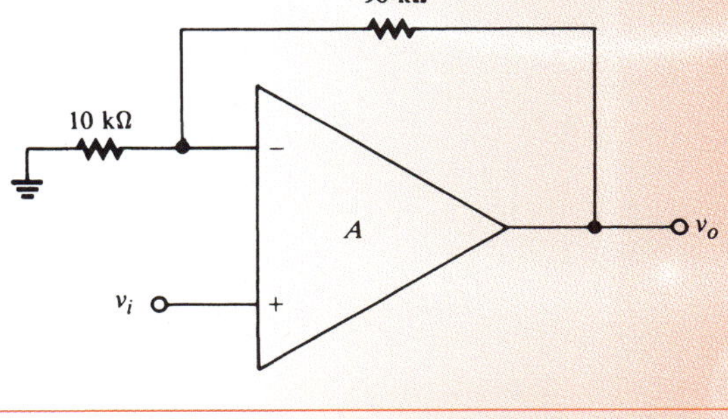

2. Using equation 10–4, the closed-loop gain when $A = 10^6$ is

$$\frac{v_O}{v_i} = \frac{A}{1 + A\beta} = \frac{10^6}{1 + 10^6(0.1)} = 9.99990$$

We see that v_o/v_i is for all practical purposes the same value when $A = 10^6$ as it is when $A = \infty$.

3. When $A = 10^3$,

$$\frac{v_o}{v_i} = \frac{10^3}{1 + 10^3(0.1)} = 9.90099$$

We see that reducing A to 1000 creates a discrepancy of about 1% with respect to the value of v_o/v_i when $A = \infty$.

Equation 10–4 shows that the closed-loop gain of a real amplifier also departs from that of an ideal amplifier when the value of β becomes very small. Small values of β correspond to large closed-loop gains.

Equation (10–4) can be written in terms of the ideal closed-loop gain 1/β by rearranging it in the form

$$A_{CL} = \frac{1/\beta}{1 + 1/A\beta} = \frac{1 + R_f/R_1}{1 + (R_1 + R_f)/AR_1} \qquad \textbf{(10–5)}$$

If we use the notation $\hat{A}_{CL}$ to denote ideal closed-loop gain, then equation 10–5 can be written as

$$A_{CL} = \frac{\hat{A}_{CL}}{1 + \dfrac{\hat{A}_{CL}}{A}} \qquad \textbf{(10–6)}$$

where $\hat{A}_{CL} = 1 + R_f/R_1 = 1/\beta$ is the ideal closed-loop gain. Equation 10–6 is in a convenient form for calculating the actual closed-loop gain from the corresponding ideal gain.

EXAMPLE 10–2

A noninverting amplifier employs an op-amp with open-loop gain of 10,000. Find the actual closed-loop gain for the following resistor combinations: (a) $R_f = 24$ kΩ, $R_1 = 1$ kΩ; (b) $R_f = 49$ kΩ, $R_1 = 1$ kΩ; (c) $R_f = 99$ kΩ, $R_1 = 1$ kΩ.

Solution

a. $\hat{A}_{CL} = 1 + 24/1 = 25$ (ideal)
$A_{CL} = 25/(1 + 25/10{,}000) = 24.94$ (actual)

Similarly,

b. $\hat{A}_{CL} = 50$ (ideal)
$A_{CL} = 49.75$ (actual)

c. $\hat{A}_{CL} = 100$ (ideal)
$A_{CL} = 99.01$ (actual)

The last two examples have shown that the closed-loop gain departs from the ideal value of 1/β when A is small or when β is small, that is, when A_{CL} is large. We can deduce that fact from another examination of equation 10–5.

$$A_{CL} = \frac{v_o}{v_i} = \frac{1/\beta}{1 + 1/A\beta}$$

Clearly, both A and β should be large if we want v_o/v_i to equal $1/\beta$. The product $A\beta$ is called the *loop gain* and is very useful in predicting the behavior of a feedback system. The name *loop gain* is derived from its definition as the product of the gains in the feedback model as one travels around the loop from amplifier input, through the amplifier, and through the feedback path (with the summing junction open).

Input Resistance

Negative feedback improves the performance of an amplifier in several ways. In the case of the noninverting amplifier, the input resistance seen by the signal source (looking directly into the + terminal) increases substantially with feedback as we will examine next.

Figure 10–5 shows a noninverting amplifier where the op-amp is replaced by its model. Our goal is to obtain an expression for the input resistance by solving for the ratio v_i/i_i. First, let us express the current i_i as

$$i_i = \frac{v_i - \beta v_o}{r_{id}} \qquad (10\text{–}7)$$

where we again assume that r_{id} does not affect the voltage divider. Neglecting the voltage drop across the relatively small r_o, v_o can be written as $v_o = Av_d = Ai_ir_{id}$. Substituting into (10–7), we obtain

$$i_i = \frac{v_i - \beta A i_i r_{id}}{r_{id}} \quad \text{or} \quad i_i r_{id}(1 + A\beta) = v_i$$

which yields

$$\frac{v_i}{i_i} = r_i = r_{id}(1 + A\beta) \qquad (10\text{–}8)$$

This equation shows that the input resistance is r_{id} multiplied by the factor $1 + A\beta$, which is usually much greater than 1 and is approximately equal to the loop gain $A\beta$. For example, if $r_{id} = 200$ kΩ, $A = 10^5$, and $\beta = 0.01$, then $r_i \approx (200\text{ k})(10^5)(0.01) = 200$ MΩ, a very respectable value. In the case of the voltage follower, $\beta = 1$, and r_{id} is multiplied by the full value of A, which accounts for the extremely large input resistance it can provide in buffer applications.

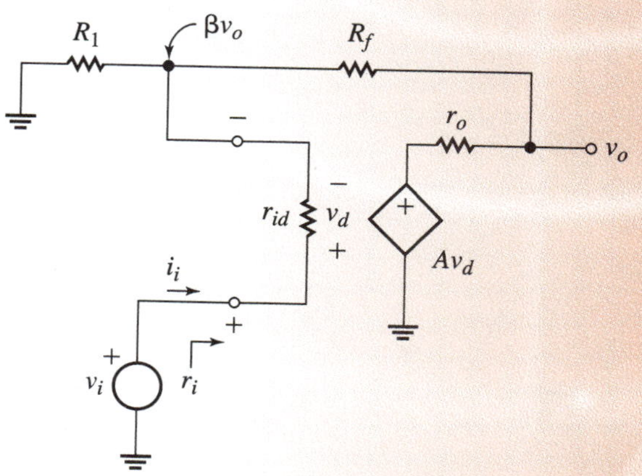

FIGURE 10–5 Obtaining the closed-loop input resistance in a noninverting amplifier

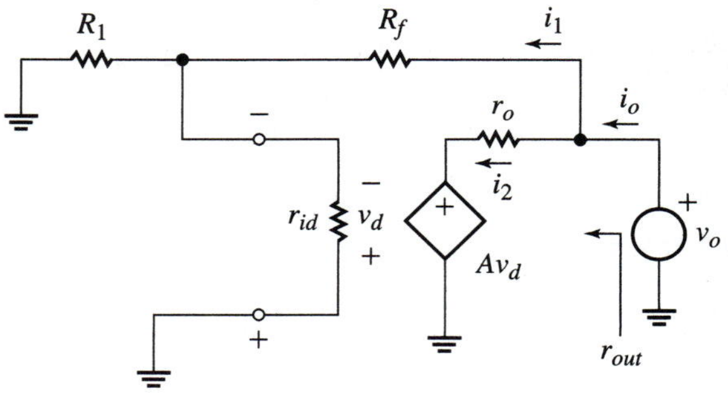

FIGURE 10–6 An external source v_o driving the output to obtain the closed-loop output resistance

Output Resistance

The closed-loop output resistance of the noninverting amplifier is also improved by negative feedback. To obtain the output resistance, we drive the output terminal with an external source, as shown in Figure 10–6.

Using KCL we write

$$i_o = i_1 + i_2$$

where $\quad i_1 = \dfrac{v_o}{R_f + r_{id} \| R_1} \quad$ and $\quad i_2 = \dfrac{v_o - Av_d}{r_o}$

But since $R_1 \ll r_{id}$, i_o can be closely approximated by

$$i_o = \frac{v_o}{R_f + R_1} + \frac{v_o - Av_d}{r_o} \tag{10–9}$$

substituting $v_d \approx -\beta v_o$, we obtain

$$i_o = \frac{v_o}{R_f + R_1} + \frac{v_o(1 + A\beta)}{r_o} \tag{10–10}$$

where it can clearly be seen that the first term in equation 10–10 is much smaller than the second and can be neglected. Solving for v_o/i_o, we finally obtain

$$\frac{v_o}{i_o} = r_{out} = \frac{r_o}{1 + A\beta} \tag{10–11}$$

Equation 10–11 shows that the output resistance is decreased by the same factor by which the input resistance is increased. A typical value for r_o is 75 Ω, so with $A = 10^5$ and $\beta = 0.01$, we have $r_{out} \approx 75/10^3 = 0.075$ Ω, which is very close to the ideal value of zero.

Finally, negative feedback *reduces distortion* caused by the amplifier itself. We will study this property later, in connection with large-signal amplifiers.

Feedback in the Inverting Amplifier

To investigate the effect of open-loop gain A and feedback ratio β on the closed-loop gain of the inverting amplifier, let us recall equations 8–6 and 8–7 from Section 8–1:

$$\frac{v_i}{R_1} - \frac{v_i^-}{R_1} = \frac{v_i^-}{R_f} - \frac{v_o}{R_f} \tag{10–12}$$

$$v_i^- = -v_o/A \tag{10–13}$$

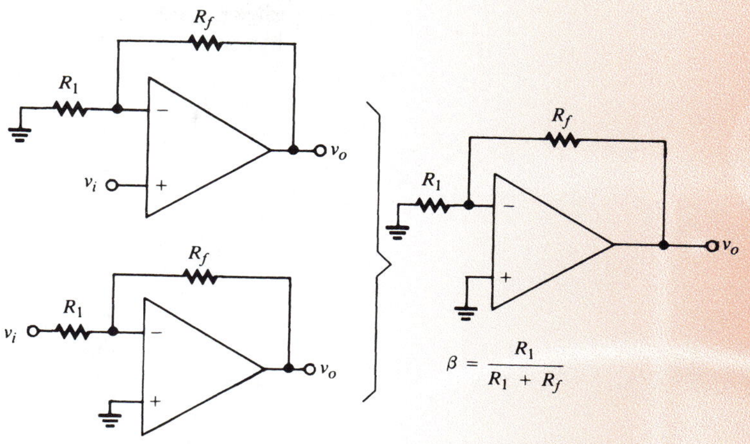

FIGURE 10–7 When v_i is grounded in both the inverting and noninverting amplifiers, it can be seen that the feedback paths are identical

Substituting (10–13) into (10–12) gives

$$\frac{v_i}{R_1} + \frac{v_o}{AR_1} = \frac{-v_o}{AR_f} - \frac{v_o}{R_f} \tag{10–14}$$

Exercise 10–7 at the end of this chapter is included to show that (10–14) can be solved for v_o/v_i with the result

$$\frac{v_o}{v_i} = \frac{-R_f/R_1}{1 + (R_1 + R_f)/AR_1} \tag{10–15}$$

Once again we see that the closed-loop gain reduces to the ideal amplifier value, $-R_f/R_1$, when $A = \infty$. Notice that the denominator of (10–15) is the same as that of (10–5), the equation for the closed-loop gain of the noninverting amplifier. Furthermore, the quantity $R_1/(R_1 + R_f)$ is also the feedback ratio β for the inverting amplifier. This fact is illustrated in Figure 10–7, which shows the feedback paths of both configurations when their signal inputs are grounded. Think of the amplifier output as a source that generates the feedback signal. By the superposition principle, we can analyze the contribution of the feedback source by grounding all other signal sources. When this is done, as shown in Figure 10–7, we see that the feedback voltage in both configurations is developed across the $R_1 - R_f$ voltage divider, and $\beta = R_1/(R_1 + R_f)$ in both cases. In view of this fact, we can write (10–15) as

$$A_{CL} = \frac{v_o}{v_i} = \frac{-R_f/R_1}{1 + 1/A\beta} \tag{10–16}$$

To obtain a block diagram of the inverting amplifier, let us make use of the principle of superposition and the voltage-division rule to express the voltage at the inverting input as

$$v_i^- = \frac{R_f}{R_1 + R_f} v_i + \frac{R_1}{R_1 + R_f} v_o \tag{10–17}$$

Note that $R_1/(R_1 + R_f) = \beta$ and hence $R_f/(R_1 + R_f) = 1 - \beta$. Therefore,

$$v_i^- = (1 - \beta)v_i + \beta v_o$$

Replacing v_i^- with $-v_o/A$, we obtain

$$v_o = -A[(1 - \beta)v_i + \beta v_o] \tag{10–18}$$

Recalling that in a block *output equals input times block function*, we recognize that the term in brackets represents the input to the $-A$ block. We also observe that the bracket term is the *sum* of the terms $(1 - \beta)v_i$ and βv_o.

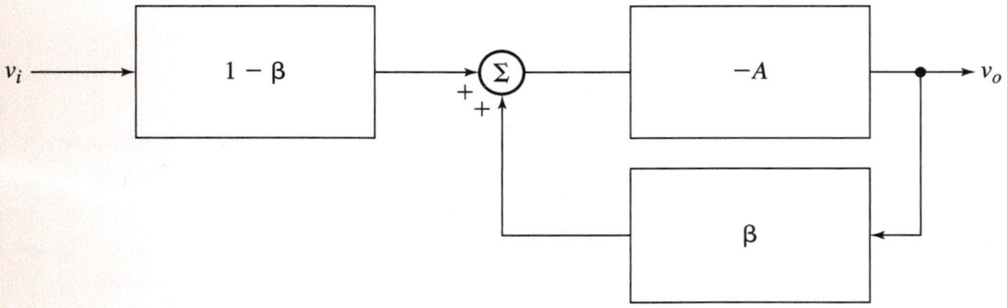

FIGURE 10–8 Block diagram representation of the inverting amplifier

With these facts, we can easily construct the block diagram shown in Figure 10–8 that satisfies this relationship. The reader can also verify that substituting $R_1/(R_1 + R_f)$ for β in equation 10–18 and solving for v_o/v_i yields equation 10–15.

Using again the notation $\hat{A}_{CL}$ for ideal gain, we can write an expression for the inverting closed-loop gain as

$$A_{CL} = \frac{\hat{A}_{CL}}{1 + \dfrac{1 - \hat{A}_{CL}}{A}} \quad \text{(inverting)} \qquad \textbf{(10–19)}$$

where $\hat{A}_{CL} = -R_f/R_1$ is the ideal inverting gain. Note that the term $1 - \hat{A}_{CL}$ in equation 10–19 is a positive quantity. For example, if $R_f = 100$ k, $R_1 = 2.5$ k, and $A = 10{,}000$, the actual closed-loop gain for an inverting amplifier would be $A_{CL} = -40/(1 + 41/10{,}000) = -39.84$. (Actual gain is less than one-half percent less in magnitude than the ideal value.)

As can be seen in Figure 10–8, the loop gain for the inverting amplifier is $A\beta$, the same as that for the noninverting amplifier. From equation 10–16, it is apparent that the greater the loop gain, the closer the closed-loop gain is to its value in the ideal inverting amplifier, $-R_f/R_1$.

Inverting Input Resistance

In an inverting amplifier the input source is connected to resistor R_1 while the noninverting (+) input is connected to ground either directly or by means of a bias-compensating resistor R_C. Figure 10–9 shows the equivalent inverting amplifier for which we want to obtain an expression for the input resistance r_{in} seen by the input source. Clearly, this resistance is R_1 plus the resistance looking into the inverting input (indicated as r_i' in the circuit), which in turn is the ratio v_i^-/i_i. To obtain this ratio, let us write the KCL equation

FIGURE 10–9 Obtaining the closed-loop input resistance for the inverting amplifier

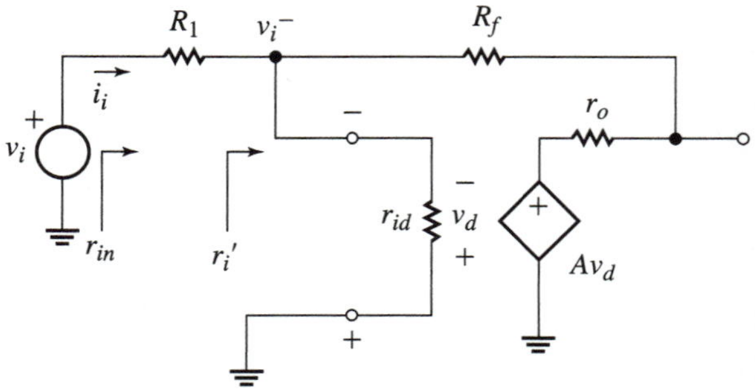

$$i_i = \frac{v_i^-}{r_{id}} + \frac{v_i^- - Av_d}{R_f + r_o} \quad (10\text{–}20)$$

Substituting $v_d = -\,v_i^-$ and simplifying, we obtain

$$i_i = \frac{v_i^-}{r_{id}} + \frac{(1 + A)v_i^-}{R_f + r_o} \approx \frac{Av_i^-}{R_f}$$

from which

$$\frac{v_i^-}{i_i} = r_i' \approx \frac{R_f}{A} \quad (10\text{–}21)$$

and finally,

$$r_{in} \approx R_1 + \frac{R_f}{A} \quad (10\text{–}22)$$

This equation confirms that the input resistance is R_1 for the ideal inverting amplifier, where $A = \infty$. It also shows that the term R_f/A, being small, makes the summing-junction node a "virtual ground."

Inverting Output Resistance

As with the noninverting amplifier, the output resistance of the inverting amplifier is decreased by negative feedback. In fact, the relationship between output resistance and loop gain is the same for both since the circuits are the same when v_i is reduced to zero and the output terminal is driven externally. Therefore,

$$r_{out} = \frac{r_o}{1 + A\beta} \approx \frac{r_o}{A\beta}$$

EXAMPLE 10–3

The operational amplifier shown in Figure 10–10 has open-loop gain equal to 2500 and open-loop output resistance 100 Ω. Find

1. the magnitude of the loop gain,
2. the closed-loop gain,
3. the input resistance seen by v_i, and
4. the closed-loop output resistance.

Solution

1. $\beta = R_1/(R_1 + R_f) = (1.5\ \text{k}\Omega)/[(1.5\ \text{k}\Omega) + (150\ \text{k}\Omega)] = 9.901 \times 10^{-3}$
   loop gain $= A\beta = (2.5 \times 10^3)(9.901 \times 10^{-3}) = 24.75$

FIGURE 10–10 (Example 10–3)

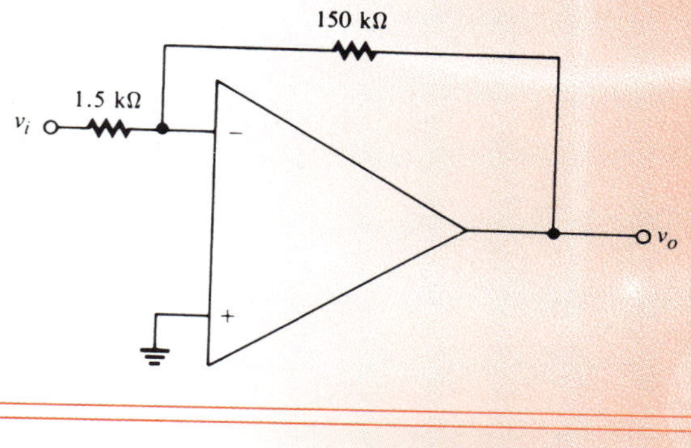

2. From equation 10–16,

$$\frac{v_o}{v_i} = \frac{-R_f/R_1}{1 + 1/A\beta} = \frac{-(150\ \text{k}\Omega)/(1.5\ \text{k}\Omega)}{1 + 1/24.75} = -96.12$$

Note that this value is about 4% less than $-R_f/R_1 = -100$.

3. From equation 10–22,

$$r_{in} = R_1 + \frac{R_f}{A} = (1.5\ \text{k}\Omega) + \frac{150\ \text{k}}{2500} = 1560\ \Omega$$

4. From equation 10–11,

$$r_{out} = \frac{r_o}{1 + A\beta} = \frac{100}{1 + 24.75} = 3.88\ \Omega$$

EXAMPLE 10–4

SPICE

Use SPICE to find the closed-loop voltage gain, the input resistance seen by v_i, and the output resistance of the inverting amplifier in Example 10–3. Use $r_{id} = 1\ \text{M}\Omega$.

Solution

Figure 10–11 shows how we can use a voltage-controlled voltage source (EOP) to model an operational amplifier in SPICE. The voltage source is controlled by the voltage between nodes 2 and 0 (NC− and NC+, respectively). Thus,

FIGURE 10–11 (Example 10–4)

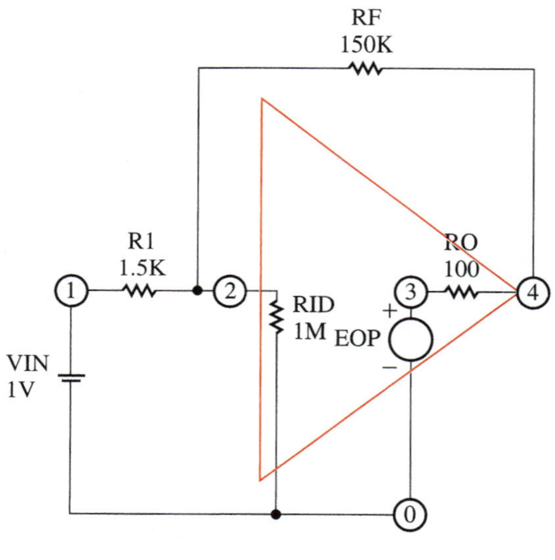


```
EXAMPLE 10-4
VIN   1   0   1V
R1   1   2   1.5K
RID   2   0   1MEG
RF   2   4   150K
EOP   3   0   0   2   2500
RO   3   4   100
VIN   1   0   1
.OP
.END
```

node 2 corresponds to the inverting input of the amplifier in Figure 10–10. Notice that the simulated amplifier has a 1 MΩ input impedance.

Since VIN = 1 V, the output voltage at node 4, V(4), is numerically equal to the closed-loop voltage gain. The results of a program run reveal that V(4) = −96.11, in close agreement with the gain calculated in Example 10–3. SPICE computes the magnitude of I(VIN) to be 0.641 mA, so the input resistance is

$$r_{in} = \frac{v_i}{i_i} = \frac{\text{VIN}}{|\text{I(VIN)}|} = \frac{1\text{ V}}{0.641\text{ mA}} = 1560\ \Omega$$

To find the output resistance of the amplifier, we must compute the current that flows through a short circuit connected to the output, since $r_o = v_L$ (open circuit)/i_L (short circuit). A zero-valued dummy voltage source, VDUM, connected between nodes 4 and 0 effectively shorts the output to ground. The current in VDUM is then the short-circuit output current. SPICE computes this value to be I(VDUM) = 24.7 A. Thus,

$$r_o = \frac{\text{V(4)(open circuit)}}{\text{I(VDUM)(short circuit)}} = \frac{96.1\text{ V}}{24.7\text{ A}} = 3.89\ \Omega$$

This value is in close agreement with that calculated in Example 10–3.

In closing our discussion of feedback theory, we should note once again that the same relationship between actual and ideal closed-loop gain applies to inverting and noninverting amplifiers. This relationship is

$$\text{actual } \frac{v_0}{v_i} = \frac{\hat{A}_{CL}}{1 + 1/A\beta} \qquad \textbf{(10–23)}$$

where $\hat{A}_{CL}$ is the closed-loop gain v_o/v_i that would result if the amplifier were ideal. We saw this relationship in equations 10–5 and 10–16, repeated here:

$$A_{CL} = \frac{v_0}{v_i} = \frac{1 + R_f/R_1}{1 + 1/A\beta} \quad \text{(noninverting amplifier)}$$

$$A_{CL} = \frac{v_0}{v_i} = \frac{-R_f/R_1}{1 + 1/A\beta} \quad \text{(inverting amplifier)}$$

In both cases, the numerator is the closed-loop gain that would result if the amplifier were ideal. Also in both cases, the greater the value of the loop gain $A\beta$, the closer the actual closed-loop gain is to the ideal closed-loop gain.

Although we have demonstrated this relationship only for inverting and noninverting amplifiers, it is a fact that equation 10–23 applies to a wide variety of amplifier configurations, many of which we shall be examining in future discussions.

Finally, it is important to emphasize the fact that all operational amplifiers have a very large open-loop gain (in excess of 100,000), which results in an actual closed-loop gain extremely close to the ideal value. In view of this, the designer should always use the ideal gain equation for calculating resistor values for two practical reasons: first, because the actual value of the open-loop gain would not be known, and second, because the resistors used in the design will have a specific tolerance. In other words, any small error on the actual closed-loop gain will be caused mostly by the resistor tolerance and only to a negligible extent by the (unknown) finite value of the open-loop gain. As long as the open-loop gain remains very large, the actual gain will be practically established by the values of R_f and R_1. In the next section, we will examine the effect that signal frequency has on open-loop and closed-loop gains.

10–3 FREQUENCY RESPONSE

In the previous section we learned some of the nonideal characteristics of operational amplifiers in inverting and noninverting configurations. The analysis, however, was done without regard for the type of input voltage or signal used in the different circuits, and hence the results hold for dc circuits only. In this section, we will examine the same circuits operating with ac input voltages at different frequencies. An ideal operational amplifier, besides having infinite gain, infinite input resistance, and zero output resistance, would also have an infinite bandwidth, meaning that all its parameters remain ideal no matter how high the operating frequency is. It is precisely in this regard that op-amps, in general, deviate substantially from being ideal.

We saw that input resistance in noninverting amplifiers is improved by feedback to levels that, in practical terms, can be considered ideal (on the order of thousands of megohms). Output resistance, although relatively low in op-amps without feedback (open-loop), reduces to practical ideal levels (on the order of milliohms) when feedback is present. We could pose the question: Can frequency response in op-amp circuits be improved by the use of feedback? The answer is yes. We will see that the *closed-loop bandwidth* of an inverting or noninverting amplifier is directly related to the feedback ratio β.

Stability

When the word *stability* is used in connection with a high-gain amplifier, it usually means the property of behaving like an amplifier rather than like an *oscillator*. An oscillator is a device that spontaneously generates an ac signal because of *positive* feedback. We will study oscillator theory in Chapter 12, but for now it is sufficient to know that oscillations are easily induced in high-gain, wide-bandwidth amplifiers due to positive feedback that occurs through reactive elements. As we know, an operational amplifier has very high gain, so precautions must be taken in its design to ensure that it does not oscillate, i.e., to ensure that it remains stable. Large gains at high frequencies tend to make an amplifier unstable because those properties enable positive feedback through stray capacitance.

To ensure stable operation, most operational amplifiers have internal *compensation* circuitry that causes the open-loop gain to diminish with increasing frequency. This reduction in gain is called *rolling-off* the amplifier. Sometimes it is necessary to connect external roll-off networks to reduce high-frequency gain even more rapidly. Because the dc and low-frequency open-loop gain of an operational amplifier is so great, the gain roll-off must begin at relatively low frequencies. As a consequence, the open-loop bandwidth of an operational amplifier is generally rather small.

In most operational amplifiers, the gain over the usable frequency range rolls off at the rate of -20 dB/decade, or -6 dB/octave. Recall from Chapter 9 that this rate of gain reduction is the same as that of a single low-pass RC network. Any device whose gain falls off like that of a single RC network is said to have a *single-pole* frequency response, a name derived from advanced circuit theory. Beyond a certain very high frequency, the frequency response of an operational amplifier exhibits further break frequencies, meaning that the gain falls off at greater rates, but for most practical analysis purposes we can treat the amplifier as if it had a single-pole response.

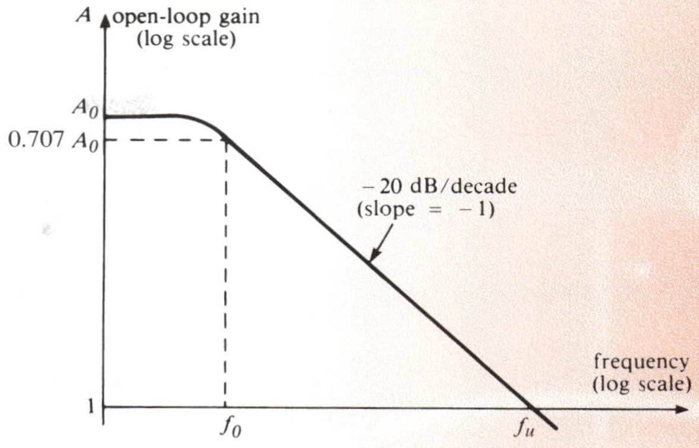

FIGURE 10–12 Frequency response of the open-loop gain of an operational amplifier; A_0 = dc gain, f_0 = cutoff frequency, f_u = unity-gain frequency

The Gain–Bandwidth Product

Figure 10–12 shows a typical frequency response characteristic for the open-loop gain of an operational amplifier having a single-pole frequency response, plotted on log-log scales. We use f_0 to denote the cutoff frequency, which is the frequency at which the gain A falls to 0.707 times its low-frequency, or dc, value (A_0). Recall that the slope of the single-pole response is -1 on log-log scaling. In Chapter 9, we studied a very similar frequency response: that of the beta of a bipolar junction transistor. Using the fact that the slope is -1, we showed that the frequency f_T at which the β falls to the value 1 (unity) is given by $f_T = \beta_m f_\beta$, where β_m is the low- and mid-frequency β and f_β is the β cutoff frequency. Using exactly the same approach, we can deduce that *the frequency at which the amplifier gain falls to value 1 equals the product of the cutoff frequency and the low-frequency gain* A_0:

$$f_u = A_0 f_0 \tag{10–24}$$

where f_u = the *unity-gain frequency*, the frequency at which the gain equals 1
A_0 = the low-frequency, or dc, value of the open-loop gain
f_0 = the cutoff frequency, or 3-dB frequency, of the open-loop gain

Since the amplifier is dc (lower cutoff frequency = 0), the bandwidth equals f_0. The term $A_0 f_0$ in equation 10–24 called the *gain–bandwidth* product, or GBP. In specifications, a value may be quoted either for the gain–bandwidth product or for its equivalent, the unity-gain frequency f_u.

The significance of the gain–bandwidth product is that it makes it possible for us to compute the bandwidth of an amplifier when it is operated in one of the more useful closed-loop configurations. Obviously, a knowledge of the upper-frequency limitation of a certain amplifier configuration is vital information when designing it for a particular application. The relationship between closed-loop bandwidth (BW_{CL}) and the gain–bandwidth product is closely approximated by

$$BW_{CL} = f_u\beta = A_0 f_0 \beta = \beta GBP \tag{10–25}$$

where β is the feedback ratio. See Appendix C for the detailed derivation of this relationship.

Another important fact regarding the gain–bandwidth product is that any point along the *sloped* portion of the open-loop gain plot satisfies the relationship **gain × frequency = GBP**. To illustrate the point, let f_1 and f_2 be two arbitrary frequencies in the sloped portion of the open-loop gain plot

FIGURE 10–13 (Example 10–5)

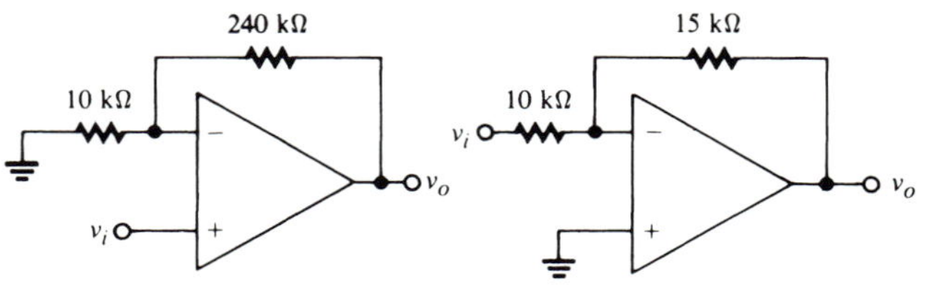

with A_1 and A_2 being their corresponding open-loop gains. The following relationship will be satisfied:

$$A_1 f_1 = A_2 f_2 = A_0 f_0 = 1 f_u = \text{GBP}$$

EXAMPLE 10–5

Each of the amplifiers shown in Figure 10–13 has an open-loop, gain–bandwidth product equal to 1×10^6 Hz. Find the cutoff frequencies in the closed-loop configurations shown.

Solution

1. In Figure 10–13(a), $\beta = R_1/(R_1 + R_f) = (10\text{ k}\Omega)/[(10\text{ k}\Omega) + (240\text{ k}\Omega)] = 0.04$. From equation 10–25, $BW_{CL} = f_u\beta = (10^6)(0.04) = 40$ kHz.
   Since the amplifier is dc, the closed-loop cutoff frequency is the same as the closed-loop bandwidth, 40 kHz.
2. In Figure 10–13(b), $\beta = R_1/(R_1 + R_f) = (10\text{ k}\Omega)/[(10\text{ k}\Omega) + (15\text{ k}\Omega)] = 0.4$. Then $BW_{CL} = 10^6(0.4) = 400$ kHz.

Let us clarify a few things regarding the frequency response of an operational amplifier. First, remember that the dc open-loop gain is not generally known unless we bother to measure it. Manufacturers simply give minimum guaranteed values, and actual values can vary substantially from device to device. The open-loop 3-dB frequency f_0 is not known either. However, the product $A_0 f_0$, which is the gain–bandwidth product (GBP) or unity-gain frequency f_u, is always provided by the manufacturer. As long as we know GBP, knowledge of A_0 and f_0 is totally irrelevant.

Going back to the closed-loop gain, it is important to understand that its magnitude is a frequency-dependent function that when plotted in a log-log graph will have two asymptotes and a corner or break point at the 3-dB frequency f_c (or BW_{CL}). Figure 10–14 shows a plot of the closed-loop gain for *noninverting* and *inverting* amplifiers, with the open-loop frequency response shown with a dotted line. Notice that the noninverting response coincides in the sloped region with the open-loop response. Appendix C demonstrates that the *magnitude* of the closed-loop gain as a function of frequency is given by

$$|A_{CL}| = \frac{A_{CLO}}{\sqrt{1 + \left(\dfrac{f}{f_c}\right)^2}} \qquad \textbf{(10–26)}$$

where $A_{CLO} = 1 + R_f/R_1$ (noninverting)
or $A_{CLO} = R_f/R_1$ (inverting)
and $f_c = \beta\ \text{GBP}$

It is worthwhile noting that in the case of the *noninverting* amplifier, the fact that the ideal closed-loop gain is $1/\beta$ makes equation 10–25 equivalent to

$$BW_{CL} = \frac{\text{GBP}}{\hat{A}_{CL}} \qquad \textbf{(10–27)}$$

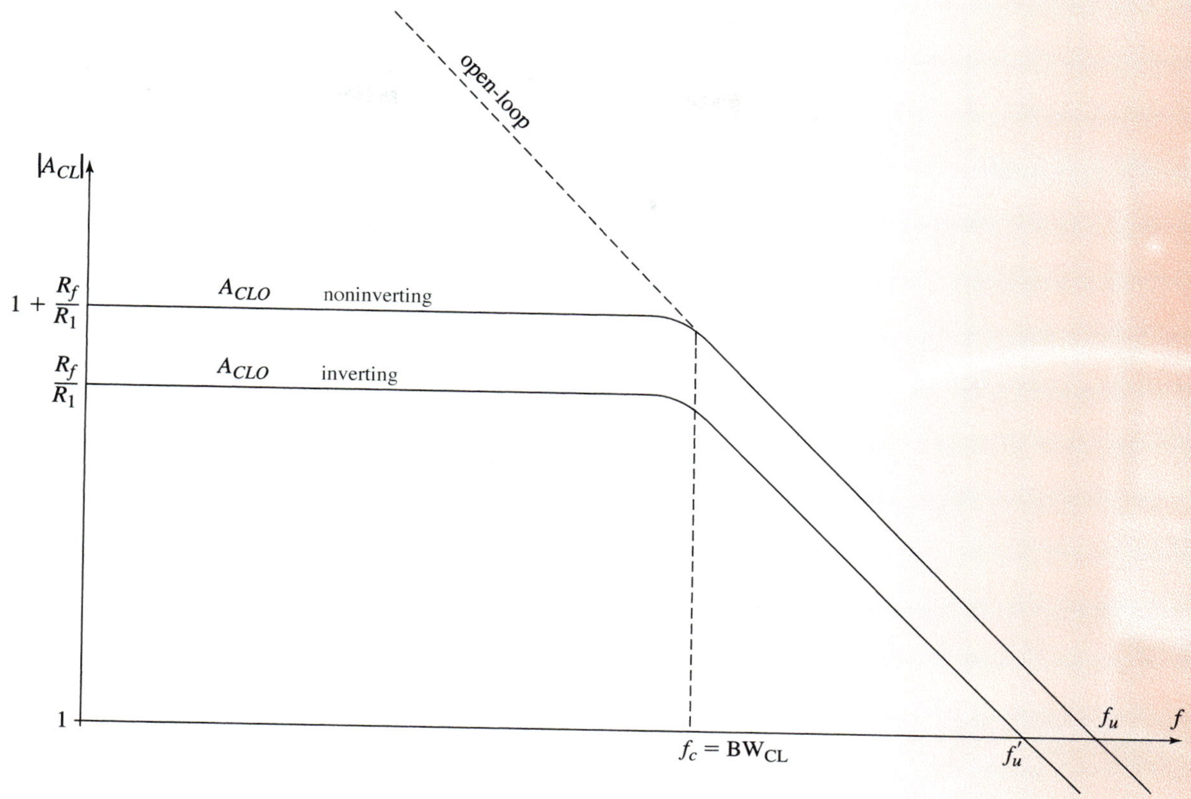

FIGURE 10–14 Closed-loop gain vs. frequency for noninverting and inverting amplifiers

or (ideal closed-loop gain) × (closed-loop bandwidth) = gain–bandwidth product. To illustrate the validity of this expression, refer to part 1 of Example 10–5. Here, the ideal closed-loop gain is $(R_1 + R_f)/R_1 = (250\ \text{k}\Omega)/(10\ \text{k}\Omega) = 25$, so 25 × (closed-loop bandwidth) = 10^6, which yields

$$\text{closed-loop bandwidth} = \text{BW}_{CL} = 10^6/25 = 40\ \text{kHz} \qquad \text{(correct)}$$

Equation 10–27 is *not* valid for the inverting amplifier. In part 2 of Example 10–5, we have

$$\text{ideal closed-loop gain} = R_f/R_1 = (15\ \text{k}\Omega)/(10\ \text{k}\Omega) = 1.5$$

If we now apply equation 10–27, we obtain

$$\text{BW}_{CL} = 10^6/1.5 = 666.6\ \text{kHz} \qquad \text{(incorrect)}$$

Although some authors interpret the gain–bandwidth product to be the product of closed-loop gain and closed-loop bandwidth regardless of configuration, we have seen that this interpretation yields a bandwidth for the inverting amplifier that is larger than its actual value. At large values of closed-loop gain, the bandwidths of the inverting and noninverting amplifiers are comparable, but at low gains the noninverting amplifier has a larger bandwidth. For example, when the closed-loop gain is 1, the bandwidth of the noninverting amplifier is equal to the GBP, whereas that of the inverting amplifier is half the GBP. See Appendix C for derivation details. In general, to avoid errors, we should always use $\text{BW}_{CL} = \beta\text{GBP}$ regardless of configuration. Appendix C also shows that the unity-gain frequency for the inverting amplifier (f_u') is given by $f_u' = (1 - \beta)\text{GBP}$. Note in Figure 10–14 that the sloped portion of the inverting response does not coincide with the open-loop response.

At frequencies below the corner frequency, f_c, the closed-loop gain is approximately equal to the dc and low-frequency gain A_{CLO}, whereas at frequencies above the corner, the gain drops at the rate of -20 dB/dec (slope $= -1$). In this sloped region, the noninverting closed-loop gain can be approximated by GBP/f, using the gain-frequency product property described for the open-loop frequency response. Obviously, the closed-loop gain at $f = f_c$ is $A_{CLO}/\sqrt{2} = 0.707A_{CLO}$, or 3 dB below A_{CLO}.

Since the actual plot of the gain deviates significantly from the asymptotes at frequencies in the vicinity of the break point, we should always use equation 10–26 when calculating the closed-loop gain at frequencies relatively close to the 3-dB frequency. In general, we can safely use the approximations when the frequency of interest is a decade or more *below or above* the corner or 3-dB frequency f_c. Otherwise, we should use the exact gain formula. In summary,

$$\begin{aligned} |A_{CL}| &\approx A_{CLO} && (f \ll f_c) \\ |A_{CL}| &= 0.707A_{CLO} && (f = f_c) \\ |A_{CL}| &\approx \text{GBP}/f && (f \gg f_c) \\ |A_{CL}| &\approx f_u'/f && (f \gg f_C, \text{inverting}) \end{aligned} \qquad \text{(10–28)}$$

Remember from Chapter 9 that another important piece of information regarding frequency response is the phase angle between input and output voltages. Appendix C shows that the phase of the closed-loop amplifier is

$$\phi = -\tan^{-1}(f/f_c) \qquad \text{(noninverting)} \qquad \text{(10–29)}$$

$$\phi = 180° - \tan^{-1}(f/f_c) \qquad \text{(inverting)} \qquad \text{(10–30)}$$

EXAMPLE 10–6

With reference to the operational amplifier whose open-loop frequency response is shown in Figure 10–15, find

1. the unity-gain frequency,
2. the open-loop 3-dB frequency,
3. the bandwidth when the feedback ratio is 0.02, and
4. the closed-loop gain at 0.4 MHz when the feedback ratio is 0.04.

Solution

1. In Figure 10–15, it is apparent that the open-loop gain equals 1 when the frequency is 1 MHz. Thus, $f_u = 1$ MHz.
2. $f_0 = \dfrac{f_u}{A_0} = 1\text{ M}/(2 \times 10^5) = 5$ Hz
3. From equation 10–25, $\text{BW}_{CL} = f_u\beta = 10^6(0.02) = 20$ kHz.
4. $\text{BW}_{CL} = f_u\beta = 10^6(0.04) = 40$ kHz. Thus, the closed-loop cutoff frequency is 40 kHz. Since the amplifier is noninverting, the closed-loop gain is $1/\beta = 25$. Since 0.4 MHz is 1 decade above the cutoff frequency, the gain can be found from GBP/$f = 10^6/400\text{ k} = 2.5$.

User-Compensated Amplifiers

As noted earlier, many commercially available amplifiers have internal compensation circuitry to make the frequency response roll off at 6 dB/octave (20 dB/decade) over the entire frequency range from f_o to f_u (Figure 10–12). Some amplifiers do not have such circuitry and must be compensated by

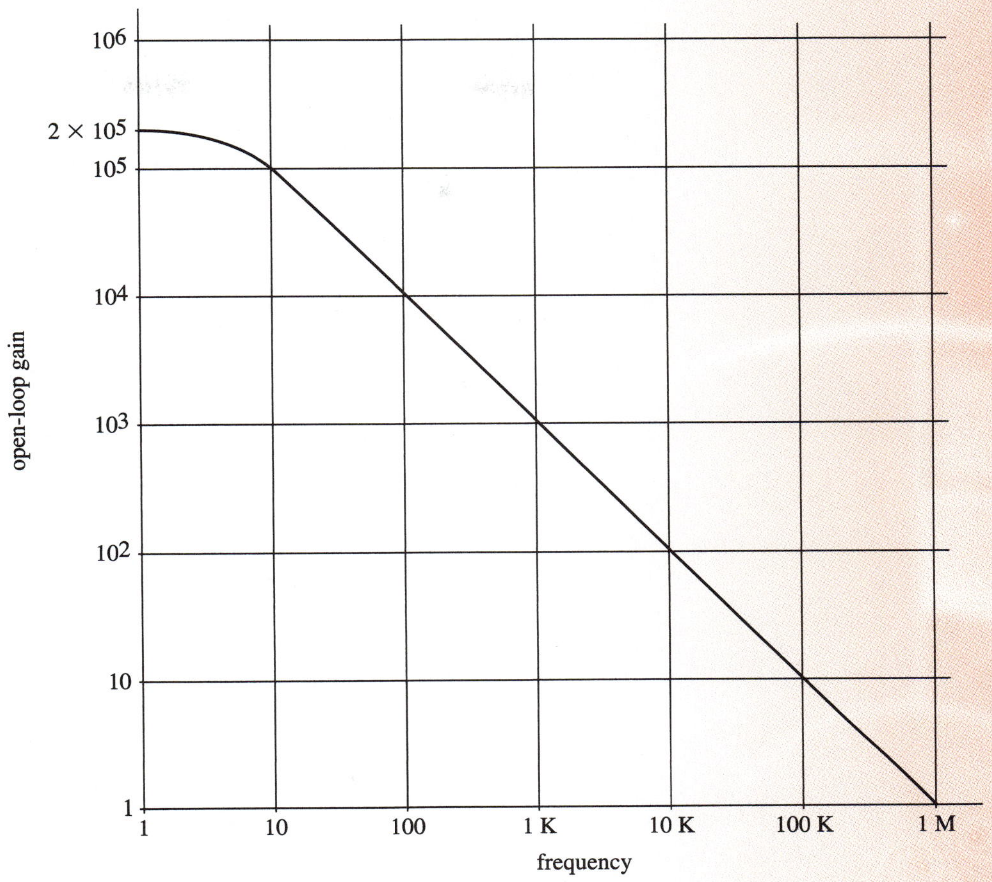

FIGURE 10–15 Open-loop frequency response for the op-amp in Example 10–6

connecting external roll-off networks. These networks, typically RC circuits, are selected by the user to ensure that the frequency response is 6 dB/octave at the closed-loop gain at which the amplifier is to be operated. Manufacturers' specifications usually include equations for determining the values of the components of external roll-off networks, based on the closed-loop gain desired.

Figure 10–16(a) shows a typical frequency response of an uncompensated amplifier. Compensation is particularly critical when the amplifier is to be operated with a small closed-loop gain, because the bandwidth is then very large and the amplifier's roll-off rate may be 12 or 18 dB/octave, rates that jeopardize stability. Figure 10–16(a) also shows an example of a response that has been compensated to roll off at 6 dB/octave when a particular value of closed-loop gain, A_{CL}, is desired. Note that the roll-off rate would be 12 dB/octave if the uncompensated amplifier were used at that value of closed-loop gain. It is clear that compensation reduces the bandwidth of the amplifier. However, it is generally true that user compensation results in a wider bandwidth than can be achieved with an internally compensated amplifier that rolls off at 6 dB/octave over its entire range. Figure 10–16(b) shows a typical RC network used for external compensation. Note that this network is usually connected to an internal stage of the amplifier at an external terminal that may be identified as "roll-off," "phase," or "frequency compensation."

The external compensation shown in Figure 10–16(b) is called *lag phase* compensation. Figure 10–16(c) shows an example of *lead* compensation, used to offset the effects of input and stray capacitance. The feedback capacitor

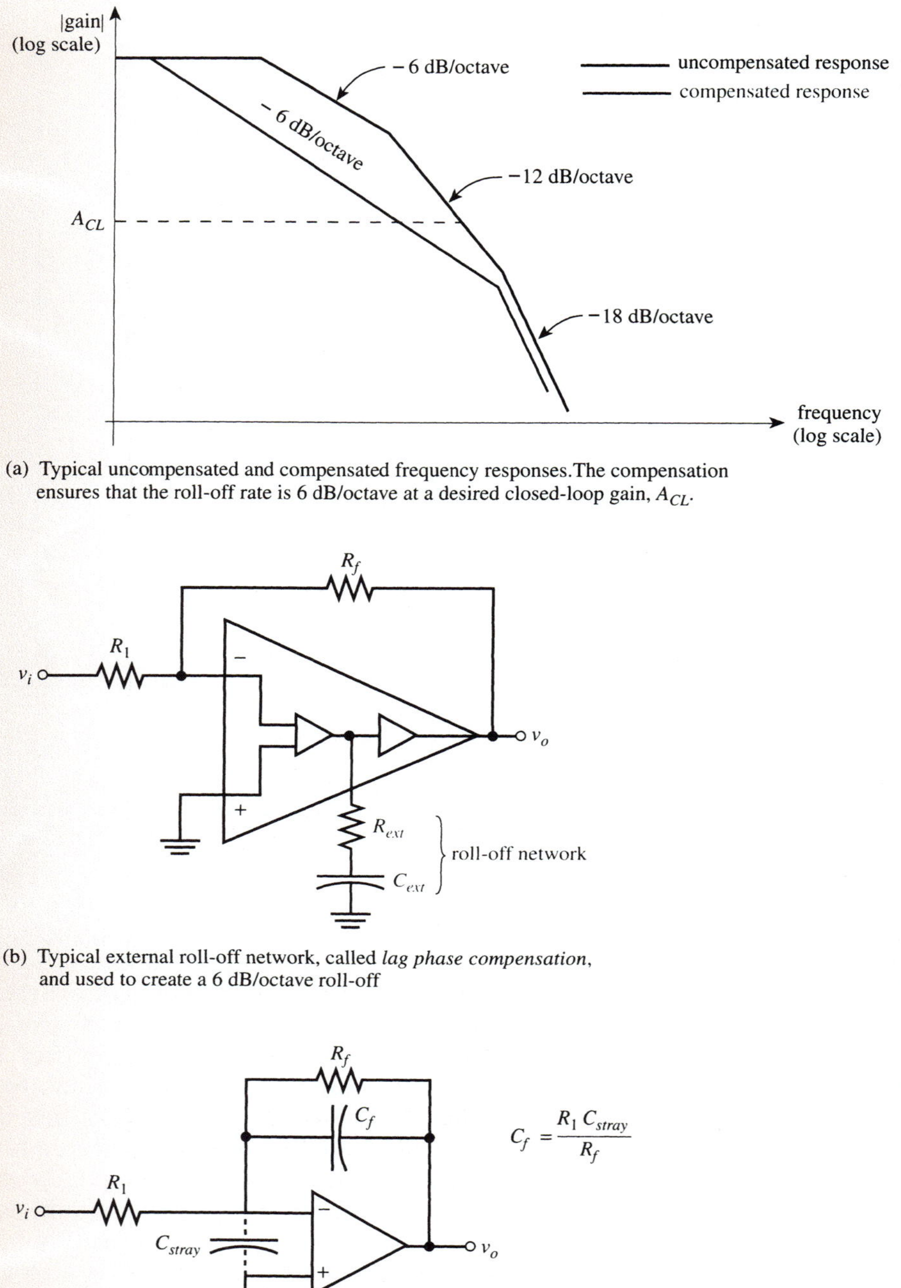

(a) Typical uncompensated and compensated frequency responses. The compensation ensures that the roll-off rate is 6 dB/octave at a desired closed-loop gain, A_{CL}.

(b) Typical external roll-off network, called *lag phase compensation*, and used to create a 6 dB/octave roll-off

(c) Use of feedback capacitance C_f to compensate for shunt capacitance at the input (lead compensation)

FIGURE 10–16 User-compensated amplifiers

C_f is selected so that the break frequency due to the combination of R_f and the input shunt capacitance equals the break frequency due to R_f and C_f:

$$\frac{1}{2\pi R_1 C_{stray}} = \frac{1}{2\pi R_f C_f}$$

$$C_f = \frac{R_1 C_{stray}}{R_f}$$

10–4 SLEW RATE AND RISE TIME

We have discussed the fact that internal compensation circuitry used to ensure amplifier stability also affects the frequency response and places a limit on the maximum operating frequency. The capacitor(s) in this compensation circuitry limit amplifier performance in still another way. When the amplifier is driven by a step or pulse-type signal, the capacitance must charge and discharge rapidly in order for the output to "keep up with," or track, the input. Because the voltage across a capacitor cannot be changed instantaneously, there is an inherent limit on the *rate* at which the output voltage can change. The maximum possible rate at which an amplifier's output voltage can change, in volts per second, is called its *slew rate*.

It is not possible for *any* waveform, input or output, to change from one level to another in *zero* time. An instantaneous change corresponds to an *infinite rate of change*, which is not realizable in any physical system. Therefore, in our investigation of performance limitations imposed by an amplifier's slew rate, we need concern ourselves only with inputs that undergo a total change in voltage, ΔV, over some nonzero time interval, Δt. For simplicity, we will assume that the change is linear with respect to time; that is, it is a *ramp*-type waveform, as illustrated in Figure 10–17. The rate of change of this kind of waveform is the change in voltage divided by the length of time that it takes for the change to occur:

$$\text{rate of change} = \frac{V_2 - V_1}{t_2 - t_1} = \frac{\Delta V}{\Delta t} \quad \text{volts/second}$$

The value specified for the slew rate of an amplifier is the maximum rate at which its output can change, so we cannot drive the amplifier with any kind of input waveform that would require the output to exceed that rate. For example, if the slew rate is 10^6 V/s (a typical value), we could not drive an amplifier having unity gain with a signal that changes from -5 V to $+5$ V in 0.1 μs because that would require the output to change at the rate $\Delta V/\Delta t = (10\text{ V})/(10^{-7}\text{ s}) = 10^8$ V/s. Similarly, we could not drive an amplifier having a gain of 10 with an input that changes from 0 V to 1 V in 1 μs because that would require the output to change from 0 V to 10 V in 1 μs, giving $\Delta V/\Delta t = 10/10^{-6} = 10^7$ V/s. When we say we "could not" drive the amplifier with these inputs, we simply mean that we could not do so and still expect the output to be a faithful replica of the input.

In specifications, the slew rate is often quoted in the units volts per microsecond. Of course, 1 V/μs is the same as 10^6 V/s: $(1\text{ V})/(10^{-6}\text{ s}) = 10^6$ V/s.

EXAMPLE 10–7

The operational amplifier in Figure 10–18 has a slew-rate specification of 0.5 V/μs. If the input is the ramp waveform shown, what is the maximum closed-loop gain that the amplifier can have without exceeding its slew rate?

FIGURE 10–17 The rate of change of a linear, or ramp, signal is the change in voltage divided by the change in time

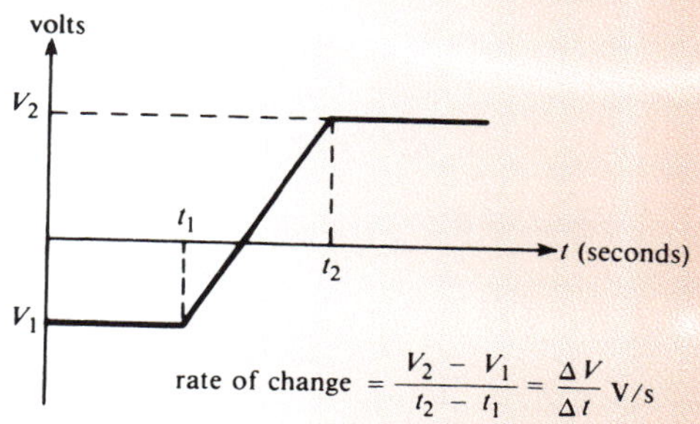

FIGURE 10–18 (Example 10–7)

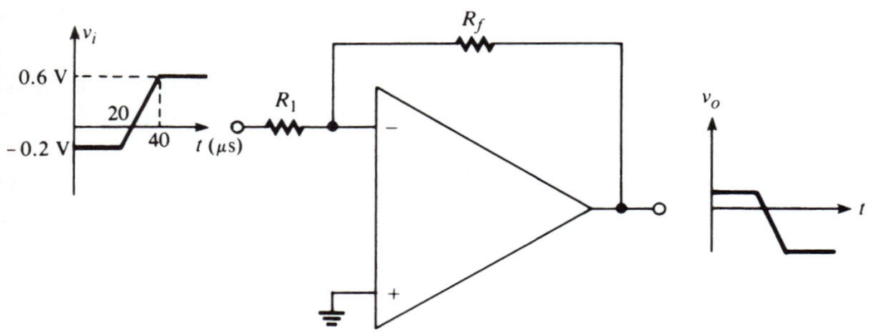

Solution

The rate of change of the input is

$$\frac{\Delta V}{\Delta t} = \frac{V_2 - V_1}{t_2 - t_1} = \frac{0.6\text{ V} - (-0.2\text{V})}{(40 - 20) \times 10^{-6}\text{ s}} = 4 \times 10^4 \text{ V/s}$$

Since the slew rate is 0.5 V/μs $= 5 \times 10^5$ V/s, the maximum permissible gain is

$$\frac{5 \times 10^5 \text{ V}}{4 \times 10^4 \text{ V}} = 12.5$$

Notice that the amplifier is connected in an inverting configuration, so the output changes from positive to negative. The inversion is of no consequence as far as slew rate is concerned. With a gain of -12.5, the output will change from $(-12.5)(-0.2\text{ V}) = +2.5$ V to $(-12.5)(0.6\text{ V}) = -7.5$ V in 20 μs, giving

$$\frac{\Delta V}{\Delta t} = \frac{10\text{ V}}{20\ \mu\text{s}} = 0.5 \text{ V}/\mu\text{s}$$

the specified slew rate.

Slew rate is a performance specification used primarily in applications where the waveforms are large-signal pulses or steps that cause the output to swing through a substantial part of its total range ($\pm V_{CC}$ volts). *However*, the slew rate imposes a limitation on output rate of change regardless of the nature of the signal waveform. In particular, if the signal is sinusoidal, or a complex waveform containing many different frequencies, we must be certain that no large-amplitude, high-frequency component will require the output to exceed the slew rate. High-frequency signals change (continuously) at rapid rates, and if their amplitudes are so large that the slew-rate specification is exceeded, distortion will result. It is especially important to realize that a frequency component may be within the bandwidth of the amplifier, as determined in Section 10–3, but may have such a large amplitude that it must be excluded because of slew-rate limitations. The converse is also true: A high-frequency signal that does not exceed the slew rate may have to be excluded because it is outside the amplifier bandwidth. In other words, the maximum frequency at which an amplifier can be operated depends on both the bandwidth and the slew rate, the latter being a function of amplitude as well as frequency. In a later discussion, we will summarize the criteria for determining the operating frequency range of an amplifier based on both slew rate and bandwidth limitations.

When the output of an amplifier is the sine-wave voltage $v_o(t) = V_P \sin \omega t$, it can be shown using calculus (differentiating with respect to t) that the signal has a maximum rate of change given by

$$\text{rate of change (max)} = V_P\,\omega \text{ volts/second}$$

where V_P is the peak amplitude of the sine wave, in volts, and ω is the angular frequency, in radians/second. The relationship clearly shows that the rate of change is proportional to both the amplitude and the frequency of the signal. If S is the specified slew rate of an amplifier, then we must have

$$V_P\,\omega \le S \quad \text{or} \quad 2\pi f V_P \le S \tag{10–31}$$

This inequality allows us to solve for the maximum frequency, f_s(max), that the slew-rate limitation permits at the output of an amplifier:

$$f_s(\text{max}) = \frac{S}{2\pi V_P} \qquad \text{or} \qquad \omega_s(\text{max}) = \frac{S}{V_P} \tag{10–32}$$

We emphasize again that f_s(max) is the frequency limit imposed by the slew rate *alone*, i.e., disregarding bandwidth limitations. Also, equations 10–31 and 10–32 apply to sinusoidal signals only. When dealing with complex waveforms containing many different frequency components, the slew rate should be at *least* as great as that necessary to satisfy equation 10–32 for the highest frequency component. Depending on phase relations, maximum rates of change may actually be additive.

In a design situation, the inequality given by equation 10–31 is used for determining the minimum slew rate required from an operational amplifier to satisfy a particular application. In that case, the condition is expressed as

$$S \ge 2\pi f_{max} V_P(\text{max}) \tag{10–33}$$

where f_{max} is the highest frequency contained in the signal and V_P(max) the maximum expected peak *output* voltage.

EXAMPLE 10–8

The operational amplifier in Figure 10–19 has a slew rate of 0.5 V/μs. The amplifier must be capable of amplifying the following input signals: $v_1 = 0.01 \sin(10^6 t)$, $v_2 = 0.05 \sin(350 \times 10^3 t)$, $v_3 = 0.1 \sin(200 \times 10^3 t)$, and $v_4 = 0.2 \sin(50 \times 10^3 t)$.

1. Determine whether the output will be distorted due to slew-rate limitations on any input.
2. If so, find a remedy (other than changing the input signals).

Solution

1. We must check each frequency to verify that $\omega \le \omega_S(\text{max}) = S/V_P$ rad/s. Note that V_P is the peak amplitude at the *output* of the amplifier, so each input amplitude must be multiplied by the closed-loop gain before the check is performed. Assuming that the closed-loop gain is ideal, we have $v_o/v_i = -R_f/R_1 = -(330\text{ k}\Omega)/(10\text{ k}\Omega) = -33$. Thus, the upper limit

FIGURE 10–19 (Example 10–8)

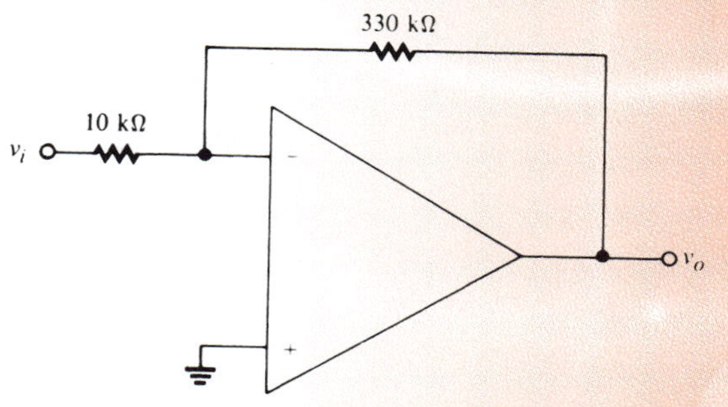

on the ω of each signal component will be $S/(33\,V_i)$, where V_i is the peak amplitude of the input.

$$v_1:\ \omega_S(\text{max}) = S/(33V_i) = 0.5 \times 10^6/(33)(0.01) = 1.515 \times 10^6$$
$$\omega = 10^6 < 1.515 \times 10^6 \qquad \text{(OK)}$$

$$v_2:\ \omega_S(\text{max}) = S/(33V_i) = 0.5 \times 10^6/(33)(0.05) = 303.03 \times 10^3$$
$$\omega = 350 \times 10^3 > 303.03 \times 10^3 \qquad \text{(not OK)}$$

$$v_3:\ \omega_S(\text{max}) = S/(33V_i) = 0.5 \times 10^6/(33)(0.1) = 151.5 \times 10^3$$
$$\omega = 200 \times 10^3 > 151.5 \times 10^3 \qquad \text{(not OK)}$$

$$v_4:\ \omega_S(\text{max}) = S/(33V_i) = 0.5 \times 10^6/(33)(0.2) = 75.75 \times 10^3$$
$$\omega = 50 \times 10^3 > 75.75 \times 10^3 \qquad \text{(OK)}$$

We see that v_2 and v_3 would both cause the slew rate specification of the amplifier to be exceeded. Consequently, the output will be distorted.

2. Since we cannot change the input signal amplitudes or frequencies, there are only two remedies: (a) find an amplifier with a greater slew rate, or (b) reduce the closed-loop gain of the present amplifier. We will investigate both remedies.

   a. The slew rate of a new amplifier must satisfy *both* $S/(33)(0.05) \ge 350 \times 10^3$ (for v_2) and $S/(33)(0.1) \ge 200 \times 10^3$ (for v_3). These inequalities are equivalent to

   $$S \ge 0.5775 \times 10^6\ \text{V/s} \quad \text{and} \quad S \ge 0.66 \times 10^6\ \text{V/s}$$

   Therefore, we must use an amplifier with a slew rate of at least 0.66×10^6 V/s, or 0.66 V/μs.

   b. If we use the present amplifier, we must reduce its closed-loop gain G so that it satisfies *both* $0.5 \times 10^6/0.05G \ge 350 \times 10^3$ (for v_2) and $0.5 \times 10^6/0.1G \ge 200 \times 10^3$ (for v_3). These inequalities are equivalent to

   $$G \le 28.57 \quad \text{and} \quad G \le 25$$

   Therefore, the maximum closed-loop gain is 25. This limit can be achieved by changing the 330-kΩ resistor in Figure 10–19 to 250 kΩ.

To ensure that an operational-amplifier circuit will not distort a signal component having frequency f, we require that *both* of the following conditions be satisfied:

$$f \le \text{BW}_{\text{CL}} \tag{10–34}$$
$$f \le S/2\pi V_P \tag{10–35}$$

If the signal is a complex waveform containing multiple frequency components, the highest frequency in the signal should satisfy both conditions. Note that both conditions depend on closed-loop gain: Large gains reduce BW_{CL} and increase the value of V_P, so the greater the closed-loop gain, the more severe the restrictions.

EXAMPLE 10–9

The operational amplifier in Figure 10–20 has a unity-gain frequency of 1 MHz and a slew rate of 1 V/μs. Find the maximum frequency of a 0.1-V-peak sine-wave input that can be amplified without slew-rate distortion.

FIGURE 10–20 (Example 10–9)

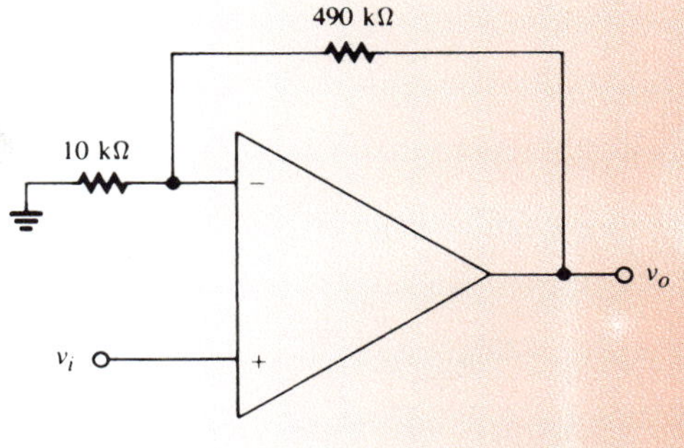

Solution

The feedback ratio is

$$\beta = \frac{R_1}{R_1 + R_f} = \frac{10\ \text{k}\Omega}{(10\ \text{k}\Omega) + (490\ \text{k}\Omega)} = 0.02$$

By equation 10–25, $BW_{CL} = f_u\beta = (1\ \text{MHz})(0.02) = 20$ kHz. The closed-loop gain for the noninverting configuration is $v_o/v_i = 1/\beta = 1/0.02 = 50$. Therefore, the peak value of the output is $V_P = 50(0.1) = 5$ V. Then $f_s(\text{max}) = S/2\pi V_P = 10^6/(2\pi)(5) = 31.83$ kHz. Since we require that f satisfy both $f \le 20$ kHz and $f \le 31.83$ kHz, we see that the maximum permissible frequency is 20 kHz. In this case, the bandwidth sets the upper limit.

EXAMPLE 10–10

DESIGN

1. Derive a design equation that imposes a limit on the closed-loop gain (A_{CL}) of an *inverting* amplifier based on bandwidth and slew-rate limitations of the operational amplifier. The known values that can be used in the equation are slew rate S, gain–bandwidth product GBP, maximum sinusoidal operating frequency f_{max}, and maximum expected peak input voltage V_i(pk).
2. Use the design equation to find the limit on A_{CL} when the op-amp has $S = 1$ V/μs, GBP $= 1$ MHz, and the input signal has maximum frequency and peak voltage of 15 kHz and 500 mV, respectively.

Solution

1. Since the amplifier is inverting, the closed-loop bandwidth is

$$BW_{CL} = \text{GBP}/(1 + |A_{CL}|)$$

which implies $f_{max} \le \text{GBP}/(1 + |A_{CL}|)$. Solving for the inverting closed-loop gain, we obtain

$$|A_{CL}| \le (\text{GBP}/f_{max}) - 1$$

According to equation (10–35), $f \le S/(2\pi V_P)$. But $V_P = |A_{CL}|V_i(\text{pk})$. Therefore, the design equation based on the slew rate becomes

$$|A_{CL}| \le \frac{S}{2\pi f_{max} V_i(\text{pk})}$$

Because $|A_{CL}|$ must be less than both the limits we have found, it must be less than the smaller of the two:

$$|A_{CL}| = \min\left(\frac{\text{GBP}}{f_{max}} - 1, \frac{S}{2\pi f_{max} V_i(\text{pk})}\right)$$

2. Substituting values into the design formula, we obtain

$$|A_{CL}| = \min(65.7, 21.2) \quad \text{or} \quad |A_{CL}| \leq 21.2$$

Another practical limit that is not considered in the foregoing is the maximum permissible output voltage of the op-amp, which depends directly on the supply voltage, $\pm V_{CC}$. If we used the gain of 65.7 in the present example, then the peak output voltage, would be 65.7(0.5 V pk) = 32.9 V pk, which is too large for many, if not most, commercial operational amplifiers.

We have seen that an amplifier's slew rate affects its ability to track, or follow, a rapidly changing input pulse. When the output voltage must change through ΔV volts, the minimum possible time in which that change can occur is

$$\Delta t = \frac{\Delta V}{S}$$

where ΔV is the total *output* voltage change. In terms of input quantities, the minimum time allowed for an input voltage change of ΔV_i volts is

$$\Delta t = \frac{(A_{CL})\Delta V_i}{S} \tag{10–36}$$

where A_{CL} is the closed-loop gain.

An amplifier's bandwidth also affects the time required for its output to change in response to a pulse input. Recall from Chapter 9 that the *rise time* t_r of a single-pole system is

$$t_r = \frac{0.35}{\text{BW}}$$

We defined rise time to be the time required for a voltage to change from 10% of its final value to 90% of its final value in response to a step input, such as a square waveform.

It is important at this point to emphasize that the rise-time concept is associated with the transient response of a single-pole system when driven by a step function. A step function is ideally a sudden change of voltage in zero time. Of course, this is not physically possible, but if the actual time is much smaller than the calculated rise time, the relationship $t_r = 0.35/\text{BW}$ will be valid. Another issue regarding rise time is that the response of the single-pole system is exponential in nature. We have to understand that a rise-time measurement or calculation based on the preceding formula is valid only if the output voltage is exponential. However, because of slew-rate limitation, the output voltage could be significantly affected and the rise-time calculation no longer valid. In fact, if the slew-rate limitation is severe, the voltage rise will have no trace of being exponential, but be just a straight line.

It is erroneous to assume that a step-like input voltage will always produce an output whose slope will be given by the slew rate. Slew-rate limitation occurs only when the maximum slope of the "intended" exponential output signal exceeds the slew-rate value. This slope depends directly on the closed-loop gain and the gain–bandwidth product. The mathematical expression of the amplifier's output voltage in response to a step input has a form identical to that for the charging voltage across a capacitor in a series RC circuit. That form is

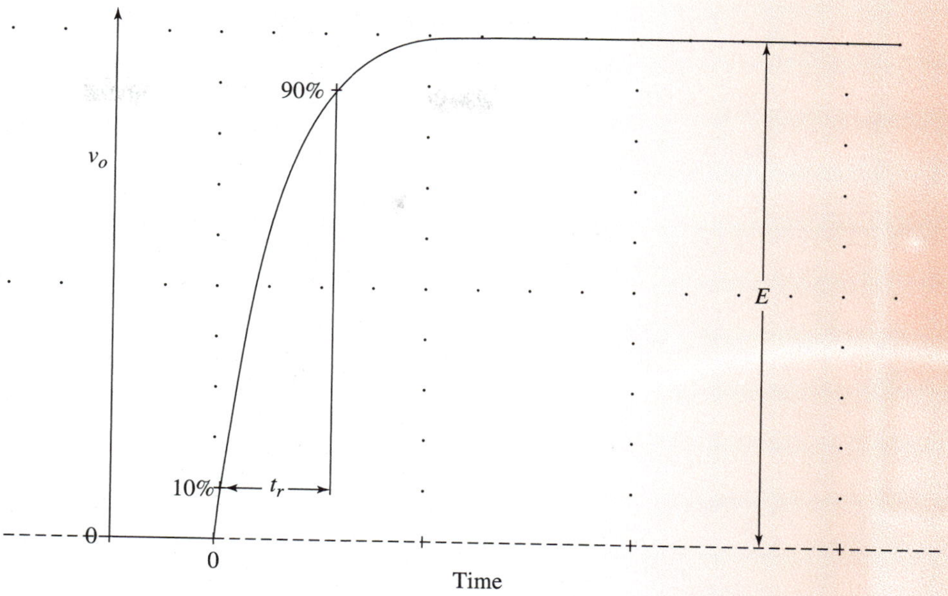

FIGURE 10–21 Exponential rise of the output voltage in response to a step when no slew-rate limitation takes place

$$v_C(t) = V(1 - e^{-t/RC})$$

where V is the applied dc voltage across the RC circuit and v_C is the capacitor's voltage.

Using this analogy, and recalling that $1/RC$ is also the 3-dB frequency in rad/s of the RC circuit, we can write the expression for the amplifier's output voltage as

$$v_o(t) = E(1 - e^{-2\pi f_c t}) \quad (10\text{–}37)$$

where E is the size of the output step (i.e., input step × closed-loop gain) and f_c is the closed-loop 3-dB frequency (BW_{CL}). The shape of this waveform is shown in Figure 10–21. It is apparent by simple observation that maximum slope occurs at $t = 0$.

If the slew rate is large enough, the output will be a clean exponential rise given by equation 10–37. In this case, we say that the output has been limited by the bandwidth only. This is a desirable situation as long as the rise time is kept small. The designer should always keep the amplifier from running into slew-rate limitation for any type of input signal. This can be accomplished by calculating the bandwidth necessary to achieve a particular rise time and then determining the minimum slew rate necessary in order to keep the amplifier from "slewing" at the expected output voltage. Therefore, the solution is to find an operational amplifier that *meets or exceeds* the minimum requirements for both gain–bandwidth product and slew rate. Let us look into that process.

To determine the bandwidth required for *step-like signals*, we simply use the rise time formula, solving for BW_{CL} in terms of the desired rise time, that is,

$$BW_{CL} = 0.35/t_r \quad (10\text{–}38)$$

Next, based on the closed-loop gain used, we calculate the required gain–bandwidth product from

$$GBP = A_{CL}BW_{CL} \quad (10\text{–}39)$$

To determine the required slew rate, we use calculus to differentiate equation 10–37 in order to obtain the maximum slope of the output voltage. The result is

$$\left.\frac{dv_o}{dt}\right|_{\max} = 2\pi f_c E \qquad \text{(volts/s)}$$

If the slew rate is larger than the maximum slope, no slew-rate limitation will occur. Therefore, the requirement for the slew rate can be determined from

$$S \geq 2\pi f_c E \tag{10–40}$$

We also need to be aware that if the actual slew rate is slightly less than the minimum given by this inequality, the limitation will be practically unnoticeable, at least as far as we could tell from observation of an amplified square waveform with an oscilloscope. This issue is an interesting topic to experiment with in the laboratory. Also, if we increase the input voltage, slew-rate limitation will be produced at some point. If we keep increasing the voltage, we will eventually not see any curving on the rising and falling edges of the output waveform. At this point, the shape of the output waveform will look trapezoidal. It follows, then, that we can have different levels of slew-rate limitation, from mild to severe. However, the designer should always aim for a condition free of slew-rate limitation.

An important issue to stress is that bandwidth limitation is always present if slew-rate limitation is not present. The larger the bandwidth, however, the smaller the resulting rise time and the less noticeable the bandwidth limitation. The rise time must be specified as much smaller than the time it takes for the input to change through ΔV_i. This will impose a certain minimum value for the GBP. In practical reality, however, the pulse or square waveform to be processed by the amplifier is very likely to have very fast transitions, maybe on the order of tens of nanoseconds. So pretending that an op-amp could yield a rise time on the order of 1 ns to minimize bandwidth limitation is in all certainty a practical impossibility. The good news is that the presence of bandwidth limitation does not matter if the resulting rise time is small compared to the period of a square waveform or to the duration of a pulse. For instance, a 5-V, 10-μs pulse with a 200-ns rise time will look "very square" on the oscilloscope. But this same pulse showing a climbing rate of 5 V over a 5-μs lapse due to slew-rate limitation is not going to look very square at all.

In conclusion, bandwidth and slew rate affect the shape of step-like waveforms such as square waves or pulse trains when passed through an amplifier. Their effect can be minimized by properly selecting an op-amp whose GBP and slew rate are appropriate for the specified performance. An example that further illustrates these concepts follows.

EXAMPLE 10–11

DESIGN

A train of 1-V pulses needs to be amplified to a 5-V level with a noninverting amplifier and is to be free from slew-rate limitation. Assume the pulse transitions are very fast. The rise time at the output should not be more than 0.8 μs (800 ns). Determine the op-amp's minimum GBP and slew rate required to satisfy this requirement.

Solution

Find the required closed-loop bandwidth of the amplifier:

$$BW_{CL} = 0.35/t_r = 0.35/0.8\ \mu s = 438\ \text{kHz}$$

Determine the GBP. Since the pulse train needs to be amplified from 1 to 5 V, the closed-loop gain must be 5. And since GBP = $BW_{CL} A_{CL}$,

$$GBP = (438 \text{ kHz})(5) = 2.2 \text{ MHz}$$

From equation 10–40,

$$S \geq 2\pi(438 \text{ k})5 = 13.8 \times 10^6 \text{ V/s} = 13.8 \text{ V/}\mu\text{s}$$

The op-amp selected must meet or exceed these values.

Let us now look at the issues pertaining to sine or complex waveforms such as audio signals. The bandwidth of the amplifier must be large enough to accommodate the maximum frequency content of the signal, and the slew rate must be greater than the minimum required to avoid slew-rate limitation. Recall that the closed-loop bandwidth, or BW_{CL}, is also the 3-dB (or corner) frequency f_c. However, it is important to remember that the gain at this point is about 30% less than at medium and low frequencies. Depending on the application, this could be a significant loss. A design requirement could be, for example, that the gain not drop more than 1 dB at the expected maximum operating frequency. Obviously, in that case, the 3-dB bandwidth must extend farther than the maximum operating frequency in order to yield such "gain flatness."

By using equation 10–26, we can solve for f_c given the minimum required gain at the maximum operating frequency implied by a particular restriction. Although this restriction is very likely to be expressed in terms of decibels, it can also be expressed by other means, such as percentage. An example describing the procedure follows.

EXAMPLE 10–12

DESIGN

Design a noninverting ultrasound amplifier with a gain of 50 for signal frequencies from dc to 40 kHz. The gain should not drop more than 1 dB at the highest frequency. Maximum input amplitude is 200 mV (pk), and the output is to be free of slew-rate limitation. Find the minimum GBP and slew rate required from an op-amp to satisfy these requirements.

Solution

In order to solve for f_c from equation 10–26, we must first determine the value of the closed-loop gain 1 dB below 50. We can either convert the magnitude of 50 to dB, subtract 1 dB, and then convert back to magnitude, or solve for the gain from $20 \log(A_{CL}/50) = -1$. The result is $A_{CL} = 44.56$. This is the minimum gain required at the maximum operating frequency.

Substituting into equation 10–26, we write

$$44.56 = \frac{50}{\sqrt{1 + \left(\dfrac{40 \text{ k}}{f_c}\right)^2}}$$

Using the calculator to isolate f_c by reversing the operations indicated by the expression, we obtain $f_c = 78.6$ kHz. The GBP is then GBP = 50(78.6 k) = 3.93 MHz.

Equation 10–33 is now used for determining the required slew rate to maintain operation free of slew-rate distortion.

$$S \geq 2\pi f_{max} V_P = 2\pi(40 \text{ k})(50 \times 0.2 \text{ V}) = 2.52 \times 10^6 \text{ V/s} \quad \text{or} \quad 2.52 \text{ V/}\mu\text{s}$$

The op-amp selected must meet or exceed the values we just obtained for GBP and S.

10–5 OFFSET CURRENTS AND VOLTAGES

Recall from Chapter 8 that one of the characteristics of an ideal operational amplifier is that it has zero output voltage when both inputs are 0 V (grounded). This characteristic is particularly important in applications where dc or low-frequency signals are involved. If the output is not zero when the inputs are zero, then the output will not be at its correct dc level when the input is a dc level other than zero.

The actual value of the output voltage when the inputs are zero is called the *output offset voltage.* Output offset is very much like a dc bias level in the output of a conventional amplifier in that it is added to whatever signal variation occurs there. If an operational amplifier is used only for ac signals, it can be capacitor-coupled if necessary or desirable to block the dc component represented by the offset. However, the capacitors may have to be impractically large if low frequencies and small impedance levels are involved. Also, a dc path must always be present between each input and ground to allow bias currents to flow. Small offsets, on the order of a few millivolts, can often be ignored if the signal variations are large by comparison. On the other hand, a frequent application of operational amplifiers is in precise, high-accuracy signal processing at low levels and low frequencies, and in these situations, very small offsets are crucial.

Manufacturers do not generally specify output offset because, as we shall see, the offset level depends on the closed-loop gain that a user designs through choice of external component values. Instead, *input* offsets are specified, and the designer can use these values to compute the output offset that results in a particular application. Output offset voltages are the result of two distinct input phenomena: input bias currents and input offset voltage. We will use the superposition principle to determine the contribution of each of these input effects to the output offset voltage.

Input Offset Current

In our discussion of differential amplifier circuits in Chapter 10, we ignored base currents because they had negligible effects on the kinds of computations that held our interest then. We know that some dc base current must flow when a transistor is properly biased, and, although small, this current flowing through the external resistors in an amplifier circuit produces a dc input voltage that in turn creates an output offset. To reduce the effect of bias currents, a *compensating resistor,* R_c, is connected in series with the noninverting (+) terminal of the amplifier. (R_c must provide a dc path to ground, so if a signal is capacitor-coupled to the + input, R_c must be connected between the + input and ground.) We will presently show that proper choice of the value of R_c will minimize the output offset voltage due to bias current. Figure 10–22 shows the bias currents I_B^+ and I_B^- flowing into

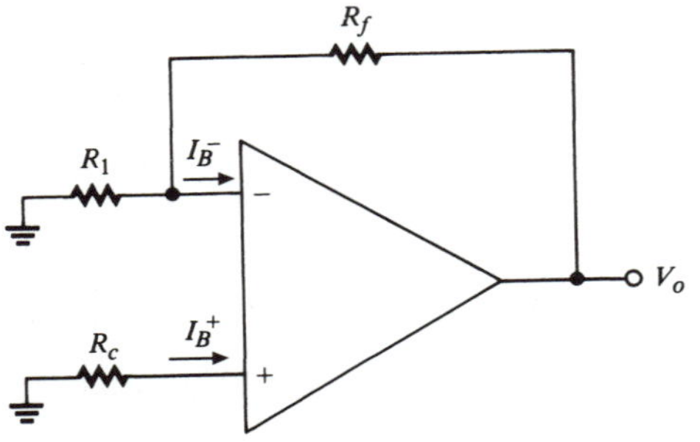

FIGURE 10–22 Input bias currents I_B^+ and I_B^- that flow when both signal inputs are grounded. R_c is a compensating resistor used to reduce the effect of bias current on output offset.

the + and − terminals of an operational amplifier when the signal inputs are grounded. Although the bias currents may actually flow into or out of the terminals, depending on the type of input circuitry, we will, for the sake of convenience, assume that the directions are as shown and that the values are always positive. These assumptions will not affect our ultimate conclusions. The figure also shows the compensating resistor R_c connected in series with the + terminal. Note that this circuit applies to both the inverting and noninverting configurations.

Figure 10–23(a) shows the equivalent circuit of Figure 10–22. Here, the bias currents are represented by current sources having resistances R_1 and R_c. Figure 10–23(b) shows the same circuit when the current sources are replaced by their Thévenin-equivalent voltage sources.

Using Figure 10–23(b), we can apply the superposition principle to determine the output offset voltage due to each input source acting alone. As illustrated in Figure 10–24(a), the amplifier acts as an inverter when the source connected to the + terminal is shorted to ground, so the output due to $I_B^- R_1$ is

$$V_{o1} = -I_B^- R_1\left(\frac{-R_f}{R_1}\right) = I_B^- R_f \qquad \text{(10–41)}$$

When the source connected to the − terminal is shorted to ground, the amplifier is in a noninverting configuration, so the output due to $I_B^+ R_c$ is

$$V_{o2} = -I_B^+ R_c\left(1 + \frac{R_f}{R_1}\right) \qquad \text{(10–42)}$$

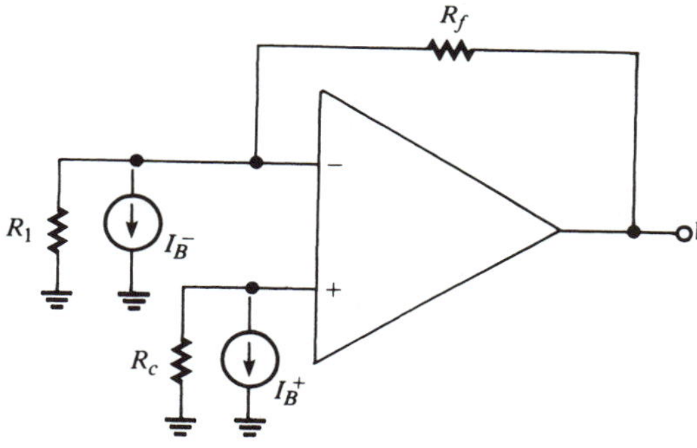

(a) The equivalent circuit of Figure 10–22.

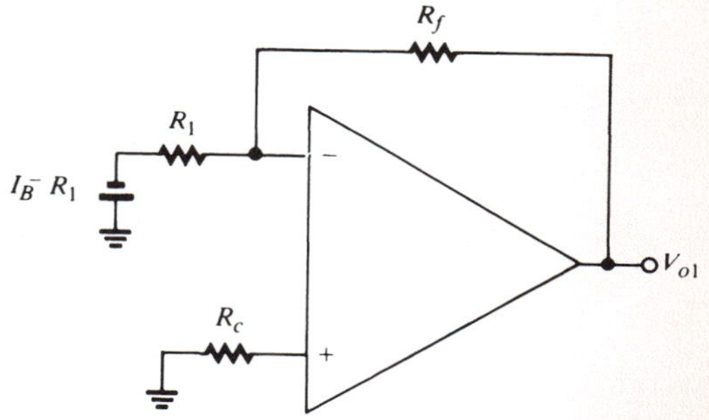

(a) When the noninverting input is grounded, the amplifier inverts and has gain $-R_f/R_1$.

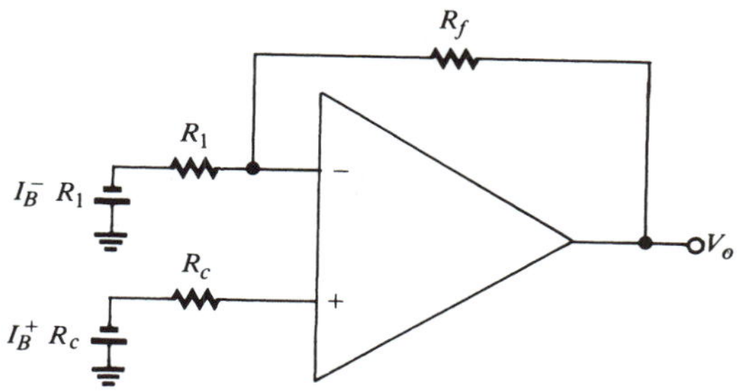

(b) The circuit equivalent to (a) when the current sources are replaced by their Thévenin equivalents

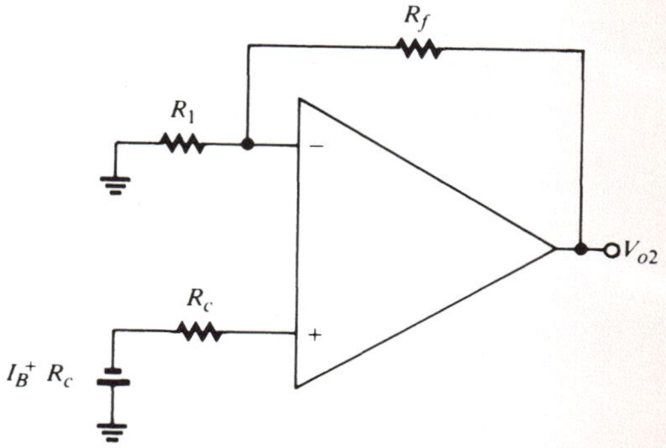

(b) When the inverting input is grounded, the noninverting amplifier has gain $1 + \frac{R_f}{R_1}$

FIGURE 10–23 Circuits equivalent to Figure 10–22

FIGURE 10–24 Applying superposition to determine the output offset voltage due to each source in Figure 10–23(b)

Combining (10–41) and (10–42), we obtain the total output offset voltage due to bias current, which we designate by $V_{OS}(I_B)$, as

$$V_{OS}(I_B) = I_B^- R_f - \left(1 + \frac{R_f}{R_1}\right) I_B^+ R_C \qquad \textbf{(10–43)}$$

Depending on which of the terms on the right side of equation 10–43 is greater, $V_{OS}(I_B)$ may be positive or negative. However, the sign of $V_{OS}(I_B)$ is of little interest, because negative offset voltage is just as undesirable as positive offset voltage. Our real interest is in finding a way to minimize the *magnitude* of $V_{OS}(I_B)$. Toward that end, let us make the reasonable assumption that the two inputs are closely matched and that, as a consequence, they have equal bias currents: $I_B^+ = I_B^- = I_{BB}$. Substituting I_{BB} for I_B^- and I_B^+ in equation 10–43 gives

$$V_{OS}(I_B) = I_{BB}\left[R_f - \left(1 + \frac{R_f}{R_1}\right)R_c\right] \qquad \textbf{(10–44)}$$

If the expression enclosed by the brackets in (10–44) were equal to zero, we would have zero offset voltage. To find a value of R_c that accomplishes that goal, we set the bracketed expression equal to 0 and solve for R_c:

$$R_f - \left(1 + \frac{R_f}{R_1}\right)R_c = 0$$

$$R_c = \frac{R_f R_1}{R_f + R_1} = R_f \| R_1 \qquad \textbf{(10–45)}$$

Equation 10–45 reveals the very important result that *output offset due to input bias currents can be minimized by connecting a resistor R_c having value $R_1 \| R_f$ in series with the noninverting input.* This method of offset compensation is valid for both inverting and noninverting configurations (see Section 8–1). Notice that we say the offset can be *minimized* using this remedy, rather than being made exactly zero, because the remedy is based on the assumption that $I_B^+ = I_B^-$, which may not be entirely valid. We can compute the exact value of $V_{OS}(I_B)$ when $R_c = R_1 \| R_f$ by substituting this value of R_c back into (10–43), where the assumption is not in force:

$$\begin{aligned} V_{OS}(I_B) &= I_B^- R_f - I_B^+\left(\frac{R_1 R_f}{R_1 + R_f}\right)\left(\frac{R_1 + R_f}{R_1}\right) \\ &= (I_B^- - I_B^+)R_f \end{aligned} \qquad \textbf{(10–46)}$$

Equation 10–46 shows that the offset voltage is proportional to the *difference* between I_B^+ and I_B^- when $R_c = R_1 \| R_f$. Because the inputs are usually reasonably well matched, the difference between I_B^+ and I_B^- is quite small. The equation confirms the fact that V_{OS} is 0 if I_B^+ exactly equals I_B^-. $V_{OS}(I_B)$ may be either positive or negative, depending on whether $I_B^+ > I_B^-$ or vice versa. Unless actual measurements are made, we rarely know which current is larger, so a more useful form of (10–46) is

$$|V_{OS}(I_B)| = I_{io} R_f \qquad \text{when } R_c = R_1 \| R_f \qquad \textbf{(10–47)}$$

where

$$I_{io} = |I_B^+ - I_B^-|$$

is the *input offset current* given in manufacturer's specifications.

Equation 10–47 shows that the output offset is directly proportional to the value of the feedback resistor R_f. For that reason, small resistance values should be used when offset is a critical consideration. However, to achieve large voltage gains when R_f is small may require impractically small values of R_1, to the extent that the amplifier may load the signal source driving it. In any event, large closed-loop gains are detrimental to another aspect of output offset, as we will see in a forthcoming discussion.

Another common manufacturers' specification is called simply *input bias current*, I_B. It refers to either or both I_B^+ and I_B^-. By convention, I_B is taken as the *average* of I_B^+ and I_B^-:

$$I_B = \frac{I_B^+ + I_B^-}{2}$$

I_B is typically much larger than I_{io} because I_B is on the same order of magnitude as I_B^+ and I_B^-, while I_{io} is the difference between the two.

It is important to know that I_{io} is normally given by manufacturers as a maximum value, meaning that it can actually be between zero and the published value. Therefore, the term $|V_{OS}(I_B)| = I_{io}R_f$ should be taken as a maximum (worst-case) value.

EXAMPLE 10–13

The specifications for the operational amplifier in Figure 10–25 state that the typical input bias current is 80 nA and that the input offset current is 10 nA, maximum.

1. Find the optimum value for R_c.
2. Find the maximum output offset voltage due to bias offset current when R_c equals its optimum value.
3. Assuming that the actual I_{io} is 10 nA and $I_B^+ > I_B^-$, find the magnitude of the output offset voltage when $R_c = 0$.

Solution

1. From equation 10–45, $R_c = R_1 \| R_f = (10\ \text{k}\Omega) \| (100\ \text{k}\Omega) = 9.09\ \text{k}\Omega$.
2. From equation 10–47, $|V_{OS}(I_B)| = I_{io}R_f = (10 \times 10^{-9})(100 \times 10^3) = 1\ \text{mV}$.
3. When $R_c = 0$, equation 10–43 becomes $V_{OS}(I_B) = I_B^- R_f$. From the average and difference of I_B^+ and I_B^-, we determine $I_B^+ = 85$ nA and $I_B^- = 75$ nA. Therefore, the magnitude of the offset voltage when $R_c = 0$ is $|V_{OS}(I_B)| = (75 \times 10^{-9})(100 \times 10^3) = 7.5\ \text{mV}$. We see that omission of the compensating resistance significantly increases the offset voltage.

Input Offset Voltage

Another input phenomenon that contributes to output offset voltage is an internally generated potential difference that exists because of imperfect matching of the input transistors. This potential may be due, for example, to

FIGURE 10–25 (Example 10–13)

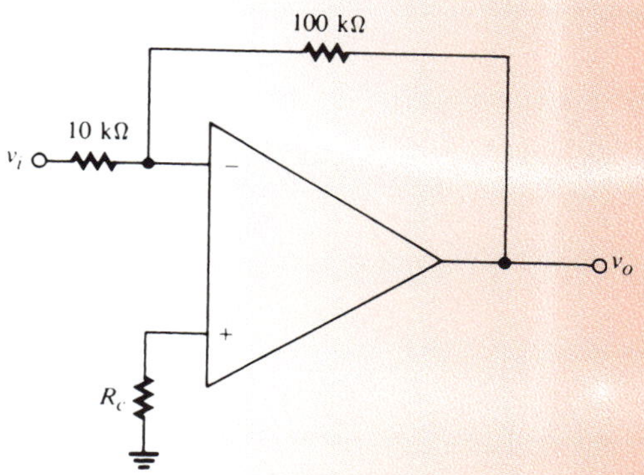

FIGURE 10–26 The effect of input offset voltage, V_{io}, is the same as if a dc source were connected in series with one of the inputs

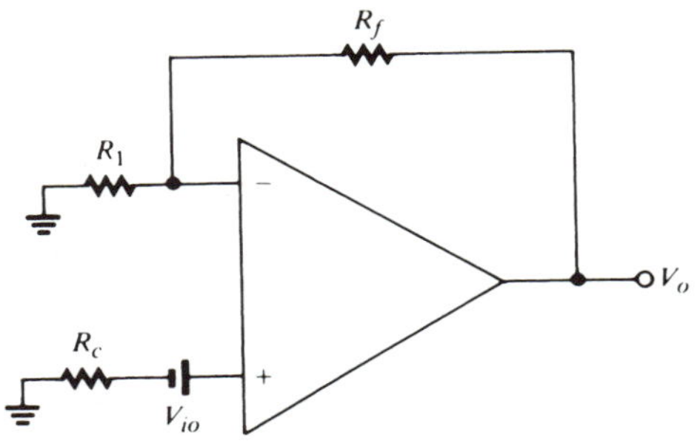

a difference between the V_{BE} drops of the transistors in the input differential stage of a BJT amplifier. Called *input offset voltage*, the net effect of this potential difference is the same as if a small dc voltage source were connected to one of the inputs. Figure 10–26 shows the equivalent circuit of an amplifier having its signal inputs grounded and its input offset voltage, V_{io}, represented as a dc source in series with the noninverting input. The effect is the same whether it is connected to the inverting or noninverting input. The polarity of the source is arbitrary, because input and output offsets may be either positive or negative. Once again, it is the magnitude of the offset that concerns us.

From Figure 10–26, it is apparent that the output voltage when the input is V_{io} is given by

$$V_{OS}(V_{io}) = V_{io}\left(1 + \frac{R_f}{R_1}\right) \qquad \textbf{(10–48)}$$

where $V_{OS}(V_{io})$ is the output offset voltage due to V_{io}. Note that the compensating resistor R_c is shown in Figure 10–26 for completeness' sake, but it has no effect on the output offset due to V_{io}. Equation 10–48 shows that input offset is magnified at the output by a factor equal to the closed-loop gain of the noninverting amplifier, as we would expect. If the amplifier is operated open-loop, the very large open-loop gain acting on the input offset voltage may well drive the amplifier to one of its output voltage limits. It is therefore important to have an extremely small V_{io} in any application or measurement that requires an open-loop amplifier.

Equation 10–48 is also valid for an amplifier in an inverting configuration. In fact, for a wide variety of amplifier configurations, it is true that

$$V_{OS}(V_{io}) = V_{io}/\beta \qquad \textbf{(10–49)}$$

where β is the feedback ratio.

EXAMPLE 10–14

The specifications for the amplifier in Example 10–13 state that the input offset voltage is 0.8 mV. Find the output offset due to this input offset.

Solution

From equation 10–48,

$$V_{OS}(V_{io}) = V_{io}\left(1 + \frac{R_f}{R_1}\right) = (0.8 \times 10^{-3}\ \text{V})\left(1 + \frac{100\ \text{k}}{10\ \text{k}}\right) = 8.8\ \text{mV}$$

The Total Output Offset Voltage

We have seen that output offset voltage is a function of two distinct input characteristics: input bias currents and input offset voltage. It may be that the polarities of the offsets caused by these two characteristics are such that they tend to cancel each other out. Of course, we cannot depend on that happy circumstance, so it is good design practice to assume a *worst-case* situation, in which the two offsets have the same polarity and reinforce each other. We can invoke the principle of superposition and conclude that the total output offset voltage is the sum of the offsets caused by the individual input phenomena, but for the worst-case situation, we assume that the total offset is the sum of the respective *magnitudes:*

$$|V_{OS}| = |V_{OS}(I_B)| + |V_{OS}(V_{io})| \text{ (worst case)} \tag{10–50}$$

where V_{OS} is the total output offset voltage, and $V_{OS}(I_B)$ is given by equation 10–43. When R_c has the optimum value $R_1 \| R_f$, the total output offset voltage is expressed as

$$|V_{OS}| = I_{io}R_f + |V_{OS}(V_{io})| \tag{10–51}$$

EXAMPLE 10–15

The operational amplifier in Figure 10–27 has the following specifications: input bias current = 100 nA; input offset current = 20 nA; input offset voltage = 0.5 mV. Find the worst-case output offset voltage. (Consider the two possibilities $I_B^+ > I_B^-$ and vice versa.)

Solution

We first check to see if the 10-kΩ resistor in series with the noninverting input has the optimum value of a compensating resistor: $R_1 \| R_f = (15\text{ k}\Omega) \| (75\text{ k}\Omega) = 12.5\text{ k}\Omega$. $R_c = 10\text{ k}\Omega$ is not optimum, and we will have to use equation 10–43 to find $V_{OS}(I_B)$. Assuming first that $I_B^+ > I_B^-$, we have

$$I_B^+ = I_B + 0.5I_{io} = (100\text{ nA}) + 0.5(20\text{ nA}) = 110\text{ nA}$$
$$I_B^- = I_B - 0.5I_{io} = (100\text{ nA}) - 0.5(20\text{ nA}) = 90\text{ nA}$$

Therefore, by equation 10–43,

$$\begin{aligned} V_{OS}(I_B) &= I_B^- R_f - \left(1 + \frac{R_f}{R_1}\right) I_B^+ R_c \\ &= (90 \times 10^{-9})(75\text{ k}) - (6)(110 \times 10^{-9})(10\text{ k}) \\ &= 0.15\text{ mV} \end{aligned}$$

If $I_B^- > I_B^+$, then $I_B^+ = 90$ nA and $I_B^- = 110$ nA. In that case,

$$\begin{aligned} V_{OS}(I_B) &= (110 \times 10^{-9})(75\text{ k}) - 6(90 \times 10^{-9})(10\text{ k}) \\ &= 2.85\text{ mV} \end{aligned}$$

We see that the worst case occurs for $I_B^- > I_B^+$, and therefore we assume that $|V_{OS}(I_B)| = 2.85$ mV.

FIGURE 10–27 (Example 10–15)

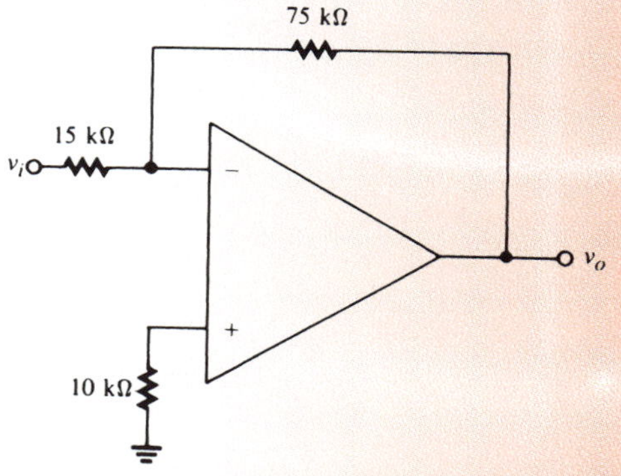

By equation 10–48,

$$V_{OS}(V_{io}) = V_{io}\left(1 + \frac{R_f}{R_1}\right) = (0.5\text{ mV})(6) = 3\text{ mV}$$

Therefore, the worst-case offset is $V_{OS} = |V_{OS}(I_B)| + |V_{OS}(V_{io})| = (2.85\text{ mV}) + (3\text{ mV}) = 5.85\text{ mV}$. (Note that the "best-case" offset would be 0.15 mV.)

The values we have used for V_{io}, I_B, and I_{io} in the examples of this section are typical for general-purpose, BJT operational amplifiers. Amplifiers with much smaller input offsets are available. The bias currents in amplifiers having FET inputs are in the picoamp range.

Most operational amplifiers have two terminals across which an external potentiometer can be connected to adjust the output to zero when the inputs are grounded. This operation is called *zeroing,* or *balancing,* the amplifier. However, operational amplifiers are subject to *drift,* wherein characteristics change with time and, particularly, with temperature. For applications in which extremely small offsets are required and in which the effects of drift must be minimized, *chopper-stabilized* amplifiers are available. Internal choppers convert the dc offset to an ac signal, amplify it, and use it to adjust amplifier characteristics so that the output is automatically restored to zero.

10–6 OPERATIONAL AMPLIFIER SPECIFICATIONS

In this section we will examine and interpret a typical set of manufacturer's specifications for an operational amplifier. The specifications shown in Figure 10–28 are those of the LF353 amplifier, a popular, inexpensive, general-purpose operational amplifier that has been produced for a number of years.

Reviewing the specifications, we see that parameter values are superior to those we have used in the examples of this chapter which were representative of a general-purpose, BJT operational amplifier. Note that most entries show a typical value and a minimum or maximum value. The range of values is the manufacturer's statement of the variation that can be expected among a large number of LF353 chips. Those parameters for which a large numerical value is desirable show a minimum value, and those for which a small numerical value is desirable show a maximum value. For example, the input offset voltage at 25°C has a typical value of 5 mV but may be as great as 10 mV. Circuit designers who are using the LF353, or any other operational amplifier, in the design of a product that will be manufactured in large quantities should use the worst-case specifications.

Note that the specifications include values for the common-mode rejection ratio (CMRR) in dB. Many of the specifications vary with operating conditions such as frequency, supply voltage, and ambient temperature. Typical variations are shown in the graphs that accompany the value listings.

EXAMPLE 10–16

Assuming worst-case conditions at 25°C, determine the following, in connection with the LF353 operational-amplifier circuit shown in Figure 10–29:

1. the closed-loop bandwidth,
2. the maximum operating frequency when the input is a 0.5-V-peak sine wave,
3. the total output offset voltage $|V_{OS}|$.

National Semiconductor

August 2000

LF353 Wide Bandwidth Dual JFET Input Operational Amplifier

LF353
Wide Bandwidth Dual JFET Input Operational Amplifier

General Description

These devices are low cost, high speed, dual JFET input operational amplifiers with an internally trimmed input offset voltage (BI-FET II™ technology). They require low supply current yet maintain a large gain bandwidth product and fast slew rate. In addition, well matched high voltage JFET input devices provide very low input bias and offset currents. The LF353 is pin compatible with the standard LM1558 allowing designers to immediately upgrade the overall performance of existing LM1558 and LM358 designs.

These amplifiers may be used in applications such as high speed integrators, fast D/A converters, sample and hold circuits and many other circuits requiring low input offset voltage, low input bias current, high input impedance, high slew rate and wide bandwidth. The devices also exhibit low noise and offset voltage drift.

Features

- Internally trimmed offset voltage: 10 mV
- Low input bias current: 50pA
- Low input noise voltage: 25 nV/√Hz
- Low input noise current: 0.01 pA/√Hz
- Wide gain bandwidth: 4 MHz
- High slew rate: 13 V/μs
- Low supply current: 3.6 mA
- High input impedance: $10^{12}\Omega$
- Low total harmonic distortion : ≤0.02%
- Low 1/f noise corner: 50 Hz
- Fast settling time to 0.01%: 2 μs

Typical Connection

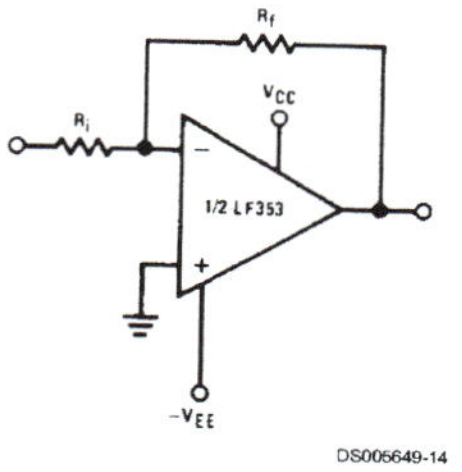

DS005649-14

Connection Diagram

Dual-In-Line Package

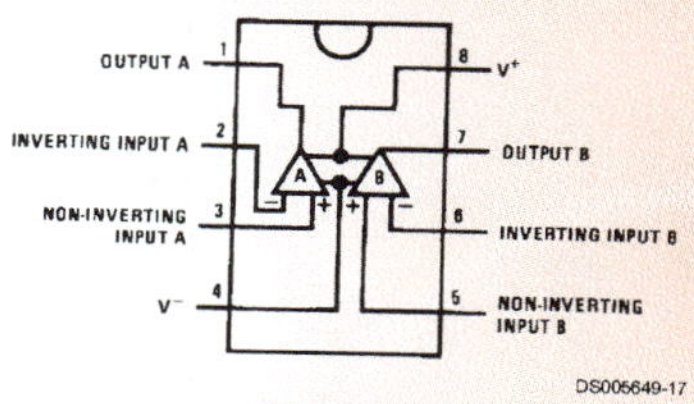

DS005649-17

Top View
Order Number LF353M, LF353MX or LF353N
See NS Package Number M08A or N08E

Simplified Schematic

1/2 Dual

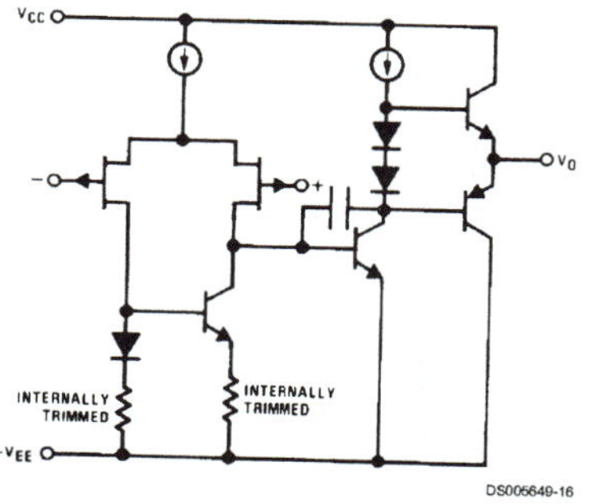

DS005649-16

BI-FET II™ is a trademark of National Semiconductor Corporation.

FIGURE 10–28 LF353 Wide Bandwidth Dual JFET Input Operational Amplifier (Reprinted with permission of National Semiconductor Corporation)

Solution

1. From equation 10–25, $BW_{CL} = f_u\beta$. The parameter labeled "Gain Bandwidth Product" in the LF353 specifications reveals a minimum value of 2.7 MHz. From Figure 10–29 (on page 412),

$$\beta = \frac{R_1}{R_1 + R_f} = \frac{12\ \text{k}\Omega}{(12\ \text{k}\Omega) + (138\ \text{k}\Omega)} = 0.08$$

Thus, $BW_{CL} = (2.7\ \text{MHz})(0.08) = 216\ \text{kHz}$.

LF353

Absolute Maximum Ratings (Note 1)

If Military/Aerospace specified devices are required, please contact the National Semiconductor Sales Office/ Distributors for availability and specifications.

Supply Voltage	±18V
Power Dissipation	(Note 2)
Operating Temperature Range	0°C to +70°C
T_j(MAX)	150°C
Differential Input Voltage	±30V
Input Voltage Range (Note 3)	±15V
Output Short Circuit Duration	Continuous
Storage Temperature Range	−65°C to +150°C
Lead Temp. (Soldering, 10 sec.)	260°C
Soldering Information Dual-In-Line Package Soldering (10 sec.)	260°C
Small Outline Package Vapor Phase (60 sec.)	215°C
Infrared (15 sec.)	220°C

See AN-450 "Surface Mounting Methods and Their Effect on Product Reliability" for other methods of soldering surface mount devices.

ESD Tolerance (Note 8)	1700V
θ_{JA} M Package	TBD

Note 1: Absolute Maximum Ratings indicate limits beyond which damage to the device may occur. Operating ratings indicate conditions for which the device is functional, but do not guarantee specific performance limits. Electrical Characteristics state DC and AC electrical specifications under particular test conditions which guarantee specific performance limits. This assumes that the device is within the Operating Ratings. Specifications are not guaranteed for parameters where no limit is given, however, the typical value is a good indication of device performance.

DC Electrical Characteristics

(Note 5)

Symbol	Parameter	Conditions	LF353 Min	LF353 Typ	LF353 Max	Units
V_{OS}	Input Offset Voltage	R_S=10kΩ, T_A=25°C		5	10	mV
		Over Temperature			13	mV
$\Delta V_{OS}/\Delta T$	Average TC of Input Offset Voltage	R_S=10 kΩ		10		μV/°C
I_{OS}	Input Offset Current	T_j=25°C, (Notes 5, 6)		25	100	pA
		T_j≤70°C			4	nA
I_B	Input Bias Current	T_j=25°C, (Notes 5, 6)		50	200	pA
		T_j≤70°C			8	nA
R_{IN}	Input Resistance	T_j=25°C		10^{12}		Ω
A_{VOL}	Large Signal Voltage Gain	V_S=±15V, T_A=25°C	25	100		V/mV
		V_O=±10V, R_L=2 kΩ				
		Over Temperature	15			V/mV
V_O	Output Voltage Swing	V_S=±15V, R_L=10kΩ	±12	±13.5		V
V_{CM}	Input Common-Mode Voltage Range	V_S=±15V	±11	+15		V
				−12		V
CMRR	Common-Mode Rejection Ratio	R_S≤ 10kΩ	70	100		dB
PSRR	Supply Voltage Rejection Ratio	(Note 7)	70	100		dB
I_S	Supply Current			3.6	6.5	mA

AC Electrical Characteristics

(Note 5)

Symbol	Parameter	Conditions	LF353 Min	LF353 Typ	LF353 Max	Units
	Amplifier to Amplifier Coupling	T_A=25°C, f=1 Hz−20 kHz (Input Referred)		−120		dB
SR	Slew Rate	V_S=±15V, T_A=25°C	8.0	13		V/μs
GBW	Gain Bandwidth Product	V_S=±15V, T_A=25°C	2.7	4		MHz
e_n	Equivalent Input Noise Voltage	T_A=25°C, R_S=100Ω, f=1000 Hz		16		nV/√Hz
i_n	Equivalent Input Noise Current	T_j=25°C, f=1000 Hz		0.01		pA/√Hz

www.national.com 2

FIGURE 10–28 *Continued*

2. By equation 10–32, the maximum operating frequency under the slew-rate limitation is

$$f_S(\max) = \frac{S}{2\pi V_P}$$

The specifications show the slew rate to be 8 V/μs, minimum. From the figure, the closed-loop gain of the inverting configuration is

AC Electrical Characteristics (Continued)

(Note 5)

Symbol	Parameter	Conditions	LF353			Units
			Min	Typ	Max	
THD	Total Harmonic Distortion	A_V=+10, RL=10k, V_O=20Vp−p, BW=20 Hz-20 kHz		<0.02		%

Note 2: For operating at elevated temperatures, the device must be derated based on a thermal resistance of 115˚C/W typ junction to ambient for the N package, and 158˚C/W typ junction to ambient for the H package.

Note 3: Unless otherwise specified the absolute maximum negative input voltage is equal to the negative power supply voltage.

Note 4: The power dissipation limit, however, cannot be exceeded.

Note 5: These specifications apply for V_S=±15V and 0˚C≤T_A≤+70˚C. V_{OS}, I_B and I_{OS} are measured at V_{CM}=0.

Note 6: The input bias currents are junction leakage currents which approximately double for every 10˚C increase in the junction temperature, T_j. Due to the limited production test time, the input bias currents measured are correlated to junction temperature. In normal operation the junction temperature rises above the ambient temperature as a result of internal power dissipation, P_D. T_j=T_A+θ_{jA} P_D where θ_{jA} is the thermal resistance from junction to ambient. Use of a heat sink is recommended if input bias current is to be kept to a minimum.

Note 7: Supply voltage rejection ratio is measured for both supply magnitudes increasing or decreasing simultaneously in accordance with common practice. V_S = ±6V to ±15V.

Note 8: Human body model, 1.5 kΩ in series with 100 pF.

Typical Performance Characteristics

Input Bias Current

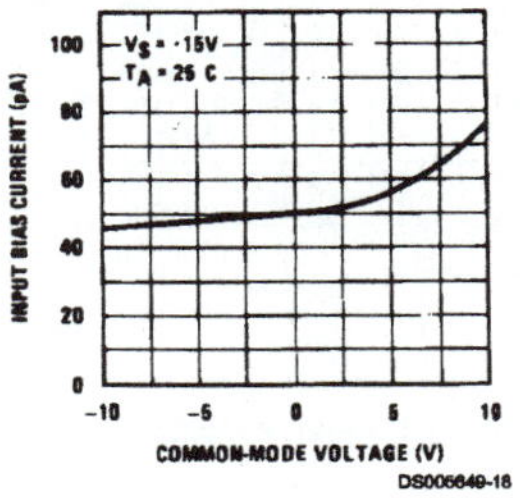

Input Bias Current

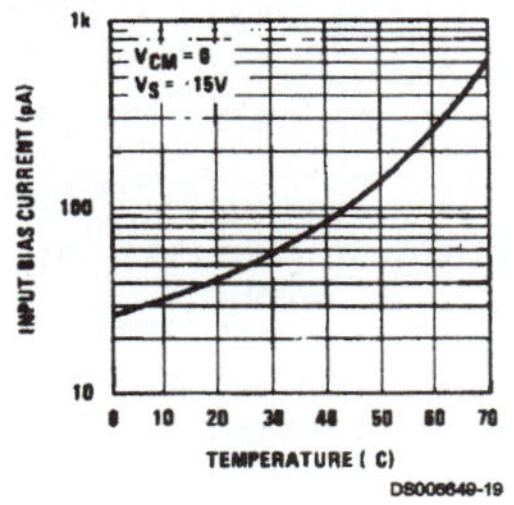

Supply Current

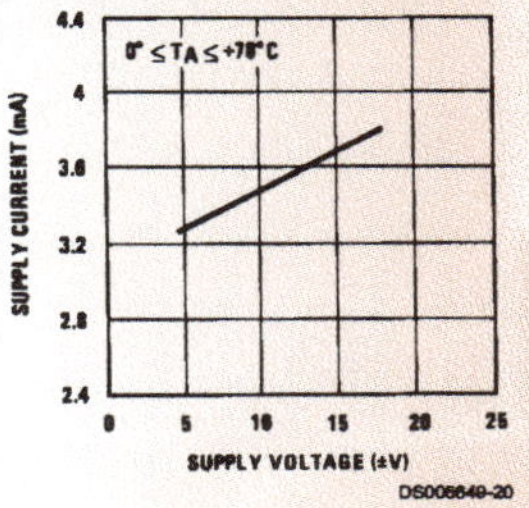

Positive Common-Mode Input Voltage Limit

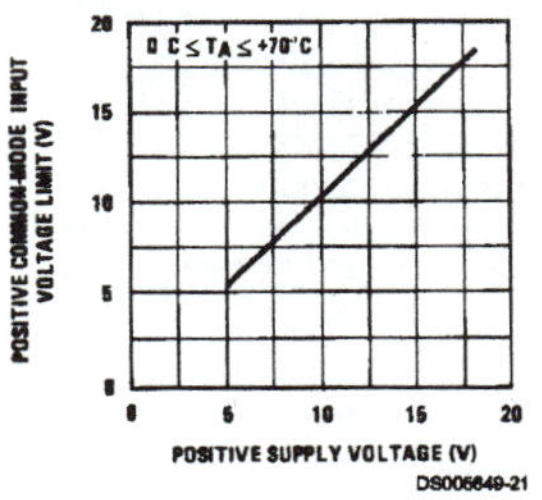

Negative Common-Mode Input Voltage Limit

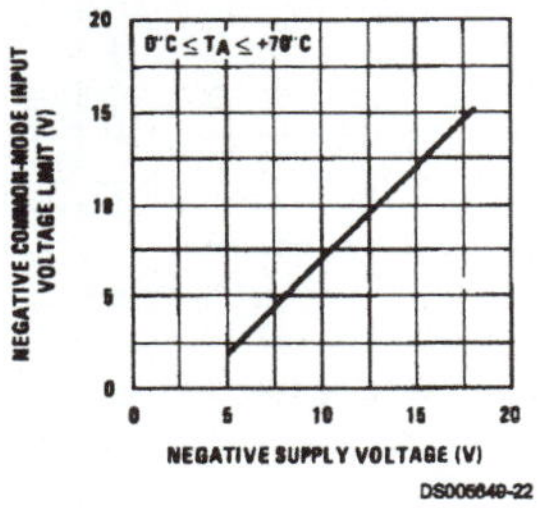

Positive Current Limit

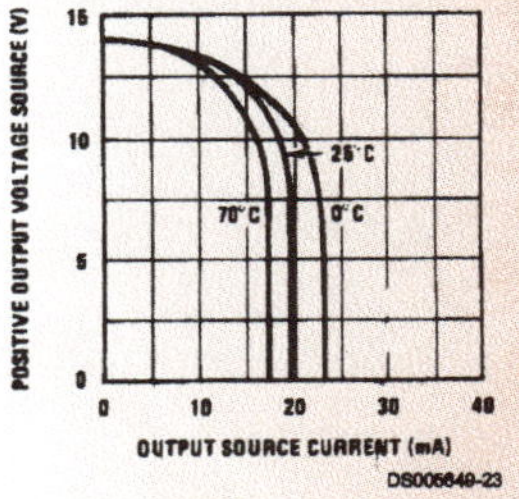

3 www.national.com

FIGURE 10–28 *Continued*

$$\frac{v_o}{v_i} = \frac{-R_f}{R_1} = \frac{-138\ \text{k}\Omega}{12\ \text{k}\Omega} = -11.5$$

Therefore, the magnitude of the peak output voltage is $V_P = (0.5)(11.5) = 5.75$ V. Thus,

$$f_S(\text{max}) = \frac{8 \times 10^6\ \text{V/s}}{2\pi(5.75)\ \text{V}} = 221.4\ \text{kHz}$$

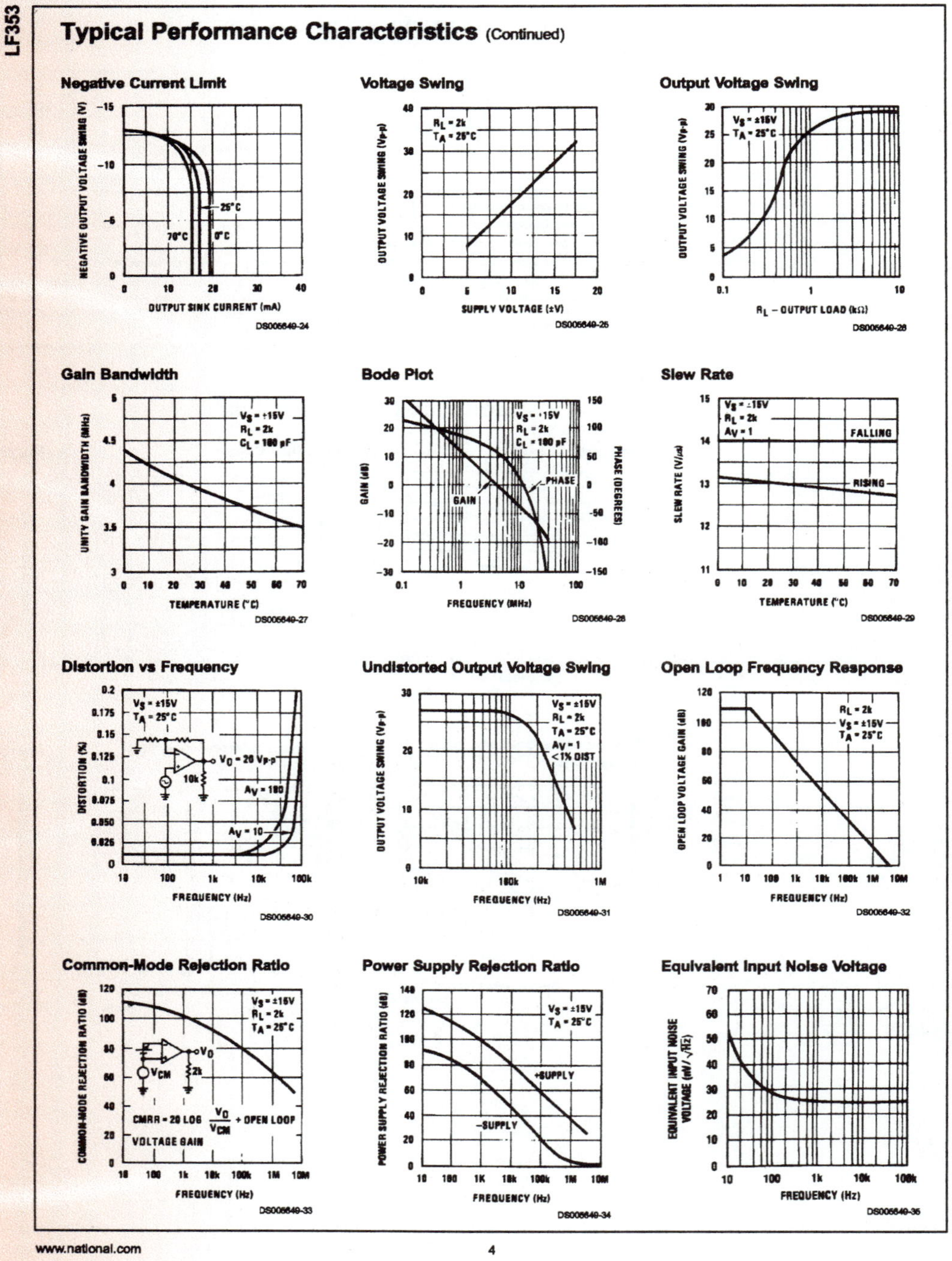

FIGURE 10–28 *Continued*

Since $f_S(\text{max}) > \text{BW}_{\text{CL}}$, the maximum operating frequency for a 0.5-V-peak input is 216 kHz.

3. $R_1 \| R_f = (12\ \text{k}\Omega) \| (138\ \text{k}\Omega) \approx 11\ \text{k}\Omega$. Therefore, the compensating resistor has its optimum value, and we can use equation 10–47 to determine the output offset due to bias currents: $|V_{OS}(I_B)| = I_{io} R_f$. The specifications list the maximum value of input offset current to be 100 pA. Therefore,

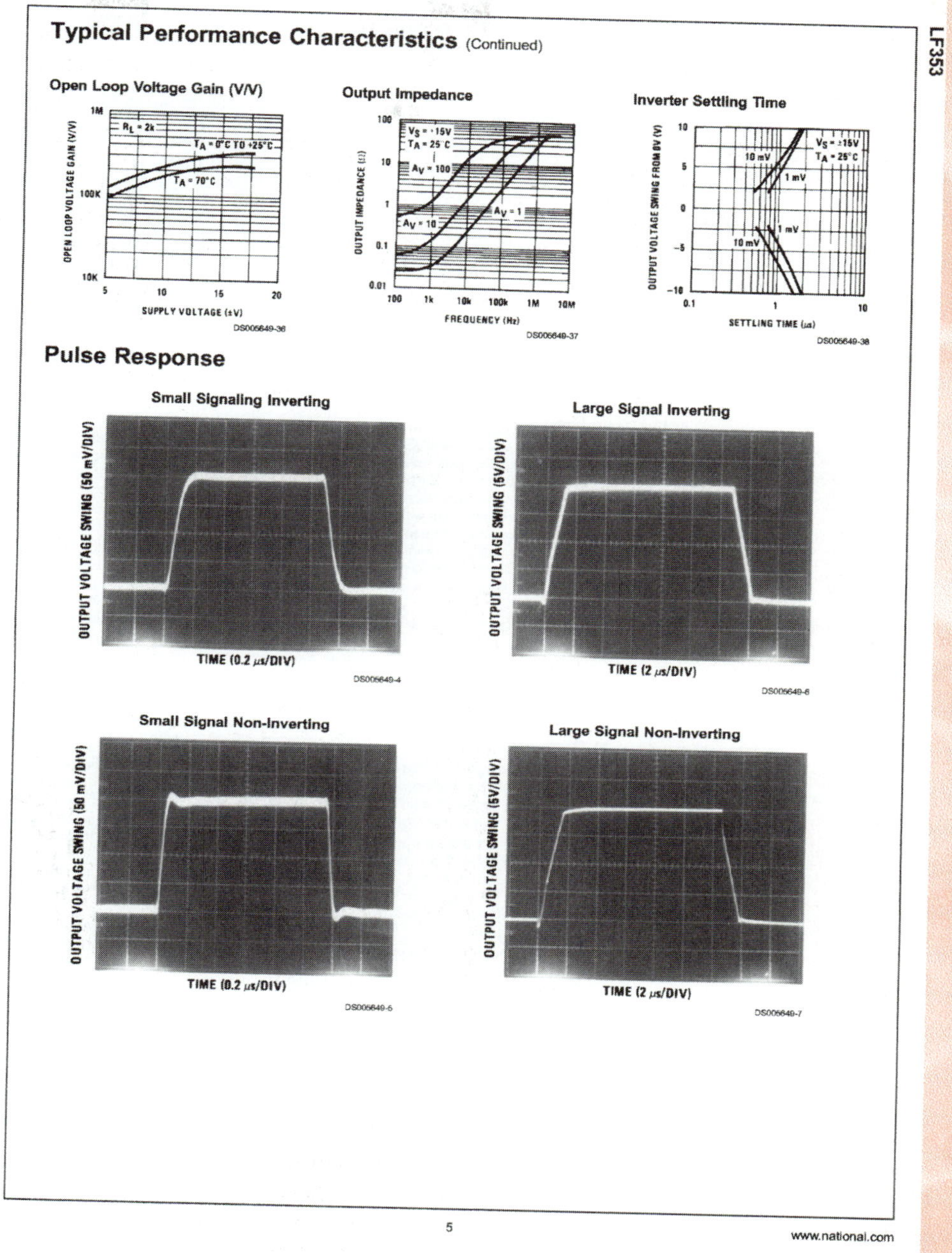

FIGURE 10–28 *Continued*

$|V_{OS}(I_B)|_{\max} = (100 \times 10^{-12})(138 \times 10^3) = 13.8\ \mu\text{V}$. The specifications list the maximum value of input offset voltage to be 10 mV. Therefore,

$$|V_{OS}(V_{io})| = V_{io}\left(\frac{R_f + R_1}{R_1}\right) = (10\text{ mV})\left[\frac{(138\text{ k}\Omega) + (12\text{ k}\Omega)}{12\text{ k}\Omega}\right] = 130\text{ mV}$$

Finally, $|V_{OS}|_{worst\ case} = (13.8\ \mu\text{V}) + (130\text{ mV}) = 130\text{ mV}$.

LF353

Pulse Response (Continued)

Current Limit (R_L = 100Ω)

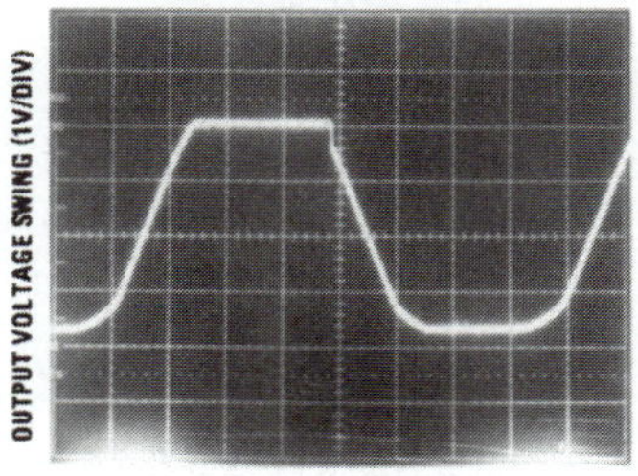

DS005649-8

Application Hints

These devices are op amps with an internally trimmed input offset voltage and JFET input devices (BI-FET II). These JFETs have large reverse breakdown voltages from gate to source and drain eliminating the need for clamps across the inputs. Therefore, large differential input voltages can easily be accommodated without a large increase in input current. The maximum differential input voltage is independent of the supply voltages. However, neither of the input voltages should be allowed to exceed the negative supply as this will cause large currents to flow which can result in a destroyed unit.

Exceeding the negative common-mode limit on either input will force the output to a high state, potentially causing a reversal of phase to the output. Exceeding the negative common-mode limit on both inputs will force the amplifier output to a high state. In neither case does a latch occur since raising the input back within the common-mode range again puts the input stage and thus the amplifier in a normal operating mode.

Exceeding the positive common-mode limit on a single input will not change the phase of the output; however, if both inputs exceed the limit, the output of the amplifier will be forced to a high state.

The amplifiers will operate with a common-mode input voltage equal to the positive supply; however, the gain bandwidth and slew rate may be decreased in this condition. When the negative common-mode voltage swings to within 3V of the negative supply, an increase in input offset voltage may occur.

Each amplifier is individually biased by a zener reference which allows normal circuit operation on ±6V power supplies. Supply voltages less than these may result in lower gain bandwidth and slew rate.

The amplifiers will drive a 2 kΩ load resistance to ±10V over the full temperature range of 0°C to +70°C. If the amplifier is forced to drive heavier load currents, however, an increase in input offset voltage may occur on the negative voltage swing and finally reach an active current limit on both positive and negative swings.

Precautions should be taken to ensure that the power supply for the integrated circuit never becomes reversed in polarity or that the unit is not inadvertently installed backwards in a socket as an unlimited current surge through the resulting forward diode within the IC could cause fusing of the internal conductors and result in a destroyed unit.

As with most amplifiers, care should be taken with lead dress, component placement and supply decoupling in order to ensure stability. For example, resistors from the output to an input should be placed with the body close to the input to minimize "pick-up" and maximize the frequency of the feedback pole by minimizing the capacitance from the input to ground.

A feedback pole is created when the feedback around any amplifier is resistive. The parallel resistance and capacitance from the input of the device (usually the inverting input) to AC ground set the frequency of the pole. In many instances the frequency of this pole is much greater than the expected 3 dB frequency of the closed loop gain and consequently there is negligible effect on stability margin. However, if the feedback pole is less than approximately 6 times the expected 3 dB frequency a lead capacitor should be placed from the output to the input of the op amp. The value of the added capacitor should be such that the RC time constant of this capacitor and the resistance it parallels is greater than or equal to the original feedback pole time constant.

www.national.com 6

FIGURE 10–28 *Continued*

EXAMPLE 10–17

PSPICE

Use PSpice and the PSpice library to obtain the value of the closed-loop bandwidth for the circuit of Example 10–16 but using the LF411 operational amplifier. Assume a 0.1-V-ac input.

Solution

The PSpice circuit and input circuit file are shown in Figure 10–30. Note that sub-circuit call X1 specifies the name LF411 for the subcircuit stored in the

Detailed Schematic

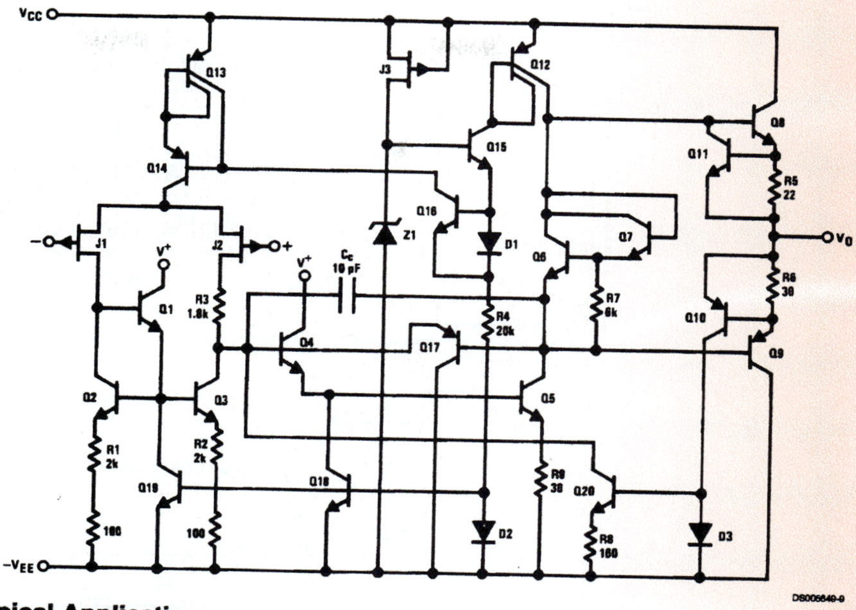

Typical Applications

Three-Band Active Tone Control

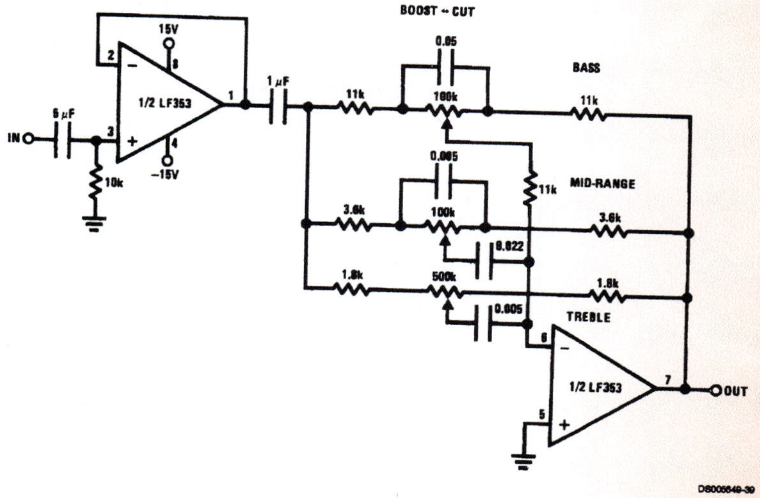

7

www.national.com

FIGURE 10–28 *Continued*

PSpice library (see Appendix Section A–17). (If the Evaluation version of PSpice is used, the .LIB statement must be written .LIB EVAL.LIB.) After execution of the program, go to Probe and obtain a plot for the closed-loop gain in dB by entering db(v(6)/v(1)) on the "Trace Expression" line. Using the cursors, you will see that the frequency at which the gain is 3 dB below its midband value is approximately 734 kHz. This op-amp has a GBP of more than 8 MHz.

LF353

Typical Applications (Continued)

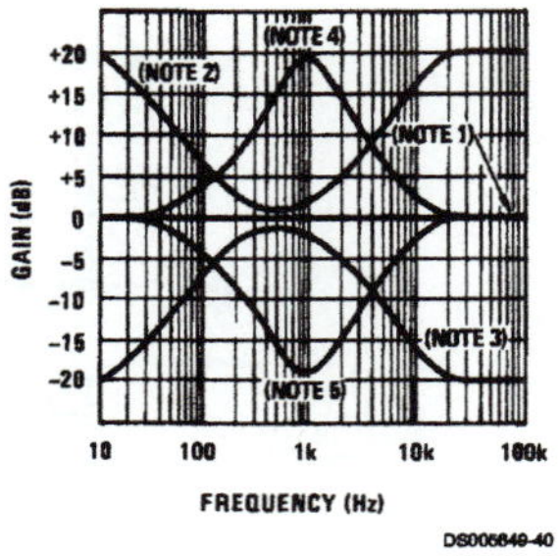

Note 1: All controls flat.
Note 2: Bass and treble boost, mid flat.
Note 3: Bass and treble cut, mid flat.
Note 4: Mid boost, bass and treble flat.
Note 5: Mid cut, bass and treble flat.

- All potentiometers are linear taper
- Use the LF347 Quad for stereo applications

Improved CMRR Instrumentation Amplifier

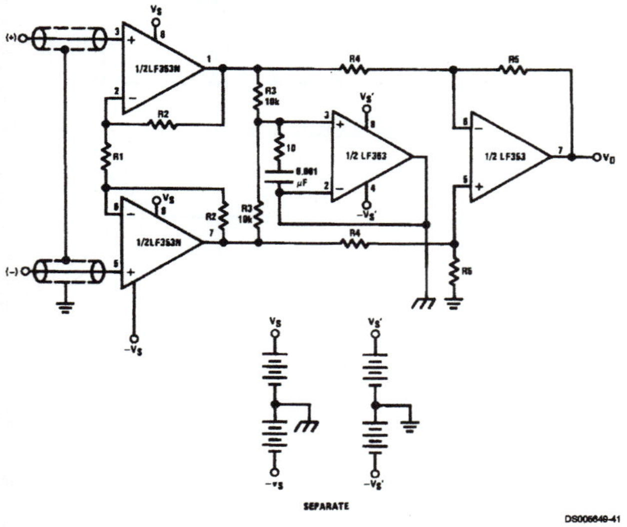

$$A_V = \left(\frac{2R2}{R1} + 1\right)\frac{R5}{R4}$$

and are separate isolated grounds
Matching of R2's, R4's and R5's control CMRR
With $A_{V_T} = 1400$, resistor matching = 0.01%: CMRR = 136 dB

- Very high input impedance
- Super high CMRR

www.national.com 8

FIGURE 10–28 *Continued*

10–7 MULTISIM EXERCISE

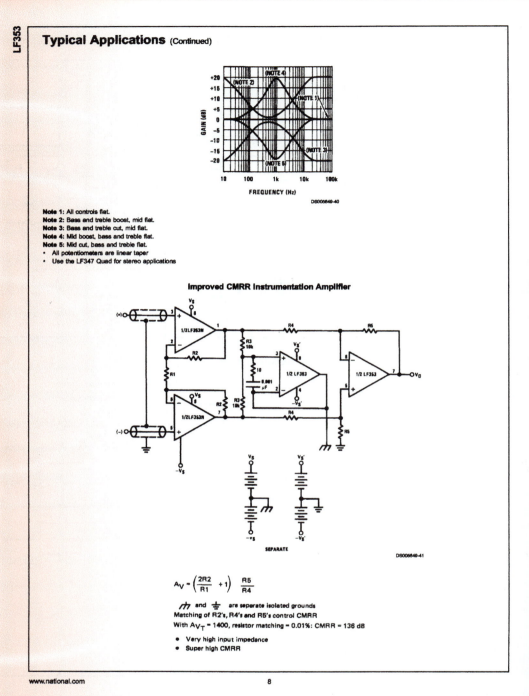

The objective of this exercise is to enhance the understanding of the effects that slew rate and bandwidth have on the transient response of an amplifier. Figure 10–31 shows the circuit schematic to be used in the simulation. As you can see, it is a noninverting amplifier using the 741 op-amp driven by a square (clock) waveform. On the first simulation, the amplitude of the input signal is 0.1 V, small enough to avoid slew-rate limitation.

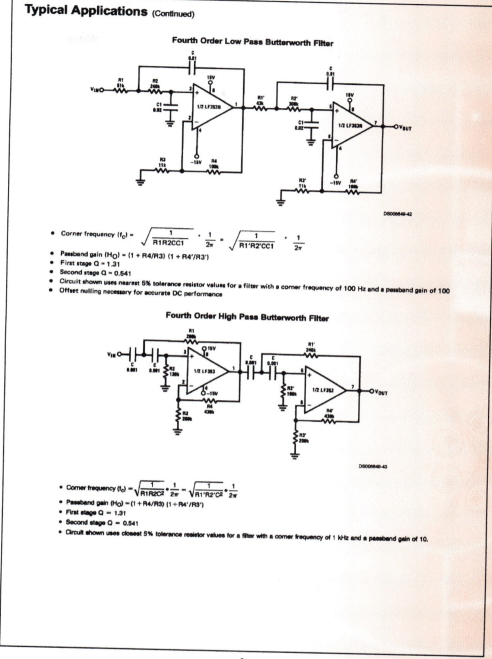

Typical Applications (Continued)

Fourth Order Low Pass Butterworth Filter

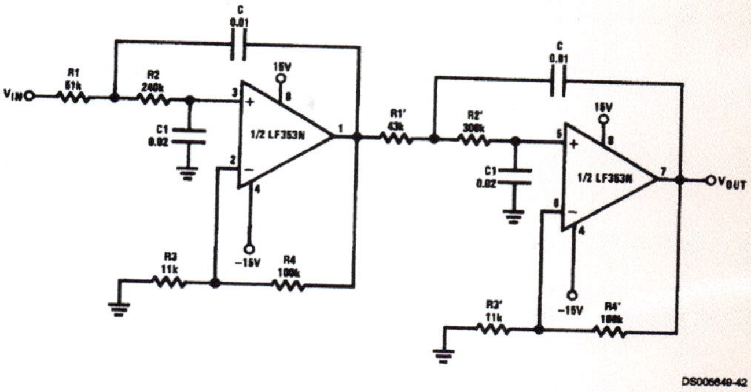

- Corner frequency $(f_c) = \sqrt{\frac{1}{R1R2CC1}} \cdot \frac{1}{2\pi} = \sqrt{\frac{1}{R1'R2'CC1}} \cdot \frac{1}{2\pi}$
- Passband gain $(H_O) = (1 + R4/R3)\ (1 + R4'/R3')$
- First stage Q = 1.31
- Second stage Q = 0.541
- Circuit shown uses nearest 5% tolerance resistor values for a filter with a corner frequency of 100 Hz and a passband gain of 100
- Offset nulling necessary for accurate DC performance

Fourth Order High Pass Butterworth Filter

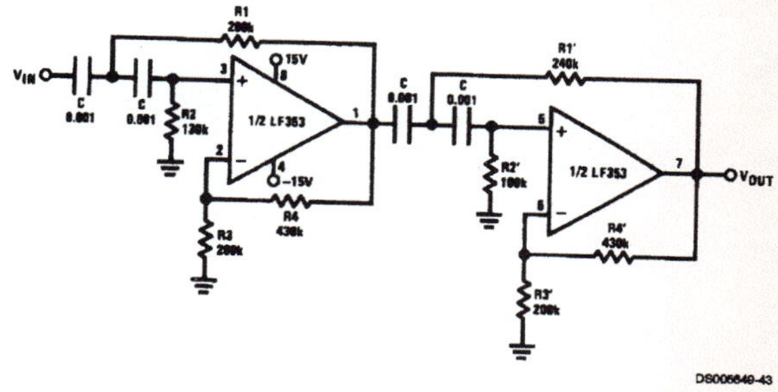

- Corner frequency $(f_c) = \sqrt{\frac{1}{R1R2C^2}} \cdot \frac{1}{2\pi} = \sqrt{\frac{1}{R1'R2'C^2}} \cdot \frac{1}{2\pi}$
- Passband gain $(H_O) = (1 + R4/R3)\ (1 + R4'/R3')$
- First stage Q = 1.31
- Second stage Q = 0.541
- Circuit shown uses closest 5% tolerance resistor values for a filter with a corner frequency of 1 kHz and a passband gain of 10.

9 www.national.com

FIGURE 10–28 *Continued*

Set up the oscilloscope to 200 mV/div and 5 μs/div for the vertical and horizontal scales, respectively. Run the simulation and observe that the rise and fall edges of the waveform are exponential curves, that is, free of slew-rate limitation. Measure the rise time using the cursors. It should be consistent with the closed-loop bandwidth of the amplifier, which is around 240 kHz.

Stop the simulation, change the voltage of the input signal to 2 V and the oscilloscope settings to 5 V/div and 20 μs/div. Restart the simulation and observe how the output signal is now severely limited by the slew rate. Using again the cursors, obtain the slew rate by computing the "rise over run" of

LF353

Typical Applications (Continued)

Ohms to Volts Converter

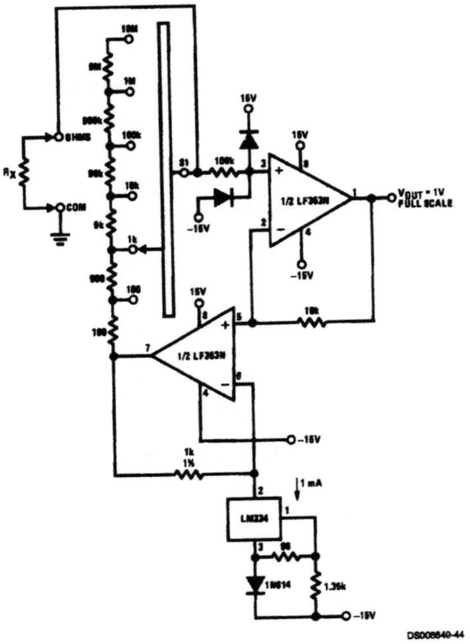

$$V_O = \frac{1V}{R_{LADDER}} \times R_X$$

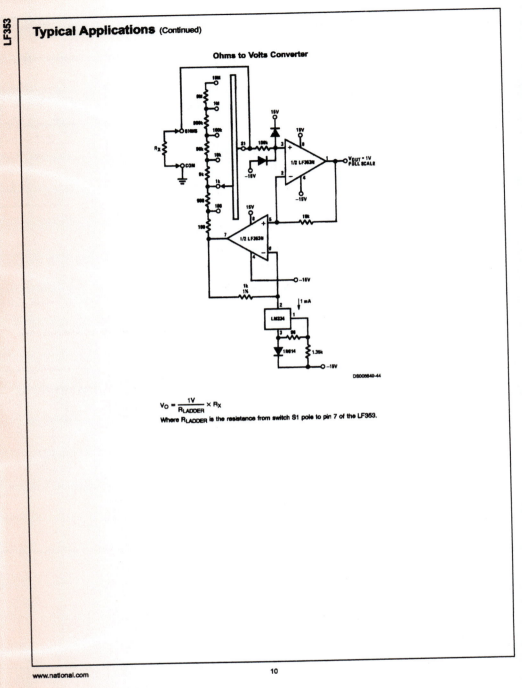

Where R_{LADDER} is the resistance from switch S1 pole to pin 7 of the LF353.

www.national.com 10

FIGURE 10–28 *Continued*

one of the waveform edges. You should obtain about 0.5 V/μs, which is the typical value for the 741.

SUMMARY

This chapter presented detailed information regarding those important nonideal characteristics in operational amplifiers that affect their performance in practical circuits. Emphasis was placed on designing op-amp circuits

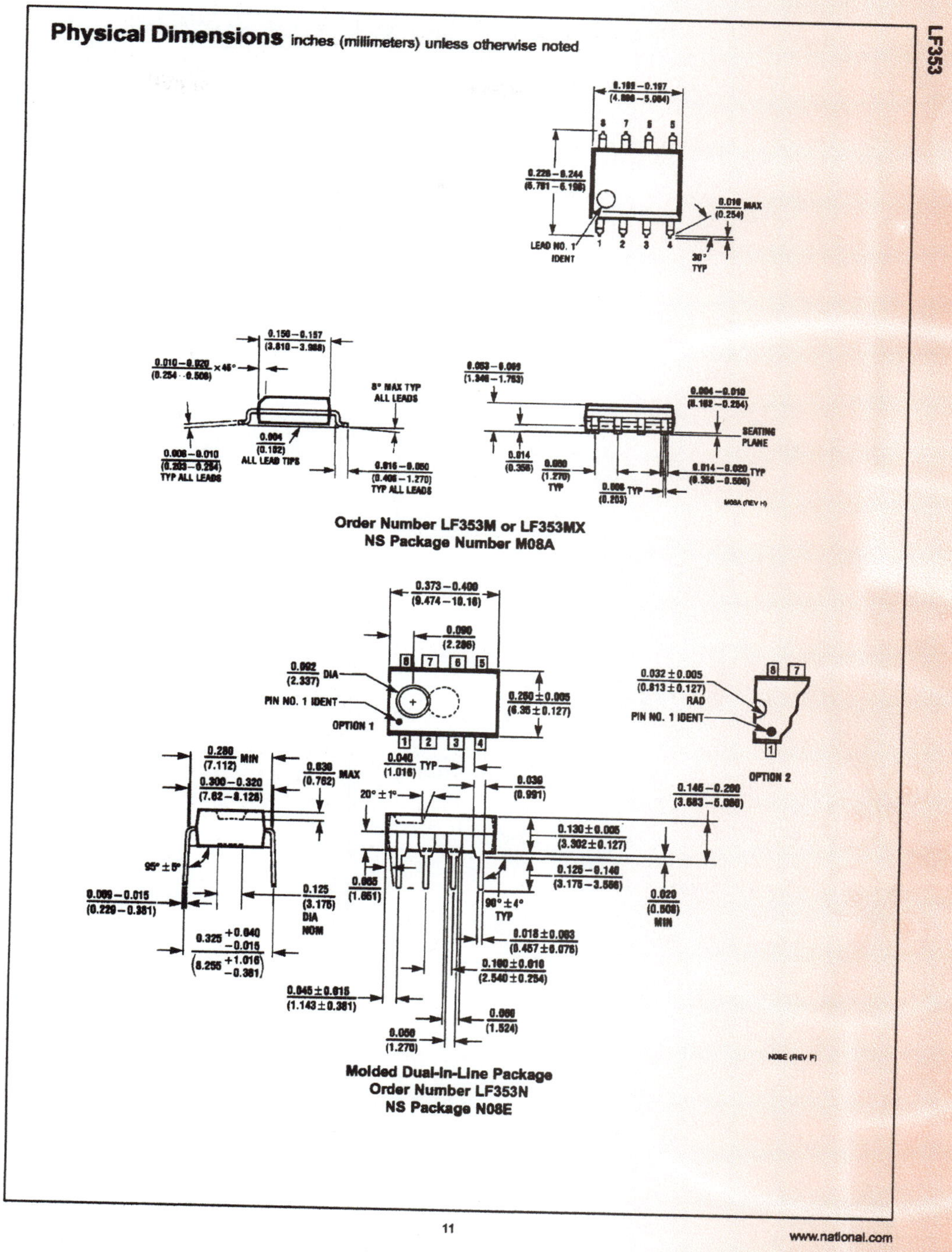

FIGURE 10–28 Continued

based on meeting or exceeding required performance characteristics. At the end of this chapter, the student should have a good understanding of the following concepts:

- Negative feedback improves the already good characteristics of operational amplifiers.
- The input resistance of a noninverting amplifier is practically infinite, whereas that for the inverting case is essentially the value of the input resistor.

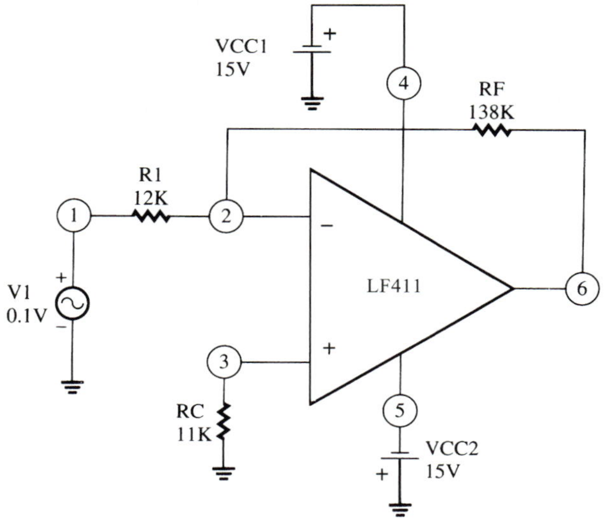

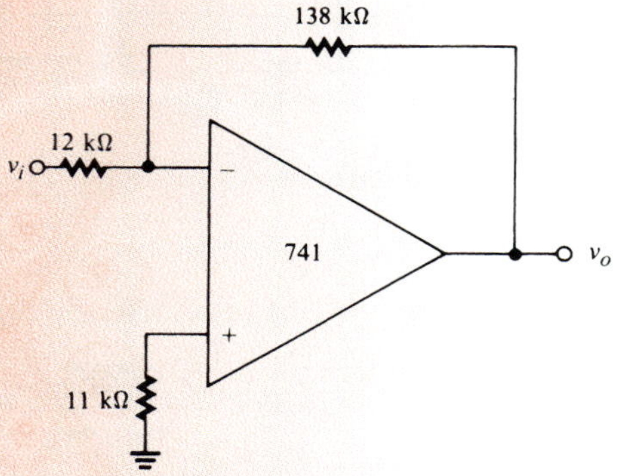


```
EXAMPLE 10-17
V1   1   0   AC   0.1V
VCC1  4   0   15V
VCC2  0   5   15V
R1   1   2   12K
RF   2   6   138K
RC   3   0   11K
X1   3   2   4   5   6   LF411
.LIB
.AC DEC 100 1K 1MEG
.PROBE
.END
```

FIGURE 10–29 (Example 10–16)

FIGURE 10–30 (Example 10–17)

FIGURE 10–31 Noninverting amplifier for observing rise time and slew rate

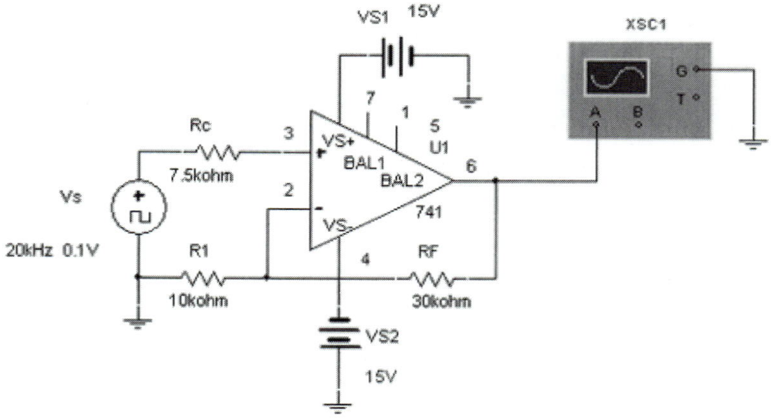

- In a closed-loop configuration, the feedback function β determines the gain and the bandwidth of the circuit.
- The slew rate of an op-amp imposes a limit on how fast the output voltage can change in a given time. It normally takes effect at relatively large output voltage swings.
- Rise time in response to a step-like input depends on the closed-loop bandwidth provided the op-amp is not under slew-rate limitation.

- Output offset voltage in operational amplifiers can be minimized by properly selecting the resistor values used in op-amp circuits.
- Data sheets for op-amps provide a wealth of information for circuit designers that helps them decide what op-amps are better suited for the application being considered.

EXERCISES

SECTION 10–2
Feedback Theory

10–1. An operational amplifier having an open-loop gain of 5000 is used in a noninverting configuration with a feedback resistor of 1 MΩ. For each of the following values of R_1, find the closed-loop gain if the amplifier were ideal and find the actual closed-loop gain.

(a) $R_1 = 5$ kΩ
(b) $R_1 = 20$ kΩ
(c) $R_1 = 100$ kΩ

10–2. Repeat Exercise 10–1 when the open-loop gain of the amplifier is increased by a factor of 2.

10–3. An operational amplifier is to be used in a noninverting configuration that has an ideal closed-loop gain of 800. What minimum value of open-loop gain should the amplifier have if the actual closed-loop gain must be at least 799?

10–4. The operational amplifier in Exercise 10–1 has a differential input resistance of 40 kΩ and an output resistance of 90 Ω. Find the closed-loop input and output resistance for each of the values of R_1 listed.

10–5. An operational amplifier has open-loop gain 10^4 and output resistance 120 Ω. It is to be used in a noninverting configuration for an application in which its closed-loop output resistance must be no greater than 1 Ω. What is the maximum closed-loop gain that the amplifier can have?

10–6. Derive equation 10–15 from equation 10–14.

10–7. An operational amplifier has an open-loop gain of 5000. It is used in an inverting configuration with a feedback ratio of 0.2. What is its closed-loop gain?

10–8. The operational amplifier in Figure 10–32 has an open-loop gain of 5×10^4. Draw a block diagram of the feedback model for the configuration shown. Label each block with the correct numerical quantity.

10–9. An operational amplifier is to be used in an inverting configuration with feedback resistance 100 kΩ and input resistance 2 kΩ. If the closed-loop gain must be no less than 49.5 in magnitude,

(a) what minimum value of loop gain should it have, and
(b) what minimum value of open-loop gain should it have? (*Hint:* Work with gain *magnitudes.*)

10–10. The operational amplifier shown in Figure 10–33 has an open-loop gain of

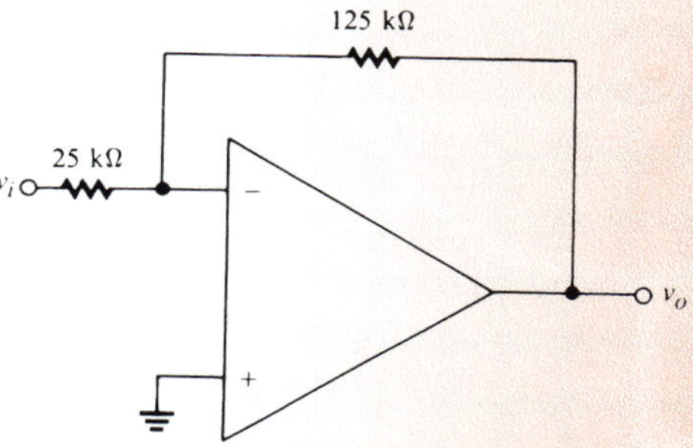

FIGURE 10–32 (Exercise 10–8)

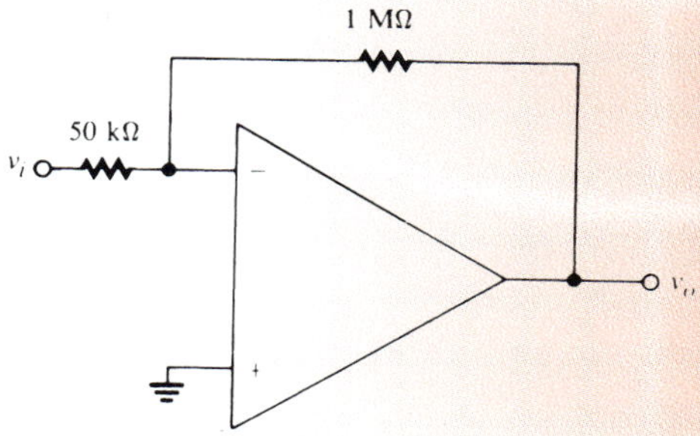

FIGURE 10–33 (Exercise 10–10)

8000 and an open-loop output resistance of 250 Ω. Find the closed-loop

(a) input resistance and

(b) output resistance.

SECTION 10–3

Frequency Response

10–11. An operational amplifier has gain-bandwidth product equal to 5×10^5 Hz and a dc, open-loop gain of 20,000. At what frequency does the open-loop gain equal 14,142?

10–12. With reference to the operational amplifier in Exercise 10–11,

(a) at what frequency is the open-loop gain equal to 0 dB, and

(b) what is the open-loop gain at 2.5 kHz?

10–13. The operational amplifier in Figure 10–34 has a unity-gain frequency of 1.2 MHz.

(a) What is the closed-loop bandwidth?

(b) What is the closed-loop gain at 600 kHz?

(c) At what frequency is the closed-loop gain equal to 0 dB?

10–14. An operational amplifier having a gain-bandwidth product of 8×10^5 Hz is to be used in a noninverting configuration as an audio amplifier (20 Hz–20 kHz). What is the maximum closed-loop gain that can be obtained from the amplifier in this application?

10–15. An operational amplifier has a dc open-loop gain of 25×10^4 and an open-loop cutoff frequency of 40 Hz. It is to be used in an inverting configuration to amplify signals up to 50 kHz. What is the maximum closed-loop gain that can be obtained from the amplifier for this application?

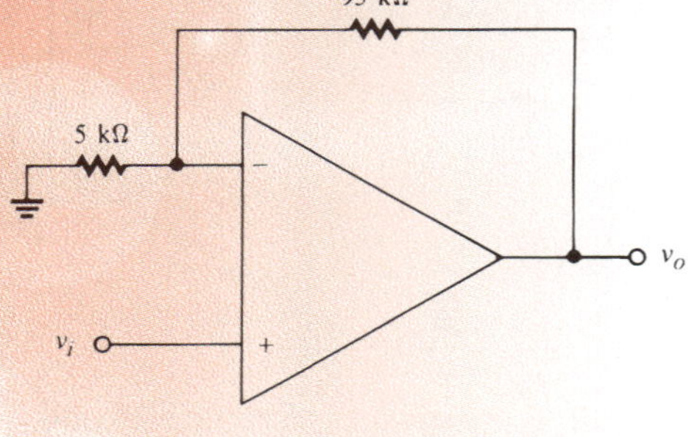

FIGURE 10–34 (Exercise 10–13)

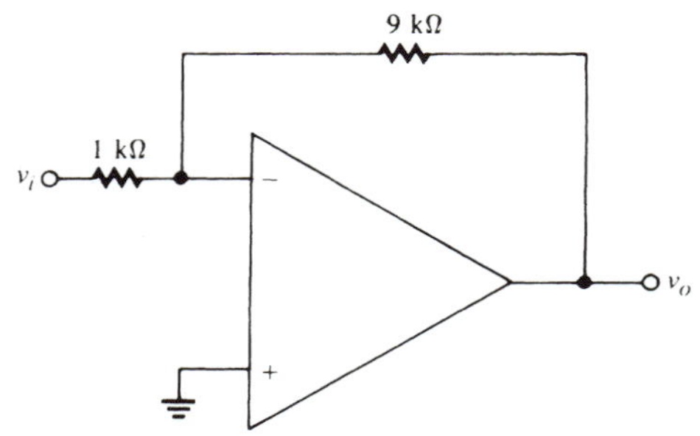

FIGURE 10–35 (Exercise 10–16)

10–16. The operational amplifier in Figure 10–35 has a unity-gain frequency of 2 MHz.

(a) What is the closed-loop bandwidth?

(b) What is the closed-loop gain at 500 kHz?

10–17. A noninverting amplifier has a closed-loop gain of 35 dB at dc and low frequencies. The operational amplifier has a GBP of 4 MHz.

(a) What is the closed-loop bandwidth of the amplifier?

(b) Determine the closed-loop gain in dB at the following frequencies: 1 kHz, 50 kHz, 1 MHz.

10–18. Each of the operational amplifiers in Figure 10–36 has a unity-gain frequency of 750 kHz. What is the approximate upper cutoff frequency of the cascaded system?

SECTION 10–4

Slew Rate and Rise Time

10–19. What minimum slew rate is necessary for a unity-gain amplifier that must pass, without distortion, the input waveform shown in Figure 10–37?

10–20. Repeat Exercise 10–19 if the amplifier is in a noninverting configuration with $R_1 = 50$ kΩ and $R_f = 100$ kΩ.

10–21. An inverting amplifier has a slew rate of 2 V/μs. What maximum closed-loop gain can it have without distorting the input waveform shown in Figure 10–37?

10–22. An operational amplifier has a slew rate of 2 V/μs. What maximum peak amplitude can a 1-MHz input signal have without exceeding the slew rate, when the amplifier is used in a voltage-follower circuit?

10–23. What minimum slew rate is required for an operational amplifier whose output

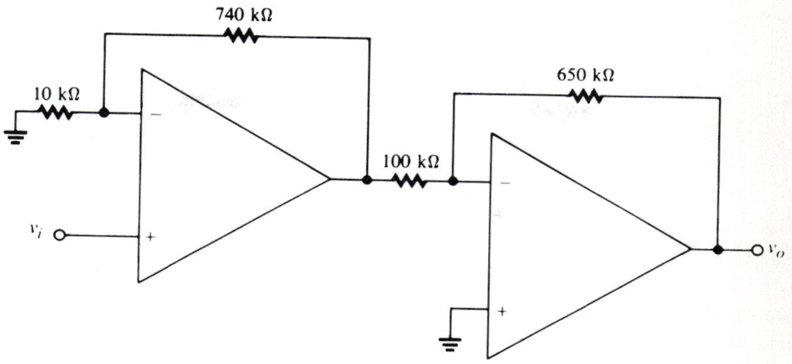

FIGURE 10–36 (Exercise 10–18)

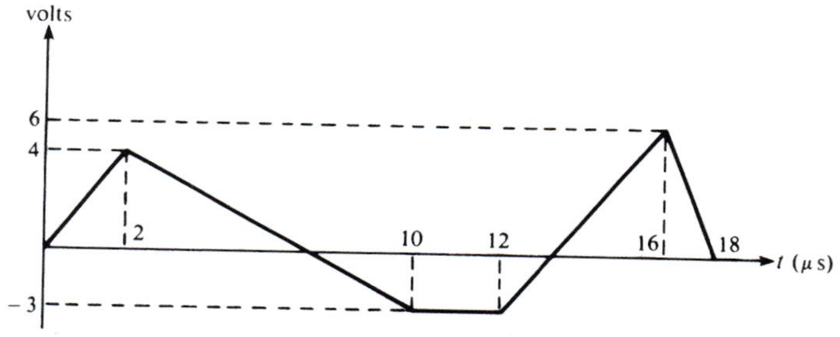

FIGURE 10–37 (Exercise 10–19)

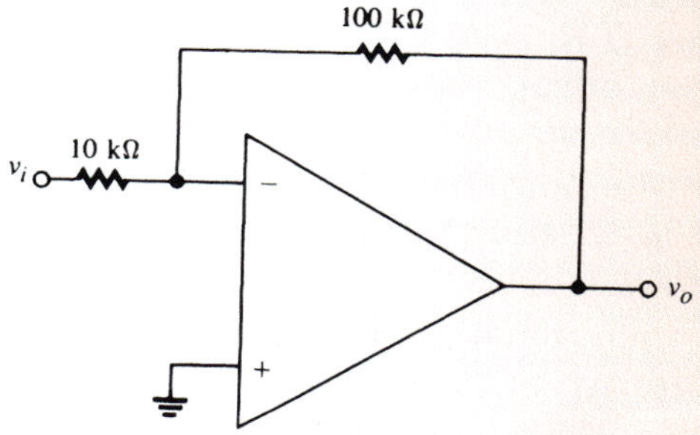

FIGURE 10–39 (Exercise 10–26)

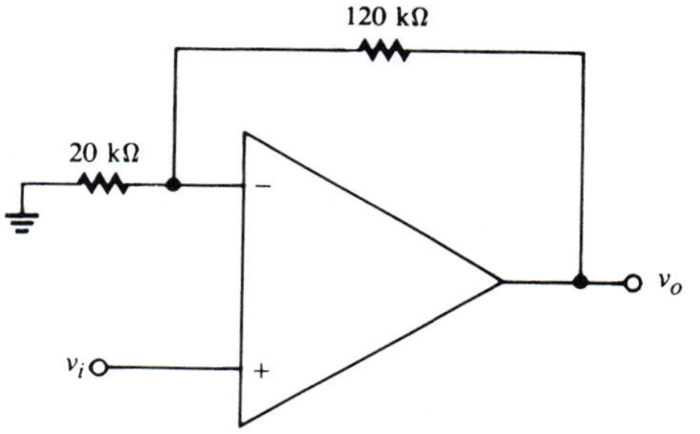

FIGURE 10–38 (Exercise 10–24)

must be at least 2 V rms over the audio frequency range?

10–24. The operational amplifier shown in Figure 10–38 has a slew rate of 1.2 V/μs. Determine whether the output will be distorted due to the slew-rate limitation when the input is any one of the following sinusoidal signals: v_1 = 0.7 V rms at 30 kHz; v_2 = 1.0 V rms at 15 kHz; v_3 = 0.5 V rms at 40 kHz; v_4 = 0.1 V rms at 20 kHz. If distortion occurs, determine the signals that are responsible.

10–25. If distortion occurs in Exercise 10–24, find remedies other than changing the input signals.

10–26. The operational amplifier shown in Figure 10–39 has a slew rate of 0.5 V/μs and a unity-gain frequency of 1 MHz. Find the maximum frequency of a 0.2-V-rms sine-wave input that can be amplified without distortion.

10–27. To what value could the feedback resistor in Exercise 10–26 be changed if it were necessary that the maximum frequency be 50 kHz?

10–28. If the input to the amplifier in Exercise 10–26 is a triangular wave that varies between −1 V and +1 V peak and has frequency 1 kHz, determine whether distortion will occur.

10–29. Find minimum gain-bandwidth product and minimum slew rate required from an op-amp in a noninverting configuration to amplify without distortion a sinusoidal signal with frequency components from 10 Hz to 22 kHz. The desired closed-loop gain is 60 and should not drop more than 5% at the maximum operating frequency. The maximum expected input voltage is 200 mV pk.

10–30. The input to an inverting amplifier is a 50-kHz square waveform of 1 V p–p and centered at 0 V. The operational amplifier has GBP = 1.8 MHz and S = 10 V/μs. Resistors used are: R_1 = 12 KΩ, R_f = 60 KΩ, R_c = 10 KΩ. Determine the kind of limitation that the output waveform will have and draw a sketch of it indicating time and voltage values.

10–31. Repeat Problem 10–30 if the slew rate is 1 V/μs.

SECTION 10–5

Offset Currents and Voltages

10–32. The operational amplifier shown in Figure 10–40 has I_B^+ = 100 nA and I_B^- = 80 nA. If $R_1 = R_f = R_c$ = 20 kΩ, find the output offset voltage due to the input bias currents.

10–33. Find the optimum value for R_c in Exercise 10–32 and repeat the exercise using that value.

10–34. An operational amplifier has an input offset current of 50 nA. It is to be used in an inverting-amplifier application where the output offset voltage due to bias currents cannot exceed 10 mV. If the amplifier must provide an impedance of at least 40 kΩ to the signal source driving it, what is the maximum permissible closed-loop gain of the amplifier?

10–35. An operational amplifier having an input offset current of 80 nA is to be used in a noninverting-amplifier application where the output offset voltage due to bias currents cannot exceed 5 mV. The value of R_1 is 10 kΩ and the amplifier must have the maximum possible closed-loop gain. Design the circuit and find its voltage gain.

10–36. Given $I_B = (I_B^+ + I_B^-)/2$ and $I_{io} = |I_B^+ - I_B^-|$, solve these equations simultaneously to show that

(a) when $I_B^+ > I_B^-$, $I_B^+ = I_B + 0.5\, I_{io}$ and $I_B^- = I_B - 0.5\, I_{io}$; and

(b) when $I_B^+ < I_B^-$, $I_B^+ = I_B - 0.5\, I_{io}$ and $I_B^- = I_B + 0.5\, I_{io}$. (*Hint*: When $I_B^+ > I_B^-$, $|I_B^+ - I_B^-| = I_B^+ - I_B^-$.)

10–37. The operational amplifier in Figure 10–41 has I_B = 100 nA and $|I_{io}|$ = 50 nA. What is the maximum possible value of $|V_{OS}(I_B)|$?

10–38. An operational amplifier has an input offset voltage of 1.2 mV. It is to be used in a noninverting-amplifier application where the output offset due to V_{io} cannot exceed 6 mV. If the feedback resistor is 20 kΩ, what is the minimum permissible value of R_1?

10–39. The input offset voltage of an operational amplifier used in a voltage-follower circuit is 0.5 mV. What is the value of $|V_{OS}(V_{io})|$?

10–40. The operational amplifier shown in Figure 10–42 has $|I_{io}|$ = 120 nA and $|V_{io}|$ = 1.5 mV. Find the worst-case output offset voltage.

10–41. The operational amplifier in Figure 10–43 has I_B = 100 nA, I_{io} = 30 nA, and $|V_{io}|$ = 2 mV. Find the worst-case value of $|V_{OS}|$.

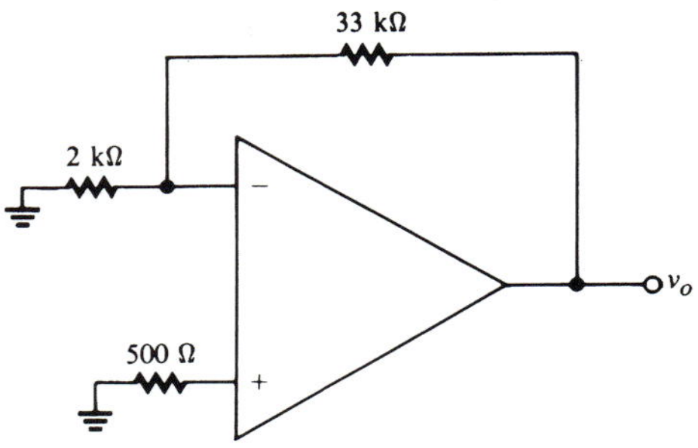

FIGURE 10–41 (Exercise 10–37)

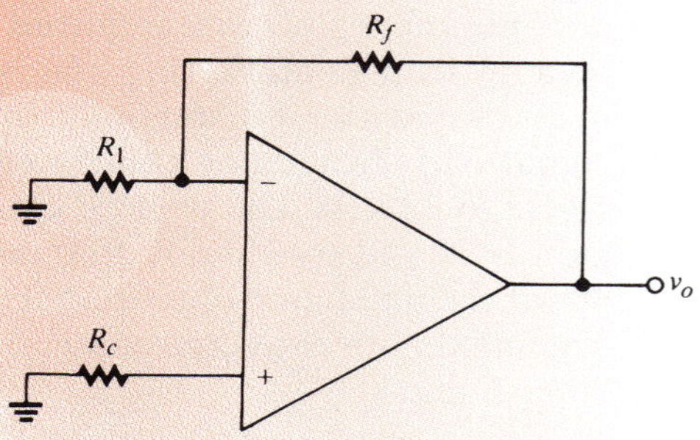

FIGURE 10–40 (Exercise 10–32)

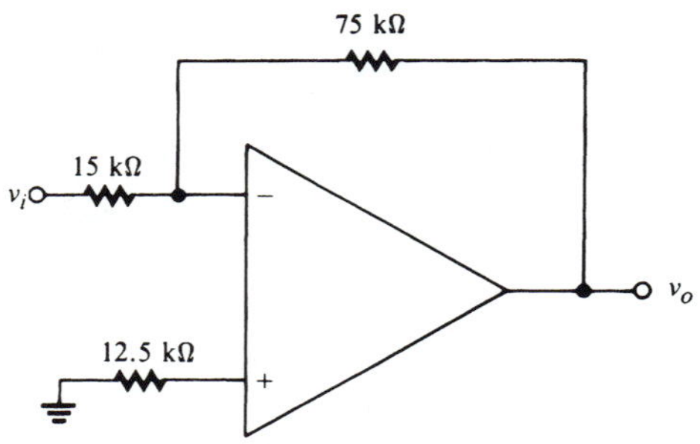

FIGURE 10–42 (Exercise 10–40)

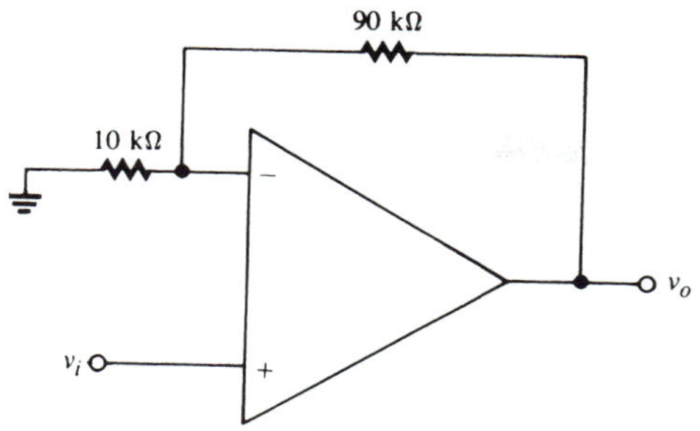

FIGURE 10–43 (Exercise 10–41)

SECTION 10–6

Operational Amplifier Specifications

10–42. Using the LF353 operational amplifier specifications, determine the following:

(a) the minimum slew rate at 25°C,

(b) the output voltage swing at 70°C when the supply voltage is 15 V,

(c) the input bias current at 60°C ambient temperature, and

(d) the input offset current at 0°C ambient temperature.

10–43. Using the LF353 operational amplifier specifications, determine the following:

(a) the maximum input offset voltage at 25°C,

(b) the CMRR at 10 kHz,

(c) the input offset current at 25°C when the supply voltage is 10 V, and

(d) the closed-loop bandwidth at 100°C if its closed-loop bandwidth at 20°C is 40 kHz.

10–44. Using the LF353, design an inverting operational-amplifier circuit that will meet the following criteria at 25°C:

(a) maximum possible closed-loop voltage gain,

(b) closed-loop bandwidth of at least 50 kHz,

(c) worst-case output offset voltage of 100 mV, and

(d) minimum input resistance 10 kΩ.

10–45. Using the LF353, design a noninverting operational-amplifier circuit that will meet the following criteria at 25°C:

(a) maximum possible closed-loop voltage gain,

(b) no distortion with an input ramp voltage that changes from −1 V to +1 V in 10 μs,

(c) worst-case output offset voltage of 200 mV, and

(d) R_1 of at least 1 kΩ.

SPICE EXERCISES

10–46. The operational amplifier in Figure 10–42 has $v_i = 2\sin(2\pi \times 10^3 t)$ V. Use SPICE to obtain a plot of the output waveform versus time over two full periods. Use the results to determine the closed-loop voltage gain and compare that with the theoretical value for an ideal amplifier. Use the LM324 op-amp.

10–47. An operational amplifier is connected in a noninverting configuration with R_1 = 10 kΩ and R_f = 90 kΩ. The open-loop voltage gain of the amplifier is 1×10^3 and its open-loop output resistance is 150 Ω. If the input impedance is 200 kΩ, use SPICE to find the closed-loop voltage gain, output resistance, and input resistance. Compare your results with the theoretical values.

CHAPTER 11

ADVANCED OPERATIONAL AMPLIFIER APPLICATIONS

OUTLINE

OBJECTIVES

- Learn the principles and limitations of practical op-amp circuits for performing integration and differentiation of signals.
- Analyze and understand phase-lag and phase-lead circuits using operational amplifiers.
- Develop an understanding of the theory and applications of instrumentation amplifiers, with emphasis to their noise-cancelling property.
- Implement different filter structures based on design tables with emphasis to performing scaling on normalized filters.
- Understand the theory and application of logarithmic amplifiers.

11–1 ELECTRONIC INTEGRATION

An *electronic integrator* is a device that produces an output waveform whose value at any instant of time equals the total *area under the input* waveform up to that point in time. (For those familiar with mathematical integration, the process produces the time-varying function $\int_0^t v_i(t)\,dt$.) Hereafter, we will use the abbreviated symbol $\int$ to represent integration. To illustrate this concept, suppose the input to an electronic integrator is the dc level E volts, which is first connected to the integrator at an instant of time we will call $t = 0$. Refer to Figure 11–1. The plot of the dc "waveform" versus time is simply a horizontal line at level E volts, because the dc voltage is constant. The more time that we allow to pass, the greater the area that accumulates under the dc waveform. At any time-point t, the total area under the input waveform between time 0 and time t is (height) $\times$ (width) $= Et$ volts, as illustrated in the figure. For example, if $E = 5$ V dc, then the output will be 5 V at $t = 1$ s, 10 V at $t = 2$ s, 15 V at $t = 3$ s, and so forth. We see that the output is the *ramp* voltage $v_o(t) = Et$.

When the input to an integrator is a dc level, the output will rise linearly with time, as shown in Figure 11–1, and will eventually reach the maximum possible output voltage of the amplifier. Of course, the integration process ceases at that time. If the input voltage goes negative for a certain interval of time, the total area during that interval is *negative* and subtracts from whatever positive area had previously accumulated, thus reducing the output voltage. Therefore, the input must periodically go positive and negative to prevent the output of an integrator from reaching its positive or negative limit. We will explore this process in greater detail in a later discussion on waveshaping.

Figure 11–2 shows how an electronic integrator is constructed using an operational amplifier. Note that the component in the feedback path is capacitor C and that the amplifier is operated in an inverting configuration. Besides the usual ideal-amplifier assumptions, we are assuming *zero input offset,* because any input dc level would be integrated as shown in Figure 11–1 and would eventually cause the amplifier to saturate. Thus, we show an *ideal-integrator* circuit. Using the standard symbol $\int_0^t v\,dt$ to represent integration

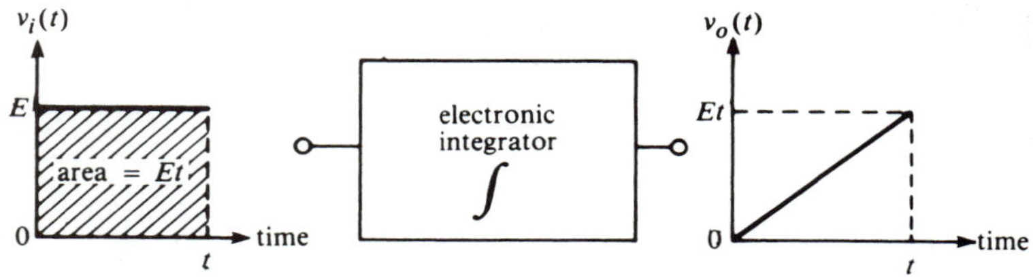

FIGURE 11–1 The output of the integrator at t seconds is the area Et under the input waveform

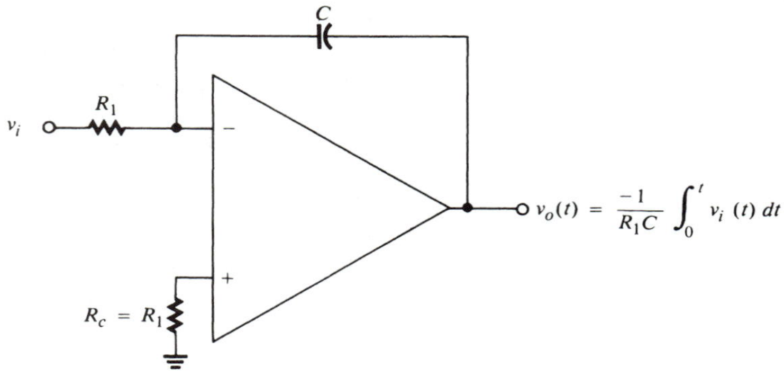

FIGURE 11–2 An ideal electronic integrator

of the voltage v between time 0 and time t, we can show that the output of this circuit is

$$v_o(t) = \frac{-1}{R_1C}\int_0^t v_i\,dt \tag{11–1}$$

This equation shows that the output is the (inverted) integral of the input, multiplied by the constant $1/R_1C$. If this circuit were used to integrate the dc waveform shown in Figure 11–1, the output would be a negative-going ramp ($v_o = -Et/R_1C$).

Readers unfamiliar with calculus can skip the present paragraph, in which we demonstrate why the circuit of Figure 11–2 performs integration. Because the current into the − input is 0, we have, from Kirchhoff's current law,

$$i_1 + i_C = 0 \tag{11–2}$$

where i_1 is the input current through R_1 and i_C is the feedback current through the capacitor. Since $v = 0$, the current in the capacitor is

$$i_C = C\frac{dv_o}{dt} \tag{11–3}$$

Thus

$$\frac{v_i}{R_1} + C\frac{dv_o}{dt} = 0 \tag{11–4}$$

or

$$\frac{dv_o}{dt} = \frac{-1}{R_1C}v_i \tag{11–5}$$

Integrating both sides with respect to t, we obtain

$$v_o = \frac{-1}{R_1C}\int_0^t v_i\,dt \tag{11–6}$$

It can be shown, using calculus, that the mathematical integral of the sine wave $A \sin \omega t$ is

$$\int (A \sin \omega t)\,dt = \frac{-A}{\omega}\sin(\omega t + 90°) = \frac{-A}{\omega}\cos(\omega t)$$

Therefore, when the input to the inverting integrator in Figure 11–2 is $v_i = A \sin \omega t$, the output is

$$v_o = \frac{-1}{R_1 C}\int (A \sin \omega t)\, dt = \frac{-A}{\omega R_1 C}(-\cos \omega t) \qquad \textbf{(11–7)}$$

$$= \frac{A}{\omega R_1 C}\cos \omega t$$

The most important fact revealed by equation 11–7 is that the output of an integrator with sinusoidal input is a sinusoidal waveform whose *amplitude is inversely proportional to its frequency.* This observation follows from the presence of ω ($= 2\pi f$) in the denominator of (11–7). For example, if a 100-Hz input sine wave produces an output with peak value 10 V, then, all else being equal, a 200-Hz sine wave will produce an output with peak value 5 V.

Note also that the output *leads* the input by 90°, regardless of frequency, since $\cos \omega t = \sin(\omega t + 90°)$. However, as shown by equation 11–7, this phase relation is due to both integration and the inverting action of the amplifier. In many practical applications, noninverting integrator circuits are used to introduce phase *lag* into a system:

$$\int (\sin x)\, dx = -\cos x = \sin(x - 90°).$$

EXAMPLE 11–1

1. Find the peak value of the output of the ideal integrator shown in Figure 11–3. The input is $v_i = 0.5 \sin(100t)$ V.
2. Repeat, when $v_i = 0.5 \sin(10^3 t)$ V.

Solution

1. From equation 11–7,

$$v_o = \frac{A}{\omega R_1 C}\cos(\omega t) = \frac{0.5}{100(10^5)(10^{-8})}\cos(100t)$$

$$= 5 \cos(100t) \text{ V} \qquad \text{peak value} = 5 \text{ V}$$

2. From equation 11–7,

$$v_o = \frac{0.5}{1000(10^5)(10^{-8})}\cos(1000t)$$

$$= 0.5 \cos(1000t) \text{ V} \quad \text{peak value} = 0.5 \text{ V}$$

FIGURE 11–3 (Example 11–1)

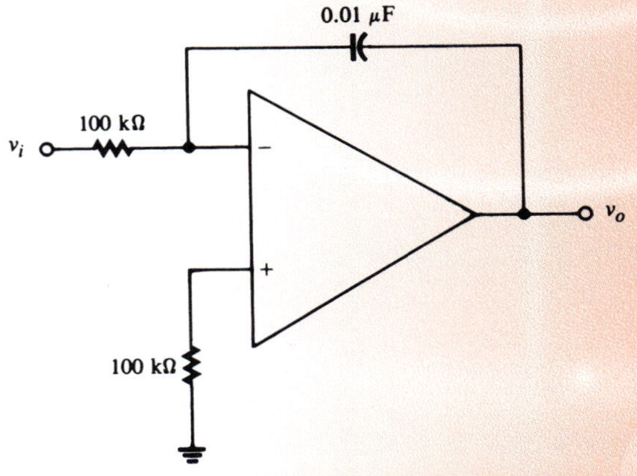

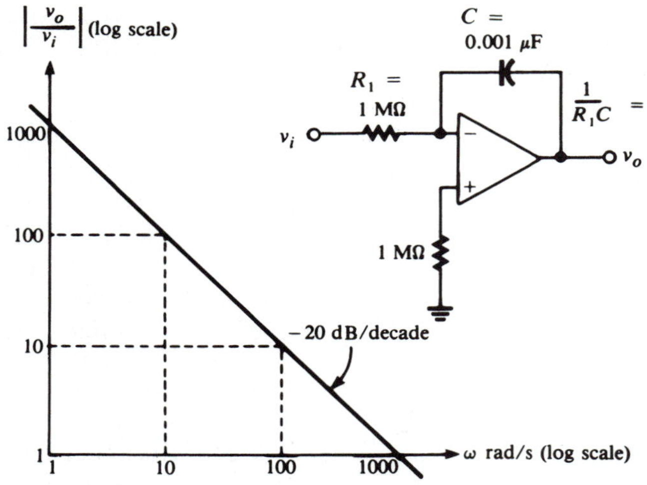

FIGURE 11–4 Bode plot of the gain of an ideal integrator for the case $R_1C = 0.001$

Example 11–1 shows that increasing the frequency by a factor of 10 causes a decrease in output amplitude by a factor of 10. This familiar relationship implies that a Bode plot for the gain of an ideal integrator will have slope −20 dB/decade, or −6 dB/octave. Gain magnitude is the ratio of the peak value of the output to the peak value of the input:

$$\left|\frac{v_o}{v_i}\right| = \frac{\left(\dfrac{A}{\omega R_1 C}\right)}{A} = \frac{1}{\omega R_1 C} \qquad (11\text{–}8)$$

This equation clearly shows that gain is inversely proportional to frequency. A Bode plot for the case $R_1C = 0.001$ is shown in Figure 11–4.

Care must be exercised, however, in making sure that the values of R_1 and C would not put the integrator's Bode plot beyond the open-loop frequency response. Recall that the sloped portion of the open-loop frequency response crosses the unity-gain value at the frequency f_u, which is also the gain–bandwidth product (GBP) of the operational amplifier. Since the magnitude of v_o/v_i for the integrator is $1/\omega R_1 C$, it is clear that the unity-gain angular frequency for the integrator is $1/R_1C$, which is equivalent to $1/2\pi R_1 C$ in Hz. So if R_1 and C are such that this frequency is larger than the GBP, the resulting Bode plot would have to be to the right of the open-loop frequency response, which is not possible. Therefore, the allowable region for an integrator's Bode plot is the area to the left of the sloped portion of the open-loop frequency response of the operational amplifier, as indicated in Figure 11–5. Clearly, in order to accomplish this, the relationship

$$1/(2\pi R_1 C) < \text{GBP} \qquad (11\text{–}9)$$

must be satisfied.

Because the integrator's output amplitude decreases with frequency, it is a kind of low-pass filter. It is sometimes called a *smoothing* circuit, because the amplitudes of high-frequency components in a complex waveform are reduced, thus smoothing the jagged appearance of the waveform. This feature is useful for reducing high-frequency noise in a signal. Integrators are also used in *analog computers* to obtain real-time solutions to differential (calculus) equations.

Practical Integrators

Although high-quality, precision integrators are constructed as shown in Figure 11–2 for use in low-frequency applications such as analog computers,

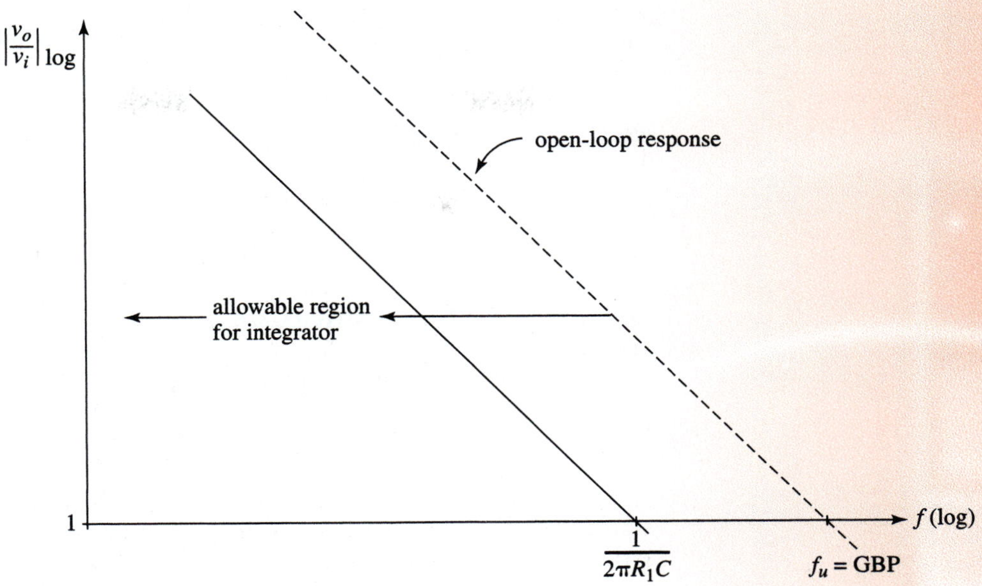

FIGURE 11–5 Allowable region of operation for an op-amp integrator

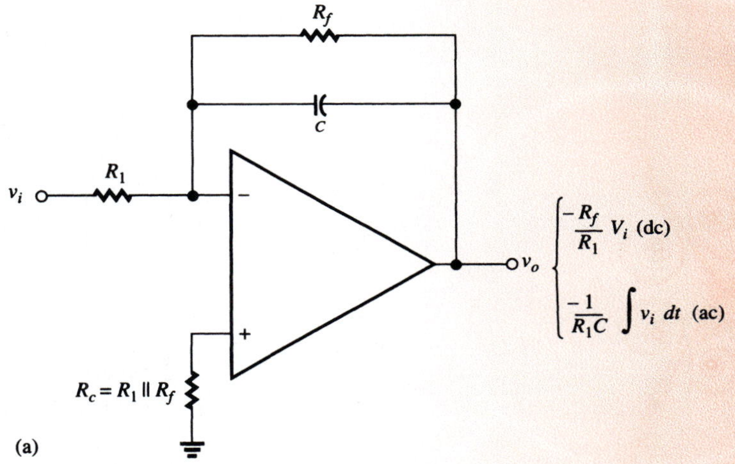

FIGURE 11–6(a) A resistor R_f connected in parallel with C causes the practical integrator to behave like an inverting amplifier to dc inputs and like an integrator to high-frequency ac inputs

$$\begin{cases} -\dfrac{R_f}{R_1}\,V_i \ \text{(dc)} \\[2ex] -\dfrac{1}{R_1C}\displaystyle\int v_i\,dt \ \text{(ac)} \end{cases}$$

(a)

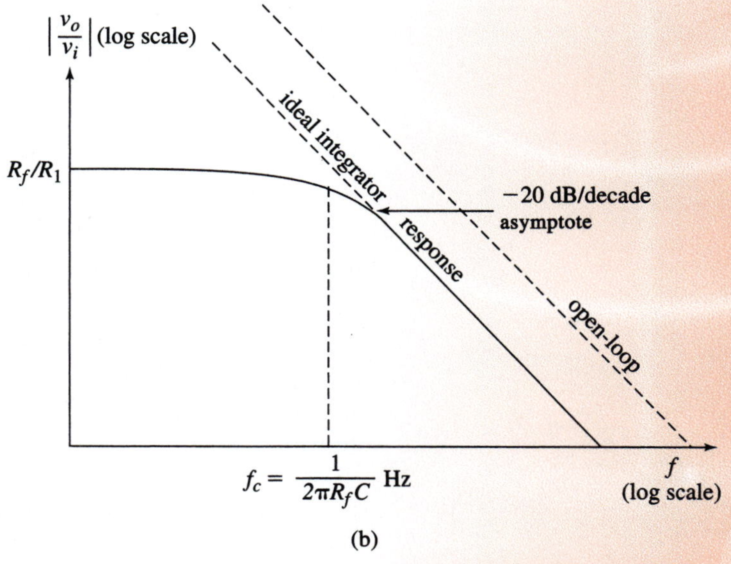

FIGURE 11–6(b) Bode plot for the practical or ac integrator, showing that integration occurs at frequencies well above $1/(2\pi R_f C)$

$$f_c = \frac{1}{2\pi R_f C} \ \text{Hz}$$

(b)

these applications require high-quality amplifiers with extremely small offset voltages or chopper stabilization. As mentioned earlier, any input offset is integrated as if it were a dc signal input and will eventually cause the amplifier to saturate. To eliminate this problem in practical integrators using general-purpose operational amplifiers, a resistor is connected in parallel with the feedback capacitor, as shown in Figure 11–6(a). Because the capacitor is an open circuit as far as dc is concerned, the integrator responds to dc inputs just as if it were an inverting amplifier. In other words, the *dc closed-loop gain* of the integrator is $-R_f/R_1$. At high frequencies, the impedance of the capacitor is much smaller than R_f, so the parallel combination of C and R_f is essentially the same as C alone, and signals are integrated as usual. These integrators are hence called *ac integrators.*

While the feedback resistor in Figure 11–6(a) prevents integration of dc inputs, it also degrades the integration of low-frequency signals. At frequencies where the capacitive reactance of C is comparable in value to R_f, the net feedback impedance is not predominantly capacitive and true integration does not occur. As a rule of thumb, we can say that satisfactory integration will occur at frequencies much greater than the frequency at which $X_C = R_f$. That is, for integrator action we want

$$X_C \ll R_f$$

$$\frac{1}{2\pi f C} \ll R_f$$

or

$$f \gg \frac{1}{2\pi R_f C} \qquad \textbf{(11–10)}$$

The frequency f_c, where X_C equals R_f,

$$f_c = \frac{1}{2\pi R_f C} \qquad \textbf{(11–11)}$$

defines a *break frequency* in the Bode plot of the practical integrator. As shown in Figure 11–6(b), at frequencies well above f_c, the gain falls off at the rate of -20 dB/decade, like that of an ideal integrator, and at frequencies below f_c, the gain approaches its dc value of R_f/R_1.

EXAMPLE 11–2

DESIGN

Design a practical integrator that

1. integrates signals with frequencies down to 100 Hz, and
2. produces a peak output of 0.1 V when the input is a 10-V-peak sine wave having frequency 10 kHz.

Find the dc component in the output when there is a +50-mV dc input.

Solution

In order to integrate frequencies down to 100 Hz, we require $f_c \ll$ 100 Hz. Let us choose f_c one decade below 100 Hz: f_c = 10 Hz. Then, from equation 11–11,

$$f_c = 10 = \frac{1}{2\pi R_f C}$$

Choose $C = 0.01\ \mu\text{F}$. Then

$$10 = \frac{1}{2\pi R_f(10^{-8})}$$

or

$$R_f = \frac{1}{2\pi(10)(10^{-8})} = 1.59\ \text{M}\Omega$$

To satisfy requirement 2, we must choose R_1 so that the gain at 10 kHz is

$$\left|\frac{v_o}{v_i}\right| = \frac{0.1\ \text{V}}{10\ \text{V}} = 0.01$$

Assuming that we can neglect R_f at this frequency (3 decades above f_c), the gain is the same as that for an ideal integrator, given by equation 11–8:

$$\left|\frac{v_o}{v_i}\right| = \frac{1}{\omega R_1 C} = 0.01$$

Thus

$$\frac{1}{2\pi \times 10^4 R_1(10^{-8})} = 0.01$$

or

$$R_1 = \frac{1}{(2\pi \times 10^4)(0.01)(10^{-8})} = 159\ \text{k}\Omega$$

The required circuit is shown in Figure 11–7. Note that $R_c = (1.59\ \text{M}\Omega) \| (159\ \text{k}\Omega) = 145\ \text{k}\Omega$.

When the input is 50 mV dc, the output is 50 mV times the dc closed-loop gain:

$$v_o = \frac{-R_f}{R_1}(50\ \text{mV}) = \frac{-1.59\ \text{M}\Omega}{159\ \text{k}\Omega}(50\ \text{mV}) = -0.5\ \text{V}$$

When we say "ac integrator" we are not implying an electronic circuit for integrating sine waves exclusively. Any periodic waveform, whether square,

FIGURE 11–7
(Example 11–2)

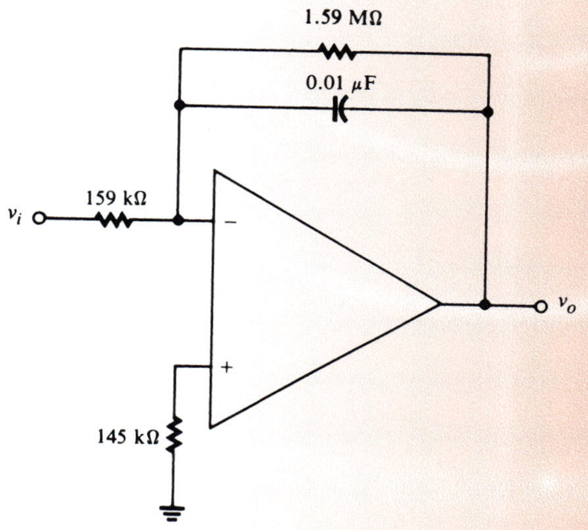

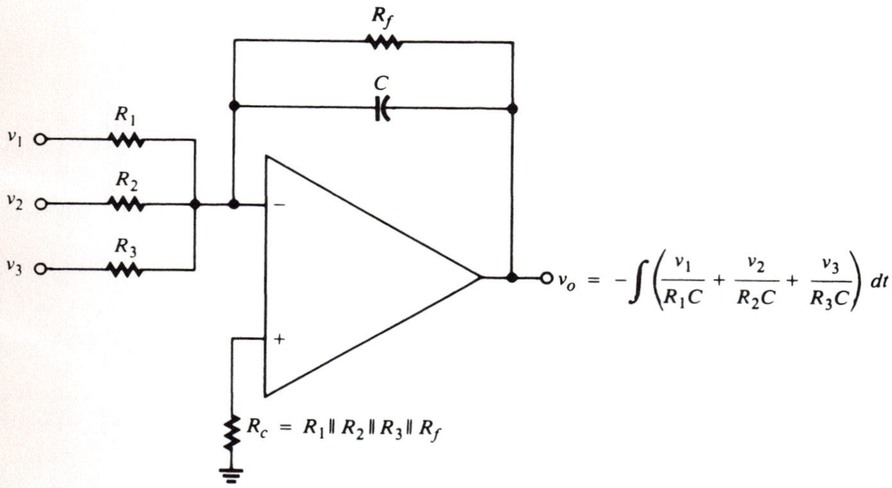

FIGURE 11–8 A three-input integrator

triangular, sawtooth, or any other, are ac signals as well. As long as the frequency of the waveform is at least a decade above the break frequency f_c, the periodic input waveform will be properly integrated. In this case, of course, we need to compute separately the ac and dc components as indicated in Figure 11–6(a). Be aware, however, that equation 11–8 should only be used with sinusoidal input.

In the next chapter, we will make use of ac integrators for obtaining a triangular wave from a square wave. We can also obtain a pseudo-sine wave by integrating a triangular waveform. We will see that in those cases it is easier to obtain the amplitude of the output waveform by computing the positive area of the ac portion of the input signal. Remember that integration of a function yields the *area under the curve.*

In closing our discussion of integrators, we should note that it is possible to scale and integrate several input signals simultaneously, using an arrangement similar to the linear combination circuit studied earlier. Figure 11–8 shows a practical, three-input integrator that performs the following operation at frequencies above f_c:

$$v_o = -\int\left(\frac{1}{R_1C}v_1 + \frac{1}{R_2C}v_2 + \frac{1}{R_3C}v_3\right) dt \tag{11–12}$$

Equation 11–12 is equivalent to

$$v_o = \frac{-1}{R_1C}\int v_1\, dt - \frac{1}{R_2C}\int v_2\, dt - \frac{1}{R_3C}\int v_3\, dt \tag{11–13}$$

If $R_1 = R_2 = R_3 = R$, then

$$v_o = \frac{-1}{RC}\int (v_1 + v_2 + v_3)\, dt \tag{11–14}$$

11–2 ELECTRONIC DIFFERENTIATION

An electronic differentiator produces an output waveform whose value at any instant of time is equal to the *rate of change* of the input at that point in time. In many respects, differentiation is just the opposite, or inverse, operation of integration. In fact, integration is also called "antidifferentiation." If we were to pass a signal through an ideal integrator in cascade with an

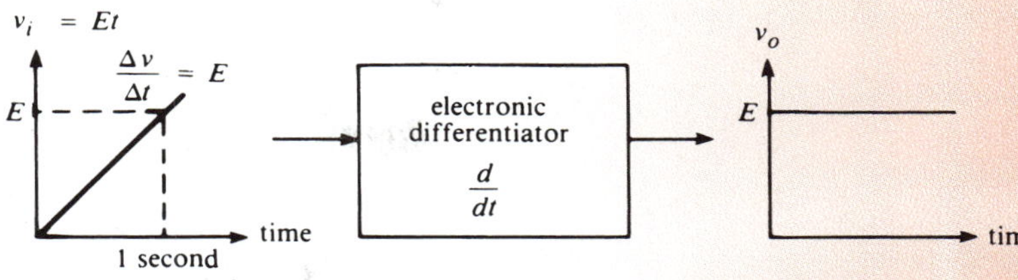

FIGURE 11–9 The ideal electronic differentiator produces an output equal to the rate of change of the input. Because the rate of change of a ramp is constant, the output in this example is a dc level.

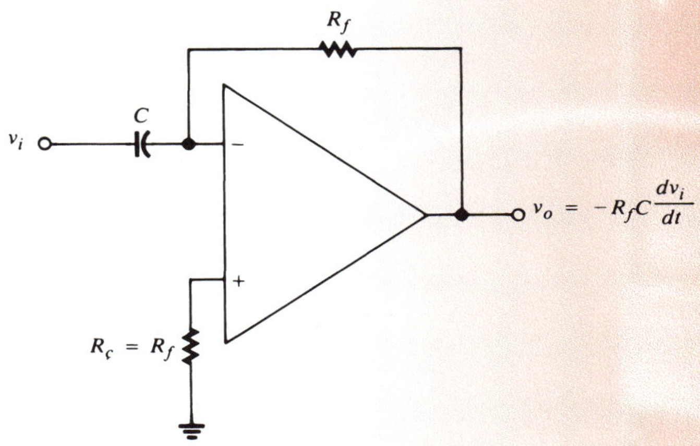

FIGURE 11–10 An ideal electronic differentiator

ideal differentiator, the final output would be exactly the same as the original input signal.

Figure 11–9 demonstrates the operation of an ideal electronic differentiator. In this example, the input is the ramp voltage $v_i = Et$. The rate of change, or slope, of this ramp is a constant E volt/second. (For every second that passes, the signal increases by an additional E volt.) Because the rate of change of the input is constant, we see that the output of the differentiator is the constant dc level E volts.

The standard symbol used to represent differentiation of a voltage v is dv/dt. (This should not be interpreted as a fraction. The dt in the denominator simply means that we are finding the rate of change of v with respect to *time*, t.) In the example shown in Figure 11–9, we would write

$$\frac{dv_i}{dt} = \frac{d(Et)}{dt} = E$$

Note that the derivative of a constant (dc level) is *zero,* because a constant does not change with time and therefore has zero rate of change.

Figure 11–10 shows how an ideal differentiator is constructed using an operational amplifier. Note that we now have a capacitive input and a resistive feedback—again, just the opposite of an integrator. It can be shown that the output of this differentiator is

$$v_o = -R_f C\frac{dv_i}{dt} \qquad \textbf{(11–15)}$$

Thus, the output voltage is the (inverted) derivative of the input, multiplied by the constant R_fC. If the ramp voltage in Figure 11–9 were applied to the input of this differentiator, the output would be a negative dc level.

Readers not familiar with calculus can skip the present paragraph, in which we show how the circuit of Figure 11–10 performs differentiation. Because the current into the – terminal is 0, we have, from Kirchhoff's current law,

$$i_C + i_f = 0 \qquad \textbf{(11–16)}$$

Since $v^- = 0$, $v_c = v_i$ and

$$i_C = C\frac{dv_i}{dt} \tag{11–17}$$

Also, $i_f = v_o/R_f$, so

$$C\frac{dv_i}{dt} + \frac{v_o}{R_f} = 0$$

or

$$v_o = -R_f C\frac{dv_i}{dt} \tag{11–18}$$

It can be shown, using calculus, that

$$\frac{d(A \sin \omega t)}{dt} = A\omega \cos \omega t \tag{11–19}$$

Therefore, when the input to the differentiator in Figure 11–10 is $v_i = A \sin \omega t$, the output is

$$v_o = -R_f C\frac{d(A \sin \omega t)}{dt} = -A\omega R_f C \cos(\omega t) = A\omega R_f C \sin(\omega t - 90°) \tag{11–20}$$

Equation 11–20 shows that when the input is sinusoidal, the *amplitude of the output of a differentiator is directly proportional to frequency* (once again, just the opposite of an integrator). Note also that the output lags the input by 90°, regardless of frequency. (In practice, noninverting differentiator circuits are used to introduce phase *lead* into a system.) The gain of the differentiator is

$$\left|\frac{v_o}{v_i}\right| = \frac{A\omega R_f C}{A} = \omega R_f C \tag{11–21}$$

Practical Differentiators

From a practical standpoint, the principal difficulty with the differentiator is that it effectively amplifies an input in direct proportion to its frequency and therefore increases the level of high-frequency noise in the output. Unlike the integrator, which "smooths" a signal by reducing the amplitude of high-frequency components, the differentiator intensifies the contamination of a signal by high-frequency noise. For this reason, differentiators are rarely used in applications requiring high precision, such as in analog computers.

In a practical differentiator, the amplification of signals in direct proportion to their frequencies cannot continue indefinitely as frequency increases, because the amplifier has a finite bandwidth. As we have already discussed in Chapter 10, there is some frequency at which the output amplitude must begin to fall off. Nevertheless, it is often desirable to design a practical differentiator so that it will have a break frequency even lower than that determined by the upper cutoff frequency of the op-amp, that is, to roll off its gain characteristic at some relatively low frequency. This action is accomplished in a practical differentiator by connecting a resistor in series with the input capacitor, as shown in Figure 11–11. We can understand how this modification achieves the stated goal by considering the net impedance of the $R_1 - C$ combination at low and high frequencies:

$$Z_i = R_1 - j/\omega C \tag{11–22}$$

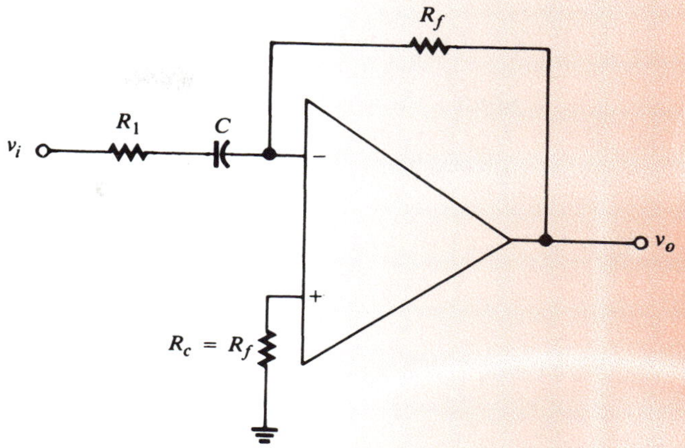

FIGURE 11–11 A practical differentiator. Differentiation occurs at low frequencies, but resistor R_1 prevents high-frequency differentiation

$$|Z_i| = \sqrt{R_1^2 + (1/\omega C)^2} \quad (11\text{–}23)$$

At very small values of ω, Z_i is dominated by the capacitive reactance component, so the combination is essentially the same as C alone, and differentiator action occurs. At very high values of ω, $1/\omega C$ is negligible, so Z_i is essentially the resistance R_1, and the circuit behaves like an ordinary inverting amplifier (with gain R_f/R_1).

The break frequency f_b beyond which differentiation no longer occurs in Figure 11–11 is the frequency at which the capacitive reactance of C equals the resistance of R_1:

$$\frac{1}{2\pi f_b C} = R_1 \quad (11\text{–}24)$$

$$f_b = \frac{1}{2\pi R_1 C}$$

In designing a practical differentiator, the break frequency should be set well above the highest frequency at which accurate differentiation is desired:

$$f_b \gg f_h \quad (11\text{–}25)$$

where f_h is the highest differentiation frequency. Figure 11–12 shows Bode plots for the gain of the ideal and practical differentiators. In the low-frequency region where differentiation occurs, note that the gain rises with frequency at the rate of 20 dB/decade. The plot shows that the gain levels off beyond the break frequency f_b and then falls off at −20 dB/decade beyond the amplifier's upper cutoff frequency. Recall that the closed-loop bandwidth, or upper cutoff frequency of the amplifier, is given by

$$f_2 = \beta f_u \quad (11\text{–}26)$$

where β in this case is $R_1/(R_1 + R_f)$.

In some applications, where very wide bandwidth operational amplifiers are used, it may be necessary or desirable to roll the frequency response off even at a lower frequency than that shown for the practical differentiator in Figure 11–12. This can be accomplished by connecting a capacitor C_f in parallel with the feedback resistor R_f. This modification will cause the response to roll off at −20 dB/decade beginning at the break frequency

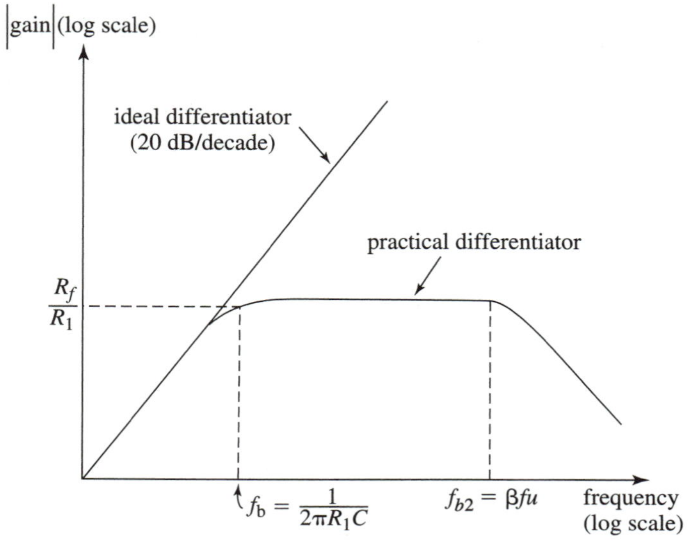

FIGURE 11–12 Bode plots for the ideal and practical differentiators. f_b is the break frequency due to the input R_1-C combination and f_2 is the upper cutoff frequency of the (closed-loop) amplifier.

$$f_{b2} = \frac{1}{2\pi R_f C_f} \tag{11–27}$$

Obviously, f_{b2} should be set higher than f_b.

EXAMPLE 11–3

DESIGN

1. Design a practical differentiator that will differentiate signals with frequencies up to 200 Hz. The gain at 10 Hz should be 0.1.
2. If the operational amplifier used in the design has a unity-gain frequency of 1 MHz, what is the upper cutoff frequency of the differentiator?

Solution

1. We must select R_1 and C to produce a break frequency f_b that is well above $f_h = 200$ Hz. Let us choose $f_b = 10\,f_h = 2$ kHz. Letting $C = 0.1\ \mu$F, we have, from equation 11–24,

$$f_b = 2 \times 10^3 = \frac{1}{2\pi R_1 C}$$

$$R_1 = \frac{1}{2\pi(2 \times 10^3)(10^{-7})} = 796\ \Omega$$

In order to achieve a gain of 0.1 at 10 Hz, we have, from equation 11–21,

$$\left|\frac{v_o}{v_i}\right| = 0.1 = \omega R_f C = (2\pi \times 10)R_f(10^{-7})$$

$$R_f = \frac{0.1}{2\pi \times 10 \times 10^{-7}} = 15.9\ \text{k}\Omega$$

The completed design is shown in Figure 11–13.

2.
$$\beta = \frac{R_1}{R_1 + R_f} = \frac{796}{796 + 15.9\text{ k}} = 0.0477$$

From equation 11–26, $f_2 = \beta f_u = (0.0477)(1\text{ MHz}) = 47.7$ kHz. The Bode plot is sketched in Figure 11–14.

FIGURE 11–13 (Example 11–3)

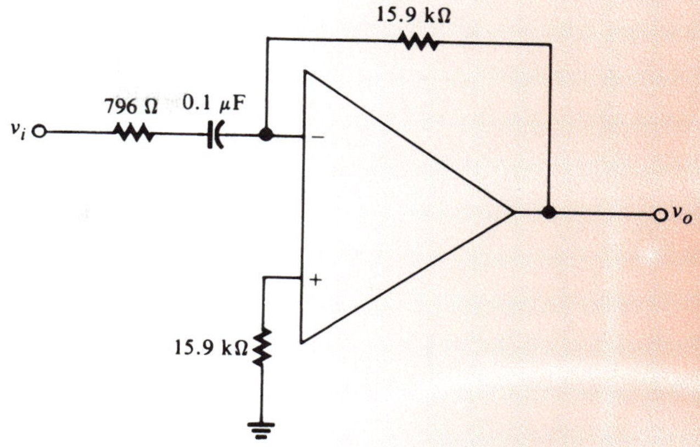

FIGURE 11–14 (Example 11–3)

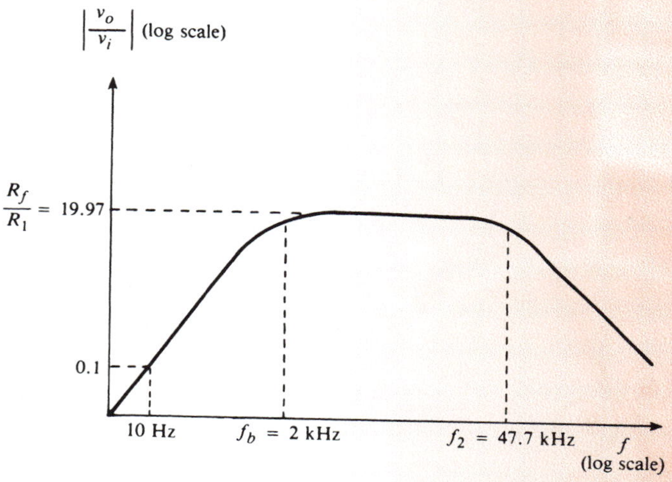

EXAMPLE 11–4

SPICE

Use SPICE to verify the design of the differentiator in Example 11–3. Assume the op-amp is ideal.

Solution

Figure 11–15(a) shows the SPICE circuit and the input data file. The inverting operational amplifier is modeled like the one in Example 10–4, except that we assume zero output resistance and a (nearly) ideal voltage gain having the very large value 1×10^9.

Figure 11–15(b) shows the frequency response computed by PSPICE over the range from 1 Hz to 100 kHz. Because the input signal has amplitude 1 V, the plot represents values of voltage gain. Notice that the gain at 10 Hz is $0.0999 \approx 0.1$, as required. The theoretical gain of the ideal differentiator at 200 Hz is

$$\omega R_f C = (2\pi \times 200)(15.9\ \text{k}\Omega)(0.1\ \mu\text{F}) = 1.998$$

The plot shows that the gain at a frequency close to 200 Hz (200.8 Hz) is 1.996, so we conclude that satisfactory differentiation occurs up to 200 Hz.

The plot shows that the high-frequency gain (at 100 kHz) is 19.97. At very high frequencies, the reactance of the 0.1-μF capacitor is negligible, so the gain, for all practical purposes, equals the ratio of R_f to R_1: 15.9 kΩ/796 Ω = 19.97. Since the break frequency f_b in the design was selected to be 2 kHz, the gain at 2 kHz should be approximately (0.707)(19.97) = 14.12. The plot

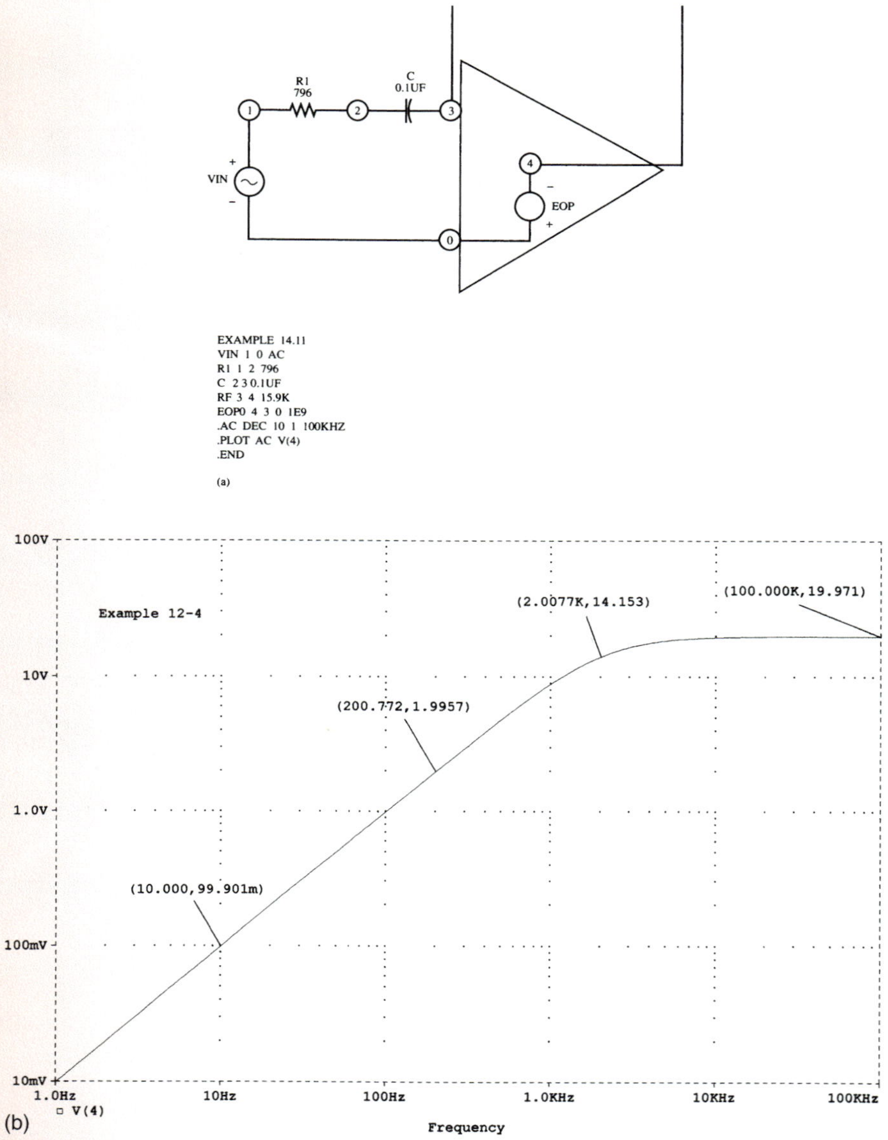

FIGURE 11–15 (Example 11–4)

shows that the gain at a frequency near 2 kHz (2.008 kHz) is 14.15, and we conclude that the design is valid. Because we assumed an ideal operational amplifier (with infinite bandwidth), the cutoff frequency f_2 does not affect the computed response. (Its presence can be simulated by connecting a 210-pF capacitor between nodes 3 and 4.)

11–3 PHASE-SHIFT CIRCUITS

We saw that integrator and differentiator circuits produce constant phase shifts, regardless of frequency, within their normal operational regions. In this section, we will look at circuits that produce phase shifts that depend on frequency but maintain a constant gain. These circuits are also called *constant-delay filters* or *all-pass filters*. Constant delay refers to the fact that

FIGURE 11–16 The phase-lag circuit

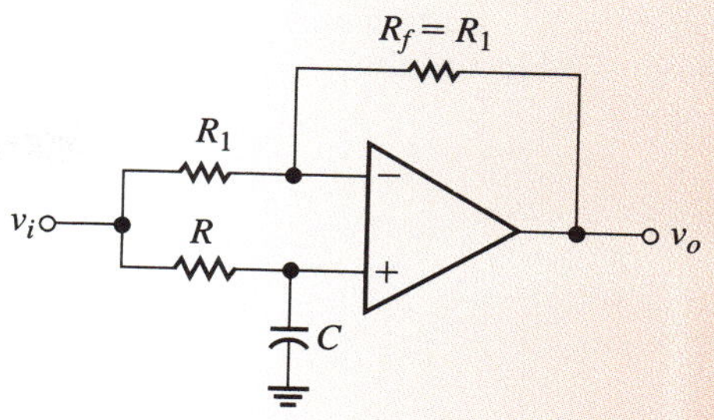

when frequency is changed, the time difference between input and output remains constant over a range of operating frequencies. The all-pass term arises from the constant gain (normally 1) that these circuits maintain at all frequencies within the operating range. We will look at two different circuits, one for producing lagging phase angles and the other for leading phase angles.

Phase-Lag Circuit

Figure 11–16 shows the phase-lag circuit built around an op-amp operating simultaneously in inverting and noninverting modes. To obtain its input-output relationship, we first observe that the input voltage drives a plain inverting amplifier on one side and a noninverting amplifier with a low-pass filter on the other. If we see this circuit as being driven by two separate, identical voltage sources, we can use the principle of superposition to derive the relationship. Note that the inverting gain is -1 and the noninverting gain, after the low-pass circuit, is $1 + R_1/R_1 = 2$. Using the standard low-pass function learned in Chapter 9, we write

$$V_o(j\omega) = -V_i + 2\frac{1}{1 + j\omega RC}V_i \tag{11–28}$$

or

$$V_o(j\omega) = V_i\left(-1 + \frac{2}{1 + j\omega RC}\right) = V_i\left(\frac{1 - j\omega RC}{1 + j\omega RC}\right)$$

The input-output relationship, as a function of jω, is then

$$\frac{V_o}{V_i}(j\omega) = \frac{1 - j\omega RC}{1 + j\omega RC} \tag{11–29}$$

Note that this relationship is complex, and therefore has both magnitude and phase. Also observe that numerator and denominator are complex conjugates. Their magnitudes are hence identical and their ratio is a constant 1. The overall phase equals the angle from the numerator term *minus* the angle from the denominator. (Recall that in a complex division angles subtract.) The phase angle θ is then

$$\theta = -\tan^{-1}(\omega RC/1) - \tan^{-1}(\omega RC/1) = -2\tan^{-1}(\omega RC) \tag{11–30}$$

Note that the phase approaches zero when ω approaches zero. On the other extreme, as ω approaches infinity, the phase angle approaches $-180°$. We can also write equation 11–30 in terms of frequency in hertz as

$$\theta = -2\tan^{-1}(f/f_o) \tag{11–31}$$

where the frequency f_o is

$$f_o = \frac{1}{2\pi RC} \tag{11–32}$$

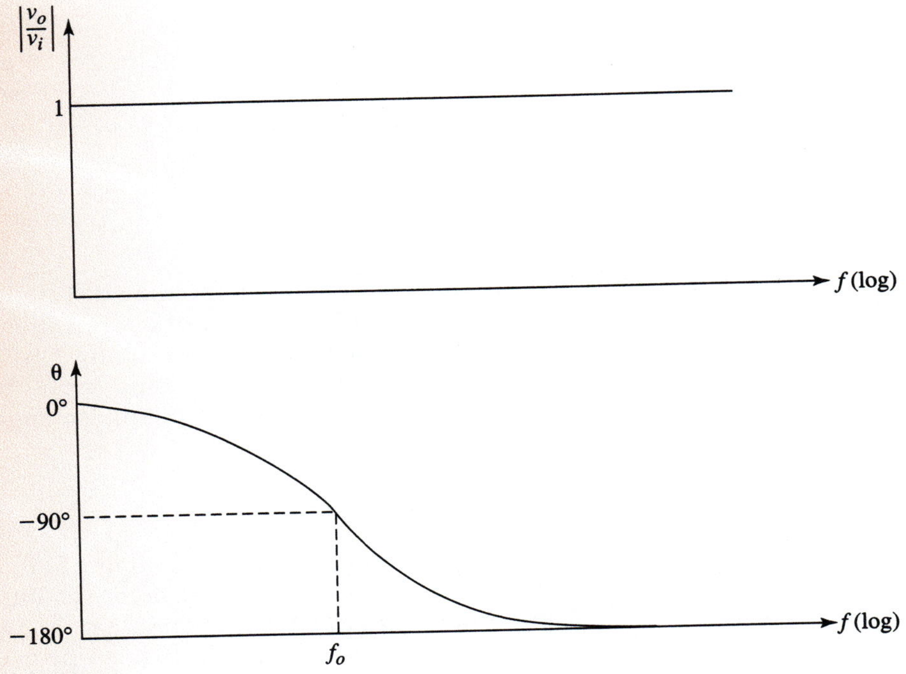

FIGURE 11–17 Bode plot for the phase-lag circuit

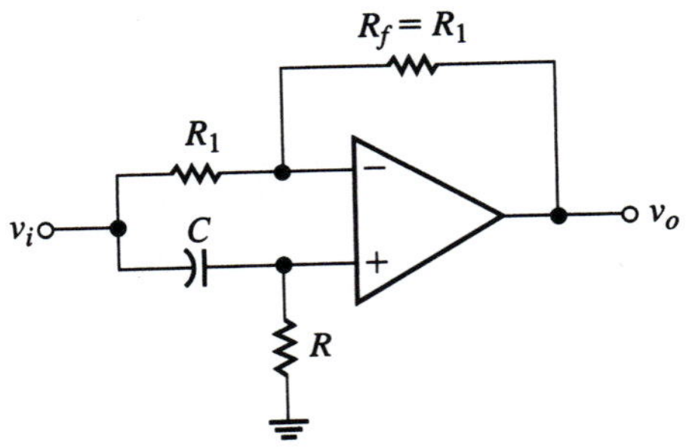

FIGURE 11–18 The phase-lead circuit

Note that when f equals f_o, the angle is $-2\tan^{-1}(1) = -90°$. Figure 11–17 shows the Bode plot for the phase-lag circuit.

Phase-Lead Circuit

Figure 11–18 shows the phase-lead circuit. Applying a similar analysis, we can obtain its input-output relationship. In this case, note that the *RC* circuit forms a high-pass network. Its function was also derived in Chapter 9. The expression for the output voltage is then

$$V_o(j\omega) = -V_i + 2\left(\frac{j\omega RC}{1 + j\omega RC}\right)V_i \qquad \textbf{(11–33)}$$

that yields

$$\frac{V_o}{V_i}(j\omega) = \frac{-1 + j\omega RC}{1 + j\omega RC} \qquad \textbf{(11–34)}$$

Here again, the ratio of magnitudes yields a constant 1. The phase is obtained in the same manner as for the previous case. Note, however, that the numerator has a negative real part making the expression for the overall phase

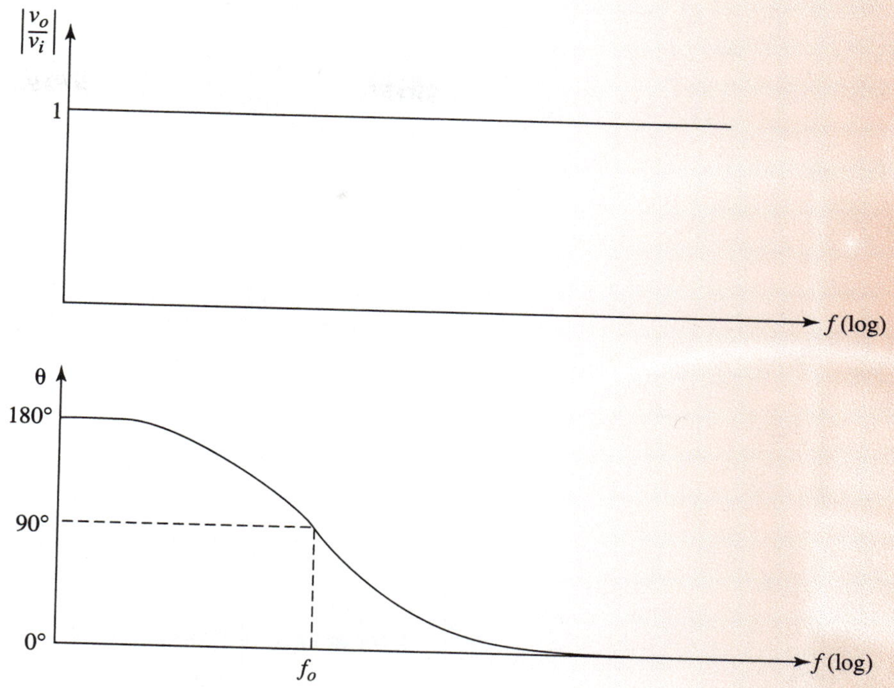

FIGURE 11–19 Bode plot for the phase-lead circuit

FIGURE 11–20 (Example 11–5)

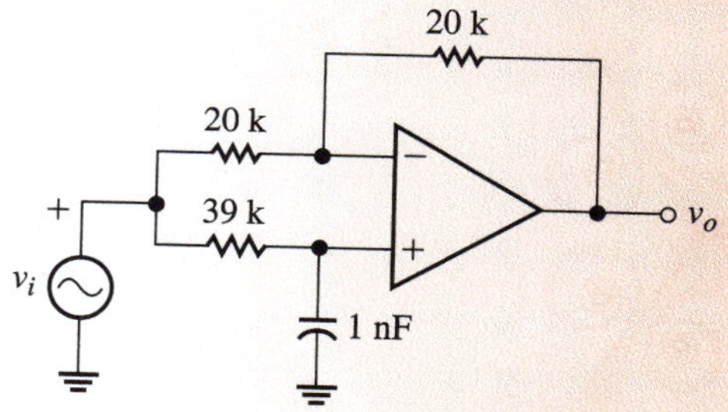

$$\theta = 180° - \tan^{-1}(\omega RC/1) - \tan^{-1}(\omega RC/1) = 180° - 2\tan^{-1}(\omega RC) \quad (11\text{–}35)$$

Observe that the phase angle in this case approaches 180° at frequencies approaching zero. As frequency is increased, the leading phase decreases and finally approaches zero at high frequencies. Equation 11–35 can also be written in terms of frequency in hertz as

$$\theta = 180° - 2\tan^{-1}(f/f_o) \quad (11\text{–}36)$$

where $f_o = \dfrac{1}{2\pi RC}$

as in the previous case. Figure 11–19 shows the Bode plot for the phase-lead circuit.

EXAMPLE 11–5

1. Find the phase angle in the circuit shown in Figure 11–20 for the following frequencies: 1 kHz, 2 kHz, 5 kHz, and 10 kHz.
2. Find the time delays for those same frequencies.

Solution

1. The circuit is a phase-lag network. Let us first calculate $f_o = 1/(2\pi RC) = 1/[2\pi(39\text{ k})(1\text{ nF})] = 4081$ Hz. From Equation 11–31, we calculate the phase angles:

   $f = 1$ kHz: $\theta = -2\tan^{-1}(1\text{ k}/4081) = -27.5°$

   $f = 2$ kHz: $\theta = -2\tan^{-1}(2\text{ k}/4081) = -52.2°$

   $f = 5$ kHz: $\theta = -2\tan^{-1}(5\text{ k}/4081) = -102°$

   $f = 10$ kHz: $\theta = -2\tan^{-1}(10\text{ k}/4081) = -136°$

2. Phase angles are directly proportional to delays. Phase angle corresponds to 360° as delay corresponds to the period, that is,

$$\theta/360° = t_d/T$$

or

$$t_d = \frac{1}{f}\frac{\theta}{360°}$$

Substituting values, the resulting delays are:

$t_d = 76.4\ \mu s$ $(f = 1\text{ kHz})$

$t_d = 72.5\ \mu s$ $(f = 2\text{ kHz})$

$t_d = 56.7\ \mu s$ $(f = 5\text{ kHz})$

$t_d = 37.8\ \mu s$ $(f = 10\text{ kHz})$

As can be seen from these results, delay is not quite constant for this particular frequency range and f_o. Observe that doubling the frequency from 1 kHz to 2 kHz produces a relatively small change as compared to the doubling from 5 kHz to 10 kHz. We conclude that if delays are to be kept fairly uniform, f_o should be larger than the upper frequency limit. Had f_o been placed at 20 kHz, that is, an octave above 10 kHz, the delays would be: 15.9, 15.8, 15.6, and 14.8 μs.

11–4 INSTRUMENTATION AMPLIFIERS

In Chapter 8 (Section 8–2) we saw a voltage subtraction circuit that could perform the general function $v_o = a_1v_1 - a_2v_2$ under a numeric restriction. We also saw that this function can be made of the form $v_o = A(v_1 - v_2)$. This function corresponds to a general differential amplifier whose gain is established by two pair of resistors, as shown in Figure 11–21. The closed-loop differential gain A_d in this circuit is given by

FIGURE 11–21 A basic differential amplifier is built with an op-amp and four resistors

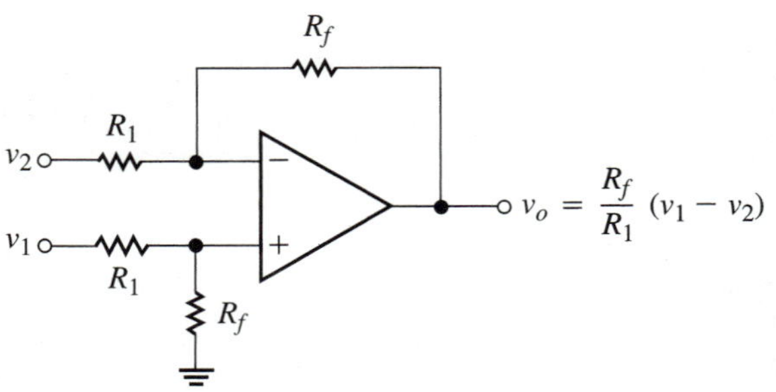

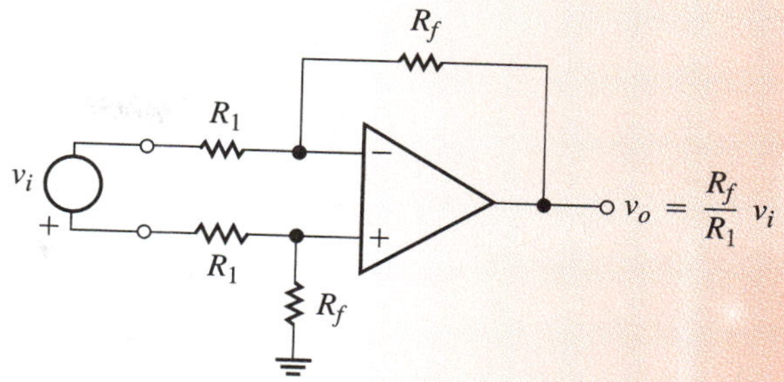

FIGURE 11–22 A floating source connected to a single op-amp differential amplifier

$$A_d = \frac{v_o}{v_1 - v_2} = \frac{R_f}{R_1} \tag{11–37}$$

where the term $v_1 - v_2$ represents the differential input voltage. Remember that an op-amp is in itself a differential amplifier. However, its huge differential gain makes it unsuitable to directly amplify a differential voltage. It is the feedback scheme formed by the four resistors that makes it a stable differential amplifier with finite gain established by the resistor values.

In most practical applications, a differential amplifier is used for amplifying a signal voltage produced by a *floating* voltage source, as shown in Figure 11–22. This scheme is particularly useful when amplifying a signal that has traveled through a long cable exposed to noise and electromagnetic interference. If both signal leads (wires) are exposed identically to the interference, the differential amplifier will eliminate it, amplifying ideally the difference, which is the desired signal. Although this basic configuration is very useful in basic, noncritical applications, it has a few drawbacks. First, the differential input resistance, which is approximately equal to $2R_1$, would not be considered high by good amplifier standards. Second, the common-mode rejection of this differential amplifier would be lower than the CMRR of the op-amp alone due to mismatch of the resistors. It is true that by inserting a voltage follower in each input line the differential input resistance would be made very large, but this does not cure the reduction in CMRR.

Figure 11–23 shows an improved configuration for producing an output proportional to the difference between two inputs. Notice that the circuit is basically the difference amplifier just discussed, with the addition of two input stages. Each input signal is connected directly to the noninverting terminal of an operational amplifier, so each signal source sees a very large input resistance. This circuit arrangement is so commonly used that it is called an *instrumentation amplifier* (IA) and is commercially available by that name in single-package units. These devices use closely matched, high-quality amplifiers and have very large common-mode rejection ratios. This is typically achieved during fabrication by laser-trimming the resistors to tolerances under 0.01%.

In our analysis of the instrumentation amplifier, we will refer to Figure 11–24, which shows current and voltage relations in the circuit. We begin by noting that the usual assumption of ideal amplifiers allows us to equate v_i^+ and v_i^- at each input amplifier ($v_i^+ - v_i^- \approx 0$), with the result that input voltages v_1 and v_2 appear across adjustable resistor R_A in Figure 11–23. For analysis purposes, let us assume that $v_1 > v_2$. Then the current i through R_A is

$$i = \frac{v_1 - v_2}{R_A} = \frac{v_{o1} - v_{o2}}{R_A + 2R} \tag{11–38}$$

Voltages v_{o1} and v_{o2} are the input voltages to the output differential stage. Since the resistors connected to that stage are all equal, the differential gain will be 1, that is,

$$v_o = v_{o1} - v_{o2} \tag{11–39}$$

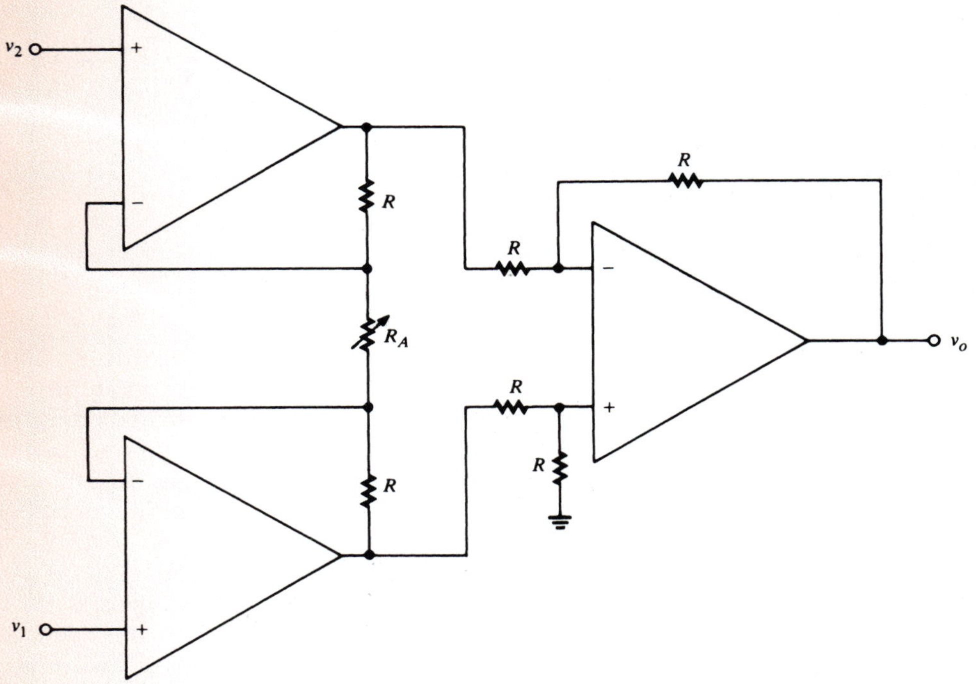

FIGURE 11–23 An instrumentation amplifier that produces an output proportional to $v_1 - v_2$. Adjustable resistor R_A is used to set the gain.

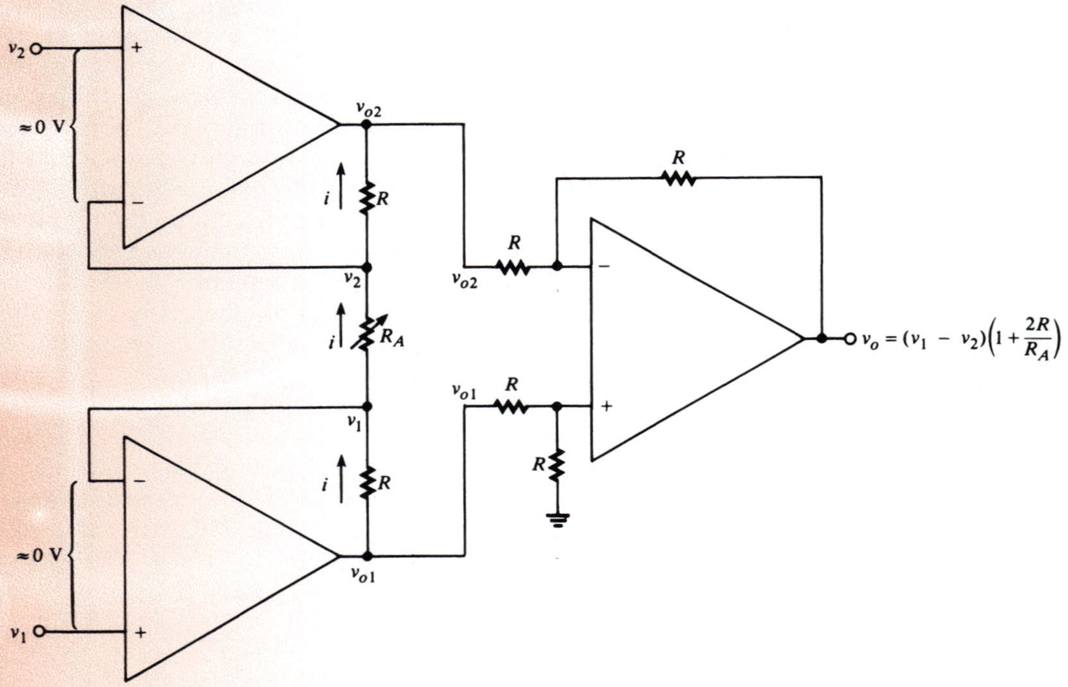

FIGURE 11–24 Voltage and current relations in the instrumentation amplifier. Note that the overall gain is inversely proportional to the value of adjustable resistor R_A.

Substituting (11–38) into (11–39) gives

$$v_o = (v_1 - v_2)(1 + 2R/R_A) \qquad (11\text{–}40)$$

Equation 11–40 shows that the output of the instrumentation amplifier is directly proportional to the difference voltage $(v_1 - v_2)$, as required. The overall closed-loop differential gain is clearly $(1 + 2R/R_A)$. R_A is made adjustable so that gain can be easily adjusted for calibration purposes. Note that the gain is inversely proportional to R_A. This resistor is the only external resistor provided by the user in a commercial IA.

The output voltage can also be expressed as

$$v_o = A_d(v_1 - v_2) \qquad (11\text{–}41)$$

where

$$A_d = 1 + \frac{2R}{R_A} \qquad (11\text{–}42)$$

is the IA's differential gain.

EXAMPLE 11–6

Assuming ideal amplifiers, find the minimum and maximum output voltage V_o of the instrumentation amplifier shown in Figure 11–25 when the 10-kΩ potentiometer R_p is adjusted through its entire range.

Solution

Referring to Figure 11–23, we see that R_A in this example is the sum of R_p and the fixed 500-Ω resistor. Assuming that R_p can be adjusted through a full range from 0 to 10 kΩ, R_A (min) = 500 Ω and R_A(max) = (500 Ω) + (10 kΩ) = 10.5 kΩ. Therefore, from equation 11–40, the minimum and maximum values of A_d are

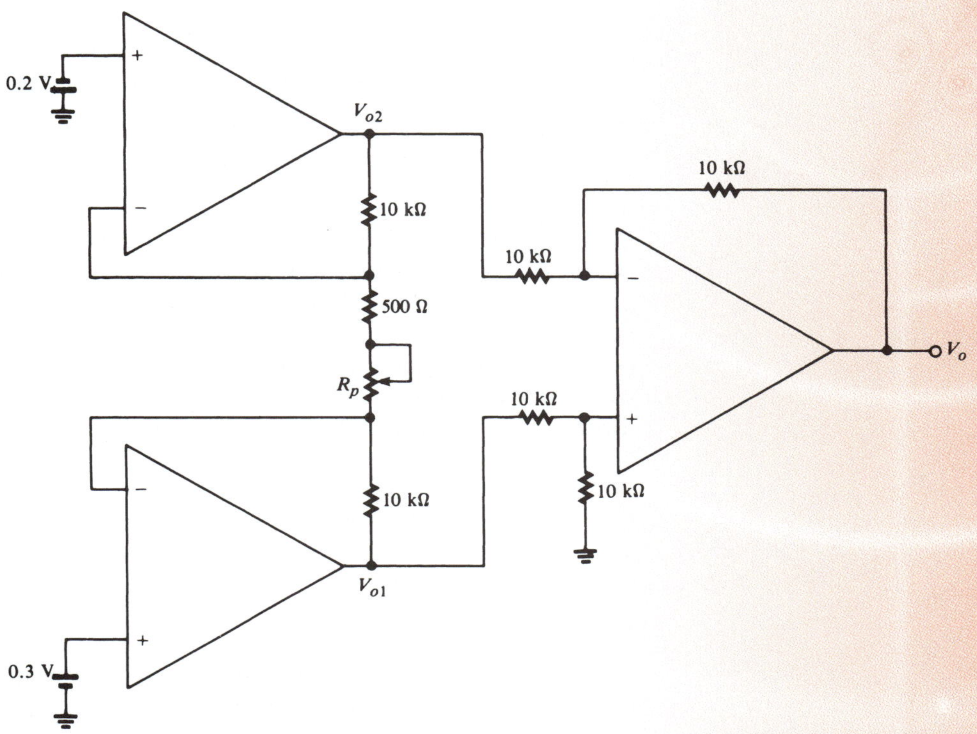

FIGURE 11–25 (Example 11–6)

$$A_d(\min) = 1 + \frac{2R}{R_A(\max)} = 1 + \frac{2(10\ \text{k}\Omega)}{10.5\ \text{k}\Omega} = 2.96$$

$$A_d(\max) = 1 + \frac{2R}{R_A(\min)} = 1 + \frac{2(10\ \text{k}\Omega)}{500\ \Omega} = 41$$

The minimum and maximum output voltages are then

$$v_o(\min) = A_d(\min)(v_1 - v_2) = 2.96[0.3 - (-0.2)] = 1.48\ \text{V}$$
$$v_o(\max) = A_d(\max)(v_1 - v_2) = 41(0.5) = 20.5\ \text{V}$$

EXAMPLE 11–7

DESIGN

An application requires the use of an instrumentation amplifier with adjustable differential gain from 20 dB to 30 dB. Determine minimum and maximum values for the external resistor R_A required to obtain such gain range. Indicate how to implement the variable-resistance circuit. Internal resistors are 20 kΩ each.

Solution

Converting the gains from dB to plain magnitude, we obtain

$$A_d(\min) = 10^{20/20} = 10 \qquad \text{and} \qquad A_d(\max) = 10^{30/20} = 31.6$$

Solving for R_A from equation (11–42), we obtain

$$R_A = \frac{2R}{A_d - 1}$$

Substituting values,

$$R_A(\min) = 1.31\ \text{k}\Omega \qquad \text{and} \qquad R_A(\max) = 4.44\ \text{k}\Omega$$

Actual implementation of the circuit for adjustable gain would require a fixed resistor of 1.31 kΩ and a potentiometer or rheostat of 3.13 kΩ connected in series. A 3.13-kΩ potentiometer would not be commercially available, so one possibility could be connecting an 8.2-kΩ resistor in parallel with a 5-kΩ potentiometer.

A very common application of instrumentation amplifiers is in the amplification of relatively weak signals contaminated by noise, such as those from thermocouples and microphones with long connecting cables. Thermocouples are industrial sensors that generate a small dc voltage (a few millivolts) that is proportional to the temperature they are sensing. They have two leads (+ and −) that connect to a shielded extension cable, which in turn connects to a central console or temperature-measuring instrument. The shield is normally grounded to reduce electromagnetic interference and other noise on the cable, which could extend over hundreds, even thousands, of feet.

In order to substantially reduce the noise picked up by the connecting leads of the signal source, the amplification must be differential, that is, the signal source must be connected to the instrumentation amplifier in a floating mode. If a noisy source is connected to a single-ended amplifier (not differential), both signal and noise will be amplified equally, so there is no improvement in the signal-to-noise ratio. Incidentally, any differential amplifier can be operated single-ended by simply grounding one of the differential inputs. The foregoing leads to one of the most important issues pertaining to IAs: their very high common-mode rejection ratio, or CMRR. CMRR is defined as the ratio of the differential gain to the common-mode gain, that is,

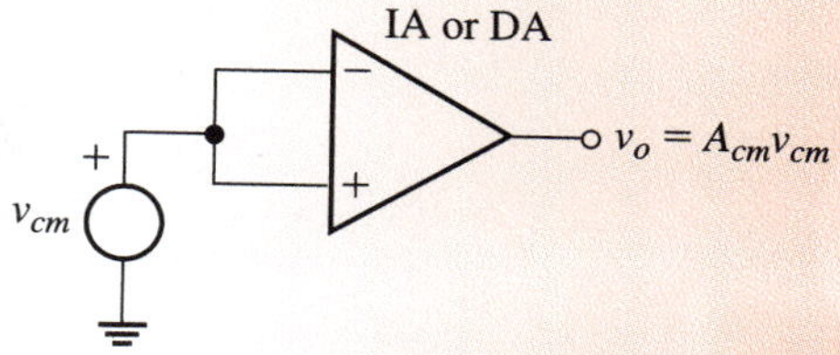

FIGURE 11–26 Measuring common-mode gain

$$\text{CMRR} = \frac{A_d}{A_{cm}} \tag{11–43}$$

or, in dB,

$$\text{CMRR}_{\text{dB}} = 20 \log \frac{A_d}{A_{cm}} \tag{11–44}$$

Ideally, a differential amplifier should have a common-mode gain of zero, which is equivalent to having an infinite CMRR. The easiest way to measure A_{cm} in a differential amplifier is by tying together both inputs and driving them with a relatively large ac voltage, as shown in Figure 11–26. The resulting output voltage, which is very small, is measured and divided by the applied input voltage to obtain the common-mode gain. By a "relatively large" input voltage we mean a peak value close to the supply voltage. For example, if $V_{CC} = \pm 15$ V, which is a typical value, the peak input voltage should be about 14 V, or 10 V rms. This is just to obtain a "readable" output voltage.

Figure 11–27 shows a typical connection for amplifying a floating signal in the differential mode. Notice that the diagram shows twisted connecting leads surrounded by a shield. By using twisted leads, both conductors get exposed equally to external disturbance. The shield should be grounded at one end only to avoid what is called a *ground loop*, which could produce an additional hum or noise voltage that would not be rejected by the IA. With knowledge of both differential and common-mode gains, we can determine how much a noisy signal can be "cleaned" by an instrumentation amplifier. Manufacturers never provide the value of A_{cm} because it depends on the value of A_d, which is normally user adjustable. They always specify, however, the CMRR in dB.

An amplified signal has two main components: signal and noise. The signal portion is the input signal voltage multiplied by the differential gain. On the other hand, the noise output level is the input noise level "attenuated" by the common-mode gain A_{cm}, which is much less than 1. Therefore, these two components can be expressed as

$$v_{os} = A_d v_{is} \tag{11–45}$$

and

$$v_{on} = A_{cm} v_{in} \tag{11–46}$$

where subscripts indicate signal (s), noise (n), input (i), and output (o). Note that equation 11–46 is better expressed as

FIGURE 11–27 A signal source connected to an IA with a shielded, twisted-pair cable

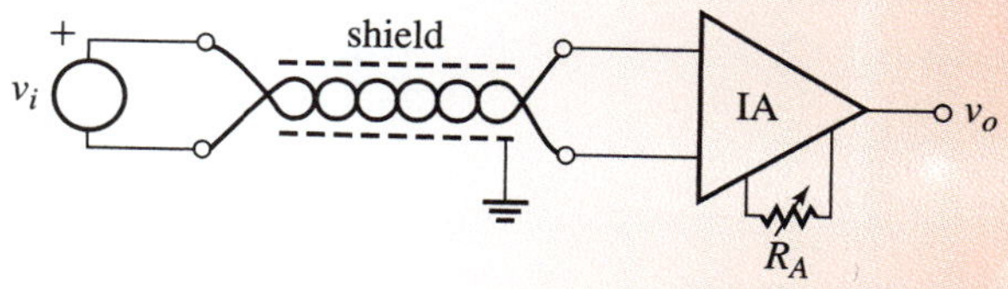

$$v_{on} = \frac{v_{in}A_d}{\text{CMRR}} \quad \textbf{(11–47)}$$

where A_{cm} was replaced with A_d/CMRR. Also note that CMRR is not in dB, which means that a conversion will be necessary if it is given in dB. If that is the case, equation 11–47 can be written as

$$v_{on} = \frac{A_d v_{in}}{10^{\text{CMRR}_{\text{dB}}/20}} \quad \textbf{(11–48)}$$

EXAMPLE 11–8

A 20-mV rms signal, contaminated by a 1.1-V rms common-mode noise voltage, is amplified in differential mode by an instrumentation amplifier with differential gain of 150. The CMRR from the data sheet is 102 dB. Determine the signal and noise output levels.

Solution

The signal output level is

$$v_{os} = A_d v_{is} = 150(20 \text{ mV}) = 3 \text{ volts}$$

The noise output level is

$$v_{on} = (A_d v_{in})/10^{\text{CMRR(dB)}/20} = (150 \times 1.1)/10^{102/20} = 1.31 \text{ mV}$$

Note the tremendous difference the IA makes: A weak signal buried in noise is amplified to a strong level of 3 V, whereas the noise is reduced to an insignificant level of just over a millivolt!

The significance of knowing the CMRR, as illustrated by the previous example, is that it allows the user to predict the noise level of the output signal, or more importantly, the output signal-to-noise ratio. In fact, we can obtain a very simple expression for predicting the output signal-to-noise ratio (SNR_o) from knowledge of the input signal-to-noise ratio (SNR_i) and the CMRR. The obvious definitions of SNR_i and SNR_o are

$$\text{SNR}_\text{i} = v_{is}/v_{in} \quad \textbf{(11–49)}$$

and

$$\text{SNR}_\text{o} = v_{os}/v_{on} \quad \textbf{(11–50)}$$

If we divide (11–50) by (11–49), we obtain

$$\frac{\text{SNR}_\text{o}}{\text{SNR}_\text{i}} = \frac{V_{os}V_{in}}{V_{is}V_{on}}$$

or

$$\frac{\text{SNR}_\text{o}}{\text{SNR}_\text{i}} = \frac{A_d}{A_{cm}} = \text{CMRR} \quad \textbf{(11–51)}$$

or

$$\text{SNR}_\text{o} = \text{SNR}_\text{i}(\text{CMRR}) \quad \textbf{(11–52)}$$

which, in dB, becomes

$$\text{SNR}_\text{o}(\text{dB}) = \text{SNR}_\text{i}(\text{dB}) + \text{CMRR}_{\text{dB}} \quad \textbf{(11–53)}$$

As you can see, the CMRR in dB represents the improvement factor that an instrumentation amplifier provides to the signal-to-noise ratio, also in dB,

FIGURE 11–28 A bridge circuit for detecting small variations of resistance

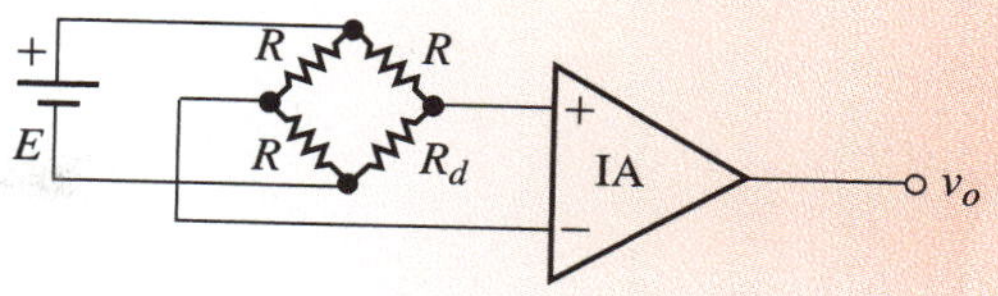

of an incoming signal. The relationship can also serve as a design equation for determining the minimum CMRR required from an IA to meet specific signal-to-noise requirements. For example, if an input signal having $SNR_i = -35$ dB needs to be "cleaned" so as to have an SNR_o of at least +60 dB, then the IA must have a CMRR of no less than 95 dB.

Another popular application of instrumentation amplifiers is in the detection of small variations of resistance in association with a Wheatstone bridge, as shown in Figure 11–28. In this circuit the IA would detect not only a small resistance change but also the direction or sign of the change. The high input impedance of the instrumentation amplifier makes it look like an open circuit to the bridge. In other words, it measures or detects without disturbing the bridge circuit. A very small resistance change, say on the order of milliohms, results in a small voltage change, maybe on the order of millivolts. The instrumentation amplifier then multiplies this small voltage to make it more readable by a voltmeter or a data acquisition system.

11–5 ACTIVE FILTERS

Basic Filter Concepts

A *filter* is a device that allows signals having frequencies in a certain range to pass through it while attenuating all other signals. An *ideal* filter has identical gain at all frequencies in its *passband* and zero gain at all frequencies outside its passband. Figure 11–29 shows the ideal frequency response for each of several commonly used filter types, along with typical responses for

FIGURE 11–29 Ideal and practical frequency responses of some commonly used filter types

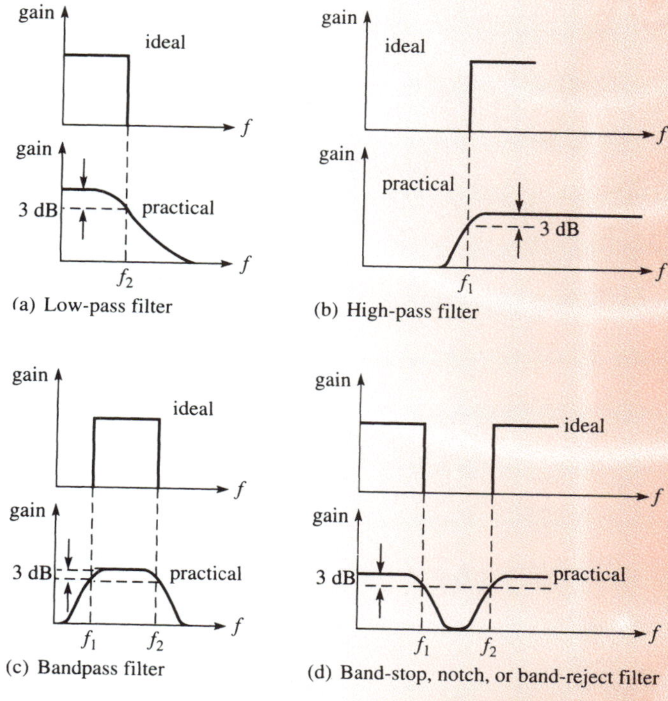

practical filters of the same types. In each case, cutoff frequencies are shown to be those frequencies at which the response is down 3 dB from its maximum value in the passband. Filters are widely used to "extract" desired frequency components from complex signals and/or reject undesired ones, such as noise. *Passive* filters are those constructed with resistors, capacitors, and inductors; *active* filters employ active components such as transistors and operational amplifiers, along with resistor–capacitor networks. Inductors are rarely used with active filters because of their bulk, expense, and lack of availability in a wide range of values.

Filters are classified by their *order*, an integer number, also called the *number of poles*. For example, a second-order filter is said to have two poles. In general, the higher the order of a filter, the more closely it approximates an ideal filter and the more complex the circuitry required to construct it. The frequency response outside the passband of a filter of order n has a slope that is asymptotic to $20n$ dB/decade or $6n$ dB/octave. Recall that a simple RC network is called a *first-order*, or *single-pole*, filter and its response falls off at 20 dB/decade for the low-pass type.

The classification of filters by their order is applicable to all the filter types shown in Figure 11–29. We should note that the response of every high-pass filter must eventually fall off at some high frequency because no physically realizable filter can have infinite bandwidth. The same is true for a band-stop filter. Therefore, every high-pass filter is, technically speaking, a bandpass filter. However, it is often the case that a filter behaves like a high-pass for all those frequencies encountered in a particular application, so for all practical purposes it can be treated as a true high-pass filter. The order of a high-pass filter affects its frequency response in the vicinity of f_1 (Figure 11–29(b)) and is not relevant to the high-frequency region, where it must eventually fall off.

Filters are also classified as belonging to one of several specific design types that, like order, affect their response characteristics within and outside of their pass bands. The two categories we will study are called the *Butterworth* and *Chebyshev* types. The gain magnitudes of low- and high-pass Butterworth filters obey the relationships

$$|G| = \frac{M}{\sqrt{1 + (f/f_2)^{2n}}} \quad \text{(low-pass)} \tag{11–54}$$

$$|G| = \frac{M}{\sqrt{1 + (f_1/f)^{2n}}} \quad \text{(high-pass)} \tag{11–55}$$

where M is a constant, n is the order of the filter, f_1 is the cutoff frequency of the high-pass filter, and f_2 is the cutoff frequency of the low-pass filter. M is the maximum gain approached by signals in the passband. Figure 11–30 shows low- and high-pass Butterworth responses for several different values of the order n. As frequency f approaches zero, note that the gain of the low-pass filter approaches M. For very large values of f, the gain of the high-pass filter approaches M. The value of M is 1, or less, for passive filters. An active filter is capable of amplifying signals having frequencies in its passband and may therefore have a value of M greater than 1. M is called the *gain* of the filter.

The Butterworth filter is also called a *maximally flat* filter because there is relatively little variation in the gain of signals having different frequencies within its passband. For a given order, the Chebyshev filter has greater variation in the passband than the Butterworth design but falls off at a steeper rate outside the passband. Figure 11–31 shows a Chebyshev frequency response of the sixth order. Note that the cutoff frequency for this filter is defined to be the frequency at the point of intersection of the response curve

FIGURE 11–30 Frequency response of low-pass and high-pass Butterworth filters with different orders

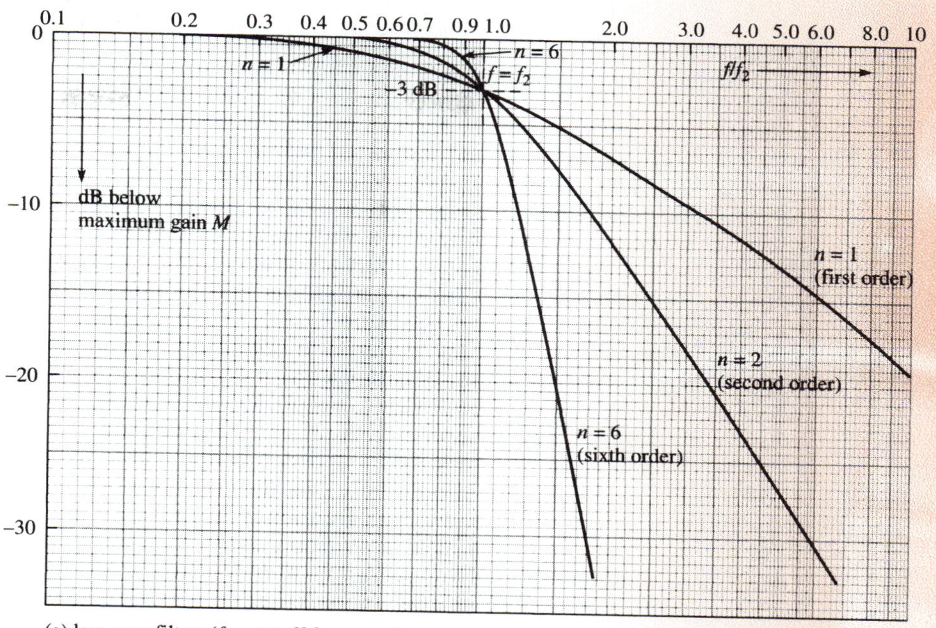

(a) low-pass filters (f_2 = cutoff frequency)

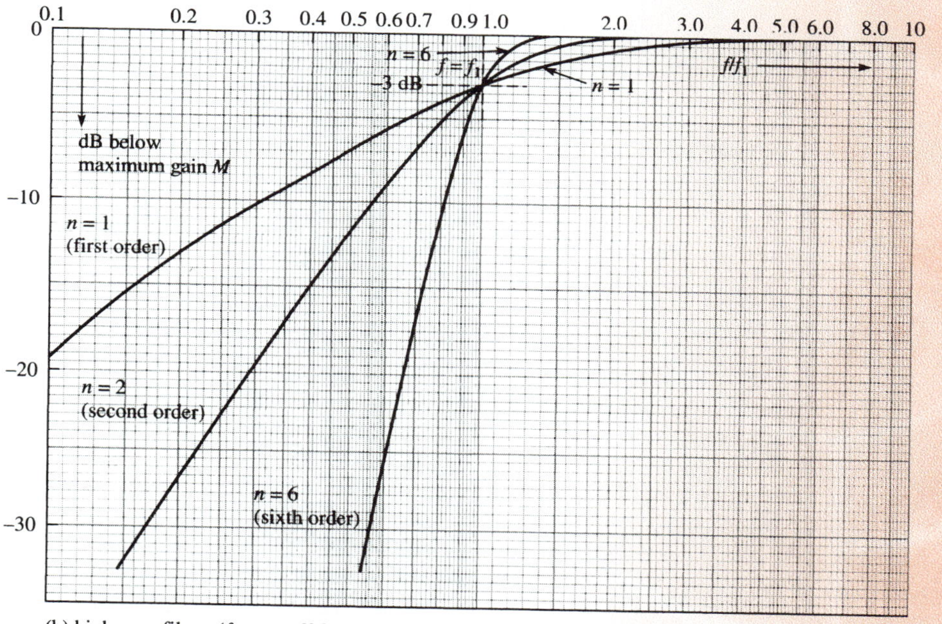

(b) high-pass filters (f_1 = cutoff frequency)

and a line drawn tangent to the lowest gain in the passband. The ripple width (RW) is the total variation in gain within the passband, usually expressed in decibels. A Chebyshev filter can be designed to have a small ripple width, but at the expense of less attenuation outside the passband.

A Chebyshev filter of the same order as a Butterworth filter has a frequency response closer to the ideal filter *outside* the passband, whereas the Butterworth is closer to ideal within the passband. This fact is illustrated in Figure 11–32, which compares the responses of second-order, low-pass Butterworth and Chebyshev filters having the same cutoff frequency. Note that both filters have responses asymptotic to 40 dB/decade outside the

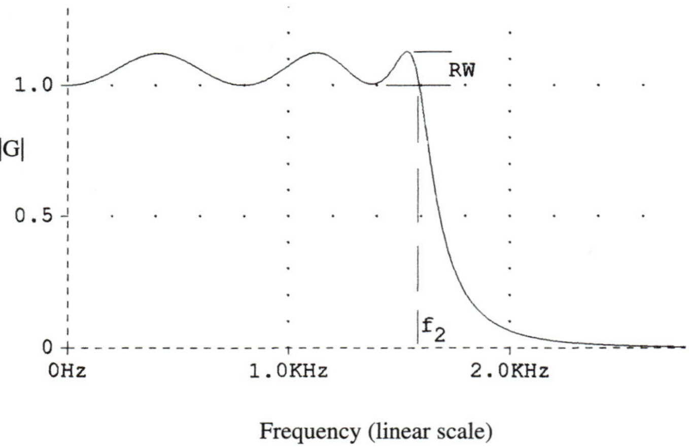

FIGURE 11–31 Chebyshev low-pass frequency response: f_2 = cutoff frequency; RW = ripple width

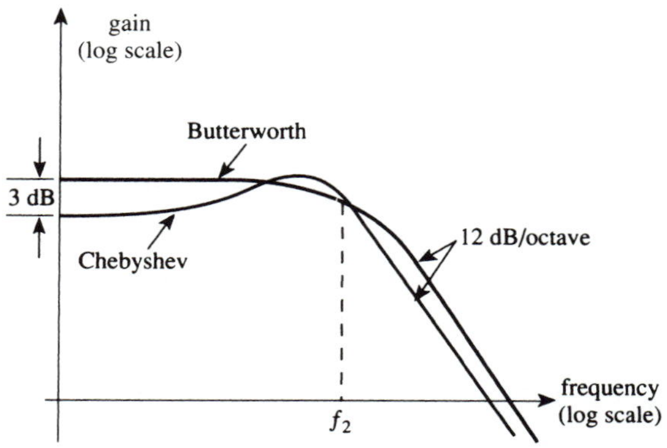

FIGURE 11–32 Comparison of the frequency responses of second-order, low-pass Butterworth and Chebyshev filters

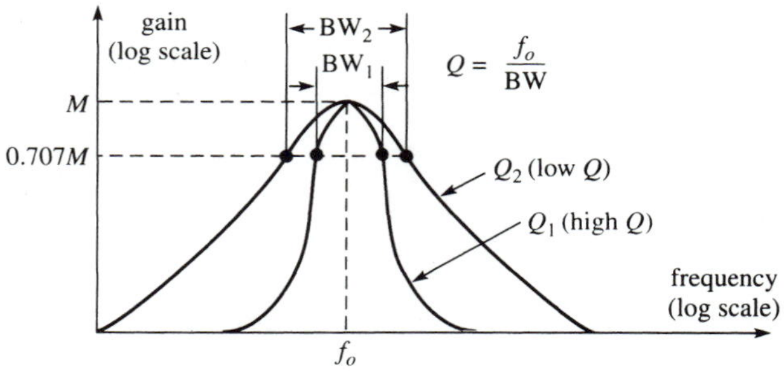

FIGURE 11–33 Comparison of the frequency responses of low-Q and high-Q bandpass filters

passband, but at any specific frequency beyond cutoff, the Chebyshev shows greater attenuation than the Butterworth.

A bandpass filter is characterized by a quantity called its Q (originating from "quality" factor), which is a measure of how rapidly its gain changes at frequencies outside its passband. Q is defined by

$$Q = \frac{f_o}{BW} \tag{11–56}$$

where f_o is the frequency in the passband where the gain is maximum, and BW is the bandwidth measured between the frequencies where the gain is 3 dB down from its maximum value. f_o is often called the *center* frequency because, for many high-Q filters ($Q \geq 10$), it is very nearly in the center of the passband. Figure 11–33 shows the frequency responses of two bandpass filters having the same maximum gain but different values of Q. It is clear that the high-Q filter has a narrower bandwidth and that its response falls off more rapidly on either side of the passband than that of the low-Q filter.

The gain magnitude of any second-order bandpass filter is the following function of frequency f:

$$|G| = \frac{M}{\sqrt{1 + Q^2(f/f_o - f_o/f)^2}} \quad \text{(11–57)}$$

where M is the maximum gain in the band. The center frequency f_o is at the *geometric* center (or geometric *mean*) of the band, defined by

$$f_o = \sqrt{f_1 f_2} \quad \text{(11–58)}$$

where f_1 and f_2 are the lower and upper cutoff frequencies. As previously mentioned, f_o is very nearly midway between the cutoff frequencies (the arithmetic mean) in high-Q filters:

$$f_o \approx f_1 + \frac{\text{BW}}{2} \approx f_2 - \frac{\text{BW}}{2} \quad \text{(11–59)}$$

which implies

$$f_1 \approx f_o - \frac{\text{BW}}{2} \text{ and } f_2 \approx f_o + \frac{\text{BW}}{2} \quad \text{(11–60)}$$

Active Filter Design

The analysis of active filters requires complex mathematical methods that are beyond the scope of this book. We will therefore concentrate on practical design methods that will allow us to construct Butterworth and Chebyshev filters of various types and orders. The discussion that follows is based on design procedures using tables constructed with information from available filter handbooks.

Figure 11–34 shows a general configuration that can be used to construct second-order high-pass and low-pass filters of both the Butterworth and Chebyshev designs. The operational amplifier is basically operated as a noninverting, voltage-controlled voltage source (see Section 8–3), and the configuration is known as the VCVS design. It is also called a *Sallen-Key* circuit. Each impedance block represents a resistance or capacitance, depending on whether a high-pass or low-pass filter is desired. Table 11–1 shows the impedance type required for each design.

FIGURE 11–34 Block diagram of a second-order, VCVS low-pass or high-pass filter

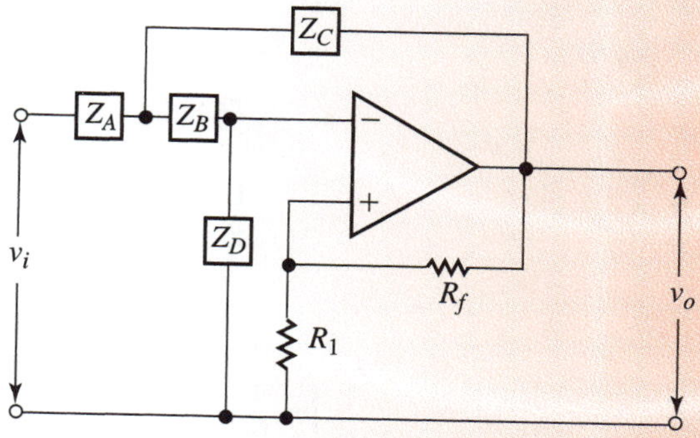

TABLE 11–1 VCVS Filter Components

	Z_A	Z_B	Z_C	Z_D
Low-Pass Filter	R	R	C	C
High-Pass Filter	C	C	R	R

Filter design can be simplified tremendously if the gain in the passband is unity. In that case, the op-amp operates simply as a voltage follower. If gain other than unity is desired, it can be implemented outside the filter block without affecting the filter characteristics. Any high-pass or low-pass filter of any order—Butterworth, Chebyshev, or other types—can be implemented by cascading unity-gain Sallen-Key stages. If the order (number of poles) of the filter is odd, an extra first-order block will be needed. For example, a sixth-order filter requires three cascaded Sallen-Key stages, whereas a seventh-order filter requires a first-order stage in addition to the three Sallen-Key blocks.

In order to obtain component values for active filters, designers can resort to a wide variety of methods: from cookbook formulas, to scaling values from normalized filters, to in-depth analysis and synthesis from scratch. The methods presented here allow the reader to quickly design and test an active filter based on unity-gain Sallen-Key blocks by either using simple formulas or by performing frequency and resistance scaling on *normalized* filters. A filter circuit or function is said to be normalized if its cutoff frequency is either 1 rad/s or 1 Hz. There are many filter handbooks that describe the different filter configurations, their mathematical functions, and different approaches for calculating component values. The reader is encouraged to browse through filter handbooks to have a better sense of the wealth of information available on filter theory and applied design.

Low-Pass Filters

Figure 11–35 shows the general structure we will use for low-pass filters. Note that the first stage, stage 0, is a first-order block and is followed by the Sallen-Key stages, which are of second order. Table 11–2 lists constants for calculating all the capacitor values for different orders, types, and desired cutoff frequency using the low-pass formulas given at the bottom of the table. All resistors, labeled R, were made identical to facilitate the calculations and avoid mistakes. Only Butterworth and Chebyshev filters are given, the latter in two ripple widths: 1 dB and 2 dB. The designer picks the resistance value and then calculates all the capacitor values. Note from the table and coefficients that only odd-ordered filters require a first-order block. Also notice that the formulas for stages 1 and up are all identical, except for the coefficients. Although Figure 11–35 shows the first-order (stage 0) block at the input, the order of the blocks in the cascade configuration is irrelevant.

High-Pass Filters

Figure 11–36 shows the high-pass general structure, which is also used in conjunction with Table 11–2. Note that all capacitors, labeled C, are identical. Using the high-pass formulas, the designer picks the capacitor value and

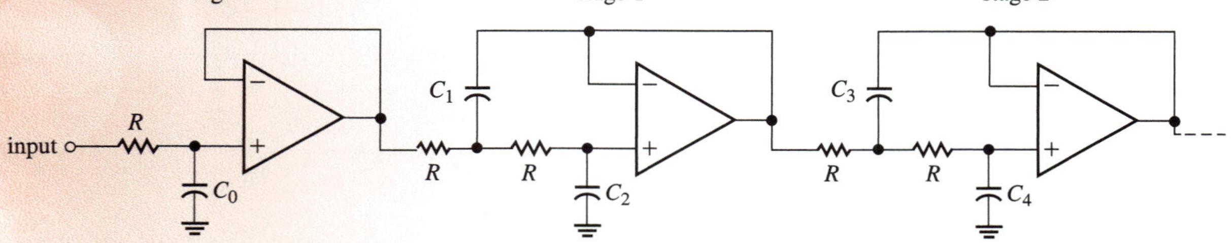

FIGURE 11–35 General low-pass filter structure; even-ordered filters do not use the first stage

TABLE 11–2 Coefficients for Butterworth and Chebyshev Filters

Butterworth	*Stage 0*	*Stage 1*		*Stage 2*		*Stage 3*		*Stage 4*	
	a_0	a_1	a_2	a_3	a_4	a_5	a_6	a_7	a_8
2-pole		1.4142	1.0000						
3-pole	1.0000	1.0000	1.0000						
4-pole		0.76537	1.0000	1.8478	1.0000				
5-pole	1.0000	0.61803	1.0000	1.6180	1.0000				
6-pole		0.51764	1.0000	1.4142	1.0000	1.9319	1.0000		
7-pole	1.0000	0.44504	1.0000	1.2470	1.0000	1.8019	1.0000		
8-pole		0.39018	1.0000	1.1111	1.0000	1.6629	1.0000	1.9616	1.0000
Chebyshev 1-dB	***Stage 0***	***Stage 1***		***Stage 2***		***Stage 3***		***Stage 4***	
	a_0	a_1	a_2	a_3	a_4	a_5	a_6	a_7	a_8
2-pole		1.0977	1.1025						
3-pole	0.49417	0.49417	0.9942						
4-pole		0.67374	0.2794	0.27907	0.98651				
5-pole	0.28949	0.46841	0.4293	0.17892	0.98832				
6-pole		0.12436	0.99073	0.33976	0.55772	0.46413	0.12471		
7-pole	0.20541	0.09142	0.99268	0.25615	0.65346	0.37014	0.23045		
8-pole		0.07002	0.99414	0.19939	0.72354	0.29841	0.34086	0.3520	0.07026
Chebyshev 2-dB	***Stage 0***	***Stage 1***		***Stage 2***		***Stage 3***		***Stage 4***	
	a_0	a_1	a_2	a_3	a_4	a_5	a_6	a_7	a_8
2-pole		0.80382	0.82306						
3-pole	0.36891	0.36891	0.8861						
4-pole		0.20977	0.92868	0.50644	0.22157				
5-pole	0.21831	0.13492	0.95217	0.35323	0.39315				
6-pole		0.09395	0.96595	0.25667	0.53294	0.35061	0.09993		
7-pole	0.15534	0.06913	0.97461	0.19371	0.63539	0.27991	0.21239		
8-pole		0.05298	0.98038	0.15089	0.70978	0.22582	0.3271	0.26637	0.0565

Low-pass Formulas (Figure 11–35)

$$C_0 = \frac{1}{2\pi a_0 f_c R}$$

$$C_1 = \frac{1}{\pi a_1 f_c R} \qquad C_2 = \frac{a_1}{4\pi a_2 f_c R}$$

$$C_3 = \frac{1}{\pi a_3 f_c R} \qquad C_4 = \frac{a_3}{4\pi a_4 f_c R}$$

$$\vdots \qquad \vdots$$

High-pass Formulas (Figure 11–36)

$$R_0 = \frac{a_0}{2\pi f_c C}$$

$$R_1 = \frac{a_1}{4\pi f_c C} \qquad R_2 = \frac{a_2}{\pi a_1 f_c C}$$

$$R_3 = \frac{a_3}{4\pi f_c C} \qquad R_4 = \frac{a_4}{\pi a_3 f_c C}$$

$$\vdots \qquad \vdots$$

then calculates resistance values. It is recommended to pick 10 nF (0.01 μF) and see how the resistors turn out. A quick modification may be necessary if the designer decides that the resulting resistor values are too low or too high. Again, first-order stages are required only for the odd-ordered filters, and the sequence of the blocks in the cascade can be altered without affecting the overall filter.

If the resulting resistor values are too low or too high, *all of them* can be changed provided all the capacitors are also changed in an opposite manner.

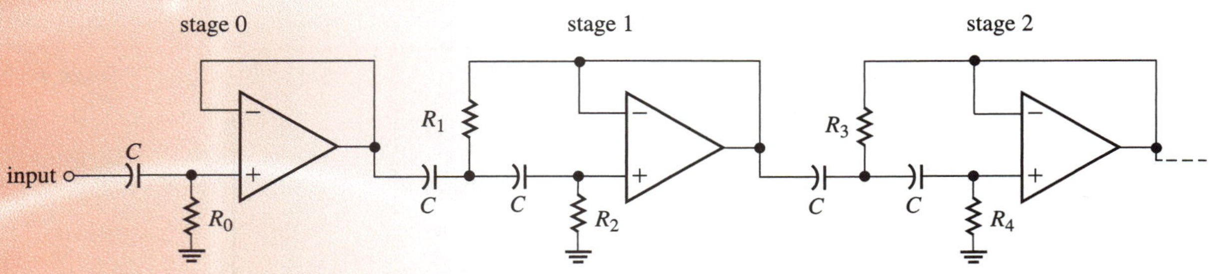

FIGURE 11–36 General high-pass filter structure; even-ordered filters do not use the first stage

For example, if all the resistors are *doubled*, then all the capacitors must be *halved* in order to maintain the same cutoff frequency.

EXAMPLE 11–9

Design a third-order, low-pass Butterworth filter for a cutoff frequency of 2.5 kHz. Select $R = 10\ \text{k}\Omega$.

Solution

From Table 11–2, $a_0 = 1$, $a_1 = 1$, $a_3 = 1$. Substituting into low-pass formulas, we obtain

$$C_0 = \frac{1}{2\pi(2.5\ \text{k})(10\ \text{k})} = 6.37\ \text{nF}$$

$$C_1 = \frac{1}{\pi(2.5\ \text{k})(10\ \text{k})} = 12.7\ \text{nF}$$

$$C_2 = \frac{1}{4\pi(2.5\ \text{k})(10\ \text{k})} = 3.18\ \text{nF}$$

EXAMPLE 11–10

Design a unity-gain, fourth-order, high-pass Chebyshev filter with 2-dB ripple for a cutoff frequency of 800 Hz. Choose $C = 100$ nF.

Solution

From Table 11–2, $a_1 = 0.20977$, $a_2 = 0.92868$, $a_3 = 0.50644$, $a_4 = 0.22157$. Substituting into high-pass formulas, we obtain

$$R_1 = \frac{0.20977}{4\pi(800)(100\ \text{nF})} = 208.7\ \Omega$$

$$R_2 = \frac{0.92868}{\pi(0.20977)(800)(100\ \text{nF})} = 17.62\ \text{k}\Omega$$

$$R_3 = \frac{0.50644}{4\pi(800)(100\ \text{nF})} = 503.8\ \Omega$$

$$R_4 = \frac{0.22157}{\pi(0.50644)(800)(100\ \text{nF})} = 1.741\ \text{k}\Omega$$

Frequency and Impedance Scaling

When the designer resorts to normalized circuits from handbooks, the component values given are for cutoff frequencies of either 1 rad/s or 1 Hz (recall $\omega = 2\pi f$). Normalized resistor and capacitor values are normally 1 Ω and 1 F. A scaling procedure is necessary to determine component values for the desired cutoff frequency. The process is relatively simple. Let us recall that in RC circuits, cutoff frequencies are inversely proportional to both R and C. So in order to change the cutoff frequency of the normalized filter from 1 Hz or 1 rad/s to a higher frequency, we must scale down C, R, or both. However, in practice, we increase the resistance in the process because the 1-ohm resistors would be impractically low. If both frequency and resistance are scaled up, it makes sense to conclude that normalized capacitor values would have to be scaled down by the *product* of the frequency and resistance *scaling factors*, that is,

$$C = \frac{C_{norm}}{K_f K_r} \quad \textbf{(11–61)}$$

where C_{norm} is a normalized capacitor value from the normalized filter, and K_f and K_r are the frequency and resistance scaling factors, respectively. For example, if the normalized filter is for a cutoff frequency of 1 Hz and we want a cutoff frequency of 12 kHz, then K_f will be 12,000 (dimensionless); if we want to scale up (multiply) all resistor values by a factor of 1000, then $K_r = 1000$.

EXAMPLE 11–11

A certain normalized low-pass filter from a handbook shows three 1-ohm resistors and three capacitors with values $C_1 = 0.564$ F, $C_2 = 0.222$ F, and $C_3 = 0.0322$ F. The normalized frequency is 1 Hz. Determine the new capacitor values required for a cutoff frequency of 5 kHz if we use 10-kΩ resistors.

Solution

The scaling factors are

$$K_f = 5 \times 10^3 \text{ and } K_r = 10^4$$

Using equation 11–61, we obtain

$$C_1 = 0.564/(5 \times 10^3 \times 10^4) = 11.3 \text{ nF}$$
$$C_2 = 0.222/5 \times 10^7 = 4.44 \text{ nF}$$
$$C_3 = 0.0322/5 \times 10^7 = 644 \text{ pF}$$

Normalized Low-Pass and High-Pass Filters

Table 11–3 gives capacitor values in farads for low-pass filters from second to eighth order. The designer needs to scale these values for the desired cutoff frequency f_c and selected resistance R. Since the table is normalized for $f_c = 1$ Hz and $R = 1\ \Omega$, the scaling factors are *numerically identical* to f_c and R. Therefore, the capacitor values can simply be obtained by dividing the normalized values from the table (see equation 11–61) by the *absolute value* of the product $f_c R$ without dimensions, that is,

$$C_k = \frac{C_{table}}{|f_c R|} \quad (k = 0, 1, 2, \ldots) \quad \textbf{(11–62)}$$

Table 11–4 is for high-pass filters. Here, the designer selects a value for the capacitance and then calculates resistor values through scaling. Note that scaling in this case is based on scaling down capacitance to a suitable value, say 10 nF, and scaling up frequency to the desired cutoff value. Because resistance is inversely proportional to both capacitance and frequency, then resistance has to be increased in the same proportion that capacitance was decreased, and it needs to be decreased in the same proportion that frequency was increased. The net result: Divide by the product of the frequency and capacitance scaling values, that is,

$$R_k = \frac{R_{table}}{K_f K_c} \quad (k = 0, 1, 2, \ldots) \quad \textbf{(11–63)}$$

TABLE 11–3 Capacitor Values in Farads for Normalized Low-pass Butterworth and Chebyshev Filters (f_c = 1 Hz, R = 1 ohm)

Butterworth	*Stage 0*	*Stage 1*		*Stage 2*		*Stage 3*		*Stage 4*	
	C_0	C_1	C_2	C_3	C_4	C_5	C_6	C_7	C_8
2-pole		0.22508	0.11254						
3-pole	0.15915	0.31831	0.07958						
4-pole		0.41589	0.06091	0.17226	0.14704				
5-pole	0.15915	0.51504	0.04918	0.19673	0.12876				
6-pole		0.61493	0.04119	0.22508	0.11254	0.16477	0.15374		
7-pole	0.15915	0.71524	0.03542	0.25526	0.09923	0.17665	0.14339		
8-pole		0.81580	0.03105	0.28648	0.08842	0.19142	0.13233	0.16227	0.15610
Chebyshev 1-dB	*Stage 0*	*Stage 1*		*Stage 2*		*Stage 3*		*Stage 4*	
	C_0	C_1	C_2	C_3	C_4	C_5	C_6	C_7	C_8
2-pole		0.28998	0.07923						
3-pole	0.32207	0.64413	0.03955						
4-pole		0.47245	0.19189	1.14061	0.02251				
5-pole	0.54978	0.67955	0.08683	1.77906	0.01441				
6-pole		2.55958	0.00999	0.93687	0.04848	0.68582	0.29616		
7-pole	0.77482	3.48184	0.00733	1.24267	0.03119	0.85997	0.12781		
8-pole		4.54599	0.00560	1.59642	0.02193	1.06669	0.06967	0.90429	0.39868
Chebyshev 2-dB	*Stage 0*	*Stage 1*		*Stage 2*		*Stage 3*		*Stage 4*	
	C_0	C_1	C_2	C_3	C_4	C_5	C_6	C_7	C_8
2-pole		0.39600	0.07772						
3-pole	0.43142	0.86284	0.03313						
4-pole		1.51742	0.01797	0.62852	0.18189				
5-pole	0.72903	2.35925	0.01128	0.90114	0.07150				
6-pole		3.38808	0.00774	1.24015	0.03833	0.90787	0.27920		
7-pole	1.02456	4.60451	0.00564	1.64323	0.02426	1.13719	0.10488		
8-pole		6.00811	0.00430	2.10955	0.01692	1.40957	0.05494	1.19499	0.37517

$$C_k = \frac{C_{table}}{K_f K_r} \qquad k = 0, 1, 2, \ldots$$

where K_f = frequency scaling factor
K_r = resistance scaling factor

where K_f and K_c are the frequency and capacitance scaling factors, respectively. Again, because the table is for f_c = 1 Hz and C = 1 F, then K_f and K_c are numerically identical to f_c and C.

If the designer wants to make scaling even easier, K_c (or C) can be selected so that the magnitude of the product f_cC is a submultiple of 10; this way, resistor values are obtained by simply shifting the decimal point to the right in the values from the table. For example, if f_c is 200 Hz, then selecting C = 50 nF will yield a factor for the resistance of $1/(200 \times 50 \times 10^{-9}) = 10^5$, which implies shifting the decimal point 5 places to the right in *all* the normalized values from the table. One practical point: This quick trick is useful when simulating the design with SPICE, but will more than likely result in a capacitance value that would not be commercially available.

TABLE 11–4 Resistor Values in Ohms for Normalized High-pass Butterworth and Chebyshev Filters ($f_c = 1$ *Hz*, $C = 1$ *farad*)

Butterworth	**Stage 0**	**Stage 1**		**Stage 2**		**Stage 3**		**Stage 4**	
	R_0	R_1	R_2	R_3	R_4	R_5	R_6	R_7	R_8
2-pole		0.11254	0.22508						
3-pole	0.15915	0.07958	0.31831						
4-pole		0.06091	0.41589	0.14704	0.17226				
5-pole	0.15915	0.04918	0.51504	0.12876	0.19673				
6-pole		0.04119	0.61493	0.11254	0.22508	0.15374	0.16477		
7-pole	0.15915	0.03542	0.71524	0.09923	0.25526	0.14339	0.17665		
8-pole		0.03105	0.81580	0.08842	0.28648	0.13233	0.19142	0.15610	0.16227
Chebyshev 1-dB	**Stage 0**	**Stage 1**		**Stage 2**		**Stage 3**		**Stage 4**	
	R_0	R_1	R_2	R_3	R_4	R_5	R_6	R_7	R_8
2-pole		0.08735	0.31970						
3-pole	0.07865	0.03932	0.64039						
4-pole		0.05361	0.13200	0.02221	1.12522				
5-pole	0.04607	0.03727	0.29173	0.01424	1.75828				
6-pole		0.00990	2.53586	0.02704	0.52251	0.03693	0.08553		
7-pole	0.03269	0.00727	3.45635	0.02038	0.81204	0.02945	0.19818		
8-pole		0.00557	4.51935	0.01587	1.15507	0.02375	0.36359	0.028101	0.06354
Chebyshev 2-dB	**Stage 0**	**Stage 1**		**Stage 2**		**Stage 3**		**Stage 4**	
	R_0	R_1	R_2	R_3	R_4	R_5	R_6	R_7	R_8
2-pole		0.06397	0.32593						
3-pole	0.05871	0.02936	0.76456						
4-pole		0.01669	1.40920	0.04030	0.13926				
5-pole	0.03475	0.01074	2.24641	0.02811	0.35428				
6-pole		0.00748	3.27271	0.02043	0.66093	0.02790	0.09072		
7-pole	0.02472	0.00550	4.48760	0.01541	1.04409	0.02227	0.24153		
8-pole		0.00422	5.89023	0.01201	1.49732	0.01797	0.46107	0.02120	0.06752

$$R_k = \frac{R_{table}}{K_f K_c} \qquad k = 0, 1, 2, \ldots$$

where K_f = frequency scaling factor
K_c = capacitance scaling factor

EXAMPLE 11–12

Design a third-order, low-pass Butterworth filter for a cutoff frequency of 4 kHz. Select $R = 8.2\ \text{k}\Omega$.

Solution

According to Table 11–3, two stages are needed, including the first-order stage. Using scaling, we obtain (equation 11–61 or 11–62),

$$C_0 = 0.1592/(8200 \times 4000) = 4.85\ \text{nF}$$
$$C_1 = 0.3183/32.8 \times 10^6 = 9.7\ \text{nF}$$
$$C_2 = 0.07958/32.8 \times 10^6 = 2.43\ \text{nF}$$

EXAMPLE 11–13

Design a fourth-order, high-pass Butterworth filter for a cutoff frequency of 150 Hz. Select $C = 10$ nF.

Solution

Two second-order stages are needed, and we need to calculate four resistor values through scaling using the normalized values from Table 11–4. Using equation 11–63, we obtain,

$$R_1 = 0.0609/150 \times 10 \times 10^{-9} = 0.0609/1.5 \times 10^{-6} = 40.6\ \text{k}\Omega$$
$$R_2 = 0.4159/1.5 \times 10^{-6} = 277.3\ \text{k}\Omega$$
$$R_3 = 0.1470/1.5 \times 10^{-6} = 98.0\ \text{k}\Omega$$
$$\text{R}_4 = 0.1723/1.5 \times 10^{-6} = 114.9\ \text{k}\Omega$$

Simple Bandpass Filters

Another configuration that can be used to construct low-pass, high-pass, and bandpass filters is called the *infinite-gain multiple-feedback* (IGMF) design. Figure 11–37 shows the component arrangement used to obtain a second-order bandpass filter. One advantage of the IGMF design is that it requires only a few components. It is also popular because it has good stability and low output impedance. The IGMF filter inverts signals in its passband.

Analysis of this circuit yields the following relationships:

$$\text{BW} = \frac{1}{\pi R_3 C} \tag{11–64}$$

$$G_o = \frac{R_3}{2R_1} \tag{11–65}$$

$$f_o = \frac{1}{2\pi C\sqrt{(R_1 \| R_2)R_3}} \tag{11–66}$$

where f_o is the center frequency, BW is the 3-dB bandwidth, and G_o is the filter gain at f_o. Recall that the "Q" of the circuit equals f_o/BW and that the geometric mean of the cutoff frequencies is f_o. Design can be done easily by using equations 11–64 through 11–66. The designer picks a suitable value for C and then determines R_3 according to desired bandwidth, then R_1 according to the gain, and finally $R_2 \| R_1$ (and hence R_2) to satisfy the center frequency.

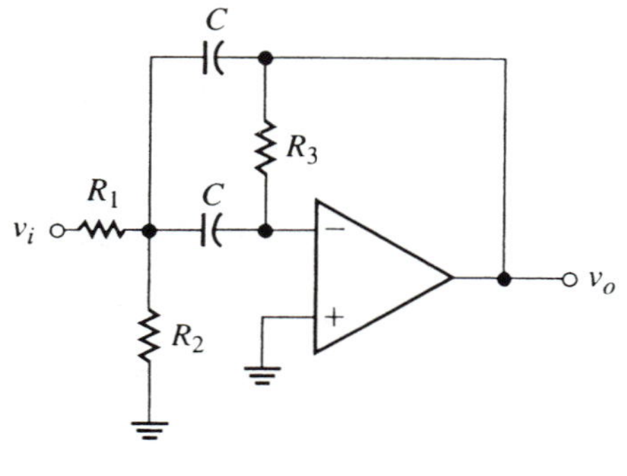

FIGURE 11–37 The IGMF second-order bandpass filter

FIGURE 11–38 (Example 11–14)

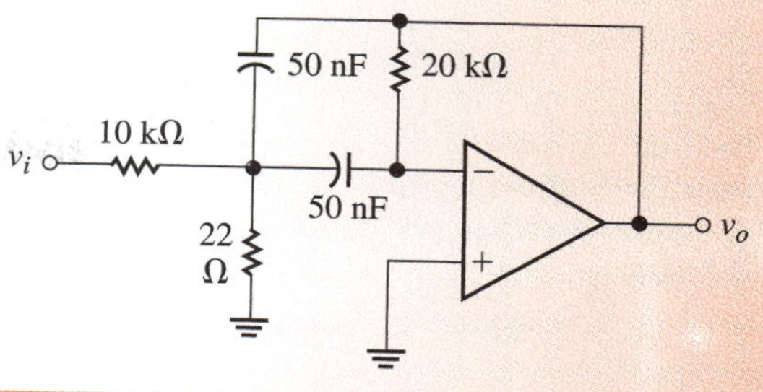

EXAMPLE 11–14

Characterize the bandpass filter shown in Figure 11–38.

Solution

Using equations 11–64 through 11–66, we obtain the following:

$$BW = \frac{1}{\pi(20\text{ k})(50\text{ nF})} = 318\text{ Hz}$$

$$G_o = \frac{20\text{ k}}{2(10\text{ k})} = 1$$

$$f_o = \frac{1}{2\pi(50\text{ nF})\sqrt{(10\text{ k}\,\|\,22)20\text{ k}}} = 4.804\text{ kHz}$$

The Q of this filter is $Q = 4804/318 = 15.1$, which is high. The cutoff frequencies are then given by $f_o \pm BW/2$, that is,

$$f_1 = 4645\text{ Hz} \quad \text{and} \quad f_2 = 4963\text{ Hz}$$

Some applications require bandpass filters with wide passbands but with steep rejections on both sides outside the passband. These filters can be easily implemented by cascading overlapping low-pass and high-pass filters, as shown in Figure 11–39. Each stage is designed separately with any of the methods described previously. Note that the lower cutoff frequency is determined by the high-pass filter, and vice versa.

Band-Stop Filters

Although a band-stop filter can be implemented using the multiple-feedback structure by itself, it can also be obtained by performing a simple

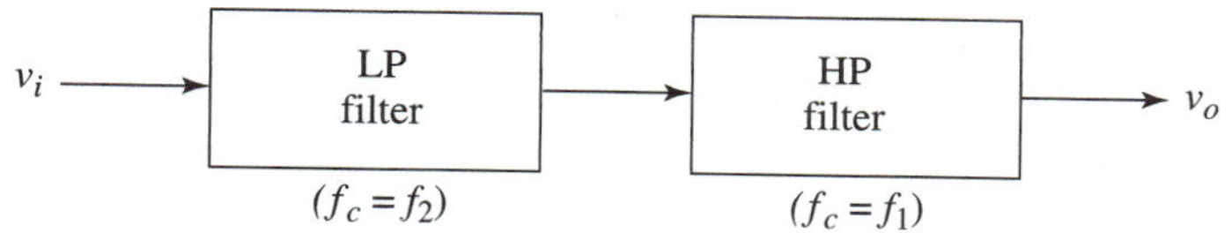

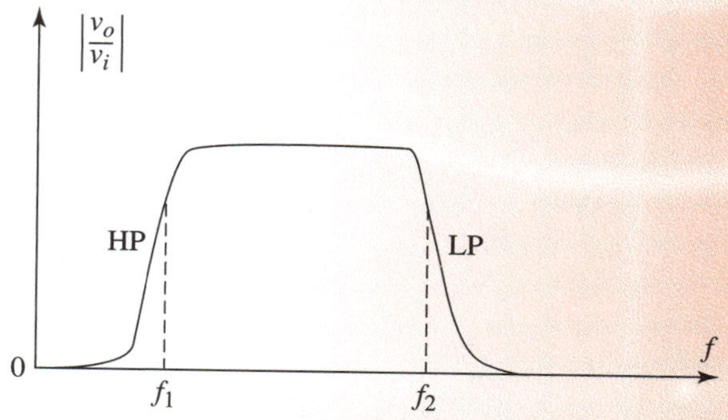

FIGURE 11–39 A wideband bandpass filter obtained by cascading overlapping low-pass and high-pass filters

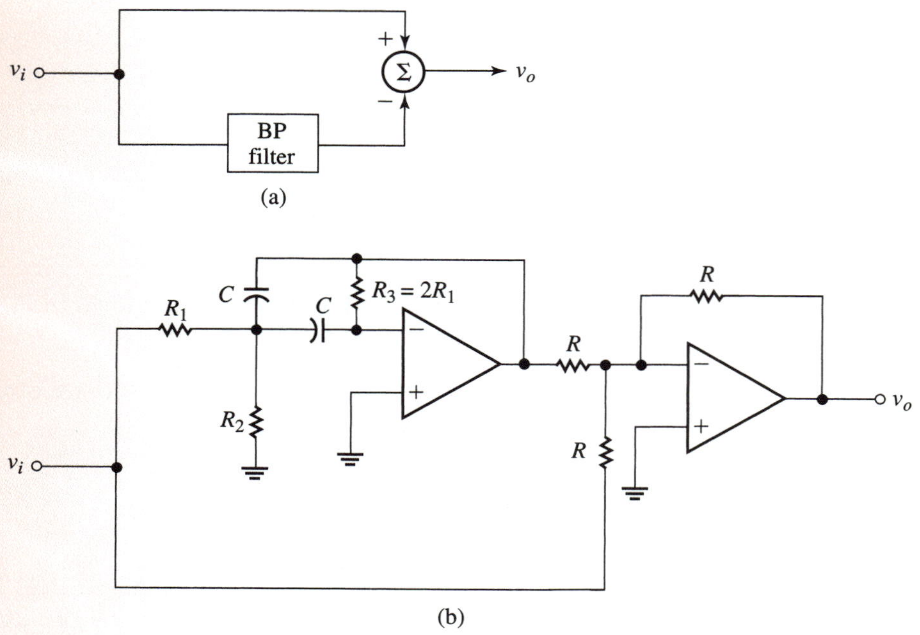

FIGURE 11–40 (a) Block diagram of a band-stop filter obtained from a unity-gain bandpass filter. (b) A possible implementation using the multiple-feedback BP filter

subtracting operation on a bandpass filter. A band-stop filter is just the opposite of a bandpass filter: It passes all frequencies but a certain bandwidth. Or, equivalently, its function is "all of the input minus the bandpass function." Figure 11–40(a) shows the block diagram of a band-stop filter obtained from a unity-gain bandpass filter. Notice that the output summing junction performs subtraction, according to the signs. This structure can also make use of wideband bandpass filters when wideband rejection (stopping) is needed. A possible implementation is shown in Figure 11–40(b) for a narrowband band-stop filter. Note that the output circuit is an inverting *summing* amplifier, not a subtracting circuit, since the bandpass filter is already inverting the signal. A noninverting summing amplifier will do the job just as well.

Wideband band-stop filters can also be implemented by "summing" *nonoverlapping* low-pass and high-pass functions, as shown in Figure 11–41. Notice that in this case the lower cutoff frequency is determined by the low-pass filter, just the opposite as in the cascade structure discussed for bandpass filters. The summing stage can be inverting or noninverting and could add gain to the circuit, if desired.

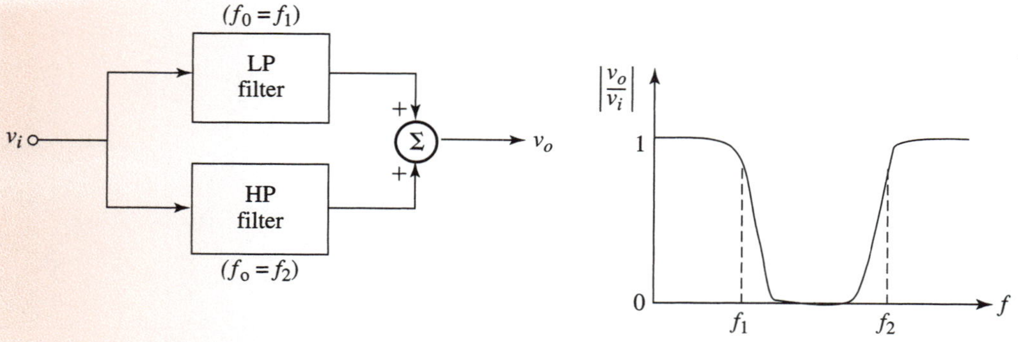

FIGURE 11–41 Obtaining a wideband band-stop filter from nonoverlapping LP and HP filters

EXAMPLE 11–15

Design a band-stop filter with center frequency of 1 kHz and a 3-dB rejection band of 150 Hz. Use the circuit from Figure 11–40(b) with unity gain.

Solution

Let $R_3 = 20\ \text{k}\Omega$ and determine C using equation 11–64.

$$C = \frac{1}{\pi \text{BW} R_3} = \frac{1}{\pi(150)20\ \text{k}} = 106\ \text{nF}$$

For unity gain, $R_1 = R_3/2 = 10\ \text{k}\Omega$. From equation 11–66, we write

$$1\ \text{kHz} = \frac{1}{2\pi(106\ \text{nF})\sqrt{(R_1 \| R_2)20\ \text{k}}}$$

to obtain $R_1 \| R_2 = 112.7\ \Omega$. Finally, R_2 is obtained from the calculated parallel combination, yielding

$$\frac{1}{R_2} = \frac{1}{112.7} - \frac{1}{10\ \text{k}} \quad \text{or} \quad R_2 = 114\ \Omega$$

The three resistors in the summing amplifier are not critical as long as their values are identical; several kilohms will be adequate.

EXAMPLE 11–16

PSPICE

Simulate fourth- and fifth-order low-pass Chebyshev filters with 1-dB ripple for a cutoff frequency of 5 kHz. Request linear ac sweep. Compare the two responses and observe where the ripples appear with respect to the unity-gain line.

Solution

The two .CIR files are shown below. Note the .AC statement with "lin" indicating linear sweep with 2000 points of data to be obtained between 10 Hz and 10 kHz.

```
4TH-ORDER 1DB CHEBYSHEV
*FILTER fc = 5 kHz
VI 1 0 AC 1
R1 1 2 10K
R2 2 3 10K
C1 2 4 9.449NF
C2 3 0 3.838NF
X1 3 4 4 opamp
R3 4 5 10K
R4 5 6 10K
C3 5 7 22.81NF
C4 6 0 450.2PF
X2 6 7 7 opamp
.LIB LINEAR.LIB
.AC LIN 2000 10 10K
.PROBE
.END
```

```
5th Order LP 1-dB Chebyshev
*Filter fc = 5 kHz
vi 1 0 ac 1
ro 1 2 10k
co 2 0 11nf
x1 2 3 3 opamp
r1 3 4 10k
r2 4 5 10k
c1 4 6 13.59nf
c2 5 0 1.737nf
c2 5 0 1.737nf
x2 5 6 6 opamp
r3 6 7 10k
r4 7 8 10k
c3 7 9 35.58nf
c3 7 9 35.58nf
c4 8 0 288.2pf
x3 8 9 9 opamp
.lib linear.lib
.ac lin 2000 10 10k
.probe
.end
```

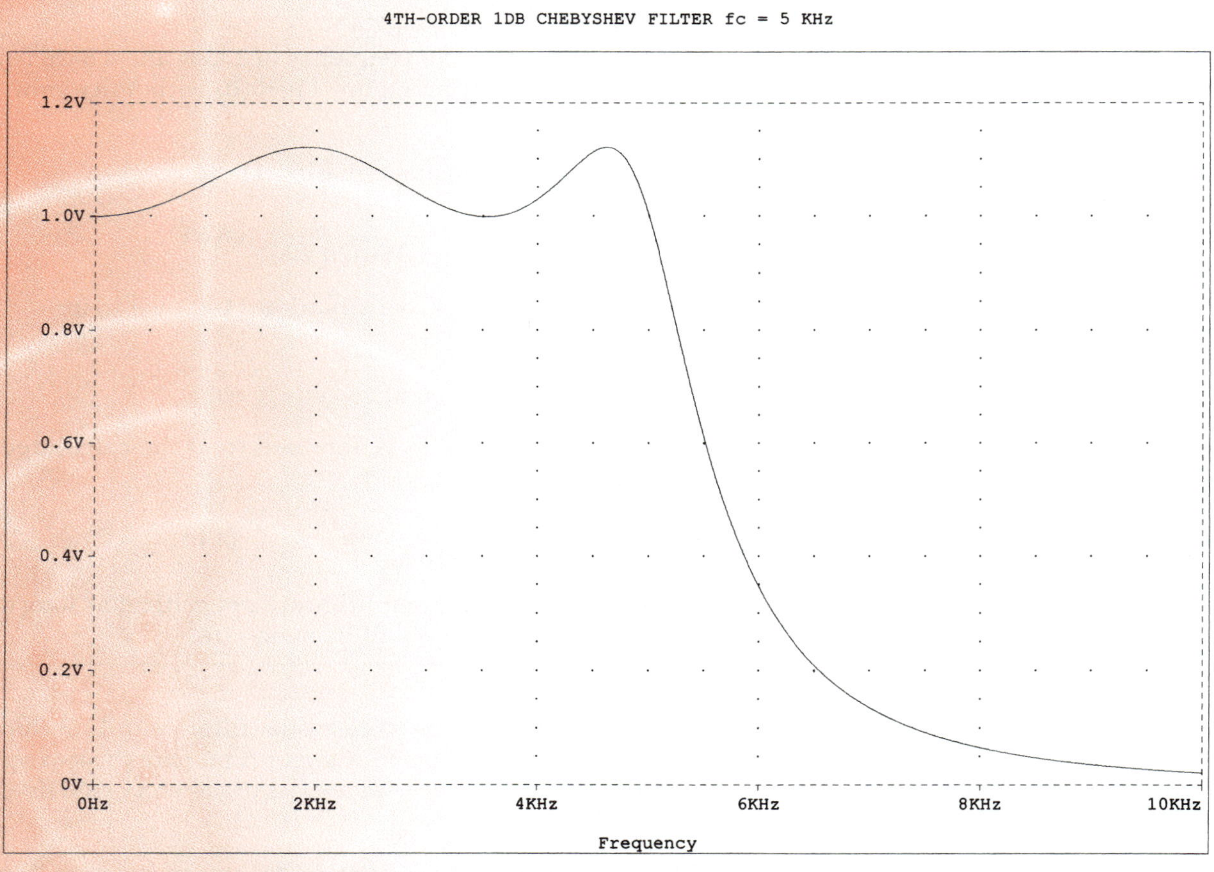

FIGURE 11–42 4th-order, low-pass, 1-dB Chebyshev filter with $f_c = 5$ kHz

Note: "opamp" is a semi-ideal op-amp model residing in LINEAR.LIB, not part of the PSpice package. Users can create their own subcircuit libraries. Figures 11–42 and 11–43 show the plots for these two filters. Note that both responses approach a magnitude of 1 when frequency approaches zero. However, the fourth-order response ripples above the magnitude of 1, whereas the fifth-order filter does it right below. Note also that in Chebyshev filters the "cutoff" frequency is actually the point where the line tangent to the bottom of the ripple crosses the rightmost part of the curve. For even-ordered filters, this crossing point is at unity-gain. For odd-ordered filters, the point is RW dB below unity, where RW is the ripple width.

11–6 LOGARITHMIC AMPLIFIERS

A logarithmic (log) amplifier produces an output that is proportional to the logarithm of its input. Since the log function is nonlinear, it is clear that a log amplifier is not linear in the sense discussed in Chapter 3. A logarithmic transfer characteristic is shown in Figure 11–44. We see that the slope of the characteristic, $\Delta V_o/V_i$, and hence the voltage gain, is small for large values of V_i, and large for small values of V_i. Because the gain decreases with increasing input signal level, the amplifier is said to *compress* signals.

One important application of log amplifiers is in the amplification of signals having a wide *dynamic range:* signals that may be very small as well as very large. Suppose, for example, that a temperature sensor generates a few millivolts at very low temperatures and a few volts at very high temperatures.

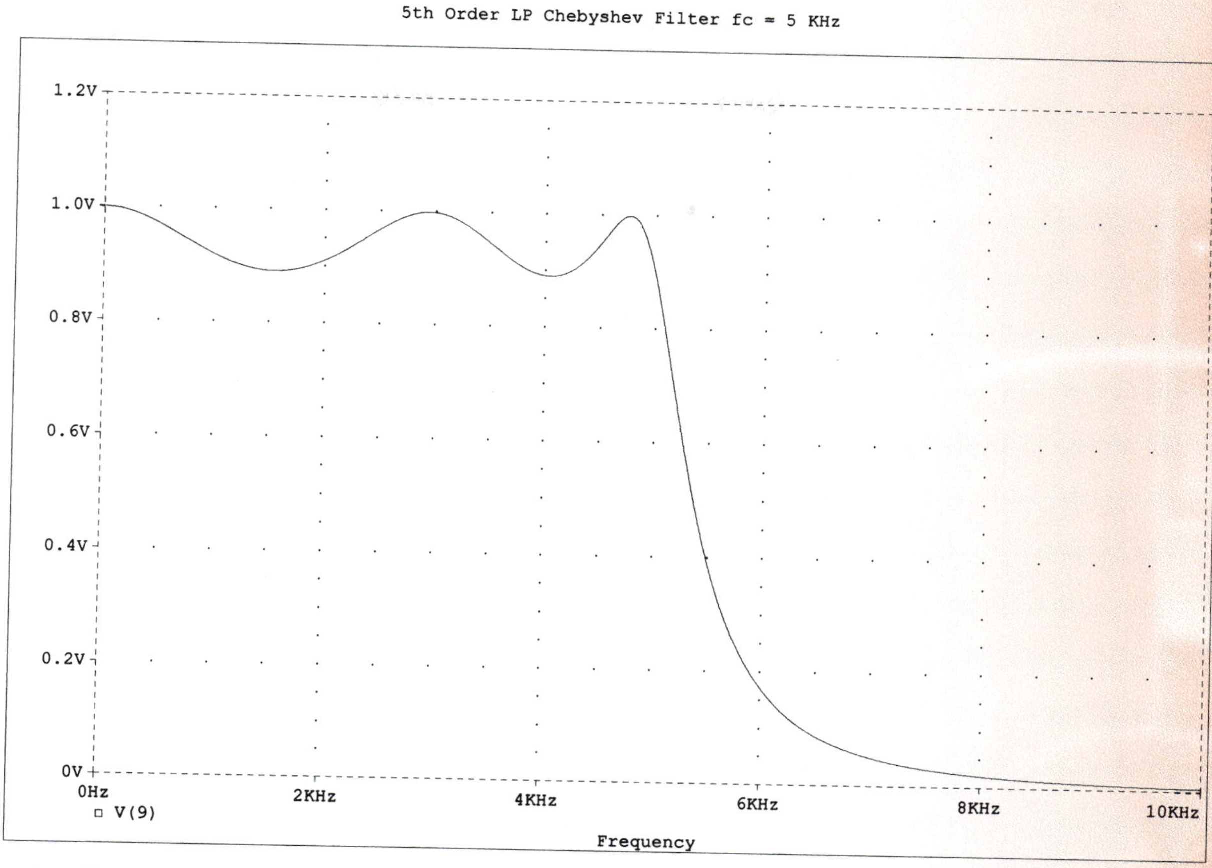

FIGURE 11–43 5th-order, low-pass, 1-dB Chebyshev filter with $f_c = 5$ kHz

To obtain good resolution, we would like the small signals to undergo significant amplification. The same amplification applied to the large signals, as would occur in a linear amplifier, would overdrive the amplifier and create clipping and distortion. The log amplifier eliminates this problem.

Of course, the nonlinear characteristic of the log amplifier creates output waveforms that are distorted versions of input waveforms. If necessary for a particular application, the distortion can be removed by an *antilogarithmic* (inverse log, or exponential) amplifier, which has a transfer characteristic that is exactly the opposite of the log amplifier. On the other hand, in some applications the antilog operation is not necessary, as, for example, when it is desired to create a display of signal magnitudes on a logarithmic scale. An

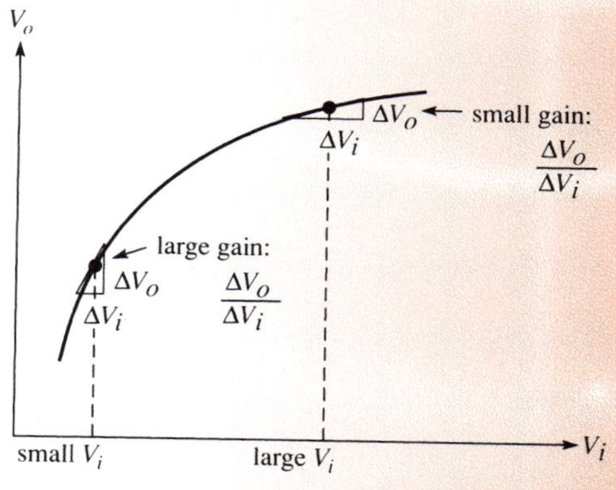

FIGURE 11–44 The transfer characteristic of a log amplifier shows that its voltage gain decreases with increasing values of V_i

example is a *spectrum analyzer*, in which the frequency content of a complex signal is displayed as a plot of decibel voltage levels versus frequency. Another application of log amplifiers is in *analog computation*, where signal voltages must be multiplied or divided. For example, if we wished to generate the product voltage v_1v_2, we could sum the outputs of two log amplifiers to obtain $\log v_1 + \log v_2 = \log v_1v_2$. The output of an antilog amplifier whose input is $\log v_1v_2$ would then be a voltage proportional to v_1v_2.

The logarithmic characteristic of a log amplifier stems from the relationship between the collector current and base-to-emitter voltage of a BJT, which is similar to the diode equation (equation 2–2):

$$I_C = I_s(e^{V_{BE}/V_T} - 1) \qquad \textbf{(11–67)}$$

where I_s is the reverse saturation current of the base-emitter diode and V_T is the thermal voltage, $V_T = kT/q$. When V_{BE} is a few tenths of a volt, $e^{V_{BE}/V_T} \gg 1$, and (11–67) becomes

$$I_C = I_s e^{V_{BE}/V_T}$$

Figure 11–45 shows the basic configuration of a log amplifier, in which a BJT is connected in the feedback path of an inverting operational amplifier. Taking the logarithm of both sides of the above and solving for V_{BE}, we find

$$V_{BE} = V_T \ln\left(\frac{I_C}{I_s}\right) \qquad \textbf{(11–68)}$$

In Figure 11–45, note that the collector is at virtual ground ($\approx$0 V), so

$$I_C = \frac{V_i}{R_1} \qquad \textbf{(11–69)}$$

Also note that $V_{BE} = -V_o$. Substituting these relationships in (11–68), we obtain

$$V_o = -V_T \ln\left(\frac{V_i}{I_s R_1}\right) \qquad \textbf{(11–70)}$$

We see that V_o is a logarithmic function of V_i. Since the common logarithm (base 10) is related to the natural logarithm by $\ln x = 2.303 \log_{10} x$, equation 11–70 can also be written in terms of $\log_{10}$ as

$$V_o = -2.303\, V_T \log_{10}\left(\frac{V_i}{I_s R_1}\right) \qquad \textbf{(11–71)}$$

At room temperature, $V_T \approx 0.0257$ V, for which equations 11–70 and 11–71 become

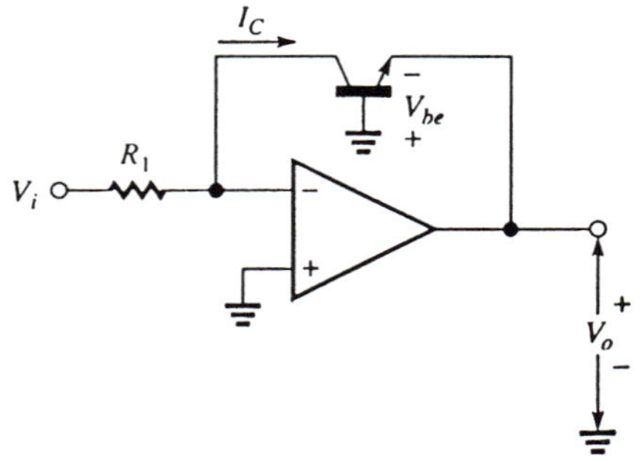

FIGURE 11–45 Basic configuration of a log amplifier

FIGURE 11–46 A practical log amplifier designed to compensate for the variability of I_s

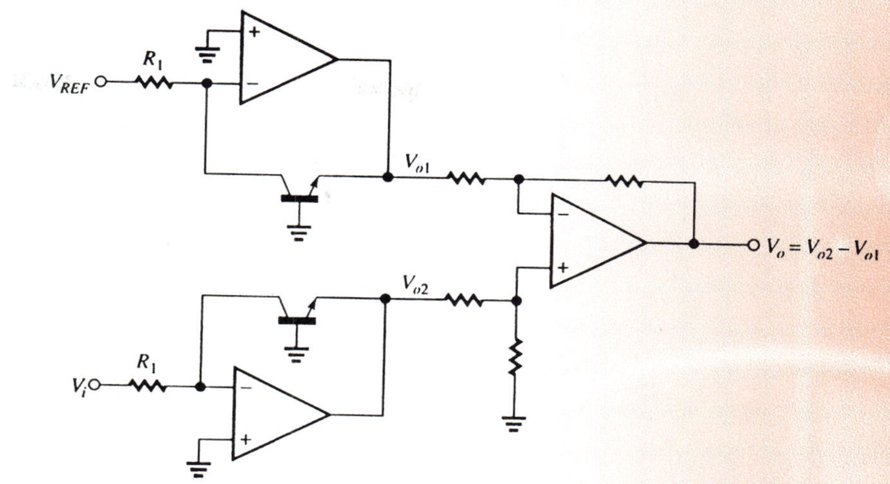

$$V_o = -0.0257 \ln\left(\frac{V_i}{I_s R_1}\right) = -0.0592 \log_{10}\left(\frac{V_i}{I_s R_1}\right) \tag{11–72}$$

The practical difficulty of the configuration we have described is that the value of I_s cannot usually be predicted accurately and is, in any event, highly sensitive to temperature variations. To overcome this problem, a practical log amplifier is constructed as shown in Figure 11–46. The two transistors are closely matched in an integrated circuit, so their values of I_s are essentially equal and change equally with temperature. Following the same procedure we used to obtain equation 11–72, we find

$$V_{o1} = -0.0257 \ln\left(\frac{V_{REF}}{I_s R_1}\right) \tag{11–73}$$

and

$$V_{o2} = -0.0257 \ln\left(\frac{V_i}{I_s R_1}\right) \tag{11–74}$$

As discussed in Section 11–4, the output amplifier forms the difference voltage $V_{o2} - V_{o1}$:

$$\begin{aligned} V_o = V_{o2} - V_{o1} &= -0.0257\left(\ln\frac{V_i}{I_sR_1} - \ln\frac{V_{REF}}{I_sR_1}\right) \\ &= -0.0257 \ln\left[\frac{(V_i/I_sR_1)}{(V_{REF}/I_sR_1)}\right] \\ &= -0.0257 \ln\left(\frac{V_i}{V_{REF}}\right) \end{aligned} \tag{11–75}$$

Here we see that the output is no longer dependent on the value of I_s. Also, the external voltage V_{REF} can be adjusted to control the overall sensitivity of the amplifier (as can the gain of the output difference amplifier).

EXAMPLE 11–17

When V_i in Figure 11–46 is 1 V, it is desired that $V_o = 50$ mV. What value of V_{REF} should be used? Assume the difference amplifier has unity gain.

Solution

From equation 11–75,

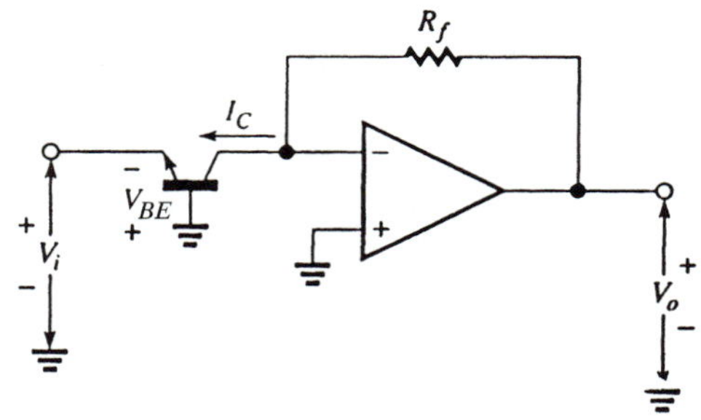

FIGURE 11–47 Basic configuration of an antilog amplifier

$$50 \text{ mV} = -0.0257 \ln\left(\frac{1 \text{ V}}{V_{REF}}\right)$$

$$\frac{50 \times 10^{-3}}{-0.0257} = -1.9456 = \ln\left(\frac{1}{V_{REF}}\right)$$

Taking the inverse logarithm of both sides,

$$0.143 = \frac{1}{V_{REF}}$$

$$V_{REF} = 7.0 \text{ V}$$

Figure 11–47 shows the basic configuration of an antilogarithmic (inverse log) amplifier. Note that the transistor in this case is connected to the input of an inverting operational amplifier. As before,

$$I_C = I_s e^{V_{BE}/V_T} \tag{11–76}$$

Since I_C flows through R_f and the collector is at virtual ground, we have

$$V_o = R_f I_C = R_f I_s e^{V_{BE}/V_T} \tag{11–77}$$

Since $V_{BE} = -V_i$, we obtain

$$V_o = R_f I_s e^{-V_i/V_T} \tag{11–78}$$

Equation 11–78 is equivalent to

$$V_o = R_f I_s \text{ antilog } (-V_i/V_T) \tag{11–79}$$

Note that values of V_i must be negative so that $-V_i/V_T$ is always positive. As in practical log amplifiers, practical antilog amplifiers use two transistors and two operational amplifiers to compensate for the variability of I_s.

Examples of integrated-circuit log and antilog amplifiers are the ICL 8048 (log) and ICL 8049 (antilog) amplifiers, manufactured by Intersil. These devices can be used over a 60-dB (1000 to 1) dynamic voltage range, with output voltages up to 14 V. The 8048 does not incorporate a difference amplifier to compensate for I_s, but arranges two matched transistors and two operational amplifiers in such a way that the output of the second amplifier is proportional to the logarithm of the input difference voltage, which accomplishes the same goal.

11–7 MULTISIM EXERCISE

The purpose of this exercise is to observe how a clock waveform can be integrated to produce a triangular waveform. The clock waveform, not being centered at zero volts, has a dc component which equals one-half of its am-

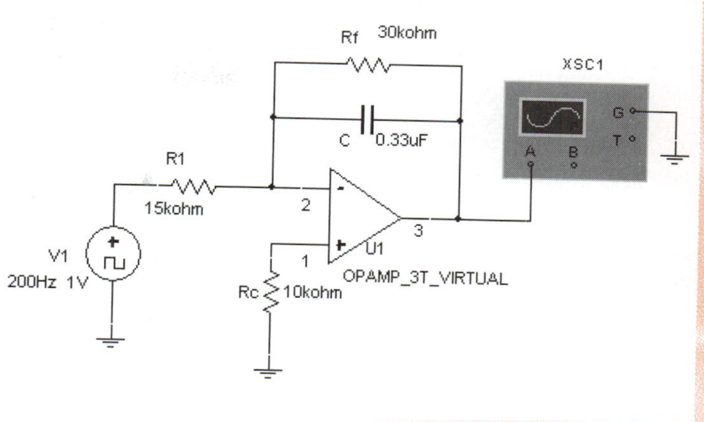

FIGURE 11–48 A practical integrator driven by a 1-V clock source

plitude when the duty cycle is 50%. We will then observe both the ac and dc components on the output and verify that they agree with the theoretical values.

Figure 11–48 shows the circuit schematic as drawn with the Multisim EWB program. Make sure you specify all the component values and supply voltages correctly. Adjust the vertical scale of the oscilloscope to 0.5 V/div and the horizontal scale to 5 ms/div. Verify that the input mode is at "DC."

Run the simulation and make adjustments to the oscilloscope settings as needed to view the output waveform in its entirety. Observe that the output is not centered at zero volts. Its dc component should be –1 volt because the dc gain is –2 and the dc input component is 0.5 V. Once you have verified the dc component level, you can select the "AC" mode to remove the dc component, and then change the vertical scale to a smaller value to enlarge the ac portion of the signal. Can you tell that the waveform is not a perfect triangular waveform?

SUMMARY

This chapter has presented a number of selected op-amp applications that are representative of circuits which require more rigorous analysis, but at the same time provide the student with a broader knowledge of operational amplifiers. At the end of this chapter, the reader should understand the following concepts:

- An ideal integrator that employs a standard operational amplifier will not be stable; its output voltage will drift off until it reaches the saturation limit.
- A practical or ac integrator will not drift off, but it is not suitable for integrating signals at low frequencies.
- Electronic phase-shift circuits are useful for providing a constant delay for signals at a specific frequency range.
- A Butterworth filter provides maximally flat response in the passband.

- A Chebyshev filter provides more attenuation in the stopband than a Butterworth filter of the same order. However, the Chebyshev filter has a rippled passband; that is, its gain is not constant in the passband.
- Frequency, resistance, and capacitance scaling in filters is a simple process that allows the designer to change the cutoff frequency or frequencies of a given filter through very simple calculations.

EXERCISES

SECTION 11–1

Electronic Integration

11–1. The input to an ideal electronic integrator is 0.25 V dc. Assume that the integrator inverts and multiplies by the constant 20.

(a) What is the output 2 s after the input is connected?

(b) How long would it take the output to reach −15 V?

11–2. Using an ideal operational amplifier, design an ideal integrator whose output will reach 5 V 200 ms after a −0.1-V-dc input is connected.

11–3. The input to the circuit shown in Figure 11–49 is $v_i = 6\sin(500t - 30°)$ V. Write a mathematical expression for the output voltage.

11–4. The input to the integrator in Exercise 11–3 is a 100-Hz sine wave with peak value 5 V.

(a) What is the peak value of the output?

(b) What is the peak value of the output if the capacitance in the feedback is halved?

(c) What is the peak value of the output (with the original capacitance) if the frequency is halved?

(d) What is the peak value of the output (with original capacitance and original frequency) if the input resistance is halved?

(e) Name three ways to double the closed-loop gain of an integrator.

11–5. Using an ideal operational amplifier, design an ideal integrator that will produce the output $v_o = 0.04\cos(2 \times 10^3 t)$ when the input is $v_i = 8\sin(2 \times 10^3 t)$.

11–6. What is the closed-loop gain, in decibels, of the integrator designed in Exercise 11–5 when the angular frequency is 500 rad/s?

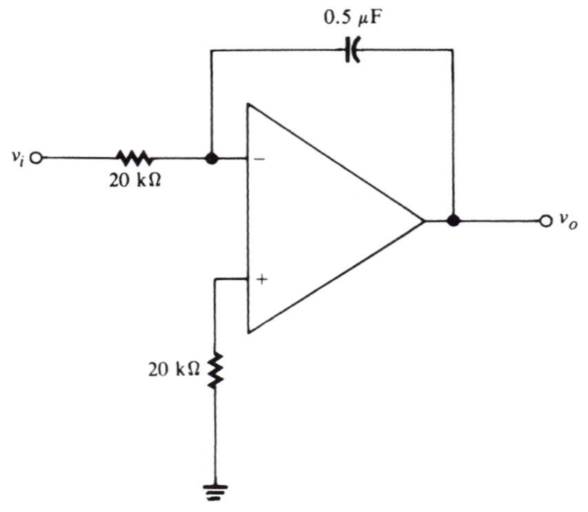

FIGURE 11–49 (Exercise 11–3)

11–7. Design a practical integrator that will integrate signals with frequencies down to 500 Hz and that will provide unity gain to dc inputs.

11–8. (a) Design a practical integrator that will integrate signals with frequencies down to 500 Hz and that will provide a closed-loop gain of 0.005 at $f = 20$ kHz.

(b) What will be the output of the integrator when the input is −10 mV dc?

11–9. (a) Write an expression for the output v_o of the system shown in Figure 11–50.

(b) Show how the same output could be obtained using a single amplifier.

SECTION 11–2
Electronic Differentiation

11–10. An ideal differentiator has a closed-loop gain of 2 when the input is a 50-Hz signal. What is its closed-loop gain when the signal frequency is changed to 2 kHz?

11–11. The differentiator shown in Figure 11–11 has $R_1 = 2.2$ kΩ, $C = 0.015$ μF, and $R_f =$ 27 kΩ.

(a) Assuming that the circuit performs satisfactory differentiation at frequencies up to 1 decade below its first break frequency, find the frequency range of satisfactory differentiation.

(b) Find the magnitude of the voltage gain at 120 Hz.

(c) Sketch a Bode plot of the gain, as it would appear if plotted on log-log graph paper. Label all break frequencies. Assume that the operational amplifier has a gain–bandwidth product of 2 MHz.

11–12. Design a practical differentiator that will perform satisfactory differentiation of signals up to 1 kHz. The *maximum* closed-loop gain of the differentiator (at any frequency) should be 60. (You may assume an ideal operational amplifier that has a wide bandwidth.)

11–13. Sketch the Bode plot for the gain of the circuit shown in Figure 11–51 the way it would appear if plotted on log-log graph

FIGURE 11–50
(Exercise 11–9)

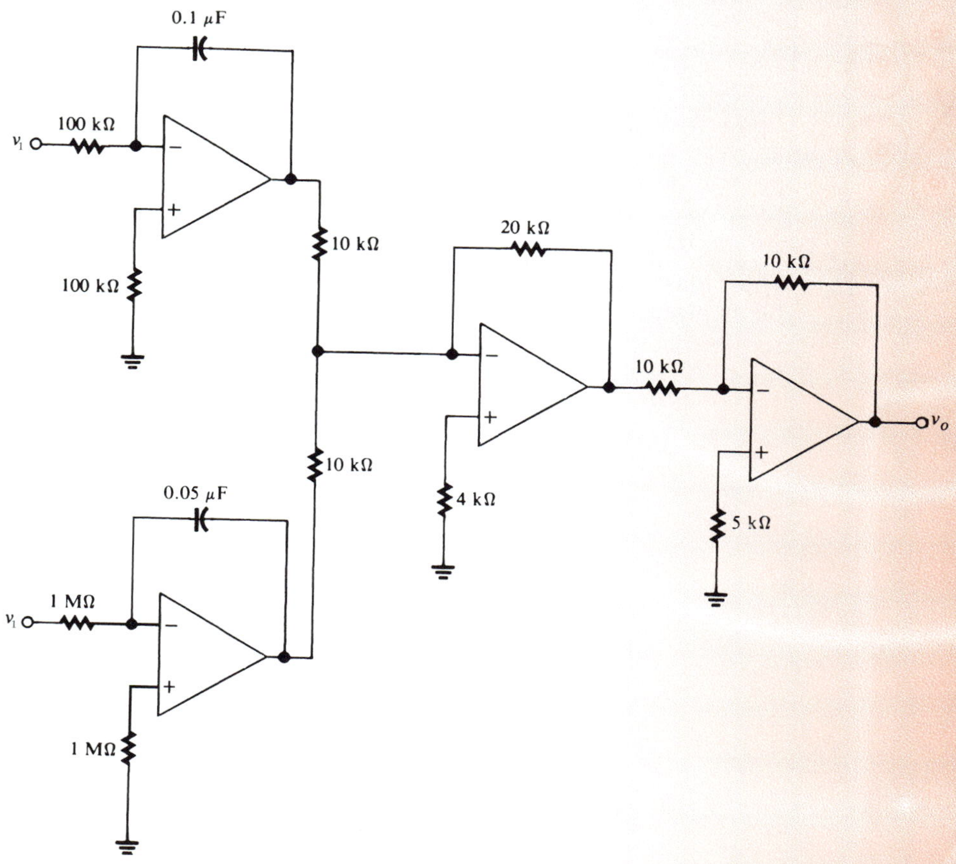

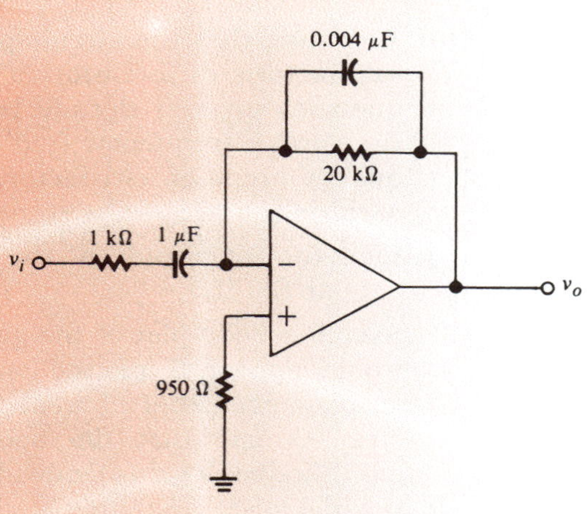

FIGURE 11–51 (Exercise 11–13)

paper. Assume that the amplifier has a unity-gain frequency of 1 MHz. On your sketch, label

(a) all break frequencies, in Hz,

(b) the slopes of all gain asymptotes, in dB/decade, and

(c) the value of the maximum closed-loop gain.

11–14. An integrator having $R_1 = 10\text{ k}\Omega$ and $C = 0.02\ \mu\text{F}$ is to be used to generate a triangular wave from a square-wave input. The operational amplifier has a slew rate of 10^5 V/s. Assume that the input has zero dc component.

(a) What maximum positive level can the square wave have when its frequency is 500 Hz?

(b) Repeat when the square-wave frequency is 5 kHz.

11–15. Sketch the output of the integrator shown in Figure 11–52. Label the peak positive and peak negative output voltages.

11–16. Design a triangular-waveform generator whose output alternates between +9 V pk and −9 V pk when the input is a 10-Hz square wave that alternates between +1.5 V and −1.5 V. The input resistance to the generator must be at least 15 kΩ. You may assume that there is no dc component or offset in the input.

11–17. Sketch the output of an ideal differentiator having $R_f = 40\text{ k}\Omega$ and $C = 0.5\ \mu\text{F}$ when the input is each of the waveforms shown in Figure 11–53. Label maximum and minimum values on your sketches.

11–18. Sketch the output of an ideal differentiator having $R_f = 60\text{ k}\Omega$ and $C = 0.5\ \mu\text{F}$ when the input is each of the waveforms shown in Figure 11–54. Label minimum and maximum values on your sketches.

SECTION 11–3

Phase-Shift Circuits

11–19. (a) Determine the RC product required for a phase-lead circuit to provide a phase shift of 120° when the frequency of the applied input signal is 3.35 kHz.

(b) If we select $R = 12\text{ k}\Omega$, what is the required capacitance C?

11–20. Determine the ratio of the maximum frequency to the minimum frequency required for a phase-lag circuit to cover the phase range −45° to −135°.

FIGURE 11–52 (Exercise 11–15)

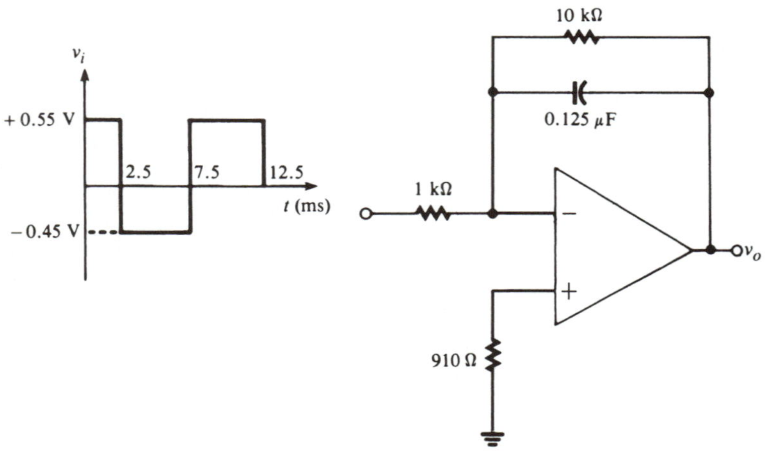

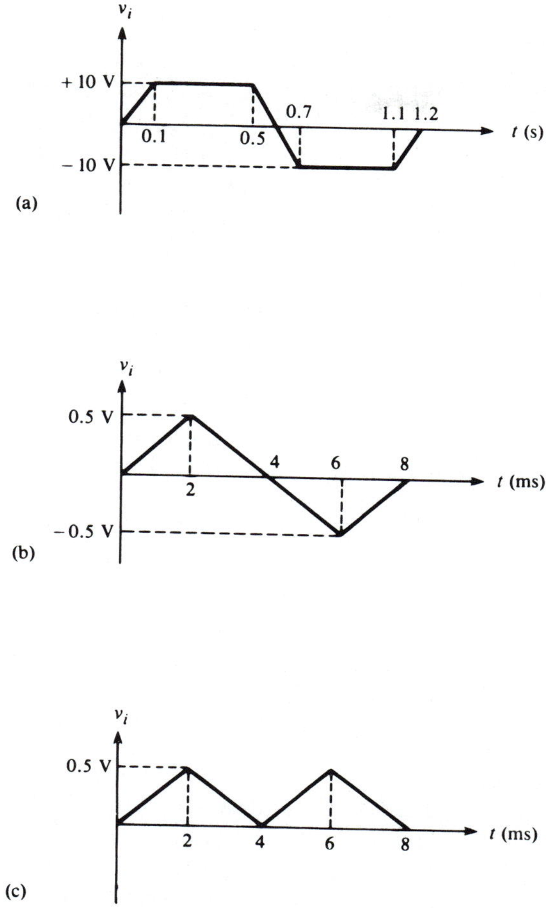

FIGURE 11–53 (Exercise 11–17)

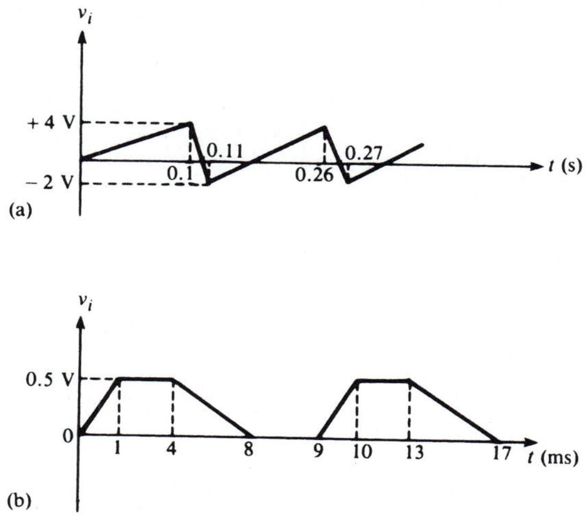

FIGURE 11–54 (Exercise 11–18)

11–21. Design a phase-shift circuit that provides a nominal delay of 50 μs over a frequency range of 100 Hz to 1 kHz. (*Hint:* Work with the larger frequency and check the delay at the lower one.)

SECTION 11–4

Instrumentation Amplifiers

11–22. (a) Assuming that the amplifiers shown in Figure 11–55 are ideal, find V_{o1}, V_{o2}, I_1, I_2, I_3, and V_o.

(b) Repeat when the inputs V_1 and V_2 are interchanged.

11–23. (a) The instrumentation amplifier in Figure 11–23 is required to have gain $|v_o/(v_1 - v_2)|$ equal to 25. Assuming that $R = 5\ \text{k}\Omega$, what should be the value of R_A?

(b) Verify that the amplifier will perform satisfactorily if the maximum permissible output voltage of each operational amplifier is 23 V. v_1 varies from 0.3 V to 1.2 V and v_2 varies from 0.5 V to 0.8 V.

11–24. A 25-mV rms signal contaminated by noise is to be amplified with an instrumentation amplifier to a level of 4 V rms. The noise level of the signal before amplification is 780 mV. The IA has a CMRR of 106 dB.

(a) Determine the value of R_A required if the internal resistors are 20 kΩ.

(b) Find the noise level at the output.

(c) Find the input and output signal-to-noise ratios.

11–25. A 10-V rms sine wave is applied to both inputs tied together in an IA. The differential gain was adjusted to 45 dB. The output voltage as measured with an oscilloscope is 90 mV p–p.

(a) What is the common-mode gain of the IA in dB?

(b) What is its common-mode rejection ratio in dB?

SECTION 11–5

Active Filters

11–26. A low-pass Butterworth filter must have an attenuation of at least −20 dB at 1 octave above its cutoff frequency. What minimum order must the filter have?

11–27. A high-pass Butterworth filter having a maximum high-frequency gain of 5 and a cutoff frequency of 500 Hz must have a gain no less than 4.9 at 2 kHz and no greater than 0.1 at 100 Hz. What minimum order must the filter have?

FIGURE 11–55 (Exercise 11–22)

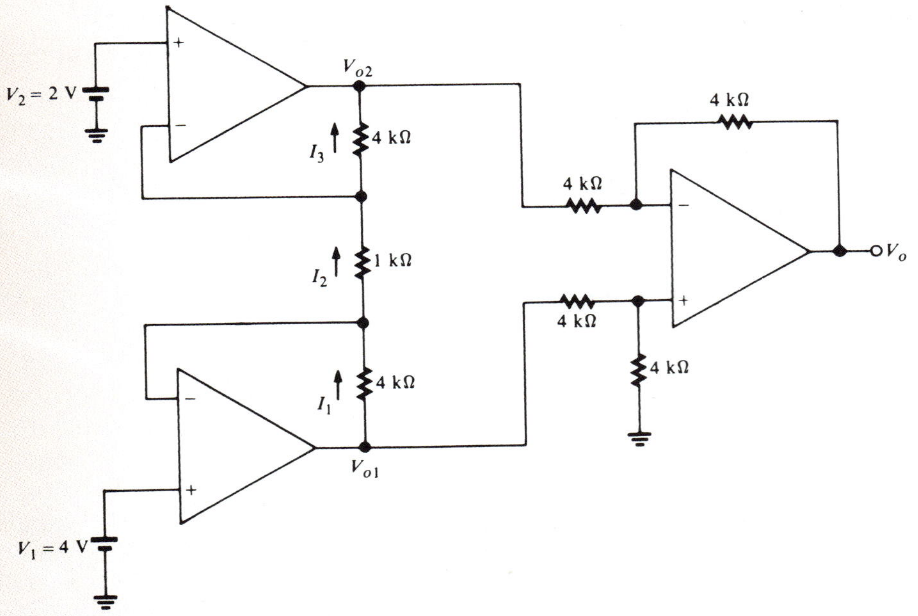

11–28. The gain of a bandpass filter at its upper cutoff frequency is 42. The lower cutoff frequency is 10.8 kHz and the Q is 50. Assume that the center frequency is midway between the cutoff frequencies.

(a) What is the gain at the center frequency?

(b) What is the center frequency?

(c) What is the bandwidth?

11–29. Design a third-order, high-pass, Chebyshev filter having a cutoff frequency of 8 kHz and a ripple width of 2 dB.

11–30. Design a second-order, low-pass Butterworth filter with cutoff frequency 1 kHz. If the input to this filter has a 500-Hz component with amplitude 1.2 V and a 4-kHz component with amplitude 5 V, what are the amplitudes of these components in the output?

11–31. Design a second-order, IGMF bandpass filter with a center frequency of 20 kHz and a bandwidth of 4 kHz. The gain should be 2 at 20 kHz.

11–32. The input to the filter in Exercise 11–31 has the following components:

(a) v_1: 6 V rms at 2 kHz

(b) v_2: 2 V rms at 20 kHz

(c) v_3: 10 V rms at 100 kHz

What are the rms values of these components in the filter's output?

11–33. Design a bandpass filter with cutoff frequencies at 200 Hz and 1 kHz. The passband is to be very flat and the attenuation 1 octave below and above the cutoff frequencies is required to be at least 15 dB.

11–34. Design a wideband band-stop filter with the following characteristics: very flat passband with unity gain, 3-dB cutoff frequencies at 300 Hz and 2 kHz, required attenuation at 600 Hz and 1 kHz at least 24 dB.

SECTION 11–6

Logarithmic Amplifiers

11–35. The log amplifier in Figure 11–46 has $V_{REF} = 5$ V. It is necessary for the output voltage to be 0.25 V when the input is 0.5 V. What should be the voltage gain of the output difference amplifier?

11–36. Assuming the output difference amplifier in Figure 11–46 has unit voltage gain, what is the change in the output voltage of the log amplifier when V_i doubles in value?

PSPICE EXERCISES

11–37. Simulate an ac integrator with $R_1 = 20\ \text{k}\Omega$, $C = 0.1\ \mu\text{F}$, and $R_f = 50\ \text{k}\Omega$ driven by a 10-V p–p square wave at a frequency of (a) 1 kHz and (b) 200 Hz. Using Probe, compare the shapes of the waveforms obtained. Verify the expected output amplitude.

11–38. Connect a second integrator in cascade with the one from the previous exercise and plot the output waveform for $f = 1$ kHz. You should obtain a sine-like wave. Verify if the amplitude of the output signal is as expected.

11–39. Add a unity-gain, multiple-feedback bandpass filter centered at 1 kHz with $Q = 20$ following the circuit from the previous exercise. The "parabolic" waveform from the second integrator should become a fairly "clean" sine wave after passing through the filter.

11–40. Design an operational-amplifier integrator that integrates signals with frequencies down to 200 Hz. The magnitude of the gain at 10 kHz should be 0.02. Verify your design using SPICE. Obtain a plot of the frequency response and compare the gain at 20 Hz, 200 Hz, and 10 kHz with theoretical values.

11–41. Design an IGMF bandpass filter having center frequency 1 kHz, $Q = 5$, and gain 1. Use SPICE to verify your design. Obtain a plot of the frequency response extending from 2 decades below the center frequency to 1 decade above the center frequency. Using SPICE to determine the bandwidth of the filter, verify that its Q is 5.

CHAPTER 12

WAVE GENERATION AND SHAPING

OUTLINE

OBJECTIVES

- Learn the operation of voltage comparators and their use in voltage-level detectors.
- Analyze and design integrator circuits for wave shaping.
- Understand the underlying principle for producing sustained oscillations.
- Analyze and design basic oscillators using op-amps.
- Develop an understanding of op-amp circuits with diodes for clipping and clamping of signals.

12–1 VOLTAGE COMPARATORS

We start the topic of wave generation by learning about a very important electronic device: the voltage comparator. These devices are widely used in the electronics industry in a vast number of applications such as signal detection, alarm systems, control, and multivibrators, to name a few. Voltage comparators can be seen as the simplest form of interfacing between the analog and digital worlds. They have a high-gain differential input stage, very similar to those found in operational amplifiers. The output stage, however, is designed so that only two output voltage levels are possible; that is, the output is digital in nature. A general description of voltage comparators will be given next, followed by basic applications including the multivibrator, which is used for generating a square wave.

As the name implies, a *voltage comparator* is a device used to compare two voltage levels. The output of the comparator indicates which of its two inputs is larger, so it is basically a switching device, producing a high output when one input is the larger, and switching to a low output if the other input becomes larger. An operational amplifier can be used as a voltage comparator by operating it open-loop (no feedback) and by connecting the two voltages to be compared to the inverting and noninverting inputs. Because it has a very large open-loop gain, the amplifier's output is driven all the way to one of its output voltage limits when there is a very small difference between the input levels. For example, if the + input voltage is slightly greater than the − input voltage, the amplifier quickly switches to its maximum positive output, and when the − input voltage is slightly greater than the + input voltage, the amplifier switches to its maximum negative output.* This behavior is illustrated in Figure 12–1.

Note that Figure 12–1(b) is a transfer characteristic showing output voltage versus *differential* input voltage $v^+ - v^-$. It can be seen that the output switches when $v^+ - v^-$ passes through 0.

*Many authors use the term *saturation voltage* to describe the minimum or maximum output voltage of the amplifier. We are avoiding this use only to prevent confusion with the saturation voltage of a transistor.

FIGURE 12–1 The operational amplifier used as a voltage comparator

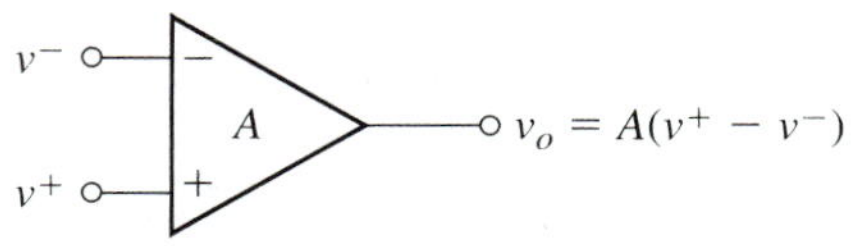

(a) Open-loop operation as a voltage comparator

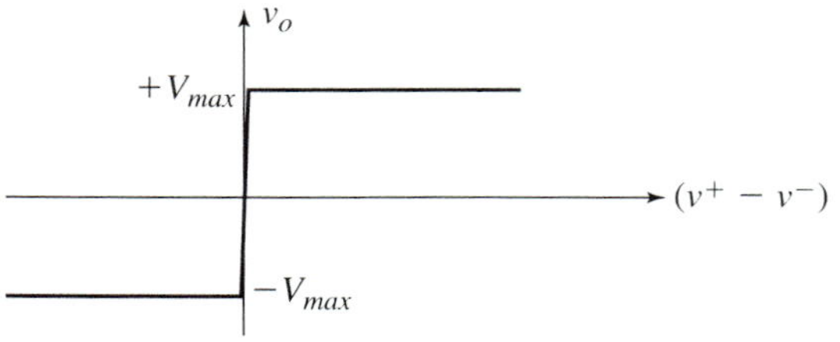

(b) Transfer characteristic of the voltage comparator. When $v^+ > v^-$, the output is at its maximum positive limit, and when $v^+ < v^-$, the output switches to its maximum negative limit.

FIGURE 12–2 The comparator output switches to $+V_{max}$ when $v^+ - v^- > 0$ V, which corresponds to the time points where v^+ rises through +6 V. The output remains high as long as $v^+ - v^- > 0$, or $v^+ > 6$ V.

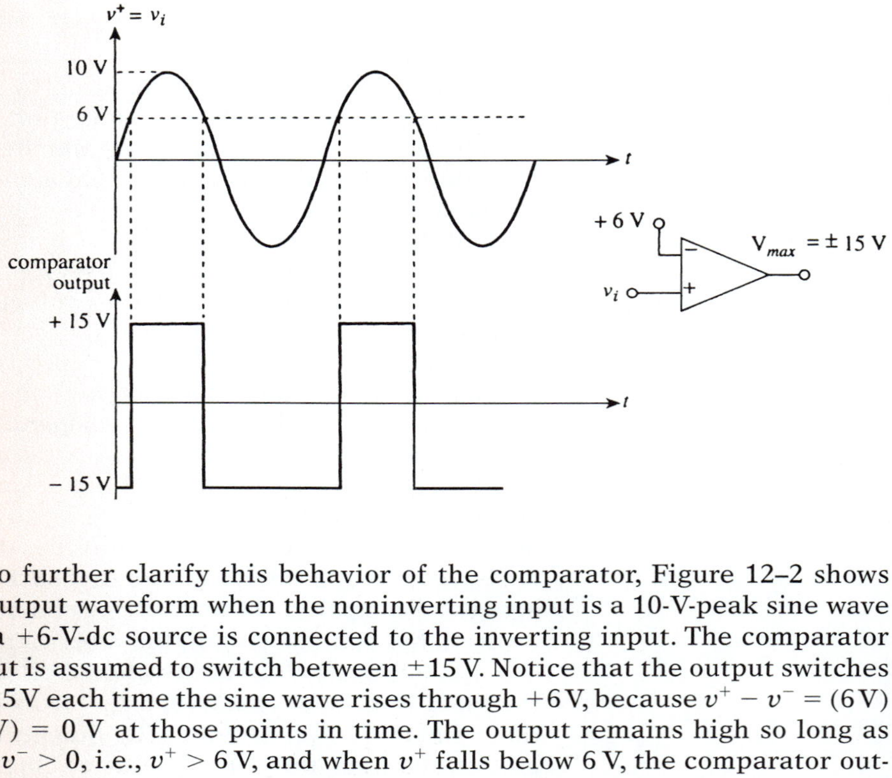

To further clarify this behavior of the comparator, Figure 12–2 shows the output waveform when the noninverting input is a 10-V-peak sine wave and a +6-V-dc source is connected to the inverting input. The comparator output is assumed to switch between ±15 V. Notice that the output switches to +15 V each time the sine wave rises through +6 V, because $v^+ - v^- = (6\text{ V}) - (6\text{ V}) = 0$ V at those points in time. The output remains high so long as $v^+ - v^- > 0$, i.e., $v^+ > 6$ V, and when v^+ falls below 6 V, the comparator output switches to −15 V. As an exercise, plot the output when the sine wave is connected to the inverting input and the +6-V-dc source is connected to the noninverting input.

In some applications, either the inverting or noninverting input is grounded, so the comparator is effectively a zero-crossing detector. It switches output states when the ungrounded input passes through 0. For example, if the inverting input is grounded, the output switches to its maximum positive voltage when v^+ is slightly positive and to its maximum negative voltage when v^+ is slightly negative. The reverse action occurs if the noninverting input is grounded. The transfer characteristics for these two cases are shown in Figure 12–3.

FIGURE 12–3 Operation of the voltage comparator as a zero-crossing detector

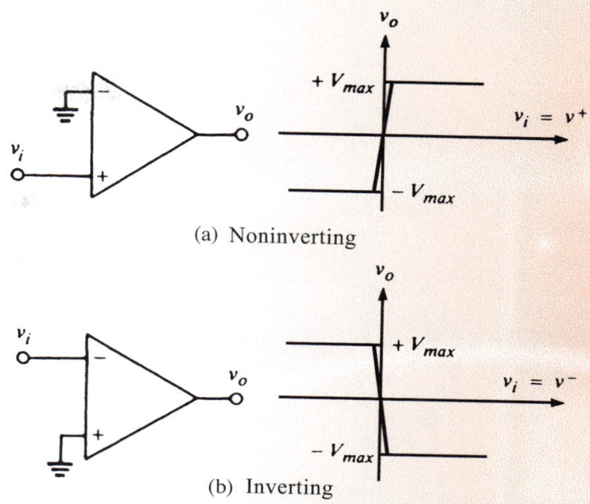

Even though operational amplifiers can be used as voltage comparators, their lack of switching speed due to slew rate makes them unsuitable in applications requiring precision and high speed. Dedicated voltage comparators, commercially available from many vendors, share all the same basic principles of operation:

$$Y = \begin{cases} \text{High} & v^+ > v^- \\ \text{Low} & v^+ < v^- \end{cases} \tag{12–1}$$

where Y is the logical output that can take on two values, high and low. All dedicated voltage comparators have provisions for the user to establish the "high" and "low" voltage levels. Let V_o be the output voltage with V_H being the high level and V_L the low level. The output from the comparator, as a voltage, can then be expressed as

$$V_o = \begin{cases} V_H & v^+ > v^- \\ V_L & v^+ < v^- \end{cases} \tag{12–2}$$

Most voltage comparators are of the *open-collector* type, which allows the user to configure the comparator for driving LEDs, relays, etc. Additionally, they provide an *output return* terminal for establishing the "low" output voltage level V_L or for using the output transistor as an emitter follower. Figure 12–4 shows examples of typical output configurations.

Two other important characteristics of a voltage comparator are its *response time* and *rise time*, illustrated in Figure 12–5. The response time is the delay between the time a step input is applied and the time the output begins to change state. It is measured from the edge of the step input to the time point where the output reaches a fixed percentage of its final value, such as 10% of $+V_{max}$. (For clarity, Figure 12–5 shows the output switching from 0 V toward $+V_{max}$; in many applications, one of the output levels actually is 0 V.) Response time is strongly dependent on the amount of *overdrive* in the input: the voltage in excess of that required to cause switching to occur. The greater the overdrive, the shorter the response time. Rise time is defined in the usual way: the time required for the output to change from 10% of its final value to 90% of its final value.

Although general-purpose operational amplifiers can be, and are, used as voltage comparators in the way we have described, their use should be limited to applications where high speed is not important and when the output levels are not required to be other than $\pm V_{max}$.

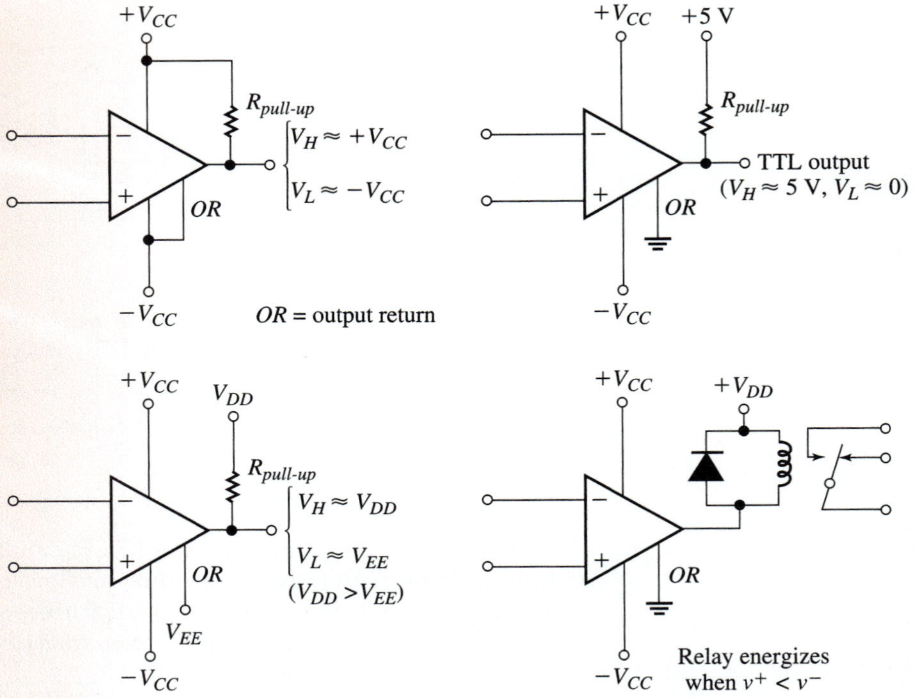

FIGURE 12–4 Typical output configurations for voltage comparators

FIGURE 12–5 Response time and rise time of a voltage comparator

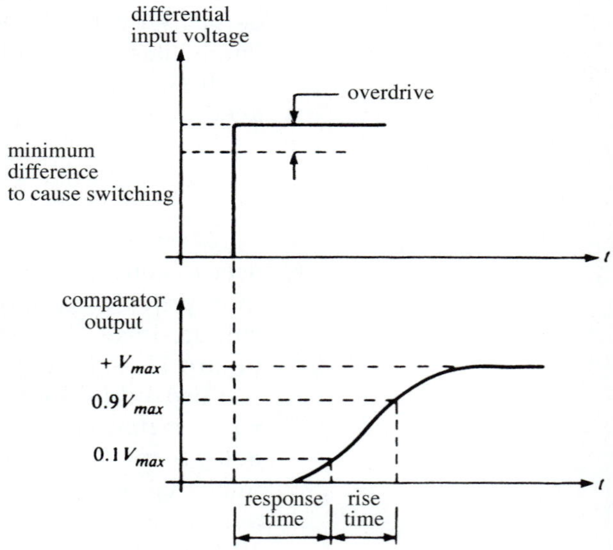

Hysteresis and Schmitt Triggers

In its most general sense, *hysteresis* is a property that means a device behaves differently when its input is increasing from the way it behaves when its input is decreasing. In the context of a voltage comparator, hysteresis means that the output will switch when the input increases to one level but will not switch back until the input falls below a *different* level. In some applications, hysteresis is a desirable characteristic because it prevents the

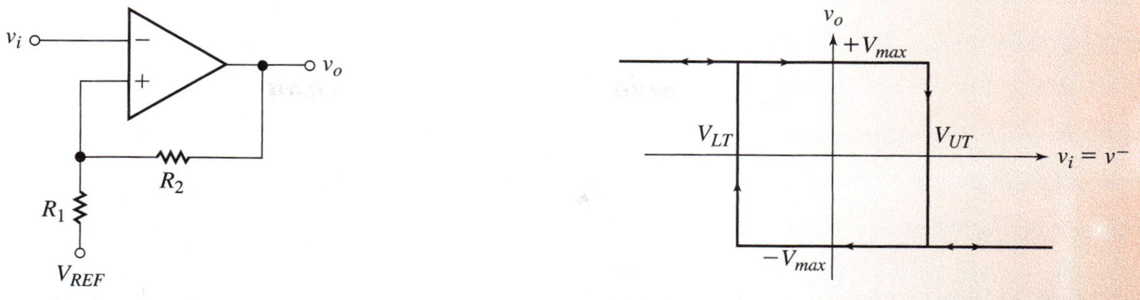

(a) Circuitry used to introduce hysteresis (V_{REF} may be 0.)

(b) Transfer characteristic of (a). The arrowheads show the portions of the characteristic followed when v_i is increasing (arrows pointing right) and when v_i is decreasing (arrows pointing left). Double-headed arrows mean that the output remains on that portion of the characteristic whether v_i is increasing or decreasing.

FIGURE 12–6 A voltage comparator with hysteresis (Schmitt trigger)

comparator from switching back and forth in response to random noise fluctuations in the input. For example, if $v^+ - v^-$ is near 0 V, and if the minimum difference is 1 mV, then noise voltages on the order of 1 mV will cause random switching of the comparator output. On the other hand, if the output will switch to one state only when the input rises past +1 V, and will switch to the other state only when the input falls below −1 V, then only a very large (2-V) noise voltage will cause it to switch states when the input is in the vicinity of 0 V.

Figure 12–6(a) shows how hysteresis can be introduced into comparator operation. In this case, the input is connected to the inverting terminal and a voltage divider is connected across the noninverting terminal between v_o and a fixed reference voltage V_{REF} (which may be 0). Figure 12–6(b) shows the resulting transfer characteristic (called a *hysteresis loop*). This characteristic shows that the output switches to V_H when v_i falls below a lower trigger level (V_{LT}) but will not switch to V_L unless v_i *rises* past an upper trigger level (V_{UT}). The arrows indicate the portions of the characteristic followed when the input is increasing (upper line) and when it is decreasing (lower line). A comparator having this characteristic is called a *Schmitt trigger*.

We can derive expressions for V_{UT} and V_{LT} using the superposition principle. Suppose first that the comparator output is shorted to ground. Then

$$v^+ = \frac{R_2}{R_1 + R_2} V_{REF} \qquad (V_0 = 0) \tag{12–3}$$

When V_{REF} is 0, we find

$$v^+ = \frac{R_1}{R_1 + R_2} V_o \qquad (V_{REF} = 0) \tag{12–4}$$

Therefore,

$$v^+ = \frac{R_2}{R_1 + R_2} V_{REF} + \frac{R_1}{R_1 + R_2}(V_o) \tag{12–5}$$

Because V_o has two possible values, V_H and V_L, so does v^+. These two values for v^+ establish the two trigger levels V_{UT} and V_{LT} that v_i must cross in order to produce switching on the output. Therefore,

$$V_{LT} = \frac{R_2}{R_1 + R_2}V_{REF} + \frac{R_1}{R_1 + R_2}(V_L) \qquad \textbf{(12–6)}$$

and

$$V_{UT} = \frac{R_2}{R_1 + R_2}V_{REF} + \frac{R_1}{R_1 + R_2}(V_H) \qquad \textbf{(12–7)}$$

Quantitatively, the hysteresis voltage (V_{hys}) of a Schmitt trigger is defined as the difference between the input trigger levels. Then,

$$V_{hys} = V_{UT} - V_{LT} \qquad \textbf{(12–8)}$$

Subtracting (12–6) from (12–7), we obtain

$$V_{UT} - V_{LT} = \left(\frac{R_1}{R_1 + R_2}\right)(V_H) - \left(\frac{R_1}{R_1 + R_2}\right)(V_L) \qquad \textbf{(12–9)}$$

or

$$V_{hys} = \frac{R_1}{R_1 + R_2}(V_H - V_L) \qquad \textbf{(12–10)}$$

EXAMPLE 12–1

1. Find the upper and lower trigger levels and the hysteresis of the Schmitt trigger shown in Figure 12–7. Sketch the hysteresis loop. The output switches between ±15 V.
2. Repeat (1) if $V_{REF} = 0$ V.
3. Repeat (1) if $V_{REF} = 0$ V and the output switches between 0 V and +6 V.

Solution

1. From equations 12–6 and 12–7,

$$V_{LT} = \left[\frac{10\text{ k}\Omega}{(5\text{ k}\Omega) + (10\text{ k}\Omega)}\right](6\text{ V}) + \left[\frac{5\text{ k}\Omega}{(5\text{ k}\Omega) + (10\text{ k}\Omega)}\right](-15\text{ V}) = -1\text{ V}$$

$$V_{UT} = \left[\frac{10\text{ k}\Omega}{(5\text{ k}\Omega) + (10\text{ k}\Omega)}\right](6\text{ V}) + \left[\frac{5\text{ k}\Omega}{(5\text{ k}\Omega) + (10\text{ k}\Omega)}\right](15\text{ V}) = +9\text{ V}$$

$$V_{hys} = (9\text{ V}) - (-1\text{ V}) = 10\text{ V}$$

(Note also, from equation 12–10, $V_{hys} = (5\text{ k}\Omega)(30\text{ V})/(15\text{ k}\Omega) = 10$ V.)

2. Since $V_{REF} = 0$,

$$V_{LT} = \left(\frac{R_1}{R_1 + R_2}\right)(V_L) = \left[\frac{5\text{ k}\Omega}{(5\text{ k}\Omega) + (10\text{ k}\Omega)}\right](-15\text{ V}) = -5\text{ V}$$

$$V_{UT} = \left(\frac{R_1}{R_1 + R_2}\right)(V_H) = \left[\frac{5\text{ k}\Omega}{(5\text{ k}\Omega) + (10\text{ k}\Omega)}\right](15\text{ V}) = +5\text{ V}$$

FIGURE 12–7 (Example 12–1)

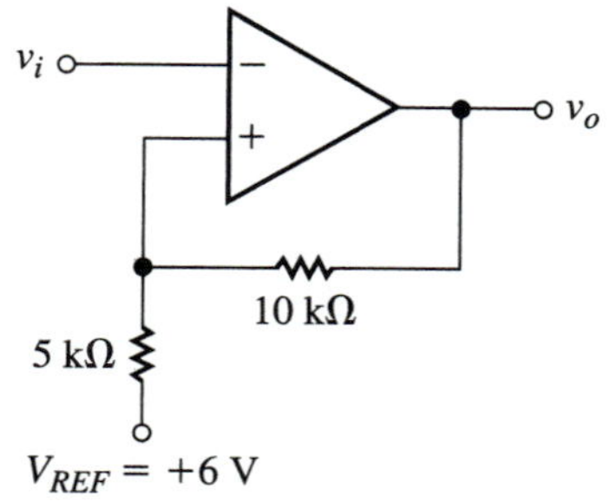

FIGURE 12–8 (Example 12–1)

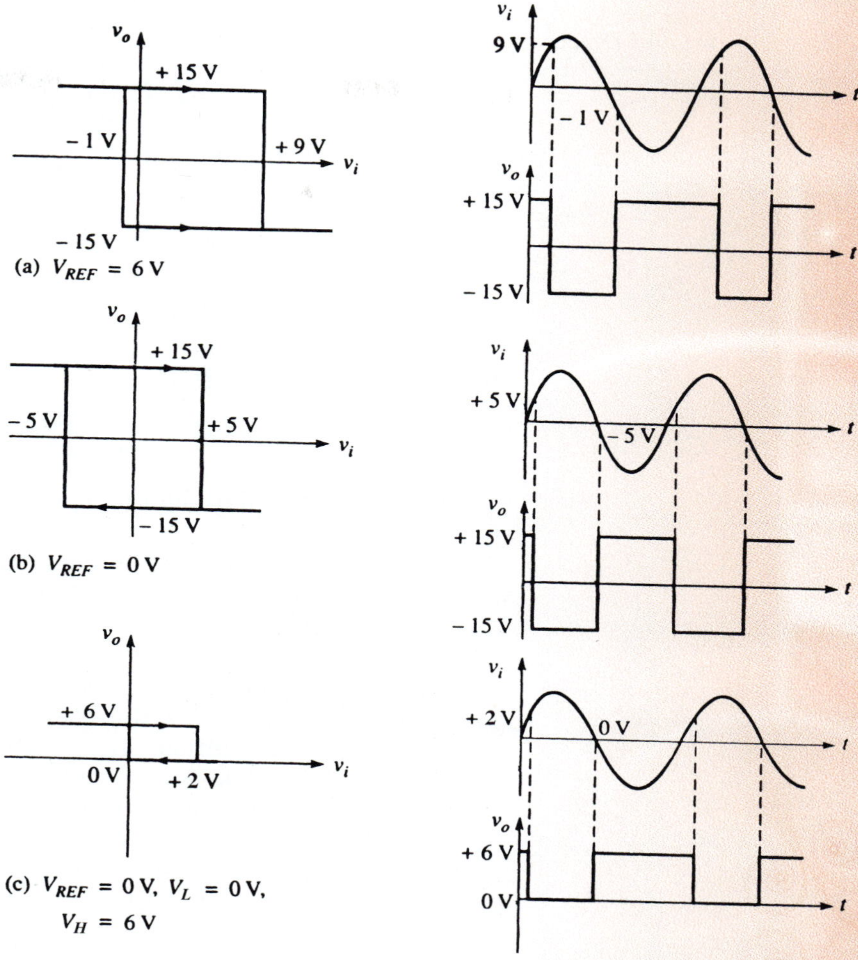

$$V_{hys} = (5\text{ V}) - (-5\text{ V}) = 10\text{ V}$$

(Again, equation 12–10 may be used.)

3. Since the output switches between 0 V and + 6 V, we must use $V_L = 0$ and $V_H = 6$ V in the trigger-level equations:

$$V_{LT} = \left(\frac{R_1}{R_1 + R_2}\right)(V_L) = 0\text{ V}$$

$$V_{UT} = \left(\frac{R_1}{R_1 + R_2}\right)(V_H) = +2\text{ V}$$

$$V_{hys} = (2\text{ V}) - (0\text{ V}) = 2\text{ V}$$

(Equation 12–10 yields the same result.)

Figure 12–8 shows the hysteresis loops for these cases, along with the output waveforms that result when v_i is a 10-V-peak sine wave.

The comparator we have discussed is called an *inverting* Schmitt trigger because the output is high when the input is low, and vice versa, as can be seen in Figure 12–8. Figure 12–9 shows a noninverting Schmitt trigger. For this circuit, the lower and upper trigger levels are

$$V_{UT} = \left(1 + \frac{R_1}{R_2}\right)V_{REF} - \frac{R_1}{R_2}V_L \qquad \textbf{(12–11)}$$

FIGURE 12–9 The noninverting Schmitt trigger

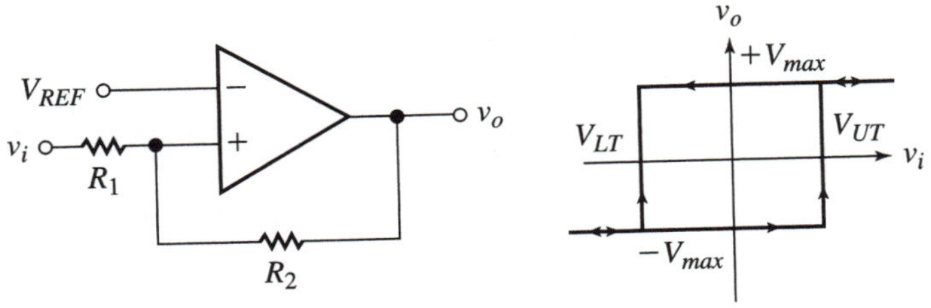

$$V_{LT} = \left(1 + \frac{R_1}{R_2}\right) V_{REF} - \frac{R_1}{R_2} V_H \tag{12–12}$$

and the hysteresis voltage is

$$V_{hys} = \frac{R_1}{R_2}(V_H - V_L) \tag{12–13}$$

The derivation of equations 12–11, 12–12, and 12–13 is left as an exercise at the end of this chapter.

Comparator circuits like those described in the foregoing discussion are, in general, known as *voltage-level detectors* or *voltage-crossing detectors*. They are further subdivided into inverting or noninverting types, and they may or may not include hysteresis or *noise immunity* in their operation. As explained earlier, hysteresis can prevent multiple transitions or "glitching" on the output when the input is noisy. This is accomplished by calculating R_1 and R_2 so that the hysteresis voltage V_{hys} is somewhat larger than the maximum expected peak-to-peak noise voltage. A safety margin of 20% or 25% is adequate. Larger margins will further decrease the probability of glitching but at the cost of increasing the error on the detection level.

To further illustrate the use of hysteresis for noise immunity in voltage-level detectors, consider a VLD required to trigger upon the crossing of +2 V by the input voltage v_i. If the maximum expected noise voltage is 100 mV, then the hysteresis voltage should be established at about 120 mV. To spread equally the error on the detection level, the actual triggering points can be set at 2 V ± 60 mV, or $V_{UT} = 2.06$ V and $V_{LT} = 1.94$ V. This is easily accomplished by calculating the required values for R_1, R_2, and V_{REF} using the equations derived for inverting and noninverting Schmitt triggers.

EXAMPLE 12–2

Design a voltage-level detector with noise immunity that indicates when an input signal crosses the nominal threshold of −2.5 V. The output is to switch from high to low when the signal crosses the threshold in the positive direction, and vice versa. Noise level expected is 0.2 V p–p, maximum. Assume the output levels are $V_H = 10$ V and $V_L = 0$ V.

Solution

From the triggering action described, we conclude that an inverting configuration is required. Let us specify the hysteresis voltage as 20% larger than the maximum p–p noise voltage, that is, $V_{hys} = 0.24$ V. The upper and lower trigger levels are then −2.5 ± 0.12, or

$$V_{UT} = -2.38\ \text{V} \quad \text{and} \quad V_{LT} = -2.62\ \text{V}$$

From equation 12–10, we obtain

$$\frac{V_H - V_L}{V_{hys}} = 1 + \frac{R_2}{R_1}$$

FIGURE 12–10 A possible final design for Example 12–2

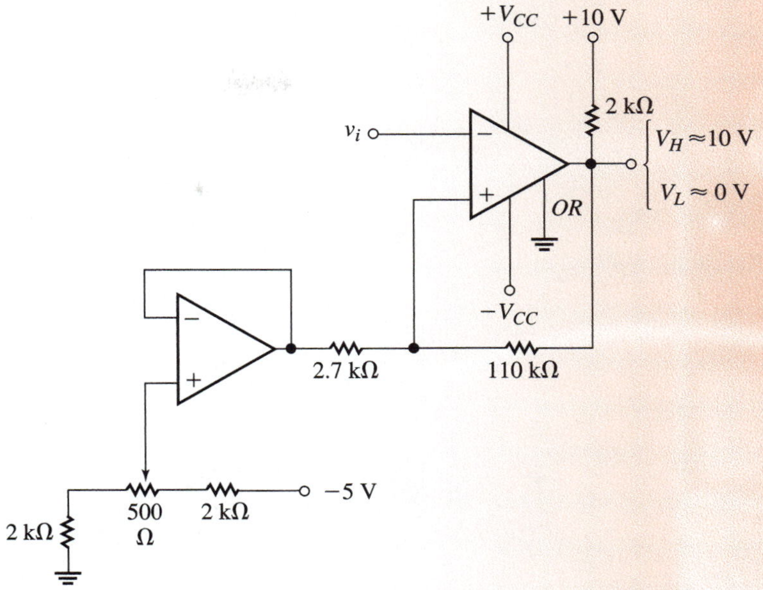

and

$$\frac{R_2}{R_1} = \frac{10 - 0}{0.24} - 1 = 40.7$$

With $V_L = 0$ and solving for V_{REF} from equation 12–6, we obtain

$$V_{REF} = (1 + R_1/R_2)\, V_{LT} = (1 + 1/40.7)\,(-2.62) = -2.68\text{ V}$$

Any values for R_2 and R_1 that satisfy the calculated ratio of 40.7 will do the job. It is a good practice to have more than 100 kΩ for the sum of R_1 and R_2 and 1 kΩ to 3 kΩ for the pull-up resistor on the output. The circuit shown in Figure 12–10 shows a possible final design. The potentiometer serves as a fine adjustment for V_{REF}, while the voltage follower makes V_{REF} to appear as an almost ideal voltage source.

12–2 MULTIVIBRATORS AND WAVESHAPING

Within the context of signal generation, voltage comparators play an important role in circuits designed to produce square waveforms from which other waveforms can be derived. For example, if a square waveform is fed into an electronic integrator, the result will be a triangular waveform. If the triangular waveform is fed into a second integrator, the output will be a pseudo-sinusoidal waveform consisting of interlaced parabolas. This processing of waveforms is known as *waveshaping* and will also be a topic in this chapter. Let us start by looking at a very simple comparator circuit capable of generating such a square waveform.

An Astable Multivibrator

The word *astable* means "unstable," and, like other unstable devices, an astable multivibrator is a (square-wave) oscillator. (A *bi*stable multivibrator, also called a *flip-flop*, is a digital device with two stable states; a *mono*stable multivibrator has one stable state; and an astable multivibrator has zero

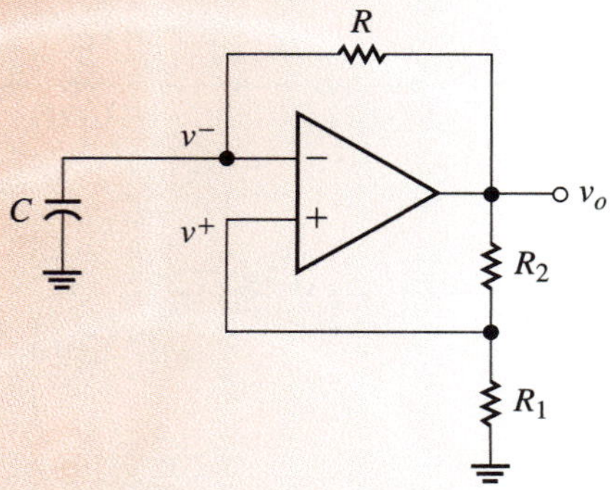

FIGURE 12–11 An astable multivibrator

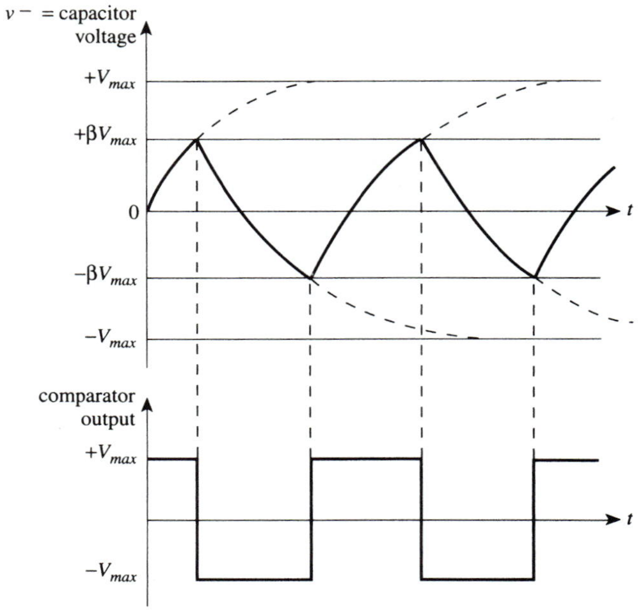

FIGURE 12–12 The waveforms on the capacitor and at the output of the astable multivibrator shown in Figure 12–11

stable states.) An astable multivibrator can be constructed by using a voltage comparator in a circuit like that shown in Figure 12–11. This circuit is an example of a *relaxation* oscillator, one whose operation depends on the repetitive charging and discharging of a capacitor.

For analysis purposes, let us assume that the output voltages of the comparator are equal in magnitude and opposite in polarity: $\pm V_{max}$. Figure 12–12 shows the voltage across capacitor C and the output waveform produced by the comparator. We begin by assuming that the output is at $+V_{max}$. Then, the voltage fed back to the noninverting input is

$$v^+ = \frac{R_1}{R_1 + R_2} V_{max} = +\beta V_{max} \quad \textbf{(12–14)}$$

Notice that v^- equals the voltage across the capacitor. The capacitor begins to charge through R *toward* a final voltage of $+V_{max}$. However, as soon as the capacitor voltage reaches a voltage equal to v^+, the comparator switches state. In other words, switching occurs at the point in time where $v^- = v^+ = +\beta V_{max}$. After the comparator switches state, its output is $-V_{max}$, and the voltage fed back to the noninverting input becomes

$$v^+ = \frac{R_1}{R_1 + R_2} (-V_{max}) = -\beta V_{max} \quad \textbf{(12–15)}$$

Because the comparator output is now negative, the capacitor begins to discharge through R toward $-V_{max}$. But, when that voltage falls to $-\beta V_{max}$, we once again have $v^+ = v^-$, and the comparator switches back to $+V_{max}$. This cycle repeats continuously, as shown in Figure 12–12, with the result that the output is a square wave that alternates between $\pm V_{max}$ volts.

It can be shown that the period of the multivibrator oscillation is

$$T = 2RC \ln\left(\frac{1 + \beta}{1 - \beta}\right) \quad \textbf{(12–16)}$$

or

$$T = 2RC \ln\left(1 + \frac{2R_1}{R_2}\right) \qquad (12\text{–}17)$$

Although the analysis of the astable multivibrator was done under the assumption that the output voltage oscillates between $+V_{max}$ and $-V_{max}$, the voltage comparator can be configured for any other output voltage levels (V_L and V_H) provided resistor R_1 is connected to a reference voltage, V_{REF}, whose value is exactly the middle point between V_L and V_H. This way, the charging and discharging voltage across the capacitor will be symmetrical and equation 12–17 will still hold. For example, if the square waveform is to be TTL (transistor transistor logic) compatible, it must oscillate between 0 V and 5 V. In that case, R_1 must be connected to a reference voltage of +2.5 V instead of being grounded.

EXAMPLE 12–3

Characterize the astable multivibrator shown in Figure 12–13. Establish the frequency range.

Solution

By observing the way the output section is connected, we conclude that the output voltage oscillates between $V_L = -5$ V and $V_H = +5$ V. To calculate the oscillation frequency range, we use the two extreme values for R_1 and R_2, which can be obtained from the two extreme positions of the potentiometer. Thus, when the wiper of the potentiometer is at its leftmost position, $R_1 = 10$ kΩ and $R_2 = 120$ kΩ. At the other extreme, $R_1 = 110$ kΩ and $R_2 = 20$ kΩ.

Substituting values into equation 12–17, we obtain

$$T = 2(15\text{ k})(47\text{ nF}) \ln\left(1 + \frac{20\text{ k}}{120\text{ k}}\right) = 217.4\ \mu\text{s}$$

$$T = 2(15\text{ k})(47\text{ nF}) \ln\left(1 + \frac{220\text{ k}}{20\text{ k}}\right) = 3.504\text{ ms}$$

Therefore,

$$f_{min} = 285\text{ Hz} \quad \text{and} \quad f_{max} = 4.6\text{ kHz}$$

FIGURE 12–13 (Example 12–3)

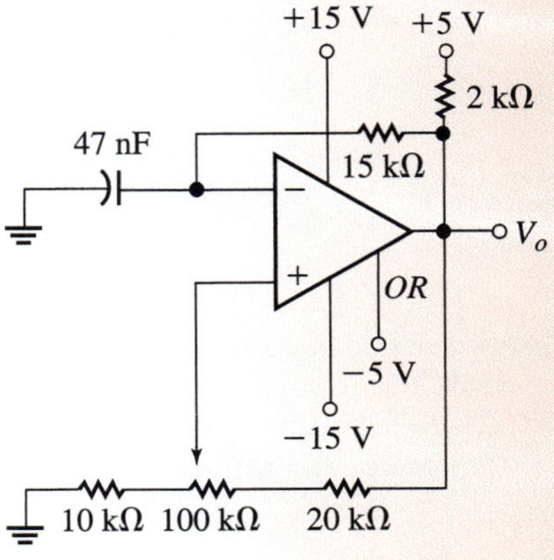

Waveshaping

Waveshaping is the process of altering the shape of a waveform in some prescribed manner to produce a new waveform having a desired shape. Examples include altering a triangular wave to produce a square wave, and vice versa, altering a square wave to produce a series of narrow pulses, altering a square or triangular wave to produce a sine wave, and altering a sine wave to produce a square wave. Waveshaping techniques are widely used in function generators, frequency synthesizers, and synchronization circuits that require different waveforms having precisely the same frequency. In this section we will discuss just two waveshaping techniques, one employing an integrator and one employing a differentiator.

Recall that the output of an ideal integrator is a voltage ramp when its input is a positive dc level. Positive area accumulates under the input waveform as time passes, and the output rises, as shown in Figure 12–14. If the dc level is suddenly made negative, and remains negative for an interval of time, then negative area accumulates, so the *net* area decreases and the output voltage begins to fall. At the point in time where the total accumulated negative area equals the previously accumulated positive area, the output reaches 0, as shown in Figure 12–14.

Figure 12–14 shows the basis for generating a triangular waveform using an integrator. When the input is a square wave that continually alternates between positive and negative levels, the output rises and falls in synchronism. Note that the *slope* of the output alternates between $+E$ and $-E$ volts/second. Figure 12–15(a) shows how a triangular wave is generated using a practical integrator. Figure 12–15(b) shows the actual waveform generated. Notice that the figure reflects the *phase inversion* caused by the amplifier in that the output decreases when the input is positive and increases when the input is negative. The *average* level (dc component) of the output is 0, assuming no input offset. Note also that the slopes of the triangular wave are $\pm E/R_1C$ volts/second, since the integrator gain is $1/R_1C$. It is important to be able to predict these slopes to ensure that they will be within the specified slew rate of the amplifier. In the figure, T represents the period of the square wave, and it can be seen that the peak value of the triangular wave is

$$|V_{pk}| = \frac{ET}{4R_1C} \qquad \textbf{(12–18)}$$

Equation 12–18 shows that the amplitude of the triangular wave decreases with increasing frequency, because the period T decreases with frequency. On the other hand, the slopes $\Delta V/\Delta t$ are (contrary to expectation) independent of frequency. This observation is explained by the fact that while an increase in frequency reduces the time Δt over which the voltage changes, it also reduces the total change in voltage, ΔV. The phenomenon is illustrated in Figure 12–16.

Equation 12–18 is a special case of a general relationship that can be used to find the integrator's peak output when the input is any periodic waveform symmetric about the horizontal axis:

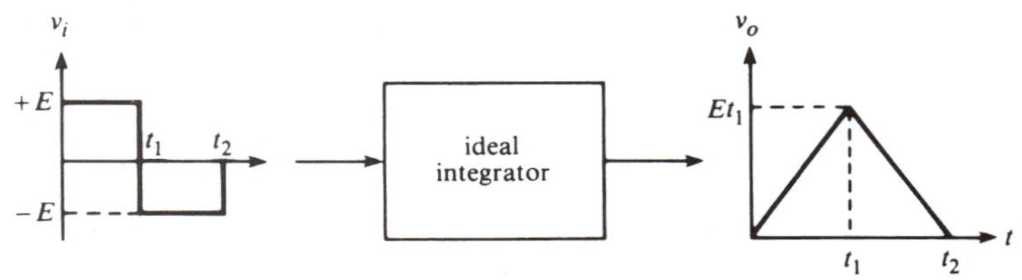

FIGURE 12–14 The integrator output rises to a maximum of Et_1 while the input is positive and then falls to 0 at time t_2 when the net area under the input is 0

FIGURE 12–15 The practical triangular-waveform generator

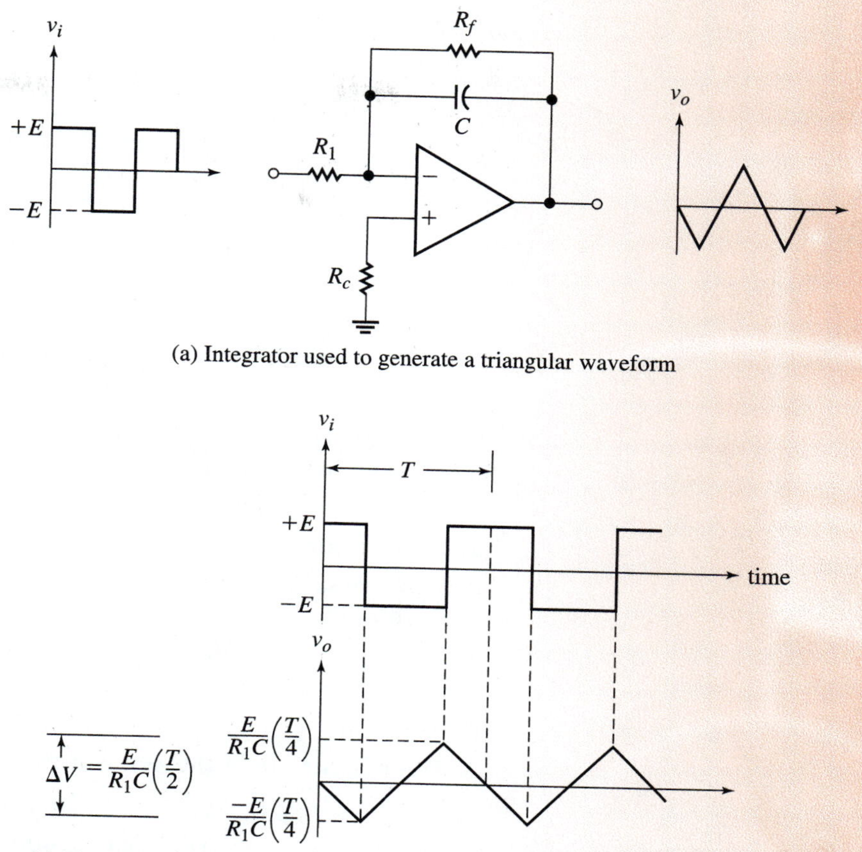

(a) Integrator used to generate a triangular waveform

(b) The triangular wave generated by the circuit in (a), showing voltage and time relations. T is the period of the square wave.

FIGURE 12–16 When the frequency of the triangular wave produced by an integrator is doubled, the amplitude is halved because the slopes remain the same

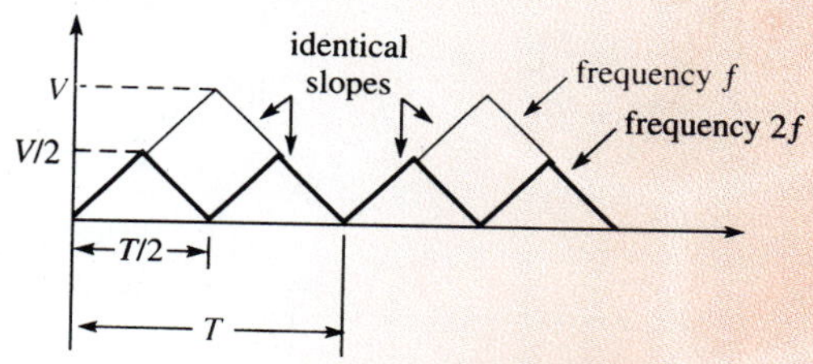

$$V_{pk} = \frac{\text{positive area in } \frac{1}{2} \text{ cycle of input}}{2R_1C} \tag{12–19}$$

For example, note that the positive area in one-half cycle of the square wave in Figure 12–15 is $E(T/2)$, so (12–19) reduces to (12–18) in that case.

Equations 12–18 and 12–19 are based on the assumption that there is *no input offset level or dc level in the input waveform.* When there is an input dc component, there will be an output dc component given by

$$V_o(\text{dc}) = \frac{-R_f}{R_1} V_i(\text{dc}) \tag{12–20}$$

Thus, a triangular output will be shifted up or down by an amount equal to $V_o(\text{dc})$.

FIGURE 12–17 (Example 12–4)

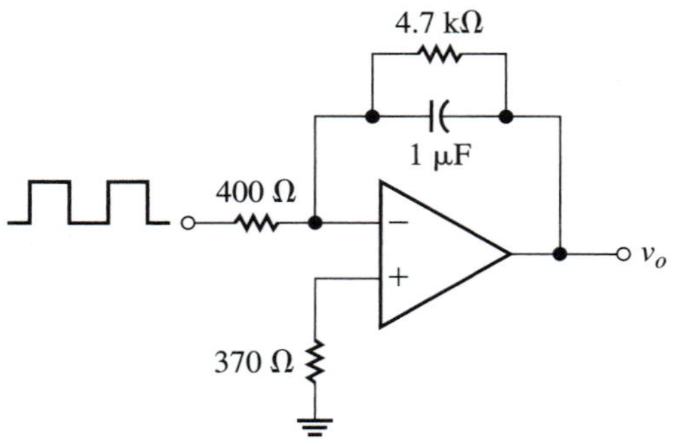

EXAMPLE 12–4

The integrator in Figure 12–17 is to be used to generate a triangular waveform from a 500-Hz square wave connected to its input. Suppose the square wave alternates between ±12 V.

1. What minimum slew rate should the amplifier have?
2. What maximum output voltage should the amplifier be capable of developing?
3. Repeat (2) if the dc component in the input is −0.2 V. (The square wave alternates between +11.8 V and −12.2 V.)

Solution

1. The magnitude of the slope of the triangular waveform is

$$\frac{\Delta V}{\Delta t} = \frac{E}{R_1 C} = \frac{12}{400(10^{-6})} = 3 \times 10^4 \text{ V/s}$$

Thus, we must have slew rate $S \geq 3 \times 10^4$ V/s.

2. The period T of the 500-Hz square wave is $T = 1/500 = 2 \times 10^{-3}$ s. By equation 12–18,

$$|V_{pk}| = \frac{ET}{4R_1 C} = \frac{12(2 \times 10^{-3})}{4(400)(10^{-6})} = 15 \text{ V}$$

The triangular wave alternates between peak values of +15 V and −15 V.

3. By equation 12–20, the dc component in the output is

$$v_o(\text{dc}) = \frac{-R_f}{R_1} V_i(\text{dc}) = \frac{-4700}{400}(-0.2 \text{ V}) = 2.35 \text{ V}$$

Therefore, the triangular output is shifted up by 2.35 V and alternates between peak values of 15 + 2.35 = 17.35 V and −15 + 2.35 = −12.65 V. The amplifier must be capable of producing a 17.35-V output to avoid distortion.

12–3 OSCILLATORS

An *oscillator* is a device that generates a periodic, ac output signal without any form of input signal required. The term is generally used in the context of a sine-wave signal generator; a square-wave generator is usually called a *multivibrator*. A *function generator* is a laboratory instrument that a user can set to produce sine, square, or triangular waves, with amplitudes and frequencies that can be adjusted at will. Desirable features of a sine-wave oscillator include the ability to produce a low distortion ("pure") sinusoidal waveform, and, in many applications, the capability of being easily adjusted so that the user can vary the frequency over some reasonable range.

Oscillation is a form of instability caused by feedback that *regenerates*, or reinforces, a signal that would otherwise die out due to energy losses. In order for the feedback to be regenerative, it must satisfy certain amplitude and phase relations that we will discuss shortly. Oscillation often plagues designers and users of high-gain amplifiers because of unintentional feedback paths that create signal regeneration at one or more frequencies. By contrast, an oscillator is designed to have a feedback path with known characteristics, so that a predictable oscillation will occur at a predetermined frequency.

The Barkhausen Criterion

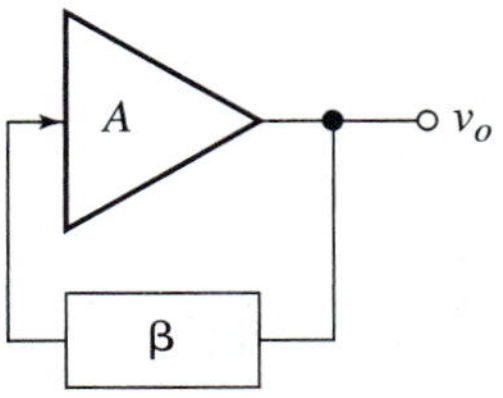

FIGURE 12–18
Block diagram of an oscillator

We have stated that an oscillator has no input per se, so the reader may wonder what we mean by "feedback"—feedback to where? In reality, it makes no difference where, because we have a closed loop with no summing junction at which any external input is added. Thus, we could start anywhere in the loop and call that point both the "input" and the "output"; in other words, we could think of the "feedback" path as the entire path through which signal flows in going completely around the loop. However, it is customary and convenient to take the output of an amplifier as a reference point and to regard the feedback path as that portion of the loop that lies between amplifier output and amplifier input. This viewpoint is illustrated in Figure 12–18, where we show an amplifier having gain A and a feedback path having gain β. β is the usual feedback ratio that specifies the portion of amplifier output voltage fed back to amplifier input. Every oscillator must have an amplifier, or equivalent device, that supplies energy (from the dc supply) to replenish resistive losses and thus sustain oscillation.

In order for the system shown in Figure 12–18 to oscillate, the loop gain $A\beta$ must satisfy the *Barkhausen criterion*, namely,

$$A\beta = 1 \qquad (12\text{–}21)$$

Imagine a small variation in signal level occurring at the input to the amplifier, perhaps due to noise. The essence of the Barkhausen criterion is that this variation will be reinforced and signal regeneration will occur only if the net gain around the loop, beginning and ending at the point where the variation occurred, is unity. It is important to realize that unity gain means not only a gain magnitude of 1 but also an *in-phase* signal reinforcement. Negative feedback causes signal cancellation because the feedback voltage is out of phase. By contrast, the unity loop-gain criterion for oscillation is often called *positive feedback*.

To understand and apply the Barkhausen criterion, we must regard both the gain and the phase shift of $A\beta$ as *functions of frequency*. Reactive elements, capacitance in particular, contained in the amplifier and/or feedback cause the gain magnitude and phase shift to change with frequency. In general, there will be only one frequency at which the gain magnitude is unity and at which, simultaneously, the total phase shift is equivalent to 0° (in phase—a multiple of 360°). *The system will oscillate at the frequency that satisfies those conditions.* Designing an oscillator amounts to selecting reactive components and incorporating them into circuitry in such a way that the conditions will be satisfied at a predetermined frequency.

To show the dependence of the loop gain $A\beta$ on frequency, we write $A\beta(j\omega)$, a complex phasor that can be expressed in both polar and rectangular form:

$$A\beta(j\omega) = |A\beta|\angle\theta = |A\beta| \cos\theta + j\,|A\beta| \sin\theta \tag{12–22}$$

where $|A\beta|$ is the loop gain magnitude, a function of frequency, and θ is the phase shift, also a function of frequency. The Barkhausen criterion requires that

$$|A\beta| = 1 \tag{12–23}$$

and

$$\theta = \pm 360°n \tag{12–24}$$

where n is any integer, including 0. In polar and rectangular forms, the Barkhausen criterion is expressed as

$$A\beta(j\omega) = 1\underline{/\pm 360°n} = 1 + j0 \tag{12–25}$$

EXAMPLE 12–5

The gain of a certain amplifier as a function of frequency is $A(j\omega) = -16 \times 10^6/j\omega$ A feedback path connected around it has $\beta(j\omega) = 10^3/(2 \times 10^3 + j\omega)^2$. Will the system oscillate? If so, at what frequency?

Solution

The loop gain is

$$A\beta = \left(\frac{-16 \times 10^6}{j\omega}\right)\left[\frac{10^3}{(2 \times 10^3 + j\omega)^2}\right] = \frac{-16 \times 10^9}{j\omega(2 \times 10^3 + j\omega)^2}$$

To determine if the system will oscillate, we will first determine the frequency, if any, at which the phase angle of $A\beta(\theta = \underline{/A\beta})$ equals 0 or a multiple of 360°. Using phasor algebra, we have

$$\theta = \underline{/A\beta} = \underline{\Big/\frac{-16 \times 10^9}{j\omega(2 \times 10^3 + j\omega)^2}} = \underline{/-16 \times 10^9} + \underline{/1/j\omega} + \underline{\Big/\frac{1}{(2 \times 10^3 + j\omega)^2}}$$

$$= -180° - 90° - 2\arctan(\omega/2 \times 10^3)$$

This expression will equal $-360°$ if $2\arctan(\omega/2 \times 10^3) = 90°$, or

$$\arctan(\omega/2 \times 10^3) = 45°$$

$$\omega/2 \times 10^3 = 1$$

$$\omega = 2 \times 10^3 \text{ rad/s}$$

Thus, the phase shift around the loop is $-360°$ at $\omega = 2000$ rad/s. We must now check to see if the gain magnitude $|A\beta| = 1$ at $\omega = 2 \times 10^3$. The gain magnitude is

$$|A\beta| = \left|\frac{-16 \times 10^9}{j\omega(2 \times 10^3 + j\omega)^2}\right| = \frac{|-16 \times 10^9|}{|j\omega||(2 \times 10^3 + j\omega)|^2}$$

$$= \frac{16 \times 10^9}{\omega[(2 \times 10^3)^2 + \omega^2]}$$

Substituting $\omega = 2 \times 10^3$, we find

$$|A\beta| = \frac{16 \times 10^9}{2 \times 10^3(4 \times 10^6 + 4 \times 10^6)} = 1$$

Thus, the Barkhausen criterion is satisfied at $\omega = 2 \times 10^3$ rad/s and oscillation occurs at that frequency ($2 \times 10^3/2\pi = 318.3$ Hz).

Example 12–5 illustrated an application of the *polar* form of the Barkhausen criterion, since we solved for $\angle A\beta$ and then determined the frequency at which that angle equals $-360°$. It is instructive to demonstrate how the same result can be obtained using the *rectangular* form of the criterion: $A\beta = 1 + j0$. Toward that end, we first expand the denominator:

$$A\beta = \frac{-16 \times 10^9}{j\omega(2 \times 10^3 + j\omega)^2} = \frac{-16 \times 10^9}{j\omega(4 \times 10^6 + j4 \times 10^3\omega - \omega^2)}$$

$$= \frac{-16 \times 10^9}{j\omega[(4 \times 10^6 - \omega^2) + j4 \times 10^3\omega]} = \frac{16 \times 10^9}{4 \times 10^3\omega^2 - j\omega(4 \times 10^6 - \omega^2)}$$

To satisfy the Barkhausen criterion, this expression for $A\beta$ must equal 1. We therefore set it equal to 1 and simplify:

$$1 = \frac{16 \times 10^9}{4 \times 10^3\omega^2 - j\omega(4 \times 10^6 - \omega^2)}$$

$$4 \times 10^3\omega^2 - j\omega(4 \times 10^6 - \omega^2) = 16 \times 10^9$$

$$(4 \times 10^3\omega^2 - 16 \times 10^9) - j\omega(4 \times 10^6 - \omega^2) = 0$$

In order for this expression to equal 0, *both the real and imaginary parts must equal 0*. Setting either part equal to 0 and solving for ω will give us the same result we obtained before:

$$4 \times 10^3\omega^2 - 16 \times 10^9 = 0 \Rightarrow \omega = 2 \times 10^3$$

$$4 \times 10^6 - \omega^2 = 0 \Rightarrow \omega = 2 \times 10^3$$

This result was obtained with somewhat more algebraic effort than previously. In some applications, it is easier to work with the polar form than the rectangular form, and in others, the reverse is true.

The RC Phase-Shift Oscillator

One of the simplest kinds of oscillators incorporating an operational amplifier can be constructed as shown in Figure 12–19. Here we see that the amplifier is connected in an inverting configuration and drives three cascaded (high-pass) RC sections. The arrangement is called an *RC phase-shift oscillator*. The inverting amplifier causes a 180° phase shift in the signal

FIGURE 12–19 An RC phase-shift oscillator

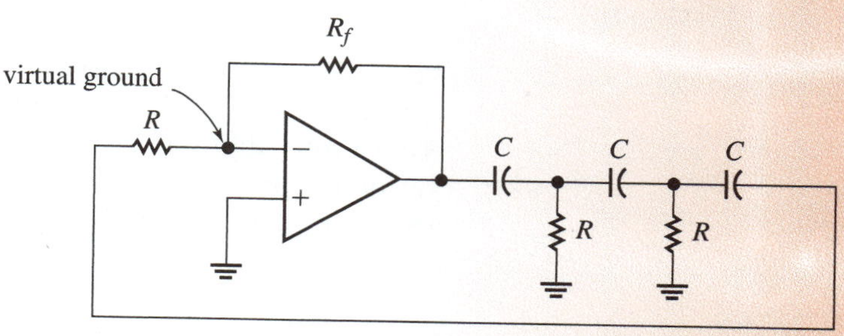

passing through it, and the purpose of the cascaded RC sections is to introduce an additional 180° at some frequency. Recall that the output of a single, high-pass RC network leads its input by a phase angle that depends on the signal frequency. When the signal passes through all three RC sections, there will be some frequency at which the cumulative phase shift is 180°. When the signal having that frequency is fed back to the inverting amplifier, as shown in the figure, the total phase shift around the loop will equal 180° + 180° = 360° (or, equivalently, −180° + 180° = 0°) and oscillation will occur at that frequency, provided the loop gain is 1. The gain necessary to overcome the loss in the RC cascade and bring the loop gain up to 1 is supplied by the amplifier ($v_o/v_i = -R_f/R$). Note that the input resistor to the inverting amplifier is also the last resistor of the RC cascade.

With considerable algebraic effort, it can be shown (Exercise 12–15) that the feedback ratio determined by the RC cascade is

$$\beta = \frac{R^3}{(R^3 - 5RX_C^2) + j(X_C^3 - 6R^2X_C)} \tag{12–26}$$

In order for oscillation to occur, the cascade must shift the phase of the signal by 180°, which means the angle of β must be 180°. When the angle of β is 180°, β is a purely real number. In that case, the imaginary part of the denominator of equation 12–26 is 0. Therefore, we can find the oscillation frequency by finding the value of ω that makes the imaginary part equal 0. Setting it equal to 0 and solving for ω, we find

$$X_C^3 - 6R^2X_C = 0$$

$$X_C^3 = 6R^2X_C$$

$$X_C^2 = 6R^2$$

$$\frac{1}{(\omega C)^2} = 6R^2$$

$$\omega_o = \frac{1}{RC\sqrt{6}} \tag{12–27}$$

where ω_o is the oscillation angular frequency, or

$$f_o = \frac{1}{2\pi\sqrt{6}RC} \tag{12–28}$$

We can find the gain that the amplifier must supply by finding the reduction in gain caused by the RC cascade. This we find by evaluating the magnitude of β at the oscillation frequency: $\omega = 1/(\sqrt{6}RC)$. At that frequency, the imaginary term in equation 12–26 is 0 and β is the real number

$$\beta|_{\omega_o} = \frac{R^3}{R^3 - 5RX_C^2} = \frac{R^3}{R^3 - 5R\left(\frac{\sqrt{6}RC}{C}\right)^2} = \frac{R^3}{R^3 - 30R^3} = -1/29 \tag{12–29}$$

The minus sign confirms that the cascade inverts the feedback at the oscillation frequency. We see that the amplifier must supply a gain of −29 to make the loop gain $A\beta = 1$. Thus, we require

$$\frac{R_f}{R} = 29 \quad \text{or} \quad R_f = 29R \tag{12–30}$$

In practice, the feedback resistor is made adjustable to allow for small differences in component values.

EXAMPLE 12–6

DESIGN

Design an RC phase-shift oscillator that will oscillate at 100 Hz.

Solution

From equation 12–28

$$f = 100\text{ Hz} = \frac{1}{2\pi\sqrt{6}RC}$$

Let $C = 0.5\ \mu$F. Then,

$$R = \frac{1}{100(2\pi)\sqrt{6}(0.5 \times 10^{-6})} = 1300\ \Omega$$

From equation 12–30, $R_f = 29R = 29(1300\ \Omega) = 37.7\text{ k}\Omega$. The completed circuit is shown in Figure 12–20. R_f is made adjustable so the loop gain can be set precisely to 1.

The Wien-Bridge Oscillator

Figure 12–21 shows a widely used type of oscillator called a *Wien bridge*. The operational amplifier is used in a noninverting configuration, and the impedance blocks labeled Z_1 and Z_2 form a voltage divider that determines the feedback ratio. Note that a portion of the output voltage is fed back through this impedance divider to the + input of the amplifier. Resistors R_g and R_f determine the amplifier gain and are selected to make the magnitude of the loop gain equal to 1. If the feedback impedances are chosen properly, there will be some frequency at which there is zero phase shift in the signal fed back to the amplifier input (v^+). Because the amplifier is noninverting, it also contributes zero phase shift, so the total phase shift around the loop is 0 at that frequency, as required for oscillation.

In the most common version of the Wien-bridge oscillator, Z_1 is a series RC combination and Z_2 is a parallel RC combination, as shown in Figure 12–22. For this configuration,

$$Z_1 = R_1 - jX_{C_1}$$

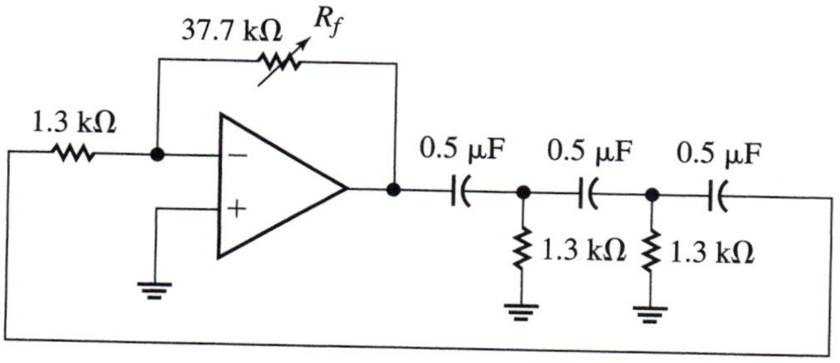

FIGURE 12–20 (Example 12–6)

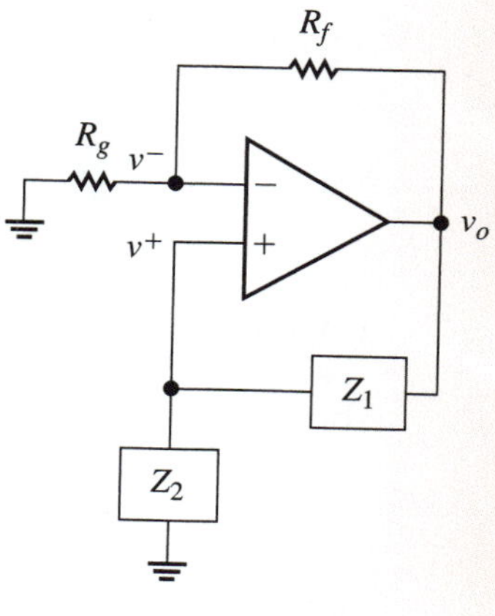

FIGURE 12–21 The Wien-bridge oscillator. Z_1 and Z_2 determine the feedback ratio to the noninverting input. R_f and R_g control the magnitude of the loop gain

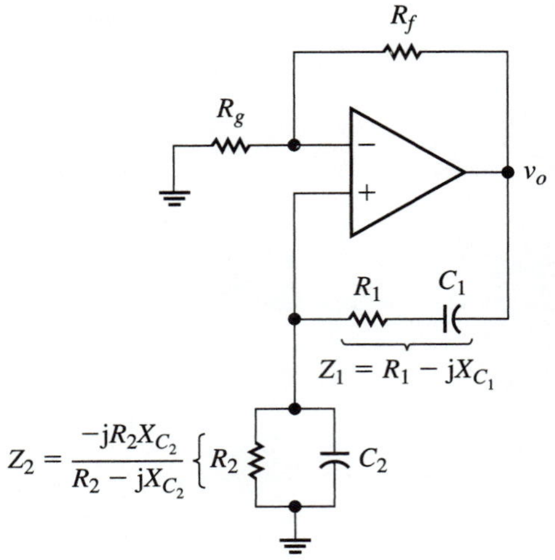

FIGURE 12–22 The Wien-bridge oscillator showing the RC networks that form Z_1 and Z_2

and

$$Z_2 = R_2 \| -jX_{C_2} = \frac{-jR_2X_{C_2}}{R_2 - jX_{C_2}}$$

The feedback ratio is then

$$\beta = \frac{v^+}{v_o} = \frac{Z_2}{Z_1 + Z_2} = \frac{-jR_2X_{C_2}/(R_2 - jX_{C_2})}{R_1 - jX_{C_1} - jR_2X_{C_2}/(R_2 - jX_{C_2})} \qquad \textbf{(12–31)}$$

which, upon simplification, becomes

$$\frac{v^+}{v_o} = \frac{R_2X_{C_2}}{(R_1X_{C_2} + R_2X_{C_1} + R_2X_{C_2}) + j(R_1R_2 - X_{C_1}X_{C_2})} \qquad \textbf{(12–32)}$$

In order for v^+ to have the same phase as v_o, this ratio must be a purely real number. Therefore, the imaginary term in (12–32) must be 0. Setting the imaginary term equal to 0 and solving for ω gives us the oscillation frequency ω_o:

$$R_1R_2 - X_{C_1}X_{C_2} = 0$$

$$R_1R_2 = \left(\frac{1}{\omega C_1}\right)\left(\frac{1}{\omega C_2}\right)$$

$$\omega_o^2 = \frac{1}{R_1R_2C_1C_2}$$

$$\omega_o = \frac{1}{\sqrt{R_1R_2C_1C_2}} \qquad \textbf{(12–33)}$$

In most applications, the resistors are made equal and so are the capacitors: $R_1 = R_2 = R$ and $C_1 = C_2 = C$. In this case, the oscillation frequency becomes

$$\omega_o = \frac{1}{\sqrt{R^2C^2}} = \frac{1}{RC} \qquad \textbf{(12–34)}$$

or

$$f_o = \frac{1}{2\pi RC} \qquad \textbf{(12–35)}$$

When $R_1 = R_2 = R$ and $C_1 = C_2 = C$, the capacitive reactance of each capacitor at the oscillation frequency is

$$X_{C_1} = X_{C_2} = \frac{1}{\omega C} = \frac{1}{\left(\frac{1}{RC}\right)C} = R$$

Substituting $X_{C_1} = X_{C_2} = R = R_1 = R_2$ in equation 12–32, we find that the feedback ratio at the oscillation frequency is

$$\frac{v^+}{v_o} = \frac{R^2}{2R^2 + j0} = \frac{1}{3}$$

Therefore, the amplifier must provide a gain of 3 to make the magnitude of the loop gain unity and sustain oscillation. Since the amplifier gain is $1 + R_f/R_g$, we require

$$1 + \frac{R_f}{R_g} = 3 \Rightarrow \frac{R_f}{R_g} = 2 \quad (12\text{–}36)$$

(Another way of reaching this same conclusion is to recognize that the operational amplifier maintains $v^+ \approx v^-$. But, from Figure 12–22, we see that $v^- = v_oR_g/(R_g + R_f)$, so $v^-/v_o = (R_g + R_f)/R_g = v^+/v_o = 1/3$.)

EXAMPLE 12–7

DESIGN

Design a Wien-bridge oscillator that oscillates at 25 kHz.

Solution

Let $C_1 = C_2 = 0.001\ \mu$F. Then, from equation 12–35,

$$f = 25 \times 10^3 \text{ Hz} = \frac{1}{2\pi R(10^{-9}\text{ F})}$$

$$R = \frac{1}{2\pi(25 \times 10^3 \text{ Hz})(10^{-9}\text{ F})} = 6366\ \Omega$$

Let $R_g = 10$ kΩ. Then, from equation 12–36,

$$\frac{R_f}{R_g} = 2 \Rightarrow R_f = 20\text{ k}\Omega$$

In practical Wien-bridge oscillators, R_f is not equal to exactly $2R_g$ because component tolerances prevent R_1 from being exactly equal to R_2 and C_1 from being exactly equal to C_2. Furthermore, the amplifier is not ideal, so v^- is not exactly equal to v^+. In a circuit constructed for laboratory experimentation, R_f should be made adjustable so that the loop gain can be set as necessary to sustain oscillation. Practical oscillators incorporate a nonlinear device in the R_f–R_g feedback loop to provide automatic adjustment of the loop gain as necessary to sustain oscillation. This arrangement is a form of *automatic gain control* (AGC), whereby a reduction in signal level changes the resistance of the nonlinear device in a way that restores gain.

The Colpitts Oscillator

In the Colpitts oscillator, the impedance in the feedback circuit is a low-pass LC network. See Figure 12–23(a). The frequency of oscillation is the frequency at which the phase shift through the network is 180°. At that

FIGURE 12–23 The Colpitts oscillator

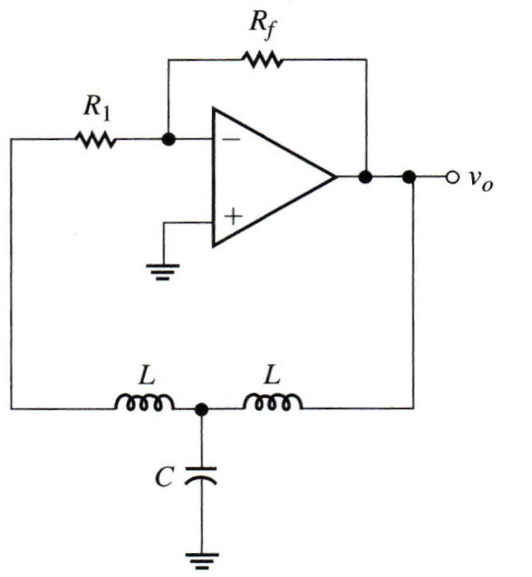

(a) The Colpitts oscillator

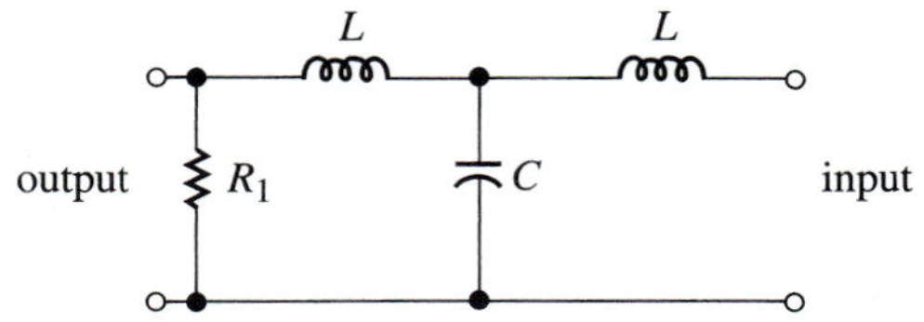

(b) The feedback network

frequency, the β function is a real number. Figure 12–23(b) shows the feedback network with R_1 as a load (recall the virtual ground in inverting amplifiers). It can be shown that the transfer characteristic function of the β network is

$$\beta = \frac{1}{(1 - \omega^2 LC) + j\left(\frac{2\omega L - \omega^3 L^2 C}{R}\right)} \quad \text{(12–37)}$$

In order for β to equal a real number, the imaginary term must equal 0. Thus,

$$2\omega L = \omega^3 L^2 C \Rightarrow \omega_o^2 LC = 2 \quad \text{(12–38)}$$

or

$$f_o = \frac{1}{2\pi\sqrt{LC/2}} \quad \text{(12–39)}$$

where f_o is the oscillation frequency in hertz. With the condition $\omega^2 LC = 2$, it follows that $\beta = -1$, thus verifying the fact that there is a 180° phase shift through the β network. It follows also that the gain from the amplifier must be −1, that is, $R_f = R_1$.

A practical design consideration is that the component values should be such that the "Q" of the β network, given by

$$Q = R_1\sqrt{C/L}$$

is not too low or too high (2 to 10 is very adequate). The reader is encouraged to simulate the β network under different R_1 values and observe the peak of the low-pass response as R_1 is increased and decreased. Note that Q is directly proportional to R_1.

EXAMPLE 12–8

DESIGN

Design a Colpitts oscillator that will oscillate at 100 kHz.

Solution

Let us choose $R_1 = R_f = 5\ \text{k}\Omega$ and $C = 0.001\ \mu\text{F}$. From equation 12–39,

$$L = \frac{2}{(2\pi)^2 f_o^2 C} = \frac{2}{(2\pi)^2(100 \times 10^3\ \text{Hz})^2(0.001 \times 10^{-6}\ \text{F})} \approx 5\ \text{mH}$$

Check on Q:

$$Q = 5\text{k}\sqrt{\frac{1\ \text{nF}}{5\ \text{mH}}} \approx 2.2\ \ (\text{OK})$$

The Hartley Oscillator

The β network in the feedback of a Hartley oscillator is a high-pass LC network consisting of two capacitors and one inductor. See Figure 12–24. As in the Colpitts oscillator, oscillation occurs at the frequency at which the β function is real. We can show (Exercise 12–19) that the oscillation frequency for the Colpitts oscillator is

$$f_o = \frac{1}{2\pi\sqrt{2LC}} \tag{12–40}$$

and that β at the oscillation frequency is, as would be expected, −1. Again, we should check on the "Q" of this circuit, which is identical to that of the Colpitts oscillator.

It should be noted that any LC low-pass or high-pass function can be used for obtaining lagging or leading phase reversal required for oscillation in conjunction with an inverting amplifier, such as with the Colpitts and Hartley circuits. In particular, a third-order, low-pass Butterworth filter, which would require different inductor values, will produce an oscillation equal to $\sqrt{2}$ times its cutoff frequency, provided the gain of the amplifier is −3 and resistor R_1 is calculated as part of the filter. Similarly, a high-pass Butterworth filter yields an oscillation frequency of 0.707 (or $1/\sqrt{2}$) times the cutoff frequency along with the gain of −3. The reader is encouraged to simulate these possibilities.

As a final comment regarding the simulation of oscillators using ideal amplifier models, because these models do not use supply voltages, one has to provide the means for "starting up" the oscillator. One simple way is to apply a large and narrow voltage pulse from a voltage source connected in

FIGURE 12–24 The Hartley oscillator

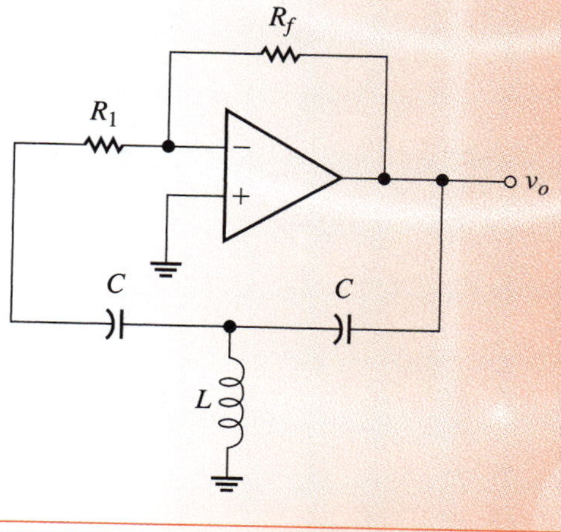

series with an inductor. The width of the pulse should be much smaller than the period of the waveform. Another way is by applying a current pulse in parallel with a capacitor. Refer to the PULSE function for voltage and current sources in SPICE or PSpice.

12–4 THE 8038 INTEGRATED-CIRCUIT FUNCTION GENERATOR

Function generators capable of producing sinusoidal, triangular, and rectangular waveforms are commercially available in integrated-circuit form. An example is the 8038, manufactured by Intersil as the ICL8038 and also available from other manufacturers. This versatile circuit is capable of generating the aforementioned waveforms simultaneously (at three separate output terminals) over a frequency range from 0.0001 Hz to 1 MHz. Frequency is determined by externally connected resistor-capacitor combinations and can also be controlled by an external voltage. In the latter mode of operation, the generator serves as a *voltage-controlled oscillator* (VCO). (A VCO is also called a *voltage-to-frequency converter*.)

The 8038 is basically a relaxation oscillator that generates a triangular waveform. The triangular waveform is converted internally to a rectangular waveform using voltage comparators and a digital-type storage device (flip-flop). Sixteen internal transistors are used to shape the triangular wave into an approximation of a sine wave. Under ideal conditions (using external circuit connections specified in the manufacturer's literature), the total harmonic distortion of the sine-wave output can be reduced to less than 1%.

Figure 12–25(a) shows a pin diagram of the 8038 and identifies the function of each pin. Pins 1 and 12, "sine wave adjust," are used for connecting external resistors to minimize sine-wave distortion, as previously mentioned. Pins 4 and 5, "duty cycle and frequency adjust," are used for connecting external resistors that, in conjunction with an external capacitor connected to pin 10, "timing capacitor," determine the *duty cycle* and frequency of the outputs. The duty cycle of an output is the time it is high, expressed as a percentage of the period of oscillation. For example, a square wave, which is high for one-half of a period and low for the other half, has a 50% duty cycle. Pin 8, "FM sweep input," is the input to which an external voltage may be connected to adjust the frequency of oscillation when the 8038 is operated as a VCO. (FM refers to *frequency modulation:* As the input voltage is adjusted, the output frequency changes in direct proportion to the change in voltage.)

Figure 12–25(b) shows the simplest possible circuit connections for using the 8038 as a fixed-frequency, 50%–duty-cycle function generator. In the configuration shown, the frequency of oscillation is

$$f_o = \frac{0.15}{RC} \qquad (12\text{–}41)$$

Manufacturer's product literature can be consulted for external circuit connections required when the 8038 is used as a VCO and for duty cycles other than 50%.

12–5 CLIPPING AND RECTIFYING CIRCUITS

Clipping Circuits

In Chapter 7, we referred to *clipping* as the undesirable result of overdriving an amplifier. We have seen that any attempt to push an output voltage beyond the limits through which it can "swing" causes the tops and/or bottoms of a

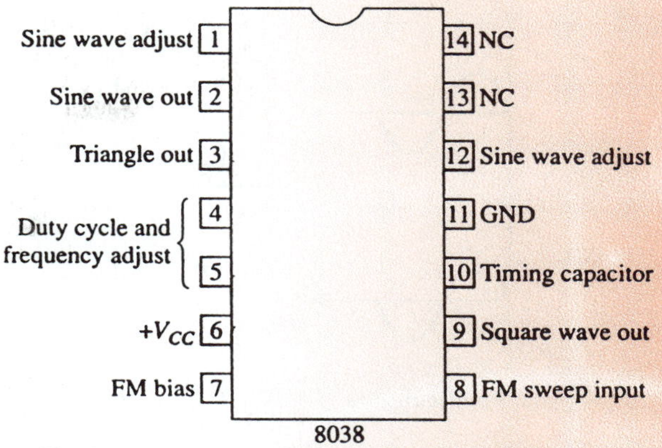

(a) Pin diagram (NC = no connection)

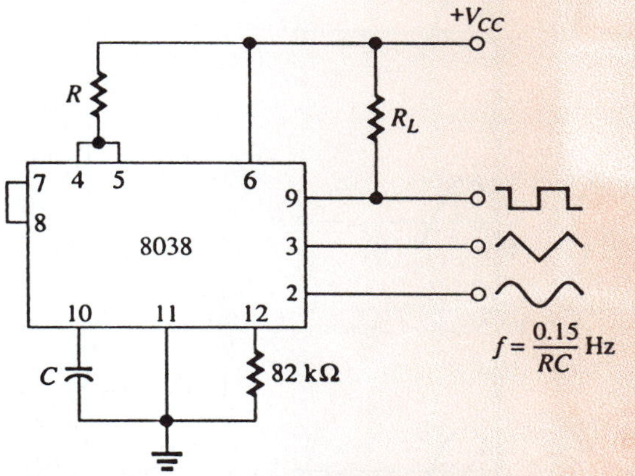

(b) Connection for operation as a 50%–duty-cycle, fixed-frequency function generator

FIGURE 12–25 The 8038 function generator

waveform to be "clipped" off, resulting in distortion. However, in numerous practical applications, including waveshaping and nonlinear function generation, waveforms are *intentionally* clipped.

Figure 12–26 shows how the transfer characteristic of a device is modified to reflect the fact that its output is clipped at certain levels. In each of the examples shown, note that the characteristic becomes horizontal at the output level, where clipping occurs. The horizontal line means that the output remains constant regardless of the input level in that region. Outside the clipping region, the transfer characteristic is simply a line whose slope equals the gain of the device. This is the region of normal, *linear* operation. In these examples, the devices are assumed to have unity gain, so the slope of each line in the linear region is 1.

Figure 12–27 illustrates a somewhat different kind of clipping action. Instead of the positive or negative peaks being "chopped off," the output follows the input when the signal is above or below a certain level. The transfer characteristics show that linear operation occurs only when certain signal levels are reached and that the output remains constant below those levels. This form of clipping can also be thought of as a special case of that shown in Figure 12–26. Imagine, for example, that the clipping level in Figure 12–26(b) is raised to a positive value; then the result is the same as Figure 12–27(a).

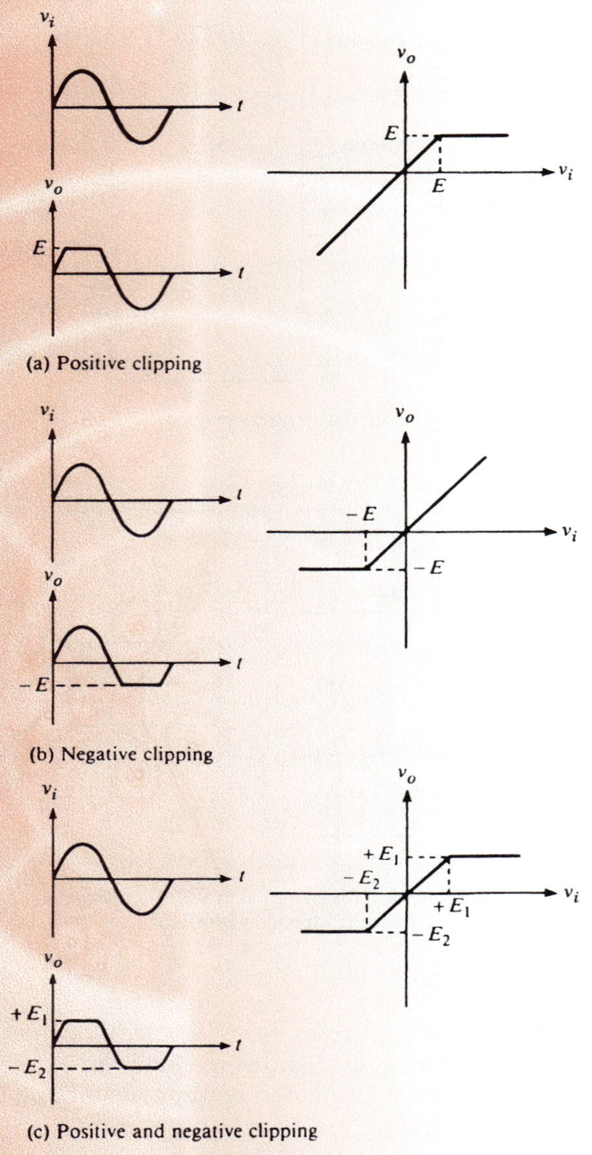

FIGURE 12–26 Waveforms and transfer characteristics of clipping circuits

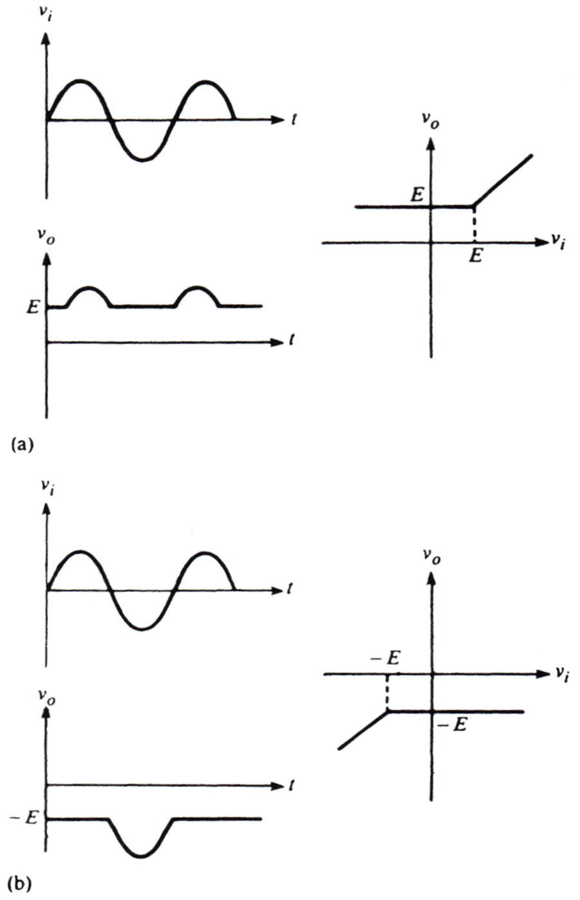

FIGURE 12–27 Another form of clipping. Compare with Figure 12–26.

Clipping can be accomplished using *biased diodes*, a technique that is more efficient than overdriving an amplifier. Clipping circuits rely on the fact that diodes have very low impedances when they are forward biased and are essentially open circuits when reverse biased. If a certain point in a circuit, such as the output of an amplifier, is connected through a very small impedance to a *constant* voltage, then the voltage at the circuit point cannot differ significantly from the constant voltage. We say in this case that the point is *clamped to* the fixed voltage. (However, we will reserve the term *clamping circuit* for a special application to be discussed later.) An ideal, forward-biased diode is like a closed switch, so if it is connected between a point in a circuit and a fixed voltage source, the diode very effectively holds the point to the fixed voltage. Diodes can be connected in operational-amplifier circuits, as well as other circuits, so that they become forward biased when a signal reaches a certain

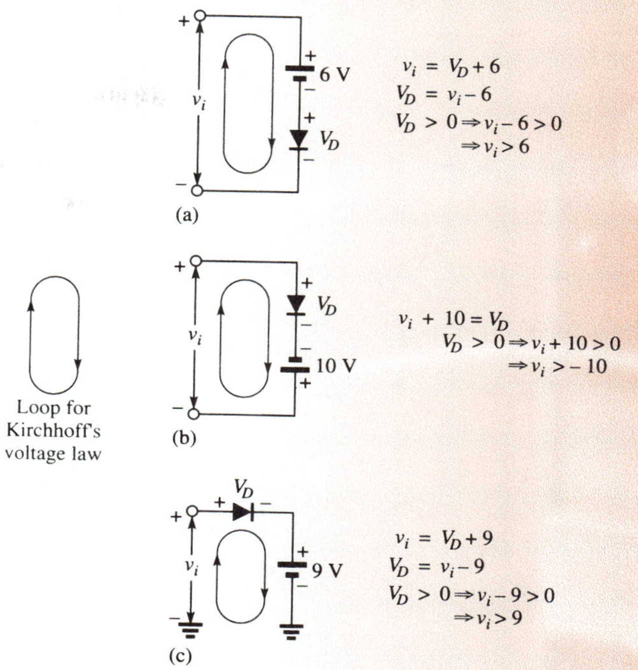

FIGURE 12–28 Examples of biased diodes and the signal voltages v_i required to forward bias them. (Ideal diodes are assumed.) In each case, we solve for the value of v_i that is necessary to make $V_D > 0$

voltage. When the forward-biasing level is reached, the diode serves to hold the output to a fixed voltage and thereby establishes a clipping level.

A biased diode is simply a diode connected to a fixed voltage source. The value and polarity of the voltage source determine what value of total voltage across the combination is necessary to forward bias the diode. Figure 12–28 shows several examples. (In practice, a series resistor would be connected in each circuit to limit current flow when the diode is forward biased.) In each part of the figure, we can write Kirchhoff's voltage law around the loop to determine the value of input voltage v_i that is necessary to forward bias the diode. Assuming that the diodes are ideal (neglecting their forward voltage drops), we determine the value of v_i necessary to forward bias each diode by determining the value of v_i necessary to make $V_D > 0$. Whenever v_i reaches the voltage necessary to make $V_D > 0$, the diode becomes forward biased and the signal source is forced to, or held at, the dc source voltage. If the forward voltage drop across the diode is not neglected, the clipping level is found by determining the value of v_i necessary to make V_D greater than that forward drop (e.g., $V_D > 0.7$ V for a silicon diode). Although these conditions can be determined through formal application of Kirchhoff's voltage law, as shown in the figure, the reader is urged to develop a mental image of circuit behavior in each case. For example, in (a), think of the diode as being reverse biased by 6 V, so the input must "overcome" that reverse bias by reaching +6 V to forward bias the diode.

Figure 12–29 shows three examples of clipping circuits using ideal biased diodes and the waveforms that result when each is driven by a sine-wave input. In each case, note that the output equals the dc source voltage when the input reaches the value necessary to forward bias the diode. Note also that the type of clipping we showed in Figure 12–26 occurs when the fixed bias voltage tends to *reverse* bias the diode, and the type shown in Figure 12–27 occurs when the fixed voltage tends to *forward* bias the diode. When the diode is reverse biased by the input signal, it is like an open circuit that disconnects the dc source, and the output follows the input. These circuits are called

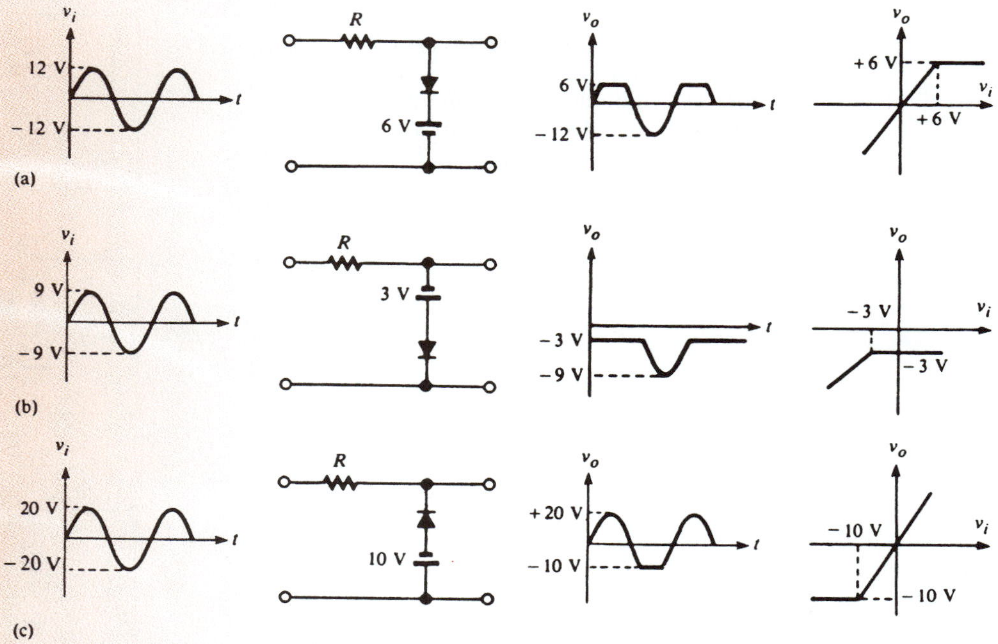

FIGURE 12–29 Examples of parallel clipping circuits

parallel clippers because the biased diode is in parallel with the output. Although the circuits behave the same way whether or not one side of the dc voltage source is connected to the common (low) side of the input and output, the connections shown in Figure 12–29(a) and (c) are preferred to that in (b) because the latter uses a floating source.

Figure 12–30(a) shows a biased diode connected in the feedback path of an inverting operational amplifier. The diode is in parallel with the feedback resistor and forms a parallel clipping circuit like that shown in Figure 12–29(a). Since v^- is at virtual ground, the voltage across R_f is the same as the output voltage v_o. Therefore, when the output voltage reaches the bias voltage E, the output is held at E volts. Figure 12–30(b) illustrates this fact for a sinusoidal input. As long as the diode is reverse biased, it acts like an open circuit and the amplifier behaves like a conventional inverting amplifier. Notice that output clipping occurs at *input* voltage $-(R_1/R_f)E$, because the amplifier inverts and has closed-loop gain magnitude R_f/R_1. The resulting transfer characteristic is shown in Figure 12–30(c). This circuit is often called a *limiting* circuit because it limits the output to the dc level clamped by the diode. (In this and future circuits in this section, we are omitting the bias compensation resistor, R_c, for clarity; it should normally be included, following the guidelines of Chapter 8.)

In practice, the biased diode shown in the feedback of Figure 12–30(a) is often replaced by a *zener* diode in series with a conventional diode. This arrangement eliminates the need for a floating voltage source. Figure 12–31 shows two operational-amplifier clipping circuits using zener diodes. The zener diode conducts like a conventional diode when it is forward biased, so it is necessary to connect a reversed diode in series with it to prevent shorting of R_f. When the reverse voltage across the zener diode reaches V_Z, the diode breaks down and conducts heavily, while maintaining an essentially constant voltage, V_Z, across it. Under those conditions, the *total* voltage across R_f, i.e., v_o, equals V_Z plus the forward drop, V_D, across the conventional diode.

FIGURE 12–30 An operational-amplifier limiting circuit

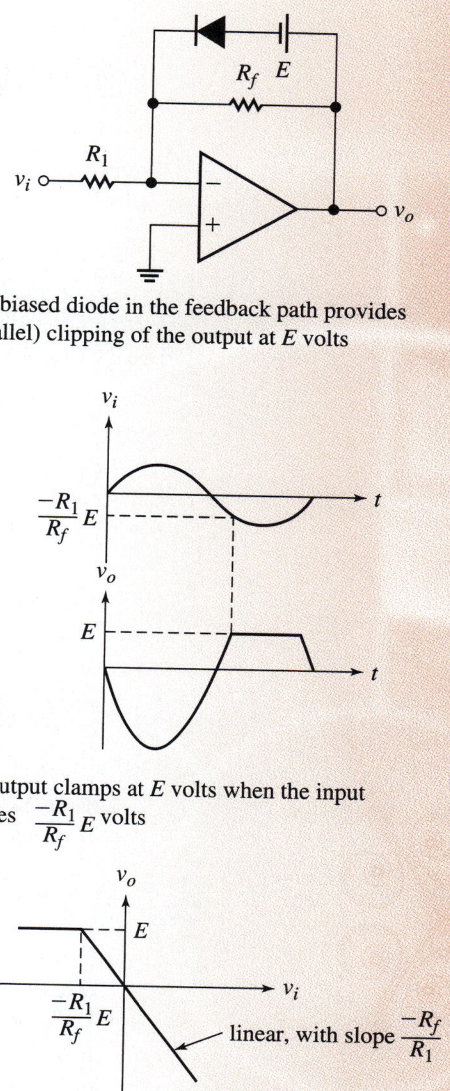

(a) The biased diode in the feedback path provides (parallel) clipping of the output at E volts

(b) The output clamps at E volts when the input reaches $\frac{-R_1}{R_f}E$ volts

(c) Transfer characteristic

Figure 12–32 shows *double-ended* limiting circuits, in which both positive and negative peaks of the output waveform are clipped. Figure 12–32(a) shows the conventional parallel clipping circuit and (b) shows how double-ended limiting is accomplished in an operational-amplifier circuit. In each circuit, note that no more than one diode is forward biased at any given time and that both diodes are reverse biased for $-E_1 < v_o < E_2$, the linear region.

Figure 12–33 shows a double-ended limiting circuit using *back-to-back* zener diodes. Operation is similar to that shown in Figure 12–31, but no conventional diode is required. Note that diode D_1 is conducting in a forward direction when D_2 conducts in its reverse breakdown (zener) region, while D_2 is forward biased when D_1 is conducting in its reverse breakdown region. Neither diode conducts when $-(V_{Z2} + 0.7) < v_o < (V_{Z1} + 0.7)$, which is the region of linear amplifier operation.

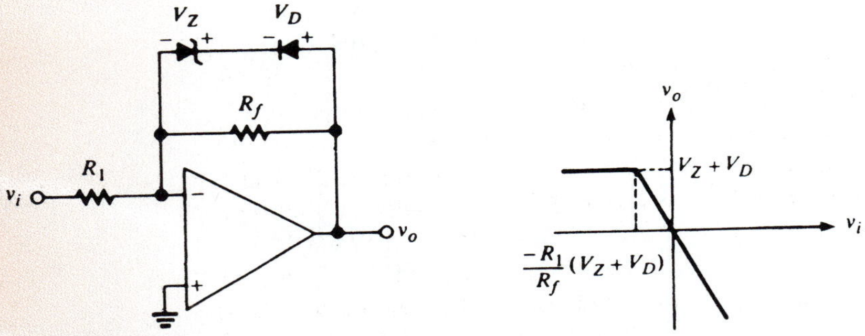

(a) Positive limiting

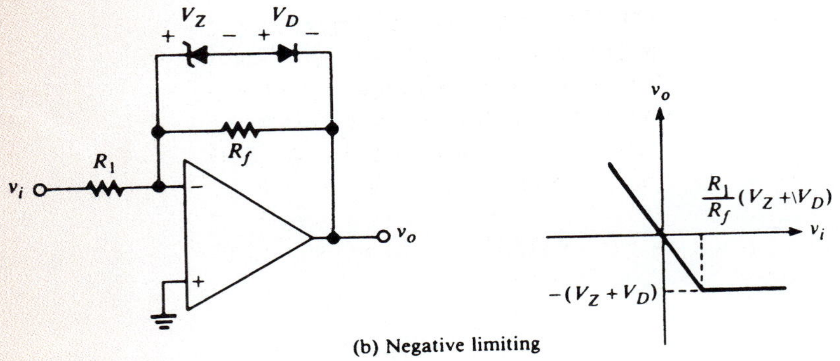

(b) Negative limiting

FIGURE 12–31 Operational-amplifier limiting circuits using zener diodes

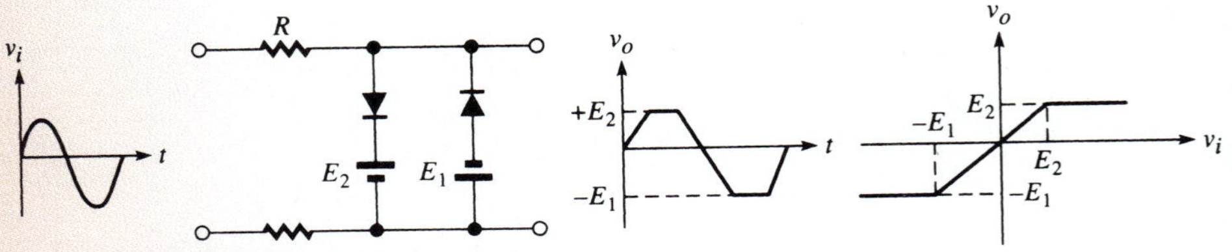

(a) Double-ended parallel clipper

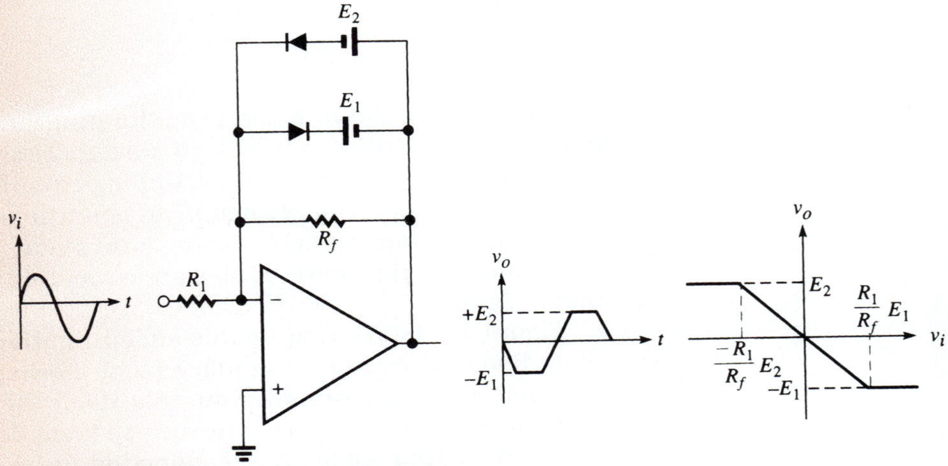

(b) Operational amplifier with double-ended clipping

FIGURE 12–32 Double-ended clipping, or limiting

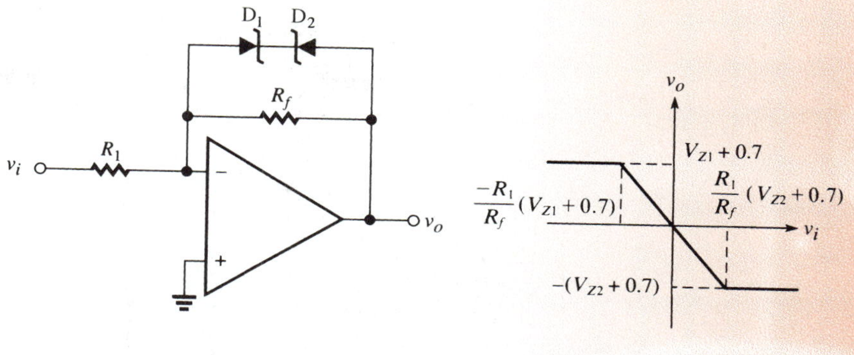

FIGURE 12–33 A double-ended limiting circuit using zener diodes

Precision Rectifying Circuits

Recall that a *rectifier* is a device that allows current to pass through it in one direction only (see Figure 3–14). A diode can serve as a rectifier because it permits generous current flow in only one direction—the direction of forward bias. Rectification is the same as clipping at the 0-V level: All of the waveform below (or above) the zero-axis is eliminated. Recall, however, that a diode rectifier has certain intervals of nonconduction and produces resulting "gaps" at the zero-crossing points of the output voltage, due to the fact that the input must overcome the diode drop (0.7 V for silicon) before conduction begins. See Figures 3–16 and 3–17. In power supply applications, where input voltages are quite large, these gaps are of no concern. However, in many other applications, especially in instrumentation, the 0.7-V drop can be a significant portion of the total input voltage swing and can seriously affect circuit performance. For example, most ac instruments rectify ac inputs so they can be measured by a device that responds to dc levels. It is obvious that small ac signals could not be measured if it were always necessary for them to reach 0.7 V before rectification could begin. For these applications, *precision* rectifiers are necessary.

Figure 12–34 shows one way to obtain precision rectification using an operational amplifier and a diode. The circuit is essentially a noninverting voltage follower when the diode is forward biased. When v_i is positive, the output of the amplifier, v_o, is positive, the diode is forward biased, and a low-resistance path is established between v_o and v^-, as necessary for a voltage follower. The load voltage, v_L, then follows the positive variations of $v_i = v^+$. Note that even a very small positive value of v_i will cause this result because of the large differential gain of the amplifier; that is, the large gain and the action of the feedback cause the usual result that $v^+ \approx v^-$. Note also that the drop across the diode does not appear in v_L.

When the input goes negative, v_o becomes negative, and the diode is reverse biased. This effectively opens the feedback loop, so v_L no longer follows v_i. The amplifier itself, now operating open-loop, is quickly driven to its maximum negative output, thus holding the diode well into reverse bias.

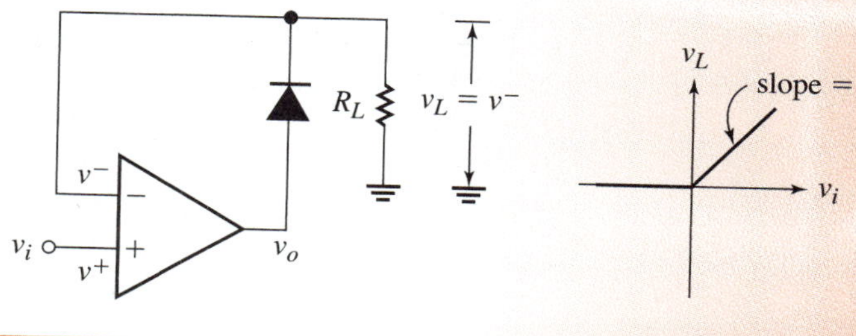

FIGURE 12–34 A precision rectifier. When v_i is positive, the diode is forward biased, and the amplifier behaves like a voltage follower, maintaining $v^+ \approx v^- = v_L$

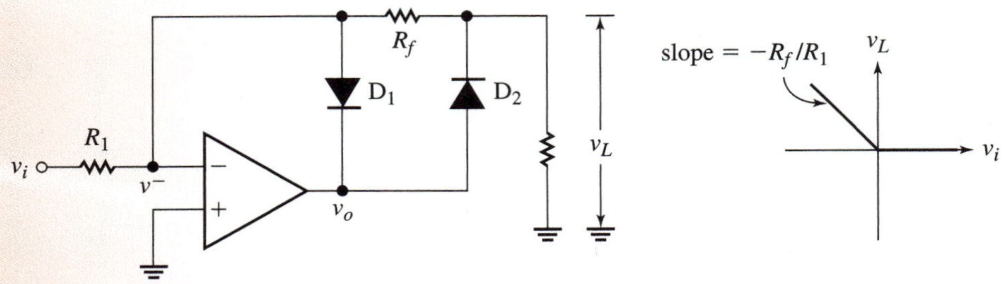

FIGURE 12–35 A precision rectifier circuit that amplifies and inverts the negative variations in the input voltage

Another precision rectifier circuit is shown in Figure 12–35. In this circuit, the load voltage is an amplified and inverted version of the *negative* variations in the input signal and is 0 when the input is positive. Also, in contrast with the previous circuit, the amplifier in this rectifier is not driven to one of its output extremes. When v_i is negative, the amplifier output, v_o, is positive, so diode D_1 is reverse biased and diode D_2 is forward biased. D_1 is open and D_2 connects the amplifier output through R_f to v^-. Thus, the circuit behaves like an ordinary inverting amplifier with gain $-R_f/R_1$. The load voltage is an amplified and inverted (positive) version of the negative variations in v_i. When v_i becomes positive, v_o is negative, D_1 is forward biased, and D_2 is reverse biased. D_1 shorts the output v_o to v^-, which is held at virtual ground, so v_L is 0. It is an exercise at the end of the chapter to analyze this circuit when the diodes are turned around.

12–6 CLAMPING CIRCUITS

Clamping circuits are used to shift an ac waveform up or down by adding a dc level equal to the positive or negative peak value of the ac signal. In the authors' opinion, *clamping* is not a particularly good term for this operation: *Level shifting* is more descriptive. Clamping circuits are also called dc *level restorers* because they are used in systems (television, for example) where the original dc level is lost in capacitor-coupled amplifier stages. It is important to recognize that the amount of dc-level shift required in these applications *varies* as the peak value of the ac signal varies over a period of time. In other words, it is not possible to simply add a fixed dc level to the ac signal using a summing amplifier. To illustrate, Figure 12–36 shows the outputs required from a clamping circuit for two different inputs. Note in both cases that the *peak-to-peak* value of the output is the same as the peak-to-peak value of the input and that the output is shifted up (in this case) by an amount equal to the negative peak of the input.

Figure 12–37(a) shows a clamping circuit constructed from passive components. When the input first goes negative, the diode is forward biased and the capacitor charges rapidly to the peak negative input voltage, V_1. The charging time-constant is very small because the forward resistance of the diode is small. The capacitor voltage, V_1, then has the polarity shown in the figure. Assuming that the capacitor does not discharge appreciably through R_L, the total load voltage as v_i begins to increase is

$$v_L = V_1 + v_i \tag{12–42}$$

(Verify this equation by writing Kirchhoff's voltage law around the loop.) Notice that the polarity of the capacitor voltage keeps the diode reverse biased, so it is like an open circuit during this time and does not discharge

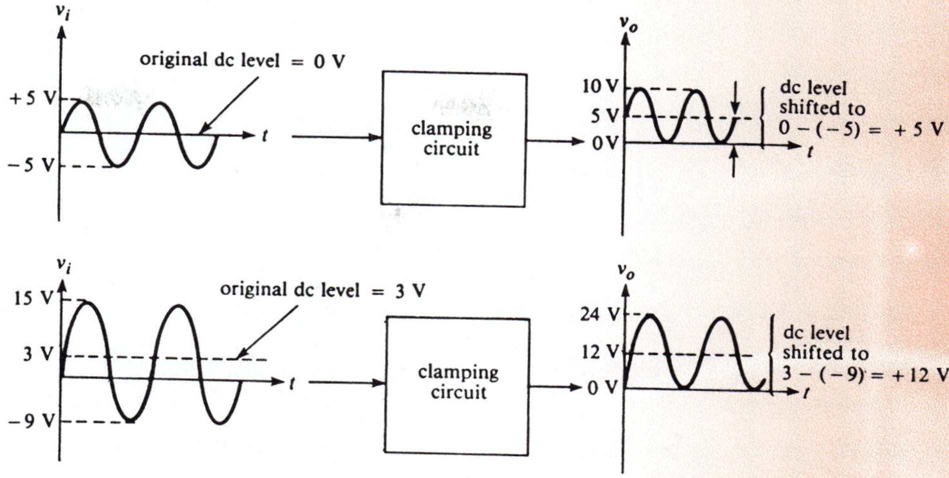

FIGURE 12–36 A clamping circuit that shifts a waveform up by an amount equal to the negative peak value

FIGURE 12–37 A clamping circuit consisting of a diode and a capacitor

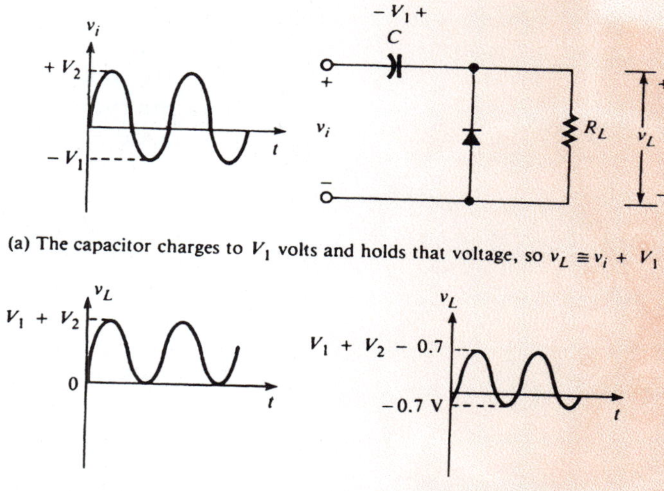

(a) The capacitor charges to V_1 volts and holds that voltage, so $v_L \cong v_i + V_1$

(b) The load voltage if the diode were ideal

(c) Actual load voltage: $v_L = v_i + V_1 - 0.7$

the capacitor. Equation 12–42 shows that the load voltage equals the input voltage shifted up by an amount equal to V_1, as required. When the input again reaches its negative peak, the capacitor may have to recharge slightly to make up for any decay that occurred during the cycle. For proper circuit performance, the discharge time-constant, R_LC, must be much greater than the period of the input. If the diode connections are reversed, the waveform is shifted down by an amount equal to the positive peak voltage, V_2. If the diode is biased by a fixed voltage, the waveform can be shifted up or down by an amount equal to a peak value plus or minus the bias voltage. Examples are given in the exercises at the end of the chapter.

Figure 12–37(b) shows the load voltage that results if the diode is assumed to have zero voltage drop. In reality, because the capacitor charges through the diode, the voltage across the capacitor reaches only V_1 minus the diode drop. Consequently,

$$v_L = v_i + V_1 - 0.7 \qquad (12\text{–}43)$$

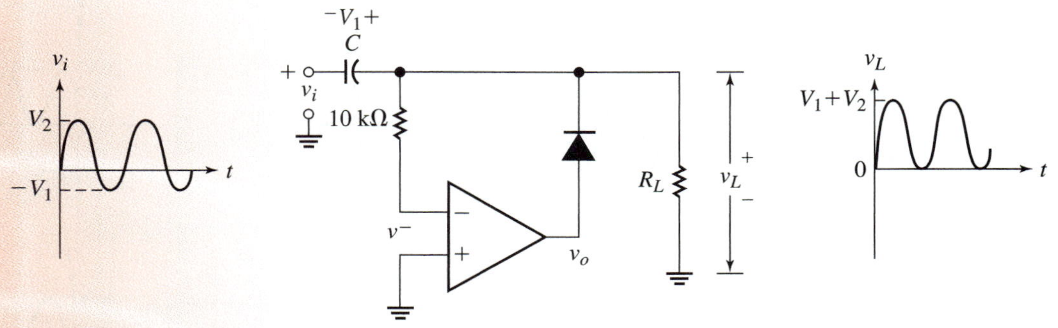

FIGURE 12–38 A precision clamping circuit

The waveform that results is shown in Figure 12–37(c). If the input voltage is large, this offset due to an imperfect diode can be neglected, as is the case in many practical circuits.

If precision clamping is required, the operational-amplifier circuit shown in Figure 12–38 can be used. When v_i in Figure 12–38 first goes negative, the amplifier output, v_o, is positive and the diode is forward biased. The capacitor quickly charges to V_1, with the polarity shown. Notice that $v_L = v_i + V_1$ and that the drop across the diode does not appear in v_L. With the capacitor voltage having the polarity shown, v^- becomes positive, and remains positive throughout the cycle, so the amplifier output is negative. Therefore, the diode is reverse biased and the feedback loop is opened. The amplifier is driven to its maximum negative output level and the diode remains reverse biased. During one cycle of the input, the capacitor may discharge somewhat into the load, causing its voltage to fall below V_1. If so, then when v_i once again reaches its maximum negative voltage, v^- will once again be negative, v_o will be positive, and the capacitor will be allowed to recharge to V_1 volts, as before.

12–7 MULTISIM EXERCISE

In this exercise, we will set up an astable multivibrator that employs the LM311N voltage comparator, as shown in Figure 12–39. Before you start the simulation, double-click on the oscilloscope and set it up to 2 ms/div and 5 V/div for the horizontal and vertical scales, respectively.

Start the simulation, pause it or stop it, and measure the period of the waveform. Compute the theoretical value according to circuit component values and Equation 12–17 and compare.

SUMMARY

This chapter has introduced the reader to basic circuits for generating and shaping signals. A new device, the voltage comparator, was introduced. Two broad types of oscillators were introduced: the astable multivibrator, based on a voltage comparator, for generating a square wave; and sinusoidal oscillators based on operational amplifiers.

At the end of the chapter, the reader should have a good understanding of the following concepts:

- Voltage comparators represent the simplest form of interfacing between the analog and digital worlds. Their basic function is to indicate if a signal level is above or below a specified voltage reference.

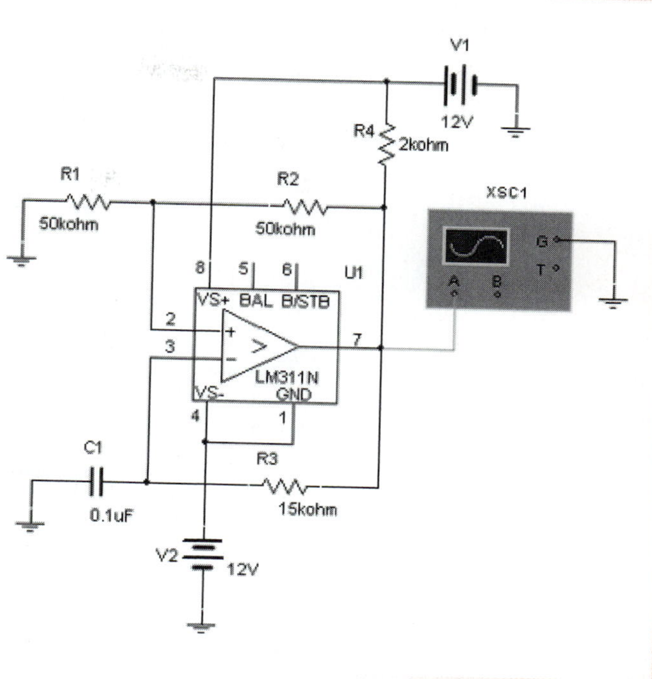

FIGURE 12–39 Astable multivibrator with the LM311N voltage comparator

- An astable multivibrator is a square-wave generator.
- An amplifier with positive feedback will oscillate if it meets the Barkhausen criterion: in-phase, unity-gain around the feedback loop.
- The feedback β-network basically distinguishes the different types of oscillators.
- A precision rectifier does not insert the diode drop that a normal rectifier does.

EXERCISES

SECTION 12–1
Voltage Comparators

12–1. The maximum output voltages of each of the voltage comparators shown in Figure 12–40 are ±15 V. Sketch the output waveforms for each when v_i is a 10-V-pk sine wave. (In each case, show v_i and v_o on your sketch and label z-voltage levels where switching occurs.)

12–2. Repeat Exercise 12–1 when each of the v^+ and v^- inputs are interchanged.

12–3. The output of the comparator shown in Figure 12–41 switches between +10 V and −10 V.

(a) Find the lower and upper trigger levels.

(b) Find the hysteresis voltage.

(c) Sketch the hysteresis loop.

(d) Sketch the output when v_i is a 10-V-pk sine wave.

12–4. Repeat Exercise 12–3 when V_{REF} = +4 V, under the assumption that the comparator output switches between +10 V and −5 V.

12–5. Design a Schmitt trigger circuit whose output switches to a high level when the input falls through −1 V and switches to a low level when the input rises through +1 V. Assume that the output switches between ±10 V.

12–6. Derive equations 12–11 and 12–12 for the lower and upper trigger levels of a noninverting Schmitt trigger.

12–7. Design a Schmitt trigger that switches to a high level when the input rises through +2 V and switches to a low level when the

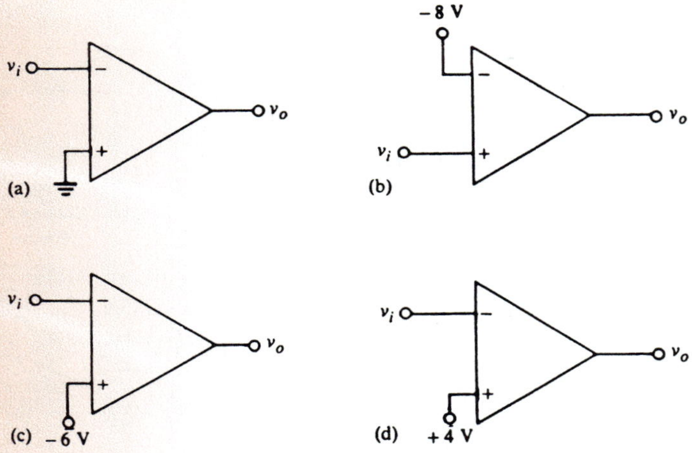

FIGURE 12–40 (Exercise 12–1)

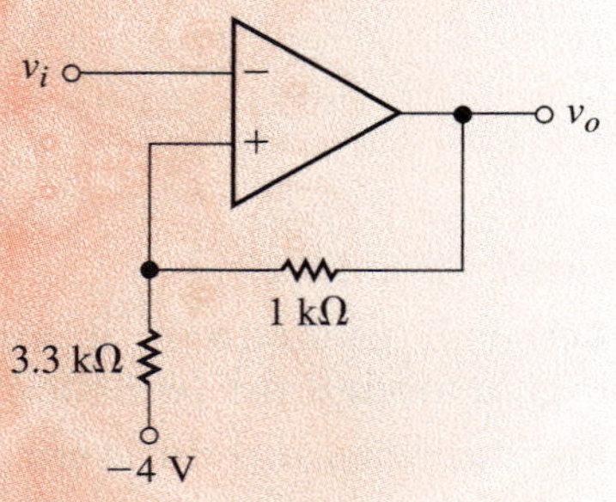

FIGURE 12–41 (Exercise 12–3)

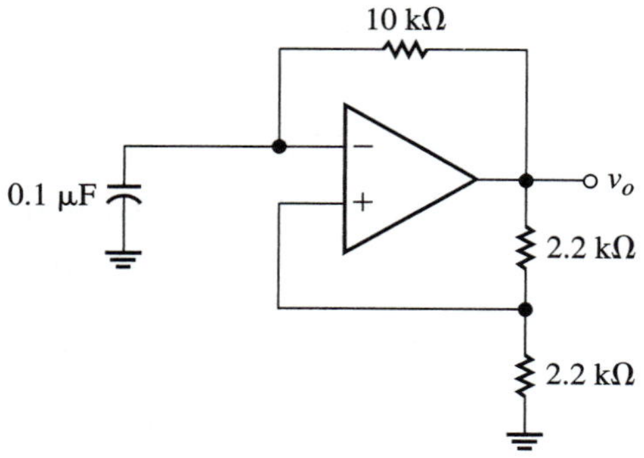

FIGURE 12–42 (Exercise 12–8)

input falls through −1 V. You may choose the high and low output levels of the comparator. Sketch the output when the input is a 5-V-pk sine wave.

SECTION 12–2

Multivibrators and Waveshaping

12–8. The output of the voltage comparator in Figure 12–42 switches between ±10 V.

(a) What maximum and minimum values does the capacitor voltage reach?

(b) What is the frequency of the output oscillation?

12–9. Design an astable multivibrator that produces a 1-kHz square wave.

12–10. Design an astable multivibrator that produces a 5-kHz, TTL square wave.

12–11. An integrator having R_1 = 10 kΩ and C = 0.01 μF is to be used to generate a pseudo-sine wave from a triangular-wave input of 1 V pk. Assume that the input has zero dc component. What minimum frequency can the triangular wave have if $V_{max} = \pm 14$ V?

12–12. Sketch the output of the integrator shown in Figure 12–43. Label the peak positive and peak negative output voltages.

12–13. Design a triangular-waveform generator whose output alternates between +5 V peak and −5 V peak when the input is a 50-Hz square wave that alternates between +1 V and −1 V. The input resistance to the generator must be at least 20 kΩ. You may assume that there is no dc component or offset in the input.

SECTION 12–3

Oscillators

12–14. Design an RC phase-shift oscillator that will oscillate at 1.5 kHz.

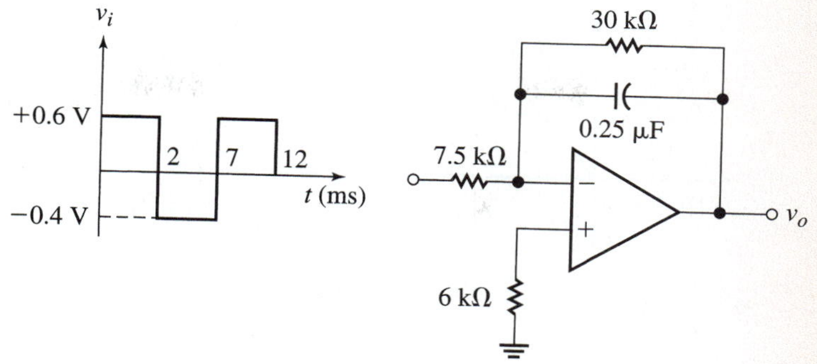

FIGURE 12–43 (Exercise 12–12)

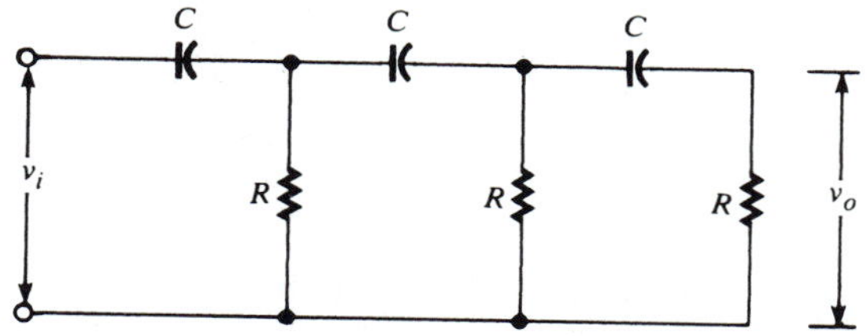

FIGURE 12–44 (Exercise 12–15)

12–15. Figure 12–44 shows the cascaded RC sections that form the feedback network for the RC phase-shift oscillator. Show that the feedback ratio is

$$\beta = \frac{v_o}{v_i} = \frac{R^3}{(R^3 - 5RX_C^2) + j(X_C^3 - 6R^2X_C)}$$

(*Hint:* Write three loop equations and solve for i_3, the current in the rightmost loop. Then $v_o = i_3R$. Solve this equation for v_o/v_i.)

12–16. Design a Wien-bridge oscillator that oscillates at 180 kHz.

12–17. (a) Determine the oscillation frequency in the Colpitts oscillator from Figure 12–23 when $L = 500\ \mu$H, $C = 2.2$ nF, and $R_1 = R_f = 4.7$ kΩ.

(b) Are the values for R_1 and R_f good choices in terms of the circuit Q?

(c) What minimum slew-rate value would be required from the operational amplifier if the output voltage were 10 V pk?

12–18. Design a Hartley oscillator based on Figure 12–24 for an oscillation frequency of 10 kHz. Only a 5-mH inductor is available. Using commercial values, specify R_1 and R_f so that the Q of the β network does not exceed 15.

12–19. Derive equation 12–40 for the oscillation frequency of the Hartley oscillator. Also show that $\beta = -1$ at the oscillation frequency.

SECTION 12–4

The 8038 Integrated-Circuit Function Generator

12–20. A 0.1-μF timing capacitor is to be used with an 8038 function generator to provide output frequencies in the range from 200 Hz to 10 kHz. Through what range should external timing resistor R be adjustable?

SECTION 12–5

Clipping and Rectifying Circuits

12–21. Assume that the input to each device whose transfer characteristic is shown in Figure 12–45 is a 10-V-pk sine wave. Sketch input and output waveforms for each device. Label voltage values on your sketches.

12–22. Sketch the output waveform for the input shown to each circuit in Figure 12–46. Label voltage values on your sketches. Assume ideal diodes.

12–23. Design an operational-amplifier clipping circuit whose output is that shown in Figure 12–47(b) when its input is as shown in 12–47(a). Sketch the transfer characteristic.

12–24. Design an operational amplifier having the following characteristics:

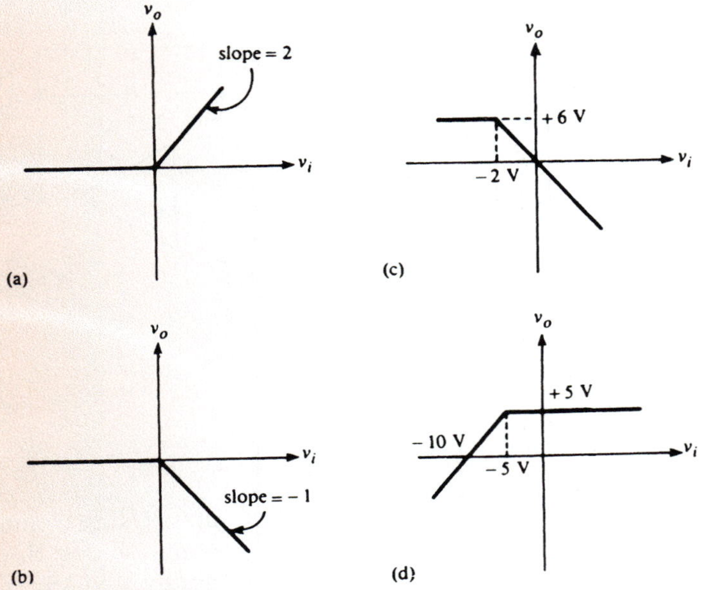

FIGURE 12–45 (Exercise 12–21)

(a) When the input is +0.4 V, the output is −2 V.

(b) When the input is a 6-V-pk sine wave, the output is clipped at +10 V and at −5 V.

(c) The input resistance is at least 10 kΩ.

12–25. Using zener diodes, design an operational-amplifier clipping circuit that clips its output at +8.3 V and −4.8 V. (Specify the zener voltages in your design.) The gain in the linear region should be −4, and the input resistance to the circuit must be at least 25 kΩ. Sketch the transfer characteristic.

12–26. Sketch v_L in the circuit shown in Figure 12–48 when the input is a sine wave that alternates between +10 V peak and −6 V peak. Sketch the transfer characteristic of the circuit.

12–27. Repeat Exercise 12–26 for the circuit shown in Figure 12–49.

SECTION 12–6

Clamping Circuits

12–28. Sketch the load-voltage waveforms in each of the circuits shown in Figure 12–50. Assume ideal diodes, and label voltage values on your sketches.

12–29. Sketch the load-voltage waveforms in each of the circuits shown in Figure 12–51. Assume ideal diodes, and label voltage values on your sketches.

12–30. Sketch the load-voltage waveform in the circuit shown in Figure 12–52. Label voltage values on your sketch.

PSPICE PROBLEMS

12–31. Simulate an astable multivibrator that employs the LM111 voltage comparator available in the EVAL.LIB library. Components are: $R_1 = R_2 = 100$ kΩ, $C = 0.1$ μF, $R_{pull\text{-}up} =$ 1 kΩ. Use a supply voltage of ±15 V that turns on at $t = 0$ by means of the PULSE function. Connect the output return terminal to the negative supply voltage. Note that the output return terminal in the LM111 subcircuit is described as "output ground" in the EVAL.LIB documentation. To call the LM111 from the circuit file, use the subcircuit call "x<id> " followed by the node numbers and "LM111". The LM111 terminals are, from left to right: noninverting input, inverting input, positive supply, negative supply, open-collector output, output return.

```
Example: x1 2 3 7 9 5 9 LM111
```

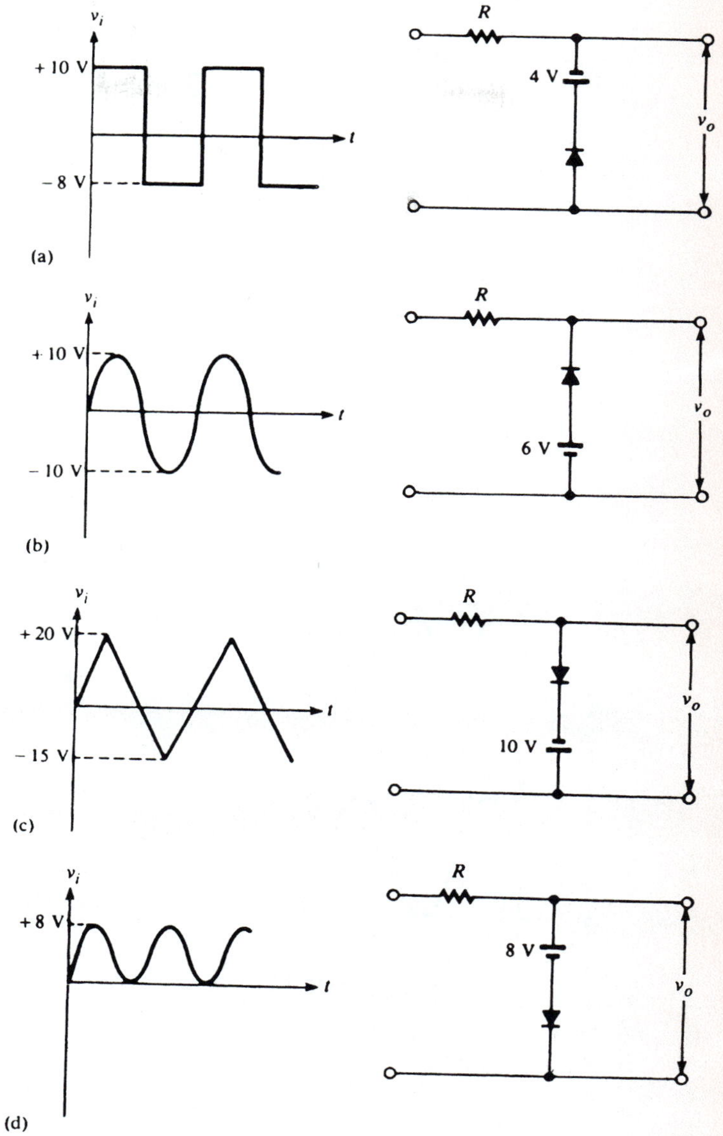

FIGURE 12–46 (Exercise 12–22)

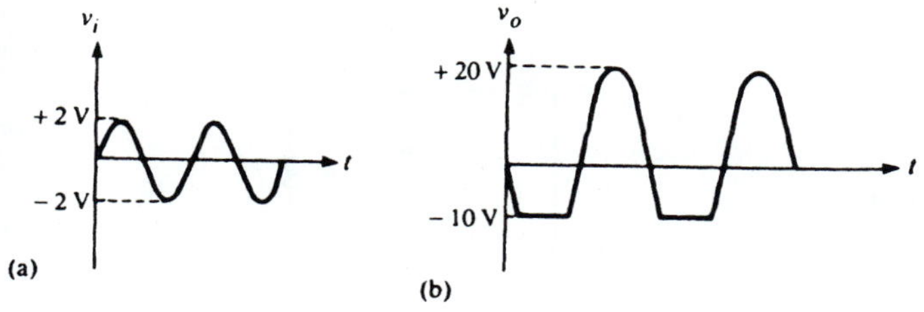

FIGURE 12–47 (Exercise 12–23)

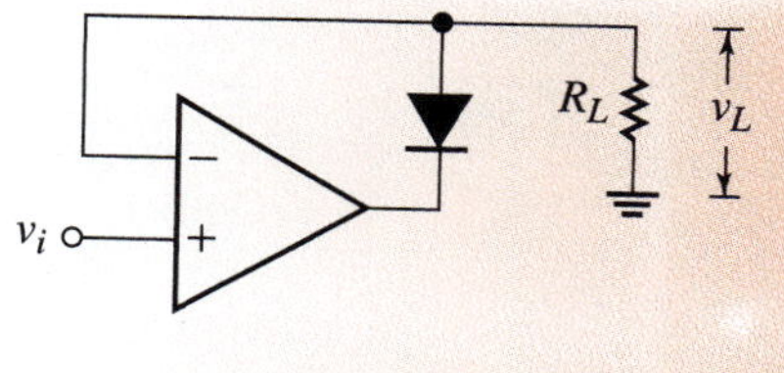

FIGURE 12–48 (Exercise 12–26)

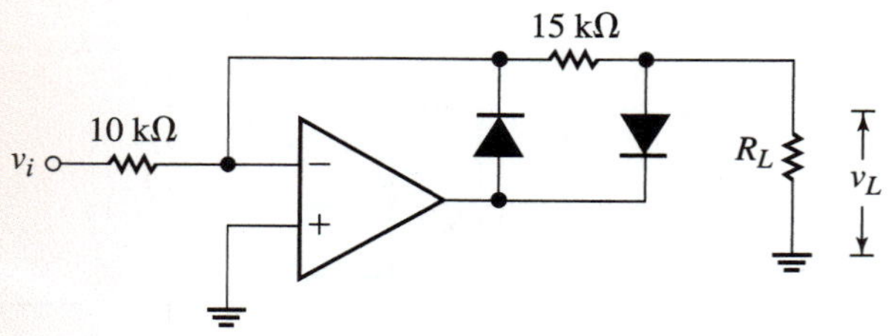

FIGURE 12–49 (Exercise 12–27)

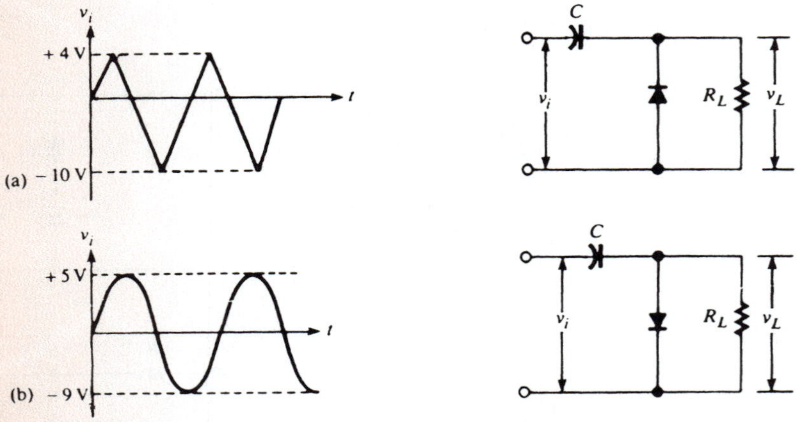

FIGURE 12–50 (Exercise 12–28)

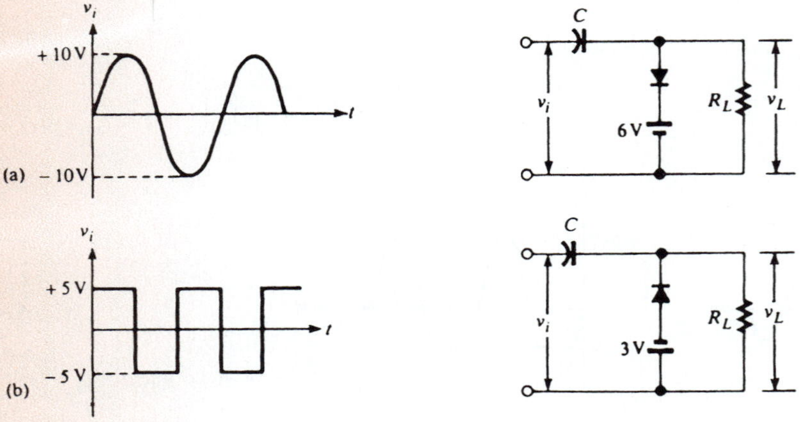

FIGURE 12–51 (Exercise 12–29)

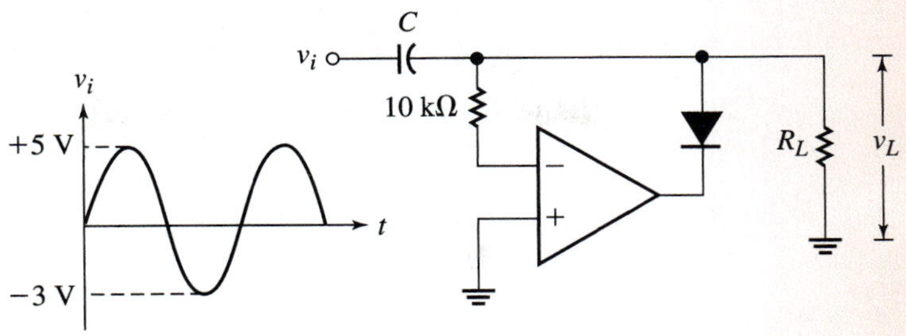

FIGURE 12–52 (Exercise 12–30)

12–32. Simulate a Hartley oscillator using an ideal operational amplifier and the following components: $L = 4$ mH, $C = 680$ pF, $R_1 = R_2 = 8.2$ kΩ. Include a pulse-generating voltage source in series with the inductor to start up the oscillator. Using the interactive Probe screen, measure the period to obtain the oscillation frequency and compare it to the theoretical value. Example for the pulsing source:

```
vpulse 7 8 pulse(0 50 0 1p 1p 1m)
```

CHAPTER 13

REGULATED AND SWITCHING POWER SUPPLIES

OUTLINE

OBJECTIVES

- Introduce the student to the basics of regulated and switching power supplies.
- Examine the basic circuits used for voltage regulation and current limiting.
- Investigate how switching regulator circuits works.
- Examine the basic concept of a dc-to-dc converter.

13–1 INTRODUCTION

The term *power supply* generally refers to a source of dc power that is itself operated from a source of ac power, such as a 120-V, 60-Hz line. A power supply of this type can therefore be regarded as an *ac-to-dc converter.* A supply that produces constant-frequency, constant-amplitude, *ac* power from a dc source is called an *inverter.* Some power supplies are designed to operate from a dc source and to produce power at a different dc level; these are called *dc-to-dc converters.*

A dc power supply operated from an ac source consists of one or more of the following fundamental components:

1. a *rectifier* that converts an ac voltage to a pulsating dc voltage and permits current to flow in one direction only,
2. a low-pass *filter* that suppresses the pulsations in the rectified waveform and passes its dc (average value) component, and
3. a *voltage regulator* that maintains a substantially constant output voltage under variations in the load current drawn from the supply and under variations in line voltage.

The extent to which each of the foregoing components is required in a given supply and the complexity of its design depend entirely on the application for which the supply is designed. For example, a dc supply designed exclusively for service as a battery charger can consist simply of a half-wave rectifier. No filter or regulator is required, since it is necessary to supply only pulsating, unidirectional current to recharge a battery. Special-purpose, fixed-voltage supplies subjected to little or no variation in current demand can be constructed with a rectifier and a large capacitive filter, and require no regulator. On the other hand, devices such as operational amplifiers and output power amplifiers require elaborate, well-regulated supplies.

Power supplies are classified as regulated or unregulated and as adjustable or fixed. Adjustable supplies are generally well regulated and are used in laboratory applications for general-purpose experimental and developmental work. The most demanding requirement imposed on these supplies is that they maintain an extremely constant output voltage over a wide range of loads and over a continuously adjustable range of voltages. Some fixed-voltage supplies are also designed so that their output can be adjusted, but over a relatively small range, to permit periodic recalibration.

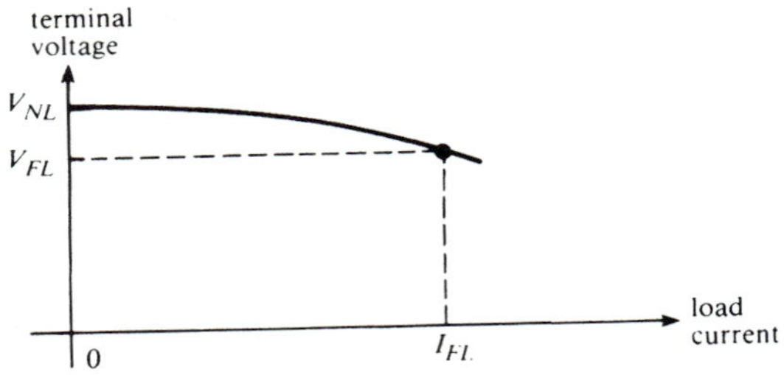

FIGURE 13–1 No-load and full-load conditions in a power supply

13–2 VOLTAGE REGULATION

An *ideal* power supply maintains a constant voltage at its output terminals no matter what current is drawn from it. The output voltage of a practical power supply changes with load current, generally dropping as load current increases. Power-supply specifications include a *full-load current* (I_{FL}) rating, which is the maximum current that can be drawn from the supply. The terminal voltage when full-load current is drawn is called the *full-load voltage* (V_{FL}). The *no-load voltage* (V_{NL}) is the terminal voltage when zero current is drawn from the supply, that is, the open-circuit terminal voltage. Figure 13–1 illustrates these terms.

One measure of power-supply performance, in terms of how well the power supply is able to maintain a constant voltage between no-load and full-load conditions, is called its *percent voltage regulation:*

$$\text{VR} = \frac{V_{NL} - V_{FL}}{V_{FL}} \times 100\% \tag{13–1}$$

More precisely, equation 13–1 defines the percent *output,* or *load,* voltage regulation, because it is based on changes that occur due to changes in load conditions, all other factors (including input voltage) remaining constant. It is clear that the numerator of equation 13–1 is the total change in output voltage between no-load and full-load and that the ideal supply therefore has *zero* percent voltage regulation.

Figure 13–2 shows the Thévenin equivalent circuit of a power supply. The Thévenin voltage is the no-load voltage V_{NL}, and the Thévenin equivalent resistance is called the *output resistance,* R_o, of the supply. Many power-supply manufacturers specify output resistance rather than percent voltage regulation. We will show that one can be obtained from the other if the full-load–voltage and full-load–current ratings are known. Let the full-load resistance be designated

$$R_{FL} = \frac{V_{FL}}{I_{FL}} \tag{13–2}$$

When R_L in Figure 13–2 equals R_{FL}, we have, by voltage-divider action,

$$V_{FL} = \left(\frac{R_{FL}}{R_{FL} + R_o}\right) V_{NL} \tag{13–3}$$

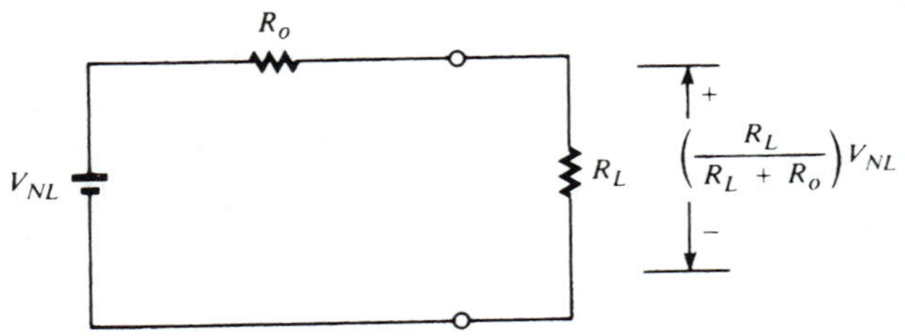

FIGURE 13–2 Thévenin equivalent circuit of a power supply

Substituting (13–3) into (13–1) (and omitting the percent factor) gives

$$\text{VR} = \frac{V_{NL} - \left(\dfrac{R_{FL}}{R_{FL} + R_o}\right) V_{NL}}{\left(\dfrac{R_{FL}}{R_{FL} + R_o}\right) V_{NL}} = \frac{R_o}{R_{FL}} \qquad (13\text{–}4)$$

or

$$\text{VR} = R_o\left(\frac{I_{FL}}{V_{FL}}\right) \qquad (13\text{–}5)$$

It is clear that the ideal supply has zero output resistance, corresponding to zero percent voltage regulation. Like the output resistance we have studied in earlier chapters, R_o can also be determined as the slope of a plot of load voltage versus load current: $R_o = \Delta V_L/\Delta I_L$.

EXAMPLE 13–1

A power supply having output resistance 1.5 Ω supplies a full-load current of 500 mA to a 50-Ω load.

1. What is the percent voltage regulation of the supply?
2. What is the no-load output voltage of the supply?

Solution

1. $V_{FL} = I_{FL} R_{FL} = (500 \text{ mA})(50\ \Omega) = 25$ V. From equation 13–5,

$$\text{VR}(\%) = R_o\left(\frac{I_{FL}}{V_{FL}}\right) \times 100\% = 1.5\left(\frac{0.5 \text{ A}}{25 \text{ V}}\right) \times 100\% = 3.0\%$$

(Note also that $\text{VR} = R_o/R_{FL} = 1.5\ \Omega/50\ \Omega = 0.03$.)

2. From equation 13–3,

$$V_{NL} = \frac{V_{FL}(R_{FL} + R_o)}{R_{FL}} = \frac{25\,(50 + 1.5)}{50} = 25.75 \text{ V}$$

EXAMPLE 13–2

Assuming that the transformer and forward-biased–diode resistances in Figure 13–3 are negligible, find the percent voltage regulation of the power supply. The full-load current is 2 A at a full-load voltage of 15 V.

Solution

Since the transformer and diode resistances can be neglected, the dc output resistance looking back into the load terminals is $R_o = 8\ \Omega$. Therefore, by equation 13–5,

FIGURE 13–3 (Example 13–2)

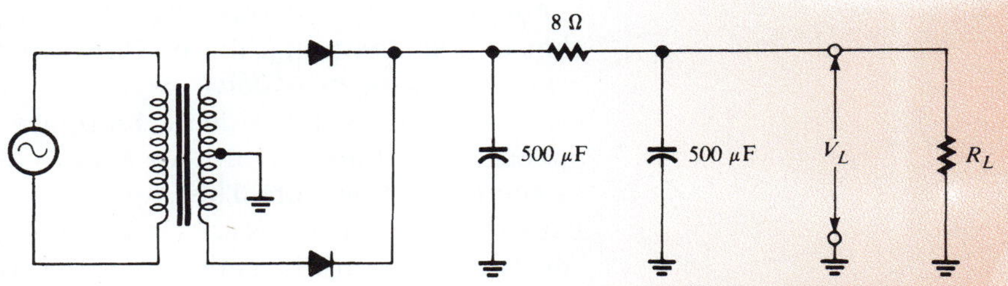

$$\mathrm{VR}(\%) = R_o\left(\frac{I_{FL}}{V_{FL}}\right) \times 100\% = 8\left(\frac{2\text{ A}}{15\text{ V}}\right) \times 100\% = 106.6\%$$

This example illustrates that the series resistance used in an RC π filter to reduce ripple can seriously degrade voltage regulation. Note that the no-load voltage in this example is, from equation 13–1, $V_{NL} = \mathrm{VR}(V_{FL}) + V_{FL} = (1.066)15 + 15 = 31$ V, so there is a 16-V change in output voltage between no-load and full-load! This is an example of an *unregulated* supply, because there is no circuitry designed to correct, or compensate for, the effect of load variations. It would be unacceptable for most applications in which the load could vary over such a wide range.

Line Regulation

Percent *line regulation* is another measure of the ability of a power supply to maintain a constant output voltage. In this case, it is a measure of how sensitive the output is to changes in *input,* or line, voltage rather than to changes in load. The specification is usually expressed as the percent change in output voltage that occurs per volt change in input voltage, with the load R_L assumed constant. For example, a line regulation of 1%/V means that the output voltage will change 1% for each 1-V change in input voltage. If the input voltage to a 20-V supply having that specification were to change by 5 V, then the output could be expected to change by (5 V)(1%/V) = 5%, or 0.05(20) = 1 V.

$$\%\text{ line regulation} = \frac{(\Delta V_o / V_o) \times 100\%}{\Delta V_i} \qquad (13\text{–}6)$$

13–3 SERIES AND SHUNT VOLTAGE REGULATORS

A voltage regulator is a device, or combination of devices, designed to maintain the output voltage of a power supply as nearly constant as possible. It can be regarded as a *closed-loop control system* because it monitors output voltage and generates *feedback* that automatically increases or decreases the supply voltage, as necessary, to compensate for any tendency of the output to change. Thus, the purpose of a regulator is to eliminate any output voltage variation that might otherwise occur because of changes in load, changes in line (input) voltage, or changes in temperature. Regulators are constructed in numerous circuit configurations, ranging from one or two discrete components to elaborate integrated circuits.

The Zener Diode as a Voltage Reference

One of the simplest discrete regulators consists of only a resistor and a *zener diode.* We studied that circuit in detail in Chapter 3, but since zener diodes are also used in the more complex regulators that are the subject of our present discussion, we will review the principal feature that makes them useful as *voltage references.* Figure 13–4 shows the *I–V* characteristic of a typical zener diode. Notice that the forward-biased characteristic is the same as that of a conventional diode. However, the zener is normally operated in its reverse-biased, *breakdown* region, which is the nearly vertical line drawn downward on the left side of the figure. The diode is manufactured to have a specific breakdown voltage, called the *zener voltage,* V_Z, which may be as small as a few volts. The important point to notice is that the nearly vertical breakdown characteristic implies that the voltage V_Z across the diode remains substantially constant as the (reverse) current through it varies over

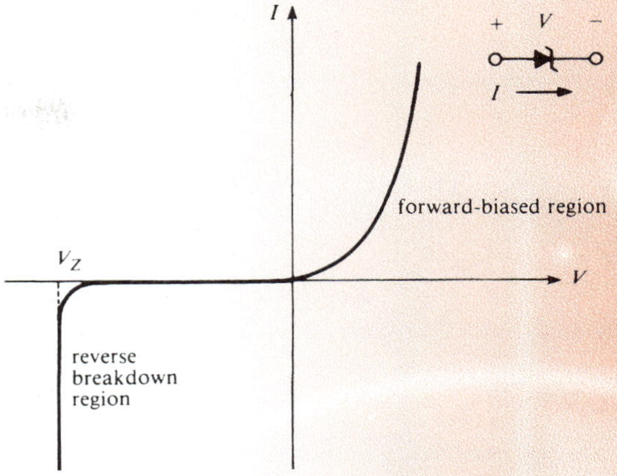

FIGURE 13–4 The *I–V* characteristic of a zener diode

a wide range. It therefore behaves much like an ideal voltage source that maintains a constant voltage independent of current. For this reason, the zener diode can be operated as an essentially constant voltage reference in regulator circuits, provided it remains in breakdown.

Series Regulators

Voltage regulators may be classified as being of either a *series* or a *shunt* design. Figure 13–5 is a *functional* block diagram of the series-type regulator. Although the individual components shown in the diagram may not be readily identifiable in every series regulator circuit, the functional block diagram serves as a useful model for understanding the underlying principles of series regulation. V_i is an unregulated dc input, such as might be obtained from a rectifier with a capacitor filter, and V_o is the regulated output voltage. Notice that the *control element,* which is a device whose operating state adjusts as necessary to maintain a constant V_o, is in a series path between V_i and V_o. A *sampling circuit* produces a feedback voltage proportional to V_o and this voltage is compared to a reference voltage. The output of the comparator circuit is the control signal that adjusts the operating state of the control element. If V_o decreases, due, for example, to an increased load, then the comparator produces an output that causes the control element to increase V_o. In other words, V_o is automatically raised until the comparator circuit no longer detects any difference between the reference and the feedback. Similarly, any tendency of V_o to increase results in a signal that causes the control element to reduce V_o.

Figure 13–6 shows a simple transistor voltage regulator of the series type. Here, the control element is the *npn* transistor, often called the *pass transistor* because it conducts, or passes, all the load current through the regulator. It is usually a power transistor, and it may be mounted on a heat

FIGURE 13–5 Block diagram of a series voltage regulator

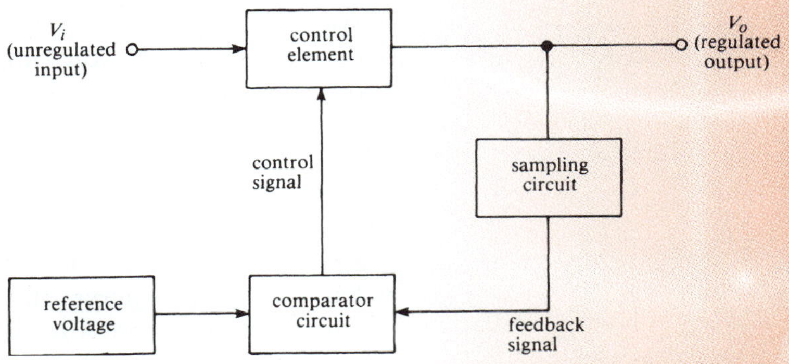

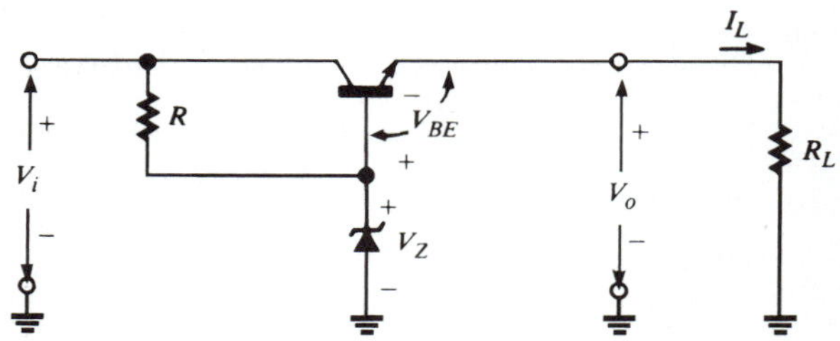

FIGURE 13–6 A transistor series voltage regulator

sink in a heavy-duty power supply that delivers substantial current. The zener diode provides the voltage reference, and the base-to-emitter voltage of the transistor is the control voltage. In this case, there is no identifiable sampling circuit, because the entire output voltage level V_o is used for feedback.

In Figure 13–6, notice that the zener diode is reverse biased and that reverse current is furnished to it through resistor R. Although V_i is unregulated, it must remain sufficiently large, and R must be sufficiently small, to keep the zener in its reverse breakdown region, as shown in Figure 13–4. Thus, as the unregulated input voltage varies, V_Z remains essentially constant. Writing Kirchhoff's voltage law around the output loop, we find

$$V_{BE} = V_Z - V_o \quad (13\text{–}7)$$

We now treat V_Z as perfectly constant. Because (13–7) is valid at all times, any change in V_o must cause a change in V_{BE} to maintain the equality. For example, if V_o decreases, V_{BE} must increase since V_Z is constant. Similarly, if V_o increases, V_{BE} must decrease.

The behavior we have just described accounts for the ability of the circuit to provide voltage regulation. When V_o decreases, V_{BE} increases, which causes the *npn* transistor to conduct more heavily and to produce more load current. The increase in load current causes an increase in V_o since $V_o = I_L R_L$. For example, suppose the regulator supplies 1 A to a 10-Ω load, so V_o = (1 A)(10 Ω) = 10 V. Now suppose the load resistance is reduced to 8 Ω. The output voltage would then fall to (1 A)(8 Ω) = 8 V. However, this reduction in V_o causes the transistor to conduct more heavily, and the load current increases to 1.25 A. Thus, the output voltage is restored to V_o = (1.25 A)(8 Ω) = 10 V. The regulating action is similar when V_o increases: V_{BE} decreases, transistor conduction is reduced, load current decreases, and V_o is reduced.

Notice that the transistor is used essentially as an emitter follower. The load is connected to the emitter and the emitter follows the base, which is the constant zener voltage.

EXAMPLE 13–3

In Figure 13–6, V_i = 20 V, R = 200 Ω, and V_Z = 12 V. If V_{BE} = 0.65 V, find

1. V_o,
2. the collector-to-emitter voltage of the pass transistor, and
3. the current in the 200-Ω resistor.

Solution

1. From equation 13–7, $V_o = V_Z - V_{BE}$ = 12 V − 0.65 V = 11.35 V.
2. By writing Kirchhoff's voltage law around the loop that includes V_i, V_{CE}, and V_o, we find $V_{CE} = V_i - V_o$ = 20 V − 11.35 V = 8.65 V.
3. The voltage drop across the 200-Ω resistor is $V_i - V_Z$ = 20 V − 12 V = 8 V. Therefore, the current in the resistor is I = (8 V)/(200 Ω) = 0.04 A.

FIGURE 13–7 An improved series regulator that incorporates transistor Q_2 to increase sensitivity

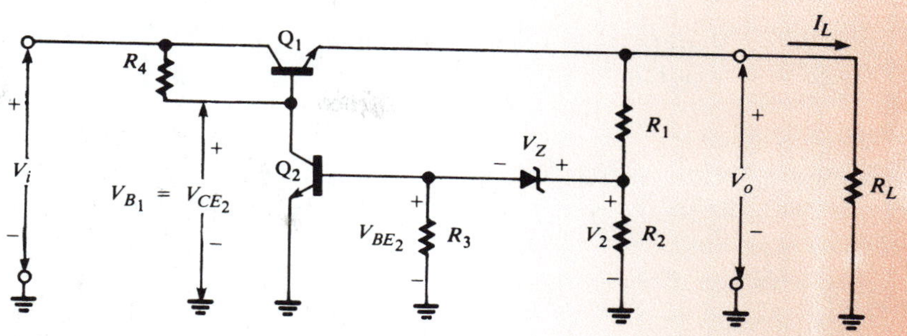

For successful regulator operation, the pass transistor must remain in its active region, V_i must not drop to a level so small that the zener diode is no longer in its breakdown region, and the zener voltage V_Z should be highly independent of both current and temperature. Also, the feedback *loop gain*, from output back through the control element, should be very large, so that small changes in output voltage can be detected and rapidly corrected. Figure 13–7 shows an improved series regulator that incorporates an additional transistor in the feedback path to increase gain.

Note that resistors R_1 and R_2 form a voltage divider across V_o and serve as the sampling circuit that produces voltage V_2 proportional to V_o. Assuming that the divider is designed so that the zener-diode current does not load it appreciably,

$$V_2 = \frac{R_2}{R_1 + R_2} V_o \tag{13–8}$$

Writing Kirchhoff's voltage law around the loop containing R_2, R_3, and the zener diode, we have

$$V_{BE_2} + V_Z = V_2 \tag{13–9}$$

Any decrease in V_o will cause a decrease in V_2 and, assuming that V_Z remains constant, V_{BE_2} must therefore decrease to maintain equality in (13–9). A decrease in V_{BE_2} cause V_{CE_2} to increase, since Q_2 is an inverting common-emitter stage. Since the base voltage V_{B_1} on Q_1 experiences this same increase, Q_1 conducts more heavily, produces more load current, and increases V_o, as previously described. Another way of viewing this feedback action is to recognize that a decrease in V_{BE_2} causes Q_2 to conduct less, and therefore allows more base current to flow into Q_1. Of course, any *increase* in V_o causes actions that are just the opposite of those we have described and results in less load current.

EXAMPLE 13–4

In the regulator shown in Figure 13–7, $R_1 = 50\ \text{k}\Omega$, $R_2 = 43.75\ \text{k}\Omega$ and VZ = 6.3 V. If the 15-V output drops 0.1 V, find the change in V_{BE_2} that results.

Solution

From equation 13–8, when $V_o = 15$ V,

$$V_2 = \frac{R_2}{R_1 + R_2} V_o = \left[\frac{43.75\ \text{k}\Omega}{(50\ \text{k}\Omega) + (43.75\ \text{k}\Omega)}\right](15\ \text{V}) = 7\ \text{V}$$

Therefore, by (13–9), $V_{BE_2} = V_2 - V_Z = 7\ \text{V} - 6.3\ \text{V} = 0.7\ \text{V}$. When $V_o = 15\ \text{V} - 0.1\ \text{V} = 14.9\ \text{V}$, V_2 becomes

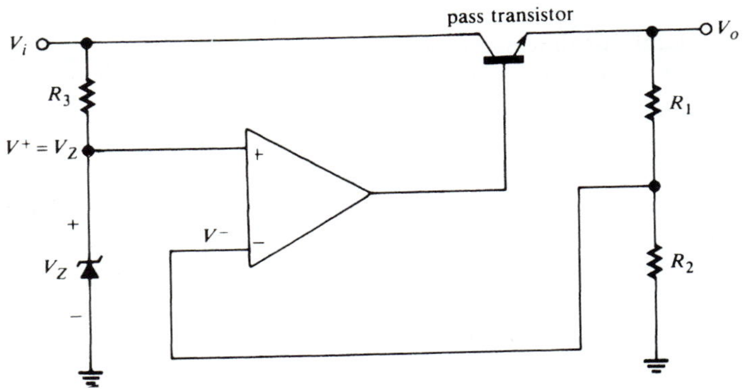

FIGURE 13–8 An operational amplifier used in a series voltage regulator

$$V_2' = \left[\frac{43.75\ \text{k}\Omega}{(50\ \text{k}\Omega) + (43.75\ \text{k}\Omega)}\right](14.9\ \text{V}) = 6.953\ \text{V}$$

Therefore,

$$V'_{BE_2} = V_2' - V_Z = 6.953\ \text{V} - 6.3\ \text{V} = 0.653\ \text{V}$$

$$\Delta V_{BE_2} = V_{BE_2} - V'_{BE_2} = 0.7\ \text{V} - 0.653\ \text{V} = 0.047\ \text{V}$$

Figure 13–8 shows how an operational amplifier is used in a series voltage regulator. Resistors R_1 and R_2 form a voltage divider that feeds a voltage proportional to V_o back to the inverting input: $V^- = V_oR_2/(R_1 + R_2)$. The zener voltage V_Z on the noninverting input is greater than V^-, so the amplifier output is positive and is proportionate to $V_Z - V^-$. If V_o decreases, V^- decreases, and the amplifier output increases. The greater voltage applied to the base of the pass transistor causes it to conduct more heavily, and V_o increases. Similarly, an increase in V_o causes V^- to increase, $V_Z - V^-$ to decrease, and the amplifier output to decrease. The pass transistor conducts less heavily and V_o decreases.

The operational amplifier in Figure 13–8 can be regarded as a noninverting configuration with input V_Z and gain

$$A = 1 + \frac{R_1}{R_2} \tag{13–10}$$

Therefore, neglecting the base-to-emitter drop of the pass transistor, the regulated output voltage is

$$V_o = \left(1 + \frac{R_1}{R_2}\right)V_Z \tag{13–11}$$

R_3 is chosen to ensure that sufficient reverse current flows through the zener diode to keep it in breakdown.

EXAMPLE 13–5

DESIGN

Design a series voltage regulator using an operational amplifier and a 6-V zener diode to maintain a regulated output of 18 V. Assume that the unregulated input varies between 20 V and 30 V and that the current through the zener diode must be at least 20 mA to keep it in its breakdown region.

Solution

From equation 13–11, 18 V = $(1 + R_1/R_2)$6 V. Thus, $(1 + R_1/R_2) = 3$. Let $R_1 = 20\ \text{k}\Omega$. Then

$$\left(1 + \frac{20\ \text{k}\Omega}{R_2}\right) = 3$$

$$R_2 = 10\ \text{k}\Omega$$

The current through R_3 is

$$I = \frac{V_i - V_Z}{R_3} \tag{13–12}$$

Since the current into the noninverting input is negligibly small, the current in R_3 is the same as the current in the zener diode. This current must be at least 20 mA, so in equation 13–12 I must be 20 mA for the smallest possible value of V_i (20 V):

$$20\ \text{mA} = \frac{(20\ \text{V}) - (6\ \text{V})}{R_3}$$

$$R_3 = \frac{14\ \text{V}}{20\ \text{mA}} = 700\ \Omega$$

Current Limiting

Many general-purpose power supplies are equipped with short-circuit or overload protection. One form of protection is called *current limiting*, whereby specially designed circuitry limits the current that can be drawn from the supply to a certain specific maximum, even if the output terminals are short-circuited. Figure 13–9 shows a popular current-limiting circuit incorporated into the operational-amplifier regulator. As load current increases, the voltage drop across resistor R_{SC} increases. Notice that R_{SC} is in parallel with the base-to-emitter junction of transistor Q_2. If the load current becomes great enough to create a drop of about 0.7 V across R_{SC}, then Q_2 begins to conduct a substantial collector current. As a consequence, current that would otherwise enter the base of Q_1 is diverted through Q_2. In this way, the pass transistor is prevented from supplying additional load current. The maximum (short-circuit) current that can be drawn from the supply is the current necessary to create the 0.7-V drop across R_{SC}:

$$I_L(\text{max}) = \frac{0.7\ \text{V}}{R_{SC}} \tag{13–13}$$

EXAMPLE 13–6

The voltage regulator in Figure 13–9 maintains an output voltage of 25 V.

1. What value of R_{SC} should be used to limit the maximum current to 0.5 A?
2. With the value of R_{SC} found in (1), what will be the output voltage when $R_L = 100\ \Omega$? When $R_L = 10\ \Omega$?

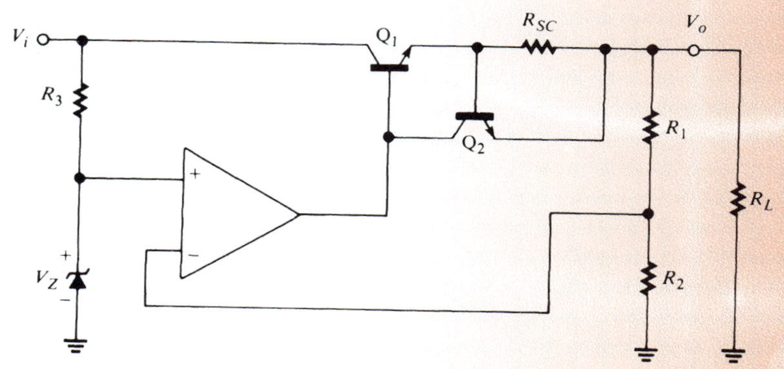

FIGURE 13–9 Q_2 and R_{SC} are used to limit the maximum current that can be drawn from the supply

Solution

1. From equation 13–13,

$$R_{SC} = \frac{0.7\text{ V}}{I_L(\text{max})} = \frac{0.7\text{ V}}{0.5\text{ A}} = 1.4\ \Omega$$

2. We must determine whether a 100-Ω load would attempt to draw more than the current limit of 0.5 A at the regulated output voltage of 25 V:

$$I_L = \frac{25\text{ V}}{100\ \Omega} = 0.25\text{ A}$$

Since 0.25 A < I_L(max) = 0.5 A, the output voltage will remain at 25 V.

If $R_L = 10\ \Omega$, the (unlimited) current would be

$$I_L = \frac{25\text{ V}}{10\ \Omega} = 2.5\text{ A}$$

Since 2.5 A > I_L(max) = 0.5 A, current limiting will occur, and the output voltage will be $V_o = I_L(\text{max})R_L = (0.5\text{ A})(10\ \Omega) = 5\text{ V}$.

Example 13–6 demonstrates that the output voltage of a current-limited regulator decreases if the load resistance is made smaller than what would draw maximum current at the regulated output voltage. Figure 13–10 shows a typical load-voltage–load-current characteristic for a current-limited regulator. The characteristic shows that load current may increase slightly beyond I_L(max) as the output approaches a short-circuit condition ($V_L = 0$). Note that Q_2 in Figure 13–9 supplies a small amount of additional current to the load once current limiting takes place.

Foldback Limiting

Another form of overcurrent protection, called *foldback limiting,* is used to reduce both the output current and the output voltage if the load resistance is made smaller than what would draw a specified maximum current, I_L(max). We have already seen that the output voltage of a current-limited regulator decreases as the load resistance is made smaller (Example 13–6). In foldback limiting, this decrease in output voltage is sensed and is used to further decrease the amount of current that can flow to the load. Thus, as load resistance decreases beyond a certain minimum, both load voltage and load current decrease. If the load becomes a short circuit, output voltage and output current both approach 0. The foldback characteristic is shown in

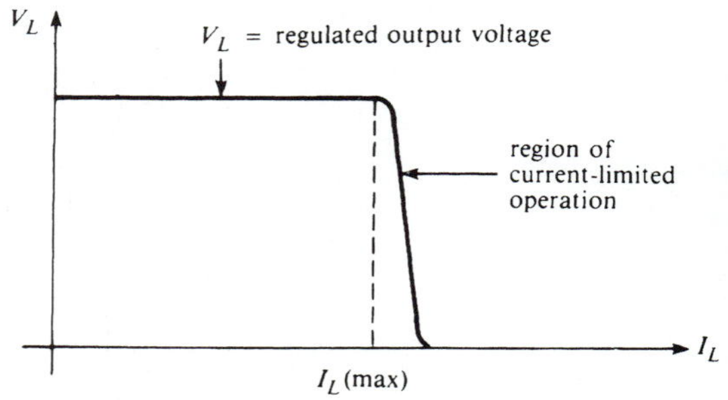

FIGURE 13–10 A typical load-voltage–load-current characteristic for a current-limited regulator

Figure 13–11. The principal purpose of foldback is to protect a load from overcurrent, as well as to protect the regulator itself.

Figure 13–12 shows one method used to add foldback limiting to the basic current-limited regulator. Notice the similarity of this circuit to the current-limited regulator circuit shown in Figure 13–9. The only difference is that the base of Q_2 is now connected to the $R_3 - R_4$ voltage divider. Writing Kirchhoff's voltage law around the loop, we find

$$V_{BE} = V_{RSC} - V_{R3} \tag{13–14}$$

Notice that V_{R3} will increase or decrease if the load voltage increases or decreases. When the load current increases to its maximum permissible value, V_{RSC} becomes large enough to make V_{BE} approximately 0.7 V; that is, V_{RSC} becomes large enough to exceed the drop across V_{R3} by about 0.7 V: $0.7 \approx V_{BE} = V_{RSC} - V_{R3}$. At this point, current limiting occurs, just as it does in the current-limited regulator. If the load resistance is now made smaller, the load voltage will drop, as we have seen previously. But when the load voltage drops, V_{R3} drops. Consequently, by equation 13–14, a smaller value of V_{RSC} is required to maintain $V_{BE} = 0.7$ V. Because V_{BE} remains essentially constant at 0.7 V, a smaller load current must flow to produce the smaller drop across R_3. Further decreases in load resistance produce further drops in load voltage and a further reduction in load current, so foldback limiting occurs. If the load resistance is restored to its normal operating value, the circuit resumes normal regulator action.

FIGURE 13–11 Foldback limiting (compare with Figure 13–10)

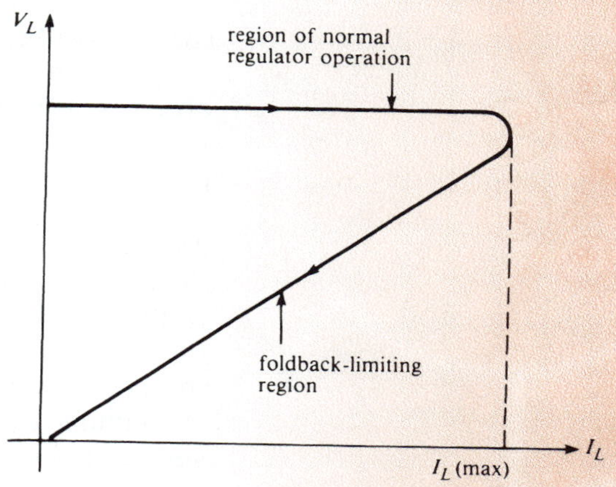

FIGURE 13–12 A current-limited regulator modified to provide foldback limiting

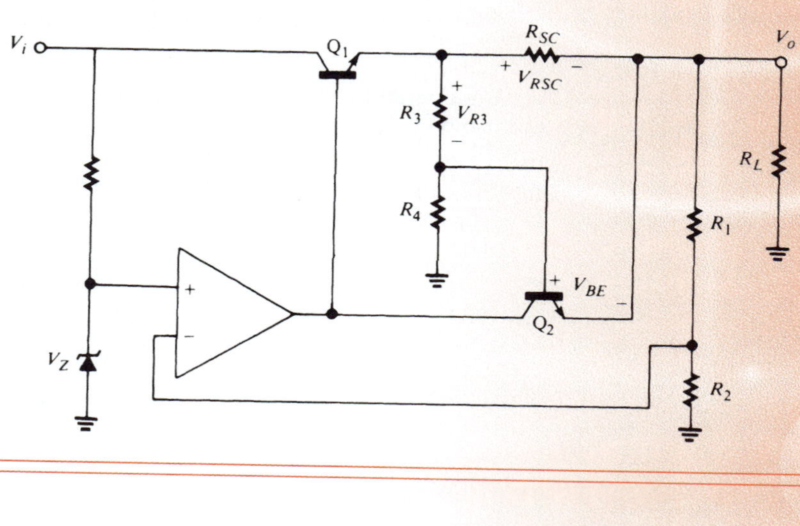

EXAMPLE 13–7

The circuit in Figure 13–12 maintains a regulated output of 6 V. If $R_3 = 1\ \text{k}\Omega$ and $R_4 = 9\ \text{k}\Omega$, what should be the value of R_{SC} to impose a maximum current limit of 1 A?

Solution

At the onset of current limiting,

$$V_{R3} = \left(\frac{R_3}{R_3 + R_4}\right)V_o = \left[\frac{1\ \text{k}\Omega}{(1\ \text{k}\Omega) + (9\ \text{k}\Omega)}\right](6\text{V}) = 0.6\ \text{V}$$

From equation 13–14,

$$V_{BE} = 0.7\ \text{V} = V_{RSC} - 0.6\ \text{V}$$

$$V_{RSC} = 1.3\ \text{V}$$

Thus,

$$R_{SC} = \frac{V_{RSC}}{I_L(\text{max})} = \frac{1.3\ \text{V}}{1\ \text{A}} = 1.3\ \Omega$$

Shunt Regulators

Figure 13–13 is a functional block diagram of the shunt-type regulator. Each of the components shown in the figure performs the same function as its counterpart in the series regulator (Figure 13–5), but notice in this case that the control element is in parallel with the load. The control element maintains a constant load voltage by shunting more or less current from the load.

It is convenient to think of the control element in Figure 13–13 as a variable resistance. When the load voltage decreases, the resistance of the control element is made to increase, so less current is diverted from the load, and the load voltage rises. Conversely, when the load voltage increases, the resistance of the control element decreases, and more current is shunted away from the load. From another viewpoint, the source resistance R_S on the unregulated side of Figure 13–13 forms a voltage divider with the parallel combination of the control element and R_L. Thus, when the resistance of the control element increases, the resistance of the parallel combination increases, and, by voltage-divider action, the load voltage increases.

Figure 13–14 shows a discrete shunt regulator in which transistor Q_1 serves as the shunt control element. Since V_Z is constant, any change in output voltage creates a proportionate change in the voltage across R_1. Thus, if V_o decreases, the voltage across R_1 decreases, as does the base voltage of Q_2. Q_2 conducts less heavily and the current into the base of Q_1 is reduced. Q_1

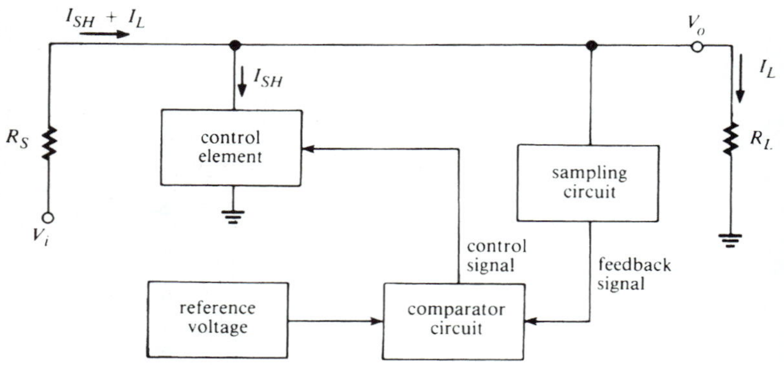

FIGURE 13–13 Functional block diagram of a shunt-type voltage regulator

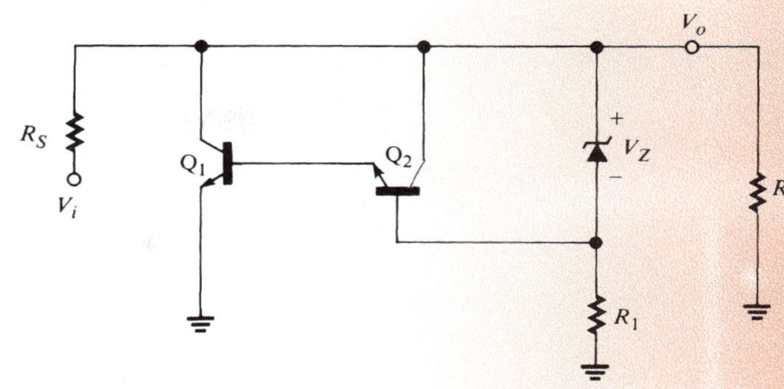

FIGURE 13–14 A discrete shunt regulator

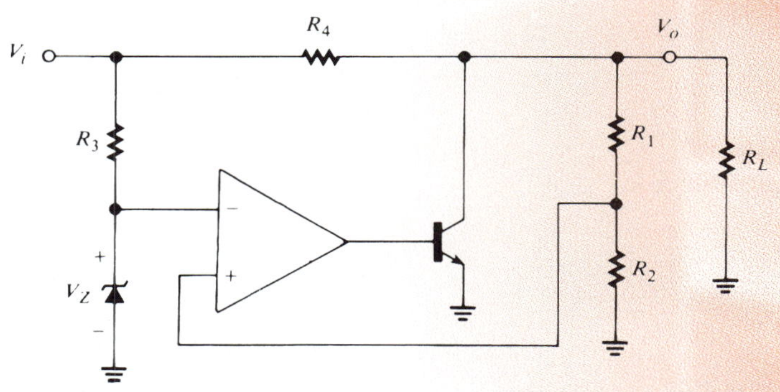

FIGURE 13–15 A shunt voltage regulator incorporating an operational amplifier

then conducts less heavily and shunts less current from the load, allowing the load voltage to rise. Conversely, an increase in V_o causes both Q_1 and Q_2 to conduct more heavily, and more current is diverted from the load.

Figure 13–15 shows a shunt regulator incorporating an operational amplifier. Resistors R_1 and R_2 form a voltage divider that feeds a voltage proportional to V_o back to the noninverting input. This voltage is greater than the reference voltage V_Z applied to the inverting input, so the output of the amplifier is a positive voltage proportional to $V_o - V_Z$. If V_o decreases, the amplifier output decreases and Q_1 conducts less heavily. Thus, less current is diverted from the load and the output voltage rises.

One advantage of the shunt-regulator circuit shown in Figure 13–15 is that it has inherent current limiting. It is clear that the load current cannot exceed V_i/R_4, which is the current that would flow through R_4 if the output were short-circuited. Because load current must flow through R_4, the power dissipation in the resistor may be quite large, particularly under short-circuit conditions.

13–4 SWITCHING REGULATORS

A principal disadvantage of the series and shunt regulators discussed in Section 13–3 is the fact that the control element in each type must dissipate a large amount of power. As a consequence, they have poor *efficiency;* that is, the ratio of load power to total input power is relatively small. Because the control element, such as a pass transistor, is operated in its active region, power dissipation can be quite large, particularly when there is a large voltage difference between V_i and V_o and when substantial load current flows. When efficiency is a major concern, the switching-type regulator is often used. In this design, the control element is switched on and off at a rapid rate and is therefore either in saturation or cutoff most of the time. As is the case with other digital switching devices, this mode of operation results in very low power consumption.

FIGURE 13–16 Operation of a pulse-width modulator

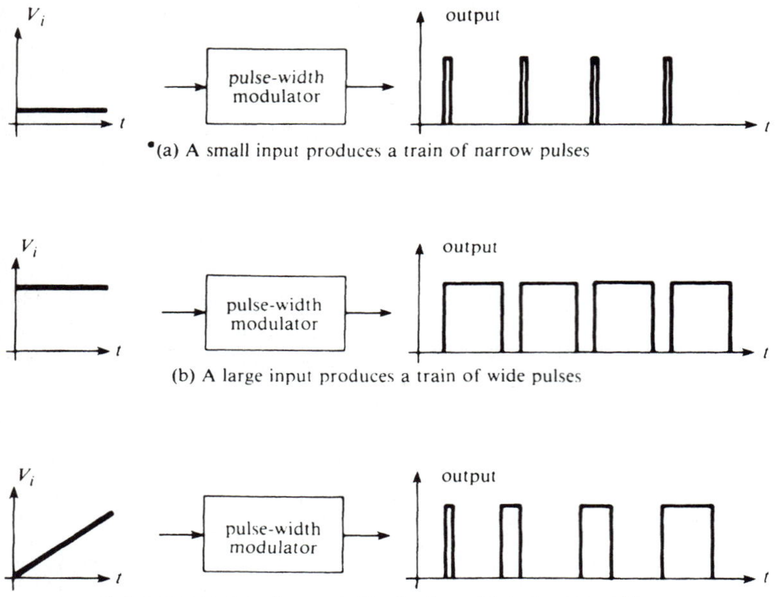

(a) A small input produces a train of narrow pulses

(b) A large input produces a train of wide pulses

(c) A ramp input produces a train of pulses with increasing widths

The fundamental component of a switching regulator is a *pulse-width modulator,* illustrated in Figure 13–16. This device produces a train of rectangular pulses having widths that are proportional to the device's input. The figure illustrates the pulse output corresponding to a small dc input, the output corresponding to a large dc input, and that corresponding to a ramp-type input. To produce a periodic sequence of pulses, the pulse-width modulator must employ some form of oscillator. Practical modulators used in switching regulators usually have inherent oscillation capabilities incorporated into their circuit designs. Figure 13–17 shows how a pulse-width modulator (used in power amplifiers) can be constructed using a voltage comparator driven by a sawtooth waveform.

FIGURE 13–17 The dc value of a pulse train is directly proportionate to its duty cycle

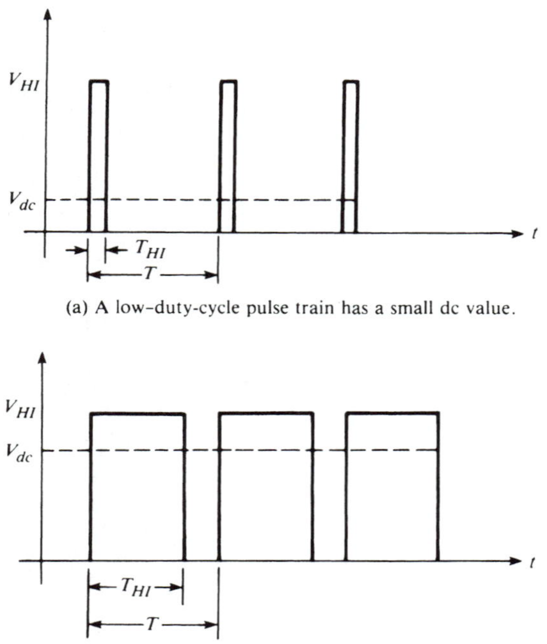

(a) A low-duty-cycle pulse train has a small dc value.

(b) A high-duty-cycle pulse train has a large dc value.

The *duty cycle* of a pulse train is the proportion of the period during which the pulse is present, i.e., the fractional part of the period during which the output is high. For example, a square wave, which is high for one half-cycle and low for the other half-cycle, has a duty cycle of 0.5, or 50%. By definition

$$\text{duty cycle} = T_{HI}/T \tag{13–15}$$

where T_{HI} is the total time during period T that the waveform is high. As illustrated in Figure 13–17, the dc, or average, value of a pulse train is directly proportionate to its duty cycle:

$$V_{dc} = V_{HI}\left(\frac{T_{HI}}{T}\right) \tag{13–16}$$

where V_{HI} is the high, or peak, value of the pulse. Equation 13–46 is valid for pulses that alternate between 0 volts (low) and V_{HI} volts.

A switching regulator uses a pulse-width modulator to produce a pulse train whose duty cycle is automatically adjusted as necessary to increase or decrease the dc value of the train. The basic circuit is shown in Figure 13–18. Note that the output of the pulse-width modulator drives pass transistor Q_1. When the drive pulse is high, Q_1 is saturated and maximum current flows through it. When the drive pulse is low, Q_1 is cut off and no current flows. Thus, the current through Q_1 is a series of pulses having the same duty cycle as the modulator output. As in the regulator circuits discussed earlier, the zener diode provides a reference voltage that is compared in the operational amplifier to the feedback voltage obtained from the R_1–R_2 voltage divider. If the load voltage V_o tends to fall, then the output of the operational amplifier increases and a larger voltage is applied to the pulse-width modulator. It therefore produces a pulse train having a larger duty cycle. The pulse train switches Q_1 on and off with a greater duty cycle, so the dc value of the current through Q_1 is increased. The increased current raises the load voltage to compensate for its initial decrease.

Series inductor L and shunt capacitor C in Figure 13–18 form a low-pass filter that recovers the dc value of the pulse waveform supplied to it by Q_1. The fundamental frequency of the pulse train and its harmonics are suppressed by the filter. Diode D_1 is used to suppress the negative voltage transient generated by the inductor when Q_1 turns off. It is called a *catcher* diode, or *free-wheeling* diode. Modern switching regulators are operated at frequencies in the kilohertz range, usually less than 100 kHz. High frequencies are desirable because the LC filter components can then be smaller, less bulky, and less expensive. However, high-frequency switching of heavy currents creates large magnetic fields that induce noise voltages in surrounding conductors. This generation of *electromagnetic interference* (EMI) is a principal limitation of switching regulators, and they require careful shielding. Also, switching frequencies are limited by the inability of the

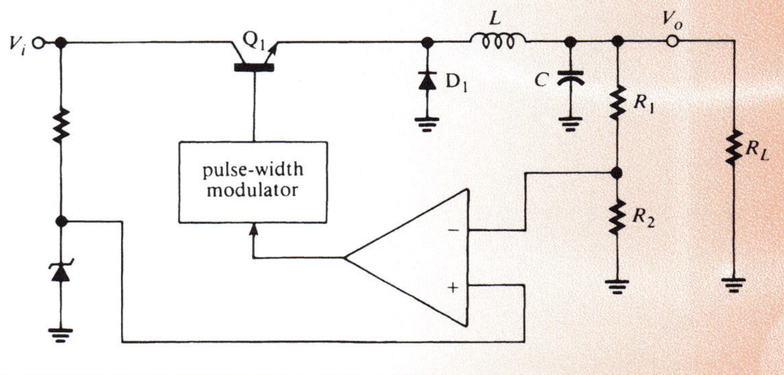

FIGURE 13–18 A basic switching-type regulator circuit

power (pass) transistor to turn on and off at high speeds. Many modern regulators use power MOSFETs of the VMOS design instead of bipolars because of their superior ability to switch heavy loads at high frequencies (see Section 16–11 on MOSFET power amplifiers).

EXAMPLE 13–8

The switching regulator in Figure 13–18 is designed to maintain a 12-V-dc output when the unregulated input voltage varies from 15 V to 24 V. When pass transistor Q_1 is conducting, its collector-to-emitter saturation voltage is 0.5 V. Assuming that the load is constant and the LC filter is ideal (so its output is the dc value of its input), find the minimum and maximum duty cycles of the pulse-width modulator.

Solution

When V_i has its minimum value of 15 V, the high voltage of the pulse train at the input to the filter is $15 - V_{CE(sat)} = 15 - 0.5 = 14.5$ V. The duty cycle of this pulse train must be sufficient to produce a dc value of 12 V. From equation 13–16, $T_{HI}/T = V_{dc}/V_{HI} = 12/14.5 = 0.828$. Thus, the *maximum* duty cycle, corresponding to the smallest value of V_i, is 0.828, or 82.8%. When V_i has its maximum value of 24 V, we find the minimum duty cycle to be $T_{HI}/T = 12/(24 - 0.5) = 0.51$.

Switching Regulator Data Sheets

A partial set of data sheets for a commercial switching regulator has been provided. The data sheet for a National Semiconductor LM3352 buck-boost switched capacitor dc/dc converter is shown in Figure 13–19. This circuit provides both boost (step-up voltage) and buck (step-down voltage) for an input dc voltage range from 2.5 to 5.5 V while outputting standard output voltages of 2.5 V, 3.0 V, and 3.3 V. This device is designed for use in battery-powered applications. Regulation of the output voltage is provided by gating the clock to the switch, which controls the output charge. Typical clock frequency is 1 MHz.

Filter capacitor selection is critical for the charge-pump portion of the LM3352 circuit. Recommendations are provided by the manufacturer regarding the choice of tantalum, ceramic, or polymer electrolytic capacitors. (*Note:* Capacitor selection in general is also addressed in Appendix D.) This information is very important for the beginning analog circuit designer. The assumption that all capacitors perform equally well for all frequencies and operating conditions is invalid.

13–5 ADJUSTABLE INTEGRATED-CIRCUIT REGULATORS

As the name implies, an *adjustable* voltage regulator is one that can be set to maintain any output voltage that is within some prescribed range. Unlike three-terminal regulators, adjustable IC regulators must have external components connected to them to perform voltage regulation.

The 723 integrated-circuit regulator is an example of a popular and very versatile adjustable regulator. It can be connected to produce positive or negative outputs over the range from 2 to 37 V, provide either current limiting or foldback limiting, be used with an external pass transistor to handle load currents up to 10 A, and be used as a switching regulator. Figure 13–20 (p. 534) is a block diagram of the regulator. Note the terminal labeled V_{REF} at the output of the voltage reference amplifier. This is an internally gener-

National Semiconductor

September 1999

LM3352
Regulated 200 mA Buck-Boost Switched Capacitor DC/DC Converter

LM3352 Regulated 200 mA Buck-Boost Switched Capacitor DC/DC Converter

General Description

The LM3352 is a CMOS switched capacitor DC/DC converter that produces a regulated output voltage by automatically stepping up (boost) or stepping down (buck) the input voltage. It accepts an input voltage between 2.5V and 5.5V. The LM3352 is available in three standard output voltage versions: 2.5V, 3.0V and 3.3V. If other output voltage options between 1.8V and 4.0V are desired, please contact your National Semiconductor representative.

The LM3352's proprietary buck-boost architecture enables up to 200 mA of load current at an average efficiency greater than 80%. Typical operating current is only 400 µA and the typical shutdown current is only 2.5 µA.

The LM3352 is available in a 16-pin TSSOP package. This package has a maximum height of only 1.1 mm.

The high efficiency of the LM3352, low operating and shutdown currents, small package size, and the small size of the overall solution make this device ideal for battery powered, portable, and hand-held applications.

Features

- Regulated V_{OUT} with ±3% accuracy
- Standard output voltage options: 2.5V, 3.0V and 3.3V
- Custom output voltages available from 1.8V to 4.0V in 100 mV increments
- 2.5V to 5.5V input voltage
- Up to 200 mA output current
- >80% average efficiency
- Uses few, low-cost external components
- Very small solution size
- 400 µA typical operating current
- 2.5 µA typical shutdown current
- 1 MHz switching frequency (typical)
- Architecture and control methods provide high load current and good efficiency
- TSSOP-16 package
- Over-temperature protection

Applications

- 1-cell LiIon battery-operated equipment including PDAs, hand-held PCs, cellular phones
- Flat panel displays
- Hand-held instruments
- NiCd, NiMH, or alkaline battery powered systems
- 3.3V to 2.5V and 5.0V to 3.3V conversion

Typical Operating Circuit

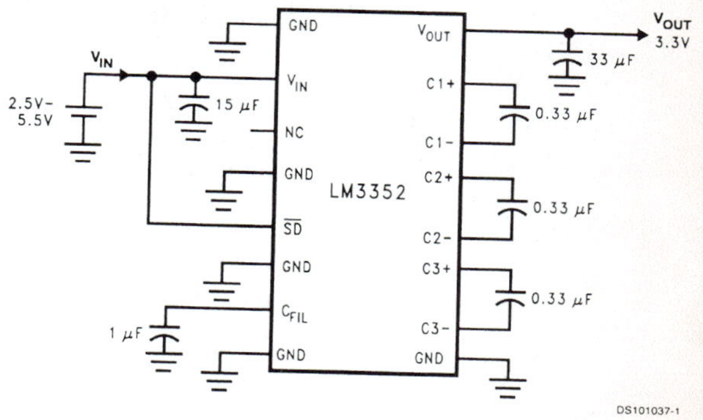

www.national.com

FIGURE 13–19 The LM3352 data sheet (Reprinted with permission of National Semiconductor Corporation)

ated voltage of approximately 7 V that is available at an external pin. To set a desired regulator output voltage, the user connects this 7-V output, or an externally divided-down portion of it, to one of the inputs of the error amplifier. The error amplifier is a comparator that compares the externally connected reference to a voltage proportionate to V_o. Depending on whether the reference is connected to the noninverting or the inverting input, the

Absolute Maximum Ratings (Note 1)

If Military/Aerospace specified devices are required, please contact the National Semiconductor Sales Office/ Distributors for availability and specifications.

V_{OUT} Pin	−0.5V to 4.5V
All Other Pins	−0.5V to 5.6V
Power Dissipation (T_A = 25°C) (Note 2)	700 mW
T_{JMAX} (Note 2)	150°C
θ_{JA} (Note 2)	150°C/W
Storage Temperature	−65°C to +150°C
Lead Temperature (Soldering, 5 sec.)	260°C
ESD Rating (Note 3)	
human body model	2 kV
machine model	100V

Operating Ratings

Input Voltage (V_{IN})	2.5V to 5.5V
Output Voltage (V_{OUT})	1.8V to 4.0V
Ambient Temperature (T_A) (Note 2)	−40°C to +85°C
Junction Temperature (T_J) (Note 2)	−40°C to +125°C

Electrical Characteristics

Limits in standard typeface are for T_J = 25°C, and limits in **boldface** type apply over the full operating temperature range. Unless otherwise specified: $C_1 = C_2 = C_3$ = 0.33 µF; C_{IN} = 15 µF; C_{OUT} = 33 µF; V_{IN} = 3.5V.

Parameter	Conditions	Min	Typ	Max	Units
LM3352-2.5					
Output Voltage (V_{OUT})	V_{IN} = 3.5V; I_{LOAD} = 100 mA	2.463	2.5	2.537	V
	2.8V < V_{IN} < 5.5V; 1 mA < I_{LOAD} < 100 mA	2.425/**2.400**	2.5	2.575/**2.600**	
	3.6V < V_{IN} < 4.9V; 1 mA < I_{LOAD} < 200 mA	2.425/**2.400**	2.5	2.575/**2.600**	
	4.9V < V_{IN} < 5.5V; 1 mA < I_{LOAD} < 175 mA	2.425/**2.400**	2.5	2.575/**2.600**	
Efficiency	I_{LOAD} = 15 mA		85		%
	I_{LOAD} = 150 mA, V_{IN} = 4.0V		75		
Output Voltage Ripple (Peak-to-Peak)	I_{LOAD} = 50 mA; C_{OUT} = 33 µF tantalum		75		mV_{P-P}
LM3352-3.0					
Output Voltage (V_{OUT})	V_{IN} = 3.5V; I_{LOAD} = 100 mA	2.955	3.0	3.045	V
	2.5V < V_{IN} < 5.5V; 1 mA < I_{LOAD} < 100 mA	2.910/**2.880**	3.0	3.090/**3.120**	
	3.8V < V_{IN} < 5.5V; 1 mA < I_{LOAD} < 200 mA	2.910/**2.880**	3.0	3.090/**3.120**	
Efficiency	I_{LOAD} = 15 mA		80		%
	I_{LOAD} = 150 mA, V_{IN} = 4.0V		75		
Output Voltage Ripple (Peak-to-Peak)	I_{LOAD} = 50 mA; C_{OUT} = 33 µF tantalum		75		mV_{P-P}
LM3352-3.3					
Output Voltage (V_{OUT})	V_{IN} = 3.5V; I_{LOAD} = 100 mA	3.251	3.3	3.349	V
	2.5V < V_{IN} < 5.5V; 1 mA < I_{LOAD} < 100 mA	3.201/**3.168**	3.3	3.399/**3.432**	
	4.0V < V_{IN} < 5.5V; 1 mA < I_{LOAD} < 200 mA	3.201/**3.168**	3.3	3.399/**3.432**	
Efficiency	I_{LOAD} = 15 mA		90		%
	I_{LOAD} = 150 mA, V_{IN} = 4.0V		80		
Output Voltage Ripple (Peak-to-Peak)	I_{LOAD} = 50 mA; C_{OUT} = 33 µF tantalum		75		mV_{P-P}
LM3352-ALL OUTPUT VOLTAGE VERSIONS					
Operating Quiescent Current	Measured at Pin V_{IN}; I_{LOAD} = 0A (Note 4)		400	**500**	µA
Shutdown Quiescent Current	SD Pin at 0V (Note 5)		2.5	**5**	µA
Switching Frequency		**0.65**	1	**1.35**	MHz
SD Input Threshold Low	2.5V < V_{IN} < 5.5V			**0.2 V_{IN}**	V
SD Input Threshold High	2.5V < V_{IN} < 5.5V	**0.8 V_{IN}**			V

3 www.national.com

FIGURE 13–19 Continued

regulated output is either positive or negative. For normal positive voltage regulation, the unregulated input is connected to the terminals labeled V^+ and V_C, and V^- is connected to ground. Note the transistor labeled "current limiter" in Figure 13–20. By making external resistor connections to the CL and CS terminals, either the current-limiting circuit of Figure 13–9 or the fold-back circuit of Figure 13–12 can be implemented. The current limiter performs the function of Q_2 in each of those figures. Of course, the terminals can be left open if no limiting is desired.

Typical Performance Characteristics Unless otherwise specified T_A = 25°C. (Continued)

V_{OUT} Ripple vs. C_{OUT}

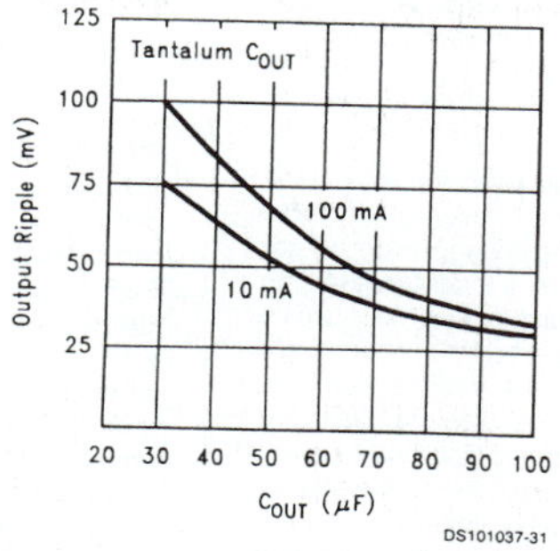

V_{OUT} Ripple vs. C_{OUT}

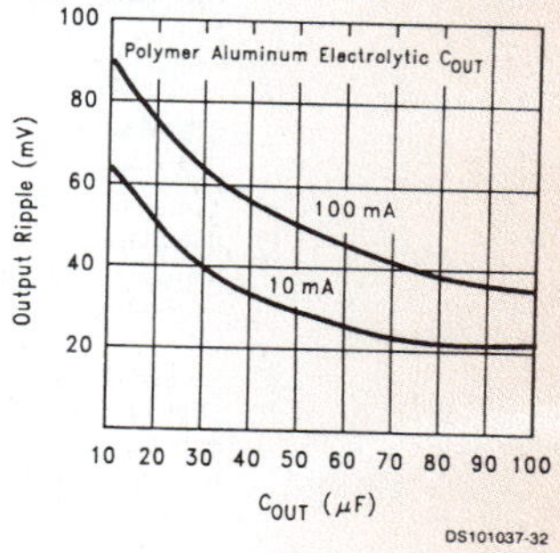

Applications Information

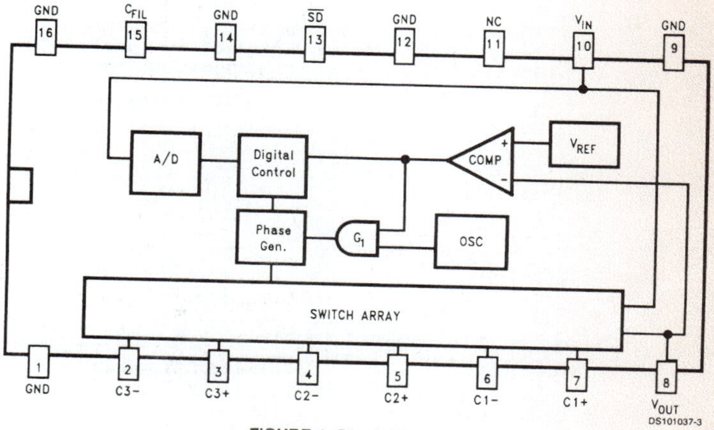

FIGURE 1. Block Diagram

Operating Principle

The LM3352 is designed to provide a step-up/step-down voltage regulation in battery powered systems. It combines switched capacitor circuitry, reference, comparator, and shutdown logic in a single 16-pin TSSOP package. The LM3352 can provide a regulated voltage between 1.8V and 4V from an input voltage between 2.5V and 5.5V. It can supply a load current up to 200 mA.

As shown in *Figure 1*, the LM3352 employs two feedback loops to provide regulation in the most efficient manner possible. The first loop is from V_{OUT} through the comparator COMP, the AND gate G_1, the phase generator, and the switch array. The comparator's output is high when V_{OUT} is less than the reference V_{REF}. Regulation is provided by gating the clock to the switch array. In this manner, charge is transferred to the output only when needed. The second loop controls the gain configuration of the switch array. This loop consists of the comparator, the digital control block, the phase generator, and the switch array. The digital control block computes the most efficient gain from a set of seven gains based on inputs from the A/D and the comparator. The gain signal is sent to the phase generator which then sends the appropriate timing and configuration signals to the switch array. This dual loop provides regulation over a wide range of loads efficiently.

Since efficiency is automatically optimized, the curves for V_{OUT} vs. V_{IN} and Efficiency vs. V_{IN} in the Typical Performance Characteristics section exhibit small variations. The reason is that as input voltage or output load changes, the digital control loops are making decisions on how to optimize efficiency. As the switch array is reconfigured, small variations in output voltage and efficiency result. In all cases where these small variations are observed, the part is operating correctly; minimizing output voltage changes and optimizing efficiency.

Charge Pump Capacitor Selection

A 0.33 μF ceramic capacitor is suggested for C1, C2 and C3. To ensure proper operation over temperature variations, an X7R dielectric material is recommended.

7

www.national.com

FIGURE 13–19 Continued

Figure 13–21 shows manufacturer's specifications for the 723 regulator, along with some typical applications. Figure 13–22 shows the 723 regulator connected to maintain its output at any voltage between +2 V and +7 V. The output is determined from

$$V_o = V_{REF}\left(\frac{R_2}{R_1 + R_2}\right) \tag{13–17}$$

Filter Capacitor Selection

a) CAPACITOR TECHNOLOGIES

The three major technologies of capacitors that can be used as filter capacitors for LM3352 are: i) tantalum, ii) ceramic and iii) polymer electrolytic technologies.

i) Tantalum

Tantalum capacitors are widely used in switching regulators. Tantalum capacitors have the highest CV rating of any technology; as a result, high values of capacitance can be obtained in relatively small package sizes. It is also possible to obtain high value tantalum capacitors in very low profile (<1.2 mm) packages. This makes the tantalums attractive for low-profile, small size applications. Tantalums also possess very good temperature stability; i.e., the change in the capacitance value, and impedance over temperature is relatively small. However, the tantalum capacitors have relatively high ESR values which can lead to higher voltage ripple and their frequency stability (variation over frequency) is not very good, especially at high frequencies (>1 MHz).

ii) Ceramic

Ceramic capacitors have the lowest ESR of the three technologies and their frequency stability is exceptionally good. These characteristics make the ceramics an attractive choice for low ripple, high frequency applications. However, the temperature stability of the ceramics is bad, except for the X7R and X5R dielectric types. High capacitance values (>1 μF) are achievable from companies such as Taiyo-yuden which are suitable for use with regulators. Ceramics are taller and larger than the tantalums of the same capacitance value.

iii) Polymer Electrolytic

Polymer electrolytic is a third suitable technology. Polymer capacitors provide some of the best features of both the ceramic and the tantalum technologies. They provide very low ESR values while still achieving high capacitance values. However, their ESR is still higher than the ceramics, and their capacitance value is lower than the tantalums of the same size. Polymers offer good frequency stability (comparable to ceramics) and good temperature stability (comparable to tantalums). The Aluminum Polymer Electrolytics offered by Cornell-Dubilier and Panasonic, and the POSCAPs offered by Sanyo fall under this category.

Table 1 compares the features of the three capacitor technologies.

TABLE 1. Comparison of Capacitor Technologies

	Ceramic	Tantalum	Polymer Electrolytic
ESR	Lowest	High	Low
Relative Height	Low for Small Values (<10 μF); Taller for Higher Values	Lowest	Low
Relative Footprint	Large	Small	Largest
Temperature Stability	X7R/X5R-Acceptable	Good	Good
Frequency Stability	Good	Acceptable	Good
V_{OUT} Ripple Magnitude @ <50 mA	Low	High	Low
V_{OUT} Ripple Magnitude @ >100 mA	Low	Slightly Higher	Low
dv/dt of V_{OUT} Ripple @ All Loads	Lowest	High	Low

b) CAPACITOR SELECTION

i) Output Capacitor (C_{OUT})

The output capacitor C_{OUT} directly affects the magnitude of the output ripple voltage so C_{OUT} should be carefully selected. The graphs titled V_{OUT} Ripple vs. C_{OUT} in the Typical Performance Characteristics section show how the ripple voltage magnitude is affected by the C_{OUT} value and the capacitor technology. These graphs are taken at the gain at which worst case ripple is observed. In general, the higher the value of C_{OUT}, the lower the output ripple magnitude. At lighter loads, the low ESR ceramics offer a much lower V_{OUT} ripple than the higher ESR tantalums of the same value. At higher loads, the ceramics offer a slightly lower V_{OUT} ripple magnitude than the tantalums of the same value. However, the dv/dt of the V_{OUT} ripple with the ceramics and polymer electrolytics is much lower than the tantalums under all load conditions. The tantalums are suggested for very low profile, small size applications. The ceramics and polymer electrolytics are a good choice for low ripple, low noise applications where size is less of a concern.

ii) Input Capacitor (C_{IN})

The input capacitor C_{IN} directly affects the magnitude of the input ripple voltage, and to a lesser degree the V_{OUT} ripple. A higher value C_{IN} will give a lower V_{IN} ripple. To optimize low input and output ripple as well as size a 15 μF polymer electrolytic, 22 μF ceramic, or 33 μF tantalum capacitor is recommended. This will ensure low input ripple at 200 mA load current. If lower currents will be used or higher input ripple can be tolerated then a smaller capacitor may be used to reduce the overall size of the circuit. The lower ESR ceramics and polymer electrolytics achieve a lower V_{IN} ripple than the higher ESR tantalums of the same value. Tantalums make a good choice for small size, very low profile applications. The ceramics and polymer electrolytics are a good choice for low ripple, low noise applications where size is less of a concern. The 15 μF polymer electrolytics are physically much larger than the 33 μF tantalums and 22 μF ceramics.

www.national.com 8

FIGURE 13–19 Continued

From the specifications in Figure 13–21, we see that V_{REF} may be between 6.95 V and 7.35 V. Therefore, the actual value produced by a given device should be measured before selecting values for R_1 and R_2, if a very accurate output voltage is required. Notice that the full (undivided) output voltage V_o is fed back to the inverting input (INV) through R_3. For maximum thermal stability, R_3 should be set equal to $R_1 \| R_2$.

The circuit shown in Figure 13–22 is connected to provide current limiting, where

Filter Capacitor Selection (Continued)

iii) C_{FIL}

A 1 μF, XR7 ceramic capacitor should be connected to pin C_{FIL}. This capacitor provides the filtering needed for the internal supply rail of the LM3352.

Of the different capacitor technologies, a sample of vendors that have been verified as suitable for use with the LM3352 are shown in *Table 2.*

TABLE 2. Capacitor Vendor Information

	Manufacturer	Tel	Fax	Website
Ceramic	Taiyo-yuden	(408) 573-4150	(408) 573-4159	www.t-yuden.com
	AVX	(803) 448-9411	(803) 448-1943	www.avxcorp.com
Tantalum	Sprague/Vishay	(207) 324-4140	(207) 324-7223	www.vishay.com
	Nichicon	(847) 843-7500	(847) 843-2798	www.nichicon.com
Polymer Electrolytic	Cornell-Dubilier (ESRD)	(508) 996-8561	(508) 996-3830	www.cornell-dubilier.com
	Sanyo (POSCAP)	(619) 661-6322	(619) 661-1055	www.sanyovideo.com

Maximum Available Output Current

The LM3352 cannot provide 200 mA under all V_{IN} and V_{OUT} conditions. The V_{OUT} vs V_{IN} graphs in the Typical Performance Characteristics section show the minimum V_{IN} at which the LM3352 is capable of providing different load currents while maintaining V_{OUT} regulation. Refer to the Electrical Characteristics for guaranteed conditions.

Maximum Load Under Start-Up

Due to the LM3352's unique start-up sequence, it is not able to start up under all load conditions. Starting with 45 mA or less will allow the part to start correctly under any temperature or input voltage conditions. After the output is in regulation, any load up to the maximum as specified in the Electrical Characteristics may be applied. Using a Power On Reset circuit, such as the LP3470, is recommended if greater start up loads are expected. Under certain conditions the LM3352 can start up with greater load currents without the use of a Power On Reset Circuit (*See application note AN-1144: Maximizing Startup Loads with the LM3352 Regulated Buck/Boost Switched Capacitor Converter*).

Thermal Protection

During output short circuit conditions, the LM3352 will draw high currents causing a rise in the junction temperature. On-chip thermal protection circuitry disables the charge pump action once the junction temperature exceeds the thermal trip point, and re-enables the charge pump when the junction temperature falls back to a safe operating point.

Typical Application Circuits

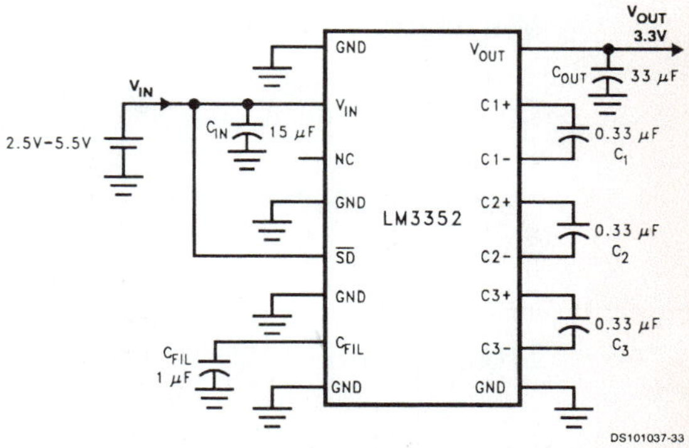

FIGURE 2. Basic Buck/Boost Regulator

9 www.national.com

FIGURE 13–19 Continued

$$I_L(\max) \approx \frac{0.7\ \text{V}}{R_{SC}} \tag{13–18}$$

The 100-pF capacitor shown in the figure is used to ensure circuit stability. When the circuit is connected to provide foldback limiting, a voltage divider (identified as resistors R_3 and R_4 in Figure 13–12) is connected across V_{OUT} in Figure 13–22. The CL terminal on the 723 regulator is then connected to the middle of the divider, instead of to V_{OUT}.

FIGURE 13–20 Block diagram of the 723 adjustable voltage regulator

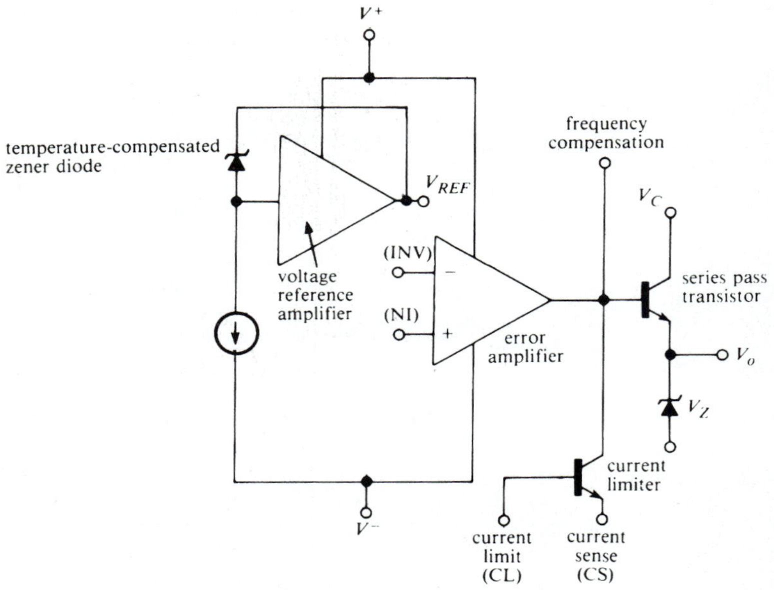

EXAMPLE 13–9

DESIGN

Design a 723-regulator circuit that will maintain an output voltage of +5 V and that will provide current limiting at 0.1 A. Assume that $V_{REF} = 7.0\text{ V}$.

Solution

We arbitrarily choose $R_2 = 1\text{ k}\Omega$. Then, from equation 13–47,

$$V_o = 5\text{ V} = (7\text{ V})\left[\frac{1\text{ k}\Omega}{R_1 + (1\text{ k}\Omega)}\right]$$

Solving for R_1, we find $R_1 = 400\ \Omega$. The optimum value of R_3 is

$$R_3 = R_1 \| R_2 = \frac{400(1000)}{1400} = 285.7\ \Omega$$

From equation 13–18,

$$I_L(\text{max}) = 0.1\text{ A} = \frac{0.7\text{ V}}{R_{SC}}$$

$$R_{SC} = 7\ \Omega$$

13–6 VOLTAGE REGULATOR CIRCUIT ANALYSIS WITH ELECTRONICS WORKBENCH MULTISIM

This section examines the use of EWB to simulate a voltage regulator circuit.

To begin the exercise, open the file **Ch13–EWB.msm** that is found in the Electronics Workbench Multisim 2001 CD-ROM packaged with the text. The circuit being demonstrated is shown in Figure 13–23. This is an LM7805 voltage regulator.

The purpose of this exercise is to demonstrate the limits of voltage regulation. Commercial voltage regulators do an excellent job of providing a fixed voltage as long as the minimum and maximum input voltages are not exceeded. Double-click on the ac signal generator and set the voltage offset to 9.0 V. Set the voltage amplitude to 3 V. The settings are shown in Figure 13–24.

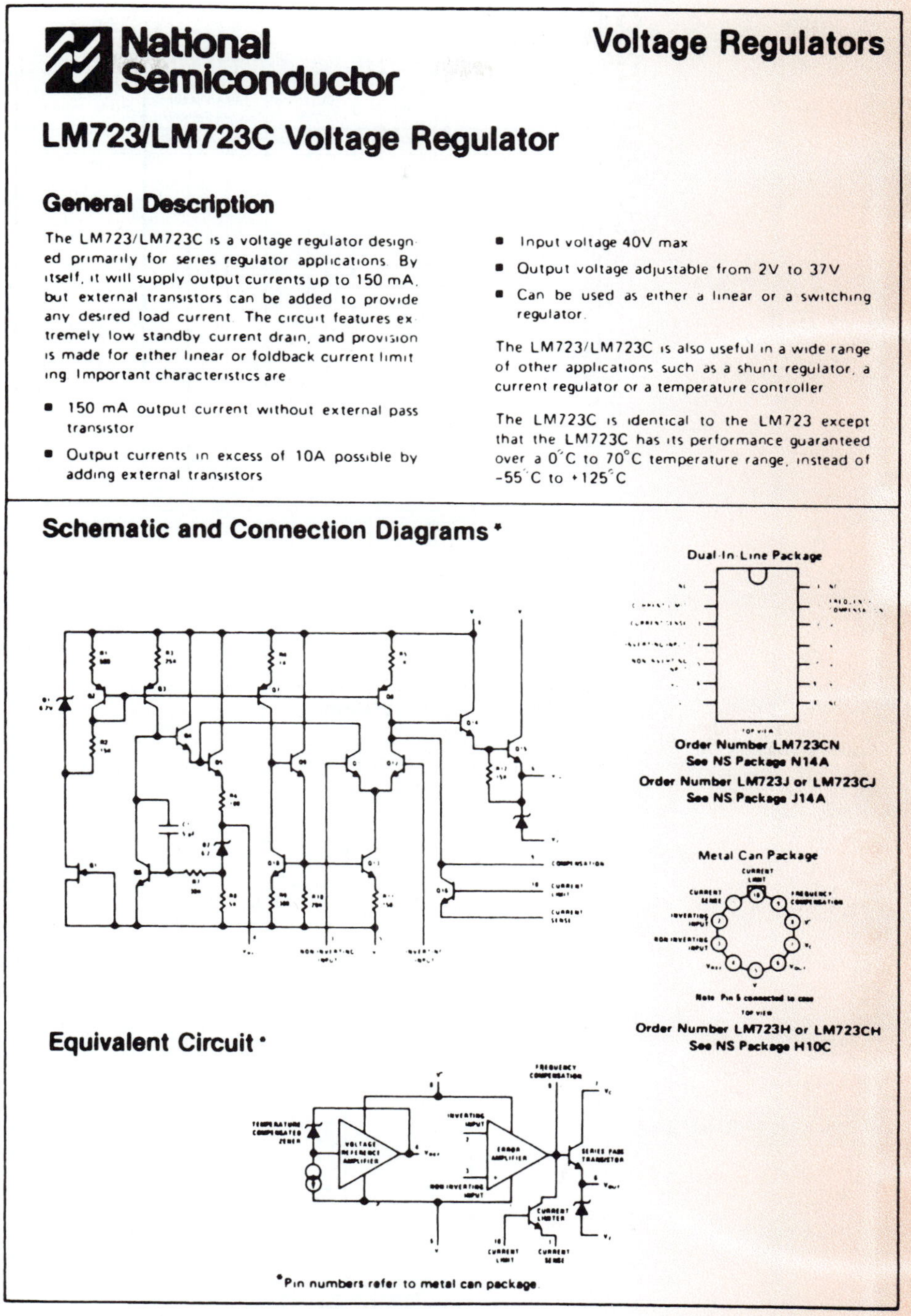

National Semiconductor

Voltage Regulators

LM723/LM723C

LM723/LM723C Voltage Regulator

General Description

The LM723/LM723C is a voltage regulator designed primarily for series regulator applications. By itself, it will supply output currents up to 150 mA, but external transistors can be added to provide any desired load current. The circuit features extremely low standby current drain, and provision is made for either linear or foldback current limiting. Important characteristics are:

- 150 mA output current without external pass transistor
- Output currents in excess of 10A possible by adding external transistors
- Input voltage 40V max
- Output voltage adjustable from 2V to 37V
- Can be used as either a linear or a switching regulator.

The LM723/LM723C is also useful in a wide range of other applications such as a shunt regulator, a current regulator or a temperature controller.

The LM723C is identical to the LM723 except that the LM723C has its performance guaranteed over a 0°C to 70°C temperature range, instead of −55°C to +125°C.

Schematic and Connection Diagrams*

Equivalent Circuit*

*Pin numbers refer to metal can package.

FIGURE 13–21 723–voltage-regulator specifications (Reprinted with permission of National Semiconductor Corporation)

This simulates an input ac signal with a dc offset of 9.0 V and a peak-to-peak ripple voltage of 6 V. Input this poor unregulated signal into the voltage regulator. Use the EWB oscilloscope to monitor the input and output voltages. You should see a result similar to that shown in Figure 13–25.

Set the signal voltage to 7 V dc offset and leave the amplitude voltage at an amplitude of 3 V. Start the simulation and view the results with the oscilloscope. You should observe a result similar to that shown in Figure

LM723/LM723C

TABLE I RESISTOR VALUES (kΩ) FOR STANDARD OUTPUT VOLTAGE

POSITIVE OUTPUT VOLTAGE	APPLICABLE FIGURES	FIXED OUTPUT ±5%		OUTPUT ADJUSTABLE ±10% (Note 5)			NEGATIVE OUTPUT VOLTAGE	APPLICABLE FIGURES	FIXED OUTPUT ±5%		5% OUTPUT ADJUSTABLE ±10%		
	(Note 4)	R1	R2	R1	P1	R2			R1	R2	R1	P1	R2
+3.0	1, 5, 6, 9, 12 (4)	4.12	3.01	1.8	0.5	1.2	+100	7	3.57	102	2.2	10	91
+3.6	1, 5, 6, 9, 12 (4)	3.57	3.65	1.5	0.5	1.5	+250	7	3.57	255	2.2	10	240
+5.0	1, 5, 6, 9, 12 (4)	2.15	4.99	.75	0.5	2.2	−6 (Note 6)	3, (10)	3.57	2.43	1.2	0.5	.75
+6.0	1, 5, 6, 9, 12 (4)	1.15	6.04	0.5	0.5	2.7	−9	3, 10	3.48	5.36	1.2	0.5	2.0
+9.0	2, 4, (5, 6, 12, 9)	1.87	7.15	.75	1.0	2.7	−12	3, 10	3.57	8.45	1.2	0.5	3.3
+12	2, 4, (5, 6, 9, 12)	4.87	7.15	2.0	1.0	3.0	−15	3, 10	3.65	11.5	1.2	0.5	4.3
+15	2, 4, (5, 6, 9, 12)	7.87	7.15	3.3	1.0	3.0	−28	3, 10	3.57	24.3	1.2	0.5	10
+28	2, 4, (5, 6, 9, 12)	21.0	7.15	5.6	1.0	2.0	−45	8	3.57	41.2	2.2	10	33
+45	7	3.57	48.7	2.2	10	39	−100	8	3.57	97.6	2.2	10	91
+75	7	3.57	78.7	2.2	10	68	−250	8	3.57	249	2.2	10	240

TABLE II FORMULAE FOR INTERMEDIATE OUTPUT VOLTAGES

Outputs from +2 to +7 volts (Figures 1, 5, 6, 9, 12, (4)) $V_{OUT} = \left(V_{REF} \times \frac{R2}{R1 + R2}\right)$	Outputs from +4 to +250 volts (Figure 7) $V_{OUT} = \frac{V_{REF}}{2} \times \frac{R2 - R1}{R1}$; $R3 = R4$	Current Limiting $I_{LIMIT} = \frac{V_{SENSE}}{R_{SC}}$
Outputs from +7 to +37 volts (Figures 2, 4, (5, 6, 9, 12)) $V_{OUT} = V_{REF} \times \frac{R1 + R2}{R2}$	Outputs from −6 to −250 volts (Figures 3, 8, 10) $V_{OUT} = -\left(\frac{V_{REF}}{2} \times \frac{R1 + R2}{R1}\right)$; $R3 = R4$	Foldback Current Limiting $I_{KNEE} = \left(\frac{V_{OUT}\,R3}{R_{SC}\,R4} + \frac{V_{SENSE}\,(R3 + R4)}{R_{SC}\,R4}\right)$ $I_{SHORT\,CKT} = \left(\frac{V_{SENSE}}{R_{SC}} \times \frac{R3 + R4}{R4}\right)$

Typical Applications

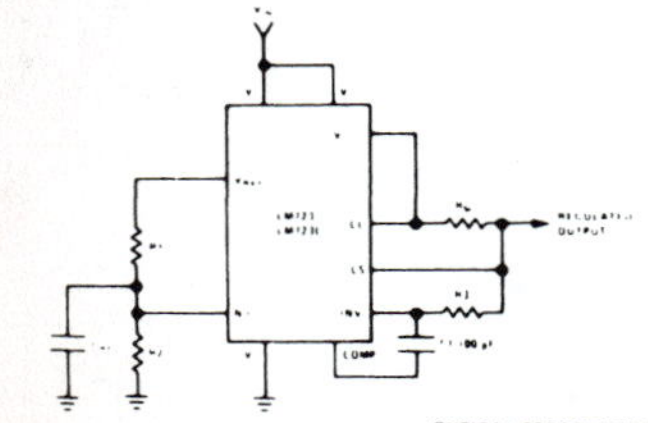

Note $R3 = \frac{R1\,R2}{R1 + R2}$ for minimum temperature drift

TYPICAL PERFORMANCE

Regulated Output Voltage	5V
Line Regulation (ΔV_{IN} = 3V)	0.5 mV
Load Regulation (ΔI_L = 50 mA)	1.5 mV

FIGURE 1. Basic Low Voltage Regulator (V_{OUT} = 2 to 7 Volts)

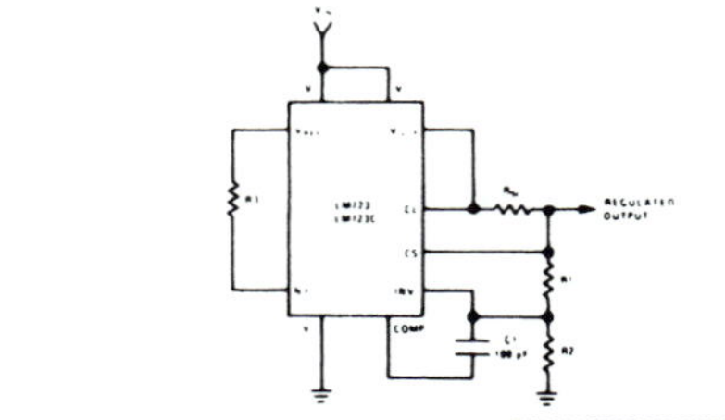

Note $R3 = \frac{R1\,R2}{R1 + R2}$ for minimum temperature drift.
R3 may be eliminated for minimum component count

TYPICAL PERFORMANCE

Regulated Output Voltage	15V
Line Regulation (ΔV_{IN} = 3V)	1.5 mV
Load Regulation (ΔI_L = 50 mA)	4.5 mV

FIGURE 2. Basic High Voltage Regulator (V_{OUT} = 7 to 37 Volts)

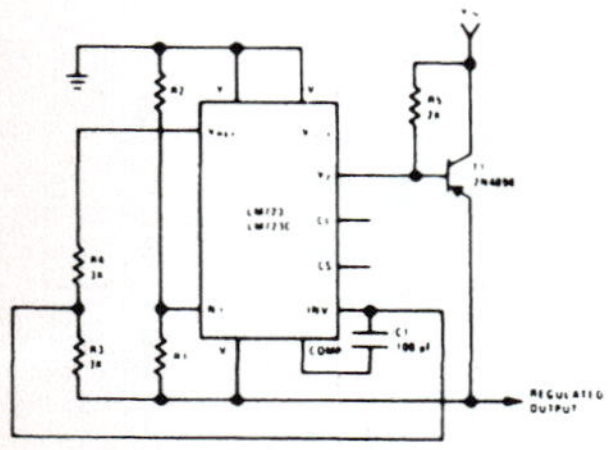

TYPICAL PERFORMANCE

Regulated Output Voltage	−15V
Line Regulation (ΔV_{IN} = 3V)	1 mV
Load Regulation (ΔI_L = 100 mA)	2 mV

FIGURE 3. Negative Voltage Regulator

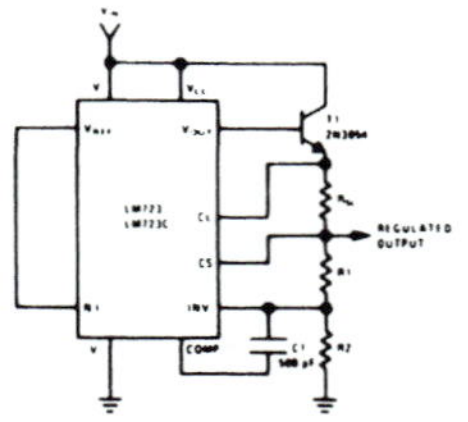

TYPICAL PERFORMANCE

Regulated Output Voltage	+15V
Line Regulation (ΔV_{IN} = 3V)	1.5 mV
Load Regulation (ΔI_L = 1A)	15 mV

FIGURE 4. Positive Voltage Regulator (External NPN Pass Transistor)

FIGURE 13–21 Continued

13–26. The oscilloscope trace shows that the output drops out of regulation when the minimum input voltage for the regulator is exceeded.

Double-click on the EWB LM7805 voltage regulator to examine how the circuit is modeled. Select "edit model," and the parameters will be displayed

LM723/LM723C

Absolute Maximum Ratings

Pulse Voltage from V⁺ to V⁻ (50 ms)	50V
Continuous Voltage from V⁺ to V⁻	40V
Input Output Voltage Differential	40V
Maximum Amplifier Input Voltage (Either Input)	7 5V
Maximum Amplifier Input Voltage (Differential)	5V
Current from V_Z	25 mA
Current from V_{REF}	15 mA
Internal Power Dissipation Metal Can (Note 1)	800 mW
Cavity DIP (Note 1)	900 mW
Molded DIP (Note 1)	660 mW
Operating Temperature Range LM723	55 C to +125 C
LM723C	0 C to +70 C
Storage Temperature Range Metal Can	-65 C to +150 C
DIP	-55 C to +125 C
Lead Temperature (Soldering, 10 sec)	300 C

Electrical Characteristics (Note 2)

PARAMETER	CONDITIONS	LM723			LM723C			UNITS
		MIN	TYP	MAX	MIN	TYP	MAX	
Line Regulation	V_{IN} = 12V to V_{IN} = 15V		01	0 1		01	0 1	% V_{OUT}
	-55°C < T_A < +125°C			0 3				% V_{OUT}
	0°C < T_A < +70°C						0 3	% V_{OUT}
	V_{IN} 12V to V_{IN} = 40V		02	0 2		0 1	0 5	% V_{OUT}
Load Regulation	I_L = 1 mA to I_L = 50 mA		03	0 15		03	0 2	% V_{OUT}
	-55°C < T_A < +125°C			0 6				% V_{OUT}
	0°C < T_A < +70°C						0 6	% V_{OUT}
Ripple Rejection	f = 50 Hz to 10 kHz, C_{REF} = 0		74			74		dB
	f 50 Hz to 10 kHz, C_{REF} 5 μF		86			86		dB
Average Temperature Coefficient of Output Voltage	-55°C < T_A < +125°C		002	015				%/°C
	0°C < T_A < +70°C					003	015	%/°C
Short Circuit Current Limit	R_{SC} = 10Ω, V_{OUT} 0		65			65		mA
Reference Voltage		6 95	7 15	7 35	6 80	7 15	7 50	V
Output Noise Voltage	BW = 100 Hz to 10 kHz, C_{REF} 0		20			20		μVrms
	BW = 100 Hz to 10 kHz, C_{REF} 5 μF		2 5			2 5		μVrms
Long Term Stability			0 1			0 1		%/1000 hrs
Standby Current Drain	I_L 0, V_{IN} 30V		1 3	3 5		1 3	4 0	mA
Input Voltage Range		9 5		40	9 5		40	V
Output Voltage Range		2 0		37	2 0		37	V
Input Output Voltage Differential		3 0		38	3 0		38	V

Note 1: See derating curves for maximum power rating above 25°C.

Note 2: Unless otherwise specified, T_A = 25°C, V_{IN} = V^+ = V_C = 12V, V^- = 0, V_{OUT} = 5V, I_L = 1 mA, R_{SC} = 0, C_1 = 100 pF, C_{REF} = 0 and divider impedance as seen by error amplifier ≤ 10 kΩ connected as shown in Figure 1. Line and load regulation specifications are given for the condition of constant chip temperature. Temperature drifts must be taken into account separately for high dissipation conditions.

Note 3: L_1 is 40 turns of No. 20 enameled copper wire wound on Ferroxcube P36/22-3B7 pot core or equivalent with 0.009 in. air gap.

Note 4: Figures in parentheses may be used if R1/R2 divider is placed on opposite input of error amp.

Note 5: Replace R1/R2 in figures with divider shown in Figure 13.

Note 6: V^+ must be connected to a +3V or greater supply.

Note 7: For metal can applications where V_Z is required, an external 6.2 volt zener diode should be connected in series with V_{OUT}.

FIGURE 13–21 Continued

for the regulator. The following is the text listed for the LM7805 model. The schematic for the model of the voltage regulator is provided in Figure 13–27. Keep in mind that this is just a model, not the circuitry for the actual integrated voltage regulator (see next page).

FIGURE 13–22 The 723 regulator connected to provide output voltages between +2 V and +7 V

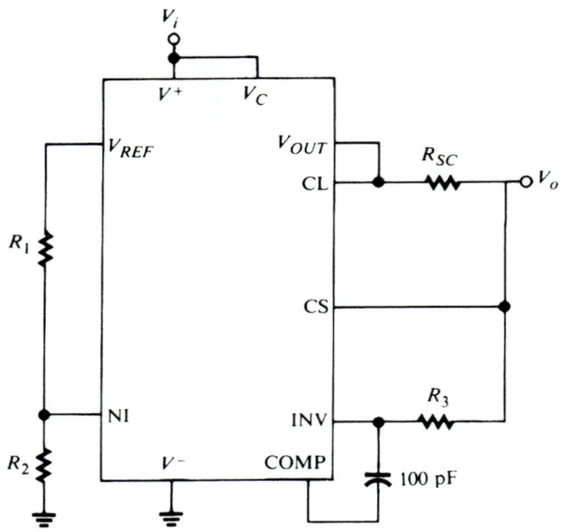

FIGURE 13–23 The voltage regulator circuit

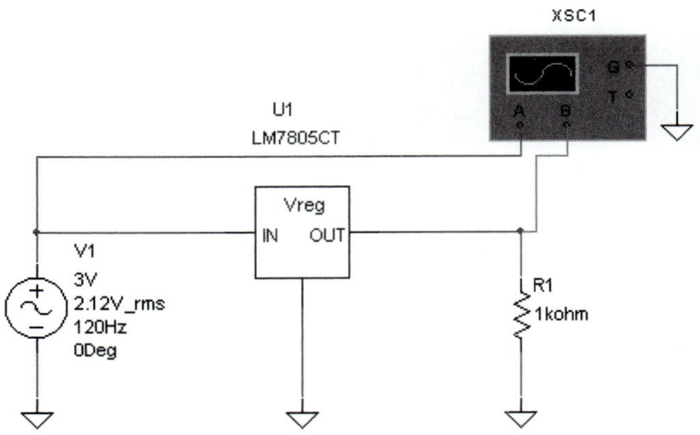


```
*POSITIVE 3 TERMINAL VOLTAGE REGULATOR
*PIN 1 = VOLTAGE INPUT
*PIN 2 = VOLTAGE REF (USUALLY GND)
*PIN 3 = VOLTAGE OUTPUT
*OUTPUT VOLTAGE = V1 - 1.3v
R1 5 3 50
R2 1 4 10000
R3 3 2 10000
V1 4 2 dc 6.3
Q2 1 4 5 CNTL_NPN
Q1 1 5 3 POWER_NPN
*
.MODEL CNTL_NPN NPN(Is=6.734f Xti=3 Eg=1.11 Vaf=74.03 Bf=416.4 Ne=1.259
+       Ise=6.734f Ikf=66.78m Xtb=1.5 Br=.7371 Nc=2 Isc=0 Ikr=0 Rc=1
+       Cjc=3.638p Mjc=.3085 Vjc=.75 Fc= .5 Cje=4.493p Mje=.2593 Vje=.75
+       Tr=239.5n Tf=301.2p Itf=.4 Vtf=4 Xtf=2 Rb=10)
.MODEL POWER_NPN NPN
+IS=3.41639e-13 BF=405.26 NF=1.03507 VAF=120.176
+IKF=10 ISE=8.44498e-12 NE=3.47601 BR=0.1
+NR=1.04536 VAR=1.10398 IKR=2.5029 ISC=5.50017e-13
+NC=3.93748 RB=1.80955 IRB=6.50293 RBM=1.80955
+RE=0.0001 RC=0.13777 XTB=0.1 XTI=1
+EG=1.05 CJE=6.36056e-10 VJE=0.4 MJE=0.479146
+TF=7.91835e-08 XTF=463.212 VTF=62107.5 ITF=1000
+CJC=2.70642e-10 VJC=0.95 MJC=0.406578 XCJC=0.128859
+FC=0.8 CJS=0 VJS=0.75 MJS=0.5
+TR=8.186e-05 PTF=0 KF=0 AF=1
```

FIGURE 13–24 The windows for setting the input ac signal parameters

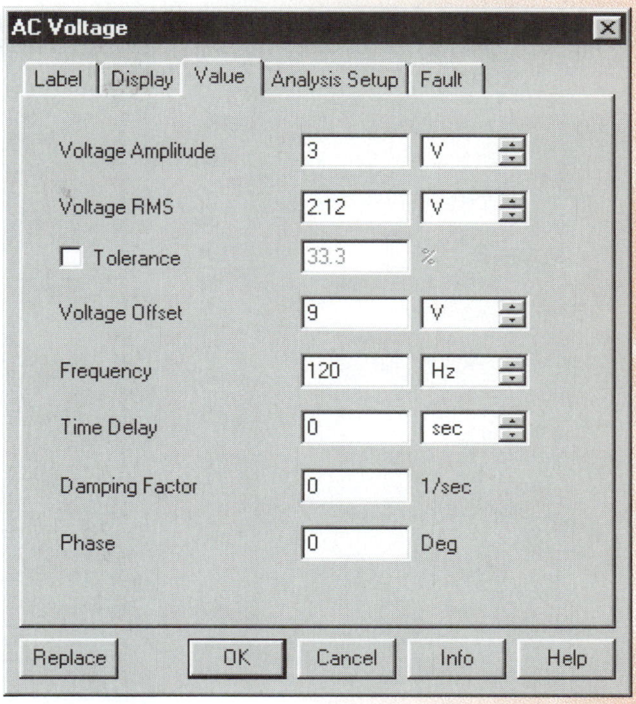

FIGURE 13–25 The oscilloscope display for the input ac signal and regulated output for the voltage regulator

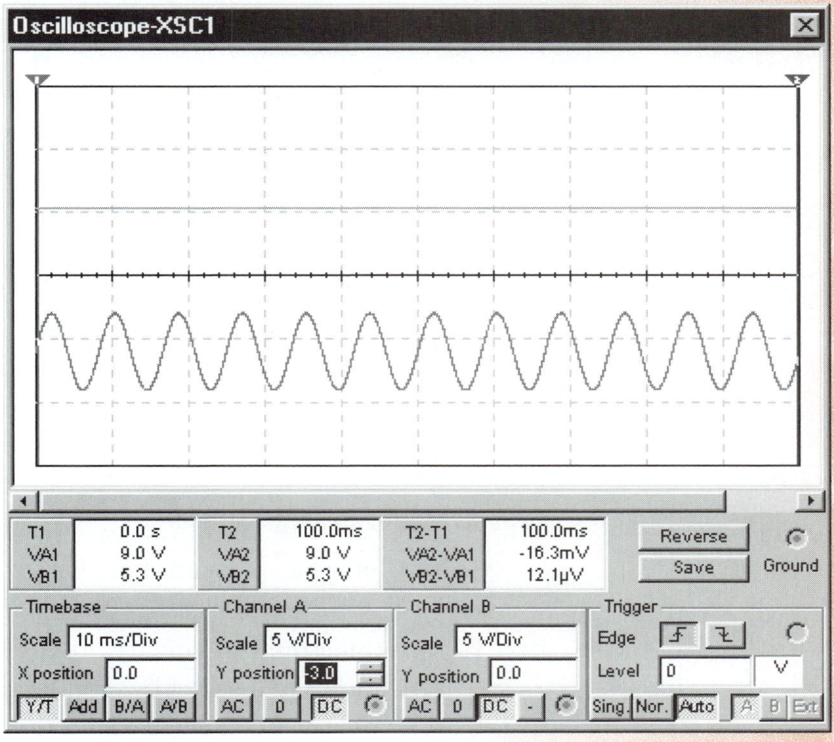

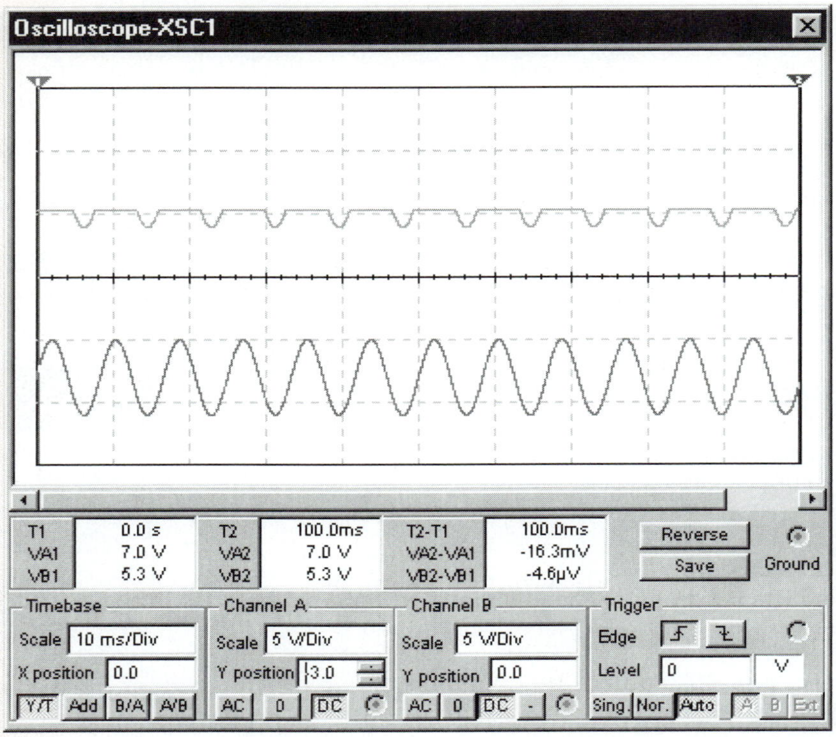

FIGURE 13–26 The oscilloscope output showing the output dropping out of regulation

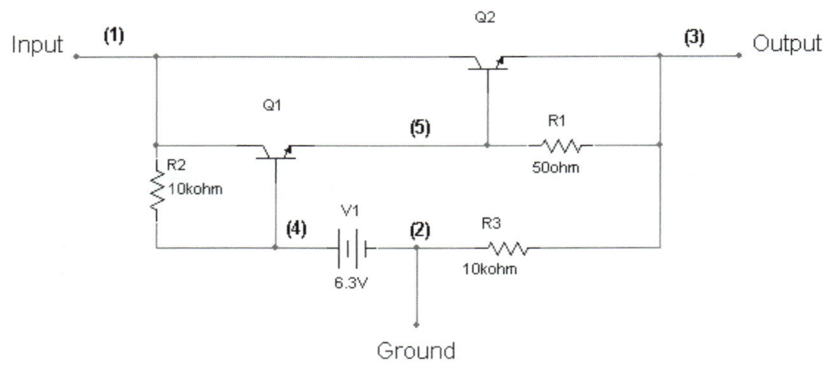

FIGURE 13–27 The schematic for the model of the voltage regulator

The model, shown in Figure 13–27 produces similar results to that obtained using the EWB LM7805 model. The difference is that ideal transistor models were used for Q_1 and Q_2.

This exercise has demonstrated the use of the EWB-Multisim voltage regulator. The student should be able to relate this material to the concepts presented in the chapter.

SUMMARY

This chapter has presented the basics of regulated and switching power supplies. The concepts the student should understand are:

- How electronic circuits regulate the output voltage.
- The basics of a switching regulator.
- The function of a dc-dc converter.

EXERCISES

SECTION 13–2

Voltage Regulation

13–1. A power supply has 4% voltage regulation and an open-circuit output voltage of 48 V dc.

(a) What is the full-load voltage of the supply?

(b) If a 120-Ω resistor draws full-load current from the supply, what is its output resistance?

13–2. One way to determine the output resistance of a power supply is to vary its load resistance while measuring load voltage. The output resistance of the supply equals the value of load resistance that is found to make the load voltage equal to one-half of the open-circuit voltage of the supply. Using an appropriate equation, explain why this method is valid.

13–3. A 50-V power supply has line regulation 0.2%/V. How large would the 75-V input voltage to the supply have to become for the output voltage to rise to 52 V?

13–4. The specifications for a certain 24-V power supply state that the output voltage increases 18 mV per volt increase in the input voltage. What is the percent line regulation of the supply?

SECTION 13–3

Series and Shunt Voltage Regulators

13–5. The base-to-emitter voltage of the transistor in Figure 13–28 is 0.7 V. V_i can vary from 12 V to 24 V.

(a) What breakdown voltage should the zener diode have if the load voltage is to be maintained at 9 V?

(b) If the zener diode must conduct 10 mA of reverse current to remain in breakdown, what maximum value should R have?

(c) With the value of R found in (b), what is the maximum power dissipated in the zener diode?

13–6. The base-to-emitter voltage of the transistor shown in Figure 13–23 is 0.7 V. V_i can vary from 18 V to 30 V. If $V_Z = 10$ V, $R = 500\ \Omega$, and $R_L = 1\ \text{k}\Omega$, find

(a) the minimum collector-to-emitter voltage of the transistor,

(b) the maximum power dissipated in the transistor, and

(c) the maximum current supplied by V_i.

13–7. Draw a schematic diagram of a series voltage regulator that employs one transistor and one zener diode and that can be operated from an unregulated negative voltage to produce a regulated negative voltage. It is not necessary to show component values.

13–8. In the regulator circuit shown in Figure 13–6, V_{BE2} is 0.68 V when V_o is maintained at 24 V. The zener diode has $V_Z = 9$ V. Find values for R_1 and R_2.

13–9. In the regulator circuit shown in Figure 13–8, $R_1 = 4\ \text{k}\Omega$, $R_2 = 1\ \text{k}\Omega$, and $R_3 = 2\text{k}\Omega$, V_i can vary from 30 V to 50 V.

(a) What breakdown voltage should the zener diode have so that the regulated output voltage is 25 V?

(b) Using the V_Z found in (a), what is the maximum current in the zener diode?

13–10. The potentiometer in Figure 13–29 has a total resistance of 10 kΩ.

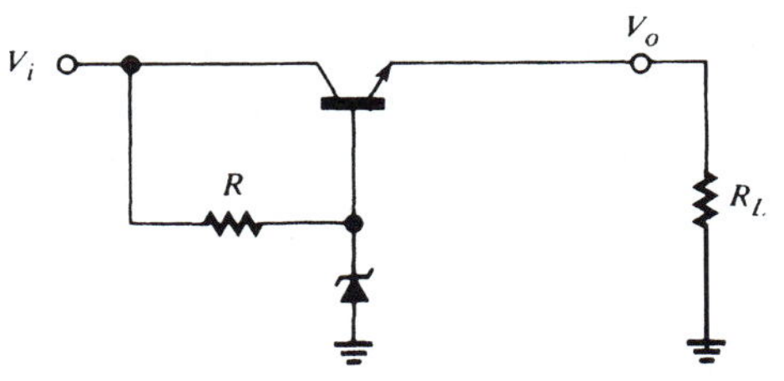

FIGURE 13–28 (Exercises 13–5 and 13–6)

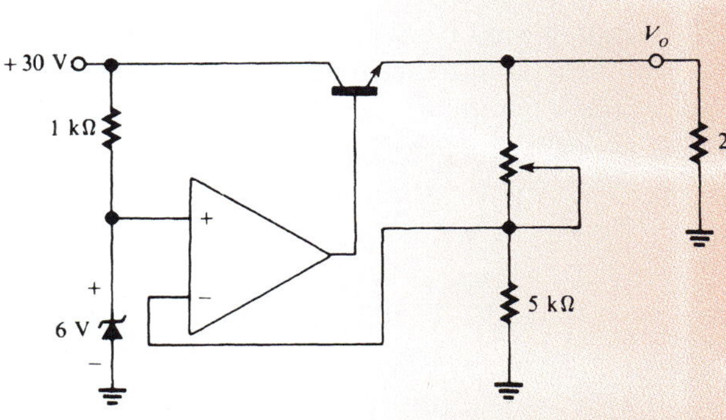

FIGURE 13–29 (Exercise 13–10)

(a) What range of regulated output voltages can be obtained by adjusting the potentiometer through its entire range?

(b) What power is dissipated in the pass transistor when the potentiometer is set for maximum resistance?

13–11. In the circuit shown in Figure 13–9, $R_1 = R_2 = 5$ kΩ, $V_Z = 6$ V, and $R_{SC} = 2$ Ω. Find V_o

(a) when $R_L = 100$ Ω, and

(b) when $R_L = 30$ Ω

13–12. Design a current-limited, series voltage regulator that includes an operational amplifier and a 10-V zener diode. The regulated output voltage should be 36 V and the output current should be limited to 700 mA. The zener current should be 10 mA when $V_i = 40$ V.

13–13. In Figure 13–12, $R_1 = 1.5$ kΩ, $R_2 = 1$ kΩ, $R_3 = 820$ Ω, $R_4 = 8.2$ kΩ, $R_{SC} = 2$ Ω and $V_Z = 6.3$ V.

(a) Find the regulated output voltage.

(b) Find the maximum load current.

13–14. When the regulated output voltage in Figure 13–14 is 18 V, V_i supplies a total current of 0.5 A. If $R_L = 100$ Ω, $V_Z = 6$ V, $R_1 = 200$Ω, and the collector current in Q_2 is negligible, find the power dissipated by Q_1.

13–15. In Figure 13–15, the unregulated input voltage varies from 6 to 9 V and $R_4 = 15$ Ω.

(a) What is the maximum (short-circuit) current that can be drawn from the regulator?

(b) What is the minimum power dissipation rating that R_4 should have?

SECTION 13–4

Switching Regulators

13–16. What should be the duty cycle of a train of pulses that alternate between 0 V and +5 V if the dc value of the train must be 1.2 V?

13–17. (a) In Figure 13–30(a), $T_1 = 2$ ms, $T = 5$ ms, $V_1 = 0.5$ V, and $V_2 = 6$ V. Find the dc value.

(b) In Figure 13–30(b), $T_1 = 2$ ms, $T = 5$ ms, $V_1 = -0.5$ V, and $V_2 = 6$ V. Find the dc value.

13–18. Derive one general expression involving V_1, V_2, T_1, and T that can be used to determine the dc value of either of the pulse trains in Figure 13–25. (Note that V_1 would be entered in the expression as a positive number for Figure 13–25(a) and as a negative number for 13–25(b).) Use your expression to solve Exercise 13–17.

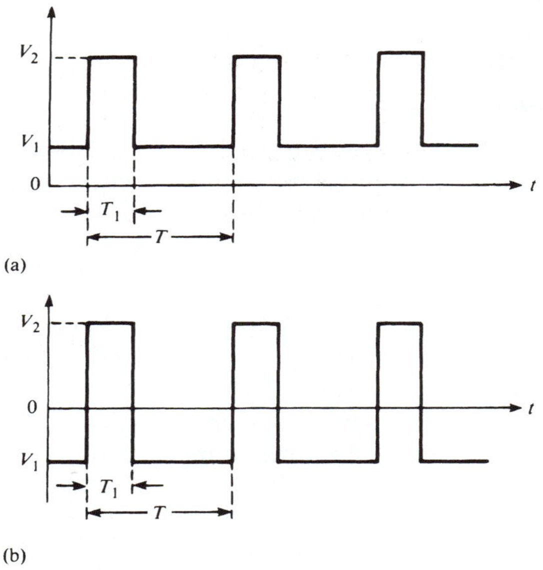

FIGURE 13–30 (Exercise 13–17)

13–19. The pulse train produced by the pulse-width modulator in Figure 13–18 has a 50% duty cycle when $V_i = 20$ V. What is its duty cycle when $V_i = 30$ V? When $V_i = 15$ V?

SECTION 13–5

Adjustable Integrated-Circuit Regulators

13–20. In Figure 13–22, $R_1 = 330$ Ω, $R_2 = 4.7$ kΩ, and $R_{SC} = 3$ Ω. Assuming that $V_{REF} = 7$ V,

(a) find the value of the regulated output voltage,

(b) find the optimum value of R_3, and

(c) find the maximum load current.

13–21. Design a 723-regulator circuit whose output voltage can be varied from 3 V to 6 V by adjusting a potentiometer connected between R_1 and R_2 in Figure 13–22. Assume that $V_{REF} = 7$ V.

13–22. Using the reference material given in the 723 regulator specifications (Figure 13–21), design a +20-V regulator circuit. Assume that $V_{REF} = 7$ V.

SPICE EXERCISES

13–23. The voltage regulator in Figure 13–7 has $R_1 = 50\ \text{k}\Omega$, $R_2 = 43.75\ \text{k}\Omega$, $R_3 = 47\ \text{k}\Omega$, and $R_4 = 1\ \text{k}\Omega$. The zener diode has a breakdown voltage of 6.3 V and the input voltage to the regulator is 20 V dc. The regulator supplies full-load current when the load resistance is 200 Ω. Use SPICE to find the percent voltage regulation of the regulator. Model the zener diode by specifying a conventional diode having a reverse breakdown voltage of 6.3 V.

CHAPTER 14

DIGITAL-TO-ANALOG AND ANALOG-TO-DIGITAL CONVERTERS

OUTLINE

OBJECTIVES

- Introduce the student to the basic operations of digital-to-analog and analog-to-digital conversion.
- Examine the key structures used in the A/D and D/A conversion process.
- Investigate performance specifications for the DAC and ADC.
- Introduce the student to the use of EWB simulation for performing A/D and D/A conversion.

14–1 OVERVIEW

Analog and Digital Voltages

An analog voltage is one that may vary *continuously* throughout some range. For example, the output of an audio amplifier is an analog voltage that may have any value (of an infinite number of values) between its minimum and maximum voltage limits. Most of the devices and circuitry we have studied in this book are analog in nature. In contrast, a digital voltage has only two useful values: a "low" and a "high" voltage, such as 0 V and +5 V. Digital voltages are used to represent numerical values in the binary number system, which has just the two digits 0 and 1. For example, four digital voltages having the values 0 V, +5 V, 0 V, and +5 V would represent the binary number 0101, which equals 5 in decimal. Digital computers perform computations using binary numbers represented by digital voltages.

Most physical variables in our environment are analog in nature. Quantities such as temperature, pressure, velocity, weight, etc., can have any of an infinite number of values. However, because of the high speed and accuracy of digital systems, such as computers, it is frequently the case that the transmission of data representing such quantities and computations performed on them are operations that are best accomplished using digital equivalents of analog values. Two modern examples of functions that have traditionally been analog in nature and that are now performed digitally include communications systems using fiber-optic cables and music reproduction using compact disks. Converting an analog value to a digital equivalent (binary number) is called *digitizing* the value. An analog-to-digital converter (ADC, or A/D converter) performs this function. Figure 14–1(a) illustrates the operation of an ADC. After a digital system has transmitted, analyzed, or otherwise processed digital data, it is often necessary to convert the results of such operations back to analog values. This function is performed by a digital-to-analog converter (DAC, or D/A converter). Figure 14–1(b) illustrates the operation of a DAC. Figure 14–1(c) shows a practical example of a system that incorporates both an ADC and a DAC. In this example, an accelerometer mounted on a vibration table (used to test components for

FIGURE 14–1 The analog-to-digital converter (ADC) and digital-to-analog converter (DAC)

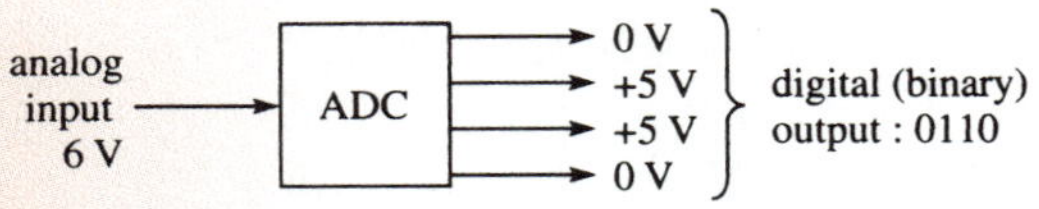

(a) The ADC converts a 6-V analog input to an equivalent digital output. (The binary number 0110 is equivalent to decimal 6.)

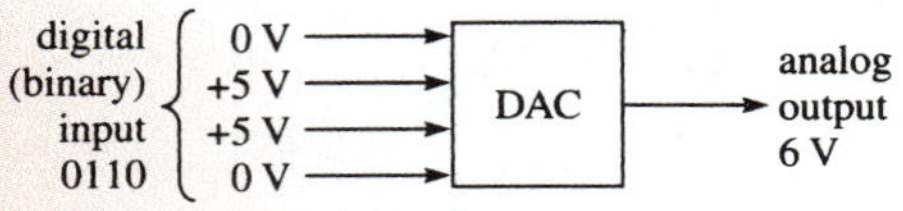

(b) The DAC converts 0110 to 6 V.

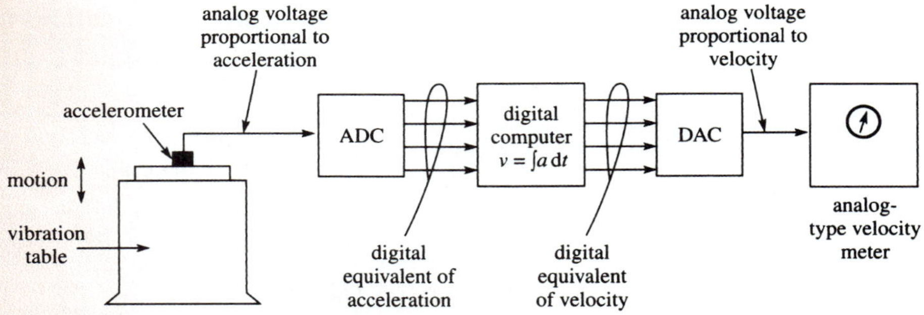

(c) Example of an instrumentation system that uses an analog-to-digital converter and a digital-to-analog converter.

vibration damage) produces an analog voltage proportionate to the instantaneous acceleration of the table. The analog voltage is converted to digital form and transmitted to a computer, which calculates the instantaneous velocity of the table (by mathematical integration: $v = \int a\,dt$). The digital quantity representing the computed velocity is then converted by a DAC to an analog voltage, which becomes the input to an analog-type velocity meter.

Converting Binary Numbers to Decimal Equivalents

As an aid in understanding subsequent discussions on digital-to-analog and analog-to-digital converters, we present here for review and reference purposes a brief summary of the mathematics used to convert binary numbers to their decimal equivalents.

As in the decimal number system, the position of each digit in a binary number carries a certain *weight.* In the decimal number system, the weights are powers of 10 (the units position: $10^0 = 1$, the tens position: $10^1 = 10$, the hundreds position: $10^2 = 100$, and so forth). In the binary number system, each position carries a weight equal to a power of 2: $2^0 = 1$, $2^1 = 2$, $2^2 = 4$, $2^3 = 8$, and so forth. A binary digit is called a *bit,* so a binary number consists of a sequence of bits, each of which can be either 0 or 1. The rightmost bit in a binary number carries the least weight and is called the *least significant bit* (LSB). If the binary number is an integer (whole number), the LSB carries weight $2^0 = 1$, and each bit to the left of the LSB has a weight equal to a power of 2 that is one greater than the weight of the bit to its right. To convert a binary number to its decimal equivalent, we compute the sum of all the weights of the positions where binary 1s occur. The following is an example:

binary number: 1 0 1 0 0 1 1

$$1 \times 2^0 = 1$$
$$1 \times 2^1 = 2$$
$$0 \times 2^2 = 0$$
$$0 \times 2^3 = 0$$
$$1 \times 2^4 = 16$$
$$0 \times 2^5 = 0$$
$$1 \times 2^6 = 64$$

$$\text{decimal equivalent} = 1 + 2 + 16 + 64 = 83$$

The leftmost bit in a binary number is called the *most significant bit* (MSB). An easy way to convert a binary number to its decimal equivalent is to write the values of the powers of 2 (1, 2, 4, 8, . . .) above each bit, beginning with the LSB and proceeding through the MSB. Then we simply add those powers where 1s occur. For example:

power of 2:	16 8 4 2 1
binary number:	1 0 1 1 0
	16 + 4 + 2 = 22 = decimal equivalent

If a binary number has a fractional part, then a *binary point* (like a decimal point) separates the integer part from the fractional part. The powers of 2 become negative and descend as we move right past the binary point. The following is an example:

					2^{-1}	2^{-2}
					‖	‖
power of 2:	8	4	2	1	0.5	0.25
binary number:	1	0	0	1.	0	1
	8 + 1 + 0.25 = 9.25					

Some Digital Terminology

Binary 0 and 1 are called logical 0 and logical 1 to distinguish them from the voltages used to represent them in a digital system. The voltages are called *logic levels.* For example, a common set of logic levels is 0 V (ground) for logical 0 and +5 V for logical 1.

A *binary counter* is a device whose binary output is numerically equal to the number of pulses that have occurred at its input. For example, the output of a 4-bit binary counter would be 0110 after 6 pulses had occurred at its input. The input pulses often occur at a fixed frequency from a signal called the system *clock*. The largest binary number that a 4-bit counter can contain is 1111, or decimal 15. (The counter *resets* to 0000 after the sixteenth pulse.) Thus, the total number of binary numbers that a 4-bit counter can produce is (counting 0000) 16, or 2^4. The total number of binary numbers that can be represented by n bits is 2^n. The counter we have just described is called an *up-counter* because its binary output increases by 1 (increments) each time a new clock pulse occurs. In a *down-counter,* the binary output decreases by 1 (decrements) each time a new clock pulse occurs. Thus, the sequence of outputs from a 4-bit down-counter would be 1111, 1110, 1101, . . . , 0000, 1111, ...

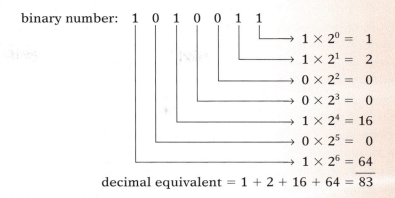

A *latch* is a digital device that stores the value (0 or 1) of a binary input. It is especially useful when a binary input is changing and we want to "latch onto" its value at a particular instant of time—as, for example, when performing a digital-to-analog conversion of a digital quantity that is continuously changing in value. A *register* is a set of latches used to store all the bits of a digital quantity, one latch for each bit.

Some analog-to-digital converters produce outputs that are in 8-4-2-1 *binary-coded–decimal* (BCD) form, rather than true binary. In 8-4-2-1 BCD, *each* decimal digit is represented by 4 bits. For example, if the input to an ADC having this type of output were 14 V, then the output would be

analog input: 14
8-4-2-1 output: (0001) (0100)

Note that this type of output is quite different from true binary. (The number 14 in binary is 1110.)

Resolution

A digital-to-analog converter having a 4-bit binary input produces only $2^4 = 16$ different analog output voltages, corresponding to the 16 different values that can be represented by the 4-bit input. The output of the converter is, therefore, not truly analog, in the sense that it cannot have an infinite number of values. If the input were a 5-bit binary number, the output could have $2^5 = 32$ different values. In short, the greater the number of input bits, the greater the number of output values and the closer the output resembles a true analog quantity. *Resolution* is a measure of this property. The greater the resolution of the DAC, the finer the increments between output voltage levels.

Similarly, an analog-to-digital converter having a 4-bit binary output produces only 16 different binary outputs, so it can convert only 16 different analog inputs to digital form. Since an analog input has an infinite number of values, the ADC does not truly convert (every) analog input to equivalent digital form. The greater the number of output bits, the greater the number of analog inputs the ADC can convert and the greater the resolution of the device. Resolution is clearly an important performance specification for both DACs and ADCs, and we will examine quantitative measures for it in later discussions.

14–2 THE *R*-2*R* LADDER DAC

The most popular method for converting a digital input to an analog output incorporates a ladder network containing series-parallel combinations of two resistor values: R and $2R$. Figure 14–2 shows an example of an R-$2R$ ladder having a 4-bit digital input.

To understand the operation of the R-$2R$ ladder, let us first determine the output voltage in Figure 14–2 when the input is 1000. We will assume that a logical-1 input is E volts and that logical 0 is 0 V (ground). Figure 14–3(a) shows the circuit with input D_3 connected to E and all other inputs connected

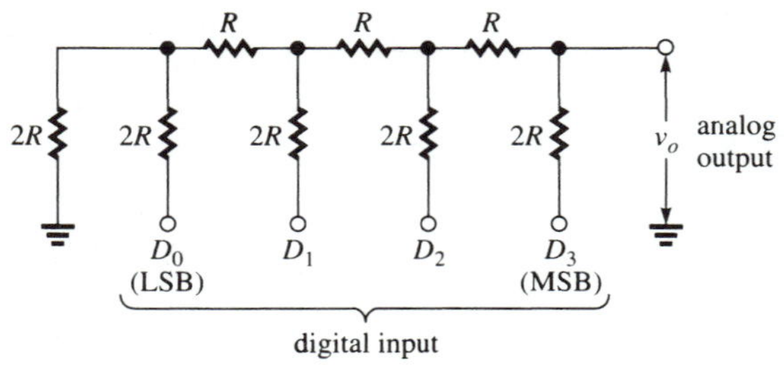

FIGURE 14–2 A 4-bit R-$2R$ ladder network used for digital-to-analog conversion

FIGURE 14–3 Calculating the output of the *R*-2*R* ladder when the input is 1000

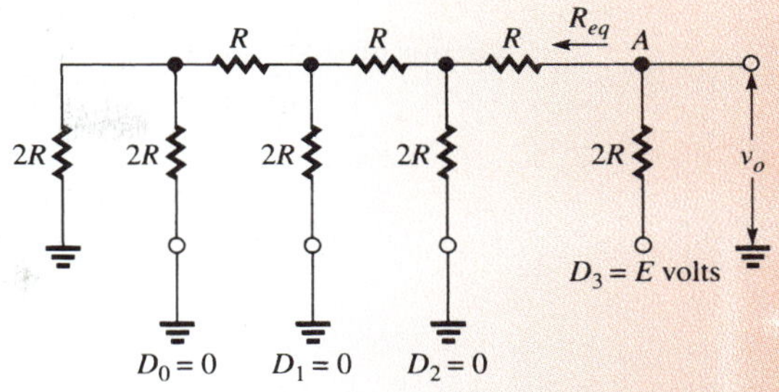

(a) When the input is 1000, D_0, D_1, and D_2 are grounded (0 V) and D_3 is *E* volts.

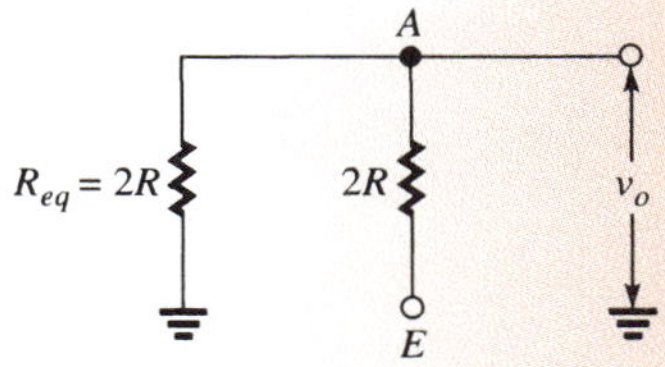

(b) The circuit equivalent to (a) when the network to the left of node *A* is replaced by its equivalent resistance, R_{eq}.

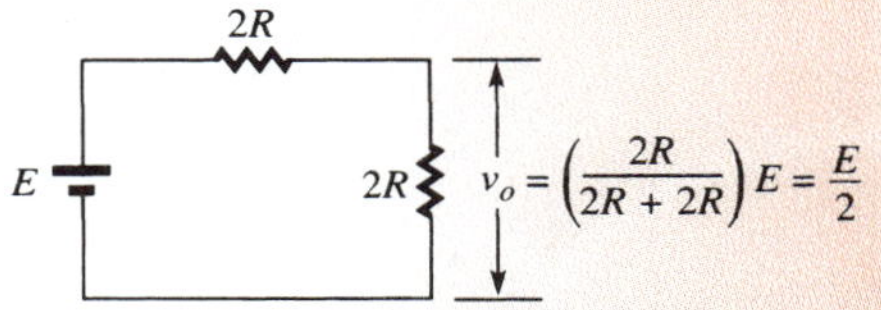

(c) Calculation of v_o using the voltage-divider rule. (Note that v_o in (b) is the voltage across $R_{eq} = 2R$.)

to ground, corresponding to the binary input 1000. We wish to find the total equivalent resistance, R_{eq}, looking to the left from node *A*. At the left end of the ladder, we see that 2*R* is in parallel with 2*R*, so that combination is equivalent to *R*. That *R* is in series with another *R*, giving 2*R*. That 2*R* is in parallel with another 2*R*, which is equivalent to *R* once again. Continuing in this manner, we ultimately find that $R_{eq} = 2R$. In fact, we see that the equivalent resistance looking to the left from every node is 2*R*. Figure 14–3(b) shows the circuit when it is redrawn with R_{eq} replacing all of the network to the left of node *A*. Figure 14–3(c) shows an equivalent way to draw the circuit in (b), and it is now readily apparent that $v_o = E/2$.

Let us now find v_o when the input is 0100. Figure 14–4(a) shows the circuit, with D_2 connected to *E* and all other inputs grounded. As demonstrated earlier, the equivalent resistance looking to the left from node *B* is 2*R*. The equivalent circuit with R_{eq} replacing the network to the left of *B* is shown in Figure 14–4(b). We proceed with the analysis by finding the (Thévenin) equivalent circuit to the left of node *A*, as indicated by the bracketed arrows. The Thévenin equivalent resistance (with *E* shorted to ground) is $R_{TH} = R + 2R \| 2R = R + R = 2R$. The Thévenin equivalent voltage is $[2R/(2R + 2R)]\,E = E/2$. The Thévenin equivalent circuit is shown in Figure 14–4(c). Figure 14–4(d) shows the ladder with the Thévenin equivalent circuit replacing everything to the left of node *A*. It is now apparent that $v_o = E/4$.

By an analysis similar to the foregoing, we find that the output of the ladder is *E*/8 when the input is 0010 and that the output is *E*/16 when the

FIGURE 14–4 Calculating the output of the *R*-2*R* ladder when the input is 0100

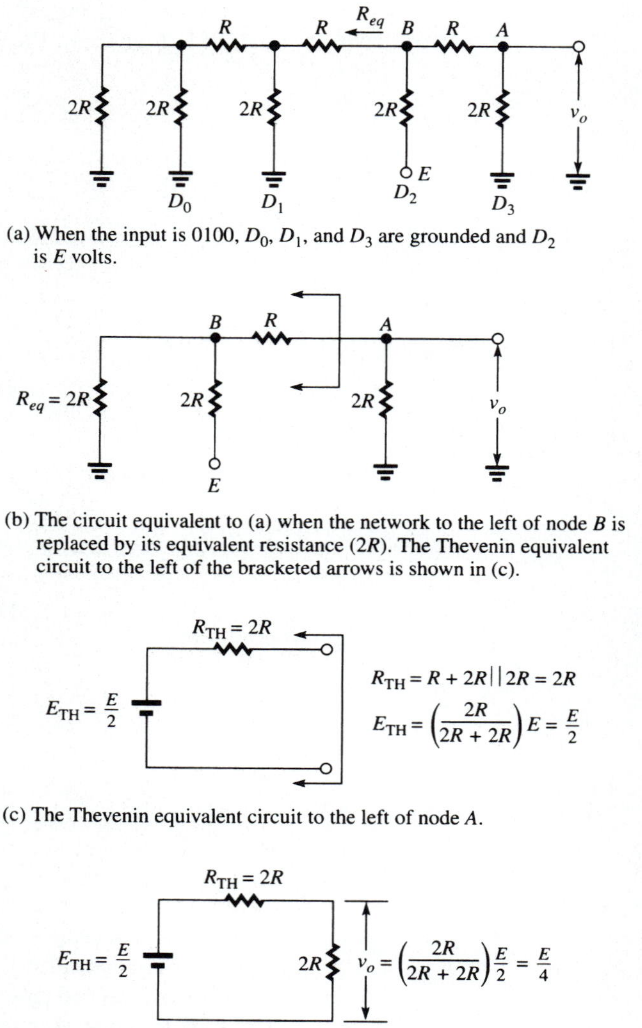

(a) When the input is 0100, D_0, D_1, and D_3 are grounded and D_2 is *E* volts.

(b) The circuit equivalent to (a) when the network to the left of node *B* is replaced by its equivalent resistance (2*R*). The Thevenin equivalent circuit to the left of the bracketed arrows is shown in (c).

$$R_{TH} = R + 2R\,||\,2R = 2R$$
$$E_{TH} = \left(\frac{2R}{2R + 2R}\right)E = \frac{E}{2}$$

(c) The Thevenin equivalent circuit to the left of node *A*.

$$v_o = \left(\frac{2R}{2R + 2R}\right)\frac{E}{2} = \frac{E}{4}$$

(d) Calculation of v_o using the voltage-divider rule.

input is 0001. In general, when the D_n input is 1 and all other inputs are 0, the output is

$$v_o = \frac{E}{2^{N-n}} \tag{14-1}$$

where *N* is the total number of binary inputs. For example, if $E = 5\,\text{V}$, the output voltage when the input is 0010 is $(5\,\text{V})/(2^{4-1}) = (5\,\text{V})/8 = 0.625\,\text{V}$.

To find the output voltage corresponding to *any* input combination, we can invoke the principle of superposition and simply add the voltages produced by the inputs where 1s are applied. For example, when the input is 1100, the output is $E/2 + E/4 = 3E/4$. The next example illustrates these computations.

EXAMPLE 14–1

The logic levels used in a 4-bit *R*-2*R* ladder DAC are $1 = +5\,\text{V}$ and $0 = 0\,\text{V}$.

1. Find the output voltage when the input is 0001 and when it is 1010.

2. Sketch the output when the inputs are driven from a 4-bit binary up counter.

FIGURE 14–5 (Example 14–1)

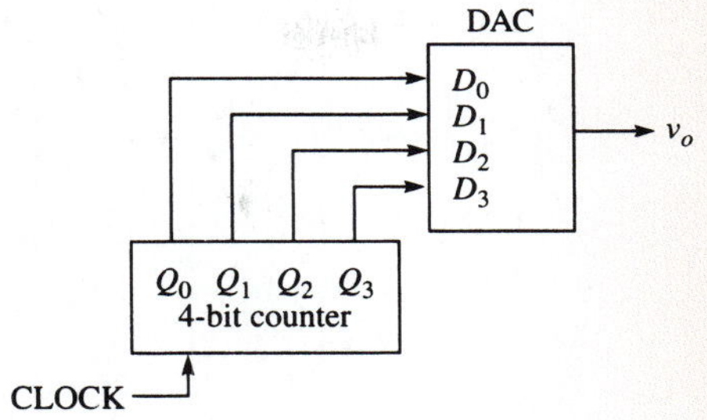

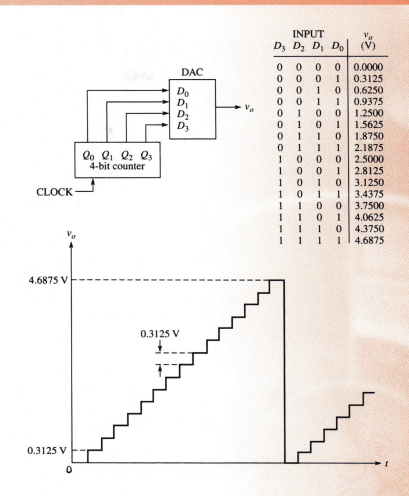

INPUT D_3	D_2	D_1	D_0	v_o (V)
0	0	0	0	0.0000
0	0	0	1	0.3125
0	0	1	0	0.6250
0	0	1	1	0.9375
0	1	0	0	1.2500
0	1	0	1	1.5625
0	1	1	0	1.8750
0	1	1	1	2.1875
1	0	0	0	2.5000
1	0	0	1	2.8125
1	0	1	0	3.1250
1	0	1	1	3.4375
1	1	0	0	3.7500
1	1	0	1	4.0625
1	1	1	0	4.3750
1	1	1	1	4.6875

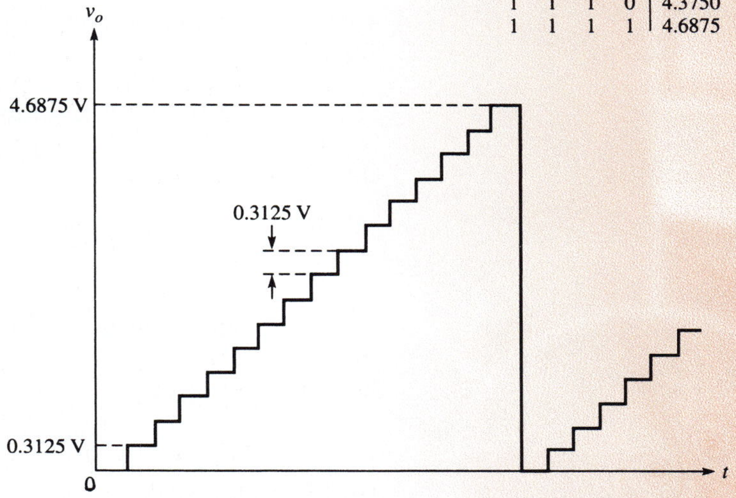

Solution

1. By equation 14–1, the output when the input is 0001 is

$$v_o = \frac{5\text{ V}}{2^{4-0}} = \frac{5\text{ V}}{16} = 0.3125\text{ V}$$

   When the input is 1000, the output is (5 V)/2 = 2.5 V, and when the input is 0010, the output is (5 V)/8 = 0.625 V. Therefore, when the input is 1010, the output is 2.5 V + 0.625 V = 3.125 V.

2. Figure 14–5 shows a table of the output voltages corresponding to every input combination, calculated using the method illustrated in part (1) of this example. As the counter counts up, the output voltage increases by 0.3125 V at each new count. Thus, the output is the *staircase* waveform shown in the figure. The voltage steps from 0 V to 4.6875 V each time the counter counts from 0000 to 1111.

Typical values for R and $2R$ are 10 kΩ and 20 kΩ. For accurate conversion, the output voltage from the R-$2R$ ladder should be connected to a high impedance to prevent loading. Figure 14–6 shows how an operational amplifier can be used for that purpose. The output of the ladder is connected to a unity-gain voltage follower, whose input impedance is extremely large and whose output voltage is the same as its input voltage.

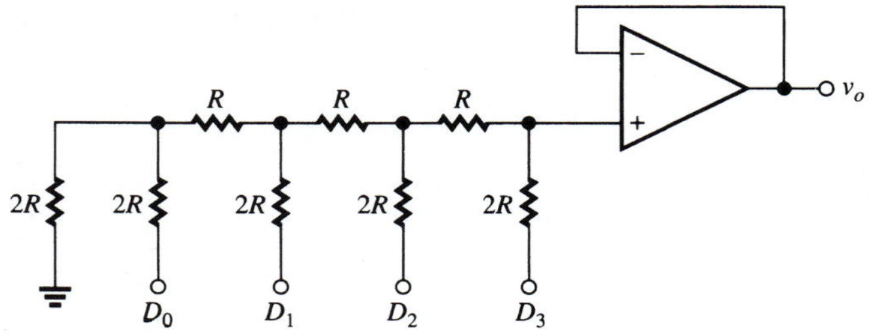

FIGURE 14–6 Using a unity-gain voltage follower to provide a high impedance for the *R*-2*R* ladder

14–3 A WEIGHTED-RESISTOR DAC

Figure 14–7 illustrates another approach to digital-to-analog conversion. The operational amplifier is used to produce a *weighted sum* of the digital inputs, where the weights are proportionate to the weights of the bit positions of the inputs. Each input is amplified by a factor equal to the ratio of the feedback resistance divided by the input resistance to which it is connected. Thus, D_3, the most significant bit, is amplified by R_f/R, D_2 by $R_f/2R = 1/2(R_f/R)$, D_1 by $R_f/4R = 1/4(R_f/R)$, and D_o by $R_f/8R = 1/8(R_f/R)$. Since the amplifier sums and inverts, the output is

$$v_o = -\left(D_3 + \frac{1}{2}D_2 + \frac{1}{4}D_1 + \frac{1}{8}D_o\right)\frac{R_f}{R} \tag{14–2}$$

The principal disadvantage of this type of converter is that a different-valued precision resistor must be used for each digital input. In contrast, the *R*-2*R* ladder network uses only two values of resistance.

EXAMPLE 14–2

Design a 4-bit, weighted-resistor DAC whose full-scale output voltage is −10 V. Logic levels are 1 = + 5 V and 0 = 0 V. What is the output voltage when the input is 1010?

Solution

The full-scale output voltage is the output voltage when the input is maximum: 1111. In that case, from equation 14–2, we require

$$\left(5\text{ V} + \frac{5\text{ V}}{2} + \frac{5\text{ V}}{4} + \frac{5\text{ V}}{8}\right)\frac{R_f}{R} = 10\text{ V}$$

or

$$9.375\frac{R_f}{R} = 10$$

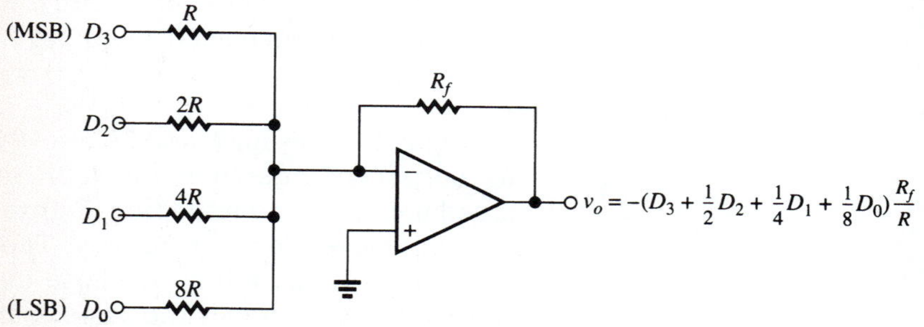

FIGURE 14–7 A weighted-resistor DAC using an inverting operational amplifier

Let us choose $R_f = 10\ \text{k}\Omega$. Then

$$R = \frac{9.375(10\ \text{k}\Omega)}{10} = 9.375\ \text{k}\Omega$$
$$2R = 18.75\ \text{k}\Omega$$
$$4R = 37.50\ \text{k}\Omega$$
$$8R = 75\ \text{k}\Omega$$

When the input is 1010, the output voltage is

$$V_o = -\left(5\ \text{V} + \frac{0\ \text{V}}{2} + \frac{5\ \text{V}}{4} + \frac{0\ \text{V}}{8}\right)\frac{10\ \text{k}\Omega}{9.375\ \text{k}\Omega} = -6.667\ \text{V}$$

14–4 THE SWITCHED CURRENT-SOURCE DAC

The D/A converters we have discussed so far can be regarded as switched voltage-source converters: When a binary input goes high, the high voltage is effectively switched into the circuit, where it is summed with other input voltages. Because of the technology used to construct integrated-circuit DACs, currents can be switched in and out of a circuit faster than voltages can. For that reason, most integrated-circuit DACs utilize some form of current switching, where the binary inputs are used to open and close switches that connect and disconnect internally generated currents. The currents are weighted according to the bit positions they represent and are summed in an operational amplifier. Figure 14–8 shows an example. Note that an R-$2R$ ladder is connected to a voltage source identified as E_{REF}. The current that flows in each $2R$ resistor is

$$I_n = \left(\frac{E_{\text{REF}}}{R}\right)\frac{1}{2^{N-n}} \tag{14–3}$$

when $n = 0, 1, \ldots, N-1$ is the subscript for the current created by input D_n, and N is the total number of inputs. Thus, each current is weighted according to the bit position it represents. For example, the current in the $2R$ resistor at the D_1 input in the figure ($n = 1$ and $N = 4$) is $(E_{REF}/R)(1/2)^3$. The binary inputs control switches that connect the currents either to ground or to the input of the amplifier. The amplifier sums all currents whose corresponding binary inputs are high. The amplifier also serves as a current-to-voltage converter. It is connected in an inverting configuration and its output is

$$v_o = -I_T R \tag{14–4}$$

FIGURE 14–8 A 4-bit switched current-source DAC

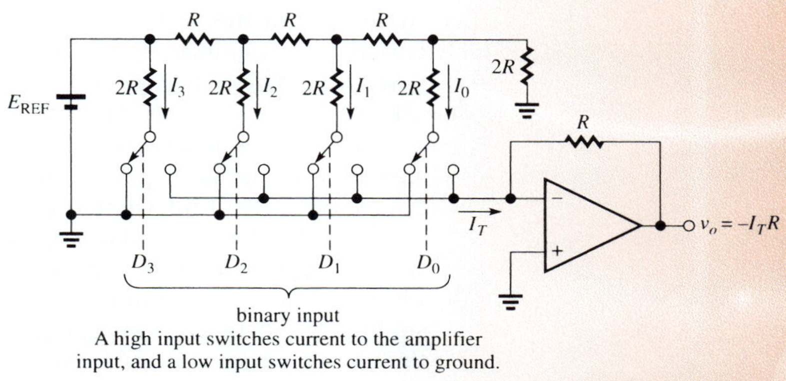

where I_T is the sum of the currents that have been switched to its input. For example, if the input is 1001, then

$$I_T = \frac{E_{\text{REF}}}{R}\left(\frac{1}{2}\right) + \frac{E_{\text{REF}}}{R}\left(\frac{1}{16}\right) = \frac{E_{\text{REF}}}{R}\left(\frac{7}{16}\right)$$

and

$$v_o = -\left(\frac{E_{\text{REF}}}{R}\right)\left(\frac{7}{16}\right)R = -\frac{7}{16}E_{\text{REF}}$$

EXAMPLE 14–3

The switched current-source DAC in Figure 14–8 has $R = 10\ \text{k}\Omega$ and $E_{\text{REF}} = 10\ \text{V}$. Find the total current delivered to the amplifier and the output voltage when the binary input is 1010.

Solution

From equation (14–3),

$$I_3 = \left(\frac{10\ \text{V}}{10\ \text{k}\Omega}\right)\frac{1}{2^{4-3}} = 0.5\ \text{mA}$$

and

$$I_1 = \left(\frac{10\ \text{V}}{10\ \text{k}\Omega}\right)\frac{1}{2^{4-1}} = 0.125\ \text{mA}$$

Therefore, $I_T = I_3 + I_1 = 0.5\ \text{mA} + 0.125\ \text{mA} = 0.625\ \text{mA}$. From equation (14–4),

$$v_o = -I_TR = -(0.625\ \text{mA})\,(10\ \text{k}\Omega) = -6.25\ \text{V}$$

The reference voltage E_{REF} in Figure 14–8 may be fixed—as, for example, when it is generated internally in an integrated circuit—or it may be externally variable. When it is externally variable, the output of the DAC is proportionate to the *product* of the variable E_{REF} and the variable binary input. In that case, the circuit is called a *multiplying* D/A converter, and the output represents the product of an analog input (E_{REF}) and a digital input.

In most integrated-circuit DACs utilizing current-source switching, the output is the total current, I_T, produced in the R-$2R$ ladder. The user may then connect a variety of operational amplifier configurations at the output to perform magnitude scaling and/or phase inversion. If E_{REF} can be both positive and negative and if the binary input always represents a positive number, then the output of the DAC is both positive and negative. In that case, the DAC is called a *two-quadrant multiplier.* A DAC whose output can be both positive and negative is said to be *bipolar,* and one whose output is only positive or only negative is *unipolar.* A four-quadrant multiplier is one in which both inputs can be either negative or positive and in which the output (product) has the correct sign for every case. Negative binary inputs can be represented using the *offset binary code,* the 4-bit version of which is shown in Table 14–1. Note that the numbers +7 through −8 are represented by 0000 through 1111, respectively, with 0_{10} represented by 1000. The table also shows the analog outputs that are produced by a multiplying D/A converter. In general, for an N-bit DAC, the offset binary code represents decimal numbers from $+2^{N-1} - 1$ through -2^{N-1} and the analog outputs range from

TABLE 14–1 The offset binary code used to represent positive and negative binary inputs to a D/A converter

Decimal Number	Offset Binary Code	Analog Output
+7	1111	$+(7/8)E_{REF}$
+6	1110	$+(6/8)E_{REF}$
+5	1101	$+(5/8)E_{REF}$
+4	1100	$+(4/8)E_{REF}$
+3	1011	$+(3/8)E_{REF}$
+2	1010	$+(2/8)E_{REF}$
+1	1001	$+(1/8)E_{REF}$
0	1000	0
−1	0111	$-(1/8)E_{REF}$
−2	0110	$-(2/8)E_{REF}$
−3	0101	$-(3/8)E_{REF}$
−4	0100	$-(4/8)E_{REF}$
−5	0011	$-(5/8)E_{REF}$
−6	0010	$-(6/8)E_{REF}$
−7	0001	$-(7/8)E_{REF}$
−8	0000	$-(8/8)E_{REF}$

$$+\left(\frac{2^{N-1}-1}{2^N-1}\right)E_{REF} \text{ through } -E_{REF}$$

in steps of size $(1/2^{N-1})\,E_{REF}$.

14–5 SWITCHED-CAPACITOR DACS

The newest technology used to construct D/A converters employs weighted capacitors instead of resistors. In this method, charged capacitors form a capacitive voltage divider whose output is proportionate to the sum of the binary inputs.

As an aid in understanding the method, let us review the theory of capacitive voltage dividers. Figure 14–9 shows a two-capacitor example. The total equivalent capacitance of the two series-connected capacitors is

$$C_T = \frac{C_1C_2}{C_1 + C_2} \tag{14–5}$$

Therefore, the total charge delivered to the circuit, which is the same as the charge on both C_1 and C_2, is

$$Q_1 = Q_2 = Q_T = C_TE = \left(\frac{C_1C_2}{C_1 + C_2}\right)E \tag{14–6}$$

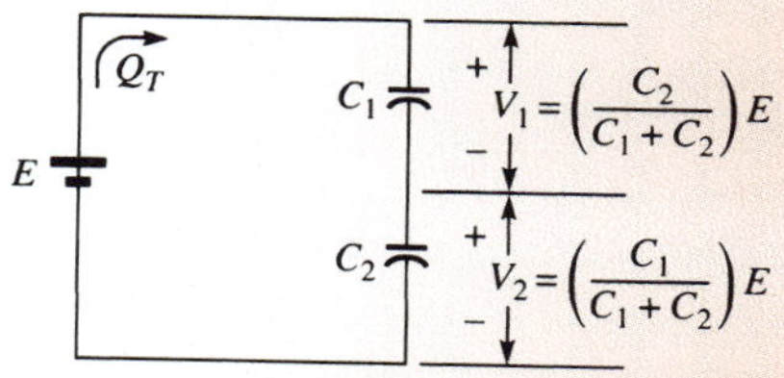

FIGURE 14–9 The capacitive voltage divider

The voltage across C_2 is

$$V_2 = \frac{Q_2}{C_2} = \frac{Q_T}{C_2} = \frac{\left(\frac{C_1C_2}{C_1 + C_2}\right)E}{C_2} = \left(\frac{C_1}{C_1 + C_2}\right)E \tag{14–7}$$

Similarly,

$$V_1 = \left(\frac{C_2}{C_1 + C_2}\right)E \tag{14–8}$$

Figure 14–10(a) shows an example of a 4-bit switched-capacitor DAC. Note that the capacitance values have binary weights. A *two-phase* clock is used to control switching of the capacitors. The two-phase clock consists of clock signals ϕ_1 and ϕ_2; ϕ_1 goes high while ϕ_2 is low, and ϕ_2 goes high while ϕ_1 is low. When ϕ_1 goes high, all capacitors are switched to ground and discharged. When ϕ_2 goes high, those capacitors where the digital input is high are switched to E_{REF}, whereas those whose inputs are low remain grounded. Figure 14–10(b) shows the equivalent circuit when ϕ_2 is high and the digital input is 1010. We see that the two capacitors whose digital inputs are 1 are in parallel, as are the two capacitors whose digital inputs are 0. The circuit is redrawn in Figure 14–10(c) with the parallel capacitors replaced by their equivalents (sums). The output of the capacitive voltage divider is

FIGURE 14–10 The switched-capacitor D/A converter

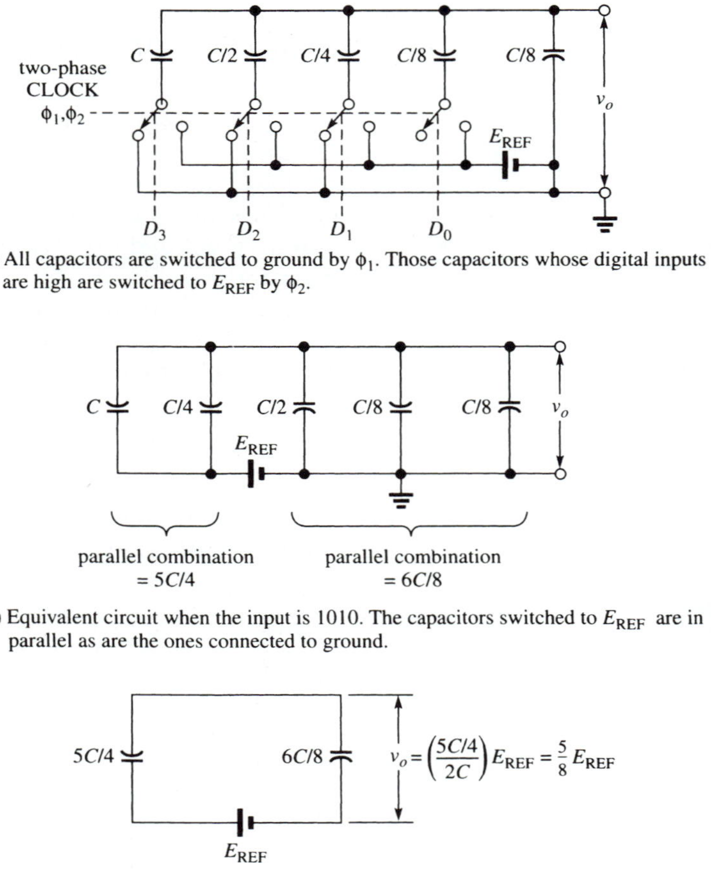

(a) All capacitors are switched to ground by ϕ_1. Those capacitors whose digital inputs are high are switched to E_{REF} by ϕ_2.

(b) Equivalent circuit when the input is 1010. The capacitors switched to E_{REF} are in parallel as are the ones connected to ground.

(c) Circuit equivalent to (b). The output is determined by a capacitive voltage divider.

TABLE 14–2 Output voltages produced by a 4-bit switched-capacitor DAC

Binary Input $D_3D_2D_1D_0$	V_o
0000	0
0001	$(1/16)E_{REF}$
0010	$(1/8)E_{REF}$
0011	$(3/16)E_{REF}$
0100	$(1/4)E_{REF}$
0101	$(5/16)E_{REF}$
0110	$(3/8)E_{REF}$
0111	$(7/16)E_{REF}$
1000	$(1/2)E_{REF}$
1001	$(9/16)E_{REF}$
1010	$(5/8)E_{REF}$
1011	$(11/16)E_{REF}$
1100	$(3/4)E_{REF}$
1101	$(13/16)E_{REF}$
1110	$(7/8)E_{REF}$
1111	$(15/16)E_{REF}$

$$v_o = \left(\frac{5C/4}{5C/4 + 6C/8}\right)E_{REF} = \left(\frac{5C/4}{2C}\right)E_{REF} = \frac{5}{8}E_{REF} \qquad \textbf{(14–9)}$$

The denominator in (14–9) will always be $2C$, the sum of all the capacitance values in the circuit. From the foregoing analysis, we see that the output of the circuit in the general case is

$$v_o = \left(\frac{C_{eq}}{2C}\right)E_{REF} \qquad \textbf{(14–10)}$$

where C_{eq} is the equivalent (sum) of all the capacitors whose digital inputs are high. Table 14–2 shows the outputs for every possible input combination, and it is apparent that the analog output is proportionate to the digital input. For the case where the input is 0000, note that the positive terminal of E_{REF} in Figure 14–10 is effectively open-circuited, so the output is 0 V.

Switched-capacitor technology evolved as a means for implementing analog functions in integrated circuits, particularly MOS circuits. It has been used to construct filters, amplifiers, and many other special devices. The principal advantage of the technology is that small capacitors, on the order of a few picofarads, can be constructed in the integrated circuits to perform the function of the much larger capacitors that are normally needed in low-frequency analog circuits. When capacitors are switched at a high enough frequency, they can be effectively "transformed" into other components, including resistors. The transformations are studied from the standpoint of sampled-data theory, which is beyond the scope of this book.

14–6 DAC PERFORMANCE SPECIFICATIONS

When specifying a D/A converter, there are two areas that are considered to be the most important. These two specifications are for resolution and accuracy, where accuracy is defined in terms of offset, gain, and nonlinearity errors. This section first examines resolution.

As discussed in Section 14–1, the resolution of a D/A converter is a measure of the fineness of the increments between output values. Given a fixed output voltage range, say, 0 to 10 V, a DAC that divides that range into 1024 distinct output values has greater resolution than one that divides it into 512 values. Because the output increment is directly dependent on the number of input bits, the resolution is often quoted as simply that total number of bits. The most commonly available integrated-circuit converters have resolutions of 8, 10, 12, or 16 bits. Resolution is also expressed as the reciprocal of the total number of output voltages, often in terms of a percentage. For example, the resolution of an 8-bit DAC may be specified as

$$\left(\frac{1}{2^8}\right) \times 100\% = 0.39\%$$

Some DAC specifications are quoted with reference to 1 or to one-half LSB (least significant bit). In this context, an LSB is simply the increment between successive output voltages. Since an n-bit converter has $2^n - 1$ such increments,

$$\text{LSB} = \frac{\text{FSR}}{2^n - 1} \tag{14–11}$$

where FSR is the full-scale range of the output voltage.

When the output of a DAC changes from one value to another, it typically overshoots the new value and may oscillate briefly around that new value before it settles to a constant voltage. The *settling time* of a D/A converter is the total time between the instant that the digital input changes and the time that the output enters a specified error band for the last time, usually ±1/2 LSB around the final value. Figure 14–11 illustrates the specification. Settling times of typical integrated-circuit converters range from 50 ns to several microseconds. Settling time may depend on the magnitude of the change at the input and is often specified for a prescribed input change.

The next important specification for a D/A converter is for accuracy. Accuracy is measured in terms of the DACs offset error, gain error, and linearity issues. Offset and gain errors are easily shown via graphs. Figure 14–12 shows the output plot of a 3-bit DAC with offset error. Notice that there is a DC offset for the signal for all digital data input values. Figure 14–13 shows the output

FIGURE 14–11 The settling time, t_s, of a D/A converter

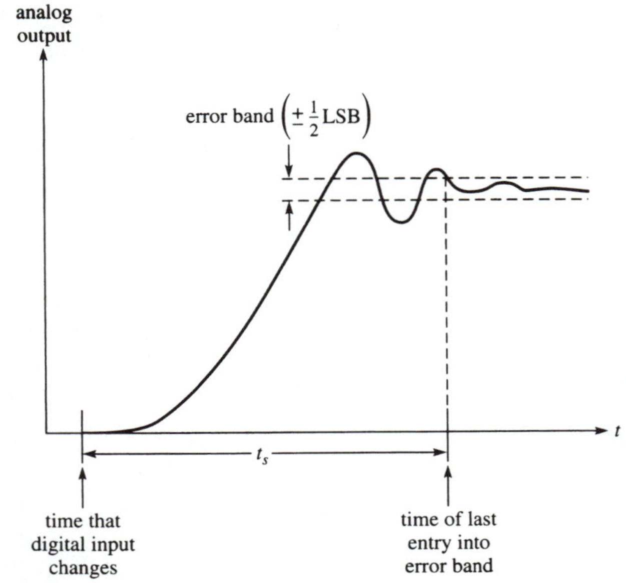

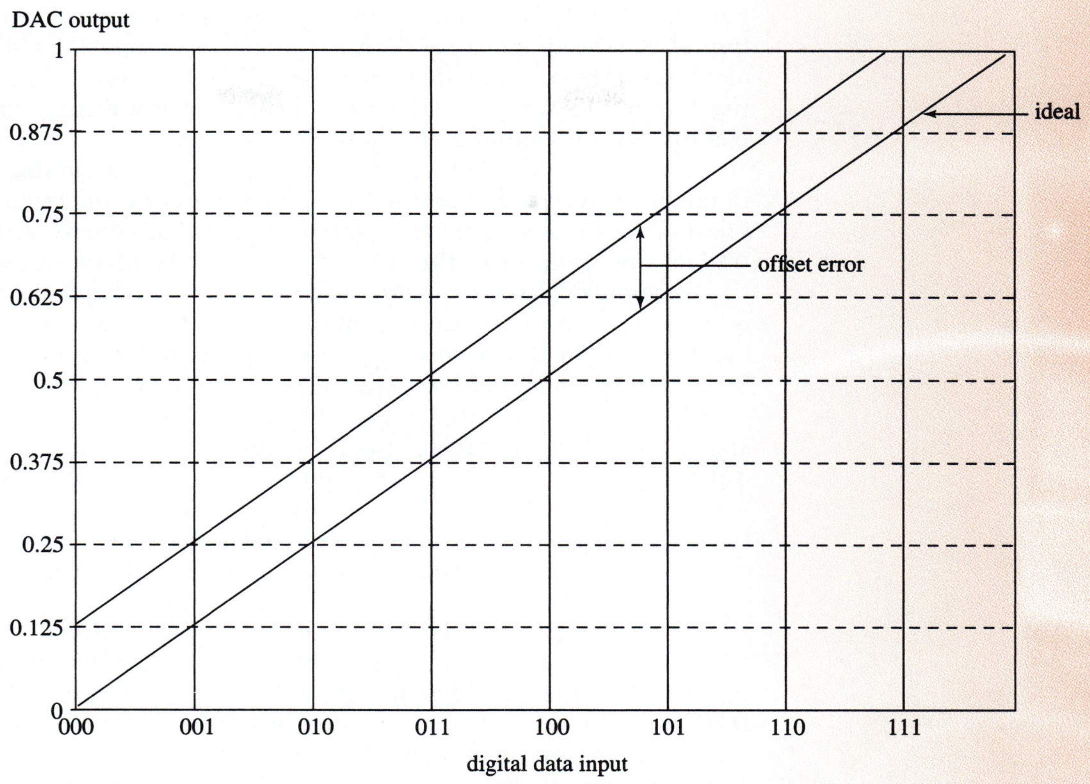

FIGURE 14–12 A plot of a D/A converter output with offset error

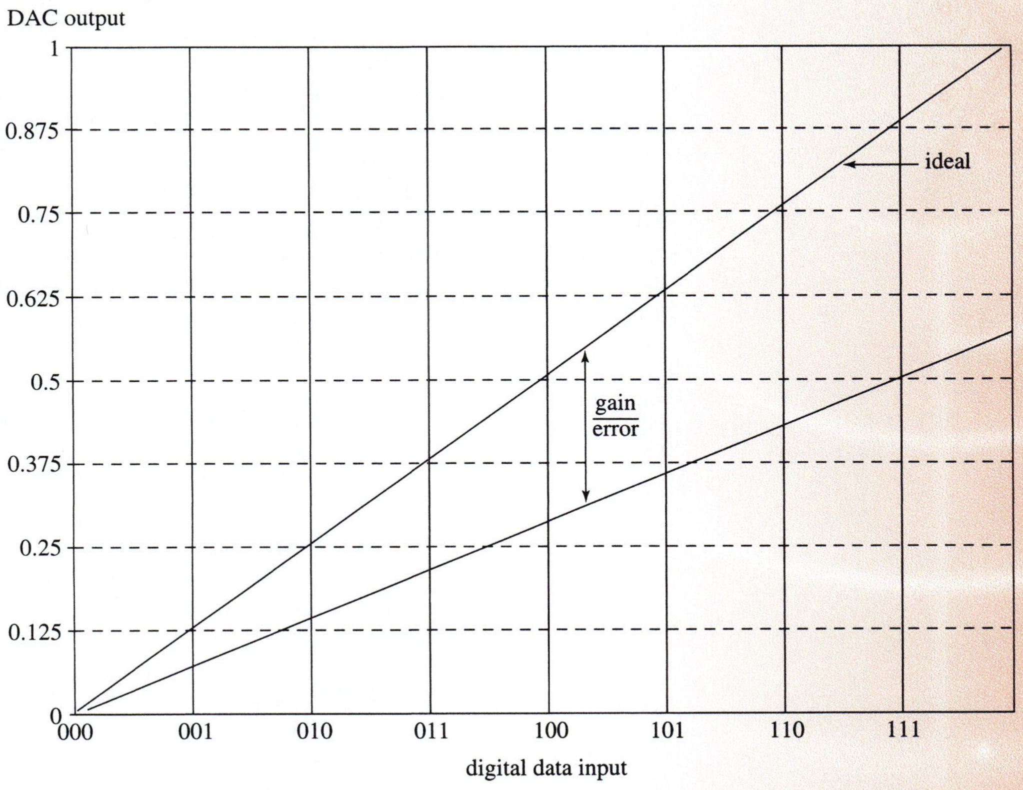

FIGURE 14–13 A plot of a D/A converter output with gain error

plot of a 3-bit DAC with gain error. In this case, the output error increases as the digital data input increases (D/A conversion is larger). For a digital input value of 000, there is minimal gain error; however, there is a large error when the larger 111 binary number is converted. Gain errors can be negative or positive. In this example, the gain error is negative.

Linearity error is the maximum deviation of the analog output from the ideal output. Since the output is ideally in direct proportion to the input, the ideal output is a straight line drawn from 0 V. Linearity error may be specified as a percentage of the full-scale range or in terms of an LSB.

Differential linearity error is the difference between the ideal output increment (1 LSB) and the actual increment. For example, each output increment of an 8-bit DAC whose full-scale range is 10 V should be $(10\text{ V})/(2^8 - 1) = 39.22$ mV. If any one increment between two successive values is, say, 30 mV, then there is a differential linearity error of 9.22 mV. This error is also specified as a percentage of the full-scale range or in terms of an LSB. If the differential linearity error is greater than 1 LSB, it is possible for the output voltage to *decrease* when there is an increase in the value of the digital input or to increase when the input decreases. Such behavior is said to be *nonmonotonic.* In a monotonic DAC, the output always increases when the input increases and decreases when the input decreases.

The input of a DAC is said to undergo a *major change* when every input bit changes, as, for example, from 01111111 to 10000000. If the switches in Figure 14–8 open faster than they close or vice versa, the output of the DAC will momentarily go to 0 or to full scale when a major change occurs, thus creating an output *glitch. Glitch area* is the total area of an output voltage glitch in volt-seconds or of an output current glitch in ampere-seconds. Commercial units are often equipped with *deglitchers* to minimize glitch area.

An Integrated-Circuit DAC

The AD7524 CMOS integrated-circuit DAC, manufactured by Analog Devices and available from other manufacturers, is an example of an 8-bit, multiplying D/A converter. Figure 14–14 shows a functional block diagram. Note that the digital input is latched under the control of $\overline{CS}$ and $\overline{WR}$. When both of these control inputs are low, the output of the DAC responds directly to the digital inputs, with no latching occurring. If either control input goes high, the digital input is latched and the analog output remains at the level corresponding to the latched data, independent of any changes

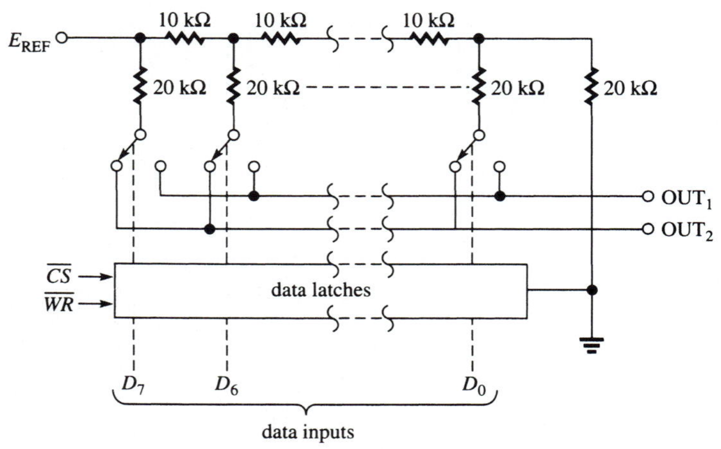

FIGURE 14–14 Functional block diagram of the AD7524 D/A converter

in the digital input. The device is then said to be in a HOLD mode, with the data bus *locked out*. The OUT2 output is normally grounded. Maximum settling time to a ±1/2 LSB error band for the Texas Instruments version is 100 ns, and maximum linearity error is ±0.2% of the full-scale range. The device can be used as a 2- or 4-quadrant multiplier, and E_{REF} can vary ±25 V.

14–7 THE COUNTER-TYPE ADC

The simplest type of A/D converter employs a binary counter, a voltage comparator, and a D/A converter, as shown in Figure 14–15(a). The output of a voltage comparator is high as long as its v^+ input is greater than its v^- input. Notice that the analog input is the v^+ input to the comparator. As long as it is greater than the v^- input, the AND gate is enabled and clock pulses are passed to the counter. The digital output of the counter is converted to an analog voltage by the DAC, and that voltage is the other input to the comparator. Thus, the counter counts up until its output has a value equal to the analog input. At that time, the comparator switches low, inhibiting the clock

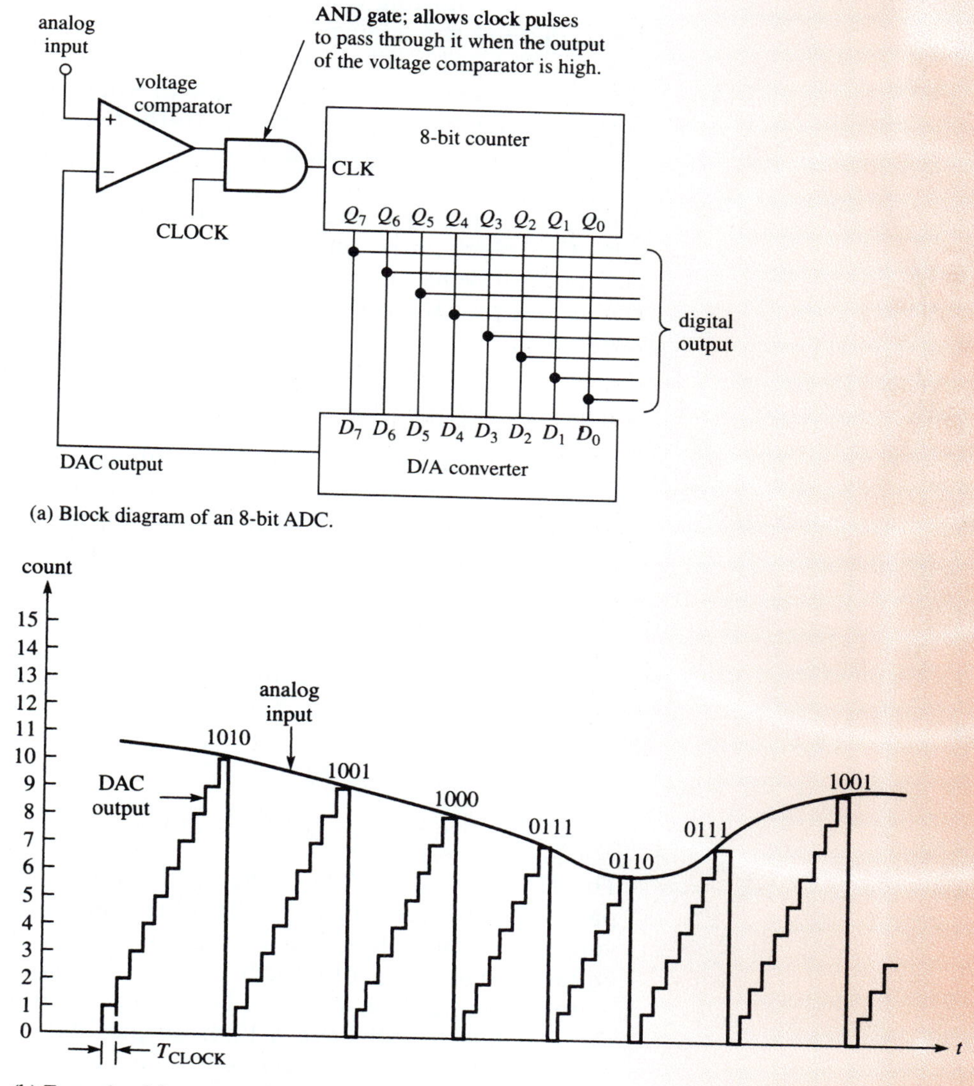

(a) Block diagram of an 8-bit ADC.

(b) Example of the output of the DAC in a 4-bit ADC.

FIGURE 14–15 The counter-type A/D converter

FIGURE 14–16 The tracking counter-type A/D converter. The counter counts up or down from its last count to reach its next count rather than resetting to 0 between counts

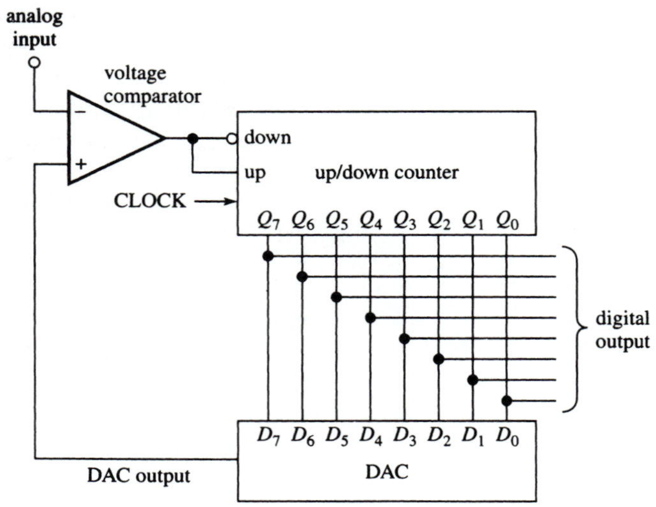

pulses, and counting ceases. The count it reached is the digital output proportionate to the analog input. Control circuitry, not shown, is used to latch the output and reset the counter. The cycle is repeated, with the counter reaching a new count proportionate to whatever new value the analog input has acquired. Figure 14–15(b) illustrates the output of a 4-bit DAC in an ADC over several counting cycles when the analog input is a slowly varying voltage. The principal disadvantage of this type of converter is that the conversion time depends on the magnitude of the analog input: the larger the input, the more clock pulses must pass to reach the proper count. An 8-bit converter could require as many as 255 clock pulses to perform a conversion, so the counter-type ADC is considered quite slow in comparison to other types we will study.

Tracking A/D Converter

To reduce conversion times of the counter-type ADC, the up counter can be replaced by an up/down counter, as illustrated in Figure 14–16. In this design, the counter is not reset after each conversion. Instead, it counts up or down from its last count to its new count. Thus, the total number of clock pulses required to perform a conversion is proportionate to the *change* in the analog input between counts rather than to its magnitude. Since the count more or less keeps up with the changing analog input, this type of ADC is called a *tracking converter.* A disadvantage of the design is that the count may oscillate up and down from a fixed count when the analog input is constant.

14–8 FLASH A/D CONVERTERS

The fastest type of A/D converter is called the *flash* (or simultaneous, or parallel) type. As shown for the 3-bit example in Figure 14–17, a reference voltage is connected to a voltage divider that divides it into 7 ($2^n - 1$) equal-increment levels. Each level is compared to the analog input by a voltage comparator. For any given analog input, one comparator and all those below it will have a high output. All comparator outputs are connected to a *priority encoder.* A priority encoder produces a binary output corresponding to the input having the highest priority, in this case, the one

FIGURE 14–17 A 3-bit flash A/D converter

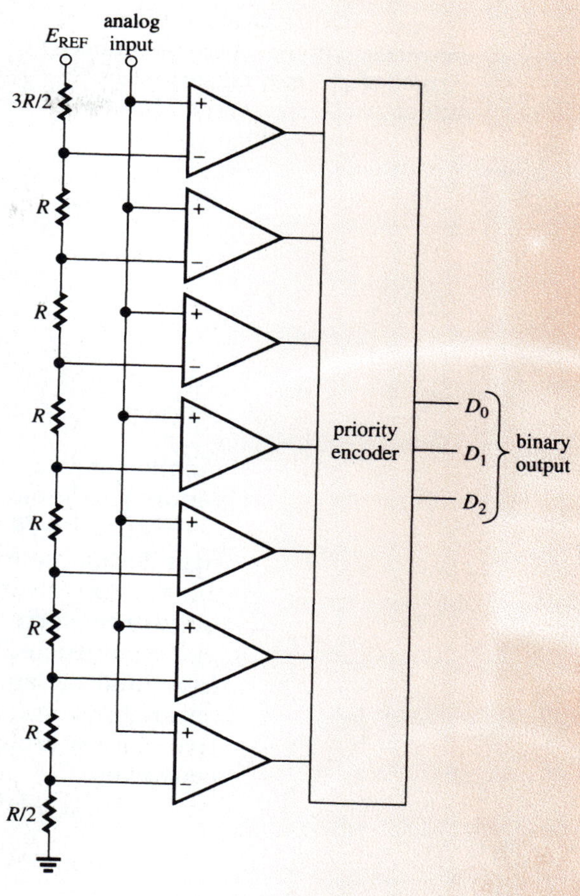

representing the largest voltage level equal to or less than the analog input. Thus, the binary output represents the voltage that is closest in value to the analog input.

The voltage applied to the v^- input of the uppermost comparator in Figure 14–17 is, by voltage-divider action,

$$\left(\frac{6R + R/2}{6R + R/2 + 3R/2}\right)E_{REF} = \frac{13R/2}{16R/2}E_{REF} = \frac{13}{16}E_{REF} \tag{14–12}$$

Similarly, the voltage applied to the v^- input of the second comparator is $(11/16)E_{REF}$, that applied to the third is $(9/16)E_{REF}$, and so forth. The increment between voltages is $(2/16)E_{REF}$, or $(1/8)E_{REF}$. An n-bit flash comparator has $n - 2R$-valued resistors, and the increment between voltages is

$$\Delta V = \frac{1}{2^n}E_{REF} \tag{14–13}$$

The voltage levels range from

$$\left(\frac{2^{n+1} - 3}{2^{n+1}}\right)E_{REF} \quad \text{through} \quad \left(\frac{1}{2^{n+1}}\right)E_{REF}$$

The flash converter is fast because the only delays in the conversion are in the comparators and the priority encoder. Under the control of a clock, a new conversion can be performed very soon after one conversion is complete. The principal disadvantage of the flash converter is the need for a large

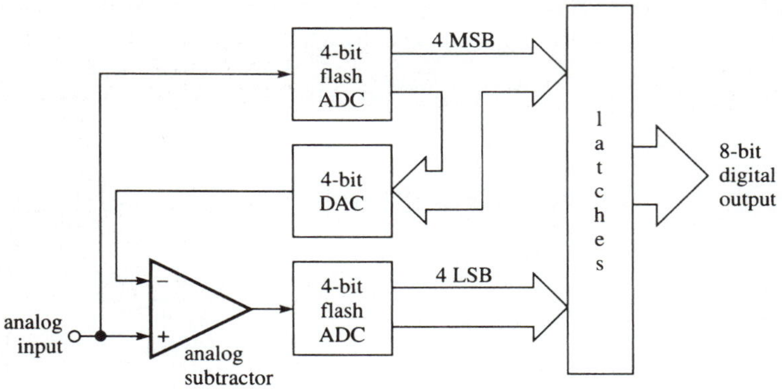

FIGURE 14–18 Modified flash converter that uses 30 comparators instead of 255 to produce an 8-bit output

number of voltage comparators ($2^n - 1$). For example, an 8-bit flash ADC requires 255 comparators.

Figure 14–18 shows a block diagram of a modified flash technique that uses 30 comparators instead of 255 to perform an 8-bit A/D conversion. One 4-bit flash converter is used to produce the 4 most significant bits. Those 4 bits are converted back to an analog voltage by a D/A converter, and the voltage is subtracted from the analog input. The difference between the analog input and the analog voltage corresponding to the 4 most significant bits is an analog voltage corresponding to the 4 least significant bits. Therefore, that voltage is converted to the 4 least significant bits by another 4-bit flash converter.

14–9 THE DUAL-SLOPE (INTEGRATING) ADC

A dual-slope ADC uses an operational amplifier to integrate the analog input. Recall from Chapter 12 that the output of an integrator is a ramp when the input is a fixed level (Figure 11–1). The slope of the ramp is $\pm E_{in}/R_1C$, where E_{in} is the input voltage that is integrated and R_1 and C are the fixed components of the integrating operational amplifier. Because R_1 and C are fixed, the slope of the ramp is directly dependent on the value of E_{in}. If the ramp is allowed to continue for a fixed time, the voltage it reaches in that time depends on the slope of the ramp and hence on the value of E_{in}. The basic principle of the integrating ADC is that the voltage reached by the ramp controls the length of time that a binary counter is allowed to count. Thus, a binary number proportionate to the value of E_{in} is obtained. In the dual-slope ADC, two integrations are performed, as described next.

Figure 14–19(a) shows a functional block diagram of the dual-slope ADC. Recall that the integrating operational amplifier inverts, so a positive input generates a negative-going ramp and vice versa. A conversion begins with the switch connected to the analog input. Assume that the input is negative, so a positive-going ramp is generated by the integrator. As discussed earlier, the ramp is allowed to continue for a fixed time, and the voltage it reaches in that time is directly dependent on the analog input. The fixed time is controlled by sensing the time when the counter reaches a specific count. At that time, the counter is reset and control circuitry causes the switch to be connected to a reference voltage having a polarity *opposite* to that of the analog input—in this case, a positive voltage. As a consequence, the output of the integrator becomes a negative-going ramp, beginning from the positive value it reached during the first integration. Because the reference voltage is fixed, so is the slope of the negative-going ramp. When the negative-going ramp reaches 0 V,

FIGURE 14–19 The dual-slope integrating A/D converter

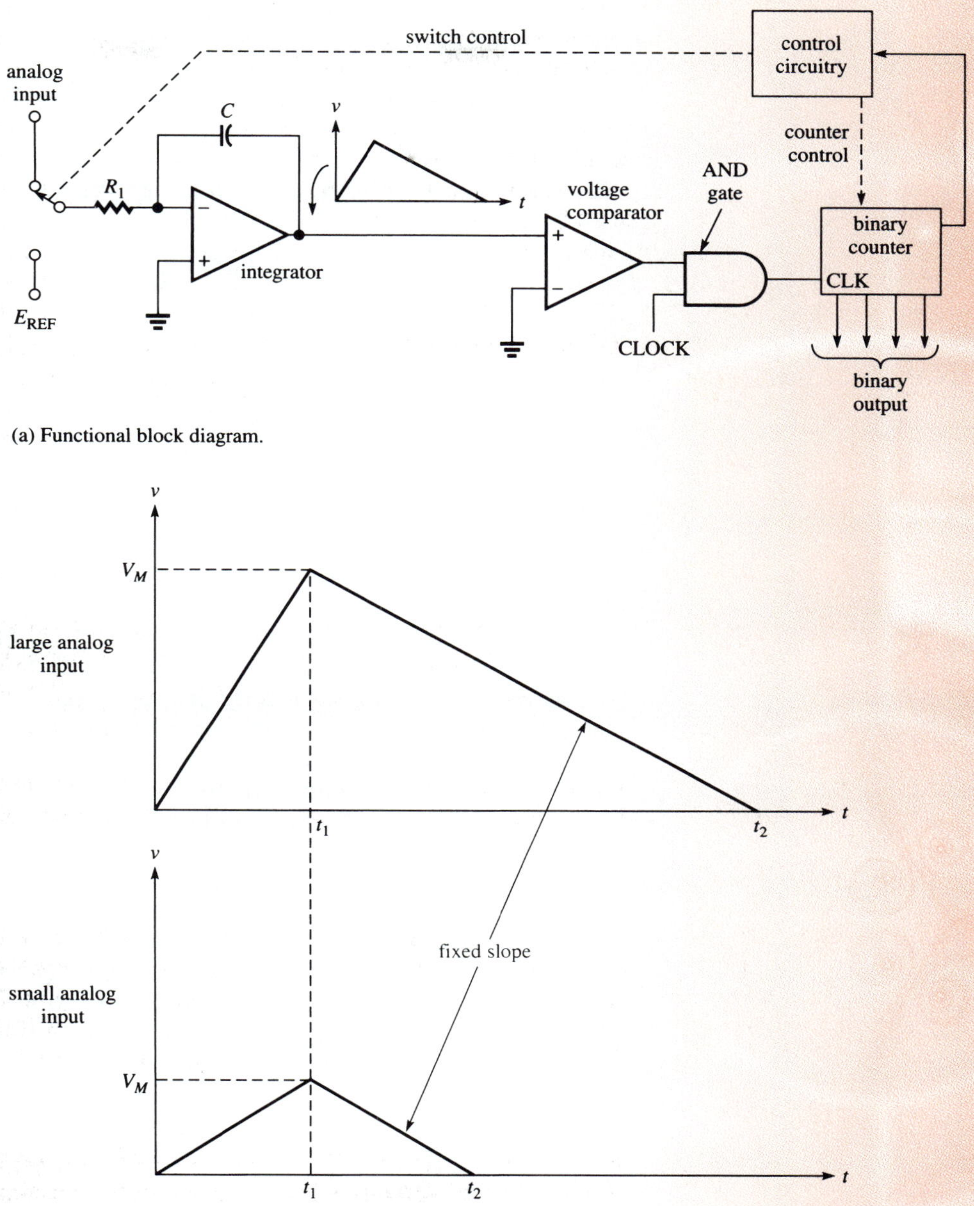

(a) Functional block diagram.

(b) Ramps resulting from large and small (negative) analog inputs.

the voltage comparator switches, the clock pulses are inhibited, and the counter ceases to count. The count it contains at that time is proportionate to the time required for the negative-going ramp to reach 0 V, which is proportionate to the positive voltage reached in the first integration. Thus, the binary count is proportionate to the value of the analog input. Figure 14–19(b) shows examples of the ramp waveforms generated by a small analog input and by a large analog input. Note that the slope of the positive-going ramp is variable (depending on E_{in}), and the slope of the negative-going ramp is fixed. The origin of the name *dual-slope* is now apparent.

One advantage of the dual-slope converter is that its accuracy depends neither on the values of the integrator components R_1 and C nor upon any long-term changes that may occur in them. This fact is demonstrated by examining the equations governing the times required for the two integrations. Since the slope of the positive-going ramp is E_{in}/R_1C, the maximum voltage V_M reached by the ramp in time t_1 is

$$V_M = \frac{|E_{in}|t_1}{R_1C} \qquad (14\text{–}14)$$

The magnitude of the slope of the negative-going ramp is $|E_{REF}|/R_1C$, so

$$V_M = \frac{|E_{REF}|}{R_1C}(t_2 - t_1) \qquad (14\text{–}15)$$

Equating (14–14) and (14–15),

$$\frac{|E_{in}|}{R_1C}t_1 = \frac{|E_{REF}|}{R_1C}(t_2 - t_1) \qquad (14\text{–}16)$$

Cancelling R_1C on both sides and solving for $t_2 - t_1$ gives

$$t_2 - t_1 = \frac{|E_{in}|}{|E_{REF}|}t_1 \qquad (14\text{–}17)$$

Since the counter contains a count proportionate to $t_2 - t_1$ (the time required for the negative-going ramp to reach 0 V) and t_1 is fixed, equation (14–17) shows that the count is directly proportionate to E_{in}, the analog input. Note that this expression does not contain R_1 or C; those quantities cancelled out in (14–16). Thus, accuracy does not depend on their values. Furthermore, accuracy does not depend on the frequency of the clock. Equation (14–17) shows that accuracy does depend on E_{REF}, so the reference voltage should be very precise.

An important advantage of the dual-slope A/D converter is that the integrator suppresses noise. Recall from equation (11–7) that the output of an integrator has amplitude inversely proportional to frequency. Thus, high-frequency noise components in the analog input are attenuated. This property makes it useful for instrumentation systems, and it is widely used for applications such as digital voltmeters. However, the integrating ADC is not particularly fast, so its use is restricted to signals having low to medium frequencies.

14–10 THE SUCCESSIVE-APPROXIMATION ADC

The method called *successive approximation* is the most popular technique used to construct A/D converters, and, with the exception of flash converters, successive-approximation converters are the fastest of those we have discussed. Figure 14–20(a) shows a block diagram of a 4-bit version. The method is best explained by way of an example. For simplifying purposes, let us assume that the output of the D/A converter ranges from 0 V through 15 V as its binary input ranges from 0000 through 1111, with 0000 producing 0 V, 0001 producing 1 V, and so forth. Suppose the "unknown" analog input is 13 V. On the first clock pulse, the output register is loaded with 1000, which is converted by the DAC to 8 V. The voltage comparator determines that 8 V is less than the analog input (13 V), so on the next clock pulse, the control circuitry causes the output register to be loaded with 1100. The output of the DAC is now 12 V, which the comparator again determines is less than the analog input. Consequently, the register is loaded with 1110 on the next clock pulse. The output of the DAC is 14 V, which the comparator now determines is larger

FIGURE 14–20 The successive-approximation A/D converter

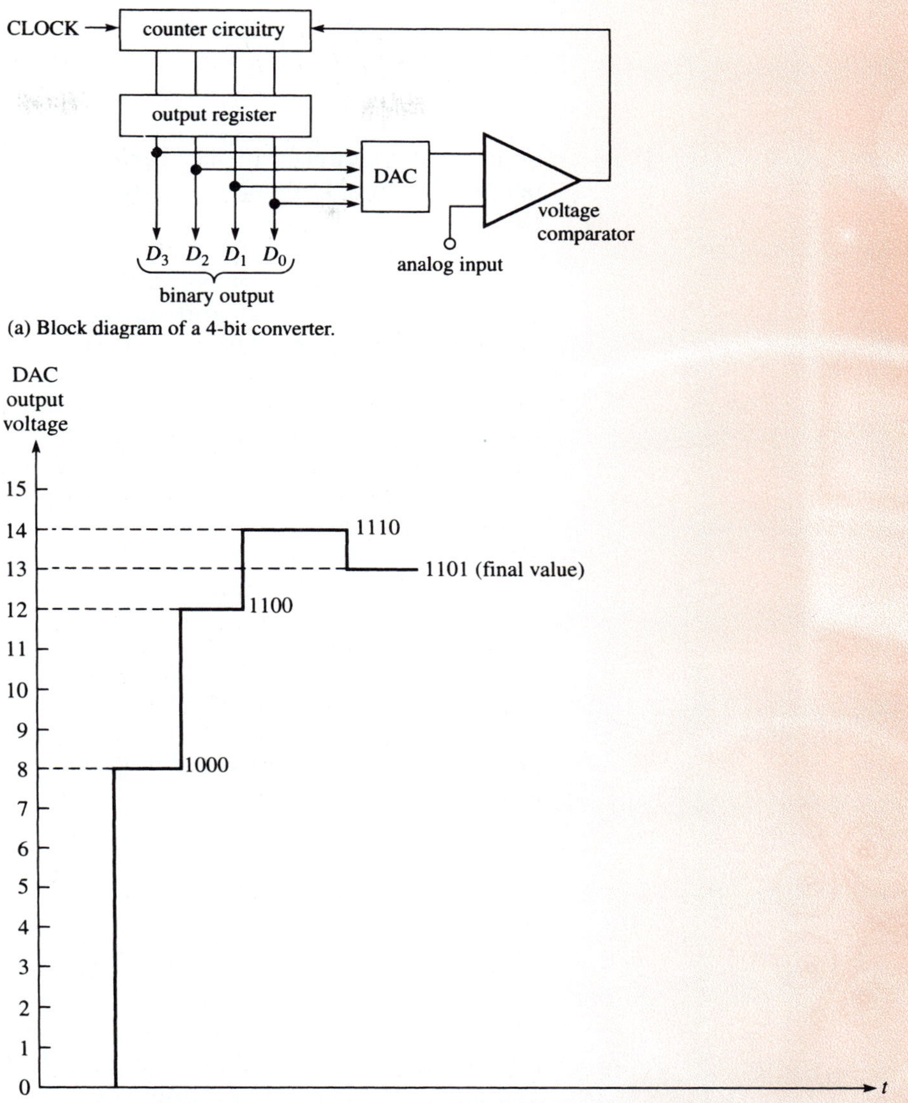

(a) Block diagram of a 4-bit converter.

(b) Output of the DAC and contents of the output register, showing successive approximations when a 13-V analog input is converted

than the 13-V analog input. Therefore, the last 1 that was loaded into the register is replaced with a 0, and a 1 is loaded into the LSB. This time, the output of the DAC is 13 V, which equals the analog input, so the conversion is complete. The output register contains 1101. We see that the method of successive approximation amounts to testing a sequence of trial values, each of which is adjusted to produce a number closer in value to the input than the previous value. This "homing-in" on the correct value is illustrated in part (b) of the figure. Note that an n-bit conversion requires n clock pulses. As another example, the following is the sequence of binary numbers that would appear in the output register of an 8-bit successive-approximation converter when the analog input is a voltage that is ultimately converted to 01101001:

```
10000000
01000000
01100000
01110000
```

```
01101000
01101100
01101010
01101001
```

Some modern successive-approximation converters have been constructed using the switched-capacitor technology discussed in connection with DACs.

The primary component affecting the accuracy of a successive-approximation converter is the D/A converter. Consequently, the reference voltage connected to it and its ladder network must be very precise for accurate conversions. Also, the analog input should remain fixed during the conversion time. Some units employ a *sample-and-hold* circuit to ensure that the input voltage being compared at the voltage comparator does not vary during the conversion. A sample-and-hold is the analog counterpart of a digital data latch. It is constructed using electronic switches, an operational amplifier, and a capacitor that charges to and holds a particular voltage level.

14–11 ADC PERFORMANCE SPECIFICATIONS

The resolution of an A/D converter is the smallest change that can be distinguished in the analog input. As in DACs, resolution depends directly on the number of bits, so it is often quoted as simply the total number of output bits. The actual value depends on the full-scale range (FSR) of the analog input:

$$\text{resolution} = \frac{\text{FSR}}{2^n} \qquad \textbf{(14–18)}$$

Some A/D converters have 8-4-2-1 BCD outputs rather than straight binary. This is especially true of dual-slope types designed for use in digital instruments. The BCD outputs facilitate driving numerical displays. The resolution of a BCD converter is quoted as the number of (decimal) digits available at the output, where each digit is represented by 4 bits. In this context, the term $\frac{1}{2}$ *digit* is used to refer to a single binary output. For example, a $1\frac{1}{2}$-digit output is represented by 5 bits. The $\frac{1}{2}$ digit is used as the most significant bit, and its presence doubles the number of decimal values that can be represented by the 4 BCD bits. Some BCD converters employ multiplexers to expand the number of output digits. An example is Texas Instruments' TLC135C $4\frac{1}{2}$-digit A/D converter. Expressed as a voltage, the resolution of a BCD A/D converter is FSR/10^d, where d is the number of output digits.

The time required to convert a single analog input to a digital output is called the *conversion time* of an A/D converter. Conversion time may be quoted as including any other delays, such as access time, associated with acquiring and converting an analog input. In that case, the total number of conversions that can be performed each second is the reciprocal of the conversion time. The reason this specification is important is that it imposes a limit on the rate at which the analog input can be allowed to change. In effect, the A/D converter *samples* the changing analog input when it performs a sequence of conversions. The *Shannon sampling theorem* states that sampled data can be used to faithfully reproduce a time-varying signal provided that the sampling rate is at least *twice* the frequency of the highest-frequency component in the signal. Thus, for the sequence of digital outputs to be a valid representation of the analog input, the A/D converter must perform conversions at a rate equal to at least twice the frequency of the highest component of the input.

EXAMPLE 14–4

What maximum conversion time can an A/D converter have if it is to be used to convert *audio* input signals? (The audio frequency range is considered to be 20 Hz to 20 kHz.)

Solution

Because the highest frequency in the input may be 20 kHz, conversions should be performed at a rate of at least 40×10^3 conversions/s. The maximum allowable conversion time is therefore equal to

$$\frac{1}{40 \times 10^3} = 25\ \mu\text{s}$$

One LSB for an A/D converter is defined in the same way it is for a D/A converter:

$$\text{LSB} = \frac{\text{FSR}}{2^{n-1}} \qquad (14\text{–}19)$$

where FSR is the full-scale range of the analog input. Other A/D converter specifications may be quoted in terms of an LSB or as a percentage of FSR.

Gain and offset error specifications, while important for A/D converters, can usually be calibrated out so that their error contribution can be minimized or removed. Other important A/D converter performance specifications are linearity or nonlinearity issues. Differential nonlinearity results in missing digital data output values in the A/D conversion process. Integral nonlinearity refers to the deviation of the output from an ideal straight line.

Integrated-Circuit A/D Converters

A wide variety of A/D converters of all the types we have discussed are available in integrated circuits. Most have additional features, such as latched *three-state outputs* (0, 1, and open circuit), that make them compatible with microprocessor systems. Some have *differential* inputs, wherein the voltage converted to a digital output is the difference between two analog inputs. Differential inputs are valuable in reducing the effects of noise, because any noise signal common to both inputs is "differenced out." Recall that this property is called *common-mode rejection.* Differential inputs also allow the user to add or subtract a fixed voltage to the analog input, thereby offsetting the values converted. Conventional (single-ended) inputs can be accommodated simply by grounding one of the differential inputs.

The advances in MOS technology that have made high-density memory circuits possible have also made it possible to incorporate many special functions into a single-integrated circuit containing an ADC. Examples include sample-and-hold circuitry and analog multiplexers. With these functions, and versatile control circuitry, complete microprocessor-compatible *data-acquisition* systems have become available in a single-integrated circuit. The microprocessor can be programmed to control the multiplexer so that analog data from many different sources (up to 19 in some versions) can be sampled in a desired sequence. The analog data from an instrumentation system, for example, is sampled, converted, and transmitted directly to the microprocessor for storing in memory or further processing.

14–12 A/D CIRCUIT ANALYSIS WITH ELECTRONICS WORKBENCH MULTISIM

This section examines the use of EWB to assemble a simple analog-to-digital converter using a simple resistive network and an operational amplifier.

The objectives of this section are as follows:

- Use EWB to demonstrate how analog-to-digital conversion is done inside an integrated circuit.
- Develop an understanding of voltage reference.
- Understand the artifacts generated in the reconstructed analog signal.

To begin this exercise open the EWB Multisim circuit **Ch14_EWB.msm** that is found in the Electronics Workbench CD-ROM packaged with the text. This file contains the A/D (analog-to-digital) and D/A (digital-to-analog) circuits shown in Figure 14–21. The input to the A/D is the function generator XFG1. The square wave generator V2 is the clock used to sample the input signal. Probes (X0–X7) are connected to the output of the A/D to provide an indicator of the logic level. The A/D voltage reference is set to 3 V. This is the maximum input signal voltage that can be accurately converted to a digital value. The output enable pin (OE) has been tied high so that the outputs are always active.

A VDAC (voltage D/A converter) is provided to convert the digital signal back to analog. The reference to the D/A is set to 3 V to match the A/D. The sampled digital values are connected to the D/A inputs (D0–D7). The 1-kΩ resistor (R1) is being used to represent a load on the D/A.

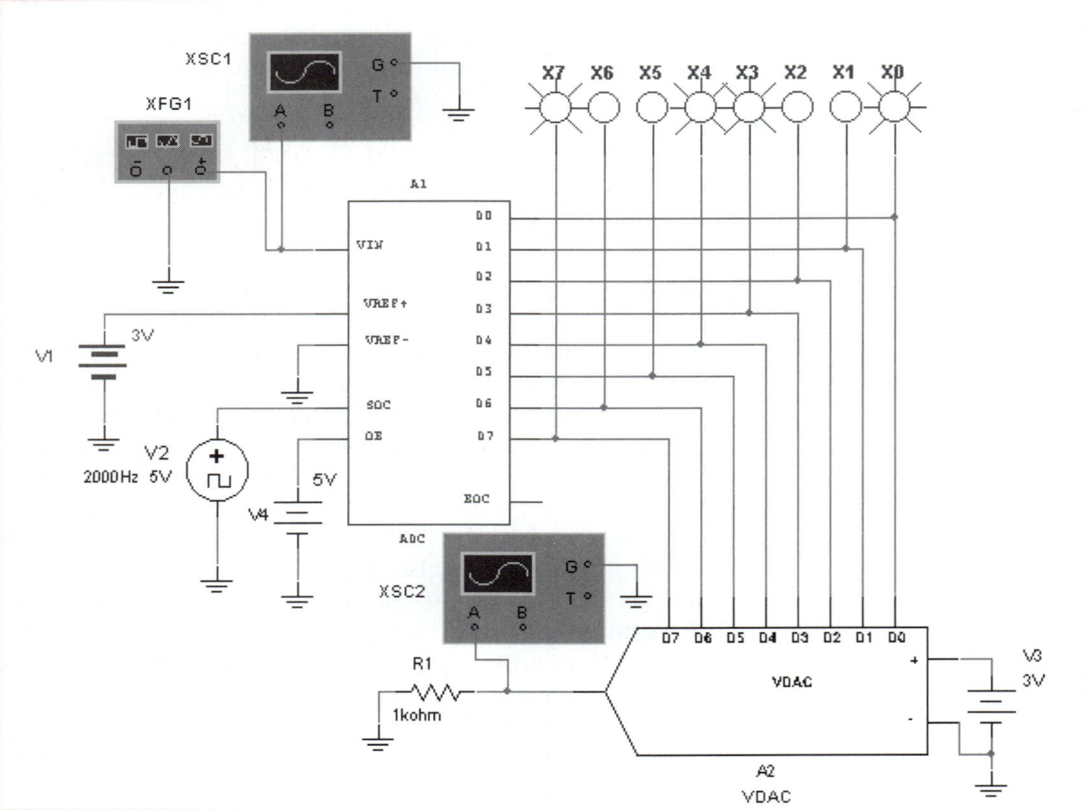

FIGURE 14–21 The A/D and D/A EWB Multisim Circuit

The input to the A/D being used for this exercise is the triangle wave shown in Figure 14–22. Start the simulation and open the EWB oscilloscope XSC2. You should see the waveform shown in Figure 14–23. (*Note:* It takes a minute for this waveform to be created.) Notice the poor resemblance of this waveform to the original input triangle wave. Why is the signal not the same?

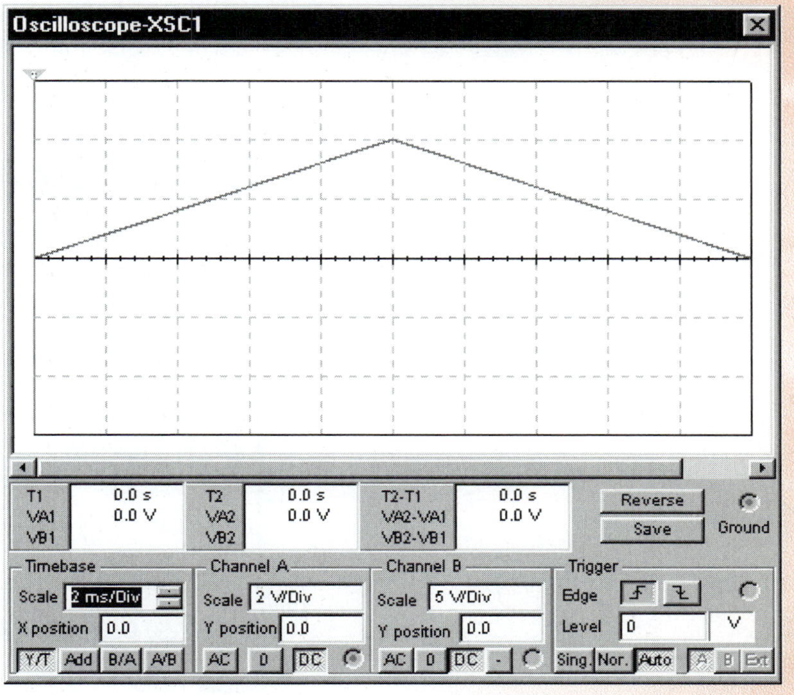

FIGURE 14–22 The input triangle wave to the A/D

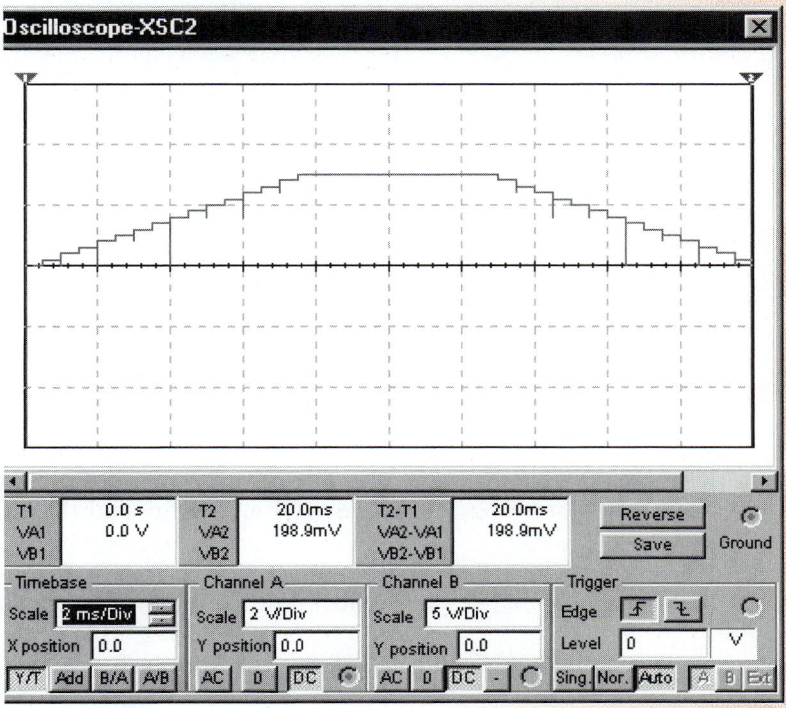

FIGURE 14–23 The reconstructed D/A signal off the output of the VDAC

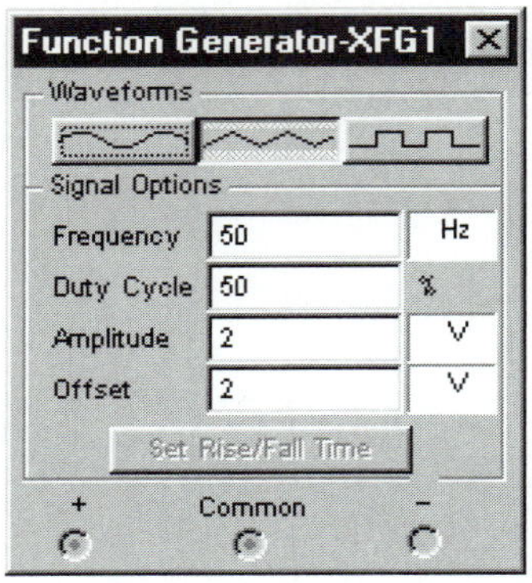

FIGURE 14–24 The original settings for the function generator

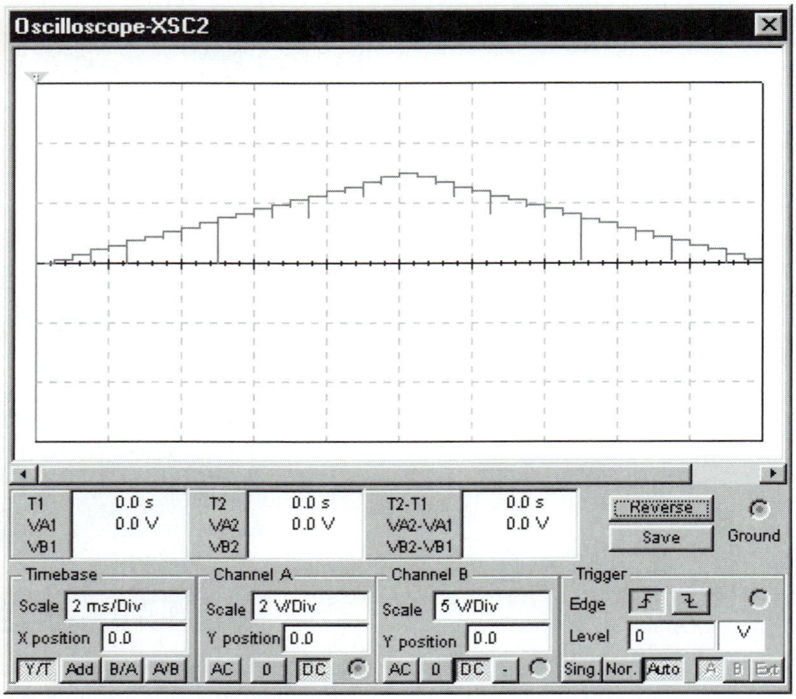

FIGURE 14–25 The VDAC output with the corrected input signal level

The signal is obviously noisy but even more significant is the flat top on the reproduced triangle wave. Why did this happen? A good step is to go back and examine the input signal from the function generator. The settings are shown in Figure 14–24. Notice the amplitude is 2 V and the offset voltage is also 2 V. This produces the 4 Vp-p triangle wave shown in Figure 14–22. It was stated earlier that the maximum input voltage to the A/D is 3 V (this is set by the voltage reference). Therefore the amplitude of the input signal is too large. The input signal was dropped to 3 Vp-p (1.5-V amplitude and 1.5-V offset), and the simulation was rerun. The new D/A signal is shown in Figure 14–25.

The VDAC output with the corrected input signal level shows correct signal level and most importantly (at this point) shows the full triangle signal. The signal is still extremely noisy and there is an obvious stair-step pattern to the reconstructed signal. Most of the noise on the reconstructed signal can be removed by adding an output filter on the VDAC. For this example, a simple RC filter was added. This is shown in Figure 14–26.

Running the simulation with the addition of the filter on the VDAC produces a much cleaner signal. This is shown in Figure 14–27.

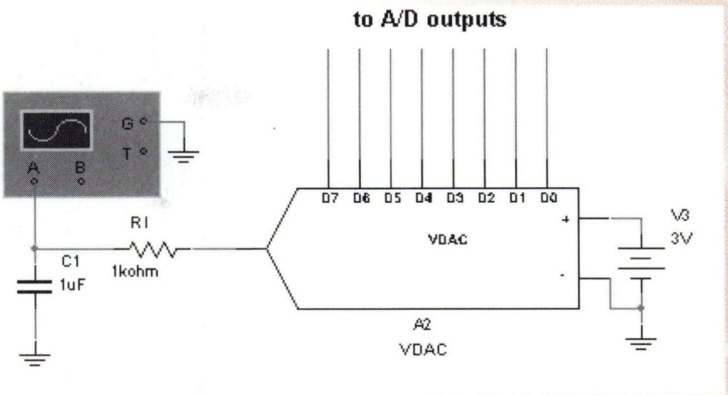

FIGURE 14–26 The addition of the RC filter to the output of the VDAC

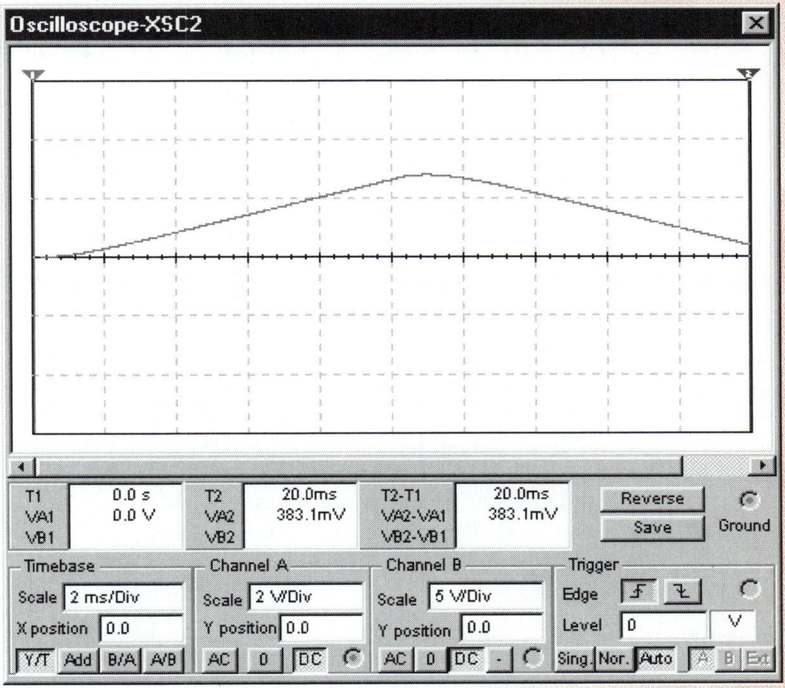

FIGURE 14–27 The filtered output of the VDAC

This exercise has demonstrated the use of EWB Multisim to perform A/D and D/A conversions as well as how the A/D voltage reference comes into play and how to clean up the output signal.

SUMMARY

This chapter has presented an overview of the key circuit structures used in the analog-to-digital and digital-to-analog conversion process. The concepts the student should understand are:

- The structure of the *R*-2*R* ladder.
- The weighted-resistor DAC.
- Switched capacitor circuits.
- Flash ADC.
- Integrating ADC.
- How to use EWB to simulate A/D and D/A circuits.

EXERCISES

SECTION 14–2

The R-2R Ladder DAC

14–1. The logic levels used in an 8-bit *R*-2*R* ladder DAC are 1 = 5 V and 0 = 0 V. Find the output voltage for each input:

(a) 00100000

(b) 10100100

14–2. The logic levels used in a 6-bit *R*-2*R* ladder DAC are 1 = +5 V and 0 = 0 V. What is the binary input when the analog output is 3.28125 V?

SECTION 14–3

A Weighted-Resistor DAC

14–3. Design a 5-bit weighted-resistor DAC whose full-scale output voltage is −15 V. Logic levels are 1 = +5 V and 0 = 0 V. What is the output voltage when the input is 01010?

14–4. Design an 8-bit weighted-resistor DAC using operational amplifiers and two *identical* 4-bit DACs of the design shown in Figure 14–7. The output should be +10 V when the input is 11110000. (*Hint:* Sum the outputs of two amplifiers in a third amplifier, using appropriate voltage gains in the summation.) What is the full-scale output of your design?

SECTION 14–4

The Switched Current-Source DAC

14–5. In the switched current-source DAC shown in Figure 14–8, R = 10 kΩ and E_{REF} = 20 V. Find the current in each 20-kΩ resistor.

14–6. An 8-bit switched current-source DAC of the design shown in Figure 14–8 has R = 10 kΩ and E_{REF} = 15 V. Find the total current I_T delivered to the amplifier and the output voltage when the input is 01101100.

14–7. An 8-bit switched current-source DAC is operated as a 2-quadrant multiplier. The binary input is positive, and the reference voltage can range from −10 V to +10 V. If R = 10 kΩ, what is the total range of the output voltage?

14–8. A 4-bit switched current-source DAC is operated as a 4-quadrant multiplier. The input is offset binary code. Find the output voltage in each case:

(a) The reference voltage is −10 V and the input is 0011.

(b) The reference voltage is +5 V and the input is 1010.

(c) The reference voltage is +10 V and the input is 0001.

SECTION 14–5

Switched-Capacitor DACs

14–9. The binary input to a 4-bit switched-capacitor DAC having C = 2 pF is 0101.

(a) What is the total capacitance connected to E_{REF} when the ϕ_2 clock is high?

(b) What is the total capacitance connected to ground when the ϕ_2 clock is high?

(c) If E_{REF} = 8 V, what is the output voltage?

14–10. Draw a schematic diagram of an 8-bit switched-capacitor DAC. Label all capacitor values in terms of capacitance, *C*. What is the total capacitance when all capacitors are in parallel?

SECTION 14–6

DAC Performance Specifications

14–11. A 12-bit D/A converter has a full-scale range of 15 V. Its maximum differential linearity error is specified to be $\pm(\frac{1}{2})$LSB.

(a) What is its percentage resolution?

(b) What are the minimum and maximum possible values of the increment in its output voltage?

14–12. The LSB of a 10-bit D/A converter is 20 mV.

(a) What is its percentage resolution?

(b) What is its full-scale range?

(c) A differential linearity error greater than what percentage of FSR could make its output nonmonotonic?

SECTION 14–7

The Counter-Type ADC

14–13. The A/D converter in Figure 14–15(a) is clocked at 1 MHz. What is the maximum possible time that could be required to perform a conversion?

14–14. The minimum conversion time of a tracking type A/D converter is 400 ns. At what frequency is it clocked?

SECTION 14–8

Flash A/D Converters

14–15. A flash-type 5-bit A/D converter has a reference voltage of 10 V.

(a) How many voltage comparators does it have?

(b) What is the increment between the fixed voltages applied to the comparators?

14–16. The largest fixed voltage applied to a comparator in a flash-type A/D converter is 14.824218 V when the reference voltage is 15 V. What is the number of bits in the digital output?

SECTION 14–9

The Dual-Slope (Integrating) ADC

14–17. In an 8-bit, dual-slope A/D converter, $R_1 = 20\ \text{k}\Omega$ and $C = 0.001\ \mu\text{F}$. An analog input of -0.25 V is integrated for $t_1 = 160\ \mu\text{s}$.

(a) What is the maximum voltage reached in the integration?

(b) If the input to the integrator is then switched to a reference voltage of +5 V, how long does it take the output to reach 0 V?

(c) If the counter is clocked at 3.125 MHz, what is the digital output after the conversion?

14–18. An 8-bit dual-slope A/D converter integrates analog inputs for $t_1 = 50\ \mu\text{s}$. What should be the magnitude of the reference voltage if an input of 25 V is to produce a binary output of 11111111 when the clock frequency is 1 MHz?

SECTION 14–10

The Successive-Approximation ADC

14–19. List the sequence of binary numbers that would appear in the output register of a 4-bit successive-approximation A/D converter when the analog input has a value that is ultimately converted to 1011.

14–20. Sketch the output of the DAC in an 8-bit successive-approximation A/D converter when the analog input is a voltage that is ultimately converted to 10101011. Label each step of the DAC output with the decimal number corresponding to the binary value it represents.

SECTION 14–11

ADC Performance Specifications

14–21. List four types of A/D converters in descending order of speed (fastest converter type first).

14–22. List the principal advantage of each of four types of A/D converters.

14–23. The analog input of an 8-bit A/D converter can range from 0 V to 10 V. Find its resolution in volts and as a percentage of full-scale range.

14–24. The resolution of a 12-bit A/D converter is 7 mV. What is its full-scale range?

14–25. The frequency components of the analog input to an A/D converter range from 50 Hz to 10 kHz. What maximum total conversion time should the converter have?

14–26. The analog input to an A/D converter consists of a 500-Hz fundamental waveform and its harmonics. If the converter has a total conversion time of 40 μs, what is the highest-order harmonic that should be in the input?

CHAPTER 15

SPECIAL ELECTRONIC DEVICES

OUTLINE

OBJECTIVES

- Introduce the student to the basics of SCR operation.
- Examine the operation of optoelectronic devices including the LCD, LED, phototransistor, photodiodes, and photoresistors.
- Introduce the basic operation of the tunnel diode.
- Examine the operation of the phase-locked loop.
- Investigate how to use EWB Multisim to analyze SCR operation.

15–1 FOUR-LAYER DEVICES

Four-layer devices, also called *thyristors*, comprise a class of semiconductor components whose structure is characterized by alternating layers of *p* and *n* material. By altering the terminal configuration and geometry of the basic structure, it is possible to construct a variety of useful devices, including silicon-controlled rectifiers, silicon-controlled switches, diacs, triacs, and Shockley diodes. These are widely used in high-power switching applications where the control of hundreds of amperes and thousands of watts is not unusual. They are especially useful in the control of ac power delivered to heavy loads such as electric motors and lighting systems.

Silicon Controlled Rectifiers (SCRs)

Figure 15–1 shows the layered *pnpn* structure of a silicon controlled rectifier (SCR). Note that there are three *pn* junctions in this structure. The figure also shows the schematic symbol for an SCR and its three terminals: anode (A), gate (G), and cathode (K). In applications, the anode is made positive with respect to the cathode. As the name and symbol imply, the SCR behaves like a diode (rectifier), which is normally forward biased but conducts from anode to cathode only if a *control* signal is applied to the gate.

As an aid in understanding SCR operation, Figure 15–2 shows a two-transistor equivalent of the four-layer device. Note that the *n* and *p* layers in the center of the structure are "shared" by the transistors: the *n* layer is the base of Q_1 and the collector of Q_2, while the *p* layer is the collector of Q_1 and the base of Q_2.

In our analysis of SCR operation, we will first consider the case where the gate is open and a positive voltage V_A is connected between anode and cathode, as shown in Figure 15–3. The positive voltage forward biases the emitter–base junction of Q_1. However, the base current in Q_1 is only a small leakage current because the base of Q_1 is the collector of Q_2, and the collector–base junction of Q_2 is reverse biased. The only base current in Q_2 is a small leakage current from Q_1. Thus, both the base and collector currents in Q_2 are small leakage currents. Since the emitter current in Q_2 is the same as the total current flowing through the SCR, and since $I_{E2} = I_{B2} + I_{C2}$, we see that the total SCR current is the sum of two small leakage currents, which we

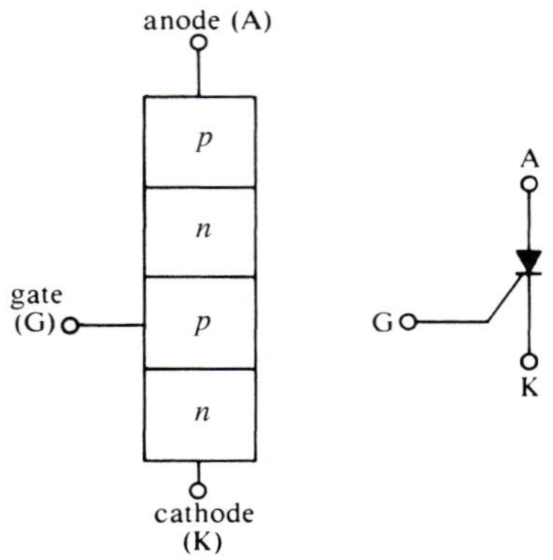

FIGURE 15–1 Structure and symbol of a silicon-controlled rectifier (SCR)

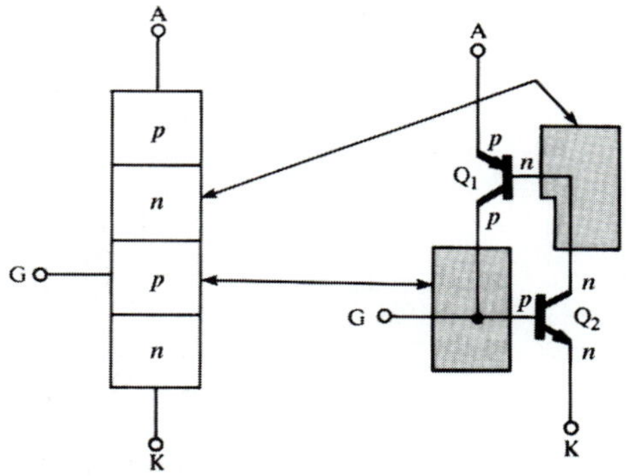

FIGURE 15–2 Two-transistor equivalent of an SCR

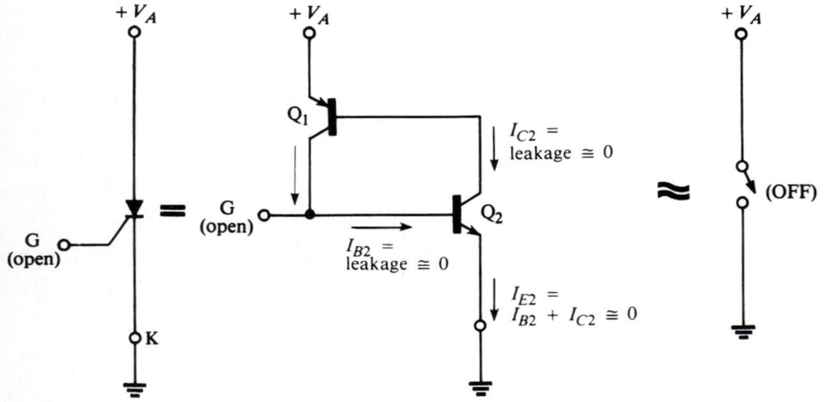

FIGURE 15–3 The SCR in its off, or forward-blocking, state

may regard as negligibly small. We say that the SCR itself is off (essentially nonconducting), or in a *forward-blocking* state: forward biased, but blocking the flow of current.

SCR operation is based on the fundamental principle that the α and β of a transistor remain small so long as its collector current is small. As we saw in Chapter 4, collector current must be increased to a certain level before those parameters acquire the normal, essentially constant values they have in the active region. When a transistor is cut off, α and β are quite small, but they rise rapidly as the transistor comes out of cutoff.

In our analysis of Figure 15–3, we assumed that the collector currents (leakage currents) were so small that the α and β of each transistor remained small. If the value of V_A is increased, the leakage currents will increase correspondingly, and so will the values of α and β. Notice that the collector current in Q_2 has a component equal to $\beta_2 I_{B2}$, where β_2 is the beta of Q_2. Thus, as β_2 increases, so does I_{C2}. Likewise, the collector current in Q_1 increases as β_1 increases. Furthermore, as I_{C2} increases, so does base current I_{B1}, since they are identical currents. But as I_{B1} increases, so does $I_{C1} = \beta_1 I_{B1}$. This, in turn, increases I_{B2}, which further increases I_{C2}, which further increases I_{C1}, and so forth. As the currents rise, so do the β-values, which in turn increase the current values. The upshot of all this is that if V_A is made sufficiently large, the transistor α- and β-values increase enough to cause significant current flow. When that point is reached, the larger collector current in Q_2 turns Q_1 fully on, so Q_1 supplies a large base current to Q_2, which drives Q_2 fully on; in other words, both transistors saturate, and the current flowing through each ensures that the other remains in saturation. The SCR is said to be *latched*. The action we have described is an example of *regenerative* switching, where one device, as it begins to conduct, causes another device to conduct, which in turn drives the first further into conduction. Once regeneration is initiated, it quickly causes both devices to saturate. In SCRs the whole process takes a few microseconds or less.

Figure 15–4 shows the SCR after V_A has been increased sufficiently to cause regenerative switching. Because both transistors are saturated, a heavy flow of current is now possible from anode to cathode. As with a forward-biased diode, external resistance is required to limit that current. The anode-to-cathode voltage that is required to initiate regenerative switching to the on state is called the *forward breakover voltage*, $V_{BR(F)}$.

Figure 15–5(a) shows an external resistance connected to the SCR to limit current flow when breakover occurs, and (b) shows the current-voltage characteristic. When V_{AK} is small and the SCR is off, the leakage current is small, so there is only a small voltage drop across R. Most of the supply voltage, V_A, appears across the SCR as V_{AK}, and the SCR is in its forward blocking region. When V_{AK} reaches $V_{BR(F)}$, the SCR suddenly turns on, so, the SCR being like a closed switch, the voltage drop across it suddenly becomes small. Notice that the characteristic loops back from the point where $V_{AK} = V_{BR(F)}$ and I_A is small to the region where V_{AK} is small and I_A is large. As an aid in understanding this behavior, imagine that you are measuring the current through and voltage across an open switch that is in series with a resistor and a battery; then close the switch and observe the measured values. Once the SCR switches on, the characteristic becomes very much like that of a forward-biased silicon diode. The dc voltage drop V_{AK} across the SCR remains at around 0.9 to 1 V, depending on current. The plot also shows that if V_A is made

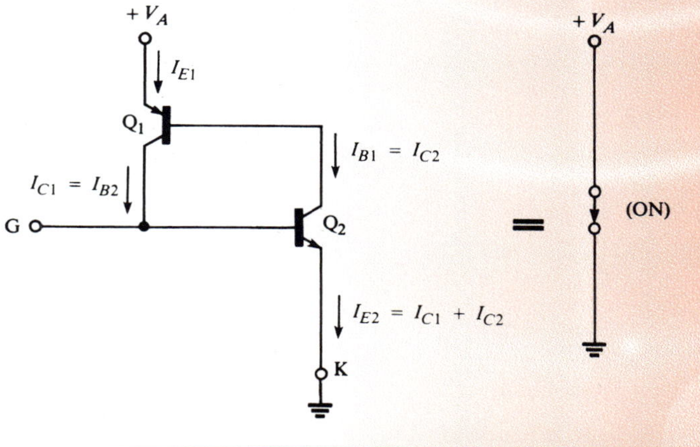

FIGURE 15–4 The SCR is like a closed switch when the anode-to-cathode voltage reaches the forward breakover voltage

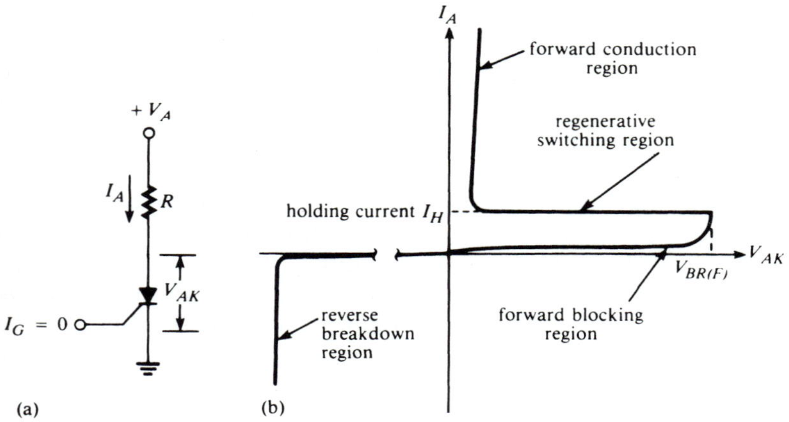

FIGURE 15–5 The current-voltage characteristic of an SCR with $I_G = 0$

sufficiently negative, a reverse breakdown voltage is reached, and reverse conduction occurs by the avalanching mechanism.

Once the SCR has switched on, it will remain on provided the current through it remains sufficiently large. If the current is reduced (by reducing V_A, for example) to the point that I_A falls below a certain *holding current*, then the SCR will revert to its off state. The holding current, I_H, is identified in Figure 15–5(b) as the flat portion of the characteristic corresponding to the regenerative switching region. Note that the SCR cannot remain in this region for any length of time; the horizontal line merely shows the path followed when the SCR switches on and is often drawn as a dashed line for that reason.

Shockley Diodes

In all our analysis to this point, we have assumed that the gate terminal is open, so I_G remained 0 at all times. Certain four-layer devices are manufactured without gate terminals, so they behave exactly like an SCR with an open gate. These two-terminal devices are called *Shockley* diodes, or four-layer diodes. Figure 15–6 shows the schematic symbol for a Shockley diode and an application called a *relaxation oscillator*. Capacitor C charges through R to the breakover voltage, at which time the diode switches on and quickly discharges the capacitor. The diode then turns off and the

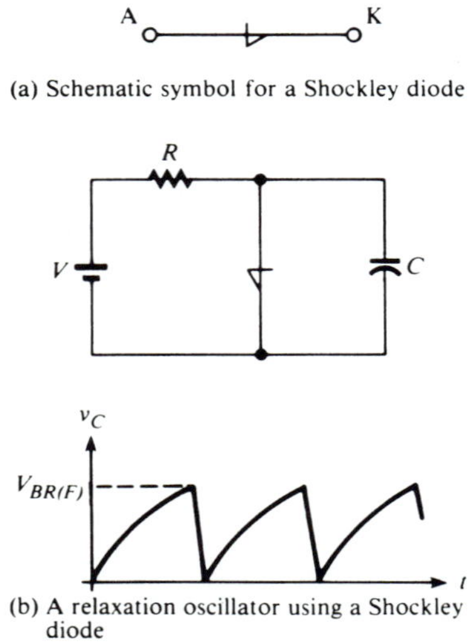

FIGURE 15–6 Schematic symbol and a typical application of a Shockley diode

capacitor begins charging again. The frequency of oscillation is controlled by the *RC* time constant.

SCR Triggering

SCRs are normally operated with an external circuit connected to the gate. The purpose of the circuit is to inject current into the base of Q_2 in Figure 15–4 and thereby "trigger" the SCR into a regenerative breakover. The additional current supplied by the gate causes I_{C2} to increase, which increases I_{C1}, and so forth, so the SCR switches to its on state, as before. Thus, the gate serves as a control input that can be used to switch the SCR from off to on when desired. It is important to realize that it is necessary to supply only a short *pulse* of current through the gate to cause the switching action; once the regeneration is initiated, latching quickly occurs and gate current is no longer required. By the same token, the SCR cannot be turned off by simply reducing gate current to 0. As described earlier, it can be turned off only by reducing the anode-to-cathode current to a value below the holding current.

Figure 15–7 is a typical family of SCR characteristics showing I_A versus V_{AK} for different values of gate current. Notice that the larger the gate current, the smaller the breakover voltage. When gate current is large, it is not necessary for V_{AK} to become large enough to induce breakover through the increase of leakage current alone. Also note that holding current decreases as gate current increases. Of course, if gate current is removed after breakover has occurred, then the holding current becomes the same as that for $I_G = 0$.

EXAMPLE 15–1

The voltage drop across the SCR in Figure 15–8 is 1 V when it is conducting. It has a holding current of 2 mA when $I_G = 0$. If the SCR is triggered on by a momentary pulse of gate current, to what value must V_A be reduced to turn the SCR off?

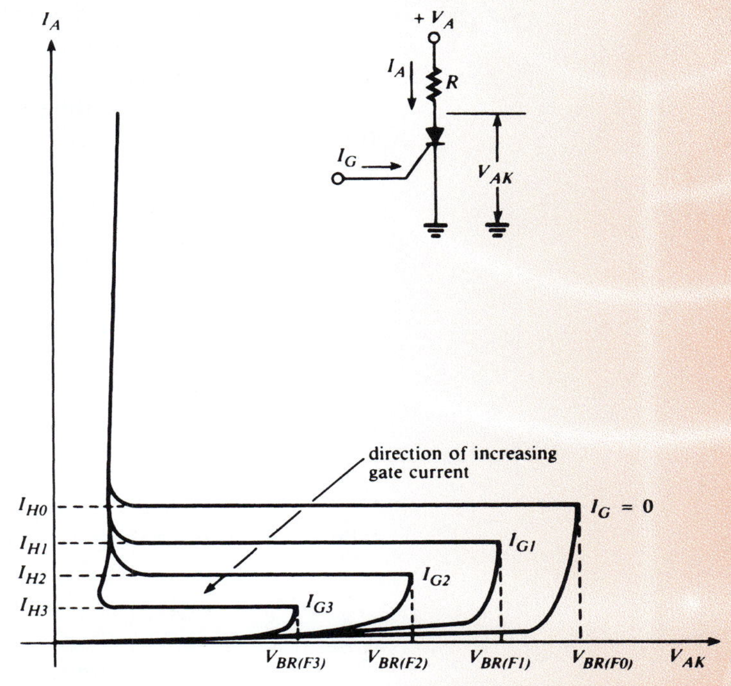

FIGURE 15–7 A family of SCR characteristics. As gate current increases, the breakover voltage and the holding current decrease.

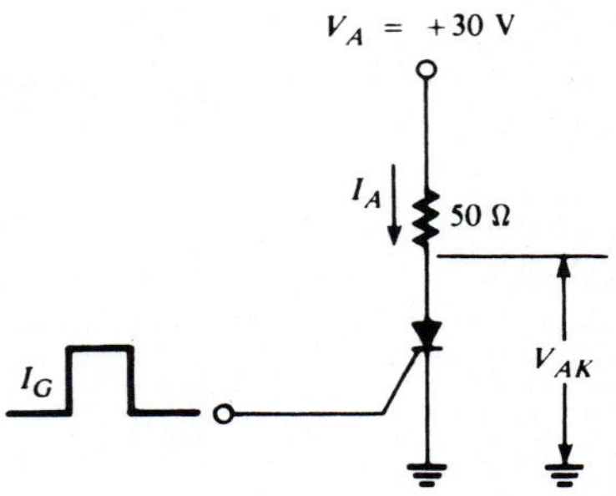

FIGURE 15–8 (Example 15–1)

Solution

The current in the 50-Ω resistor is

$$I_A = \frac{V_A - V_{AK}}{50}$$

When the SCR is conducting, $V_{AK} = 1$ V, and

$$I_A = \frac{V_A - 1}{50}$$

Assuming that V_{AK} remains constant at 1 V as V_A and I_A are reduced (actually, V_{AK} decreases slightly), the value of V_A necessary to make $I_A = I_H = 2$ mA is found from

$$2 \text{ mA} = \frac{V_A - 1}{50} \qquad \text{or} \qquad V_A = 1.1 \text{ V}.$$

This example shows that V_A must be reduced to a very small value to turn the SCR off. In practice, turn-off is accomplished either by switching the supply voltage off entirely or by short-circuiting the SCR.

We will now derive a general expression for the SCR current I_A in terms of gate current I_G and leakage currents I_{CBO1} and I_{CBO2}. The result will demonstrate vividly how the magnitudes of α_1 and α_2 affect the SCR current below and at the onset of breakover. In our derivation, we will use the following two relations, which are easily obtained from the standard transistor equations given in Chapter 4 (4–4 and 4–9):

$$I_C = \alpha I_E + I_{CBO} \qquad \text{or} \qquad I_E = \frac{I_C - I_{CBO}}{\alpha} \tag{15–1}$$

$$I_B = \frac{1-\alpha}{\alpha} I_C - \frac{I_{CBO}}{\alpha} \tag{15–2}$$

Refer to Figure 15–9. Writing Kirchhoff's current law for the external currents, as shown in Figure 15–9(a), we find

$$I_A = I_K - I_G \tag{15–3}$$

where $I_K = I_{E2}$ in Figure15–9(b). From equation 15–1,

$$I_K = I_{E2} = \frac{I_{C2} - I_{CBO2}}{\alpha_2} \tag{15–4}$$

Note that $I_{C2} = I_{B1}$. From equation 15–2,

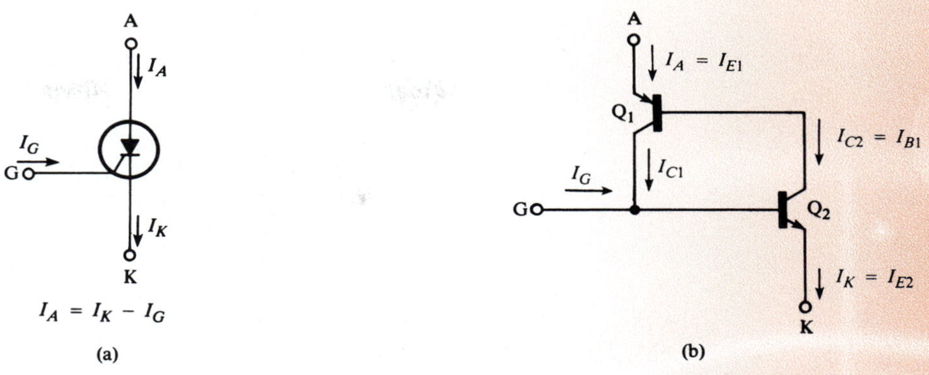

FIGURE 15–9 Current relations used to derive equation 15-8

$$I_{C2} = I_{B1} = \frac{1-\alpha_1}{\alpha_1} I_{C1} - \frac{I_{CBO1}}{\alpha_1} \quad (15\text{–}5)$$

From (15–1),

$$\begin{aligned} I_{C1} &= \alpha I_{E1} + I_{CBO1} \\ &= \alpha_1 I_A + I_{CBO1} \end{aligned} \quad (15\text{–}6)$$

Substituting (15–6) into (15–5) and the result into (15–4) gives

$$I_K = \frac{(1-\alpha_1)I_A + \dfrac{(1-\alpha_1)}{\alpha_1} I_{CBO1} - \dfrac{I_{CBO1}}{\alpha_1} - I_{CBO2}}{\alpha_2} \quad (15\text{–}7)$$

Substituting (15–7) into (15–3) and solving for I_A gives

$$I_A = \frac{\alpha_2 I_G + I_{CBO1} + I_{CBO2}}{1 - (\alpha_1 + \alpha_2)} \quad (15\text{–}8)$$

Equation 15–8 shows that if α_1 and α_2 are small, the current I_A is essentially the sum of two small leakage currents. However, if this current, plus the component of gate current, increases enough to raise the values of α_1 and α_2 to where $\alpha_1 + \alpha_2 = 1$, the denominator of (15–8) becomes 0. In that case, I_A is theoretically infinite, limited only by external resistance.

EXAMPLE 15–2

An SCR in its forward-blocking region has $V_A = 30\,\text{V}$, $I_G = 1\,\mu\text{A}$, and $I_{CBO1} = I_{CBO2} = 50\text{ nA}$. If $\alpha_1 = 0.5$ and $\alpha_2 = 0.3$ under those conditions, find the anode current I_A and the dc resistance of the SCR.

Solution

By equation 15–8,

$$I_A = \frac{\alpha_2 I_G + I_{CBO1} + I_{CBO2}}{1 - (\alpha_1 + \alpha_2)} = \frac{0.3(1\,\mu\text{A}) + (100\text{ nA})}{1 - 0.8} = 2\,\mu\text{A}$$

$$R_{dc} = \frac{30\text{ V}}{2\,\mu\text{A}} = 15\text{ M}\Omega$$

Half-Wave Power Control Using SCRs

SCRs are commonly used to adjust, or control, the average power delivered to a load from an ac source. In these applications, the SCR is periodically switched on and off, and the average power delivered to the load is controlled by adjusting the total length of time the SCR conducts during each ac cycle. Figure 15–10 shows a simple half-wave power controller based on this principle.

When the ac voltage in Figure 15–10 is positive and the SCR is conducting, current flows through load resistance R_L. As shown in (b) and (c), the total length of time the SCR conducts depends on the point in each cycle at which the SCR switches on, or *fires*. After firing, the SCR continues to conduct until the ac voltage drops to near 0, causing the SCR current to fall below the holding current. It then remains off until firing again at the same point in the next positive half-cycle. The angle of the ac waveform at which the SCR switches on is called the *firing angle*, θ_f, or the *delay angle*. The firing angle is controlled by adjusting R_p. When R_p is large, the ac voltage must reach a large value to generate enough gate current to fire the SCR. Consequently, the firing angle is large, as shown in Figure 15–10(c). Conversely, when R_p is small, a relatively small voltage will cause the SCR to fire, and θ_f will be small. In the circuit shown, θ_f can be adjusted from 0° to 90°. The diode is used to protect the gate from the negative portion of the ac voltage.

When the SCR conducts, the peak load current I_P is

$$I_P = \frac{V_P - V_{AK}}{R_L} \tag{15–9}$$

where V_P is the peak ac voltage and V_{AK} is the drop across the SCR, about 1 V, when it is conducting. It can be shown that the average and rms values of the current are

FIGURE 15–10 An SCR half-wave power controller and load-current waveforms. Current flows in the load during the shaded intervals shown on the waveforms.

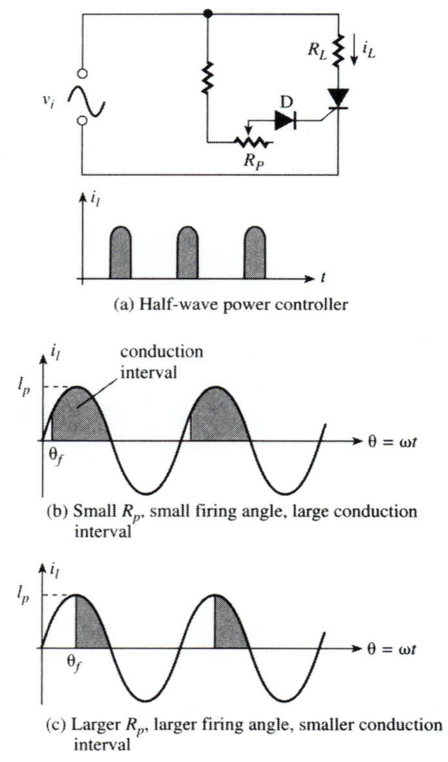

$$I_{AVG} = \frac{I_P}{2\pi}(1 + \cos\theta_f)$$

$$I_{rms} = \frac{I_P}{2}\sqrt{\left(1 - \frac{\theta_f}{180°}\right) + \frac{\sin 2\theta_f}{2\pi}} \quad (15\text{–}10)$$

where θ_f is in degrees. When $\theta_f = 0$, the current is the same as that in a half-wave rectifier, and equation 15–10 shows that $I_{AVG} = I_P/\pi$ and $I_{rms} = I_P/2$. When $\theta_f = 90°$, $I_{AVG} = I_P/2\pi$ and $I_{rms} = I_P/\sqrt{2}$.

One problem encountered in practical applications of SCRs, particularly high-power devices, is a tendency to "false trigger" if there is a large rate of change of anode voltage. This tendency is attributable to the large junction capacitance associated with the large junction areas that are necessary to dissipate heavy amounts of power. Recall that capacitor current is proportionate to the rate of change of capacitor voltage. The large charging current that results when the anode voltage changes at a high rate may induce regenerative breakover. To suppress a false triggering, a shunting capacitor is sometimes connected between anode and cathode.

EXAMPLE 15–3

In Figure 15–10(a), v_i is a 120-V-rms ac voltage and $R_L = 40\ \Omega$. Assume that the drop across the SCR is 1 V when it is conducting.

1. What should be the firing angle if it is desired to deliver an average current of 1 A to the load?
2. What is the average power delivered to the load under the conditions of (1)?

Solution

1. $V_P = \sqrt{2}(120) = 169.7$ V. From equation 15–9,

$$I_P = \frac{V_P - V_{AK}}{R_L} = \frac{169.7 - 1}{40} = 4.2175 \text{ A}$$

From equations 15–10,

$$1 = \frac{4.2175}{2\pi}(1 + \cos\theta_f)$$

$$\cos\theta_f = 0.4898$$

$$\theta_f = 60.67°$$

2. From (15–10)

$$I_{rms} = \frac{4.2175}{2}\sqrt{\left(1 - \frac{60.67}{180}\right) + \frac{\sin 2(60.67)}{2\pi}} = 1.885 \text{ A rms}$$

$$P_{AVG} = I_{rms}^2 R_L = (1.885)^2 40 = 142.1 \text{ W}$$

Figure 15–11 shows a typical set of SCR specifications. Notice the variety of case styles and reverse voltage ratings available. The same manufacturer supplies SCRs with forward current ratings up to 55 A rms.

On-State (RMS) Current									
8.0 AMPS		12 AMPS				12.5 AMPS	15 AMPS		
T_C = 83°C		T_C = 90°C	T_C = 85°C			T_C = 80°C	T_C = 80°C		

Case 86-01 Style 1	Case 87L-02 Style 1	Case 221A-02 TO-220AB Style 3	Case 86-01 Style 1	Case 221A-02 TO-220AB Style 3	Case 342-01 Style 1	Case 54-05 Style 2	Case 221A-02 TO-220AB Style 3		TO-220 Style 3
2N4167	2N4183		MCR67-1	MCR68-1	MCR568-1			V_{DRM} V_{RRM}	25 V
2N4168	2N4184	2N6394	MCR67-2	MCR68-2	MCR568-2				50 V
2N4169	2N4185	2N6395	MCR67-3	MCR68-3	MCR568-3	2N3668			100 V
2N4170	2N4186	2N6396				2N3669	MCR2150-4,A4		200 V
2N4171	2N4187	MCR220-5					MCR2150-5,A5		300 V
2N4172	2N4188	2N6397	MCR67-6	MCR68-6	MCR568-6	2N3670	MCR2150-6,A6		400 V
2N4173	2N4189	MCR220-7				2N4103	MCR2150-7,A7		500 V
2N4174	2N4190	2N6398					MCR2150-8,A8		600 V
		MCR220-9					MCR2150-9,A9		700 V
		2N6399					MCR2150-10,A10		800 V
100			300(1)			200	160	MAXIMUM ELECTRICAL CHARACTERISTICS	I_{TSM} (Amps) 60 Hz
30						40	50		I_{GT} (mA)
1.5						2.0	2.5		V_{GT} (V)
−40 to +100		−40 to +125				−40 to +100	−40 to +125		T_J Operating Range (°C)

FIGURE 15–11 SCR specifications (Courtesy of Motorola, Inc.)

Silicon Controlled Switches (SCSs)

A silicon controlled switch (SCS) is essentially an SCR equipped with a second gate terminal. Like a conventional switch, and unlike an SCR, an SCS can be switched either on or off through external gate control. Figure 15–12 shows the structure, symbol, and two-transistor equivalent of an SCS. Note that the gate terminal at the base of Q_2 is now called the *cathode gate*, and the second gate, at the base of Q_1, is called the *anode gate*.

The silicon controlled switch can be turned on in exactly the same way as an SCR: by supplying a positive pulse of current to the cathode gate, which drives Q_2 on and initiates regenerative breakover. It can also be turned on by supplying a *negative* current pulse to the anode gate. Notice that negative current turns on *pnp* transistor Q_1 so Q_1 initiates the regeneration in this case. The SCS can be turned off either by supplying negative current to the cathode gate, which turns off Q_2, or by supplying positive current to the anode gate, which turns off Q_1. Figure 15–13 summarizes the methods by which an SCS can be turned on and off. Silicon controlled switches can be turned on and off faster than SCRs, but they are not available with the high-power, high-current ratings of SCRs.

DIACs and TRIACs

A *DIAC* is a four-layer device whose "top" and "bottom" layers contain both *n* and *p* material, as shown in Figure 15–14. The right side of the stack can be regarded as a *pnpn* structure with the same characteristics as a gateless SCR, while the left side is an inverted SCR having an *npnp* structure. A four-transistor equivalent circuit of a DIAC, resembling an SCR in parallel with

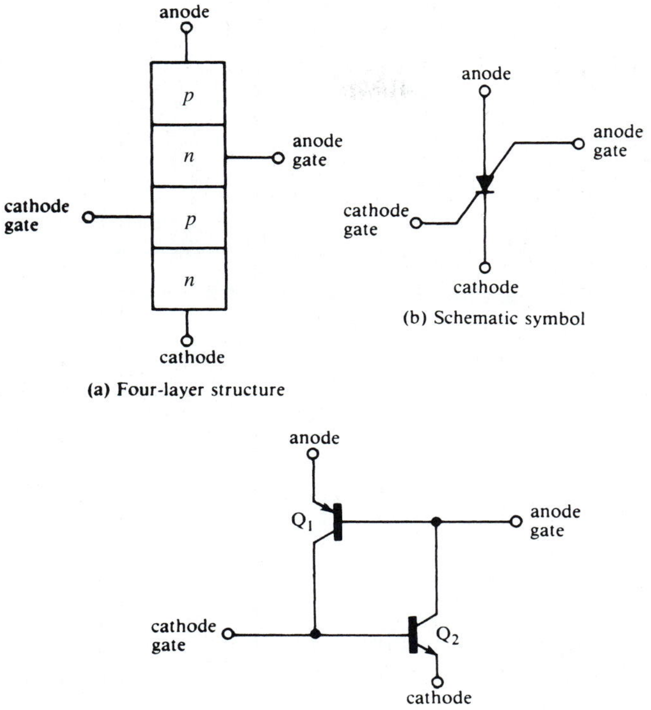

FIGURE 15–12 The silicon controlled switch (SCS)

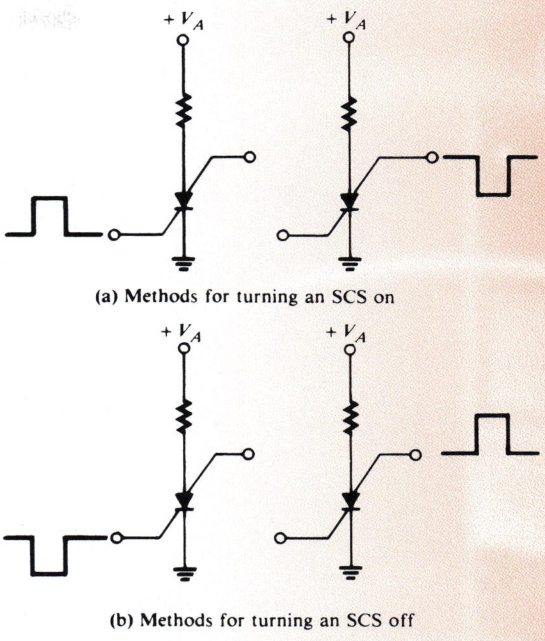

FIGURE 15–13 Turning an SCS on and off

FIGURE 15–14 DIAC construction and operation

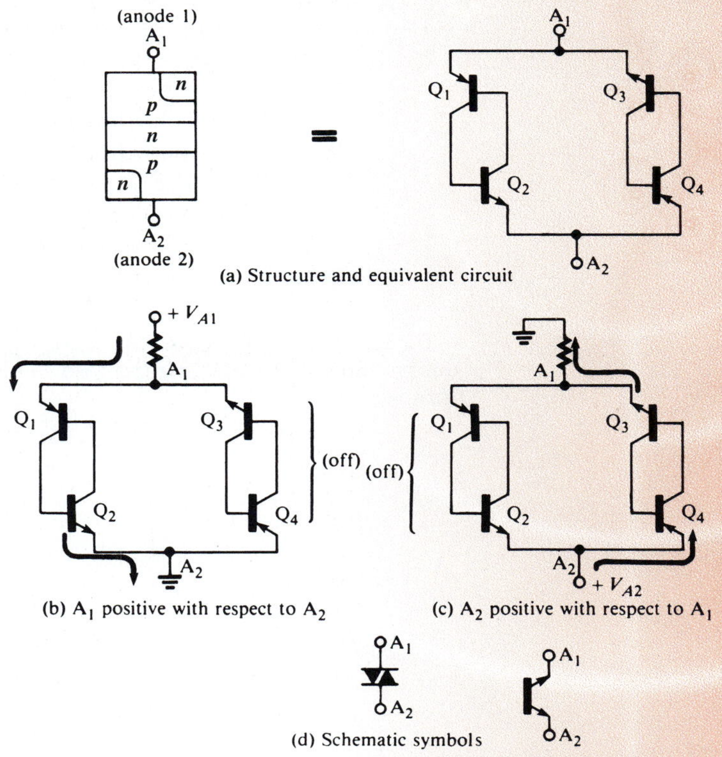

an inverted SCR, is shown in the figure. Note that the terminals are labeled A_1 (anode 1) and A_2 (anode 2). There are no gates. The DIAC can conduct current in *either* direction: from A_1 to A_2 through Q_1 and Q_2, or from A_2 to A_1 through Q_3 and Q_4. If A_1 is made sufficiently positive with respect to A_2 to induce breakover in Q_1 and Q_2, then conduction occurs in that path, while Q_3

FIGURE 15–15 The I–V characteristic of a DIAC

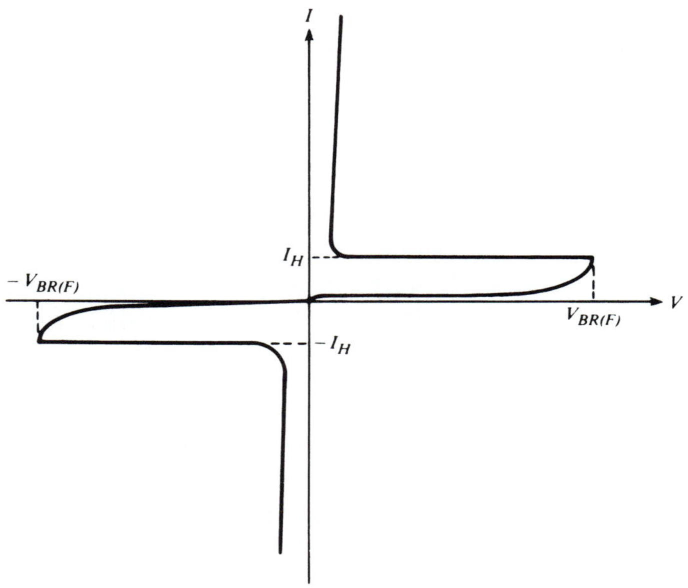

and Q_4 remain off. Similarly, if A_2 is sufficiently positive with respect to A_1, then Q_3 and Q_4 conduct while Q_1 and Q_2 remain off. The two cases are illustrated in Figure 15–14(b) and (c). Part (d) shows the schematic symbols.

Figure 15–15 shows the $I-V$ characteristic of a DIAC. We may arbitrarily assume one direction through the DIAC as positive (from A_1 to A_2, for example), and the opposite direction as negative. The positive portion of the characteristic is the same as that of an SCR with zero gate current. The negative portion shows that breakover occurs when the reverse voltage reaches $-V_{BR(F)}$, at which time a large reverse current flows. To turn the DIAC off, the current must be reduced below the positive holding current I_H if conducting in the forward direction, or below the negative holding current $-I_H$ if conducting in the reverse direction.

A TRIAC is equivalent to a DIAC having a gate terminal, as shown in Figure 15–16. Note that the gate is connected to the base of Q_2 and to the base

FIGURE 15–16 A TRIAC equivalent to a DIAC having a gate terminal

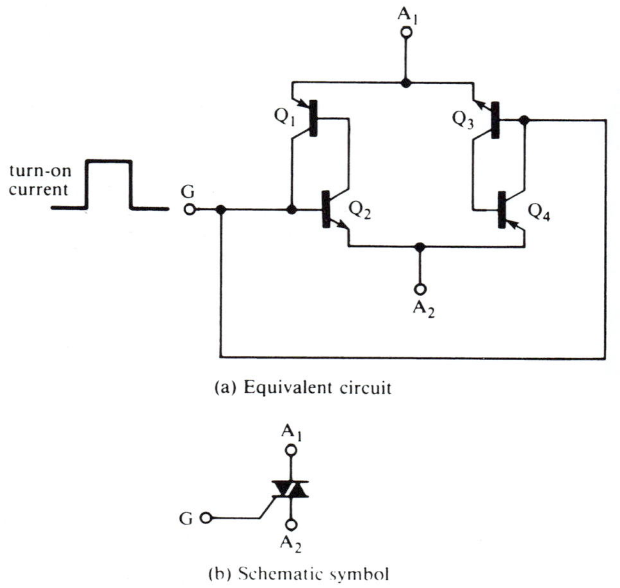

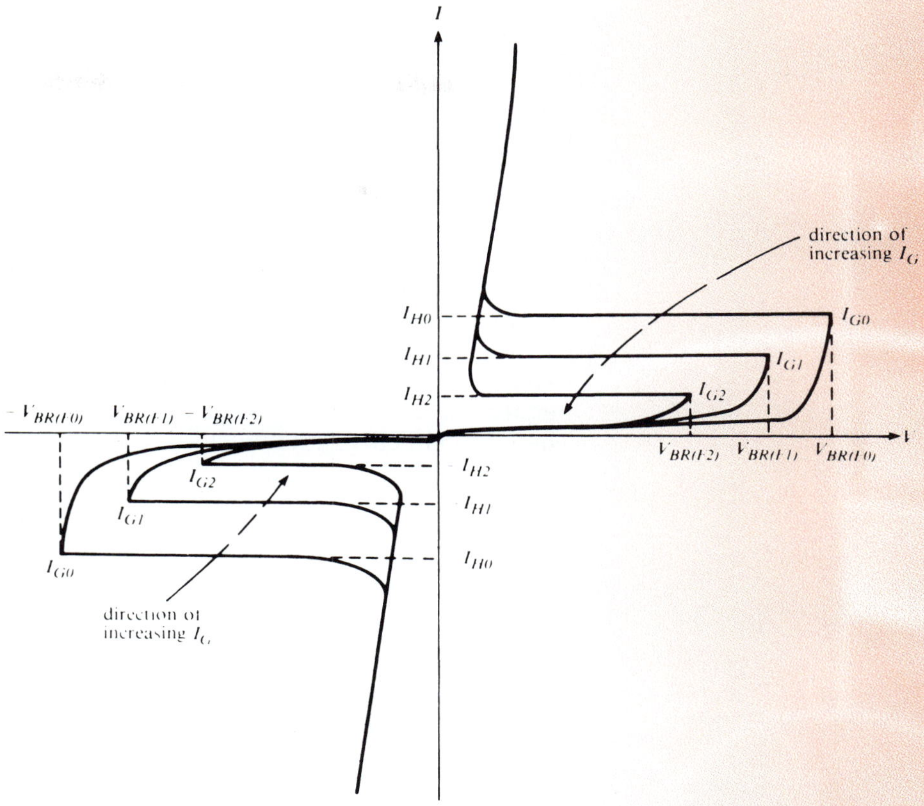

FIGURE 15–17 *I–V* characteristics of a TRIAC

of Q_3, both of which are *npn* transistors. Therefore, a pulse of current flowing into the gate will induce current flow through Q_1 and Q_2 if A_1 is positive with respect to A_2 or will induce current flow through Q_3 and Q_4 if A_2 is positive with respect to A_1. The gate must be made positive with respect to A_2 to turn on Q_1 and Q_2 and must be made positive with respect to A_1 to turn on Q_3 and Q_4. Modern TRIACs have an additional *n* layer connected to one gate. This layer serves as the emitter of an *npn* transistor that increases gate sensitivity on one side, but negative gate current is required to trigger that side on.

Figure 15–17 shows a family of characteristic curves for a TRIAC. Like the SCR, the magnitudes of the breakover voltage and holding current become smaller as the values of gate current increase. Whatever the direction of the current through the TRIAC, its value must be reduced below the holding current to turn the TRIAC off.

Figure 15–18 shows a power controller utilizing a TRIAC that is triggered by positive gate current when A_1 is positive with respect to A_2 and by

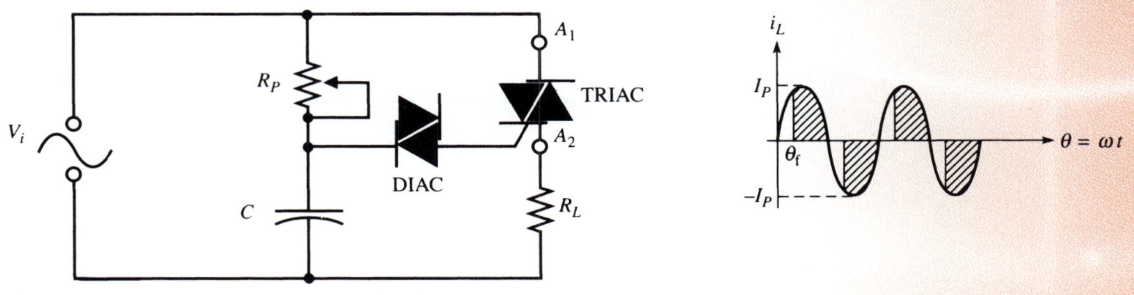

FIGURE 15–18 A TRIAC power controller

negative gate current when A_1 is negative with respect to A_2. Notice that a DIAC is connected between the gate and an RC circuit. When the capacitor voltage rises far enough to overcome the sum of the breakover voltage of the DIAC and the forward drop of the gate, the DIAC fires, the capacitor discharges rapidly into the gate, and the TRIAC is triggered into conduction. Because R_P controls the time constant at which the capacitor charges, it is used to control the firing angle, θ_f: Large values of R_P delay the charging and therefore increase the firing angle. When A_1 is negative with respect to A_2, the polarities of all voltages are reversed, and the TRIAC is triggered into conduction during a portion of the negative half-cycle. The firing angle can be adjusted in practical circuits from near 0° to near 180°, so load current can be made to flow for nearly an entire cycle of input or, at the other extreme, for a very small portion of an input cycle. Because the positive and negative areas of the current waveform are equal, the average current is 0. The rms load current is

$$I_{rms} = \frac{I_P}{\sqrt{2}}\sqrt{\left(1 - \frac{\theta_f}{180°}\right) + \left(\frac{\sin 2\theta_f}{2\pi}\right)} \qquad \textbf{(15–11)}$$

where θ_f is in degrees.

EXAMPLE 15–4

In the circuit shown in Figure 15–18, v_i is 60 V rms and $R_l = 15\ \Omega$. Neglecting the drop across the TRIAC when it is conducting, find

1. the maximum possible power that can be delivered to the load, and
2. the percent of maximum load power delivered when $\theta_f = 45°$.

Solution

1. Maximum power is delivered when $\theta_f = 0°$, in which case the current is a full sine wave and the load voltage is 60 V rms:

$$P_{max} = \frac{V_{rms}^2}{R_L} = \frac{60^2}{15} = 240 \text{ W}$$

2.

$$V_P = \sqrt{2}(60) = 84.85 \text{ V}$$

$$I_P = \frac{V_P}{R_L} = \frac{84.85}{15} = 5.66 \text{ A}$$

From equation 15–11,

$$I_{rms} = \frac{5.66}{\sqrt{2}}\sqrt{\left(1 - \frac{45}{180}\right) + \frac{\sin 2(45)}{2\pi}} = 3.82 \text{ A rms}$$

$$P = I_{rms}^2 R_L = (3.82)^2(15) = 218.8 \text{ W}$$

$$\frac{P}{P_{max}} \times 100\% = \frac{218.8}{240} \times 100\% = 91.2\%$$

15–2 OPTOELECTRONIC DEVICES

Optoelectronic devices are those whose characteristics are changed or controlled by light or those that create and/or modify light. In recent years, important technological advances in the design and fabrication of solid-state optoelectronic devices have been responsible for what are now widespread applications in areas such as visual displays, alarm systems, fiber-optic communications, optical readers, lasers, and solar energy converters.

Light is a form of electromagnetic radiation and, as such, is characterized by its frequency or, equivalently, by its wavelength. The fundamental relationship between frequency and wavelength is

$$f = \frac{c}{\lambda} \tag{15–12}$$

where f is frequency, in hertz; λ is wavelength, in meters; and c is the speed of light, 3×10^8 m/s.

Another commonly used unit of wavelength is the angstrom (Å), where

$$1\ \text{Å} = 10^{-10}\ \text{m} \tag{15–13}$$

Visible light occurs at frequencies in the range from about 4.3×10^{14} Hz (7000 Å) to 7.5×10^{14} Hz (4000 Å). The lowest frequency appears red and the highest appears violet. Frequencies less than that of red, down to 10^{12} Hz, are called *infrared*, and those greater than that of violet, up to 5×10^{17} Hz, are called *ultraviolet*. *White* light is a mixture, or composite, of all frequencies in the visible range.

A light *spectrum* is a plot of light energy versus frequency or wavelength. An *emission* spectrum is a spectrum of the light produced, or emitted, by a light source, such as an incandescent bulb or a light-emitting diode (LED). A *spectral response* shows how one characteristic of a light-sensitive device varies with the frequency or wavelength of the light to which it is exposed. Spectral response is similar in concept to the frequency response of an electronic system. For example, the spectral response of the human eye shows that it is most sensitive to yellow light, at wavelengths around 5700 Å, and least sensitive to the frequencies at opposite ends of the visible band, i.e., red and violet.

Luminous flux is the *rate* at which visible light energy is produced by a light source, or at which it is received on a surface, and is measured in *lumens* (lm), where

$$1\ \text{lm} = 1.496\ \text{mW} \tag{15–14}$$

Thus, a surface receiving visible light energy at the rate of 1.496×10^{-3} J/s is said to have a luminous flux of 1 lumen. Luminous *intensity* is luminous flux per unit area. For example, if the luminous flux on a 0.2-m^2 surface is 1.5×10^3 lm, then the intensity is $1.5 \times 10^3\ \text{lm}/(0.2\ \text{m}^2) = 75 \times 10^3\ \text{lm/m}^2$. One lumen per square foot equals the older unit, one footcandle. Manufacturers of light-sensitive electronic devices also use the units *milliwatts per square centimeter* to specify light intensity in their product literature. Care must be exercised in the interpretation of light-intensity figures, because measured values depend on the spectral response of the instrument used to measure them.

EXAMPLE 15–5

Find the horsepower equivalent of 1×10^6 lm/m^2 (the typical intensity of visible sunlight at noon on a 10-m^2 surface).

Solution

$$P = (10^6\ \text{lm/m}^2)(10\ \text{m}^2) = 10^7\ \text{lm}$$

$$(10^7 \text{ lm})(1.496 \times 10^{-3} \text{ W/lm}) = 1.496 \times 10^4 \text{ W}$$

$$(1.496 \times 10^4 \text{ W})\left(\frac{1 \text{ hp}}{746 \text{ W}}\right) = 20 \text{ hp}$$

Photoconductive Cells

A photoconductive cell is a passive device composed of semiconductor material whose surface is exposed to light and whose electrical resistance decreases with increasing light intensity. Recall that electrons in a semiconductor become *free* electrons if they acquire enough energy to break the bonds that hold them in a covalent structure. In conventional diodes and transistors, electrons are freed by heat energy. In a photoconductive cell, free electrons are similarly created by light energy; the greater the light intensity, the greater the number of free electrons. Since the conductivity of the material increases as the number of free electrons increases, the resistance of the material decreases with increasing light intensity. Photoconductive cells are also called *photoresistive* cells, or *photoresistors*. They are the simplest and least expensive of components belonging to the general class of *photoconductive* devices: those whose conductance changes with light intensity.

Figure 15–19 shows the typical geometry of a photoconductive cell and its schematic symbol. The light-sensitive semiconductor material is arranged in a zigzag strip whose ends are attached to external pins. A glass or transparent plastic cover is attached for protection. The light-sensitive material used in photoconductive cells is either *cadmium sulfide* (CdS) or *cadmium selenide* (CdSe). The resistance of each of these materials changes rather slowly with changes in light intensity, the CdSe cell having a response time of about 10 ms, and that of CdS being about 100 ms. However, the slower CdS cell is much less sensitive to temperature changes than is the CdSe cell. The spectral response of CdS, i.e., its sensitivity as a function of the wavelength of incident light, is similar to that of the human eye, so it responds to visible light. The response of CdSe is greater at red and infrared frequencies.

Figure 15–20 shows a plot of resistance versus light intensity for a typical photoconductive cell. Note that the plot is linear on logarithmic axes. The

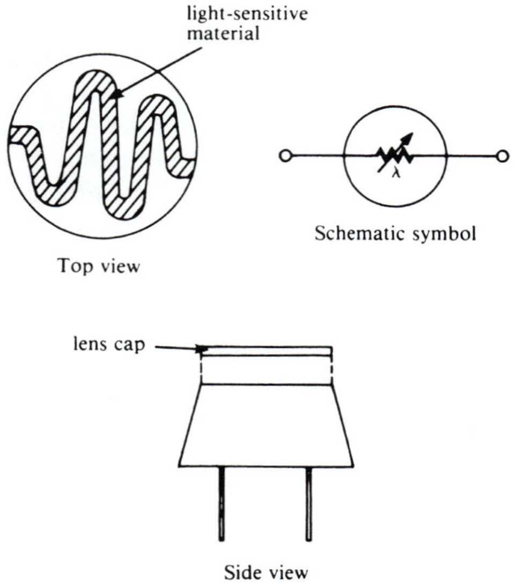

FIGURE 15–19 Structure and symbol for a photoconductive cell

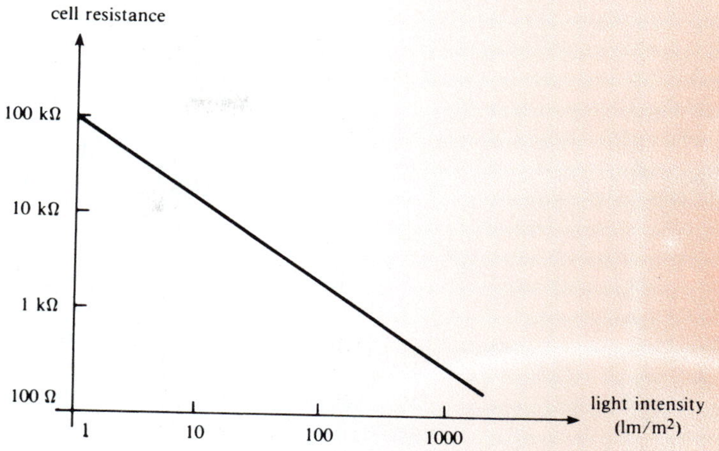

FIGURE 15–20 Resistance versus light intensity for a typical photoconductive cell

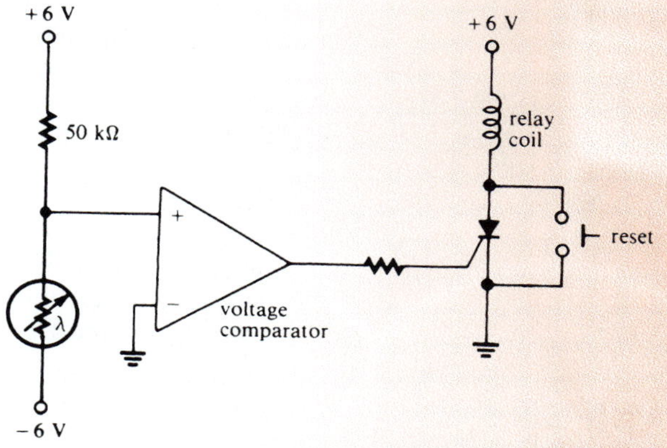

FIGURE 15–21 (Example 15–6)

plot demonstrates that the resistance of a typical cell can change over a very wide range, from about 100 kΩ to a few hundred ohms. The large resistance of the cell when it is not illuminated is called its *dark resistance*.

Photoconductive cells are used in light meters, lighting controls, automatic door openers, and intrusion detectors. In the latter applications, the change in resistance that occurs when incident light is interrupted causes a circuit in which the cell is connected to energize a switch or relay that in turn generates an audible or visual signal. The next example demonstrates such an application.

EXAMPLE 15–6

When the photoconductive cell in Figure 15–21 is illuminated by a light beam, it has a resistance of 20 kΩ. The dark resistance is 100 kΩ. Show that the relay is de-energized when the cell is illuminated and energized when the light beam is interrupted.

Solution

When the cell is illuminated and has resistance 20 kΩ, we find V^+ using superposition:

$$V^+ = \left[\frac{50\text{ k}\Omega}{(20\text{ k}\Omega) + (50\text{ k}\Omega)}\right](-6\text{ V}) + \left[\frac{20\text{ k}\Omega}{(20\text{ k}\Omega) + (50\text{ k}\Omega)}\right](+6\text{ V})$$
$$= -2.57\text{ V}$$

Because V^+ is negative, the output of the comparator is negative and the SCR is off. No current flows in the relay coil, so it is de-energized.

When the light beam is interrupted, the cell has a dark resistance of 100 kΩ, and

$$V^+ = \left[\frac{50\ \text{k}\Omega}{(100\ \text{k}\Omega) + (50\ \text{k}\Omega)}\right](-6\ \text{V}) + \left[\frac{100\ \text{k}\Omega}{(100\ \text{k}\Omega) + (50\ \text{k}\Omega)}\right](+6\ \text{V})$$
$$= +2\ \text{V}$$

The output of the comparator is therefore positive, the SCR turns on, and the relay is energized. To de-energize the relay, the reset pushbutton is depressed, thus shorting the SCR and dropping its current below the holding current.

Note that a latching device such as an SCR is required in this application if the circuit is to respond to *momentary* interruptions of the light beam. Also, cell response time may be critical in detecting short-duration interruptions. The circuit can be made more sensitive by increasing the values of the positive and negative voltages applied to the comparator, but *photodiodes*, discussed in the next section, have much faster response times and are therefore more appropriate for detecting very short-duration interruptions of light.

Photodiodes

A photodiode is a *pn* junction constructed so that it can be exposed to light. When reverse biased, it behaves as a *photoconductive* device because its resistance changes with light intensity. In this case, the change in resistance manifests itself as a change in reverse leakage current. Recall that reverse leakage current in a conventional diode is due to thermally generated minority carriers that are swept through the depletion region by the barrier voltage. In the photodiode, additional minority carriers are generated by light energy, so the greater the light intensity, the greater the reverse current and the smaller the effective resistance.

Figure 15–22 shows a family of characteristic curves for a typical photodiode. For a fixed value of reverse bias (along a vertical line), the magnitude

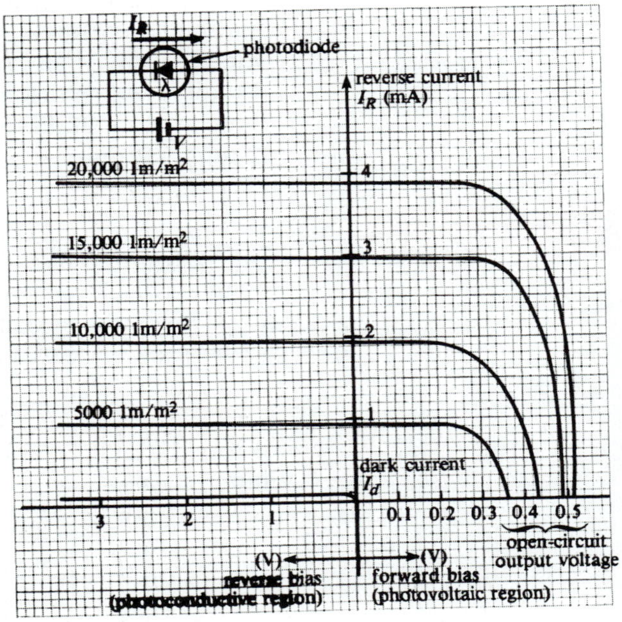

FIGURE 15–22 Characteristic curves of a photodiode (note that the horizontal scale differs along the forward and reverse axes)

of the reverse current increases with increasing light intensity. Along a line of constant light intensity, there is relatively little change in reverse current with increasing reverse voltage.

EXAMPLE 15–7

The photodiode whose characteristics are shown in Figure 15–22 has a reverse bias of 2 V. Find its resistance when the light intensity is 5000 lm/m^2 and when it is 20,000 lm/m^2.

Solution

Constructing a vertical line through the 2-V reverse-bias voltage in Figure 15–22, we find that it intersects the 5000-lm/m^2 characteristic at approximately $I_R = 1$ mA. It intersects the 20,000-lm/m^2 characteristic at approximately $I_R = 3.9$ mA. Thus,

$$R\,(\text{at } 5000\text{ lm/m}^2) = \frac{V_R}{I_R} = \frac{2\text{ V}}{1\text{ mA}} = 2\text{ k}\Omega$$

$$R\,(\text{at } 20{,}000\text{ lm/m}^2) = \frac{V_R}{I_R} = \frac{2\text{ V}}{3.9\text{ mA}} = 512.8\ \Omega$$

These results confirm that the resistance of the photodiode decreases with increasing light intensity.

In the preceding example, it is worth noting that a fourfold increase in light intensity resulted in a very nearly fourfold decrease in resistance, indicating that the photodiode has good *linearity.* This property is apparent in the way the characteristic curves are nearly equally spaced, implying that the change in reverse current is proportionate to the change in light intensity. But, from an applications standpoint, perhaps the most desirable feature of a photodiode is its very fast response time, on the order of nanoseconds. The ability to change resistance very quickly when there is a sudden change in light level makes the photodiode useful in high-speed digital applications, such as optical readers.

The *dark current* in the photodiode is the small reverse current that flows due to thermally generated carriers alone. Using the notation I_d for dark current, its equation is the same as that presented in Chapter 2 for the ideal diode under reverse bias*:

$$I_d = I_s(1 - e^{-V_R/\eta V_T}) \qquad \textbf{(15–15)}$$

Letting I_p represent the reverse current produced due to carriers generated by light energy, the total reverse current is the sum of the thermal and light-generated components:

$$I_R = I_d + I_p \qquad \textbf{(15–16)}$$

When a small forward-biasing voltage is connected across the photodiode, it is found that reverse current continues to flow. This phenomenon can be seen in the forward-bias characteristics of Figure 15–22 and is attributable to the *excess carriers* produced by the light energy. If the forward bias is made sufficiently large, the reverse current must eventually become 0. As can be seen in the forward characteristics, the greater the light intensity, the greater

*See equation 2–13. Because the reverse current is negative, we multiply both sides of (2–13) by −1 so we can represent I_d as a positive quantity in (15–15).

the value of forward bias required to reduce the current to 0. The implication of this result is that light energy creates an internal voltage across the terminals of the photodiode, positive on the anode side and negative on the cathode side, because an external voltage of equal magnitude and opposite polarity is necessary to stop the flow of current. When the terminals are left *open* and the photodiode is exposed to light, that internal voltage appears across the terminals and can be measured. The value increases with increasing light intensity and is identified as the *open-circuit output voltage* in Figure 15–22. When operated in this manner, as a voltage generator, the photodiode is said to be a *photovoltaic* device, rather than a photoconductive device.

When the photodiode terminals are open-circuited, $I_R = 0$ in equation 15–16. Also, letting V_{oc} be the open-circuit output voltage, we substitute $-V_R = V_{oc}$ in equation 15–15 and obtain

$$0 = I_p + I_d$$

$$0 = I_p + I_s(1 - e^{V_{oc}/\eta V_T}) \qquad \textbf{(15–17)}$$

Solving for V_{oc} gives

$$V_{oc} = \eta V_T \ln\left(1 + \frac{I_p}{I_s}\right) \qquad \textbf{(15–18)}$$

Unless the diode is operated at very low light levels, the saturation current, I_s, is much smaller than I_p. Therefore, under at least moderate light levels, $(1 + I_p/I_s) \approx I_p/I_s$ and (15–18) becomes

$$V_{oc} = \eta V_T \ln(I_p/I_s) \qquad \textbf{(15–19)}$$

Equation 15–19 shows that the open-circuit output voltage is a logarithmic function of I_p and therefore a logarithmic function of light intensity.

EXAMPLE 15–8

A photodiode operated in the photovoltaic mode under moderate light intensity has an open-circuit output voltage of 0.4 V. Assuming that the component of reverse current produced by light energy is directly proportionate to light intensity, find the output voltage when the intensity is doubled. Assume that $\eta = 1$ and $V_T = 0.026$ V.

Solution

From equation 15–19, $V_{oc} = 0.4 = (1)(0.026)\ln(I_p/I_s)$. By the linearity assumption, doubling the light intensity doubles I_p, so we must find V_{oc} from

$$V_{oc} = 0.026 \ln(2I_p/I_s) = 0.026 \ln 2 + 0.026 \ln(I_p/I_s)$$

$$= 0.018 + 0.4 = 0.418 \text{ V}$$

Phototransistors

Like a photodiode, a phototransistor has a reverse-biased *pn* junction that is exposed to light. In this case, the junction is the collector–base junction of a bipolar transistor. Figure 15–23(a) shows the currents that flow in an *npn* transistor when the base is open. Recall from Chapter 4 that reverse leakage current I_{CBO} flows across the reverse-biased collector–base junction due to thermally generated minority carriers. Also recall that the collector current in this situation is

$$I_C = I_{CEO} = \frac{I_{CBO}}{1 - \alpha} = (\beta + 1)I_{CBO} \qquad \textbf{(15–20)}$$

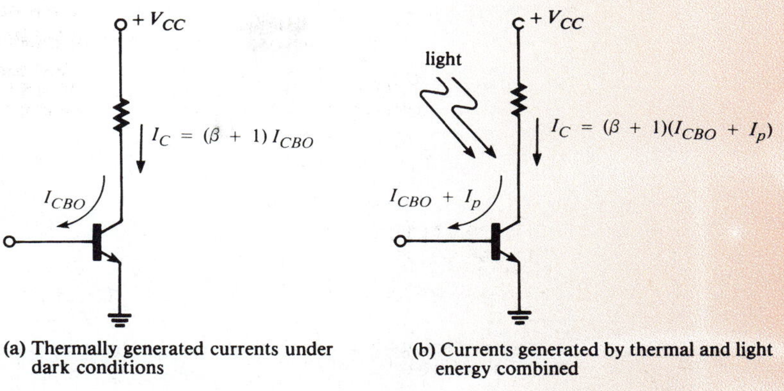

FIGURE 15–23 Currents in a phototransistor

When the junction is exposed to light, an additional component of reverse current, I_p, is generated due to minority carriers produced by light energy, as in the photodiode. Thus, the total collector current is

$$I_C = (\beta + 1)\,(I_{CBO} + I_p) \tag{15–20a}$$

The thermally generated collector current $(\beta + 1)I_{CBO}$ is the dark current. In most applications, it is much smaller than the collector current created by light energy, so

$$I_c \approx (\beta + 1)\, I_p \approx \beta I_p \tag{15–20b}$$

Equation 15–20b shows that the collector current in a phototransistor is directly proportionate to I_p, which, as in the photodiode, is proportionate to light intensity. The advantage of the phototransistor is that it provides current gain and is therefore more *sensitive* to light than the photodiode. It is used in many of the same kinds of applications as photodiodes, but it has a slower response time, on the order of microseconds, compared to the nanosecond response of a photodiode. The phototransistor is constructed with a lens that focuses incident light on the collector–base junction. Some are constructed with no externally accessible base terminal, and others have a base connection that can be used for external bias purposes.

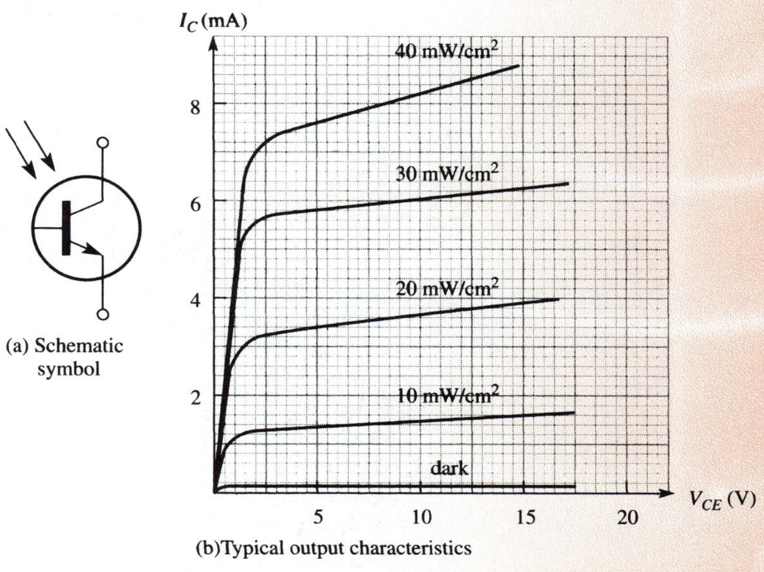

FIGURE 15–24 Phototransistor symbol and characteristics

FIGURE 15–25 (Example 15–9)

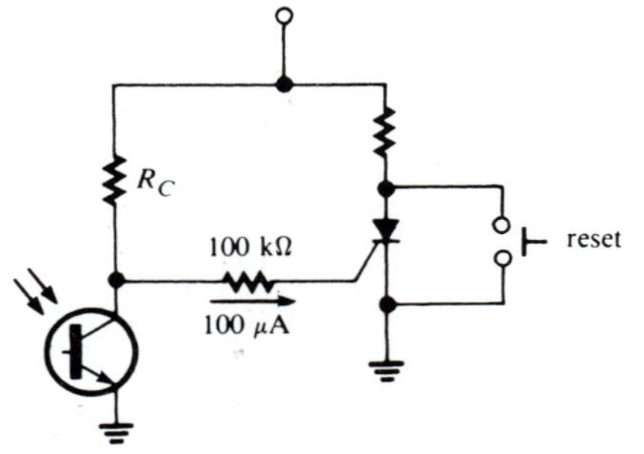

(a) When the light intensity falls, the collector voltage of the phototransistor rises and supplies the 100-μA gate current necessary to fire the SCR

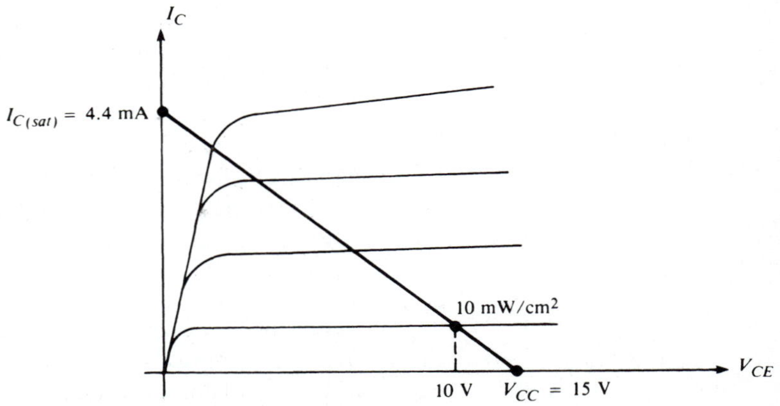

(b) The value required for R_C is found by constructing a load line on the output characteristics

Figure 15–24 shows the schematic symbol and a typical set of output (collector) characteristics for a phototransistor. Notice the similarity of these characteristics to those of a conventional transistor. Light intensity in milliwatts per cubic centimeter serves as the control parameter instead of base current.

EXAMPLE 15–9

The phototransistor whose characteristics are shown in Figure 15–24 is to be used in the detector circuit shown in Figure 15–25(a). When the light intensity falls below a certain level, the collector voltage rises far enough to supply the 100 μA of gate current that is necessary to turn on the SCR. Find the value of R_C that should be used if the SCR must fire when the light intensity falls to 10 mW/cm^2.

Solution

To fire the SCR, the collector voltage must rise to $V_{CE} = (100\ \mu\text{A})(100\ \text{k}\Omega) = 10\ \text{V}$. Therefore, the dc load line for the phototransistor must intersect the 10-mW/cm^2 characteristic at $V_{CE} = 10$ V. As shown in Figure 15–25(b), the load line is drawn between that point and the point where it must intersect the V_{CE}-axis at $V_{CC} = 15$ V. (Use the output characteristic in Figure 15–24 to construct an accurate plot of the load line.) The load line is then extended to its intersection with the I_C-axis, where $I_{C(sat)} \approx 4.4$ mA. Thus,

$$R_C = \frac{V_{CC}}{I_{C(sat)}} = \frac{15\ \text{V}}{4.4\ \text{mA}} = 3410\ \Omega$$

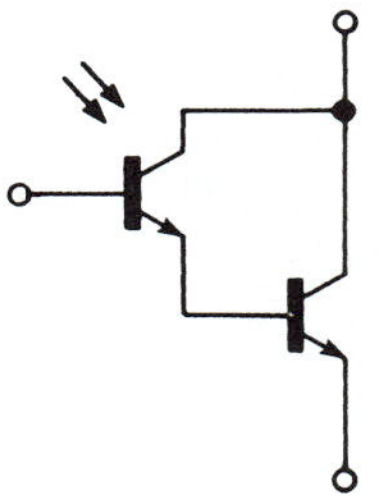

FIGURE 15–26 The photodarlington

A *photodarlington* is a phototransistor packaged with another transistor connected in a Darlington configuration, as studied in Chapter 7. Figure 15–26 shows the configuration. With its large current gain, the photodarlington can produce greater output current than either the photodiode or phototransistor, and it is therefore a more light-sensitive device. However, the photodarlington has a slower response time than either of the other two devices.

Figure 15–27 shows typical manufacturer's specifications for photodiodes, phototransistors, and photodarlingtons. Compare the response times and sensitivities for these devices. Note, for example, that the response times of the photodiodes are all 1 ns, compared to rise and fall times on the order of 2 μs for the phototransistors and 15 μs for the photodarlingtons. On the other hand, the light currents range from only 2 μA at 5 mW/cm^2 for a photodiode to 25 mA at 0.5 mW/cm^2 for a photodarlington. Note also the very small dark currents of each, in comparison to the values of light-generated current.

Solar Cells

A solar cell is simply a large photodiode operated in its photovoltaic mode. In its most common application, it is used to convert light energy to electrical energy, and it thereby serves as a source of dc power. Unlike the photodiode, the solar energy converter is usually constructed in the form of a circular wafer with a very thin layer of *p* material at its surface. Light penetrates the thin *p* layer to the *pn* junction, where excess minority carriers are generated, as in the photodiode. Silicon is the semiconductor material most commonly used in solar-cell construction. Figure 15–28 shows the solar-cell symbol and its connection to a fixed load resistance, R_L. Remember that the positive load current I_L shown here is actually the reverse current through the *pn* junction.

The characteristic curves of a solar cell have the same appearance as photodiode characteristics in the photovoltaic region, shown in Figure 15–22. With reference to I_L and V_L in Figure 15–28, these curves are simply plots of the photodiode equation

$$I_L = I_p + I_s(1 - e^{V_L/\eta V_T}) \qquad \textbf{(15–21)}$$

for different values of I_p, i.e., for different values of light intensity. A typical set of curves for a 3-in.-diameter cell, with axes relabeled as I_L and V_L, is shown in Figure 15–29. A nominal value for the solar intensity at noon on the earth's surface on a clear day is 100 mW/cm^2, which is often referred to as *one sun* in solar-power technology.

The intersection of each curve with the I_L-axis in Figure 15–29 is the *short-circuit current,* I_{sc}, corresponding to a given light intensity. Of course, zero power is delivered to the load when the load is a short. The intersection of each curve with the V_L-axis in Figure 15–29 is the *open-circuit voltage,* V_{oc}, corresponding to a given light intensity. Again, zero power is delivered when the load is open. On each curve of constant light intensity, there is a point where the power, $P_L = V_L I_L$, is maximum. This point is on the knee of the curve, and maximum load power is delivered when the value of R_L is such that the line

$$V_L = I_L R_L \qquad \textbf{(15–22)}$$

intersects that point of maximum power. In Figure 15–29 it can be seen that a load resistance of 0.67 Ω results in maximum power when the light intensity

Silicon Photodetectors

A variety of silicon photodetectors are available, varying from simple PN diodes to complex, single chip 400 volt triac drivers. They are available in packages offering choices of viewing angle and size in either economical plastic cases or rugged, hermetic metal cans. They are spectrally matched for use with Motorola infrared emitting diodes.

Photodiodes

Package	Type Number	Light Current µA Typ	@ V_R Volts	H mW/cm²	$V_{(BR)R}$ Volts Min	Dark Current nA Max	@ Volts	Response Time ns Typ
Case 209-02 Metal Convex Lens (Actual Size)	MRD500	9.0	20	5.0	100	2.0	20	1.0
Case 210-01 Metal Flat Lens (Actual Size)	MRD510	2.0	20	5.0	100	2.0	20	1.0
Case 349-01 Plastic, Style 1 (Actual Size)	MRD721	4.0	20	5.0	100	10	20	1.0

Phototransistors

Package	Type Number	Light Current mA Typ	@ V_{CC}	H mW/cm²	$V_{(BR)CEO}$ Volts Min	t_r, t_f/t_{on}, t_{off}* Typ µs	@ V_{CC}	I_L µA
Case 173-01 Plastic, Style 1 (Actual Size)	MRD150	2.2	20	5.0	40	2.5 40	20	1000
Case 82-05 Metal, Style 1 (Actual Size)	MRD310	3.5	20	5.0	50	2.0 2.5	20	1000
	MRD300	8.0	20	5.0	50	2.0 2.5	20	1000
	MRD3050	0.1 Min	20	5.0	[illegible]	2.0 2.5	20	1000
	MRD3051	0.2 Min	20	5.0	30	2.0 2.5	20	1000
	MRD3054	0.5 Min	20	5.0	30	2.0 2.5	20	1000
	MRD3055	1.5 Min	20	5.0	30	2.0 2.5	20	1000
	MRD3056	2.0 Min	20	5.0	30	2.0 2.5	20	1000
Case 81A-06 Metal, Style 1* / Case 81A-07 Metal, Style 1 (Actual Size)	MRD601, 611*	1.5	5.0	20	50	1.5 15	30	800
	MRD602, 612*	3.5	5.0	20	50	1.5 15	30	800
	MRD603, 613*	6.0	5.0	20	50	1.5 15	30	800
	MRD604, 614*	8.5	5.0	20	50	1.5 15	30	800
Case 349-01 Plastic, Style 2 (Actual Size)	MRD701	0.5	5.0	0.5	30	10 60*	5.0*	

Photodarlingtons

Package	Type Number	Light Current mA Typ	@ V_{CC}	H mW/cm²	$V_{(BR)CEO}$ Volts Min	t_r, t_f/t_{on}, t_{off}* Typ µs	@ V_{CC}	I_L µA
Case 82-05 Metal, Style 1 (Actual Size)	MRD370	10	5.0	0.5	40	15 40	5.0	1.0
	MRD360	20	5.0	0.5	40	15 65	5.0	1.0
Case 349-01 Plastic, Style 2 (Actual Size)	MRD711(1)	25	5.0	0.5	60	125 150*	5.0*	

(1)Photodarlington with integral base-emitter resistor

FIGURE 15–27 Typical specifications for light-sensitive devices (Courtesy of Motorola, Inc.)

FIGURE 15–28 Solar cell with load connected

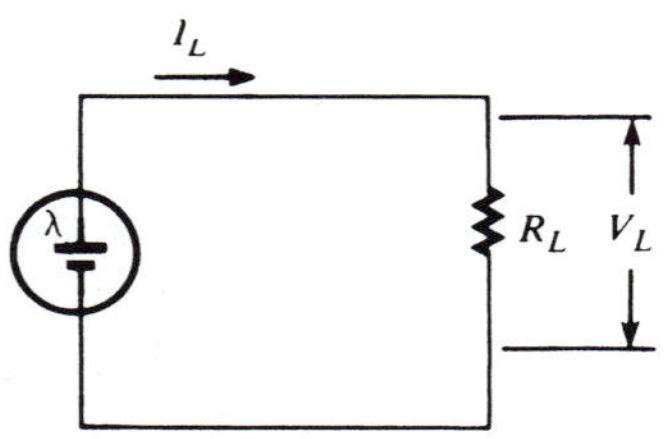

FIGURE 15–29 Typical characteristics of a 3-inch solar cell

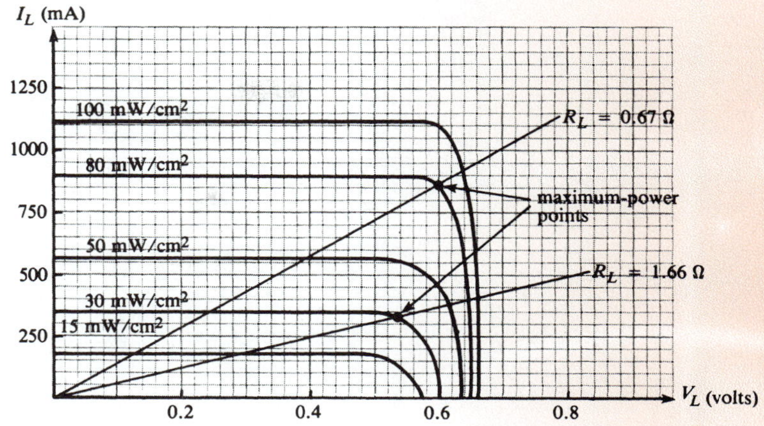

is 80 mW/cm^2. However, a load resistance of 1.66 Ω is required to achieve maximum power at 15 mW/cm^2.

In solar-power applications, solar cells are connected in series/parallel arrays. Parallel connections increase the total current that can be supplied by the array, and series connections increase the voltage.

EXAMPLE 15–10

An array of solar cells is to be constructed so it will deliver 4 A at 6 V under a certain light intensity. If each cell delivers 0.8 A at 0.6 V under the given intensity, how many cells are required and how should they be connected?

Solution

To obtain a 6-V output from cells that produce 0.6 V each, we require 6/0.6 = 10 series-connected cells.

The current in a series string of cells is the same as the current in any one cell: 0.8 A. To obtain 4 A, we require 4/0.8 = 5 parallel paths, each of which contributes 0.8 A.

Thus, the total number of cells required is 10 × 5 = 50. These are arranged in 5 parallel branches, each consisting of 10 series-connected cells, as shown in Figure 15–30.

Figure 15–31 shows specifications for a commercially available array of 36 series-connected solar cells, the Solavolt MSVM4010. Notice the dependence of the characteristics on temperature as well as solar intensity ("irradiance").

Light-Activated SCR (LASCR)

A light-activated SCR (LASCR) is constructed like a conventional SCR except that the reverse-biased *np* junction in the center of the structure is exposed to light (see Figure 15–1). Light energy causes reverse current to flow across the junction in the same manner as it does in a photo-diode. Recall that the SCR will regeneratively switch to its on state if the leakage current becomes sufficiently large. Thus, the LASCR can be triggered on if a sufficiently intense light falls on the exposed junction. Since the leakage current also increases with temperature, the LASCR turns on at lower light levels when temperature increases. Most LASCRs have an accessible gate terminal so they can be controlled conventionally by an external circuit.

FIGURE 15–30 (Example 15–10)

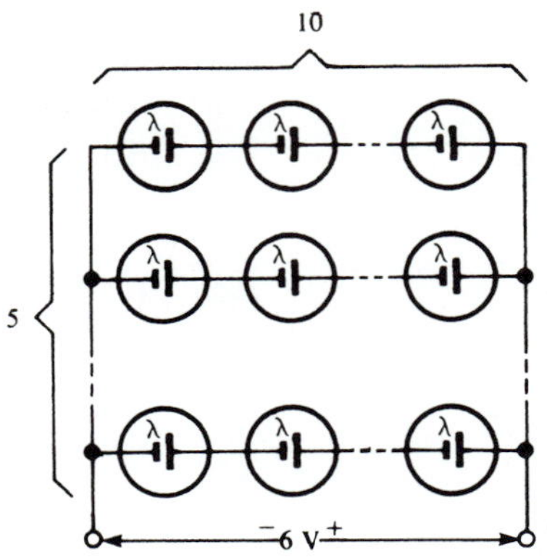

ELECTRICAL SPECIFICATIONS — all tests conducted with an air mass (AM) of 1.5

Specification	Symbol	TEST CONDITIONS: Cell temp (°C) 25	Irradiance (mW/cm²) 100	Cell temp (°C) 48(NOCT)*	Irradiance (mW/cm²) 100	Cell temp (°C) 48(NOCT)*	Irradiance (mW/cm²) 80	Unit
Maximum Power Point Voltage	Vmp	16.5 ± .75		14.7 ± .75		14.7 ± .75		Vdc
Maximum Power Point Current	Imp	2.45 ± .25		2.50 ± .25		2.00 ± .20		Adc
Maximum Power (Vmp x Imp)	Pmp (typ)	40.4		36.7		29.4		W
Open Circuit Voltage	Voc (typ)	20.6		18.8		18.8		Vdc
Short Circuit Current	Isc (typ)	2.74		2.80		2.24		Adc

*Normal Operating Cell Temperature (NOCT) = 48°C (118°F). This is cell temperature with 20°C air temperature, 80 mW/cm² irradiance, 1 m/s wind velocity and module tilted at 45° to the horizontal.

FIGURE 2 — I-V CHARACTERISTIC AS A FUNCTION OF IRRADIANCE AT T_C=25°C

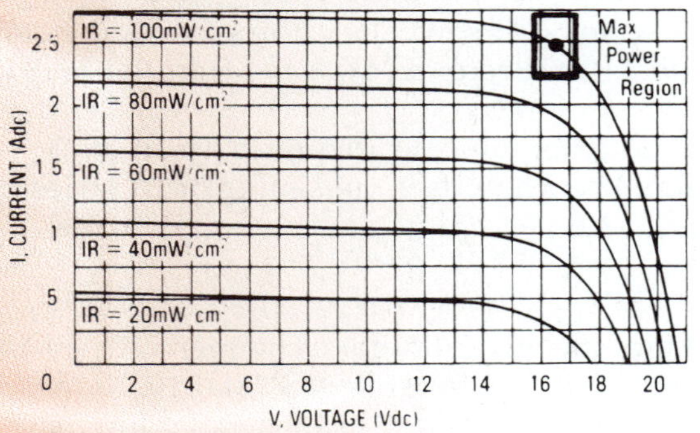

FIGURE 3 — I-V CHARACTERISTIC AS A FUNCTION OF TEMPERATURE

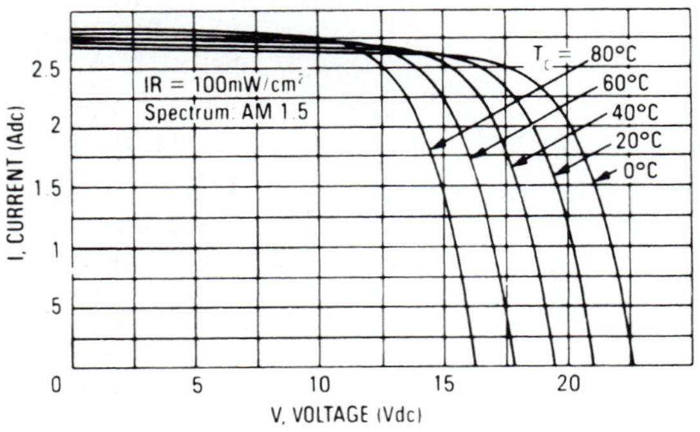

FIGURE 15–31 Specifications for the Solavolt model MSVM4010 solar cell module (Courtesy of Solavolt International)

Figure 15–32 shows the schematic symbol for a LASCR and a typical application. When switch S_1 is closed, the lamp illuminates the LASCR and triggers it on, thus allowing current to flow in the load. Momentary push-button switch S_2 is used to reset the LASCR by reducing its current below the holding current. The principal advantage of this arrangement is the complete electrical isolation it provides between the control circuit and the load.

Light-Emitting Diodes (LEDs)

When current flows through a forward-biased *pn* junction, free electrons cross from the *n* side and recombine with holes on the *p* side. Free electrons

FIGURE 15–32 The schematic symbol and an application for the light-activated SCR (LASCR)

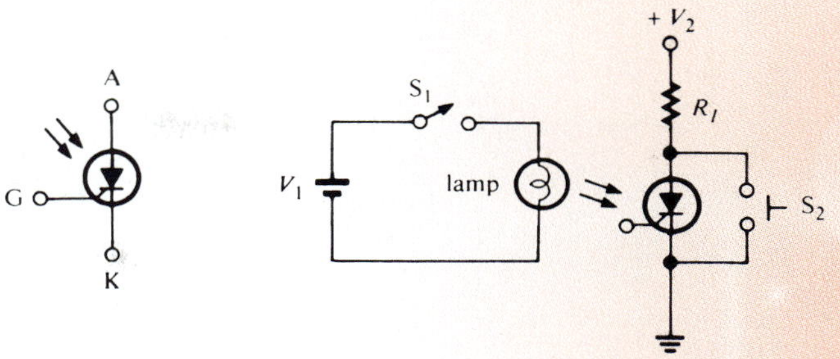

are in the *conduction* band and therefore have greater energy than holes, which are in the *valence* band. When an electron in the conduction band recombines with a hole in the valence band, it releases energy as it falls into that lower energy state. The energy is released in the form of heat and light. In some materials, such as silicon, most of the energy is converted to heat, while in others it is in the form of light. If the material is translucent and if the light energy released is visible, then a *pn* junction having those properties is called a *light-emitting diode* (LED). This conversion of electrical energy to light energy is an example of the phenomenon called *electroluminescence*.

The difference in energy between the conduction band and the valence band is called the *energy gap*, and its value, which depends on the material, governs the wavelength of the emitted light:

$$\lambda = \frac{1.24 \times 10^{-6}}{\text{E.G.}} \tag{15–23}$$

where E.G. is the energy gap in electron volts (eV). In silicon, E.G. ≈ 1.1 eV, and

$$\lambda = \frac{1.24 \times 10^{-6}}{1.1} = 1.13 \times 10^{-6}\ \text{m}$$

This wavelength is in the infrared region of the light spectrum. Because infrared is not visible and because most energy is released as heat, silicon is not used in the fabrication of light-emitting diodes. For the same reasons, germanium is not used. Instead, *gallium arsenide* (GaAs), *gallium phosphide* (GaP), and *gallium arsenide phosphide* (GaAsP) are commonly used. The color of the light emitted by these materials can be controlled by the type and degree of doping. Red, green, and yellow LEDs are commonly available.

Figure 15–33 shows typical manufacturer's specifications for a line of GaP LEDs. Note that the forward voltage drop across an LED is considerably greater than that across a silicon or germanium diode. The drop is typically between 2 and 3 V, depending on forward current. Note also that the maximum permissible reverse voltage is relatively small, 5 V in this case.

EXAMPLE 15–11

Figure 15–34 shows an LED *driver* circuit designed to illuminate the LED when the transistor is turned on (saturated) by a 5-V positive input at its base. Find the values of R_B and R_C that should be used if the LED is to be illuminated by 20 mA of forward current. Assume that the forward drop across the LED is 2.5 V and that the silicon transistor has $\beta = 50$.

Solution

When the transistor is saturated,

$$I_C = 20\ \text{mA} = \frac{V_{CC} - V_D}{R_C} = \frac{(5\ \text{V}) - (2.5\ \text{V})}{R_C} = \frac{2.5\ \text{V}}{R_C}$$

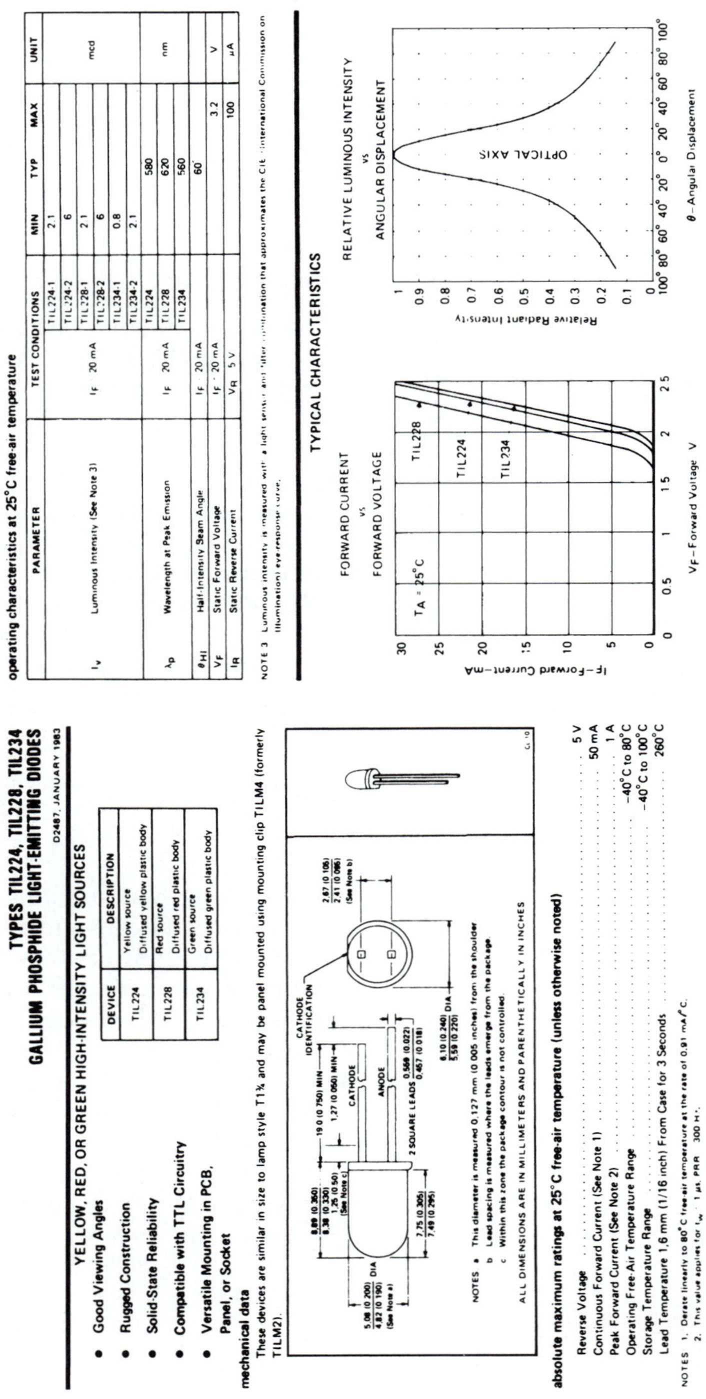

TYPES TIL224, TIL228, TIL234
GALLIUM PHOSPHIDE LIGHT-EMITTING DIODES

D2487, JANUARY 1983

YELLOW, RED, OR GREEN HIGH-INTENSITY LIGHT SOURCES

- **Good Viewing Angles**
- **Rugged Construction**
- **Solid-State Reliability**
- **Compatible with TTL Circuitry**
- **Versatile Mounting in PCB, Panel, or Socket**

DEVICE	DESCRIPTION
TIL224	Yellow source Diffused yellow plastic body
TIL228	Red source Diffused red plastic body
TIL234	Green source Diffused green plastic body

mechanical data

These devices are similar in size to lamp style T1¾ and may be panel mounted using mounting clip TILM4 (formerly TILM2).

NOTES a. This diameter is measured 0,127 mm (0.005 inches) from the shoulder.
b. Lead spacing is measured where the leads emerge from the package.
c. Within this zone the package contour is not controlled.

ALL DIMENSIONS ARE IN MILLIMETERS AND PARENTHETICALLY IN INCHES

absolute maximum ratings at 25°C free-air temperature (unless otherwise noted)

Reverse Voltage	5 V
Continuous Forward Current (See Note 1)	50 mA
Peak Forward Current (See Note 2)	1 A
Operating Free-Air Temperature Range	−40°C to 80°C
Storage Temperature Range	−40°C to 100°C
Lead Temperature 1,6 mm (1/16 inch) From Case for 3 Seconds	260°C

NOTES 1. Derate linearly to 80°C free-air temperature at the rate of 0.91 mA/°C.
2. This value applies for t_w ≤ 1 μs, PRR ≤ 300 Hz.

operating characteristics at 25°C free-air temperature

	PARAMETER	TEST CONDITIONS		MIN	TYP	MAX	UNIT
I_v	Luminous Intensity (See Note 3)	I_F = 20 mA	TIL224-1	2.1			mcd
			TIL224-2	6			
			TIL228-1	2.1			
			TIL228-2	6			
			TIL234-1	0.8			
			TIL234-2	2.1			
λ_p	Wavelength at Peak Emission	I_F = 20 mA	TIL224		580		nm
			TIL228		620		
			TIL234		560		
θ_{HI}	Half-Intensity Beam Angle	I_F = 20 mA			60°		
V_F	Static Forward Voltage	I_F = 20 mA				3.2	V
I_R	Static Reverse Current	V_R = 5 V				100	μA

NOTE 3. Luminous intensity is measured with a light sensor and filter combination that approximates the CIE (International Commission on Illumination) eye-response curve.

TYPICAL CHARACTERISTICS

FIGURE 15–33 Light-emitting diode (LED) specifications (Courtesy of Texas Instruments, Inc.)

FIGURE 15–34 (Example 13–11)

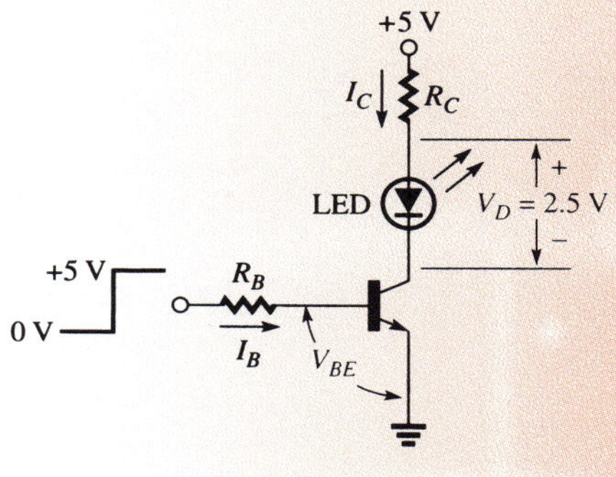

$$R_C = \frac{2.5\text{ V}}{20\text{ mA}} = 125\ \Omega$$

$$I_B = I_C/\beta = 20\text{ mA}/50 = 0.4\text{ mA}$$

$$R_B = \frac{5 - V_{BE}}{I_B} = \frac{(5 - 0.7)\text{V}}{0.4\text{ mA}} = 10.75\text{ k}\Omega$$

LEDs are used in visual displays of all kinds. The principal disadvantage of the LED is that it draws considerable current in comparison to the types of low-power circuits with which it is typically used. For that reason, LEDs are no longer widely used in such low-power devices as calculators and watches. In some applications, power is conserved by pulsing the LEDs on and off at a rapid rate, rather than supplying them with a steady drive current. The LEDs appear to be continuously illuminated because of the eye's *persistence*, that is, its maintenance of the perception of light during the short intervals during which the LEDs are off.

A very popular use for LEDs is in the construction of *seven-segment displays*, such as those illustrated in the specification sheet shown in Figure 15–35. Notice that each of the seven segments is an LED identified by a letter from A through G. By illuminating selected segments, any numeral from 0 through 9 can be displayed. For example, the numeral 3 is displayed by illuminating only those LEDs corresponding to segments A, B, C, D, and G. The seven-segment display must be driven from logic circuitry (called a *decoder*) that supplies current to the correct combination of segments, depending on the numeral to be displayed. In some displays, the decoder circuitry is built-in. In Figure 15–35, note that both *common-anode* (TIL312) and *common-cathode* (TIL313) configurations are available. In the TIL312, a positive voltage is connected to the common anode, so selected LEDs are illuminated by making their respective cathodes low (0 V). In the TIL313, the common cathode is held low and LEDs are illuminated by making their respective anodes high. Other LED configurations are available for *hexadecimal* displays (0 through 9 and A through F) and for *alphanumeric* displays (numerals and all alphabetic characters).

Optocouplers

An optocoupler combines a light-emitting device with a light-sensitive device in one package. One of the simplest examples is an LED packaged with a phototransistor. The LED is illuminated by an input circuit, and the phototransistor, responding to the light, drives an output circuit. Thus, the input and output circuits are *coupled* by light energy alone. The principal advantage of this arrangement is the excellent electrical isolation it provides between the input and output. In fact, these devices are often called *optoisolators*. Other examples of optocouplers include LEDs packaged with photodarlingtons and LASCRs.

TYPES TIL312 THRU TIL315, TIL327, TIL328, TIL333 THRU TIL335, TIL339 THRU TIL341 NUMERIC DISPLAYS

D1924, SEPTEMBER 1981 - REVISED DECEMBER 1982

SOLID-STATE DISPLAYS WITH RED, GREEN, OR YELLOW CHARACTERS

- 7.62-mm (0.300-inch) Character Height
- Continuous Uniform Segments
- Wide Viewing Angle
- High Contrast
- Yellow and Green Displays are Categorized for Uniformity of Luminous Intensity and Wavelength among Units within Each Category

	SEVEN SEGMENTS WITH RIGHT AND LEFT DECIMALS, COMMON ANODE	SEVEN SEGMENTS WITH RIGHT DECIMAL, COMMON CATHODE	PULSE/MINUS ONE WITH LEFT DECIMAL
RED	TIL312	TIL313	TIL327
GREEN	TIL314	TIL315	TIL328
RED+	TIL333	TIL334	TIL335
YELLOW	TIL339	TIL340	TIL341

Red+ stands for high efficiency red

mechanical data

NOTES a The true position spacing (T P) between centerlines is 2.54 mm (0.100 inch) Each pin centerline is within 0.26 mm (0.010 inch) of its true longitudinal position relative to pin 1

b All dimensions associated with segments and decimal points are nominal

c Left decimal point of TIL313 TIL315 TIL334 and TIL340 is not operational

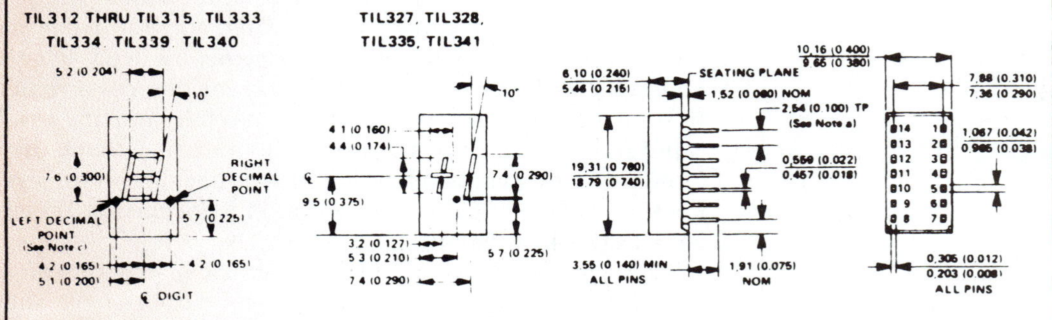

ALL DIMENSIONS ARE IN MILLIMETERS AND PARENTHETICALLY IN INCHES

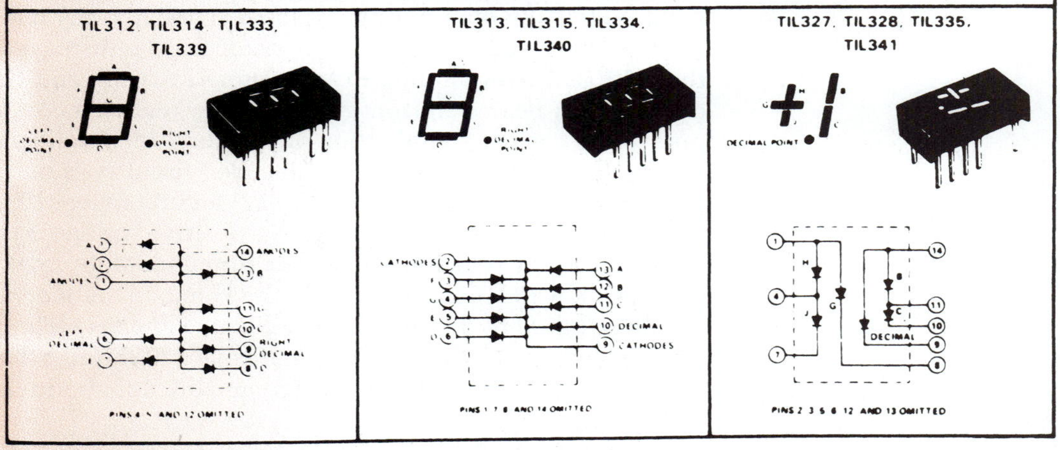

FIGURE 15–35 Seven-segment display specifications (Courtesy of Texas Instruments, Inc.)

Figure 15–36 shows the circuit symbol and two methods used to construct an optocoupler consisting of an LED and a phototransistor. In each structure, the LED is actually a GaAs *infrared* emitter producing light with wavelengths in the vicinity of 0.9×10^{-6} m. In one device, the coupling medium is a special "infrared glass," while the other uses an air gap for better electrical isolation.

An important specification for an optocoupler is its *current-transfer ratio:* the ratio of output (phototransistor) current to input (LED) current, often

FIGURE 15–36 Optocoupler symbol and two construction methods (Courtesy of Texas Instruments, Inc.)

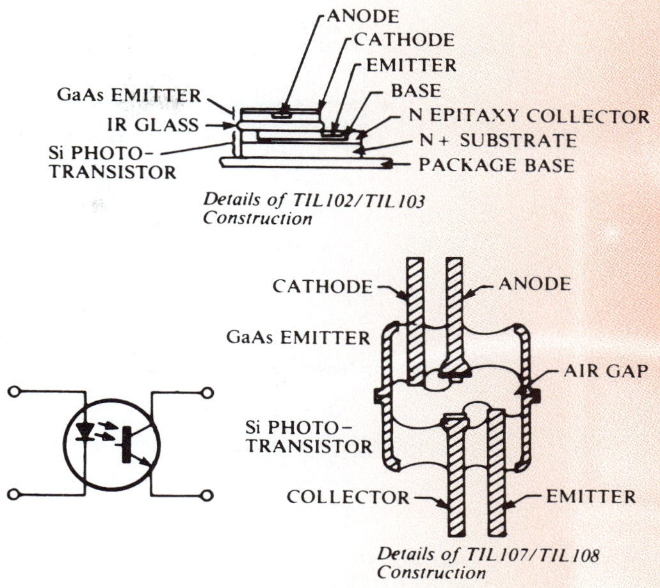

expressed as a percentage. The ratio varies with the value of LED current and can range from 0.1 or less to several hundred in photodarlington devices. Figure 15–37 shows a set of specification sheets for the TIL111, -114, -116, and -117 series of optocouplers. Note the high voltage isolation these devices provide: up to 2500 V. Note also that the base terminal of the phototransistor is externally accessible, so the coupler can be operated in a photodiode mode by taking the output between the base and collector terminals, as shown in Test Circuit B. The specifications for switching characteristics show that the response time is improved by a factor of 5 when in photodiode operation.

EXAMPLE 15–12

Find the current-transfer ratio (CTR) of the TIL117 optocoupler when the diode current is 20 mA and the collector-to-emitter voltage of the phototransistor is 10 V. Assume that $I_B = 0$.

Solution

Referring to the log-log plot of "collector current versus input-diode forward current," shown in the "typical characteristics" section of the specifications in Figure 15–37, we see that $I_C = 20$ mA when $I_F = 20$ mA (from the TIL117 characteristic). Note that the characteristic applies to the case when $V_{CE} =$ 10 V and $I_B = 0$. Thus,

$$\text{CTR}(\%) = \frac{I_C}{I_F} \times 100\% = \frac{20\text{ mA}}{20\text{ mA}} \times 100\% = 100\%$$

Liquid-Crystal Displays (LCDs)

Liquid crystals have been called "the fourth state of matter" (after solids, liquids, and gases) because they have certain crystal properties normally found in solids, yet flow like liquids. In recent years, they have found wide application in the construction of low-power visual displays, such as those found in watches and calculators. Unlike LEDs and other electroluminescent devices, LCDs do not generate light energy, but simply alter or control existing light to make selected areas appear bright or dark.

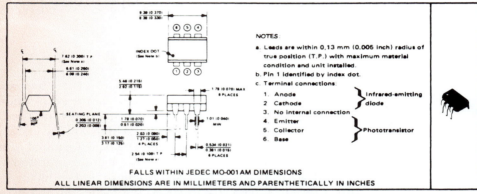

TYPES TIL111, TIL114, TIL116, TIL117
OPTOCOUPLERS

D1607, NOVEMBER 1973—REVISED FEBRUARY 1983

COMPATIBLE WITH STANDARD TTL INTEGRATED CIRCUITS

- **Gallium Arsenide Diode Infrared Source Optically Coupled to a Silicon N-P-N Phototransistor**
- **High Direct-Current Transfer Ratio**
- **High-Voltage Electrical Isolation . . . 1.5-kV or 2.5-kV Rating**
- **Plastic Dual-In-Line Package**
- **High-Speed Switching: t_r = 5 μs, t_f = 5 μs Typical**

mechanical data

The package consists of a gallium arsenide infrared-emitting diode and an n-p-n silicon phototransistor mounted on a 6-lead frame encapsulated within an electrically nonconductive plastic compound. The case will withstand soldering temperature with no deformation and device performance characteristics remain stable when operated in high-humidity conditions. Unit weight is approximately 0.52 grams.

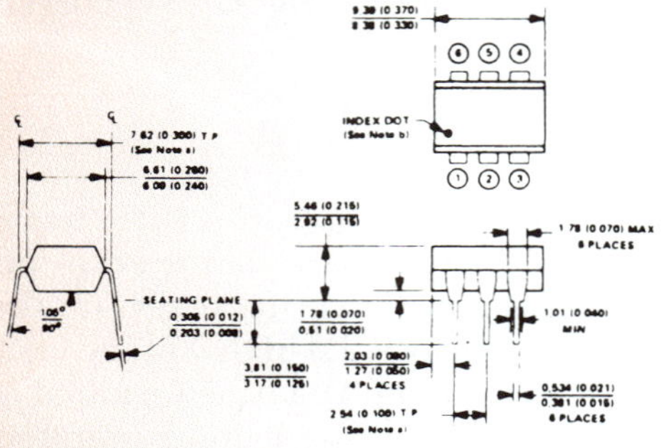

NOTES:

a. Leads are within 0.13 mm (0.005 inch) radius of true position (T.P.) with maximum material condition and unit installed.

b. Pin 1 identified by index dot.

c. Terminal connections:

1. Anode } Infrared-emitting diode
2. Cathode
3. No internal connection
4. Emitter } Phototransistor
5. Collector
6. Base

FALLS WITHIN JEDEC MO-001AM DIMENSIONS

ALL LINEAR DIMENSIONS ARE IN MILLIMETERS AND PARENTHETICALLY IN INCHES

7 OPTOCOUPLERS

absolute maximum ratings at 25°C free-air temperature (unless otherwise noted)

Input-to-Output Voltage: TIL111	±1.5 kV
TIL114, TIL116, TIL117	±2.5 kV
Collector-Base Voltage	70 V
Collector-Emitter Voltage (See Note 1)	30 V
Emitter-Collector Voltage	7 V
Emitter-Base Voltage	7 V
Input-Diode Reverse Voltage	3 V
Input-Diode Continuous Forward Current at (or below) 25°C Free-Air Temperature (See Note 2)	100 mA
Continuous Power Dissipation at (or below) 25°C Free-Air Temperature:	
Infrared-Emitting Diode (See Note 3)	150 mW
Phototransistor (See Note 4)	150 mW
Total, Infrared-Emitting Diode plus Phototransistor (See Note 5)	250 mW
Storage Temperature Range	−55°C to 150°C
Lead Temperature 1,6 mm (1/16 Inch) from Case for 10 Seconds	260°C

NOTES 1. This value applies when the base-emitter diode is open-circuited.
2. Derate linearly to 100°C free-air temperature at the rate of 1.33 mA/°C.
3. Derate linearly to 100°C free-air temperature at the rate of 2 mW/°C.
4. Derate linearly to 100°C free-air temperature at the rate of 2 mW/°C.
5. Derate linearly to 100°C free-air temperature at the rate of 3.33 mW/°C.

TEXAS INSTRUMENTS
INCORPORATED
POST OFFICE BOX 225012 • DALLAS, TEXAS 75265

FIGURE 15–37 Optocoupler specifications (Courtesy of Texas Instruments, Inc.)

There are two fundamental ways in which liquid crystals are used to control properties of light and thereby alter its appearance. In the *dynamic-scattering* method, the molecules of the liquid crystal acquire a random orientation by virtue of an externally applied electric potential. As a result, light passing through the material is reflected in many different directions and has a bright, frosty appearance as it emerges. In the *absorption* method,

TYPES TIL111, TIL114, TIL116, TIL117
OPTOCOUPLERS

electrical characteristics at 25°C free-air temperature

PARAMETER			TEST CONDITIONS	TIL111 TIL114			TIL116			TIL117			UNIT
				MIN	TYP	MAX	MIN	TYP	MAX	MIN	TYP	MAX	
$V_{(BR)CBO}$	Collector-Base Breakdown Voltage		$I_C = 10\ \mu A$, $I_E = 0$, $I_F = 0$	70			70			70			V
$V_{(BR)CEO}$	Collector-Emitter Breakdown Voltage		$I_C = 1\ mA$, $I_B = 0$, $I_F = 0$	30			30			30			V
$V_{(BR)EBO}$	Emitter-Base Breakdown Voltage		$I_E = 10\ \mu A$, $I_C = 0$, $I_F = 0$	7			7			7			V
I_R	Input Diode Static Reverse Current		$V_R = 3\ V$			10			10			10	μA
$I_{C(on)}$	On-State Collector Current	Phototransistor Operation	$V_{CE} = 0.4\ V$, $I_F = 16\ mA$, $I_B = 0$	2	7								mA
			$V_{CE} = 10\ V$, $I_F = 10\ mA$, $I_B = 0$				2	5		5	9		mA
		Photodiode Operation	$V_{CB} = 0.4\ V$, $I_F = 16\ mA$, $I_E = 0$	7	20		7	20		7	20		μA
$I_{C(off)}$	Off-State Collector Current	Phototransistor Operation	$V_{CE} = 10\ V$, $I_F = 0$, $I_B = 0$		1	50		1	50		1	50	nA
		Photodiode Operation	$V_{CB} = 10\ V$, $I_F = 0$, $I_E = 0$		0.1	20		0.1	20		0.1	20	nA
h_{FE}	Transistor Static Forward Current Transfer Ratio		$V_{CE} = 5\ V$, $I_C = 10\ mA$, $I_F = 0$	100	300					200	550		
			$V_{CE} = 5\ V$, $I_C = 100\ \mu A$, $I_F = 0$				100	300					
V_F	Input Diode Static Forward Voltage		$I_F = 16\ mA$		1.2	1.4					1.2	1.4	V
			$I_F = 60\ mA$					1.25	1.5				V
$V_{CE(sat)}$	Collector-Emitter Saturation Voltage		$I_C = 2\ mA$, $I_F = 16\ mA$, $I_B = 0$		0.25	0.4							V
			$I_C = 2.2\ mA$, $I_F = 15\ mA$, $I_B = 0$					0.25	0.4				V
			$I_C = 0.5\ mA$, $I_F = 10\ mA$, $I_B = 0$								0.25	0.4	V
r_{IO}	Input-to-Output Internal Resistance		$V_{in\text{-}out} = \pm 1.5\ kV$ for TIL111, $\pm 2.5\ kV$ for all others, See Note 6	10^{11}			10^{11}			10^{11}			Ω
C_{io}	Input-to-Output Capacitance		$V_{in\text{-}out} = 0$, $f = 1\ MHz$, See Note 6		1	1.3		1	1.3		1	1.3	pF

NOTE 6 These parameters are measured between both input diode leads shorted together and all the phototransistor leads shorted together

switching characteristics at 25°C free-air temperature

PARAMETER			TEST CONDITIONS	TIL111 TIL114			TIL116			TIL117			UNIT
				MIN	TYP	MAX	MIN	TYP	MAX	MIN	TYP	MAX	
t_r	Rise Time	Phototransistor Operation	$V_{CC} = 10\ V$, $I_{C(on)} = 2\ mA$, $R_L = 100\ \Omega$, See Test Circuit A of Figure 1		5	10		5	10		5	10	μs
t_f	Fall Time				5	10		5	10		5	10	μs
t_r	Rise Time	Photodiode Operation	$V_{CC} = 10\ V$, $I_{C(on)} = 20\ \mu A$, $R_L = 1\ k\Omega$, See Test Circuit B of Figure 1		1			1			1		μs
t_f	Fall Time				1			1			1		μs

TEXAS INSTRUMENTS
INCORPORATED
POST OFFICE BOX 225012 • DALLAS, TEXAS 75265

FIGURE 15–37 Continued

the molecules are oriented in such a way that they alter the *polarization* of light passing through the material. Polarizing filters are used to absorb or pass the light, depending on the polarization it has been given, so light is visible only in those regions where it can emerge from the filter.

The basic construction of a liquid-crystal display is shown in Figure 15–38. A thin layer of liquid crystal, 10–20 μm thick, is encapsulated

TYPES TIL111, TIL114, TIL116, TIL117
OPTOCOUPLERS

PARAMETER MEASUREMENT INFORMATION

Adjust amplitude of input pulse for
$I_{C(on)}$ = 2 mA (Test Circuit A) or
$I_{C(on)}$ = 20 μA (Test Circuit B)

TEST CIRCUIT A
PHOTOTRANSISTOR OPERATION

VOLTAGE WAVEFORMS

TEST CIRCUIT B
PHOTODIODE OPERATION

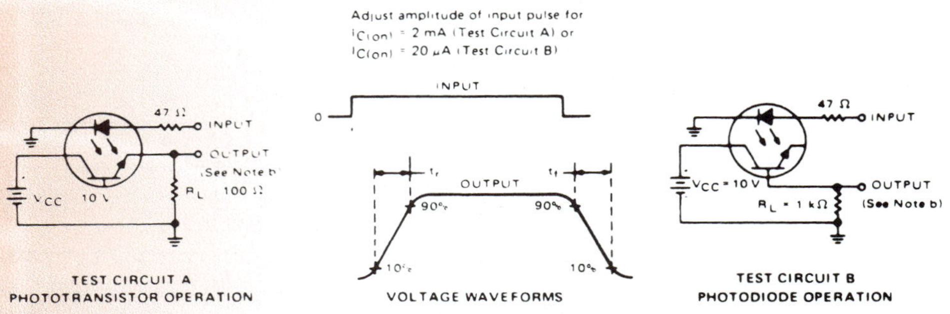

NOTES: a. The input waveform is supplied by a generator with the following characteristics: Z_{out} = 50 Ω, t_r ≤ 15 ns, duty cycle ≈ 1%, t_w = 100 μs.
b. The output waveform is monitored on an oscilloscope with the following characteristics: t_r ≤ 12 ns, R_{in} ≥ 1 MΩ, C_{in} ≤ 20 pF.

FIGURE 1—SWITCHING TIMES

TYPICAL CHARACTERISTICS

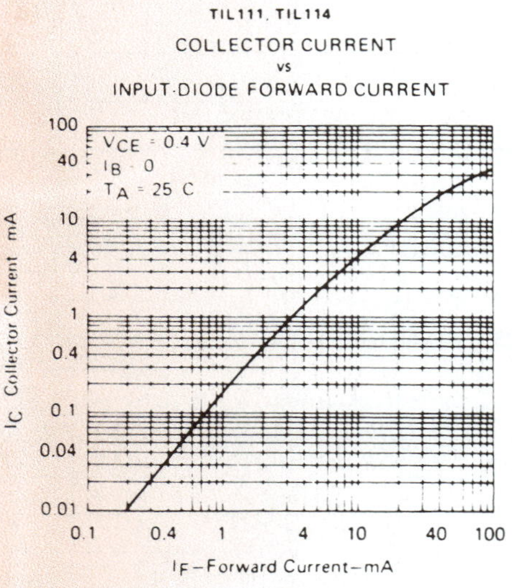

TIL111, TIL114
COLLECTOR CURRENT
vs
INPUT-DIODE FORWARD CURRENT

FIGURE 2

TIL116, TIL117
COLLECTOR CURRENT
vs
INPUT-DIODE FORWARD CURRENT

FIGURE 3

TEXAS INSTRUMENTS
INCORPORATED
POST OFFICE BOX 225012 ● DALLAS, TEXAS 75265

FIGURE 15–37　Continued

between two glass sheets. To enable application of an electric field across the crystal, a transparent conducting material is deposited on the inside surface of each glass sheet. The conducting material forms the *electrodes* to which an external voltage is connected when it is desired to alter the molecular structure of the liquid crystal. The electrodes are etched in patterns or in individually accessible segments that can be selectively

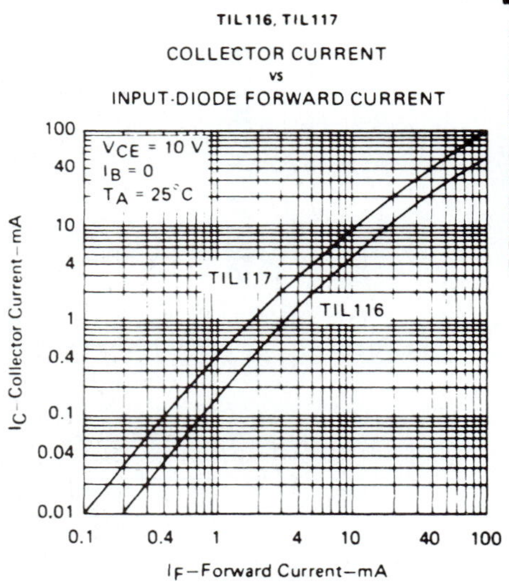

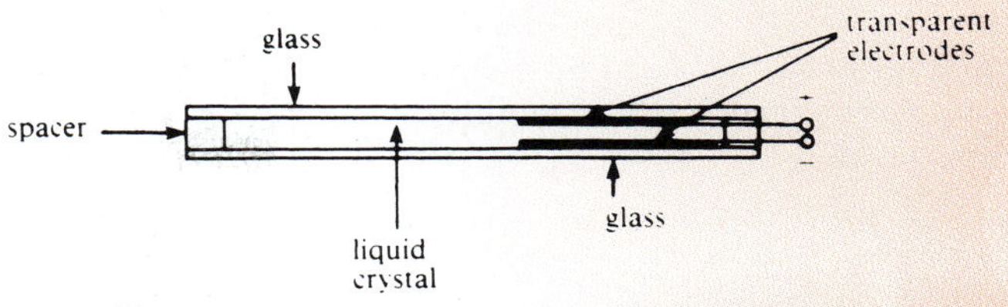

FIGURE 15–38 Basic construction of a liquid-crystal display

energized to create a desired display, similar to a seven-segment LED display. Materials used to form the electrodes include *stannic oxide* (SnO_2) and *indium oxide* (In_2O_3).

Some liquid-crystal displays are designed so that light passes completely through them, while others have a mirrored surface that reflects light back to the viewer. In the first instance, called the *transmissive* mode of operation, light originates from a light source on one side of the crystal and is altered in the desired pattern as it passes through to emerge on the other side. In the *reflective* mode, light enters one side, is altered within the material, and is reflected by a mirror to emerge from the same side it entered. This mode is particularly useful for small displays such as watches, where ambient light is present on one side only.

Figure 15–39 shows a dynamic-scattering LCD operated in the transmissive mode. In the region where no electric field is applied, the rod-shaped molecules have the common alignment shown, and in the region that has been activated by an externally applied field, the molecules have random orientations. The light source is displaced in such a way that the unactivated region appears dark. In the activated region, the molecules cause light to be reflected in many directions (scattered), and it escapes with a bright, diffuse appearance. Thus, the pattern appears bright on a dark background. A special light-control film can be placed on the illuminated glass to permit the use of a diffuse light source.

A dynamic-scattering LCD operated in the reflective mode has essentially the same construction shown in Figure 15–39, except that a mirrored surface replaces or is added behind one of the glass sheets. However, unwanted reflections limit the readability of displays of this type, which must therefore be fitted with special *retroreflectors* in practical applications.

Liquid-crystal displays that depend on light absorption rather than light scattering are constructed using so-called *twisted-nematic* crystals. The electrodes in these displays are coated so that the molecules of the liquid crystal

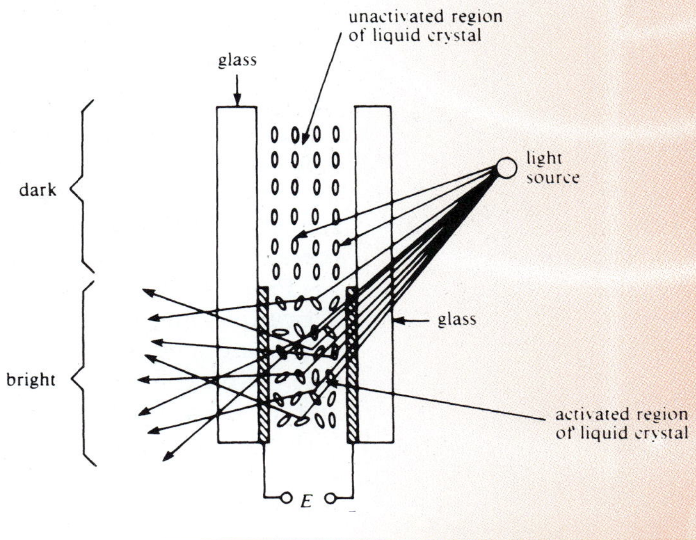

FIGURE 15–39 Dynamic-scattering LCD; transmissive mode

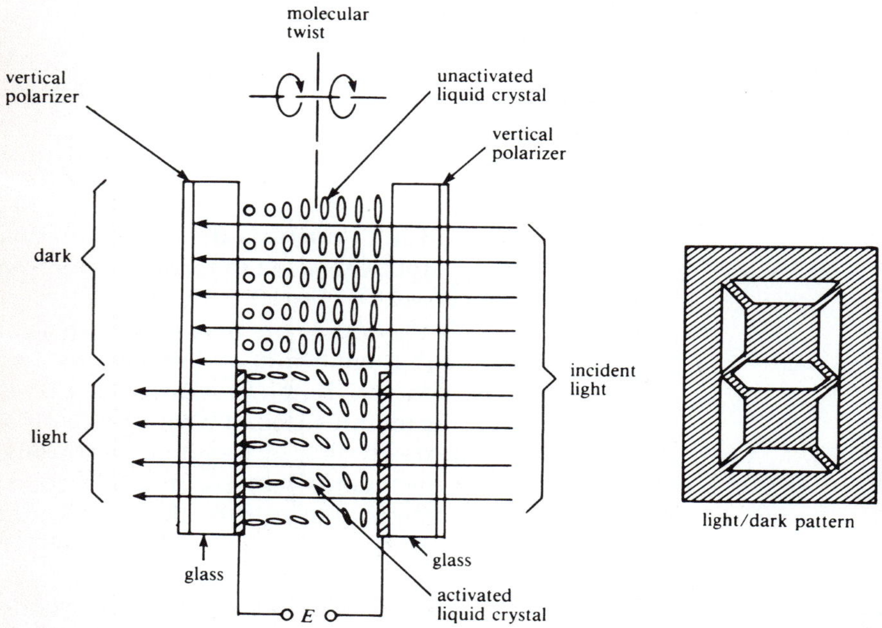

FIGURE 15–40 Twisted-nematic, field-effect LCD based on light absorption; transmissive-mode operation

undergo a 90° rotation from one side of the cell to the other. As a result, light passing through the material undergoes a 90° change in polarization. However, if the crystal is activated by an externally applied electric field, the molecules are reoriented in such a way that the change in polarization does not occur. The absorption-type LCD operates on the principle that horizontally polarized light is not visible through a vertically polarizing filter, i.e., it is absorbed. As shown in Figure 15–40, light enters the (transmissive) LCD through a vertical polarizer and can exit through another vertical polarizer provided the liquid crystal is activated. In the region where the crystal is not activated, the vertically polarized light becomes horizontally polarized and is absorbed in the vertical filter. Thus, the unactivated region appears dark and the activated region appears bright. Since it is the electric field resulting from an externally applied voltage that reorients the molecule, the twisted-nematic display is often called a *field-effect* type.

Figure 15–41 shows a light-absorbing, field-effect LCD operated in the reflective mode. As in the transmissive LCD, incident light is vertically polarized, but in this case a horizontal polarizer is placed between one of the glass sheets and a reflector. The vertically polarized light becomes horizontally polarized when it passes through an unactivated region of the crystal. This light can pass through the horizontal polarizer and is reflected back into the crystal. The unactivated crystal then shifts the polarization back to vertical, so the light emerges through the vertical polarizer where it entered. The activated region of the crystal does not alter the vertical polarization, so light passing through it cannot pass the horizontal polarizer and is not reflected. As a consequence, the pattern created in the activated region appears dark on a light background.

Practical liquid-crystal displays are activated by ac waveforms because dc activation causes a plating effect that shortens their lifetime. The dynamic-scattering type requires a relatively high voltage, 25–30 V at 50–60 Hz. However, LCDs consume very little power in comparison to LED displays. One method used to activate the crystal is to drive each of the opposing electrodes with a square wave. If the square waves are in phase, then the net voltage across the crystal is zero, and it remains unactivated. When the driver circuitry causes the square waves to be 180° out of phase, the net voltage across

FIGURE 15–41 Twisted-nematic, field-effect LCD based on light absorption; reflective-mode operation

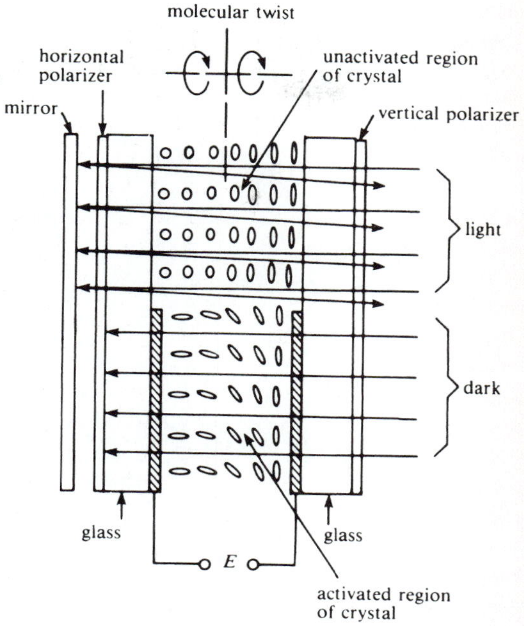

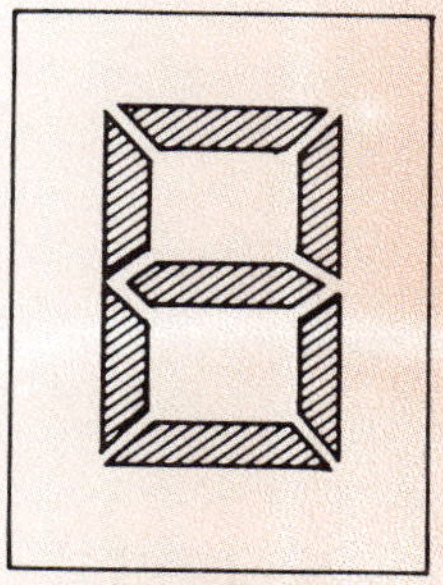

TABLE 15–1 Typical characteristics of liquid-crystal displays

Characteristic	Dynamic-Scattering Display (4 digits)	Twisted-Nematic Display ($4\frac{1}{2}$ digits)
Voltage	15–30 V	3 V
Frequency	50–60 Hz	30–1500 Hz
Operating temperature	−5°C to +65° C	−15°C to +60° C
Total current	60 μ A	6 μA
Total capacitance	750 pF	2 nF
Rise time	25 ms	100 ms
Contrast ratio	20 : 1	50 : 1

the electrodes is twice the peak-to-peak value of either wave, and the crystal is activated.

One disadvantage of a liquid-crystal display is its relatively long response time, i.e., the time required for the display to appear or disappear. LCDs are much slower in that respect than LED displays. However, in practical applications such as watches, the long response time does not impose a limitation on their usefulness. Table 15–1 shows a comparison of the principal characteristics of dynamic-scattering and twisted-nematic LCDs. Note that the twisted-nematic type requires about one-tenth the voltage and current of the dynamic-scattering type and has better contrast, but is about four times slower.

15–3 UNIJUNCTION TRANSISTORS

As shown in Figure 15–42, a unijunction transistor consists of a bar of lightly doped (high-resistivity) *n* material to which a heavily doped, *p*-type rod is attached on one side. Ohmic contacts are made at opposing ends of the *n*-type bar, which are called *base 1* (B_1) and *base 2* (B_2) of the transistor. The *p*-type rod is called the *emitter* and forms a conventional *pn* junction with

FIGURE 15–42 Construction and schematic symbol of the unijunction transistor (UJT)

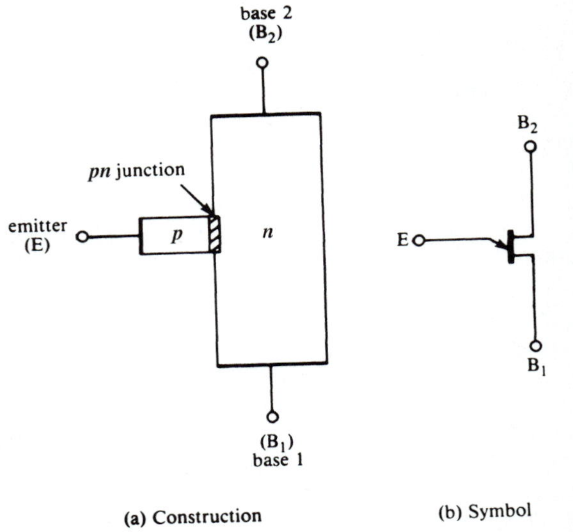

FIGURE 15–43 External circuit connections to a UJT

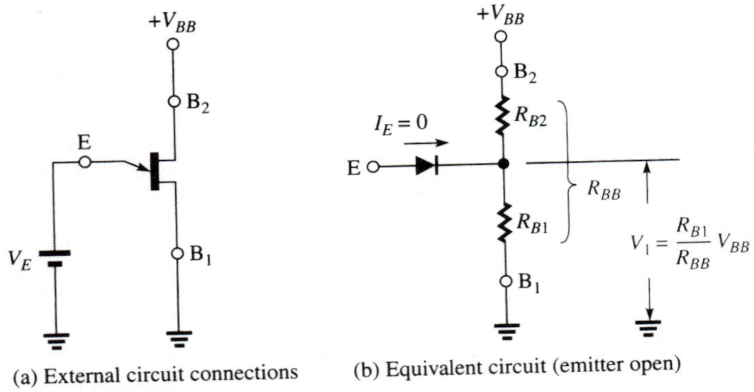

the bar. Because this device has only one such junction, it behaves quite differently from a conventional bipolar transistor. The single junction accounts for the name *uni*junction transistor, and it is also called a *double-base diode.* Notice that the schematic symbol for a unijunction transistor (UJT) has an arrow pointing in the direction of conventional forward current through the junction. UJTs are also manufactured with *n*-type emitters and *p*-type bars.

Figure 15–43 shows external circuit connections to the UJT and its equivalent circuit. Here, the *pn* junction is represented by an equivalent diode. Note that base 1 is the common (ground) for both the emitter input circuit and the supply voltage, V_{BB}. The *interbase resistance,* R_{BB}, is defined to be the total resistance between B_1 and B_2 when the emitter circuit is open ($I_E = 0$). R_{BB} is therefore the sum of the resistances R_{B1} and R_{B2} between each base and the emitter, when $I_E = 0$:

$$R_{BB} = (R_{B1} + R_{B2})\Big|_{I_E = 0} \tag{15–24}$$

By the voltage-divider rule, the voltage V_1 across R_{B1} is (again, for $I_E = 0$)

$$V_1 = \left(\frac{R_{B1}}{R_{B1} + R_{B2}}\right)V_{BB} = \frac{R_{B1}}{R_{BB}}V_{BB} \tag{15–25}$$

The voltage-divider ratio in (15-25) is called the *intrinsic standoff ratio* η:

$$\eta = \frac{R_{B1}}{R_{BB}}\Big|_{I_E = 0} \tag{15–26}$$

FIGURE 15–44 When V_E is increased to $V_P = \eta V_{BB} + V_D$ there is a sudden drop in the value of R_{B1}

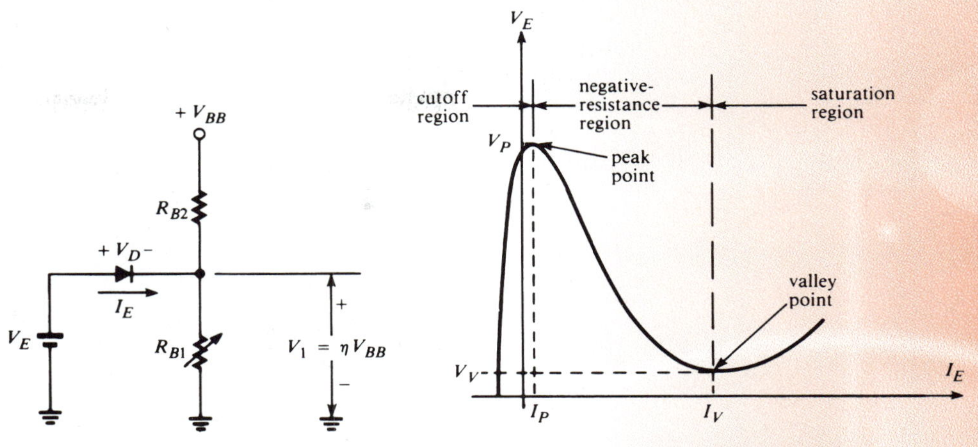

The value of η clearly depends on how close the emitter terminal is to base 2, since R_{B1} becomes larger and R_{B2} becomes smaller as the emitter is moved closer to B_2. η is typically in the range from 0.50 to 0.85, while R_{BB} may range from 4 kΩ to 12 kΩ.

Figure 15–44 shows the equivalent circuit with external emitter voltage V_E connected. Notice that R_{B1} in this case is shown as a variable resistance. If V_E is less than the positive voltage V_1 across R_{B1}, then the diode is reverse biased, and R_{B1} has essentially the same resistance as when $I_E = 0$. However, if V_E is increased to the onset of forward bias, a small amount of forward current I_E begins to flow through the emitter and into the base 1 region, causing the resistance of that region to decrease. The decrease in R_{B1} is attributable to the presence of additional charge carriers resulting from the forward current flow. If the forward bias is increased slightly, there is a sudden and dramatic reduction in R_{B1}. This phenomenon occurs because the increase in current reduces R_{B1}, which further increases the current, which further reduces R_{B1}, and so forth. In other words, a *regenerative* action occurs. The value of emitter voltage at which regeneration is initiated is called the *peak voltage*, V_P. As can be seen in Figure 15–44(a), the value that the emitter voltage must reach is V_1 plus the forward drop V_D across the diode. Since $V_1 = \eta V_{BB}$, we have

$$V_P = \eta V_{BB} + V_D \tag{15–27}$$

Figure 15–44(b) shows a characteristic curve for the unijunction transistor. Once the emitter voltage has reached V_P, emitter current increases even as V_E is made smaller. This fact is conveyed by the negative slope of the characteristic in the region to the right of the peak point (decreasing voltage accompanied by increasing current). The region is therefore appropriately referred to as the *negative-resistance* portion of the characteristic. This region is said to be *unstable* because the UJT cannot actually be operated there. The curve simply shows the combinations of V_E and I_E that the UJT undergoes during the regenerative transition. Beyond the *valley point* shown in the figure, the resistance R_{B1} has reached its minimum possible value, called the *saturation resistance.* In this saturation region, emitter current can be increased further only by again increasing V_E. The characteristic in this region is similar to that of a conventional forward-biased diode. Note that the region to the left of the peak point is called the *cutoff* region.

In applications, the UJT is used like a voltage-controlled switch. When the input voltage is raised to V_P, the UJT "fires," or turns on, allowing a generous flow of current from V_{BB} to ground. To turn the device off, i.e., to return it to the cutoff region, the emitter current must be reduced below the valley

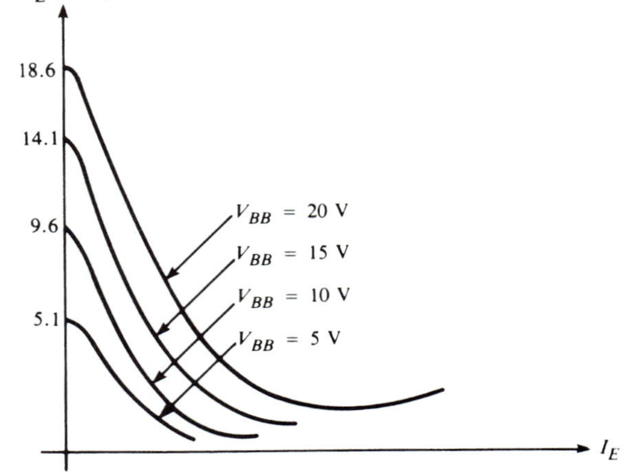

FIGURE 15–45 A typical family of characteristic curves for a UJT

current I_V shown in Figure 15–44(b). It must be remembered that V_P is a function of V_{BB}, as shown by equation 15–27. Thus, if V_{BB} is increased, the emitter voltage must be raised to a higher value to switch the UJT on. Figure 15–45 shows a typical family of characteristic curves, corresponding to different values of V_{BB}. (The cutoff regions are not shown.) The intersection of each characteristic curve with the V_E-axis is (approximately) the value of V_P corresponding to a particular V_{BB}, and it can be seen that V_P increases with increasing V_{BB}.

Figure 15–46(a) shows a common application of a UJT in a *relaxation* oscillator circuit. Capacitor C charges through R_E until the capacitor voltage reaches V_P. At that time, the UJT switches on, and the capacitor discharges through the emitter. The resulting surge of current through external resistor R develops a sharp voltage "spike," as shown in Figure 15–46(b). Once the capacitor has discharged, the UJT switches back to its off state, provided the emitter current drops below I_V. The capacitor then begins to recharge and the cycle is repeated. The resulting capacitor-voltage and output-voltage waveforms are shown in (b). This type of oscillator is used to generate trigger and timing pulses in a wide variety of control, synchronization, and nonsinusoidal oscillator circuits.

The period T of the oscillation in Figure 15–46 is given by

$$T = R_E C \ln\left(\frac{V_{BB} - V_V}{V_{BB} - V_P}\right) \tag{15–28}$$

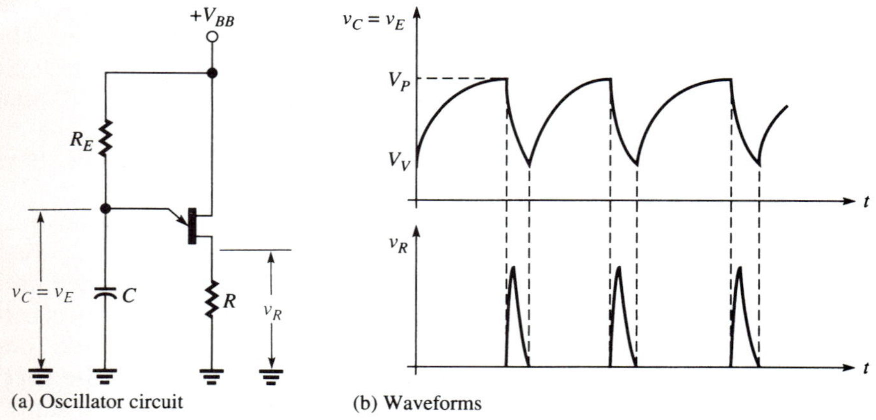

FIGURE 15–46 A UJT relaxation oscillator

The choice of value for R_E in this equation is not entirely arbitrary. R_E must be small enough to permit the emitter current to reach I_P when the capacitor voltage reaches V_P, and yet be large enough to make the emitter current less than I_V when the capacitor discharges to V_V. These constraints are necessary to ensure that the UJT switches on when V_E reaches V_P and to ensure that it switches off when V_E falls to V_V. Thus, R_E must satisfy the following two inequalities:

$$R_E < R_E(\max) = \frac{V_{BB} - V_P}{I_P} \tag{15–29}$$

$$R_E > R_E(\min) = \frac{V_{BB} - V_V}{I_V} \tag{15–30}$$

If $V_V \ll V_{BB}$, then equation 15–28 can be approximated by

$$T \approx R_E C \ln\left(\frac{V_{BB}}{V_{BB} - V_P}\right) \tag{15–31}$$

If $\eta V_{BB} \gg V_D$, then $V_P \approx \eta V_{BB}$, and (15–31) can be further approximated by

$$T \approx R_E C \ln\left(\frac{V_{BB}}{V_{BB} - \eta V_{BB}}\right) \tag{15–32}$$

or

$$T \approx R_E C \ln\left(\frac{1}{1 - \eta}\right) \tag{15–33}$$

Notice that R in Figure 15–46 is in series with R_{B1} and R_{B2} in the UJT. Its value should therefore be small in comparison to R_{BB}, to prevent it from altering the value of V_P.

Figure 15–47 shows typical specifications for a number of different unijunction transistors. Notice the broad range in the value of η that a transistor of a given type may have. The specification for I_{EB20} is the emitter reverse current.

EXAMPLE 15–13

DESIGN

Design a 1-kHz relaxation oscillator using an MU4892 UJT with a 20-V-dc supply. Use the specifications given in Figure 15–47. The MU4892 has V_D = 0.56 V. Assume that $V_V = 2$ V.

Solution

We will use a value of η equal to the average of the minimum and maximum values given in the specifications: $\eta = (0.51 + 0.69)/2 = 0.60$. From equation 16–27, $V_P = \eta V_{BB} + V_D = (0.6)(20) + 0.56 = 12.56$ V. Since inequality 15–29 gives the maximum permissible value of R_E and I_P appears in the denominator, we should use the maximum possible value of I_P in the calculation. Similarly, since (15–30) gives the minimum permissible value of R_E, we should use the minimum specified value for I_V. From the specifications, $I_P(\max) = 2\ \mu$A and $I_V(\min) = 2$ mA. Thus, from (15–29) and (15–30),

$$R_E < \frac{V_{BB} - V_P}{I_P} = \frac{(20\text{ V}) - (12.56\text{ V})}{2\ \mu\text{A}} = 3.72\text{ M}\Omega$$

$$R_E > \frac{V_{BB} - V_P}{I_V} = \frac{(20\text{ V}) - (2\text{ V})}{2\text{ mA}} = 9\text{ k}\Omega$$

Thyristors – Trigger Devices

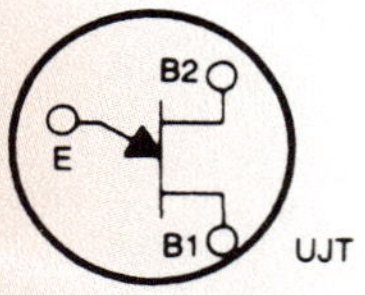

Unijunction Transistors — UJT

Highly stable devices for general-purpose trigger applications and as pulse generators (oscillators) and timing circuits. Useful at frequencies ranging (generally) from 1.0 Hz to 1.0 MHz. Available in low-cost plastic package TO-226AA (TO-92) and in hermetically sealed metal package (Case 22A).

Device Type	η Min	η Max	I_P µA Max	I_{EB20} µA Max	I_V mA Min
Plastic TO-92					
MU10	0.50	0.85	5.0	1.0	1.0
2N4870	0.56	0.75	5.0	1.0	2.0
2N4871	0.70	0.85	5.0	1.0	4.0
MU2646	0.56	0.75	5.0	12	4.0
MU4891	0.55	0.82	5.0	0.01	2.0
MU4892	0.51	0.69	2.0	0.01	2.0
MU4893	0.55	0.82	2.0	0.01	2.0
MU4894	0.74	0.86	1.0	0.01	2.0
Metal TO-18					
MU20	0.50	0.85	5.0	1.0	1.0
2N2646	0.56	0.75	5.0	12	4.0
2N2647	0.68	0.82	2.0	0.2	8.0
2N3980	0.68	0.82	2.0	0.01	1.0
2N4851	0.56	0.75	2.0	0.1	2.0
2N4852	0.70	0.85	2.0	0.1	4.0
2N4853	0.70	0.85	0.4	0.05	6.0
2N4948*	0.55	0.82	2.0	0.01	2.0
2N4949*	0.74	0.86	1.0	0.01	2.0
2N5431*	0.72	0.80	0.4	0.01	2.0

*Also available as JAN and JANTX devices

FIGURE 15–47 Unijunction transistor specifications (Courtesy of Motorola, Inc.)

We see that we have a wide range of choices for R_E. Both inequalities are satisfied if we choose $R_E = 10\ \text{k}\Omega$

Since $T = 1/f = 1/(1\text{ kHz}) = 10^{-3}$ s, we have, from equation 15–28,

$$10^{-3} = 10^4 C \ln\left(\frac{20 - 2}{20 - 12.56}\right)$$

which gives $C = 0.113\ \mu\text{F}$.

A typical value for resistance R in Figure 15–46 is 47 Ω.

Checking the accuracy of approximation 15–33, we find

$$T \approx R_E C \ln\left(\frac{1}{1-\eta}\right) = (10^4)(0.113 \times 10^{-6}) \ln\left(\frac{1}{1-0.6}\right) \approx 1.04 \text{ ms}$$

The approximation gives a value slightly higher than the design value of 1 ms, but, considering the variation that is possible in the parameter values, the approximation is quite good. For accurate frequency control, R_E should be made adjustable.

Unijunction transistors are often used in SCR trigger circuits, a typical example of which is shown in Figure 15–48. When the UJT switches on, a pulse of gate current is supplied to the SCR to turn it on. In this ac power-control circuit, the SCR switches off during every negative half-cycle, so the UJT must retrigger

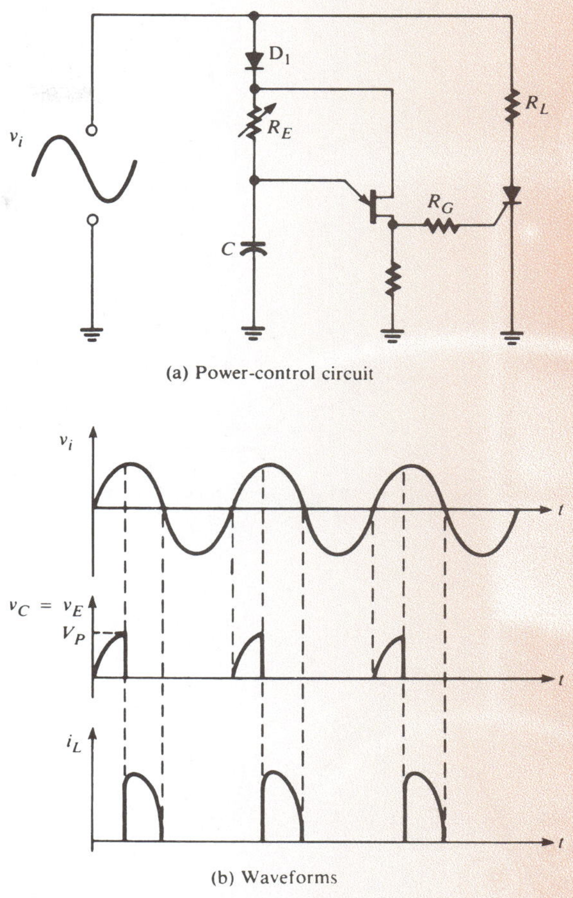

FIGURE 15–48 Using a UJT to trigger an SCR in an ac power-control circuit

it during each positive half-cycle. Adjusting the value of R_E controls the rate at which the capacitor charges and therefore controls the point in time at which the emitter voltage reaches V_P. A small value of R_E allows the capacitor to charge quickly, which makes the UJT and SCR switch on early in the half-cycle and allows load current to flow for a significant portion of the half-cycle. A large value of R_E causes a greater delay in switching and allows load current to flow during a smaller time interval. Diode D_1 isolates the UJT from negative half-cycles.

Programmable UJTs (PUTs)

A programmable UJT (PUT) is not a unijunction transistor at all, but a four-layer device. Its name is derived from the fact that its characteristic curve and many of its applications are similar to those of a UJT. Figure 15–49 shows the construction and schematic symbol of a PUT, which resembles an SCR more than a UJT. Like an SCR, its three terminals are designated anode (A), cathode (K), and gate (G). However, note that the gate is connected to the *n* region below the uppermost *pn* junction, like the anode gate in an SCS (Figure 15–12).

In applications, the gate of the PUT is biased positive with respect to the cathode. If the anode is then made about 0.7 V more positive than the gate, the device regeneratively switches on. Thus, the PUT is "programmed" by its gate-to-cathode bias voltage. Figure 15–50(a) shows a typical bias arrangement, where the gate voltage is obtained from a resistive voltage divider. Note that V_{AK} and V_{BB} must be separate voltage sources, because V_G, which is derived from V_{BB}, must remain constant, while V_{AK} varies. Figure 15–50(b) shows the characteristic curve. Note the resemblance of the characteristic

FIGURE 15–49 Construction and symbol for a programmable UJT (PUT)

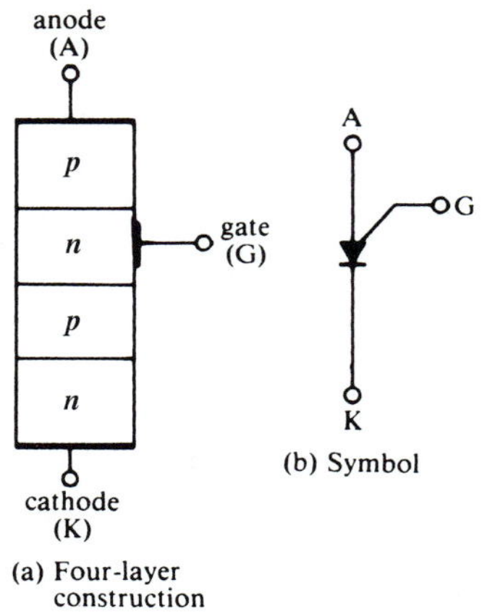

FIGURE 15–50 PUT biasing and characteristic curve

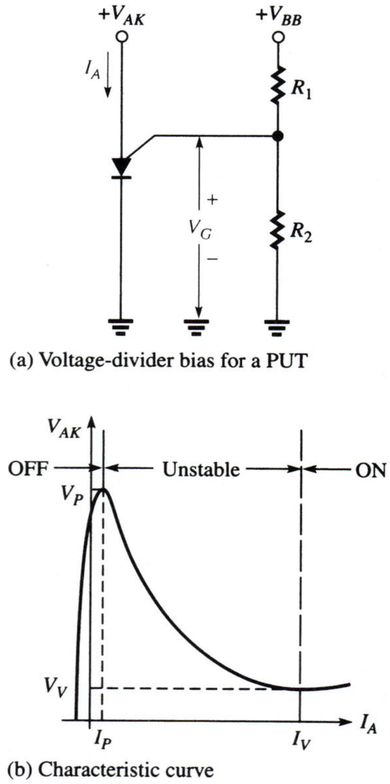

to that of a UJT. The peak and valley points have the same significance as in a UJT. The advantage of a PUT is that it has a smaller on resistance than a UJT and can handle heavier currents.

A notation for PUTs that is consistent with that for UJTs can be developed by defining

$$\eta = \frac{R_2}{R_1 + R_2} \quad \textbf{(15–34)}$$

where R_1 and R_2 are the resistors in the voltage divider shown in Figure 15–50(a). Then it is clear that

$$V_G = \eta V_{BB} \quad \textbf{(15–35)}$$

The peak-point value of V_{AK}, V_P, at which the PUT switches on, can then be written as

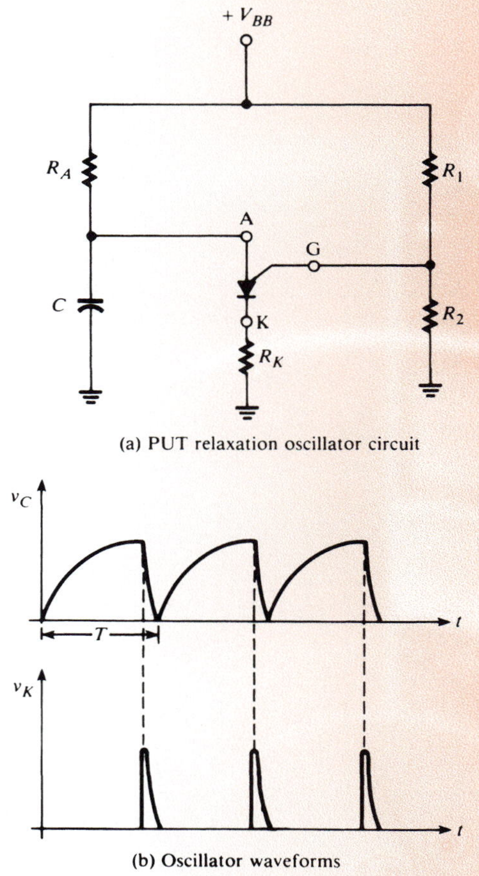

FIGURE 15–51 PUT relaxation oscillator

$$V_P = V_G + V_D = \eta V_{BB} + V_D \tag{15–36}$$

where $V_D \approx 0.7$ V. Equation 15–36 is completely consistent with equation 15–27 for the UJT.

Figure 15–51(a) shows a PUT relaxation oscillator. The capacitor charges through R_A until V_{AK} reaches V_P, at which time the PUT switches on and the capacitor discharges. Note that the anode-to-cathode voltage V_{AK} when the PUT is *off* is approximately the same as the anode-to-ground voltage V_A, which is the same as the capacitor voltage V_C, because there is very little drop across R_K. Thus, the PUT switches on when $V_C \approx V_P$. The surge of current through the PUT when it turns on creates a voltage pulse across R_K, as shown in Figure 15–51(b). When the current through the PUT falls below the holding current (I_V in Figure 15–50), the device switches back to its off state and the cycle repeats itself.

As in the UJT oscillator, R_A must be small enough to permit I_P to flow when V_C equals V_P yet large enough to ensure that the device switches back to its off state when the capacitor discharges. The range of values of R_A that ensure oscillation is found in the same way as for the UJT oscillator:

$$R_A < R_A(\text{max}) = \frac{V_{BB} - V_P}{I_P} \tag{15–37}$$

$$R_A > R_A(\text{min}) = \frac{V_{BB} - V_V}{I_V} \tag{15–38}$$

Also, the period of the oscillation can be determined using the same approximation that applies to the UJT oscillator:

$$T \approx R_A C \ln\left(\frac{1}{1-\eta}\right) \qquad (15\text{–}39)$$

Remember that η is an externally adjusted resistance ratio in this case rather than a device parameter.

15–4 TUNNEL DIODES

A tunnel diode, also called an *Esaki* diode after its inventor, is constructed with very heavily doped *p* and *n* materials, usually germanium or gallium arsenide. The doping level in these highly conductive regions may be from 100 to several thousand times that of a conventional diode. As a consequence, the depletion region at the *pn* junction is extremely narrow. Recall that forward conduction in a conventional diode occurs only if the forward bias is sufficient to give charge carriers the energy necessary to overcome the barrier potential opposing their passage through the depletion region. When a tunnel diode is only slightly forward biased, many carriers are able to cross through the very narrow depletion region without acquiring that energy. These carriers are said to *tunnel* through the depletion region because their energy level is lower than the level of those that overcome the barrier potential.

Because of the tunneling phenomenon, there is a substantial flow of current through a tunnel diode at the onset of forward bias. The same is true when the diode is reverse biased. This behavior is evident in the *I–V* characteristic of a tunnel diode, shown in Figure 15–52. Notice the steep slope (small resistance) of the characteristic in the region around the origin. For comparison purposes, the dashed line shows the characteristic of a conventional diode. With increasing forward bias, the tunneling phenomenon continues, until it reaches a maximum at I_P in the figure. The forward bias at this point is V_P. Because of higher energies acquired by carriers on the *n* side, current due to tunneling begins to diminish rapidly with a further increase in forward bias. As a result, the characteristic displays a

FIGURE 15–52 *I–V* characteristic of a tunnel diode

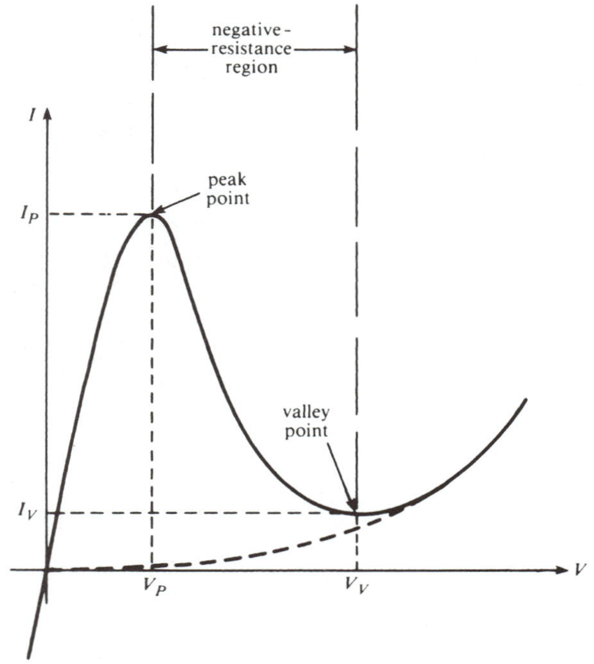

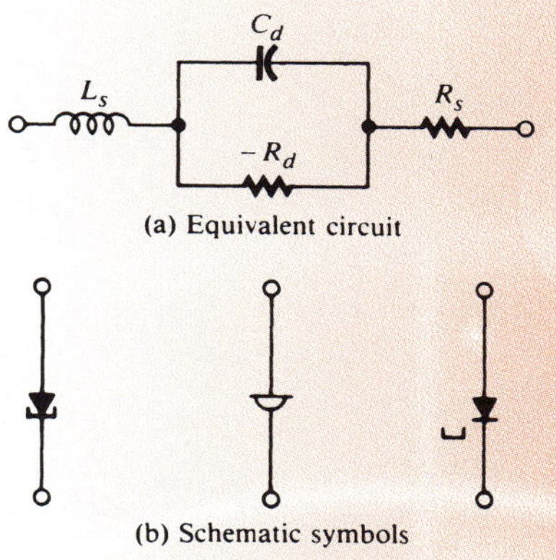

FIGURE 15–53 Equivalent circuit and schematic symbols for a tunnel diode

negative-resistance region, similar to that of a UJT, where increasing voltage is accompanied by decreasing current. When the forward bias becomes large enough to supply carriers with the energy required to surmount the barrier potential, forward current begins to increase again. As shown in the figure, the forward characteristic beyond that point coincides with the characteristic of a conventional diode.

Figure 15–53 shows the equivalent circuit of a tunnel diode in its negative-resistance region and several commonly used schematic symbols. The device is most often used in high-frequency or high-speed switching circuits, so the junction capacitance, C_d, and the inductance of the connecting leads, L_s, are important device parameters. C_d may range from 5 to 100 pF, while L_s is on the order of a few nanohenries. Note the negative resistance labeled $-R_d$, typically $-10\ \Omega$ to $-200\ \Omega$. R_s is the resistance of the leads and semiconductor material, on the order of $1\ \Omega$.

The negative-resistance characteristic of a tunnel diode has fostered some applications that are unusual among two-terminal devices, including oscillators, amplifiers, and high-speed electronic switches. In the latter application, the tunnel diode is switched between its peak and valley points with nano- or picosecond switching times. The peak voltage is typically rather small, less than 200 mV, but the peak current may range from 1 to 100 mA.

15–5 PHASE-LOCKED LOOPS

A phase-locked loop (PLL) is a set of components that, through the use of feedback, generate a signal whose frequency tracks that of another externally connected input signal. The term *phase-locked* is derived from the fact that the (apparent) phase difference between two signals of different frequencies is used to control, or maintain, the frequency of the output, as will be discussed in more detail presently. Entire books have been written on the many applications of PLLs, particularly in signal-processing and communications equipment, but we will discuss just two of the most common: FM demodulation and frequency synthesis.

Figure 15–54 shows a block diagram of a phase-locked loop. Note that the voltage-controlled oscillator (VCO), which was discussed in Chapter 12, produces a signal whose frequency is proportionate to input voltage. The output of the VCO is connected to a *phase comparator*, which generates an output voltage proportionate to the difference in phase between the VCO signal and the externally connected input signal, v_i. When the frequency of the input signal fluctuates, the output of the phase comparator is a fluctuating voltage that is, in effect, an *error* voltage proportionate to the difference in *frequency* between the two inputs to the comparator. The

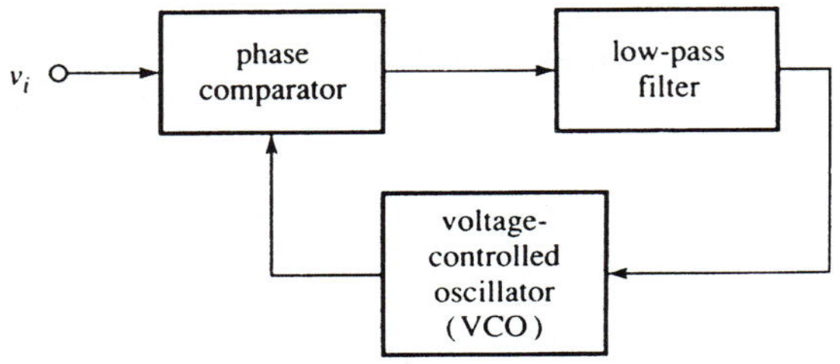

FIGURE 15–54 Block diagram of a phase-locked loop

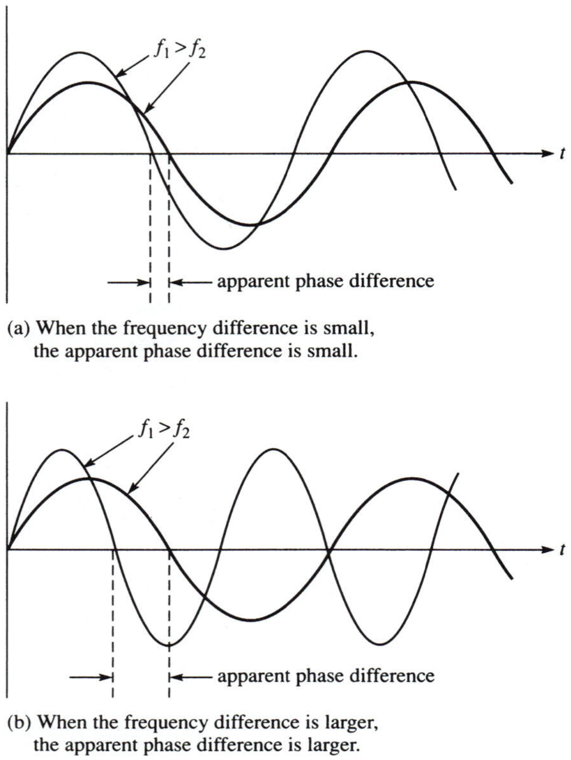

$f_1 > f_2$
t
apparent phase difference
(a) When the frequency difference is small, the apparent phase difference is small.
$f_1 > f_2$
t
apparent phase difference
(b) When the frequency difference is larger, the apparent phase difference is larger.

FIGURE 15–55 Apparent phase difference between signals of different frequencies

fluctuating error voltage is smoothed by the low-pass filter and applied to the input of the VCO. The system is designed so that the error voltage causes the VCO to adjust its frequency to match the frequency of v_i. When the VCO frequency and the frequency of v_i are equal, the loop is "locked" at that frequency.

Strictly speaking, phase difference is not defined for two signals having different frequencies. However, the phase comparator "sees" an apparent phase difference when the frequency of one of its inputs changes, as illustrated in Figure 15–55. We see that a small frequency difference creates a small (apparent) phase difference and a larger frequency difference creates a larger apparent phase difference. By considering how these diagrams would appear if one frequency were very much larger than the other, we can understand why a practical PLL may not achieve lock in such a situation without some design modifications.

FM Demodulation

Recall that frequency modulation (FM) is the process in which the frequency of one signal is controlled by the magnitude of another signal. A VCO generates an FM signal, as illustrated in Figure 15–56. In this example, the input

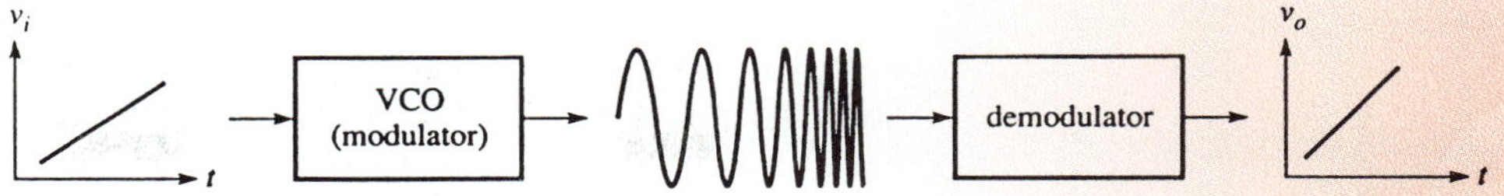

FIGURE 15–56 Frequency modulation and demodulation

to the VCO (the *modulating* signal) is a ramp voltage whose amplitude increases with time. Consequently, the output of the VCO (the *modulated* signal) is a signal whose frequency increases with time. In a communications system, the modulated signal is transmitted to a receiver, where it must be *demodulated*; that is, the original modulating signal must be recovered from the modulated signal. As shown in the example of Figure 15–56, the FM demodulator performs this function. The input to the demodulator is the FM signal and the output is the ramp voltage that created it. An FM demodulator is also called a *frequency-to-voltage converter*.

A phase-locked loop can serve as an FM demodulator by connecting the FM signal to the input and taking the output from the low-pass filter. As the input frequency changes, the output of the phase comparator and filter changes the same way. For example, an increasing input frequency (originally produced as the result of an increasing modulating signal) will create an increasing error voltage because the frequency of the VCO must be increased to track the input. Thus, the error voltage duplicates the original modulating signal.

Frequency Synthesizers

A frequency synthesizer is simply a signal generator whose frequency can be readily adjusted. A phase-locked loop can be used as a frequency synthesizer by inserting a frequency divider (counter) in the feedback path to the phase comparator and taking the output from the VCO. This arrangement is shown in Figure 15–57. Note that a low-pass filter is not required. The counter is a digital-type device that produces a signal having frequency f/N, where N is an integer, when the input to the counter has frequency f. As can be seen in the figure, a signal with fixed reference frequency, f_{REF}, is compared in the phase comparator to the output of the counter. Consider what happens when this arrangement is first connected: The signal fed back to the phase comparator has a frequency that is initially smaller (by the factor $1/N$) than f_{REF}. Consequently, the phase comparator produces an output that causes the VCO to increase its frequency, in the comparator's usual attempt to bring the two frequencies to equality. The VCO increases its frequency until the signal fed back to the comparator has frequency f_{REF} and lock is achieved. But, for the output of the counter to have frequency f_{REF}, the VCO must have reached frequency Nf_{REF}, because the counter divides its input frequency by N. Thus, the output of the phase-locked loop (the output of the VCO) is a signal whose frequency is the multiple Nf_{REF} of the reference frequency.

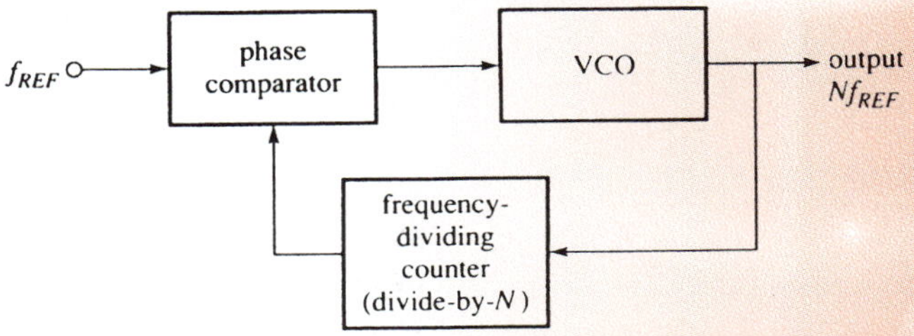

FIGURE 15–57 Block diagram of a PLL used as a frequency multiplier (synthesizer)

One advantage of a frequency synthesizer using a PLL is that the signal providing the reference frequency can be very stable, as, for example, from a crystal-controlled oscillator. The higher frequency, Nf_{REF}, will then also be very stable. In some high-frequency applications, it is not possible to achieve such crystal-controlled stability at high frequencies without using such a scheme. We should note that a frequency-dividing counter can also be connected between the f_{REF} input and the phase comparator. This connection has just the opposite effect of connecting a divider in the feedback: The frequency of the VCO is *divided* by the same factor that the input divider provides.

The 565 Integrated-Circuit PLL

An example of a commercially available, integrated-circuit PLL is the LM565 manufactured by National Semiconductor. The inputs and outputs of the phase comparator and VCO are accessible at separate pins so that external components, such as filters and frequency dividers, can be inserted as required for different applications. Some important specifications of a PLL, and their values for the LM565, are the maximum VCO frequency (500 kHz), the demodulated output voltage (300 mV for a 10% change in input frequency), and the phase comparator sensitivity (0.68 V/radian). A portion of the data sheets for National Semiconductor's LM565 are provided in Figure 15–58.

15–6 SCR CIRCUIT ANALYSIS WITH ELECTRONICS WORKBENCH MULTISIM

This section uses EWB Multisim to examine the basic operation of the SCR. Chapter 15 introduced the basic theory behind the operation of the SCR. The student has learned that once turned on, the SCR will remain on until the hold current is removed. The objective of this EWB exercise is to:

- Use EWB to examine the hold operation for the SCR.
- Develop an understanding of how to apply the use of an SCR to other circuits.

To begin this exercise, open the circuit **Ch15_EWB.msm**, that is found in the Electronics Workbench CD-ROM packaged with the text. This file is the SCR circuit shown in Figure 15–59. The circuit uses a 2N1599 SCR. The gate on the SCR is connected to the 1-kΩ potentiometer R2. R2 is used to control the gate voltage. In Figure 15–59 the gate voltage is 0.45 V, the SCR is off and the voltage across the SCR is approximately 5.0 V. Multimeter XMM1 is being used to measure the current flowing through the gate.

The SCR can be turned on by adjusting potentiometer R2 until the SCT reaches $V_{BR(F)}$, the forward breakdown voltage. Press "A" to increase the resistance value of the potentiometer and thus increase the gate current. The SCR begins to turn on at about 18%, and the SCR fully turns on when the potentiometer is set to 21%. This is shown in Figure 15–60.

Notice that the potentiometer setting is at 21%, the voltage across the SCR now reads 0.576 V, and the current meter reads 2.106 mA. How can the SCR be turned back off? This requires that the hold current be made sufficiently low by adjusting the potentiometer to about 9% or 10%. The next step is to determine how to use the SCR in a clocked operation. The SCR circuit was modified to include a clocking circuit on the gate. The modified circuit is shown in Figure 15–61.

National Semiconductor

May 1999

LM565/LM565C Phase Locked Loop

General Description

The LM565 and LM565C are general purpose phase locked loops containing a stable, highly linear voltage controlled oscillator for low distortion FM demodulation, and a double balanced phase detector with good carrier suppression. The VCO frequency is set with an external resistor and capacitor, and a tuning range of 10:1 can be obtained with the same capacitor. The characteristics of the closed loop system — bandwidth, response speed, capture and pull in range — may be adjusted over a wide range with an external resistor and capacitor. The loop may be broken between the VCO and the phase detector for insertion of a digital frequency divider to obtain frequency multiplication.

The LM565H is specified for operation over the −55˚C to +125˚C military temperature range. The LM565CN is specified for operation over the 0˚C to +70˚C temperature range.

Features

- 200 ppm/˚C frequency stability of the VCO
- Power supply range of ±5 to ±12 volts with 100 ppm/% typical
- 0.2% linearity of demodulated output
- Linear triangle wave with in phase zero crossings available
- TTL and DTL compatible phase detector input and square wave output
- Adjustable hold in range from ±1% to > ±60%

Applications

- Data and tape synchronization
- Modems
- FSK demodulation
- FM demodulation
- Frequency synthesizer
- Tone decoding
- Frequency multiplication and division
- SCA demodulators
- Telemetry receivers
- Signal regeneration
- Coherent demodulators

Connection Diagrams

Metal Can Package

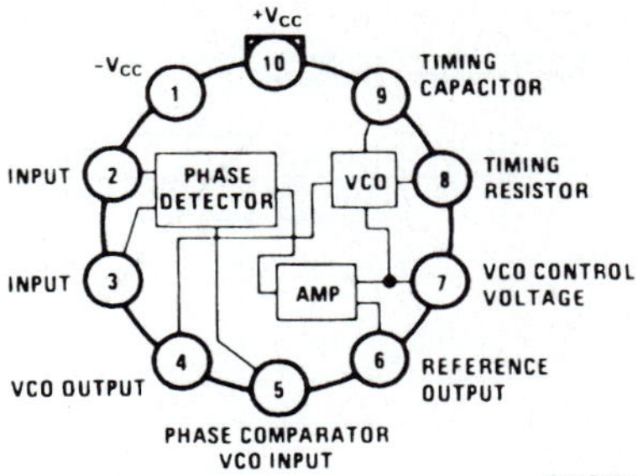

Order Number LM565H
See NS Package Number H10C

Dual-in-Line Package

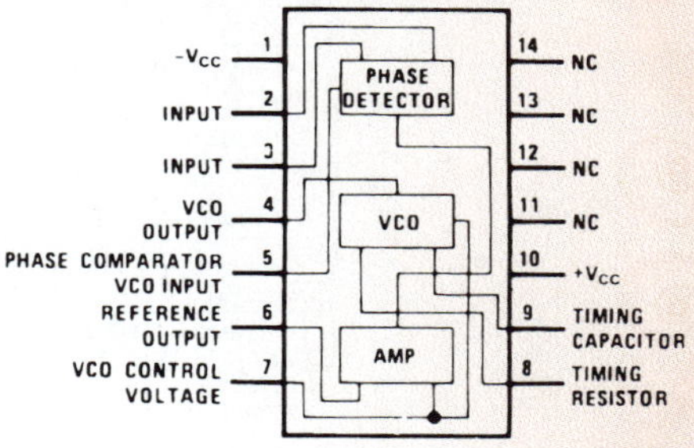

Order Number LM565CN
See NS Package Number N14A

FIGURE 15–58 The data sheets for the LM565 phase locked loop (Reprinted with permission of National Semiconductor Corporation)

The settings for the waveform generator driving the SCR's gate are shown in Figure 15–62. The clock frequency is set to 60 Hz, the amplitude is set to 5 V and the offset is set to 0 V. This means that the input clock is switching between ±5 V. The settings for the rise and fall time are shown. The rise/fall time are set to 1 ms to minimize any problems with the simulator. *Note*: Fast rise and fall times can result in a simulator error. This change reduces the potential for simulator analysis problems.

Absolute Maximum Ratings (Note 1)

If Military/Aerospace specified devices are required, please contact the National Semiconductor Sales Office/ Distributors for availability and specifications.

Supply Voltage	±12V
Power Dissipation (Note 2)	1400 mW
Differential Input Voltage	±1V
Operating Temperature Range	
LM565H	−55°C to +125°C
LM565CN	0°C to +70°C
Storage Temperature Range	−65°C to +150°C
Lead Temperature (Soldering, 10 sec.)	260°C

Electrical Characteristics

AC Test Circuit, T_A = 25°C, V_{CC} = ±6V

Parameter	Conditions	LM565			LM565C			Units
		Min	Typ	Max	Min	Typ	Max	
Power Supply Current			8.0	12.5		8.0	12.5	mA
Input Impedance (Pins 2, 3)	−4V < V_2, V_3 < 0V	7	10			5		kΩ
VCO Maximum Operating Frequency	C_o = 2.7 pF	300	500		250	500		kHz
VCO Free-Running Frequency	C_o = 1.5 nF R_o = 20 kΩ f_o = 10 kHz	−10	0	+10	−30	0	+30	%
Operating Frequency Temperature Coefficient			−100			−200		ppm/°C
Frequency Drift with Supply Voltage			0.1	1.0		0.2	1.5	%/V
Triangle Wave Output Voltage		2	2.4	3	2	2.4	3	V_{p-p}
Triangle Wave Output Linearity			0.2			0.5		%
Square Wave Output Level		4.7	5.4		4.7	5.4		V_{p-p}
Output Impedance (Pin 4)			5			5		kΩ
Square Wave Duty Cycle		45	50	55	40	50	60	%
Square Wave Rise Time			20			20		ns
Square Wave Fall Time			50			50		ns
Output Current Sink (Pin 4)		0.6	1		0.6	1		mA
VCO Sensitivity	f_o = 10 kHz		6600			6600		Hz/V
Demodulated Output Voltage (Pin 7)	±10% Frequency Deviation	250	300	400	200	300	450	mV_{p-p}
Total Harmonic Distortion	±10% Frequency Deviation		0.2	0.75		0.2	1.5	%
Output Impedance (Pin 7)			3.5			3.5		kΩ
DC Level (Pin 7)		4.25	4.5	4.75	4.0	4.5	5.0	V
Output Offset Voltage $\|V_7 - V_6\|$			30	100		50	200	mV
Temperature Drift of $\|V_7 - V_6\|$			500			500		μV/°C
AM Rejection		30	40			40		dB
Phase Detector Sensitivity K_D			0.68			0.68		V/radian

Note 1: Absolute Maximum Ratings indicate limits beyond which damage to the device may occur. Operating Ratings indicate conditions for which the device is functional, but do not guarantee specific performance limits. Electrical Characteristics state DC and AC electrical specifications under particular test conditions which guarantee specific performance limits. This assumes that the device is within the Operating Ratings. Specifications are not guaranteed for parameters where no limit is given, however, the typical value is a good indication of device performance.

Note 2: The maximum junction temperature of the LM565 and LM565C is +150°C. For operation at elevated temperatures, devices in the TO-5 package must be derated based on a thermal resistance of +150°C/W junction to ambient or +45°C/W junction to case. Thermal resistance of the dual-in-line package is +85°C/W.

www.national.com 2

FIGURE 15–58 Continued

The switching of the SCR was observed using the oscilloscope. The trace is shown in Figure 15–63. Notice that the output is switching between 5 V and about 0.6 V. This is similar to the result obtained in the first part of this exercise.

This section has examined the basic concept of SCR operation. The student should understand how the SCR is turned on and what causes the SCR to turn off. The student should also understand how to clock the SCR so that it turns on and off.

FIGURE 15–59 The SCR circuit shown in the off state

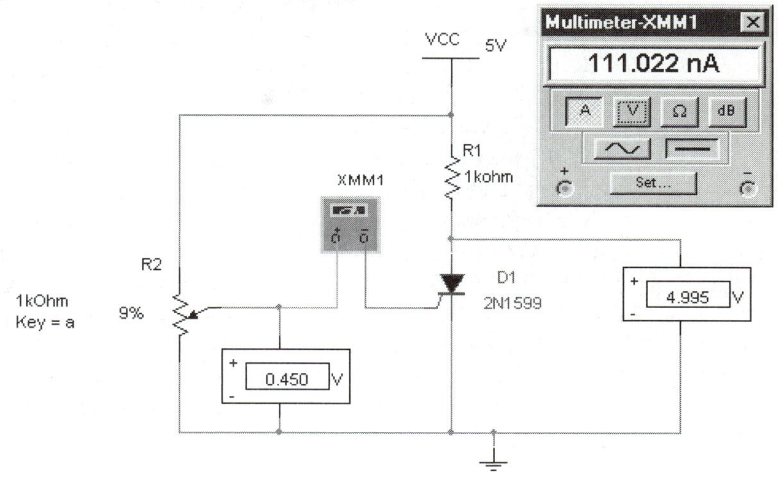

FIGURE 15–60 The SCR fully turned on

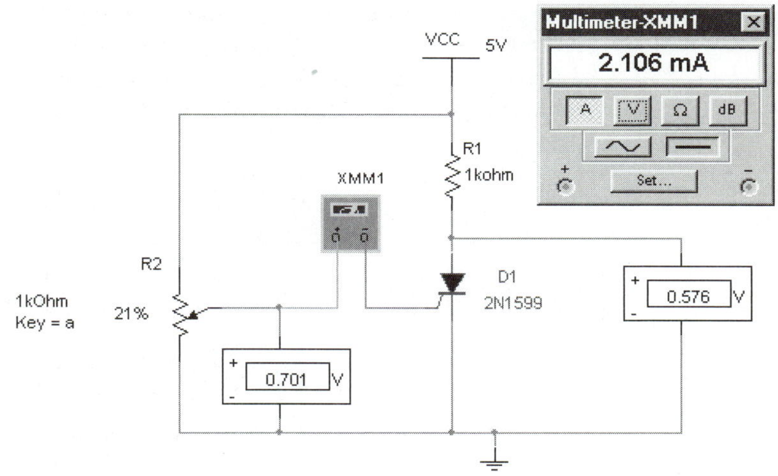

FIGURE 15–61 The SCR with a clocked input to the gate

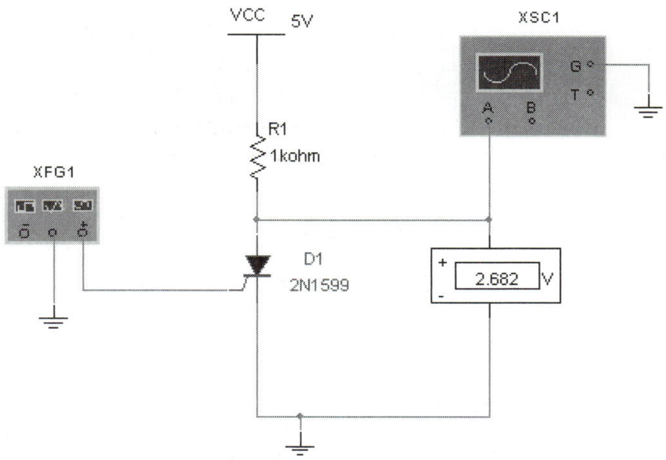

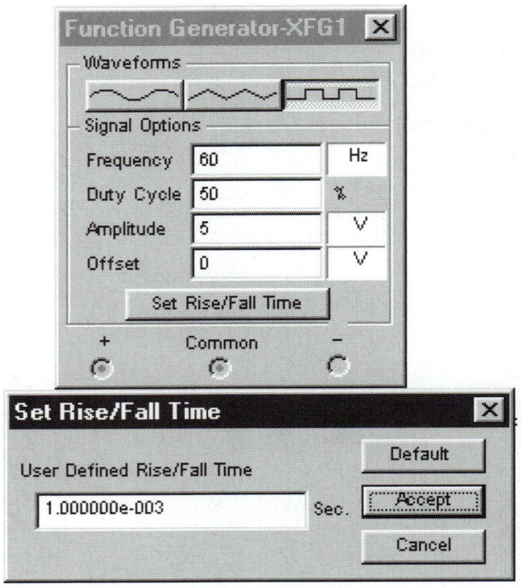

FIGURE 15–62 The settings for the clock to the SCR's gate

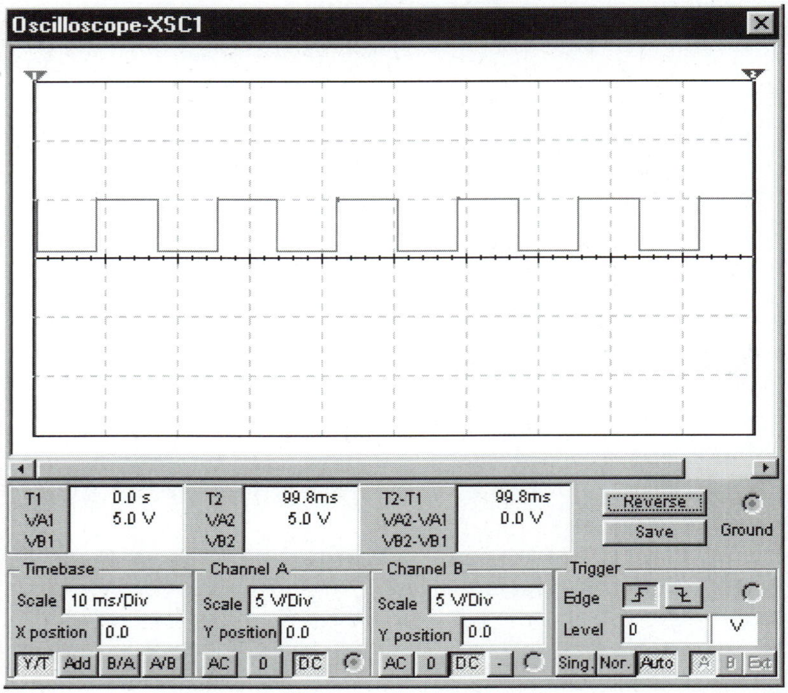

FIGURE 15–63 The output of the SCR with the clocked gate

SUMMARY

This chapter has presented an overview of some of the special devices used in electronics. Each section presented a different topic that has a different application in electronics. The concepts the student should understand are:

- How electronic devices are used in power control.
- Basic optoelectronic devices including the LCD and LED.
- The basic operation of the phase-locked loop.
- How to examine the operation of the SCR using EWB Multisim.

EXERCISES

SECTION 15–1

Four-Layer Devices

15–1. Figure 15–64 shows a Shockley diode used in a time-delay circuit to turn on an SCR. After the switch is closed, the capacitor charges until its voltage reaches the forward breakover voltage of the Shockley diode. The diode then conducts gate current to the SCR and turns it on. The adjustable resistance R is used to vary the RC time constant and thus vary the time delay between switch closure and SCR turn-on. If the forward breakover voltage of the diode is 30 V, find the time delay when $R = 100$ kΩ.

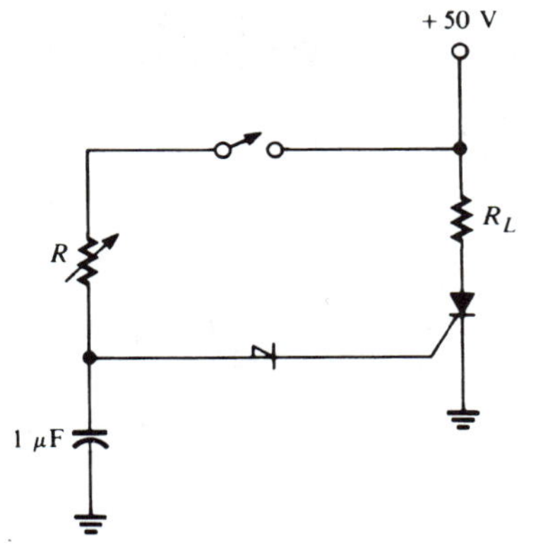

FIGURE 15–64 (Exercise 15–1)

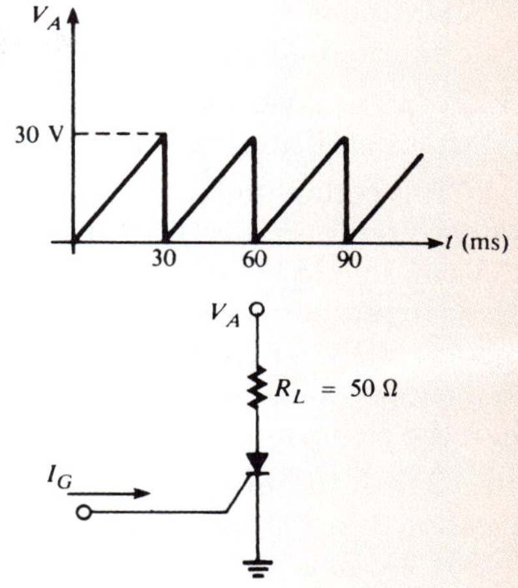

FIGURE 15–65 (Exercise 15–2)

15–2. The SCR in Figure 15–65 has $V_{BR(F)} = 10$ V. When it is conducting, the voltage drop V_{AK} is 1 V. If the anode voltage is the sawtooth waveform shown, find the average current in R_L.

15–3. Find the average voltage across the SCR in Exercise 15–2.

15–4. In the experimental circuit shown in Figure 15–66, V_A was set to +50 V and V_{GG} was adjusted until the SCR turned on. The value of V_G at that point was recorded. The value of V_A was then reduced until the SCR turned off. The value of V_A at that point was recorded. V_A was then restored to +50 V, and the measurements were repeated with a new setting for V_{GG}. The procedure was repeated several times, and the following data were obtained:

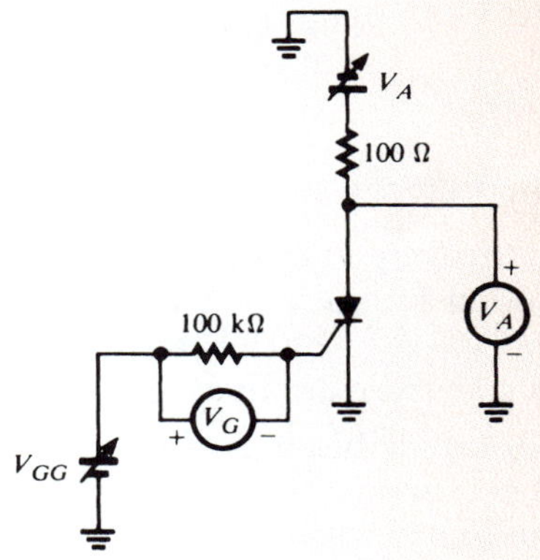

FIGURE 15–66 (Exercise 15–4)

V_G *(volts)*	V_A *(volts)*
0	1.15
0.2	1.075
0.4	1.00
1.0	0.95

Make a table showing holding current versus gate current for the SCR. Assume that the voltage drop V_{AK} across the SCR is 0.9 V when it is conducting.

15–5. Derive equation 15–7 from (15–4), (15–5), and (15–6). Then derive equation 15–8 from (15–7) and (15–3).

15–6. An SCR is connected in a half-wave power-control circuit (Figure 15–10) in which $v_i = 100$ V rms and $R_L = 80$ Ω. The forward voltage drop across the SCR when it is conducting is 1.0 V.

(a) Find the firing angle necessary to deliver two-thirds of the theoretical maximum average current that the circuit could deliver to R_L.

(b) Find the average load power under that condition.

15–7. An SCR is connected in a half-wave power-control circuit (Figure 15–10) in which v_i = 125 V rms and R_L = 50 Ω. Assuming that the forward voltage drop across the SCR is a constant 0.9 V when it is conducting, find the average power dissipated by the SCR when the firing angle is 30°.

15–8. The TRIAC in the power-controller circuit shown in Figure 15–18 has a voltage drop of 1 V between its anodes when it is conducting. If R_L = 30 Ω, find the rms value of v_i that is necessary if it is required to deliver an average power of 100 W to R_L when the firing angle is 45°.

SECTION 15–2

Optoelectronic Devices

15–9. Find the frequency range corresponding to radiation having wavelengths in the range from 1500 Å to 2000 Å. In what part of the light spectrum are these frequencies?

15–10. The total visible light energy received on a 1-cm^2 surface in 1 min is 0.042 J. What is the light intensity in $1m/m^2$?

15–11. The resistance of a certain photoconductive cell plotted versus light intensity is a straight line on log-log axes, as in Figure 15–20. If the cell has a resistance of 40 kΩ when the light intensity is 8 lm/m^2, and 30 kΩ when the intensity is 15 lm/m^2, what is the resistance when the intensity is 100 lm/m^2?

15–12. Using the photoconductive cell whose characteristics are described in Example 15–6, redesign the circuit in Figure 15–21 so that the relay coil remains de-energized as long as the cell is dark and energizes when the cell is illuminated. Demonstrate by circuit calculations that your design works.

15–13. Assuming that the photodiode characteristics shown in Figure 15–22 are perfectly linear for light intensities between 5000 and 10,000 $1m/m^2$, calculate the light intensity required to make the effective resistance 1250 Ω when the reverse voltage is 2 V.

15–14. A photodiode has $\eta = 1$, $V_T = 0.026$ V, and I_s = 0.5 nA. How much light-generated reverse current is necessary to produce an open-circuit output of 0.50 V when the diode is operated as a photovoltaic device?

15–15. What light intensity will fire the SCR in Example 15–9 (Figure 15–25) if the 100-kΩ gate resistor is changed to 35 kΩ ?

15–16. The phototransistor in Figure 15–67 has the characteristics shown in Figure 15–24. To what intensity must the incident light rise to cause the comparator output to switch low?

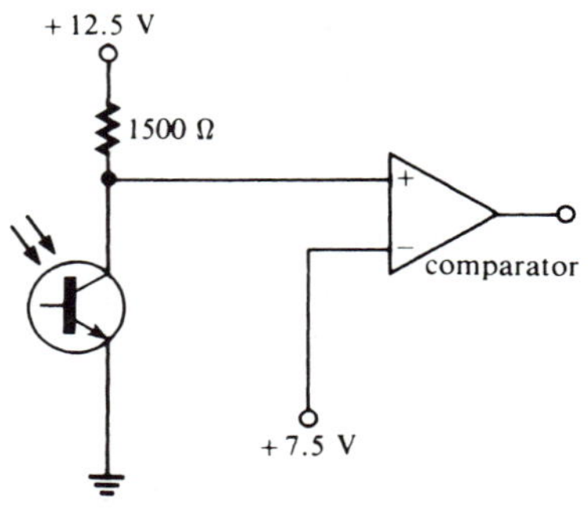

FIGURE 15–67 (Exercise 15–16)

15–17. What is the maximum power that can be delivered to a load by the solar cell whose characteristics are shown in Figure 15–29 when the light intensity is 80 mW/cm^2? When the light intensity is 15 mW/cm^2?

15–18. A series/parallel array of solar cells is used to provide power to a certain load. The array consists of 12 parallel branches, each of which has 8 cells in series. If each cell produces 900 mA at 0.55 V, how much power is delivered to the load?

15–19. The energy gaps for gallium arsenide and gallium phosphide are 1.37 eV and 2.25 eV, respectively. Find the wavelengths, in Å, of the light produced by LEDs fabricated from each of these materials.

15–20. If the maximum permissible forward current in the LED shown in Figure 15–34 (Example 15–11) is 40 mA, find the smallest permissible value of R_C. Assume that the transistor and LED parameters are the same as those given in the example.

15–21. The LED driver circuit shown in Figure 15–68 is designed to turn the LED off when 5 V is applied to R_B. Assuming that

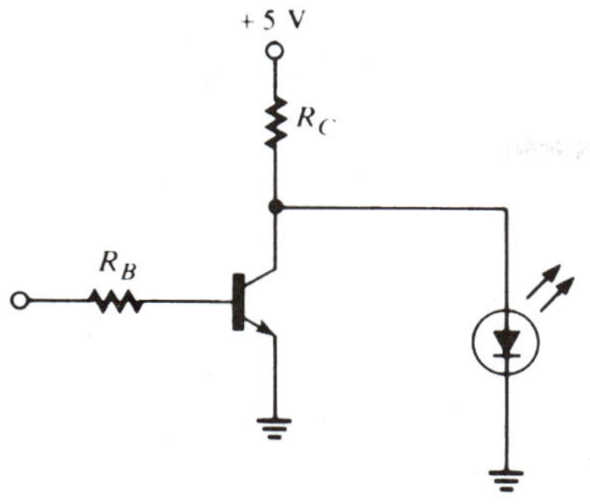

FIGURE 15–68 (Exercise 15-21)

the current in the LED should be 15 mA when it is on and that the LED has a forward voltage drop of 2 V, find values for R_B and R_C. The silicon transistor has $\beta = 100$.

15–22. Each LED in a certain seven-segment display draws 20 mA and has a 1.5-V drop when it is illuminated.

(a) Which numeral (from 0 through 9) requires the most power to display? The least power?

(b) How much power is required to display the numeral that requires the most power?

15–23. Figure 15–69 shows an optocoupler used as a logic inverter. When a positive pulse (high) is applied to the input, the LED illuminates and the transistor saturates, thus making the output low. To operate properly, the LED must be supplied with 15 mA, and it has a 1.2-V drop when it is illuminated. The optocoupler has an 80% current-transfer ratio. Find values for R_D and R_C. The high input level is +5 V.

15–24. Name the two methods used to alter light properties in liquid-crystal displays. In what two modes can an LCD be designed to operate using each of the light-altering methods?

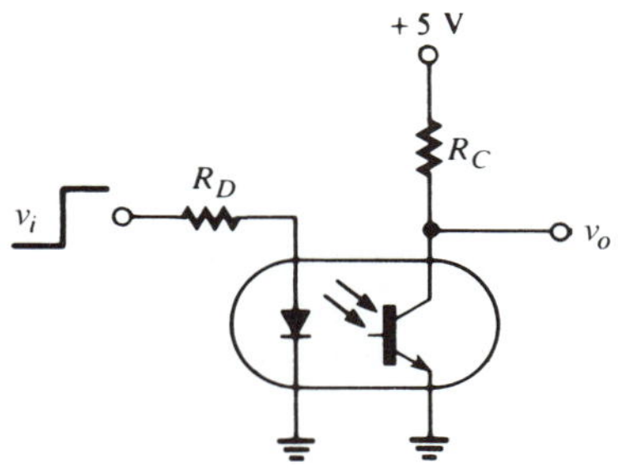

FIGURE 15–69 (Exercise 15-23)

15–25. What method of light alteration and what mode of operation are used in a field-effect (twisted-nematic) LCD to create a dark display on a light background?

15–26. What is the principal advantage of an LCD in comparison to an LED display? What are the principal limitations?

SECTION 15–3

Unijunction Transistors

15–27. In an experiment with a certain UJT, it was found that with V_{BB} set to 8 V, it was necessary to increase the emitter voltage to 5 V to cause it to switch on. When V_{BB} was raised to 15 V, it was necessary to raise the emitter voltage to 8.85 V to cause it to switch on. What is the intrinsic standoff ratio of the UJT?

15–28. A UJT relaxation oscillator of the design shown in Figure 15–46 must oscillate somewhere in the range from 750 Hz to 1.5 kHz, no matter what UJT is used. If $R_E = 68\ \text{k}\Omega$ and $C = 0.01\ \mu\text{F}$, use approximation 15–33 to determine the permissible range of values of η that UJTs used in the circuit can have.

15–29. Design a bias circuit for a programmable unijunction transistor so that it will have a peak voltage of 16 V. Use a 20-V-dc supply and assume that $V_D = 0.7$ V. Draw the schematic diagram for your design.

15–30. The PUT relaxation oscillator in Figure 15–51(a) has $R_A = 10\ \text{k}\Omega$, $C = 0.1\ \mu\text{F}$, and $R_2 = 15\ \text{k}\Omega$. Through what range of values (approximately) should it be possible to adjust R_1 if it is desired to adjust the oscillation frequency from 700 Hz to 1700 Hz?

SECTION 15–4

Tunnel Diodes

15–31. Describe the tunneling phenomenon and explain how it accounts for the appearance of the characteristic curve of a tunnel diode in the region of small forward bias.

15–32. The negative-resistance region of a certain tunnel diode occurs for forward voltages in the range from 0.2 V to 0.4 V. Is the statement that follows correct? Why or why not? If the diode has negative resistance −200 Ω when the voltage is 0.3 V, then the current under that amount of bias is $I = 0.3\ \text{V}/(-200\ \Omega) = -1.5$ mA.

CHAPTER 16

POWER AMPLIFIERS

OUTLINE

OBJECTIVES

- Evolve from small-signal amplifiers to large-signal or power amplifiers.
- Relate transistor power dissipation to heat transfer and thermal effects.
- Distinguish the different types and classes of amplifiers.
- Understand nonlinearties and distortion in power amplifiers.
- Develop a knowledge of complementary and quasi-complementary configurations for BJT power amplifiers.
- Understand the operation of power amplifiers in Class C and the generation of AM signals.

16–1 DEFINITIONS, APPLICATIONS, AND TYPES OF POWER AMPLIFIERS

As the name implies, a *power amplifier* is designed to deliver a large amount of power to a load. To perform this function, a power amplifier must itself be capable of *dissipating* large amounts of power; that is, it must be designed so that the heat generated when it is operated at high current and voltage levels is released into the surroundings at a rate fast enough to prevent destructive temperature buildup. Consequently, power amplifiers typically contain bulky components having large surface areas to enhance heat transfer to the environment. A *power transistor* is a discrete device with a large surface area and a metal case, characteristics that make it suitable for incorporation into a power amplifier.

A power amplifier is often the last, or *output,* stage of an amplifier system. The preceding stages may be designed to provide voltage amplification, to provide buffering to a high-impedance signal source, or to modify signal characteristics in some predictable way, functions that are collectively referred to as *signal conditioning.* The output of the signal-conditioning stages drives the power amplifier, which in turn drives the load. Some amplifiers are constructed with signal-conditioning stages and the output power stage all in one integrated circuit. Others, especially those designed to deliver very large amounts of power, have a *hybrid* structure, in the sense that the signal-conditioning stages are integrated and the power stage is discrete.

Power amplifiers are widely used in audio components—radio and television receivers, phonographs and tape players, stereo and high-fidelity systems, recording-studio equipment, and public address systems. The load in these applications is most often a loudspeaker ("speaker"), which requires considerable power to convert electrical signals to sound waves. Power amplifiers are also used in electromechanical *control systems* to drive electric motors. Examples include computer disk and tape drives, robotic manipulators, autopilots, antenna rotators, pumps and motorized valves, and manufacturing and process controllers of all kinds.

Large-Signal Operation

Because a power amplifier is required to produce large voltage and current variations in a load, it is designed so that at least one of its semiconductor components, typically a power transistor, can be operated over substantially the *entire* range of its output characteristics, from saturation to cutoff. This mode of operation is called *large-signal operation.* Recall that small-signal operation occurs when the range of current and voltage variation is small enough that there is no appreciable change in device parameters, such as β and r_e'. By contrast, the parameters of a large-signal amplifier at one output voltage may be considerably different from those at another output voltage. There are two important consequences of this fact:

1. Signal distortion occurs because of the change in amplifier characteristics with signal level. *Harmonic distortion* always results from such *nonlinear* behavior of an amplifier (see Figure 7–11). Compensating techniques, such as negative feedback, must be incorporated into a power amplifier if low-distortion, high-level outputs are required.
2. Many of the equations we have developed for small-signal analysis of amplifiers are no longer valid. Those equations were based on the assumption that device parameters did not change, contrary to fact in large-signal amplifiers. Rough approximations of large-signal–amplifier behavior can be obtained by using average parameter values and applying small-signal analysis techniques. However, graphical methods are used more frequently in large-signal amplifier design.

As a final note on terminology, we should mention that the term *large-signal operation* is also applied to devices used in digital switching circuits. In these applications, the output level switches between "high" and "low" (cutoff and saturation) but remains in those states *most of the time.* Power dissipation is therefore not a problem. Either the output voltage or the output current is near 0 when a digital device is in an ON or an OFF state, so power, which is the product of voltage and current, is near 0 except during the short time when the device switches from one state to the other. On the other hand, the variations in the output level of a power amplifier occur in the active region, *between* the two extremes of saturation and cutoff, so a substantial amount of power is dissipated.

16–2 TRANSISTOR POWER DISSIPATION

Recall that power is, by definition, the *rate* at which energy is consumed or dissipated (1 W = 1 J/s). If the rate at which heat energy is dissipated in a device is less than the rate at which it is generated, the temperature of the device must rise. In electronic devices, electrical energy is converted to heat energy at a rate given by $P = VI$ watts, and temperature rises when this heat energy is not removed at a comparable rate. Because semiconductor material is irreversibly damaged when subjected to temperatures beyond a certain limit, temperature is the parameter that ultimately limits the amount of power a semiconductor device can handle.

Transistor manufacturers specify the maximum permissible junction temperature and the maximum permissible power dissipation that a transistor can withstand. As we shall presently see, the maximum permissible power dissipation is specified as a function of *ambient temperature* (temperature of the surroundings) because the rate at which heat can be liberated from the device depends on the temperature of the region to which the heat must be transferred. Most of the conversion of electrical energy to heat energy in a bipolar

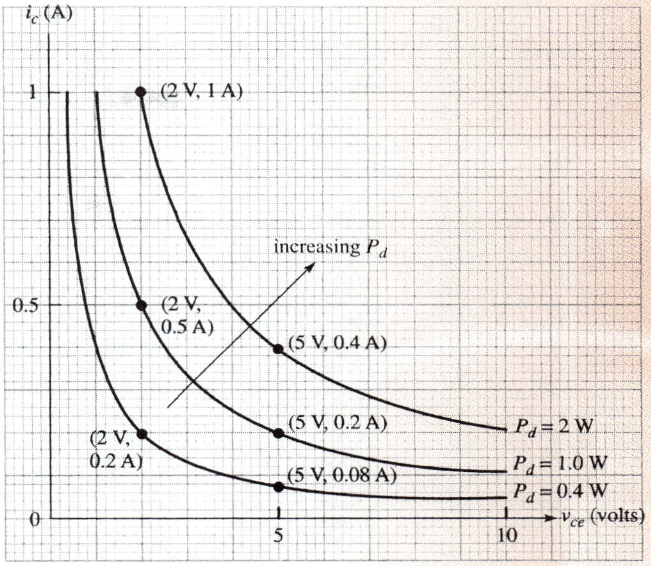

FIGURE 16–1 A family of graphs (hyperbolas) corresponding to $P_d = V_{CE}I_C$ for different values of P_d. Each hyperbola represents all combinations of collector voltage and collector current that result in a specific power dissipation (value of P_d)

junction transistor occurs at the junctions. Because power is the product of voltage and current, and because collector and emitter currents are approximately equal, the greatest power dissipation occurs at the junction where the voltage is greatest. In normal transistor operation, the collector–base junction is reverse biased and has, on average, a large voltage across it, while the base-emitter junction has a small forward-biasing voltage. Consequently, most of the heat generated in a transistor is produced at the collector–base junction. The total power dissipated at the junctions is

$$P_d = V_{CB}I_C + V_{BE}I_E \approx (V_{CB} + V_{BE})I_C = V_{CE}I_C \qquad (16\text{–}1)$$

Equation 16–1 gives, for all practical purposes, the total power dissipation of the transistor.

For a fixed value of P_d, the graph of Equation 16–1 is a *hyperbola* when plotted on $I_C - V_{CE}$ axes. Larger values of P_d correspond to hyperbolas that move outward from the axes, as illustrated in Figure 16–1. Each hyperbola represents all possible combinations of V_{CE} and I_C that give a product equal to the same value of P_d. The figure shows sample coordinate values (I_C, V_{CE}) at points on each hyperbola. Note that the product of the coordinate values is the same for points on the same hyperbola and that the product equals the power dissipation corresponding to the hyperbola.

Figure 16–2 shows a simple common-emitter amplifier and its dc load line plotted on $I_C - V_{CE}$ axes. Also shown are a set of hyperbolas corresponding to different values of power dissipation. Recall that the load line represents all possible combinations of I_C and V_{CE} corresponding to a particular value of R_C. As the amplifier output changes in response to an input signal, the collector current and voltage undergo variations along the load line and intersect different hyperbolas of power dissipation. It is clear that the power dissipation changes as the amplifier output changes. For safe operation, the load line must lie below and to the left of the hyperbola corresponding to the maximum permissible power dissipation. When the load line meets this requirement, there is no possible combination of collector voltage and current that results in a power dissipation exceeding the rated maximum.

It can be shown that the point of maximum power dissipation occurs at the *center* of the load line, where $V_{CE} = V_{CC}/2$ and $I_C = V_{CC}/2R_C$ (see Figure 16–2).

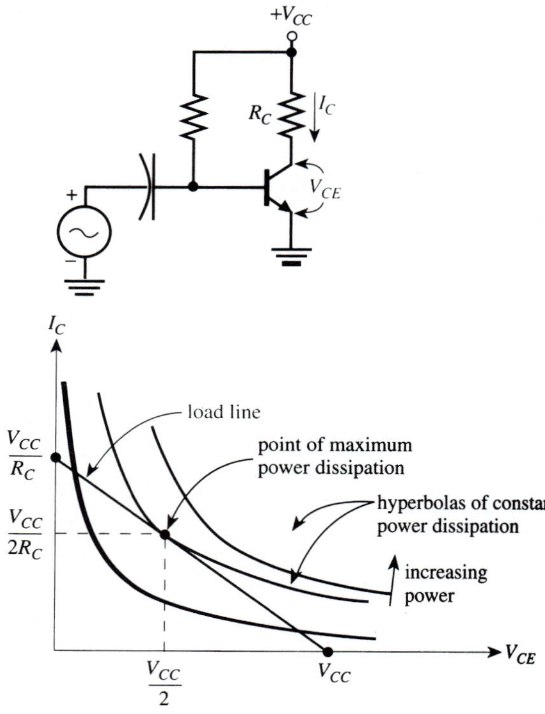

FIGURE 16–2 Load lines and hyperbolas of constant power dissipation

Therefore, the maximum power dissipation is

$$P_d(\text{max}) = \left(\frac{V_{CC}}{2}\right)\left(\frac{V_{CC}}{2R_C}\right) = \frac{V_{CC}^2}{4R_C} \qquad \textbf{(16–2)}$$

To ensure that the load line lies below the hyperbola of maximum dissipation, we therefore require that

$$\frac{V_{CC}^2}{4R_C} < P_d\,(\text{max}) \qquad \textbf{(16–3)}$$

where P_d(max) is the manufacturer's specified maximum dissipation at a specified ambient temperature. Inequality 16–3 enables us to find the maximum permissible value of R_C for a given V_{CC} and a given P_d(max):

$$R_C > \frac{V_{CC}^2}{4P_d\,(\text{max})} \qquad \textbf{(16–4)}$$

EXAMPLE 16–1

The amplifier shown in Figure 16–2 is to be operated with $V_{CC} = 20$ V and $R_C = 1$ kΩ.

1. What maximum power dissipation rating should the transistor have?
2. If an increase in ambient temperature reduces the maximum rating found in (1) by a factor of 2, what new value of R_C should be used to ensure safe operation?

Solution

1. From inequality 16–3,

$$P_d(\text{max}) > \frac{V_{CC}^2}{4R_C} = \frac{(20\text{ V})^2}{4(10^3\ \Omega)} = 0.1\text{ W}$$

2. When the dissipation rating is decreased to 0.05 W, the minimum permissible value of R_C is, from (16–4),

$$R_C > \frac{V_{CC}^2}{4P_d(\max)} = \frac{(20\ \text{V})^2}{4(0.05\ \text{W})} = 2\ \text{k}\Omega$$

16–3 HEAT TRANSFER IN SEMICONDUCTOR DEVICES

Conduction, Radiation, and Convection

Heat transfer takes place by one or more of three fundamental mechanisms: *conduction, radiation,* and *convection,* as illustrated in Figure 16–3. Conduction occurs when the energy that the atoms of a material acquire through heating is transferred to adjacent, less energetic atoms in a cooler region of the material. For example, conduction is the process by which heat flows from the heated end of a metal rod toward the cooler end, and the process by which heat is transferred from the heating element of a stove to a pan, and from the pan to the water it contains. Heat radiation is similar to other forms of radiation because no physical medium is required between the heat source and its destination. For example, heat is transferred by radiation through space from the sun to the earth. Convection occurs when a physical medium that has been heated by conduction or radiation *moves* away from the source of heat, i.e., is displaced by a cooler medium, so more conduction or radiation can occur. Examples include *forced* convection, where a fan is used to push air past a heated surface, and *natural* convection, where heated air rises and is replaced by cooler air from below.

Semiconductor devices are cooled by all three heat-transfer mechanisms. Heat is conducted from a junction, through the semiconductor material, through the case, and into the surrounding air. It is radiated from the surface of the case. Heating of the surrounding air creates airflow around the device, so convection cooling occurs. Various measures are taken to enhance heat transfer by each mechanism. To improve conduction, good physical contact is made between a junction and a device's metal case, or enclosure. In fact, in many power transistors the collector is in direct contact with the case, so the two are electrically the same as well as being at approximately the same temperature. To improve the conduction and radiation of heat from the case to the surrounding air, power devices are often equipped with *heat sinks.* These are attached to the device to be cooled and conduct heat outward to metal *fins* that increase the total surface area from which conduction and radiation into the air can take place (see Figure 16–4). Finally, convection cooling is enhanced by the use of fans that blow cool air past the surface of the case and/or the heat sink.

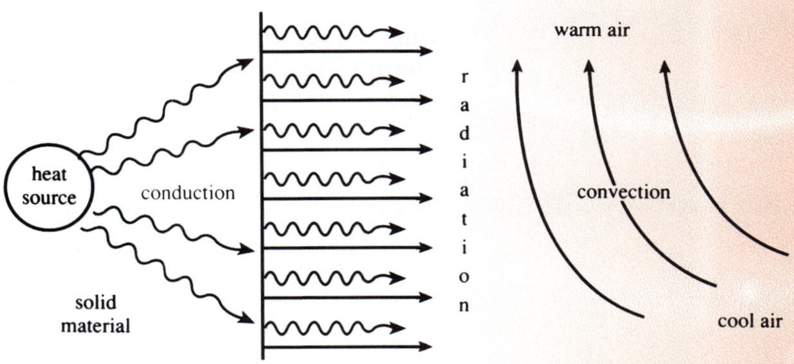

FIGURE 16–3 Examples illustrating heat transfer by conduction, radiation, and convection

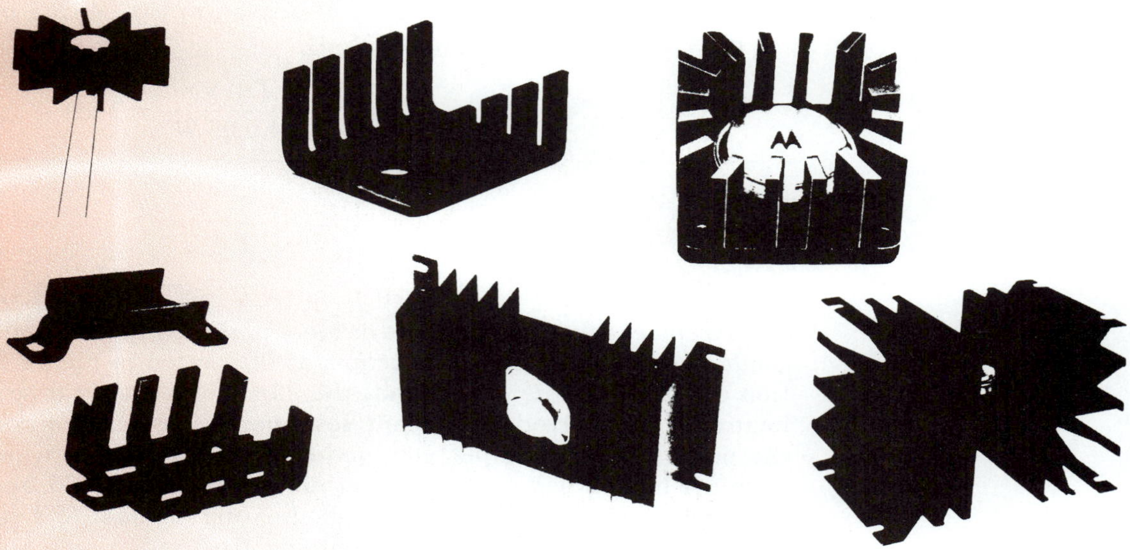

FIGURE 16–4 Typical heat sinks (Courtesy of EG & G Wakefield Engineering)

Thermal Resistance

Heat flow by conduction is very much like the conduction of electrical charge, i.e., current. Recall that current is the *rate* of flow of charge and is proportional to the difference in voltage across a resistance. Similarly, power is the rate of flow of heat energy and is proportionate to the difference in *temperature* across the region through which heat is conducted. We can regard any impediment to heat flow as *thermal resistance*, θ. Thus, the rate of flow of heat (i.e., power) is directly proportionate to temperature difference and inversely proportionate to thermal resistance:

$$P = \frac{\Delta T}{\theta} = \frac{T_2 - T_1}{\theta} \quad \text{J/s, or W} \tag{16–5}$$

Insulating materials, like wool, have large thermal resistances and metals have small thermal resistances. Note the similarity of equation 16–5 to Ohm's law: $I = (V_2 - V_1)/R$. Solving for θ in (16–5), we obtain the following equivalent relation that allows us to determine the units of thermal resistance:

$$\theta = \frac{T_2 - T_1}{P} \quad \text{degrees/W} \tag{16–6}$$

In a power amplifier, heat is transferred through different types of materials and across boundaries of dissimilar materials, each of which presents different values of thermal resistance. These are usually treated like series electrical circuits, so thermal resistances are added when computing the total heat flow through the system or the temperatures at various points in the system.

EXAMPLE 16–2

The collector–base junction of a certain transistor dissipates 2 W. The thermal resistance from junction to case is 8°C/W, and the thermal resistance from case to air is 20°C/W. The free-air temperature (ambient temperature) is 25°C.

1. What is the junction temperature?
2. What is the case temperature?

Solution

1. The total thermal resistance between junction and ambient is the sum of that from junction to case and that from case to ambient: $\theta_T = \theta_{JC} + \theta_{CA} = 8°C/W + 20°C/W = 28°C/W$. Therefore, by equation 16–5, with $T_2 = T_J =$ junction temperature and $T_1 = T_A =$ ambient temperature,

$$T_J - T_A = P\theta_T$$

$$T_J = P\theta_T + T_A = (2\text{ W})(28°\text{C/W}) + 25°\text{C} = 81°\text{C}$$

2. The rate of heat flow, P, is constant through the series configuration, so

$$T_J - T_C = P\theta_{JC}$$

$$T_C = T_J - P\theta_{JC} = 81°\text{C} - (2\text{ W})(8°\text{C/W}) = 65°\text{C}$$

To reduce thermal resistance and improve heat conductivity, a special silicone grease is often used between contact surfaces, such as between the case and heat sink. Mica washers are used to isolate the case and heat sink *electrically* when the case is electrically common to one of the device terminals, such as a collector. The washer creates additional thermal resistance in the heat-flow path between the case and heat sink and may have to be taken into account in heat-flow computations.

The values of the thermal resistance of various types of heat sinks, washers, and metals can be determined from published manufacturers' data. The data are often given in terms of thermal *resistivity* ρ (°C · in./W or °C · m/W), and the thermal resistance is computed by

$$\theta = \frac{\rho t}{A} \quad °\text{C/W} \tag{16–7}$$

where
ρ = resistivity of material
t = thickness of material
A = area of material

For example, a typical value of ρ for mica is 66°C · in./W, so a 0.002-in.-thick mica washer having a surface area of 0.6 in.^{2} will have a thermal resistance of

$$\theta = \frac{(66°\text{C}\cdot\text{in./W})(2 \times 10^{-3}\text{ in.})}{0.6\text{ in.}^2} = 0.22°\text{C/W}$$

EXAMPLE 16–3

The maximum permissible junction temperature of a certain power transistor is 150°C. It is desired to operate the transistor with a power dissipation of 15 W in an ambient temperature of 40°C. The thermal resistance are as follows:

$$\theta_{JC} = 0.5°\text{C/W} \qquad \text{(junction to case)}$$

$$\theta_{CA} = 10°\text{C/W} \qquad \text{(case to ambient)}$$

1. Determine whether a heat sink is required for this application.
2. If a heat sink is required, determine the maximum thermal resistance it can have. Assume that a mica washer having thermal resistance $\theta_W = 0.5°C/W$ must be used between case and heat sink.

Solution

1. $$\theta_T = \theta_{JC} + \theta_{CA} = 0.5°\text{C/W} + 10°\text{C/W} = 10.5°\text{C/W}$$

$$T_J - T_A = \theta_T P = (10.5°\text{C/W})(15\ \text{W}) = 157.5°\text{C}$$

$$T_J = 157.5°\text{C} + T_A = 157.5°\text{C} + 40°\text{C} = 197.5°\text{C}$$

Because the junction temperature, 197.5°C, will exceed the maximum permissible value of 150°C under the given conditions, a heat sink must be used to reduce the total thermal resistance.

2. Setting T_J equal to its maximum permissible value of 150°C, we can solve for the maximum total thermal resistance:

$$T_J - T_A = \theta_T P$$

$$150 - 40 = (\theta_T)(15)$$

$$\theta_T = \frac{110°\text{C}}{15\ \text{W}} = 7.33°\text{C/W}$$

The thermal resistance of the heat sink, θ_H, will replace the thermal resistance θ_{CA} between case and ambient that was present in (1). Thus, taking into account the thermal resistance of the washer, we have $\theta_T = 7.33°\text{C/W} = \theta_{JC} + \theta_W + \theta_H = 0.5°\text{C/W} + 0.5°\text{C/W} + \theta_H$, or $\theta_H = (7.33 - 1.0)°\text{C/W} = 6.33°\text{C/W}$.

Derating

The maximum permissible power dissipation of a semiconductor device is specified by the manufacturer at a certain temperature, either case temperature or ambient temperature. For example, the maximum power dissipation of a power transistor may be given as 10 W at 25°C ambient, and that of another device as 20 W at 50°C case temperature. These ratings mean that each device can dissipate the specified power at temperatures up to and including the given temperature, but not at greater temperatures. The maximum permissible dissipation *decreases* as a function of temperature when temperature increases beyond the given value. The decrease in permissible power dissipation at elevated temperatures is called *derating,* and the manufacturer specifies a *derating factor,* in W/°C, that is used to find the *decrease* in dissipation rating beyond a certain temperature.

To illustrate derating, suppose a certain device has a maximum rated dissipation of 20 W at 25°C ambient and a derating factor of 100 mW/°C at temperatures above 25°C. Then the maximum permissible dissipation at 100°C is

$$P_d(\text{max}) = (\text{rated dissipation at } T_1) - (T_2 - T_1)(\text{derating factor}) \quad (16\text{–}8)$$

$$= (20\ \text{W}) - (100°\text{C} - 25°\text{C})(100\ \text{mW/°C})$$

$$= (20\ \text{W}) - (75°\text{C})(100\ \text{mW/°C}) = (20\ \text{W}) - (7.5\ \text{W}) = 12.5\ \text{W}$$

EXAMPLE 16–4

A certain semiconductor device has a maximum rated dissipation of 5 W at 50°C case temperature and must be derated above 50°C case temperature. The thermal resistance from case to ambient is 5°C/W.

1. Can the device be operated at 5 W of dissipation without auxiliary cooling (heat sink or fan) when the ambient temperature is 40°C?
2. If not, what is the maximum permissible dissipation, with no auxiliary cooling, at 40°C ambient?
3. What derating factor should be applied to this device, in watts per ambient degree?

Solution

1. We determine the case temperature when the ambient temperature is 40°C:

$$T_C - T_A = \theta_{CA}P$$
$$T_C - 40°\text{C} = (5°\text{C/W})(5\text{ W})$$
$$T_C = 40 + 25 = 65°\text{C}$$

Because the case temperature exceeds 50°C, the full 5 W of power dissipation is not permitted when the ambient temperature is 40°C.

2. The derated dissipation at $T_A = 40°\text{C}$ is

$$P_d(\max) = (T_C - T_A)/\theta = \frac{(50 - 40)°\text{C}}{5°\text{C/W}} = 2\text{ W}$$

3. We must first find the maximum ambient temperature at which the device can dissipate 5 W when the case temperature is 50°C:

$$5\text{ W} = \frac{(50°\text{C} - T_A)}{5°\text{C/ W}}$$
$$T_A = 50 - 25 = 25°\text{C}$$

Thus, derating will be necessary when the ambient temperature exceeds 25°C. Let D be the derating factor. Then $P_d(\max) = (5\text{ W}) - (T_A - 25)D$. Substituting the results of (2), we find

$$2\text{ W} = (5\text{ W}) - (40 - 25)D$$
$$D = 0.2\text{ W/°C}$$

Power Dissipation in Integrated Circuits

All the heat transfer concepts we have described in connection with power devices apply equally to integrated circuits, including derating. Of course, a maximum junction temperature is not specified for an integrated circuit, but a maximum device temperature or case temperature is usually given. Many integrated circuits are available in a variety of case types, and the rated power dissipation may depend on the case used. For example, the specifications given in Chapter 10 for the LF353 operational amplifier show that the maximum dissipation is 500 mW for the metal package, 310 mW for the DIP, and 570 mW for the flatpak. Note that these specifications require the use of derating factors above 70°C ambient and that the factors differ, depending on case type.

The total power dissipation of an integrated circuit can be calculated by measuring the current drawn from each supply voltage used in a particular application and forming the product of each current with the respective voltage. For example, if an operational amplifier employing ±15-V-dc power supplies draws 10 mA from the positive supply and 8 mA from the negative supply, the total dissipation is $P_d = (15\text{ V})(10\text{ mA}) + (15\text{ V})(8\text{ mA}) = 270\text{ mW}$.

16–4 AMPLIFIER CLASSES AND EFFICIENCY

Class-A Amplifiers

All the small-signal amplifiers we have studied in this book have been designed so that output voltage can vary in response to both positive and negative inputs; that is, the amplifiers are *biased* so that under normal operation the output never saturates or cuts off. An amplifier that has that property is called a *class-A* amplifier. More precisely, an amplifier is class A if its output remains in the active region during a complete cycle (one full period) of a sine-wave input signal.

Figure 16–5 shows a typical class-A amplifier and its input and output waveforms. In this case, the transistor is biased at $V_{CE} = V_{CC}/2$, which is midway between saturation and cutoff, and which permits maximum output voltage swing. Note that the output can vary through (approximately) a full V_{CC} volts, peak-to-peak. The output is in the transistor's active region during a full cycle (360°) of the input sine wave.

Efficiency

The efficiency of a power amplifier is defined to be

$$\eta = \frac{\text{average signal power delivered to load}}{\text{average power drawn from dc source(s)}} \tag{16–9}$$

Note that the numerator of (16–9) is average *signal* power, that is, average *ac* power, excluding any dc or bias components in the load. Recall that when voltages and currents are sinusoidal, average ac power can be calculated using any of the following relations:

$$P = V_{rms} I_{rms} = \frac{V_P I_P}{2} = \frac{V_{PP} I_{PP}}{8} \tag{16–10}$$

$$P = I_{rms}^2 R = \frac{I_P^2 R}{2} = \frac{I_{PP}^2 R}{8} \tag{16–11}$$

$$P = \frac{V_{rms}^2}{R} = \frac{V_P^2}{2R} = \frac{V_{PP}^2}{8R} \tag{16–12}$$

where the subscripts *P* and *PP* refer to peak and peak-to-peak, respectively.

As a consequence of the definition of efficiency (equation 16–9), the efficiency of a class-A amplifier is 0 when no signal is present. (The amplifier is said to be in *standby* when no signal is applied to its input.) We will now derive a general expression for the efficiency of the class-A amplifier shown in Figure 16–5. In doing so, we will not consider the small power consumed in the base-biasing circuit, i.e., the power at the input side: $I_B^2 R_B + v_{be} i_b$. Figure 16–6 shows the voltages and currents used in our analysis. Notice that resistance *R* is considered to be the load. We will re-

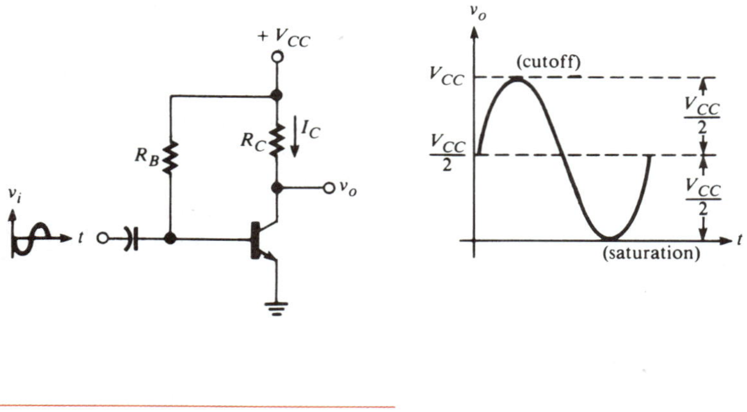

FIGURE 16–5 The output of a class-A amplifier remains in the active region during a full period (360°) of the input sine wave. In this example, the transistor output is biased midway between saturation and cutoff

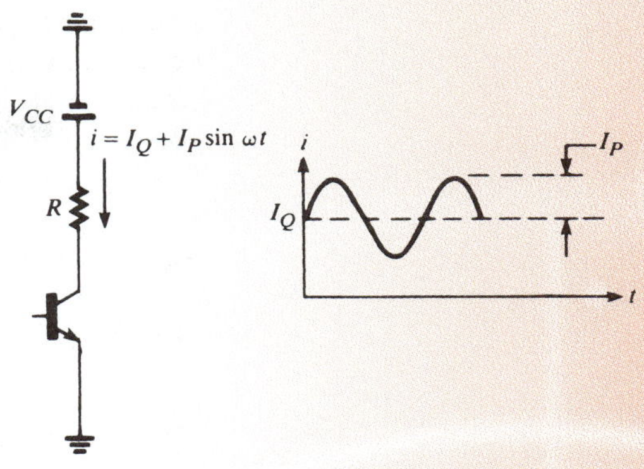

FIGURE 16–6 Voltages and currents used in the derivation of an expression for the efficiency of a series-fed, class-A amplifier

fer to this configuration as a *series-fed* class-A amplifier, and we will consider capacitor- and transformer-coupled loads in a later discussion.

The instantaneous power from the dc supply is

$$p_S(t) = V_{CC}i = V_{CC}(I_Q + I_P \sin \omega t) = V_{CC}I_Q + V_{CC}I_P \sin \omega t \qquad (16\text{–}13)$$

Since the average value of the sine term is 0, the average power from the dc supply is

$$P_S = V_{CC}I_Q \qquad (16\text{–}14)$$

The average *signal* power in load resistor R is, from equation 16–11,

$$P_R = \frac{I_P^2 R}{2} \qquad (16\text{–}15)$$

Therefore, by equation 16–9,

$$\eta = \frac{P_R}{P_S} = \frac{I_P^2 R}{2V_{CC}I_Q} \qquad (16\text{–}16)$$

We see again that the efficiency is 0 under no-signal conditions ($I_P = 0$) and that efficiency rises as the peak signal level I_P increases. The maximum possible efficiency occurs when I_P has its maximum possible value without distortion. When the bias point is at the center of the load line, as shown in Figure 16–2, the quiescent current is one-half the saturation current, and the output current can swing through the full range from 0 to V_{CC}/R A without distorting (clipping). Thus, the maximum undistorted peak current is also one-half the saturation current:

$$I_Q = I_P(\text{max}) = \frac{V_{CC}}{2R} \qquad (16\text{–}17)$$

Substituting (16–17) into (16–16), we find the maximum possible efficiency of the series-fed, class-A amplifier:

$$\eta(\text{max}) = \frac{(V_{CC}/2R)^2 R}{2V_{CC}(V_{CC}/2R)} = 0.25 \qquad (16\text{–}18)$$

This result shows that the best possible efficiency of a series-fed, class-A amplifier is undesirably small: Only 1/4 of the total power consumed by the circuit is delivered to the load, under optimum conditions. For that reason, this type of amplifier is not widely used in heavy power applications. The principal advantage of the class-A amplifier is that it generally produces less signal distortion than some of the other, more efficient classes that we will consider later.

Another type of efficiency used to characterize power amplifiers relates signal power to total power dissipated at the collector. Called *collector efficiency,* its practical significance stems from the fact that a major part of the cost and bulk of a power amplifier is invested in the output device itself and the means used to cool it. Therefore, it is desirable to maximize the ratio of signal power in the load to power consumed by the device. Collector efficiency η_c is defined by

$$\eta_c = \frac{\text{average signal power delivered to load}}{\text{average power dissipated at collector}} \tag{16–19}$$

The average power P_C dissipated at the collector of the class-A amplifier in Figure 16–6 is the product of the dc (quiescent) voltage and current:

$$P_C = V_Q I_Q = (V_{CC} - I_Q R)I_Q \tag{16–20}$$

Therefore,

$$\eta_c = \frac{I_P^2 R/2}{(V_{CC} - I_Q R)I_Q} \tag{16–21}$$

The maximum value of η_c occurs when I_P is maximum, $I_P = I_Q = V_{CC}/2R$, as previously discussed. Substituting these values into (16–21) gives

$$\eta_c(\text{max}) = \frac{(V_{CC}/2R)^2 R/2}{(V_{CC} - V_{CC}/2)V_{CC}/2R} = 0.5 \tag{16–22}$$

Figure 16–7 shows the output side of a class-A amplifier with capacitor-coupled load R_L. Also shown are the dc and ac load lines that result. In this case, the average power delivered to the load is

$$P_L = I_{PL}^2 R_L/2 \tag{16–23}$$

where I_{PL} is the peak ac load current. The average power from the dc source is computed in the same way as for the series-fed amplifier: $P_S = V_{CC}I_Q$, so the efficiency is

FIGURE 16–7 A class-A amplifier with capacitor-coupled load R_L

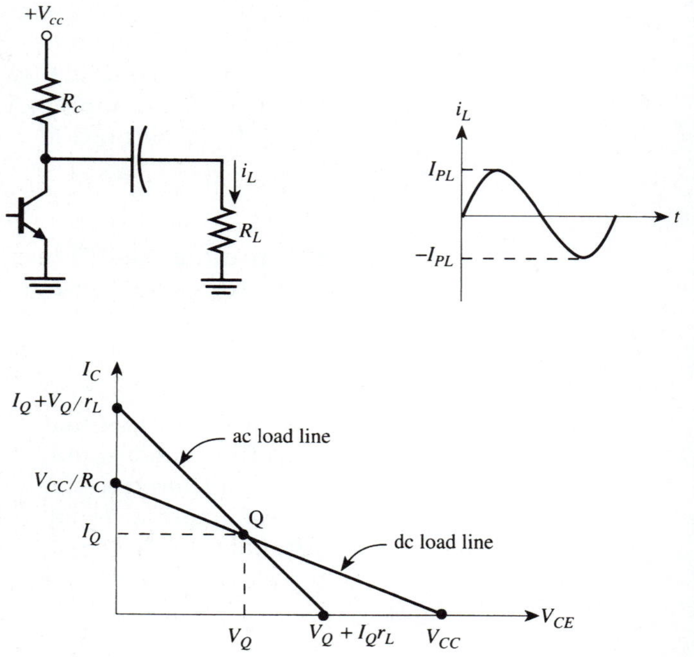

$$\eta = \frac{I_{PL}^2 R_L}{2V_{CC}I_Q} \tag{16–24}$$

As in the case of the series-fed amplifier, the efficiency is 0 under no-signal (standby) conditions and increases with load current I_{PL}.

Recall that maximum output swing can be achieved by setting the Q-point in the center of the ac load line at

$$I_Q = \frac{V_{CC}}{R_C + r_L} \tag{16–25}$$

The peak collector current under those circumstances is $V_{CC}/(R_C + r_L)$. Neglecting the transistor output resistance, the portion of the collector current that flows in R_L is, by the current-divider rule,

$$I_{PL} = \left(\frac{V_{CC}}{R_C + r_L}\right)\left(\frac{R_C}{R_C + R_L}\right) \tag{16–26}$$

The average ac power in the load resistance R_L is then

$$P_L = \frac{I_{PL}^2 R_L}{2} = \left[\left(\frac{V_{CC}}{R_C + r_L}\right)\left(\frac{R_C}{R_C + R_L}\right)\right]^2 (R_L/2) \tag{16–27}$$

The average power supplied from the dc source is

$$P_S = V_{CC}I_Q = \frac{V_{CC}^2}{R_C + r_L} \tag{16–28}$$

Therefore, the efficiency under the conditions of maximum possible undistorted output is

$$\eta = \frac{P_L}{P_S} = \frac{\left[\left(\frac{V_{CC}}{R_C + r_L}\right)\left(\frac{R_C}{R_C + R_L}\right)\right]^2 R_L}{2\left(\frac{V_{CC}^2}{R_C + r_L}\right)} \tag{16–29}$$

Algebraic simplification of (16–29) leads to

$$\eta = \frac{R_C R_L}{2(R_C + 2R_L)(R_C + R_L)} = \frac{r_L}{2(R_C + 2R_L)} \tag{16–30}$$

Equation 16–30 shows that the efficiency depends on both R_C and R_L. In practice, R_L is a fixed and known value of load resistance, while the value of R_C is selected by the designer. If R_C is fixed and R_L can be selected, it can be shown using calculus (differentiating equation 16–30 with respect to R_C), that η is maximized by setting

$$R_L = \frac{R_C}{\sqrt{2}} \tag{16–31}$$

With this value of R_C, the maximum efficiency is

$$\eta(\text{max}) = 0.0858 \tag{16–32}$$

Another criterion for choosing R_L is to select its value so that maximum power is transferred to the load. Under the constraint of maintaining maximum output swing, it can be shown that maximum power transfer occurs when $R_L = R_C/2$. Under that circumstance, by substituting into (16–30), we find the maximum possible efficiency with maximum power transfer to be

$$\eta(\text{max}) = 0.0833 \text{ (max power transfer)} \tag{16–33}$$

It is interesting to note that the efficiency under maximum power transfer (0.0833) is somewhat less than what can be achieved (0.0858) without regard to power transfer. In either case, the maximum efficiency is substantially less than that attainable in the series-fed class-A amplifier.

EXAMPLE 16–5

The class-A amplifier shown in Figure 16–8 is biased at $V_{CE} = 12\text{ V}$. The output voltage is the maximum possible without distortion. Find

1. the average power from the dc supply,
2. the average power delivered to the load,
3. the efficiency, and
4. the collector efficiency.

Solution

1. $I_Q = (V_{CC} - V_Q)/R_C = (24 - 12)/50 = 0.24\text{ A}$

   $P_S = V_{CC}I_Q = (24\text{ V})(0.24\text{ A}) = 5.76\text{ W}$

2. As can be seen in Figure 16–7, the maximum value of the peak output voltage is the smaller of V_Q and $I_Q r_L$. In this case, $V_Q = 12\text{ V}$ and $I_Q r_L = (0.24\text{ A})(50 \| 50) = 6\text{ V}$. Thus, the peak undistorted output voltage is 6 V and the ac power delivered to the 50-Ω load is

$$P_L = \frac{V_{PL}^2}{2R_L} = \frac{6^2}{100} = 0.36\text{ W}$$

$$\eta = \frac{P_L}{P_S} = \frac{0.36\text{ W}}{5.76\text{ W}} = 0.0625$$

3. The efficiency is less than the theoretical maximum because the bias point permits only a ±6-V swing. As an exercise, find the quiescent value of V_{CE} that maximizes the swing and calculate the efficiency under that condition.
4. The average power dissipated at the collector is $P_C = V_Q I_Q = (12)(0.24) = 2.88\text{ W}$. Therefore,

$$\eta_c = \frac{P_L}{P_C} = \frac{0.36}{2.88} = 0.125$$

Transformer-Coupled Class-A Amplifiers

Transformers are used to couple power amplifiers to their loads and, in those applications, are called *output transformers*. As in other coupling applications, the advantages of a transformer are that it provides an opportunity to

FIGURE 16–8 (Example 16–5)

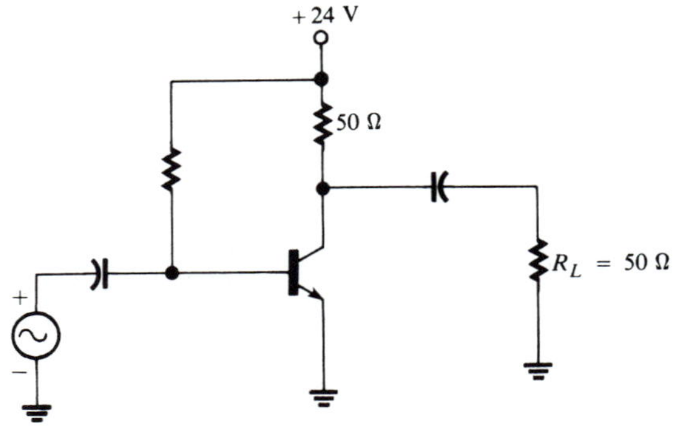

achieve impedance matching for maximum power transfer and that it blocks the flow of dc current in a load.

Figure 16–9 shows a transformer used to couple the output of a transistor to load R_L. Also shown are the dc and ac load lines for the amplifier. Here we assume that the dc resistance of the primary winding is negligibly small, so the dc load line is vertical (slope $= -1/R_{dc} = -\infty$). Recall that the ac resistance r_L reflected to the primary side is

$$r_L = (N_p/N_s)^2 R_L \tag{16–34}$$

where N_p and N_s are the numbers of turns on the primary and secondary windings, respectively. As shown in the figure, the slope of the ac load line is $-1/r_L$.

Since we are assuming that there is negligible resistance in the primary winding, there is no dc voltage drop across the winding, and the quiescent collector voltage is therefore V_{CC} volts, as shown in Figure 16–9. Conventional base-bias circuitry (not shown in the figure) is used to set the quiescent collector current I_Q. The Q-point is the point on the dc load line at which the collector current equals I_Q.

Since V_{CE} cannot be negative, the maximum permissible decrease in V_{CE} below its quiescent value is $V_Q = V_{CC}$ volts. Thus, the maximum possible peak value of V_{CE} is V_{CC} volts. To achieve maximum peak-to-peak output variation, the intercept of the ac load line on the V_{CE}-axis should therefore be $2V_{CC}$ volts, as shown in Figure 16–9. The quiescent current I_Q is selected so that the ac load line, a line having slope $-1/r_L$, intersects the V_{CE}-axis at $2V_{CC}$ volts.

The ac load line intersects the I_C-axis in Figure 16–9 at $I_C(\text{max})$. Note that there is no theoretical limit to the value that I_C may have, because there is no limiting resistance in the collector circuit. However, in practice, I_C must not exceed the maximum permissible collector current for the transistor and it must not be so great that the magnetic flux of the transformer saturates. When the transformer saturates, it can no longer induce current in the secondary winding and signal distortion results. When I_Q is set for maximum signal swing (so that $V_{CE}(\text{max}) = 2V_{CC}$), I_Q is one-half $I_C(\text{max})$; that is, $I_C(\text{max}) = 2I_Q$, as shown in Figure 16–9. Thus, the maximum values of the peak primary voltage and peak primary current are V_{CC} and I_Q, respectively. Since the ac output can vary through this range, below and above the quiescent point, the amplifier is of the class-A type.

Note that, unlike the case of the capacitor-coupled or series-fed amplifier, the collector voltage can exceed the supply voltage. A transistor having a collector breakdown voltage equal to at least twice the supply voltage should be used in this application.

The ac power delivered to load resistance R_L in Figure 16–9 is

$$P_L = \frac{V_s^2}{2R_L} = \frac{V_{PL}^2}{2R_L} \tag{16–35}$$

where $V_s = V_{PL}$ is the peak value of the secondary, or load, voltage. The average power from the dc supply is

$$P_S = V_{CC} I_Q \tag{16–36}$$

Therefore, the efficiency is

$$\eta = \frac{P_L}{P_S} = \frac{V_{PL}^2}{2R_L V_{CC} I_Q} \tag{16–37}$$

Under maximum signal conditions, the peak primary voltage is V_{CC} volts, so the peak load voltage is

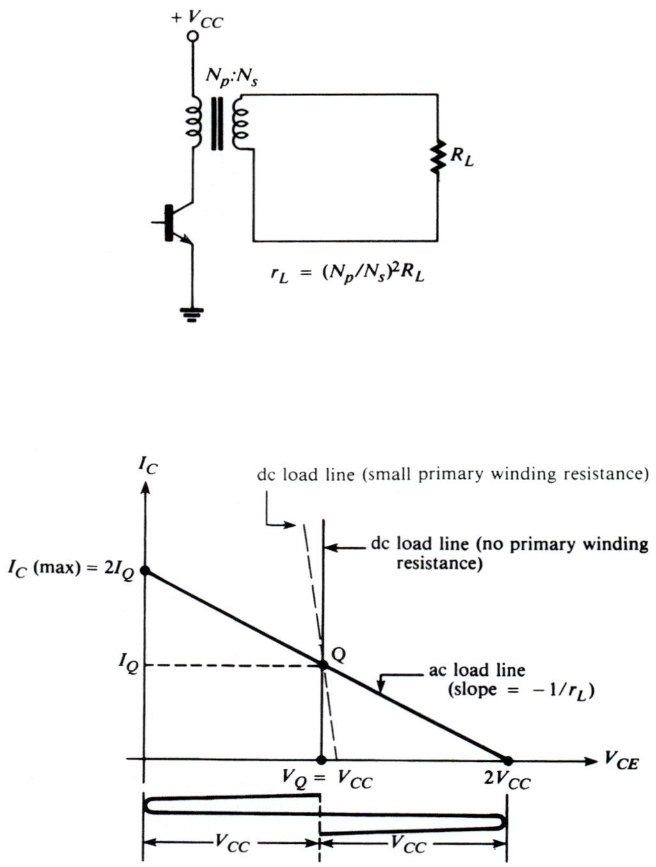

FIGURE 16–9 Transformer-coupled, class-A amplifier and load lines. The amplifier is biased for maximum peak-to-peak variation in V_{CE}.

$$V_{PL} = (N_s/N_p)V_{CC} \tag{16–38}$$

Also, since the slope of the ac load line is $-1/r_L$, we have

$$\frac{|\Delta I_C|}{|\Delta V_{CE}|} = \frac{1}{r_L} = \frac{I_Q}{V_Q}$$

or

$$I_Q = \frac{V_Q}{r_L} = \frac{V_{CC}}{(N_p/N_s)^2 R_L} \tag{16–39}$$

Substituting (16–38) and (16–39) into (16–37), we find the maximum possible efficiency of the class-A, transformer-coupled amplifier:

$$\eta(\max) = \frac{(N_s/N_p)^2 V_{CC}^2}{(2R_L V_{CC})\dfrac{V_{CC}}{(N_p/N_s)^2 R_L}} = 0.5 \tag{16–40}$$

We see that the maximum efficiency is twice that of the series-fed class-A amplifier and six times that of the capacitor-coupled class-A amplifier. This improvement in efficiency is attributable to the absence of external collector resistance that would otherwise consume dc power. Note that the collector efficiency of the transformer-coupled class-A amplifier is the same as the overall amplifier efficiency, because the average power from the dc supply is the same as the collector dissipation:

$$P_S = V_{CC}I_Q = V_Q I_Q = P_C \tag{16–41}$$

In practice, a full output voltage swing of $2V_{CC}$ volts cannot be achieved in a power transistor. The device is prevented from cutting off entirely by virtue of a relatively large leakage current, and it cannot be driven all the way into saturation ($I_C = I_C(\text{max})$) without creating excessive distortion. These points are illustrated in the next example.

EXAMPLE 16–6

The transistor in the power amplifier shown in Figure 16–10 has the output characteristics shown in Figure 16–11. Assume that the transformer has zero resistance.

1. Construct the (ideal) dc and ac load lines necessary to achieve maximum output voltage swing. What quiescent values of collector and base current are necessary to realize the ac load line?
2. What is the smallest value of $I_C(\text{max})$ for which the transistor should be rated?
3. What is the maximum peak-to-peak collector voltage, and what peak-to-peak base current is required to achieve it? Assume that the base current cannot go negative and that, to minimize distortion, the collector should not be driven below 2.5 V in the saturation region.
4. Find the average power delivered to the load under the maximum signal conditions of (3).
5. Find the power dissipated in the transistor under no-signal conditions (standby).
6. Find the efficiency.

Solution

1. The vertical dc load line intersects the V_{CE}-axis at $V_{CC} = 15$ V, as shown in Figure 16–11. To find the slope of the ac load line, we must find r_L. From equation 16–34, $r_L = (N_p/N_s)^2 R_L = (14.6/8)^2(10) = 33.3\ \Omega$. Thus, the slope of the ac load line is $-1/r_L = -1/33.3 = -0.03$. To achieve the ideal maximum output swing, we want the ac load line to intercept the V_{CE}-axis at $2V_{CC} = 30$ V. Because the slope of that line has magnitude 0.03, it will intercept the I_C-axis at $I_C = (0.03\text{ A/V})(30\text{ V}) = 0.9$ A. The ac load line is then drawn between the two intercepts (0 A, 30 V) and (0.9 A, 0 V), as shown in Figure 16–11.

   The ac load line intersects the dc load line at the Q-point. The quiescent collector current at that point is seen to be $I_Q = 0.45$ A. The corresponding

FIGURE 16–10 (Example 16–6)

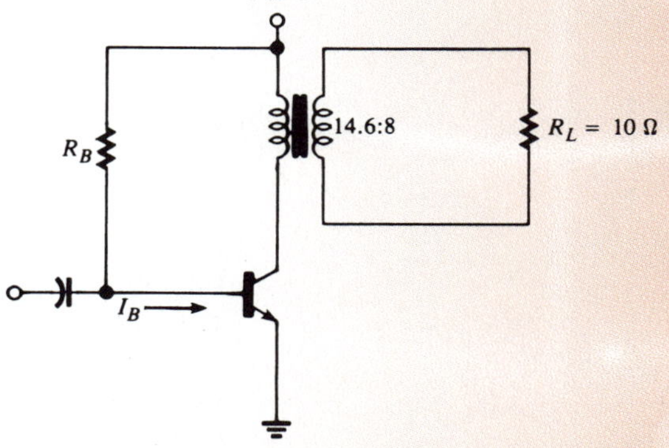

FIGURE 16–11 (Example 16–6)

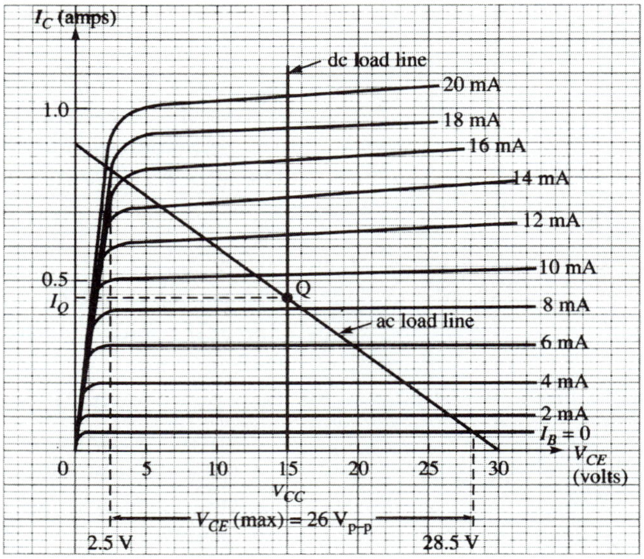

base current is approximately halfway between $I_B = 8$ mA and $I_B = 10$ mA, so the quiescent base current must be 9 mA.

2. The maximum collector current is $I_C(\text{max}) = 0.9$ A, at the intercept of the ac load line on the I_C-axis. Actually, we will not operate the transistor that far into saturation, because we do not allow V_{CE} to fall below 2.5 V. However, a maximum rating of 0.9 A (or 1 A) will provide us with a margin of safety.
3. The maximum value of V_{CE} occurs on the ac load line at the point where $I_B = 0$. As shown in Figure 16–11, this value is 28.5 V. Because the minimum permissible value of V_{CE} is 2.5 V, the maximum peak-to-peak voltage swing is $28.5 - 2.5 = 26$ V p–p. As can be seen on the characteristic curves, the base current must vary from $I_B = 0$ to $I_B = 18$ mA, or 18 mA peak-to-peak, to achieve that voltage swing.
4. The peak primary voltage in the transformer is (1/2)(26 V) = 13 V. Therefore, the peak secondary, or load, voltage is $V_{PL} = (N_s/N_p)(13\text{ V}) = (8/14.6)13 = 7.12$ V. The average load power is then

$$P_L = \frac{V_{PL}^2}{2R_L} = \frac{(7.12)^2}{20} = 2.53\text{ W}$$

5. The standby power dissipation is $P_d = V_Q I_Q = V_{CC} I_Q = (15\text{ V})(0.45\text{ A}) = 6.75$ W.
6. The standby power dissipation found in (5) is the same as the average power supplied from the dc source, so

$$\eta = \frac{2.53\text{ W}}{6.75\text{ W}} = 0.375$$

(Why is this value less than the theoretical maximum of 0.5?)

Class-B Amplifiers

Transistor operation is said to be *class B* when output current varies during only one half-cycle of a sine-wave input. In other words, the transistor is in its

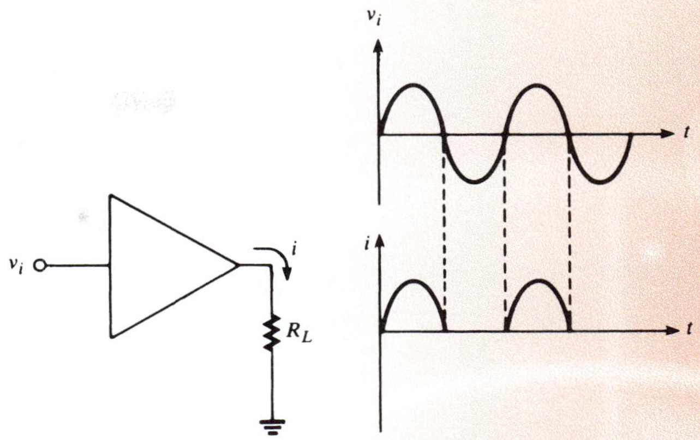

FIGURE 16–12 In class-B operation, output current variations occur only during each positive or each negative half-cycle of input. In the example shown, output current flows during positive half-cycles of the input, and the amplifier is cut off during negative half cycles.

active region, responding to signal input, only during a positive half-cycle or only during a negative half-cycle of the input. This operation is illustrated in Figure 16–12.

It is clear that class-B operation produces an output waveform that is severely clipped (half-wave rectified) and therefore highly distorted. The waveform *by itself* is not suitable for audio applications. However, in practical amplifiers, *two* transistors are operated as class B: one to amplify positive signal variations and the other to amplify negative signal variations. The amplifier output is the composite waveform obtained by combining the waveforms produced by each class-B transistor. We will study these amplifiers in detail in the next section because they are more efficient and more widely used in power applications than are class-A amplifiers. An amplifier utilizing transistors that are operated as class B is called a *class-B* amplifier.

16–5 PUSH-PULL AMPLIFIER PRINCIPLES

A *push-pull* amplifier uses two output devices to drive a load. The name is derived from the fact that one device is primarily (or entirely) responsible for driving current through the load in one direction (pushing), while the other device drives current through the load in the opposite direction (pulling). The output devices are typically two transistors, each operated as class B, one of which conducts only when the input is positive, and the other of which conducts only when the input is negative. This arrangement is called a *class-B, push-pull* amplifier and its principle is illustrated in Figure 16–13.

Note in Figure 16–13 that amplifying devices 1 and 2 are driven by equal-amplitude, *out-of-phase* input signals. The signals are identical except for phase. Here we assume that each device conducts only when its input is positive and is

FIGURE 16–13 The principle of class-B, push-pull operation. Output amplifying device 1 drives current i_L through the load in one direction while 2 is cut off, and device 2 drives current in the opposite direction while 1 is cut off.

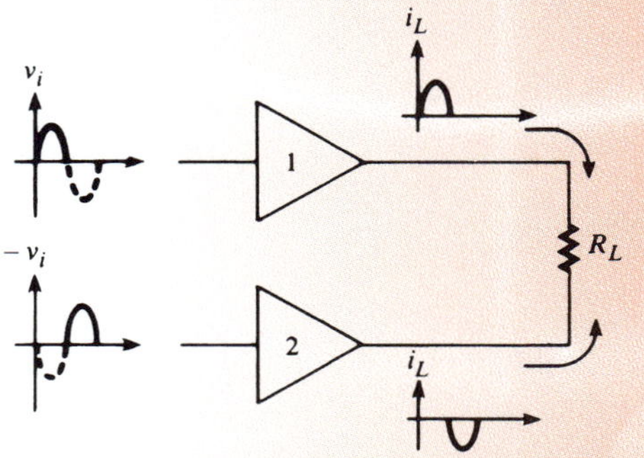

cut off when its input is negative. The net effect is that device 1 produces load current when the input is positive and device 2 produces load current, in the opposite direction, when the input is negative. An example of a device that produces output (collector) current only when its input (base-to-emitter) voltage is positive is an *npn* transistor having no base-biasing circuitry, that is, one that is biased at cutoff. As we shall see, *npn* transistors can be used as the output amplifying devices in push-pull amplifiers. However, the circuitry must be somewhat more elaborate than that diagrammed in Figure 16–13 because we must make provisions for load current to flow through a *complete circuit,* regardless of current direction. Obviously, when amplifying device 1 in Figure 16–13 is cut off, it cannot conduct current produced by device 2, and vice versa.

Push-Pull Amplifiers with Output Transformers

Figure 16–14 shows a push-pull arrangement that permits current to flow in both directions through a load, even though one or the other of the amplifying devices (*npn* transistors) is always cut off. The *output transformer* shown in the figure is the key component. Note that the primary winding is connected between the transistor collectors and that its *center tap* is connected to the dc supply, V_{CC}. The center tap is simply an electrical connection made at the center of the winding, so there are an equal number of turns between each end of the winding and the center tap. The figure does not show the push-pull *driver* circuitry, which must produce out-of-phase signals on the bases of Q_1 and Q_2. We will discuss that circuitry later.

Figure 16–15 shows how current flows through the amplifier during a positive half-cycle of input and during a negative half-cycle of input. In 16–15(a), the input to Q_1 is the positive half-cycle of the signal, and because the input to Q_2 is out-of-phase with that to Q_1, Q_2 is driven by a negative half-cycle. Notice that neither class-B transistor is biased. Consequently, the positive base voltage on Q_1 causes it to turn on and conduct current in the counterclockwise path shown. The negative base voltage on Q_2 keeps that transistor cut off. Current flowing in the upper half of the transformer's primary induces current in the secondary, and current flows through the load. In 16–15(b), the input signal on the base of Q_1 has gone negative, so its inverse on the base of Q_2 is positive. Therefore, Q_2 conducts current in the clockwise path shown, and Q_1 is cut off. Current induced in the secondary winding is in the direction opposite that shown in Figure 16–15(a). The upshot is that current flows through the load in one direction when the input signal is positive and in the opposite direction when the input signal is negative, just as it should in an ac amplifier.

Figure 16–16 displays current flow in the push-pull amplifier in the form of a *timing diagram.* Here the complete current waveforms are shown over

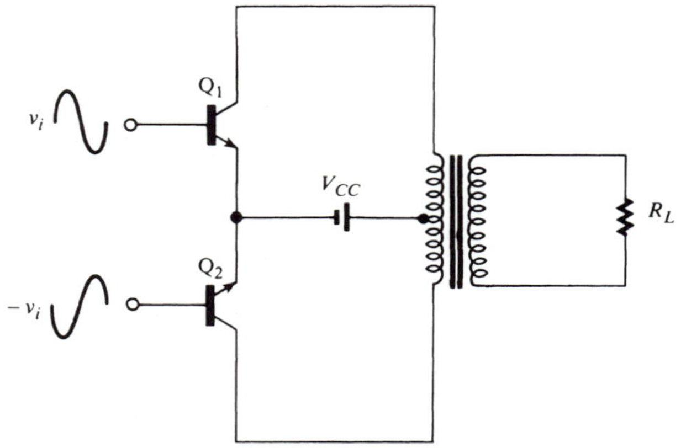

FIGURE 16–14 A center-tapped output transformer used in a push-pull amplifier

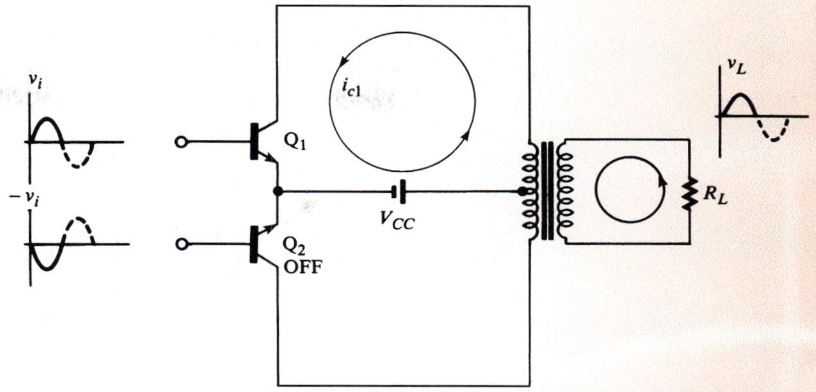

FIGURE 16–15 Current flow in a push-pull amplifier with output transformer. (Currents are those that flow during the portion of the input shown as a solid line)

(a) When v_i is positive, Q_1 conducts and Q_2 is cut off. A counterclockwise current is induced in the load.

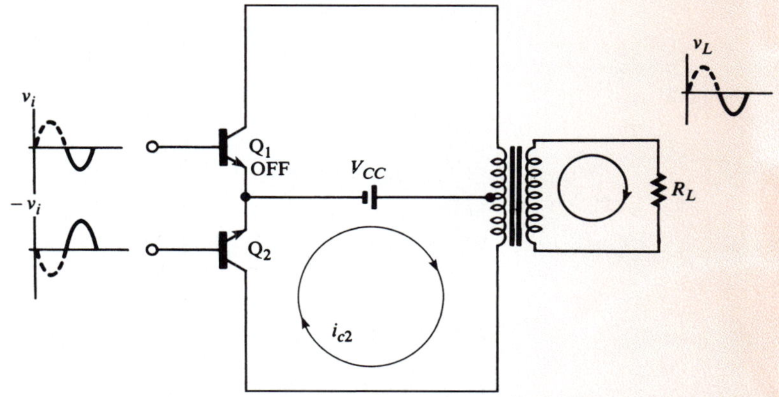

(b) When v_i is negative, Q_2 conducts and Q_1 is cut off. A clockwise current is induced in the load.

two full cycles of input. For purposes of this illustration, counterclockwise current (in Figure 16–15) is arbitrarily assumed to be positive and clockwise current is therefore negative. Note that, as far as the load is concerned, current flows during the full 360° of input signal. Figure 16–16(e) shows that the current i_S from the power supply varies from 0 to the peak value I_P every half-cycle. Because the current variation is so large, the power supply used in a push-pull amplifier must be particularly well regulated—that is, it must maintain a constant voltage, independent of current demand.

Class-B Efficiency

The principal advantage of using a class-B power amplifier is that it is possible to achieve an efficiency greater than that attainable in a class-A amplifier. The improvement in efficiency stems from the fact that no power is dissipated in a transistor during the time intervals that it is cut off. Also, like the transformer-coupled class-A amplifier, there is no external collector resistance that would otherwise consume power.

We will derive an expression for the maximum efficiency of a class-B push-pull amplifier assuming ideal conditions: perfectly matched transistors and zero resistance in the transformer windings. The current supplied by each transistor is a half-wave–rectified waveform, as shown in Figure 16–16. Let I_P represent the peak value of each. Then, the peak value of the current in the secondary winding, which is the same as the peak load current, is

$$I_{PL} = (N_p/N_s)I_P \quad (16\text{–}42)$$

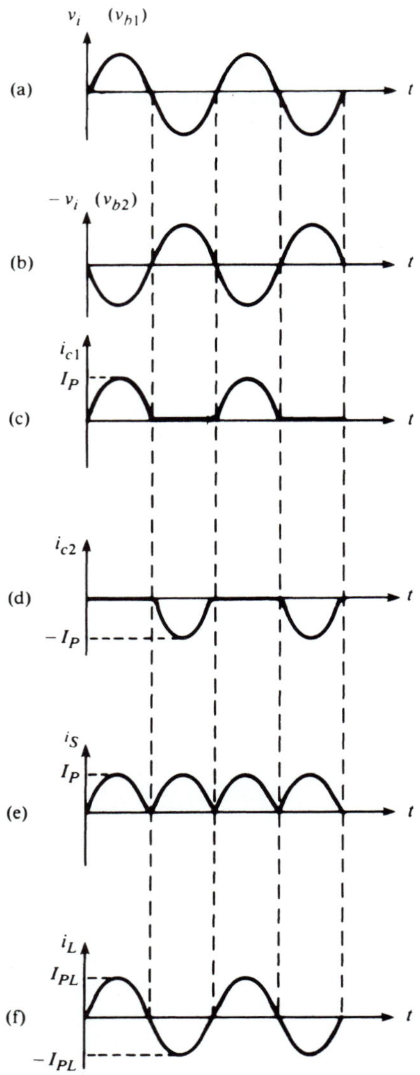

FIGURE 16–16 A timing diagram showing currents in the push-pull amplifier of Figure 16–15 over two full cycles of input

where N_p/N_s is the turns ratio between one-half the primary winding and the secondary winding. (Note that only those primary turns between one end of the winding and its center tap are used to induce current in the secondary.) Similarly, the peak value of the load voltage is

$$V_{PL} = (N_s/N_p)V_P \tag{16–43}$$

where V_P is the peak value of the primary (collector) voltage. Because the load voltage and load current are sinusoidal, the average power delivered to the load is, from equation 16–10,

$$P_L = \frac{V_{PL}I_{PL}}{2} = \frac{(N_s/N_p)V_P(N_p/N_s)I_P}{2} = \frac{V_PI_P}{2} \tag{16–44}$$

As shown in Figure 16–16(e), the power supply current is a full-wave–rectified waveform having peak value I_P. The dc, or average, value of such a waveform is known to be $2I_P/\pi$. Therefore, the average power delivered to the circuit by the dc supply is

$$P_S = \frac{2I_PV_{CC}}{\pi} \tag{16–45}$$

The efficiency is then

$$\eta = \frac{P_L}{P_S} = \frac{V_P I_P/2}{2I_P V_{CC}/\pi} = \frac{\pi V_P}{4V_{CC}} \qquad (16\text{–}46)$$

Under maximum signal conditions, $V_P = V_{CC}$, and 16–46 becomes

$$\eta(\text{max}) = \frac{\pi}{4} = 0.785 \qquad (16\text{–}47)$$

Equation 16–47 shows that a class-B push-pull amplifier can be operated with a much higher efficiency than the class-A amplifiers studied earlier. Furthermore, unlike the case of class-A amplifiers, the power dissipated in the transistors is 0 under standby (zero-signal) conditions because both transistors are cut off. A general expression for the total power dissipated in the transistors can be obtained by realizing that it equals the difference between the total power supplied by the dc source and the total power delivered to the load:

$$P_d = \frac{2I_P V_{CC}}{\pi} - \frac{V_P I_P}{2} \qquad (16\text{–}48)$$

Using calculus, it can be shown that P_d is maximum when $V_P = 2V_{CC}/\pi = 0.636V_{CC}$. We conclude that maximum transistor dissipation does *not* occur when maximum load power is delivered ($V_P = V_{CC}$), but at the intermediate level $V_P = 0.636V_{CC}$.

EXAMPLE 16–7

The push-pull amplifier in Figure 16–14 has $V_{CC} = 20$ V and $R_L = 10\ \Omega$. The *total* number of turns on the primary winding is 100, and the secondary winding has 50 turns. Assume that the transformer has zero resistance.

1. Find the maximum power that can be delivered to the load.
2. Find the power dissipated in each transistor when maximum power is delivered to the load.
3. Find the power delivered to the load and the power dissipated in each transistor when transistor power dissipation is maximum.

Solution

1. The turns ratio between each half of the primary and the secondary is $N_p/N_s = (100/2){:}50 = 50{:}50$. Therefore, the peak values of primary and secondary voltages are equal, as are the peak values of primary and secondary current.

$$V_P(\text{max}) = V_{PL}(\text{max}) = V_{CC} = 20\text{ V}$$

$$I_P(\text{max}) = I_{PL}(\text{max}) = \frac{V_{CC}}{R_L} = \frac{20\text{ V}}{10\ \Omega} = 2\text{ A}$$

Therefore,

$$P_L(\text{max}) = \frac{V_P(\text{max})I_P(\text{max})}{2} = \frac{(20\text{ V})(2\text{ A})}{2} = 20\text{ W}$$

2. From equation 16–48,

$$P_d = \frac{2I_P V_{CC}}{\pi} - \frac{V_P I_P}{2} = \frac{2(2\text{ A})(20\text{ V})}{\pi} - 20\text{ W} = 5.46\text{ W}$$

Since 5.46 W is the total power dissipated by both transistors, each dissipates one-half that amount, or 2.73 W.

3. Transistor power dissipation is maximum when $V_P = 0.636V_{CC} = (0.636)(20) = 12.72$ V. Then,

$$I_P = I_{PL} = \frac{12.72\text{ V}}{10\ \Omega} = 1.272\text{ A}$$

and

$$P_L = \frac{V_P I_P}{2} = \frac{(12.72\text{ V})(1.272\text{ A})}{2} = 8.09\text{ W}$$

From equation 16–48,

$$P_d(\max) = \frac{2I_P V_{CC}}{\pi} - \frac{V_P I_P}{2} = \frac{2(1.272\text{ A})(20\text{ V})}{\pi} - 8.09\text{ W} = 8.09\text{ W}$$

The maximum power dissipation in each transistor is then 8.09/2 = 4.05 W.

The preceding example demonstrates some results that are true in general for push-pull amplifiers: (1) When transistor power dissipation is maximum, its value equals the power delivered to the load; and (2) the maximum total transistor power dissipation equals approximately 40% of the maximum power that can be delivered to the load.

16–6 PUSH-PULL DRIVERS

We have seen that the push-pull amplifier described earlier must be driven by out-of-phase input signals. Figure 16–17 shows how a transformer can be used to provide the required drive signals. Here, the *secondary* winding has a grounded center tap that effectively splits the secondary voltage into two out-of-phase signals, each having one-half the peak value of the total secondary voltage. The input signal is applied across the primary winding, and a voltage is developed across secondary terminals A–B in the usual transformer fashion. To understand the phase-splitting action, consider the instant at which the voltage across A–B is +6 V, as shown in the figure. Then, since the center point is at ground, the voltage from A to ground must be +3 V and that from B to ground must be −3 V: $V_{AB} = V_A - V_B = 3 - (-3) = +6$ V. The same logic applied at every instant throughout a complete cycle shows that v_B with respect to ground is always the negative of v_A with respect to ground; in other words, v_A and v_B are equal-amplitude, out-of-phase driver signals, as required.

A specially designed amplifier, called a *phase-splitter,* can be used instead of a driver transformer to produce equal amplitude, out-of-phase drive signals.

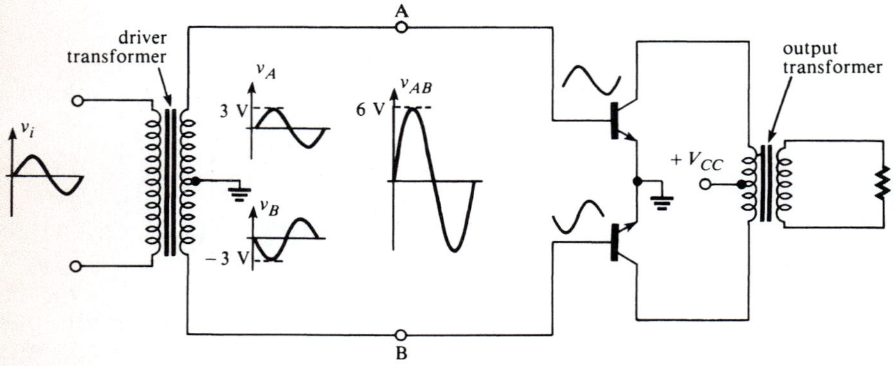

FIGURE 16–17 Using a driver transformer to create equal-amplitude, out-of-phase drive signals (v_A and v_B) for a push-pull amplifier

FIGURE 16–18 Phase-splitting methods for obtaining push-pull drive signals

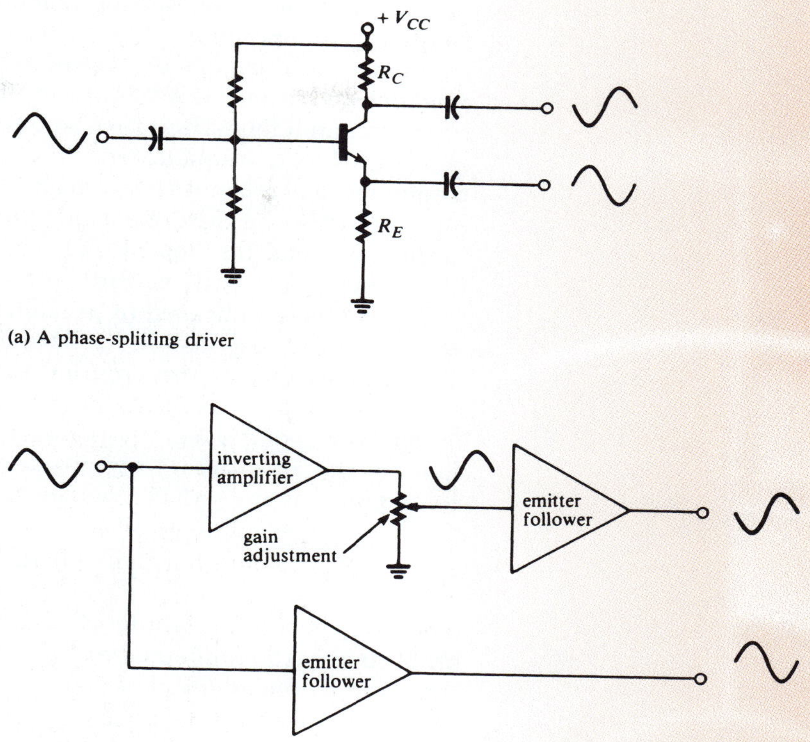

(a) A phase-splitting driver

(b) An improved method for obtaining equal-amplitude, out-of-phase signals

Figure 16–18 shows two possible designs. Figure 16–18(a) is a conventional amplifier circuit with outputs taken at the collector and at the emitter. The collector output is out of phase with the input and the emitter output is in phase with the input, so the two outputs are out of phase with each other. With no load connected to either output, the output signals will have approximately equal amplitudes if $R_C = R_E$. However, the output impedance at the collector is significantly greater than that at the emitter, so when loads are connected, each output will be affected differently. As a consequence, the signal amplitudes will no longer be equal, and gain adjustments will be required. Furthermore, the nonlinear nature of the large-signal load (the output transistors) means that the gain of the collector output may vary appreciably with signal level. Intolerable distortion can be created as a result. A better way to drive the output transistors is from two low-impedance signal sources, as shown in Figure 16–18(b). Here, the output from an inverting amplifier is buffered by an emitter-follower stage, as is the original signal.

16–7 HARMONIC DISTORTION AND FEEDBACK

Harmonic Distortion

Recall that any periodic waveform, sinusoidal or otherwise, can be represented as the sum of an infinite number of sine waves having different amplitudes, frequencies, and phase relations. The mathematical technique called *Fourier analysis* is concerned with finding the exact amplitudes, frequencies, and phase angles of the sine-wave components that reproduce a given waveform when they are added together. Of course, if the waveform is itself a pure sine wave, then all other frequency components have zero amplitudes.

Apart from a possible dc component (the average value), the lowest frequency component in the infinite sum is a sine wave having the same frequency as the periodic waveform itself. This component is called the *fundamental* and usually has an amplitude greater than that of any other frequency component. The other frequency components are called *harmonics*, and each has a frequency that is an *integer multiple* of the fundamental frequency. For example, the second harmonic of a 3-kHz waveform has frequency 6 kHz, the third harmonic has frequency 9 kHz, and so forth. Not all harmonics need be present. For example, a square wave contains only odd harmonics (third, fifth, seventh, etc.).

As we have indicated in previous discussions, nonlinear amplifier characteristics are responsible for harmonic distortion of an output signal. This distortion is in fact the creation of harmonic frequencies that would not otherwise be present in the output when the input is a pure sine wave. The extent to which a particular harmonic component distorts a signal is specified by the ratio of its amplitude to the amplitude of the fundamental component, expressed as a percentage:

$$\%\ n\text{th harmonic distortion} = \%D_n = \frac{A_n}{A_1} \times 100\% \qquad (16\text{–}49)$$

where A_n is the amplitude of the nth harmonic component and A_1 is the amplitude of the fundamental. For example, if a 1-V-peak, 10-kHz sine wave is distorted by the addition of a 0.1-V-peak, 20-kHz sine wave and a 0.05-V-peak, 30-kHz sine wave, then it has

$$\%D_2 = \frac{0.1}{1} \times 100\% = 10\% \text{ second harmonic distortion}$$

and

$$\%D_3 = \frac{0.05}{1} \times 100\% = 5\% \text{ third harmonic distortion}$$

The *total harmonic distortion* (THD) is the square root of the sum of the squares of all the individual harmonic distortions:

$$\%\text{THD} = \sqrt{D_2^2 + D_3^2 + \cdots} \times 100\% \qquad (16\text{–}50)$$

Special instruments, called *distortion analyzers*, are available for measuring the total harmonic distortion in a waveform.

EXAMPLE 16–8

The principal harmonics in a certain 15-kHz signal having a 10-V pk fundamental are the second and fourth. All other harmonics are negligibly small. If the THD is 12% and the amplitude of the second harmonic is 0.5 V pk, what is the amplitude of the 60-kHz harmonic?

Solution

$$\text{THD} = 0.12 = \sqrt{D_2^2 + D_4^2}$$

$$0.0144 = D_2^2 + D_4^2$$

$$D_2 = 0.5/10 = 0.05 \quad (5\%)$$

$$0.0144 = (0.05)^2 + D_4^2$$

$$D_4 = \sqrt{0.0144 - 0.0025} = 0.1091 \qquad (10.91\%)$$

$$A_4 = D_4 A_1 = (0.1091)(10) = 1.091 \text{ V}$$

Using Negative Feedback to Reduce Distortion

We have mentioned in several previous discussions that one of the important benefits of negative feedback is that it reduces distortion caused by amplifier nonlinearities. Having defined a quantitative measure of distortion, we can now undertake a quantitative investigation of the degree to which feedback affects harmonic distortion in the output of an amplifier. Let us begin by recognizing that a nonlinear amplifier is essentially an amplifier whose gain changes with signal level. Figure 16–19(a) shows the *transfer characteristic* of an ideal, distortionless amplifier, in which the gain, i.e., the slope of the characteristic, $\Delta V_o/\Delta V_i = 50$, is constant. Figure 16–19(b) shows a nonlinear transfer characteristic in which the slope, and hence the gain, increases with increasing signal level. We see that the gain at $V_i = 0.2$ V is 50, while the gain at 0.4 V is 100. There is a 100% increase in gain over that range, so serious output distortion is to be expected.

In earlier discussions of negative feedback, we showed that the closed-loop gain A_{CL} of an amplifier having open-loop gain A and feedback ratio β can be found from

$$A_{CL} = \frac{A}{1 + A\beta} \tag{16–51}$$

Let us suppose we introduce negative feedback into the amplifier with the nonlinear characteristic shown in Figure 16–19(b). Assume that $\beta = 0.05$. Then the gain at $v_i = 0.2$ V becomes

$$A_{CL} = \frac{50}{1 + 50(0.05)} = 14.29$$

and the gain at $v_i = 0.4$ V becomes

$$A_{CL} = \frac{100}{1 + 100(0.05)} = 16.67$$

Notice that the *change* in gain (16.7%) is now much less than it was without feedback (100%). The effect of feedback has been to "linearize" the transfer characteristics somewhat, as shown in Figure 16–19(c). We conclude that less-severe distortion will result. Of course, the penalty we pay for this improved performance is an overall reduction in gain.

To compute the reduction in harmonic distortion caused by negative feedback, consider the models shown in Figure 16–20. Figure 16–20(a) shows how amplifier distortion (without feedback) can be represented as a distortionless amplifier having a distortion component d added to its output. For example, d could represent the level of a second-harmonic sine wave added to an otherwise undistorted fundamental. The output of the amplifier is

$$v_o = Av_i + d \tag{16–52}$$

so the harmonic distortion is

$$D = \frac{d}{Av_i} \tag{16–53}$$

Figure 16–20(b) shows the amplifier when negative feedback is connected. We realize that the negative feedback will reduce the closed-loop gain, so the output level will be reduced if v_i remains the same as in (a). But, for comparison purposes, we want the output levels in both cases to be the same since the amount of distortion depends on that level. Accordingly, we assume that v_i in (b) is increased to v_i' as necessary to make the distortion component d equal to its value in (a). As shown in the figure,

$$v_o = A\,(v_i' - \beta v_o) + d \tag{16–54}$$

Solving for v_o, we find

$$v_o = \frac{Av_i'}{1 + A\beta} + \frac{d}{1 + A\beta} \tag{16–55}$$

FIGURE 16–19 Transfer characteristics and the effect of negative feedback

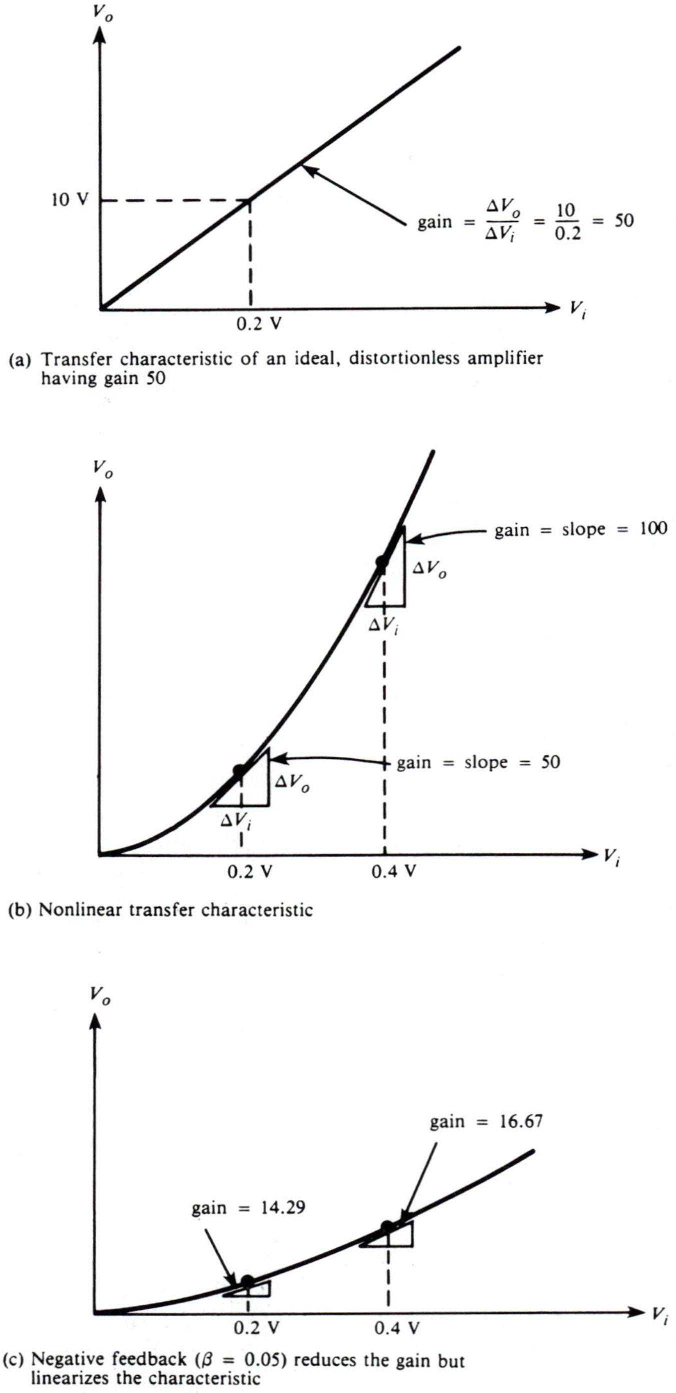

(a) Transfer characteristic of an ideal, distortionless amplifier having gain 50

(b) Nonlinear transfer characteristic

(c) Negative feedback (β = 0.05) reduces the gain but linearizes the characteristic

FIGURE 16–20 Amplifier distortion models

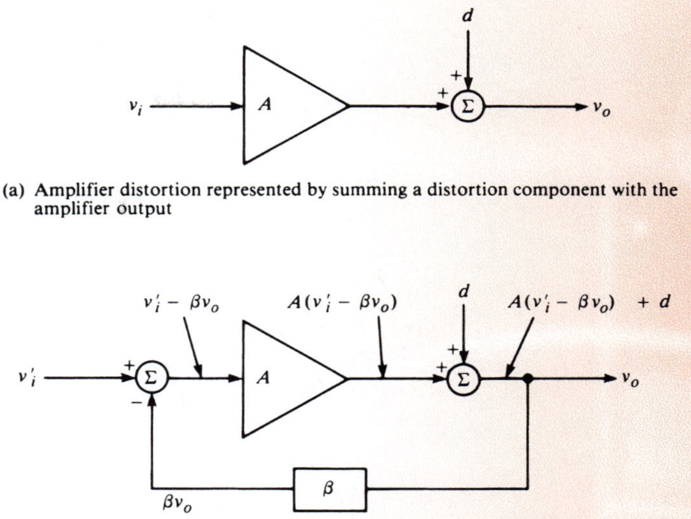

(a) Amplifier distortion represented by summing a distortion component with the amplifier output

(b) Model for an amplifier with distortion component d and negative feedback ratio β

When we adjusted v_i to v_i', we did it so that the amplifier outputs in (a) and (b) of the figure would be the same:

$$Av_i = \frac{A}{1 + A\beta} v_i' \tag{16–56}$$

or

$$v_i' = (1 + A\beta)v_i \tag{16–57}$$

Substituting (16–57) into (16–55), we find

$$v_o = Av_i + \frac{d}{1 + A\beta} \tag{16–58}$$

The harmonic distortion is now

$$D = \frac{d/(1 + A\beta)}{Av_i} = \frac{d}{(1 + A\beta)Av_i} \tag{16–59}$$

Comparing with equation 16–53, we see that the distortion has been reduced by the factor $1/(1 + A\beta)$.

It is important to remember that it was necessary to increase the input level in our analysis to achieve the improved performance. From a practical standpoint, that means that a relatively distortion-free *preamplifier* must be used to compensate for the loss in gain caused by negative feedback around the output amplifier. Fortunately, it is generally easier to achieve low distortion in small-signal preamplifiers than in large-signal, high-power output amplifiers.

EXAMPLE 16–9

An output amplifier has a voltage gain of 120 and generates 20% harmonic distortion with no feedback.

1. How much negative feedback should be used if it is desired to reduce the distortion to 2%?
2. How much preamplifier gain will have to be provided in cascade with the output amplifier to maintain an overall gain equal to that without feedback?

Solution

1. From equation 16–59,

$$D = \frac{d}{(1 + A\beta)Av_i}$$

where $d/Av_i = 0.2$ is the distortion without feedback. Then,

$$0.02 = \frac{0.2}{1 + A\beta} = \frac{0.2}{1 + 120\beta}$$

or $\beta = 0.075$. Thus, the negative feedback voltage must be 7.5% of the output voltage.

2. The closed-loop gain with 7.5% feedback is

$$A_{CL} = \frac{A}{1 + A\beta} = \frac{120}{1 + 120(0.075)} = 12$$

To maintain an overall gain of 120, the preamplifier gain, A_p, must be such that $12A_p = 120$, or $A_p = 10$.

16–8 DISTORTION IN PUSH-PULL AMPLIFIERS

Cancellation of Even Harmonics

Recall that push-pull operation effectively produces in a load a waveform proportionate to the *difference* between two input signals. Under normal operation, the signals are out of phase, so their waveform is reproduced in the load. If the signals were in phase, cancellation would occur. It is instructive to view a push-pull output as the difference between two distorted (half-wave rectified) sine waves that are out of phase with each other. This viewpoint is illustrated in Figure 16–21. It can be shown that a half-wave–rectified sine wave contains only the fundamental and all *even* harmonics. Figure 16–21(a) shows the two out-of-phase, half-wave–rectified sine waves that drive the load, and 16–21(b) shows the fundamental and second-harmonic components of each. Notice that the fundamental components are out of phase. Therefore, the fundamental component is reproduced in the load, as we have already seen (Figure 16–16). However, the second-harmonic components are in phase, and therefore cancel in the load. Although not shown in this figure, the fourth and all other even harmonics are also in phase and therefore also cancel. Our conclusion is an important property of push-pull amplifiers: *Even harmonics are canceled in push-pull operation.*

The cancellation of even harmonics is an important factor in reducing distortion in push-pull amplifiers. However, perfect cancellation would occur only if the two sides were perfectly matched and perfectly balanced: identical transistors, identical drivers, and a perfectly center-tapped transformer. Of course, this is not the case in practice, but even imperfect push-pull operation reduces even harmonic distortion. Odd harmonics are out of phase, so cancellation of those components does not occur.

Crossover Distortion

Recall that a forward-biasing voltage applied across a *pn* junction must be raised to a certain level (about 0.7 V for silicon) before the junction will conduct any significant current. Similarly, the voltage across the

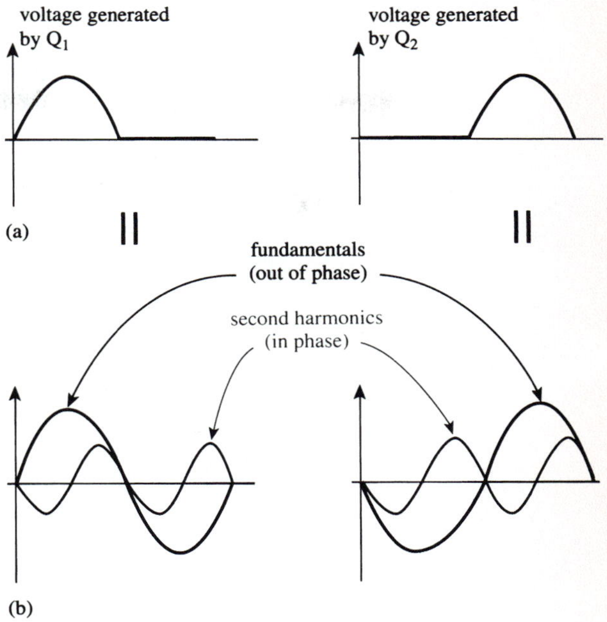

FIGURE 16–21 The fundamental components of the half-wave–rectified waveforms are out of phase, but the second harmonics are in phase. Therefore, second-harmonic distortion is canceled in push-pull operation

base–emitter junction of a transistor must reach that level before any appreciable base current, and hence collector current, can flow. As a consequence, the drive signal applied to a class-B transistor must reach a certain minimum level before its collector current is properly in the active region. This fact is the principal source of distortion in a class-B, push-pull amplifier, as illustrated in Figure 16–22. Figure 16–22(a) shows that the initial rise of collector current in a class-B transistor lags the initial rise of input voltage for the reason we have described. Also, collector current prematurely drops to 0 when the input voltage approaches 0. Figure 16–22(b) shows the voltage waveform that is produced in the load of a push-pull amplifier when the distortion generated during each half-cycle by each class-B transistor is combined. This distortion is called *crossover* distortion because it occurs where the composite waveform crosses the zero-voltage axis. Clearly, the effect of crossover distortion becomes more serious as the signal level becomes smaller.

Class-AB Operation

Crossover distortion can be reduced or eliminated in a push-pull amplifier by biasing each transistor slightly into conduction. When a small forward-biasing voltage is applied across each base–emitter junction, and a small base current flows under no-signal conditions, it is not necessary for the base drive signal to overcome the built-in junction potential before active operation can occur. A simple voltage-divider bias network can be connected across each base for this purpose, as shown in Figure 16–23. Figure 16–23(a) shows how two resistors can provide bias for both transistors when a driver transformer is used. Figure 16–23(b) shows the use of two voltage dividers when the drive signals are capacitor-coupled. Typically, the base–emitter junctions are biased to about 0.5 V for silicon transistors, or so that the collector current under no-signal conditions is about 1% of its peak signal value.

When a transistor is biased slightly into conduction, output current will flow during more than one-half cycle of a sine-wave input, as illustrated in

FIGURE 16–22 Crossover distortion

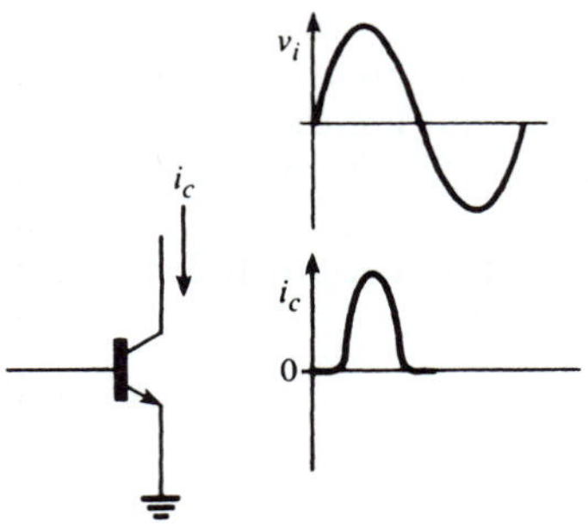

(a) Collector current in a class-B transistor does not follow input voltage in the regions near 0 (crossovers)

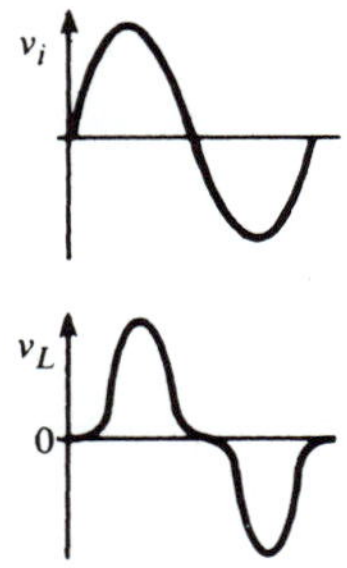

(b) The load voltage in a class-B push-pull amplifier, showing the combined effects of the distortion generated during each half-cycle of input

FIGURE 16–23 Methods for providing a slight forward bias for push-pull transistors to reduce crossover distortion

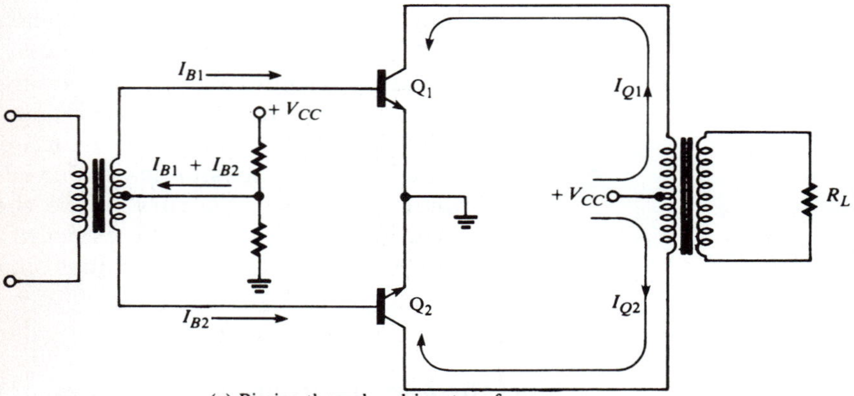

(a) Biasing through a driver transformer

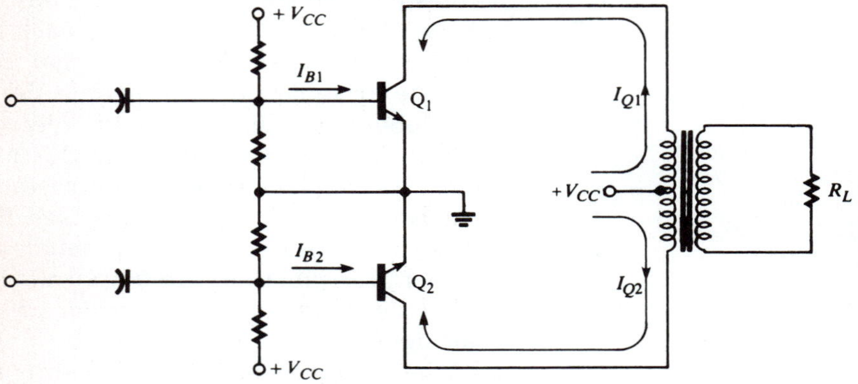

(b) Biasing with capacity-coupled drivers

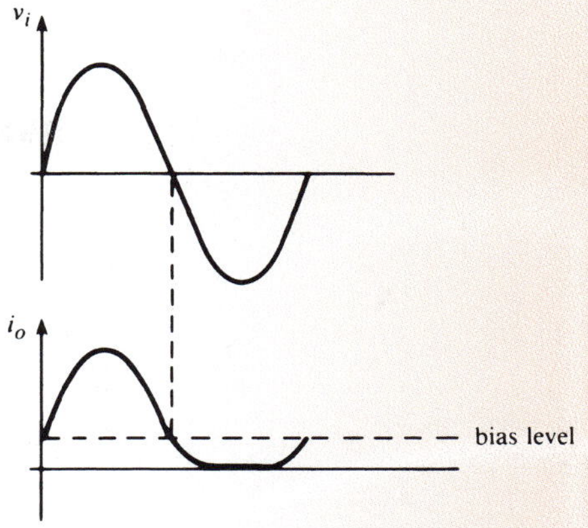

FIGURE 16–24 Class-AB operation. Output current i_o flows during more than one-half but less than a full cycle of input

Figure 16–24. As can be seen in the figure, conduction occurs for more than one-half but less than a full cycle of input. This operation, which is neither class A nor class B, is called *class-AB* operation.

Although class-AB operation reduces crossover distortion in a push-pull amplifier, it has the disadvantage of reducing amplifier efficiency. The fact that bias current is always present means that there is continuous power dissipation in both transistors, including the time intervals during which one of the transistors would be cut off if the operation were class B. The extent to which efficiency is reduced depends directly on how heavily the transistors are biased, and the maximum achievable efficiency is somewhere between that which can be obtained in class-A operation (0.5) and that attainable in class-B operation (0.785).

In Figure 16–23, notice that the quiescent collector currents I_{Q1} and I_{Q2} flow in opposite directions through the primary of the transformer. Thus, the magnetic flux created in the transformer by one dc current opposes that created by the other, and the net flux is 0. This is an advantageous situation, in comparison with the class-A transformer-coupled amplifier, because it means that transformer current can swing positive and negative through a maximum range. If the transformer flux had a bias component, the signal swing would be limited in one direction by the onset of magnetic saturation.

16–9 TRANSFORMERLESS PUSH-PULL AMPLIFIERS

Complementary Push-Pull Amplifiers

The principal disadvantage of the push-pull amplifier circuits we have discussed so far is the cost and bulk of their output transformers. High-power amplifiers in particular are encumbered by the need for very large transformers capable of conducting large currents without saturating.Figure 16–25(a) shows a popular design using *complementary* (*pnp* and *npn*) output transistors to eliminate the need for an output transformer in push-pull operation. This design also eliminates the need for a driver transformer or any other drive circuitry producing out-of-phase signals.

Figure 16–25(b) shows that current flows in a counterclockwise path through the load when the input signal on the base of *npn* transistor Q_1 is positive. The positive input simultaneously appears on the base of *pnp* transistor Q_2 and keeps it cut off. When the input is negative, Q_1 is cut off and Q_2 conducts current through the load in the opposite direction, as shown in 16-25(c).

Note that each transistor in Figure 16–25 drives the load in an emitter-follower configuration. The advantageous consequence is that low-impedance

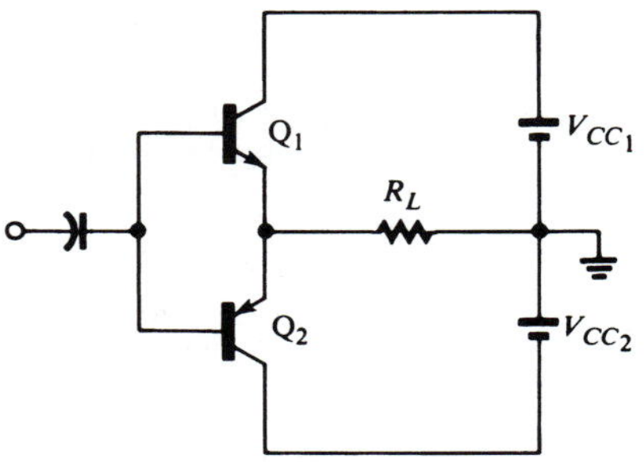

(a) Complementary push-pull amplifier

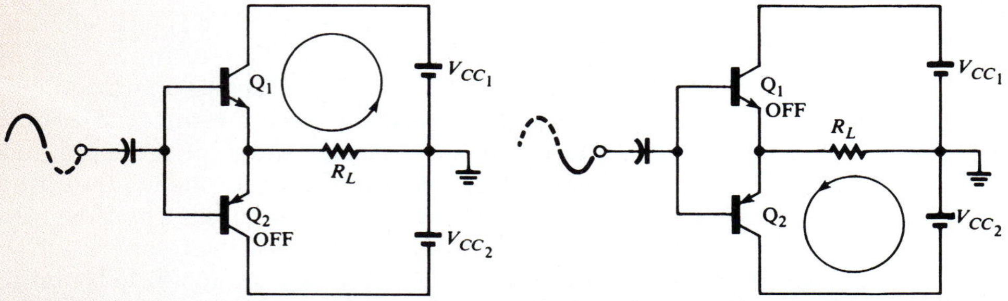

(b) When the input is positive, Q_1 conducts and Q_2 is cut off.

(c) When the input is negative, Q_2 conducts and Q_1 is cut off.

FIGURE 16–25 Push-pull amplification using complementary transistors. Note that the direction of current through R_L alternates each half-cycle, as required

loads can be driven from a high-impedance source. Also, the large negative feedback inherent in emitter-follower operation reduces the problem of output distortion. However, as is the case in all emitter followers, voltage gains greater than unity cannot be realized. The maximum positive voltage swing is V_{CC1} and the maximum negative swing is V_{CC2}. Normally, $|V_{CC1}| = |V_{CC2}| = V_{CC}$, so the maximum peak-to-peak swing is $2V_{CC}$ volts. Since the voltage gain is near unity, the input must also swing through $2V_{CC}$ volts to realize maximum output swing. Notice that under conditions of maximum swing, the cutoff transistor experiences a maximum reverse-biasing collector-to-base voltage of $2V_{CC}$ volts. For example, when Q_1 is off, its collector is at $+V_{CC}$ and its base voltage (the input signal) swings to $-V_{CC}$. Thus, each transistor must have a rated breakdown voltage of at least $2V_{CC}$.

EXAMPLE 16–10

The amplifier in Figure 16–25 must deliver 30 W to a 15-Ω load under maximum drive.

1. What is the minimum value required for each supply voltage?
2. What minimum collector-to-base breakdown-voltage rating should each transistor have?

Solution

1.
$$P_L(\text{max}) = \frac{V_P^2(\text{max})}{2R_L}$$

$$V_p(\text{max}) = \sqrt{2R_L P_L(\text{max})} = \sqrt{2(15)(30)} = 30 \text{ V}$$

Therefore, for maximum swing, $V_{CC} = V_P(\max) = 30$ V; i.e., $V_{CC1} = +30$ V and $V_{CC2} = -30$ V.

2. The minimum rated breakdown is $2V_{CC} = 60$ V.

EXAMPLE 16–11

PSPICE

The push-pull amplifier in Figure 16–25 has $|V_{CC1}| = |V_{CC2}| = 30$ V and $R_L = 15\ \Omega$. The input coupling capacitor is 100 μF. Assuming the transistors have their default parameters, use PSpice to obtain a plot of one full cycle of output when the input is

1. a 1-kHz sine wave with peak value 30 V,
2. a 1-kHz sine wave with peak value 2 V.

Solution

Figure 16–26(a) shows the PSpice circuit and input data file for the case where the input signal has peak value 30 V. The resulting plot is shown in Figure 16–26(b). We see that the computed output swings from −29 to +28.98 V. The output that results when the peak value of the input is changed to 2 V is shown in Figure 16–26(c). Crossover distortion is now clearly apparent in this low-amplitude case.

Figure 16–27 shows two variations on the basic complementary push-pull amplifier. In (a) the transistors are replaced by emitter-follower Darlington pairs. Because power transistors tend to have low betas, particularly at high current levels, the Darlington pair improves the drive capabilities and the current gain of the amplifier, as discussed in Chapter 10. These devices are available in matched complementary sets with current ratings up to 20 A. Figure 16–27(b) shows how emitter-follower transistors can be operated in parallel to increase the overall current-handling capability of the amplifier. In this variation, the parallel transistors must be matched closely to prevent "current hogging," wherein one device carries most of the load, thus subverting the intention of load sharing. Small emitter resistors, R_E, shown in the figure introduce negative feedback and help prevent current hogging, at the expense of efficiency. Amplifiers capable of dissipating several hundred watts have been constructed using this arrangement.

FIGURE 16–26
(Example 16-11)

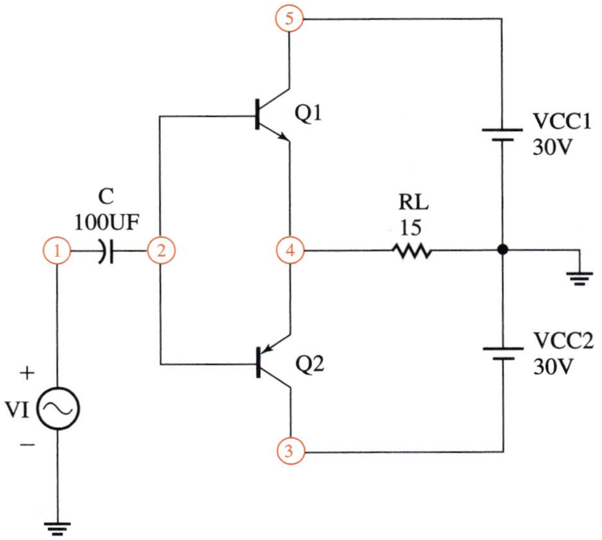

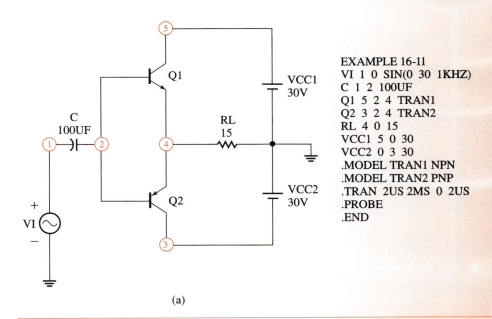

```
EXAMPLE 16-11
VI 1 0 SIN(0 30 1KHZ)
C 1 2 100UF
Q1 5 2 4 TRAN1
Q2 3 2 4 TRAN2
RL 4 0 15
VCC1 5 0 30
VCC2 0 3 30
.MODEL TRAN1 NPN
.MODEL TRAN2 PNP
.TRAN 2US 2MS 0 2US
.PROBE
.END
```

(a)

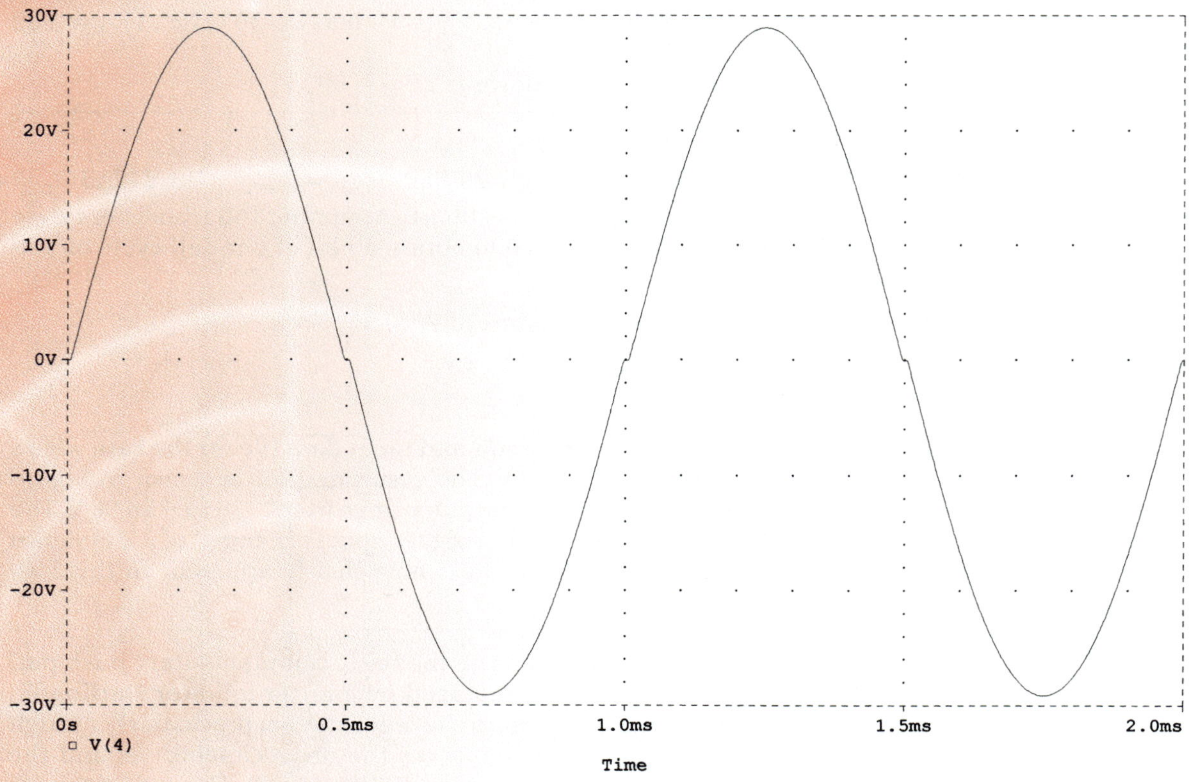

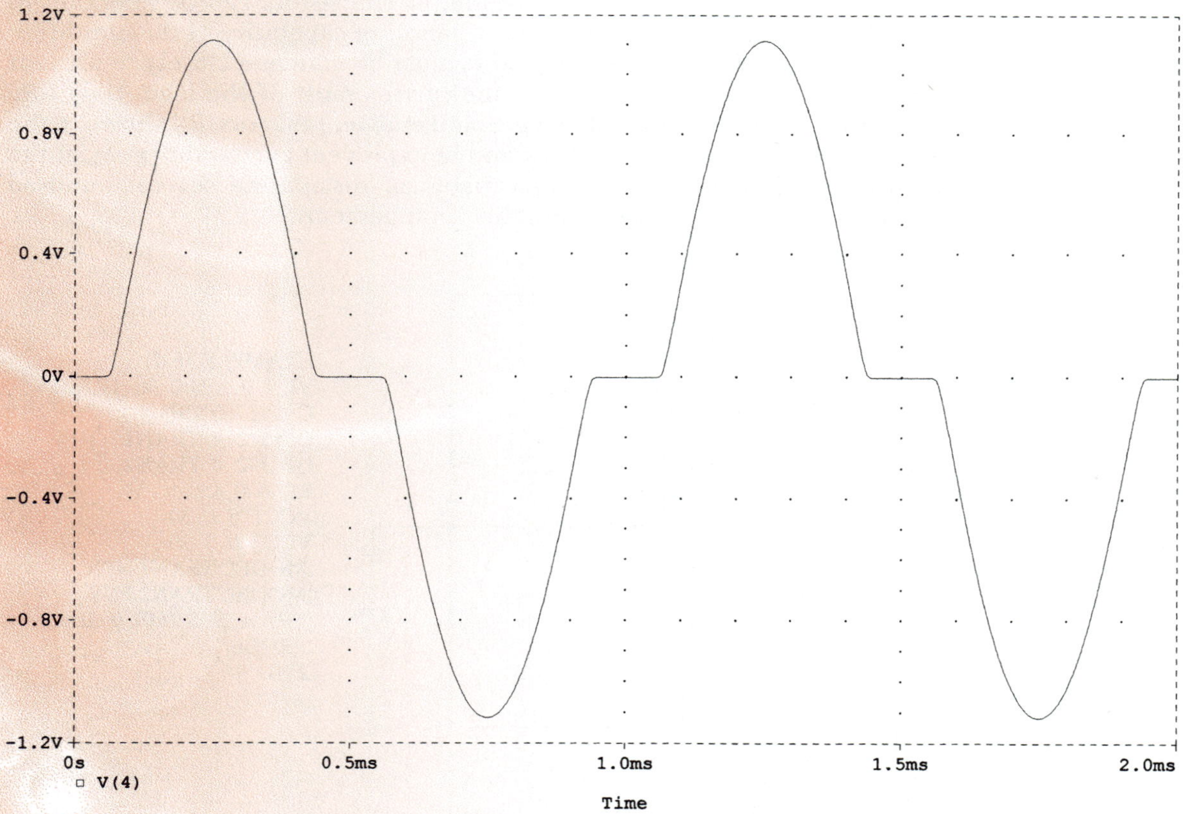

FIGURE 16–26 Continued

FIGURE 16–27 Variations on the basic complementary push-pull amplifier

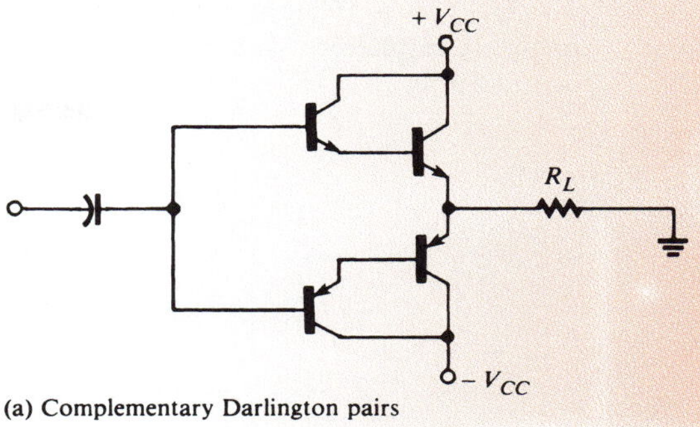

(a) Complementary Darlington pairs

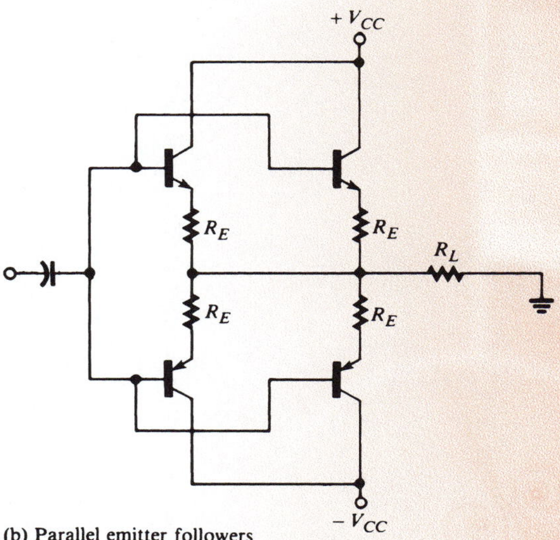

(b) Parallel emitter followers

One disadvantage of the complementary push-pull amplifier is the need for two power supplies. Also, like the transformer-coupled push-pull amplifier, the complementary class-B amplifier produces crossover distortion in its output. Figure 16–28(a) shows another version of the complementary amplifier that eliminates these problems and that incorporates some additional features.

The complementary amplifier in Figure 16–28 can be operated with a single power supply because the output, v_o, is biased at half the supply voltage and is capacitor-coupled to the load. The resistor–diode network connected across the transistor bases is used to bias each transistor near the threshold of conduction. Crossover distortion can be reduced or eliminated by inserting another resistor (not shown in the figure) in series with the diodes to bias the transistors further into AB operation. Assuming that all components are perfectly matched, the supply voltage will divide equally across each half of the amplifier, as shown in Figure 16–28(b). (In practice, one of the resistors R can be made adjustable for balance purposes.) Resistors R_E provide bias stability to prevent thermal runaway but are made as small as possible because they adversely affect efficiency. Since each half of the amplifier has $V_{CC}/2$ V across it, the forward-biased diode drops appear across the base–emitter junctions with the proper polarity to bias each transistor toward conduction. The diodes are selected so that their characteristics track the base–emitter junctions under temperature changes and thus ensure bias stability. The diodes are typically mounted on the same heat sinks as the transistors so that both change temperature in the same way.

FIGURE 16–28 Complementary push-pull amplifier using a single power supply

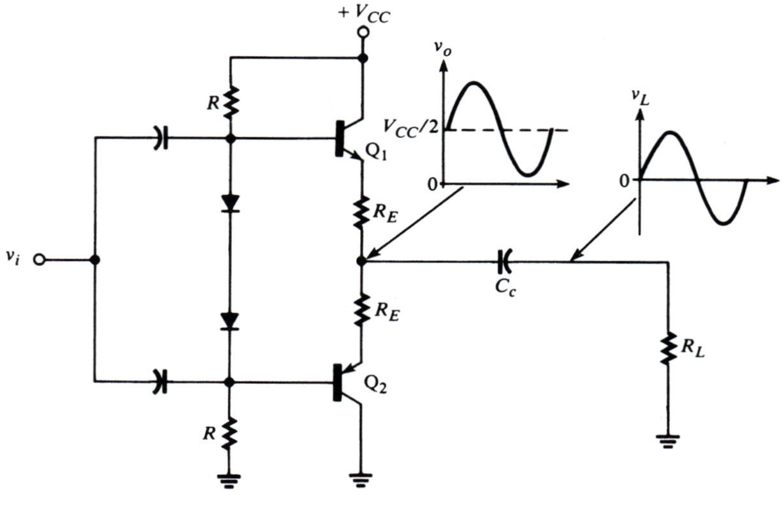

(a) The output is biased at $V_{CC}/2$ V

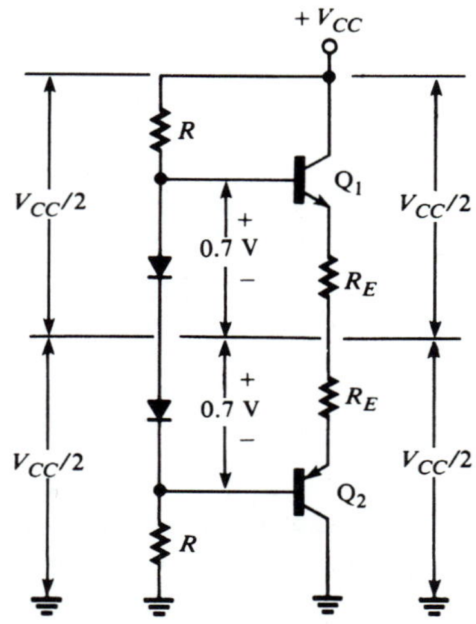

(b) dc bias voltages (all components are assumed to be perfectly matched)

In ac operation, when input v_i is positive and Q_1 is conducting, current is drawn from the power supply and flows through Q_1 to the load. When Q_1 is cut off by a negative input, no current can flow from the supply. At those times, Q_2 is conducting and capacitor C_c discharges through that transistor. Thus, current flows from the load, through C_c, and through Q_2 to ground whenever the input is negative. The R_LC_c time constant must be much greater than the period of the lowest signal frequency. The lower cutoff frequency due to C_c is given by

$$f_L = \frac{1}{2\pi(R_L + R_E)C_c} \tag{16–60}$$

The peak load current is the peak input voltage V_P divided by $R_L + R_E$:

$$I_{PL} = \frac{V_P}{R_L + R_E} \tag{16–61}$$

Therefore, the average ac power delivered to the load is

$$P_L = \frac{I_{PL}^2 R_L}{2} = \frac{V_P^2 R_L}{2(R_L + R_E)^2} \tag{16–62}$$

Because current is drawn from the power supply only during positive half-cycles of input, the supply-current waveform is half-wave rectified, with peak value $V_P/(R_L + R_E)$ amperes. Therefore, the average value of the supply current is

$$I_S(\text{avg}) = \frac{V_P}{\pi(R_L + R_E)} \qquad (16\text{–}63)$$

and the average power from the supply is

$$P_S = V_{CC}I_S(\text{avg}) = \frac{V_{CC}V_P}{\pi(R_L + R_E)} \qquad (16\text{–}64)$$

Dividing (16–62) by (16–64), we find the efficiency to be

$$\eta = \frac{P_L}{P_S} = \frac{\pi}{2}\left(\frac{R_L}{R_L + R_E}\right)\left(\frac{V_P}{V_{CC}}\right) \qquad (16\text{–}65)$$

The efficiency is maximum when the peak voltage V_P has its maximum possible value, $V_{CC}/2$. In that case,

$$\eta(\text{max}) = \frac{\pi}{2}\left(\frac{R_L}{R_L + R_E}\right)\left(\frac{V_{CC}/2}{V_{CC}}\right) = \frac{\pi}{4}\left(\frac{R_L}{R_L + R_E}\right) \qquad (16\text{–}66)$$

Equations 16–65 and 16–66 show that efficiency decreases, as expected, when R_E is increased. If $R_E = 0$, then the maximum possible efficiency becomes $\eta(\text{max}) = \pi/4 = 0.785$, the theoretical maximum for a class-B amplifier.

EXAMPLE 16–12

Assuming that all components in Figure 16–29 are perfectly matched, find

1. the base-to-ground voltages V_{B1} and V_{B2} of each transistor,
2. the power delivered to the load under maximum signal conditions,
3. the efficiency under maximum signal conditions, and
4. the value of capacitor C_c if the amplifier is to be used at signal frequencies down to 20 Hz.

Solution

1. Because all components are matched, the supply voltage divides equally across the resistor–diode network, as shown in Figure 16–30. Then, as can be seen from the figure, $V_{B1} = 10 + 0.7 = 10.7$ V and $V_{B2} = V_{B1} - 1.4 = 9.3$ V.
2. From equation 16–62, with $V_P(\text{max}) = V_{CC}/2 = 10$ V,

$$P_L(\text{max}) = \frac{V_P^2(\text{max})R_L}{2(R_L + R_E)^2} = \frac{10^2(10)}{2(10 + 1)^2} = 4.132 \text{ W}$$

3. From equation 16–66,

$$\eta(\text{max}) = \frac{\pi}{4}\left(\frac{R_L}{R_L + R_E}\right) = \frac{\pi}{4}\left(\frac{10}{11}\right) = 0.714$$

4. From equation 16–60,

$$C_c = \frac{1}{2\pi(R_L + R_E)f_L} = \frac{1}{2\pi(11)(20)} = 723\ \mu\text{F}$$

FIGURE 16–29 (Example 16–12)

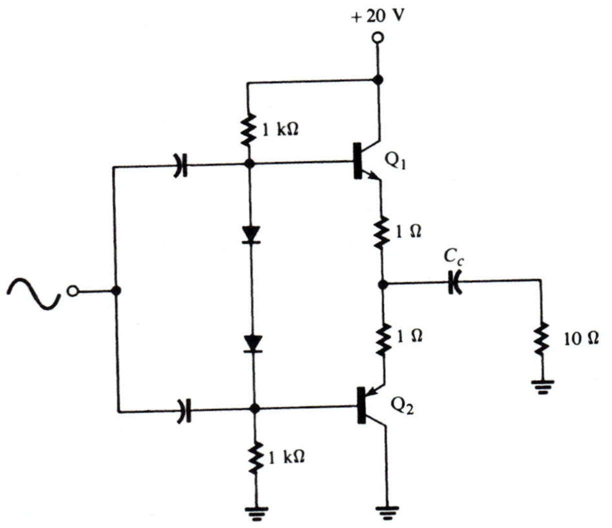

FIGURE 16–30 (Example 16–12)

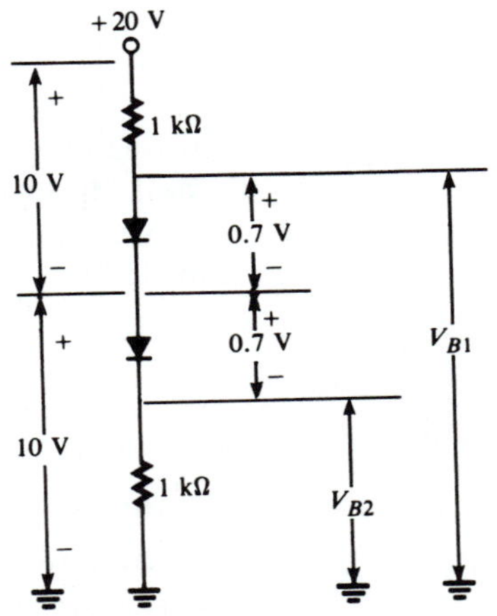

This result demonstrates a principal disadvantage of the complementary, single-supply, push-pull amplifier: The coupling capacitor must be quite large to drive a low-impedance load at a low frequency. These operating requirements are typical for audio power amplifiers.

Quasi-Complementary Push-Pull Amplifiers

As has been discussed in previous chapters, modern semiconductor technology is such that *npn* transistors are generally superior to *pnp* types. Figure 16–31 shows a popular push-pull amplifier design that uses *npn* transistors for both output devices. This design is called *quasi-complementary* because transistors Q_3 and Q_4 together perform the same function as the *pnp* transistor in a complementary push-pull amplifier. When the input signal is positive, *pnp* transistor Q_3 is cut off, so *npn* transistor Q_4 receives no base current and is also cut off. When the input is negative, Q_3 conducts and supplies base current to Q_4, which can then conduct load current. Transistors Q_1 and Q_2 form an *npn* Darlington pair, so Q_1 and Q_2 provide emitter-follower action to the load when the input is positive, and Q_3 and Q_4 perform that function when the input is negative. The entire configuration thus performs push-pull operation in the same manner as the complementary push-pull amplifier. The current gain of the Q_3Q_4 combination is

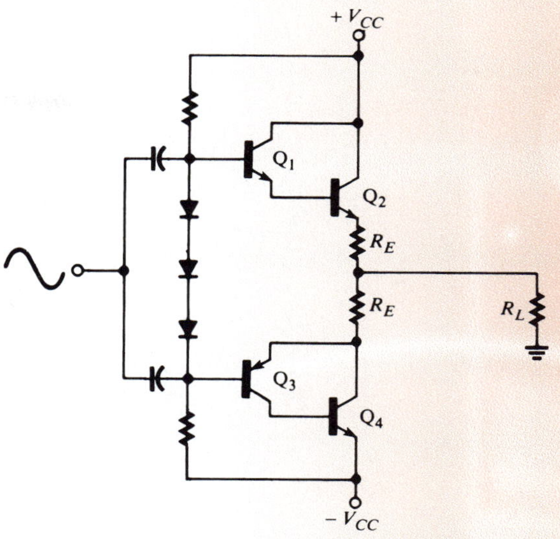

FIGURE 16–31 A quasi-complementary push-pull amplifier. Transistors Q_3 and Q_4 conduct during negative half-cycles of input.

$$\frac{I_E}{I_B} = \beta_3\beta_4 + (1 + \beta_3) \tag{16–67}$$

This gain is very nearly that of a Darlington pair. Transistors Q_1 and Q_2 are connected as a Darlington pair to ensure that both sides of the amplifier have similar gain.

Integrated-Circuit Power Amplifiers

Low- to medium-power audio amplifiers (in the 1- to 20-W range) are available in integrated-circuit form. Many have differential inputs and quasi-complementary outputs. Figure 16–32 shows an example, the LM380 2.5-W audio amplifier manufactured by National Semiconductor. Transistors Q_1 and Q_2 are input emitter followers that drive the differential pair consisting of Q_3 and Q_4. Transistors Q_5 and Q_6 are active loads, and transistors Q_{10} and Q_{11} form a current mirror supplying bias current. The output of the differential stage, at the collector of Q_4, drives a common-emitter stage (Q_{12}), which in turn drives a quasi-complementary output stage. The 10-pF capacitor provides internal frequency compensation.

One feature of the LM380 is that internal resistor networks (R_1 and R_2) are used to set the output bias level automatically at one-half the supply voltage, V_S. The output is normally capacitor-coupled to the load. The voltage gain of the amplifier is fixed at 50 (34 dB). Manufacturer's specifications state that the bandwidth is 100 kHz and that the total harmonic distortion is typically 0.2%. The supply voltage can be set from 10 to 22 V.

16–10 CLASS-C AMPLIFIERS

A class-C amplifier is one whose output conducts load current during *less* than one-half cycle of an input sine wave. Figure 16–33 shows a typical class-C current waveform, and it is apparent that the total angle during which current flows is less than 180°. This angle is called the *conduction angle*, θ_c.

Of course, the output of a class-C amplifier is a highly distorted version of its input. It could not be used in an application requiring high fidelity, such as an audio amplifier. Class-C amplifiers are used primarily in high-power, high-frequency applications, such as radio-frequency transmitters. In these

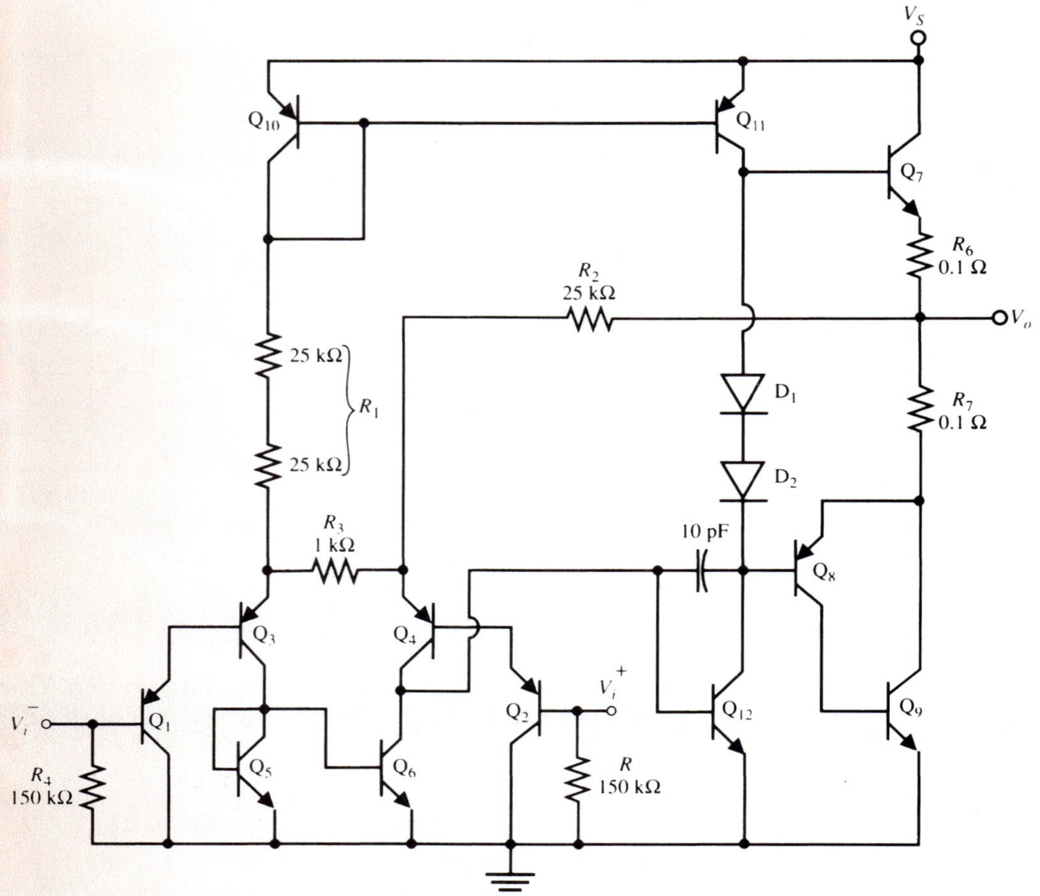

FIGURE 16–32 Simplified schematic diagram of the LM380 integrated-circuit amplifier

FIGURE 16–33 Output current in a class-C amplifier

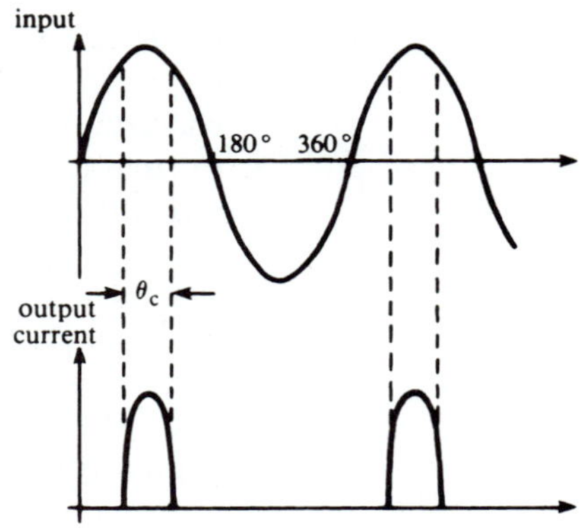

applications, the high-frequency pulses handled by the amplifier are not themselves the signal but constitute what is called the *carrier* for the signal. The signal is transmitted by varying the amplitude of the carrier, using the process called *amplitude modulation* (AM). The signal is ultimately recovered in a *receiver* by filtering out the carrier frequency. The principal advantage of a class-C amplifier is that it has a very high efficiency, as we shall presently demonstrate.

Figure 16–34 shows a simple class-C amplifier with a resistive load. Note that the base of the *npn* transistor is biased by a *negative* voltage, $-V_{BB}$,

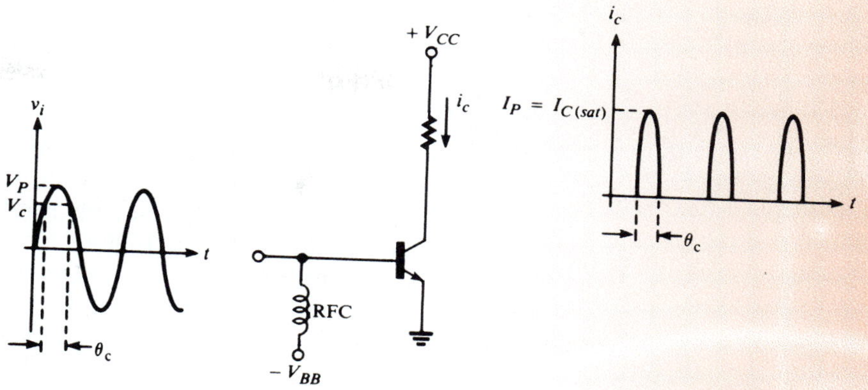

FIGURE 16–34 A class-C amplifier with resistive load. The transistor conducts when $v_i \geq |V_{BB}| + 0.7$

connected through a coil labeled *RFC*. The RFC is a *radio-frequency choke* whose inductance presents a high impedance to the high-frequency input and thereby prevents the dc source from shorting the ac input. For the transistor to begin conducting, the input must reach a level sufficient to overcome both the negative bias and the V_{BE} drop of about 0.7 V:

$$V_c = |V_{BB}| + 0.7 \tag{16–68}$$

where V_c is the input voltage at which the transistor begins to conduct. As shown in the figure, the transistor is cut off until v_i reaches V_c, then it conducts, and then it cuts off again when v_i falls below V_c. Clearly, the more negative the value of V_{BB}, the shorter the conduction interval. In most class-C applications, the amplifier is designed so that the peak value of the input, V_P, is just sufficient to drive the transistor into saturation, as shown in the figure.

The conduction angle θ_c in Figure 16–34 can be found from

$$\theta_c = 2 \arccos\left(\frac{V_c}{V_P}\right) \tag{16–69}$$

where V_P is the peak input voltage that drives the transistor to saturation. If the peak input only just reaches V_c, then $\theta_c = 2 \arccos(1) = 0°$. At the other extreme, if $V_{BB} = 0$, then $V_c = 0.7$, $(V_c/V_P) \approx 0$, and $\theta_c = 2 \arccos(0) = 180°$, which corresponds to class-B operation.

Figure 16–35 shows the class-C amplifier as it is normally operated, with an LC *tank* network in the collector circuit. Recall that the tank is a *resonant* network whose center frequency, assuming small coil resistance, is closely approximated by

FIGURE 16–35 A tuned class-C amplifier with an LC tank circuit as load

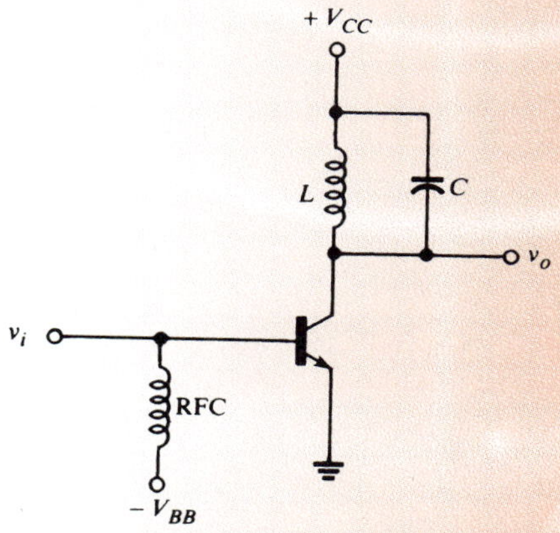

$$f_c \approx \frac{1}{2\pi\sqrt{LC}} \tag{16–70}$$

The purpose of the tank is to produce the fundamental component of the pulsed, class-C waveform, which has the same frequency as v_i. The configuration is called a *tuned* amplifier, and the center frequency of the tank is set equal to (tuned to) the input frequency. There are several ways to view its behavior as an aid in understanding how it recovers the fundamental frequency. We may regard the tank as a highly selective (high-Q) filter that suppresses the harmonics in the class-C waveform and passes its fundamental. We may also recall that the voltage gain of the transistor equals the impedance in the collector circuit divided by the emitter resistance. Because the impedance of the tank is very large at its center frequency, the gain is correspondingly large at that frequency, while the impedance and the gain at harmonic frequencies are much smaller.

The amplitude of the fundamental component of a class-C waveform depends on the conduction angle θ_c. The greater the conduction angle, the greater the ratio of the amplitude of the fundamental component to the amplitude of the total waveform. Let r_1 be the ratio of the peak value of the fundamental component to the peak value of the class-C waveform. The value of r_1 is closely approximated by

$$r_1 \approx (-3.54 + 4.1\theta_c - 0.0072\theta_c^2) \times 10^{-3} \tag{16–71}$$

where $0° \le \theta_c \le 180°$. The values of r_1 vary from 0 to 0.5 as θ_c varies from 0° to 180°.

Let r_0 be the ratio of the dc value of the class-C waveform to its peak value. The value of r_0 can be found from

$$r_0 = \frac{\text{dc value}}{\text{peak value}} = \frac{\theta_c}{\pi(180°)} \tag{16–72}$$

where $0° \le \theta_c \le 180°$. The values of r_0 vary from 0 to $1/\pi$ as θ_c varies from 0° to 180°.

The efficiency of a class-C amplifier is large because very little power is dissipated when the transistor is cut off, and it is cut off during most of every full cycle of input. The output power at the fundamental frequency under maximum drive conditions is

$$P_o = \frac{(r_1 I_P)V_{CC}}{2} \tag{16–73}$$

where I_P, is the peak output (collector) current. The average power supplied by the dc source is V_{CC} times the average current drawn from the source. Because current flows only when the transistor is conducting, this current waveform is the same as the class-C collector-current waveform having peak value I_P. Therefore,

$$P_S = (r_0 I_P) V_{CC} \tag{16–74}$$

The efficiency is then

$$\eta = \frac{P_o}{P_S} = \frac{r_1 I_P V_{CC}}{2 r_0 I_P V_{CC}} = \frac{r_1}{2r_0} \tag{16–75}$$

EXAMPLE 16–13

A class-C amplifier has a base bias voltage of −5 V and V_{CC} = 30 V. It is determined that a peak input voltage of 9.8 V at 1 MHz is required to drive the transistor to its saturation current of 1.8 A.

1. Find the conduction angle.

2. Find the output power at 1 MHz.
3. Find the efficiency.
4. If an LC tank having $C = 200$ pF is connected in the collector circuit, find the inductance necessary to tune the amplifier.

Solution

1. From equation 16–68, $V_c = |V_{BB}| + 0.7 = 5 + 0.7 = 5.7$ V. From equation 16–69,

$$\theta_c = 2 \arccos\left(\frac{V_c}{V_P}\right) = 2 \arccos\left(\frac{5.7}{9.8}\right) = 108.9°$$

2. From equation 16–71, $r_1 \approx [-3.54 + 4.1(108.9) - 0.0072(108.9)^2] \times 10^{-3} = 0.357$. From equation 16–73,

$$P_o = \frac{(r_1 I_P)V_{CC}}{2} = \frac{(0.357)(1.8\text{ A})(30\text{ V})}{2} = 9.64\text{ W}$$

3. From equation 16–72,

$$r_0 = \frac{\theta}{\pi(180°)} = \frac{108.9}{\pi(180°)} = 0.193$$

From equation 16–75,

$$\eta = \frac{r_1}{2r_0} = \frac{0.357}{2(0.193)} = 0.925 \text{ (or } 92.5\%)$$

This result demonstrates that a very high efficiency can be achieved in a class-C amplifier.

4. From equation 16–70,

$$L = \frac{1}{(2\pi f_o)^2 C} = \frac{1}{(2\pi \times 10^6)^2(200 \times 10^{-12})} = 0.127\text{ mH}$$

Amplitude Modulation

We have mentioned that amplitude modulation is a means used to transmit signals by varying the amplitude of a high-frequency carrier. In a typical application, the signal is a low-frequency audio waveform and the amplitude of a high-frequency (radio-frequency, or rf) carrier is made to increase and decrease as the audio signal increases and decreases. Figure 16–36 shows the waveforms that are generated by an amplitude modulator when the signal input is a small dc value, a large dc value, and a low-frequency sinusoidal wave. Notice that the carrier input to the modulator is a constant-amplitude, constant-frequency rf sine wave. It can be seen that the audio waveform is reproduced in the variations of the positive and negative peaks of the output. The audio waveform superimposed on the high-frequency peaks is called the *envelope* of the modulated wave.

It is important to note that amplitude modulation is achieved by *multiplying* two waveforms (signal × carrier), which is a *nonlinear* process. As such, the modulated output contains frequency components that are not present in either the signal or the carrier. Amplitude modulation cannot be achieved by simply adding two waveforms. Unfortunately, the term *mixing* is

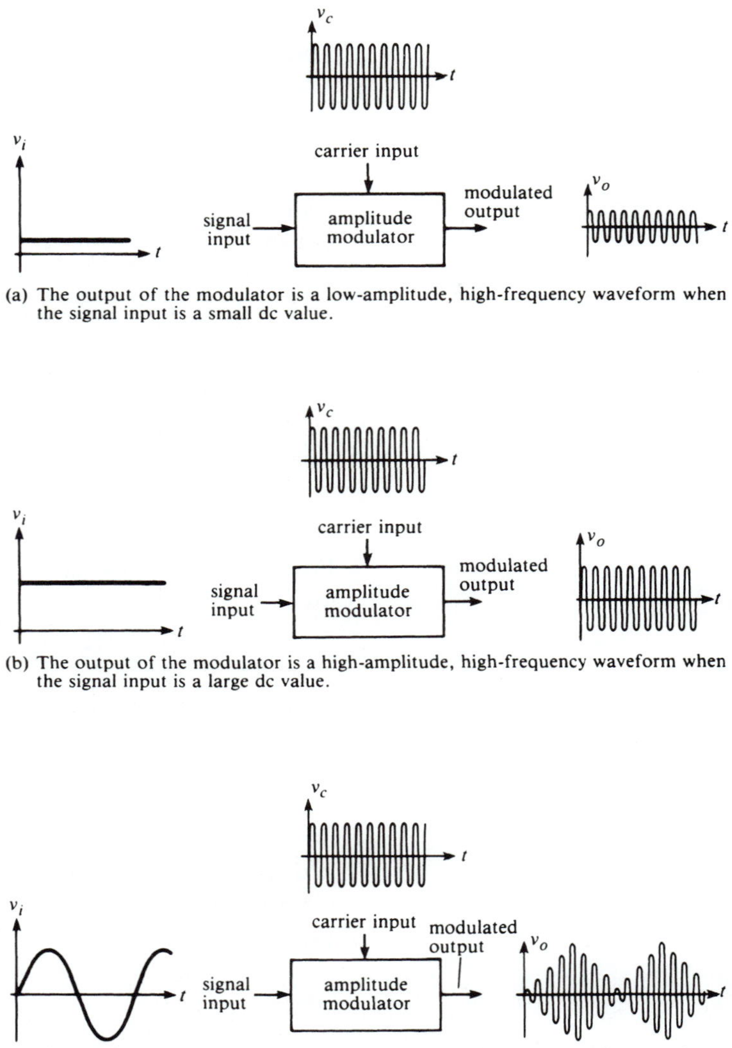

(a) The output of the modulator is a low-amplitude, high-frequency waveform when the signal input is a small dc value.

(b) The output of the modulator is a high-amplitude, high-frequency waveform when the signal input is a large dc value.

(c) The amplitude of the output varies in the same way that the signal input varies.

FIGURE 16–36 Amplitude-modulated (AM) waveforms

used in the broadcast industry to mean both the summation of signals and the multiplication of signals (some modulators are called *mixers*), but these are two very distinct processes.

We will now demonstrate that amplitude modulation creates new frequency components. Let the input signal be a pure sine wave designated

$$v_s(t) = A_s \sin \omega_s t \tag{16–76}$$

and let the carrier input be designated

$$v_c(t) = A_c \sin \omega_c t \tag{16–77}$$

where A_s and A_c are the peak values of the signal and the carrier, respectively. The signal and carrier frequencies are $\omega_s = 2\pi f_s$ rad/s and $\omega_c = 2\pi f_c$ rad/s, respectively. The envelope of the modulated output is the time-varying signal voltage added to the peak carrier voltage:

$$\text{envelope} = e(t) = v_s(t) + A_c = A_s \sin \omega_s t + A_c \tag{16–78}$$

The modulated AM waveform has frequency ω_c and has a (time-varying) peak value equal to the envelope:

$$v_o(t) = e(t)\sin \omega_c t$$

$$\begin{aligned}\text{envelope} &= \text{peak value}\\ &= (A_s \sin \omega_s t + A_c)\sin \omega_c t\\ &= A_s(\sin \omega_c t)(\sin \omega_s t) + A_c \sin \omega_c t\end{aligned} \quad (16\text{–}79)$$

Equation 16–79 makes it apparent now that amplitude modulation involves the multiplication of two sine waves. Using a trigonometric identity, equation 16–79 may be expressed as (Exercise 16–34)

$$v_o(t) = A_c \sin \omega_c t + \tfrac{1}{2}A_s \cos(\omega_c - \omega_s)t - \tfrac{1}{2}A_s \cos(\omega_c + \omega_s)t \quad (16\text{–}80)$$

Equation 16–80 shows that the modulated waveform contains the new frequency components $\omega_c + \omega_s$ and $\omega_c - \omega_s$, called the *sum and difference frequencies*, as well as a component at the carrier frequency, ω_c. Note that there is *no* frequency component equal to ω_s. In practice, the input signal will consist of a complex waveform containing many different frequencies. Therefore, the AM output will contain many different sum and difference frequencies. The band of difference frequencies is called the *lower sideband* because its frequencies are all less than the carrier frequency, and the band of sum frequencies is called the *upper sideband*, each of its frequencies being greater than the carrier frequency. The significance of this result is that it allows us to determine the bandwidth that an amplifier must have to pass an AM waveform. The bandwidth must extend from the smallest difference frequency to the largest sum frequency.

EXAMPLE 16–14

An amplitude modulator is driven by a 580-kHz carrier and has an audio signal input containing frequency components between 200 Hz and 9.5 kHz.

1. What range of frequencies must be included in the passband of an amplifier that will be used to amplify the modulated signal?
2. What frequency components are in the lower sideband of the AM output? In the upper sideband?

Solution

1. The smallest difference frequency is $2\pi(580 \times 10^3) - 2\pi(9.5 \times 10^3)$ rad/s, or (580 kHz) − (9.5 kHz) = 570.5 kHz, and the largest sum frequency is (580 kHz) + (9.5 kHz) = 589.5 kHz. Therefore, the amplifier must pass frequencies in the range from 570.5 kHz to 589.5 kHz. An amplifier having only this range would be considered a *narrowband* amplifier.
2. The lowest sideband extends from the smallest difference frequency to the largest difference frequency: [(580 kHz) − (9.5 kHz)] to [(580 kHz) − (200 Hz)], or 570.5 kHz to 579.8 kHz. The upper sideband extends from the smallest sum frequency to the largest sum frequency: [(580 kHz) + (200 Hz)] to [(580 kHz) + (9.5 kHz)], or 580.2 kHz to 589.5 kHz.

Figure 16–37 shows how a class-C amplifier can be used to produce amplitude modulation. The circuit is similar to Figure 16–35, except that the coil in the collector circuit has been replaced by the secondary winding of a transformer. The low-frequency signal input is connected to the transformer primary. The voltage at the collector is then the sum of V_{CC} and a signal proportionate to v_s. As v_s increases and decreases, so does the collector voltage. In effect, we are varying the supply voltage on the collector. When the carrier signal on the base drives the transistor to saturation, the peak collector

FIGURE 16–37 A class-C amplitude modulator

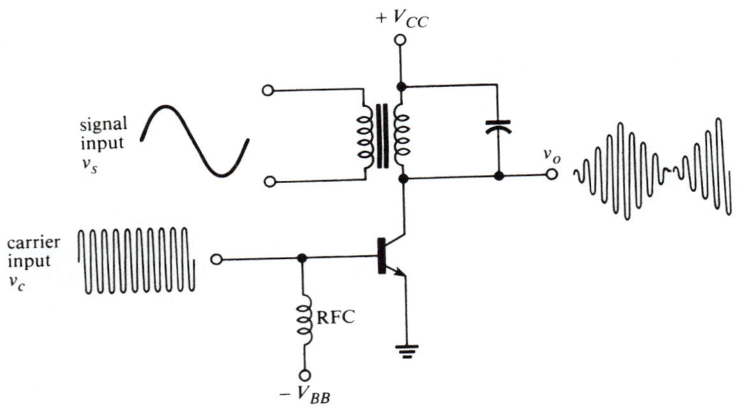

current $I_{c(sat)}$ that flows depends directly on the supply voltage: The greater the supply voltage, the greater the peak current. Since the supply voltage varies with the signal input, so does the peak collector current. As a consequence, the collector current is amplitude modulated by v_s, and the tank circuit produces an AM output voltage, as shown.

16–11 MOSFET AND CLASS-D POWER AMPLIFIERS

MOSFET Amplifiers

MOSFET devices of the VMOS design (Figure 5–45) are becoming a popular choice for power amplifiers, particularly those designed to switch large currents on and off. Examples include line drivers for digital switching circuits, switching-mode voltage regulators (discussed in Chapter 13), and class-D amplifiers, which we will discuss presently. One advantage of a MOSFET switch is the fact that turn-off time is not delayed by minority-carrier storage, as it is in a bipolar switch that has been driven deeply into saturation. Recall that current in field-effect transistors is due to the flow of majority carriers only. Also, MOSFETs are not susceptible to thermal runaway like bipolar transistors. Finally, a MOSFET has a very large input impedance, which simplifies the design of driver circuits.

Figure 16–38 shows a simple, class-A audio amplifier that uses a VN64GA VMOS transistor to drive an output transformer. Notice that the driver for the output stage is a J108 JFET transistor connected in a common-source configuration. Also notice the feedback path between the transformer secondary and the input to the JFET driver. This negative feedback is incorporated to reduce distortion, as discussed earlier. The amplifier will reportedly deliver between 3 and 4 W to an 8-Ω speaker with less than 2% distortion up to 15 kHz.

MOSFET power amplifiers are also constructed for class-B and class-AB operation. Like their bipolar counterparts, these amplifiers can be designed without the need for an output transformer, since power MOSFETs are available in complementary pairs (*n*-channel and *p*-channel). Figure 16–39 shows a typical push-pull MOSFET amplifier using complementary output devices.

Class-D Amplifiers

A class-D amplifier is one whose output is switched on and off, that is, one whose output is in its linear range for essentially *zero* time during each cycle of an input sine wave. As we progress through the letters designating the various classes of operation, A, B, C, and D, we see that linear operation occurs for shorter and shorter intervals of time, and in class D we reach the limiting case where no linear operation occurs at all. The only time that a class-D output

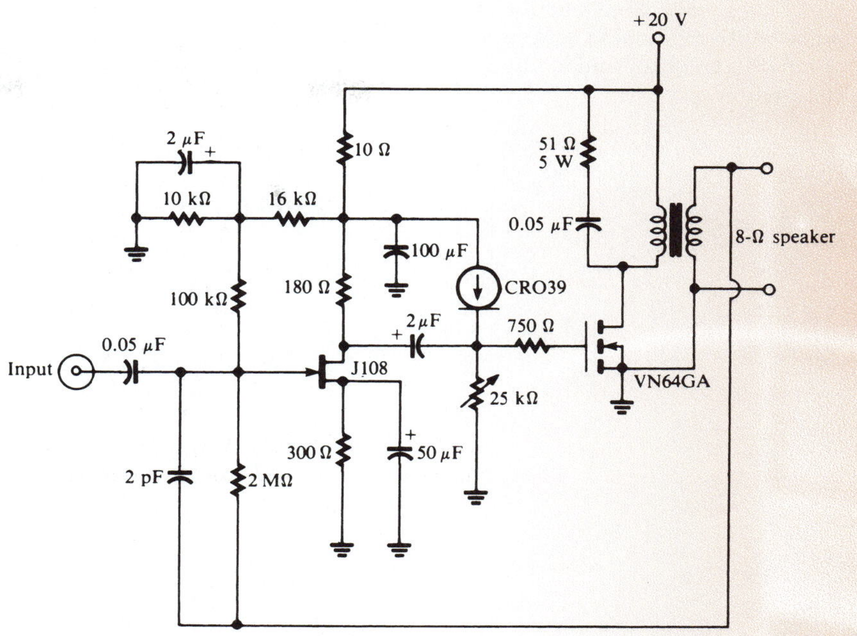

FIGURE 16–38 A two-stage, class-A audio amplifier with a VMOS output transistor (Courtesy of Siliconix Incorporated)

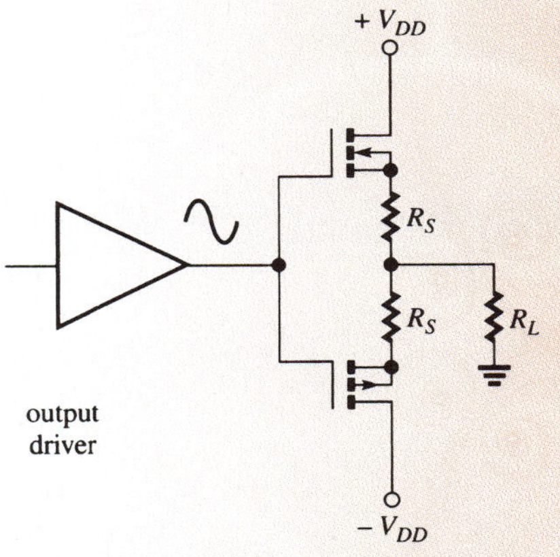

FIGURE 16–39 A complementary MOSFET amplifier

device is in its linear region is during that short interval required to switch from saturation to cutoff, or vice versa. In other words, the output device is a digital power switch, an application ideally suited for VMOS transistors.

A fundamental component of a class-D amplifier is a *pulse-width modulator*, which produces a train of pulses having widths that are proportionate to the level of the amplifier's input signal. When the signal level is small, a series of narrow pulses is generated, and when the input level is large, a series of wide pulses is generated. (See Figure 13–16, which shows some typical outputs of a pulse-width modulator used in a voltage-regulator application.) As the input signal increases and decreases, the pulse widths increase and decrease in direct proportion.

Figure 16–40 shows how a pulse-width modulator can be constructed using a *sawtooth generator* and a voltage comparator. A sawtooth waveform is one that rises linearly and then quickly switches back to its low level, to begin another linear rise, as illustrated in the figure. When the sawtooth voltage is greater than v_i, the output of the comparator is low, and when the sawtooth falls below v_i, the comparator switches to its high output. Notice that the comparator must switch high each time the sawtooth makes its vertical descent. These time points mark the beginning of each new pulse. The comparator output remains

FIGURE 16–40 Construction of a pulse-width modulator using a sawtooth generator and a voltage comparator

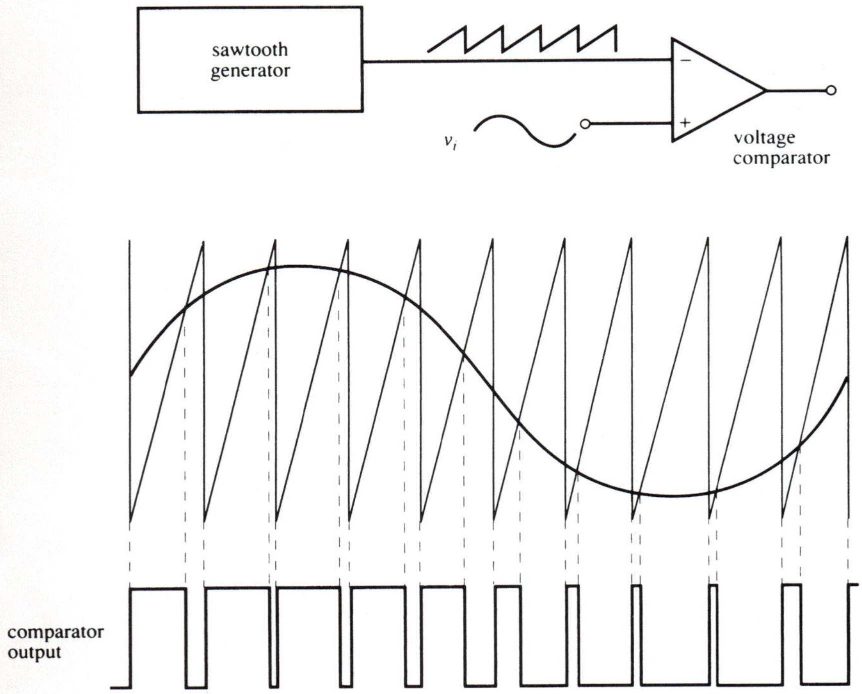

high until the sawtooth rises back to the value of v_i, at which time the comparator output switches low. Thus the width of the high pulse is directly proportionate to the length of time it takes the sawtooth to rise to v_i, which is directly proportionate to the level of v_i. As shown in the figure, the result is a series of pulses whose widths are proportionate to the level of v_i. Notice that the peak-to-peak voltage of the sawtooth must exceed the largest peak-to-peak input voltage for successful operation. Also, the frequency of the sawtooth should be at least 10 times as great as the highest frequency component of v_i.

The pulse-width modulator drives the output stage of the class-D amplifier, causing it to switch on and off as the pulses switch between high and low. Figure 16–41 shows a popular switching circuit, called a *totem pole,* used to drive heavy loads. The MOSFET version of the totem pole is shown in the figure, because MOSFETs are generally used in class-D amplifiers. Bipolar versions of the totem pole are also widely used in digital logic circuits, particularly in the integrated-circuit family called *TTL* (transistor-transistor logic). The totem pole shown in the figure *inverts,* in the sense that the output is low when the input is high, and vice versa.

When the input in Figure 16–41 is high, Q_1 and Q_3 are on, so the output is low (R_L is effectively connected to ground through Q_3). Because Q_1 is on, the gate of Q_2 is low and Q_2 is held off. Thus, when the input is high, Q_2 is like an open switch and Q_3 is like a closed switch. When the input is low, Q_1 is off and a high voltage (V_{DD}) is applied through R to the gate of Q_2, turning it on. Thus, R_L is connected through Q_2 to the high level V_{DD}. Because Q_3 is also off, a low input makes Q_3 an open switch and Q_2 a closed switch. The advantage of this arrangement is that the load is driven from a low-impedance signal source, a turned-on MOSFET, both when the output is low and when the output is high.

Like the class-C amplifier, a class-D amplifier must have a *filter* to extract, or recover, the signal from the pulsed waveform. However, in this case the signal may have many frequency components, so a tank circuit, which resonates at a single frequency, cannot be used. Instead, a low-pass filter having a cutoff frequency near the highest signal frequency is used. The low-pass filter suppresses

FIGURE 16–41 A MOSFET totem pole, used to switch heavy load currents

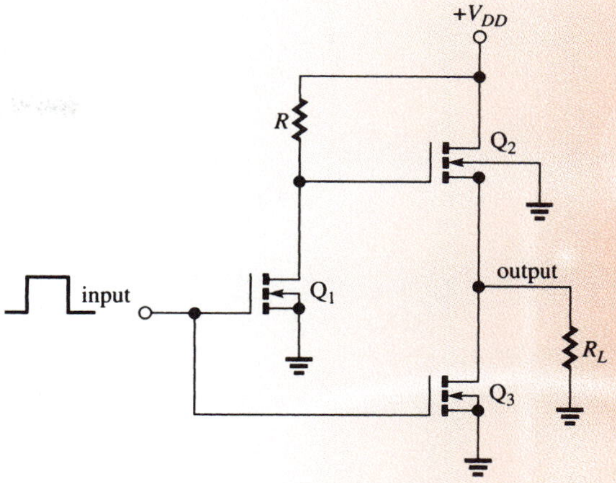

FIGURE 16–42 Block diagram of a class-D amplifier

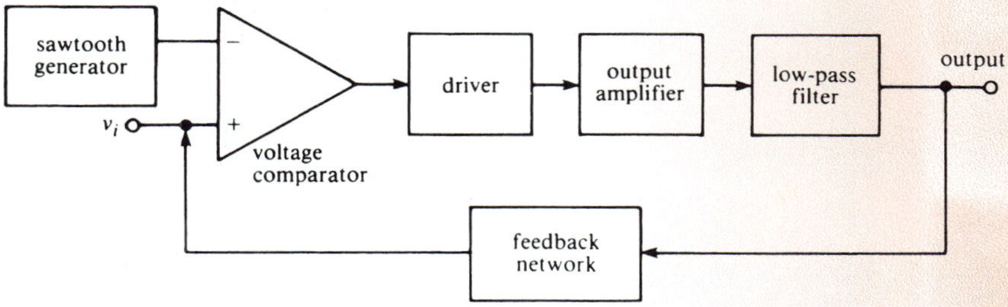

the high-frequency components of the pulse train and, in effect, recovers the *average value* of the pulse train. Because the average value of the pulses depends on the pulse widths, the output of the filter is a waveform that increases and decreases as the pulse widths increase and decrease, that is, a waveform that duplicates the input signal, v_i. Figure 16–42 shows a block diagram of a complete class-D amplifier, including negative feedback to reduce distortion.

The principal advantage of a class-D amplifier is that it may have a very high efficiency, approaching 100%. Like that of a class-C amplifier, the high efficiency is due to the fact that the output device spends very little time in its active region, so power dissipation is minimal. The principal disadvantages are the need for a very good low-pass filter and the fact that high-speed switching of heavy currents generates noise through electromagnetic coupling, called *electromagnetic interference,* or *EMI.*

Figure 16–43 shows a 100-W class-D amplifier using a MOSFET totem pole with a bipolar driver.

EXAMPLE 16–15

A class-D audio amplifier is to be driven by a signal that varies between ± 5 V. The output is a MOSFET totem pole with V_{DD} = 30 V.

1. What minimum frequency should the sawtooth waveform have?
2. What minimum peak-to-peak voltage should the sawtooth waveform have, assuming that the input is connected directly to the voltage comparator?
3. Assume that the amplifier is 100% efficient and is operated at the frequency found in (1). What average current is delivered to an 8-Ω load when the pulse-width modulator produces pulses that are high for 2.5 μs?

Solution

1. The nominal audio-frequency range is 20 Hz to 20 kHz, and the sawtooth waveform should have a frequency equal to at least 10 times the highest input frequency, i.e., 10(20 kHz) = 200 kHz.

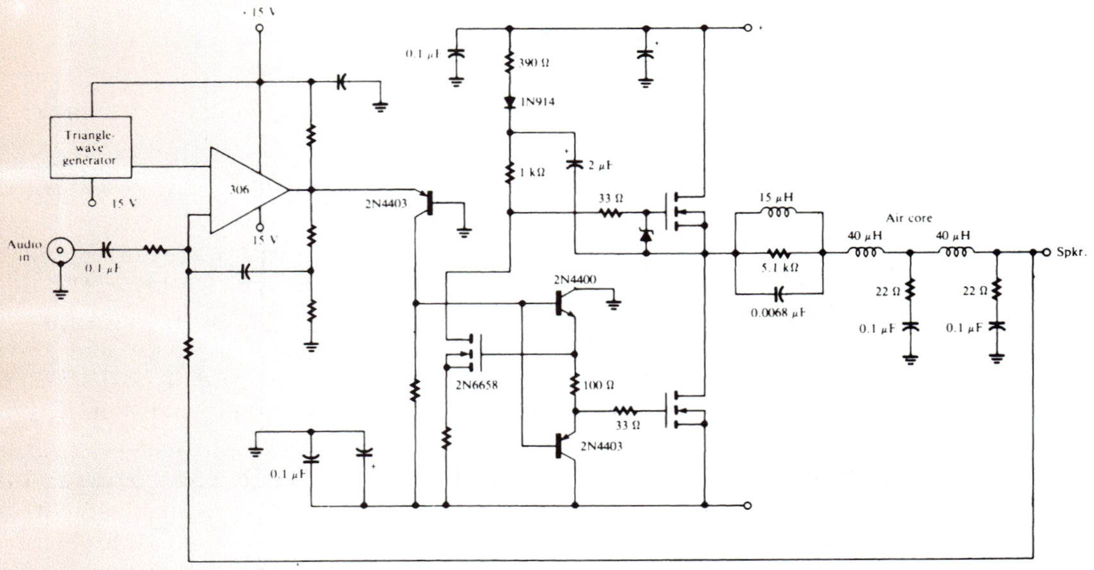

FIGURE 16–43 A 100-W class-D amplifier (Reprinted with permission from Power FETs and Their Applications by Edwin S. Oxner, Prentice-Hall, 1982)

2. The peak-to-peak sawtooth voltage should at least equal the maximum peak-to-peak input voltage, which is 10 V.
3. The period of the 200-kHz sawtooth is

$$T = \frac{1}{200 \times 10^3} = 5\ \mu\text{s}$$

Therefore, the pulse train is high for (2.5 μs)/(5 μs) = one-half the period. The totem-pole output therefore switches between 0 V and 30 V with pulse widths equal to one-half the period. Thus, the average voltage is

$$V_{avg} = \frac{1}{2}(30) = 15\ \text{V}$$

and the average current is

$$I_{avg} = \frac{V_{avg}}{R_L} = \frac{15\ \text{V}}{8\ \Omega} = 1.875\ \text{A}$$

SUMMARY

This chapter has introduced the reader to the basic principles of power amplification and related topics such as power dissipation, heat transfer, and power derating. At the end of the chapter, the student should have a good understanding of the following concepts:

- Power amplification is the process of boosting voltage and current necessary to drive low-impedance loads.
- Power transistors dissipate significant amounts of power. More efficient amplifiers deliver a larger portion of their dissipating power to the load.
- Push-pull amplifiers are more efficient and produce less distortion than single-ended amplifiers.
- Class-C amplifiers cannot be used as linear amplifiers, but they are used extensively in tuned, high-efficiency RF amplifiers in conjunction with amplitude modulators.

EXERCISES

SECTION 16-2

Transistor Power Dissipation

16–1. The transistor in Figure 16–44 is biased at $V_{CE} = 12$ V. If an increase in temperature causes the quiescent collector current to increase, will the power dissipation at the Q-point increase or decrease? Explain. What is the power dissipation at the point of maximum power dissipation?

16–2. Repeat Exercise 16–1 if the supply voltage is changed to 20 V and the Q-point is set at $V_{CE} = 13$ V.

16–3. The amplifier in Figure 16–45 is to be operated with a 40-V-dc supply. If the maximum permissible power dissipation in the transistor is 1 W, what is the minimum permissible value of R_C?

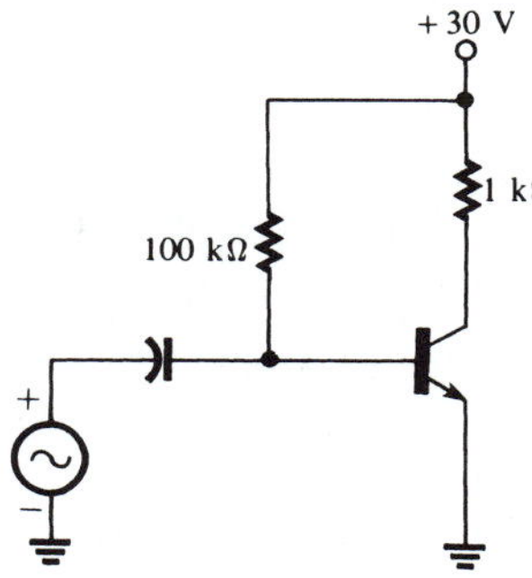

FIGURE 16–44 (Exercise 16–1)

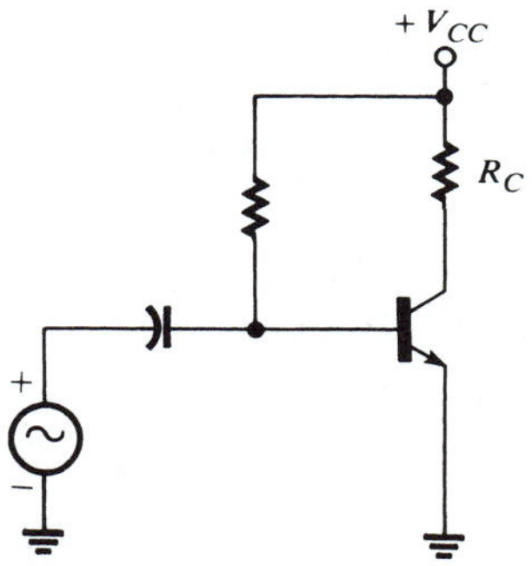

FIGURE 16–45 (Exercise 16–3)

16–4. If the amplifier in Exercise 16–3 has $R_C = 100\ \Omega$, what is the maximum permissible value of the supply voltage that can be used?

SECTION 16–3

Heat Transfer in Semiconductor Devices

16–5. The collector–base junction of a transistor conducts 0.4 A when it is reverse biased by 21 V. At what rate, in J/s, must heat be removed from the junction to prevent its temperature from rising?

16–6. A semiconductor junction at temperature 64°C releases heat into a 25°C ambient at the rate of 7.8 J/s. What is the thermal resistance between junction and ambient?

16–7. The thermal resistance between a semiconductor device and its case is 0.8°C/W. It is used with a heat sink whose thermal resistance to ambient is 0.5°C/W. If the device dissipates 10 W and the ambient temperature is 47°C, what is the maximum permissible thermal resistance between case and heat sink if the device temperature cannot exceed 70°C?

16–8. A semiconductor dissipates 25 W through its case and its heat sink to the surrounding air. The thermal resistances are the following: device-to-case, 1°C/W; case-to-heat sink, 1.2°C/W; and heat sink-to-air, 0.7°C/W. In what maximum air temperature can the device be operated if its temperature cannot exceed 100°C?

16–9. The mica washer shown in Figure 16–46 has a thermal resistivity of 66°C· in./W. Find its thermal resistance.

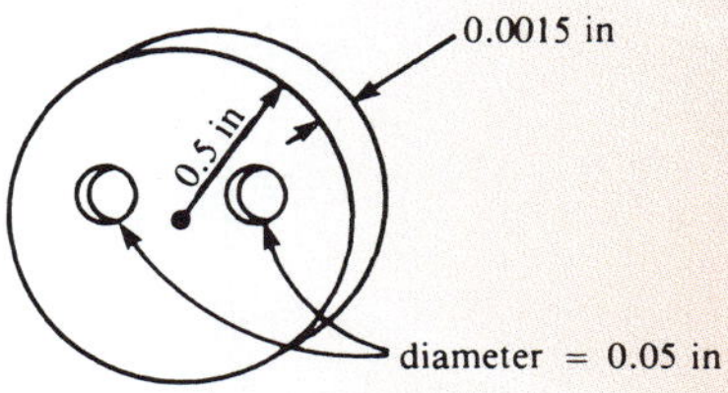

FIGURE 16–46 (Exercise 16–9)

16–10. A transistor has a maximum dissipation rating of 10 W at ambient temperatures up to 40°C and is derated at 10 mW/°C at higher temperatures. At what ambient temperature is the dissipation rating one-half its low-temperature value?

16–11. The temperature of a certain semiconductor device is 100°C when the device dissipates 1.2 W. The device temperature cannot exceed 100°C. If the total thermal resistance from device to ambient is

50°C/W, above what ambient temperature should the dissipation be derated?

SECTION 16–4
Amplifier Classes and Efficiency

16–12. The amplifier shown in Figure 16–47 is biased at $I_Q = 20$ mA. Find
- (a) the ac power in the load resistance when the voltage swing is the maximum possible without distortion,
- (b) the amplifier efficiency under the conditions of (a), and
- (c) the collector efficiency under the conditions of (a).

16–13. (a) Find the value of quiescent current in the amplifier of Exercise 16–12 that will maximize the efficiency under maximum output voltage swing.
- (b) Find the power delivered to the load resistance under the conditions of (a).

16–14. The amplifier shown in Figure 16–48 is biased at $I_Q = 0.2$ A. Find
- (a) the ac power in the load resistance when the voltage swing is the maximum possible without distortion,
- (b) the amplifier efficiency under the conditions of (a), and
- (c) the collector efficiency under the conditions of (a).

16–15. (a) Find the value of quiescent voltage in the amplifier of Exercise 16–14 that will maximize the efficiency under maximum output voltage swing.
- (b) Find the efficiency under the conditions of (a).
- (c) Why is the efficiency less than the theoretical maximum (0.0858)?

16–16. The class-A amplifier in Figure 16–49 is biased at $I_C = 0.2$ A. The transformer resistance is negligible.
- (a) What is the slope of the ac load line?
- (b) At what value does the ac load line intersect the V_{CE}-axis?
- (c) What is the maximum peak value of the collector voltage without distortion?
- (d) What is the maximum power delivered to the load under the conditions of (c)?
- (e) What is the amplifier efficiency under the conditions of (c)?

16–17. The transistor in Exercise 16–16 has a beta of 40.
- (a) Find the value of base current necessary to set a new Q-point that will permit maximum possible peak-to-peak output voltage.
- (b) What power can be delivered to the load under the conditions of (a)?
- (c) What are the amplifier and collector efficiencies under the conditions of (a)?

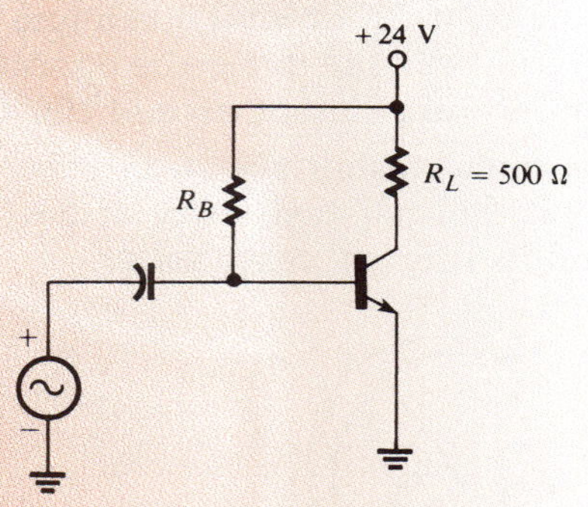

FIGURE 16–47 (Exercise 16–12)

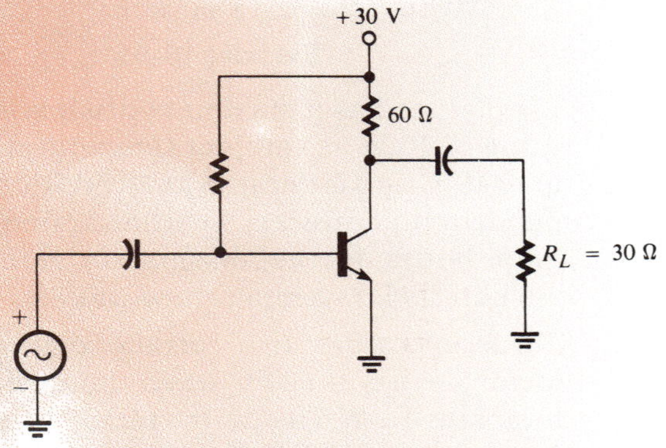

FIGURE 16–48 (Exercise 16–14)

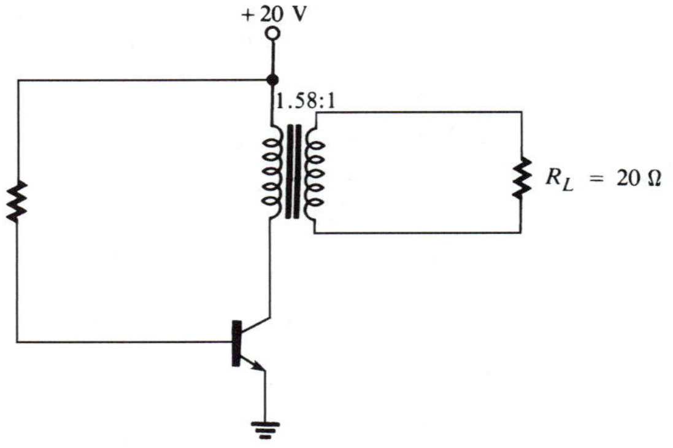

FIGURE 16–49 (Exercise 16–16)

SECTION 16–5

Push-Pull Amplifier Principles

16–18. Draw a schematic diagram of a class-B, push-pull, transformer-coupled amplifier using *pnp* transistors. Draw a loop showing current direction through the primary when the input is positive and another showing current when the input is negative.

16–19. The peak collector current and voltage in each transistor in Figure 16–50 are 4 A and 12 V, respectively. Assuming that the transformer has negligible resistance, find

(a) the average power delivered to the load,

(b) the average power supplied by the dc source,

(c) the average power dissipated by each transistor, and

(d) the efficiency.

16-20. Repeat Exercise 16–19 when the drive signals are increased so that the peak collector voltage in each transistor is the maximum possible. Assume that the peak current increases proportionately.

16-21. A push-pull amplifier utilizes a transformer whose primary has a total of 160 turns and whose secondary has 40 turns. It must be capable of delivering 40 W to an 8-Ω load under maximum power conditions. What is the minimum possible value of V_{CC}?

SECTION 16-7

Harmonic Distortion and Feedback

16–22. A square wave having peak value V_P volts and frequency ω rad/s can be expressed as the sum of an infinite number of sine waves as follows: $v(t) = 4V_P/\pi\,[\sin \omega t + (1/3)\sin 3\omega t + (1/5)\sin 5\omega t + \cdots]$. Assuming that the square wave represents a distorted sine wave whose frequency is the same as the fundamental, find its total harmonic distortion. Neglect harmonics beyond the ninth.

16–23. An amplifier has 1% distortion with negative feedback and 10% distortion without feedback. If the feedback ratio is 0.02, what is the open-loop gain of the amplifier?

16–24. An amplifier has a voltage gain of 80 when no feedback is connected. When negative feedback is connected, the amplifier gain is 20 and its harmonic distortion is 10%. What is the harmonic distortion without feedback?

SECTION 16-8

Distortion in Push-Pull Amplifiers

16–25. Draw schematic diagrams of class-AB amplifiers using *pnp* transistors and an output transformer and having

(a) an input coupling transformer, and

(b) capacitor-coupled inputs.

It is not necessary to show component values, but label the polarities of all power supplies used. In each diagram, draw arrows showing the directions of the quiescent base currents and the dc currents in the primary of the output transformer.

SECTION 16-9

Transformerless Push-Pull Amplifiers

16–26. (a) What is the maximum permissible peak value of input v_i in the amplifier shown in Figure 16–51?

(b) What is the maximum power that can be delivered to the load?

(c) What should be the breakdown voltage rating of each transistor?

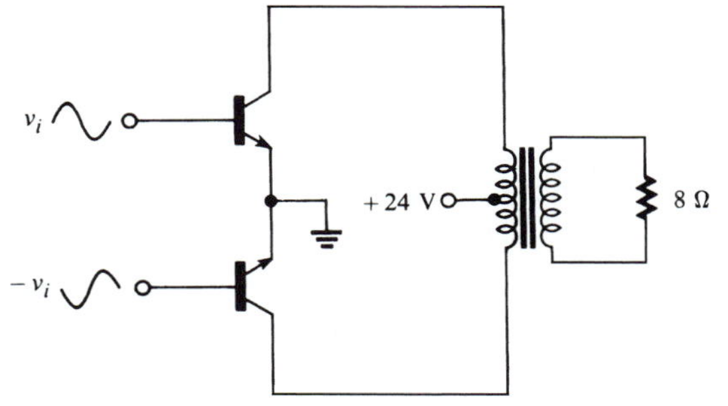

FIGURE 16–50 (Exercise 16–19)

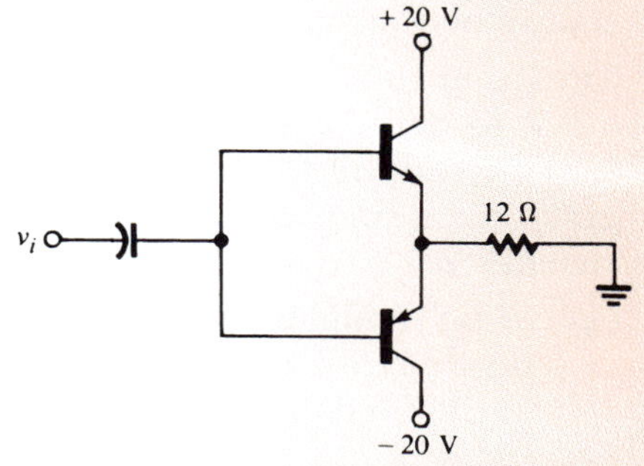

FIGURE 16–51 (Exercise 16–26)

16–27. If the amplifier in Exercise 16-26 must be redesigned so that it can deliver 36% more power to the load, by what percentage must the supply voltages be increased? What should be the minimum breakdown voltage of the transistors used in the new design?

16–28. Assuming that all the components in Figure 16–52 are perfectly matched, find
(a) the current in each 680-Ω resistor,
(b) the base-to-ground voltage of each transistor,
(c) the efficiency when the peak voltage V_P is 10 V, and
(d) the lower cutoff frequency.

16–29. (a) What new value of load resistance is required in the amplifier of Exercise 16–28 if the lower cutoff frequency must be 15 Hz?
(b) What is the maximum efficiency with the new load resistance determined in (a)?

SECTION 16-10

Class-C Amplifiers

16–30. The peak input voltage required to saturate the transistor in Figure 16–34 is 5.5 V. What should be the value of V_{BB} in order to achieve a conduction angle of 90°?

16–31. A peak input of 8 V is required to drive the transistor in Figure 16–34 to its 1.2-A saturation current. It is desired to produce a collector current that has a dc (average) value of 0.3 A. What should be the value of V_{BB}?

16–32. A peak input of 9 V is required to drive the transistor in Figure 16–34 to its 2.0-A saturation current. It is desired to produce a collector current whose fundamental has a peak value of 0.5 A. What should be the value of V_{BB}?

16–33. The input to a class-C amplifier must be 6.8 V to drive the output to its saturation current of 1.0 A. The amplifier begins to conduct when the input voltage reaches 2.6 V.
(a) What should be the value of V_{CC} if it is desired to obtain 5 W of output power at the fundamental frequency?
(b) What is the efficiency of the amplifier under the conditions of (a)?

16–34. Derive equation 16–80 from equation 16–79. (*Hint:* Use the trigonometric identity for (sin *A*) (sin *B*).)

16–35. An amplitude-modulated waveform has a carrier frequency 810 kHz and an upper sideband that extends from 812.5 kHz to 829 kHz. What is the frequency range of the lower side-band?

16–36. An amplitude modulator is driven by an 8-V-rms, 10-kHz carrier and has a 2-V-rms, 10-kHz signal input. Write the mathematical expression for the modulated output.

SECTION 16-11

MOSFET and Class-D Power Amplifiers

16–37. The pulse-width modulator in Figure 16–40 is driven by a 100-kHz sawtooth. When v_i is +2 V, the modulator output is a series of pulses having widths equal to 1 μs. What are the pulse widths when $v_i = +3.5$ V?

16–38. The voltage comparator in Figure 16–40 switches between 0 V and +10 V. A 1-kHz sawtooth that varies between 0 V and +5 V is applied to the inverting input. A ramp voltage that rises from 0 V to 5 V in 5 ms is applied to the noninverting input. Sketch the input and output waveforms. Label the time points where the comparator output switches from low to high.

16–39. Repeat Exercise 16–38 if the inputs to the comparator are reversed. Label time points where the comparator switches from high to low instead of low to high. What would be the purpose of this input reversal, assuming that the comparator output was connected directly to a MOSFET totem-pole output stage?

16–40. Draw a schematic diagram of a MOSFET totem pole using *p*-channel devices. Describe an input signal that would be appropriate for driving this totem pole. Describe the operation of the totem pole, including its output levels, for each input level.

16–41. A class-D amplifier is to be designed to deliver a maximum average current of 2 A to a 12-Ω load. A 50-kHz sawtooth is to be used for the pulse-width modulator. Under maximum drive, the totem-

FIGURE 16–52 (Exercise 16-28)

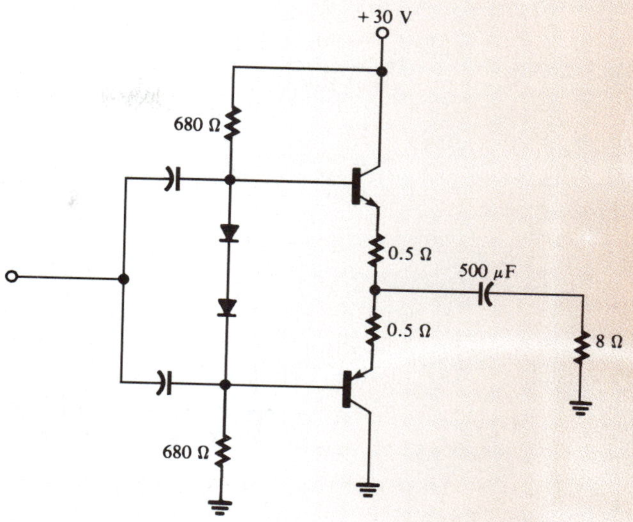

pole output produces pulses that are 15 μs wide.

(a) What should be the maximum input signal frequency?

(b) Assuming 100% efficiency, what should be the value of V_{DD} used in the totem pole?

PSPICE EXERCISES

16–42. Use PSpice to obtain a plot of the load voltage in the class-B amplifier shown in Figure 16–27(a). The supply voltages are ±15 V dc, the coupling capacitor is 100 μF, and the load resistance is 100 Ω. The input signal is a sine wave having frequency 1 kHz. Obtain a plot of one full cycle of output when

(a) the peak input is 12.5 V. What is the (approximate) power delivered to the load, based on the PSpice results?

(b) the peak input is 2.5 V. Why is the crossover distortion so much more severe than in Example 16–11?

16–43. The amplifier in Figure 16–27(b) has R_E = 1 Ω, V_{CC} = ±50 V, R_L = 20 Ω, and a 100 μF coupling capacitor. Use PSpice to obtain a plot of one full cycle of load voltage when the input is a 40-V peak sine wave having frequency 1 kHz. Also obtain plots of the collector current in each of transistors Q_3 and Q_4. Use the PSpice results to determine the power delivered to the load.

16–44. The input coupling capacitors in the amplifier in Figure 16–29 are each 100 μF and the output coupling capacitor, C_c, is 723 μF. The input signal is a 200-Hz sine wave having peak value 10 V. Use PSpice to obtain a plot of the load voltage over one full cycle and a plot of the current drawn from the power supply over one full cycle. Use the PSpice results to answer the following questions:

(a) What is the dc base-to-ground voltage of each transistor? Compare with the values calculated in Example 16–12.

(b) What is the power delivered to the load?

(c) What is the voltage gain of the amplifier?

(d) What is the peak current drawn from the power supply? Explain the appearance of the current waveform. (Why is it not sinusoidal?)

16–45. The quasi-complementary amplifier in Figure 16–31 has R_E = 1 Ω, V_{CC} = ±20 V, R_L = 10 Ω and 100-μF-input coupling capacitors. The input signal is a 12-V-pk sine wave having frequency 1 kHz. Use PSpice to obtain a plot of the load voltage over one full cycle. Using the results, determine the peak-to-peak load current, the voltage gain of the amplifier, and the power delivered to the load.

CHAPTER 17

TRANSISTOR ANALOG CIRCUIT BUILDING BLOCKS

OUTLINE

OBJECTIVES

- Investigate the building blocks for analog circuits.
- Explore the use of computer-based analysis to investigate a circuit's performance.
- Develop an understanding of the fabrication model parameters for MOSFETS and BJTs.
- Investigate how to use computer simulation to simplify the analog circuit design process.
- Develop an appreciation of the internal structure of an operational amplifier so that the user can achieve predictable results.

17–1 INTRODUCTION

The material in Chapter 17 brings together the basic concepts for building analog circuits using transistors. The techniques presented are suitable for both discrete and Very Large Scale Integration (VLSI) design applications. The methodology does require the use of some advanced concepts but the techniques can be mastered by novice analog circuit designers. The examples presented in each section demonstrate the use of the techniques and provide sufficient background for the student to undertake new design tasks. Examples for each building block are provided using BJTs and MOSFET transistors.

Section 17–2 reinvestigates the BJT and MOSFET transistor model. A good understanding of this model helps the designer better understand the limitations of each transistor device. The concept of using a transistor as a current source is presented in Section 17–3. The current mirror is explored in Section 17–4 and the gain stage is presented in 17–5. An important building block in analog circuits is the differential amplifier, which is explored in Section 17–6. These building blocks provide an excellent starting point for understanding basic analog circuit design. In fact, simple operational amplifier integrated circuits such as the Miller op-amp are constructed with these fundamental building blocks. The Miller op-amp is discussed in Section 17–7.

17–2 THE BJT AND MOSFET SMALL-SIGNAL EQUIVALENT CIRCUITS

The small-signal equivalent circuit is used by a circuit designer to better understand a transistor's operation and limitation in the circuit. SPICE simulation has minimized the need for the designer to obtain exact solutions by hand. However, these equivalent circuits are critical for obtaining a "ballpark" estimate of the circuit's performance. We will examine the high-frequency BJT and MOSFET models. All circuits have some frequency limitation, and these models help the designer to better understand the transistor's limitations in a specific application.

The BJT Model

The small-signal equivalent circuit of a BJT is shown in Figure 17–1. This circuit is also called the *high-frequency hybrid-π* model. The base, collector, and emitter terminals are shown as B, C, and E, respectively. The model shows the transistor parasitic capacitances, which affect the performance of the device. The resistance r_π is used to describe the input resistance to the transistor. C_π describes the forward-biased base–emitter junction capacitance, and C_μ and r_μ describe the parasitic capacitance that exists between the collector and base. C_μ, the reverse-biased base–collector junction capacitance, is an extremely important parameter when designing a transistor amplifier that requires a large bandwidth and large gain. It will be shown that the value of C_μ increases with gain and thereby limits the bandwidth. C_{CS} describes the reverse-biased collector–substrate junction capacitance. The resistance r_o describes the common-emitter output resistance of the transistor.

The values provided in this model can be used to determine the transition frequency (f_T) of a transistor, which is the frequency where the short-circuit current gain of a common-emitter amplifier (β or h_{fe}) falls to unity. The equation for f_T is

$$f_T = \frac{1}{2\pi}\frac{g_m}{C_\pi + C_\mu} \tag{17–1}$$

The values from the model can also be used to find the 3-dB frequency of the current gain β. The 3-dB frequency is called f_β and can be calculated using equation 17–2:

$$f_\beta = \frac{1}{2\pi\beta}\frac{g_m}{C_\pi + C_\mu} = \frac{1}{2\pi r_\pi(C_\pi + C_\mu)} \tag{17–2}$$

Each of the components in the model can be quantified using information derived from the quiescent bias collector current, often referred to as I_{CQ} or I_C. Table 17–1 lists the parameters in the BJT transistor model, the equations for calculating the parameters, and typical values for components in the model. The calculations for the capacitor values have been omitted because these parameters are dependent on the physical layout. The other terms are bias-current dependent. Using the equations and typical values given in Table 17–1, the designer can estimate the performance of the BJT in the circuit.

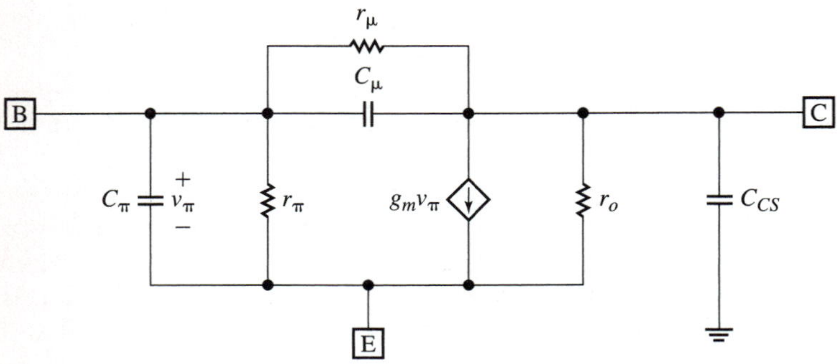

FIGURE 17–1 The small-signal equivalent circuit for a BJT transistor

TABLE 17–1 Equations and typical values for the BJT equivalent circuit

Component	Equation	Typical Value (I_c = 1 mA)
r_π	β/g_m	2.6 kΩ
r_μ	[5 to 10] × (βr_o)	100 MΩ
r_o	V_A/I_{C_Q}	100 kΩ
g_m	I_{C_Q}/V_T	40 mA/V
C_π	—	1 pF
C_μ	—	0.3 pF
C_{CS}	—	3 pF

The MOSFET Model

The high-frequency model of the MOSFET transistor is shown in Figure 17–2. Note that there is not a dc input resistance into the device as there was with the BJT. This is because the MOSFET has a capacitive input. A dc current does not flow into the gate of the device; hence the input resistance is considered to be infinite. This is a very desirable characteristic for many amplifiers. Tables 17–2 and 17–3 provide a brief review of transistor input resistances for the BJT, JFET, and the MOSFET transistor gain stages (CE- and CS-biased amplifiers). Notice that there are five significant parasitic capacitors in the MOSFET model. The feedback capacitance from drain-to-gate, C_{gd}, is important because its Miller-effect value can dominate the frequency response of a gain circuit. The transition frequency, f_T, for a MOSFET can be calculated using equation 17–3:

$$f_T = \frac{1}{2\pi}\frac{g_m}{C_{gs} + C_{gd}} \tag{17–3}$$

FIGURE 17–2 The high frequency model of the MOSFET transistor

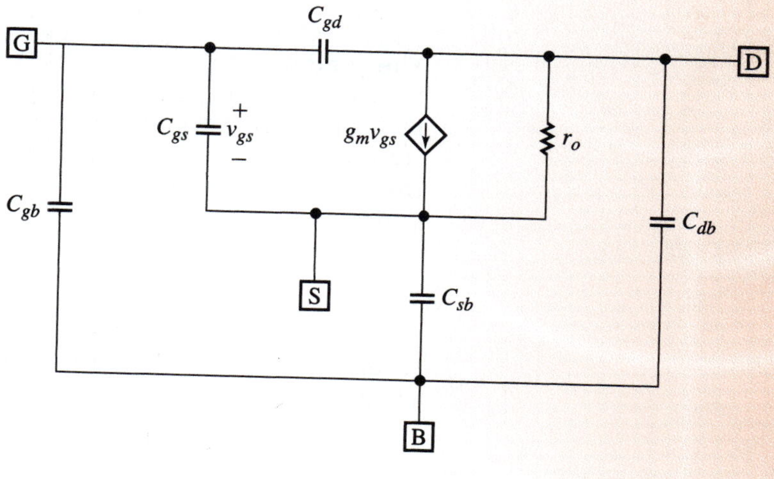

TABLE 17–2 Approximate input resistance values for BJT, JFET, and MOSFET gain stages

Device	Characteristics (Comments)
BJT	low to moderate input resistance (input is a forward-biased *pn* junction)
JFET	very high (input is a reverse-biased *pn* junction)
MOSFET	open circuit (capacitive input)

TABLE 17–3 Equations and typical values for the MOSFET equivalent circuit

Component	Equation	Typical Value
r_o	$\frac{1}{\lambda I_D}$	$\lambda \approx 0.02$ to 0.04
g_m	$\beta(V_{GS} - V_t)$ or $\sqrt{2\beta I_D}$ $2I_D/(V_{gs} - V_t)$	$g_{mMOS} \ll g_{mBJT}$
C_{gd}, C_{gs}, C_{db} C_{gb}, C_{sb}	—	device layout dependent

Notice the similarity to the BJT model. In fact, the design methods presented in the following sections demonstrate the similarity in the design methodology for a BJT and a MOSFET.

17–3 THE CURRENT SOURCE/SINK

In the previous chapters, several techniques were presented for using resistors for biasing the transistor to a particular Q-point. In some applications, it is not practical to bias transistors using resistor networks. This is especially true in integrated circuits designs. Integrated circuit designers will use transistors in many configurations to provide the bias for transistor amplifiers. They also take advantage of the fact that transistors, fabricated on the same chip, are well matched, meaning that they will have comparable operating characteristics. Additionally, the designer will lay out the device so that random errors, created in the fabrication process, can be minimized. For example, MOS threshold voltages (ΔV_t) and BJT beta variations ($\Delta\beta$) can be minimized by layout techniques that address processing errors.

The BJT Current Source

The fundamental bias circuit for an IC is a dc constant-current source, a transistor circuit that provides a nearly constant current to a load. For a BJT current source, a constant bias voltage is applied to the base of the transistor. The bias voltage should be sufficient to place the transistor in the "active" mode. The current flowing in the transistor is described by the equation

$$I_C = I_S e^{\frac{|V_{BE}|}{V_t}} \left(1 + \frac{V_{CE}}{V_A}\right) \quad (17\text{–}4)$$

where I_S is the reverse saturation current (~10^{-13}A), V_{BE} is the magnitude of the base–emitter voltage (0.5 V to 0.7 V), and V_T is the thermal voltage (~26 mV at room temperature). The $(1 + V_{CE}/V_A)$ describes the effect that the collector–emitter voltage has on I_C. V_A is the Early voltage, typical values of which are 50 to 150 volts, although some transistors will have a smaller Early voltage. For practical purposes, $(1 + V_{CE}/V_A)$ simplifies to 1 and I_C can be written as

$$I_C = I_S e^{\frac{|V_{BE}|}{V_t}} \quad (17\text{–}5)$$

This equation enables the designer to determine I_C given a specific value for V_{BE} and can be modified so that it can be solved for a required base bias voltage (V_{bias}) to source or to sink the required current. This equation is written as

$$V_{BE} = \ln\frac{I_C}{I_S} \cdot V_T \qquad (17\text{–}6)$$

where
$$V_{bias} = V_{BE} \ (npn)$$
$$V_{bias} = +V - |V_{BE}| \ (pnp)$$

A BJT current source is shown in Figure 17–3(a), and a BJT current sink is shown in Figure 17–3(b). R_L represents the load to the current sink/source. The load is typically another transistor configured as a gain stage, a differential amplifier, or many of the other possible transistor circuits. Example 17–1 demonstrates how equations 17–5 and 17–6 are used to calculate the bias currents and bias voltages in a BJT current sink or current source.

EXAMPLE 17–1

1. Determine the current I_C for the circuit shown in Figure 17–3(b). The bias voltage (V_{bias}) equals 0.62 V, I_S = 1 E–13 A, and the temperature is 27°C.

Solution

The temperature 27°C needs to be converted to Kelvin so that the thermal voltage V_T can be calculated.

$$\text{Kelvin} = 273 + 27°\text{C} = 300 \qquad kT/q = (1.38 \times 10^{-23})(300)/(1.6 \times 10^{-19}) = 0.0258\ \text{V}$$

Use equation 17–5 to solve the collector current I_C.

$$I_C = I_S e^{\frac{|V_{BE}|}{V_t}} = 1 \times 10^{-13} e^{\frac{0.62}{0.0258}} = 2.73\ \text{mA}$$

2. Determine the required base bias voltage required to source 1 mA of current given that I_S = 1 E–13 A and the temperature is 27°C.

Solution

Use equation 17–6 to find V_{BE}.

$$V_{BE} = (0.0258)\ \ln\frac{1\ \text{mA}}{1 \times 10^{-13}} = 0.594\ \text{V}$$

FIGURE 17–3 The BJT current source (a) and sink (b)

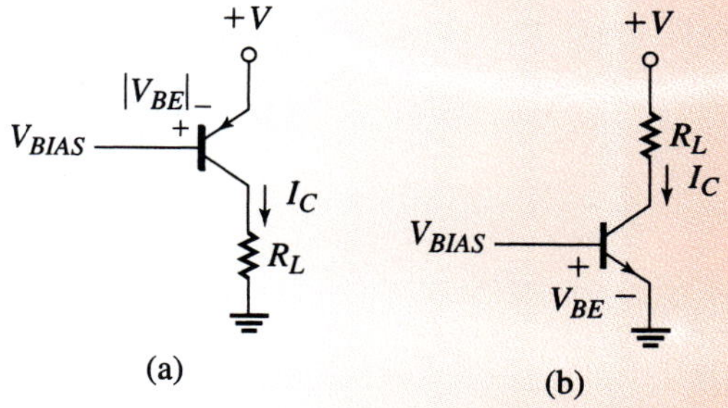

The MOSFET Current Source

For a MOSFET current source, a constant bias voltage is applied to the gate of the transistor. The bias voltage should be sufficient to place the transistor in the "saturation" mode. The current flowing in the transistor is defined by the equation:

$$I_D = \frac{\beta}{2}(|V_{GS}| - |V_t|)^2(1 + \lambda V_{DS}) \quad \textbf{(17–7)}$$

where β is a complex quantity defining the gain of a MOSFET transistor. $\beta = KP \times W/L$, where *KP* is a transconductance parameter with units (A/V^2) and *W/L* are layout parameters that define the width and length of the MOSFET transistor's gate. *KP* is fabrication specific and must be provided for each transistor. V_{GS} is the magnitude of the gate-to-source voltage for the transistor (voltage range is ~0.7 V to $V_{DD} - V_t$). V_t is the magnitude of the threshold voltage for the transistor. Typical threshold voltages for MOSFETs are 0.7 V to 0.85 V. Lambda, λ in the $1 + \lambda V_{DS}$ term, is called the *channel-length modulation* parameter, which is multiplied by the V_{DS} voltage. Typical values for lambda are small, on the order of 0.01 to 0.04. For practical purposes, this term can be ignored and equation 17–8 will yield a satisfactory answer for I_D.

$$I_D = \frac{\beta}{2}(|V_{GS}| - |V_t|)^2 \quad \textbf{(17–8)}$$

As with the BJT, the equation for the drain current can be modified so that the designer can solve for the required gate bias voltage (v_{bias}) to source or sink the required current. This equation is written as

$$V_{GS} = \sqrt{\frac{2I_D}{\beta}} + |V_t| \quad \textbf{(17–9)}$$

where $V_{bias} = V_{GS}$ (*NMOS*)

$V_{bias} = +V - |V_{GS}|$ (*PMOS*)

A MOSFET current source is shown in Figure 17–4(a) and a MOSFET current sink is shown in Figure 17–4(b). R_L represents the load to the current sink/source. As with the BJT, the load is typically another transistor configured as a gain stage, a differential amplifier, or many of the other possible transistor circuits.

Example 17–2 demonstrates how equations 17–8 and 17–9 are used to calculate the bias currents and bias voltages in a MOSFET current sink or current source.

EXAMPLE 17–2

1. Determine the current I_D for the circuit shown in Figure 17–4(b). The bias voltage (V_{bias}) equals 1.2 V, $\beta_n = 4 \times 10^{-4}$ A/V², $V_{in} = 0.7$ V, $R_L = 100\ \Omega$, $V_{DD} = 1.5$ V.

Solution

Use equation 17–8 to find the value for I_D.

$$I_D = \frac{4 \times 10^{-4}}{2}(1.2 - 0.7)^2 = 50\ \mu A$$

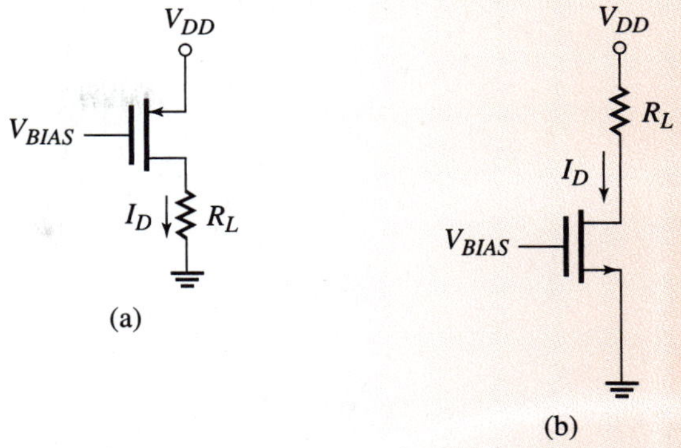

FIGURE 17–4 The MOSFET current source (a) and sink (b)

2. Determine the bias voltage required to source an I_D current of 100 μA for the circuit shown in Figure 17–4(a). $I_D = 100\ \mu A$, $\beta_p = 3.9 \times 10^{-4}\ A/V^2$, $V_{tp} = -0.82\ V$, $R_L = 100\ \Omega$, $V_{DD} = 2.0\ V$.

Solution

Use equation 17–9 to find the required input bias voltage, V_{GS}.

$$V_{bias} = V_{GS} = \sqrt{\frac{2I_D}{\beta}} + |V_t| = \sqrt{\frac{(2)(100\ \mu A)}{3.9 \times 10^{-4}}} + |0.82| = 1.536\ V$$

The answers obtained in parts (1) and (2) are close approximations to the exact solution. The exact answer may be off as much as 3 to 4 μA due to the effects of the $1 + \lambda V_{DS}$ term.

17–4 THE CURRENT MIRROR

The BJT Current Mirror

In Section 17–3, a technique was developed for designing a circuit to sink or source a specific current. This section explores a technique for distributing the current to other circuits using a transistor configuration called a *current mirror.* This circuit is shown in Figure 17–5. In the figure, a current source, I_{bias}, is shown connected to transistor Q_1 whose collector is shorted to its base. This is called a *diode-connected* transistor. The collector–base short forces the device to operate in the active region as long as current is flowing. The connection creates the condition where the

FIGURE 17–5 A BJT current mirror

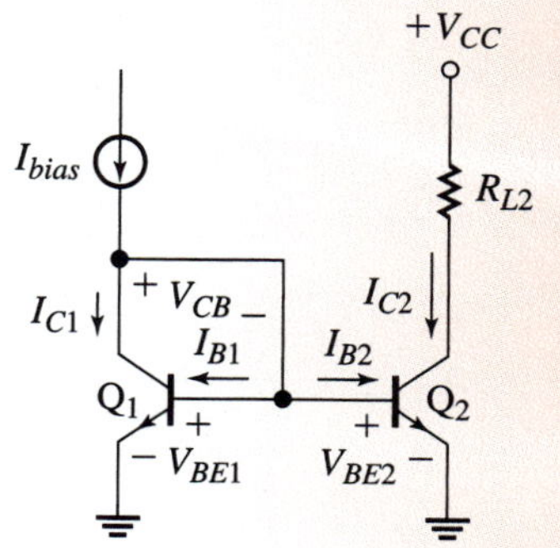

collector–base junction, V_{CB}, is never forward biased. Recall that the collector–base junction must be reverse biased for the transistor to operate in the active region.

It will be shown that the current I_{bias} is essentially mirrored (or reflected) in the collector of Q_2. In most cases, it can be assumed that transistors Q_1 and Q_2 are fabricated on the same chip and are therefore closely matched. Q_1 and Q_2 could also be a "matched" pair of discrete transistors. The base–emitter junctions for each are in parallel and the equal values of V_{BE} will yield equivalent values of I_C. For practical purposes, $I_{B1} = I_{B2}$. I_B is a small current and can be ignored for developing "ballpark" estimates. Once again, SPICE will provide an accurate answer if the proper fabrication model parameters are used and an exact answer is required. The equation for the current I_{C2} is developed as follows:

$$I_{C1} = \beta_1 I_{B1} \quad \text{and} \quad I_{C2} = \beta_2 I_{B2} \tag{17–10}$$

Rewriting these equations to express the relationship of the output current I_{C2} to I_{C1} yields

$$\frac{I_{C2}}{I_{C1}} = \frac{\beta_2 I_{B2}}{\beta_1 I_{B1}}$$

where $I_{B_2} = I_{B_1}$ for matched pairs; therefore,

$$I_{C2} = \frac{\beta_2}{\beta_1} \times I_{C1} \tag{17–11}$$

Equation 17–11 shows that the output current, I_{C2}, for a current mirror is primarily dependent on the β ratios of the BJT transistors multiplied by the bias current.

EXAMPLE 17–3

1. Assuming that $\beta_1 = \beta_2 = 150$, find the approximate value of I_{C2} and V_{CE} in the circuit shown in Figure 17–6.

Solution

$$\beta_1 = \beta_2\text{; therefore, } I_{C2} \approx \beta_2/\beta_1 \times 3.1 \text{ mA} = 3.1 \text{ mA}$$

2. Assuming that $\beta_1 = 100$ and $\beta_2 = 175$, find the approximate value of I_{C2} and V_{CE} in the circuit shown in Figure 17–6.

Solution

$$I_{C2} \approx \beta_2/\beta_1 \times 3.1 \text{ mA} = 175/100 \times 3.1 \text{ mA} = 5.425 \text{ mA}$$

FIGURE 17–6 The circuit for Example 17–3

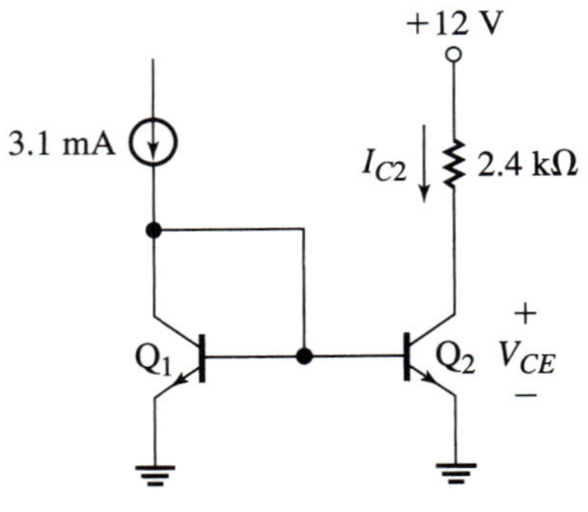

A single current source can be used with a current mirror to set the bias for numerous amplifiers in the same integrated circuit. This is demonstrated in Example 17–4. When current mirroring is used to bias several BJT transistors from a single current source, each collector current is actually somewhat smaller due to the loss of I_B currents for each transistor. The greater the number of transistors, the greater the reduction in collector current for each stage.

Because of the extensive use of current mirrors in integrated circuits, a different convention is sometimes used in schematic diagrams to show transistors having common-base terminals. This is shown in Figure 17–7. The horizontal line representing the connection to the base terminal is extended through the vertical bar of the BJT schematic symbol, as shown.

EXAMPLE 17–4

1. Find the ideal value for I_{C2}, I_{C3}, and I_{C4} for the current mirror circuit shown in Figure 17–7. Assume that $\beta_1 = 100$, $\beta_2 = 100$, $\beta_3 = 200$, and $\beta_4 = 150$. $I_{bias} = 2$ mA.
2. Verify your answer using SPICE.

Solution

1. Neglecting I_B,

$$I_{C2} \approx \beta_2/\beta_1 \times 2\text{ mA} = 100/100 \times 2\text{ mA} = 2\text{ mA}$$

$$I_{C3} \approx \beta_3/\beta_1 \times 2\text{ mA} = 200/100 \times 2\text{ mA} = 4\text{ mA}$$

$$I_{C4} \approx \beta_4/\beta_1 \times 2\text{ mA} = 150/100 \times 2\text{ mA} = 3\text{ mA}$$

2. The SPICE .cir file for Figure 17–7 follows. Notice that after each BJT transistor description in the .cir file, a number appears (1, 1, 2, 1.5) following the reference link to the model. For example, transistor Q_3 is described as follows:

```
Q3 3 1 0 tran 2
```

The Q indicates a BJT transistor. The collector–base–emitter nodes are next defined (3 1 0), followed by the reference link (tran) to the .Model. The last number is the area value for a BJT. This parameter is necessary to accurately model changes in the I_C current. In an integrated circuit layout, the layout areas of the emitter are used to scale the current for each device. The β value for each transistor is scaled using the area value. The

FIGURE 17–7 The circuit for Example 17–4

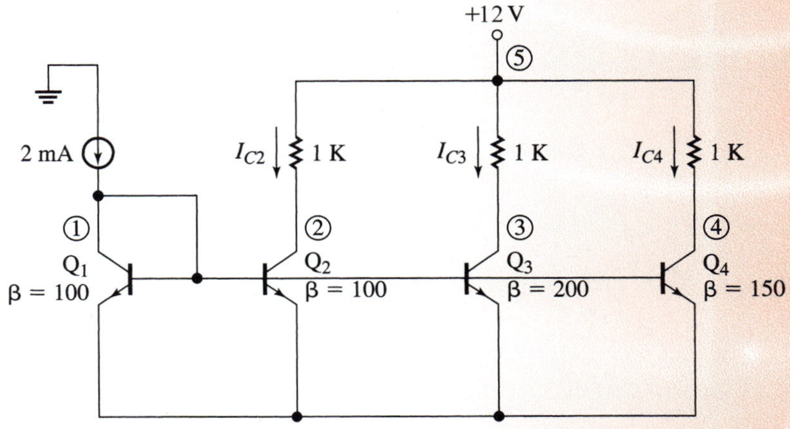

reference β, defined by the .Model statement, is BF= 100. Therefore, Q_1 and Q_2 have a relative ratio of 1, Q_3 has twice the reference β of 100, so its area value is 2, and Q_4 has a β of 150, which is 1.5 times the reference β.

Mathematically, the saturation current I_S is multiplied by the scaling factor specified by the area.

```
Example 17-4
Ibias 0 1 2MA
Vcc 5 0 12.0V
Q1 1 1 0 tran 1
Q2 2 1 0 tran 1
Q3 3 1 0 tran 2
Q4 4 1 0 tran 1.5
R2 5 2 1k
R3 5 3 1k
R4 5 4 1k
.model tran NPN BF = 100
.TRAN 1ms 100ms
.PROBE
.END
```

The SPICE simulations yield the following results:

$I_{C1} = 1.896$ mA
$I_{C2} = 1.896$ mA
$I_{C3} = 3.79$ mA
$I_{C4} = 2.84$ mA

These answers are close to the ideal answers but our assumption of neglecting I_B does introduce some error. However, for approximation purposes, our ideal answers are suitable.

The MOSFET Current Mirror

The MOSFET current mirror operates very much like the BJT, but the equations defining the current differ. For a MOSFET, the current I_D is defined as follows:

$$I_D \approx \frac{\beta}{2}(|V_{GS}| - |V_t|)^2 \qquad \textbf{(17–12)}$$

The circuit for a MOSFET current mirror is shown in Figure 17–8. The PMOS transistors, M_1 and M_2, form the current mirror. M_1 is a diode-connected device (the drain is shorted to the gate). The drain–gate short will force the V_{DS} voltage to always be greater than the $(V_{GS} - V_t)$ voltage; therefore, the transistor will always operate in the saturation mode as long as current is flowing in the device. MOSFET M_3 is a current sink, which was described in Section 17–3. M_3 provides the bias current (I_{D1}) for the circuit. The drain current I_{D2} is the reflected current, and its value is obtained as follows. For matched devices, it can be assumed that $V_{t1} = V_{t2}$, and by inspection $V_{GS1} = V_{GS2}$. Therefore, equation 17–12 can be reduced to the following relationship expressing the output current I_{D2} to the bias current I_{D1}:

$$\frac{I_{D2}}{I_{D1}} = \frac{\beta_2}{\beta_1}$$

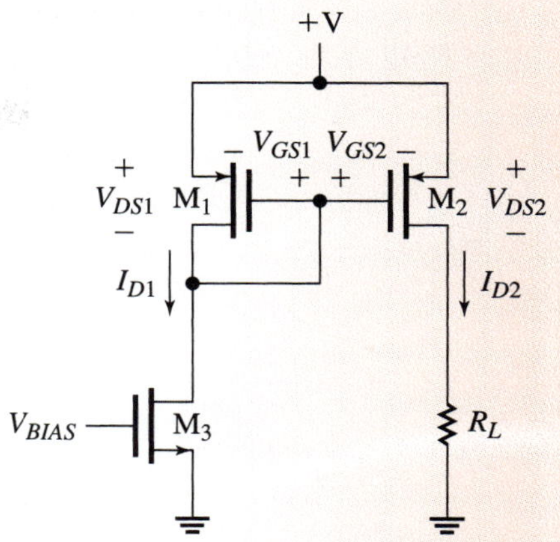

FIGURE 17–8 The MOSFET current mirror

yielding our design equation of

$$I_{D2} = \frac{\beta_2}{\beta_1} \times I_{D1} \tag{17–13}$$

Example 17–5 demonstrates how equation 17–13 is used in the design process.

EXAMPLE 17–5

Given the MOSFET transistor circuit shown in Figure 17–8, if the bias current I_{D1} is set to 50 μA, $\beta_1 = 4 \times 10^{-4}\ \text{A/V}^2$, and $\beta_2 = 8 \times 10^{-4}\ \text{A/V}^2$, determine the output current I_{D2}.

Solution

Use equation 17–13 to solve for I_{D2}.

$$I_{D2} = \frac{\beta_2}{\beta_1} \times I_{D1} = \frac{8 \times 10^{-4}}{4 \times 10^{-4}}(50\ \mu\text{A}) = 100\ \mu\text{A}$$

We showed that with a BJT current mirror, the collector current ratio was controlled by the areas of the emitter. For a MOSFET device, the drain currents are also controlled by an integrated circuit layout parameter. This control is the width/length (W/L) ratio of the gate. For a MOSFET transistor, β is defined to be

$$\beta = KP \cdot \frac{W}{L} \tag{17–14}$$

where KP is the transconductance parameter (specified in the fabrication model parameters for the NMOS and PMOS transistors) and W/L is the gate width/length ratio. This layout ratio is controlled by the VLSI designer and is selected so that the circuit will meet a desired operational specification.

In the next example, we will demonstrate that a transistor can be used to supply the bias current to the current mirror and that the currents can be mirrored to two circuit paths, each with a different W/L ratio, each, therefore, with a different current.

EXAMPLE 17–6

For the current mirror circuit shown in Figure 17–9, PMOS transistors provide the mirror current provided by M_1. The W/L (gate width/length) ratios are specified for each transistor. NMOS transistor M_5 is used to bias the current mirror. A bias voltage of 1.2 volts is applied to the gate of the M_1 transistor. $KP_n = 8.16 \times 10^{-5}\ \text{A/V}^2$ and $KP_p = 2.83 \times 10^{-5}\ \text{A/V}^2$.

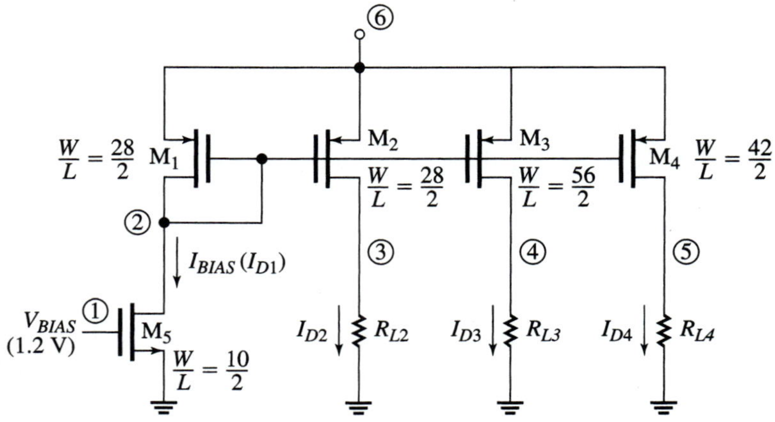

FIGURE 17–9 A MOSFET current mirror with three reflected currents

1. Determine the bias current, I_{D1}, and the ideal currents for I_{D2}, I_{D3}, and I_{D4}. $V_{tn} = 0.7\text{ V}$, $V_{tp} = -0.82\text{ V}$.
2. Verify your answers using SPICE.

Solution

1. The βs for each transistor need to be calculated using equation 17–14.

$$\beta_1 = KP_p \cdot \frac{W_1}{L_1} = 2.83 \times 10^{-5} \cdot \frac{28}{2} = 3.96 \times 10^{-4}$$

$$\beta_2 = KP_p \cdot \frac{W_2}{L_2} = 2.83 \times 10^{-5} \cdot \frac{28}{2} = 3.96 \times 10^{-4}$$

$$\beta_3 = KP_p \cdot \frac{W_3}{L_3} = 2.83 \times 10^{-5} \cdot \frac{56}{2} = 7.92 \times 10^{-4}$$

$$\beta_4 = KP_p \cdot \frac{W_4}{L_4} = 2.83 \times 10^{-5} \cdot \frac{42}{2} = 5.974 \times 10^{-4}$$

$$\beta_5 = KP_p \cdot \frac{W_5}{L_5} = 8.16 \times 10^{-5} \cdot \frac{10}{2} = 4.08 \times 10^{-4}$$

Next, calculate the bias current.

$$I_{bias} = I_{D1} = \frac{\beta_5}{2}(V_{GS5} - V_{t5})^2$$

$$I_{D1} = \frac{4.08 \times 10^{-4}}{2}(1.2 - 0.7)^2 = 50\ \mu\text{A}$$

Using the value calculated for I_{D1}, calculate the reflected currents for I_{D2}, I_{D3}, and I_{D4}.

$$I_{D2} = \beta_2/\beta_1 \times I_{D1} = 3.96 \times 10^{-4}/3.96 \times 10^{-4} \times 50\ \mu\text{A} = 50\text{ uA}$$

$$I_{D3} = \beta_3/\beta_1 \times I_{D1} = 7.92 \times 10^{-4}/3.96 \times 10^{-4} \times 50\ \mu\text{A} = 100\text{ uA}$$

$$I_{D4} = \beta_4/\beta_1 \times I_{D1} = 5.974 \times 10^{-4}/3.96 \times 10^{-4} \times 50\ \mu\text{A} = 75\text{ uA}$$

2. To verify the answers using SPICE, we created the following .cir file.

```
* A CMOS Inverter Using 2 Micron Channel Lengths
* D G S B
M1 2 2 6 6 CMOSP W=28.0U L=2.0U AS=84P AD=84P
M2 3 2 6 6 CMOSP W=28.0U L=2.0U AS=84P AD=84P
M3 4 2 6 6 CMOSP W=56.0U L=2.0U AS=84P AD=84P
M4 5 2 6 6 CMOSP W=42.0U L=2.0U AS=84P AD=84P
M5 2 1 0 0 CMOSN W=10.0U L=2.0U AS=30P AD=30P
Vbias 1 0 DC 1.2
VDD 6 0 2.0
RL2 3 0 100
RL3 4 0 100
RL4 5 0 100
*
*The following are fabrication parameters obtained from the MOSIS service.
.Model CMOSN NMOS LEVEL=2 LD=0.121440U TOX=410.000E-10
+ NSUB=2.355991E+16 VTO=0.7 KP=8.165352E-05 GAMMA=1.05002
+ PHI=0.6 UO=969.492 UEXP=0.308914 UCRIT=40000
+ DELTA=0.262772 VMAX=71977.5 XJ=0.300000U LAMBDA=3.937849E-02
+ NFS=1.00000E+12 NEFF=1.001 NSS=0 TPG=1.000000
+ RSH=33.290002 CGDO=1.022762E-10 CGSO=1.022762E-10
+ CGBO=5.053170E-11 CJ=1.368000E-04
+ MJ=0.492500 CJSW=5.222000E-10 MJSW=0.235800 PB=0.490000
* Weff= Wdrawn - Delta_W
* The suggested Delta_W is 0.06 um
*
.Model CMOSP PMOS LEVEL=2 LD=0.180003U TOX=410.000E-10
+ NSUB=1.000000E+16 VTO=-0.821429 KP=2.83164E-05 GAMMA=0.684084
+ PHI=0.6 UO=336.208 UEXP=0.351755 UCRIT=30000
+ DELTA=1.000000E-06 VMAX=94306.1 XJ=0.300000U
+ LAMBDA=4.861781E-02
+ NFS=2.248211E+12 NEFF=1.001 NSS=1.000000E+12 TPG=-1.000000
+ RSH=119.500003 CGDO=1.515977E-10 CGSO=1.515977E-10
+ CGBO=2.273927E-10 CJ=2.517000E-04 MJ=0.528100
+ CJSW=3.378000E-10
+ MJSW=0.246600 PB=0.480000
* Weff= Wdrawn - Delta_W
* The suggested Delta_W is 0.27 um
* .PLOT TRAN V(3) V(2) V(1)
.Tran .1n 250n
.Probe
.End
```

The SPICE analysis provides the following values for the currents:

$$I_{D1} = I_{bias} = 49.96\ \mu\text{A}$$
$$I_{D2} = 53.3\ \mu\text{A}$$
$$I_{D3} = 103.6\ \mu\text{A}$$
$$I_{D4} = 78.9\ \mu\text{A}$$

These results are close to the values calculated using approximations. SPICE model parameters describing the NMOS and PMOS transistors were used in the simulation and provided by the MOSIS service. The model parameters reflect the parameters obtained for a particular fabrication run. The geometry of each transistor gate is described by the W = and L = values. The purpose of the AS and AD (area of the source and drain) values will be explored later in Chapter 18.

17–5 THE GAIN STAGE

This section addresses the use of the transistor as an amplifier. The concepts behind this material were first presented in Chapter 6, and although we will take a VLSI design perspective here, the concepts directly relate to the discrete transistor concepts we have previously presented. The term *gain stage* is just another way of labeling an analog building block that amplifies. This section will explore the design, analysis, and applications of a gain stage for both the BJT and MOSFET transistors. Concepts from Sections 17–2, 17–3, and 17–4 will be incorporated as needed.

The BJT Gain Stage

We will first address the equations for calculating the component values in the simplified BJT transistor model shown in Figure 17–10. BJT transistors have an input resistance, r_π, which is defined in equation 17–15 where r_π is expressed as the reciprocal of the change in the base current relative to the change in the base–emitter voltage.

$$r_\pi = \left(\frac{\Delta I_B}{\Delta V_{BE}}\right)^{-1} = \frac{\beta}{g_m} \qquad (17\text{–}15)$$

A BJT transistor has a finite output resistance, r_o, which is defined as

$$r_o = \left(\frac{\Delta I_C}{\Delta V_{CE}}\right)^{-1} \approx \frac{V_A}{I_C} \qquad (17\text{–}16)$$

The value g_m is defined to be equal to

$$g_m = \frac{\Delta I_C}{\Delta V_{BE}} = \frac{I_C}{V_T} \qquad (17\text{–}17)$$

In Chapter 4, the output current was defined in terms of β. The dependent current source was written as $i_C = \beta i_B$. In Figure 17–10, the dependent current source is written as $g_m v_{be}$, and we will show that βi_B and $g_m v_{be}$ are equivalent. By inspection,

$$g_m v_{be} = g_m i_b r_\pi$$

where $r_\pi = \beta/g_m$

$$\therefore g_m i_b r_\pi = g_m i_b\, \beta/g_m = \beta i_b$$

Voltage gain is symbolized as A_V, which is the ratio of the output voltage to the input base–emitter voltage, v_L/v_{be}. The output resistance has been defined as $r_o = V_A/I_C$. V_A is the Early voltage, whose typical value range is 50 to 150 volts. Typical I_C values are on the order of 1 to 2 mA. Therefore, the

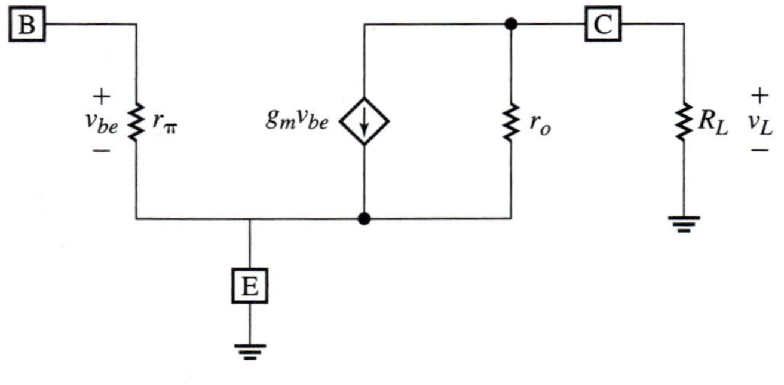

FIGURE 17–10 The simplified BJT model

value of r_o will be in the range of 50 kΩ to 150 kΩ. In other words, it is a very large resistance value, which can usually be ignored for typical values of R_L. To find the equation for A_v, write the equation defining the voltage, v_L:

$$v_L = -g_m v_{be} r_L$$

$$\frac{v_L}{v_{be}} = -g_m r_L \qquad (17\text{–}18)$$

Equation 17–18 shows that the gain of the transistor is simply $-g_m r_L$. Recall from Chapter 6 that the gain for a simple CE amplifier was shown to be $-r_L/r_e'$. This is the same result as provided in equation 17–17.

$$\frac{V_L}{V_{be}} = -g_m r_L \qquad \text{where } g_m = \frac{I_C}{V_T}$$

and

$$\frac{1}{g_m} = \frac{1}{\frac{I_C}{V_T}} \qquad \text{where } \frac{I_C}{V_T} = \frac{1}{r_e'}$$

therefore,

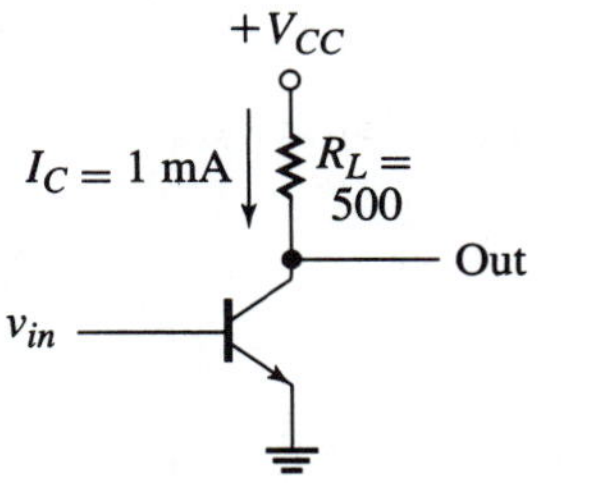

FIGURE 17–11 The gain stage for Example 17–7

$$A_V = -g_m r_L = \frac{-r_L}{r_e'} \qquad (17\text{–}19)$$

If $\beta = g_m v_{be}$ then why not use β instead? As demonstrated in equations 17–15, 17–16, and 17–17, the transistor parameters are expressed in terms of either I_C or g_m. In other words, these parameters are all dependent on the bias current, a design specification. The use of g_m also allows for consistent use of the A_V expression for the BJT ($-g_m r_L$), JFET ($-g_m r_L$), and the MOSFET ($-g_m r_L$) transistors. Example 17–7 demonstrates how to apply the relationships developed in this section for a BJT gain stage.

EXAMPLE 17–7

Given the circuit shown in Figure 17–11, determine the gain A_V given a bias current $I_C = 1$ mA. Assume room temperature conditions and $r_L = R_C = 500\Omega$.

Solution

A_V is defined to be $-g_m r_L = -r_L/r_e'$, where $g_m = 1$ mA/26 mV and $r_e' = \frac{1}{g_m} = 26\ \Omega$. Therefore, $A_V = -r_L/r_e' = -500/26 = -19.23$.

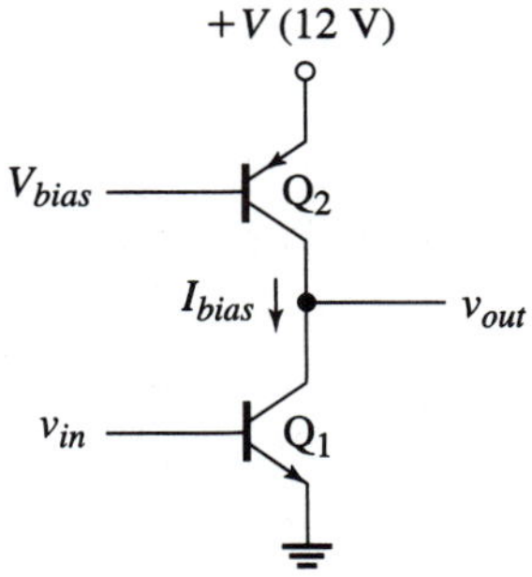

FIGURE 17–12 A BJT gain stage with an active BJT load

ACTIVE BJT LOADS How could a transistor be used to replace R_L in Figure 17–11? How could the circuit in Example 17–7 be modified so that a specific bias current is supplied by a current source? How will this affect the gain of the gain stage? It was previously mentioned that integrated circuits rarely use resistors for biasing. The bias current is almost always provided by active components. The concept of a BJT current source was developed in Section 17–2.

Figure 17–12 shows what the gain stage will look like if the resistor is replaced with an active load (a transistor). The gain (v_{out}/v_s) for this gain stage is still $-g_m r_L$, where g_m is still based on the bias current, I_{bias}. The major change is that r_L is now equal to $r_{o2} \| r_{o1}$, the parallel resistance of the output resistances for each transistor. The output resistance r_o was not considered previously because typical transistor output resistances are much larger than R_L. This means that A_V for a gain stage with an active load will be quite large.

In fact, in integrated circuits, maximum gain for a gain stage is obtained using an active load. The bias current is provided to the circuit by setting the bias voltage (V_{bias}) as described in Section 17–2 and by setting the dc bias point to the input transistor at the same point. The use of the current source is demonstrated in Example 17–8.

EXAMPLE 17–8

Design a BJT gain stage using an active load. The V_{BE} voltage is 0.66 V and v_{in} is a 1-mV sinusoid at 1 kHz sitting on a dc offset voltage of 0.66 V. Assume a temperature of 27°C ($V_T = 0.0258$ V).

1. Find the following for the circuit shown in Figure 17–12: $I(Q_1)$, A_V for the gain stage, and V_{bias} for the current source Q_2.
2. Compare your answers for the bias current and gain to SPICE results using device model parameters for a 2N3904 and a 2N3906.

Given from the device model parameters:

Parameter	2N3904 (Q_1)	2N3906 (Q_2)
Saturation Current (I_S)	6.734×10^{-15}	1.41×10^{-15}
Early Voltage	74.03	18.7

Solution

1. First determine the bias current for a $V_{BE} = 0.66$ V. Using equation 17–5, $I_C = 6.734 \times 10^{-15} e^{(0.66/.0258)} = 0.867$ mA. This is the required bias current for the circuit.

   To determine the gain $(A_V = -g_{m1} r_{o1} \| r_{o2})$,

$$g_{m1} = I_C/V_T = 0.867 \text{ mA}/0.0258 \text{ V} = 33.6 \times 10^{-3}$$

$$r_{o1} = V_{A1}/I_C = 74.03/0.867 \text{ mA} = 85.38 \text{ k}\Omega$$

$$r_{o2} = V_{A2}/I_C = 18.7/0.867 \text{ mA} = 21.57 \text{ k}\Omega$$

   Therefore, $A_V = -g_{m1} r_{o1} \| r_{o2} = (-33.6 \times 10^{-3})(85.38 \text{ k} \| 21.57 \text{ k}) = 578.6$.

   $V_{bias} = \ln(0.867 \text{ mA}/1.41 \times 10^{-15}) \times (0.0258) = 0.7$ V (set V_{bias} to Q_2 to 11.3 V).

2. The SPICE simulation results are

$$I_C = 0.891 \text{ mA}$$
$$A_V = -639.8$$

The following is the .cir file used to simulate the circuit. The nodes for each transistor correspond to Figure 17–12.

```
A BJT Gain Stage Using An Active Load
*
Q1 3 1 0 Q2N3904
Q2 3 2 4 Q2N3906
Vbias 2 0 11.3
Vin 1 0 AC 1 sin(0.66 .001 1kHz)
```

```
VCC 4 0 12V
.model Q2N3904 NPN(Is=6.734f Xti=3 Eg=1.11 Vaf=74.03 Bf=416.4 Ne=1.259
+ Ise=6.734f Ikf=66.78m Xtb=1.5 Br=.7371 Nc=2 Isc=0 Ikr=0 Rc=1
+ Cjc=3.638p Mjc=.3085 Vjc=.75 Fc=.5 Cje=4.493p Mje=.2593 Vje=.75
+ Tr=239.5n Tf=301.2p Itf=.4 Vtf=4 Xtf=2 Rb=10)
.model Q2N3906 PNP(Is=1.41f Xti=3 Eg=1.11 Vaf=18.7 Bf=180.7 Ne=1.5 Ise=0
+ Ikf=80m Xtb=1.5 Br=4.977 Nc=2 Isc=0 Ikr=0 Rc=2.5 Cjc=9.728p
+ Mjc=.5776 Vjc=.75 Fc=.5 Cje=8.063p Mje=.3677 Vje=.75 Tr=33.42n
+ Tf=179.3p Itf=.4 Vtf=4 Xtf=6 Rb=10)
.tran 1us 10ms
.ac dec 10 1 100MEG
.probe
.end
```

This circuit has a large gain and as a result is not very stable. Slight changes in the dc bias for v_{in} or V_{bias} will result in the transistor saturating or entering the cutoff region. This does not mean that this circuit cannot be used, but stability is an issue with large gain. Feedback will improve the stability in a gain stage.

The MOSFET Gain Stage

The simplified small-signal model of the MOSFET transistor is provided in Figure 17–13. The gate–source is an open circuit ($r_\pi = \infty$). The dependent current source is written as $g_m v_{gs}$, which identifies that the output current is controlled by the v_{gs} voltage. For a MOSFET device, the transconductance g_m is defined as shown in equation 17–20.

$$g_m = \frac{\Delta I_D}{\Delta V_{GS}} = \beta(V_{GS} - V_t) = \frac{2I_D}{V_{GS} - V_t} \qquad (17\text{–}20)$$

Equation 17–20 shows that g_m can be expressed in two different ways. The two relationships are helpful when designing MOSFET circuits. The output resistance of a MOSFET transistor is given by the expression

$$r_o = \left(\frac{\Delta I_D}{\Delta v_{DS}}\right)^{-1} = \frac{1}{\lambda I_D} \qquad (17\text{–}21)$$

As with the BJT, the output resistance for a MOSFET transistor is large.

The gain of a MOSFET gain stage is written as $A_V = -g_m r_L$. Notice the similarity of the equation to the BJT (equation 17–19). Example 17–9 demonstrates how to apply these equations to a simple MOSFET gain stage.

FIGURE 17–13 The simplified small signal model of the MOSFET transistor

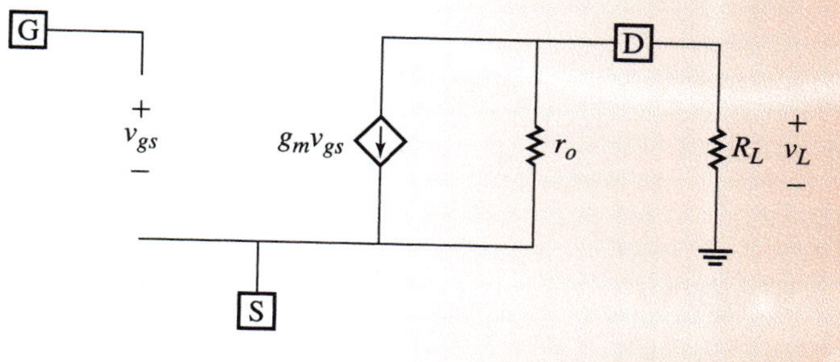

FIGURE 17–14 The MOSFET gain stage for Example 17–9

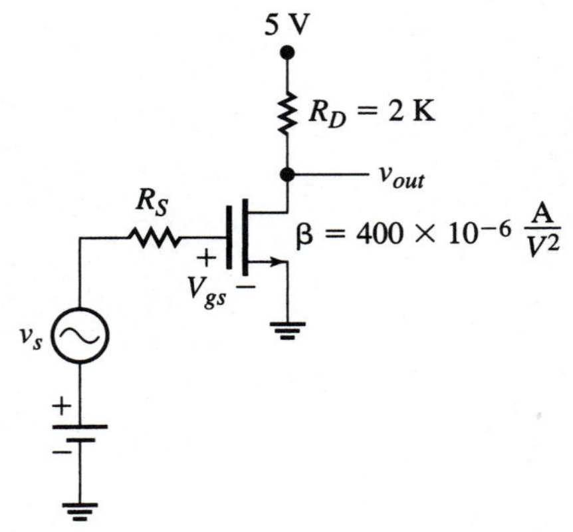

EXAMPLE 17–9

Given the circuit shown in Figure 17–14, determine the voltage gain A_V when $\beta = 400$ E-06 A/V^2, $V_{GS} = 2.0$ V, $V_t = 0.8$ V, and $R_D = 10$ k.

Solution

$$A_V = -g_m r_L = -g_m R_D$$

where $g_m = \beta(V_{GS} - V_t) = (400 \times 10^{-6})(2.0 - 0.8) = 4.8 \times 10^{-4}$

Therefore, $A_V = -(4.8 \times 10^{-4})$ (10k) $= -4.8$ (*Note:* MOSFET gain stages typically have lower gains because $g_{mBJT} \gg g_{mMOS}$.)

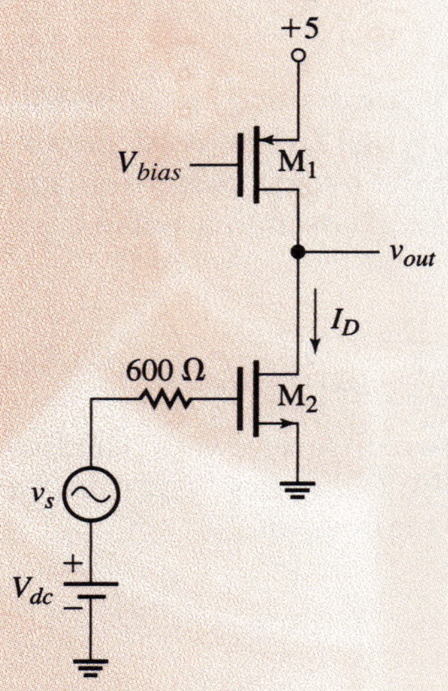

FIGURE 17–15 A MOSFET gain stage with an active load

ACTIVE MOSFET LOADS As with the BJT, a MOSFET transistor can be used as an active load to replace the resistor R_D. Figure 17–15 shows what a typical MOSFET gain stage with an active load looks like. The voltage source is v_S with a dc offset voltage of V_{dc}. The dc offset voltage is required so that the gain transistor M_2 is properly biased. Insufficient dc offset voltage will cause a decrease in the bias current and result in a decrease in the gain. The relation for determining the magnitude of the required dc offset (or bias) voltage for the gain transistor M_2 is given by equation 17–22.

$$|V_{GS}| = \sqrt{\frac{2I_D}{\beta}} + V_t \qquad (17\text{–}22)$$

which is derived from the basic current equation for a MOSFET transistor (see equation 17–8). Figure 17–15 also shows that the signal source is real, with a source resistance of 600 Ω. V_{bias} is the dc bias voltage applied to the gate of current source M_1. The desired bias current, I_D, is set using equation 17–8 from Section 17–3. Example 17–10 demonstrates how to apply these equations to the design of a MOSFET gain stage with an active load.

EXAMPLE 17–10

Determine the gain A_V for the circuit shown in Figure 17–15 given the following conditions:

$$KP_p = 2.3 \times 10^{-5}\ \text{A/V}^2 \qquad KP_n = 5.5 \times 10^{-5}\ \text{A/V}^2$$
$$VTO_p = -0.8\ \text{V} \qquad VTO_n = 0.9\ \text{V}$$
$$\lambda_p = 5.6 \times 10^{-2} \qquad \lambda_n = 1.7 \times 10^{-2}$$

$$(W/L)_p = \frac{6}{2} \qquad (W/L)_n = \frac{4}{2}$$

$$V_{bias} = 3.5\text{ V}$$

Solution

The solution to this problem requires multiple steps. Each step in the solution is listed sequentially to aid the reader. The bias current is set by the current source M_1 and the bias voltage V_{bias}.

Step 1. Determine the bias current I_D.

$$I_D = \frac{KP_p}{2} \cdot \left(\frac{W}{L}\right)_p \cdot (|V_{GS}| - |V_t|)^2 = \frac{2.3 \times 10^{-5}}{2} \times \frac{6}{2}(1.5 - 0.8)^2 = 16.9\ \mu\text{A}$$

Step 2. Calculate the output resistance, r_o, for M_1.

$$r_{oM1} = \frac{1}{\lambda_p I_D} = \frac{1}{(5.6 \times 10^{-2})(16.9\ \mu\text{A})} = 1.056 \times 10^6\ \Omega$$

Next make the necessary calculations for the gain device, in this case, M_2.

Step 3. Calculate β_n for M_2.

$$\beta_n = KP_n \times \left(\frac{W}{L}\right)_n = (5.5 \times 10^{-5})\left(\frac{4}{2}\right) = 11 \times 10^{-5}\text{ A/V}^2$$

Step 4. Use the bias current value calculated in step 1 to determine the V_{GS} required to properly bias M_2.

$$V_{GS} = \sqrt{\frac{2I_D}{\beta_n}} + V_{tn} = \sqrt{\frac{2(16.9\ \mu\text{A})}{11 \times 10^{-5}}} + 0.9 = 1.4543\text{ V}$$

Step 5. Calculate g_{m2}.

$$g_{m2} = \beta_n (V_{GS} - V_{tn}) = (11 \times 10^{-5})(1.4543 - 0.9) = 6.09 \times 10^{-5}\text{ A/V}$$

Note: An alternative way to calculate g_m is by using the following expression:

$$g_m = \sqrt{2 \cdot KP \cdot \frac{W}{L} \cdot I_D} \qquad (17\text{–}23)$$

$$= \sqrt{(2)(5.5 \times 10^{-5})\left(\frac{4}{2}\right)(16.9\ \mu\text{A})} = 6.09 \times 10^{-5}\text{ A/V}$$

The solution for g_m is presented for both methods to show equivalency.

Step 6. Calculate the output resistance, r_o, for M_2.

$$r_{o2} = \frac{1}{\lambda_n I_D} = \frac{1}{(1.7 \times 10^{-2})(16.9 \times 10^{-6})} = 3.48 \times 10^6\ \Omega$$

Step 7. Use the values obtained in steps 1–6 to determine the voltage gain A_V for the amplifier.

$$A_V = -g_{m2}[r_{o1} \| r_{o2}] = -(6.09 \times 10^{-5})(1.056 \times 10^6 \| 3.48 \times 10^6)$$

$$= -49$$

The reader's initial examination of Example 17–10 might be a little overwhelming. The math is not extremely difficult, but the work is very tedious and time-consuming. It is also very easy to make a mistake. Let's review the steps in Example 17–10 to make sure the procedure is understood.

Step 1. The bias current is provided by the current source, PMOS transistor M_1. A designer will usually have a desired bias current specified based on considerations such as g_m and output resistance. How do you know which transistor to use for calculating I_D? Identify which transistor is the current source and which is the gain transistor. The gain transistor will usually have the input voltage applied to it in some manner. M_1 only has a dc bias voltage applied to it. Will the gain device always be NMOS? No, the gain device and the active loads can be either NMOS or PMOS.

Step 2. In step 2, the output resistance of the active load is being calculated. Remember, this device replaces a resistor. For small bias currents, you should expect a large output resistance.

Step 3. The gain, β, for the NMOS device is required for several calculations. All of the information for making this calculation is provided in the problem statement.

Step 4. A bias current of 16.9 μA was calculated in step 1. NMOS transistor M_2 should also have this value for optimal performance. The value for V_{GS} is determined using the calculated and given values. V_{GS} is the dc offset, or bias voltage. This voltage places the device in the saturation region. This concept is further examined later in this chapter.

Step 5. The transconductance, g_m, of M_2 is required so that the gain, A_V, can be determined. Recall that $A_V = -g_m r_L$. An alternative way to calculate g_m is shown, which can be used when certain parameters are specified. The second equation requires knowledge of the *W/L* layout dimensions, whereas the first equation requires only that β be known.

Step 6. The output resistance for the gain stage is made up of the parallel output resistances of the transistors.

Step 7. The voltage gain, A_V, can be determined using the values calculated in steps 1–6.

Simplifying the Design Process

Once the reader has gained a little confidence with Example 17–10, the design process for a gain stage and other building blocks can be greatly simplified by using SPICE as an analytical tool. You can let the computer do a lot of the work as long as you are able to

Identify whether the SPICE solution is reasonable!

This may seem silly but understanding what to expect from a circuit is a big portion of the battle. The following discussion demonstrates how SPICE can be used as an analytical tool for solving the design problem.

A good place to begin is with the basic schematic of the circuit and a SPICE file. The SPICE .cir file is presented here as a generic example. Notice in Figure 17–16 that the signal source contains a source resistance (r_S) of 600 Ω. Although r_S is sometimes omitted in text examples, you should avoid omitting it in circuit simulations to avoid errors in your results. The *W*

FIGURE 17–16 A MOSFET gain stage and an active MOSFET load

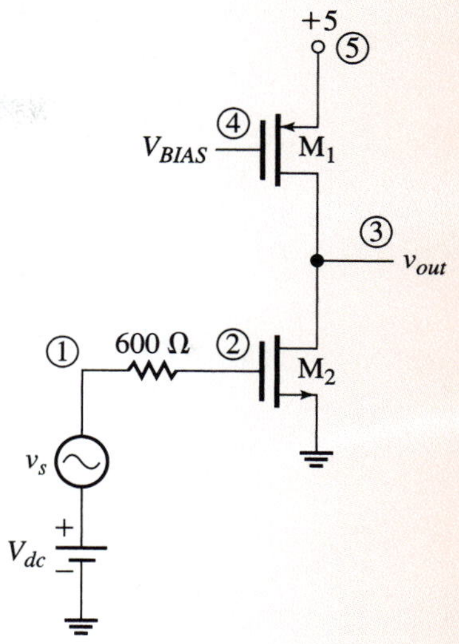

(width) and L (length) values have been included in the simulation. These values should be used if the specifications are available. The models, referenced by CMOSP and CMOSN, have been greatly simplified to reflect only the device data specified in Example 17–10. Normally, these files are quite detailed. Simplifying the model files will make it easier for us to perform calculations that will be comparable to the simulation results.

This .cir file also contains the command .OP, which instructs SPICE to print the bias point in the .out file. This command will instruct SPICE to print currents and power dissipation values for all devices. An additional factor is that the .OP statement will cause SPICE to print the small-signal values for all semiconductor and nonlinear controlled sources.

Let's begin our SPICE analysis by conducting a DC analysis of the circuit by stepping the input of the circuit through the entire voltage range using the following command:

```
.DC VS 0 5 .01
```

This steps the voltage source VS from 0 to 5 volts in 0.01-V increments. The small step-voltage increments will make it easier to examine the results.

```
MOSFET Gain Stage with Active Load
****************************************************************
M1 3 4 5 5 CMOSP W=6u L=2u
M2 3 2 0 0 CMOSN W=4U L=2U
****************************************************************
VS 1 0 SIN(1.45 5m .2k)
RS 1 2 600
VBIAS 4 0 3.5v
VDD 5 0 5V
.MODEL CMOSN NMOS LEVEL=2 VTO=0.9 KP=5.5E−05 LAMBDA=1.7E−02
.MODEL CMOSP PMOS LEVEL=2 VTO=−0.8 KP=2.3E−05 LAMBDA=5.6E−02
.OP
.DC VS 0 5 .01
.TRAN 0.01NS 10mS
.PROBE
.END
```

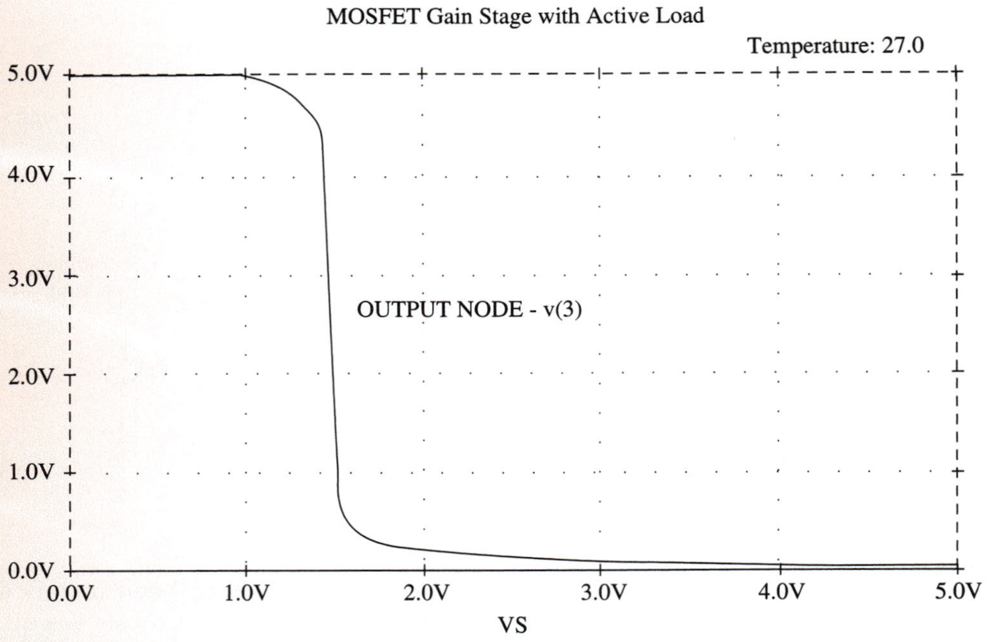

FIGURE 17–17 The DC sweep of the MOSFET gain stage

The first simulation result is a DC sweep of VS with a VBIAS voltage set to 3.5 V. This bias voltage was calculated in Example 17–10 for an I_D current of 16.9 μA. You can see from the transfer characteristics in Figure 17–17 that there is a large gain when VS is set to ~1.5 V. You see in the .cir listing that when V_{GS} was calculated for the circuit, a DC offset voltage of 1.4543 V was calculated. Our SPICE analysis gave us a comparable result graphically.

The transconductance value, g_m, for the NMOS transistor can be obtained from SPICE if the .OP statement is included. For this problem, SPICE calculated g_m for each transistor to equal

```
GM 5.10E-05      6.50E-05
```

The calculated value of g_m for M_2, the NMOS gain transistor, was determined to be equal to 6.09 E-05. Our calculated value is comparable to SPICE's value. Why the difference? SPICE takes many factors into consideration that our simple approximations omit. If our model parameters were more complete, we would also see a larger discrepancy in the SPICE solution and our calculated value. This certainly underscores the importance of incorporating accurate model parameters in the simulation.

The .OP command will also cause the bias current to be printed. For this problem, the operating point information provides an I_D of 1.79E-05 (17.9 μA), which is very close to the 16.9 μA calculated value.

What about the gain? This information can also be obtained from the DC sweep. Figure 17–18 shows an expanded plot of the DC transfer characteristics of the 1.0-V to 2.0-V region. Using the cursor mode in SPICE, we obtain a gain of $-2/0.046 = -43.47$. This is close to the calculated value of -49. The discrepancies are once again due to the assumptions used in the calculations and the limited model parameters incorporated into the simulation.

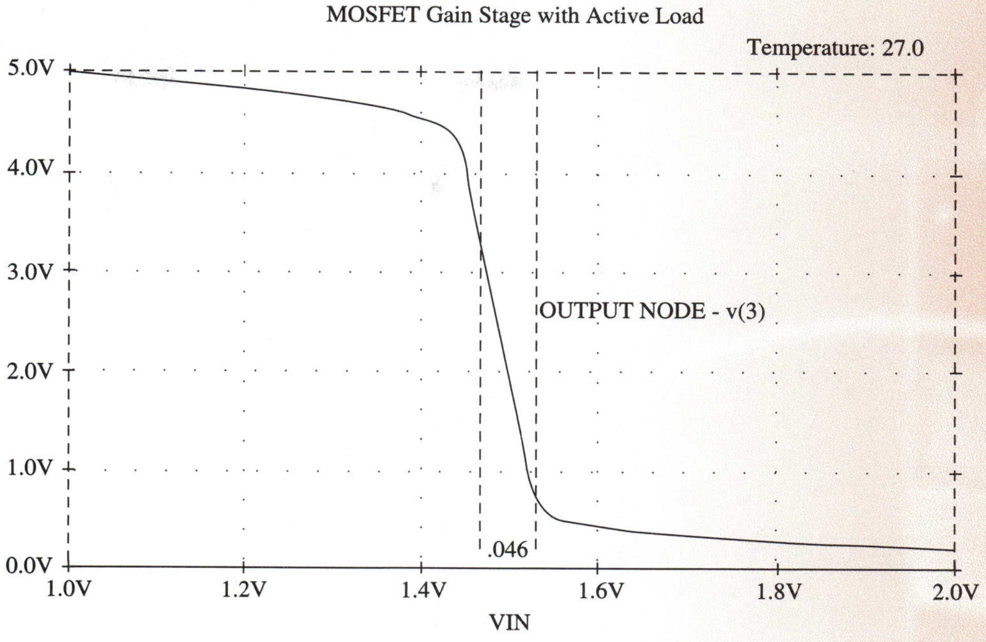

FIGURE 17–18 The DC transfer characteristic expanded so the gain can be graphically measured

This has shown how SPICE can be used to solve the design problem. It also showed that the SPICE values were comparable to the calculated values. This will be helpful as the circuits get more complicated and we begin to rely on the computer simulations to verify our designs, but never forget to **"identify if the SPICE solution is reasonable!"**

SPICE can also be used to experiment with the circuit in a quick manner. Figure 17–19(a) shows the amplified sinusoid output with the DC offset voltage at 1.45 V. This produces an amplified sinusoid with a DC offset of approximately 4.2 V. Changing the DC offset voltage to 1.5 should cause the output DC offset to shift closer to ground. Examine Figure 17–18, which is the expanded plot of the DC transfer characteristics. When the DC offset voltage for VS equals 1.5 V, the output node v(3) will have a DC offset of ~1.9 V. This result is shown in Figure 17–19(b).

This discussion shows that it is possible to easily shift the output offset voltage, giving the designer better control of the signal. Lastly, the dc transfer characteristics show that the gain is constant within a small range (1.4 V–1.53 V). If the V_{GS} voltage on the input exceeds this range, the gain will decrease dramatically. Always examine the dc transfer characteristics of your amplifier design to verify its operational range.

Diode-Connected Transistors

Diode-connected transistors are used in gain stages to simulate the low value of resistances, unlike the active load, which was shown to exhibit high resistance values. Diode-connected transistors can also be used to construct voltage divider circuits in integrated circuits and in discrete applications. This section will address both applications for diode-connected transistors.

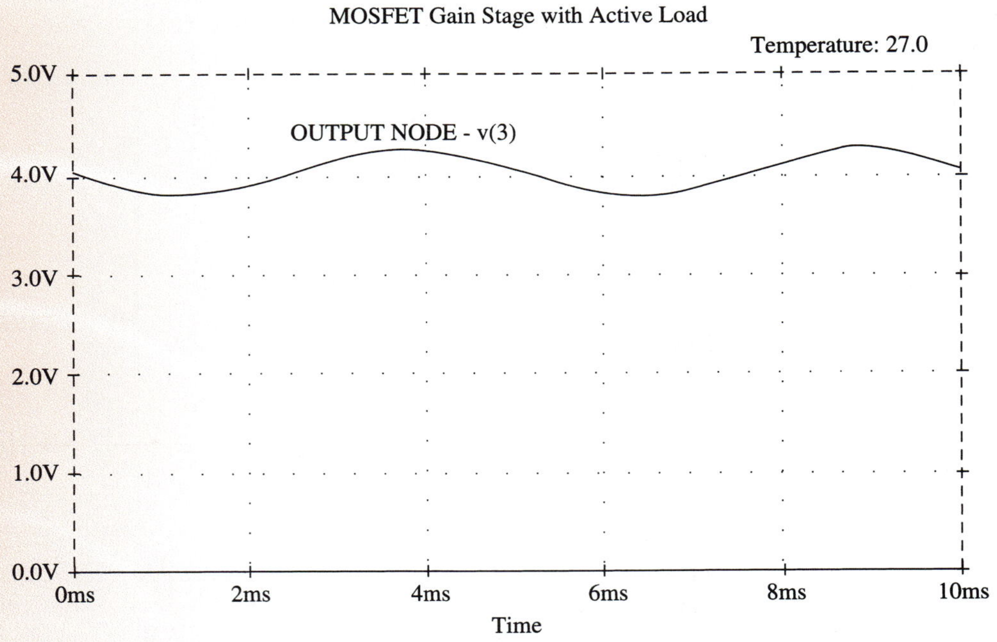

(a)

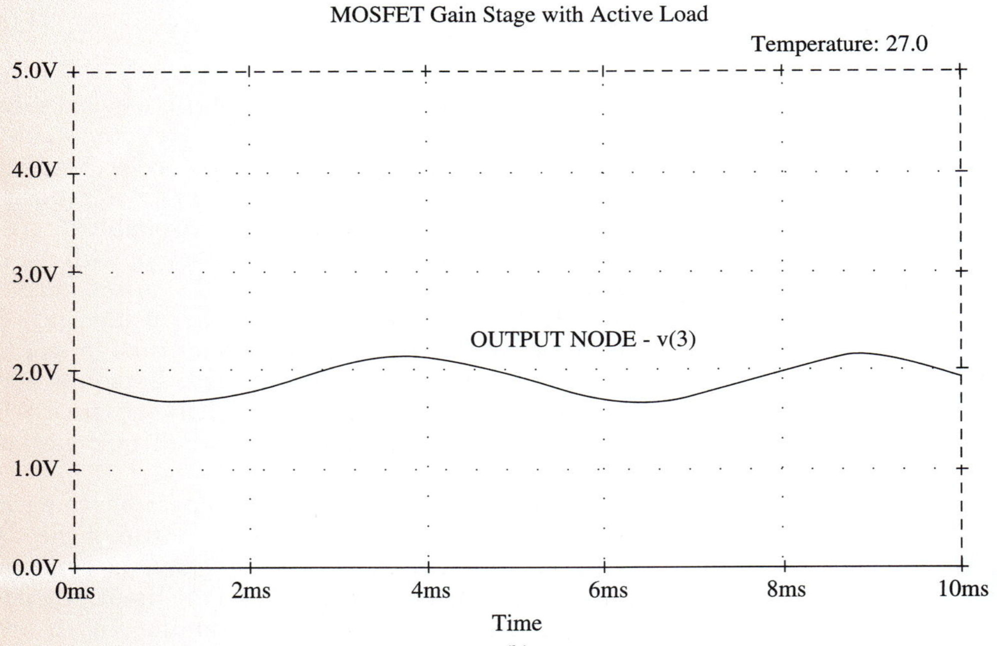

(b)

FIGURE 17–19 (a) The sinusoid output with a DC offset of 4.2 V, (b) the output sinusoid with a DC offset of 1.9 V

THE BJT DIODE-CONNECTED TRANSISTOR Figure 17–20(a) shows the schematic symbol for a diode-connected BJT. The low-frequency model for diode-connected BJT is provided in Figure 17–20(b), and the simplified model is given in Figure 17–20(c). The result in Figure 17–20(c) reduces down to a two-terminal device with a resistance of approximately $1/g_m$, which is considered a low impedance. A diode-connected transistor is the lowest possible resistance value ($1/g_m$) of a given technology, BJT or MOSFET.

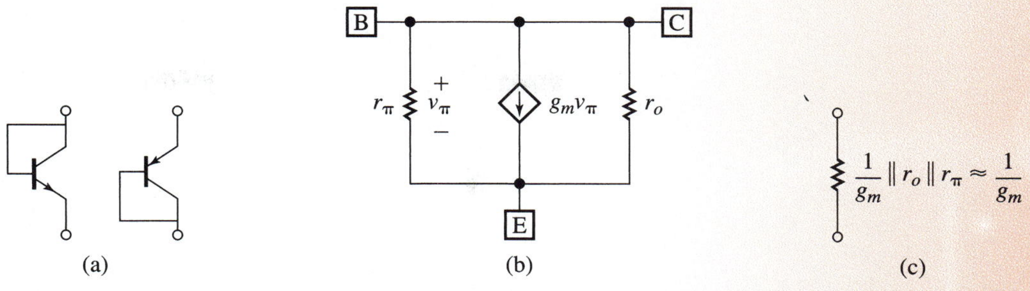

FIGURE 17–20 The diode-connected BJT transistor: (a) schematic symbols, (b) the low-frequency model, (c) the simplified two-terminal device

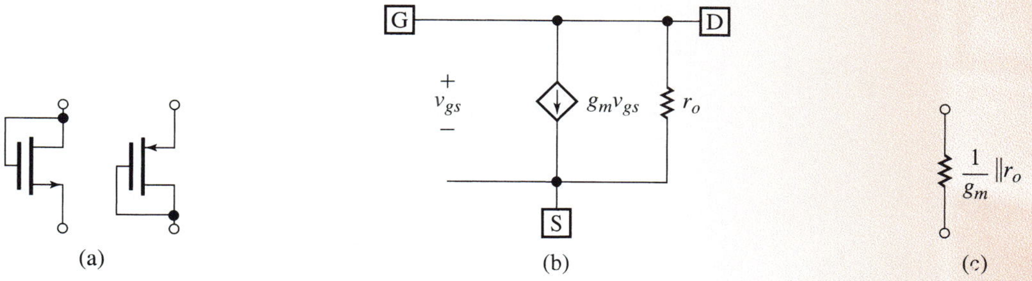

FIGURE 17–21 The diode-connected MOSFET transistor: (a) schematic symbols, (b) the low-frequency model, (c) the simplified two-terminal device

THE MOSFET DIODE-CONNECTED TRANSISTOR Figure 17–21(a) shows the schematic symbol for a diode-connected MOSFET. The low-frequency model for a diode-connected MOSFET is provided in Figure 17–21(b), and the simplified model is given in Figure 17–21(c). The result in Figure 17–21(c) reduces down to a two-terminal device with a resistance of approximately $1/g_m$, which is considered a low impedance.

Diode Transistor Applications

This part of Section 17–5 will address two applications of diode-connected transistors: (1) diode-connected loads in gain stages, and (2) the voltage divider.

DIODE-CONNECTED LOADS IN GAIN STAGES Figures 17–20 and 17–21 show that the diode-connected transistor reduces down to a two-terminal device with a resistance value of $1/g_m$. It has been stated many times that the voltage gain (A_V) of a BJT or a MOSFET is $-g_m r_L$. Figures 17–20 and 17–21 show that for a diode-connected transistor, $r_L = 1/g_m$. It has been shown previously that for a BJT, $g_m = I_C/V_T$ and for a MOSFET transistor, $g_m = \beta(V_{GS} - V_t)$. Example 17–11 provides an example of both a BJT and a MOSFET gain stage with a diode-connected load. Remember that for both the BJT and the MOSFET, $A_V = -g_m r_L$, where $r_L = 1/g_m$ for a diode-connected transistor.

EXAMPLE 17–11

1. Determine the voltage gain of the BJT gain stage shown in Figure 17–22. Assume $V_T = 0.026$ V and $I_C = 1$ mA.

Solution

$$g_{m1} = g_{m2} = I_C/V_T$$

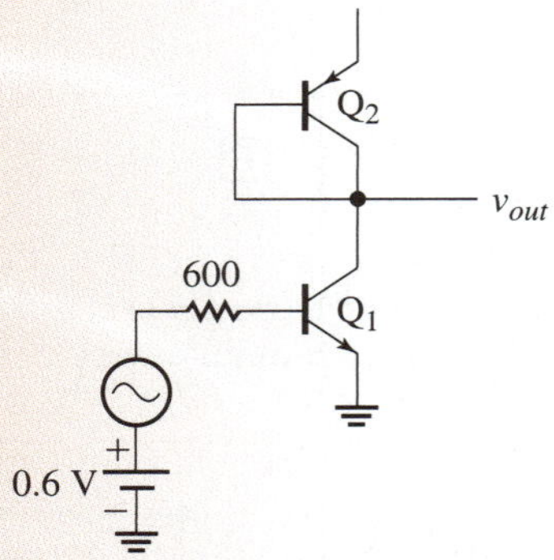

FIGURE 17–22 The circuit for Example 17–11 part 1

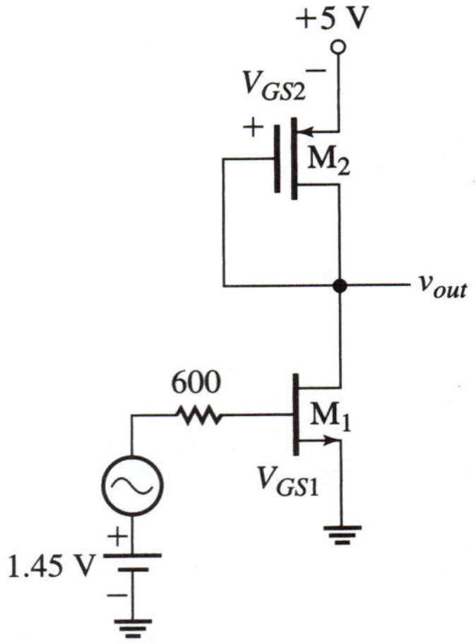

FIGURE 17–23 The circuit for Example 17–11 part 2

Therefore,

$$A_V = -g_m r_L = -g_{m1}\,(1/g_{m2}) = -1$$

Both transistors must have the same I_C and will have the same V_T; therefore, g_m is the same for both devices.

2. Determine the voltage gain for the MOSFET gain stage shown in Figure 17–23. Assume $V_{tn} = |V_{tp}| = 0.7$ V, $V_{GS1} = 1.45$ V, $\beta_n = 27.5\text{E-}05$ A/V^2, $\beta_p = 11.5\text{E-}05$ A/V^2, and $|V_{GS2}| = 1.83$ V. Determine g_m for both transistors.

Solution

$$g_{m1} = \beta_1\,(V_{GS_1} - V_{t1}) = (27.5\ \text{E}-05)(1.45 - 0.7) = 20.625\ \text{E-}05$$

$$g_{m2} = \beta_2(|V_{GS}| - |V_{t1}|) = (11.5\ \text{E}-05)(|1.83| - |0.7|) = 12.995\ \text{E-}05$$

$$A_V = -g_{m1}/g_{m2} = 20.625/12.995 = -1.587$$

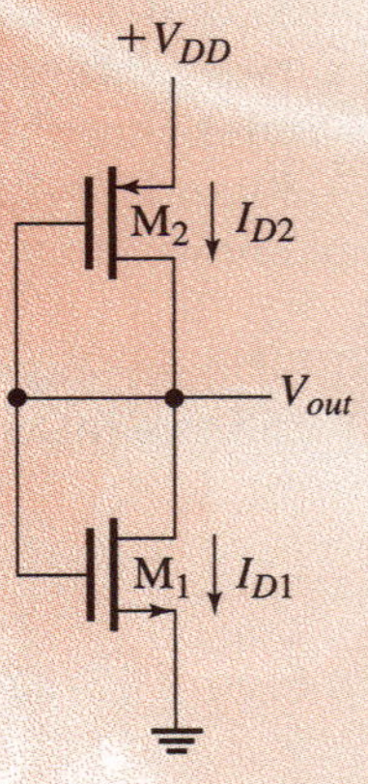

FIGURE 17–24 A MOSFET transistor diode-connected voltage divider

THE VOLTAGE DIVIDER CIRCUIT In Example 17–10, a 3.5 dc bias voltage was applied to the PMOS current source. It has been easy for us to just specify 3.5 V without worrying where 3.5 V will come from. How is this bias voltage created in integrated circuits? One might assume that an external bias voltage might be applied to the IC, but this is not a popular or sophisticated technique. Most ICs require a limited number of external input voltages; requiring many external bias-voltage pins is inefficient and awkward. Additionally, a designer does not want to be limited to only the 3.5-V dc bias voltage. This section explores a technique that enables the designer to create any voltage that falls within the power supply range.

A very useful application of diode-connected MOSFET transistors is the voltage divider. This technique uses a diode-connected PMOS and NMOS transistor to form a voltage divider. The circuit is shown in Figure 17–24.

By inspection, I_{D1} will equal I_{D2}. This relationship can be used to develop an equation for V_{out}. Equation 17–24 shows the result for V_{out}:

$$V_{out} = \frac{\sqrt{\frac{\beta_2}{\beta_1}}(V_{DD} - |V_{t_2}|) + V_{t_1}}{1 + \sqrt{\frac{\beta_2}{\beta_1}}} \qquad (17\text{–}24)$$

Equation 17–24 provides us with a design equation for obtaining a specified bias voltage. Note that when $V_{t_1} = V_{t_2}$ and $\beta_1 = \beta_2$, $V_{out} = V_{DD}/2$. This equation can be rewritten in a form that enables the designer to obtain the β ratio of the NMOS and PMOS transistors given the power supply voltage, V_{DD}, V_{tn}, and V_{tp}. This relationship is given in equation 17–25:

$$\sqrt{\frac{\beta_2}{\beta_1}} = \frac{V_{out} - V_{t_1}}{V_{DD} - |V_{t_2}| - V_{out}} \qquad (17\text{–}25)$$

Example 17–12 demonstrates how to use equation 17–25 when designing a voltage divider.

EXAMPLE 17–12

Design a MOSFET voltage divider that will provide a V_{out} of (1) 1.5 V, (2) 2.5 V, and (3) 3.5 V. Assume a KP_n = 5.6 E-05, KP_p = 2.24 E-05, $L = 2\ \mu$, $V_{tn} = |V_{tp}| = 0.75$ V. The circuit is shown in Figure 17–24. Verify your answers with SPICE.

Solution

Use equation 17–25 to solve for the β_2/β_1 ratio required to provide the specified output voltage. The SPICE simulation width values and V_{out} are shown in ( ).

1. For $V_{out} = 1.5$ V,

$$\sqrt{\frac{\beta_2}{\beta_1}} = \frac{1.5 - 0.75}{5 - |0.75| - 1.5} = 0.2727$$

$$\frac{\beta_2}{\beta_1} = 0.2727^2 = 0.0743$$

$$\frac{\beta_2}{\beta_1} = 0.0743 = \frac{KP_P \times (W/L)_P}{KP_n \times (W/L)_n} = 0.4\,\frac{W_P}{W_n}$$

$$W_n = 5.38\ W_P \quad \text{(width ratio)}$$

(For $W_p = 20\ \mu$, $W_n = 107\ \mu$; $V_{out} = 1.558$ V.)

2. For $V_{out} = 2.5$ V,

$$\sqrt{\frac{\beta_2}{\beta_1}} = \frac{2.5 - 0.75}{5 - |0.75| - 2.5} = 1$$

$$\frac{\beta_2}{\beta_1} = 1^2 = 1 = \frac{KP_P \times \left(\frac{W}{L}\right)_P}{KP_n \times \left(\frac{W}{L}\right)_n} = 0.4\,\frac{W_P}{W_n}$$

$$W_P = 2.5\ W_n$$

(For $W_n = 10\ \mu$, $W_p = 25\ \mu$; $V_{out} = 2.545$ V.)

3. For $V_{out} = 3.5\text{ V}$,

$$\sqrt{\frac{\beta_2}{\beta_1}} = \frac{3.5 - 0.75}{5 - |0.75| - 3.5} = 3.67$$

$$\frac{\beta_2}{\beta_1} = 3.67^2 = 13.44$$

$$13.44 = \frac{KP_P \times \left(\dfrac{W}{L}\right)_P}{KP_n \times \left(\dfrac{W}{L}\right)_n} = 0.4\,\frac{W_P}{W_n}$$

$$W_P = 33.6\,W_n$$

(For $W_n = 20\ \mu$, $W_p = 672\ \mu$; $V_{out} = 3.507\text{ V}$.)

The SPICE. cir file for the circuit shown in Example 17–12 is provided for the 3.5-V reference. The other voltages are obtained by using the specified W = values for the PMOS and NMOS transistors. For example, the 2.5-V reference will have $W_n = 10\ \mu$ and $W_p = 25\ \mu$.

```
MOSFET Voltage Divider (3.5 V)
*****************************************************************
M1 2 2 5 5 CMOSP W=672u L=2u
M2 2 2 0 0 CMOSN W=20U L=2U
*****************************************************************
VDD 5 0 5V
.MODEL CMOSN NMOS LEVEL=2 VTO=0.75 KP=5.6E-05 LAMBDA=1.7E-02
.MODEL CMOSP PMOS LEVEL=2 VTO=-0.75 KP=2.24E-05 LAMBDA=5.6E-02
.OP
.TRAN 0.01NS 10mS
.PROBE
.END
```

The SPICE model parameters have been minimized to simplify the discussion. Full model parameters would have produced a slightly different answer but your answer should be close to the calculated value.

17–6 DIFFERENTIAL AMPLIFIERS

Section 17–6 explores the use of transistors connected as a differential pair. This concept is first investigated from a BJT point of view followed by examples of JFET and MOSFET transistors configured as differential amplifiers.

Differential amplifiers are widely used in linear integrated circuits. They are a fundamental component of every *operational amplifier*, which, as we shall learn, is an extremely versatile device with a broad range of practical applications. We will study the circuit theory of differential amplifiers in some detail, in preparation for a more comprehensive investigation of the capabilities (and limitations) of operational amplifiers.

Difference Voltages

A differential amplifier is also called a *difference* amplifier because it amplifies the difference between two signal voltages. Let us refine the notion of a *difference voltage* by reviewing some simple examples. We have already encountered difference voltages in our study of transistor amplifiers. Recall, for example, that the collector-to-emitter voltage of a BJT is the difference between the collector-to-ground voltage and the emitter-to-ground voltage:

$$V_{CE} = V_C - V_E \quad (17\text{–}26)$$

The basic idea here is that a difference voltage is the mathematical difference between two other voltages, each of whose values is given with respect to ground. Suppose the voltage at point A in a circuit is 12 V with respect to ground, and the voltage at point B is 3 V with respect to ground. The notation V_{AB} for the difference voltage means the voltage that would be measured if the positive side of a voltmeter were connected to point A and the negative side to point B; in this case, $V_{AB} = V_A - V_B = 12 - 3 = 9$ V. If the voltmeter connections were reversed, we would measure $V_{BA} = V_B - V_A = 3 - 12 = -9$ V. Thus, $V_{BA} = -V_{AB}$.

To help get used to thinking in terms of difference voltages, consider the system shown in Figure 17–25, where two identical amplifiers are driven by two different signal voltages. Although a differential amplifier does not behave in exactly the same way as this amplifier arrangement, the concepts of input and output difference voltages are similar. The two signal input voltages, v_1 and v_2, are shown as sine waves, one greater in amplitude than the other. For illustrative purposes, their peak values are 3 V and 1 V, respectively. If the voltage gain of each amplifier is A, then the amplifier outputs are Av_1 and Av_2. The input difference voltage, $v_{12} = v_1 - v_2$, is a sine wave and the output difference voltage, $Av_1 - Av_2 = A(v_1 - v_2)$, is seen to be an amplified version of the input difference voltage. In our illustration, the gain A is 10 and the input difference voltage is (3 V pk) − (1 V pk) = 2 V pk. The output difference voltage is $A(v_1 - v_2) = 10(3\text{ V} - 1\text{ V}) = 10(2\text{ V}) = 20$ V pk.

The Ideal Differential Amplifier

The principal feature that distinguishes a differential amplifier from the configuration shown in Figure 17–25 is that a signal applied to one input of a differential amplifier induces a voltage with respect to ground on the amplifier's other output. This fact will become clear in our study of the voltage and current relations in the amplifier.

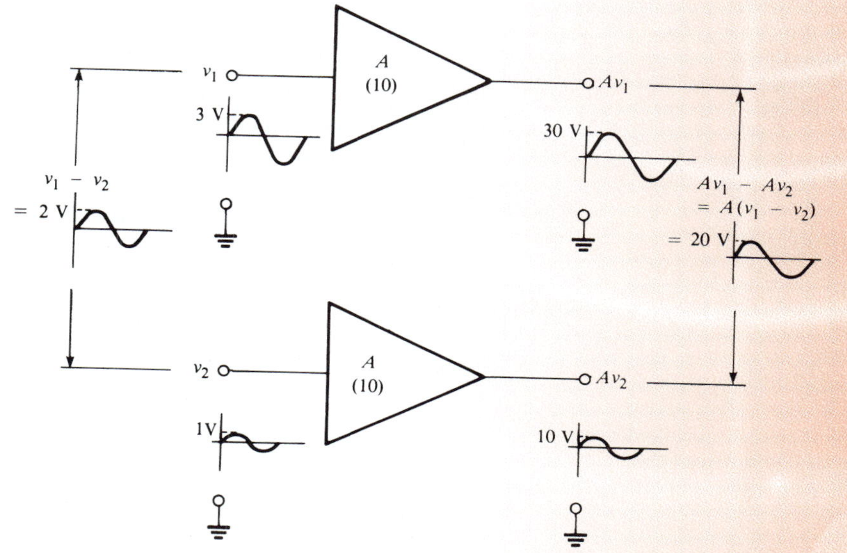

FIGURE 17–25 The amplification of difference voltages

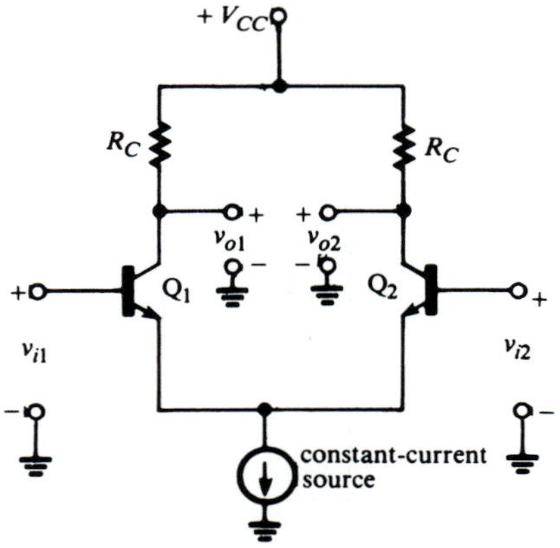

FIGURE 17–26 The basic BJT differential amplifier. The two transistors can be regarded as CE amplifiers having a common connection at their emitters. The base terminals are the inputs to the differential amplifier, and the collectors are the outputs.

Figure 17–26 shows the basic BJT version of a differential amplifier. Two transistors are joined at their emitter terminals, where a constant-current source is connected to supply bias current to each. The current source is typically one of the transistor constant-current sources that we studied in Section 17–3, but for now we will represent it by an ideal current source. Note that each transistor is basically in a common-emitter configuration, with an input supplied to its base and an output taken from its collector. The two base terminals are the two signal inputs to the differential amplifier, v_{i1} and v_{i2}, and the two collectors are the two outputs, v_{o1} and v_{o2}, of the differential amplifier. Thus, the differential input voltage is $v_{i1} - v_{i2}$, and the differential output voltage is $v_{o1} - v_{o2}$.

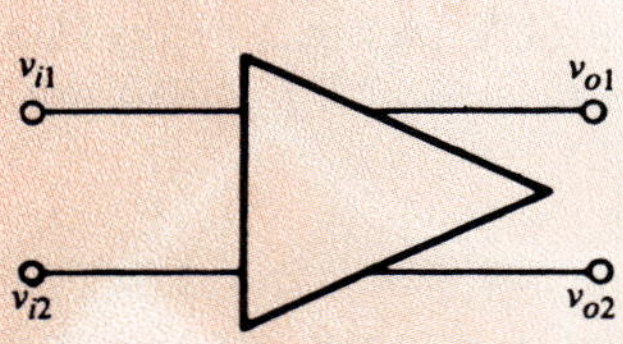

FIGURE 17–27 Schematic symbol for the differential amplifier

Figure 17–27 shows the schematic symbol for the differential amplifier. Because there are two inputs and two outputs, the amplifier is said to have a *double-ended* (or double-sided) input and a double-ended output.

We will postpone, temporarily, our analysis of the dc bias levels in the amplifier and focus on its behavior as a small-signal amplifier. Toward that end, we will determine the output voltage at each collector due to each input voltage acting *alone,* that is, with the opposite input grounded, and then apply the superposition principle to determine the outputs due to both inputs acting simultaneously. Figure 17–28 shows the amplifier with input 2 grounded ($v_{i2} = 0$) and a small signal applied to input 1. The ideal current source presents an infinite impedance (open circuit) to an ac signal, so we need not consider its presence in our small-signal analysis. We also assume the ideal situation of perfectly matched transistors, so Q_1 and Q_2 have identical values of β, r'_e, etc. Because Q_1 is essentially a common-emitter amplifier, the voltage at its collector (v_{o1}) is an amplified and *inverted* version of its input, v_{i1}. Note that there is also an ac voltage, v_{e1}, developed at the emitter of Q_1. This voltage is in phase with v_{i1} and exists because of emitter-follower action across the base-emitter junction of Q_1.

Now, the voltage v_{e1} is developed across the emitter resistance r'_e looking into the emitter of Q_2 (in parallel with the infinite resistance of the current source). Therefore, as far as the emitter-follower action of Q_1 is concerned, the load resistance seen by Q_1 is r'_e. Because the emitter resistance of Q_1 is itself r'_e, it follows from equation 17–19 that the emitter-follower gain is 1/2:

$$A_v = \frac{r_L}{r_L + r'_e} = \frac{r'_e}{r'_e + r'_e} = 0.5$$

Therefore, v_{e1} is in phase with, and one-half the magnitude of, v_{i1}. Now, it is clear that v_{e1} is the emitter-to-ground voltage of *both* transistors. *When* v_{e1}

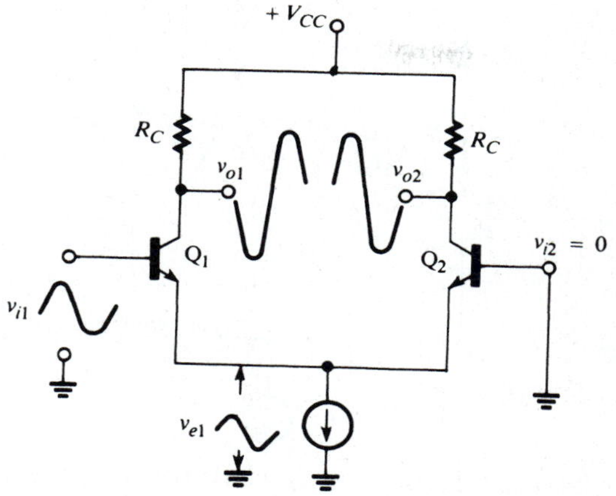

FIGURE 17–28 The small-signal voltages in a differential amplifier when one input is grounded. Note that v_{e1} is in phase with v_{i1} and that v_{o1} is out of phase with v_{o2}

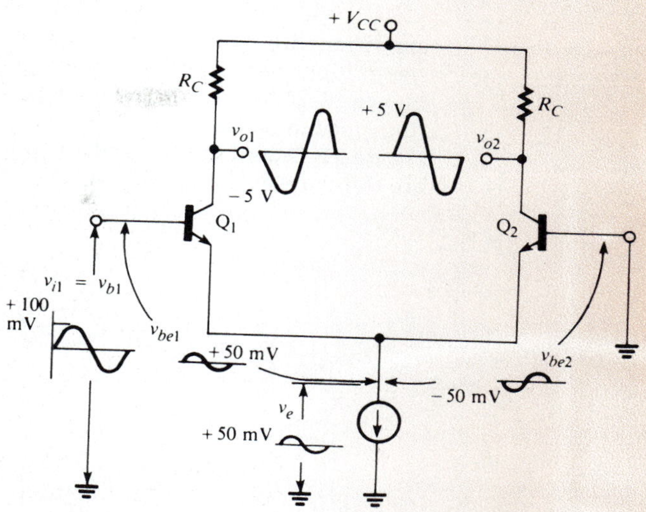

FIGURE 17–29 Each transistor has identical voltage gain −100, and the outputs at the collectors are −100 times their respective base-to-emitter voltages

goes positive, the base-to-emitter voltage of Q_2 *goes negative by the same amount.* In other words, $v_{be2} = v_{b2} - v_{e1} = 0 - v_{e1}$. (Because the base of Q_2 is grounded, its base-to-emitter voltage is the same as the negative of its emitter-to-ground voltage.) We see that even though the base of Q_2 is grounded, there exists an ac base-to-emitter voltage on Q_2 that is out of phase with v_{e1} and therefore out of phase with v_{i1}. Consequently, there is an ac output voltage v_{o2} produced at the collector of Q_2 and it is out of phase with v_{o1}.

Because both transistors are identical, they have equal gain and the output v_{o2} has the same magnitude as v_{o1}. To verify this last assertion, and to help solidify all the important ideas we have presented so far, let us study the specific example illustrated in Figure 17–29. We assume that v_{i1} (which is the base-to-ground voltage of Q_1) is a 100-mV-pk sine wave and that each transistor has voltage gain −100, where, as usual, the minus sign denotes phase inversion. By "transistor voltage gain," we mean the collector voltage divided by the *base-to-emitter* voltage.

Since the emitter-follower gain of Q_1 is 0.5, v_e is a 0.5(100 mV) = 50-mV-pk sine wave. The peak value of v_{be1} is therefore $v_{b1} - v_{e1} = (100\text{ mV}) - (50\text{ mV}) = 50\text{ mV}$. When v_{be1} is at this 50-mV peak, v_{o1} is $-100(50\text{ mV}) = -5\text{ V}$, that is, an inverted 5-V-pk sine wave. At this same point in time, where v_e is at its 50-mV peak, the base-to-emitter voltage of Q_2 is $0 - (50\text{ mV}) = -50\text{ mV}$ pk. Therefore, v_{o2} is $(-100)(-50\text{ mV}) = +5\text{ V}$ pk; that is, v_{o2} is a 5-V-pk sine wave in phase with v_{i1} and out of phase with v_{o1}.

Note in Figure 17–29 that the input difference voltage is $v_{i1} - v_{i2} = (100\text{ mV}) - 0 = 100\text{ mV}$ pk, and the output difference voltage is 10 V pk, since v_{o1} and v_{o2} are out of phase. Therefore, the magnitude of the *difference voltage gain* $(v_{o1} - v_{o2})/(v_{i1} - v_{i2})$ is 100. So, while the voltage gain v_o/v_i for each side is only 50, the difference voltage gain is the same as the gain v_c/v_{be} of each transistor. We will refine and generalize this idea later, when we finish our small-signal analysis using superposition.

In many applications, the two inputs of a differential amplifier are driven by signals that are equal in magnitude and out of phase: $v_{i2} = -v_{i1}$. Continuing our analysis of the amplifier, let us now ground input 1 ($v_{i1} = 0$) and assume that there is a signal applied to input 2 equal to and out of phase with the v_{i1} signal we previously assumed. Because the transistors

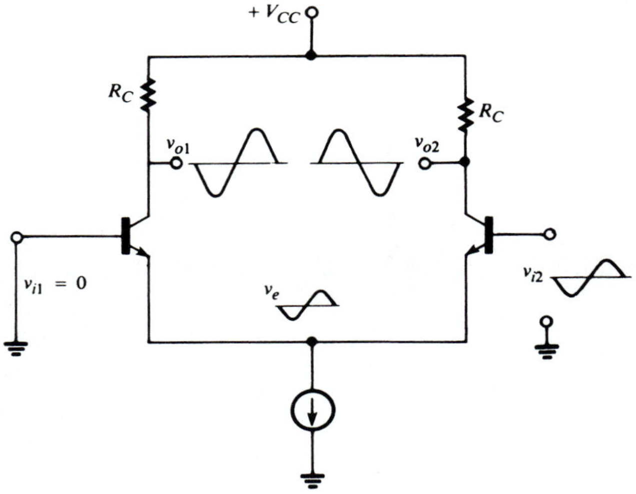

FIGURE 17–30 The differential amplifier with v_{i1} grounded and a signal input v_{i2}. Compare with Figure 17–28. Note that v_{i2} here is the opposite phase from v_{i1} in Figure 17–28 and that v_{o1} and v_{o2} are the same as in Figure 17–28.

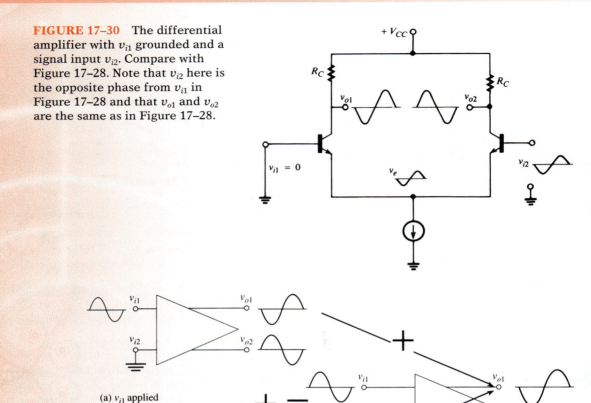

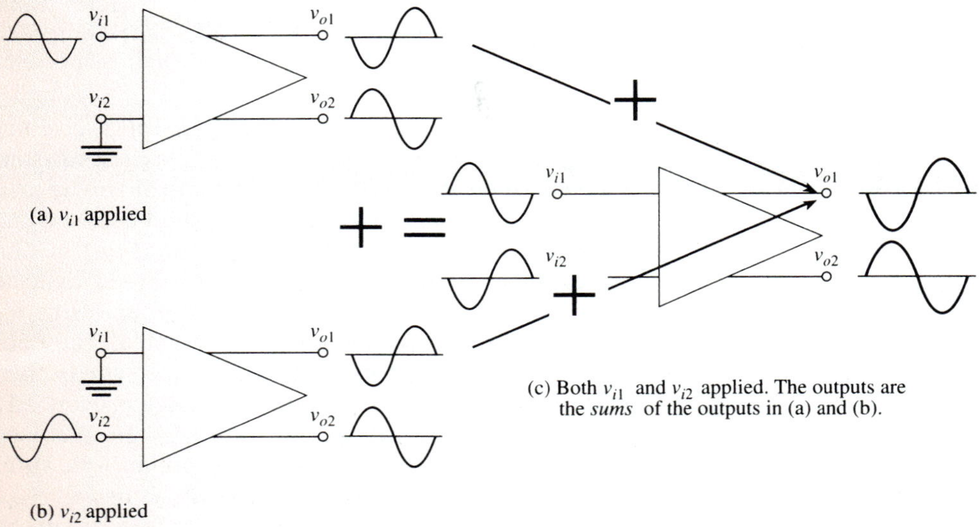

FIGURE 17–31 By the superposition principle, the output v_{o1} when both inputs are applied is the sum of the v_{o1} outputs due to each signal acting alone. Likewise for v_{o2}.

are identical and the circuit is completely symmetrical, the outputs have exactly the same relationships to the inputs as they had before: v_{o2} is out of phase with v_{i2}, and v_{o1} is in phase with v_{i2}. These relationships are illustrated in Figure 17–30.

When we compare Figure 17–30 with Figure 17–28, we note that the v_{o1} outputs are identical, as are the v_{o2} outputs. In other words, driving the two inputs with equal but out-of-phase signals reinforces, or duplicates, the signals at the two outputs. By superposition, each output is the sum of the voltages resulting from each input acting alone, so the outputs are exactly twice the level they would be if only one input signal were present. These ideas are summarized in Figure 17–31.

In many applications, the output of a differential amplifier is taken from just one of the transistor collectors—v_{o1}, for example. In this case, the input is a difference voltage and the output is a voltage with respect to ground. This use of the amplifier is called *single-ended output* operation and the voltage gain in that mode is

$$A_{v\,(\text{single-ended output})} = \frac{v_{o1}}{v_{i1} - v_{i2}} \qquad (17\text{–}27)$$

The next example demonstrates that the single-ended output gain is one-half the difference voltage gain. To distinguish between these terms, we will hereafter refer to $(v_{o1} - v_{o2})/(v_{i1} - v_{i2})$ as the *double-ended voltage gain.*

EXAMPLE 17–13

The magnitude of the voltage gain (v_c/v_{be}) for each transistor in Figure 17–26 is 100. If v_{i1} and v_{i2} are out of phase and 100-mV-pk signals are applied simultaneously to the inputs, find

1. the peak values of v_{o1} and v_{o2},
2. the magnitude of the double-ended voltage gain $(v_{o1} - v_{o2})/(v_{i1} - v_{i2})$, and
3. the magnitude of the single-ended output gain $v_{o1}/(v_{i1} - v_{i2})$.

Solution

1. As demonstrated in Figure 17–29, the peak value of each output is 5 V when one input is driven and the other is grounded. Because the outputs are doubled when the inputs are equal and out of phase, each output is 10 V pk.
2. Since $v_{i1} = -v_{i2}$, the input difference voltage is $v_{i1} - v_{i2} = 2v_{i1} = 200$ mV pk. Similarly, $v_{o1} = -v_{o2}$, so the output difference voltage is $v_{o1} - v_{o2} = 2v_{o1} = 20$ V pk. Therefore, the magnitude of $(v_{o1} - v_{o2})/(v_{i1} - v_{i2})$ is (20 V)/(200 mV) = 100.
3. The magnitude of the single-ended output gain is

$$|A_{v\,(\text{single-ended output})}| = \left|\frac{v_{o1}}{v_{i1} - v_{i2}}\right| = \frac{10\text{ V}}{200\text{ mV}} = 50$$

Because v_{o1} is out of phase with $(v_{i1} - v_{i2})$, the correct specification for the single-ended output gain is −50. If the single-ended output is taken from the *other* side (v_{o2}), which is out of phase with v_{o1}, then the gain $v_{o2}/(v_{i1} - v_{i2})$ is +50.

Note once again that the double-ended voltage gain is the same as the voltage gain v_c/v_{be} for each transistor. *Note also that the single-ended output gain is one-half the double-ended gain.* Because the output difference voltage $v_{o1} - v_{o2}$ is out of phase with the input difference voltage $v_{i1} - v_{i2}$, the correct specification for the double-ended voltage gain is −100.

It should now be clear that if the two inputs are driven by equal *in-phase* signals, the output at each collector will be exactly 0, and the output difference voltage will be 0. Of course, in this case, the input difference voltage is also 0. These ideas are illustrated in Figure 17–32.

We can now derive general expressions for the double-ended and single-ended output voltage gains in terms of the circuit parameters. Figure 17–33 shows one side of the differential amplifier with the other side replaced by its emitter resistance, r_e. Recall that this is the resistance in series with the emitter of Q_1 when the input to Q_2 is grounded. We are again assuming that the current source has infinite resistance.

Neglecting the output resistance r_o at the collector of Q_1 we can use the familiar approximation for the voltage gain of the transistor:

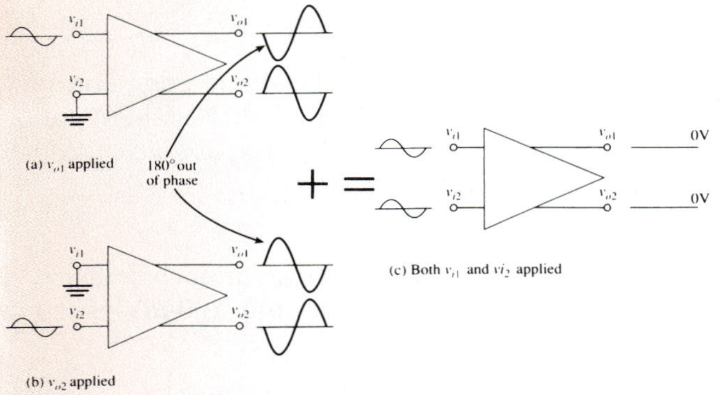

FIGURE 17–32 The outputs of the differential amplifier are 0 when the two inputs are equal and in phase

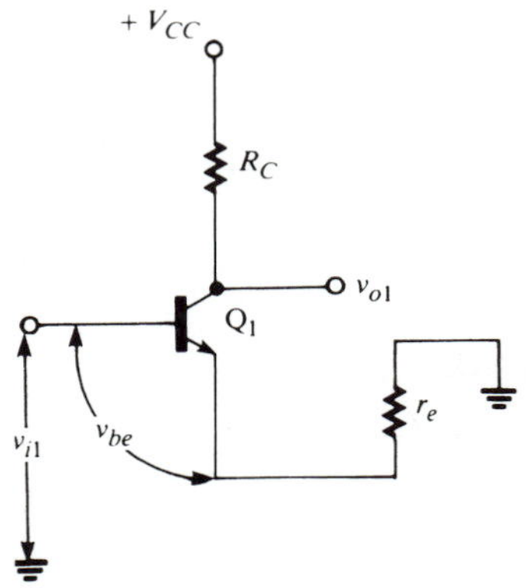

FIGURE 17–33 When the input to Q_2 is grounded, there is resistance r_e in series with the emitter of Q_1

$$\frac{v_{o1}}{v_{be1}} \approx \frac{-R_C}{r'_e} \tag{17–28}$$

where r'_e is the emitter resistance of Q_1. It is clear from Figure 17–33 that the voltage gain v_{o1}/v_{i1} is

$$\frac{v_{o1}}{v_{i1}} \approx \frac{-R_C}{2r_e} \tag{17–29}$$

where the quantity $2r'_e$ is in the denominator because we assume that the emitter resistances of Q_1 and Q_2 are equal. Equations 17–28 and 17–29 confirm our previous conclusion that the transistor voltage gain is twice the value of the gain, v_{o1}/v_{i1}. Also, we have already shown that the double-ended (difference) voltage gain equals the transistor gain and that the single-ended output gain is one-half that value. Therefore, we conclude that

$$\frac{v_{o1} - v_{o2}}{v_{i1} - v_{i2}} \approx \frac{-R_C}{r'_e} \tag{17–30}$$

and

$$\frac{v_{o1}}{v_{i1} - v_{i2}} \approx \frac{-R_C}{2r'_e} \tag{17–31}$$

We should note that these gain relations are valid regardless of the magnitudes and phase relations of the two inputs v_{i1} and v_{i2}. We have considered only the two special cases where v_{i1} and v_{i2} are equal and in phase and where they are equal and out of phase, but equations 17–30 and 17–31 hold under any circumstances. Note also that v_{o1} and v_{o2} will always have the same amplitude and be out of phase with each other. Thus,

$$\frac{v_{o2}}{v_{i1} - v_{i2}} \approx \frac{R_C}{2r'_e} \tag{17–32}$$

The small-signal *differential input resistance* is defined to be the input difference voltage divided by the total input current. Imagine a signal source connected *across* the input terminals, so the same current that flows out of the source into one input of the amplifier flows out of the other input and returns to the source. The signal-source voltage, which is the input difference voltage, divided by the signal-source current, is the differential input resistance. Since the total small-signal resistance in the path from one input through both emitters to the other input is $2r_e'$, the differential input resistance is

$$r_{id} = 2(\beta + 1)r_e' \qquad (17\text{–}33)$$

Figure 17–34 shows the dc voltages and currents in the ideal differential amplifier. Because the transistors are identical, the source current I divides equally between them, and the emitter current in each is, therefore,

$$I_E = I_{E1} = I_{E2} = I/2 \qquad (17\text{–}34)$$

The dc output voltage at the collector of each transistor is

$$V_{o1} = V_{CC} - I_{C1}R_C$$

$$V_{o2} = V_{CC} - I_{C2}R_C$$

Because $I_C \approx I_E = I/2$ in each transistor, we have

$$V_{o1} = V_{o2} \approx V_{CC} - (I/2)R_C \qquad (17\text{–}35)$$

To determine the ac emitter resistance of each transistor, we can use equation 17–34 and the familiar approximation $r_e' \approx 0.026/I_E$ to obtain

$$r_{e1}' = r_{e2}' = r_e' \approx \frac{0.026}{I_E} = \frac{0.026}{I/2} \qquad (17\text{–}36)$$

EXAMPLE 17–14

For the ideal differential amplifier shown in Figure 17–35, find

1. the dc output voltages V_{o1} and V_{o2},
2. the single-ended output gain $v_{o1}/(v_{i1} - v_{i2})$, and
3. the double-ended gain $(v_{o1} - v_{o2})/(v_{i1} - v_{i2})$.

Solution

1. The emitter current in each transistor is $I_E = I/2 = (2\text{ mA})/2 = 1\text{ mA} \approx I_C$. Therefore, $V_{o1} = V_{o2} = V_{CC} - I_C R_C = 15 - (1\text{ mA})(6\text{ k}\Omega) = 9\text{ V}$.

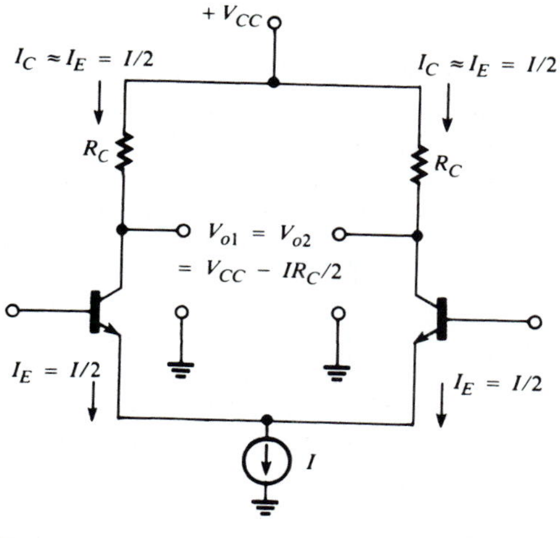

FIGURE 17–34 dc voltages and currents in an ideal differential amplifier

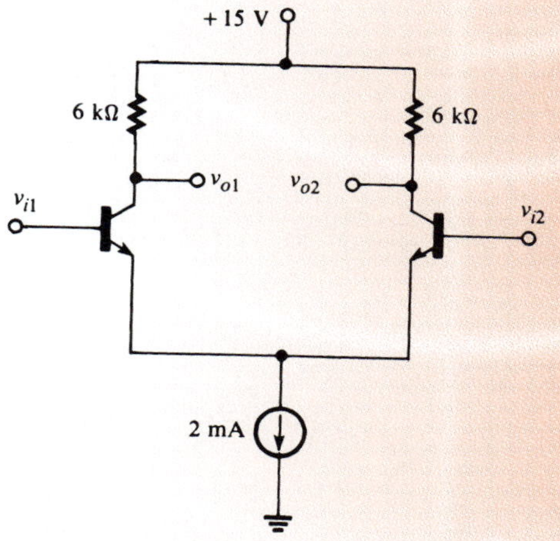

FIGURE 17–35 (Example 17–14)

2. The emitter resistance of each transistor is

$$r_e \approx \frac{0.026}{I_E} = \frac{0.026}{1\text{ mA}} = 26\ \Omega$$

Therefore, from equation 17–31.

$$\frac{v_{o1}}{v_{i1} - v_{i2}} \approx \frac{-R_C}{2r_e} = \frac{-6\text{ k}\Omega}{52\ \Omega} = -115.4$$

3. From equation 17–30,

$$\frac{v_{o1} - v_{o2}}{v_{i1} - v_{i2}} = \frac{-R_C}{r_e'} = \frac{-6\text{ k}\Omega}{26\ \Omega} = -230.8$$

The JFET Differential Amplifier

Many differential amplifiers are constructed using field-effect transistors because of the large impedance they present to input signals. This property is exceptionally important in many applications, including operational amplifiers, instrument amplifiers, and charge amplifiers. A large voltage gain is also important in these applications, and although the FET does not produce much gain, an FET differential amplifier is often the first stage in a multistage amplifier whose overall gain is large. Because FETs are easily fabricated in integrated-circuit form, FET differential amplifiers are commonly found in linear integrated circuits.

Figure 17–36 shows a JFET differential amplifier, and it can be seen that it is basically the same configuration as its BJT counterpart. The two JFETs operate as common-source amplifiers with their source terminals joined. A constant-current source provides bias current.

The derivations of the gain equations for the JFET amplifier are completely parallel to those for the BJT version. A source-to-ground voltage is developed at the common-source connection by source-follower action. With one input grounded, the output resistance and load resistance of the source follower are both equal to $1/g_m$ (assuming matched devices), so the source-follower gain is 0.5. Therefore, as shown in Figure 17–37, one-half the input voltage is developed across $1/g_m$, and the current is

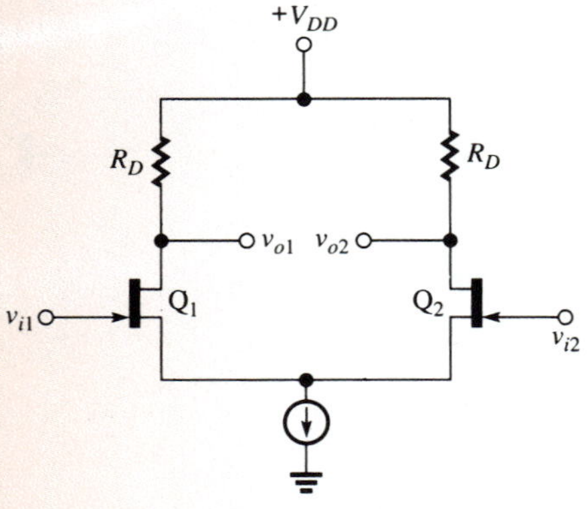

FIGURE 17–36 The JFET differential amplifier

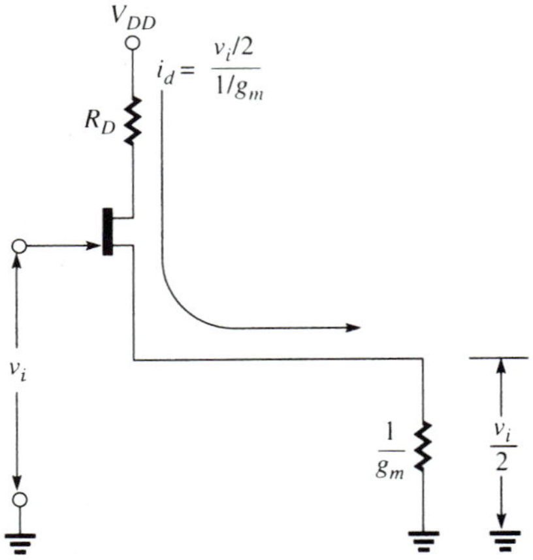

FIGURE 17–37 Source-follower action with gain 0.5 results in one-half the input voltage developed across $(1/g_m)$ ohms

$$i_d = \frac{v_i/2}{1/g_m} = \frac{v_i g_m}{2}$$

The output voltage is then

$$v_o = -i_d R_D = \frac{-v_i g_m R_D}{2} \tag{17–37}$$

from which we find the voltage gain

$$\frac{v_o}{v_i} = \frac{-g_m R_D}{2} \tag{17–38}$$

Applying the superposition principle in the same way we did for the BJT version, we readily find that

$$\frac{v_{o1}}{v_{i1} - v_{i2}} = \frac{-g_m R_D}{2} \tag{17–39}$$

and

$$\frac{v_{o1} - v_{o2}}{v_{i1} - v_{i2}} = -g_m R_D \tag{17–40}$$

Like the gain equations for the BJT differential amplifier, equations 17–39 and 17–40 show that the double-ended (difference) voltage gain is the same as the gain of one transistor, and the single-ended output gain is one-half that value.

EXAMPLE 17–15

The matched transistors in Figure 17–38 have $I_{DSS} = 12$ mA and $V_p = -2.5$ V. Find

1. the dc output voltages v_{o1} and v_{o2},
2. the single-ended output gain $v_{o1}/(v_{i1} - v_{i2})$, and
3. the double-ended gain $(v_{o1} - v_{o2})/(v_{i1} - v_{i2})$.

Solution

1. The dc current in each JFET is $I_D = (1/2)(6\text{ mA}) = 3$ mA. Therefore, $V_{o1} = V_{o2} = V_{DD} - I_D R_D = 15 - (3\text{ mA})(3\text{ k}\Omega) = 6$ V.

2. $$g_m = \frac{2I_{DSS}}{|V_p|}\sqrt{\frac{I_D}{I_{DSS}}} = \frac{2(12\text{ mA})}{2.5\text{ V}}\sqrt{\frac{3}{12}} = 4.8\text{ mS}$$

FIGURE 17–38 (Example 17–15)

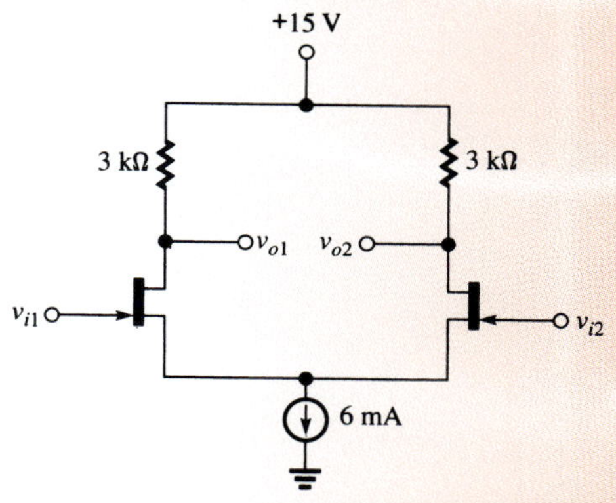

From equation 17–39,

$$\frac{v_{o1}}{v_{i1} - v_{i2}} = \frac{-g_m R_D}{2} = \frac{-(4.8\text{ mS})(3\text{ k}\Omega)}{2} = -7.2$$

3. From equation 17–40,

$$\frac{v_{o1} - v_{o2}}{v_{i1} - v_{i2}} = -g_m R_D = -(4.8\text{ mS})(3\text{ k}\Omega) = -14.4$$

The MOSFET Differential Amplifier

Many modern operational amplifier systems are incorporating MOSFET transistors in the design. This is especially true in low-power CMOS operational-amplifier analog applications. The MOSFET differential amplifier amplifies the difference in the V^+ and V^- inputs in a way similar to the BJT amplifier, and the equations are very similar to the JFET. This section explores the basic structure of a MOS differential amplifier, including the building blocks used to design the device. SPICE analysis and analytical treatments are both presented.

A MOSFET differential amplifier is shown in Figure 17–39. Rather than jump into the analysis, let's discuss the operation of the amplifier from a general point of view. First, let's answer a few basic questions about the amplifier. Which input, V_1 or V_2, is the inverting or noninverting input? How can you verify your assumption? Let's begin by making sure we recognize all of the building blocks in the circuit. Devices M_1 and M_2 form a differential pair. Devices M_3 and M_4 form a current mirror, and device M_5 is a current sink, which provides the bias current to the amplifier.

The inverting and noninverting inputs can be identified by examining the currents flowing in the circuit. The procedure is as follows. Begin the functional analysis by setting $V_1 = V_2$. Assume that M_5 is properly biased so that there is a bias current, I_5. Also assume that the transistors are matched so that their g_ms are equivalent.

Observations:

$$\boldsymbol{V_1 = V_2}$$

$$I_{D1} + I_{D2} = I_{D5}$$

$$I_{D1} = I_{D2} \qquad \text{since } V_1 = V_2 (V_{GS1} = V_{GS2})$$

$$I_{D3} = I_{D1}$$

Assume $\beta_3 = \beta_4 \therefore I_{D4} = I_{D3}$,
where

$$I_{D3} = I_{D1} = I_{D4}$$

Therefore, $I_{D4} = I_{D2}$ and $I_{out} = 0$.

$$\boldsymbol{V_1 > V_2}$$

$$I_{D1} + I_{D2} = I_{D5}, I_{D1} > I_{D2} \qquad \text{since } V_{GS1} > V_{GS2}$$

$$I_{D3} = I_{D1}$$

$$I_{D4} = I_{D3}$$

where

$$I_{D3} = I_{D1} = I_{D4}$$

$$\Sigma I_{\text{enter}} = \Sigma I_{\text{leaving}} \qquad \text{for the output node}$$

$$I_{D4} > I_{D2} \qquad \text{since } I_{D1} > I_{D2}$$

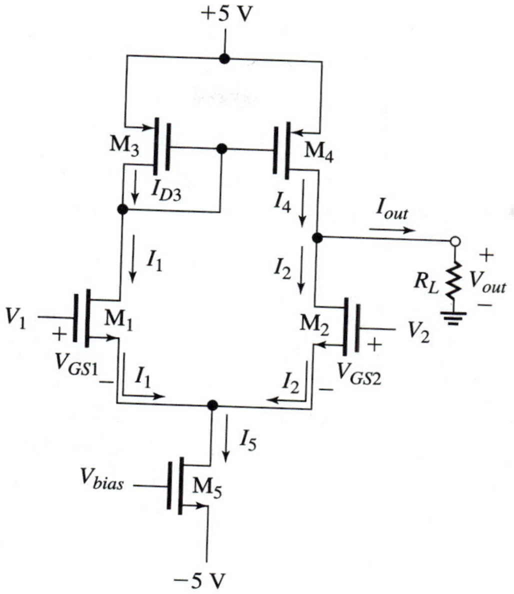

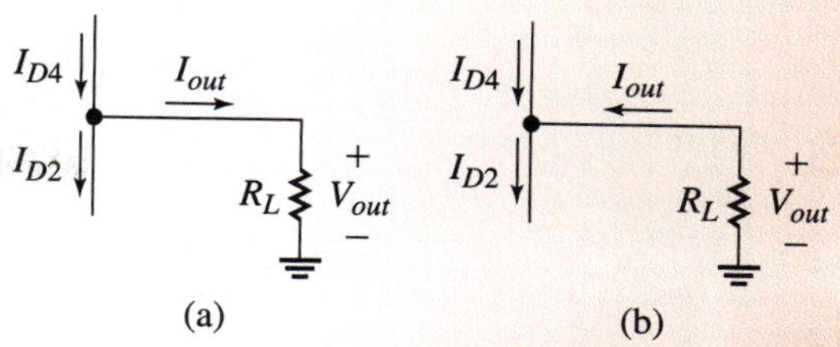

FIGURE 17–40 The load current direction for (a) $V(+) > V(-)$ where $V_1 = V(+)$ (b) $V(+) < V(-)$ where $V_2 = V(-)$

FIGURE 17–39 A MOSFET differential amplifier

Therefore, I_{out} is positive, current is leaving the node.

$$\mathbf{V_1 < V_2}$$

$$I_{D1} + I_{D2} = I_{D5}, I_{D1} < I_{D2} \quad \text{since } V_{GS1} < V_{GS2}$$

$$I_{D3} = I_{D1}$$

$$I_{D4} = I_{D3}$$

where

$$I_{D3} = I_{D1} = I_{D4}$$
$$I_{D4} < I_{D2} \qquad \text{since } I_{D1} < I_{D2}$$

Therefore, I_{out} is negative, current is entering the node.

When $V_1 = V_2, I_{out} = 0$; therefore, the output voltage is zero. When $V_1 > V_2$, the current I_{out} is leaving the output node, as shown in Figure 17–40(a). This indicates that V_{out} is positive and designates V_1 as the positive (+) input. When $V_1 < V_2$ ($V_2 > V_1$), the current I_{out} is entering the output node, as shown in Figure 17–40(b). This indicates that V_{out} is negative and designates V_2 as the negative (−) input.

Devices M_1 and M_2 are considered to be matched devices, so it can be assumed that each transistor has comparable operating properties such as the same transconductance value g_m. The MOSFET differential amplifier shown in Figure 17–39 is a single-ended output amplifier. Therefore, by inspection, the voltage gain of the differential amplifier is

$$A_V = \frac{V_{out}}{V_1 - V_2} = \frac{-g_{m2}(r_{o2} \| r_{o4})}{2} \qquad (17\text{–}41)$$

where $V_1 - V_2$ is the differential input voltage, and the load resistance R_L has been ignored. This is a valid assumption when the differential amplifier is driving a MOSFET input.

Equation 17–41 can also be written in terms of the conductance g_{ds}, replacing the r_o value with g_{ds}. This is a convenient form because SPICE prints the GDS value for each MOSFET transistor when the .OP option is specified.

TABLE 17–4 Sample values generated by the .OP command in SPICE

Name Model	M1 CMOSN	M2 CMOSN	M3 CMOSP	M4 CMOSP	M5 CMOSN
ID	3.41E−04	3.41E−04	−3.41E−04	−3.41E−04	6.83E−04
VGS	2.49E+00	2.49E+00	−4.54E+00	−4.54E+00	1.50E+00
GM	3.83E−04	3.83E−04	1.78E−04	1.78E−04	1.71E−03
GDS	3.52E−06	3.52E−06	7.51E−06	7.51E−06	7.00E−06

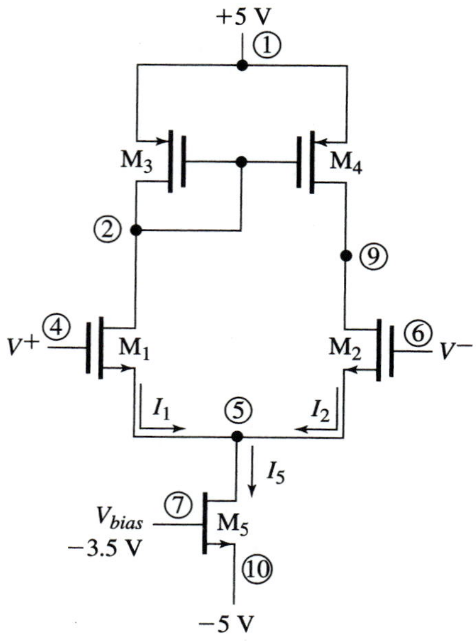

FIGURE 17–41 The MOSFET differential amplifier for Example 17–16

$$A_V = -\frac{g_m}{2} \cdot \frac{1}{(g_{ds2} + g_{ds4})} \tag{17–42}$$

SPICE also lists the g_m and g_{ds} values for each MOSFET transistor with the .OP command, so in most circumstances it is more convenient to use equation 17–42 to calculate the voltage gain A_V when working with the SPICE values. A sample SPICE .out file listing GM and GDS as generated by the .OP command in SPICE is provided in Table 17–4. These values were generated for a SPICE run for the differential amplifier shown in Figure 17–41.

Example 17–16 demonstrates how to determine the voltage gain for a differential amplifier.

EXAMPLE 17–16

Determine the voltage gain for the differential amplifier shown in Figure 17–41. Verify your answers (1) analytically and (2) with SPICE. Incorporate the following fabrication parameters in your analysis.

NMOS	PMOS
$KP_n = 5.2\text{E}-05$	$KP_p = 2.1\text{E}-05$
$VTO = 0.7$	$VTO = -0.7$
Lambda = 0.01	Lambda = 0.02
$(W/L)_{1,2} = 40\ \mu/10\ \mu$	$(W/L)_{3,4} = 20\ \mu/10\ \mu$
$(W/L)_5 = 80\ \mu/2\ \mu$	

Solution

1. Start your analysis by first determining the bias current I_5.

$$V_{bias} = 3.5 \text{ V} \qquad V_{SS} = -5 \text{ V} \qquad |V_{GS}| = 1.5 \text{ V}$$

Using Equation 17–8,

$$I_{D5} = \frac{\beta_5}{2}(|V_{GS_5}| - |V_{t_5}|)^2$$

where

$$\beta_5 = KP_n \times \left(\frac{W}{L}\right)_5 = (5.2\text{E-}05)\left(\frac{80}{2}\right) = 2.08 \times 10^{-3}$$

Therefore,

$$I_{D5} = \frac{2.08 \times 10^{-3}}{2}(1.5 - 0.7)^2 = 6.656 \times 10^{-4}$$

and

$$I_{D2} = \frac{I_{D5}}{2} = 3.328 \times 10^{-4}$$

$$g_{m2} = \sqrt{2 \cdot KP_n \cdot \left(\frac{W}{L}\right)_2 \cdot I_{D2}}$$

$$= \sqrt{(2)(5.2\text{E-}05)\left(\frac{40}{10}\right)(3.328 \times 10^{-4})}$$

$$= 3.721 \times 10^{-4}$$

$$r_{o2} = \frac{1}{\lambda_n I_{D2}} = \frac{1}{(0.01)(3.328 \times 10^{-4})} = 3 \times 10^5$$

$$r_{o4} = \frac{1}{\lambda_p I_{D2}} = \frac{1}{(0.02)(3.328 \times 10^{-4})} = 1.5 \times 10^5$$

$$\text{Voltage gain } A_V = \frac{(3.721 \times 10^{-4})}{2}(3 \times 10^5 \| 1.5 \times 10^5)$$

$$= -18.62$$

2. SPICE analysis for the circuits was performed using the following .cir file:

```
MOSFET DIFFERENTIAL AMPLIFIER-DIODE CONNECTED LOAD
****************************************************************
M1 2 4 5 10 CMOSN W=40u L=10u
M2 9 6 5 10 CMOSN W=40u L=10u
M3 2 2 1 1 CMOSP W=20u L=10u
M4 9 2 1 1 CMOSP W=20u L=10u
M5 5 7 10 10 CMOSN W=80U L=2U
****************************************************************
VMINUS 6 0 SIN(0 .1 100K)
VPLUS 4 0 DC 0
VBIAS 7 0 -3.5v
VDD 1 0 5V
VSS 10 0 -5
.MODEL CMOSN NMOS LEVEL=2 VTO=0.7 KP=5.2E-05 LAMBDA=.01
```

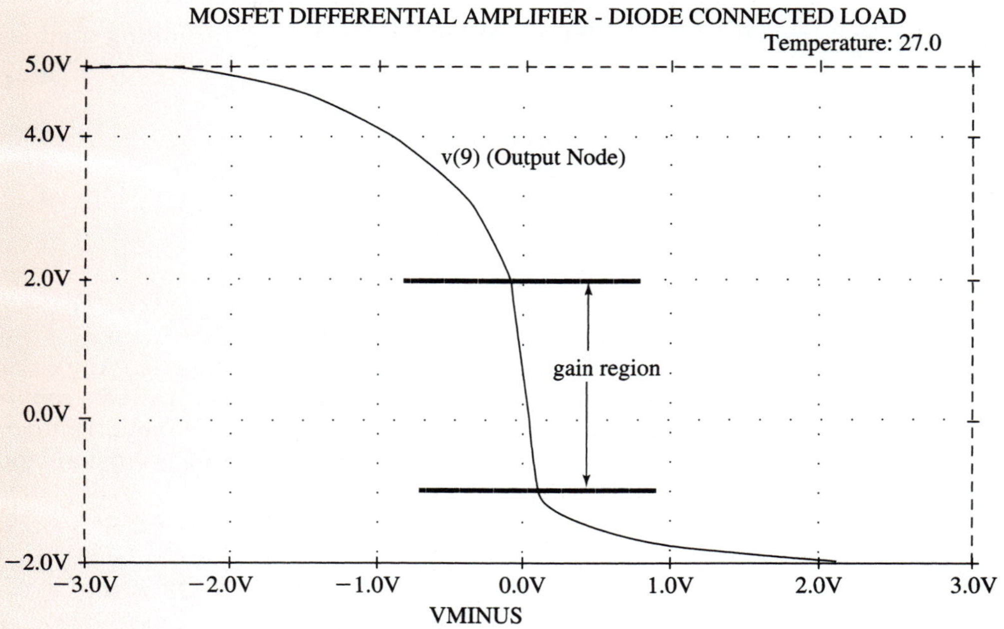

FIGURE 17–42 The SPICE analysis for Example 17–16

```
.MODEL CMOSP PMOS LEVEL=2 VTO=−0.7 KP=2.1E-05 LAMBDA=.02
*************************************************************
.DC VMINUS − 5 5 .1
.TRAN 0.01NS 100uS
.OP
.PROBE
.END
```

A graphical result for the gain was obtained using SPICE analysis. The analysis was obtained by performing a DC sweep of the circuit for V^- and measuring the gain graphically, as shown in Figure 17–42. Figure 17–42 shows that the output node v(9) changes −2.86 V for an input change of 0.185 V, which translates to a gain of −15.46. This is comparable to the value of −18.62 that we calculated.

EXAMPLE 17–17

Compare the results for I_D, g_{m2}, r_{o2}, r_{o4}, and A_V obtained in Example 17–16 using the SPICE values shown in Table 17–4 to determine SPICE calculated results.

Solution

For I_D: SPICE 6.83×10^{-4} A
Example 17–16 6.656×10^{-4} A

For g_{m2}: SPICE 3.83×10^{-4} A/V
Example 17–16 3.721×10^{-4} A/V

For r_{o2}: SPICE $g_{ds2} = 3.52 \times 10^{-6}$ $(2.84 \times 10^5\ \Omega)$
Example 17–16 $r_{o2} = 3 \times 10^5\ \Omega$

For r_{o4}: SPICE $g_{ds4} = 7.51 \times 10^{-6}$ $(1.33 \times 10^5\ \Omega)$
Example 17–15 $r_{o4} = 1.5 \times 10^5\ \Omega$

$$\text{For } A_V\text{: } A_V = \frac{-3.83 \times 10^{-4}}{2} \cdot \frac{1}{3.52 \times 10^{-6} + 7.51 \times 10^{-6}} = -18.62$$

This result should equal −18.62 which agrees with Example 17–16.

This section has explored basic circuit analysis and SPICE simulation techniques for a MOS differential amplifier. The calculations performed in this section provide a basic understanding of the building blocks for analog systems.

Common-Mode Parameters

One attractive feature of a differential amplifier is its ability to reject signals that are *common* to both inputs. Because the outputs are amplified versions of the difference between the inputs, any voltage component that appears identically in both signal inputs will be "differenced out," that is, will have zero level in the outputs. (We have already seen that the outputs are exactly 0 when both inputs are identical, in-phase signals.) Any dc or ac voltage that appears simultaneously in both signal inputs is called a *common-mode* signal. The ability of an amplifier to suppress, or zero-out, common-mode signals is called *common-mode rejection*. An example of a common-mode signal whose rejection is desirable is electrical *noise* induced in both signal lines, a frequent occurrence when the lines are routed together over long paths. Another example is a dc level common to both inputs, or common dc fluctuations caused by power-supply variations.

In the ideal differential amplifier, any common-mode signal will be completely cancelled out and therefore have no effect on the output signals. In practical amplifiers, which we will discuss in the next section, mismatched components and certain other nonideal conditions result in imperfect cancellation of common-mode signals. Figure 17–43 shows a differential amplifier in which a common-mode signal v_{cm} is applied to both inputs. Ideally, the output voltages should be 0, but in fact some small component of v_{cm} may appear. The differential *common-mode gain*, A_{cm}, is defined to be the ratio of the output difference voltage caused by the common-mode signal to the common-mode signal itself:

$$A_{cm} = \frac{(v_{o1} - v_{o2})_{cm}}{v_{cm}} \tag{17–43}$$

We can also define a single-ended common-mode gain as the ratio of $(v_{o1})_{cm}$ or $(v_{o2})_{cm}$ to v_{cm}. Obviously, the ideal amplifier has common-mode gain equal to 0.

A widely used specification and figure of merit for a differential amplifier is its *common-mode rejection ratio* (CMRR), defined to be the ratio of the magnitude of its differential (difference-mode) gain A_d to the magnitude of its common-mode gain:

$$\text{CMRR} = \frac{|A_d|}{|A_{cm}|} \tag{17–44}$$

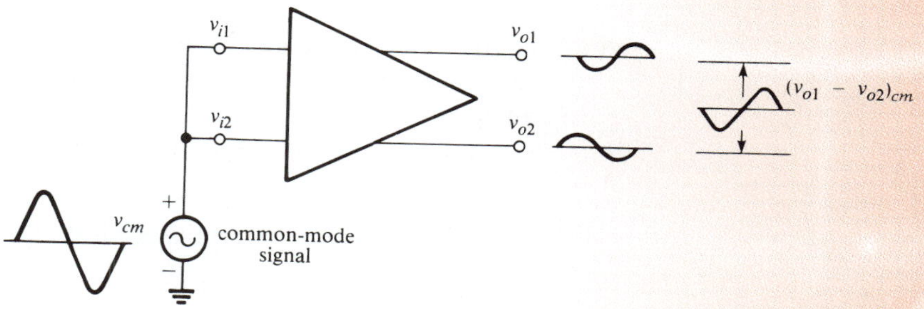

FIGURE 17–43 If the differential amplifier were ideal, both outputs would be 0 when the inputs have the same (common-mode) signal. In reality, there is a small common-mode output, as shown

The value of the CMRR is often given in decibels:

$$\text{CMRR} = 20\log_{10}\left|\frac{A_d}{A_{cm}}\right| \tag{17–45}$$

EXAMPLE 17–18

When the inputs to a certain differential amplifier are $v_{i1} = 0.1 \sin \omega t$ and $v_{i2} = -0.1 \sin \omega t$, it is found that the outputs are $v_{o1} = -5 \sin \omega t$ and $v_{o2} = 5 \sin \omega t$. When both inputs are $2 \sin \omega t$, the outputs are $v_{o1} = -0.05 \sin \omega t$ and $v_{o2} = 0.05 \sin \omega t$. Find the CMRR in dB.

Solution

We will use the peak values of the various signals for our gain computations, but note carefully how the minus signs are used to preserve phase relations. The difference-mode gain is

$$A_d = \frac{v_{o1} - v_{o2}}{v_{i1} - v_{i2}} = \frac{-5 - 5}{0.1 - (-0.1)} = \frac{-10}{0.2} = -50$$

The common-mode gain is

$$A_{cm} = \frac{(v_{o1} - v_{o2})_{cm}}{v_{cm}} = \frac{-0.05 - 0.05}{2} = \frac{-0.1}{2} = -0.05$$

The common-mode rejection ratio is

$$\text{CMRR} = \frac{|A_d|}{|A_{cm}|} = \frac{50}{0.05} = 1000$$

Expressing this result in dB, we have CMRR $= 20\log_{10}(1000) = 60$ dB.

17–7 THE MILLER OPERATIONAL AMPLIFIER

Sections 17–1 to 17–6 have provided a basic understanding of the building blocks used in analog systems. Frequency analysis of analog amplifiers was presented in Chapter 9 so this concept has not been included here. This chapter has addressed the following analog building blocks:

- current source/sink
- current mirror
- gain stages
- differential amplifiers

The output buffer building block was presented in Chapter 6.

These building blocks are used in analog systems, including the operational amplifier (op-amp). Op-amps are typically very complex from a transistor point of view. However, you will be able to identify some of the transistor building blocks in the example presented in this section, which is very simple but meets all of the requirements for an op-amp. The following is a list of some ideal operating characteristics for an operational amplifier:

- infinite input impedance
- a differential structure so that common noise is not amplified (i.e., good common-mode rejection)

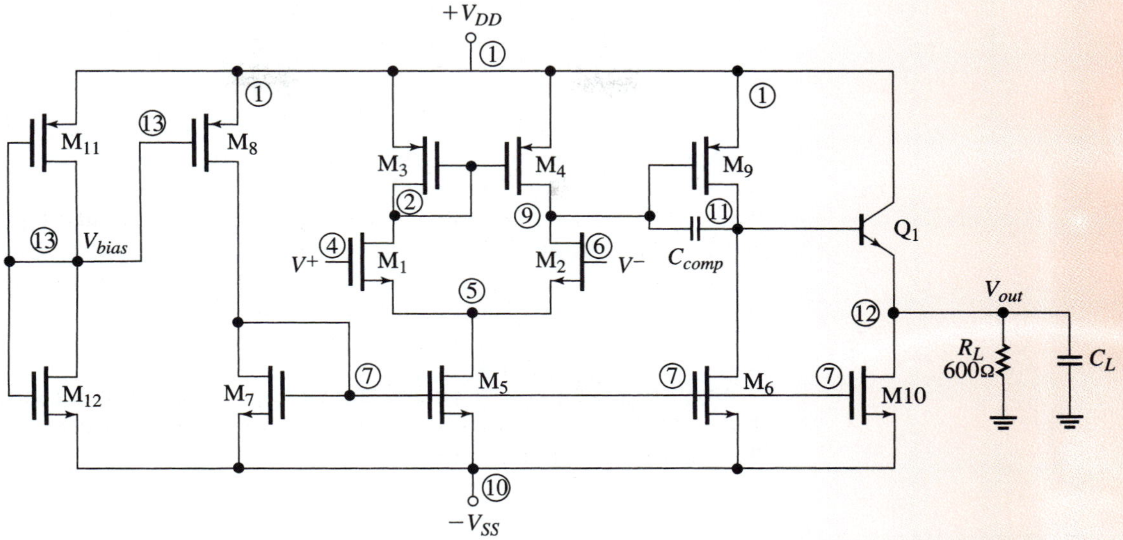

FIGURE 17–44 A simple BiCMOS Miller operational amplifier

- high gain (i.e., infinite)
- ability to drive heavy loads (low output resistance—ideally 0)

Real amplifiers cannot satisfy these requirements, but modern operational amplifiers come close.

The most important concepts the reader can extract from this section are what the basic building blocks for the operational amplifier are. We present a Miller op-amp constructed with MOSFET transistors and incorporating a bipolar output driver. This op-amp is very simple but demonstrates the fundamentals of the internal building blocks. Modern op-amps are much more complex but resemble this structure in concept. Each analog building block will be identified and briefly discussed.

Referring to Figure 17–44, the inputs to the amplifier are the gates of M_1 (V^+) and M_2 (V^-). The bias to the M_1/M_2 differential pair is provided by the current sink (M_5), which is biased by the current mirror formed by M_7 and M_5. The loading for the differential amplifier is provided by the current mirror made up of MOS transistors M_3 and M_4. Node (9) connects the output of the differential amplifier block to the gain-stage stage made up of M_9 and M_6. NMOS transistor M_6 is the active load, which is biased by the current mirror made up of M_7 and M_6. Notice that M_7 provides the mirror current for both M_5 and M_6 and that a capacitor connects the output (node 11) of the gain stage back to the input (node 9). This is called a *Miller capacitor* and is sometimes used in op-amps to provide stability through what is called a *dominant pole.* In more practical terms, this capacitor helps to add stability to an op-amp that has high open-loop gain. Without the capacitor, the op-amp could potentially have oscillations. The capacitor adds stability by moving the 3-dB cutoff frequency (refer to Chapter 9) to a much lower value.

The output of the gain stage (node 11) feeds the base of the BJT transistor Q_1. Q_1 and NMOS transistor M_{10} form an emitter-follower output buffer. The combination of BJTs and CMOS transistors, on the same IC, is called a *BiCMOS* device. These devices incorporate the benefits of each transistor, such as the low output impedance of the emitter follower and its current-drive capability, which allows the op-amp to drive loads containing small resistance values and moderate capacitance values with minimal distortion to the signal. NMOS transistor M_{10} is used as an active load for the emitter follower. The advantage of the MOSFET device as an active load is the control of the current afforded by the *W/L* ratio. Control of the current provides control of the output resistance ($r_o = 1/\lambda I_D$). M_{10} is biased by the current mirror formed by M_7 and reflected to M_{10}. Notice that M_7 is used to reflect the current to the entire op-amp (M_5, M_6, and M_{10}). Control of the current is provided by the *W/L* ratios for each transistor in the current mirror. (This concept was presented in Section 17–4.) The current source feeding the current mirror is provided by M_8, which is biased by the voltage divider formed by transistors M_{11}/M_{12}.

We include a SPICE .cir file here so that you can experiment with the transistor-level implementation of an op-amp.

```
*************************************************************
* Miller OPAMP Demonstration
*************************************************************
M1 2 4 5 5 CMOSN W=100u L=50u AD=375P AS=375P PD=160U PS=160U
M2 9 6 5 5 CMOSN W=100u L=50u AD=375P AS=375P PD=160U PS=160U
M3 2 2 1 1 CMOSP W=450u L=2u AD=375P AS=375P PD=160U PS=160U
M4 9 2 1 1 CMOSP W=450u L=2u AD=375P AS=375P PD=160U PS=160U
M5 5 7 10 10 CMOSN W=360U L=2U AD=925P AS=925P PD=380U PS=380U
M6 11 7 10 10 CMOSN W=45U L=2U AD=925PAS=925P PD=380U PS=380U
M7 7 7 10 10 CMOSN W=180U L=2U AD=925P AS=925P PD=380U PS=380U
M8 7 13 1 1 CMOSP W=180u L=2u AD=375P AS=375P PD=160U PS=160U
M9 11 9 1 1 CMOSP W=180u L=2u AD=375P AS=375P PD=160U PS=160U
M10 12 7 10 10 CMOSN W=380U L=2U AD=925P AS=925P PD=380U PS=380U
M11 13 13 1 1 CMOSP W=820u L=2u AD=3280P AS=3280P PD=1648U PS=1648U
M12 13 13 10 10 CMOSN W=5U L=2U AD=20P AS=20P PD=18U PS=18U
Q1 1 11 12 qn1
CCOMP 9 11 6pF
RL 12 0 600
CL 12 0 10pf
Vminus 6 0 AC 1.0 SIN(0 .01 2500)
Vplus 4 0 DC 0.0
VDD 1 0 5V
VSS 10 0 -5
.model qn1 npn VJE=.7
.MODEL CMOSN NMOS LEVEL=2 LD=0.25U TOX=398.00008E-10
+ NSUB=2.374951E+16 VTO=0.954612 KP=5.371E-05 GAMMA=1.0234
+ PHI=0.6 UO=619.676 UEXP=0.20018 UCRIT=97000.7
+ DELTA=4.5376 VMAX=74815.1 XJ=0.250000U LAMBDA=1.900757E-02
+ NFS=3.91E+11 NEFF=1 NSS=1.000000E+10 TPG=1.00000
+ RSH=27.67 CGDO=3.25390E-10 CGSO=3.25390E-10 CGBO=4.31921E-10
+ CJ=3.9278E-04 MJ=0.427343 CJSW=5.4554E-10 MJSW=0.371244 PB=0.800
+ NUSB=2.374951E+16 VTO=0.954612 KP=5.371E-05 GAMMA=1.0234
+ PHI=0.6 UO=619.676 UEXP=0.20018 UCRIT=97000.7
+ DELTA=4.5376 VMAX=74815.1 XJ=0.250000U LAMBDA=1.900757E-02
+ NFS=3.91E+11 NEFF=1 NSS=1.000000E+10 TPG=1.00000
+ RSH=27.67 CGDO=3.25390E-10 CGSO=3.25390E-10 CGBO=4.31921E-10
+ CJ=3.9278E-04 MJ=0.427343 CJSW=5.4554E-10 MJSW=0.371244 PB=0.800
* Weff=wDRAWN - dELTA w
```

```
* The suggested Delta W is 0.53 um
.MODEL CMOSP PMOS LEVEL=2 LD=0.249723U TOX=398.00E−10
+ NSUB=5.799569E+15 VTO=−0.821283 KP=2.160E−05 GAMMA=0.5057
+ PHI=0.6 UO=249 UEXP= 0.217968 UCRIT=19160.6
+ DELTA=1.55841 VMAX=42259.5 XJ=0.2500U LAMBDA=4.495901E−02
+ NFS=3.23E+11 NEFF=1.001 NSS=1.0000E+10 TPG=−1.000
+ RSH=72.45 CGDO=3.249985E−10 CGSO=3.249985E−10 CGBO=4.093809E−10
+ CJ=2.06930E−04 MJ=0.462054 CJSW=2.2334E−10 MJSW=0.117772 PB=0.700
* Weff=wDRAWN − dELTA w
* The suggested Delta W is − 0.39 um
*****************************************************************
.AC DEC 10 1 10MEG
.DC VMINUS −5 5 .1
.TRAN 0.01NS 2mS
.PROBE
.END
```

SPICE simulation results for frequency response and the transient response for a sinusoid input are provided in Figures 17–45(a), (b), and (c). Figure 17–45(a) shows the frequency response of the op-amp with the Miller compensation capacitor (CCOMP) included. The 3-dB corner frequency is at approximately 60 kHz. If CCOMP is removed (see Figure 17–45(b)), the bandwidth is greatly increased. In this case, the new 3-dB corner frequency is 3.48 MHz. Remember, the Miller compensation capacitor is added to the op-amp for stability purposes. Figure 17–45(c) shows the transient simulation result for a 10-mV peak sinusoid input. The peak output voltage is approximately 0.94 V for a voltage gain of 94. This agrees with the result of the frequency analysis, which shows a gain of 94. Note that the output sinusoid has a slight dc offset of approximately 16 mV. This can be seen at the start of the sinusoid, which begins slightly above the 0-axis.

We have presented this example to reinforce your understanding of the use of analog building blocks. Op-amps are not the only application of these circuits, but they are an important part of an introductory text in circuits because they require the necessary basic understanding of transistor circuits.

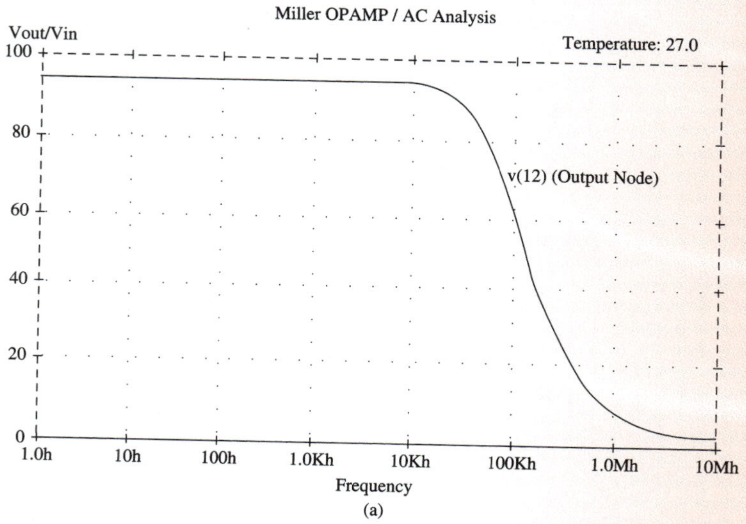

FIGURE 17–45 (a) The frequency analysis for the Miller op-amp demonstration with the Miller feedback capacitance (b) The Miller op-amp with no Miller feedback capacitance (c) Transient analysis for the Miller op-amp demonstration

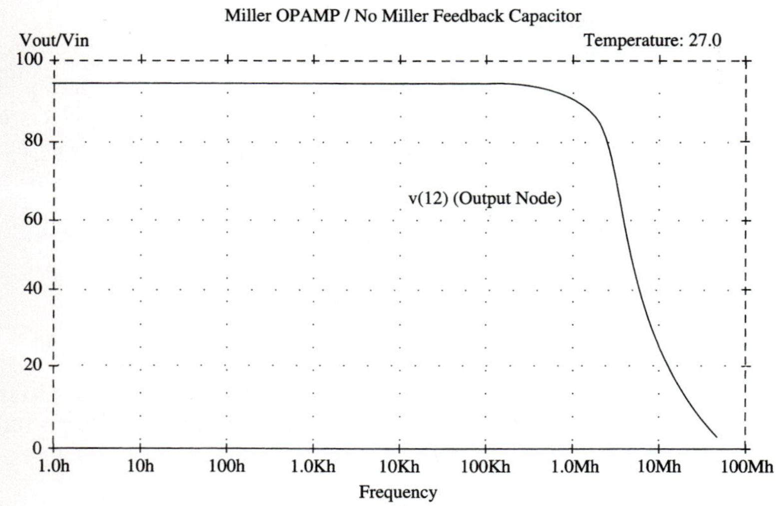

(b)

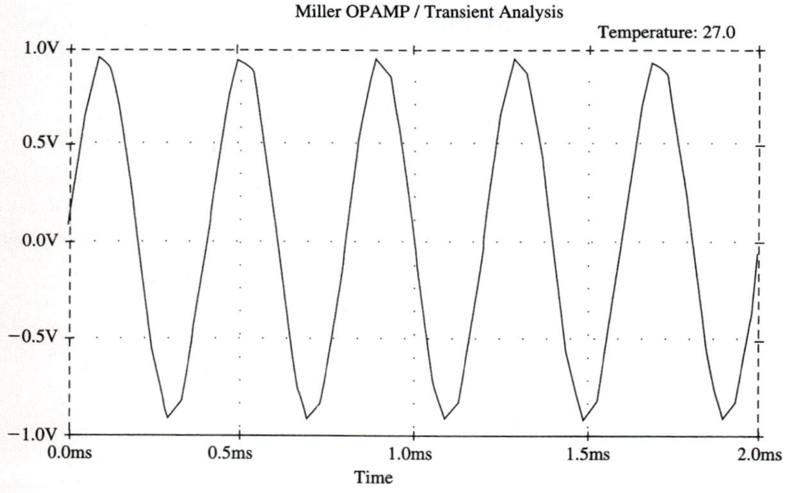

(c)

FIGURE 17–45 Continued

17–8 BJT GAIN STAGE CIRCUIT ANALYSIS WITH ELECTRONICS WORKBENCH MULTISIM

This section examines the use of EWB to determine the proper biasing point for a BJT gain stage (a common-emitter amplifier). The BJT gain stage is a major analog circuit building block and can be represented in many different styles. The circuit examined in this section is a simple BJT gain stage. In fact, this circuit is seldom used in discrete form because of many instability issues; however, this circuit can be used to demonstrate how to determine the proper bias point for a transistor amplifier.

Chapter 17 introduced the basic analog circuit building blocks. The student has learned how to use Pspice analysis to design and verify various BJT and MOSFET circuits. This exercise demonstrates how to use EWB Multisim to perform similar analysis. The objective of this EWB exercise is to:

- Use EWB to determine the proper bias point for the simple BJT gain stage.
- Develop an understanding on how to use the EWB postprocessor to determine the proper bias voltage to the BJT gain stage.

To begin this exercise, open the circuit, **Ch17_EWB.msm** that is found in the Electronics Workbench CD-ROM packaged with the text. This file is the simple gain stage shown in Figure 17–46. The circuit is a simple BJT amplifier. The input signal is a 1-kHz sine wave with a peak amplitude of 50 mV. The resistor R2 (1 kΩ) is modeling the internal resistance of the signal source. The BJT being used is a 2N2222 and a collector resistance of 1 kΩ (R1). The V_{CC} voltage for the circuit is 9 V.

The first task is to determine the offset voltage needed to place the amplifier in the proper mode of operation. The V_{CC} voltage is 9 V; therefore the V_{CE} voltage should be about 4.5 V to provide maximum dynamic range for the swing of the output signal. The V_{CE} voltage can be adjusted by varying the dc bias on the transistor. Increasing the dc bias voltage increases the current I_C. This increases the voltage drop across the collector resistor R1 and as a result, decreases the V_{CE} voltage. The objective of this task is to determine the dc bias voltage required to place the transistor at the proper bias point ($V_{CE} = 4.5$ V). In this case, the dc bias voltage will be provided by the dc offset voltage from the input signal source V_{in}. The analysis function dc sweep can be used to find the proper bias point. Click on Simulate and Analyses as shown in Figure 17–47. Select the menu option DC Sweep.

The menu for the DC sweep is shown in Figure 17–48.

The source for the dc sweep is VIN, the input sinusoid. The dc start voltage is 0 V, the stop voltage is 1 V, and the increment is .01 V. You must also select the output variables to be plotted. This can be done by

FIGURE 17–46 The BJT gain stage used in the EWB exercise

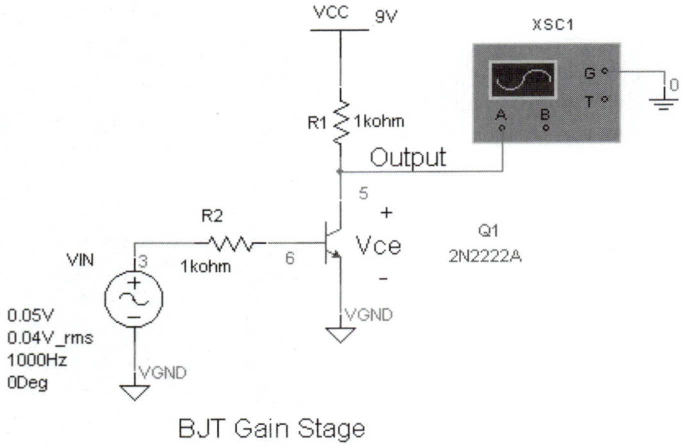

FIGURE 17–47 The menu for selecting the EWB analyses

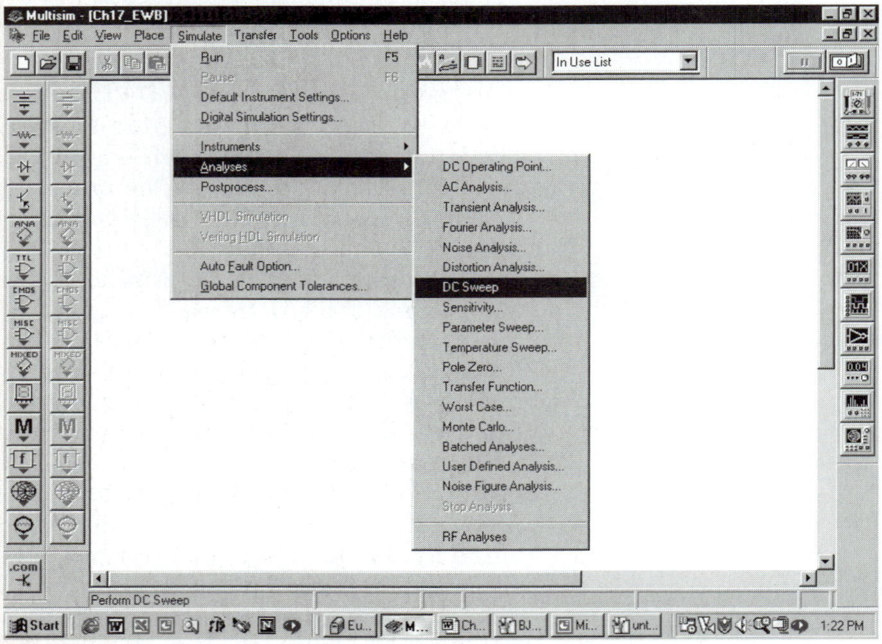

clicking on the Output variables tab. This menu is shown in Figure 17–49. The VCE voltage (node 5) is selected for the simulation analysis. (*Note:* The node numbers in EWB Multisim can be displayed by clicking on Options in the main EWB Multisim window, selecting Preferences and then selecting the Circuit tab, click on Show node names.) Click on Simulate to start the simulation.

The window shown in Figure 17–50 should appear, showing a plot of the output voltage V(5) previously specified on the y-axis. The x-axis shows the sweep of the VIN voltage from 0 to 1 V. Notice that the output swings from 9 to approximately 0 V.

The objective is to determine the V_{IN} voltage that places the V_{CE} voltage at $V_{CC}/2$ (4.5 V). EWB Multisim has cursors that can be used to deter-

FIGURE 17–48 The menu for the DC Sweep Analysis

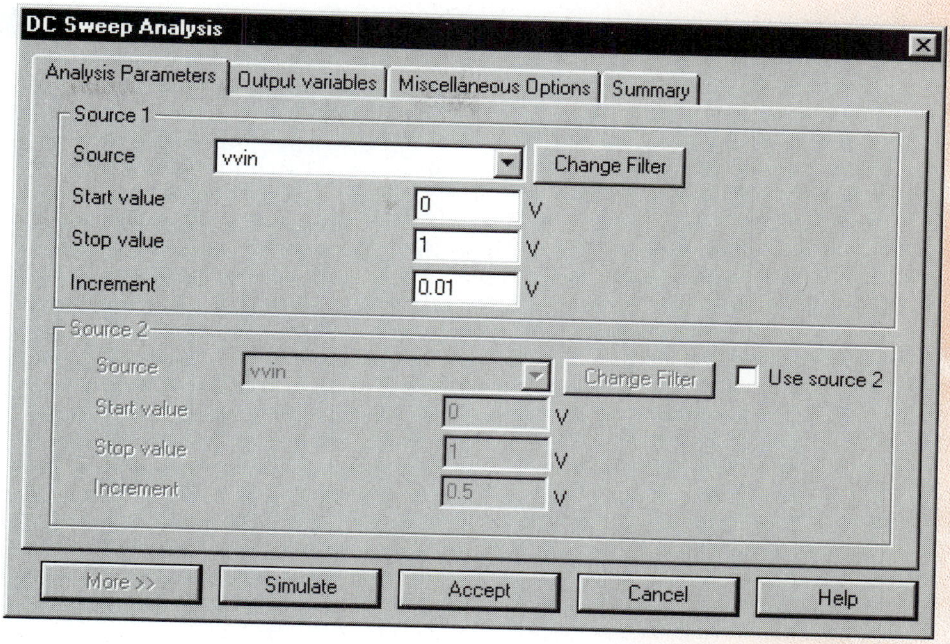

FIGURE 17–49 The menu for selecting the variables for plotting during the DC sweep simulation

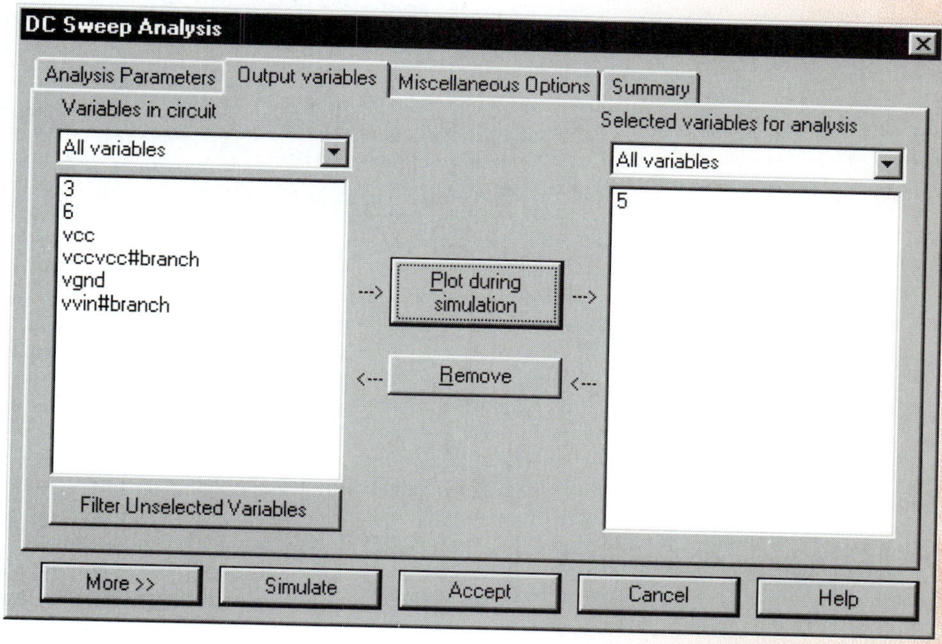

mine the analysis values. The button for selecting the cursors is shown in Figure 17–51. The x and y cursor values for trace 5 are also shown. Move cursor 1 until the y value changes from 9 to approximately 4.5 V. The x value at this point is the dc voltage for V_{IN} that should be used as the DC offset voltage. This result is shown in Figure 17–52. The x-axis voltage of .686 V can be used for setting the dc offset voltage for the input signal source, V_{IN}. The output sine wave will now be sitting on a 4.5 V (approximately) dc voltage as shown in Figure 17–53.

This exercise has demonstrated how to use the EWB Multisim dc sweep analysis to determine the input dc bias voltage for an amplifier.

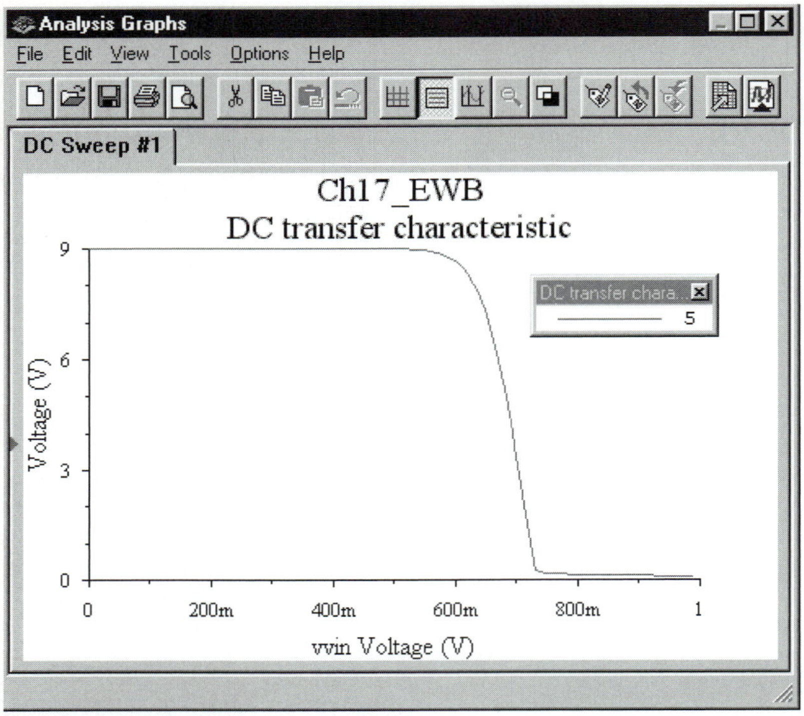

FIGURE 17–50 The analysis graph of node V5

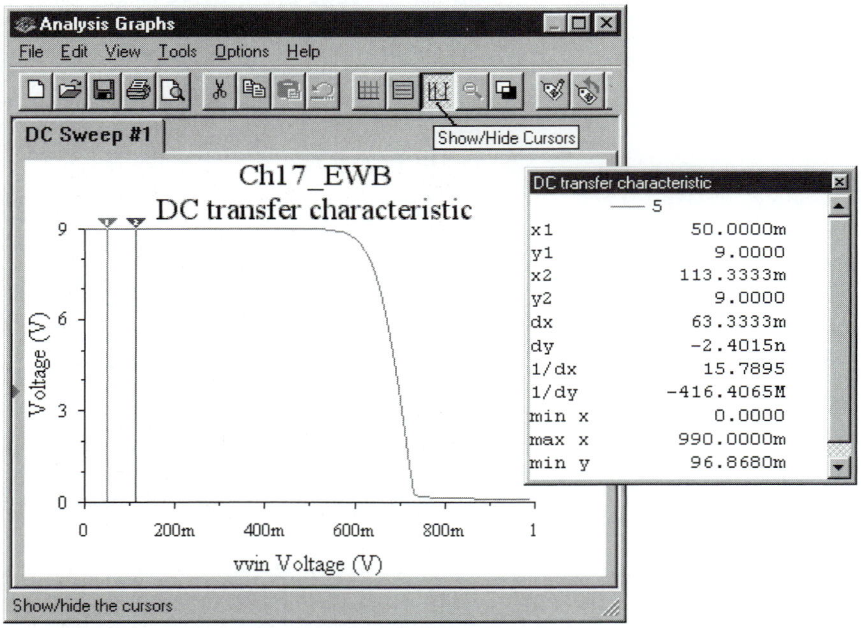

FIGURE 17–51 The EWB cursors and the x- and y-axis values for trace 5

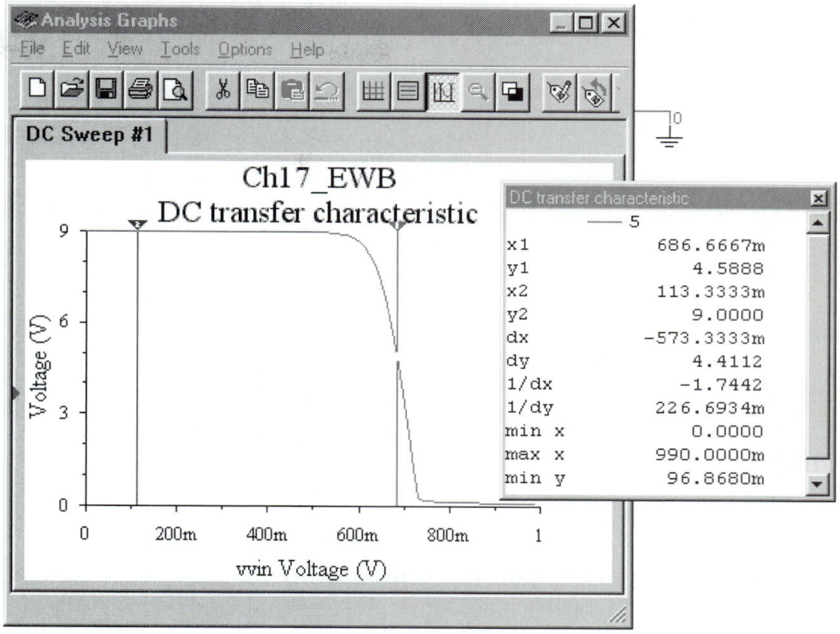

FIGURE 17–52 The cursor at the midpoint voltage of 4.588 V and the x-axis value of 686 mV

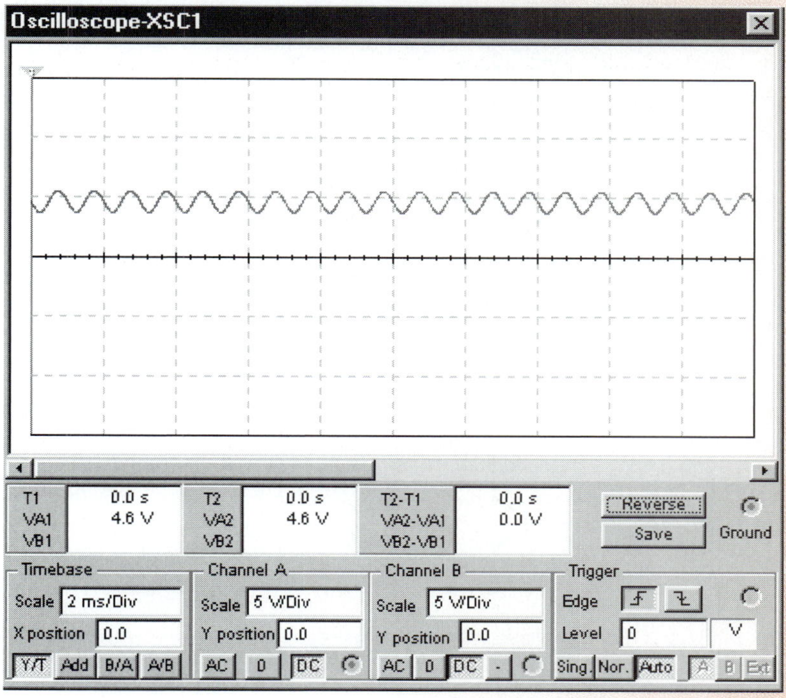

FIGURE 17–53 The output sine wave sitting on a ~4.5 V dc offset

SUMMARY

This chapter has presented an overview of the basic building blocks for analog circuits. The chapter focused on BJT and MOSFET devices. The concepts the student should understand are:

- The structure for each basic analog circuit building block.
- The key equations required to analyze each building block.
- How to use computer simulation to analyze the performance of a circuit.
- How to identify the key issues of each circuit building block (e.g. temperature effects).

EXERCISES

SECTION 17–3

The Current Source/Sink

17–1. Determine I_C for a BJT if $I_S = 1.3 \times 10^{-13}$ A, $|V_{BE}| = 0.65$ V, and the temperature is 26°C.

17–2. Determine I_C for a BJT circuit shown in Figure 17–54. $I_S = 1.3 \times 10^{-13}$ A; $V_T = 0.026$ V.

17–3. Determine I_C for a BJT given the circuit shown in Figure 17–55. $I_S = 1.3 \times 10^{-13}$ A and $V_T = 0.025$ V.

17–4. Determine the base-bias voltage required for a BJT to source 1.5 mA of current given that $I_S = 1.4 \times 10^{-13}$ A and $V_T = 0.026$ V.

17–5. Determine the base-bias voltage (V_{bias}) required to sink 1.5 mA for the circuit shown in Figure 17–56. $I_S = 1.4 \times 10^{-13}$ A and the temperature is 27°C.

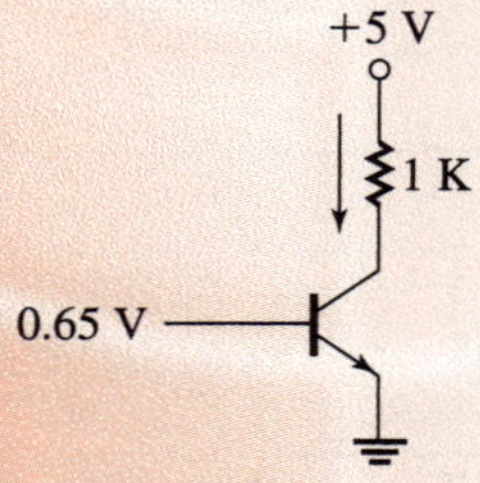

FIGURE 17–54 (Exercise 17–2)

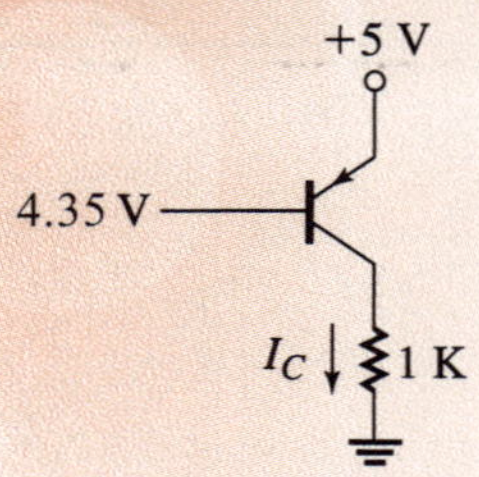

FIGURE 17–55 (Exercise 17–3)

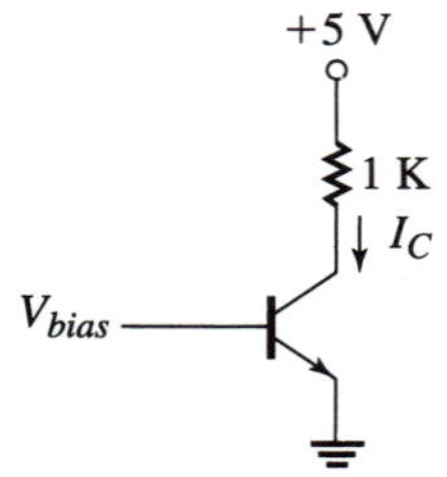

FIGURE 17–56 (Exercise 17–5)

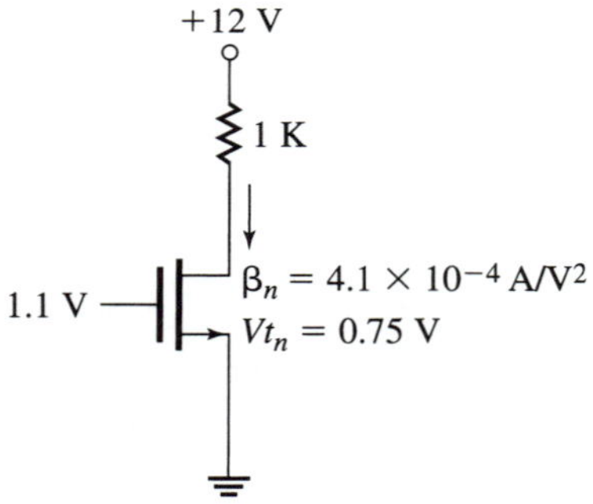

FIGURE 17–57 (Exercise 17–6)

17–6. Determine the current I_D for the circuit shown in Figure 17–57. $\beta_n = 4.1 \times 10^{-4}$ A/V^2, $Vt_n = 0.75$ V.

17–7. Determine the current I_D for the circuit shown in Figure 17–58. $\beta_p = 3.8 \times 10^{-4}$ A/V^2, $V_{tp} = -0.82$ V.

17–8. A bias voltage of 1.45 V is applied to the current sink shown in Figure 17–59. If $I_D = 150$ μA, what is the β value of the transistor? $Vt_n = 0.72$ V.

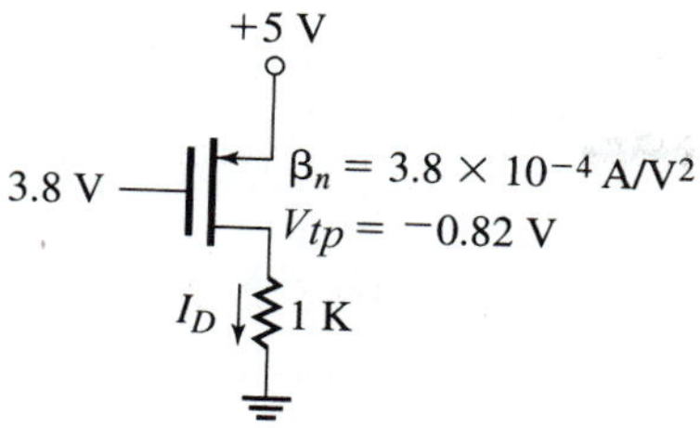

FIGURE 17–58 (Exercise 17–7)

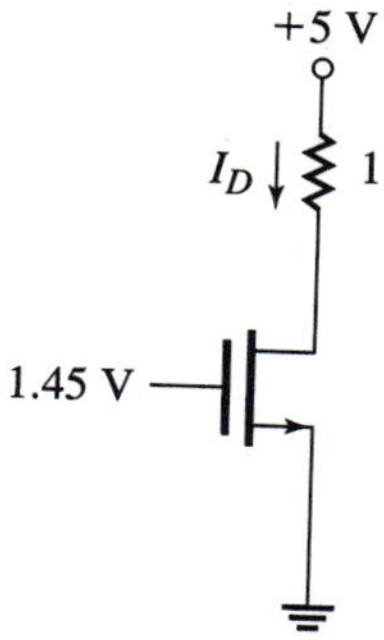

FIGURE 17–59 (Exercise 17–8)

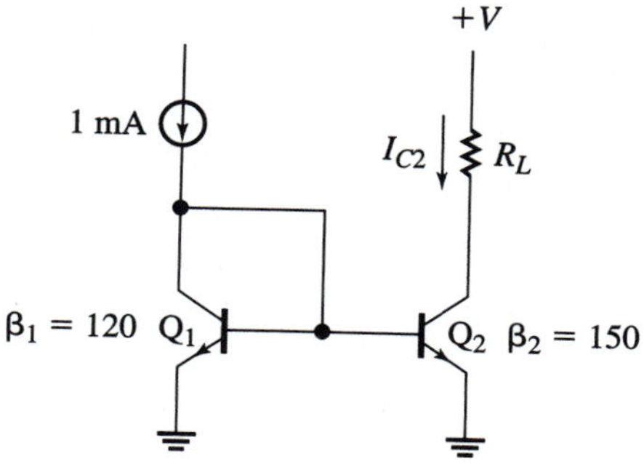

FIGURE 17–60 (Exercise 17–9)

SECTION 17–4

The Current Mirror

17–9. Determine I_{C2} for the current mirror shown in Figure 17–60.

17–10. What is the value of I_{C2} for the circuit shown in Figure 17–61?

17–11. Determine the values for I_{C2}, I_{C3}, and I_{C4} for the circuit shown in Figure 17–62.

17–12. Given $I_{D1} = 100$ μA, $\beta_1 = 4.2 \times 10^{-4}$ A/V^2, $\beta_2 = 3.8 \times 10^{-4}$ A/V^2, find I_{D2} for the circuit shown in Figure 17–63.

17–13. Determine I_{out} given $I_{bias} = 150$ μA and $KP_n = 5 \times 10^{-5}$ A/V^2 for the circuit provided in Figure 17–64. The *W/L* ratios for the transistors are specified in the circuit drawing.

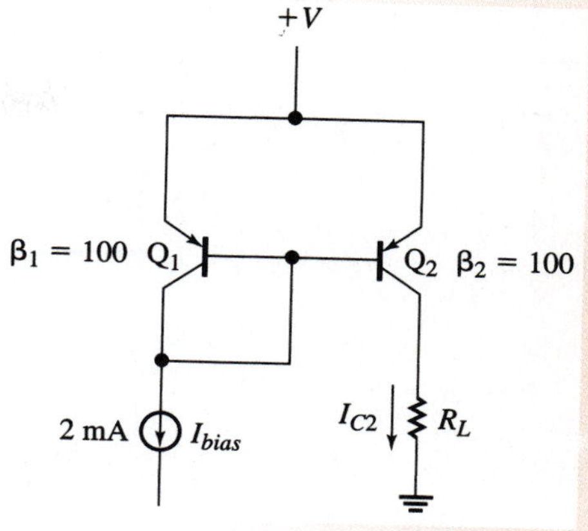

FIGURE 17–61 (Exercise 17–10)

17–14. Find I_{C2} and I_{C3} for the circuit shown in Figure 17–65 given the following information from the SPICE .cir file:

```
Q1 1 2 3 QNL 1.0
Q2 5 2 3 QNL 1.5
Q3 6 2 3 QNL 2.0
.model QNL NPN BF=100
```

17–15. Determine I_{D2} for the MOS current mirror shown in Figure 17–66 given the following fabrication parameter information. The *W/L* ratios for the transistors are specified on the circuit diagram.

NMOS	PMOS
$KP = 5.1 \times 10^{-5}$ A/V^2	$KP = 2.5 \times 10^{-5}$ A/V^2
$Vt_n = 0.7$ V	$Vt_p = -0.8$ V

SECTION 17–5

The Gain Stage

17–16. What is the approximate gain (V_{out}/V_{in}) for the circuit shown in Figure 17–67?

17–17. What is the approximate gain (V_{out}/V_{in}) for the circuit shown in Figure 17–68? $I_C = 1$ mA.

17–18. Determine the gain, V_{out}/V_{in}, for the circuit shown in Figure 17–69. The bias voltage for Q_1 is 0.65 V. $I_{S_{Q1}} = 5.8 \times 10^{-14}$, $V_{A_{Q1}} = 100$, $I_{S_{Q2}} = 1.5 \times 10^{-15}$ A and $V_{A_{Q2}} = 80$. Assume $V_T = 0.026$ V.

17–19. (a) Determine the voltage gain for the circuit shown in Figure 17–70 $Vt_n = 0.75$ V, $V_{GS} = 1.1$ V.

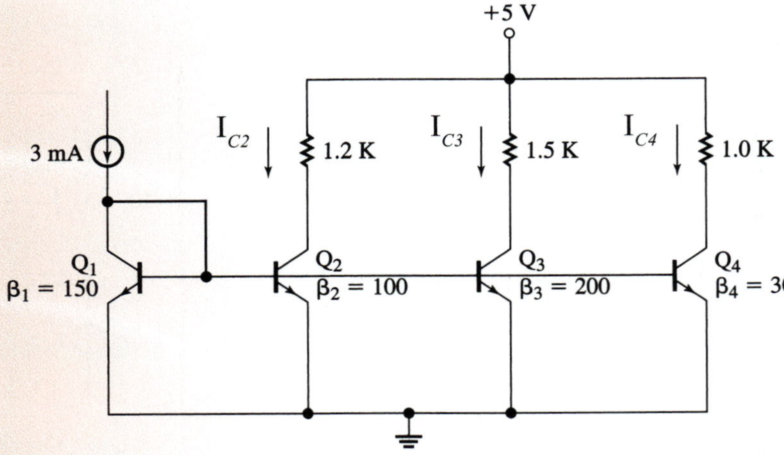

FIGURE 17–62 (Exercise 17–11)

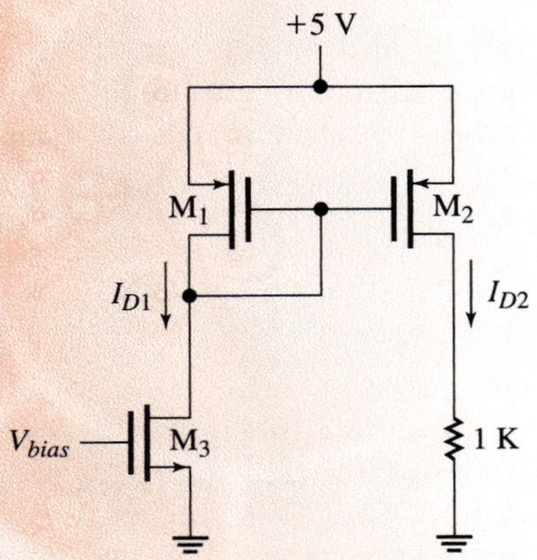

FIGURE 17–63 (Exercise 17–12)

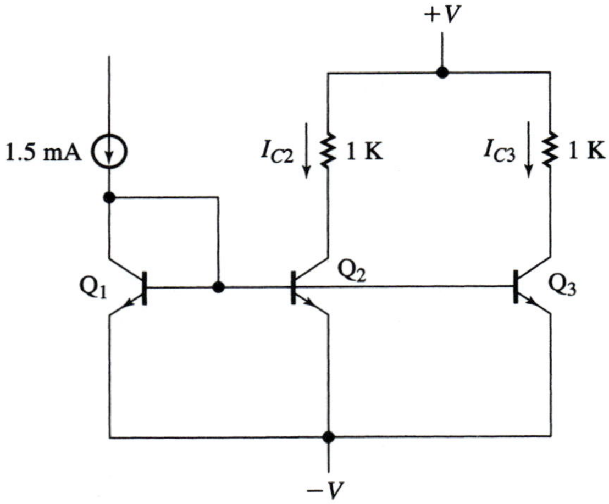

FIGURE 17–65 (Exercise 17–14)

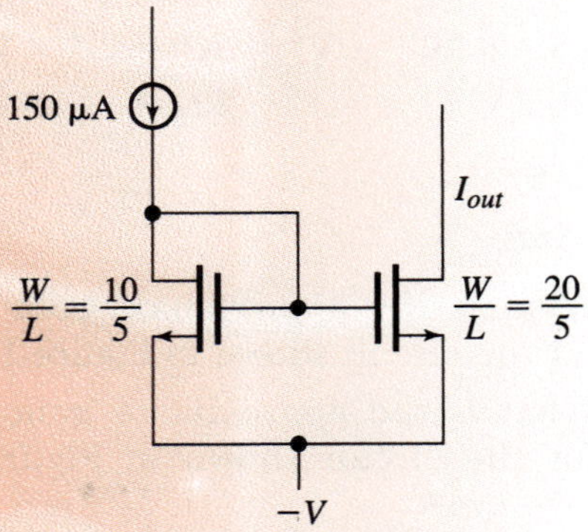

FIGURE 17–64 (Exercise 17–13)

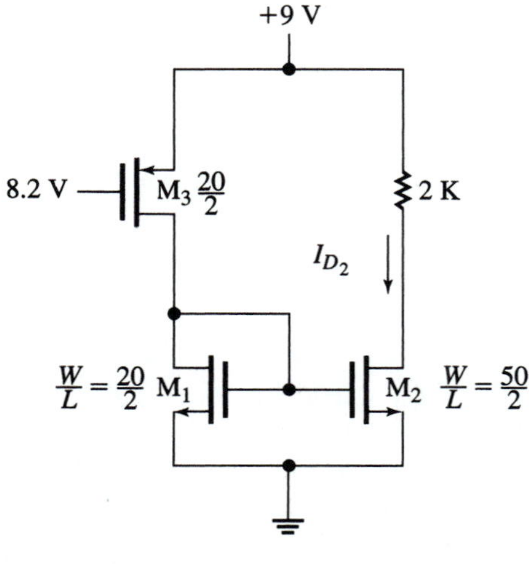

FIGURE 17–66 (Exercise 17–15)

(b) What is the voltage gain if $V_{GS} = 1.5\text{V}$?

(c) What is the voltage gain if $V_{GS} = -1.1\text{V}$?

17–20. Determine the required value for V_{GS} given the MOSFET gain stage shown in

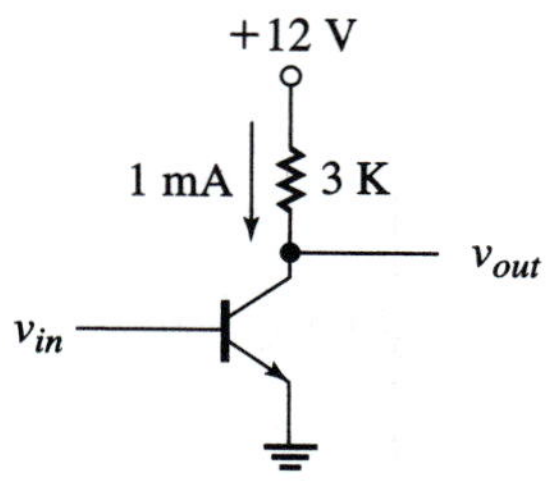

FIGURE 17–67 (Exercise 17–16)

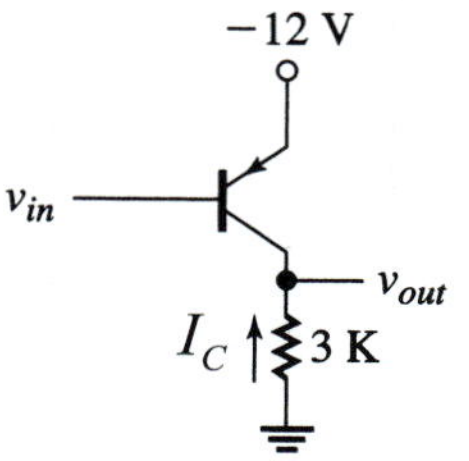

FIGURE 17–68 (Exercise 17–17)

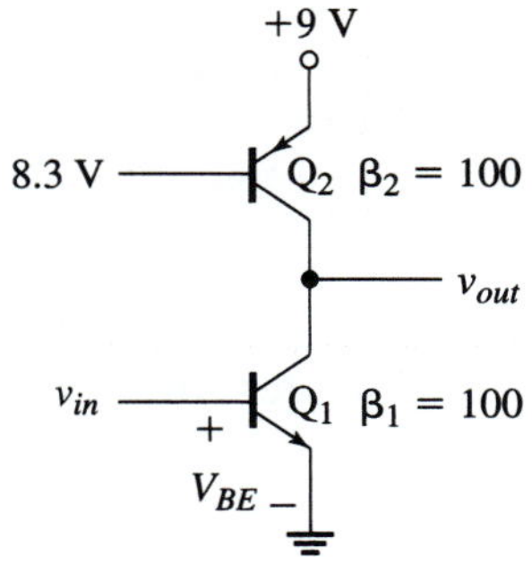

FIGURE 17–69 (Exercise 17–18)

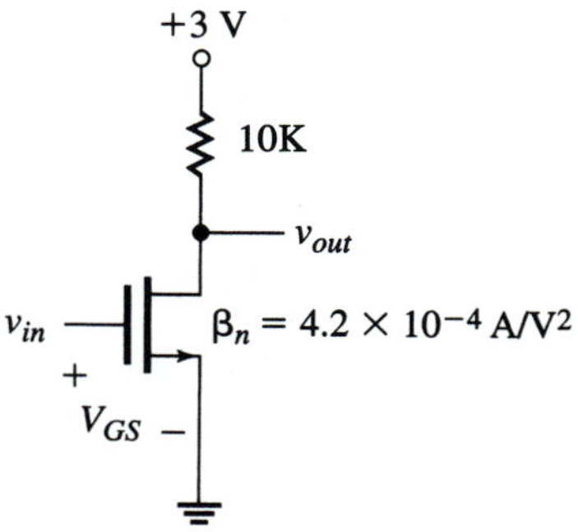

FIGURE 17–70 (Exercise 17–19)

Figure 17–71, if $I_D = 500\ \mu A$, $\beta_p = 5.5 \times 10^{-5}$ A/V^2, and $Vt_p = -0.8$ V.

17–21. Given the circuit shown in Figure 17–72 and the fabrication parameters listed, answer the following questions. $V_{bias} = 10.6$ V.

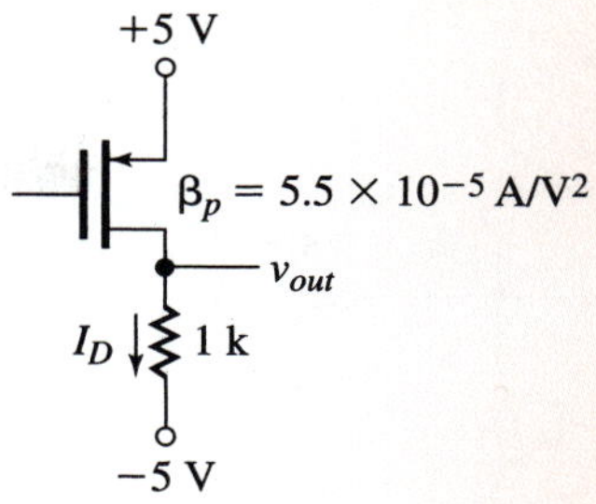

FIGURE 17–71 (Exercise 17–20)

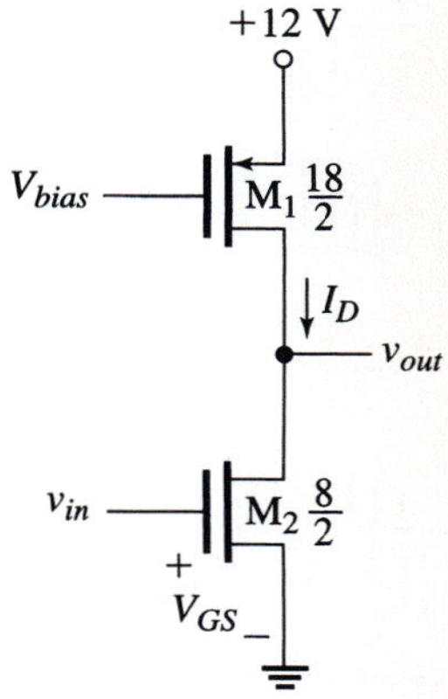

FIGURE 17–72 (Exercise 17–21)

NMOS	PMOS
$KP_n = 4.9 \times 10^{-5}$ A/V^2	$KP_p = 2.5 \times 10^{-5}$ A/V^2
$Vt_n = 0.7$ V	$Vt_p = -0.7$ V
$\lambda_n = 1.6 \times 10^{-2}$	$\lambda_p = 5.8 \times 10^{-2}$

(a) Determine I_D.

(b) Calculate the output resistance for M_1 and M_2.

(c) Determine β_n for M_2.

(d) Determine the required V_{GS} voltage for M_2 so that the current I equals the value found in part (a).

(e) Determine g_{m2}.

(f) Determine the voltage gain A_V.

17–22. Determine the voltage gain for the BJT gain stage shown in Figure 17–73. $V_T = 0.027$ V.

17–23. (a) Determine g_m for each of the MOSFET transistors for the gain stage shown in Figure 17–74. Assume $Vt_n = |Vt_p| = 0.8$ V, $V_{GS1} = 1.52$ V, $V_{GS2} = 1.72$ V, $\beta_n = 32.3$ E–05 A/V^2, and $\beta_p = 16.2$ E–05 A/V^2.

(b) Determine the voltage gain.

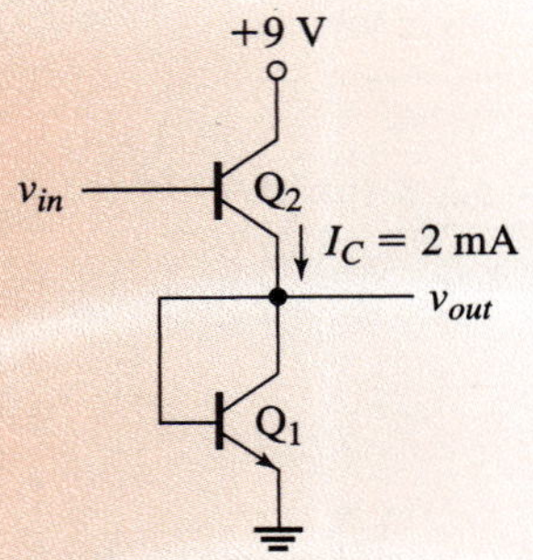

FIGURE 17–73 (Exercise 17–22)

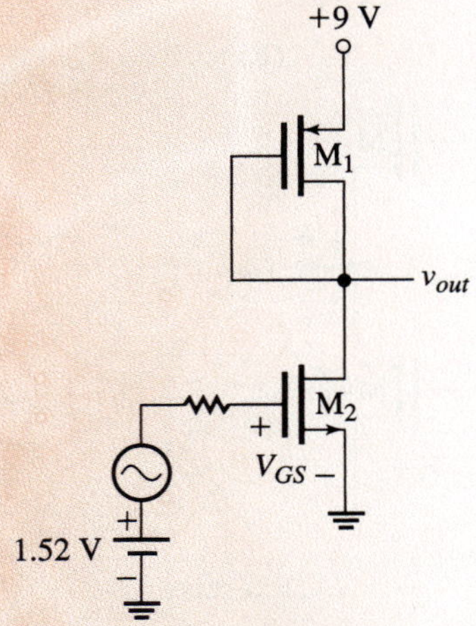

FIGURE 17–74 (Exercise 17–23)

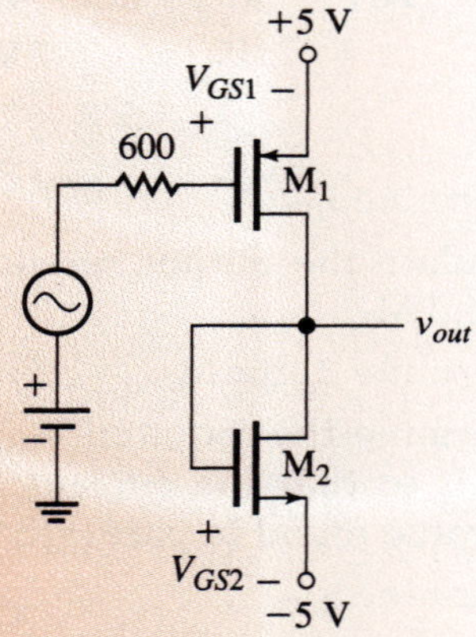

FIGURE 17–75 (Exercise 17–24)

17–24. The MOSFET gain stage in Figure 17–75 has the following parameters:

NMOS	PMOS
$Vt_n = 0.7$ V	$Vt_p = -0.8$ V
$V_{GS2} = 1.3$ V	$\lvert V_{GS1}\rvert = 1.5$ V
$\beta_n = 18.5$ E–05 A/V^2	$\beta_p = 29.2$ E–05 A/V^2

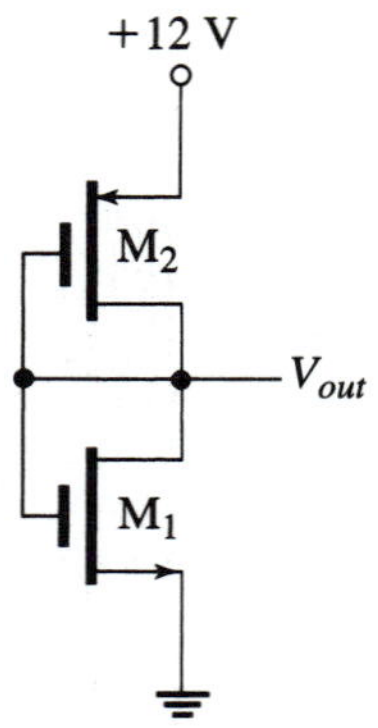

FIGURE 17–76 (Exercise 17–25)

(a) Determine g_m for each transistor.

(b) Determine the voltage gain.

17–25. For the voltage divider shown in Figure 17–76, determine V_{out}.

NMOS	PMOS
$Vt_n = 0.7$ V	$Vt_p = -0.75$ V
$\beta_n = 50$ E–05 A/V^2	$\beta_p = 100$ E–05 A/V^2

17–26. Design a MOSFET voltage divider ($V_{DD} = 5.0$ V) which provides an output voltage of

(a) 3.2 V,

(b) 1.8 V, and

(c) 2.5 V.

Given:

NMOS	PMOS
$Vt_n = 0.7$ V	$Vt_p = -0.7$ V
$KP_n = 4.9$ E-05 A/V^2	$KP_p = 2.52$ E–05 A/V^2
$L_n = 2\mu$	$L_p = 2\mu$

Specify the *W/L* ratios for the MOSFET transistors in your solution.

17–27. Determine the output voltage for the MOSFET voltage divider specified by the following SPICE .cir file:

```
M1 2 2 5 5 CMOSP W=520U L=2U
M2 2 2 0 0 CMOSN W=400U L=2U
VDD 5 0 5V
.MODEL CMOSN NMOS LEVEL=2 VTO=0.7
+ KP=5.1E-05 LAMBDA=1.6E-02
.MODEL CMOSP PMOS LEVEL=2 VTO=-0.8
+ KP=2.8E-05 LAMBDA=5.6E-02
```

SECTION 17–6

Differential Amplifiers

17–28. The voltage gain of each transistor in the ideal differential amplifier shown in Figure 17–28 is $v_o/v_{be} = -160$. If v_{i1} is a 40-mV-peak sine wave and $v_{i2} = 0$, find

(a) the peak value of v_{e1},

(b) the peak value of v_{o1},

(c) the peak value of v_{o2}, and

(d) the voltage gain v_{o1}/v_{i1}.

17–29. A 40-mV-peak sine wave that is out of phase with v_{i1} is applied to v_{i2} in the amplifier in Exercise 17–28. Find

(a) the single-ended voltage gains $v_{o1}/(v_{i1} - v_{i2})$ and $v_{o2}/(v_{i1} - v_{i2})$, and

(b) the double-ended voltage gain $(v_{o1} - v_{o2})/(v_{i1} - v_{i2})$.

17–30. Repeat Exercise 17–29 if the signal applied to v_{i2} is a 40-mV-peak sine wave that is *in phase* with v_{i1}. (Think carefully.)

17–31. The ideal BJT differential amplifier shown in Figure 17–26 is biased so that 0.75 mA flows in each emitter. If R_C = 9.2 kΩ, find

(a) the single-ended voltage gains $v_{o1}/(v_{i1} - v_{i2})$ and $v_{o2}/(v_{i1} - v_{i2})$, and

(b) the double-ended voltage gain $(v_{o1} - v_{o2})/(v_{i1} - v_{i2})$.

17–32. If the transistors in the differential amplifier of Exercise 17–31 each have $\beta = 120$, find the differential input resistance of the amplifier.

17–33. The β of each transistor in the ideal differential amplifier shown in Figure 17–77 is 100. Find

(a) the dc output voltages V_{o1} and V_{o2},

(b) the double-ended voltage gain, and

(c) the differential input resistance.

17–34. The current source in Exercise 17–33 is changed to 1 mA. If v_{i1} is a 16-mV-peak sine wave and $v_{i2} = 0$, find the peak values of v_{o1} and v_{o2}.

17–35. The FETs in the ideal differential amplifier shown in Figure 17–78 have I_{DSS} = 10 mA and $V_p = -2$ V. Find

(a) the dc output voltages V_{o1} and V_{o2};

(b) the single-ended output gain $v_{o1}/(v_{i1} - v_{i2})$; and

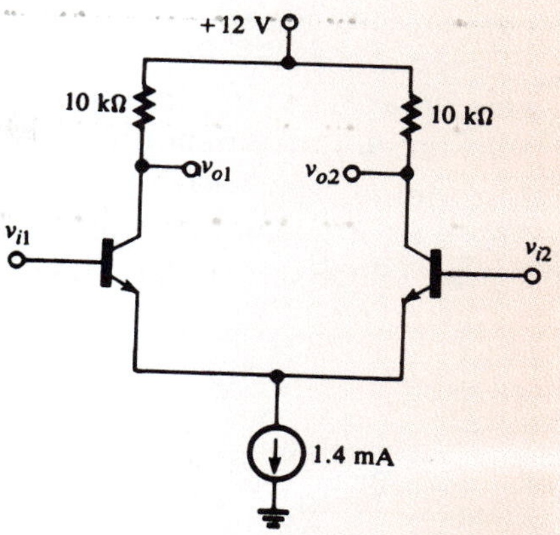

FIGURE 17–77 (Exercise 17–33)

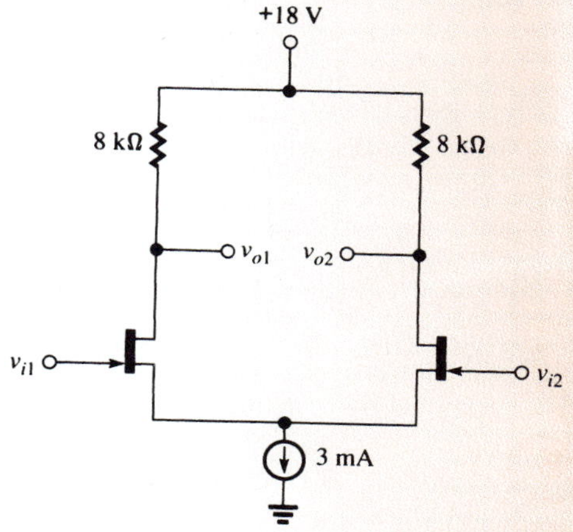

FIGURE 17–78 (Exercise 17–35)

(c) the output difference voltage when the input difference voltage is 50 mV rms.

17–36. The inputs to the differential amplifier in Exercise 17–35 are v_{i1} = 65 mV rms and v_{i2} = 10 mV rms. v_{i1} and v_{i2} are in phase. Find the rms values of v_{o1} and v_{o2}.

17–37. A differential amplifier has CMRR = 68 dB and a differential mode gain of 175. Find the rms value of the output difference voltage when the common-mode signal is 1.5 mV rms.

17–38. The noise signal common to both inputs of a differential amplifier is 2.4 mV rms. When an input difference voltage of 0.1 V rms is applied to the amplifier, each output must have 4 V rms of signal level. Assuming that the amplifier produces no noise and that the noise component in the output difference voltage must be no greater than 500 μV rms, what is the minimum CMRR, in dB, that the amplifier should have?

17–39. Determine the voltage gain for the MOSFET differential amplifier described by the SPICE .cir file provided.

```
M1 2 4 5 10 CMOSN W=45u L=10u
M2 9 6 5 10 CMOSN W=45u L=10u
M3 2 2 1 1 CMOSP W=25u L=10u
M4 9 2 1 1 CMOSP W=25u L=10u
M5 5 7 10 10 CMOSN W=75U L=2U
VMINUS 6 0 SIN(0 .1 100K)
VPLUS 4 0 DC 0
VBIAS 7 0 -4.5v
VDD 1 0 6V
VSS 10 0 -6
.MODEL CMOSN NMOS LEVEL=2 LD=0.25U
+ VTO=0.7
KP=5.2E-05
LAMBDA=.01
.MODEL CMOSP PMOS LEVEL=2
+ LD=0.249723U
VTO=-0.7 KP=2.1E-05
LAMBDA=.02
.DC Vminus -5 5 .1
.TRAN 0.01NS 100uS
.OP
.PROBE
.END
```

17–40. Find the voltage gain for the MOSFET differential amplifier shown in Figure 17–79. The *W/L* ratios are specified for each transistor; use the following fabrication parameters:

NMOS	PMOS
VTO= 0.75	VTO=−0.75
KP=5.4E–05	KP=2.0E–05
LAMBDA=.015	LAMBDA=.022

17–41. Describe the operation, in terms of current flow, for the MOSFET differential amplifier shown in Figure 17–80 for the following conditions and identify the V^+ and V^- inputs:

(a) $V_1 = V_2$,

(b) $V_1 > V_2$, and

(c) $V_1 < V_2$.

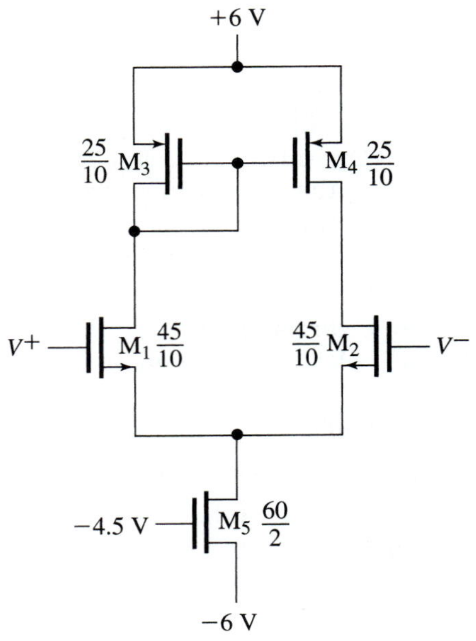

FIGURE 17–79 (Exercise 17–40)

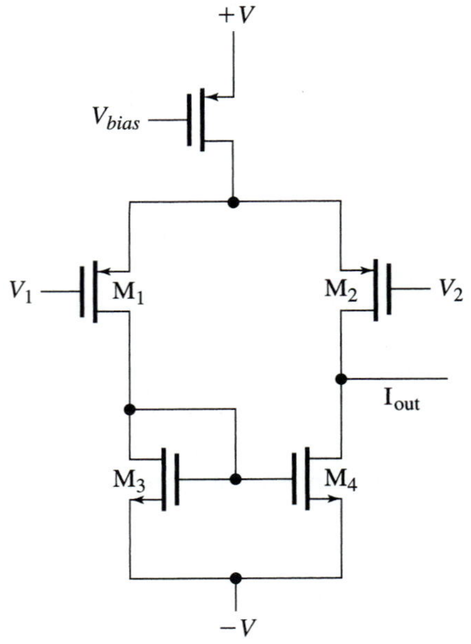

FIGURE 17–80 (Exercise 17–41)

SPICE PROBLEMS

17–42. Verify the analytical answers to Exercise 17–15 using SPICE analysis.

17–43. Simulate the circuit design in Exercise 17–20 using the specified model parameters. Verify that the value obtained for V_{GS} produces an I_D of 500 μA.

17–44. Use SPICE to determine the gain for the circuit in Exercise 17–24.

NMOS	PMOS
KP = 3.7 E–05	KP = 2.92 E–05
W = 5μ	W = 20μ
L = 2μ	L = 2μ
VTO = 0.7	VTO = −0.8

17–45. Use SPICE to verify your answers to Exercise 17–26.

17–46. Perform SPICE analysis on the Miller op-amp shown in Figure 17–44. Show that you can generate transient and ac analysis results comparable to the results shown in Section 17–7. Include the compensation capacitor in your simulations. Use the model parameters provided in Section 17–7.

CHAPTER 18

INTRODUCTION TO DIGITAL VLSI DESIGN

OUTLINE

OBJECTIVES

- Explore the use of MOSFET transistors to create modern digital logic circuits
- Demonstrate a methodology to create CMOS combinational logic circuits
- Investigate how to perform simulation analysis on complex logic circuits.
- Investigate how signals are clocked into and out of CMOS logic circuits
- Develop an understanding of memory structures using transistors
- Explore the characteristics of buffering input and output stages

18–1 INTRODUCTION

Chapters 4, 5, 7, and 17 address the use of transistors in analog applications. Why discuss digital VLSI design in an analog circuits textbook? Because more and more integrated systems incorporate a mixed-signal (analog/digital) environment. It is possible to perform digital logic design without a strong background in knowledge of transistors, but digital circuits are not just devices that turn on and off. They pass through all the operating modes of a transistor each time they transition from High to Low or Low to High. Modern digital VLSI circuits can contain millions of integrated transistors. Many people believe that digital and analog circuits incorporate two completely different design philosophies at the transistor level. Although on the surface this statement may appear to be correct, a quick comparison of the design philosophies demonstrates that the two methods are more similar than you might believe. Table 18–1 provides a simple comparison of the concerns in the two design methods. We do not mean to imply that a good analog designer will also be good at digital design, or vice versa, but the person who thoroughly understands transistor behavior will better interface with both methodologies and with future IC developments.

The examination of transistors in digital circuits will complete our introduction to the transistor. The topics in Chapter 18 address digital VLSI design issues as implemented with complementary metal-oxide semiconductors (CMOS) devices. This chapter will further develop an understanding of the use of the transistor as a switch in logic systems. The reader was briefly introduced to the use of a very useful transistor circuit, the BJT switch, in Chapter 4, where the technique was primarily based on the use of the BJT in driving slow-responding devices such as relays and solenoids. Section 18–2 examines the functionality of MOSFET transistors when they are connected in series and in parallel. These configurations are

TABLE 18–1 A quick comparison of digital and analog circuit design issues

Analog	Digital
Slew rate	Rise- and fall-time characteristics
Load considerations	Drive capability
Bandwidth	Switching speeds
Push-pull amplifier	The inverter
Problems with parasitics	Problems with parasitics

the fundamental building blocks for full CMOS digital logic circuits. The construction and operation of a CMOS transmission gate is also examined. A design procedure for creating CMOS combinational logic circuits is introduced in Section 18–3. Section 18–4 examines the transient behavior of a CMOS transistor-level structure through an in-depth PSPICE analysis. Clocked circuits are examined in Section 18–5, followed by a discussion of transistors in random-access memory (RAM) in Section 18–6. Interestingly, RAM is both digital (storage) and analog (read/write). High-speed RAM requires the use of sophisticated sensing amplifiers, which are complex analog amplifiers. Section 18–7 details how digital VLSI circuits interface to the outside world via input/output (I/O) pads, which are examined in sufficient detail to give the reader a better understanding of logic-level conversion, device protection, and drive capability.

18–2 TRANSISTOR-LEVEL IMPLEMENTATION OF CMOS COMBINATIONAL LOGIC CIRCUITS

Complementary metal-oxide semiconductor (CMOS) logic devices are the most common devices used today in the high-density, large-transistor-count circuits found in everything from complex microprocessor integrated circuits to signal processing and communication circuits. The CMOS structure is popular because of its inherently lower power requirements, high operating clock speed, and ease of implementation at the transistor level. You can gain insight into the operation of these CMOS devices through a brief introduction to constructing simple CMOS combinational logic circuits such as AND/NAND gates and OR/NOR gates. These circuits are created using both *p*- and *n*-channel metal-oxide semiconductor field effect transistors (MOSFETs) connected in complementary configurations. The simple CMOS inverter also requires one *p*-channel and one *n*-channel MOSFET transistor connected in complementary operation.

The complementary *p*-channel and *n*-channel transistor networks are used to connect the output of the logic device to either the V_{DD} or V_{SS} power-supply rails for a given input logic state. At a simplified level, MOSFET transistors can be treated as simple switches. This is adequate for our initial introduction to simple CMOS circuits. Switching speeds, propagation delays, drive capability, and rise and fall times are addressed, in Section 18–4, after you gain insight into logic device construction.

The MOSFET Transistor in Digital Circuits

Schematically, MOSFET transistors are typically identified using one of three symbols, which are shown in Figure 18–1 for both *n*-channel (NMOS) and *p*-channel (PMOS) devices. The MOSFET schematic symbols in Figure 18–1(a) show the drain (D), gate (G), source (S), and bulk (B)

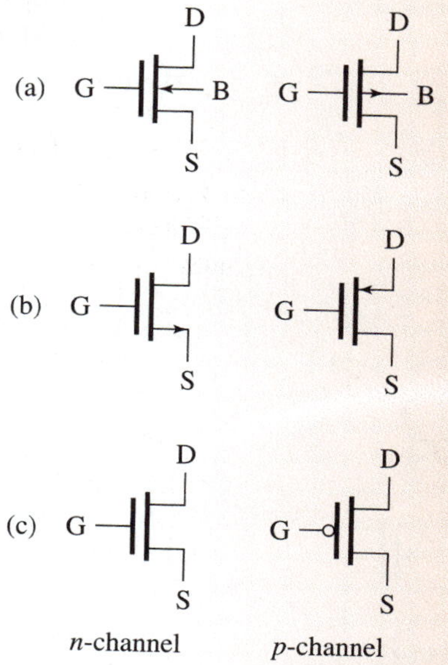

FIGURE 18–1 Schematic symbols for the MOSFET transistor

connections for the transistor. The bulk, also called the *bulk-substrate* or *substrate,* is shown unconnected in the figure, but it must be properly connected before power is applied.

> MOSFET Rule Number 1: The bulk connections for MOSFET transistors should be connected to a power supply rail when practical. *p*-channel bulk connections are typically tied to the V_{DD} rail, and *n*-channel bulk connections are typically tied to the V_{SS} rail.

The MOSFET schematic symbols in Figure 18–1(b) show the symbols for the *p*- and *n*-channel MOSFET transistors when the source–bulk connection has been shorted ($V_{SB} = 0.0$ V). These symbols are most commonly used in documenting analog CMOS circuits. The MOSFET schematic symbols shown in Figure 18–1(c) show the schematic symbols for *p*- and *n*-channel MOSFET transistors. In this case, the bulk–substrate connection is not indicated. Notice too that the gates for the *p*- and *n*-channel devices differ. The *p*-channel device is identified by a "bubble" on the gate input: the *n*-channel device does not have a "bubble." The presence or absence of a bubble on the gate input is used to signify what logic level is best used to turn on that particular transistor style. The presence of a bubble on the *p*-channel device indicates that it should have a logic Low applied to the gate input to turn on the transistor, whereas the absence of a bubble on the *n*-channel device indicates that it should have a logic High applied to the input to turn on the device. These schematic symbols are most commonly used when documenting CMOS logic circuits. The bulk–substrate connections are almost always connected to the power supply rails using MOSFET Rule Number 1.

A MOSFET transistor has three major regions (modes) of operation: *cutoff, saturation (pinch-off), and ohmic (or triode).* Some digital VLSI texts refer to the ohmic region as the *nonsaturated* region. In the ohmic region, the voltage drop across the drain–source terminals approaches zero as the magnitude of the voltage across the gate–source terminals approaches $V_{DD} - V_{SS}$. For example, in a 5-V system, the drain–source voltage approaches zero V as the magnitude of the gate–source voltage approaches

5 V. In the cutoff region, the drain-to-source current, I_{DS}, approaches zero (i.e., the drain–source resistance approaches infinity—an open circuit). Hence, the drain and source terminals of a MOSFET transistor can be treated as an ideal switch alternating between the OFF (cutoff) and ON (nonsaturated) modes of operation. However, there is a limitation on the use of MOSFET transistors as ideal switches:

> MOSFET Rule Number 2: For proper operation as an ideal switch, the *p*-channel MOSFET transistor or network must be connected to the most positive voltage rail and the *n*-channel MOSFET transistor or network must be connected to the most negative voltage rail.

The AND/OR and Inverter Structure

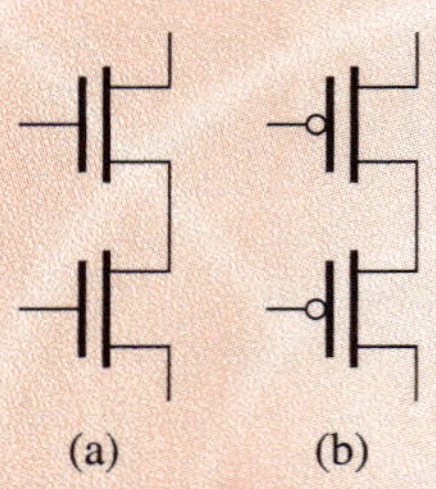

FIGURE 18–2 (a) NMOS AND structure; (b) PMOS AND structure

Creating AND and OR structures using MOSFET transistors is easily accomplished by placing the NMOS and PMOS transistors either in series (AND) or parallel (OR), as shown in Figures 18–2 and 18–3. Shown in Figure 18–2(a) and (b) are two MOSFET transistors connected in series. The singular current path in both structures defines the AND operation. Shown in Figure 18–3(a) and (b) are two MOSFET transistors connected in parallel. The parallel current paths represent the OR structure.

Figure 18–4 shows an NMOS AND structure with the source of MOSFET transistor M_1 connected to ground according to MOSFET Rule Number 2. Recall that an NMOS switch is turned ON when a logic High is applied to the gate input. The logic expression for the circuit shown in Figure 18–4 is $F = (A \cdot B)_{-L}$, meaning that the output F is Low if A and B are High. This is called the *analogous structure.* If gate inputs A and B are a logic High, then the output node of the AND structure will have a path to ground (a logic Low). If either input A or B is a logic Low, then there will not be a path to ground because both MOSFET transistors will not be turned ON. In CMOS technology, a complementary transistor structure is required to connect the output node to the opposite power supply rail (in this case, the V_{DD} rail). The expression and transistor configuration for the complementary structure is obtained by applying DeMorgan's theorem to the expression defining the analogous structure. A method for creating the full CMOS transistor structure is described in the design procedure (Section 18–2).

Creating a CMOS inverter requires only one PMOS and one NMOS transistor. The NMOS transistor provides the path to ground when the input is a logic High, and the PMOS device provides the path to the V_{DD} power-supply rail when the input to the inverter circuit is a logic Low. This is consistent with MOSFET Rule Number 2. The transistor configuration for a CMOS inverter is shown in Figure 18–5.

Another important building block in CMOS circuits is the transmission gate. The transmission gate is used to pass a signal from input to output.

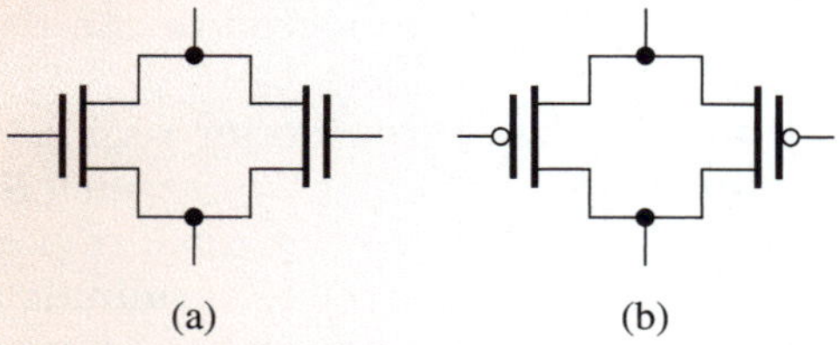

FIGURE 18–3 (a) NMOS OR structure; (b) PMOS OR structure

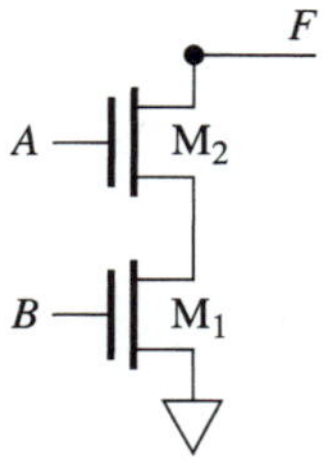

FIGURE 18–4 An NMOS transistor structure realizing the expression $F = (A \cdot B)_{-L}$

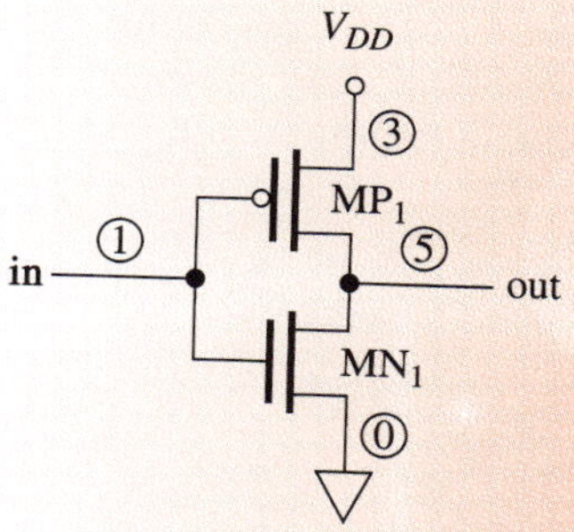

FIGURE 18–5 The transistor view of a CMOS inverter

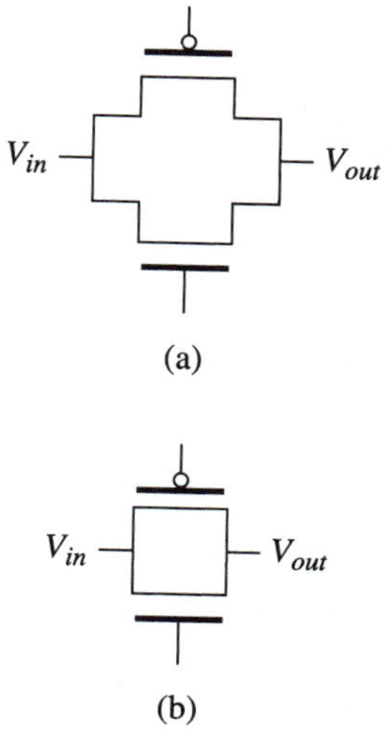

FIGURE 18–6 (a) The transistor view of the CMOS transmission gate, (b) An alternative schematic symbol for the CMOS transmission gate

Digital designers often use the CD4016 or CD4066 IC analog switches or similar packages, which are constructed of transmission gates. The transistor view for the transmission gate is shown in Figure 18–6(a); an alternative symbol for the transmission gate is shown in Figure 18–6(b).

CMOS transmission gates require the parallel connection of the PMOS and NMOS transistors so that a rail-to-rail input voltage can be passed. For a 5.0-V system, the CMOS transmission gate will pass voltages from 0.0 V to 5.0 V. When used individually as a pass transistor, MOSFET transistors will not pass a voltage from rail to rail. NMOS transistors will pass a maximum voltage of $V_{DD} - V_{tn}$, and PMOS transistors will pass a minimum voltage of $|V_{tp}|$. This is due to the limitation in the V_{GS} voltage required for establishing a channel from drain to source in the MOSFET transistor. For practical purposes, the CMOS transmission gate can be treated as an ideal switch capable of switching rail-to-rail voltages.

EXAMPLE 18–1

Design a 2-input multiplexer that selects input *A* when CLK is High and input *B* when CLK is Low.

Solution

The circuit is created using two CMOS transmission gates. The CLK lines are connected so that only one transmission gate is enabled at one time. The solution is provided in Figure 18–7.

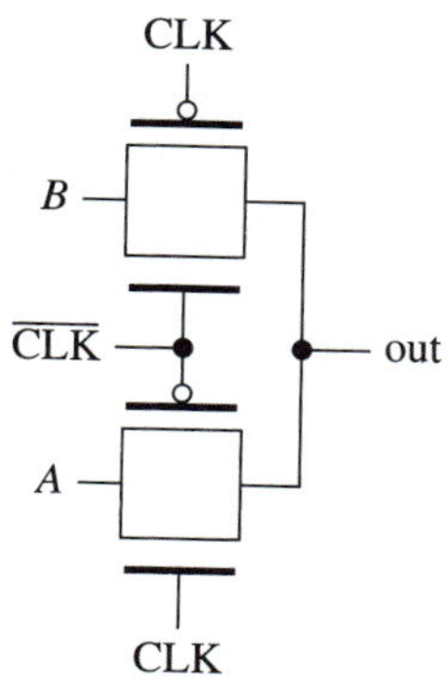

FIGURE 18–7 The 2-input multiplexer for Example 18–1

18–3 A DESIGN PROCEDURE FOR CREATING CMOS COMBINATIONAL LOGIC CIRCUITS

The following design process provides a method for obtaining an optimal CMOS combinational transistor structure given a functional (Boolean) expression. The method is based on the use of mixed logic concepts. The input variables should have a designated assertion level (i.e., assert Low or assert High).

In CMOS designs, two transistor structures (one PMOS and one NMOS) are required for implementing the functional expression. In logic systems, the analogous expression defines what is required to generate the required output assertion level. The complementary expression, obtained by applying DeMorgan's theorem to the functional expression, defines the complementary structure. These two expressions, the analogous and the complementary, are then used to create the transistor network for a CMOS circuit. The design procedure is described in five steps as follows:

1. Identify the "most common" input level by examining the input assertion levels. This requires that the input assertion levels be defined. An input variable containing a conflict is treated as if it has the opposite assertion

TABLE 18–2 Defining transistors and the most common input level

Most Common Input Level	Analogous Structure
Low	PMOS transistors are used to create the analogous structure
High	NMOS transistors are used to create the analogous structure

level. The most common input level will be either Low or High. This is determined by counting the number of asserted High or asserted Low inputs after adjusting for conflicted inputs.

2. The most common input level is used to specify the type of transistors to be used for implementing the analogous structure. See Table 18–2.
3. If there is not a most common input level, then select the input level to be the opposite level of the required output assertion level.
4. a. Create the analogous transistor structure directly from the functional logic expression. Use the transistor type specified in step 2 for creating the structure.

   b. The complementary structure is created by applying DeMorgan's theorem to the analogous expression. The transistor type is opposite to that used in step 4(a).
5. Assemble the analogous and complementary structures to create the full CMOS-equivalent circuit. In some cases, an inverter must be added to the output of the circuit to correct the output assertion level.

EXAMPLE 18–2

Given: $F = (A \cdot B)_{-L}$. Both inputs A and B are defined to assert High while the output is defined to assert Low. This expression reads, "The output is asserted Low when inputs A and B are both asserted."

Solution

Step 1: Determine the most common input level. Inputs A and B both assert High and neither input has a conflict; therefore, the most common input level is High. *Step 2:* NMOS transistors are to be used to create the analogous structure. Notice that this is an AND-type structure. The NMOS transistors are connected in series to ground, as shown in Figure 18–8(a). *Step 4:* Applying DeMorgan's theorem to the functional expression yields $\overline{F} = \overline{A} + \overline{B}$. In this case, PMOS transistors are used to create the complementary structure. The PMOS complementary circuit is an OR structure with the PMOS transistors providing the switch connection to the V_{DD} rail. The complementary structure is shown in Figure 18–8(b). *Step 5*: The completed CMOS circuit is shown in Figure 18–9. The output assertion level (Low) is correct.

EXAMPLE 18–3

Given: $F = (A \cdot B)_{-H}$. Inputs A and B are defined to assert High and the output is defined to assert High. This expression reads, "The output is asserted High when both inputs A and B are asserted."

Solution

Step 1: Determine the most common input level. Inputs A and B both assert High and neither input has a conflict; therefore, the "most common" input level is High. *Step 2*: NMOS transistors are to be used to create the analogous structure. Notice that this is an AND-type structure. The NMOS transistors

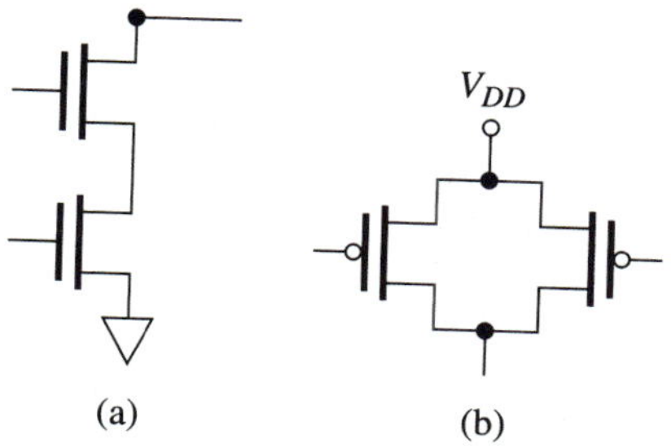

FIGURE 18–8 The (a) analogous and (b) complementary structures for Example 18–2

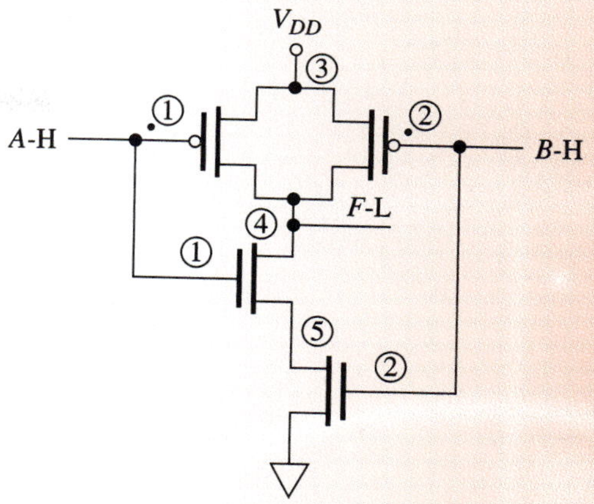

FIGURE 18–9 The complete transistor circuit for realizing the expression $F = (A \cdot B)_{-L}$

are connected in series to ground, as shown in Figure 18–10(a). *Step 4*: Applying DeMorgan's theorem to the functional expression yields $\overline{F} = \overline{A} + \overline{B}$. In this case, PMOS transistors are used to create the complementary structure. The PMOS complementary circuit is an OR structure, with the PMOS transistors providing the switch connection to the V_{DD} rail. The complementary structure is shown in Figure 18–10(b). *Step 5*: The output assertion level must be corrected by adding an inverter to the output. The completed CMOS circuit is shown in Figure 18–11.

The procedures and results for creating the transistor-equivalent circuits in Example 18–2 and Example 18–3 are the same except that the circuit in Example 18–3 required the placement of an inverter on the output to correct the assertion level. The logic circuit created in Example 18–2 is commonly called a *positive logic NAND gate*. The logic circuit created in Example 18–3 is

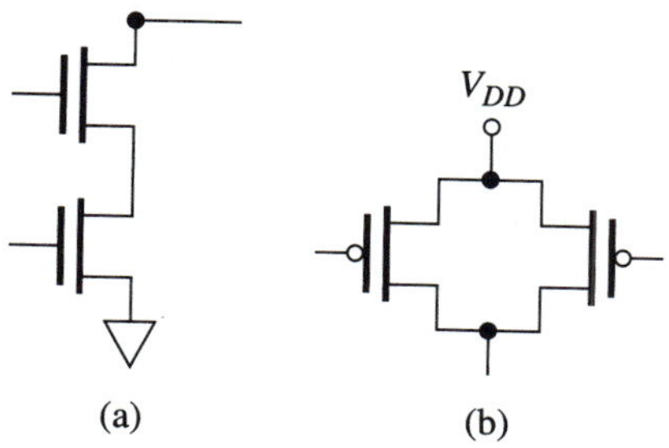

FIGURE 18–10 The (a) analogous and (b) complementary structures for Example 18–3

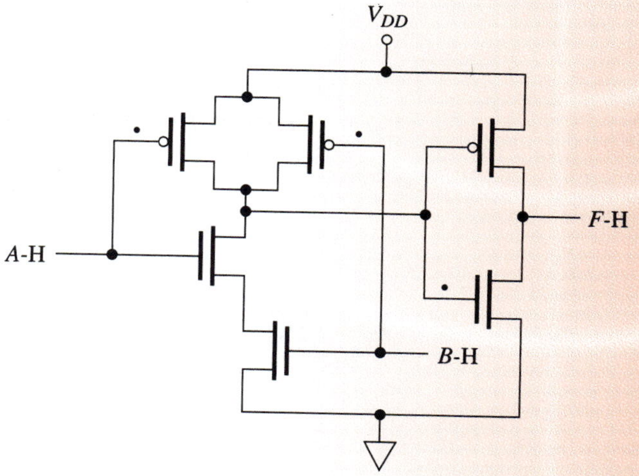

FIGURE 18–11 The complete transistor circuit for realizing the expression $F = (A \cdot B)_{-H}$

commonly called a *positive logic AND gate.* This same procedure can be easily applied to create NOR, OR, XOR, and XNOR gates.

EXAMPLE 18–4

Given: $F = [(A + B) \cdot \overline{C}]_{-H}$. Inputs A and C are defined to assert High and input B is defined to assert Low. The output is defined to assert High. This expression reads, "The output is asserted High when inputs A OR B are asserted AND C is not asserted."

Solution

Step 1: Determine the most common input level. Inputs A and C both assert High. Input C is conflicted; therefore, for the purpose of determining the most common input level, C is treated as a Low input. Input B is defined to be asserted Low and does not contain a conflict; therefore, the most common input level is Low. *Step 2:* PMOS transistors are to be used to create the analogous structure. Notice that the analogous structure contains both an AND and an OR structure. The PMOS transistors are connected to the V_{DD} rail as shown in Figure 18–12(a). Input A is defined to be asserted High, and a PMOS device requires an asserted Low input signal; therefore, the assertion level of A is changed to a Low to avoid a conflict. This is consistent with mixed-logic methods. *Step 4:* Applying DeMorgan's theorem to the functional expression yields $\overline{F} = \overline{A} \cdot \overline{B} + C$. In this case, NMOS transistors are used to create the complementary structure. The NMOS complementary circuit contains both an AND and an OR structure, with the NMOS transistors providing the switch connection to the ground rail. The complementary structure is shown in Figure 18–12(b). *Step 5:* The output assertion level will be High when the required input assertion levels are met. An inverter on the output is not required. The completed CMOS circuit is shown in Figure 18–13. The 2 inside the inverter symbol indicates that 2 transistors were required to change the assertion level of input A.

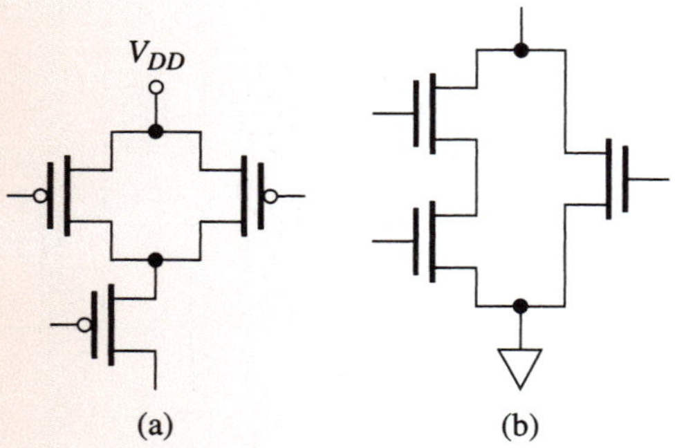

FIGURE 18–12 (a) the analogous and (b) complementary structure

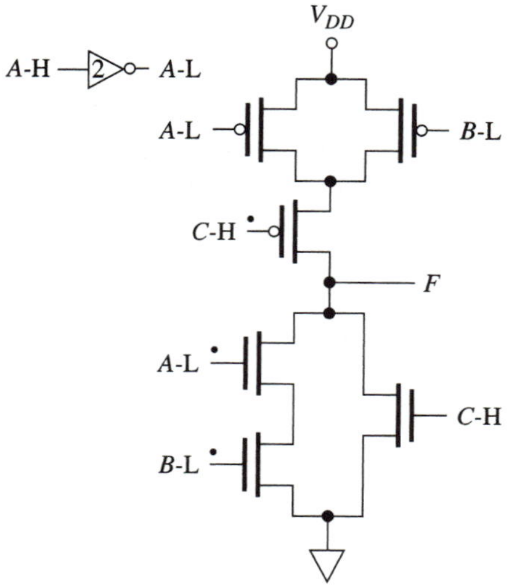

FIGURE 18–13 The complete transistor circuit for realizing the expression $F = (A + B) \cdot C_{-H}$

Constructing a CMOS Logic Circuit Using the CD4007 Transistor Array Package

Once the logic circuit is designed and verified with SPICE, a hardware circuit can be created using the CD4007 CMOS transistor array package. The CD4007 contains six transistors—three PMOS and three NMOS transistors, which include an inverter pair. The transistors are accessible via the 14-pin DIP terminals. A connection diagram and a schematic of the package are provided in Figure 18–14. Proper bulk–substrate connections are already made in the device package. A NAND gate (see Figure 18–9) can be created using the CD4007 by making the connections as shown in Figure 18–15. Notice that the gate connections are shared at pins 6 and 3. This is a convenient option for creating a CMOS combinational logic circuit.

The design technique provides a systematic method for designing and constructing any reasonably sized CMOS combinational circuit device. The technique assumes that MOSFET devices operate as ideal switches with only an ON and OFF mode. For simple circuits, the omission of the switching transient behavior is acceptable as long as information regarding the operating speed, propagation delay, and drive capability is not needed. This issue is addressed in Section 18–4.

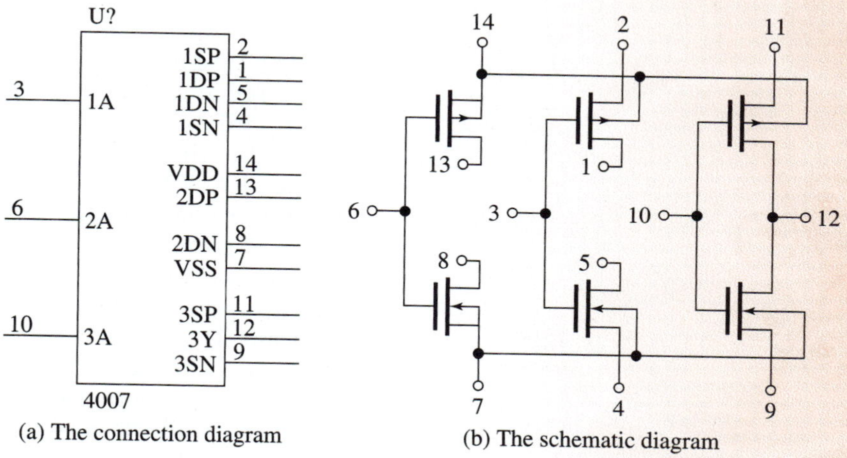

FIGURE 18–14 The CD4007 transistor array package

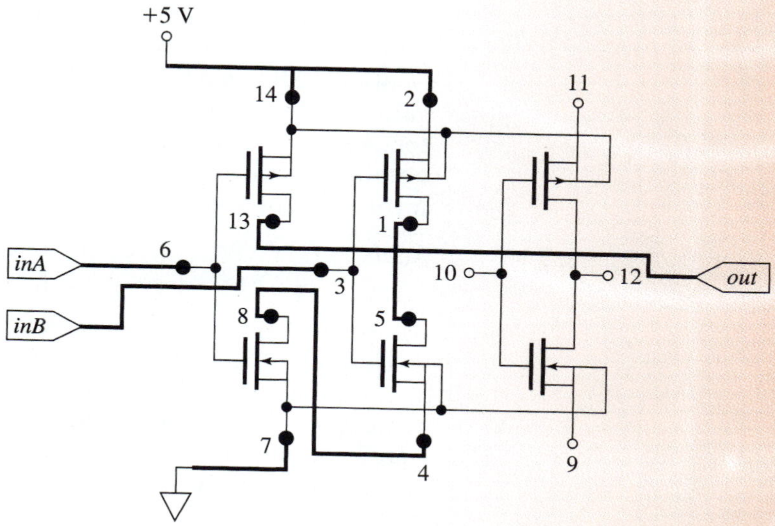

FIGURE 18–15 An implementation of a CMOS NAND gate using the CD4007

18–4 TRANSIENT BEHAVIOR OF CMOS LOGIC CIRCUITS

In this section, we examine the following transient behaviors of CMOS circuits—switching characteristics, switching speed, drive capability, and propagation delay—using SPICE simulation techniques and analytical treatments. The transient response (rise/fall time, propagation delay, etc.) of integrated circuits can be accurately modeled by incorporating proper fabrication-model parameters, and reasonable estimations of the operating behavior are possible using simplified equations. Some approximations will be used to simplify the calculations. Where appropriate, we have used MOSIS (the *M*etal-*O*xide *S*emiconductor *I*mplementation *S*ervice) fabrication parameters in the simulations. The SPICE simulation of a digital circuit is an analog simulation, which means that you will be observing behaviors comparable to those examined in analog circuits. You will observe the behavior of the transistor networks throughout the entire transition. Although this gives an accurate picture of the circuit's analog behavior, the simulation is not very fast nor practical for logic-level analysis of large-transistor-count circuits.

Transient Behavior of a CMOS Inverter

The MOSFET transistor-level implementation of a CMOS inverter will be the circuit used to examine CMOS-circuit transient behavior. This circuit is shown in Figure 18–5. We will first present SPICE simulations to describe the circuit's switching behavior. Many of the modern simulation software packages contain an excellent GUI (graphical user interface). However, we will present the text (.CIR) file to describe the functions being used in the simulation in a generic form. These functions are typically accessible graphically with the mouse and pull-down windows.

Table 18–3 is a listing of the .CIR used in the simulation of the CMOS inverter. Each transistor is described in terms of its nodal connections. For the PMOS transistor MP_1, and the NMOS transistor MN_1, the nodes are listed in terms of the drain (D), gate (G), source (S), and bulk (B) nodes. The model describing the behavior of the transistor is listed next. CMOSP for MP_1 and CMOSN for MN_1. This is an important reference if you want the simulation of the circuit to behave similarly to a fabricated part. Next follows the width (W) and channel length (L) in microns (μ), which describe the gate size of the transistor. This is a mask-level layout parameter. The channel lengths used in our simulations will be $L = 2\mu$. The channel-length value is dependent on the fabrication process, and submicron lengths are currently in use in many fabrications. AS and AD describe the area of the source and drain regions. These values are used to extract part of the parasitic capacitances for the transistor. AS = 252P comes from multiplying $28\mu \times 9\mu$, where the 9μ value is obtained from the layout of the diffusion source and drain area. The SPICE-level descriptions of the transistors are as follows:

```
MP1 5 1 3 3 CMOSP W=28.0U L=2.0U AS=252P AD=252P
MN1 5 1 0 0 CMOSN W=10.0U L=2.0U AS=90P AD=90P
```

There are other values such as the perimeter of the source (PS) and perimeter of the drain (PD) that could also be listed, but this discussion will be limited to the values just described, which will produce reasonable results for our examination. However, if maximum speed or minimum propagation issues are being examined, then all parameters should be included in the circuit description.

TABLE 18–3 A SPICE simulation of a CMOS inverter

```
* A CMOS Inverter Using 2 Micron Channel Lengths
*
* D G S B
MP1 5 1 3 3 CMOSP W=28.0U L=2.0U AS=252P AD=252P
MN1 5 1 0 0 CMOSN W=10.0U L=2.0U AS=90P AD=90P
VIN 1 0 PWL(0 0 100n 5.0 200n 0)
VDD 3 0 DC 5.0
*
* The following are fabrication parameters obtained
* from the MOSIS service.
.MODEL CMOSN NMOS LEVEL=2 LD=0.121440U TOX=410.000E-10
+ NSUB=2.355991E+16 VTO=0.7 KP=8.165352E-05 GAMMA=1.05002
+ PHI=0.6 UO=969.492 UEXP=0.308914 UCRIT=40000
+ DELTA=0.262772 VMAX=71977.5 XJ=0.300000U LAMBDA=3.937849E-02
+ NFS=1.000000E+12 NEFF=1.001 NSS==0 TPG=1.000000
+ RSH=33.290002 CGDO=1.022762E-10 CGSO=1.022762E-10
+ CGBO=5.053170E-11 CJ=1.368000E-04
+ MJ=0.492500 CJSW=5.222000E-10 MJSW=0.235800 PB=0.490000
* Weff = Wdrawn - Delta_W
* The suggested Delta_W is 0.06 um
*
.MODEL CMOSP PMOS LEVEL=2 LD=0.180003U TOX=410.000E-10
+ NSUB=1.000000E+16 VTO=-0.821429 KP=2.83164E-05 GAMMA=0.684084
+ PHI=0.6 UO=336.208 UEXP=0.351755 UCRIT=30000
+ DELTA=1.000000E-06 VMAX=94306.1 XJ=0.300000U
+ LAMBDA=4.861781E-02 NFS=2.248211E+12 NEFF=1.001 NSS=1.000000E+12
+ TPG=-1.000000 RSH=119.500003 CGDO=1.515977E-10 CGSO=1.515977E-10
+ CGBO=2.273927E-10 CJ=2.517000E-04 MJ=0.528100
+ CJSW=5.222E-10 MJSW=0.246600 PB=0.480000
* Weff= Wdrawn - Delta_W
* The suggested Delta_W is 0.27 um
* .PLOT TRAN V(3) V(2) V(1)
.TRAN .1n 250n
.PROBE
.END
```

A piecewise linear approximation is used to model a ramp to the input to the CMOS inverter using the SPICE PWL option. This is listed as

```
VIN 1 0 PWL(0 0 100n 5.0 200n 0)
```

where at time = 0 s the voltage value is 0 V, at 100 ns the voltage value is 5.0 V, and at 200 ns the value returns to 0 V. This input will enable us to observe the transient behavior of the inverter, including easy observation of the switching points of the device. The SPICE simulation is shown in Figure 18–16. The transient current behavior of the inverter is shown in Figure 18–17.

Figures 18–16 and 18–17 provide our first simulation examination of the switching behavior of MOSFET transistors as configured in a CMOS logic circuit. Figure 18–16 shows how the output of a simple CMOS inverter behaves as the input is varied from 0.0 V to 5.0 V. As the input voltage is varied, the NMOS (MN_1) transistor changes from the cutoff mode to

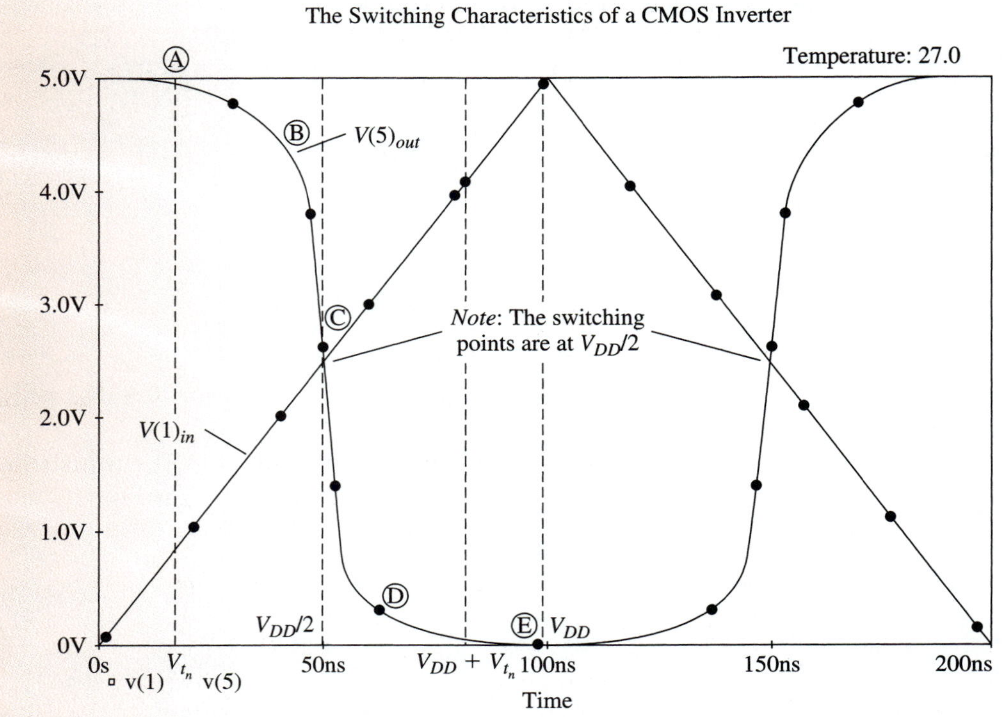

FIGURE 18–16 The SPICE simulation of the switching behavior of a CMOS inverter

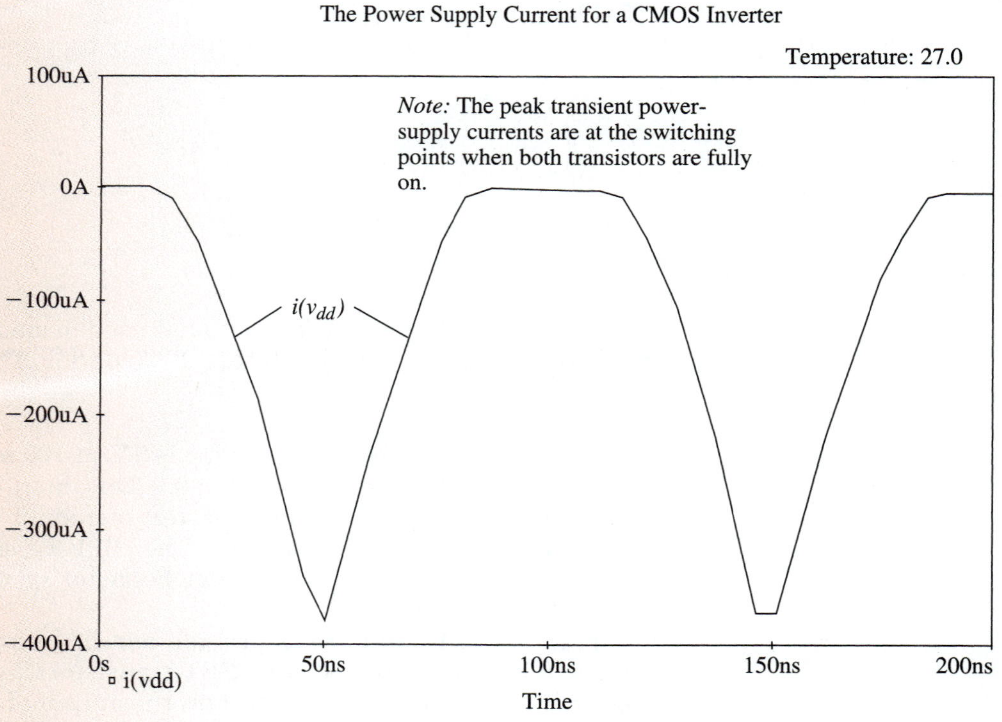

FIGURE 18–17 The SPICE simulation of the transient current behavior of the CMOS inverter

saturation and finally to the ohmic (or triode) region. The operating modes for a MOSFET transistor, the defining conditions, and the current equation for each mode are listed in the following table. The magnitudes are used in the expressions so that the same conditions work for both NMOS and PMOS transistors.

Current Equation	Condition	Mode
0	$\lvert V_{GS} - V_t \rvert \leq 0$	cutoff
$I_{DS} = \frac{\beta}{2}(V_{GS}-V_t)^2$	$0 \leq \lvert V_{GS} - V_t \rvert \leq \lvert V_{DS} \rvert$	saturation
$\beta(V_{GS}-V_t)V_{DS}-V_{DS}^2/2$	$0 \leq \lvert V_{DS} \rvert \leq \lvert V_{GS} \rvert - \lvert V_t \rvert$	ohmic (triode)

The reverse is true for the PMOS (MP_1) transistor. The various modes are indicated in Figure 18–16 as A–E. The operating modes of the switching transistors are defined in Table 18–4.

For CMOS logic circuits, both transistor structures are only ON during the switching from High to Low or Low to High. This means that in a static state there is minimal current (≈0.0A) flowing in the device. This is what makes CMOS circuits so attractive for use in low-power battery operations. This is also the reason it has been possible to pack millions of MOSFET transistors into a single package without exceeding the power-dissipation limitations of the package. Essentially, no current flows through the transistor circuit when the device is in a static state. Maximum current flows only when both transistors are in the saturation region. This is shown in Figure 18–17. The point at which maximum current is flowing is called the *switching point* of the circuit. For this example, the switching point has been set at a voltage of approximately $V_{DD}/2$, or 2.5 V. The switching point of the inverter is controlled by the gain of each transistor. This value, β, was defined in Chapter 17. In this example, $\beta_n = \beta_p$, which sets the switching point at half the supply voltage. The β for each device can be calculated using values from the SPICE .MODEL statement and .CIR listing, as follows:

$$\beta_n = KP_n \times (W/L)_n = 8.165352\text{E}{-05} \times (10/2) = 4.0827\text{E}{-04}\ \text{A/V}^2 \quad (18\text{–}1)$$

$$\beta_p = KP_p \times (W/L)_p = 2.83164\text{E}{-05} \times (28/2) = 3.9643\text{E}{-04}\ \text{A/V}^2 \quad (18\text{–}2)$$

These β values for the transistors are not exactly equal but are very close. In terms of the tolerances for fabrication parameters, these two values are essentially equal: $\beta_n = \beta_p$. The equation for the switching point can be derived by setting

$$I_{DSn} = -I_{DSp}$$

TABLE 18–4 The operating regions for the inverter

Region	PMOS (MP_1)	NMOS (MN_1)
A	ohmic	cutoff
B	ohmic	saturation
C	saturation	saturation
D	saturation	ohmic
E	cutoff	ohmic

The transistors will be in the saturation region since $V_{GS} - V_t \leq V_{DS}$. Solving the equality in terms of V_{in} yields equation 18–3:

$$V_{in} = \frac{V_{DD} + V_{tp} + V_{tn}\sqrt{\beta_n/\beta_p}}{1 + \sqrt{\dfrac{\beta_n}{\beta_p}}} \qquad (18\text{–}3)$$

Equation 18–3 provides a relationship for determining the switching point (point of maximum current) for the CMOS circuit as it switches. The equation assumes that both transistors are operating in the saturation mode ($V_{GS} - V_t < V_{DS}$). It can easily be seen that $V_{in} = V_{DD}/2$ when $\beta_n = \beta_p$. This relationship is demonstrated in Example 18–5 for the inverter described in Figure 18–5.

EXAMPLE 18–5

Determine the switching point for the CMOS inverter shown in Figure 18–5. $\beta_n = 4.0827\text{E-}04\ \text{A/V}^2$, $\beta_p = 3.9643\text{E-}04\ \text{A/V}^2$.

Solution

For the inverter shown in Figure 18–5, the switching point can be calculated using equation 18–3. It is important to remember that this equation is valid for the mode only when both transistors are operating in the saturation region.

The desired response for the circuit is when $V_{in} = 2.5\,\text{V}$, $V_{out} = 2.5\,\text{V}$ ($V_{DSn} = 2.5\,\text{V}$, $|V_{DSp}| = 2.5\ \text{V}$). For MP_1, $|V_{GS}| = 2.5\ \text{V}$ $|V_t| = 0.82\ \text{V}$ (this value is obtained from the model parameters), and $|V_{DS}| = 2.5\ \text{V}$.

$$2.5\,\text{V} - 0.82 < 2.5\,\text{V} \qquad (\text{MP}_1 \text{ is in the saturation mode})$$

For MN_1, $|V_{GS}| = 2.5\ \text{V}$, $V_t = 0.7\ \text{V}$ (this value is obtained from the model parameters), and $|V_{DS}| = 2.5\ \text{V}$.

$$2.5\,\text{V} - 0.7 < 2.5\,\text{V} \qquad (\text{MN}_1 \text{ is in the saturation mode})$$

Because both transistors are in the saturation mode, equation 18–3 can be used to solve for the switching point:

$$V_{in} = \frac{5.0 + (-0.82) + 0.7\sqrt{\dfrac{4.0827}{3.9643}}}{1 + \sqrt{\dfrac{4.0827}{3.9643}}} = 2.42$$

or approximately $V_{DD}/2$. It can be seen in Figure 18–16 that the switching point is approximately at $V_{DD}/2$.

Rise and Fall Time Characteristics

The rise and fall time characteristics of a CMOS circuit are examined in VLSI circuit design so that the designer can develop a better approximation of the circuit's behavior under load. This includes an understanding of the circuit's rise and fall time behavior. A precise analysis of the circuit's behavior requires the use of accurate fabrication-model parameters. Without these parameters, the simulation is a poor and essentially useless estimate. It is also possible for the designer to estimate the circuit's rise and fall time performance using a few calculations. These calculations provide a "ballpark" estimate, but this should not be used in place of a SPICE simulation incorporating accurate fabrication-model parameters.

We showed earlier, in Figure 18–16, that when a CMOS inverter switches from Low to High or High to Low, the transistors pass through many modes of operation. This makes it difficult to accurately estimate the rise and fall time, as the output signal varies from the 10% to 90% (or 90%–10%) value. The rise time for the CMOS inverter is primarily a function of the load capacitance and the beta of the PMOS transistor, which is identified as β_p. Equation 18–4 can be used to estimate the rise time for the inverter, but it is only a ballpark estimate:

$$t_r = \frac{4C_L}{\beta_p V_{DD}} \tag{18–4}$$

Equation 18–5 can be used to estimate the fall time for the inverter. The fall time is primarily a function of the beta of the NMOS transistor, which is identified as β_n:

$$t_f = \frac{4C_L}{\beta_n V_{DD}} \tag{18–5}$$

These equations require that the designer be able to estimate the load capacitance (C_L) that the CMOS circuit is driving. For a CMOS inverter, C_L consists of the load capacitance of the next stage (C_{load}) and the drain–bulk capacitances for the PMOS and NMOS transistors. These drain–bulk parasitic capacitances are abbreviated C_{dbp} and C_{dbn}. The capacitances affecting the inverter are shown in Figure 18–18. The procedure for calculating the β of the transistor was shown in equations 18–1 and 18–2. V_{DD} is the particular power supply voltage used in the circuit. The "4" is a multiplier that is often increased or decreased to improve the accuracy of the estimated rise or fall time. The drain–bulk capacitances for the PMOS and NMOS devices are in parallel. These capacitances can be calculated using values from the model parameters.

The device descriptions for the PMOS and NMOS devices are shown next. Notice that the PD and PS (drain and source perimeter) values are now included so that a more accurate estimate of the parasitic capacitances can be obtained.

```
MP1 5 1 3 3 CMOSP W=28.OU L=2.0U AS=252P AD=252P PS=74U PD=74U
MN1 5 1 0 0 CMOSN W=10.0U L=2.0U AS=90P AD=90P PS=38U PD=38U
```

A simplified view of a MOS transistor layout is shown in Figure 18–19. W is the width of the transistor gate and a is the extent of the diffusion.

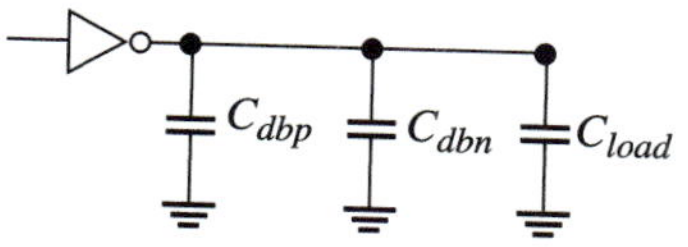

FIGURE 18–18 The capacitances making up the capacitive load for the CMOS inverter

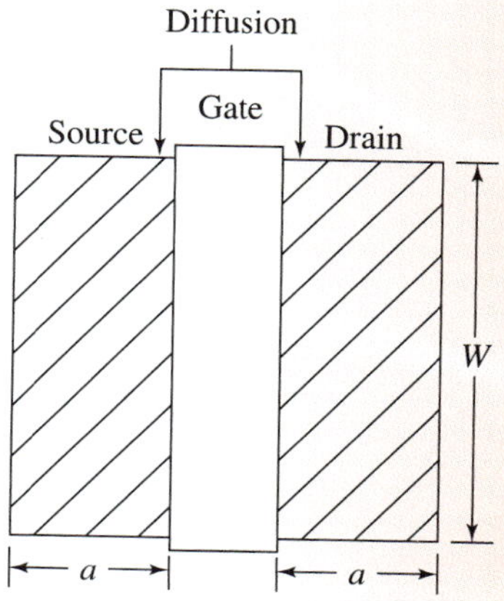

FIGURE 18–19 A simplified MOS transistor layout

The drain–bulk capacitances for the drain and source regions can be calculated using equations 18–6 and 18–7. CJ is the junction capacitance of the drain and source region. This capacitance is a result of the creation of an *n*- or *p*-region within a substrate. Recall from Chapter 2 that a junction capacitance is created when *p*- and *n*-regions are adjacent to each other. CJ has units of F/μm^2 and is obtained from the model parameters for the respective transistor. CJSW defines the sidewall capacitance of the drain–source diffusion regions. CJSW has units of F/μm and is also provided in the respective transistor's fabrication model parameters.

$$C_{dbn} = CJ_n(a \times w) + CJSW_n(2a + 2w) \qquad (18\text{–}6)$$

$$C_{dbp} = CJ_p(a \times w) + CJSW_p(2a + 2w) \qquad (18\text{–}7)$$

Or the equation can be written in terms of the values used to describe the geometry of the transistor listed in the SPICE circuit description:

$$C_{dbn} = CJ_n(AD) + CJSW_n(PD) \qquad (18\text{–}8)$$

$$C_{dbp} = CJ_p(AD) + CJSW_p(PD) \qquad (18\text{–}9)$$

Example 18–6 demonstrates how to use equations 18–8 to 18–9 for calculating the drain–bulk capacitances.

EXAMPLE 18–6

Given the following SPICE description of a CMOS inverter and using the model parameters in Table 18–3 (page 765), calculate the drain–bulk capacitances, C_{dbn} and C_{dbp}.

```
MP1 5 1 3 3 CMOSP W=28.0U L=2.0U AS=252P AD=252P PS=74U PD=74U
MN1 5 1 0 0 CMOSN W=10.0U L=2.0U AS=90P AD=90P PS=38U PD=38U
```

Solution

These drain–bulk capacitances can be calculated using equations 18–8 to 18–9 and the CJ and CJSW values in Table 18–2.

$$\begin{aligned} C_{dbn} &= 1.368 \times 10^{-4}(90\ \text{P}) + 5.222 \times 10^{-10}(38\ \mu) \\ &= 1.2312 \times 10^{-14} + 1.9844 \times 10^{-14} \\ &= 3.2156 \times 10^{-14}\ (\sim 32.2\ \text{fF}) \\ C_{dbp} &= 2.517 \times 10^{-4}\ (252\ \text{P}) + 5.222 \times 10^{-10}(74\ \mu) \\ &= 1.2312 \times 10^{-14} + 1.9844 \times 10^{-14} \\ &= 12.1498 \times 10^{-14}\ (\sim 121.5\ \text{fF}) \end{aligned}$$

The solutions to this problem were obtained by using the AD and PD values (Table 18–3) from the SPICE listing. The drain–bulk capacitance for the PMOS device is considerably larger than the C_{dbn} because the geometry for the PMOS device is approximately 2.5 times larger.

Example 18–7 demonstrates how to calculate the rise and fall time for the inverter.

EXAMPLE 18–7

Given that the load capacitance, C_{load}, for a CMOS inverter is 200 fF, calculate the rise and fall time using the SPICE parameters and the drain–bulk capacitances calculated in Example 18–6.

Solution

Rise and fall time are calculated using equations 18–4 and 18–5. C_L is equal to the sum of all the capacitances attached to the drains (the output) of the CMOS inverter. For this example,

$$C_L = C_{load} + C_{dbn} + C_{dbp} = 200\text{ fF} + 32.2\text{ fF} + 121.5\text{ fF} = 353.7\text{ fF}$$

Solving for t_r,

$$t_r = \frac{(4)(353.7 \times 10^{-15})}{(3.9643 \times 10^{-4})(5.0)} = 0.714\text{ ns}$$

Solving for t_f,

$$t_f = \frac{(4)(353.7 \times 10^{-15})}{(4.0827 \times 10^{-4})(5.0)} = 0.69\text{ ns}$$

Notice that the rise and fall times are comparable. This is due to the fact that $\beta_n \approx \beta_p$.

Examples 18–6 and 18–7 demonstrate the calculations required to estimate the circuit's rise and fall time. A simple CMOS inverter was used in the examples that required only a limited number of calculations to estimate the loading capacitance. In real CMOS systems, the circuitry is quite complex and the calculations can be tedious. This is why it is important to remember to incorporate accurate fabrication-model parameters in your SPICE simulations. It is much easier to let the simulator make these same calculations and there is less chance for error.

Propagation Delay

Propagation delay (or gate delay) of a logic gate can be easily estimated using the rise and fall time values calculated in Example 18–6. Propagation delay is a measure of the delay from the time that the input signal reaches a 50% level to the time for the output to reach a 50% level. This relationship is shown in Figure 18–20.

The propagation delay for a CMOS circuit can be calculated using equation 18–10.

$$t_{delay} = \frac{t_r + t_f}{4} \qquad \text{(18–10)}$$

EXAMPLE 18–8

Use the rise and fall time values obtained in Example 18–7 to calculate the average propagation delay.

FIGURE 18–20 The points used to measure the propagation delay

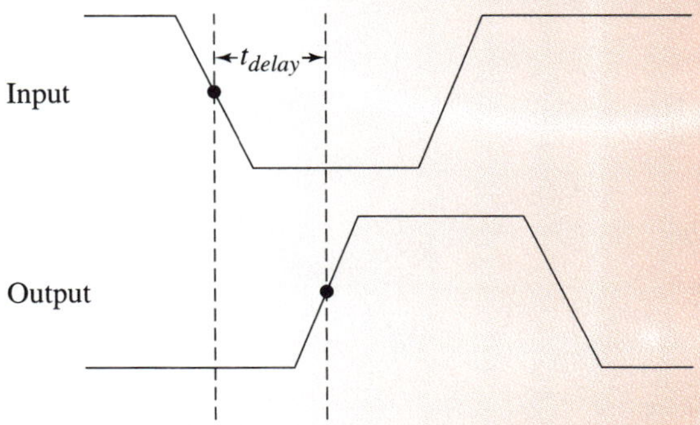

Solution

Given: $t_r = 0.714$ ns and $t_f = 0.69$ ns.

$$t_{delay} = \frac{0.714 \times 10^{-9} + 0.69 \times 10^{-9}}{4} = 0.351 \text{ ns}$$

18–5 CLOCKED CMOS VLSI CIRCUITS

The concept of creating basic combinational CMOS integrated circuits was presented in Sections 18–2 and 18–3. Section 18–4 outlined the transient behavior of CMOS transistor circuits and included a design methodology for sizing the transistors for proper switching behavior. This section expands on the use of transistors in clocked (sequential) circuits; in particular, this section will introduce a fundamental building block in VLSI systems, the D-type flip-flop (DFF), the master/slave DFF, and a nonoverlapping clock circuit.

The D-Type Flip-Flop

The D-type flip-flop is a fundamental building block in VLSI systems. The first DFF examined in this section is a basic latch constructed of a cross-coupled inverter and a simple CMOS transmission gate. The circuit for the basic DFF is shown in Figure 18–21(a). This circuit requires the use of six transistors—two transistors for each inverter and two for the CMOS transmission gate. The "2" inside the inverter symbol is a shorthand method used

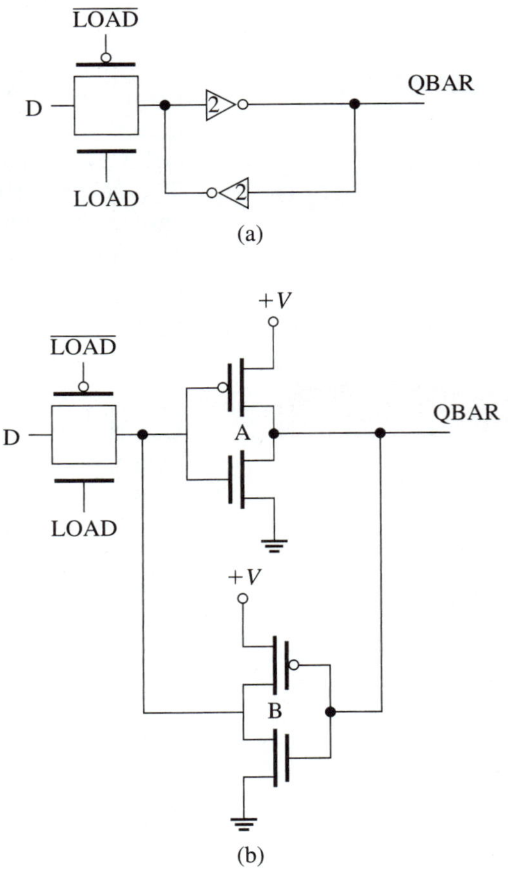

FIGURE 18–21 (a) The basic D-type flip-flop drawn with logic blocks. (b) The full transistor-equivalent circuit

to indicate that this logic block requires two transistors to create the device. The full transistor-equivalent circuit is shown in Figure 18–21(b). A logic value is input into the DFF by asserting the LOAD input (i.e., a VDD voltage). The asserted High LOAD input drives (turns ON) the NMOS pass transistor, and the LOADBAR asserted Low turns ON the PMOS pass transistor. The transmission gate will pass the D input voltage to the flip-flop latch.

Assume that the DFF circuits shown in Figure 18–21 work of a +5 V supply voltage. A logic High can be placed on the D input. The gates to inverter A will be driven High by the logic High voltage. The output of inverter A will be driven Low, thus driving Low the gates to inverter B. This causes the output of inverter B to be driven High, which, in turn, will continue driving the gate to inverter A High even when the LOAD voltage is deasserted. Hence, the logic data value is "latched" as long as power is applied to the circuit. Note that the latched output is inverted from the D input—hence it is called *QBAR*.

The DFF shown in Figure 18–21 can only change states if the D input can "drive" the cross-coupled inverters to an alternate state. In other words, the D input must be able to sink or source sufficient current to overcome the output drive of feedback inverter B. Although this can easily be done in practice, it results in a pass transistor that can be considerably larger (*W/L* ratio) as compared to the other transistors in the circuits. The larger *W/L* ratios of the CMOS pass transistors require more layout area, which is costly in terms of manufacturing. The issue of "drive" capability can be minimized by adding a clocked transmission gate in the feedback path. This change in the circuit is shown in Figure 18–22.

A simulation for the two DFF circuits is shown in Figure 18–23. Trace (A) shows the LOAD input to the CMOS transmission gate. Trace (B) is the data being input into the latch. Trace (C) is the output of the DFF shown in Figure 18–21. Trace (D) is the output of the DFF (Figure 18–22) with the feedback transmission gate. The simulation shows that both circuits latch the input data; however, the basic DFF (Figure 18–21) required a transmission gate with a *W/L* ratio approximately 10 times the size of the transmission gate used for the DFF with the transmission gate (Figure 18–22). The feedback loop CMOS transmission gate opens the feedback drive during the LOAD operation, making it easier to source/sink sufficient current to place the data into the DFF.

Master/Slave DFF Operation

The simulation shows that the basic D flip-flop will latch the data. When the LOAD input is High, the D input is transferred to the output. This means that the QBAR output follows the D input. If data on the D input are changing when CLK is asserted, then the output will also change, increasing the likelihood that the wrong data level will be latched. This is a common symptom of level-triggered latches. The DFF can be made into a master/slave operation in

FIGURE 18–22 The DFF with a transmission gate added to the feedback path

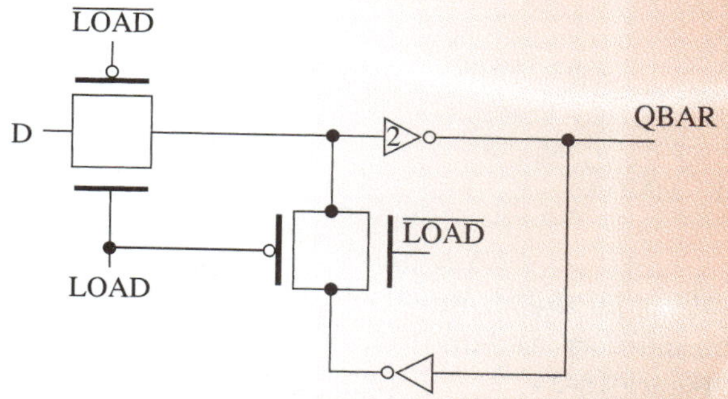

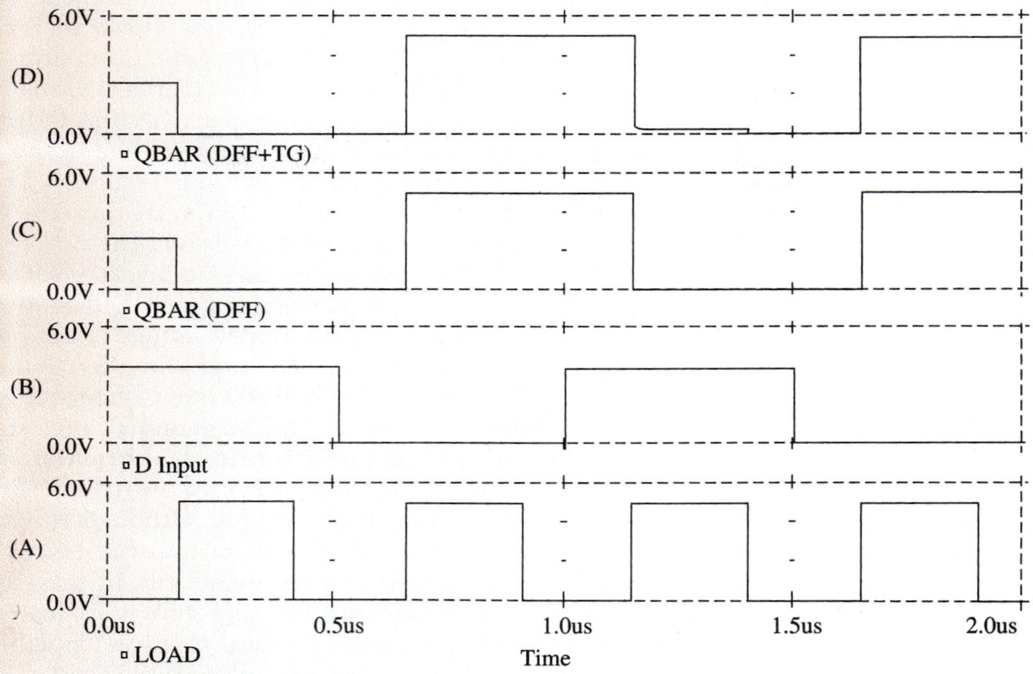

FIGURE 18–23 The simulation of the DFF circuits shown in Figures 18–21 and 18–22

which the data are loaded into the latch on one level of the CLK and transferred to the output on the opposite level. As you might suspect, this circuit requires the use of another DFF with a transmission gate. This section will address the issue of constructing a master/slave DFF with MOSFET transistors.

The transistor-level circuit for a master/slave DFF circuit is shown in Figure 18–24. This circuit requires the use of two basic DFFs. The four transmission gates are full CMOS, which means that each transmission gate incorporates the use of both an NMOS and PMOS pass transistor. All of the transistors can be of the same *W/L* size, such as the values listed in Table 18–5. The SPICE simulation of the master/slave DFF circuit is provided in Figure 18–25. Note that the data are loaded from the D input when CLK is Low and the data are output when CLK is High. Changes in the operation of the DFF are easily accomplished by adding a few transistors, which add additional control to the circuit. For example, many DFF circuits come with an asynchronous SET or RESET input. Example 18–9 demonstrates how this change is made.

EXAMPLE 18–9

Modify the DFF in Figure 18–24 so that an asynchronous SET input is provided. The Q output should be SET when the SET input is Low. We will use the term SET(L) to describe the asynchronous SET input.

Procedure

1. Develop a set of conditions and observations concerning the DFF in Figure 18–24. (*Note:* Referring to Figure 18–24, make sure that you can

TABLE 18–5 Transistor sizing for the MOSFET transistors used in Figure 18–24

PMOS Transistors	CMOSP	W=28.0U	L=2.0U	AS=252P	AD=252P
NMOS Transistors	CMOSN	W=10.0U	L=2.0U	AS=90P	AD=90P

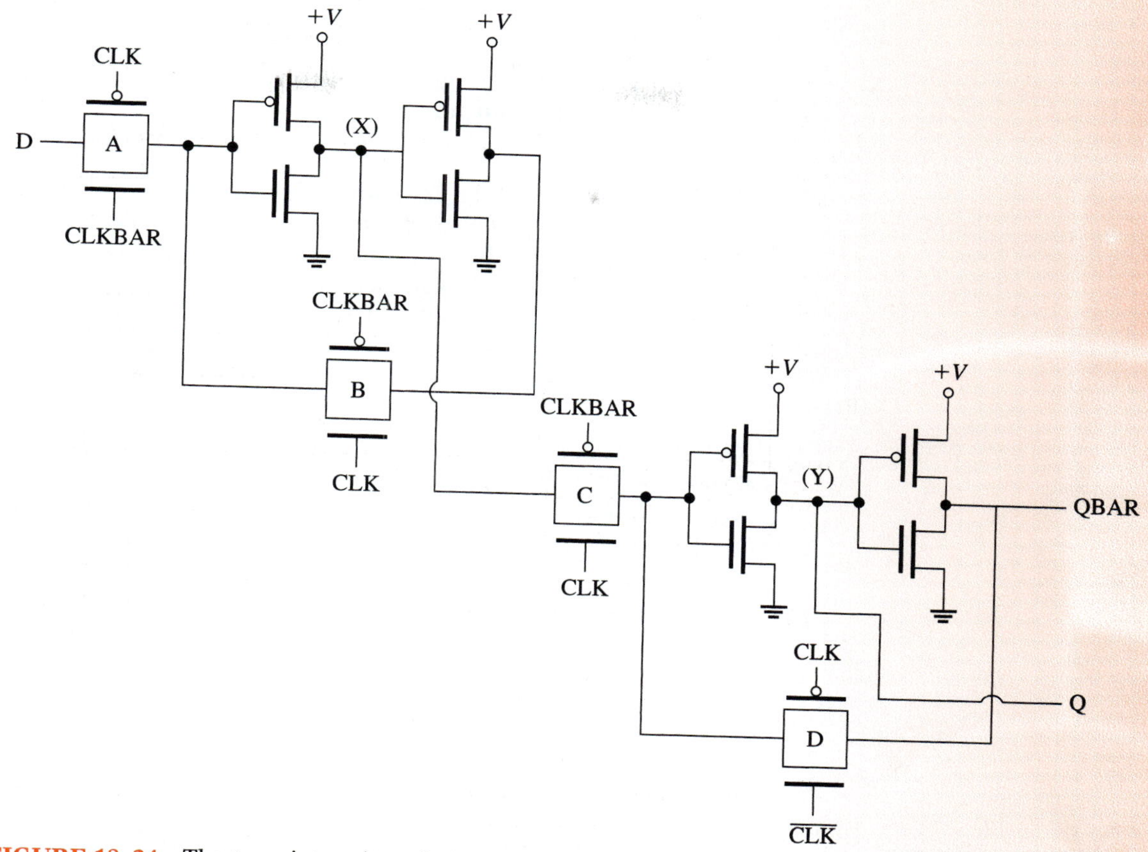

FIGURE 18–24 The transistor view of a master-slave DFF

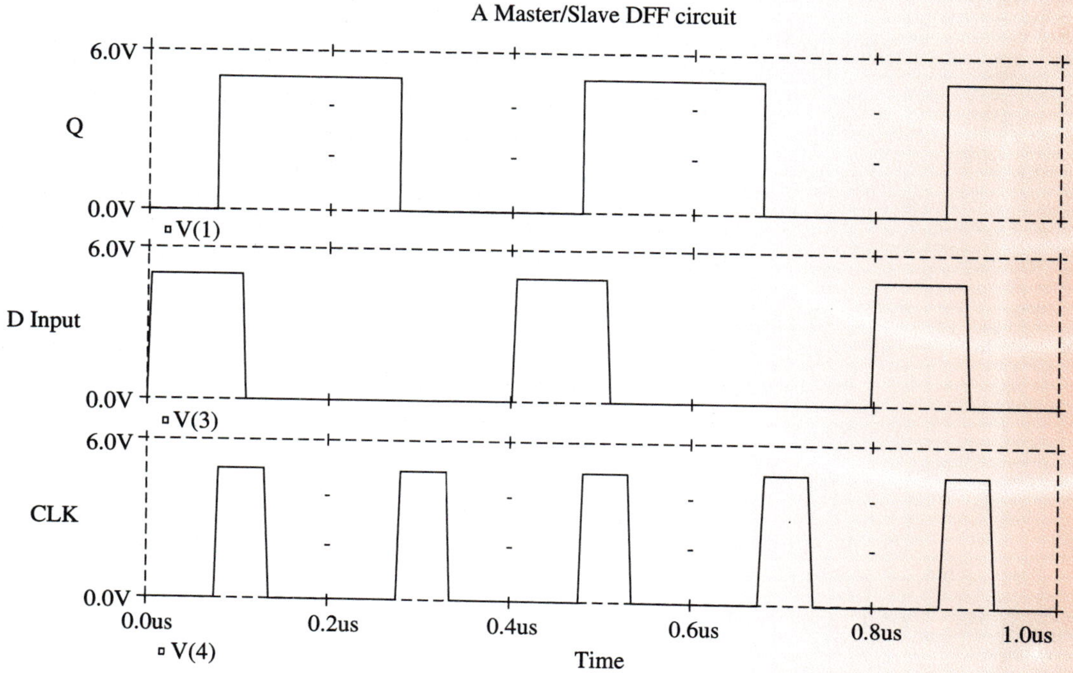

FIGURE 18–25 The SPICE simulation of a master/slave DFF

follow the output of the first DFF, which feeds the second DFF stage. The first DFF stage feeds the C transmission gate from the (X) node.)

2. Develop the required logic using the appropriate MOSFET transistors to obtain the desired output.

Solution

1. When the proposed SET(L) input is asserted Low, the Q output must go High. The only way to guarantee this logic state when you have an asynchronous input is to "force" the slave DFF to a SET condition. Second, because this is an asynchronous input, it is possible that the SET(L) input could be asserted when CLK is asserted, therefore the master DFF must be placed in the proper state, which means the master DFF output (node X in Figure 18–24), which feeds the C transmission gate, must be set Low when SET(L) is asserted. This will ensure that the slave DFF remains in a SET condition if CLK is asserted during the time when SET(L) is asserted.
2. The complete circuit is shown in Figure 18–26. The proposed changes are shown inside the dotted lines. The new circuits are both 2-input NAND gates created using two NMOS and two PMOS transistors. The simulation for the new M/S DFF with asynchronous SET(L) is shown in Figure 18–27.

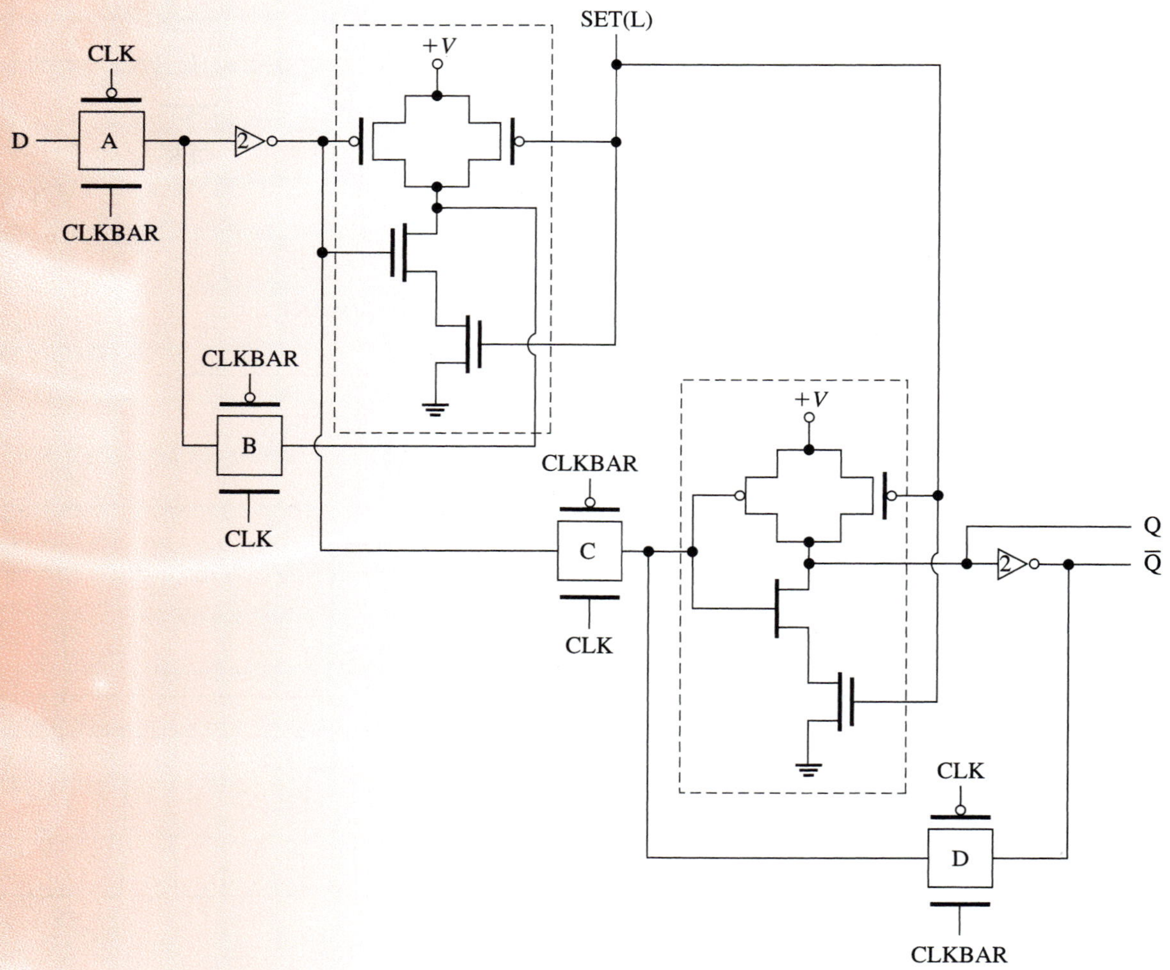

FIGURE 18–26 The M/S DFF with asynchronous SET(L) created in Example 18–9

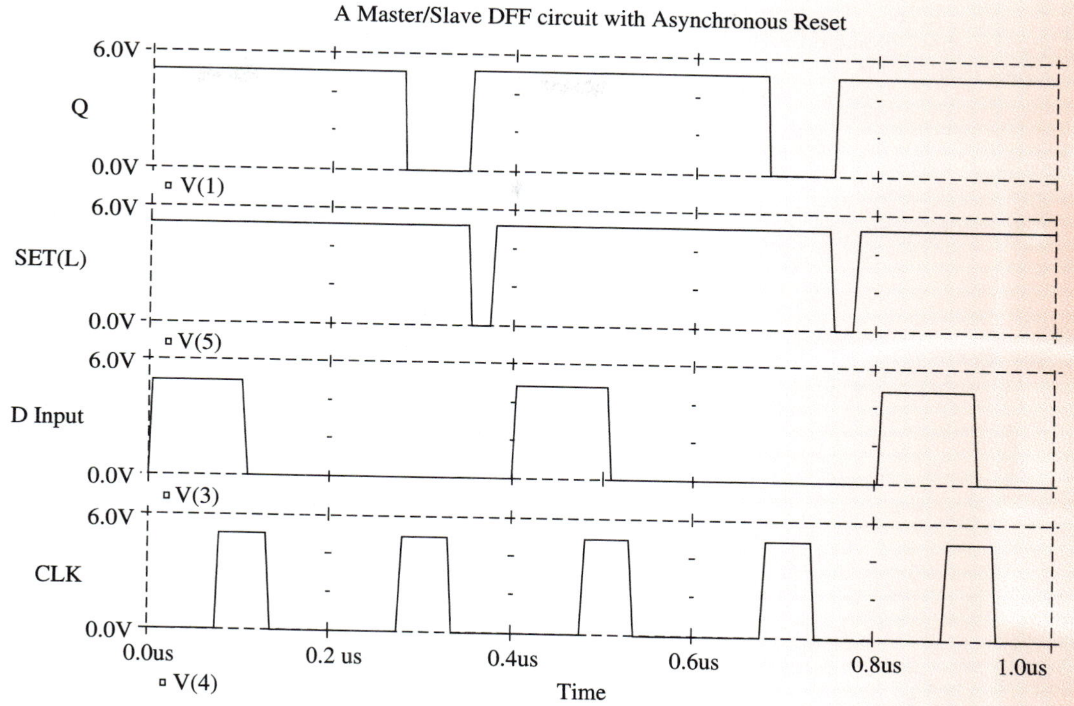

FIGURE 18–27 The SPICE simulation of the M/S DFF with asynchronous SET(L) created in Example 18–9

TABLE 18–6 Transistor sizing for the MOSFET transistors used in Figure 18–26

PMOS Transistors	CMOSP	W=28.0U	L=2.0U	AS=252P	AD=252P
NMOS Transistors	CMOSN	W=10.0U	L=2.0U	AS=90P	AD=90P

The transistor sizes for the simulation are listed in Table 18–6, and the .MODEL parameters for the NMOS and PMOS transistors are the same as provided in Table 18–5.

Two-Phase Nonoverlapping Clock Circuit

In some applications, clocking a signal through a digital or an analog system typically can require the use of a nonoverlapping clock. In other words, CLK and CLKBAR will not have coincident edges. This will facilitate the smooth transfer of the input data to the latch (or registers) and, finally, to the output without introducing racing problems. A block diagram of a 2-phase nonoverlapping CLK circuit is shown in Figure 18–28. The number of transistors required to construct each function is listed on the block diagram. This circuit uses the concept of a cross-coupled NOR gate coupled with inverters for buffering. The transistor view of the 2-phase nonoverlapping circuit is shown in Figure 18–29. The CLK and CLKBAR outputs are 180° out of phase and do

FIGURE 18–28 The block diagram view of a 2-phase nonoverlapping CLK circuit

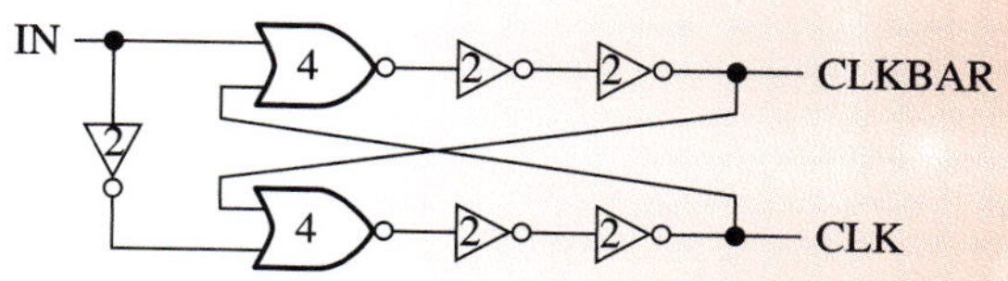

FIGURE 18–29 The transistor view of the 2-phase nonoverlapping CLK circuit

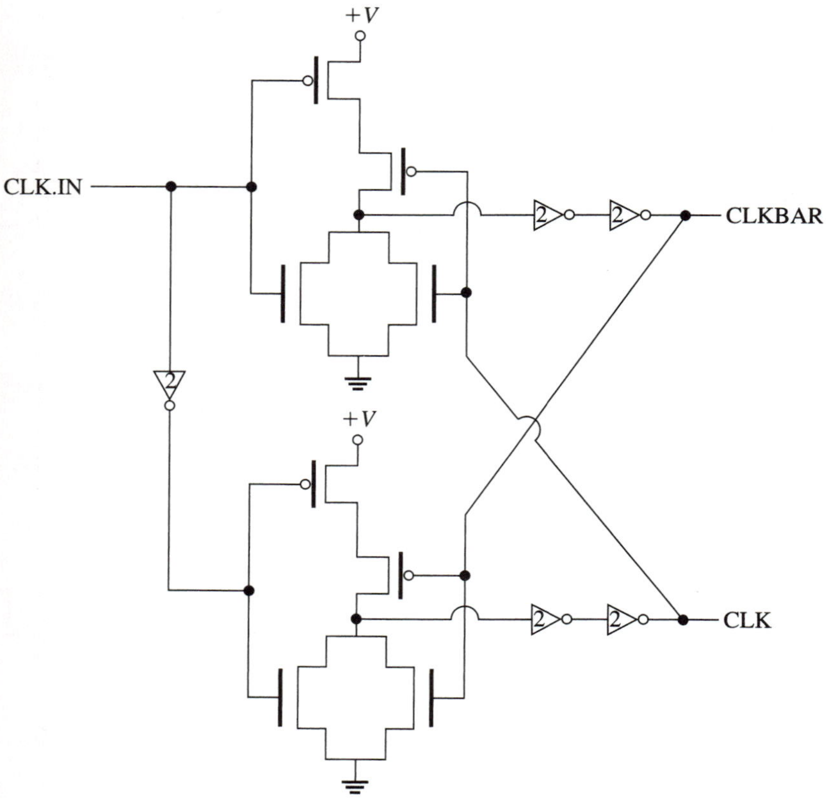

not have overlapping of the clock edge transitions. The SPICE simulation for the circuit is shown in Figure 18–30. Transistor sizes for the simulation are listed in Table 18–7 and the .MODEL parameters for the NMOS and PMOS transistors are the same as provided in Table 18–5.

18–6 RANDOM-ACCESS MEMORY AND READ-ONLY MEMORY

This section presents an overview of the use of MOSFET transistors in memory circuits. Examples of static and dynamic memory, sense amplifiers, ROM, and electrically eraseable memory using a floating gate are used to demonstrate the basic concepts of creating memory circuits with transistors. In practice, memory design—reliably addressing, reading, and writing data to millions of memory elements on an IC—is extremely complex.

The Six-Transistor Static RAM Cell

The fundamental building block for static random-access memory is the six-transistor RAM cell. The basic structure of this device was presented in Section 18–5 in the form of a D flip-flop. The transistor view of the six-transistor RAM cell is shown in Figure 18–31. M_1/M_3 and M_2/M_4 form cross-coupled CMOS inverters, which are used to latch the data. M_5/M_6 are pass transistors, which are used to couple the memory cell to the BIT(H) and BIT(L) lines for performing READ/WRITE operations. These pass transistors are not fully CMOS since they lack the complementary PMOS pass transistor. The PMOS devices are omitted for cell size considerations. The following are the basic steps for performing READ/WRITE operations in a static RAM cell:

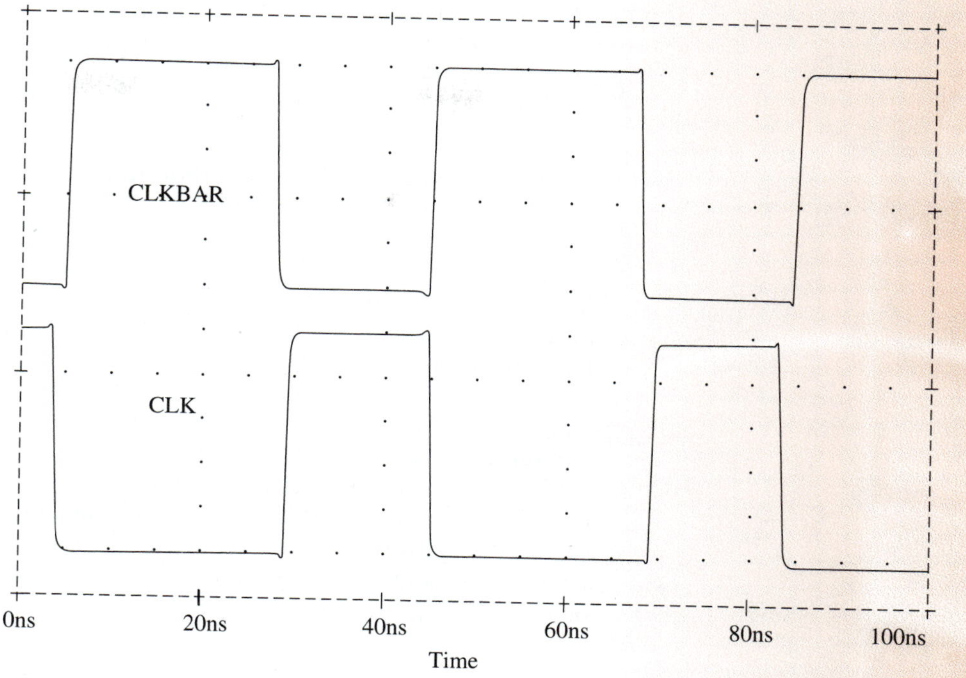

FIGURE 18–30 The SPICE simulation results for the 2-phase nonoverlapping CLK circuit

TABLE 18–7 Transistor sizing for the MOSFET transistors used in Figure 18–29

PMOS Transistors	CMOSP	W=28.0U	L=2.0U	AS=252P	AD=252P
NMOS Transistors	CMOSN	W=10.0U	L=2.0U	AS=90P	AD=90P

Write

1. Place the desired logic levels on the BIT(H) and BIT(L) lines.
2. Assert the WORD(H) lines.
3. Leave the WORD(H) lines asserted long enough for the desired data values to be latched in the RAM cell. Deassert the WORD(H) line.

Read

1. The BIT(H) and BIT(L) lines are precharged to a predefined voltage level. The precharging is stopped and the BIT lines are allowed to float.

FIGURE 18–31 The six-transistor static RAM cell

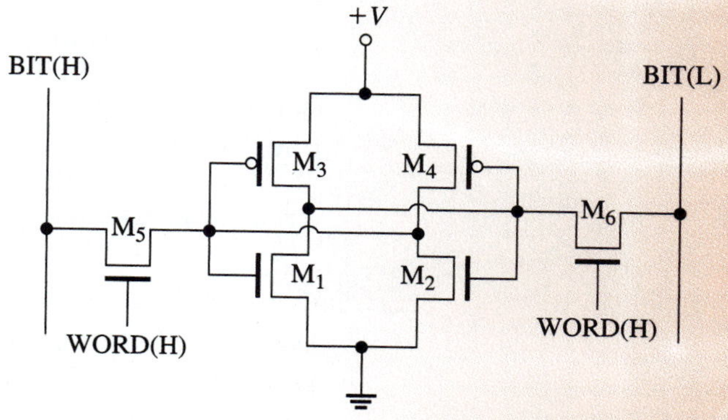

2. The WORD(H) line is asserted, coupling the stored logic values to the BIT lines. The BIT line voltages will begin to change to reflect the logical contents of the RAM cell. The voltage levels on the BIT lines are passed through a circuit called a *sense amplifier* and output for the user.

The feature sizes (gate width/length) can be minimal for the static RAM cell. However, some ratio sizing of the pass transistors may be required so that the stored data are not erased during a READ operation. One design objective is to pack many RAM cells into one IC. Looking at Figure 18–31, imagine that thousands of these same cells are connected to the same BIT(H) and BIT(L) lines. Recall that MOSFET transistors have small parasitic capacitances. In fact, there is a small drain–bulk capacitance (C_{db}) for each pass–transistor coupled to the BIT lines. This means that, in effect, a large capacitance is attached to the BIT lines due to the small parasitic drain–bulk capacitance (C_{db}) for each of the pass transistors, M_5 and M_6. Multiply the drain–bulk capacitance by the number of static RAM cells incorporated into the IC, and the result is a large capacitance on the BIT lines. This means that the response time for reading data can be lengthy, especially if the user must wait for the logic levels on the BIT lines to reach CMOS or TTL logical (High or Low) voltage levels.

The Sense Amplifier

In the READ mode, the BIT lines are precharged to set the voltage level on the BIT lines to a predefined level in order to speed up the READ operation. The BIT lines are fed to a sense amplifier, which is basically an analog circuit used to amplify the voltage levels on the BIT lines, and are precharged to the same voltage level. Referring to Figure 18–31, assume that a logic High is stored on M_2/M_4 and a logic Low is stored on M_1/M_3. When the WORD(H) input is asserted, the static RAM cell latched outputs are placed on the BIT lines. Recall that the BIT lines are precharged to a predefined level. When the static RAM output data are placed on the BIT lines, the voltage levels on the BIT lines will begin to change to reflect the stored voltage values of the static RAM cell. For example, assume that the memory circuit is being used in a 5-V logic system and that the BIT lines are precharged to 2.5 V. The BIT lines are left floating (tri-state condition) after precharging; therefore, the voltage values on the BIT lines will not corrupt the contents of the RAM cell. Second, both BIT lines are precharged to the same voltage levels; therefore, the cross-coupled inverter in the static RAM cell will not be trying to force the flip-flop to a new state. The BIT lines will begin to change to reflect the voltage values of the static RAM cell. This means that one BIT line will begin to increase in value while the other will begin to decrease. If both BIT lines, BIT(H) and BIT(L), are fed into a differential amplifier, then a small difference in BIT-line voltages can be amplified to a large voltage change. As a result, the response (READ) time of the memory circuit can be quite fast even though the BIT lines contain a relatively large capacitance.

Example 18–10 demonstrates how a differential amplifier and support circuitry can be used to speed up the READ time for a static RAM cell.

EXAMPLE 18–10

Design a sense amplifier for static RAM memory. The objective is to have valid data as soon as possible. Assume that a 5-V logic system is being used. Verify your proposed design using SPICE analysis.

Solution

A possible solution to this problem is shown in Figure 18–32, which shows MOSFET transistors M_1–M_5 for a differential amplifier whose inputs are driven by the BIT and BITBAR lines from the static RAM cell. The differential amplifier requires +5 V and −5 V, and the bias current transistor M_5 is biased with −3.985 V. The output of the differential amplifier drives the gate of M_6, a MOSFET gain stage, with M_7 being used as an active load. Note that the gain stage is connected to +5 V and ground. At this point, the sense amplifier is outputting a signal that changes within +5 V and ground. The third stage is a CMOS inverter whose βs have been altered to provide rail-to-rail switching of the sense amplifier output.

The plot of the SPICE simulation is provided in Figure 18–33. It can be seen that a small voltage change in BIT and BITBAR, 0.2 V to 0.4 V, causes the

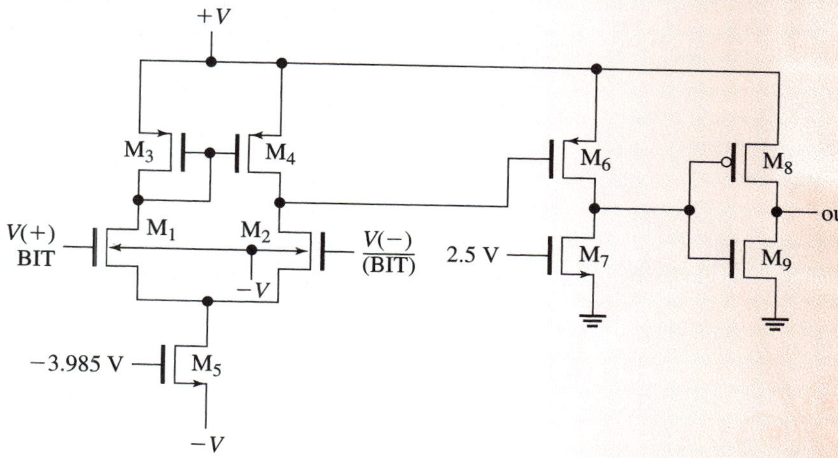

FIGURE 18–32 A differential sense amplifier circuit

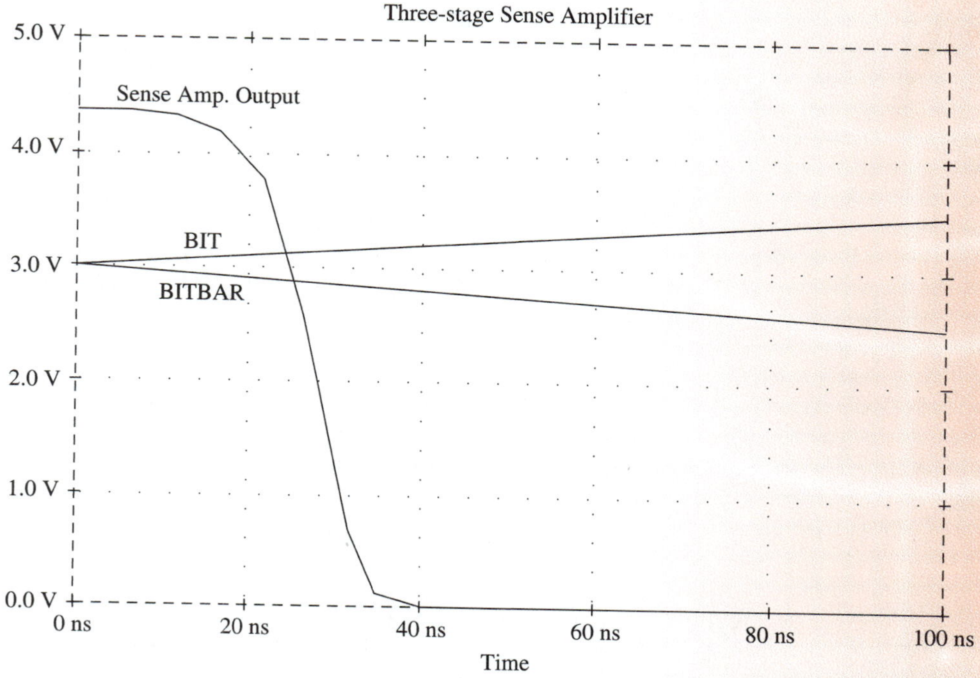

FIGURE 18–33 A differential sense amplifier circuit

sense amplifier output to change states. The .CIR file for this simulation is provided here. The layout parameters for the area of the source and drain (AS and AD) are not specified for this simulation.

```
Three-stage Sense Amplifier
M1 2 4 5 10 CMOSN W=100u L=10u
M2 9 6 5 10 CMOSN W=100u L=10u
M3 2 2 1 1 CMOSP W=20u L=10u
M4 9 2 1 1 CMOSP W=20u L=10u
M5 5 7 10 10 CMOSN W=80u L=2U
M6 110 9 1 1 CMOSP W=150u L=2u
M7 110 90 0 0 CMOSN W=20U L=2U
v7 90 0 2.5
M8 210 110 1 1 CMOSP W=45u L=2u
M9 210 110 0 0 CMOSN W=20U L=2U
******************************************************************
VMINUS 6 0 PWL(0 3.0 100n 3.5 200n 4)
VPLUS 4 0 PWL(0 3 100n 2.5 200n 2)
VBIAS 7 0 -3.985v
VDD 1 0 5V
VSS 10 0 -5
.MODEL CMOSN NMOS LEVEL=2 LD=0.25U TOX=398.00008E-10
+ NSUB=2.374951E+16 VTO=1.0 KP=20E-06 GAMMA=1.0234
+ PHI=0.6 UO=619.676 UEXP=0.20018 UCRIT=97000.7
+ DELTA=4.5376 VMAX=74815.1 XJ=0.250000U LAMBDA=.01
+ NFS=3.91E+11 NEFF=1 NSS=1.000000E+10 TPG=1.00000
+ RSH=27.67 CGDO=3.25390E-10 CGSO=3.25390E-10 CGBO=4.31921E-10
+ CJ=3.9278E-04 MJ=0.427343 CJSW=5.4554E-10 MJSW=0.371244 PB=0.800
*Weff=wDRAWN - dELTA_w
*The suggested Delta_W is 0.53 um
.MODEL CMOSP PMOS LEVEL=2 LD=0.249723U TOX=398.00E-10
+ NSUB=5.799569E+15 VTO=-1.0 KP=10E-06 GAMMA=0.5057
+ PHI=0.6 UO=249 UEXP=0.217968 UCRIT=19160.6
+ DELTA=1.55841 VMAX=42259.5 XJ=0.2500U LAMBDA=.02
+ NFS=3.23E+11 NEFF=1.001 NSS=1.0000E+10 TPG=-1.000
+ RSH=72.45 CGDO=3.249985E-10 CGSO=3.249985E-10 CGBO=4.093809E-10
+ CJ=2.06930E-04 MJ=0.462054 CJSW=2.2334E-10 MJSW=0.117772 PB=0.700
*Weff=wDRAWN - dELTA_w
*The suggested Delta_W is -0.39 um
******************************************************************
.TRAN 0.01NS 250nS .1n
.PROBE
.END
```

The sense amplifier we have presented is only an example. There are many different types of sense amplifiers, and each manufacturer will have its own method of providing fast and reliable READ/WRITE operations.

Dynamic RAM

Dynamic RAM cells use capacitance to temporarily store a charge on a capacitor. The charge reflects the desired logic value to be placed in memory. A simple 1-bit dynamic RAM cell is shown in Figure 18–34. The logic value is stored in the memory capacitor, C_M. NMOS transistor M_1 is turned ON/OFF by the WORD line and is used to couple the memory capacitor to the BIT line.

FIGURE 18–34 A simple 1-bit dynamic RAM cell

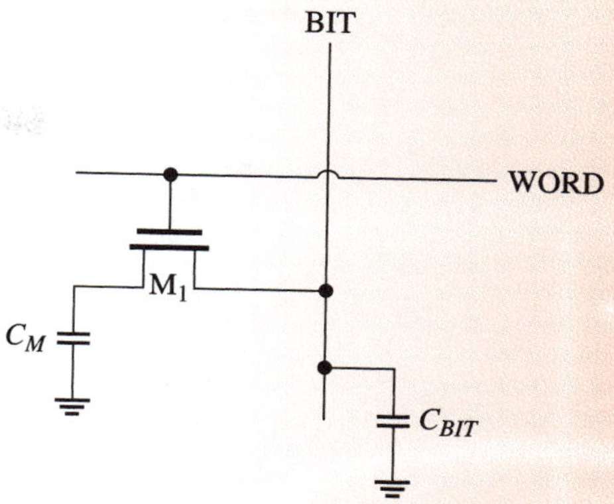

A simplified description of the READ/WRITE operations of a dynamic RAM cell follows.

Write

1. The logic value is placed in the BIT line.
2. The WORD line is asserted and C_M charges to $V_{BIT} - V_{in}$. The memory capacitor is leaky and will not hold the charge for a long time. Therefore, the memory capacitor must be refreshed periodically for the value to remain.

Read

1. The BIT line is precharged and then placed in a floating (tri-state) condition.
2. The WORD line is asserted, turning on pass transistor M_1 and connecting C_M to the BIT line.
3. The BIT charge on the BIT line will increase or decrease depending on the charge stored in the memory capacitor. A sense amplifier detects the small change and outputs the appropriate logic voltage. C_M is small relative to C_{BIT}, so only a small voltage change will be observed on the BIT line.
4. The READ operation is destructive and the data must be rewritten into the memory capacitor.

Dynamic memory requires a different type of sense amplifier. An example of a possible dynamic RAM sense amplifier is shown in Figure 18–35. The 1-bit dynamic memory cell is shown inside the dotted lines. C_{BIT} represents the capacitance on the BIT line. The BIT line is precharged to a predefined voltage. NMOS transistor M_2 is driven by a clamp voltage that is set to a threshold voltage (V_t) greater than the precharged voltage on the BIT line. This places M_2 on the edge of being turned on. PMOS transistor M_3 is driven by a clock (ϕ). M_3 is used to charge capacitor C_2 to the V_{DD} voltage, placing 5 V on node A. CLK (ϕ) is not asserted during the READ mode, only during the precharge phase.

The dynamic memory cell can have either a Low or a High stored in it. If a logic High is stored, then when the WORD line is asserted, the voltage on the BIT will increase slightly, the clamp transistor, M_2, will not turn ON, and no change should occur on node A; therefore, the voltage value on the DATA line is valid instantly. If a logic Low is stored on the memory cell and the WORD line is asserted, the BIT line voltage will drop because C_{BIT} will be charging C_M. This will cause clamp transistor M_2 to turn ON and result in a charge transfer from C_2 to C_{BIT} in an attempt to replenish the charge on the

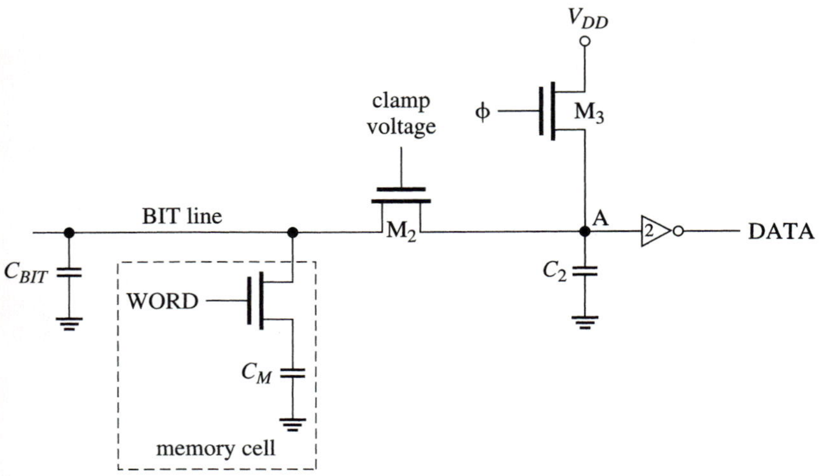

FIGURE 18–35 A charge-balance sense amplifier

BIT line. This in turn will cause a large voltage change on node A, resulting in a corresponding voltage change on the DATA line because C_2 is a small capacitor relative to the BIT line capacitance, C_{BIT}. The charge-balance sense amplifier provides for a very fast sensing in dynamic memory circuits.

Read-Only Memory

Read-only memory (ROM) cells are created using transistors to set a desired logic level on the BIT lines. An example of a simplified CMOS ROM array is shown in Figure 18–36. This is a 3 × 3-bit CMOS ROM cell that requires three BIT and three WORD lines. The NMOS transistors are used to pull down the selected BIT lines when a specific WORD line is asserted. The PMOS transistors have small *W/L* features and are used as active pull-up resistors. They are turned ON by the CLK line during the READ operation. The PMOS devices are clocked so that they are ON only in the READ mode. This feature reduces the power requirements for the device.

When the WORD line is asserted, the desired logic level is placed on the BIT lines. For example, in Figure 18–36, when $WORD_0$ is asserted, a 0 0 0 is placed on the BIT_0, BIT_1, and BIT_2 lines. For $WORD_1$, a 1 0 1 is placed on the BIT lines; for $WORD_2$, a 0 1 0 is placed on the BIT lines.

EXAMPLE 18–11

Design a 3 × 3 CMOS ROM cell that has the following 3-bit words stored in it:

	B_0	B_1	B_2
$WORD_0$	1	1	1
$WORD_1$	0	0	1
$WORD_2$	1	0	0

The solution is shown in Figure 18–37.

The Floating Gate

The floating gate is used in electrically programmable ROMs (EPROMs) and electrically eraseable programmable ROMs (EEPROMs) to store a logic state

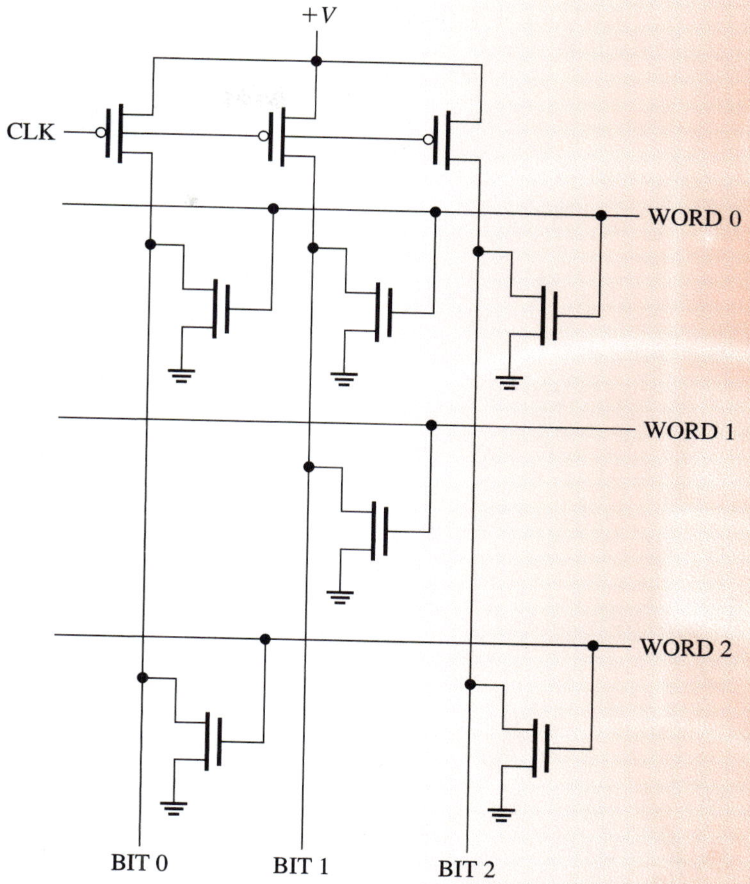

FIGURE 18–36 A 3 × 3-bit CMOS ROM cell

in a MOSFET transistor circuit. A cross-sectional block diagram of a MOS device containing a floating gate is shown in Figure 18–38.

The floating gate (polysilicon) is surrounded by a dielectric or insulating material (typically silicon dioxide, SiO_2). A charge placed on the floating gate will remain there for many years as long as the integrity of the insulating material is not compromised. The gate is programmed by injecting what are called "hot electrons" into it. Given the right voltage conditions, the electrons gain sufficient energy to jump across the silicon–silicon dioxide barrier. This breakdown is called the *Fowler-Nordheim tunneling effect*. The injected electrons placed on the floating gate will cause a shift in the threshold voltage of the MOS device.

The EPROM

A simplified EPROM array is shown in Figure 18–39. The structure is similar to the ROM circuit. There are three modes for an EPROM device: WRITE, READ, and ERASE. The modes are described as follows:

Write

a. The appropriate column line is grounded.

b. Apply a 20–25 V short-duration (≈10 ms) pulse to the appropriate Row Select line while elevating the corresponding Column line voltage to ≈18 V.

c. Columns that are not selected are set to 0 V.

FIGURE 18–37 The solution to Example 18–11

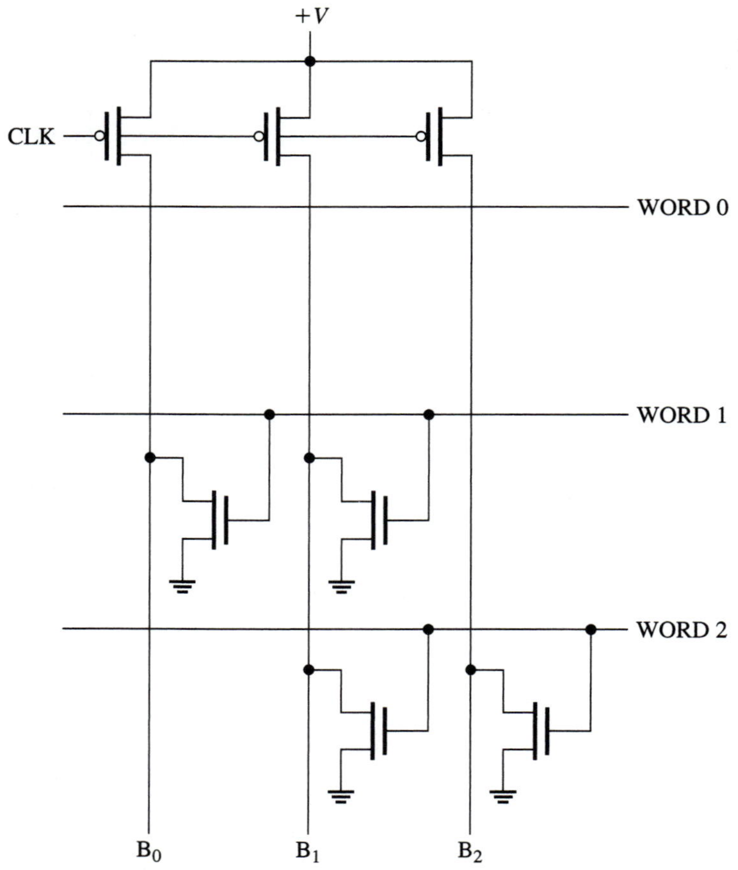

FIGURE 18–38 The floating gate for an EPROM and EEPROM MOS device

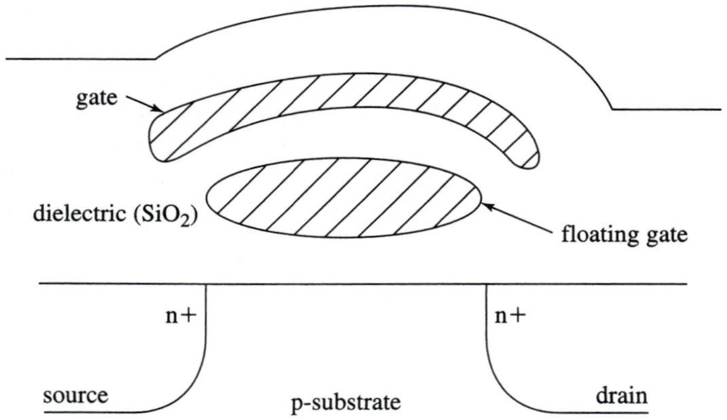

Read

Apply +5 V to the appropriate Row Select and Column lines. A cell that has been written to now has an increased threshold voltage due to the injected electrons, which prevents a 5 V READ voltage from turning ON the device. Therefore, the device outputs a logic 1.

Erase

EPROM memory cells are erased by exposing the cells to ultraviolet light. The electrons stored in the floating gate are excited by the ultraviolet light

FIGURE 18–39 A CMOS EPROM array

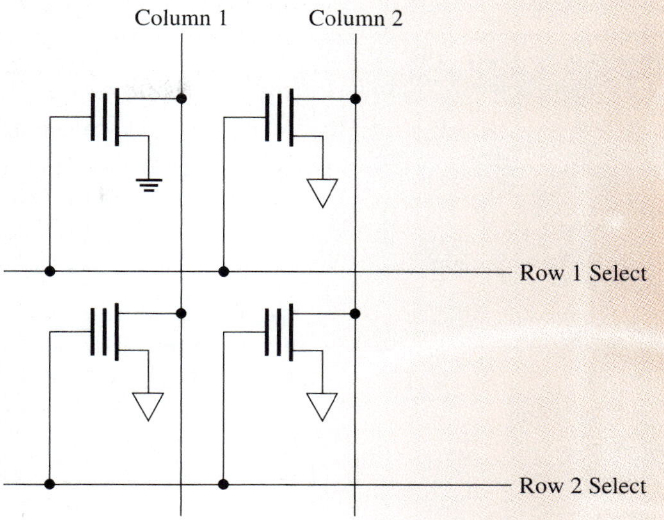

and will discharge to the substrate. This places all of the EPROM memory cells at a logic 0 state.

The EEPROM

EEPROM devices differ considerably from EPROMs in that they can be electrically erased. This is due to the design and fabrication techniques used to make the device, which provide for bidirectional tunneling of electrons to and from the floating gate. Hundreds of thousands of erase cycles are possible with EEPROM before the dielectric begins to break down.

A typical EEPROM cell is shown in Figure 18–40. There are two EEPROM cells in a device and each requires two MOSFET transistors, one with a floating gate. The WRITE, READ, and ERASE modes are defined as follows:

Write

a. 25 V is applied to the appropriate Row Select line.
b. The appropriate Column line is grounded.

FIGURE 18–40 An EEPROM memory cell

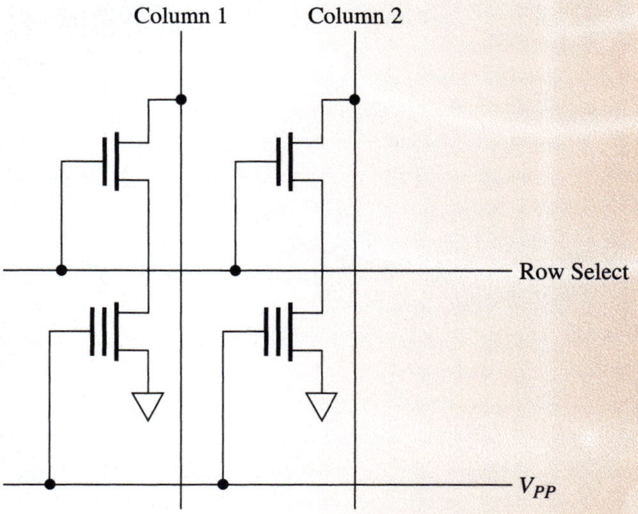

c. A 10-ms 25-V pulse is applied to the V_{PP} line, thus causing electrons to tunnel to the floating gate.

Read

a. Apply +5 V to the appropriate Row Select and Column lines.
b. Cells that have been written to will have an increased threshold voltage and will remain OFF, thus producing a logic 1 on the output. Cells that have not been written to will output a logic 0.

Erase

Apply 25 V to the appropriate Row Select and Column lines while keeping the V_{PP} line at 0 V. This enables the electrons to tunnel back from the floating gate to the drain of the device.

18–7 INPUT/OUTPUT (I/O) BUFFERS

We have examined the use of MOSFET transistors in digital circuits. In this section, we will look at the techniques for inputting or outputting a logic signal to and from the IC. On the surface, this may appear to be a trivial task, but the integrated circuit world of submicron feature sizes, nanoamp currents, and femtofarad capacitances is far different from the outside world of milliamp currents, picofarad capacitances, and large-drive requirements. The issues associated with I/O pads revolve around the following:

- I/O levels
- pad driver sizing
- power supply levels and logic levels
- electrostatic discharge (ESD) protection

These issues are introduced through the techniques used in the design of input and output pads for digital integrated circuits. Additional issues, such as ground bounce, metal migration, and latchup, are important and must be taken into account by the VLSI designer; however, these issues are not addressed in this text. We discuss first the task of inputting a logic signal into the IC.

The Input Pad

The purposes of the input pad are to (1) provide ESD protection; (2) condition the input digital signal, i.e., level conversion and removing noise; and (3) buffer the input signal. A block diagram of an input pad is shown in Figure 18–41. The bonding pad is a rectangle of metal used to connect the IC circuitry to the package.

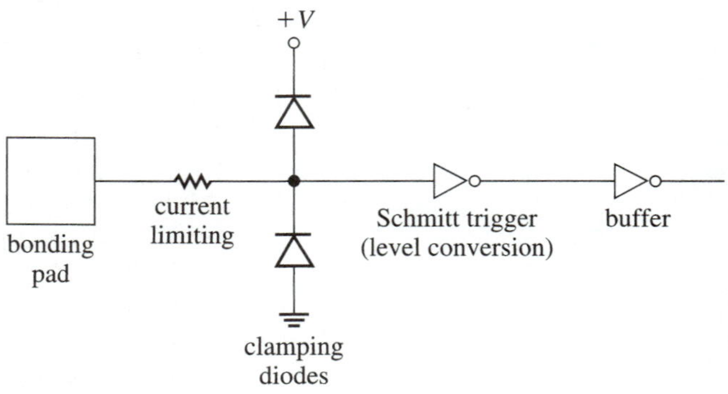

FIGURE 18–41 The block diagram of a CMOS input pad

ELECTROSTATIC DISCHARGE (ESD) PROTECTION CMOS integrated circuits are very sensitive to damage caused by an electrostatic discharge (ESD) due to the fragile physical structure of MOSFET device gates. The gate input of a MOS device is capacitive with a thin dielectric. A high voltage spike, such as that caused by ESD, can puncture the gate input, damaging the device, so some form of ESD protection is required to prevent damage. Damage prevention is provided in two ways: by safe handling and by incorporating ESD protection on the device. Input pads typically contain current-limiting resistors and clamping diodes, as shown in Figure 18–41. The resistor is typically formed from a "well" resistance. A well resistor is not a high-quality resistor as compared to a polysilicon resistor, but it has a pn junction (diode) inherent in its fabrication, as shown in Figure 18–42.

THE SCHMITT TRIGGER A Schmitt-trigger circuit is typically used to make sure that the input logic level is correct and to remove noise from the signal. For example, a TTL logic voltage is not correct for CMOS circuits. The Schmitt trigger will make sure that the voltage level is correct. In modern microprocessors, the motherboard voltage may be 3.3 V, while the internal IC voltage operates at 2.8 V. Level conversion is required for the IC to interface properly with the motherboard. Second, a Schmitt-trigger circuit has hysteresis built into it, which means that the input voltage level must exceed a minimum and maximum voltage range before the output transitions. Hysteresis helps minimize noise problems and false logic levels by preventing the logic level from switching unless the input voltage exceeds a minimum and maximum. The concept of hysteresis is explored in greater detail in Chapter 12 where op-amp comparators are discussed.

The switching characteristics of the Schmitt trigger are shown in Figure 18–43, where V^- and V^+ are the input trigger points, or the points where the output of the device begins to change state. The input voltage level must exceed V^+ for the output to change from High to Low, and conversely, the input voltage must fall below V^- for the output to change from Low to High. The transistor-equivalent circuit for a CMOS Schmitt trigger is shown in Figure 18–44.

The Schmitt-trigger circuit is basically a modified CMOS inverter. Transistors M_3 and M_6 have been added to provide control of the feedback switching point. Control of the V^+ and V^- switching points is provided by the β ratios of M_1/M_3 for V^+ and M_4/M_6 for V^-. Recall that $\beta = KP \times W/L$ for a MOSFET transistor. The equation for the V^+ trigger point is given by Equation 18–11.

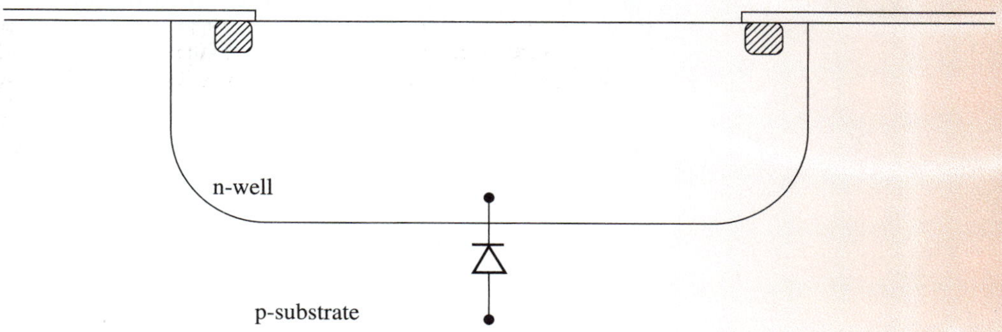

FIGURE 18–42 A cross-sectional view of the current limiting resistor and clamping diode protection provided by a well resistor

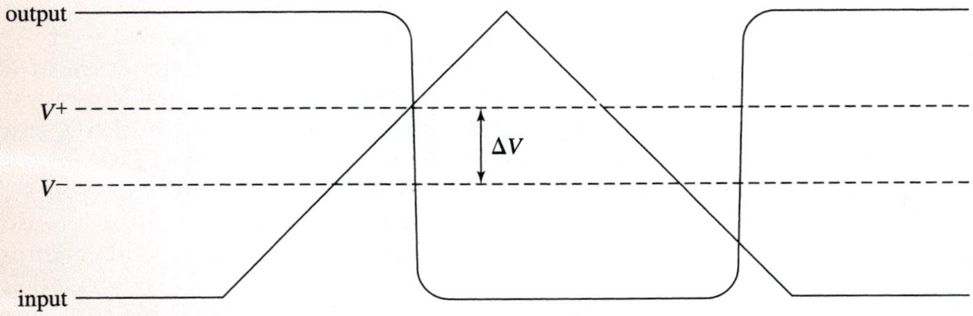

FIGURE 18–43 The switching characteristics of a CMOS Schmitt trigger

FIGURE 18–44 The transistor-equivalent circuit for a CMOS Schmitt trigger

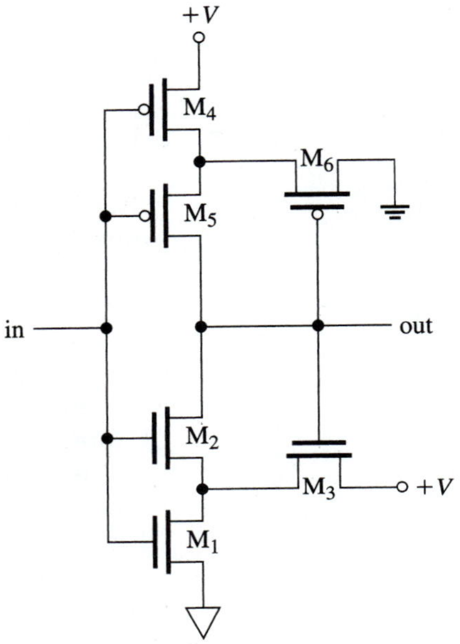

$$V_{(+)} = \frac{V_{DD} + V_{t_n} \cdot \sqrt{\frac{\beta_1}{\beta_3}}}{1 + \sqrt{\frac{\beta_1}{\beta_3}}} \tag{18–11}$$

From equation 18–11, the design equation for determining the β ratios for M_1 and M_3 can be obtained, as shown in Equation 18–12.

$$\frac{\beta_1}{\beta_3} = \frac{(V_{DD} - V^+)^2}{(V^+ - Vt_n)^2} \tag{18–12}$$

The expression for the V^- trigger point is given by equation 18–13.

$$V^- = \frac{\sqrt{\frac{\beta_4}{\beta_6}} \times (V_{DD} - |V_{t_p}|)}{1 + \sqrt{\frac{\beta_4}{\beta_6}}} \tag{18–13}$$

From equation 18–13, the design equation for determining the β ratios for M_4 and M_6 can be obtained, as shown in equation 18–14.

$$\frac{\beta_4}{\beta_6} = \frac{(V^-)^2}{(V_{DD} - V^- - |V_{t_p}|)^2} \quad (18\text{–}14)$$

Example 18–12 demonstrates how equations 18–12 and 18–14 can be used to design a Schmitt-trigger circuit to meet specific design requirements.

EXAMPLE 18–12

(a) Design a CMOS Schmitt-trigger circuit given that $V_{DD} = 5.0$ V, $V^+ = 3.0$ V, $V^- = 2.0$ V, $V_{t_n} = 0.7$ V, and $V_{t_p} = -0.821429$ V. Let $(W/L)_1 = (W/L)_2 = 20\mu/2\mu$. Let $(W/L)_4 = (W/L)_5 = 58\mu/2\mu$.

(b) Verify your answer using SPICE analysis and the model parameters given in Table 18–3.

Solution

a. Using equation 18–12,

$$\beta_1/\beta_3 = (V_{DD} - V^+)^2/(V^+ - Vt_n)^2 = (5 - 3)^2/(3 - 0.7)^2 = 0.756$$

KP_n and Vt_n are the same for M_1 and M_3; therefore, the β ratios are only dependent on the width (W) ratios.

$$\beta_1/\beta_3 = W_1/W_3$$
$$0.756 = 20/W_3$$
$$\therefore W_3 = 26\ \mu$$

Using equation 18–14,

$$\beta_4/\beta_6 = (V^-)^2/(V_{DD} - V^- - |Vt_p|^2) = (2)^2/(5 - 2 - |0.821429|^2) = 0.84278$$

KP_p and Vt_p are the same for M_4 and M_6; therefore, the β ratios are only dependent on the width (W) ratios.

$$\beta_4/\beta_6 = W_4/W_6$$
$$0.84278 = 58/W_6$$
$$\therefore W_3 = 69\ \mu$$

Feature sizes:

$$(W/L)_{M1} = (W/L)_{M2} = 20\ \mu/2\ \mu \text{ and } (W/L)_{M3} = 26\ \mu/2\ \mu$$
$$(W/L)_{M4} = (W/L)_{M5} = 58\ \mu/2\ \mu \text{ and } (W/L)_{M6} = 69\ \mu/2\ \mu$$

b. The following is the .CIR file used in the SPICE simulation; a plot of the simulation results is shown in Figure 18–45. The SPICE simulation shows that the output transitions from High to Low when the input voltage exceeds 3.0 V and switches from Low to High when the input voltage falls below 2.0 V.

```
* Schmitt Trigger simulation for Example 18-12
M4 2 1 3 3 CMOSP W=58.0U L=2.0U
M5 2 1 4 3 CMOSP W=58.0U L=2.0U
M6 2 4 0 3 CMOSP W=69.0U L=2.0U
M1 5 1 0 0 CMOSN W=20U L=2.0U
M2 4 1 5 0 CMOSN W=20U L=2.0U
M3 5 4 3 0 CMOSN W=26U L=2.0U
VIN 1 0 PWL(0 0 .1 5 .2 0)
VDD 3 0 DC 5.0
*
.MODEL CMOSN NMOS LEVEL=2 LD=0.121440U TOX=410.000E-10
```

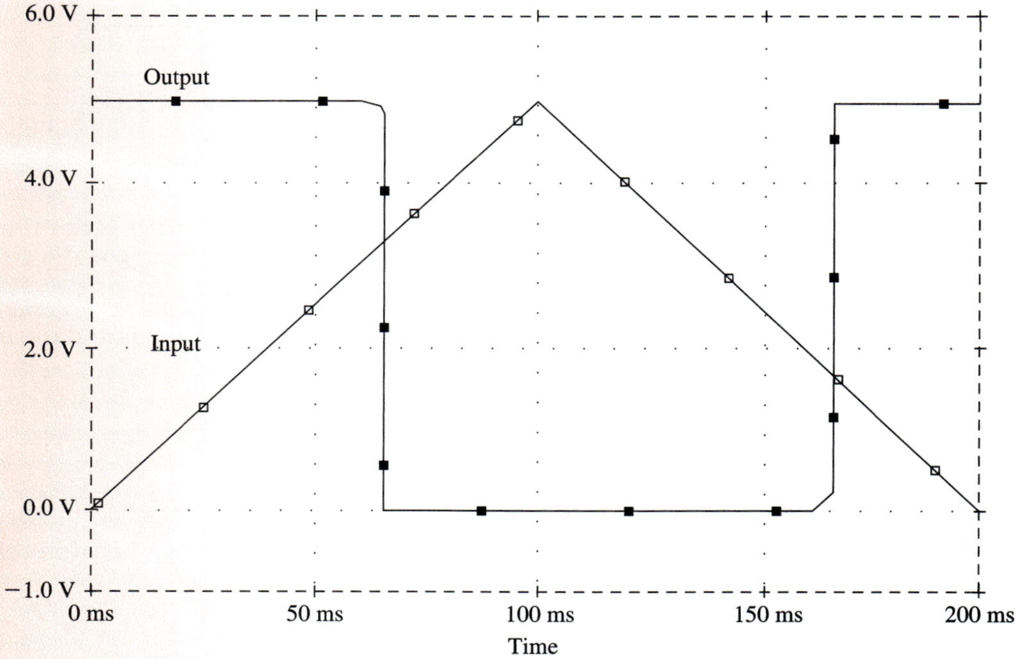

FIGURE 18–45 The input output transfer characteristics for the Schmitt trigger circuit designed in Example 18–12

```
+ NSUB=2.355991E+16 VTO=0.7 KP=8.165352E-05 GAMMA=1.05002
+ PHI=0.6 UO=969.492 UEXP=0.308914 UCRIT=40000
+ DELTA=0.262772 VMAX=71977.5 XJ=0.3000000U LAMBDA=3.3937849E-02
+ NFS=1.000000E+12 NEFF=1.001 NSS=0 TPG=1.000000
+ RSH=33.290002 CGDO=1.022762E-10 CGSO=1.022762E-10
+ CGBO=5.053170E-11 CJ=1.368000E-04
+ MJ=0.492500 CJSW=5.222000E-10 MJSW=0.235800 PB=0.490000
* WEFF=Wdrawn - Delta_W
* The suggested Delta_W is 0.06 um
*
.MODEL CMOSP PMOS LEVEL=2 LD=0.180003U TOX=410.000E-10
+ NSUB=1.000000E+16 VTO=-0.821429 KP=2.83164E-05 GAMMA=0.684084
+ PHI=0.6 UO=336.208 UEXP=0.351755 UCRIT=30000
+ DELTA=1.000000E-06 VMAX=94306.1 XJ=0.300000U LAMBDA=4.861781E-02
+ NFS=2.248211E+12 NEFF=1.001 NSS=1.000000E+12 TPG=-1.000000
+ RSH=119.500003 CGDO=1.515977E-10 CGSO=1.515977E-10
+ CGBO2=2.273927E-10 CJ=2.517000E-04 MJ=0.528100 CJSW=3.378000E-10
+ MJSW=0.246600 PB=0.480000
* Weff=Wdrawn - Delta_W
* The suggested Delta_W is 0.27 um
.TRAN 1n .21
.PROBE
.END
```

Figure 18–46 is a hysteresis plot for the Schmitt trigger with the *x*-axis being the input voltage. The output plot clearly shows that the Schmitt trigger begins to change at 2.0 and 3.0 V. The arrows show the direction of the output voltage change. When the input exceeds 3.0 V, the output begins to transition

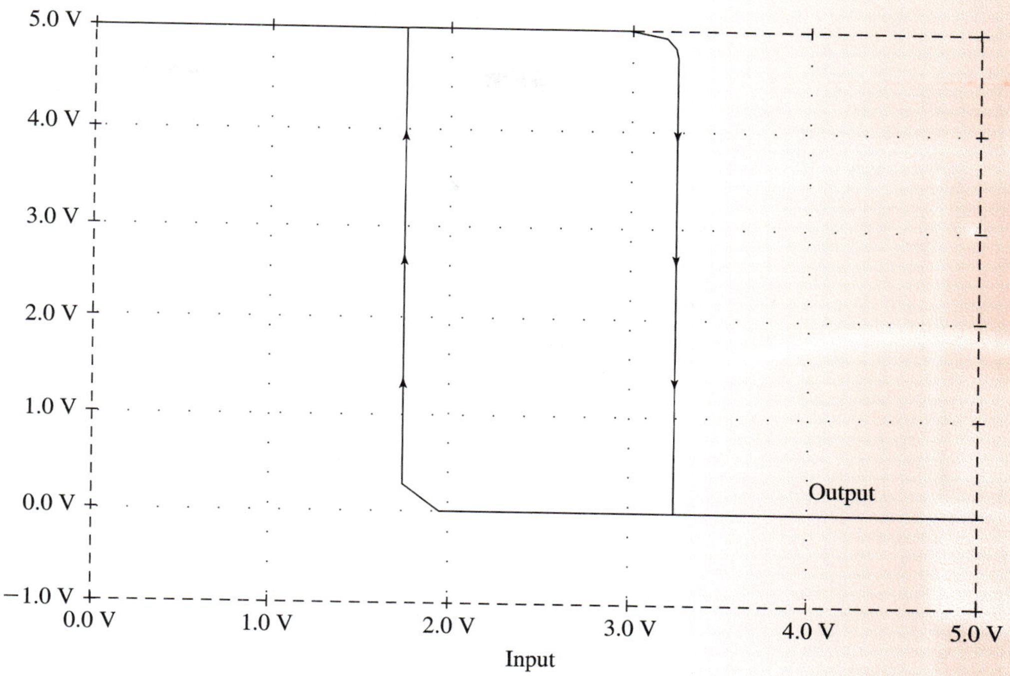

FIGURE 18–46 The hysteresis plot for the Schmitt trigger

from 5.0 V to 0.0 V. The output voltage begins to transition from 0.0 V to 5.0 V when the input voltage falls below 2.0 V.

The Output Pad

DRIVING OFF-CHIP LOADS Output buffers are used to transfer the IC's logic signals to the outside world. The two primary issues for an output driver are

- stage ratio sizing
- level conversion

A typical inverter on an IC is designed to drive femtofarad (10^{-15})-size capacitances, while the outside world uses picofarads (10^{-12}) and larger. Proper sizing of the outside-world drivers is critical to minimize propagation delay. A block diagram of a 1X inverter driving the outside world is shown in Figure 18–47. The "1X" indicates that this inverter is a typical-size device within the IC. For example, in a given technology, the transistor devices will have a specified or standard gate width and length. Referring to Figure 18–47, n represents the number of stages required to drive the load, with the driver being the last of the n stages. The load, designated C_L, represents the total load capacitance, which consists of the capacitance associated with the metal bus, the bonding pad, and the load. It was demonstrated in Section 18–4 that

FIGURE 18–47 A block diagram of the circuitry required to drive off-chip loads

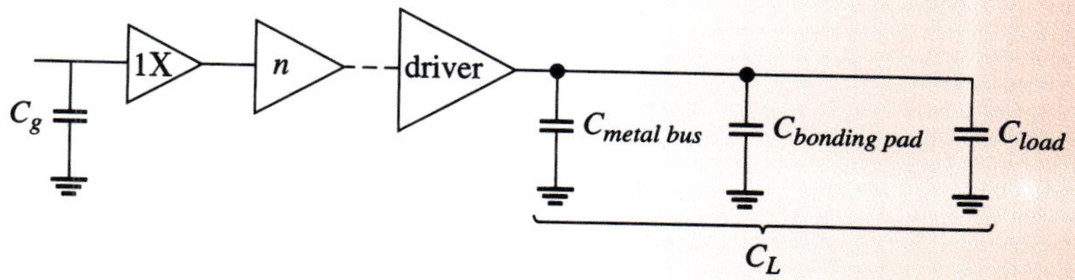

capacitive loading affects rise/fall time and propagation. The propagation delay encountered by a 1X inverter driving a several picofarad load will severely degrade the performance of a logic device. This means that some form of cascaded driver is required to compensate for the extreme difference in device sizing and to minimize propagation delay.

A general rule of thumb for driving off-chip loads is to add n buffer stages. Each buffer stage should be approximately 2.71 times as large as the previous stage. This technique, developed by Mead and Conway (*Introduction to VLSI Systems,* Addison-Wesley, 1980), yields an optimal stage ratio while minimizing layout area and power consumption. The equation for calculating n, the number of stages, is provided in equation 18–15.

$$n = \ln\frac{C_L}{C_g} \tag{18–15}$$

where C_g is the gate capacitance of the 1X device and C_L is the expected load capacitance. Example 18–13 demonstrates how to design a driver stage.

EXAMPLE 18–13

Given that $C_g = 63$ fF and total load capacitance $C_L = 10.25$ pF, calculate the number of stages required to buffer the signal. Draw a block diagram showing the result and list the W/L ratio for each stage. Assume that the W/L ratio for the 1X is 6 μ /1 μ. Plot a comparison of the 1X device and the last stage of the driver driving C_L.

Solution

Calculate the number of stages required:

$$n = \ln\frac{C_L}{C_g} = \ln\frac{10.25\text{ pF}}{63\text{ fF}} = 5.09$$

Round the number of stages to 5. The block diagram solution for the circuit driving the off-chip load is shown in Figure 18–48. Note that the gate width for each stage increases by an ideal factor of 2.71. The length remains at 1 μ. In practice, there is some room for varying the ratio sizes to a value other than 2.71. In general, ratio sizes up to 4 are suitable as long as the designer understands that this is not optimal and can affect performance. The width values, shown in Figure 18–48, are the calculated numbers, but the accuracy of each number will vary depending on the layout geometries permitted for a particular fabrication technology.

A plot comparing the drive capability of the 1X device to the last stage of the output driver is provided in Figure 18–49. It is obvious that the performance increases by incorporating the driver. The PMOS and NMOS transistor gate widths of the driver stage (last n stage) are very large (2192 μ /877 μ). In fact, widths this size would take up a large amount of chip real estate. To more

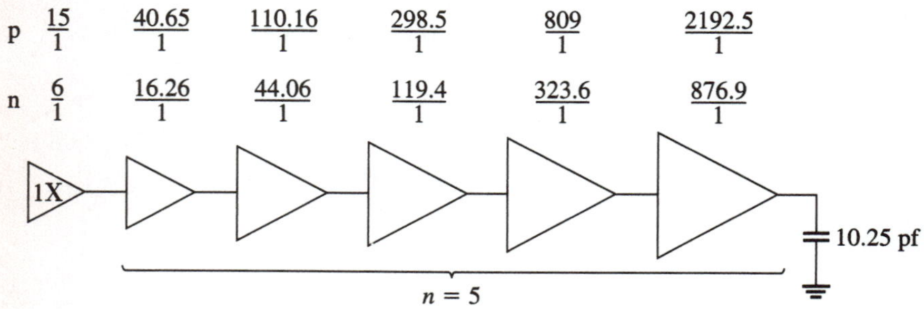

FIGURE 18–48 The block diagram solution for driving the off-chip load

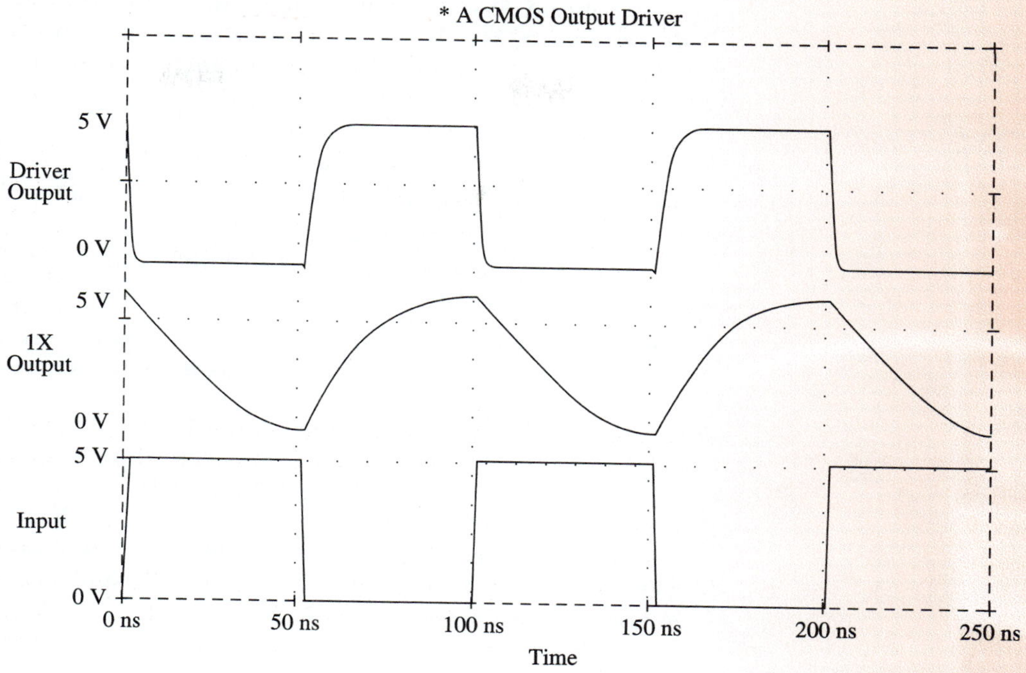

FIGURE 18–49 A comparison of the drive capability of a 1X and buffeted output driver

efficiently use limited IC real estate, moderate-size inverters are placed in parallel to make the driver more compact. The MOSFET transistor view of an output driver is shown in Figure 18–50. Note that four inverters are connected in parallel. This increases the effective gain of the buffer to 4X, or four times the gain of a 1X device. In this same manner, the last stage driver in Example 18–13 could be constructed by placing eight inverters (PMOS—270 μ/1 μ) and (NMOS—110 μ/1 μ) in parallel, which would closely approximate the desired drive capability without consuming large amounts of IC real estate.

Level Conversion

Logic-level conversion was discussed in Section 18–4, and an example of level conversion was provided in Example 18–5 using equation 18–3. Modern microprocessors use an internal logic voltage of 2.8 V, and the motherboards might use 3.3 V. This means that the output buffers must convert the voltage. When the internal logic level is 1.4 V ($V_{DD}/2$), the voltage-level output to the motherboard should be 1.65 V or $V_{DD}/2$ for the motherboard. CMOS logic circuits switch from rail to rail (V_{DD} to V_{SS} or ground); therefore, the switching point is assumed to be at $V_{DD}/2$. In some cases, the assumption that the switching point is at $V_{DD}/2$ may not be correct.

FIGURE 18–50 A 4X output driver

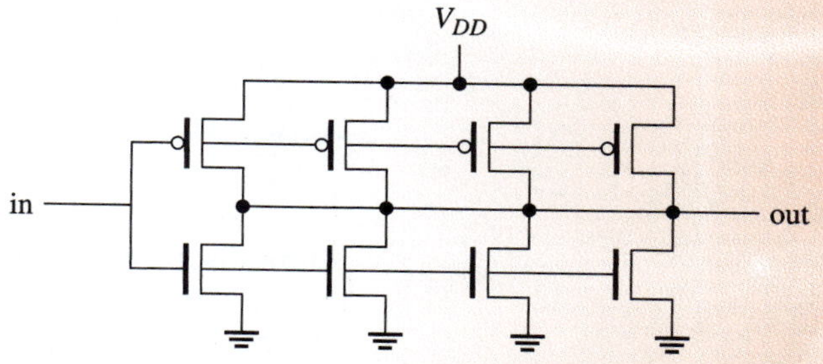

Equation 18–3 must be written in a form that enables the designer to determine the required β ratios. Equation 18–16 provides the designer with a proper format for designing a logic-level conversion.

$$\sqrt{\frac{\beta_n}{\beta_p}} = \frac{(V_{in} - V_{DD} - V_{t_p})}{(-V_{in} + V_{t_n})} \quad \textbf{(18–16)}$$

where V_{in} is the specified input voltage (1.4 V)
V_{DD} is the output voltage (3.3 V)
V_{tp} and V_{tn} are the threshold voltages for the NMOS and PMOS transistors

This equation assumes that the transistors are in the saturation mode of operation ($V_{gs} - V_t \leq V_{DS}$). Example 18–14 demonstrates how to use Equation 18–16.

EXAMPLE 18–14

Design a logic-level interface using a CMOS inverter that meets the following criteria. When the internal logic level is 1.4 V ($V_{DD}/2$), the voltage level output to the motherboard should be 1.65 V or $V_{DD}/2$ for the motherboard. The motherboard voltage is 3.3 V, $V_{t_n} = 0.7$ V and $V_{t_p} = -0.82$ V. The ratio for $KP_n/KP_p = 2.88$. Assume that $L_n = L_p = 2\ \mu$.
Perform a SPICE simulation of the circuit.

Solution

$$\sqrt{\frac{\beta_n}{\beta_p}} = \frac{(V_{in} - V_{DD} - V_{tp})}{(-V_{in} + V_{tn})} = \frac{1.4 - 3.3 - (-0.82)}{1.4 + 0.7} = 1.54$$

Therefore,

$$\beta_n/\beta_p = 1.54^2 = 2.37$$

Given that $W_n = 10\ \mu$, solve for W_p.

$\beta_n/\beta_p = 2.38 = 2.88 \times W_n/W_p = 2.88 \times 10\ \mu / W_p$ (assume that $L_n = L_p = 2\ \mu$)

Therefore,

$$W_p = 12\ \mu$$

The NMOS device feature sizes are 10 μ /2 μ, and the PMOS device feature sizes are 12 μ /2 μ. A portion of the SPICE .CIR follows, and the results of the SPICE simulation are shown in Figure 18–51.

```
* A CMOS Output Driver
M1 2 1 3 3 CMOSP W=12U L=2.OU
M2 2 1 0 0 CMOSN W=10.OU L=2.OU
VDD 3 0 3.3
Vin 1 0 PWL(0 0 1 2.8 2 0)
* the fabrication model parameters listed in Table 18-3 were used
.TRAN .ln 2.1
.PROBE
.END
```

The Tri-State Driver

Some digital operations require the use of a tri-state data bus such as the one used for memory. MOSFET transistors can be configured to provide tri-state operation by including an enable input that will turn the circuit ON/OFF.

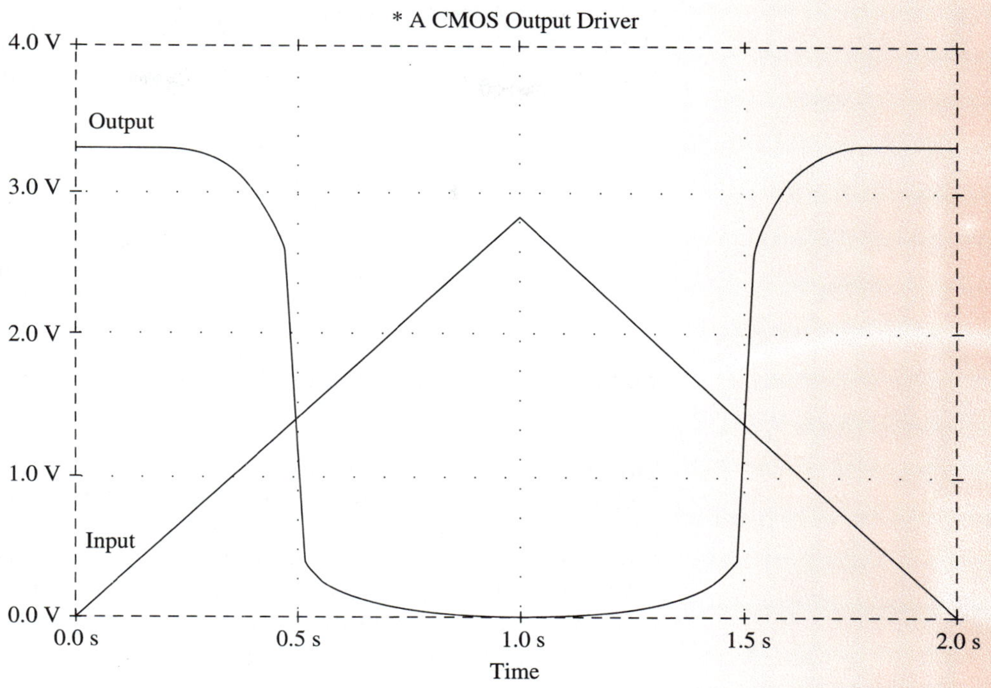

FIGURE 18–51 The results of the SPICE simulation for the logic-level converter design in Example 18–14

A simplified inverting tri-state device and its corresponding logic symbol are shown in Figure 18–52. MP_2 and MN_2 form a CMOS inverter whose paths to the V_{DD} and GND rail are provided by transistors MP_1 and MN_1. Control of MP_1/MN_1 is provided by the CMOS inverter (MP_3/MN_3). When the Enable-L line is Low, MN_1 and MP_1 are connected to the GND and V_{DD} rails, respectively, enabling the input signal (IN) to be passed to the output (OUT). A voltage table is provided in Table 18–8. High-Z is the tri-state condition.

Tri-state output drivers would look very similar to Figure 18–2 except more CMOS inverters would be in parallel. When examining MOSFET circuits,

FIGURE 18–52 A simplified inverting tri-state device

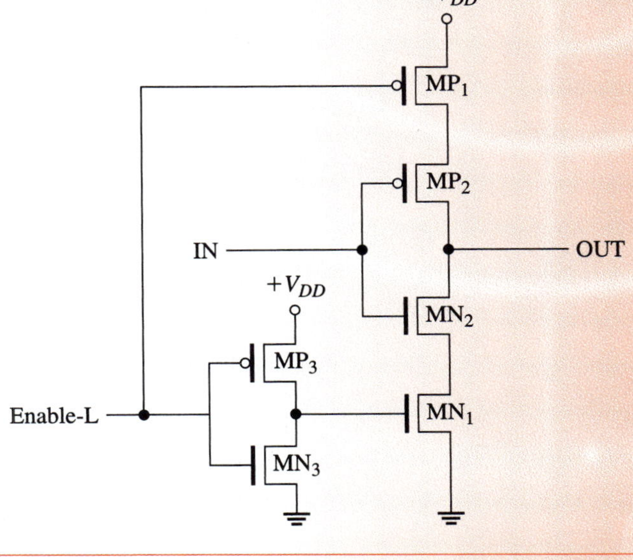

TABLE 18–8 The voltage table for the tri-state inverter

Enable-L	IN	OUT
Low	Low	High
Low	High	Low
High	Low	High-Z
High	High	High-Z

including input or output drivers, just remember to treat the transistor as an ideal switch to determine the function a particular circuit is providing.

18–8 TRANSISTOR-LEVEL LOGIC CIRCUIT ANALYSIS WITH ELECTRONICS WORKBENCH MULTISIM

This chapter examined the use of MOSFET transistors to create CMOS logic circuits. The concept was presented that the NMOS network was responsible for connecting the output to ground, while the PMOS network was responsible for connecting the output to V_{DD}. This section examines the use of EWB to simulate the transistor-level implementation of two CMOS combinational logic circuits, the CMOS Inverter and the CMOS NOR gate.

The objectives of this section are as follows:

- Use EWB to simulate transistor-level implementations of CMOS logic circuits.
- Develop an understanding of how to set the parameters on the input signals and how to verify the circuit is working properly.

The first circuit examined is a simple two-transistor CMOS Inverter. This is shown in Figure 18–53. Transistors M1 and M2 form the CMOS inverter. The 100 um 100 um next to each transistor indicate the channel length and width. This setting can be viewed by double-clicking on the transistor as shown in Figure 18–54.

The next step is to set the input signal level to the CMOS inverter. In this example, the input signal level is set to switch between 0 and 5.0 V. The settings for the function generator are shown in Figure 18–55. The function generator is set to output a square wave with an offset voltage set to 2.5 V and the amplitude set to 2.5 V. This produces the 0 V−5 V square wave. The frequency of the square wave is set to 1 kHz. The menu for the function

FIGURE 18–53 The CMOS Inverter circuit

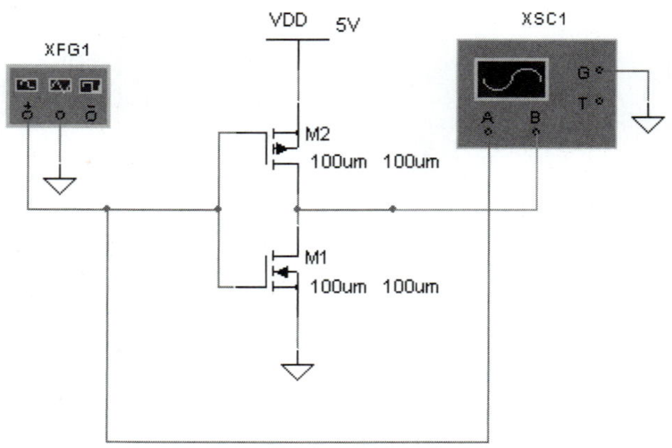

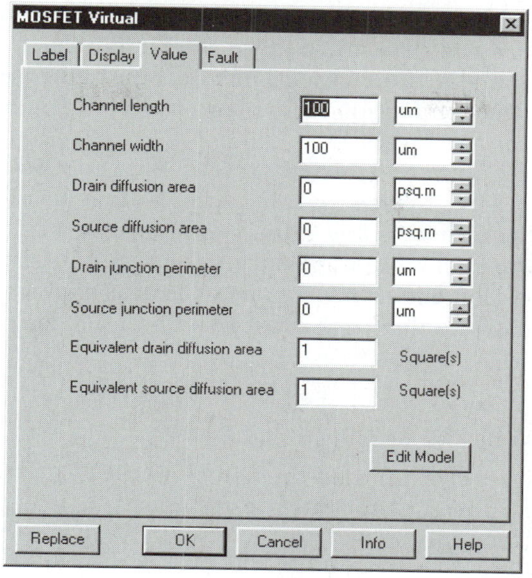

FIGURE 18–54 The menu for setting the channel length and width parameters for the MOSFET transistor

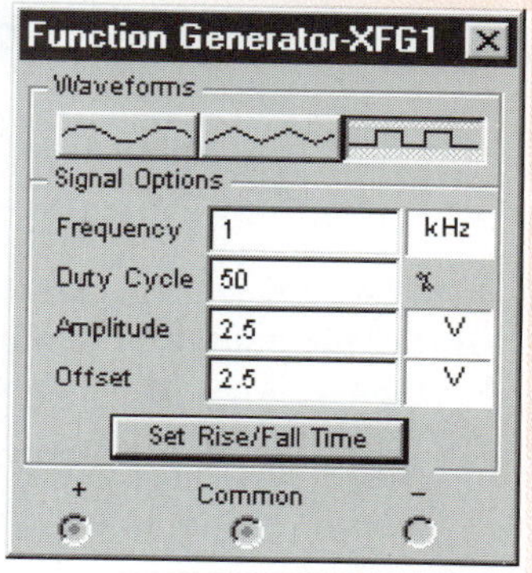

FIGURE 18–55 The settings for the function generator

generator is viewed by double-clicking on the function generator icon. The menu is displayed in Figure 18–55.

The circuit is now ready for simulation. Click on the start simulation button and double-click on oscilloscope to view the input and output signals. The results for this simulation are provided in Figure 18–56. Notice that when the input is high, the output voltage level is low, and when the input level is low, the output is high, an inverting function.

FIGURE 18–56 The input and output signals for the CMOS inverter

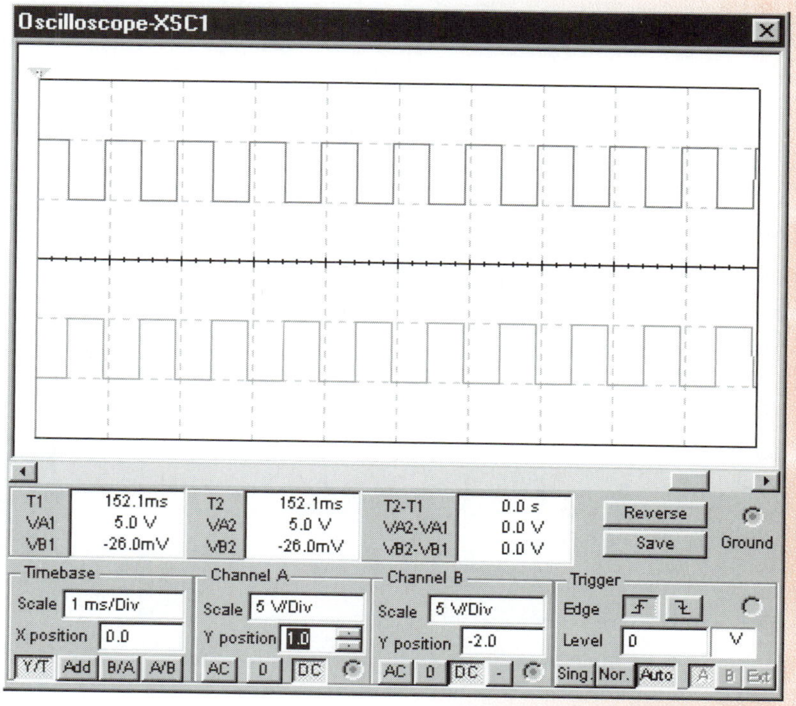

The next circuit is a CMOS NOR gate created using MOSFET transistors. The circuit is shown in Figure 18–57. Notice that the NMOS transistors are used to make the connection to ground and the PMOS transistor network makes the connection to V_{DD}. Two function generators are used to create two input voltage levels.

The input and output traces for the CMOS NOR gate are shown in Figure 18–58. Notice that when the voltage level for any of the two inputs are high, the output is low. This is what is expected for a NOR gate.

Note: It may be necessary to change the rise/fall time characteristics of the function generator signals to 1 μs as shown in Figure 18–59 to avoid simulation errors due to the fast rise and fall time of the input signals.

FIGURE 18–57 The transistor-level implementation of the CMOS NOR gate

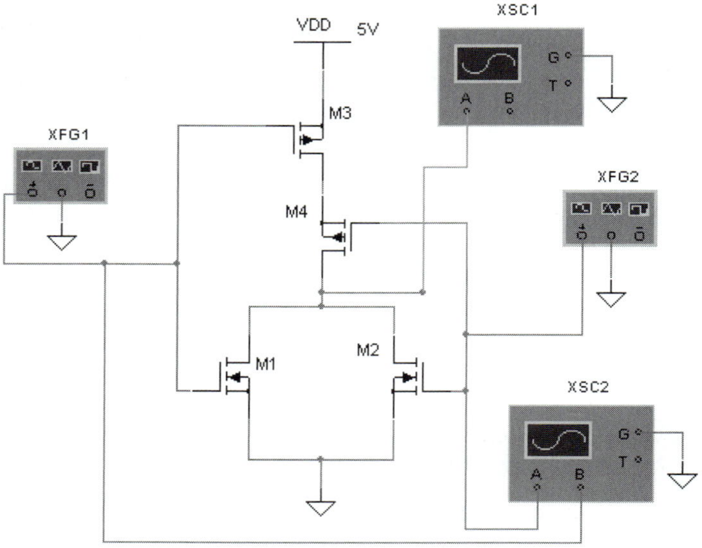

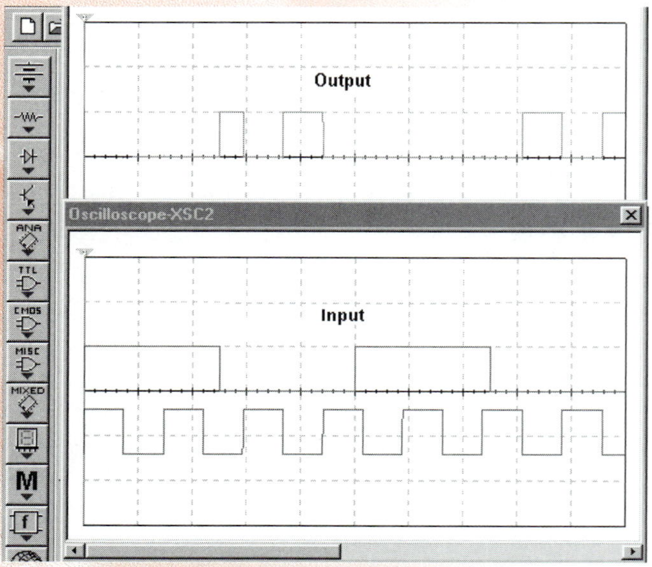

FIGURE 18–58 The input and output traces for the CMOS NOR gate

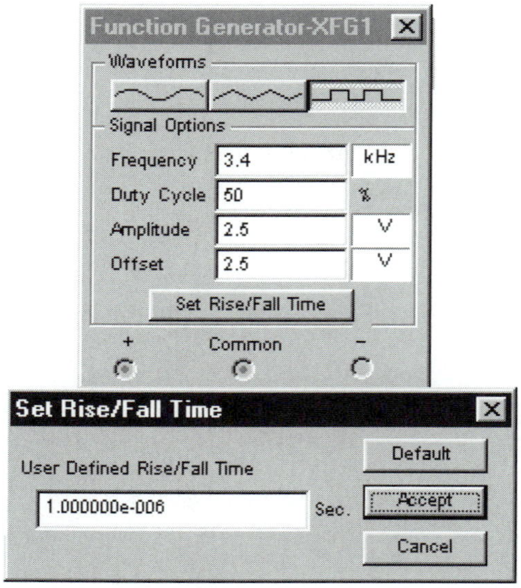

FIGURE 18–59 The menu for setting the rise/fall time characteristics of the function generator

This section has demonstrated how to use EWB Multisim to simulate CMOS logic circuits using MOSFET transistors. The student must remember that CMOS circuits have a pull-down network made up of NMOS transistors and a pull-up network made up of PMOS transistors.

SUMMARY

This chapter has addressed the basic use of MOSFET transistors in integrated CMOS circuits. VLSI design is a complex task that requires many years of study and practice to be successful. However, the reader should have gained sufficient understanding to identify and design simple CMOS logic circuits, perform some basic analytical measurements of performance, and simulate the circuit using accurate fabrication-model parameters.

The concepts the student should understand are:

- The concept of the pull-up and pull-down networks in CMOS circuits.
- How combinational logic circuits are constructed using transistors.
- How signals are clocked into and out of CMOS logic circuits.
- The analog signal issues of memory circuits.
- The purpose of the input/output buffers in CMOS integrated circuits.

EXERCISES

SECTION 18–2

Transistor-Level Implementation of CMOS Combinational Logic Circuits

18–1. Several transistor devices are shown in Figure 18–60(a–d). Which of the devices form

(a) a CMOS inverter?

(b) a CMOS transmission gate?

(c) an AND structure?

(d) an OR structure?

18–2. Identify in which one of the CMOS circuits shown in Figure 18–61 the transistors are not properly connected.

SECTION 18–3

A Design Procedure for Creating CMOS Combinational Logic Circuits

18–3. Design a transistor-equivalent circuit for a CMOS 2-input AND gate.

18–4. Design a transistor-equivalent circuit for a CMOS 2-input NOR gate.

18–5. Design a transistor-equivalent circuit for a CMOS 4-input NAND gate.

18–6. Design a transistor-equivalent circuit for a CMOS 3-input OR gate.

18–7. Given that $F = [A \oplus B]_{-L}$, create the transistor-equivalent circuit for this function.

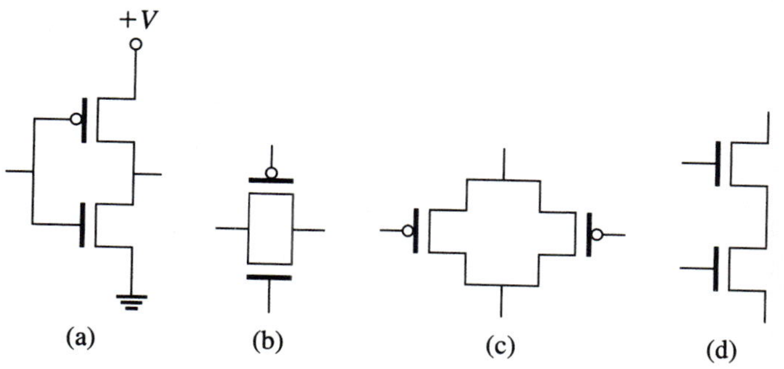

FIGURE 18–60 (Exercise 18–1)

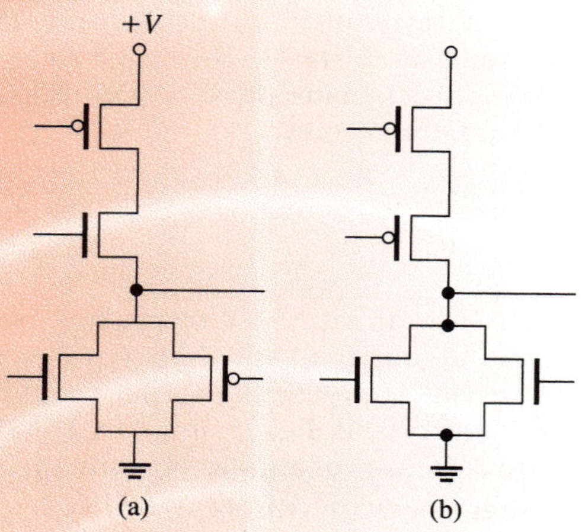

FIGURE 18–61 (Exercise 18–2)

18–8. Identify the logic function provided by the circuit shown in Figure 18–62.

18–9. Create the equivalent transistor CMOS circuit for the following function: Assume all inputs assert high.

$$F = [A \cdot \overline{B} + \overline{C} \cdot D]_{-L}$$

18–10. Create the equivalent transistor CMOS circuit for the following function: Assume all inputs assert high.

$$F = [[A \cdot \overline{B} + \overline{C} \cdot D] \cdot F]_{-L} \quad F = [A \cdot \overline{B} + \overline{C} \cdot D]_{-H}$$

18–11. Create the equivalent transistor CMOS circuit for the following function: Assume all inputs assert high.

$$F = [A \cdot \overline{B} + \overline{A} \cdot B]_{-L}$$

18–12. Create the equivalent transistor CMOS circuit for the following function. Assume that all of the inputs assert High.

$$F = [X + \overline{Y}Z]_{-H}$$

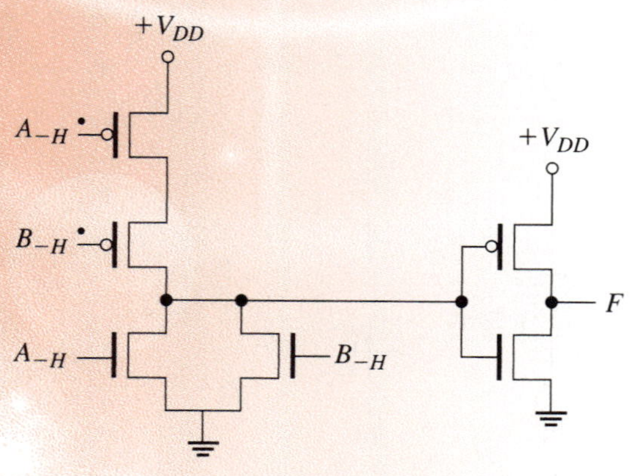

FIGURE 18–62 (Exercise 18–8)

18–13. Create the equivalent transistor CMOS circuit for the following function. Assume that X asserts Low and Y and Z assert high.

$$F = [X + \overline{Y}Z]_{-H}$$

18–14. Create the equivalent transistor CMOS circuit for the following function. Assume that X asserts High, and Y and Z assert Low.

$$F = [X + \overline{Y}Z]_{-H}$$

SECTION 18–4

Transient Behavior of CMOS Logic Circuits

18–15. Determine the operating mode for a PMOS and NMOS transistor configured as a CMOS inverter. Assume that $\beta_n = \beta_p$ and $V_{tn} = |V_{tp}| = 0.7$ V for the following conditions:

(a) $V_{in} = 1.65$ V $V_{out} = 2.5$ V
(b) $V_{in} = 1.4$ V $V_{out} = 2.1$ V
(c) $V_{in} = 4.4$ V $V_{out} = 0.65$ V

18–16. Given a CMOS inverter with $W_n = 20\ \mu$, $L_p = L_n = 2\ \mu$, $KP_p = 2.0\text{E-05 A/V}^2$, and $KP_n = 5.0\text{E-05 A/V}^2$, find the value of the gate width, W_p, so that $\beta_n = \beta_p$.

18–17. Determine the β of an NMOS transistor with a $(W/L)_n$ of 5 μ/1 μ and $Kp_n = 8.43\text{ E-05 A/V}^2$. If $\beta_n = 3\beta_p$ and $(W/L)_p = 12\ \mu/1\ \mu$, what is the value of KP_p?

18–18. Determine the switching point of a CMOS inverter if $\beta_p = 2 \times \beta_n$. Assume $V_{tn} = |V_{t_p}| = 1.0$ V and $KP_n = 2.5\ KP_p$, $V_{DD} = 5.0$ V, and $L_p = L_n = 2.0\ \mu$.

18–19. Find the switching point for the CMOS Inverter if $V_{t_n} = |V_{t_p}| = 0.7$ V and $KP_p = 2KP_p$, $V_{DD} = 3.3$ V, and $\beta_p = \beta_n$.

18–20. Determine the drain-bulk capacitance value C_{dbn} given the following line from a SPICE .CIR file. CJ = 1.21E-04, CJSW = 5.321E-10

```
MN1 4 1 0 0 CMOSN W=15U L=2U AS=75U
AD=75U PS=40U PD=40U
```

18–21. For the CMOS inverter shown in Figure 18–63, determine the rise and fall time given that $\beta_n = 25\ \mu\text{A/V}^2$ and $\beta_p = 10\ \mu\text{A/V}^2$.

18–22. Determine the rise and fall times for a CMOS inverter that is driving a total load capacitance of 1 pF. You are given the following: $V_{DD} = 3.3$ V.

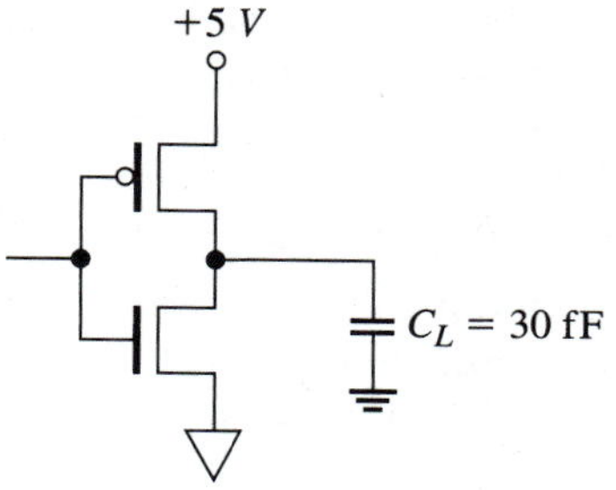

FIGURE 18–63 (Exercise 18–21)

(p-device)	KP_p = 6.7E-06	W = 56 μ	L = 2 μ
(n-device)	KP_n = 1.3E-06	W = 23 μ	L = 2 μ

18–23. The following specifications are provided for a CMOS inverter: $t_r = 0.82$ ns, $t_f = 0.88$ ns. What is the propagation delay for the circuit?

18–24. A CMOS inverter with a propagation delay of 0.351 ns is driving the circuit in Exercise 18–22. What is the total propagation delay for a signal passing through both circuits?

SECTION 18–5

Clocked CMOS VLSI Circuits

18–25. Modify the DFF in Figure 18–24 so that it has an asynchronous RESET input. The Q output should reset when the RESET input is Low.

18–26. Examine the circuit shown in Figure 18–64. Determine the function of the circuit. Identify the purpose of nodes A,B,C,D, and E.

18–27. Examine the circuit shown in Figure 18–65. Determine the function of the circuit. Identify the purpose of node A.

SECTION 18–6

Random-Access Memory and Read-Only Memory

18–28. Design a 2 × 4-bit CMOS ROM circuit that contains two 4-bit words with the following information stored in it:

	BIT_3	BIT_2	BIT_1	BIT_0
$WORD_1$	0	0	0	1
$WORD_2$	1	0	1	0

18–29. Design a 4 × 4-bit CMOS ROM circuit that contains four 4-bit words with the following information stored in it:

	BIT_3	BIT_2	BIT_1	BIT_0
Word 1	0	0	0	1
Word 2	0	0	1	1
Word 3	1	0	0	1
Word 4	1	1	1	1

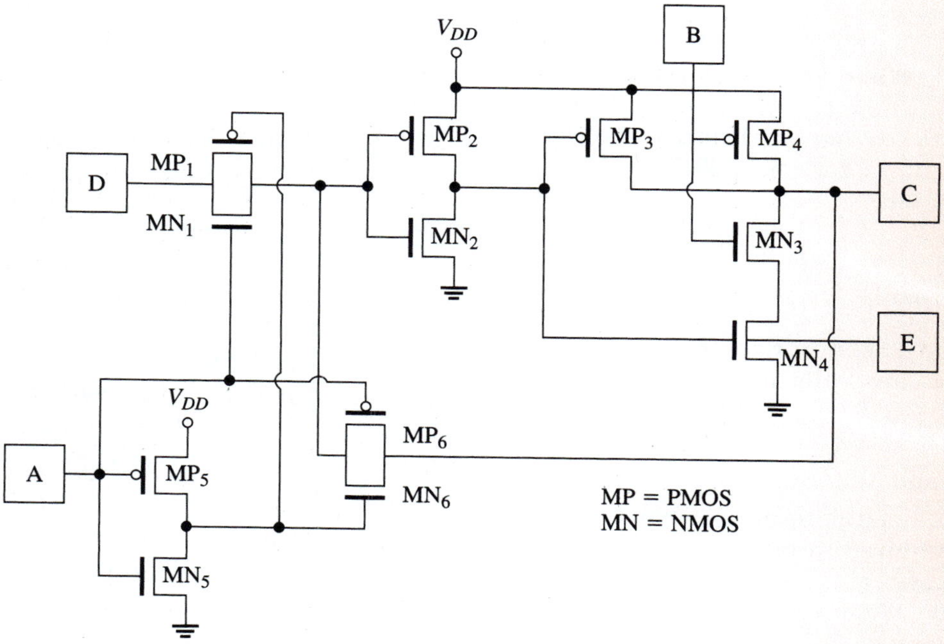

FIGURE 18–64 (Exercise 18–26)

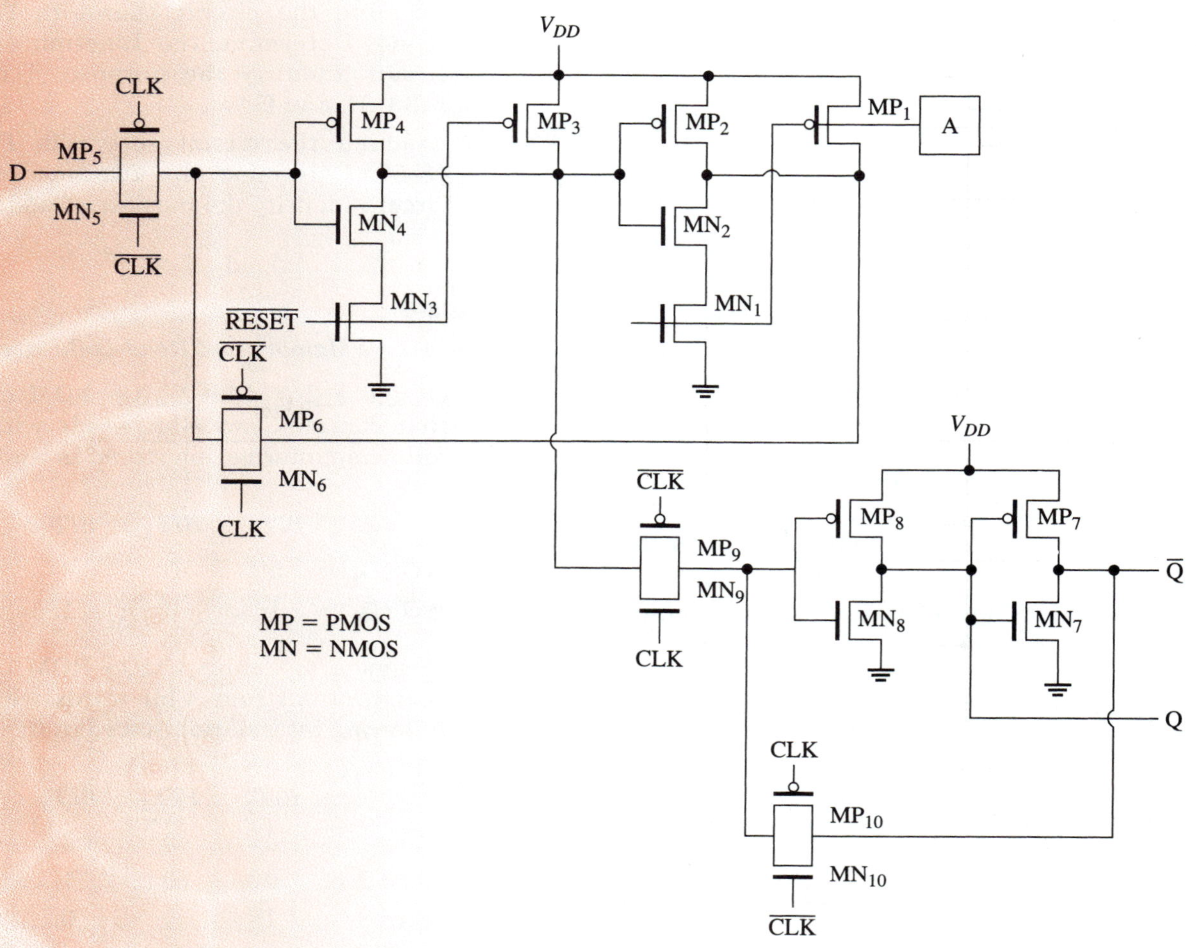

FIGURE 18–65 (Exercise 18–27)

18–30. Examine the circuit shown in Figure 18–66. What are the logic outputs of $B_{0,1,2,3}$ when the Word 1, 2, 3, or 4 is asserted? What condition of the clock ϕ is required for each WORD line to be evaluated?

SECTION 18–7

Input/Output (I/O) Buffers

18–31. Design a Schmitt-trigger circuit that provides the following switching points: $V^+ = 2.8$ V, $V^- = 1.8$ V, $V_{DD} = 5.0$ V, $V_{t_n} = 0.7$ V, $V_{t_p} = -0.7$ V. Solve for the β_1/β_3 and β_4/β_5 ratios.

18–32. It has been determined that five stages are required to buffer an output signal. The gate capacitance of the 1X device is 50 pF. What is the value of the load capacitance C_L?

18–33. Determine the number of stages required to buffer an output signal given that $C_L = 4$ pF and $C_g = 428$ fF.

18–34. Design an input logic-level interface using a CMOS inverter that meets the given criteria. Convert a TTL input with a switching point of 1.6 V to a CMOS level of 2.5 V. Express your answer in terms of the β_n/β_p ratio. When $V_{in} = 1.6$ V, $V_{out} = 2.5$ V. $V_{t_n} = |V_{t_p}| = 1.0$ V, $V_{DD} = 5.0$ V.

18–35. Calculate the switching point for a CMOS inverter if $\beta_n = 1.5\beta_p$. Assume $V_{t_n} = |V_{t_p}| = 0.7$ V, $V_{DD} = 5.0$ V, $KP_n = 2.5\,KP_p$, $L_n = L_p = 1.0\ \mu$.

18–36. Describe and identify the function of the MOSFET circuit shown in Figure 18–67.

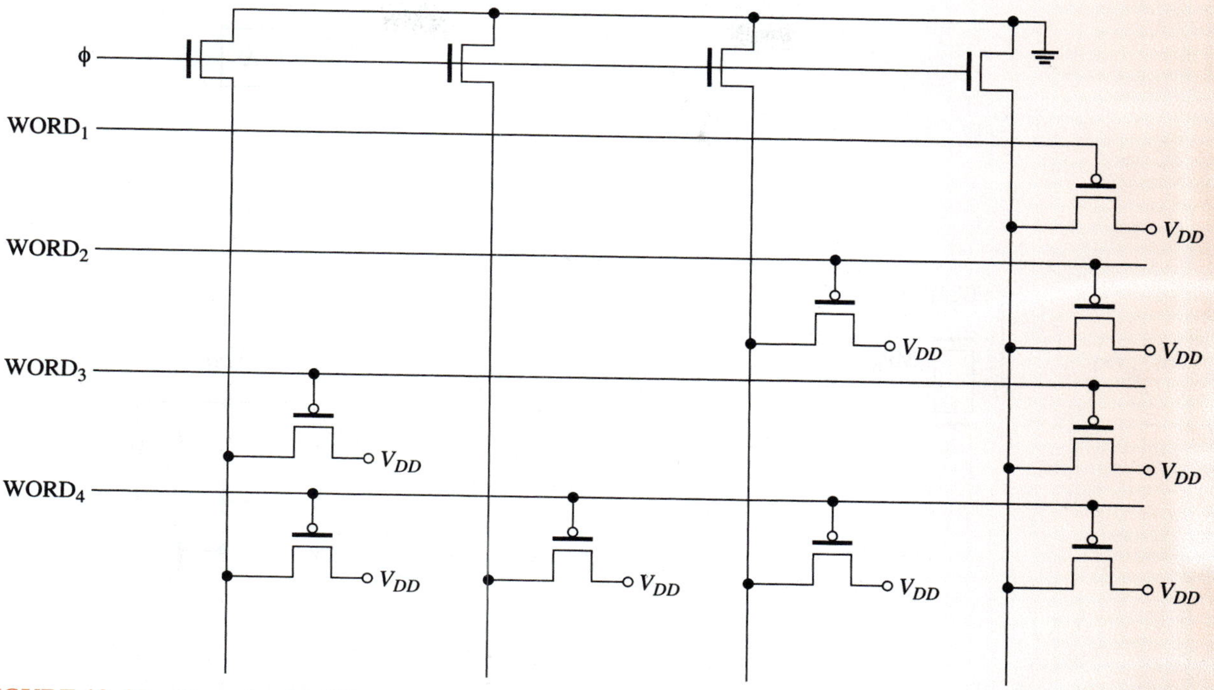

FIGURE 18–66 (Exercise 18–30)

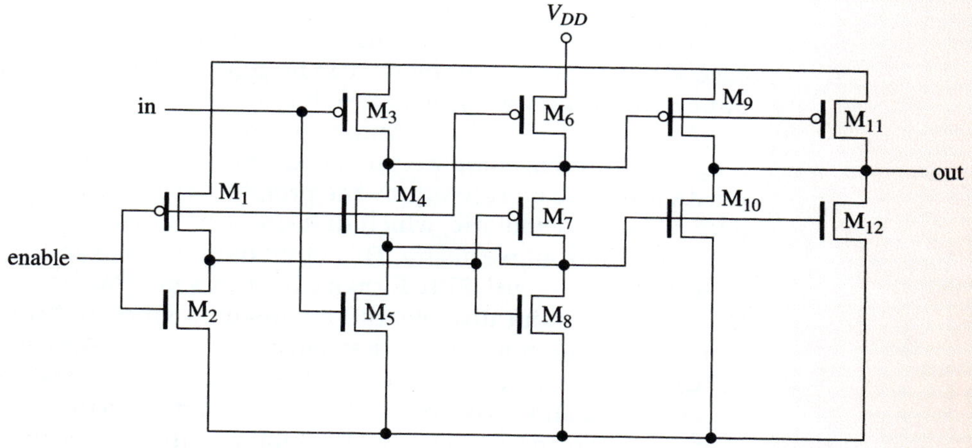

FIGURE 18–67 (Exercise 18–36)

SPICE EXERCISES

18–37. Perform a SPICE simulation to verify the Schmitt trigger designed in Exercise 18–31. Use the model parameters provided in Table 18–3.

18–38. Verify your answer in Exercise 18–34 using SPICE analysis. Use the model parameters provided in Table 18–3 in your SPICE simulation.

18–39. Perform SPICE analysis on Exercise 18–4 using the Table 18–3 model parameters in your simulation.

18–40. Perform SPICE analysis on Exercise 18–5 using the Table 18–3 model parameters in your simulation.

APPENDIX A

SPICE AND PSPICE

A–1 INTRODUCTION

SPICE—Simulation Program with Integrated Circuit Emphasis—was developed at the University of California, Berkeley, as a computer aid for designing integrated circuits. However, it is readily used to analyze discrete circuits as well and can, in fact, analyze circuits containing no semiconductor devices at all. In addition to this versatility, SPICE owes its current popularity to the ease with which a circuit model can be constructed and the wide range of output options (analysis types) available to the user.

As a brief note on terminology, observe that SPICE is a computer *program,* stored in computer memory, which we do not normally inspect or alter. As users, we merely supply the program with data, in the form of an *input data file,* or a schematic, which describes the circuit we wish to analyze and the type of output we desire. This input data file is usually supplied to SPICE by way of a keyboard. SPICE then executes a program run, using the data we have supplied, and displays the results on a video terminal or printer. The mechanisms, or *commands,* that must be used to supply the input data to SPICE and to cause it to execute a program run vary widely with the computer system and software used, so we cannot describe that procedure here. Depending on the version used, minor variations in *syntax* (the format for specifying the input data) may be encountered. The *User's Guide,* supplied with most versions, should be consulted if any difficulty is experienced with any of the programs in this book. All programs here have been run successfully using PSpice.

PSPICE

One of the most widely used versions of SPICE designed for operation on microcomputers is PSpice (a registered trademark of the MicroSim Corporation). It has numerous features and options that make it more versatile and somewhat easier to use than the original Berkeley version of SPICE. However, with a few minor exceptions, any input data file written to run on Berkeley SPICE will also run on PSpice. In the discussions that follow, features of PSpice are highlighted immediately after the corresponding capabilities or requirements of Berkeley SPICE are described.

A–2 DESCRIBING A CIRCUIT FOR A SPICE INPUT FILE

The input data file consists of successive lines, which we will hereafter refer to as *statements*, each of which serves a specific purpose, such as identifying one component in the circuit. The statements do not have to be numbered and, except for the first and last, can appear in any order.

The Title

The first statement in every input file must be a *title*. Subject only to the number of characters permitted by a particular version, the title can be anything we wish. Examples are:

AMPLIFIER

EXERCISE 2.25

A DIODE (1N54) TEST

Nodes and Component Descriptions

The first step in preparing the circuit description is to identify and number all the *nodes* in the circuit. It is good practice to draw a schematic diagram with nodes shown by circles containing the node numbers. Node numbers can be any positive integers, and one of them must be 0. (The zero node is usually, but not necessarily, the circuit common, or ground.) Node numbers can be assigned in any sequence, such as 0, 1, 2, 3, . . . or 0, 2, 4, 6, ...

Once the node numbers have been assigned, each component in the circuit is identified by a separate statement that specifies the type of component it is and the node numbers between which it is connected. The first letter appearing in the statement identifies the component type. Passive components (resistors, capacitors, and inductors) are identified by the letters *R, C,* and *L*. Any other characters can follow the first letter, but each component must have a unique designation. For example, the resistors in a circuit might be identified by R1, R2, RB, and REQUIV.

The node numbers between which the component is connected appear next in the statement, separated by one or more spaces. Except in some special cases, it does not matter which node number appears first. The component value, in ohms, farads, or henries, appears next. Resistors cannot have value 0. Following are some examples:

R25 6 0 100	(Resistor R25 is connected between nodes 6 and 0 and has value 100 Ω.)
R25 0 6 1E2	(Interpreted by SPICE the same as in the first case.)
CIN 3 5 22E–6	(Capacitor CIN is connected between nodes 3 and 5 and has value 22 μF.)
LSHUNT 12 20 0.01	(Inductor LSHUNT is connected between nodes 12 and 20 and has value 0.01 H.)

PSPICE

In the foregoing examples, all letters used in the input data files are capitalized, a requirement of Berkeley SPICE. In PSpice, either lowercase or capital letters (or both) may be used. For example, resistor R1 can be listed as r1 in one line and referred to again as R1 in another line, and PSpice will recognize both as representing the same component.

Node numbers in PSpice do not have to be integers; they can be identified by any set of alphanumeric characters (letters or numbers) up to 31 in length. However, one node must be node 0.

Specifying Numerical Values

Suffixes representing powers of 10 can be appended to value specifications (with no spaces in between). Following are the SPICE suffix designations:

T	(tera: 10^{12})	U	(micro: 10^{-6})
G	(giga: 10^{9})	N	(nano: 10^{-9})
MEG	(mega: 10^{6})	P	(pico: 10^{-12})
K	(kilo: 10^{3})	F	(femto: 10^{-15})
M	(milli: 10^{-3})		

Note that M represents *milli* (10^{-3}) and that MEG is used for 10^6. Following are some examples of equivalent ways of representing values, all of which are interpreted by SPICE in the same way:

0.002 = 2M = 2E–3 = 2000U
5000E–12 = 5000P = 5N = .005U = 0.005E–6
0.15MEG = 150K = 150E3 = 150E+3 = .15E+6

Any characters can follow a value specification, and unless the characters are one of the powers-of-10 suffixes just given, SPICE simply ignores them. Characters are often added to designate units. Following are some examples, all of which are interpreted by SPICE in the same way:

100UF = 100E–6F = 100U = 100UFARADS
56N = 56NSEC = 0.056US = 56E–9SECONDS
2.2K = 2200OHMS = 0.0022MEGOHMS = 2.2E3
0.05MV = 50UVOLTS = 50E–6V = 0.05MILLIV

Be careful not to use a unit that begins with one of the power-of-10 suffixes. For example, a 0.0001-F capacitor specified as 0.0001F would be interpreted by SPICE as 0.0001×10^{-15} F.

dc Voltage Sources

The first letter designating a voltage source, dc or ac, is V. As with passive components, any characters can follow the V, and each voltage source must have a unique designation. Examples are VI, VIN, and VSIGNAL. Following is the format for representing a dc source:

```
V******* N+ N- <DC> value
```

where ******* are arbitrary characters, N+ is the number of the node to which the positive terminal of the source is connected, and N− is the number of the node to which the negative terminal is connected. The symbol < > enclosing DC means that the specification DC is optional: If a source is not designated DC, SPICE will automatically assume that it is DC. *Value* is the source voltage, in volts. *Value* can be negative, which is equivalent to reversing the N+ and N− node numbers. The following are examples:

```
VIN 5 0 24VOLTS
VIN 0 5 DC -24
```

Both of these statements specify a 24-V-dc source designated VIN whose positive terminal is connected to node 5 and whose negative terminal is connected to node 0. Both statements are treated the same by SPICE.

Figure A–1 shows some examples of circuits and the statements that could be used to describe them in a SPICE input file. (These examples are not complete input files because we have not yet discussed *control* statements, used to specify the type of analysis and the output desired.)

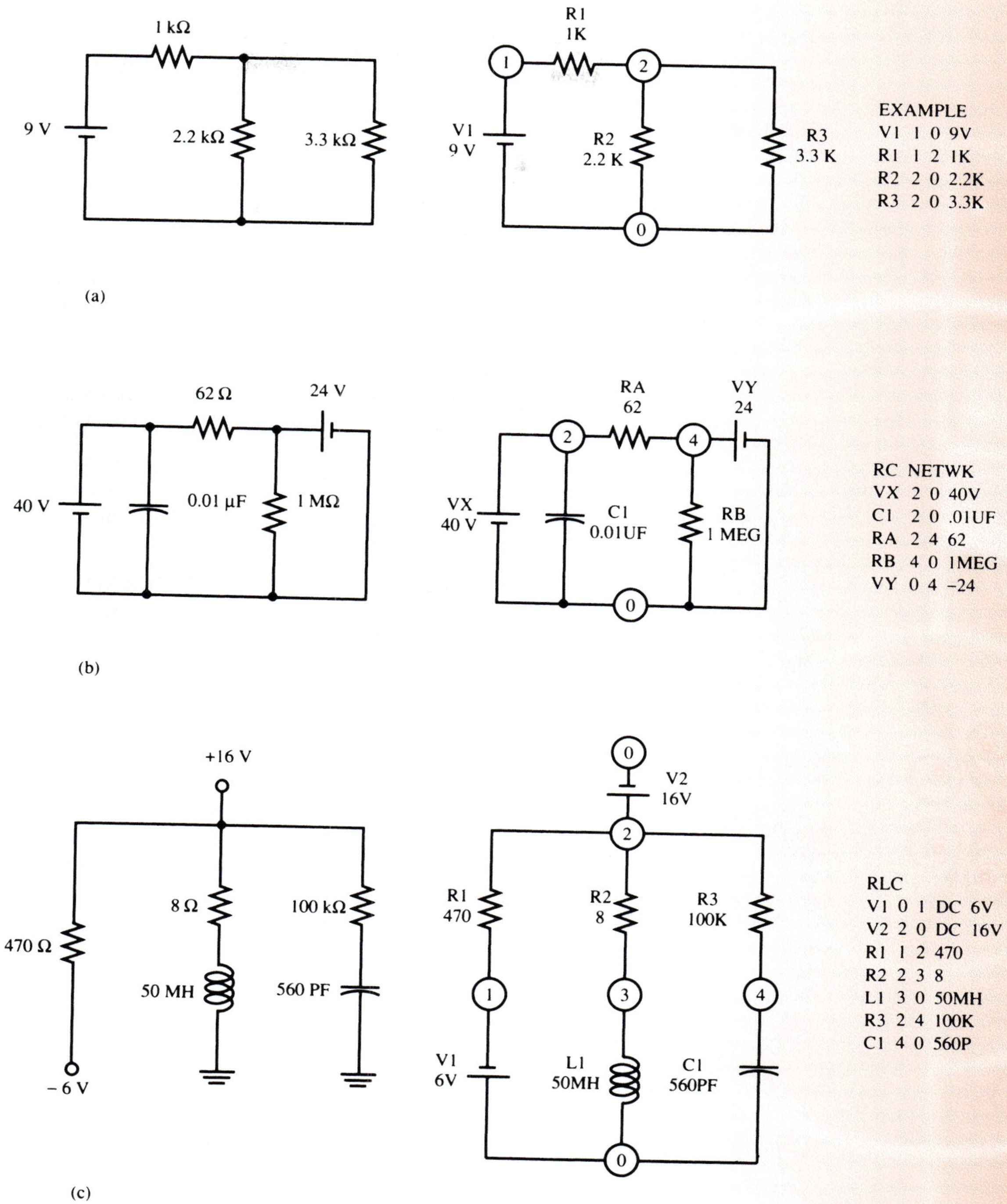

FIGURE A–1 Examples of circuit descriptions for a SPICE input data file

dc Current Sources

The format for specifying a dc current source is

```
I******* N+ N- <DC> value
```

where *value* is the value of the source in amperes and DC is optional. Note the following *unconventional* definition of N+ and N−: *N− is the number of*

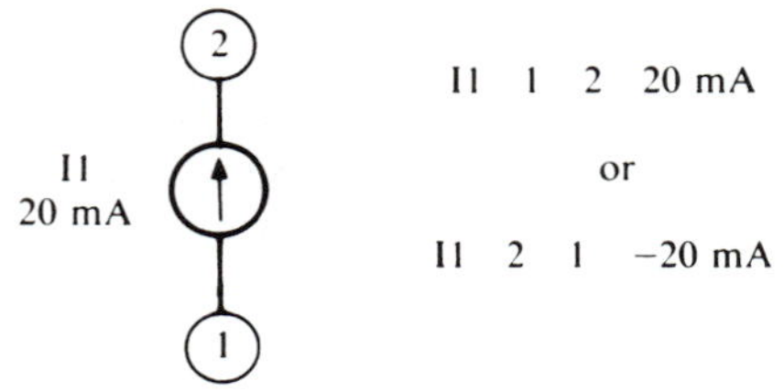

FIGURE A–2 Equivalent ways of specifying a dc current source. Note that the "negative" terminal is the one to which current is delivered ($N- = 2$)

the node to which the source delivers current. To illustrate, Figure A–2 shows two equivalent ways of specifying a 20-mA dc current source.

A–3 THE .DC AND .PRINT CONTROL STATEMENTS

A control statement is one that specifies the type of analysis to be performed or the type of output desired. Every control statement begins with a period, followed immediately by a group of characters that identifies the type of control statement it is.

The .DC Control Statement

The .DC control statement tells SPICE that a dc analysis is to be performed. This type of analysis is necessary when the user wishes to determine dc voltages and/or currents in a circuit. Although ac sources can be present in the circuit, there must be at least one dc (voltage or current) source present if a dc analysis is to be performed. The format for the .DC control statement is

```
.DC name1 start stop incr. <name2 start2 stop2 incr.2>
```

where *name1* is the name of one dc voltage or current source in the circuit and *name2* is (optionally) the name of another. The .DC control statement can be used to *step* a source through a sequence of values, a useful feature when plotting characteristic curves. For that use, *start* is the first value of voltage or current in the sequence, *stop* is the last value, and *incr.* is the value of the increment, or step, in the sequence. A second source, *name2*, can also be stepped. If analysis is desired at a *single* dc value, we set *start* and *stop* both equal to that value and arbitrarily set *incr.* equal to 1. In cases where the circuit contains several fixed-value sources, one of them (any one) *must* be specified in the .DC control statement. Following are some examples:

`.DC V1 24 24 1`	V1 is a fixed 24-V-dc voltage source.
`.DC IB 50U 50U 1`	IB is a fixed 50-μA-dc current source.
`.DC VCE 0 50 10`	VCE is a dc voltage source that is stepped from 0 to 50 V in 10-V increments.

When two sources are stepped in a .DC control statement, the first source is stepped over its entire range for each value of the second source. The following is an example:

```
.DC VCC 0 25 5 IBB 0 20U 2U
```

In this example, the dc voltage source named VCC is stepped from 0 to 25 V for *each* value of the dc current source named IBB.

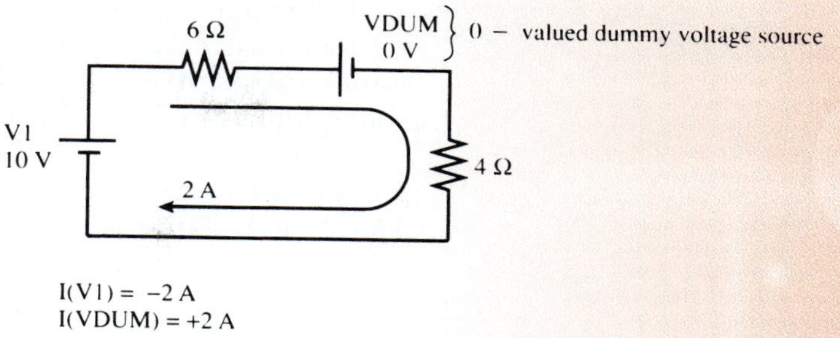

FIGURE A–3 SPICE treats conventional current flowing out of a positive terminal as negative current and current flowing into a positive terminal as positive current

Identifying Output Voltages and Currents

To tell SPICE the voltages whose values we wish to determine (the *output* voltages we desire), we must identify them in one of the following formats:

V(N+, N−) or V(N+)

In the first case, V(N+, N−) refers to the voltage at node N+ with respect to node N−. In the second case, V(N+) is the voltage at node N+ with respect to node 0. The following are examples:

V(5, 1) The voltage at node 5 with respect to node 1.

V(3) The voltage at node 3 with respect to node 0.

The only way to obtain the value of a current in Berkeley SPICE is to request the value of the current in a *voltage* source. Thus, a voltage source must be in the circuit at any point where we wish to know the value of the current. The current is identified by I(*Vname*), where *Vname* is the name of the voltage source. *We can insert zero-valued voltage sources (dummy sources) anywhere in a circuit for the purpose of obtaining a current.* These dummy voltage sources serve as ammeters for SPICE and do not in any way affect circuit behavior. Note carefully the following unconventional way that SPICE assigns polarity to the current through a voltage source: Positive current flows *into* the positive terminal (N+) of the voltage source. Figure A–3 shows an example. Here, conventional positive current flows in a clockwise direction, but a SPICE output would show I(V1) equal to −2 A. On the other hand, SPICE would show the current in the zero-valued dummy source I(VDUM) to be +2 A.

The .PRINT Control Statement

The .PRINT statement tells SPICE to print the values of voltages and/or currents resulting from an analysis. The format is

```
.PRINT type out1 <out2 out3 . . . >
```

where *type* is the type of analysis and *out1, out2, . . .* , identify the output voltages and/or currents whose values we desire. So far, DC is the only analysis type we have discussed. For example, the statement

```
.PRINT DC V(1, 2) V(3) I(VDUM)
```

tells SPICE to print the values of the voltages V(1, 2) and V(3) and the value of the current I(VDUM) resulting from a dc analysis. If the .DC control statement specifies a stepped source, the .PRINT statement will print all the output values resulting from all the stepped values. The number of output variables whose values can be requested by a single .PRINT statement may vary with the version of SPICE used (up to eight can be requested in version 2G.6). Any number of .PRINT statements can appear in a SPICE input file.

PSPICE

In PSpice, the current through any component can be requested directly, without using a dummy voltage source. For example, I(R1) is the current through resistor R1. The reference polarity of the current is from the first node number of R1 to the second node number. In other words, positive current is assumed to flow from the first node given in the description of R1 to the second node. For example, if 5 A flows from node 1 to node 2 in a circuit whose input data file contains

```
R1 1 2 10
.PRINT DC I(R1)
```

then the .PRINT statement will produce 5 A. On the other hand, if the description of the same R1 were changed to R1 2 1 10, then the same .PRINT statement would produce −5 A.

Also, voltages across components can be requested directly, as, for example, V(RX), the voltage across resistor RX. The reference polarity is from the first node listed in the description of RX to the second node. For example, if the voltage across RX from node 5 to node 6 is 12 V in a circuit whose input data file contains

```
RX 5 6 1K
.PRINT DC V(RX)
```

then the .PRINT statement will produce 12 V. On the other hand, if the description of the same resistor were changed to RX 6 5 1K, the same .PRINT statement would produce −12 V.

The .PRINT statement has its place in SPICE analysis but sees limited use in modern PSpice packages.

The .END Statement

The last statement in every SPICE input file must be .END. We have now discussed enough statements to construct a complete SPICE input file, as demonstrated in the next example.

EXAMPLE A–1

Use SPICE to determine the voltage drop across and the current through every resistor in Figure A–4(a).

Solution

Figure A–4(b) shows the circuit when redrawn and labeled for analysis by SPICE. Note that two dummy voltage sources are inserted to obtain currents in two branches. The polarities of these sources are such that positive currents will be computed. In PSpice, we can simply request I(R2) and I(R3) in the .PRINT statement. The current in R1 is the same as the current in V1, and we must simply remember that SPICE will print a negative value for that current.

Figure A–4(c) shows the SPICE input file. Note that it is not necessary to specify a voltage value in the statement defining V1, because that value is given in the .DC statement. Figure A–4(d) shows the results of a program run. The outputs appear under the heading “DC TRANSFER CURVES,” which refers to the type of output obtained when the source(s) are stepped. In our case, the heading is irrelevant. Note that the analysis is performed under the (default) assumption that the circuit temperature is 27°C (80.6°F). We will see later that we can specify different temperatures. Referring to the circuit nodes in Figure A–4(b), we see that the printed results give the following voltages and currents:

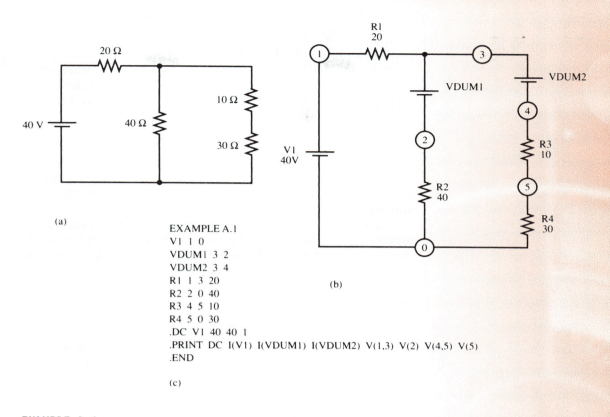

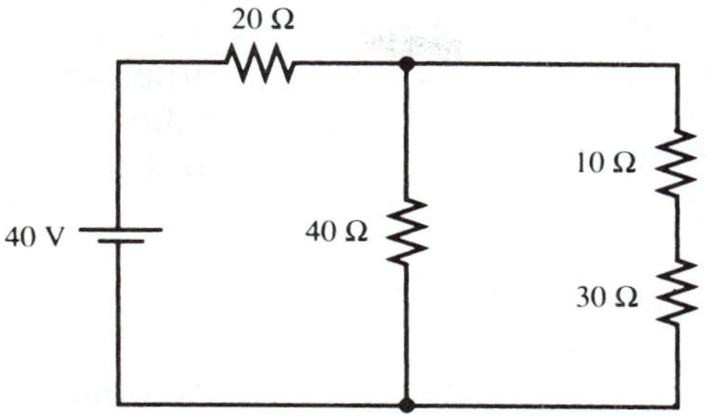

(a)

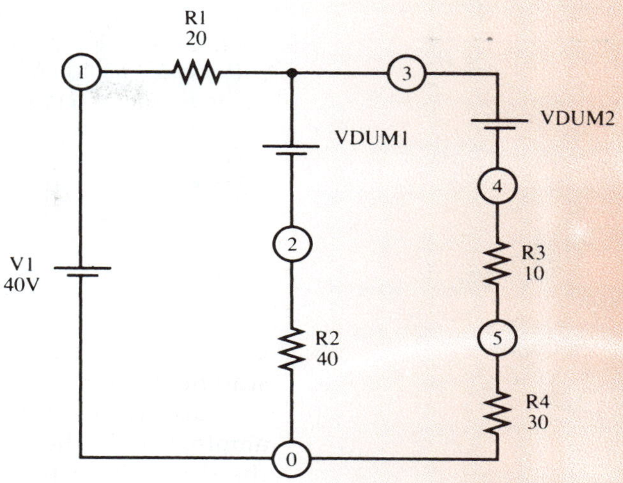

(b)

```
EXAMPLE A.1
V1 1 0
VDUM1 3 2
VDUM2 3 4
R1 1 3 20
R2 2 0 40
R3 4 5 10
R4 5 0 30
.DC V1 40 40 1
.PRINT DC I(V1) I(VDUM1) I(VDUM2) V(1,3) V(2) V(4,5) V(5)
.END
```

(c)

```
EXAMPLE A.1
****      DC TRANSFER CURVES                  TEMPERATURE =     27.000 DEG C
******************************************************************************
  V1             I(V1)         I(VDUM1)       I(VDUM2)       V(1,3)         V(2)
 4.000E+01     -1.000E+00     5.000E-01      5.000E-01      2.000E+01      2.000E+01


******************************************************************************
  V1             V(4,5)        V(5)
 4.000E+01      5.000E+00     1.500E+01
```

(d)

FIGURE A–4 (Example A–1)

	I	V
R_1	1 A	20 V
R_2	0.5 A	20 V
R_3	0.5 A	5 V
R_4	0.5 A	15 V

Circuit Restrictions

Every node in a circuit defined for SPICE must have a *dc path to ground* (node 0). A dc path to ground can be through a resistor, inductor, or voltage source but not through a capacitor or current source. Figure A–5(a) shows two examples of nodes that do not have dc paths to ground and that cannot, therefore, appear in a SPICE circuit. However, we can connect a very large resistance in parallel with a capacitor or current source to provide a dc path to ground. The resistance

FIGURE A–5 Examples of circuits that cannot be simulated in SPICE (for the reasons cited). To force a simulation, a very large resistance can be connected in parallel with a capacitor in (a) and a very small resistance can be connected in series with either V1 or L in (b)

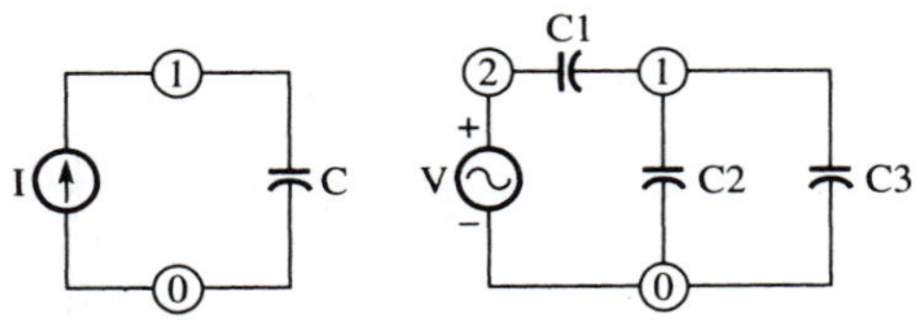

(a) Neither circuit has a dc path from node 1 to ground (node 0).

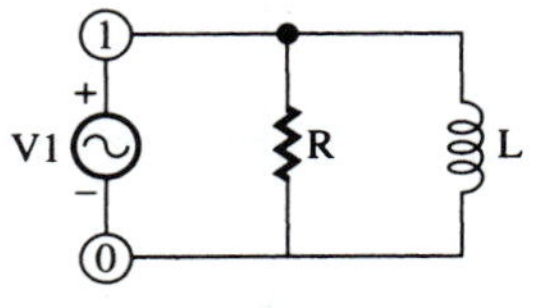

(b) Voltage source V1 and inductance L appear in a closed loop.

should be very large in comparison to other impedances in the circuit so that it will have a negligible effect on the computations. For example, if the capacitive reactances in the circuits of Figure A–5(a) are less than 1 MΩ we can specify a 10^{12}-Ω resistor, RDUM (1 million megohms), between nodes 1 and 0:

```
RDUM 1 0 1E12
```

SPICE does not permit any loop (closed circuit path) to consist exclusively of inductance(s) and voltage source(s). Figure A–5(b) shows an example. Here, inductance L appears in a closed loop with voltage source V1. To circumvent this problem, we can insert a very small resistance in such a loop. The resistance should be much smaller than the impedances of other elements in the circuit in order to have a negligible effect on the computations. For example, if the impedances of R and L in Figure A–5(b) are greater than 1 Ω, we could insert a 1-pΩ resistor, RDUM (10^{-12} Ω), in series with either L or V1.

A–4 THE .TRAN CONTROL STATEMENT

The .TRAN Control Statement

The .TRAN control statement (derived from "transient") is used when we want to obtain values of voltages or currents versus *time* (whether they are transients in the traditional sense or not). We must specify the total time interval over which we wish to obtain the time-varying values and the increment of time between each using the format

```
.TRAN STEP TSTOP <TSTART> <STEPCEILING>
```

where *STEP* is the time increment and *TSTOP* is the largest value of time at which values will be computed. Unless we optionally specify the start time, *TSTART,* SPICE assumes it to be 0. *STEPCEILING* is the maximum time increment allowed in the computations. To illustrate, the statement

```
.TRAN 5M 100M
```

will cause SPICE to produce values of the output(s) at the 21 time points 0, 5 ms, 10 ms, . . . , 100 ms. When .TRAN is used as the analysis type in a .PRINT statement, 21 values of each output variable specified in the .PRINT statement will be printed. We will show an example of a .TRAN analysis (Example A–2) after discussing a few more statement types.

A–5 THE SIN AND PULSE SOURCES

If we wish to obtain a printout or plot of an output versus time, as in a .TRAN analysis, at least one source in the circuit should itself be a time-varying voltage or current. In other words, we will not be able to observe an output *waveform* unless an input waveform is defined. It is not sufficient to indicate in a component definition that a particular source is AC instead of DC. A source designated AC is used by SPICE in an .AC analysis, to be discussed presently, and that analysis type does *not* cause SPICE to display time-varying outputs. For a .TRAN analysis, we use a different format to define the time-varying sources. The two sources that are most widely used for that purpose are the SIN (sinusoidal ac) and PULSE sources.

The SIN Source

The format for specifying a sinusoidal voltage source is

```
V******* N+ N- SIN(VO VP FREQ TD θ)
```

where ******* are arbitrary characters, *N*+ and *N*− are the node numbers of the positive and negative terminals, *VO* is the *offset* (dc, or bias level), *VP* is the peak value in volts, and *FREQ* is the frequency in hertz. *TD* and θ are special parameters related to time delay and damping, both of which are set to 0 to obtain a conventional sine wave. A sinusoidal current source is defined by using I instead of V as the first character. Note that it is not possible to specify a phase shift (other than 180°, by reversing *N*+ and *N*−). Figure A–6 shows an example of a SIN source definition. The zero values for *TD* and θ can be omitted in the specification, and SPICE will assume they are zero by default.

PSPICE

In PSpice, a phase angle can be specified for a SIN source. The format (for a sinusoidal voltage) is

```
V******* N+ N- SIN(VO VP FREQ TD θ PH)
```

where *PH* is the phase angle in degrees and all other parameters are the same as in the SIN source description in Berkeley SPICE.

The PULSE Source

The PULSE source can be used to simulate a dc source that is switched into a circuit at a particular instant of time (a *step* input) or to generate a sequence of square, trapezoidal, or triangular pulses. Figure A–7(a) shows the parameters used to define a voltage pulse or pulse-type waveform: one having time delay (*TD*) that elapses before the voltage begins to rise linearly with time, a *rise time (TR)* that represents the time required for the voltage to change from *V1* volts to *V2* volts, a *pulse width (PW)*, and a *fall time (TF)*,

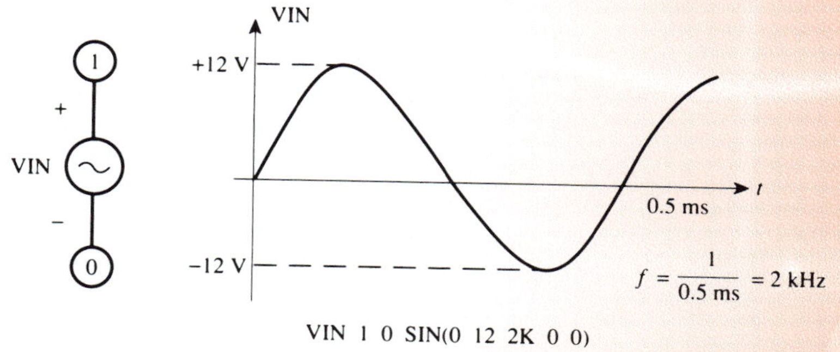

FIGURE A–6 An example of the specification of a sinusoidal voltage source

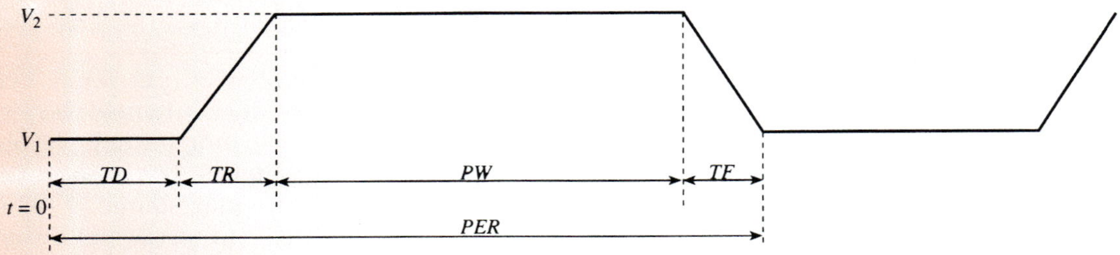

V ******* $N+$ $N-$ PULSE($V_1 V_2$ < *TD TR TF PW PER* >)
Default values: *TD* = 0, *TR* = *STEP*, *TF* = *STEP*, *PW* = *TSTOP*,
PER = *TSTOP*.

(a)

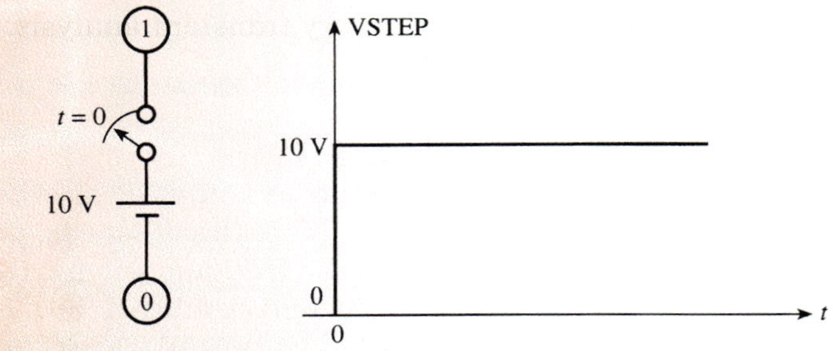

VSTEP 1 0 PULSE(0 10 0 0 0)

(b) Using the PULSE source to simulate a 10-V dc source switched into a circuit at $t = 0$. Even though *TR* and *TF* are set to 0, SPICE assigns each the default value of *STEP* specified in a .TRAN statement. *PW* and *PER* both default to the value of *TSTOP*.

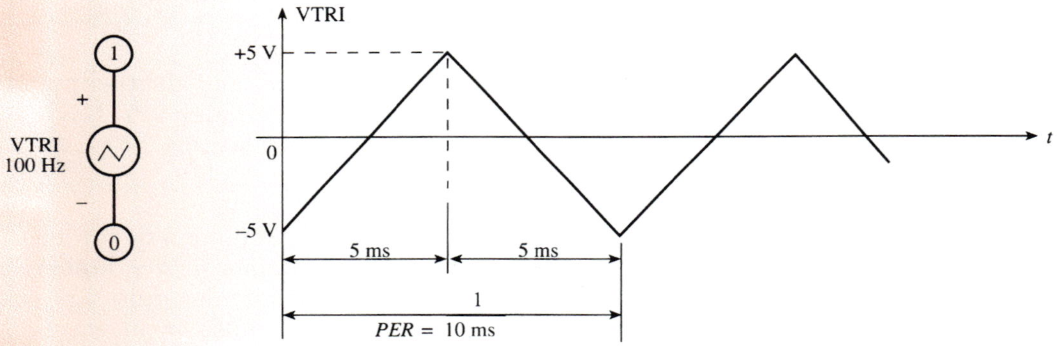

VTRI 1 0 PULSE(–5 5 0 5M 5M 1P 10M)

(c) Using the PULSE source to generate a 100-Hz triangular waveform. Note that $PW = 1$ ps $\cong 0$.

FIGURE A–7 The PULSE source

during which the voltage falls from *V2* volts to *V1* volts. If the pulse is repetitive, a value for the period of the waveform *(PER)* is also specified. If *PER* is not specified, its default value is the value of *TSTOP* in a .TRAN analysis; that is, the pulse is assumed to remain at *V2* volts for the duration of the analysis, simulating a step input. The default value for the rise and fall times is the *STEP* time specified in a .TRAN analysis. The figure shows the format for identifying a pulsed voltage source. Current pulses can be obtained by using

I instead of V as the first character. Figure A–7(b) shows an example of how the PULSE source is used to simulate a step input caused by switching a 10-V-dc source into a circuit at $t = 0$. Figure A–7(c) shows an example of how the PULSE source is used to define a triangular waveform that alternates between ± 5 V with a frequency of 100 Hz. SPICE does not accept a zero pulse width, as would be necessary to define an ideal triangular waveform, but *PW* can be made negligibly small. In this example, we set $PW = 10^{-12}$ s = 1 ps, which makes the period 10^{10} times as great as the pulse width.

A–6 THE INITIAL TRANSIENT SOLUTION

The next example demonstrates the use of the PULSE source to generate a square wave and contains some important discussion on how SPICE performs a dc analysis in conjunction with every transient analysis.

EXAMPLE A–2

Use SPICE to obtain a plot of the capacitor voltage versus time in Figure A–8(a). The plot should cover two full periods of the square-wave input.

Solution

Figure A–8(b) shows the circuit when redrawn for analysis by SPICE and the corresponding input data file. The period of the input is T = 1/(2.5 Hz) = 0.4 s, so *TSTOP* in the .TRAN statement is set to 0.8 s to obtain a plot covering two full periods. Note that *PW* in the definition of V1 is set to 0.2 s. The square wave is idealized by setting the rise and fall times, *TR* and *TF*, to 0, so the actual value assigned by SPICE to *TR* and *TF* is the value of *STEP:* 0.02 s.

Figure A–9 shows the results of a program run. When SPICE performs a .TRAN analysis, it first obtains an "initial transient solution." This solution

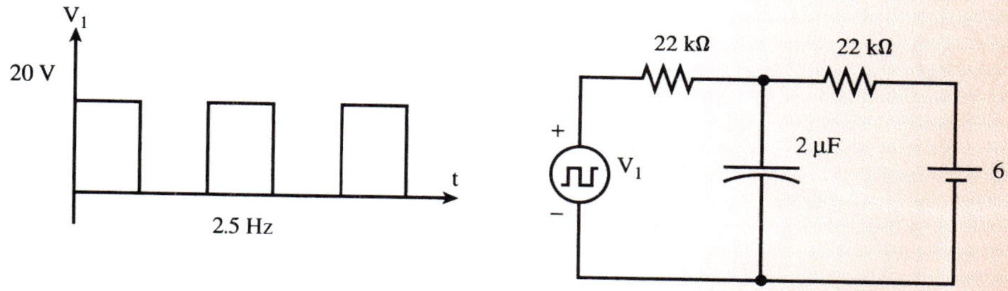

(a)

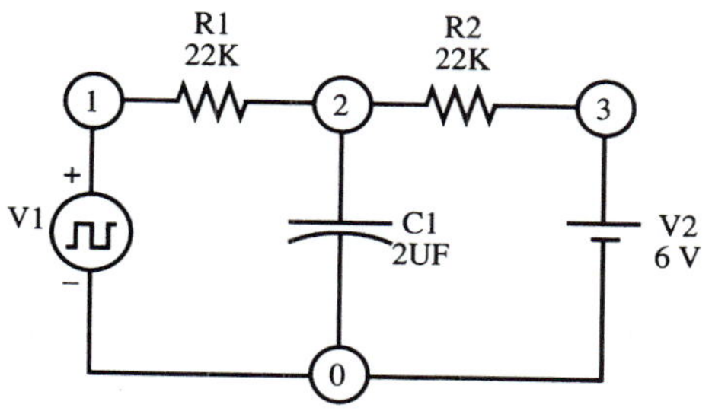


```
EXAMPLE A.2
V1 1 0 PULSE(0 20 0 0 0 0.2 0.4)
V2 3 0 6V
R1 1 2 22K
R2 2 3 22K
C1 2 0 2uF
.TRAN 0.02 0.8
.END
```

(b)

FIGURE A–8 (Example A–2)

```
EXAMPLE A.2
****      INITIAL TRANSIENT SOLUTION        TEMPERATURE =   27.000 DEG C
******************************************************************************
 NODE    VOLTAGE       NODE   VOLTAGE       NODE    VOLTAGE
(  1)       .0000     (  2)    3.0000      (  3)     6.0000
   VOLTAGE SOURCE CURRENTS
   NAME          CURRENT
   V1           1.364D-04
   V2          -1.364D-04
   TOTAL POWER DISSIPATION    8.18D-04   WATTS
```

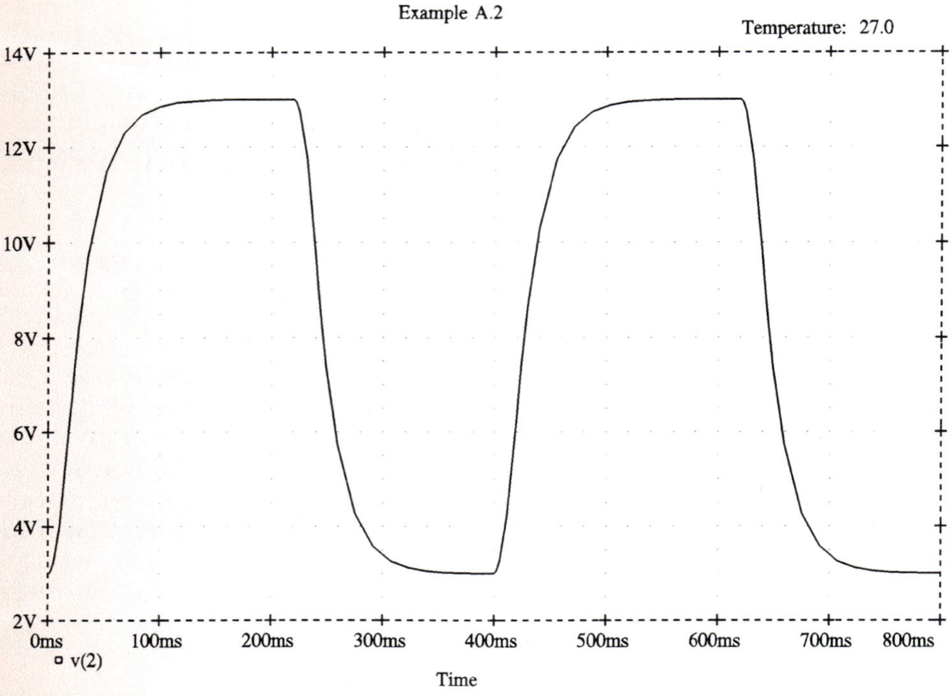

FIGURE A–9 (Example A–2)

is obtained from a dc analysis with all time-varying sources set to zero. Thus, the initial solution represents the *quiescent,* or dc, operating conditions in the circuit, useful information for determining the bias point(s) in circuits containing transistors. The actual time-varying outputs are computed with the initial voltages and currents as the starting points, so the outputs do not reflect initial transients associated with the charging of capacitors, such as coupling capacitors, in the circuit.

In our example, the initial transient solution gives the dc voltages and currents when V1, the square-wave generator, is set to 0. The figure shows that the dc voltages at all nodes are printed, as are the dc currents in all voltage sources and the total dc power dissipation in the circuit. (Note that dc current flows *into* V1 when it is set to 0 V.) Since the capacitor is charged to +3 VDC, the time-varying plot shows its voltage to begin at +3 V. The actual transient that would occur (beginning at $t = 0$) while the capacitor charged to 3 V does not appear in the output.

A–7 DIODE MODELS

When there is a semiconductor device in a circuit to be analyzed by SPICE, the SPICE input file must contain two new types of statements: one that identifies the device by name and gives its node numbers in the circuit and

FIGURE A–10 The parameter values of diodes D1 and D3 are given in the model whose name is MODA, and the parameter values of D2 and D4 are given in the model whose name is MODB

```
D1 5 6 MODA
D2 6 7 MODB
D3 4 7 MODA
D4 5 4 MODB
.MODEL MODA D IS = 0.5P
.MODEL MODB D IS = 0.1P
```

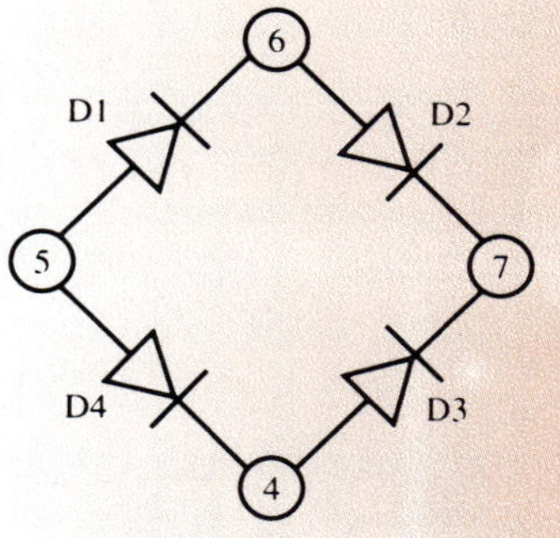

another, called a .MODEL statement, that specifies the values of the device *parameters* (saturation current and the like).

All diode names must begin with D. The format for identifying a diode in a circuit is

```
D******* NA NC Mname
```

where *NA* and *NC* are the numbers of the nodes to which the anode and cathode are connected, respectively, and *Mname* is the *model name.* The model name associates the diode with a particular .MODEL statement that specifies the parameter values of the diode:

```
.MODEL Mname D <Pval1 = n1 Pval2 = n2 . . .>
```

where D, signifying diode, *must* appear as shown, and *Pval1* = *n1*, . . . specify parameter values, to be described shortly. Note that several diodes, having different names, can all be associated with the same .MODEL statement, and other diodes can be associated with a different .MODEL statement. Figure A–10 shows an example: a diode bridge in which diodes D1 and D3 are modeled by MODA and diodes D2 and D4 are modeled by MODB. MODA specifies a diode having saturation current (IS) 0.5 pA, and MODB specifies a diode having saturation current 0.1 pA.

Table A–1 lists the diode parameters whose values can be specified in a .MODEL statement and the default values of each. The default values are

TABLE A–1 Diode parameters

Parameter	Identification	Units	Default Value
Saturation current	IS	A	1×10^{-14}
Ohmic resistance	RS	Ω	0
Emission coefficient	N	—	1
Transit time	TT	s	0
Zero-bias junction capacitance	CJO	F	0
Junction potential	VJ	V	1
Grading coefficient	M	—	0.5
Activation energy	EG	eV	1.11
Saturation current temperature exponent	XT1	—	3
Flicker noise coefficient	KF	—	0
Flicker noise exponent	AF	—	0
Coefficient for forward-bias depletion capacitance equation	FC	—	0.5
Reverse breakdown voltage	BV	V	Infinite
Current at breakdown voltage	IBV	A	1×10^{-3}

typical, or average, values and are acceptable for most electronic circuit analysis at our level of study. In some examples in the book, we specify parameter values different from the default values to illustrate certain points, but average diode behavior is adequate for most of our purposes and can be realized without knowledge of specific values. In practice, some of these diode parameters are very difficult to obtain or measure and are necessary only when a highly accurate model is essential. For example, if we were designing a new diode to have specific low-noise characteristics, we would want to know the noise parameters, KF and AF, very accurately.

PSPICE

In PSpice, the current through and/or the voltage across a diode can be printed or plotted by specifying output variables I(D*******) and V(D*******), where D******* is the diode name. For example, the following statements request the value of the dc current through diode D1 and a plot of the ac voltage across diode D2:

```
.PRINT DC I(D1)
.PLOT AC V(D2)
```

The reference polarity in each case is from the anode node to the cathode node.

A–8 BJT MODELS

The format for identifying a bipolar junction transistor in a circuit is

```
Q******* NC NB NE Mname
```

where *NC, NB,* and *NE* are the numbers of the nodes to which the collector, base, and emitter are connected and *Mname* is the name of the model that specifies the transistor parameters. The format of the .MODEL statement for a BJT is

```
.MODEL Mname type <Pval1 = n1 Pval2 = n2 . . .>
```

where *type* is either *npn* or *pnp.* Figure A–11 shows an example of how an *npn* transistor and a *pnp* transistor in a circuit are identified and modeled. The *npn* model specifies that the "ideal maximum forward beta" (BF) of the transistor is 150, and the *pnp* model allows that parameter to have its default value of 100.

As shown in Table A–2, we can specify up to 40 different parameter values for a BJT. The mathematical model used by SPICE to simulate a BJT is very complex but very accurate, provided that all parameter values are accurately known. However, to an even greater extent than in the diode model, many of the parameters are very difficult to measure, estimate, or deduce

FIGURE A–11 An example of how an npn transistor and a pnp transistor are identified and modeled

```
Q1 4 3 2 TYPE1
Q2 0 1 2 TYPE2
.MODEL TYPE1 NPN BF=150
.MODEL TYPE2 PNP
```

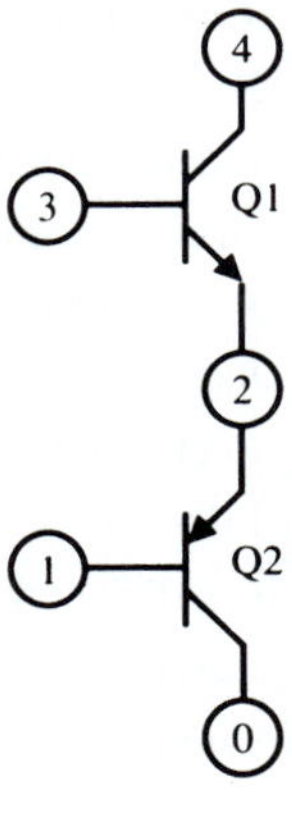

TABLE A–2 BJT parameters

Parameter	Identification	Units	Default Value
Transport saturation current	IS	A	10^{-16}
Ideal maximum forward beta	BF	—	100
Forward-current emission coefficient	NF	—	1
Forward Early voltage	VAF	V	Infinity
Corner for forward beta high-current roll-off	IKF	A	Infinity
Base–emitter leakage saturation current	ISE	A	0
Base–emitter leakage emission coefficient	NE	—	1.5
Ideal maximum reverse beta	BR	—	1
Reverse-current emission coefficient	NR	—	1
Reverse Early voltage	VAR	V	Infinity
Corner for reverse beta high-current roll-off	IKR	A	Infinity
Base–collector leakage saturation current	ISC	A	0
Base–collector leakage emission coefficient	NC	—	2
Zero-bias base resistance	RB	Ω	0
Current where base resistance falls to half of its minimum value	IRB	A	Infinity
Minimum base resistance at high currents	RBM	Ω	RB
Emitter resistance	RE	Ω	0
Collector resistance	RC	Ω	0
Base–emitter zero-bias depletion capacitance	CJE	F	0
Base–emitter built-in potential	VJE	V	0.75
Base–emitter junction exponential factor	MJE	—	0.33
Ideal forward transit time	TF	s	0
Coefficient for bias dependence of TF	XTF	—	0
Voltage describing V_{BC}-dependence of TF	VTF	V	Infinity
High-current parameter for effect on TF	ITF	A	0
Excess phase at $f = 1/(2\pi \text{ TF})$ Hz	PTF	Degrees	0
Base–collector zero-bias depletion capacitance	CJC	F	0
Base–collector built-in potential	VJC	V	0.75
Base–collector junction exponential factor	MJC	—	0.33
Fraction of base–collector depletion capacitance to internal base node	XCJC	—	1
Ideal reverse transit time	TR	s	0
Zero-bias collector–substrate capacitance	CJS	F	0
Substrate–junction built-in potential	VJS	V	0.75
Substrate–junction exponential factor	MJS	—	0
Forward and reverse beta temperature exponent	XTB	—	0
Energy gap for temperature effect on IS	EG	eV	1.11
Temperature exponent for effect on IS	XTI	—	3
Flicker-noise coefficient	KF	—	0
Flicker-noise exponent	AF	—	1
Coefficient for forward-bias depletion capacitance formula	FC	—	0.5

from other characteristics. Once again, for routine circuit analysis, it is generally sufficient to allow most of the parameters to have their default values. This approach is further justified by the fact that there is usually a wide variation in parameter values among transistors of the same type. Therefore, it is unrealistic and unnecessary in many practical applications to seek a highly accurate analysis of a circuit containing transistors. On the other hand, there are situations, such as the design and development of a new integrated circuit required to have certain properties, where an accurate determination of parameter values is warranted.

PSPICE

PSpice has a *library* option that allows access to model statements for many commonly used diodes and transistors. These model statements specify values for the many device parameters that we could not ordinarily determine from a manufacturer's specifications. The PSpice library is discussed in Section A–17.

In PSpice, the voltage across any pair of BJT terminals and/or the current into any BJT terminal can be printed using the letters B, E, and C to represent the base, emitter, and collector terminals. Following are some examples of output variables that could be specified with a .PRINT statement:

`VBE(Q1)`	The base-to-emitter voltage of transistor Q1
`IC(Q2)`	The collector current of transistor Q2
`VC(QA)`	The collector-to-ground voltage of transistor QA

A–9 THE .TEMP STATEMENT

As noted earlier, all SPICE computations are performed under the assumption that the temperature of the circuit is 27°C, unless a different temperature is specified. The .TEMP statement is used to request an analysis at one or more different temperatures:

```
.TEMP T1 <T2 T3 . . . >
```

where *T1* is the temperature in degrees Celsius at which an analysis is desired. *T2, T3,* . . . are optional additional temperatures at which SPICE will repeat the analysis, once for each temperature. Temperature is a particularly important parameter in circuits containing semiconductor devices because their characteristics are temperature-sensitive. However, SPICE will not adjust all device characteristics for temperature unless the parameters in the .MODEL statement that relate to temperature sensitivity are given specific values. A case in point is β in a BJT. The parameter BF is called the "ideal maximum forward beta." The actual value of β used in the computations depends on other factors, including the forward Early voltage (VAF) and the temperature. However, no temperature variation in the value of BF will occur unless the parameter XTB (forward and reverse beta temperature exponent) is specified to have a value other than 0 (its default value). When SPICE performs an analysis at a temperature other than 27°C, it prints a list of temperature-adjusted values of device parameters that are temperature-sensitive.

The values of resistors in a circuit are not adjusted for temperature unless either first- or second-order temperature coefficients of resistance, tc_1 and tc_2 (or both), are given values in the statements defining resistors. The format is

```
R******* N1 N2 value TC = tc1, tc2
```

The temperature-adjusted value is then computed by

$$R_T = R_{27}\,[1 + tc_1(T - 27°) + 1 + tc_2(T - 27°)]$$

where R_T is the resistance at temperature T and R_{27} is the resistance at temperature 27°C. The default values of tc_1 and tc_2 are 0.

A–10 SOURCES AND THE .AC CONTROL STATEMENT

The format for identifying an ac voltage source in a circuit is

```
V******* N+ N- AC &lt;mag&gt; &lt;phase&gt;
```

where ******* is an arbitrary sequence of characters, *N*+ and *N*− are the numbers of the nodes to which the positive and negative terminals are connected, *mag* is the magnitude of the voltage in volts, and *phase* is its phase angle in degrees. All ac sources are assumed to be sinusoidal. If *mag* is omitted, its default value is 1 V, and if *phase* is omitted, its default value is 0°. An ac current source is identified by making the first character I instead of V. Note that *mag* may be regarded as either a peak or an rms value, since SPICE output from an ac analysis does not consist of instantaneous (time-varying) values.

The .AC Control Statement

The .AC control statement is used to compute ac voltages and currents in a circuit *versus frequency.* A single frequency or a range of frequencies can be specified. The circuit must contain at least one source designated AC, and all AC sources are assumed to be sinusoidal and to have identical frequencies or to undergo the same frequency variation, if any. The format is

```
.AC vartype N fstart fstop
```

where *vartype* specifies the way frequency is to be varied in the range from *fstart* through *fstop*. *N* is a number related to the number of frequencies at which computations are to be performed, as will be discussed shortly. The *vartype* is one of DEC, OCT, or LIN (decade, octave, or linear). If analysis at a single frequency is desired, we can use any *vartype*, set *fstart* equal to *fstop*, and let $N = 1$.

When *vartype* is LIN, the frequencies at which analysis is performed vary linearly from *fstart* through *fstop*. In that case, *N* is the total number of frequencies at which the analysis is performed (counting *fstart* and *fstop*). Thus, the interval between frequencies is

$$\Delta f = \frac{fstop - fstart}{N - 1}$$

For example, the statement

```
.AC LIN 21 100 1K
```

will tell SPICE to perform an ac analysis at 21 frequencies from 100 Hz through 1 kHz. The frequency interval will be $(1000 - 100)/20 = 45$ Hz, so computations will be performed at 100 Hz, 145 Hz, 190 Hz, . . . , 1 kHz.

When the *vartype* is DEC or OCT, the analysis is performed at logarithmically spaced intervals and *N* is the total number of frequencies *per decade or per octave.* For example, the statement

```
.AC DEC 10 100 10K
```

will cause SPICE to analyze the circuit at ten frequencies in each of the decades 100 Hz to 1 kHz and 1 kHz through 10 kHz. The frequencies will be at one-tenth–decade intervals, so each interval will be different. The frequencies at one-tenth–decade intervals are 10^x, $10^{x+0.1}$, $10^{x+0.2}$, . . . , where $x = \log_{10}(fstart)$. In general, the frequencies at which SPICE performs an ac analysis using the DEC *vartype* are 10^x, $10^{x+1/N}$, $10^{x+2/N}$, . . . , where $x = \log_{10}(fstart)$. The last frequency in this sequence is not necessarily *fstop,* but SPICE will compute at frequencies in the sequence up through the first frequency that is equal to or greater than *fstop.* The frequencies at which SPICE performs an ac analysis when the *vartype* is OCT are 2^x, $2^{x+1/N}$, $2^{x+2/N}$, . . . , where $x = \log_2(fstart)$.

ac Outputs

ac voltages and currents whose values are desired from an .AC analysis are specified the same way as dc voltages and currents, using *V(N1, N2)*, *V(N1)*, or I(*Vname*) in a .PRINT statement. The values of the magnitudes of these quantities are printed or plotted. In addition, we can request certain other values, as indicated in the following list of voltage characteristics:

VR	real part
VI	imaginary part
VM	magnitude, $\lvert V\rvert$
VP	phase, degrees
VDB	$20 \log_{10}\lvert V\rvert$

The same values for ac currents can be obtained by substituting I for V. To illustrate, the statement

```
.PRINT AC V(1) VLP (2, 3) II(VDUM) IR(VX)
```

causes SPICE to print the magnitude of the ac voltage between nodes 1 and 0, the phase angle of the ac voltage VL between nodes 2 and 3, the imaginary part of the current in VDUM, and the real part of the current in VX. Note that ac must be listed as the analysis type.

PSPICE

Probe

Probe is a PSpice option that makes it possible to obtain output plots with high resolution. Probe produces a virtually solid line on a high-resolution monitor. Also, Probe has other features that make it very useful for analyzing output data. It behaves very much like a high-quality oscilloscope that allows the user to position a cursor on a trace and to obtain a direct readout of the value of the variable displayed at the position of the cursor.

If the statement

```
.PROBE
```

appears anywhere in an input data file that requests a .DC, .AC, or .TRAN analysis, then Probe automatically generates plotting data for the voltage (with respect to ground) at every node in the circuit and for the current entering every device in the circuit. Probe stores the data in an *output data file* named PROBE.DAT. To initiate a Probe run, the user enters the command PROBE directly from the keyboard. If more than one analysis type appears in the input data file (called the *circuit file* in PSpice), a "start-up menu" appears, and the user selects a single analysis type for the Probe run. If only one analysis type appeared in the circuit file, then a set of axes and a menu are displayed immediately after Probe is entered. Selecting the option ADD TRACE from this menu prompts the user to enter the variable to be plotted, using the same format used to specify outputs in PSpice [such as V(2), I(R1), etc.]. The plot is scaled automatically and displayed on the monitor. Addi-

FIGURE A–12 Example of a PSpice Probe display

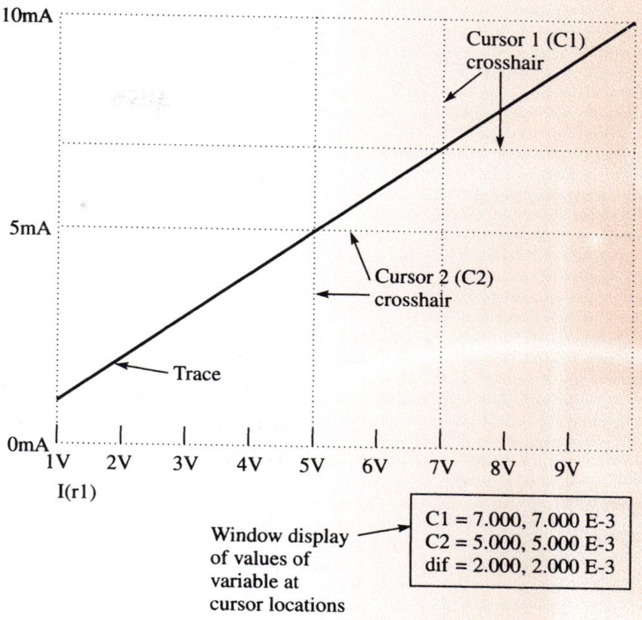

tional variables can be plotted simultaneously by repeatedly selecting the ADD TRACE option. Also, plots can be deleted by selecting the DELETE TRACE option.

Selecting the CURSOR option from the Probe menu creates two sets of cross hairs, identified as cursor 1 (C1) and cursor 2 (C2), on the display. These cross hairs can be moved along the plot using direction keys (← and →) on the keyboard. Using the direction keys alone moves C1, and using the direction keys with SHIFT depressed moves C2. The numerical values of the variable at the positions of the cross hairs on the plot are displayed in a window, along with the difference in the values. Figure A–12 shows an example.

If desired, Probe plots can be limited to specific variables by giving their names in the .PROBE statement in the circuit file. For example, the statement

```
.PROBE V(2)
```

causes Probe to create an output data file containing only plotting data for the voltage at node 2. This option is useful when memory capacity is too small to store all voltages and currents.

Another feature of Probe is that it can create plots of mathematical functions of the variables. For example, after selecting ADD TRACE, entering the expression

```
I(R1)*I(R1)*1K
```

creates a plot of the power (I^2R) dissipated by 1-kΩ resistor R1. Numerous mathematical functions, including trigonometric functions, logarithmic functions, average values, square roots, and absolute values, of the variables can also be plotted. One difference between Probe and PSpice is that suffixes in expressions written for Probe must use lowercase m to represent milli and capital M to represent meg (as opposed to M for milli and MEG for meg).

It should be noted that the high-resolution plots created by Probe are *not* the result of Probe increasing the number of output values computed in a circuit simulation. Rather, Probe *interpolates* values between the data points that a circuit file specifies are to be computed. Thus, if the output variable

TABLE A–3

Vartype	Output Requested	Frequency (Vertical) Axis	Output (Horizontal) Axis
	Magnitude	Linear	Log
	Phase	Linear	Linear
LIN	Imaginary part	Linear	Linear
	Real part	Linear	Linear
	Decibels	Linear	Linear
	Magnitude	Log	Log
DEC or OCT	Phase	Log	Linear
	Imaginary part	Log	Linear
	Real part	Log	Linear
	Decibels	Log	Linear

has a sudden change in value (such as a sharp peak) and the circuit file does not specify a fine-enough increment between data points to detect that change, Probe cannot be expected to detect it either.

ac Plots

When ac voltages or currents are obtained through Probe, we obtain a linear, semilog, or log-log plot, depending on the type of voltage or current output requested and the *vartype.* Table A–3 summarizes the types of plots produced for each combination. "Log" in the table means that the scale supplied by SPICE has logarithmically spaced values.

Small-Signal Analysis and Distortion

When a circuit contains active devices such as transistors, an .AC analysis by SPICE is assumed to be a small-signal analysis. That is, ac variations are assumed to be small enough that the values of device parameters do not change. As in a .TRAN analysis, SPICE performs an initial dc analysis to determine quiescent voltages and currents so that the values of those device parameters affected by dc levels can be computed. In the ac analysis, the device parameters are assumed to retain those initial values, regardless of the actual magnitudes of the ac variations. *Thus, in an .AC analysis, SPICE does not take into account any distortion, even clipping, that would actually occur if we were to severely overdrive a transistor by specifying very large ac inputs.* For example, if the output swing of an actual transistor circuit were limited to 10 V, this fact would not be "known" to SPICE, and by overdriving the computer-simulated circuit, we could obtain outputs of several hundred volts from an .AC analysis.

To observe the effects of distortion, such as clipping, it is necessary to perform a .TRAN analysis and obtain a plot of the output waveform versus time. There is also a .DISTO (distortion) control statement that can be used to obtain limited information on harmonic distortion, but we will not have occasion to use that statement.

The next example illustrates an .AC analysis over a range of frequencies.

EXAMPLE A–3

Use SPICE to perform an ac analysis of the transistor amplifier in Figure A–13(a) over the frequency range from 100 Hz through 100 kHz. Obtain a plot of the magnitude and phase angle of the collector-to-emitter voltage over the frequency range, with ten frequencies per decade.

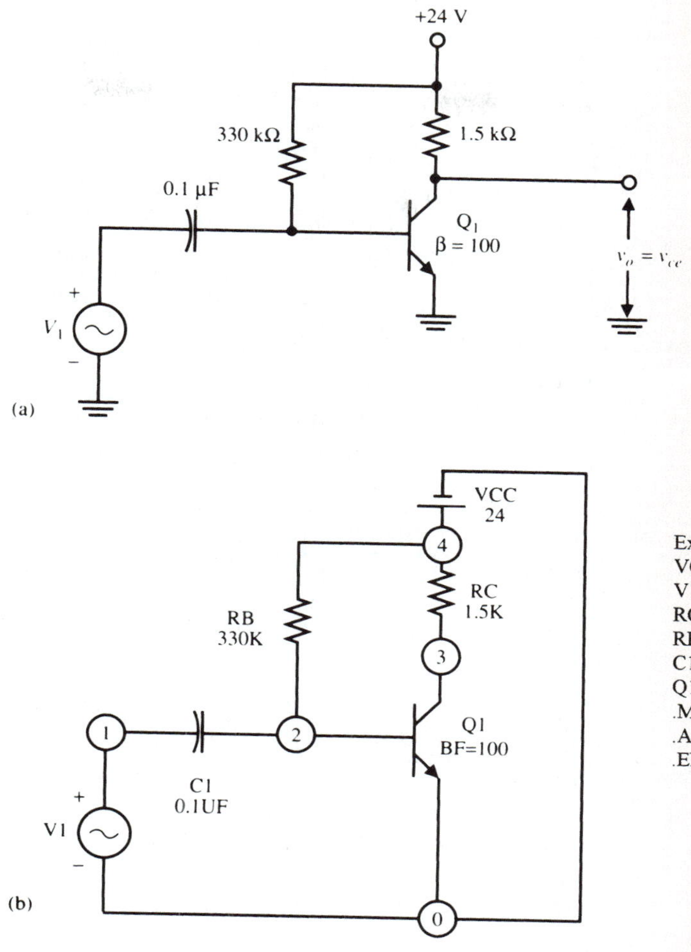


```
Example A.3
VCC 4 0 24
V1 1 0 AC 1.0
RC 4 3 1.5k
RB 4 2 330K
C1 1 2 0.1uF
Q1 3 2 0 TRANS
.MODEL TRANS NPN BF=100
.AC DEC 10 100 100k
.END
```

FIGURE A–13 (Example A-3)

Solution

Figure A–13(b) shows the circuit redrawn for analysis by SPICE and the corresponding input data file. Note that we allow the magnitude and phase of the ac source (V1) to have the default values 1 V and 0°, respectively.

Figure A–14 shows the results of a program run. Note the following points in connection with an .AC analysis of a circuit containing a transistor:

1. SPICE prints a list of the values of the transistor model parameters, which in this case are default values. (Since the default value of BF is 100, we could have omitted that specification in the .MODEL statement.)
2. SPICE obtains a "small-signal bias solution" to determine the dc voltages and currents in the circuit with the ac source set to 0. This is similar to the "initial transient solution" obtained in a .TRAN analysis.
3. Using the dc values obtained from the small-signal bias solution, SPICE computes "operating point information." This information is provided in the form of a list of important bias-dependent parameter values.
4. The two outputs are plotted using * to represent V(3), the magnitude of V_{CE}, on a log scale and + to represent VP(3), the phase angle of V_{CE} on a linear scale. The values of V(3) are printed along the vertical scale, along

```
EXAMPLE A.3
****     BJT MODEL PARAMETERS             TEMPERATURE =    27.000 DEG C
*****************************************************************************
             TRANS
TYPE         NPN
IS         1.00D-16
BF          100.000
NF            1.000
BR            1.000
NR            1.000
*******02/22/89 ********  SPICE 2G.6    3/15/83 *******13:15:17*****
EXAMPLE A.3
****     SMALL SIGNAL BIAS SOLUTION       TEMPERATURE =    27.000 DEG C
*****************************************************************************
 NODE    VOLTAGE       NODE    VOLTAGE       NODE    VOLTAGE       NODE    VOLTAGE
(  1)     .0000       (  2)     .8246       (  3)    13.4657      (  4)    24.0000
    VOLTAGE SOURCE CURRENTS
    NAME         CURRENT
    VCC        -7.093D-03
    V1          0.000D+00
    TOTAL POWER DISSIPATION    1.70D-01   WATTS
*


EXAMPLE A.3
****     OPERATING POINT INFORMATION      TEMPERATURE =    27.000 DEG C
*****************************************************************************
**** BIPOLAR JUNCTION TRANSISTORS
              Q1
MODEL       TRANS
IB          7.02E-05
IC          7.02E-03
VBE             .825
VBC          -12.641
VCE           13.466
BETADC       100.000
GM          2.72E-01
RPI         3.68E+02
RX          0.00E+00
RO          1.00E+12
CPI         0.00E+00
CMU         0.00E+00
CBX         0.00E+00
CCS         0.00E+00
BETAAC       100.000
FT          4.32E+18
*
```

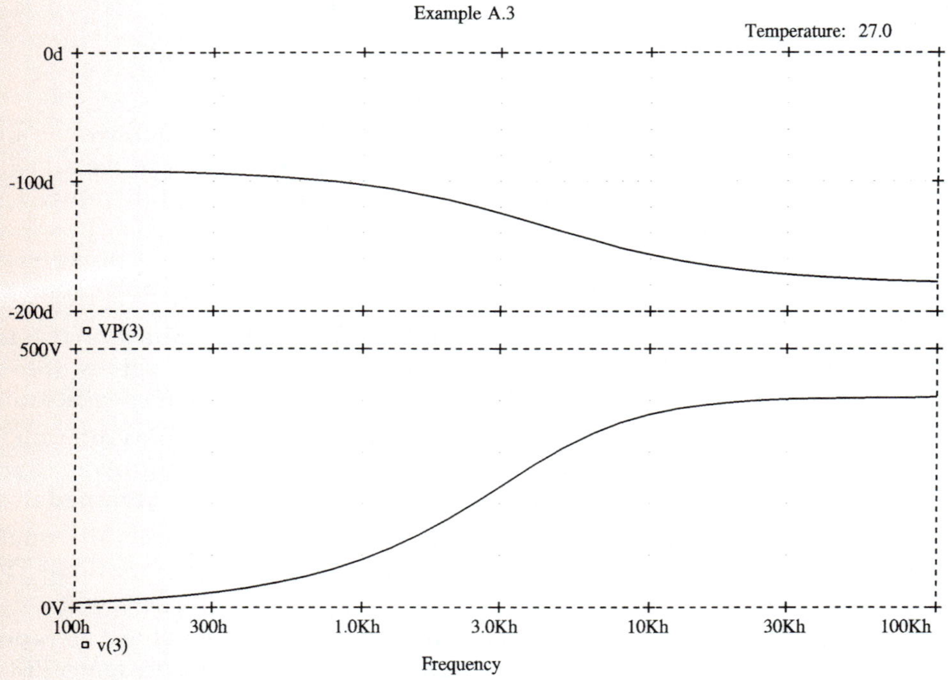

FIGURE A–14 (Example A–3)

with the logarithmically spaced frequencies. Thus, the plot is a log-log plot of voltage magnitude and a semilog plot of phase angle.

5. The plots intersect at $f = 1$ kHz, and SPICE prints an X where that occurs.
6. The maximum output voltage (at 100 kHz) is 406.9 V, which is clearly impossible in the actual circuit. As noted earlier, SPICE does not take output voltage limits (clipping) into account during an .AC analysis. Because the input signal magnitude is 1 V, the output magnitude also represents the voltage gain at each frequency.
7. At high frequencies, the phase angle approaches −180°, confirming the phase inversion caused by a common-emitter amplifier.

A–11 JUNCTION FIELD-EFFECT TRANSISTORS (JFETs)

An *n*-channel or *p*-channel JFET in a circuit to be analyzed by SPICE must be given a name that begins with J, such as JX or JFET1. The format for specifying both *p*-channel and *n*-channel devices is

```
J******* ND NG NS Mname
```

where *ND* is the node number of the drain, *NG* the node number of the gate, *NS* the node number of the source, and *Mname* is the model name. The model name appears in the .MODEL statement for a JFET using the format

```
.MODEL Mname type <Pval1 = n1 Pval2 = n2 . . . >
```

where *type* is NJF for an *n*-channel JFET and PJF for a *p*-channel JFET. Table A–4 shows the JFET parameters that can be specified in a .MODEL statement and their default values.

The two principal parameters related to the dc characteristics of a JFET are VTO (the pinch-off voltage, V_p) and BETA (the transconductance parameter, β). In the pinch-off region, these are related by

$$\beta = \frac{I_D}{(V_{GS} - V_p)^2} \quad \text{A/V}^2$$

TABLE A–4 JFET parameters

Parameter	Identification	Units	Default Value
Threshold voltage	VTO	V	−2
Transconductance parameter	BETA	A/V^2	10^{-4}
Channel-length-modulation parameter	LAMBDA	1/V	0
Drain ohmic resistance	RD	Ω	0
Source ohmic resistance	RS	Ω	0
Zero-bias gate-to-source junction capacitance	CGS	F	0
Zero-bias gate-to-drain junction capacitance	CGD	F	0
Gate–junction saturation current	IS	A	10^{-14}
Gate–junction potential	PB	V	1
Flicker-noise coefficient	KF	—	0
Flicker-noise exponent	AF	—	1
Coefficient for forward-bias depletion capacitance formula	FC	—	0.5

Because $I_D = I_{DSS}$ when $V_{GS} = 0$, we have

$$\beta = \frac{I_{DSS}}{V_p^2}$$

When modeling a JFET for which values of I_{DSS} and V_p are known, we must calculate β using this equation so that the parameter BETA can be specified in the .MODEL statement. *Note this important point: The value entered for* the *parameter VTO is always negative, whether the JFET is an n-channel or a p-channel device.*

The parameter LAMBDA (λ, the channel-length-modulation factor) controls the extent to which the characteristic curves rise for increasing values of V_{DS} in the pinch-off region. If λ is 0 (the default value), the characteristic curves are perfectly flat. A typical value for λ is 10^{-4}.

A–12 MOS FIELD-EFFECT TRANSISTORS (MOSFETs)

An *n*-channel or *p*-channel MOSFET must be identified in a SPICE data file by a name beginning with the letter M, using the format

```
M******* ND NG NS NSS Mname
```

where *ND, NG, NS,* and *NSS* are the node numbers of the drain, gate, source, and substrate, respectively, and *Mname* is the model name. The values of certain geometric parameters, such as the length and width of the channel, can be optionally specified with the MOSFET identification, but these will not concern us and we can allow all of them to default.

The MOSFET model is very complex. Like other semiconductor models in SPICE, it involves many parameters whose values are difficult to determine and many that are beyond the scope of our treatment. There are actually three built-in models, referred to as *levels* 1, 2, and 3. LEVEL is one of the parameters that can be specified in a MOSFET .MODEL statement, and its value prescribes the particular model to be used. The default level is 1, which we can assume for all purposes in this book. The format of the MOSFET .MODEL statement is

```
.MODEL Mname type<Pval1 = n1 Pval2 = n2 . . .>
```

where *type* is NMOS for an *n*-channel MOSFET and PMOS for a *p*-channel MOSFET. A MOSFET can be either the depletion-mode type or of the enhancement-mode type, as is discussed shortly.

Because the MOSFET parameters in the three-level SPICE model are so numerous and complex, we will not present a table showing their identifications, units, and default values. (Such information should be available in a user's guide furnished with the SPICE software used.) In any case, there is a significant variation among the several versions of SPICE in the number and type of MOSFET parameters that can be specified. However, there are two fundamental parameters whose values should probably be specified in every MOSFET simulation: β and V_T. In SPICE, β is called the *intrinsic transconductance parameter* and is identified in a .MODEL statement by KP = β provided the channel length and width parameters, L and W, are allowed to default; otherwise, KP = β(L/W). Its value is always positive. V_T is called the zero-bias threshold voltage and is identified by VTO. Following is an example of a .MODEL statement for an *n*-channel MOSFET having $\beta = 0.5 \times 10^{-3}$ A/V^2 and $V_T = 2$ V:

```
.MODEL M1 NMOS KP = 0.5E-3 VTO = 2
```

The threshold voltage, VTO, is positive or negative, according to the following table:

Mode	Channel	Sign of VTO
Depletion	n	−
Depletion	p	+
Enhancement	n	+
Enhancement	p	−

For example, the value specified for the VTO of an n-channel, enhancement-mode MOSFET should be positive.

PSPICE

In PSpice, the voltage across any pair of JFET or MOSFET terminals and/or the current into any JFET or MOSFET terminal can be printed or plotted using the letters D, G, and S to represent the drain, gate, and source terminals. The following are some examples of output variables that could be specified with a .PRINT or .PLOT statement:

VGS(J1)	gate-to-source voltage of the JFET named J1
ID(MXY)	drain current of the MOSFET named MXY
VD(M25)	drain-to-ground voltage of the MOSFET named M25

A–13 CONTROLLED (DEPENDENT) SOURCES

A controlled voltage source is one whose output voltage is controlled by (depends on) the value of a voltage or current elsewhere in the circuit. The simplest and most familiar example is a voltage amplifier: It is a voltage-controlled voltage source because its output voltage *depends* on its input voltage. The output voltage equals the input voltage multiplied by the gain. A current-controlled voltage source obeys the relation $v_o = ki$, where i is the controlling current and k is a constant having the units of resistance: $k = v_o/i$ volts per ampere, or ohms. In the context of a current-controlled voltage source, k is called a *transresistance.*

Similarly, a controlled current source produces a current whose value depends on a voltage or a current elsewhere in the circuit. A voltage-controlled current source obeys the relation $i_o = kv$, where v is the controlling voltage and k has the units of conductance: $k = i_o/v$ amperes per volt, or siemens. In the context of a controlled source, k is called a *transconductance.*

The four types of controlled sources—voltage-controlled voltage sources, current-controlled voltage sources, voltage-controlled current sources, and current-controlled current sources—can be modeled in SPICE. Figure A–15 shows the format used to model each type. Note that the controlling voltage in voltage-controlled sources (the voltage between $NC+$ and $NC-$ in (a) and (b) of the figure) can be the voltage between any two nodes; it is not necessary that a component be connected between those nodes. Also note that the controlling current in controlled current sources [parts (c) and (d) of the figure] is always the current in a voltage source. Thus, it may be necessary to insert a dummy voltage source in a circuit in order to specify a controlling current at a desired point in the circuit. Observe how the "plus" and "minus" nodes are defined in the figure, in connection with the polarity assumptions made by SPICE for the currents and current sources.

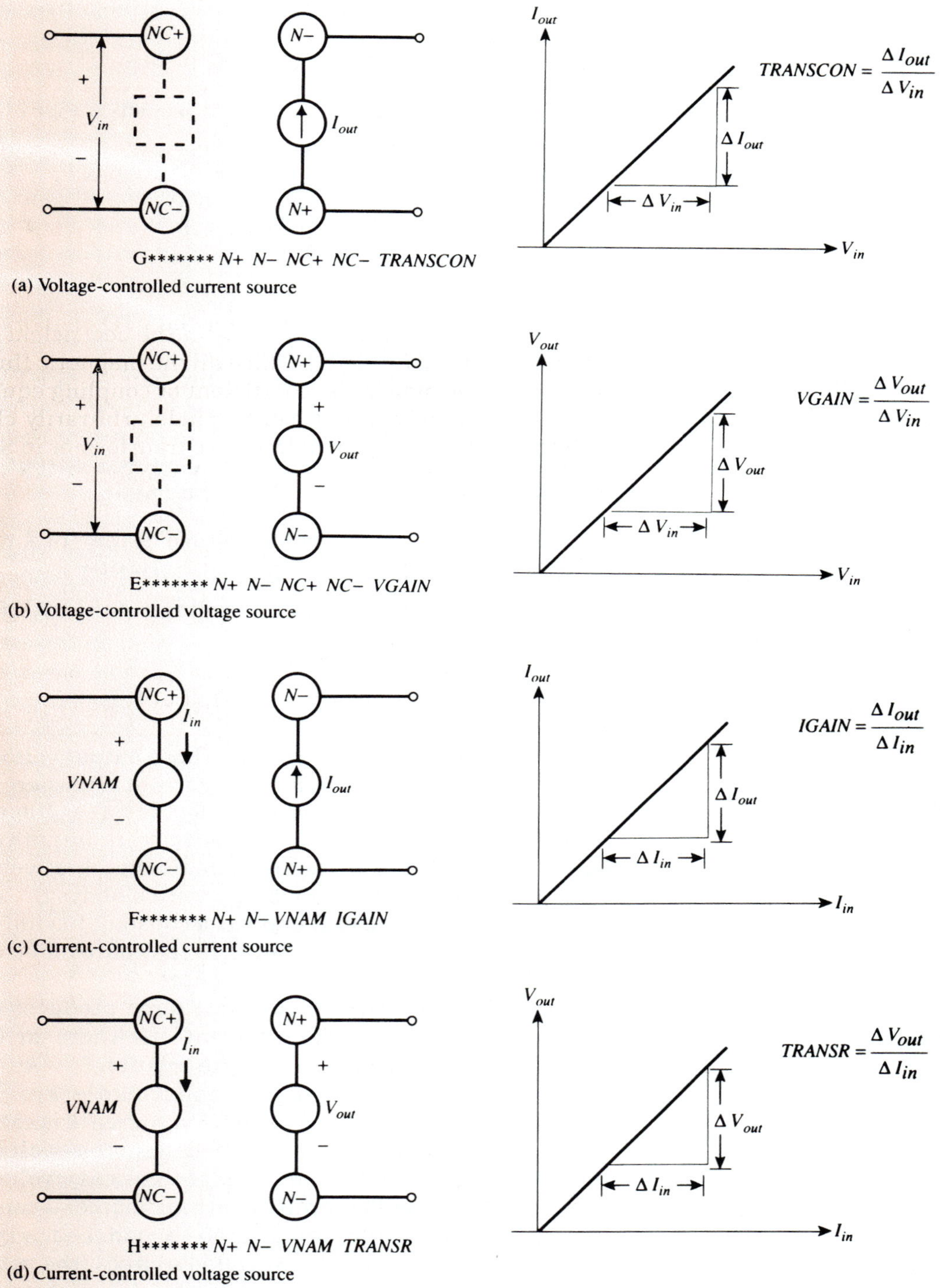

FIGURE A–15 Specification of controlled sources in SPICE

A–14 TRANSFORMERS

A transformer can be modeled in SPICE using three statements: one to specify the nodes and inductance of the primary winding, one to specify the nodes and inductance of the secondary winding, and one to specify the *coefficient of coupling* between the windings. The primary and secondary windings are specified exactly as ordinary inductors are, using an L prefix, node numbers, and the inductance (in henries). The coefficient of coupling is specified in a statement that must begin with K:

```
K******* LNAM1 LNAM2 k
```

where L*NAM1* and L*NAM2* are the inductors comprising the primary and secondary windings and k is the value of the coefficient of coupling ($0 < k < 1$). For an ideal transformer, in which all the magnetic flux in the primary is coupled to the secondary, the coefficient of coupling equals 1. (In PSpice, k cannot be set equal to 1, but can be made arbitrarily close to it, for example, 0.999.) To simulate an ideal iron-core transformer in SPICE, it is necessary to specify inductance values so that the reactance of the primary winding is much greater than the impedance of the signal source driving it and so that the reactance of the secondary winding is much greater than the load impedance. (The primary and secondary windings of an ideal transformer have infinite inductance.) Thus, it may be necessary to specify unrealistically large values for the inductances of the primary and secondary windings. (Such will be the case in all examples found in this book.) The following is an example showing the specification of a transformer (KXFRMR) having primary and secondary windings named LPRIM and LSEC:

```
LPRIM 5 0 20H
LSEC 6 0 0.2H
KXFRMR LPRIM LSEC 1
```

If operation of this transformer is to be simulated at 1 kHz, the source impedance should be much smaller than $2\pi f(\text{LPRIM}) = 2\pi(10^3\text{ Hz})(20\text{ H}) = 125.7\text{ k}\Omega$, and the load impedance should be much smaller than $2\pi f(\text{LSEC}) = 2\pi(10^3\text{ Hz})(0.2\text{ H}) = 1.257\text{ k}\Omega$. In SPICE, an inductor cannot appear in a loop isolated from ground, so it is *not* possible to isolate the primary and secondary windings from each other, as is done in a real transformer.

The turns ratio of an ideal transformer is determined by the inductance values of the primary and secondary windings, according to

$$\frac{N_p}{N_s} = \sqrt{\frac{L_p}{L_s}}$$

where N_p/N_s is the turns ratio and L_p and L_s are the primary and secondary inductances, respectively. For example, the turns ratio of the transformer in the foregoing example is

$$\frac{N_p}{N_s} = \sqrt{\frac{20\text{ H}}{0.2\text{ H}}} = \sqrt{100} = 10$$

Thus, the transformer is a step-down type whose secondary voltage is one-tenth of its primary voltage.

A–15 SUBCIRCUITS

A complex electronic circuit will often contain several components, or subsections of circuitry, that are identical to each other. Examples include active filters containing several identical operational amplifiers, multistage amplifiers consisting of identical amplifier stages, and digital logic systems containing numerous identical logic gates. When modeling such circuits in SPICE, it is a tedious and time-consuming task to write numerous sets of identical statements describing identical circuitry. Furthermore, the input data file for such a system may become so long and cumbersome that it is difficult to interpret or modify. To alleviate those kinds of problems, SPICE allows a user to create a *subcircuit:* circuitry that can be defined just once and then, effectively, inserted into a larger system (which we will call the *main* circuit) at as many places as desired. The concept is similar to that of a *subroutine* in conventional computer programming. In SPICE, it is possible to define several different subcircuits in one data file, and, in fact, one subcircuit can contain other subcircuits.

The first statement of a subcircuit is

```
.SUBCKT NAME N1 <N2 N3 . . . >
```

where *NAME* is any name chosen to identify the subcircuit and *N1, N2,* . . . are the numbers of the nodes in the subcircuit that will be joined to other nodes in the main circuit. None of these can be node 0. Components in a subcircuit are defined in exactly the same way they are in any SPICE data file, using successive statements following the .SUBCKT statement. A subcircuit can contain .MODEL statements, but it cannot contain any control statements, such as .DC, .TRAN, .PRINT, or .PLOT. The node numbers in a subcircuit do not have to be different from those in the main program. SPICE treats a subcircuit as a completely separate (isolated) entity, and a node having the same number in a subcircuit as another node in the main circuit will still be treated as a different node. The exception is node 0: If node 0 appears in a subcircuit, it is treated as the same node 0 as in the main circuit. (In the language of computer science, subcircuit nodes are said to be *local,* except node 0, which is *global.*) The last statement in a subcircuit must be

```
.ENDS<NAME>
```

If a subcircuit itself contains subcircuits, the *NAME* must be given in the .ENDS statement to specify which subcircuit definition has been ended.

In order to "insert" a subcircuit into the main circuit, we must write a subcircuit *call* statement in the main program. A different call statement is required for each location where the subcircuit is to be inserted. The format of a call statement is

```
X******* N1<N2 N3 . . . > NAME
```

where ******* are arbitrary characters that must be different for each call; *N1, N2,* . . . , are the node numbers in the main circuit that are to be joined to the subcircuit nodes specified in the .SUBCKT statement; and *NAME* is the name of the subcircuit to be inserted. The node numbers in the call statement will be joined to the nodes in the .SUBCKT statement in exactly the same order as they both appear. That is, *N1* in the subcircuit will be joined to *N1* in the main circuit, and so forth. The following is an example:

```
.SUBCKT OPAMP 1 2 3
-
-}Statements describing components in the subcircuit
-
.ENDS
```

```
X1 8 4 12 OPAMP
X2 1 4 3 OPAMP
-
-
```

In this example, the subcircuit named OPAMP is inserted at two locations in the main program. The first call statement (X1) connects subcircuit nodes 1, 2, and 3 to main-circuit nodes 8, 4, and 12, respectively. The X2 call statement connects subcircuit nodes 1, 2, and 3 to main-circuit nodes 1, 4, and 3, respectively. (In this case, some of the subcircuit and main-circuit node numbers are the same.)

The next example illustrates the use of a subcircuit to model an RC filter containing three identical stages. Although this circuit could not be considered complex enough to warrant the use of a subcircuit, it does serve to demonstrate the syntax we have described.

EXAMPLE A–4

Using a SPICE subcircuit, determine the magnitude and phase angle of the output of the RC filter in Figure A–16(a) when the input is a 1-kHz sine wave with peak value 10 V.

Solution

Figure A–16(b) shows the RC subcircuit, named STAGE, and the way that it is inserted into the main circuit in three locations. Its locations are identified by rectangles labeled X1, X2, and X3. The node numbers inside the rectangles are the subcircuit node numbers and those outside are the main-circuit node numbers. In the input data file, shown in part (c) of the figure, note that the node numbers in the three call statements are the main-circuit nodes that are connected to subcircuit nodes 1, 2, 3, and 4, in that order. The results of a program run are shown in Figure A–16(d). We see that the output voltage has peak value 3.769 V and lags the input by 71.07°.

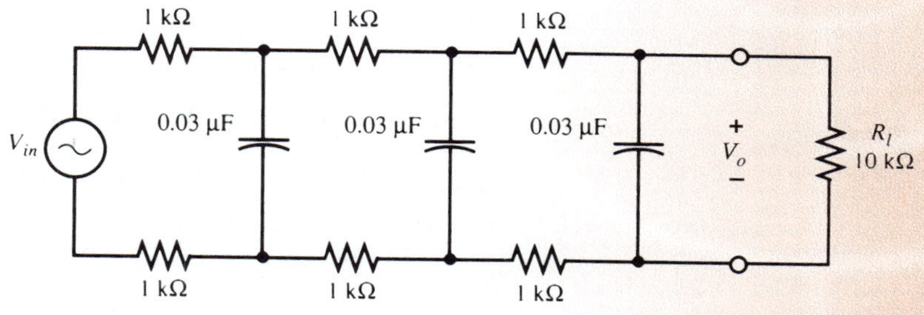

(a)

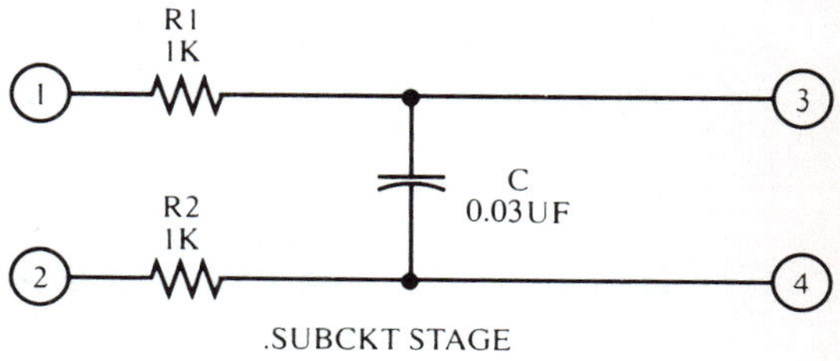

FIGURE A–16 (Example A–4)

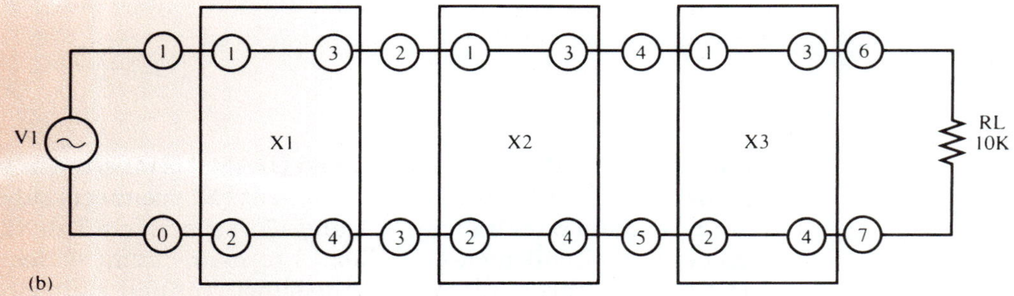

(b)

```
EXAMPLE A.4
V1 1 0 AC 10V
.SUBCKT STAGE 1 2 3 4
R1 1 3 1K
R2 2 4 1K
C 3 4 0.03UF
.ENDS
X1 1 0 2 3 STAGE
X2 2 3 4 5 STAGE
X3 4 5 6 7 STAGE
RL 6 7 10K
.AC LIN 1 1K 1K
.PRINT AC VM(6,7) VP(6,7)
.END
```

(c)

```
EXAMPLE A.4
****      AC ANALYSIS                              TEMPERATURE =    27.000 DEG C
*******************************************************************************
   FREQ         VM(6,7)       VP(6,7)
 1.000E+03       3.769E+00   -7.107E+01
```

(d)

FIGURE A–16
(Example A–4) Continued

A–16 DESIGN MANAGER

Design Manager is a PSpice option that coordinates and simplifies writing, editing, and running input circuit files and using the various options and features available in PSpice. For example, Design Manager can be used to make Probe run automatically after every circuit simulation. Options are selected by the user from various menus displayed by Design Manager.

In a system equipped with Design Manager, the first (main) menu displayed contains, among others, selections titled Text Edit, PSpice A/D, and Probe. These are the principal choices that will be used in creating and running most circuit simulations. When the user selects Text Edit, another menu is displayed with options that include File, Edit, Search, and View. Text Edit allows the user to edit an existing circuit file or to create and name a new circuit file. The circuit file must be saved before running PSpice, which is done by selecting PSpice A/D.

To perform a program run of an input circuit file that has already been created, the option PSpice A/D is selected from the main menu. This selection opens another menu that allows the user to specify the circuit file to be simulated. Because at least one analysis type is (presumably) already specified in the input circuit file, the simulation program starts running as soon as the circuit file is selected. It should be noted that if there is an error in the input circuit file, Design Manager will open a window with a list of errors.

Selecting Probe from the main menu creates another menu that allows the user to request automatic running of Probe after a PSpice simulation run. Probe can also be started from the PSpice A/D window, provided the circuit file was successfully run.

A–17 THE PSPICE LIBRARY

PSPICE

The MicroSim Corporation has created models and subcircuits for over 3500 standard analog devices, including transistors, diodes, and operational amplifiers. These can be stored in a computer system as a *library* that can be accessed by users who wish to specify any one or more of the devices in an input circuit file. It is not necessary to write the entire model or subcircuit defining a particular device in the input file; it is necessary only to refer to it by name. PSpice automatically retrieves the description from the library and uses it in the circuit simulation.

Transistors and diodes are stored in the library as model statements. If we wish to incorporate a particular transistor in an input circuit file, we specify the device name in place of the model name in the statement defining the transistor in our circuit. The statement .LIB is used to inform PSpice that we are accessing the library. The following is an example:

```
Q1 3 5 8 Q2N2222A
.LIB
```

These statements cause PSpice to use the model statement in the library for the 2N2222A transistor as the model statement for Q1. That model statement is as follows:

```
.model Q2n2222A NPN (Is=14.34f Xti=3 Eg=1.11 Vaf=74.03 Bf=255.9 Ne=1.307
Ise=14.34f Ikf=.2847 Xtb=1.5 Br=6.092 Nc=2 Isc=0 Ikr=0 Rc=1
Cjc=7.306p Mjc=.3416 Vjc=.75 Fc=.5 Cje=22.01p Mje=.377 Vje=.75
Tr=46.91n Tf=411.1p Itf=.6 Vtf=1.7 Xtf=3 Rb=10)
```

We should note that the PSpice library actually consists of a number of library files, each having a name and containing devices of a certain type. For example, DIODE.LIB is the name of a library file that contains model statements for diodes, and LINEAR.LIB contains subcircuits for operational amplifiers. PSpice can be directed to a particular library file by including the name of the file in the .LIB statement. For example,

```
.LIB LINEAR.LIB
```

directs PSpice to the linear library file. In particular, in the *evaluation version* (student version) of PSpice, the library file is called EVAL.LIB, and this must be included in the .LIB statement. Linear devices in EVAL.LIB include the following:

Q2N2222A	*npn* transistor
Q2N2907A	*pnp* transistor
Q2N3904	*npn* transistor
Q2N3906	*pnp* transistor
D1N750	zener diode

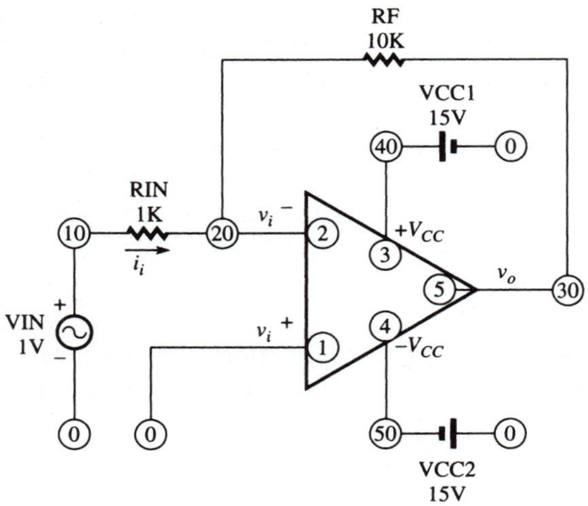

FIGURE A–17 Example of a call for an operational amplifier subcircuit from the PSpice library

```
OPAMP EXAMPLE
VIN 10 0 AC 1V
RIN 10 20 1K
RF 20 30 10K
VCC1 40 0 15V
VCC2 0 50 15V
X1 0 20 40 50 30 UA741
.LIB EVAL.LIB
.AC DEC 1 1KHZ 1KHZ
.PRINT AC V(30) VP(30) I(RIN)
.END
```

(EVAL.LIB for Evaluation version of PSpice only)

MV2201	voltage variable-capacitance diode
D1N4148	switching diode
J2N3819	*n*-channel JFET
J2N4398	*n*-channel JFET
LM324	operational amplifier
UA741	operational amplifier
LM111	voltage comparator

Because operational amplifiers are stored in the library as subcircuits, they must be accessed in the input circuit file by subcircuit *calls* (statements beginning with X; see Section A–15). Figure A–17 shows an example of an input circuit file containing a call for the 741 operational amplifier. Note that the subcircuit nodes are

1	noninverting input
2	inverting input
3	positive supply voltage
4	negative supply voltage
5	output

In this example, the amplifier is connected in an inverting configuration with voltage gain $-R_f/R_1 = -10\ \text{k}\Omega/1\ \text{k}\Omega = -10$. The input is a 1-V, 1-kHz sine wave. Execution of the program reveals that v_o = V(30) = 9.999 V, $\underline{/v_o}$ = VP(30) = 179.4°, and i_{in} = I(RIN) = 0.9999 mA.

The subcircuits for operational amplifiers are the *functional equivalents* of the devices they represent. That is, the circuits do not contain the transistors, diodes, resistors, etc., that are in the actual amplifier circuits. These functionally equivalent circuits are designed to exhibit nominal amplifier characteristics (such as bandwidth), not the worst-case values given in manufacturers' specifications. These characteristics do *not* change with temperature, as they do in actual amplifiers. Finally, we should note that component names used in library subcircuits are all *local.* In other words, PSpice will recognize that R1, for example, used in a library subcircuit is different from another resistor named R1 in the input circuit file.

STANDARD VALUES OF RESISTORS AND READING AND SELECTING CAPACITORS

B–1 STANDARD VALUES OF RESISTORS

Resistors with 5% tolerance are available in all values shown. Resistors with 10% tolerance are available only in the boldfaced values.

Ohms (Ω)					Kilohms (kΩ)		Megohms (MΩ)	
0.10	**1.0**	**10**	**100**	**1000**	**10**	**100**	**1.0**	**10.0**
0.11	1.1	11	110	1100	11	110	1.1	11.0
0.12	**1.2**	**12**	**120**	**1200**	**12**	**120**	**1.2**	**12.0**
0.13	1.3	13	130	1300	13	130	1.3	13.0
0.15	**1.5**	**15**	**150**	**1500**	**15**	**150**	**1.5**	**15.0**
0.16	1.6	16	160	1600	16	160	1.6	16.0
0.18	**1.8**	**18**	**180**	**1800**	**18**	**180**	**1.8**	**18.0**
0.20	2.0	20	200	2000	20	200	2.0	20.0
0.22	**2.2**	**22**	**220**	**2200**	**22**	**220**	**2.2**	**22.0**
0.24	2.4	24	240	2400	24	240	2.4	
0.27	**2.7**	**27**	**270**	**2700**	**27**	**270**	**2.7**	
0.30	3.0	30	300	3000	30	300	3.0	
0.33	**3.3**	**33**	**330**	**3300**	**33**	**330**	**3.3**	
0.36	3.6	36	360	3600	36	360	3.6	
0.39	**3.9**	**39**	**390**	**3900**	**39**	**390**	**3.9**	
0.43	4.3	43	430	4300	43	430	4.3	
0.47	**4.7**	**47**	**470**	**4700**	**47**	**470**	**4.7**	
0.51	5.1	51	510	5100	51	510	5.1	
0.56	**5.6**	**56**	**560**	**5600**	**56**	**560**	**5.6**	
0.62	6.2	62	620	6200	62	620	6.2	
0.68	**6.8**	**68**	**680**	**6800**	**68**	**680**	**6.8**	
0.75	7.5	75	750	7500	75	750	7.5	
0.82	**8.2**	**82**	**820**	**8200**	**82**	**820**	**8.2**	
0.91	9.1	91	910	9100	91	910	9.1	

B–2 READING AND SELECTING CAPACITORS

For the student new to electronics, reading capacitor values and selecting the appropriate capacitor can be a little confusing. This portion of Appendix B provides sufficient information for the student to choose the correct capacitor value and type with confidence.

Reading Capacitor Values

Capacitor values are marked on the device in several ways, either by a typical value expressed in microfarads (10^{-6}) or picofarads (10^{-12}). Some capacitors also have color codes but this is rare with modern capacitors. Capacitors will also typically have many initials on the package. We will only address the tolerance issues in this appendix.

The following examples demonstrate how to read some common capacitors:

G indicates that the capacitor has a tolerance value of 2%
J indicates that the capacitor has a tolerance value of 5%
K indicates that the capacitor has a tolerance value of 10%

Examples of reading capacitor values:

101 J — 1 0 plus 1 zero expressed in pF = 100 pF, ±5%
102 K — 1 0 plus 2 zeros expressed in pF = 1000 pF, ±10%
103 J — 1 0 plus 3 zeros expressed in pF = 10,000 pF, ±5%
104 K — 1 0 plus 4 zeros expressed in pF = 100,000 pF, ±10%
153 K — 1 5 plus 3 zeros expressed in pF = 15,000 pF, ±10%
223 K — 2 2 plus 3 zeros expressed in pF = 22,000 pF or .022 μF, ±10%
333 J — 3 3 plus 3 zeros expressed in pF = 33,000 pF = .033 μF, ±5%
.001 K — Anytime you see a decimal point followed by a number (e.g.,.001) you can assume that the value is expressed in μF. In this case, .001 = .001 μF or 1000 pF, ±10%.

Capacitor values greater than or equal to 1 μF are always marked with values expressed in μF.

Conversion Table of Common Capacitor Values

Marking on the Capacitor (Alternate Marking)	Capacitance Value	
101	100 pF	
151	150 pF	
221	220 pF	
331	330 pF	
471	470 pF	
681	680 pF	
102	.001 μF	(1000 pF)
152	.0015 μF	(1500 pF)
222	.0022 μF	(2200 pF)
332	.0033 μF	(3300 pF)
472	.0047 μF	(4700 pF)
682	.0068 μF	(6800 pF)

(Continued)

Conversion Table of Common Capacitor Values *(Continued)*

Marking on the Capacitor (Alternate Marking)		Capacitance Value	
.01	(103)	.01 μF	(10,000 pF)
.015	(153)	.015 μF	(15,000 pF)
.022	(223)	.022 μF	(22,000 pF)
.033	(333)	.033 μF	(33,000 pF)
.047	(473)	.047 μF	(47,000 pF)
.068	(683)	.068 μF	(68,000 pF)
.082	(823)	.082 μF	(82,000 pF)
.1	(104)	.1 μF	(100,000 pF)
.15	(154)	.15 μF	(150,000 pF)
.22	(224)	.22 μF	(220,000 pF)
.33	(334)	.33 μF	(330,000 pF)
.47	(474)	.47 μF	(470,000 pF)
.68	(684)	.68 μF	(680,000 pF)

Selecting the Capacitor

There are many issues in selecting capacitors. An important one is the useful frequency of a capacitor. The following table shows the useful ranges for the many type of capacitors.

Estimated Useful Frequency Range for Capacitors

Capacitor Type	Useful Frequency Range
Electrolytic, tantalum	up to 10 kHz
Paper	up to 100 kHz
Ceramic (High Dielectric)	1 kHz to 100 kHz
Polycarbonate, polyethylene	1 kHz to over 1 MHz
Solid tantulum	up to 10 MHz
Mica, ceramic, air, gas	50 kHz to 100 MHz
Polystyrene, polypropylene	up to 1 GHz

APPENDIX C

FREQUENCY RESPONSE DERIVATIONS

Derivation of Equation 9–23 for Finding the Lower Cutoff Frequency f_L in an Amplifier with Two Interacting Corner Frequencies, f_1 and f_2

The magnitude of the overall gain, A_{vs}, at medium and low frequencies has the (high-pass) form

$$|A_{vs}| = \frac{A_m}{\sqrt{1 + \left(\frac{f_1}{f}\right)^2}\sqrt{1 + \left(\frac{f_2}{f}\right)^2}}$$

At the cutoff frequency f_L, $|A_{vs}| = A_m/\sqrt{2}$. Substituting this value and simplifying yields

$$\left[1 + \left(\frac{f_1}{f_L}\right)^2\right]\left[1 + \left(\frac{f_2}{f_L}\right)^2\right] = 2$$

which reduces to

$$f_L^4 - (f_1^2 + f_2^2)f_L^2 - f_1^2 f_2^2 = 0$$

Solving this equation using the general quadratic formula yields f_L^2.

Derivation of Equation 9–32 for Finding the Upper Cutoff Frequency f_H in an Amplifier with Two Interacting Corner Frequencies, f_A and f_B

At medium and high frequencies, the magnitude of the gain A_{vs} has the (low-pass) form

$$|A_{vs}| = \frac{A_m}{\sqrt{1 + \left(\frac{f}{f_A}\right)^2}\sqrt{1 + \left(\frac{f}{f_B}\right)^2}}$$

cutoff frequency f_H, $|A_{vs}| = A_m/\sqrt{2}$. Substitution and simplifica-

$$f_H^4 + (f_A^2 + f_B^2)f_H^2 - f_A^2 f_B^2 = 0$$

so be solved using the general quadratic formula to obtain f_H^2.

y Response in Inverting and Noninverting Configurations

TING The closed-loop gain of the noninverting amplifier is

$$A_{CL}(jf) = \frac{A(jf)}{1 + A(jf)\beta}$$

Here, $A(jf)$ is no longer just a huge number but the open-loop gain as a function of frequency described by the single-pole (or single-corner) low-pass function

$$A(jf) = \frac{A_o}{1 + \dfrac{jf}{f_o}}$$

where A_o is the (very large) dc open-loop gain and f_o is the open-loop 3-dB (or corner) frequency. After substituting $A(jf)$ and simplifying, we obtain

$$A_{CL}(jf) = \frac{A_o}{1 + A_o\beta}\,\frac{1}{1 + \dfrac{jf}{f_o(1 + A_o\beta)}} \approx \frac{1/\beta}{1 + \dfrac{jf}{f_o A_o \beta}}$$

where the numerator $1/\beta$ is clearly recognized as the ideal noninverting closed-loop gain.

In magnitude, the closed-loop gain can be expressed as

$$|A_{CL}| = \frac{A_{CLO}}{\sqrt{1 + \left(\dfrac{f}{f_{CL}}\right)^2}}$$

where $A_{CLO} = 1/\beta = 1 + R_f/R_1$ is the closed-loop gain at dc, low, and midband frequencies; and $f_{CL} = A_o f_o \beta = \text{GBP}/A_{CLO}$ is the closed-loop 3-dB (or corner) frequency.

The two asymptotes for the closed-loop gain are

$$|A_{CL}| = A_{CLO} \qquad (f \ll f_{CL})$$

and

$$|A_{CL}| = A_{CLO} f_{CL}/f = \text{GBP}/f \qquad (f \gg f_{CL})$$

and intercept each other at the corner frequency f_{CL}. Note that the second asymptote, which has a slope of -1 on log-log scaling, crosses unity gain at $f = \text{GBP}$; that is, it coincides with the sloped portion of the open-loop frequency response of the operational amplifier.

INVERTING The closed-loop gain in this case is given by

$$A_{CL}(jf) = \frac{-A(jf)(1 - \beta)}{1 + A(jf)\beta}$$

where $A(jf)$ is the open-loop gain defined in the noninverting case. After substituting $A(jf)$, we obtain

$$A_{CL}(jf) = \frac{-A_o(1-\beta)}{1+A_o\beta}\,\frac{1}{1+\dfrac{jf}{f_o(1+A_o\beta)}} \approx \frac{1-1/\beta}{1+\dfrac{jf}{A_of_o\beta}}$$

where the numerator can easily be shown to be the ideal inverting closed-loop gain $-R_f/R_1$. Note that the denominator is identical to the noninverting case; this means that the closed-loop 3-dB frequency is the same for both configurations for the same β.

In magnitude, the inverting closed-loop gain can be expressed as

$$|A_{CL}| = \frac{1/\beta - 1}{\sqrt{1+\left(\dfrac{f}{A_of_o\beta}\right)^2}} = \frac{R_f/R_1}{\sqrt{1+\left(\dfrac{f}{f_{CL}}\right)^2}}$$

The asymptotes in this case are given by

$$|A_{CL}| = R_f/R_1 \qquad (f \ll f_{CL})$$

and

$$|A_{CL}| = (1-\beta)A_of_o/f = (1-\beta)\text{GBP}/f \qquad (f \gg f_{CL})$$

Let f_u' be the frequency at which the inverting gain becomes unity. It is clearly seen that

$$f_u' = (1-\beta)\text{GBP}$$

Therefore, the second asymptote does not coincide with the sloped portion of the open-loop frequency response, which crosses unity gain at GBP or f_u. For large gains (small β), however, the inverting unity-gain frequency f_u' is very close to GBP or f_u, as in the noninverting case.

APPENDIX D

SEMICONDUCTOR THEORY

D–1 INTRODUCTION

Semiconductor devices are the fundamental building blocks from which all types of useful electronic products are constructed—amplifiers, high-frequency communications equipment, power supplies, computers, control systems, to name only a few. It is possible to learn how these devices—diodes, transistors, integrated circuits—can be connected together to create such useful products with little or no knowledge of the semiconductor theory that explains how the devices themselves operate. However, the person who understands that theory has a greater knowledge of the capabilities and limitations of devices and is therefore able to use them in more innovative and efficient ways than the person who does not. Furthermore, it is often the case that the success or failure of a complex electronic system can be traced to a certain peculiarity or operating characteristic of a single device, and an intimate knowledge of how and why that device behaves the way it does is the key to reliable designs or to practical remedies for substandard performance.

In Sections D–2 through D–7, we present the fundamental theory underlying the flow of charged particles through semiconductor material, the material from which all modern devices are constructed. We will also learn the theory of operation of the most fundamental semiconductor device: the diode. The theory presented here includes numerous equations that allow us to compute and assign quantitative values to important atomic-level properties of semiconductors.

D–2 ATOMIC STRUCTURE

A study of modern electronic devices must begin with a study of the materials from which those devices are constructed. Knowledge of the principles of material composition, at the level of the fundamental structure of matter, is an important prerequisite in the field of study we call *electronics* because the ultimate concern of that field is predicting and controlling the flow of atomic charge. Towards developing an appreciation of how atomic structure influences the electrical properties of materials, let us begin with a review of the structure of that most fundamental of all building blocks: the atom itself.

Every chemical element is composed of atoms, and all of the atoms within a *single* element have the same structure. Each element is unique because that common structure of its atoms is unique. Each atom is itself composed of a central *nucleus* containing one or more positively charged particles called *protons.* When an atom is complete, its nucleus is surrounded by negatively charged particles, called *electrons,* equal in number to the quantity of protons in the nucleus. Since the positive charge on each proton is equal in magnitude to the negative charge on each electron, the complete atom is electrically neutral. Depending on the element that a particular atom represents, the nucleus may also contain particles called *neutrons* that carry no electrical charge. An atom from the simplest of all elements, hydrogen, contains exactly one electron and has a nucleus composed of one proton and no neutrons. All other elements have at least one neutron.

Figure D–1(a) is a diagram of the structure of an isolated atom of the element *silicon,* the material used most often in the construction of modern electronic devices. Rather than showing a cluster of individual protons and neutrons in the nucleus, the figure simply shows that the nucleus contains 14 (positively charged) protons and 14 neutrons. Notice that the atom is neutral, since it contains a total of 14 (negatively charged) electrons. These electrons are shown distributed among three distinct *orbits* around the nucleus. The electrons within a given orbit are said to occupy an electron *shell,* and it is well known that each shell in an atom can contain no more than a certain maximum number of electrons. If the first four shells are numbered in sequence beginning from the innermost shell (shell number 1 being that closest to the nucleus), then the maximum number (N_e) of electrons that shell number n can contain is

$$N_e = 2n^2 \qquad \text{(D–1)}$$

Notice in Figure D–1(a) that shell number 1 (labeled K) is filled because it contains $2 \times 1^2 = 2$ electrons and shell number 2 (labeled L) is filled because

FIGURE D–1

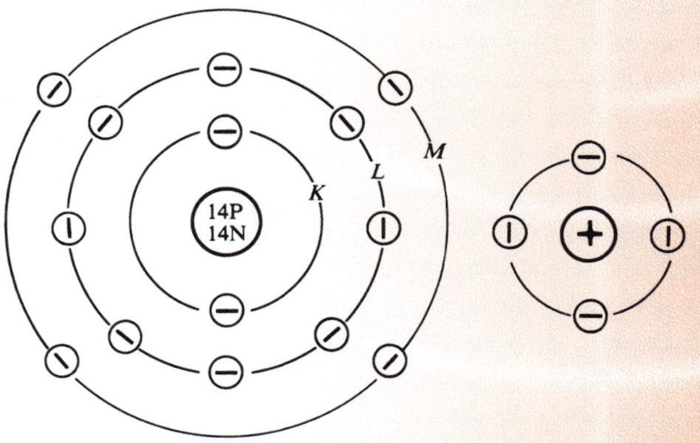

(a) A diagram of the silicon atom, showing its nucleus and electron shells. P = proton; N = neutron.

(b) An abbreviated diagram of the silicon atom, showing valence electrons only.

it contains $2 \times 2^2 = 8$ electrons. However, shell number 3 (M) is not filled because it has a capacity of $2 \times 3^2 = 18$ electrons but contains only 4.

Each shell is divided into *subshells*, the nth shell having n subshells. The first subshell in an electron shell can contain 2 electrons. If the shell has a second subshell, it can contain 4 additional, or 6 total, electrons; and if there is a third subshell, it can contain four more, or 10, electrons. The first shell of silicon has only one subshell ($n = 1$) and it is filled (2 electrons); the second shell has two ($n = 2$) filled subshells (containing 2 and 6 electrons); and the third shell has one filled subshell containing 2 electrons. The second subshell of the third shell can contain 6 electrons but has only 2. In practice, shells are given letter designations rather than numerical ones, so shells 1, 2, 3, 4, . . . , are referred to as shells K, L, M, N, . . . , as shown in Figure D–1(a). Subshells are designated s, p, d, f.

EXAMPLE D–1

The nucleus of a germanium atom has 32 protons. Assuming that each shell must be filled before a succeeding shell can contain any electrons, determine the number of electrons in each of its shells and in the subshells of each shell.

Solution

There are 32 electrons, and they fill the shells as shown in the *contents* column of the following table:

Shell	Capacity ($2n^2$)	Contents
K ($n = 1$)	2	2
L ($n = 2$)	8	8
M ($n = 3$)	18	18
N ($n = 4$)	32	4
		Total: 32

The contents of the subshells in each shell are shown in the following table:

Shell	Subshells	Capacity	Contents
K	s	2	2
L	s	2	2
	p	6	6
M	s	2	2
	p	6	6
	d	10	10
N	s	2	2
	p	6	2
	d	10	0
	f	14	0
			Total: 32

Not every electron in every atom is constrained forever to occupy a certain shell or subshell of an atomic nucleus. Although electrons tend to remain in their shells because of their force of attraction to the positively charged nucleus, some of them acquire enough energy (as, for example, from heating) to break away from their "parent" atoms and wander randomly through the material. Electrons that have escaped their shells are called *free* electrons. Conductors have a great many free electrons, while insulators have relatively few.

The outermost shell in an atom is called the *valence* shell, and the number of electrons in the valence shell has a significant influence on the electrical properties of an element. The reason that the number of valence electrons is important is that electrons in a nearly empty valence shell or subshell are more easily dislodged (freed) than electrons in a filled or nearly filled shell. Moreover, valence electrons, being farther from the nucleus than electrons in the inner shells, are the ones that experience the least force of attraction to the nucleus. Conductors are materials whose atoms have very few electrons in their valence shells (copper atoms have only one), and, in these materials, the heat energy available at room temperature (25°C) is enough to free large numbers of those loosely bound electrons. When an electrical potential is applied across the ends of a conductor, free electrons readily move from one end to the other, creating a transfer of charge through the conductor, i.e., an electrical current. Valence electrons in insulators, on the other hand, are tightly bound to their parent atoms.

Because we will be concerned primarily with the behavior of electrons in the valence shell, we will abbreviate all future diagrams of atoms so that only the nucleus and the valence electrons are shown. The silicon atom is shown in this abbreviated manner in Figure D–1(b).

D–3 SEMICONDUCTOR MATERIALS

Virtually all modern electronic devices are constructed from *semiconductor* material. As the name implies, a semiconductor is neither an electrical insulator (like rubber or plastic) nor a good conductor of electric current (like copper or aluminum). Furthermore, the mechanism by which charge flows through a semiconductor cannot be entirely explained by the process known to cause the flow of charge through other materials. In other words, a semiconductor is something more than just a conductor that does not conduct very well or an insulator that allows some charge to pass through it.

The electrical characteristics of a semiconductor stem from the way its atoms interlock with each other to form the structure of the material. Recall that conductors have nearly empty valence shells and tend to produce free electrons, while insulators tend to retain valence electrons. The valence shell of a semiconductor atom is such that it can just fill an incomplete subshell by acquiring four more electrons. For example, we have already noted that the *p* subshell of the *M* shell in silicon contains 2 electrons. Given four more, this subshell would be filled. A semiconductor atom seeks this state of stability and achieves it by *sharing* the valence electrons of four of its neighboring atoms. It in turn shares each of its own four electrons with its four neighbors and thus contributes to the filling of their subshells. Every atom duplicates this process, so every atom uses four of its own electrons and one each from four of its neighbors to fill its *p* subshell. The result is a stable, tightly bound, lattice structure called a *crystal.*

The interlocking of semiconductor atoms through electron sharing is called *covalent bonding.* It is important to be able to visualize this structure, and Figure D–2 shows a two-dimensional representation of it. Of course, a true crystal is a three-dimensional object—the covalent bonding occurs in all directions—but the figure should help clarify the concept of covalent bonding. Remember that we show only the valence electrons of each atom.

In Figure D–2, each pair of shared electrons forms a *covalent bond,* which is shown as two electrons enclosed by an oval. Note that only the center atom

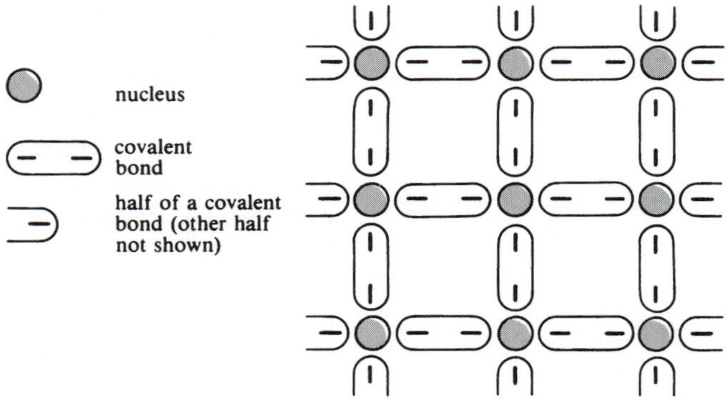

FIGURE D–2 Covalent bonding in a semiconductor crystal

is shown with a complete set of 4 covalent bonds, but all of the other atoms would be similarly interlocked with their neighbors. As previously described, the center atom uses its own four electrons and one from each of four neighbors, so it effectively has 8 electrons in its *M* shell. The atom directly above the center one similarly has four of its own electrons and one from each of four neighbors (although only three are shown), so it too has 8.

Germanium (symbol Ge) is another element whose valence shell enables it to establish covalent bonds with its neighbors and form a crystalline structure. In Example D–1, we saw that its *N* shell contains 4 electrons, 2 of which are in the *p* subshell. Therefore, it too can complete a subshell by the acquisition of 4 electrons. Like silicon (symbol Si), germanium atoms interlock and form a semiconductor material that is used to construct electronic devices. (However, because of temperature-related properties that we will discuss later, germanium is not now so widely used as silicon.) Note that the diagram in Figure D–2 applies equally to germanium and silicon, since only valence shell electrons are shown. Carbon also has a valence shell that enables it to establish covalent bonds and form a crystal, but it assumes that form only after being subjected to extreme heating and pressure. A carbon crystal is in fact a *diamond,* and it is not used in the construction of semiconductor devices.

D–4 CURRENT IN SEMICONDUCTORS

We have mentioned that the source of electrical charge available to establish current in a conductor is the large number of free electrons in the material. Recall that an electron is freed by acquiring energy, typically heat energy, that liberates it from a parent atom. Electrons are freed in semiconductor materials in the same way, but a greater amount of energy is required, on the average, because the electrons are held more tightly in covalent bonds. When enough energy is imparted to an electron to allow it to escape a bond, we say that a covalent bond has been *ruptured.*

It is instructive to view the electron-liberation process from the standpoint of the quantity of energy possessed by the electrons. The unit of energy that is conventionally used for this purpose is the *electron volt* (eV), which is the energy acquired by 1 electron if it is accelerated through a potential difference of 1 volt. One eV equals 1.602×10^{-19} joules (J). According to modern quantum theory, an electron in an isolated atom must acquire a very specific amount of energy in order to be freed, the amount depending on the kind of atom to which it belongs and the shell it occupies. Electrons in a valence shell already possess considerable energy, because a relatively small amount of additional energy will liberate them. Electrons in inner shells possess little energy, since they are strongly attracted to the nucleus and would

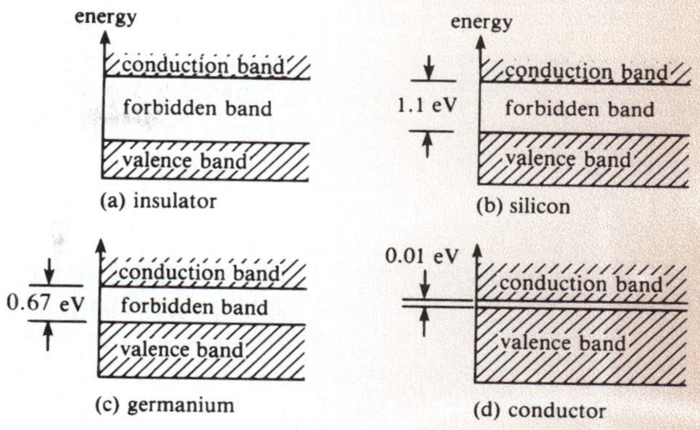

FIGURE D–3 Energy-band diagrams for several different materials

therefore need a great deal of additional energy to be freed. Electrons can also move from one shell to a more remote shell, provided they acquire the distinct amount of energy necessary to elevate them to the energy level represented by the new shell. Furthermore, electrons can *lose* energy, which is released in the form of heat or light, and thereby fall into lower-energy shells. Free electrons, too, can lose a specific amount of energy and fall back into a valence shell.

When atoms are in close proximity, as they are when they are interlocked to form a solid material, the interactions between adjoining atoms make the energy levels less distinct. In this case, it is possible to visualize a nearly continuous *energy band,* and we refer to electrons as occupying one band or another, depending on their roles in the structure. Energy-band diagrams are shown in Figure D–3. Free electrons are said to be in the *conduction band* because they are available as charge carriers for the conduction of current. *Valence-band* electrons have less energy and are shown lower in the energy diagram. The region between the valence and conduction bands is called a *forbidden band,* because quantum theory does not permit electrons to possess energies at those particular levels. The width of the forbidden band is the energy *gap* that electrons must surmount to make the transition from valence band to conduction band. Note in Figure D–3(a) that a typical insulator has a large forbidden band, meaning that valence electrons must acquire a great deal of energy to become available for conduction. For example, the energy gap for carbon is 5.4 eV. Energy gaps for semiconductors depend on temperature. As shown in Figure D–3, the room-temperature values for silicon and germanium are about 1.1 eV and 0.67 eV, respectively. The energy gap for a conductor is quite small ($\leq$0.01 eV) or nonexistent, and the conduction and valence bands are generally considered to overlap.

As might be surmised from the foregoing discussion, the number of free electrons in a material, and consequently its electrical *conductivity,* is heavily dependent on temperature. Higher temperatures mean more heat and therefore greater electron energies. At absolute zero (-273°C, or 0 K), all electrons have zero energy. But as the temperature is raised, more and more electrons acquire sufficient energy to cross the gap into the conduction band. For a semiconductor, the result is that conductivity increases with temperature (resistance decreases), which means that a semiconductor has a *negative* temperature coefficient of resistance. Although the number of conduction-band electrons in a conductor also increases with temperature, there are so many more of these than in a semiconductor that another effect predominates: Their number becomes so vast that they collide frequently and interfere with each other's progress when under the influence of an applied electric potential. The increased heat energy imparted to them at higher temperatures also makes their motion more erratic and compounds the problem. Consequently, it becomes more difficult to establish a uniform flow of charge at higher temperatures, resulting in a positive temperature coefficient of resistance for conductors.

Holes and Hole Current

What really distinguishes electrical current in a conductor from that in a semiconductor is the existence in the latter of another kind of charge flow. Whenever a covalent bond in a semiconductor is ruptured, a *hole* is left in the crystal structure by virtue of the loss of an electron. Since the atom that lost the electron now has a net positive charge (the atom becomes what is called a positive *ion*), we can regard that hole as representing a unit of positive charge. The increase in positive charge is, of course, equal to the decrease in negative charge, i.e., the charge of one electron: $q_e = 1.6 \times 10^{-19}$ coulombs (C). *If a nearby valence-band electron should now enter the hole, leaving behind a new hole, then the net effect is that a unit of positive charge has moved from the first atom to the second.* This transfer of a hole from one atom to another constitutes a flow of (positive) charge and therefore represents a component of electric current, just as electron flow contributes to current by the transfer of negative charge. We can therefore speak of *hole current* in a semiconductor as well as electron current. Figure D–4 illustrates the concept.

Figure D–4 illustrates the *single* repositioning of a hole, but it is easy to visualize still another valence electron entering the new hole (at B in the figure), causing the hole to move again, and so forth, resulting in a hole path through the crystal. Note that holes moving from left to right cause charge transfer in the same direction as electrons moving from right to left. It would be possible to analyze semiconductor current as two components of electron transfer, but it is conventional to distinguish between conduction-band electron flow and valence-band hole flow. Here is an important point that is worth repeating because it is a source of confusion to students and is not emphasized enough in most textbooks: Hole current occurs at the *valence*-band level, because valence-band electrons do not become free electrons when they simply move from atom to atom. Electron current always occurs in the *conduction* band, and involves only the flow of free electrons. If a conduction-band electron falls into a hole (which it may), this does *not* constitute current flow; indeed, such an occurrence is a cancellation of charge, and we say that a hole–electron pair has been *annihilated,* or that a *recombination* has occurred. Since charge transfer can take place by the motion of either negatively charged electrons or positively charged holes, we refer to electrons and holes collectively as charge *carriers.* Note that hole current does not occur in a conductor.

Because holes in a (pure) semiconductor are created by electrons that have been freed from their covalent bonds, the number of free electrons must equal the number of holes. This equality applies to the semiconductor materials we have studied so far because we have assumed them to be pure (in the

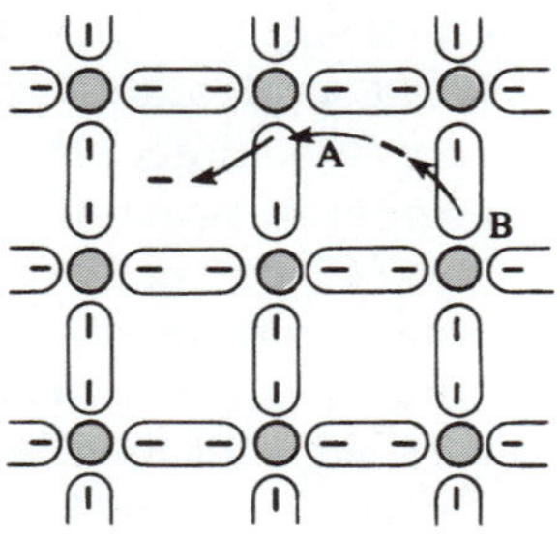

FIGURE D–4 Hole current. When the electron in position A is freed, a hole is left in its place. If the electron in position B moves into the hole at A, the hole, in effect, moves from A to B

sense that they are composed exclusively of atoms from one kind of element). Later, we will study semiconductor materials that have been made impure purposely in order to change the balance between holes and electrons. Pure semiconductor material is said to be *intrinsic*. It follows that the electron *density*, in electrons/m^3, equals the hole density, holes/m^3, in an intrinsic semiconductor. The subscript i is used to denote an intrinsic property; n_i refers to intrinsic electron density, and p_i is intrinsic hole density. Thus,

$$n_i = p_i \tag{D–2}$$

At room temperature, the charge carrier densities for germanium and silicon are approximately $n_i = p_i = 2.4 \times 10^{19}$ carriers/m^3 for germanium and $n_i = p_i = 1.5 \times 10^{16}$ carriers/m^3 for silicon. These figures seem to imply vast numbers of carriers per cubic centimeter, but consider the fact that a cubic centimeter of silicon contains more than 10^{22} atoms. Thus, there are approximately 10^{12} times as many atoms as there are carriers in silicon. Consider also that the conductor copper contains approximately 8.4 10^{28} carriers (free electrons) per cubic meter, a carrier density that is immensely greater than that of either germanium or silicon. Note that the carrier density of germanium is greater than that of silicon because the energy gap, as shown in Figure D–3, is smaller for germanium than for silicon. At a given temperature, the number of germanium electrons able to escape their bonds and enter the conduction band is greater than the number of silicon electrons that can do likewise.

Drift Current

When an electric potential is applied across a semiconductor, the electric field established in the material causes free electrons to drift in one direction and holes to drift in the other. Because the positive holes move in the opposite direction from the negative electrons, these two components of current *add* rather than cancel. The total current due to the electric field is called the *drift* current. Drift current depends, among other factors, on the ability of the charge carriers to move through the semiconductor, which in turn depends on the type of carrier and the kind of material. The measure of this ability to move is called *drift mobility* and has the symbol μ. Following are typical values for hole mobility (μ_p) and electron mobility (μ_n) in germanium and silicon:

Silicon	Germanium
$\mu_n = 0.14\ \text{m}^2/(\text{V} \cdot \text{s})$	$\mu_n = 0.38\ \text{m}^2/(\text{V} \cdot \text{s})$
$\mu_p = 0.05\ \text{m}^2/(\text{V} \cdot \text{s})$	$\mu_p = 0.18\ \text{m}^2/(\text{V} \cdot \text{s})$

Note that the units of μ are square meters per volt-second. Recall that the units of electric field intensity are V/m, so μ measures carrier velocity (m/s) per unit field intensity: (m/s)/(V/m) = m^2/(V · s). It follows that

$$v_n = \overline{E}\mu_n \quad \text{and} \quad v_p = \overline{E}\mu_p \tag{D–3}$$

where $\overline{E}$ is the electric field intensity in V/m and v_n and v_p are the electron and hole velocities in m/s. Although the value of μ depends on temperature and the actual value of the electric field intensity, the values listed above are representative of actual values at low to moderate field intensities and at room temperature.

We can use carrier mobility to compute the total *current density J* in a semiconductor when the electric field intensity is known. Current density is current per unit cross-sectional area.

$$J = J_n + J_p = \overbrace{nq_n\mu_n\overline{E}}^{J_n} + \overbrace{pq_p\mu_p\overline{E}}^{J_p}$$
$$= nq_nv_n + pq_pv_p \qquad \textbf{(D–4)}$$

where J = current density, A/m^2
n, p = electron and hole densities, carriers/m^3
$q_n = q_p$ = unit electron charge = 1.6×10^{-19} C
μ_n, μ_p = electron and hole mobilities, m^2/(V · s)
$\overline{E}$ = electric field intensity, V/m
v_n, v_p = electron and hole velocities, m/s

Equation D–4 expresses the fact that the total current density is the sum of the electron and hole components of current density, J_n and J_p. For intrinsic material, equation D–4 can be simplified as follows:

$$J = n_iq_n\overline{E}(\mu_n + \mu_p) = p_iq_p\overline{E}(\mu_n + \mu_p)$$
$$= n_iq_n(v_n + v_p) = p_iq_p(v_n + v_p)$$

An analysis of the units of equation D–4 shows that

$$\overset{n}{\left(\frac{\text{carriers}}{\text{m}^3}\right)}\overset{q}{\left(\frac{\text{coulombs}}{\text{carrier}}\right)}\overset{\overline{E}}{\left(\frac{\text{V}}{\text{m}}\right)}\overset{\mu}{\left(\frac{\text{m}^2}{\text{V}\cdot\text{s}}\right)} = \frac{\text{C}}{\text{s}\cdot\text{m}^2} = \frac{\text{A}}{\text{m}^2} = \text{current density}$$

EXAMPLE D–2

A potential difference of 12 V is applied across the ends of the intrinsic silicon bar shown in Figure D–5. Assuming that $n_i = 1.5 \times 10^{16}$ electrons/m^3 μ_n = 0.14 m^2/(V · s), and μ_p = 0.05 m^2/(V · s), find

1. the electron and hole velocities,
2. the electron and hole components of the current density,
3. the total current density, and
4. the total current in the bar.

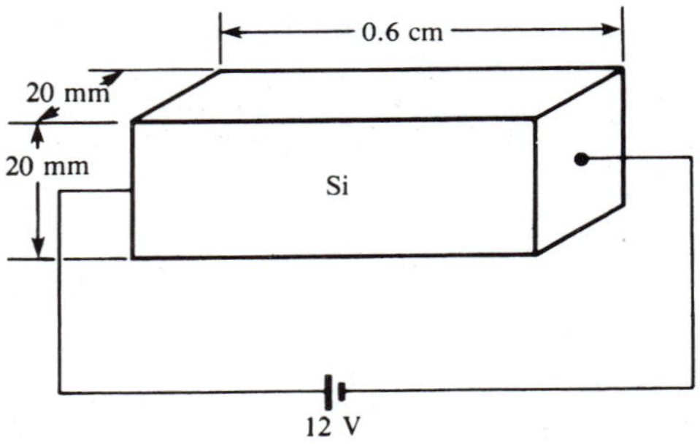

FIGURE D–5
(Example D–2)

Solution

We will assume that the electric field is established uniformly throughout the bar and that all current flow is along the horizontal axis of the bar (in the direction of the electric field).

1. $\overline{E} = (12\text{ V})/(0.6 \times 10^{-2}\text{ m}) = 2 \times 10^3$ V/m. From equation D–3.

$$v_n = \overline{E}\mu_n = (2 \times 10^3\text{ V/m})[0.14\text{ m}^2/(\text{V}\cdot\text{s})] = 2.8 \times 10^2\text{ m/s}$$
$$v_p = \overline{E}\mu_p = (2 \times 10^3\text{ V/m})[0.05\text{ m}^2/(\text{V}\cdot\text{s})] = 10^2\text{ m/s}$$

2. Since the material is intrinsic,

$$p_i = n_i = 1.5 \times 10^{16} \text{ carriers/m}^3$$

and

$$J_n = n_i q_n v_n = (1.5 \times 10^{16})(1.6 \times 10^{-19})(2.8 \times 10^2) = 0.672 \text{ A/m}^2$$
$$J_p = p_i q_p v_p = (1.5 \times 10^{16})(1.6 \times 10^{-19})(10^2) = 0.24 \text{ A/m}^2$$

3. $J = J_n + J_p = 0.672 + 0.24 = 0.912 \text{ A/m}^2$

4. The cross-sectional area A of the bar is $(20 \times 10^{-3} \text{ m}) \times (20 \times 10^{-3} \text{ m}) = 4 \times 10^{-4} \text{ m}^2$. Therefore, $I = JA = (0.912 \text{ A/m}^2)(4 \times 10^{-4} \text{ m}^2) = 0.365 \text{ mA}$.

Recall that the resistance of any body can be calculated using

$$R = \frac{\rho l}{A} \qquad \text{(D–5)}$$

where R = resistance, ohms (Ω)
ρ = *resistivity* of the material, Ω · m
l = length, m
A = cross-sectional area, m^2

Conductance, which has the units of siemens (S), is defined to be the reciprocal of resistance, and *conductivity* is the reciprocal of resistivity:

$$\sigma = \frac{1}{\rho} \qquad \text{(D–6)}$$

Thus, the units of conductivity are 1/(Ω · m), or siemens/meter (S/m).

The conductivity of a semiconductor can be computed using

$$\sigma = n\,\mu_n\,q_n + p\,\mu_p\,q_p \qquad \text{(D–7)}$$

Note that it is again possible to identify a component of conductivity due to electrons and a component due to holes.

EXAMPLE D–3

1. Compute the conductivity and resistivity of the bar of intrinsic silicon in Example D–2 (Figure D–5).
2. Use the results of (1) to find the current in the bar when the 12-V potential is applied to it.

Solution

1. $n = p = n_i = p_i = 1.5 \;\; 10^{16}/\text{m}^3$. From equation D–7,

$$\sigma = (1.5 \times 10^{16})(0.14)(1.6 \times 10^{-19}) + (1.5 \times 10^{16})(0.05)(1.6 \times 10^{-19}) = 4.56 \times 10^{-4} \text{ S/m}$$

Then $\rho = 1/\sigma = 1/(4.56 \times 10^{-4}) = 2192.98 \ \Omega \cdot \text{m}$.

2. $R = \rho l/A = (2192.98)(0.6 \times 10^{-2})/(4 \times 10^{-4}) = 32.89 \text{ k}\Omega$
$I = E/R = 12/(32.89 \times 10^3) = 0.365 \text{ mA}$

The current computed this way is the same as that computed in Example D–2.

Diffusion Current

Under certain circumstances, another kind of current besides drift current can exist in a semiconductor. Whenever there is a concentration of carriers (electrons or holes) in one region of a semiconductor and a scarcity in another, the carriers in the high-density region will migrate toward the low-density region, until their distribution becomes more or less uniform. In other words, there is a natural tendency for energetic carriers to disperse themselves to achieve a uniform concentration. Visualize, in an analogous situation, a small room crowded with people who are constantly squirming, crowding, and elbowing each other; if the room were suddenly to expand to twice its size, the occupants would tend to nudge and push each other outward as required to fill the new space. Another example of this natural expansion of energetic bodies is the phenomenon observed when a fixed quantity of gas is injected into an empty vessel: The molecules disperse to fill the confines of the container.

During the time that carriers are migrating from the region of high concentration to the one of low concentration, there is a transfer of charge taking place, and therefore an electric current. This current is called *diffusion* current, and the carriers are said to diffuse from one region to another. Diffusion is a transient (short-lived) process unless the region containing the higher concentration of charge is continually replenished. In many practical applications that we will study later, carrier replenishment does occur and diffusion current is thereby sustained.

D–5 *p*- AND *n*-TYPE SEMICONDUCTORS

Recall that intrinsic semiconductor material has the same electron density as hole density: $n_i = p_i$. In the fabrication of semiconductor materials used in practical applications, this balance between carrier densities is intentionally altered to produce materials in which the number of electrons is greater than the number of holes, or in which the number of holes is greater than the number of electrons. Such materials are called *extrinsic* (or *impure*) semiconductors. They are called *impure* because, as we will discuss presently, the desired imbalance is achieved by introducing certain impurity atoms into the crystal structure. Materials in which electrons predominate are called *n-type* materials, and those in which holes predominate are called *p-type* materials.

Let us first consider how *n*-type material is produced. Suppose that we are able, by some means, to insert into the crystal structure of a semiconductor an atom that has 5 instead of 4 electrons in its valence shell. Then 4 of those 5 electrons can (and do) participate in the same kind of covalent bonding that holds all the other atoms together. In this way, the impurity atom becomes an integral part of the structure, but it differs from the other atoms in that it has one "excess" electron. Its fifth valence electron is not needed for any covalent bond. Figure D–6 illustrates how the structure of a silicon crystal is modified by the presence of one such impurity atom. An impurity atom that produces an excess electron in this way is called a *donor* atom, because it donates an electron to the material. The nucleus of the donor atom is labeled D in Figure D–6. When a large number of donor atoms are introduced into the material, a correspondingly large number of excess electrons are created. Materials used as donor impurities in silicon include *antimony, arsenic,* and *phosphorus.*

Extrinsic semiconductor material is said to have been *doped* with impurity atoms, and the process is called *doping.* The impurity material is called a

FIGURE D–6 Structure of a silicon crystal containing a donor atom. The donor's nucleus is labeled D and the nuclei of the silicon atoms are labeled Si. The donor electrons are shown by colored dashes. Note the excess electron.

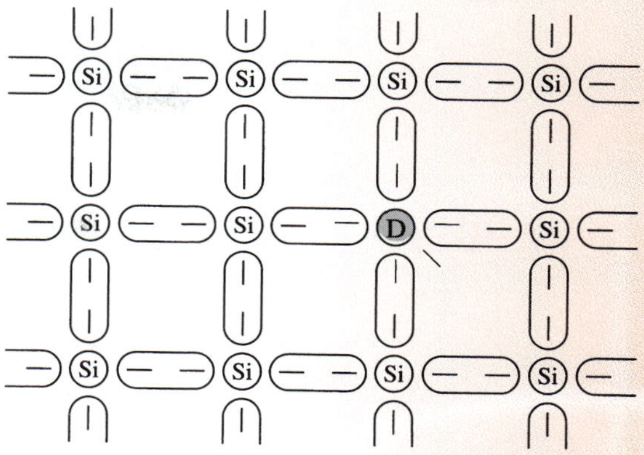

dopant. Because the silicon material illustrated in Figure D–6 has been doped with donor atoms and therefore contains an excess of electrons, it now constitutes *n*-type material. Note that the electrons are in "excess" only in the sense that there are now more electrons than holes; the material is still electrically neutral, because the number of protons in each donor nucleus still equals the total number of electrons that the donor atom brought to the material. Each excess electron is, however, very loosely bound to a donor atom and is, for all practical purposes, in the conduction band. The donor atom itself therefore becomes a positive ion. In all subsequent computations, we will assume that all impurity atoms are thus ionized.

p-type material is produced by doping a semiconductor with impurity atoms that have only three electrons in their outermost shells. When this kind of impurity atom is introduced into the crystal structure, an electron deficiency results, because the impurity atom contributes only three of the required four electrons necessary for covalent bonding. In other words, a hole is created everywhere the impurity atom appears in the crystal. Such impurity atoms are called *acceptors,* because the holes they produce can readily accept electrons. Figure D–7 shows a single acceptor atom in a silicon crystal. Materials used for doping silicon to create *p*-type material include *aluminum, boron, gallium,* and *indium.* Note once again that *p*-type material, like *n*-type, is electrically neutral because the electron deficiency exists only in the sense that there are insufficient electrons to complete all covalent bonds.

Although electrons are more numerous than holes in *n*-type material, there are still a certain number of holes present. The extent to which electrons dominate depends on the level of the doping: The more heavily the material is doped with donor atoms, the greater the degree to which the

FIGURE D–7 Structure of a silicon crystal containing an acceptor atom. The acceptor's nucleus is labeled A, and the nuclei of the silicon atoms are labeled Si. The acceptor electrons are shown by colored dashes. Note the incomplete bond and resulting hole caused by the acceptor's presence

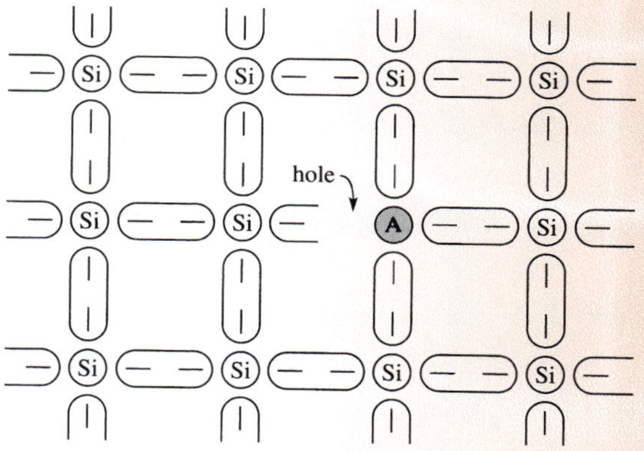

number of electrons exceeds the number of holes. In n-type material, electrons are said to be the *majority* carriers and holes the *minority* carriers. Similarly, the degree of acceptor doping controls the number of holes in p-type material. In this case, holes are the majority carriers and electrons the minority carriers.

An important relationship between the electron and hole densities in most practical semiconductor materials is given by

$$np = n_i^2 \tag{D–8}$$

where n = electron density
p = hole density
n_i = intrinsic electron density

Equation D–8 states that the product of electron and hole densities equals the square of the intrinsic electron density. Because $n_i = p_i$, equation D–8 is, of course, equivalent to $np = p_i^2 = n_i p_i$. All of the theory we have discussed so far in connection with mobility, conductivity, and current density is applicable to extrinsic as well as intrinsic semiconductors. The carrier densities used in the computations are often found using equation D–8.

EXAMPLE D–4

A bar of silicon with intrinsic electron density 1.4×10^{16} electrons/m^3 is doped with impurity atoms until the hole density is 8.5×10^{21} holes/m^3. The mobilities of the electrons and holes are $\mu_n = 0.14$ m^2/(V · s) and $\mu_p = 0.05$ m^2/(V · s).

1. Find the electron density of the extrinsic material.
2. Is the extrinsic material n-type or p-type?
3. Find the extrinsic conductivity.

Solution

1. From equation D–8,

$$n = \frac{n_i^2}{p} = \frac{(1.4 \times 10^{16})^2}{8.5 \times 10^{21}} = 2.3 \times 10^{10} \text{ electrons/m}^3$$

2. Since $p > n$, the material is p-type.
3. From equation D–7,

$$\begin{aligned}\sigma &= n\mu_n q_n + p\mu_p q_p \\ &= (2.3 \times 10^{10})(0.14)(1.6 \times 10^{-19}) + (8.5 \times 10^{21})(0.05)(1.6 \times 10^{-19}) \\ &= 5.152 \times 10^{-10} + 68 \approx 68 \text{ S/m}\end{aligned}$$

Note in the preceding example that the conductivity, 68 S/m, is for all practical purposes determined exclusively by the component of the conductivity due to holes, which are the majority carriers in this case. In practice, this is almost always the case: The conductivity essentially depends only on the majority carrier density. This result is due to a phenomenon called *minority carrier suppression*. To illustrate, suppose that the majority carriers are electrons, and that there are substantially more electrons than holes. Under these conditions, there is an increased probability that an electron—hole recombination (annihilation) will occur, thus eliminating both a free electron and a hole. Since there are very many more electrons than holes, the resultant percent decrease in electrons is much less than the percent decrease in holes. This effective suppression of minority carriers is reflected in equation D–8 and leads to the following approximations, valid in most practical cases, for computing conductivity:

$$\sigma \approx n\mu_n q_n \quad (n\text{-type material})$$
$$\sigma \approx p\mu_p q_p \quad (p\text{-type material}) \tag{D-9}$$

When one carrier type has a substantial majority, it is apparent from these equations that the conductivity of a semiconductor increases in direct proportion to the degree of doping with impurity atoms that produce the majority carriers.

D–6 THE *pn* JUNCTION

When a block of p-type material is joined to a block of n-type material, a very useful structure results. The region where the two materials are joined is called a *pn junction* and is a fundamental component of many electronic devices, including transistors. The junction is not formed by simply placing the two materials adjacent to each other, but rather through a manufacturing process that creates a transition from p to n within a single crystal. Nevertheless, it is instructive to view the formation of the junction in terms of the charge redistribution that would occur if two dissimilar materials were, in fact, suddenly brought into very close contact with each other.

Let us suppose that a block of p-type material on the left is suddenly joined to a block of n-type material on the right, as illustrated in Figure D–8(a). In the figure, the acceptor atoms and their associated "excess" holes are shown in the p region. Remember that the p region is initially neutral because each acceptor atom has the same number of electrons as protons. Similarly, the donor atoms are shown with their associated "excess" electrons in the n region, which is likewise electrically neutral. Remember also that *diffusion* current flows whenever there is a surplus of carriers in one region and a corresponding lack of carriers of the same kind in another region. Consequently, at the instant the p and n blocks are joined, electrons from the n region diffuse into the p region, and holes from the p region diffuse into the n region. (Recall that this hole current is actually the repositioning of holes due to the motion of *valence*-band electrons.)

For each electron that leaves the n region to cross the junction into the p region, a donor atom that now has a net positive charge is left behind. Similarly, for each hole that leaves the p region (that is, for each acceptor atom that captures an electron), an acceptor atom acquires a net negative charge. The upshot of this process is that negatively charged acceptor atoms begin to line the region of the junction just inside the p block, and positively charged donor atoms accumulate just inside the n region. This charge distribution is illustrated in Figure D–8(b) and is often called *space charge*.

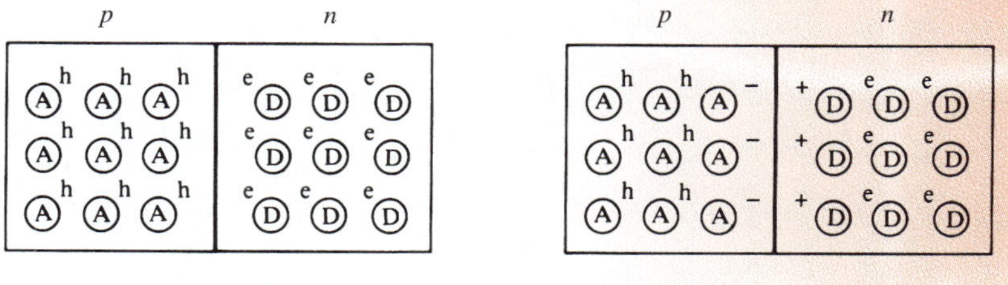

(a) Blocks of p and n materials at the instant they are joined; both blocks are initially neutral

(b) The pn junction showing charged ions after hole and electron diffusion

FIGURE D–8 Formation of a *pn* junction. A = acceptor atom; h = associated hole; D = donor atom; e = associated electron; + = positively charged ion; – = negatively charged ion.

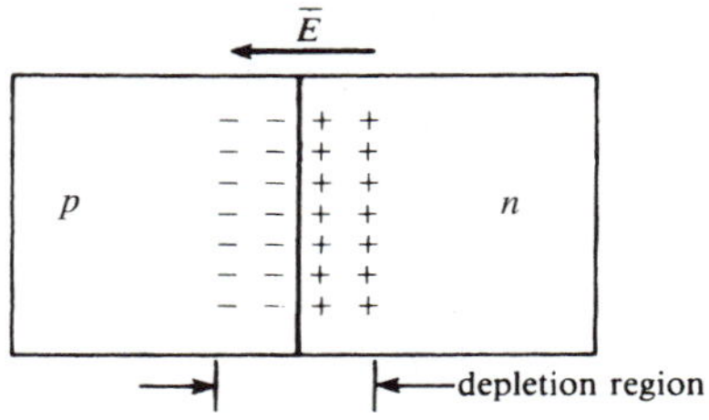

FIGURE D–9 The electric field $\overline{E}$ across a *pn* junction inhibits diffusion current from the *n* to the *p* side. There are no mobile charge carriers in the depletion region (whose width is proportionally much smaller than that shown)

It is well known that accumulations of electric charge of opposite polarities in two separated regions cause an electric field to be established between those regions. In the case of the *pn* junction, the positive ions in the *n* material and the negative ions in the *p* material constitute such accumulations of charge, and an electric field is therefore established. The direction of the field (which by convention is the direction of the force on a positive charge placed in the field) is from the positive *n* region to the negative *p* region. Figure D–9 illustrates the field $\overline{E}$ developed across a *pn* junction.

Note that the direction of the field is such that it *opposes* the flow of electrons from the *n* region into the *p* region, and the flow of holes from the *p* region into the *n* region. In other words, the positive and negative charges whose locations were *caused* by the original diffusion current across the junction are now inhibiting the further flow of current across the junction. An equivalent interpretation is that the accumulation of negative charge in the *p* region prevents additional negative charge from entering that region (like charges repel), and, similarly, the positively charged *n* region repels additional positive charge. Therefore, after the initial surge of charge across the junction, the diffusion current dwindles to a negligible amount.

The direction of the electric field across the *pn* junction enables the flow of *drift* current from the *p* to the *n* region, that is, the flow of electrons from left to right and of holes from right to left, in Figure D–9. There is therefore a small drift of *minority* carriers (electrons in the *p* material and holes in the *n* material) in the opposite direction from the diffusion current. This drift current is called *reverse* current, and when *equilibrium* conditions have been established, the small reverse drift current exactly cancels the diffusion current from *n* to *p*. The net current across the junction is therefore 0.

In the region of the junction where the charged atoms are located, there are no mobile carriers (except those that get swept immediately to the opposite side). Remember that the *p*-region holes have been annihilated by electrons, and the *n*-region electrons have migrated to the *p* side. Because all charge carriers have been depleted (removed) from this region, it is called the *depletion* region. See Figure D–9. It is also called the *barrier* region because the electric field therein acts as a barrier to further diffusion current, as we have already described. The width of the depletion region depends on how heavily the *p* and *n* materials have been doped. If both sides have been doped to have the same impurity densities (not the usual case), then the depletion region will extend an equal distance into both the *p* and *n* sides. If the doping levels are not equal, the depletion region will extend farther into the side having the smaller impurity concentration. The width of a typical depletion region is on the order of 10^{-16} m. In the practical *pn* junction, there is not necessarily the abrupt transition from *p*- to *n*-type material shown in Figure D–8. The junction may actually be formed, for example, by a gradual increase in the donor doping level of one block of p-type material, so that it gradually changes its nature from p-type to n-type with increasing distance through the block.

The electric field shown in Figure D–9 is the result of the *potential difference* that exists across the junction due to the oppositely charged sides of the junction. This potential is called the *barrier* potential because it acts as a barrier to diffusion current. (It is also called a *junction* potential, or *diffusion* potential.) The value of the barrier potential, V_0, depends on the doping levels in the p and n regions, the type of material (Si or Ge), and the temperature. Equation D–10 shows how these variables affect V_0:

$$V_0 = \frac{kT}{q} \ln\left(\frac{N_A N_D}{n_i^2}\right) \qquad \textbf{(D–10)}$$

where V_0 = barrier potential, volts
k = Boltzmann's constant = 1.38×10^{-23} J/K
T = temperature of the material in kelvin (K = 273 + °C; note that the correct SI unit of temperature is kelvin, *not* °K or degrees Kelvin.)
q = electron charge = 1.6×10^{-19} C
N_A = acceptor doping density in the p material
N_D = donor doping density in the n material
n_i = intrinsic electron density

Note that the barrier potential is directly proportionate to *temperature.* As we shall see throughout the remainder of our study of semiconductor devices, temperature plays a very important role in determining device characteristics and therefore has an important bearing on circuit design techniques. The quantity kT/q in equation D–10 has the units of volts and is called the *thermal voltage, V_T:*

$$V_T = \frac{kT}{q} \text{ volts} \qquad \textbf{(D–11)}$$

Substituting equation D–11 into equation D–10,

$$V_0 = V_T \ln\left(\frac{N_A N_D}{n_i^2}\right) \qquad \textbf{(D–12)}$$

EXAMPLE D–5

A silicon pn junction is formed from p material doped with 10^{22} acceptors/m^3 and n material doped with 1.2×10^{21} donors/m^3. Find the thermal voltage and barrier voltage at 25°C.

Solution. $T = 273 + 25 = 298$ K. From equation D–11,

$$V_T = \frac{kT}{q} = \frac{(1.38 \times 10^{-23})(298)}{1.6 \times 10^{-19}} = 25.7 \text{ mV}$$
$$n_i^2 = (1.5 \times 10^{16})^2 = 2.25 \times 10^{32}$$

From equation D–12,

$$V_0 = V_T \ln\left(\frac{N_A N_D}{n_i^2}\right)$$
$$= 0.0257 \ln\left(\frac{10^{22} \times 1.2 \times 10^{21}}{2.25 \times 10^{32}}\right)$$
$$= (0.0257)(24.6998) = 0.635 \text{ V}$$

D–7 FORWARD- AND REVERSE-BIASED JUNCTIONS

In the context of electronic circuit theory, the word *bias* refers to a dc voltage (or current) that is maintained in a device by some externally connected source. We discuss the concept of bias and its practical applications in considerable detail in Chapter 3. For now, suffice it to say that a *pn* junction can be biased by connecting a dc voltage source across its *p* and *n* sides.

Recall that the internal electric field established by the space charge across a junction acts as a *barrier* to the flow of diffusion current. When an external dc source is connected across a *pn* junction, the polarity of the connection can be such that it either opposes or reinforces the barrier. Suppose a voltage source V is connected as shown in Figure D–10, with its positive terminal attached to the *p* side of a *pn* junction and its negative terminal attached to the *n* side. With the polarity of the connections shown in the figure, the external source creates an electric field component across the junction whose direction *opposes* the internal field established by the space charge. In other words, the barrier is reduced, so diffusion current is enhanced. Therefore, current flows with relative ease through the junction, its direction being that of conventional current, from *p* to *n, as* shown in Figure D–10. With the polarity of the connections shown in the figure, the junction is said to be *forward biased.* (It is easy to remember that a junction is forward biased when the positive terminal of the external source is connected to the *p* side, and the negative terminal to the *n* side.)

When the *pn* junction is forward biased, electrons are forced into the *n* region by the external source, and holes are forced into the *p* region. As free electrons move toward the junction through the *n* material, a corresponding number of holes progresses through the *p* material. Thus, current in each region is the result of majority carrier flow. Electrons diffuse through the depletion region and recombine with holes in the *p* material. For each hole that recombines with an electron, an electron from a covalent bond leaves the *p* region and enters the positive terminal of the external source, thus maintaining the equality of current entering and leaving the source.

Since there is a reduction in the electric field barrier at the forward-biased junction, there is a corresponding reduction in the quantity of ionized acceptor and donor atoms required to maintain the field. As a result, the depletion region *narrows* under forward bias. It might be supposed that the forward-biasing voltage V could be increased to the point that the barrier field would be completely overcome, and in fact reversed in direction. This is not, however, the case. As the forward-biasing voltage is increased, the corresponding increase in current causes a larger voltage drop across the *p* and *n* material outside the depletion region, and the barrier field can never shrink to 0.

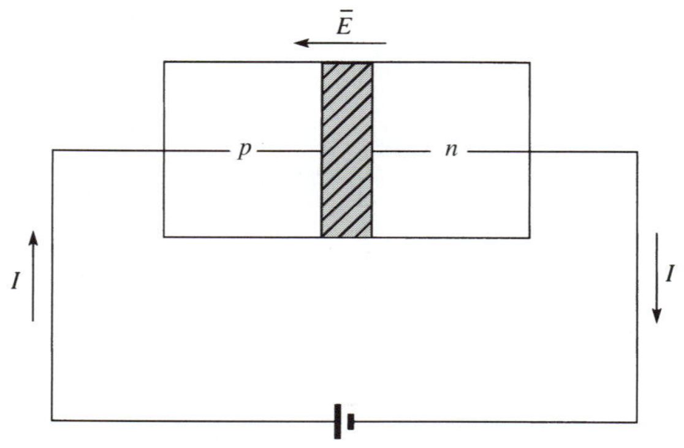

FIGURE D–10 A voltage source V connected to forward bias a *pn* junction. The depletion region (shown shaded) is narrowed.

D–8 A SEMICONDUCTOR GLOSSARY

Acceptor An impurity atom used in the doping process to create a hole in a semiconductor crystal; contains 3 electrons in its outermost valence shell.

Annihilation See *recombination.*

Anode The *p* side of a *pn* junction diode.

Avalanching Large current flow through a reverse-biased diode when it breaks down; caused by a high electric field imparting high velocities to electrons that then rupture covalent bonds.

Band, energy See *energy band.*

Barrier diode See *Schottky diode.*

Barrier potential Potential (voltage) established by the presence of layers of charge lying on opposite sides of a *pn* junction.

Barrier voltage See *barrier potential.*

Bias Connection of an external voltage source across a *pn* junction. See also *forward bias* and *reverse bias.*

Boltzmann's constant (k) Constant used in the diode equation; $k = 1.38 \times 10^{-23}$ J/K.

Breakdown voltage The reverse-biasing voltage across a diode that causes it to conduct heavily in the reverse direction (cathode to anode).

Carrier See *charge carrier.*

Carrier density The number of carriers (holes or electrons) per cubic meter of semiconductor material. See also *electron density* and *hole density.*

Cathode The *n* side of a *pn* junction diode.

Charge carrier An electron, which carries one unit of negative charge, q, or a hole, which carries one unit of positive charge.

Charge of an electron (q) $q = 1.6 \times 10^{-19}$ coulombs (C).

Conduction band The energy band of free electrons.

Conductivity (σ) A measure of the ability of a particular material to conduct current; units: S/m; the reciprocal of resistivity.

Covalent bond Shared electrons held to a parent atom in a crystal.

Current density (J) Current per unit cross-sectional area, A/m^2.

Depletion region Region in a *pn* junction where there are no mobile charge carriers.

Diffusion current Migration of carriers from a region where there are a large number of their own type to a region where there are fewer.

Diffusion potential See *barrier potential.*

Diode A *pn* junction. A discrete diode is a *pn* junction fitted with an enclosure and terminals for connection to its *p* and *n* sides (anode and cathode).

Donor An impurity atom used in the doping process to create a free electron in a semiconductor crystal; contains 5 electrons in its outermost valence shell.

Doping The process of introducing impurity atoms into a crystal to create *p*- or *n*-type material.

Drift current Current created by the motion of charge carriers under the influence of an electric field.

Electron, charge on (q) See *charge of an electron.*

Electron current The flow of (free) electrons.

Electron density (n) The number of free electrons per cubic meter of material.

Electron, free See *free electron.*

Electron-volt (eV) Unit of energy; the energy acquired by an electron when accelerated through a potential difference of 1 V: 1 eV = 1.602×10^{-19} J.

Emission coefficient (η) Coefficient used in the diode equation; its value depends on voltage and type of material; $1 \leq \eta \leq 2$.

Energy band A range of energies possessed by electrons. See also *conduction band, forbidden band,* and *valence band.*

Energy gap The width of the forbidden band; the difference in energy levels between the conduction and valence bands.

eV See *electron-volt.*

Extrinsic Impure; refers to semiconductor material that has been doped with impurity atoms.

Forbidden band The range of energy levels that electrons cannot possess in a particular type of material.

Forward bias Connection of the positive terminal of a voltage source to the *p* side of a *pn* junction and the negative terminal to the *n* side (in a diode, positive terminal to anode and negative terminal to cathode).

Forward current Current that flows from *p* to *n* in a forward-biased *pn* junction (anode to cathode in a diode).

Free electron An electron available to serve as a charge carrier in the creation of electrical current; a free electron is in the conduction energy band and is not bound to a parent atom.

Gap, energy See *energy gap.*

Ge Symbol for the element germanium.

Hole A vacant position in the structure of interlocking valence electrons in a crystal; the absence of an electron in a covalent bond; considered to have 1 unit of positive charge equal in magnitude to q.

Hole current The flow of holes, which is actually the repositioning of holes due to electrons moving from hole to hole.

Hole density (p) The number of holes per cubic meter of semiconductor material.

Impurity atom Atom introduced into a semiconductor crystal to create either *n* or *p* material; see also *donor* and *acceptor.*

Intrinsic Pure; refers to undoped semiconductor material.

I_s See *saturation current.*

J See *current density.*

Junction See *pn junction* or *MS junction.*

k See *Boltzmann's constant.*

Kelvin (K) Unit of temperature; K = 273 + °C.

Leakage current Cathode-to-anode (reverse) current that flows over the surface of a diode; not accounted for by the diode equation.

Majority carrier The carrier type whose number or density exceeds that of the other type in a particular material; electrons in *n* material and holes in *p* material.

Minority carrier The carrier type whose number or density is less than that of the other type; holes in *n* material and electrons in *p* material.

Mobility (μ) A measure of the ability of a carrier to move within a semiconductor; units: $m^2/V \cdot s$.

MS junction A metal-semiconductor junction that has diode properties; see also *Schottky diode.*

n See *electron density.*

N_A Acceptor doping density; the number of acceptor atoms per cubic meter in *p*-type material.

N_D Donor doping density; the number of donor atoms per cubic meter in *n*-type material.

n_i Electron density of intrinsic semiconductor material.

n^+ Heavily doped *n* material.

Neutron Atomic particle in the nucleus of an atom; electrically neutral.

***n* material** Semiconductor material that has been doped (with donor atoms) so that it contains more free electrons than holes.

Nucleus The center of an atom; consists of protons and neutrons.

Ohmic contact A junction of dissimilar materials that does not have diode properties; examples include aluminum and n^+ material and aluminum and *p* material.

Orbit Path occupied by electrons around the nucleus of an atom.

p See *hole density.*

p_i Hole density of intrinsic semiconductor material.

p^+ Heavily doped *p* material.

***p* material** Semiconductor material that has been doped (with acceptor atoms) so that it contains more holes than free electrons.

***pn* junction** Boundary between adjoining regions of *p* and *n* material; forms a diode.

Proton Atomic particles in the nucleus of an atom; has one unit of positive charge equal in magnitude to the charge of one electron (q).

q See *charge of an electron.*

q_n Same as q.

q_p Positive charge carried by one proton (or hole); has the same magnitude as q.

Recombination The occurrence of a free electron falling into a hole.

Resistivity (ρ) A measure of the ability of a particular material to create resistance; the resistance per unit length and cross-sectional area of the material; $\rho = 1/\sigma$.

Reverse bias Connection of the positive terminal of a voltage source to the *n* side of a *pn* junction and the negative terminal to the *p* side (in a diode, positive terminal to cathode and negative terminal to anode).

Rupture, of a covalent bond The release of an electron from the bond; creates a free electron and a hole.

Saturation current (I_s) The reverse current through a *pn* junction that flows when the junction is reverse biased by a few tenths of a volt.

Schottky diode A diode formed by an MS junction consisting of aluminum and *n*-type silicon.

Semiconductor A crystalline material that is neither a good conductor nor an insulator; current flow in a semiconductor is the motion of both electrons and holes.

Shell An orbit around the nucleus of an atom; can contain a specific maximum number of electrons.

Si Symbol for the element silicon.

Space charge A layer of charged particles lying in the *p* and *n* sides of a *pn* junction.

Subshell A subdivision of a shell; can contain a specific maximum number of electrons.

Thermal voltage (V_T) A component of the barrier potential at a *pn* junction; value depends on temperature.

Valence band Band of energies possessed by valence electrons in an atom.

Valence electron An electron in the outermost shell of an atom.

Valence shell Outermost electron shell of an atom.

V_{BR} See *breakdown voltage.*

V_T See *thermal voltage.*

Zener diode A diode designed to break down at a specific reverse voltage and to be used in the breakdown region.

Answers to Odd-Numbered Exercises

Chapter 2

2–1. (a) 4.6 mA (b) 5 mA
2–3. $I_D = 3.8$ mA, $I_D \approx 4.0$ mA, 5.26%
2–5. 0.2118 mA
2–7. $V_T = 0.025875$ V, $V = 0.6432$ V
2–9. (a) −0.0189 pA (b) −0.0199 pA (c) −0.02 pA
2–11. 150 pA
2–13. 0.5 mA
2–15. 4.39×10^{-14} A
2–17. (a) forward (b) reverse (c) forward (d) forward
2–19. (a) 0.7 V (b) 0.7 V (c) 0.7 V (d) 10.7 V
2–21. (a) −0.7 V (b) −0.7 V (c) −0.7 V (d) −5.7 V
2–23. 1.2 pF
2–25. 0.0208 μF

Chapter 3

3–1. 0.1 ms
3–3. (a) 320 Ω (b) 16 Ω (c) silicon
3–5. (a) 5.4 kΩ (b) 182.9 Ω
3–7. (a) 260 Ω (b) 7.43 Ω
3–9. 18.18 MΩ
3–11. (a) 4.23 V (b) 0.192 Ω
3–13. (a) 0.641 V (b) 12.42 Ω
3–15. $I_1 = 3.78$ mA, $I_2 = 3.26$ mA, $I_{D1} = 0.52$ mA
3–17. $I_{D3} = 9.9$ mA, $I_{D2} = 20.5$ mA, $I_{D1} = 31.8$ mA
3–19. 15.3 mA
3–21. $V_{rms} = 55.5$ V, $V_{PR} = 78.5$ V
3–23. (a) 167.4 V (b) 1.86 V (c) V_{pp} will approximately double
3–25. (a) 13.5 Ω (b) 3.4W (c) 22.6 V
3–27. (a) 7.37 V (b) 0.737 V (c) 14.14 V
3–29. −113 V (bottom of the ripple), −101.7 V (top of the ripple), $T = 10$ mS
3–31. 333.3 Ω
3–33. $I_{surge} = 5.54$ A, $V_{dc} = 166.2$ V
3–35. 1.21 mA, $V_{PR} = 121$ V
3–37. (a) 100 mA, 40 mA (b) 1.2 W, 0.48 W (c) 1.28 W
3–39. 118 Ω(max), 67 Ω(min)
3–41. (a) Use two compensating diodes (b) 13.36 V, 13.364 V (c) 0.6%
3–43. (a) 66.15 Ω (b) 53.3 Ω (decreases)
3–45. The diode rated at 200 PIV, 2 A would be the least expensive one ($3)
3–47. PIV(max) = 35 V
3–49. Diode 1N4004
3–51. $V_P = 339$, V, $I_{dc} = 3$ A, $I_{surge} = 60$ A. Bridge KBPC 604 (400 PIV, 6A, 125A surge)

Chapter 4

4–1. $I_E = 1.01$, $I_C = 12$ mA, 0.12 mA
4–3. (a) $\alpha = 0.995$ (b) $I_C = 22.1055$ mA (c) $\alpha = 0.9952$
4–7. $I_C \approx 7.6$ mA
4–9. $\alpha = 0.99375$
4–11. for α close to one
4–13. (a) $I_C \approx 0.95$ mA (b) $\beta = 95$
4–15. (a) $\beta = 156$ (b) $\beta = 168$ (c) 70 V
4–19. (a) $I_C = 1.515$ mA (b) $V_{CB} = 11.7$ V
4–21. (a) $I_C \approx 19.5$ mA, $V_{CB} = 4.2$ V (b) $I_C \approx 20$ mA, $V_{CE} = 4$ V

4–23. (a) $V_{CE} = 16.95$ V (b) $I_C = 2.55$ mA (c) $V_{CE} = 24$ V
4–25. (a) $V_{CE} = 3.8$ V (b) 1 mA (c) $I_C = 42$ mA, $V_{CE} = 3.8$ V
4–27. $R_B = 209.86$ kΩ
4–29. $V_{EC} = 6.35$ V
4–31. (a) $R_E = 4.7$ kΩ, $R_C = 5.6$ kΩ (b) $I_E(\text{min}) = 1.88$ mA, $I_E(\text{max}) = 2.08$ mA, $V_{CB}(\text{min}) = 7.77$ V, $V_{CB}(\text{max}) = 10$ V
4–33. (a) $R_E = 1.8$ kΩ, $R_B = 110$ kΩ, $V_{CE}(\text{min}) = 9.89$ V
4–35. $I_C(\text{sat}) = 4.545$ mA
4–37. $R_C = 1.97$ kΩ
4–39. 4.8 V
4–41. (a) $\beta = 20$ (b) $\beta = 80$
4–43. $I_{CBO}(\text{max}) = 0.01$ μA

Chapter 5

5–1. (a) 156.3 Ω (b) 250 Ω (c) 500 Ω
5–3. 5 V
5–5. (a) 2.6 mA (b) 4 mA
5–7. (a) 4.508 mA (b) 0.828 mA
5–9. (a) 2.5 mA 5.3V (b) 4.1 mA, 1V
5–11. (a) 1.4 valid (b) 10.97 V valid (c) 3.3 valid
5–13. 1.85 valid
5–15. (a) 4 valid (b) 4.09 valid
5–17. 500 Ω < R_L < 1.65 kΩ
5–19. 0.584 V
5–23. (a) 15 mA (b) −6 V (c) −30 V (d) 2N4420, 2N4420
5–25. 5.35 V
5–27. (a) 3.3 mA (b) 0
5–29. (a) 7 V (b) 7.06 V
5–31. (a) 9.68 V (b) 271.3 k
5–33. CMOS-Complementary Metal-Oxide Semiconductor because of the complementary NMOS and PMOS structures
5–35. 1. double-diffusion process
2. high power applications
3. parasitic diode between source and drain

Chapter 6

6–1. (a) 8.3 (b) 3 mA (c) 0.6 Vrms (d) 3.3 Vrms (e) 1.325 kΩ (f) 0.8215 W
6–3. $r_{in} \geq 3$k
6–5. 11.25 mW
6–7. 191 V
6–9. (a) −8568 (b) −11.67
6–11. $A_1 = A_2 = 10, A_3 = 5$

Chapter 7

7–1. 5.4 V, 2.85 mA
7–3. 9.4 V
7–5. $I_0 = \dfrac{V_{CC}}{R_C}$
7–7. (a) 3.4 mA (b) 162.5
7–9. (a) 14.72 Ω (b) 2.2 kΩ (c) 1.039 Vrms
7–11. 4.77 kΩ
7–13. (a) 3.08 Vpp (b) 9.92 Vpp (c) 12.8 Vpp
7–15. (a) 3.27 mA (b) 3.04 mA (c) 14.1 V
7–17. (a) 57.75 kΩ (b) 22.6 Ω (c) 0.9943 (d) 0.965 or 0.969
7–19. 0.49 V rms
7–21. (a) 305.2 kΩ (b) 0.866 (c) 3377
7–23. (a) 54.2 mS (b) 162.6 mS (c) 81.29 mS
7–25. $g_m = 92.47$ mS, $r_o = 45.59$ kΩ, $r_\pi = 1.08$ kΩ
7–27. $h_{fe} = 18.75\%$, $h_{re} = -77.3\%$, $h_{ie} = -69.4\%$, $h_{oe} = 56.25\%$
7–29. $g_m(\text{min}) = 8.57$ mS, $g_m(\text{max}) = 150$ mS, $h_{re} \approx -69.4\%$, $h_{oe} \approx 56.25\%$
7–31. 0
7–33. (a) 4.38×10^{-3} S (b) 4.375×10^{-3} S
7–35. −3.67 V
7–37. (a) 3.16 mA (b) 3.16×10^{-3} S (d) 2V
7–39. −7.32
7–41. −1.8476
7–43. 181 Ω
7–45. −3.74
7–47. 2.25×10^{-3} S

Chapter 8

8–1. (a) −0.312 Vdc (b) $-1.3 \sin \omega t$ V(c) 6.5 Vdc (d) $-10.4 + 2.6 \sin \omega t$ V (e) $-2.08 \sin (\omega t + 75°)$ V
8–3. (a) 8.06 μA (b) $-0.067 \sin \omega t$ μA (c) $(21.3 - 133.3 \sin \omega t)$ μA (d) $26.6 \sin(\omega t - 30°)$μA
8–7. (a) 4.5 Vrms (b) 0.6068 Vrms (c) 5.5 Vrms (d) 0.55 Vrms
8–9. (a) −6.8 V (b) −10.2 V
8–19. (a) 62.5 kHz (b) 33.75 kHz
8–21. (b) 2.5 kΩ
8–23. 1 mA
8–25. −1.875 V
8–27. 18 V

Chapter 9

9–1. (a) 0.88 MHz (b) 0.6 Vrms (c) 0.8 W (d) 0.4 W
9–3. 0.125 Vrms, 0.25 ms

9–5. 1 Vrms
9–7. (a) 32 dB (b) 30.46 dB
9–9. (a) 0.856 Vrms (b) 18 dB
9–11. (a) 3.6 Vrms (b) 72 Vrms (c) 0.18 Vrms (d) 360 Vrms
9–13. 4×4 log paper
9–15. (a) $f_L \approx 500$ Hz, $f_H \approx 15$ kHz (b) 0.5 Vrms (c) 63° (d) 5
9–17. (a) 159.15 Hz (b) 70.67 (c) 84.29° (d) 52.63 Hz
9–19. (a) 72.34 Hz (b) 19.38 Hz (c) 97.01
9–21. (a) 491.2 Hz (b) 170.1 (c) 1.94 :F
9–23. (a) 120.3 Hz (b) 37 dB (c) 34 dB (d) 1058 (e) 13.2 Hz
9–25. (a) 3.04 MHz (b) 157.5 (c) 43.62 (d) −73.18
9–27. 3383 Ω
9–29. (a) 271 kHz (b) 105.47
9–31. 55.4 kHz
9–33. (a) 6029 Hz (b) 58 μsec (c) decreases (d) increases
9–35. ≈ 135 Hz
9–37. 307 Hz
9–39. $C_E = 86.5$ μF, $C_1 = 17.8$ μF, $C_2 = 3.23$ μF
9–41. (a) $C_1 = 2.2$ μF, $C_2 = 10$ μF (b) 34.3 Hz (c) $f_L \approx 33.6$ Hz
9–43. $f_H \approx 1.18$ MHz
9–45. 191 Hz
9–47. 7708 Ω
9–49. 128.3 pF

Chapter 10

10–1. (a) 201, 193.23 (b) 51, 50.485 (c) 11, 10.976
10–3. $A \geq 639{,}200$
10–5. 84
10–7. −3.996
10–9. (a) $A\beta \geq 99$ (b) 5049
10–11. 25 Hz
10–13. (a) 60 kHz (b) 2
10–15. $R_f/R_1 \leq 199$
10–17. (a) 71.13 kHz (b) 35 dB, 33.26 dB (c) ≈12 dB
10–19. 3 V/μs
10–23. 355, 376 V/μs
10–25. 6.4
10–27. 56.2 kΩ
10–29. 1.66 V/μs
10–31. trapezoidal waveform of 5 Vpp
10–33. 0.4 mV
10–35. 62.5 kΩ, 7.25, 8.62 kΩ
10–37. 3.469 mV
10–39. 0.5 mV
10–41. 30.35 mV
10–43. (a) 3 mV (b) ≈ 58 dB (c) 3.5 mA (d) ≈ 34 kHz
10–45. $R_f = 18.27$ kΩ

Chapter 11

11–1. (a) −10 V (b) 3 sec
11–3. $v_o = 1.2 \cos(500\,t - 30°)$ V $= 1.2 \sin(500\,t + 60°)$ V
11–5. $R_1C = 0.1$ sec
11–7. $C = 0.01$ μF, $R_f = R_1 = 318$ kΩ
11–9. (a) $v_o = -\int(200v_1 + 40\,v_2)dt$ (b) 0.1 μF, $R_1 = 50$ kΩ, $R_2 = 250$ kΩ, $R_c = 41.67$ kΩ
11–11. (a) 482 Hz (b) 0.305 (c) 150.7 kHz
11–19. (a) $RC = 27.43$ μs (b) $C = 2.29$ nF
11–21. $C = 10$ nF, $R = 2.52$ kΩ
11–23. (a) 416.67 Ω
11–25. (a) −49.95 dB (b) 94.95 dB
11–27. $n = 3$
11–29. Stage 0: $C = 12.5$ nF, $R_0 = 587.1$ Ω
Stage 1: $C = 12.5$ nF (2 capacitors), $R_1 = 293.6$ Ω, $R_2 = 7646$ Ω
11–31. $C = 0.01$ μF, $R_1 = 1.99$ kΩ, $R_2 \approx 83$ Ω, $R_3 = 7.96$ kΩ
11–33. Low-pass section ($f_c = 1$ kHz): $R = 10$ kΩ, $C_0 = 1.592$ nF, $C_1 = 3.183$ nF, $C_2 = 796$ pF;
High-pass section ($f_c = 200$ Hz): $C = 0.5$ μF, $R_0 = 1591.5$ Ω, $R_1 = 795.8$ Ω, $R_2 = 3183.1$ Ω
11–35. $A_d = 4.22$

Chapter 12

12–3. (a) 6.74 V (b) –8.61 V (c) 15.35 V
12–5. 90 kΩ, 2 V
12–7. $V_{REF} = 1.25$ V, $R_1 = 30$ kΩ, $R_2 = 50$ kΩ
12–9. 45.5 kΩ, $R_1 = R_2 = 10$ kΩ, $C = 0.01$ μF
12–11. 179 Hz
12–13. 637 kΩ
12–17. (a) 214.6 kHz (b) 9.86 (OK) (c) 13.5 V/μs

Chapter 13

13–1. (a) 46.15 V (b) 4.8 Ω
13–3. 95 V
13–5. (a) 9.7 V (b) 230 Ω (c) 0.603 W
13–9. (a) 5 V (b) 22.5 mA
13–11. (a) $V_o = 12$ V (b) 10.5 V
13–13. (a) 15.75 V (b) 1.065 A
13–15. (a) 0.6 A (b) 5.4 W
13–17. (a) 2.7 V (b) 2.1 V
13–19. 1/3, or 33.3%; 2/3, or 66.6%
13–21. $R_1 = 333$ Ω; $R_2 = 1$ kΩ;

Chapter 14

14–1. (a) 0.625 V (b) 3.203125 V
14–3. 6.4583 kΩ, −4.8387 V
14–5. I_0 = 0.125 mA, I_1 = 0.250 mA, I_2 = 0.500 mA, I_3 = 1.00 mA
14–7. I_0 = 0.0039625 mA, I_1 = 0.0078125 mA, I_2 = 0.015625 mA, I_3 = 0.03125 mA, I_4 = 0.0625 mA, I_5 = 0.125 mA, I_6 = 0.25 mA, I_7 = 0.5 mA, I_T = 0.99609375 mA
14–9. (a) 1.25 pF (b) 2.75 pF (c) 2.5 V
14–11. (a) 0.0244% (b) 1.8315 mV, 5.4945 mV
14–13. 255 μs
14–15. (a) 31 (b) 0.3125 V
14–17. (a) 2 V (b) 8 μs (c) 00011001
14–19. 1000
1100
1010
1011
14–23. 39.06 mV, 0.3906%, 28.672 V
14–25. 50 μs

Chapter 15

15–1. 0.0916 sec
15–3. 0.667 V
15–7. 0.375 V
15–9. 1.5×10^{15} Hz→2×10^{15} Hz→ultraviolet
15–11. 12.590 Ω
15–13. 8750 1m/m^2
15–15. 20 mW/cm^2
15–17. 0.537 W, 0.176 W
15–19. 9050 Δ, 5510 Δ
15–21. R_C = 200 Ω, R_B = 17.2 kΩ
15–23. R_D = 253 Ω, R_C = 417 Ω
15–25. absorption and reflection
15–27. η = 0.55
15–29. R_1 = 3.07 kΩ, R_2 = 10 kΩ

Chapter 16

16–1. up, decreases, 225 mW
16–3. 400 Ω
16–5. 8.4 J/sec
16–7. 1°C/W
16–9. 0.127°C/W
16–11. 40°C
16–13. 24 mA, 0.144 W
16–15. (a) 7.5 V (b) 0.08333
16–17. (a) 0.01 A (b) 4 W (c) 0.5
16–19. (a) 24 W (b) 61.12 W (c) 18.56 W (d) 0.3927
16–21. $V_{CC} \geq 50.6$ V
16–23. A = 450
16–27. 16.62%, 46.66 V
16–29. (a) 20.72 Ω (b) 0.767
16–31. 1.95 V
16–33. (a) 23.9 V (b) 0.877
16–35. 791 kHz to 807.5 kHz
16–37. Directly proportional, 1.75 μs
16–39. inverts

Chapter 17

17–1. 11.48 mA
17–3. 25.45 mA
17–5. 0.5958
17–7. 27.43 μA
17–9. 1.25 mA
17–11. I_{C2} = 2 mA, I_{C3} = 4 mA, I_{C4} = 6 mA
17–13. 300 μA
17–15. I_{BIAS} = 5 μA, I_{D2} = 12.5 μA
17–17. −115.38
17–19. (a) −1.47 (b) −3.15 (c) $A_v = 0$
17–21. (a) 55.125 μA (b) 312.77 kΩ, 1.134 MΩ (c) 1.96×10^{-4} A/V^2 (d) 1.45 V (e) 1.47×10^{-4} (f) −36.038
17–23. (a) $g_{m1} = 1.166 \times 10^{-4}$ A/V, $g_{m2} = 2.97 \times 10^{-4}$ A/V (b) −2.547
17–25. 6.88 V
17–27. 2.3 V
17–29. (a) −80, 80 (b) −160
17–31. (a) −132.7, 132.7 (b) −265.4
17–33. (a) 5 V (b) −269.5 (c) 7.49 kΩ
17–35. (a) 6 V (b) −15.48 (c) 1.55 Vrms
17–37. 0.104 m Vrms
17–39. −40.8
17–41. V_1 = (+), V_2 = (−)

Chapter 18

18–1. (a) a CMOS inverter
(b) a CMOS transmission gate
(c) an "OR" structure
(d) an "AND" structure
18–3.

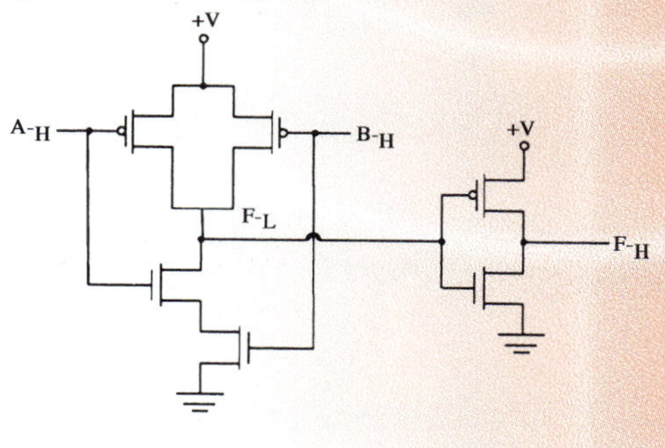

18–5.

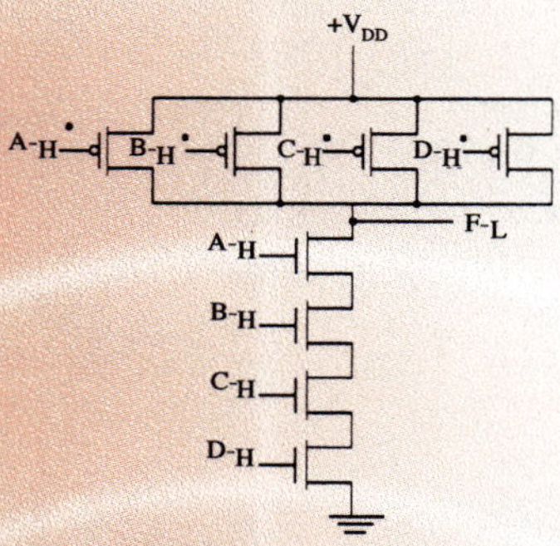

18–7.

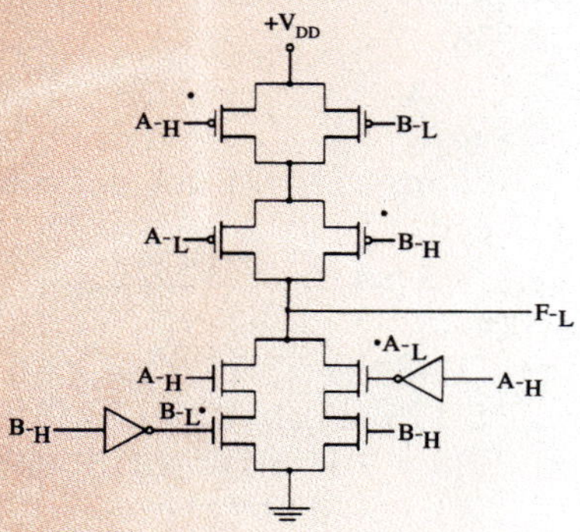

18–9.

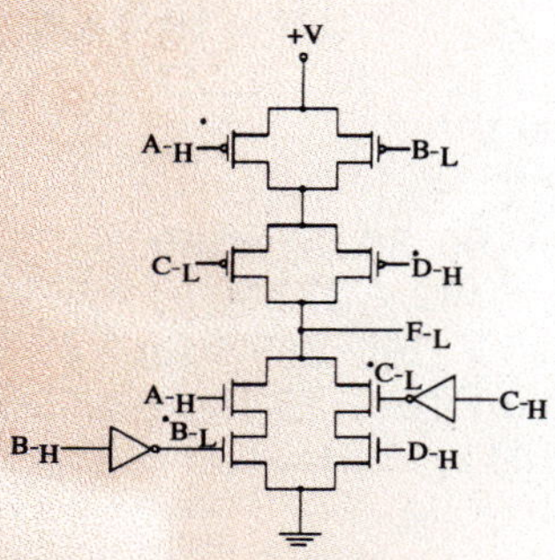

18–11.

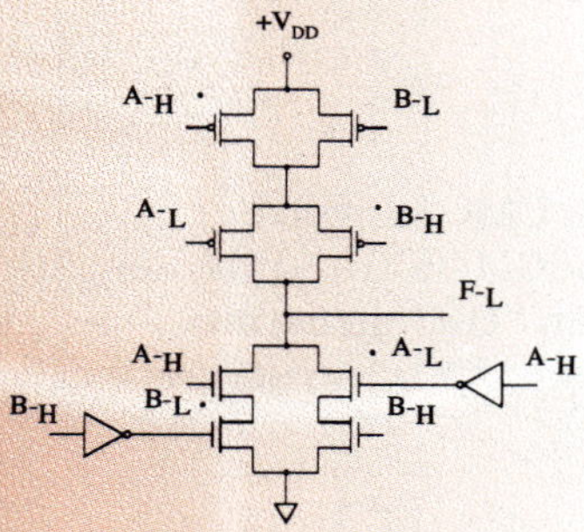

18–13.

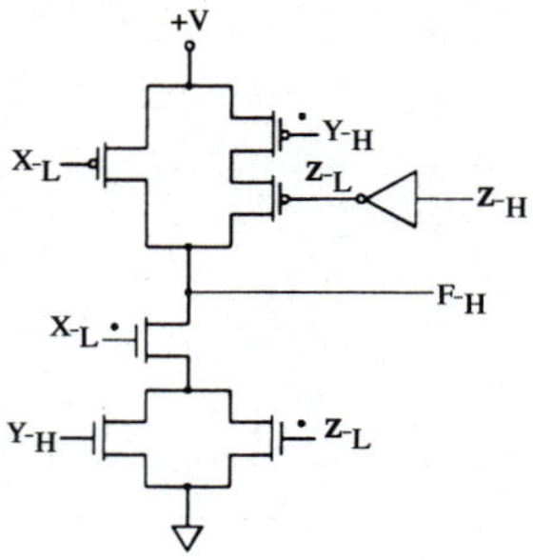

18–15. (a) $1.65 - .7 < 2.5$ saturation (b) $1.4 - .7 < 2.1$ saturation (c) $4.4 - .7 > .65$ Ohmic (triode)
18–17. 1.17×10^{-5}
18–19. 1.65
18–21. 2.4 nS, .96 nS
18–23. .425 nS

18–29.

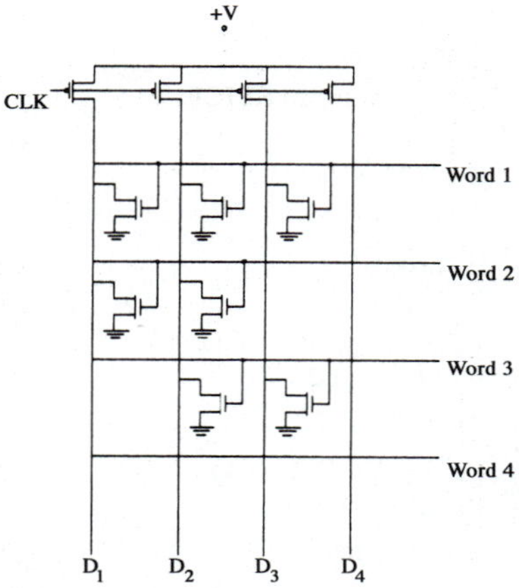

18–31. 1.097, .72
18–33. 2.24
18–35. 2.317

Index